The Evolution of Western Eurasian Neogene Mammal Faunas

The Evolution of Western Eurasian Neogene Mammal Faunas

Edited by
Raymond L. Bernor, Volker Fahlbusch, and Hans-Walter Mittmann

COLUMBIA UNIVERSITY PRESS • NEW YORK

Columbia University Press
Publishers Since 1893
New York   Chichester, West Sussex
Copyright © 1996 Columbia University Press

Library of Congress Cataloging-in-Publication Data

The evolution of western Eurasian Neogene mammal faunas / edited by
  Raymond L. Bernor, Volker Fahlbusch, and Hans-Walter Mittmann.
      p.   cm.
   Includes bibliographical references and indexes.
   ISBN 0–231–08246–0
     1. Mammals, Fossil—Europe—Congresses.   2. Mammals, Fossil—
  Middle East—Congresses.   3. Paleontology—Neogene—Congresses.
  I. Bernor, Raymond Louis. 1949–   .   II. Fahlbusch, Volker.
  III. Mittmann, Hans-Walter.
  QE881.E86   1996   96–11747
  569′.095—dc20
∞
Casebound editions of Columbia University Press books are printed on
permanent and durable acid-free paper.

Printed in the United States of America
c 10 9 8 7 6 5 4 3 2 1

Contents

PART II: MAMMALIAN SYSTEMATICS, BIOGEOGRAPHY, AND BIOCHRONOLOGY

PART III: PALEOBOTANY, PALEOBIOGEOGRAPHY, AND PALEOECOLOGY

Acknowledgments

During the course of this initiative, our dear colleague and friend Professor Dr. Heinz Tobien, Mainz University, passed away. There is no individual scientist with whom we have more enjoyed working or who has had a more positive influence on the subject we pursue here. We believe it most appropriate to dedicate this volume to his scientific legacy.

We wish to thank Professor Dr. Siegfried Rietschel, Director of the Staatliches Museum für Naturkunde, for the considerable assistance and sound advice he has provided throughout the development of this project. We thank the Volkswagen Stiftung for supporting the workshop, as well as the National Geographic Society, the L. S. B. Leakey Foundation, the Smithsonian Institution and the Joint American-Hungarian Research Fund, and the Ann and Gordon Getty Foundation for funding the related Höwenegg and Rudabánya research projects. Bernor wishes to thank the Alexander Von Humboldt Stiftung for supporting his early formulation of this project. We wish to thank the township of Immendingen and the Schloss Reisensburg for extending every courtesy to us and the workshop participants. We further wish to thank all of those colleagues who agreed to participate in this project, and to recognize the deadlines they maintained, the editorial suggestions they entertained, and the reams of correspondence, faxes, and e-mails they generated in order to make this volume possible.

The Evolution of Western Eurasian Neogene Mammal Faunas

1

The Evolution of Western Eurasian Neogene Mammal Faunas: The 1992 Schloss Reisensburg Workshop Concept

R. L. BERNOR, V. FAHLBUSCH, H.-W. MITTMANN, AND S. RIETSCHEL

Western Eurasian Neogene mammal faunas have attracted active scientific inquiry for well over a century. From early on investigators have pursued mammalian evolutionary studies largely within a stratigraphic framework. Associated with this approach have been occasional inquiries concerned with the paleoenvironmental and biogeographic contexts of faunal turnover. However, stratigraphic paleontology generally seeks relationships based on faunal similarity while biogeographically based studies seek dissimilarities among isochronous faunas. Despite abundant examples from extant and fossil faunas alike, stratigraphic paleontologists often assume synchronous faunal change over broad geographic distances in order to strike correlations. But what are the geographic limits of faunal change? What regulates the distinction of faunal provinces, if indeed they existed in the past? Do geographic limits of these provinces change? If so, what regulates change at any given time? How does one determine provinciality?

These are not new questions for the later Neogene of Western Eurasia. Indeed, the great French vertebrate paleontologist Albert Gaudry recognized these very issues in the middle of the nineteenth century. In his comparisons of the late Miocene faunas from Pikermi, Greece (1865), and Mont Lubéron, France (1873), with the late Miocene German locality of Eppelsheim, Gaudry noticed the similarity of the circum-Mediterranean mammal faunas to one another and their mutual difference from the Eppelsheim fauna. Were these differences due to temporal or paleoenvironmental distinctions? The most salient aspect of Gaudry's inquiry was that he recognized the *differences* between these faunas, in the way a biologist would, rather than the *similarities* with which a stratigraphic paleontologist is more likely to be concerned.

Gaudry's investigations of the time/biofacies problems surrounding pan-European "Pikermian" faunas paved the way for inquiries that followed, albeit sporadically, by Koenigswald (1929), Thenius (1951), and Crusafont and Villalta (1954). In a short paper delivered at the 1967 Committee on Mediterranean Neogene Stratigraphy, Bologna, Tobien most clearly articulated this particular issue and hypothesized the existence of "identical and unchanged steppe faunas in the East during the whole Pontian (i.e., Vallesian plus Pikermian) time span, and a similarly unchanged and evolutionary stable woodland fauna in the central part of Europe" (Tobien 1967:3). By the time a modern biochronology of "Greater Europe" was emerging (Thaler 1966), the concept that Western Eurasia exhibited strong provincial distinctions was a well-established idea, but a poorly demonstrated phenomenon.

Mein's (1975) initial organization of Mammal Neogene (MN) zones (or units, after Fahlbusch 1991) was truly revolutionary in organizing all the principal Western Eurasian and North African Neogene faunas into a relative biochronologic time scale. MN units exacted correlations based on faunal similarities, including characteristic evolutionary lineages, common taxonomic associations, and faunal datum events, but they did not initially address the kinds of provincial distinctions alluded to by previous investigators. Bernor's (1978) and later Bernor et al.'s (1980) attempts to strike species-level correlations between the late Miocene locality of Maragheh (Iran) and the European MN framework of Mein (1975) found very distinct and restricted geographic limits to his correlation, that later led to proposals for a provincial MN unit system (Bernor 1978, 1983, 1984). These distinctions gained support from the results of quantitative tests (Bernor 1978; Bernor and Pavlakis 1987) and found agreement with paleogeographic reconstructions by Rögl and Steininger (1983) and Bernor et al. (1988). Mein (1979, 1989) responded to these observations by adopting a more provincial segregation of faunal biochronologic sequences.

Development of the 1992 Schloss Reisensburg Workshop Concept

While several interim meetings of the Regional Committee on Mediterranean Neogene Stratigraphy (RCMNS) gave periodic updates on European Neogene mammal biochronology, the 1988 Schloss Reisensburg International Workshop on *European Neogene Mammal Chronology*, organized under the aegis of the North Atlantic Treaty Organization (NATO) by Everett H. Lindsay (USA), Volker Fahlbusch (Germany), and Pierre Mein (France), attempted to give a definitive state of the art update on the subject (Lindsay et al. 1989). Forty-seven participants were invited to bring forward new data pertinent to the subject. The participants responded with manuscripts that canvassed eight broad subjects: aspects of mammal chronology; regional treatments of faunal change; the species bases of faunal datum events; biogeographic syntheses; paleoecological aspects; application of magnetostratigraphy to continental mammal sequences; bases for correlation to East Asia, South Asia, and Africa; comparisons between the development of European MN units and American land mammal ages. Some of the most strenuously debated subjects addressed by the workshop participants were MN unit boundary synchroneity/diachroneity, the existence and definition of faunal provinces, and the forcing of evolutionary change by regional or global abiotic events. While there was no solution to these controversial issues, there was a wealth of new data, and the seed was planted for the 1992 Schloss Reisensburg workshop, which is the subject of this volume.

The International Geologic Congress was convened in Washington, D.C., during the summer of 1989. R. L. Bernor and F. F. Steininger met at this congress and developed a rough outline for a revision of the MN unit system. They agreed that a constructive step for resolving MN unit chronology could best be made by addressing a specific scientific problem that sought parallel revisions of the mammalian systematic, stratigraphic, and chronologic frameworks and tested specific scientific hypotheses. The existence of provinciality and the alternatives of global versus regional abiotic forcing of faunal turnover seemed to be two potential lines of inquiry (re: Bernor, 1983, 1984, 1985; Rögl and Steininger 1983; Vrba 1985). The most relevant geographic areas for this test were where provinciality appeared to be most striking: the middle and late Miocene of Central Europe and Southeast Europe/Southwest Asia.

Our first circular was sent in September 1989 to twenty-one European and American colleagues, seeking their advice for pursuing a follow-up workshop at the Schloss Reisensburg. Most colleagues responded positively, providing many useful suggestions on the range of topics to be addressed, and virtually unanimously agreed that a round-

table format offered the most effective forum for the workshop. Many colleagues further recommended that this workshop be more structured than previous ones, with specialists invited to provide new analyses and results on prespecified subjects, including: stratigraphy, geochronology, mammalian phylogeny, paleobotany, paleoecology, and biogeography. These presentations would develop a thoroughly updated data base for revising a critical segment of the time scale, evaluating the species-level bases for interprovincial correlations and testing selected paleogeographic and paleoecologic hypotheses to be addressed by the workshop attendees.

In order to determine the most relevant subjects, and select appropriate contributors, Bernor issued a second circular and convened an organizational workshop meeting in Mittlebergheim (Alsace), France, July 2–4, 1990, which included: S. Rietschel (Karlsruhe), H.-W. Mittmann (Karlsruhe), V. Fahlbusch (Munich), and S. Sen (Paris). This meeting determined the general course of the workshop. The geographic/chronologic focus would be the middle and late Miocene of Central Europe and Southeast Europe/Southwest Asia. However, authors would be encouraged to engage a broader geographic area and longer chronologic interval where relevant to their evolutionary reconstructions. The objectives included:

1. to give an updated summary of all faunas in the research area, including species present, stratigraphic, taphonomic, and paleoecological contexts;
2. to provide stratigraphic and chronologic updates on critical localities;
3. to strike as many species-level correlations between these geographic areas as possible to test hypotheses of provinciality;
4. to identify abiotic factors which might have forced faunal change (regional or global factors, or the coupling of the two);
5. to identify scientific problems in particular need of future work;
6. to publish the results in a single volume.

At this juncture the organizing committee subdivided responsibilities so that Fahlbusch would oversee the development of small mammal systematic, biogeographic, and biochronologic studies; Bernor would develop the equivalent for the large mammals and geochronologic framework; and Mittmann would organize and develop the faunal locality data base. A list of thirty participants was generated, and invitations extended through a third circular issued immediately following the Mittlebergheim meeting. The venue would be at two sites, the town of Immendingen and the Schloss Reisensburg. The workshop was delayed one year (from 1991 to 1992) in order to provide time to raise funding.

Over the course of the next two years three additional

circulars and hundreds of pages of correspondence were generated to further hone the workshop's objectives, establish collaborations, and set the stage for a productive round-table discussion. Funds were awarded by the Volkswagen Stiftung to support the workshop, and twenty-nine scientists from eleven countries participated, including: Austria (G. Daxner-Höck and J. Kovar-Eder); England (P. Andrews and A. Gentry); Finland (M. Fortelius); France (S. Sen and P. Mein); Germany (V. Fahlbusch, J. Franzen, K. Heissig, E. Heizmann, H.-W. Mittmann, G. Phillipi, G. Rietschel, S. Rietschel, H. Walter, and R. Ziegler); Greece (G. Koufos and P. Pavlakis); Holland (H. de Bruijn and J. Van der Made); Hungary (L. Kordos); Switzerland (T. Bolliger); Turkey (E. Ünay); and the USA (R. L. Bernor, D. Ekart, N. Solounias, C. C. Swisher III, and M. O. Woodburne).

The first two days were spent at Immendingen (Hegau), a few kilometers from the late Miocene locality of Höwenegg. Here, opening presentations included a lecture on the background to the workshop concept, and lectures on the biogeography and ecology of extant European mammals (G. Rietschel) and on vegetation communities (G. Phillipi). Time was also allocated for the large and small mammal groups to conduct informal round-table discussions to assemble data for later presentation at the Schloss Reisensburg. Finally, there was an extended field excursion to the nearby locality of Höwenegg, where an extraordinary assemblage of early late Miocene mammal skeletons were excavated in the 1950s and 1960s (Tobien 1986) and where a German-American group of scientists was pursuing resolution of various stratigraphic, chronologic, sedimentologic, and paleontologic objectives under the aegis of the Staatliches Museum für Naturkunde and with support from the National Geographic Society (re: Swisher, this volume; Woodburne et al., this volume a, b). Three and one-half days followed at the Schloss Reisensburg, where brief summaries on mammalian evolution, stratigraphy, geochronology, paleoecology, and biogeography were given and followed by extended discussions and active debate. Through this week-long round-table format the working group organized collaborations amongst attendees and a broad international group of colleagues who would assemble and analyze data and participate in preparing manuscripts. A follow-up workshop was convened by Bernor, Fahlbusch, and Mittmann in Salzburg, June 10–12, 1993, to finalize the subjects to be included in this volume and set final deadlines for the manuscripts.

The current volume includes thirty-two chapters, with published contributions by twenty-five of the twenty-nine attendees and thirty-eight additional scholars. Our results are broader ranging than initially intended, and for the most part canvass all of Western Eurasia—including Western, Central, and Southeast Europe, Southwest Asia, often North Africa and the former Western USSR. Systematic chapters most frequently extend lineage reconstructions down well into the early Miocene and up into the early Pliocene in order to give greater continuity to their records and resulting biogeographic and paleoecologic interpretations (re: Fortelius et al., this volume; Bernor et al., this volume).

This volume is subdivided into three sections: (1) Geological Background; (2) Mammalian Systematics, Biogeography, and Biochronology; and (3) Paleobotany, Paleobiogeography, and Paleoecology. All sections present new data and analysis on the subject and form the bases for the evolutionary, biogeographic, and paleoenvironmental parameters we interpret. The data base initially organized by Mittmann was based on faunal lists presented by the participants for the countries investigated. It was subsequently revised by all workshop participants at the Schloss Reisensburg and further refined by M. Fortelius and S. Viranta (Helsinki) by rebuilding revised faunal lists from mammalian systematic revisions provided in the chapters that follow here. This project was supported by the Finnish Academy Science under a data basing project entitled NOW (Neogene Old World; M. Fortelius, Principal Investigator), which is a direct outgrowth of this workshop. Also, we wish to acknowledge that our project closely parallels, and we hope compliments, two other similar projects on the evolution of Phanerozoic ecosystems (Behrensmeyer et al. 1992) and on Neogene Paleoclimate and Evolution (Vrba et al. 1995).

The authors of this volume do not present their work as the "final word" on this subject, but rather as a stimulus for further investigations on the Old World Neogene. There are no neat answers to the issues investigated throughout this volume, but the revision of the mammalian evolutionary record, placed within a more highly resolved chronologic framework, has provided a sounder basis for interpreting the complex biological phenomena we approach here.

LITERATURE CITED

Behrensmeyer, A. K., J. D. Damuth, W. A. Di Michele, R. Potts, H.-D. Sues, and S. L. Wing, eds. 1992. *Terrestrial Ecosystems Through Time.* Chicago: University of Chicago Press.

Bernor, R. L. 1978. *The Mammalian Systematics, Biostratigraphy, and Biochronology of Maragheh and Its Importance for Understanding Late Miocene Hominoid Zoogeography and Evolution.* Ph.D. diss., University of California, Los Angeles.

——. 1983. Geochronology and zoogeographic relationships of Miocene Hominoidea. In *New Interpretations of Ape and Human Ancestry,* ed. R. L. Ciochon and R. S. Corruccini, pp. 21–64. New York: Plenum.

——. 1984. A zoogeographic theater and biochronologic play: The time/biofacies phenomena of Eurasian and African Miocene mammal provinces. *Paléobiologie Continentale* 14:121–42.

——. 1985. Neogene palaeoclimatic events and continental mammalian response: Is there global synchroneity? *Sud-Afrikaanse Tydskrif-Wetenskap* 81:261.

Bernor, R. L., V. Fahlbusch, M. Fortelius, G. Daxner-Höck, F. Rögl, F. F. Steininger, and L. Werdelin. This volume. The evolution of Western Eurasian later Neogene mammal faunas: A chronologic, systematic, biogeographic, and paleoenvironmental synthesis.

Bernor, R. L., J. Kovar-Eder, D. Lipscomb, F. Rögl, S. Sen, and H. Tobien. 1988. Systematics, stratigraphic, and paleoenvironmental contexts of first-appearing Hipparion in the Vienna Basin, Austria. *Journal of Vertebrate Paleontology* 8:427–52.

Bernor, R. L. and P. P. Pavlakis. 1987. Zoogeographic relationships of the Sahabi large mammal fauna (early Pliocene, Libya). In *Neogene Geology and Paleontology of Sahabi*, ed. N. T. Boaz, J. de Heinzelin, W. Gaziry, and A. El-Arnauti, pp. 349–84. New York: Liss.

Bernor, R. L., M. O. Woodburne, and J. A. Van Couvering. 1980. A contribution to the chronology of some Old World Miocene faunas based on hipparionine horses. *Géobios* 13:25–59.

Crusafont, M. and J. Villalta. 1954. Ensayo de sintesis sobre el Mioceno de la meseta castellana. *Boletín de la Real Sociedad Española de Historia Natural (Madrid)* 1954:215–27.

Fahlbusch, V. 1991. The meaning of MN zonation: Considerations for a subdivision of the European continental Tertiary using mammals. *Newsletters on Stratigraphy* 24:159–73.

Fortelius, M., L. Werdelin, P. Andrews, R. L. Bernor, A. Gentry, H.-W. Mittmann, and S. Viranta. This volume. Provinciality, diversity, turnover, and paleoecology in land mammal faunas of the later Miocene of Western Eurasia.

Gaudry, A. 1865. *Animaux fossiles et géologie de l'Attique*. Paris.

——. 1873. *Animaux fossiles du Mont Lebéron*. Paris.

Koenigswald, G. H. R. von. 1929. Bemerkungen zur Säugetierfauna des rheinhessischen Dinotheriensandes. *Senckenbergiana* 11:267–79.

Lindsay, E. H., V. Fahlbusch, and P. Mein, eds. 1989. *European Neogene Mammalian Chronology*. New York: Plenum.

Mein, P. 1975. Résultats du Groupe de Travail des Vertébrés. In *Report on Activity of the RCMNS Working Groups (1971–1975)*, ed. J. Senes, pp. 78–81. Bratislava: SAV.

——. 1979. Rapport d'activité du Groupe de Travail des Vertébrés: Mise à jour de la biostratigraphie du Néogène basée sur les mammifères. *Annales Géologie Pays Hellénica* 1979:1367–72.

——. 1989. European mammal correlations. In E. H. Lindsay, V. Fahlbusch, and P. Mein, eds., *Topics on European Mammalian Geochronology*, pp. 73–90. New York: Plenum.

Rögl, F. and F. F. Steininger. 1983. Vom Zerfall der Tethys zur Paratethys. *Annalen des Naturhistorischen Museums, Wien* 85:135–63.

Swisher, C. C., III. This volume. New ^{40}Ar/^{39}Ar dates and their contribution toward a revised chronology for the late Miocene nonmarine of Europe and West Asia.

Thaler, L. 1966. Les Rongeurs fossiles du Bas-Languedoc dans leurs rapports avec l'histoire des faunes et la stratigraphie du Tertiary d'Europe. *Memoire Museum National Histoire Naturelle*, Series C 17:1–295.

Thenius, E. 1951. Die jungtertiäre Säugetierfauna des Wiener Beckens in ihrer Beziehung zu Stratigraphie und Ökologie. *Erdölztg* 5:52–54.

Tobien, H. 1967. Subdivision of Pontian mammal faunas. *Committee on Mediterranean Neogene Stratigraphy, Proceedings IV Session, Bologna, 1967*, pp. 1–5. Giornale di Geologia 35.

——. 1986. Die jungtertiäre Fossilgrabungsstätte Höwenegg im Hegau (Süwestdeutschland): Ein Statusbericht. *Carolinea* 44:9–34.

Vrba, E. 1985. Environment and evolution: Alternative causes of the temporal distribution of evolutionary events. *Suid-Afrikaanse Tydskrif Wetenskap* 81:229–36.

Vrba, E., G. H. Denton, T. C. Partridge, and L. H. Burckle, eds. 1995. *Paleoclimate and Evolution with Emphasis on Human Origins*. New Haven: Yale University Press.

Woodburne, M. O., R. L. Bernor, and C. C. Swisher III. This volume a. An appraisal of the stratigraphic and phylogenetic bases for the "Hipparion" Datum in the Old World.

Woodburne, M. O., G. Theobald, R. L. Bernor, H. König, C. C. Swisher III, and H. Tobien. This volume b. Advances in the geology and stratigraphy at Höwenegg, southwestern Germany.

I

GEOLOGICAL BACKGROUND

2

Circum-Mediterranean Neogene (Miocene and Pliocene) Marine—Continental Chronologic Correlations of European Mammal Units

F. F. STEININGER, W. A. BERGGREN, D. V. KENT,
R. L. BERNOR, S. SEN, AND J. AGUSTI

This paper is an update of Steininger et al. 1989. It is mainly concerned with the chronologic correlation of the Miocene and Pliocene European Mammal Biozonations, the "European Neogene Mammal Faunal Zones" (MN zones, for discussion see Steininger et al. 1989) and the "European Mammal Faunal Units" (Agenic, Orleanic, etc., of Fahlbusch 1975; for discussion see Steininger et al. 1989), with the new Geomagnetic Polarity Time Scale (Cande and Kent 1992; in press) and the revised Geomagnetic Polarity Time Scale, Cenozoic Geochronology, Chronostratigraphy, and Planktonic Foraminifera Zonation (Berggren et al., in press).

Within the last years rapid progress has been made in dating mammal-bearing Miocene and Pliocene sediments, either by radioisotopic dating (see also Swisher et al., this volume) and/or by geomagnetic dating (see also the following articles in this volume: Bernor et al.; Kappelman et al.; Sen; and Woodburne et al.). These recent results have been incorporated into this new data base. For a more recent compilation of the Pleistocene record see Agusti 1991 and Agusti et al. 1987.

This paper could not have been written without the continuous input of comments and new data by the following colleagues, who, in a more correct sense, are coauthors of this paper and are listed (in alphabetical order) gratefully below: Jean-Pierre Aguilar (Montpellier, France); Hans de Bruijn (Utrecht, Netherlands); Gudrun Daxner-Höck (Wien, Austria); Volker Fahlbusch (München, Germany); Leonard Ginsburg (Paris, France); Kurt Heissig (München, Germany); K. Kowalski (Kraków, Poland); G. D. Koufos (Thessaloniki, Greece); Pierre Mein (Lyon, France); Jacques Michaux (Montpellier, France); Michael Rasser (Wien, Austria); Fred Rögl (Wien, Austria); Danilo Torre (Firenze, Italy); Engin Ünay (Ankara, Turkey).

The Neogene (Miocene and Pliocene) Geomagnetic Polarity Time Scale, Geochronology, Chronostratigraphy, and Planktonic Foraminifera Magnetobiochronology

Neogene Chronostratigraphy

A three-fold subdivision of the Miocene is generally accepted by most stratigraphers (Berggren et al. 1985a, b). The relationship between calcareous plankton biostratigraphy and standard chronostratigraphic units was discussed at length in Berggren et al. 1985a and 1985b, and the reader is referred to that source for background information. We review here only several (minor) adjustments and/or studies that have been made since 1985.

The Paleogene/Neogene Working Group of the IUGS Neogene Subcommission (under the leadership of Fritz F. Steininger) has been focusing on the definition of a GSSP (Global Stratigraphic Section and Point) for the Neogene and the Paleogene/Neogene boundary for over a decade. It has recently been decided (Steininger et al. 1994) to recommend that the GSSP be located at the 35-m level of the Rigoroso Formation in the Lemme-Carrosio section of NE Italy corresponding to the base of Chron C6Cn.2n and the FAD of *Globorotalia kugleri* with an age estimate of 23.8 Ma (Cande and Kent, in press; Berggren et al., in press; Hodell and Woodruff 1994). For a somewhat different point of view see Srinivasan and Kennett (1983).

The Aquitanian/Burdigalian Boundary The Aquitanian/Burdigalian boundary in the Contessa section (Gubbio, Italy) has been correlated with the FAD of *Globigerinoides altiaperturus* (Iaccarino 1985), and the Burdigalian/

Langhian boundary with the FAD of *Praeorbulina sicana* (Cita and Blow 1969).

The Burdigalian/Langhian Boundary   The base of the Langhian Stage was subsequently said (Cita and Blow 1969) to correspond to the FO of *Praeorbulina bisphericus* (vel *sicana*), which is stratigraphically slightly lower in order than the FO of *Praeorbulina glomerosa* (Blow 1969; Cita and Premoli Silva 1968; Jenkins et al. 1981). We use the FO of *Praeorbulina sicana* (not *bisphericus*) to denote the base of the Langhian Stage.

The Langhian/Serravallian Boundary   The Langhian/Serravallian boundary remains somewhat controversial (see discussions in Berggren et al. 1985a, b; Iaccarino 1985). Originally defined so as to coincide with the LAD of *Praeorbulina glomerosa* (Cita and Premoli Silva 1968), which occurs within the interval of Zone N 9 (Blow 1969), it was subsequently defined to coincide with the N 10/11 boundary of Blow (1969) in Cita and Blow 1969. Iaccarino (1985) draws attention to the fact that the upper limit of the type Langhian coincides with the top of the Cessole Formation as originally defined by Cita and Premoli Silva (1960), coincident in turn, with the top of their *Orbulina suturalis* Zone, a level that is within the upper part of Zone N 9 of Blow (1969). Iaccarino (1985) accordingly equated the Langhian/Serravallian boundary with the LAD of *Praeorbulina glomerosa*, which she indicates is approximately correlative with the upper part of the *Orbulina suturalis* Zone, which is correlative, in turn, with the upper part of Zone N 9. Our data indicate that these events occur essentially simultaneously and coincident with the FAD of *Globorotalia peripheroacuta* (= Zone N 9/10 boundary of Blow 1969, 1979 and M 6/7 boundary of Berggren et al., in press), and we have thus drawn the Langhian/Serravallian boundary coincident with the Zone N 9/10 (= M 6/7) boundary. This level would appear to be consistent with, and equivalent to, the base of the Serravallian type section as (re)defined by Vervloet (1966) and correlated with the base of his *Globorotalia menardii* (including *Gt. praemenardii*) Zone. The FAD of *Gt. praemenardii* coincides essentially with the base of Zone N 10 (Bolli and Saunders, 1985).

The Serravallian/Tortonian Boundary   The upper limit of the Serravallian Stage, as redefined by Boni (1967) and Mosna and Micheletti (1968) in the Gavi section, is older than the FAD of *Globorotalia lenguaensis*, whereas the Arguello-Lequio parastratotype section of Cita and Promoli Silva (1968) includes a *Globorotalia mayeri–Gt. lenguaensis* Zone in its upper part, indicating that the upper stratigraphic limit of the Serravallian Stage is within Zone N 14 (Blow 1969) = M 11 of Berggren et al. (in press). There is thus a short gap between the top of the Serraval-

lian and the base of the Tortonian that lies within (the lower part of) Zone N 15 (= M 12) and perhaps upper N 14 (= M 11).

The Tortonian/Messinian Boundary   The lower boundary of the Messinian Stage has undergone several modifications over the past forty years (compare Gianotti 1953; Selli 1960; d'Onofrio et al. 1975; Colalongo et al. 1979); Colalongo et al. (1979) suggest that the GSSP of the Tortonian/Messinian boundary stratotype be linked with the FAD of *Globorotalia conomiozea* in the Falconara section in Sicily. In this study we follow this convention.

The Miocene/Pliocene Boundary   This boundary is equated here with the base of the Zanclean Stage as stratotypified at Capo Rosello in Sicily (Cita 1975; Hilgen 1991; Hilgen and Langereis 1993; Langereis and Hilgen 1991), which appears to be bracketed by the FADs of *Globorotalia tumida* and *Gt. sphericomiozea* (below) and the FADs of *Ceratolithus acutus* and *G. puncticulata* and the LAD of *Discocaser quinqueramus* (above). An alternative point of view suggests that in view of the Perimediterranean lithologic unconformity between the nonmarine Messinian (below) and marine Zanclean (above) sediments and the difficulty of biostratigraphically extending the lithostratigraphic base of the Zanclean Stage away from its stratotype area, a boundary stratotype section should be sought in a continuous marine section outside the Mediterranean (for further discussion see Benson and Hodell 1994; Berggren et al., in press). For the purpose of this paper we follow the commonly accepted usage of base Pliocene = base Zanclean *sensu* Cita 1975 (see also Hilgen and Langereis, 1993).

Paratethys Chronostratigraphy

Correlation of the Paratethys Neogene chronostratigraphic stage system has undergone an extensive revision by Rögl et al. (1993) and Rögl and Daxner-Höck (this volume) for upper middle Miocene, upper Miocene, and Pliocene. An extensive compilation including new radioisotopic ages for the Miocene of the Eastern Paratethys was published lately by Chumakov et al. (1984; 1992a, b) and a compilation of the geomagnetic calibration of the Sarmatian s.l. mammal localities has coincidently been published by Pevzner and Vangengeim (1993). Only the most pertinent results will be summarized here.

Central Paratethys   The base of the Sarmatian Stage sensu SUESS is recalculated by Rögl et al. (1993) and Rögl and Daxner-Höck (this volume) at 13.6 Ma; the base of the Pannonian Stage by Rögl et al. (1993) and Rögl and Daxner-Höck (this volume) at 11.5 Ma; the base of the Pontian Stage by Rögl et al. (1993) and Rögl and Daxner-

Höck (this volume) at 7.1 Ma; and the base of the Dacian Stage by Rögl et al. (1993) and Rögl and Daxner-Höck (this volume) at 5.6 Ma.

Eastern Paratethys　The confusion in correlation of middle to late Miocene between the Central and the Eastern Paratethys arises because of inappropriate use of the Sarmatian Stage in the Eastern Paratethys; its stratotypic characterization is actually in the Central Paratethys. The base of the Eastern Paratethys Sarmatian "Stage" (sensu lato) in the Eastern Paratethys has been calibrated by Rögl et al. (1993) and Rögl and Daxner-Höck (this volume) as being 13.6 Ma; however the top of the Sarmatian s.l. is calibrated by Rögl et al. (1993) and Rögl and Daxner-Höck (this volume) as being 9.5 Ma. This Sarmatian s.l. of the Eastern Paratethys is subdivided into the following stages: Volhynian Stage: base by Rögl et al. (1993) and Rögl and Daxner-Höck (this volume) at 13.6 Ma; Bessarabian Stage: base by Rögl et al. (1993) and Rögl and Daxner-Höck (this volume) at 12.2 Ma; and Khersonian Stage: base by Rögl et al. (1993) and Rögl and Daxner-Höck (this volume) at 10.2 Ma. Above the Sarmatian s.l. follows the Maeotian Stage: base by Rögl et al. (1993) and Rögl and Daxner-Höck (this volume): 9.8 Ma, followed by the Pontian Stage: base by Rögl et al. (1993) and Rögl and Daxner-Höck (this volume) at 7.1 Ma; and the Kimmerian Stage: base by Rögl et al. (1993) and Rögl and Daxner-Höck (this volume) at 5.4 Ma. With this calibration the duration of the Pontian in the Central Paratethys seems to be somewhat shorter (0.2 m.y.) than the Pontian Stage as used in the Eastern Paratethys. This calibration of Rögl et al. (1993) and Rögl and Daxner-Höck (this volume) is based on biostratigraphic correlations and in accordance with the radioisotopic ages published by Chumakov et al. (1984; 1988; 1992a, b).

Pevzner and Vangengeim (1993) have summarized lately the geomagnetic results of the middle (= Bessarabian) and late (= Khersonian) Sarmatian mammal localities of the Eastern Paratethys. In their figures 5 and 6, the boundary of the early Sarmatian = Volhynian and the middle Sarmatian = Bessarabian falls into the lower part of Chron 11 = Chron C5An.2n. The base of subchron C5An.2n according to Berggren et al. (in press) is at 12.40 Ma, and the top at 12.18 Ma. The top of the Bessarabian, and the boundary between the Bessarabian and the Khersonian (= late Sarmatian), would fall into the upper third of Chron 10 = Chron C5r. In their figures 5 and 6, Chron 10 is shown to be a completely reversed zone. However, the geomagnetic pattern of their Eldari section (= their fig. 3) shows for Chron 10 the following pattern from base to top: longer normal = N-1-short reversal = R-1-normal = N-2-longer reversal = R-2-short normal = N-3-longer reversal = R-3, followed by the long normal of Chron 9 = subchron 5n.2n. According to the GPTS of Berggren et

al. (in press) this geomagnetic pattern can be interpreted as follows: N-1 = C5An.1n; R-1 = C5r.3r; N-2 = C5r.2n; R-2 = C5r.2r; N-3 = C5r.1n; and R-3 = C5r.1r. The Bessarabian/Khersonian boundary in this section is drawn for biostratigraphic reasons below normal N-3 = C5r.1n: base at 11.09 Ma and top at 11.05 Ma. These ages are in good accordance with Rögl and Daxner-Höck (this volume), which we follow here. The Khersonian spans the upper part of Chron 10 = approximately Chron C5r.2r, C5r.1n, C5r.1r, Chron 9 = Chron C5n.2n up to the lowermost reversed part of Chron 8 = Chron C5n.1r. The Khersonian/Maeotian boundary would then fall at the Chron C5n.1r/C5n.1n boundary, with an age of 9.88 Ma, according to Berggren et al. (in press).

For the discussion concerning the proposed European Continental Chronostratigraphic Stages see discussion in Steininger et al. 1989 (pp.24 ff).

Chronology of Neogene Chronostratigraphy

The *Paleogene/Neogene*, or the *Oligocene/Miocene*, boundary, as calibrated to Chron C6Cn.2n, has an estimated age of 23.8 Ma and corresponds to the FAD of *Globorotalia kugleri* (Berggren et al. 1985; Cande and Kent 1992, 1994; Berggren et al., in press). The early Miocene is biostratigraphically bracketed by the FAD of *Gt. kugleri* (in Hodell and Woodruff 1994: slightly older: 23.99 Ma) and the FAD of *Praeorbulina sicana* (C5CN.1n: 16.49 Ma), giving the early Miocene a duration of 7.4 Ma.

The Aquitanian/Burdigalian boundary has been correlated with the LAD of *Globorotalia kugleri* (= Zone N 4/5 boundary of Blow 1969, 1979; = M 1/2 boundary of Berggren et al., in press), but it has also been correlated with the FAD of *Globigerinoides altiaperturus* in Mediterranean and Aquitaine Basin stratigraphies (see discussion in Montanari et al. 1991). The LAD of *Gt. kugleri* has been observed in Chron C6Ar (with an estimated age of 21.5 Ma: Cande and Kent, in press; Berggren et al., in press) in Hole 516F, whereas it has been recorded at the base of C6An (with an estimated age of 21.32 Ma; Cande and Kent, in press; Berggren et al., in press) in the Contessa Highway section (Montanari et al. 1990; 1991) and with an estimated age of 21.07 Ma in Hole 289 (Hodell and Woodruff 1994). The FAD of *Globigerinoides altiaperturus* has been recorded at the top of C6An (with an estimated age of 20.52 Ma [Cande and Kent, in press]) in Hole 516F, whereas it has been recorded in the older part of C6r (with an estimated age that is indistinguishable from that of the record at Hole 516F: 20.5 Ma, in view of the fact that the entire C6r is only 0.387 kyrs long) in Cande and Kent's chronology (1993 unpublished); Berggren et al. (in press) on the Contessa Highway section (Montanari et al. 1991).

Accordingly, we would recommend correlation of the Aquitanian/Burdigalian boundary with the top of Chron

C6An (with an estimated age of 20.52 Ma; (Cande and Kent 1992, 1993; Berggren et al. 1995); this procedure would facilitate regional/global correlation and would correspond with generally accepted practice in Mediterranean stratigraphies where Neogene chronostratigraphy is rooted (Van Couvering and Berggren 1977). We note, in passing, the close temporal correspondence between the estimated ages of Chron C6An in the chronologies of Cande and Kent (in press), Berggren et al. (1995), and Montanari et al. (1991). The latter have obtained an isochron age of 21.17 ± 0.23 Ma based on ^{40}Ar/^{39}Ar dating of plagioclase on an ash termed the Livello Rafaello, about a half meter below the LAD of *Gt. kugleri* and the base of C6An. Chron C6An has an estimated age of 20.518–21.320 Ma in the magnetochronology of Cande and Kent (in press).

The Burdigalian/Langhian boundary (= FAD of *Praeorbulina glomerosa*) is calibrated to the base of C5C.2n: 16.49 Ma (Cande and Kent 1992, 1994; Berggren et al., in press).

The middle Miocene is biostratigraphically bracketed by the FAD of *Praeorbulina sicana* and a level within Zone M 12 (i.e., between the LAD of *Neogloboquadrina mayeri* and the FAD of *N. acostaensis* (cf. Berggren et al. 1985a, b, in which the middle/upper Miocene boundary was pragmatically, but incorrectly, correlated with the FAD of *N. acostasensis*). The restoration of Zone M 12 = N 15 (Berggren 1993) with an estimated duration of about 0.5 m.y. between 11.4 and 10.9 Ma (Berggren et al. 1995) suggests that the middle/late Miocene (Serravallian/Tortonian) boundary is in Chron C5r.2r at 11.2 Ma.

The Langhian/Serravallian boundary is correlated here with the FAD of *Globorotalia peripheroacuta* at the top of C5Bn.1n: 14.8 Ma (Cande and Kent 1992, 1995; Berggren et al., in press). The middle Miocene has a time span of about 5.3 Ma (16.49–11.2 Ma).

The late Miocene (Tortonian and Messinian Stages) is bracketed biostratigraphically by a level within Zone M 12 (= Zone N 15, 7E11.2 Ma) to a level slightly higher/younger than the FAD of *Globorotalia tumida* (M 14/Pl 1 boundary) and/ or *G. sphericomiozea*, which lies within Chron C3r with an estimated age of 5.6 Ma (Berggren et al., in press).

The Tortonian/Messinian boundary is correlated here with the FAD of *Globorotalia conomiozea*, which has been magnetobiostratigraphically correlated in Crete with Chron C3Br.1r and has an estimated (astrochronologic) age of 7.1 Ma (Krijgsman et al. 1994). Calibration to Cande and Kent 1992 yields an age estimate of 6.92 Ma and to the chronology subsequently derived by Cande and Kent (1995) and adopted here a magnetochronologic age estimate of 7.12 Ma, essentially identical to the age estimate of Krijgsman et al. (1994). A slightly younger position for the FAD of *G. conomiozea* in Chron C3Bn has been reported by Benson and Rakic-El Bied (1991) in the Vera Basin of Spain, which would place this event at approximately 7.0 Ma in the chronology adapted here. Thus the astrochronologic and the magnetochronologic scales are seen to be coherent and concordant back to Chron C3Br, at approximately 7.0 Ma. At the same time we note that the late Miocene (Chron 6) carbon shift has been observed to start near the base of Chron 6 in Hole 588 (Hodell and Kennett 1986) = Chron C3Br.2r (7.20 Ma in Cande and Kent 1992; 7.4 Ma in Cande and Kent 1995; 7.34 Ma in Berggren et al., 1995).

An age of 7.26 ± 0.1 Ma for the Tortonian/Messinian boundary in the Northern Apennines of Romagna was recently suggested by Vai et al. (1993) based on a K/Ar (biotite) and a ^{40}Ar/^{39}Ar (plagioclase) date of 7.33 Ma on volcanogenic horizons a few meters below the FAD of *Globorotalia conomiozea* and *Gt. mediterranea* and a K/Ar (biotites) date of 7.72 ± 0.15 Ma on the stratigraphically lower FAD of *Globorotalia suterae* (which agrees closely with the magnetochronologic estimate for this datum event proposed here; cf. the estimate of 5.6 Ma and correlation to Chron C3An for the Tortonian/Messinian boundary by Langereis and Dekkers [1992]). The late Miocene thus has a span of about 6 m.y. (11.2–5.3 Ma). In a recent integrated magnetobiostratigraphic study of the Sorbas (Andalusia, Spain) and Caltanisetta (Sicily, Italy) Basins the evaporitic phase ("salinity crisis" of the Mediterranean) has been shown to be restricted to Chron C3r (Gilbert reversed) and to have a duration of approximately 0.57 m.y.—from 5.89 to about 5.32 Ma in the chronology of this paper (Gaultier et al. 1994). The implication of these results is that the pre-evaporitic Messinian (7.12 to 5.8 Ma: 1.32 m.y.) represents about two-thirds of the duration of the Messinian Age itself (7.12 to 5.32 Ma: or 1.8 m.y.), whereas the "late"/evaporitic Messinian represents but one-third the duration of the Messinian Age (5.89 to 5.32 Ma: 0.57 m.y.).

There is considerable debate regarding the adoption of an astronomically (Hilgen and Langereis 1989) versus a magnetostratigraphically (Cande and Kent 1992; 1994) based Neogene time scale, but this is beyond the scope of this paper. At the present time the two scales have been reconciled and are consistent back to the Gilbert (= C3n + C3r)/C3An boundary at 5.89 Ma. The situation has been discussed at length in Berggren et al. (1995), to which the reader is referred.

Neogene Planktonic Foraminiferal Magnetobiochronology

Miocene   About seventy planktonic foraminiferal datum levels have been identified in the Miocene, many of which were already recognized in Berggren et al. 1985a, b. We have (re)calibrated these datum levels following the revised magnetochronology of Cande and Kent (1992; 1995) (see also Berggren et al., 1995) and added several more. In some instances revisions/reinterpretations of magnetostratigraphy have resulted in revised age estimates for several datum levels as well. In general we have found our current analysis and/or evaluation of new (post 1985) date to be quite consistent with earlier interpretations and/or calibrations. The major difference between Berggren et al. 1995 and Berggren et al. 1985a, b is the more secure documentation of the regional correlation of the biogeographically overlapping zonal schemes adopted here within a more precise magnetochronologic framework.

Early Miocene   We have recognized some twenty-two datum events spanning the 7.3 Ma interval of the early Miocene (23.8–16.49 Ma), or an average of about 2.7 events/1 m.y. Zones M 1 (2.24 Ma), M 2 (2.75 Ma), and M 3 (1.50 Ma) can be contrasted with the much shorter Zones M 4 and M 5 (and indeed middle Miocene Zones M 6, M 8–10 as well), reflecting the late/early to early/middle Miocene global warming trend and concomitant flurry of speciation events in (sub)tropical environments, which allows fine-scaled biostratigraphic subdivision. Of some consternation is the (continuing) lack of a direct magnetobiostratigraphic correlation of the FAD of *Globigerinatella insueta* (which defines the base of Zone M 3).

Middle Miocene   Some twenty-five datum events have been identified in the middle Miocene (16.4–11.2 Ma) interval of 5.2 Ma, or an average of 4.6 events/1 m.y., which provides the highest degree of biostratigraphic resolution for the Cenozoic except for the Pliocene, where nearly forty-five different types of biostratigraphic events spread over the ca. 3 m.y. extent of the Pliocene provides some fifteen datum events/1 m.y. (Berggren et al. 1995). The warming trend responsible for the high degree of middle Miocene biostratigraphic resolution may be contrasted with the accelerated (and punctuated) cooling trends of the Pliocene responsible for the relatively rapid LADs of numerous (predominantly Miocene) taxa, the various biogeographic immigration/disappearance events and FADs of several taxa.

Late Miocene   Some twenty-three datum events have been recognized in the 5.9 Ma interval of the late Miocene (11.2–5.3 Ma), or an average of events of about 3.9/1 m.y. Of particular significance is the replacement of the *Globorotalia merotumida-plesiotumida*–group by *Globorotalia lenguaensis* in subdividing the upper Miocene of (sub)tropical and transitional regions, which provides more confident calibration to the GPTS. The joint occurrence of the FAD of *Globorotalia sphericomiozea* and *Globorotalia tumida* in Hole 519 has provided a means for regional correlation between (sub)tropical and transitional regions

at the Miocene/Pliocene boundary (5.2 Ma in the chronology of Cande and Kent 1992; 5.3 Ma in Cande and Kent 1995 and Berggren et al. 1995).

Plio/Pleistocene   A comprehensive compilation has recently been made of some forty-five Pliocene and eight Pleistocene planktonic foraminiferal datum events in connection with a larger review of the current status of late Neogene (Plio/Pleistocene) astro- and magnetobiochronology (Berggren et al. 1995), and the interested reader is referred to that source for additional information. We adopt the current three-fold scheme for the Pliocene: Zanclean (early), Piacenzian (middle), and Gelasian (late) (Rio et al. 1994).

Correlation and Ages of Continental European Mammalian Biostratigraphic Units: "Neogene Mammal Faunal Zones" (MN Zones) and "Neogene Mammal Faunal Units"

For a general discussion of the philosophy of these mammalian biostratigraphic units, the "Neogene Mammal Faunal Zones" (the MN zones) and the "Neogene Mammal Faunal Units" (as there are: Agenian, Orleanian, etc.), see Steininger et al. 1989:20ff.

Here we discuss only the results of the present correlation to the different chronostratigraphic units, the geomagnetic time scale, and their most probable ages implied by these correlations according to Berggren et al. 1995. The biostratigraphic definition of the Faunal Zones (MN zones) and the Faunal Units is taken in general from Bruijn et al. 1992; Agusti et. al. 1986, 1991; and Berger 1992.

In our data base we have arranged the mammal localities: (1) according to their biochronological position and ages within the specific Mammal Zone, following in general the biochronologic arrangements of mammal localities by de Bruijn et al. (1992; see also their comments on pp. 69–71) and comments of various colleagues; and (2) followed by mammal localities that cannot be specifically arranged within the particular Mammal Zone.

Agenian (MN 1 to MN 2) (fig. 2.1)

The base of the Agenian (= base of MN 1) is best estimated with an age of: 23.8 Ma and correlates with the Oligocene/Miocene = the Paleogene/Neogene boundary as proposed by Steininger et al. (1994). This correlation is based on the geomagnetic calibration of the locality Torrente del Cinca 68 (Ebro Basin, Spain). The top is best geomagnetically calibrated by the upper MN 1 localities Findreuse 3, 4, 22, 27, 31, 33 and Fornant 13 (11) (Haute-Savoie, France) to an age of approximately 22.8 Ma.

EARLY MIOCENE TIME SCALE

TIME (Ma)	CHRONS	POLARITY	EPOCH	AGE	PLANKTONIC FORAMINIFERA (SUB)TROPICAL — Berggren (this work)	Blow (1969)	CENTRAL PARATETHYS STAGES	FAUNAL UNITS	EUROPEAN FAUNAL ZONES	CHRONOSTRAT. STAGES
	C5ADn (n)			SER.	M7 — *Gl. peripheroacuta* Lin. Z	N10		ASTARACIAN	MN-6	
15	C5ADr / C5Bn 1 n/r 2n	1n / 2n	MIDDLE	LANGHIAN	M6 — *O. sutur. – Gl. peripher.* IZ	N9	BADENIAN	— 15.2	— 15.2	ARAGONIAN
16	C5Br				M5 b — *Pr. glomerosa – Orb. suturalis* ISZ; *Pr. sicana – Orb. suturalis* IZ	N8				
	C5Cn 1/2/3n	1n / 2n / 3n			M5 a — *Pr. sicana – Pr. glomerosa* ISZ		— 16.4	16.5	16.5 MN-5	
17	C5Cr				M4 b — *G. bispherica – PRSZ*; a — *Cat. dissimilis – Gl. birnageae* ISZ; *C. dissimilis – Pr. sicana* IZ	N7	KARPATIAN		— 17.0	
	C5Dn (n)			BURDIGALIAN			— 17.2		MN-4	
18	C5Dr				M3 — *Globigerinatella insueta – Catapsydrax dissimilis* Conc. RZ	N6	OTTNANGIAN — 18.0	ORLEANIAN	— 18.0	— 13.0
	C5En (n)									
19	C5Er						EGGENBURGIAN		MN-3	
	C6n (n)		MIOCENE / EARLY		M2 — *Catapsydrax dissimilis* IZ	N5				
20	C6r						— 20.5	— 20.0	— 20.0	
21	C6An 1 n/r 2n	1n / 2n								RAMBLIAN
	C6Ar			AQUITANIAN					MN-2	
22	C6AAn (n) / C6AAr 1/2/3r	1n / 2n			M1 b — *Gl. kugleri – Gl. dehiscens* Conc. RSZ; *Gl. kugleri* TRZ	N4	EGERIAN	AGENIAN		
23	C6Bn 1 2n	1n / 2n							— 22.8	
	C6Br				M1 a — *Gl. primordius* ISZ				MN-1	
24	C6Cn 1/2/3n	1n / 2n / 3n	OLIGOCENE / LATE	CHATTIAN	P22 — *Gl. ciperoensis* IZ	P22		— 23.8	— 23.8 MP-29	— 23.8 "OCTOGESIAN"
	C6Cr									

FIGURE 2.1 Early Miocene circum-Mediterranean marine-continental chronologic correlations of the European Mammal Units and Zones.

The base of MN 2 (fig. 1) is best estimated with an age of 22.8 Ma based on biostratigraphic ties and "grade dating" of Gans and Aillas (Bordeaux Basin, France); as is the top by localities Balizac, La Brete, and Laugnac (Bordeaux Basin, France), still intercalated into the type Aquitanian sediments and "grade dated" around 19.8 Ma with an age of 20.0 Ma.

Orleanian (MN 3 to MN 5) (fig. 2.1)

The base of Orleanian (= base of MN 3) is best estimated with an age of 20.0 Ma. This estimate is based mainly on unpublished strontium dates (87Sr/86Sr: 0.708581; Scharbert and Steininger, in prep.) from Maigen (Molasse Zone, Austria) and the litho-biostratigraphic correlations of Lisboa (Portugal), Estrepouy, and Laugnac (Bordeaux Basin, France) with "grade dates" circa 19 Ma. Averages for the top are best estimated at Beaulieu (France) by biostratigraphic ties and radiometric estimates of 18.7 to 17.5 Ma with an age of 18.0 Ma.

The base of MN 4 is best estimated around 18 Ma by the locality Belchatowc (Poland). The top can be estimated by the geomagnetically calibrated localities of Gemerek, Horlak 1a, 1b and 2 (Kayseri-Sivas Basin, Turkey), with an age of 17.0 Ma. However, ages younger than 17 or 16.5 Ma have to be considered in relation to the oldest well-established estimates of MN 5 and MN 6 (see below). The age of 14.1 Ma for the MN 4/5 boundary calculated by Krijgsman et al. (1994) is in contradiction with all estimated ages for this boundary so far.

The base of MN 5 is best estimated as being 17 Ma based on the localities of Gemerek, Eibiswald (Styrian Basin, Austria), Belthalów (Poland), and Teiritzberg (Korneuburg Basin, Austria).

Astaracian (MN 6 to MN 8) (figs. 2.1, 2.2)

The base of the Astaracian (= base of MN 6) depends on where the biostratigraphic position of the Devinska Nova Ves fissure fillings is placed. This locality is correlated with basal MN 6 by de Bruijn et al. (1992) and by Fejfar (pers. comm. 10/9/1994). For tectonic and paleogeographic reasons (Fejfar 1989; Rögl and Steininger 1983) those fissures could have been filled only before the Langhian (= Badenian) transgressive event. By placing the Devinska Nova Ves fissure fillings into basal MN 6, the base of the Astaracian (= base of MN 6) is estimated to be circa 16.5 Ma.

The age of the MN 6 type locality, Sansan (France), has a geomagnetically calculated age of 15.2 Ma (Sen, this volume), which Sen correlates with lowermost MN 6. The localities of Devinska Nova Ves sandhill (Slovakia), Luc sûr Orbieu (France), and Inönü I-loc. 24, 24A (Turkey; Kappelman et al., this volume) indicate ages around 15 Ma and younger for lower MN 6. The localities of Goldberg and Steinberg (Germany) have an age younger than 14.7 Ma. The age of these European lower MN 6 vertebrate localities (ca. ≤15.2 Ma) contradict the age for basal MN 6, based on tectonic and paleogeographic considerations, but are more congruent with one another than Krijgsman et al.'s (1994) 13.8 Ma determination for the MN 5/6 boundary.

As set forth in Steininger et al. (1989), it is Bernor's opinion that the base of MN 6 corresponds with the end Langhian regression ≥14.8 Ma, when an extensive biogeographic interchange occurred between Eurasia and Africa. Bernor and Tobien (1990) further recognized that MN 5 and MN 6 were distinctive by virtue of their first occurrence of immigrant species and correlated MN 5 with the terminal Burdigalian regression (ca. 16.5 Ma here) and

MN 6 yet again with the terminal Langhian regression (ca. 15 Ma). Bernor (here) still advocates these correlations and believes that Sen's estimation of Sansan's age at 15.2 Ma represents a sound age determination for the base of MN 6.

The base of MN 7 + 8 (fig. 2.2) is best estimated at 12.5 Ma by the biostratigraphically correlated localities of La Grenatiere, Santarem (Portugal), La Grive (France), St. Stephan (Austria), and Comanesti 1 (Roumania) in Volhynian sediments. The base of MN 8 is estimated by biostratigraphic ties, according to the locality C. Almirall (Spain), to have an age of 11.9 Ma. This estimated age for the MN 6/7 + 8 boundary is in good agreement with the geomagnetically calculated age of 12.5 Ma by Krijgsman et al. (1994).

Vallesian (MN 9 to MN 10) (fig. 2.2)

Based on the evolutionary history of "Hipparion" (= *Hippotherium* of Bernor et al., this volume a) in the Pannonian and Central Paratethys (= Vienna Basin; see Bernor et al. 1988, 1993a, b; Swisher, this volume; Woodburne et al., this volume) and on the correlation of the Pannonian zonation (see Rögl and Daxner-Höck, this volume), the Gaiselberg (Pannonian C, Vienna Basin, Austria) hipparion is considered to best represent the actual

MIDDLE–LATE MIOCENE TIME SCALE

TIME (Ma)	CHRONS	POLARITY	EPOCH	AGE	PLANKTONIC FORAMINIFERA (SUB)TROPICAL — Berggren (this work)	Blow (1969)	PARATETHYS-STAGES CENTRAL-	PARATETHYS-STAGES EASTERN-	FAUNAL UNITS	EUROPEAN FAUNAL ZONES	CHRONOSTRAT. STAGES
5	C3n (1, 2, 3, 4n)		PLIOCENE / EARLY	ZANCLIAN	PL1 b: Gl. cibaoensis – G. nepenthes ISZ; PL1 a: Gl. tumida – Gl. cibaoensis IRZ (G. tumida – G. nepenthes IZ)	N 19 / N18	DACIAN	KIMMERIAN	RUSCINIAN	4.2 / MN-14 / 5.3	NOT NAMED / 5.3
6	C3r; C3An (1, 2n)		MIOCENE / LATE	MESSINIAN	M14 Gl. lenguaensis - G. tumida IZ		5.6	5.4		MN-13	
7	C3Ar; C3Br				M13 b: Gl. extremus/Gl. plesiotumida - Gl. lenguaensis ISZ	N 17	PONTIAN	PONTIAN / 7.1	TUROLIAN	7.1 / MN-12	TUROLIAN
8	C4n (1, 2n); C4r (1, 2r)			TORTONIAN			7.1	MAEOTIAN		8.2 / MN-11	
9	C4An; C4Ar (1, 2, 3)				M13 a: N. acostaensis - Gl. extremus/Gl. plesiotumida ISZ	N 16	PANNONIAN			9.0 / MN-10 / 9.5	9.0
10	C5n 2n							KHERSONIAN ("SARMATIAN" s.l.)	VALLESIAN		VALLESIAN
11	C5r (2, 3r)				M12 N. mayeri – N. acostaensis IZ	N 15		9.88		MN-9	
11.5				SERRAVALLIAN	M11 G. nepenthes/N. mayeri Conc. RZ	N14	11.5	11.0	11.2	11.2	11.2
12	C5An (1, 2n)		MIDDLE		M10 Gl. f. robusta - G. nepenthes IZ; M9 b: Gl. f. robusta Tot. RZ; M9 a: Gl. f. lobata Lin. Z	N13 / N12	SARMATIAN sensu SUESS	BESSARABIAN		MN-7+8	
13	C5Ar (1, 2, 3r); C5AAn; C5AAr; C5ABn; C5ABr				M8 Gl. fohsi s.s. Lin. Z; (Gl. f. lobata - Gl. f. robusta IZ)	N11		12.2 / VOLHYNIAN / 13.6	ASTARACIAN	12.5	
14	C5ACn; C5ADn				M7 Gl. peripheroacuta Lin. Z	N 10	13.6	EASTERN PARATETHYS STAGES NOT CORRELATED		MN-6	ARAGONIAN
15	C5ADr; C5Bn			LANGHIAN	M6 O. sutur. - Gl. peripher. IZ	N9	BADENIAN		15.2	15.2	15.2
	C5Br				M5 b: Pr. glomerosa - Orb. suturalis ISZ	N8					

FIGURE 2.2 Middle and late Miocene circum-Mediterranean marine-continental chronologic correlations of the European Mammal Units and Zones.

Old World "Hipparion Datum" on combined morphologic and stratigraphic grounds. Rögl and Daxner-Höck (this volume) estimate the age of Pannonian C deposits, and the Gaiselberg hipparion, to be 11.2 Ma based on Central–Eastern Paratethys correlations, but not direct dating of Vienna Basin Pannonian C hipparions themselves. Within the Eastern Paratethys realm, the first MN 9 faunas are known from the upper Bessarabian Stage (base at 12.2, top at 11.0, for arguments see Rögl and Daxner-Höck, this volume), providing further support of this age estimation of 11.2 Ma for the base of the Vallesian.

The most precise chronologic estimates for the Old World "Hipparion" Datum are derived from the Siwaliks (ca. 10.5 Ma, Swisher, this volume; 10.45 Ma, Kappelman et al., this volume; but note Pilbeam et al. [this volume] estimate this to be 10.7 Ma) and the Sinap sequence (ca. 10.44 Ma, estimating the occurrence between Sinap localities 64 [without hipparion] and 94 [with hipparion and dated 10.38 Ma], Kappelman et al., this volume), where relatively continuous rock sequences make possible magnetostratigraphic correlations of first occurring hipparions that are directly pertinent to an Old World "Hipparion Datum" (Bernor et al., this volume; Swisher, this volume; Woodburne et al., this volume). The Höwenegg hipparions are slightly more evolutionarily derived than Pannonian C hipparions (Bernor et al., this volume; Woodburne et al., this volume) and are directly dated by ^{40}Ar/^{39}Ar at $\leq$10.3 Ma (= the maximum age of the Höwenegg fossil localities; Swisher, this volume). Bernor's position here is that until evidence is documented that a primitive hipparion is found in a magnetostratigraphically robust (long "continuous" sequence with good magnetic control) older than the lower middle portion of Chron 5's long normal interval (the interval is calibrated here as being 10.949–9.740, giving a 10.5 Ma age estimate for the lower middle portion and the "Hipparion Datum"), or in association with a precise ^{40}Ar/^{39}Ar date, it would be imprudent to accept a date older than at least the base of Chron 5n (= 10.95 Ma) and most probably lower middle portion (= 10.5 Ma). for the Old World "Hipparion Datum." Swisher (pers. comm.) and Kappelman (pers. comm.) concur with this point of view. The base of MN 10 (fig. 2.2) is best estimated to have an age of 9.3 Ma, following the geomagnetic interpretation of the locality Bou Hanifia BH 5 (Algeria) with a calibration of either 9.75 to 9.55 or 9.55 to 9.25 Ma. There are no possibilities to establish an estimate on the top of MN 10.

Turolian (MN 11 to MN 13) (fig. 2.2)

The base of the Turolian (= base of MN 11) is best estimated to have an age of 9.0 Ma by Bernor et al. (this volume). This estimate is based on the following localities: base of Maragheh (Kopran locality, Iran) fossiliferous section estimated by Bernor et al. (this volume b) as being

9.0 Ma; Kayadibi (Turkey) with radiometric dates below mammal horizon of 9.4 Ma; Prochroma 1 (Macedonia, Greece) with geomagnetic calibrations around 9.6 to 9.3 Ma.

The base of MN 12 is best estimated to have an age of 8.24 Ma for the base of the *Hipparion prostylum* zone at Maragheh (Iran). The top of MN 12 is estimated by the localities Samos Main Bone Bed levels (MBB; Greece) radioisotopically between 7.3 and 7.1 Ma.

The base of MN 13 is best estimated with an age of 7.1 Ma. It contains localities like Samos L (Greece), with a radiometric age of 6.2 Ma, and Venta del Moro, geomagnetically calibrated at 6.55 to 5.9 Ma. The top can be estimated with an age of 5.3 Ma by the localities La Alberca (Spain)—still within marine sediments of uppermost Messinian and the Brisighella (Faenza, Italy) mammal fauna overlain by lowermost marine Zanclean sediments.

Ruscinian (MN 14 to MN 15) (figs. 2.2, 2.3)

The base of the Ruscinian (= base of MN 14) is best estimated to have an age of 5.3 Ma (figs. 2.2, 2.3). An older age could be derived from the locality Celleneuve (France), geomagnetically calibrated to 5.9 to 4.2 Ma. The top is best estimated to have an age of 4.2 Ma at the locality of Villalba Alto Rio 1 (Spain). Younger ages of approximately 3.5 Ma for this boundary are indicated by the localities Mesas de Asta (Spain) and Elbistan (Turkey). However, these ages are contradicted by the ages of numerous MN 15 localities (see below).

The base of MN 15 (fig. 2.3) is best estimated to have an age of 4.2 Ma. This age is derived from the geomagnetically dated section of Villalba Alta Rio 2 (Teruel Graben, Spain), with a calibrated age of 4.2 to 2.6 Ma and an inferred age of 3.85 to 3.4 Ma. The top can be estimated to have an age of 3.4 Ma by the localities Villalba Alta Rio 5 (Teruel Graben, Spain), geomagnetically calibrated with an age of 3.8 to 3.6 Ma and an inferred age of 3.4 Ma, and Escorihuela B (Teruel Graben, Spain), geomagnetically calibrated to an age of 4.2 to 3.6 Ma and an inferred age of 3.4 Ma.

Villanyian (= "Villafranchian"; MN 16 to MN 17) (fig. 2.3)

The base of the Villanyian is best estimated with an age of 3.4 Ma by the localities Escorihuela (Teruel Graben, Spain), with an inferred age of 3.4 Ma; Triversa (Asti, Italy), geomagnetically and biostratigraphically calibrated to an age of 4.18 to 3.2 Ma; Arcille (Grosseto, Italy), biostratigraphically younger than 3.6 Ma; Vialette (France) fauna on top of a basalt flow radioisotopically dated at 3.33 Ma. The top can be best estimated to have an age of 2.6 Ma at the localities Valdeganga 14, 15 (Júcar Basin,

PLIOCENE TIME SCALE

TIME (Ma)	CHRONS	POLARITY	EPOCH	AGE	PLANKTONIC FORAMINIFERA Berggren (1973, 1977, this work) ATLANTIC	INDO-PACIFIC	PARATETHS-STAGES CENTRAL-	EASTERN-	EUROPEAN FAUNAL UNITS	FAUNAL ZONES	CHRONOSTRAT. STAGES
	C1r 2r		PLEISTOCENE								
2	C2n	n	PLIOCENE	GELASIAN LATE	PL6 *Gt. miocenica – Gl. fistulosus* IZ	*Gt. pseudomiocenica – Gl. fistulosus* IZ	STAGES NOT CORRELATED			— 1.95 —	— 1.95 —
	C2r 1 r/n, n				PL5 *D. altispira – Gt. miocenica* IZ	*D. altispira – Gt. pseudomiocenica* IZ			VILLANYIAN = "VILLA-FRANCHIAN" p.p.	MN-17 — 2.6 —	
	2r			PIACENZIAN MIDDLE	PL4 *Sph. seminulina – D. altispira* IZ					MN-16	NOT NAMED
3	C2An 1, 1n; 2 r/r, 2n; 3n				PL3 *Gt. margaritae – Sph. seminulina* IZ				— 3.4 —	— 3.4 —	
4	C2Ar			ZANCLIAN EARLY	PL2 *G. nepenthes – Gt. margaritae* IZ		ROMANIAN	AKTSCHAGYL.		MN - 15	
	C3n 1 n/r, 1n; 2 n/r, 2n; 3 n/r, 3n; 4n				PL1 b *Gt. cibaoensis – G. nepenthes* ISZ	*Gl. tumida – G. nepenthes* IZ		— 4.2 —	RUSCINIAN	— 4.2 — MN - 14	
5					PL1 a *Gt. tumida – Gt. cibaoensis* ISZ		— 4.8 — DACIAN	KIMMERIAN			
	C3r		MIOCENE MESS. LATE		M14 *Gt. lenguaensis – Gt. tumida* I Z		— 5.6 — PONTIAN	— 5.4 — PONTIAN	TUROLIAN	— 5.3 — MN - 13	— 5.3 — TUROLIAN
	C3An1n 1n										

FIGURE 2.3 Pliocene circum-Mediterranean marine-continental chronologic correlations of the European Mammal Units and Zones.

Spain), geomagnetically calibrated to an age of 3.0 to 2.6 Ma, and Stranzendorf C (Molasse Zone, Austria), geomagnetically calibrated, horizon C with an inferred age of 3.04 to 2.58 Ma.

The base of MN 17 is best estimated to have an age of 2.6 Ma at the localities of Valdeganga 7, 10 and 2, 3 and 6 (Júcar Basin, Spain), geomagnetically calibrated with an age of 2.6 to 2.25; Rocca Neyra (Italy), with radioisotopic dates of 2.5 to 2.4 Ma; and Stranzendorf D (Molasse Zone, Austria), geomagnetically calibrated to 2.58 Ma. The top can be best estimated by the locality of Stranzendorf L (Molasse Zone, Austria), geomagnetically calibrated and with an inferred age of 1.95 Ma. Therefore, the top of MN 17 is inferred to correlate with the Plio/Pleistocene boundary.

LITERATURE CITED

Adrover, R. 1978. Les rongeurs et lagomorphes (Mammalia) du Miocène inférieur continental de Navarrete del Rio (Province de Teruel, Espagne). *Document Laboratoire Géologie Science, Lyon* 72:3–47.

Aguilar, J. P. 1974. Les rongeurs du Miocène inférieur en Bas-Languedoc et les corrélations entre èchelles marine et continentale. *Géobios* 7:345–98.

——. 1981. *Evolution des rongeurs Miocènes et paléogéographie de la Méditerranée occidentale.* Ph.D. diss., Université Montpellier.

——. 1982a. Stratigraphie—Biozonation du Miocène d'Europe occidentale à l'aide des Rongeurs et corrélations avec l'échelle stratigraphique marine. *Compte Rendus Academie des Sciences, Paris* 294:49–54.

——. 1982b. Contributions a l'étude des micromammiféres du gisement Miocène supérieur de Montredon (Herault). 2. Les rongeurs. *Palaeovertébrata* 12:81–117.

——. 1987, 1989. Vide in Steininger et. al. 1987.

Aguilar, J. P., M. Calvet, and J. Michaux. 1986. Découvertes de faunes de micromammifères dans les Pyrénées-Orientales (France) de l'Oligocène supérieur au Miocène supérieur; espèces nouvelles et réflexion sur l'étalonnage des échelles marine et continentale. *Compte Rendus Academie des Sciences, Paris* 303:755–60.

Aguilar, J. P., G. Clauzon, and J. Michaux. 1989. La limite Mio-pliocènes dans le sud de la France d'après les faunes de rongeurs; état de la question et remarques sur les datations à l'aide des rongeurs. *Bolletino della Societá Paleontologia Italiana* 28:137–45.

Aguilar, J. P. and J. Michaux. 1984. Le gisement á micromammi-fères du Mont-Helene (Pyrenees-Orientales): Apports á la connaissance de l'Histoire des faunes et de environnements continentaux implications stratigraphiques pour le Pliocène du Sud de la France. *Paléobiologie Continentale* 14:19–31.

Aguirre, E., J. Agusti, C. Castillo, and F. J. Férriz. 1992. Marine-Continental correlation in the Pliocene of the Guadalquivir Basin and the Mediterranean Margin (Spain). *First Congress R.C.A.N.S. Lisboa 1992 (October 12–15)*, pp. 11–14.

Aguirre, E. and J. Morales. 1991. Villafranchian faunal record of Spain. *Quartärpaläontologie*, p. 8.

Agusti, J. 1985. Biozonación mediante Roedores (Mammalia) del tránsito Oligoceno-Mioceno en el sector sûreste de la cuenca del Ebro. *Paleontologia i Evolucio* 18:131–49.

———. 1986. Continental Mammal Units of the Plio-Pleistocene from Spain. *Memoire Societá Italiana* 31:167–73.

———. 1991. The *Allophaiomys* Complex in Southern Europe. *Géobios* 25:133–44.

Agusti, J., P. Anadón, S. Arbiol, L. Cabrera, F. Colombo, and A. Sáez. 1987. Biostratigraphical characteristics of the Oligocene sequences of North-Eastern Spain (Ebro and Campins Basins). *Münchner Geowissenschaftliche Abhandlungen* 10:35–42.

Agusti, J., X. Barberà, L. Cabrera, J. M. Parés and M. Llenas. 1994. Magnetobiostratigraphy of the Oligocene–Miocene transition in the Ebro Basin: State of the art. *Münchner Geowissenschaftliche Abhandlungen*, Reihe A, 26:161–72.

Agusti, J., L. Cabrera, P. Anadon, and S. Arbiol. 1988. A late Oligocene–early Miocene rodent biozonation from the SE Ebro Basin (NE Spain): A potential mammal stage stratotype. *Newsletters on Stratigraphy* 18:81–97.

Agusti, J., J. Gibert, and S. Moyà-Solà. 1981. Casa del Acero: Nueva fauna turoliense de Vertebrados (Mioceno superior de Fortuna, Murcia). *Boletin Instituto Paleontologos Sabadell* 13:69–87.

Agusti, J. and S. Moyà-Solà. 1991. Spanish Neogene mammal succession and its bearing on the continental biochronology. *Newsletters on Stratigraphy* 25:91–114.

Agusti, J., S. Moyà-Solà, J. Gibert, J. Guillén, and M. Labrador. 1985. Nuevos datos sobre la biostratigrafia del Neógeno continental de Murcia. *Paleontologia Evolucio* 18:83–94.

Agusti, J., S. Moyà-Solà, and E. Martin-Suárez. 1989. Review of the late Miocene–early Pliocene mammalian faunas from eastern Spain. *Bulletino Societa Paleontologica Italiana* 28:155–60.

Agusti, J., S. Moyà-Solà, and J. Pons-moyà. 1987. La sucesión de Mamiferos en el Pleistoceno inferior de Europa: Proposición de una nueva escala bioestratigráfica. *Paleontologia i Evolucio*, Memoria Especial, 1:287–95.

Alberdi, M. T. and M. F. Bonadonna. 1987. Evaluation of lower and middle Villafranchian chronostratigraphy. *Proceedings of the VIIIth RCMNS Congress*, pp. 85–91. Annales Instituti Geologici Publici Hungarici 70.

Alberdi, M. T., C. Arias, G. Bigazzi, et al. 1982. Nuevo yacimiento de moluscos y vertebrados del Villafranguiense de la Cuenca del Jucar (Albacete, Espana). *Colloque "Le Villafranchien mediterraneen,"* pp. 255–71.

Alberdi, M. T., N. López, J. Morales, C. Sesé, and D. Soria. 1981. Bioestratigrafia y biogeograpfia de la fauna de Mamiferos de los Valles de Fuentidueña (Segovia). *Estudios geologia* 37:503–11.

Alcalá, L. and P. Montoya. 1994. Las faunas de macromamíferos del Turoliense inferior español. *Paleontologia Evolucion* 27.

Alvinerie, J., R. Anglada, M. Caralp, and F. Catzigras. 1977. *Stratotype et parastratotype de l'Aquitanien.* Paris: Editeur Centre National de Reserche Scientifique, Paris.

Alvinerie, J. and J. Gayet. 1971. Sur l'importance de la coupe de Balizac (Gironde) pour la compréhension du Miocène inférieur de la région de Villandraut (feuille d'Hostens au 1/50.000). *Bulletin du Bureaux Regional de Géologie Mediterranée*, deuxième série: 47–51.

Andreescu, I. 1981. Middle-upper Neogene and early Quaternary chronostratigraphy from the Dacic Basin and correlations with the neighbouring areas. *Annales Géologiques des Pays Helléniques*, hors série, 4:130–38.

Andrews, P. 1989. Small mammal taphonomy. In *European Neogene Mammal Chronology*, ed. E. H. Lindsay, V. Fahlbusch, and P. Mein, pp. 487–94. New York: Plenum.

Andrews, P., P. J. Whybrow and C. B. Stringer. 1980. Stratigraphy and palaeontology of Miocene deposits at Yeni Eskihisar, Turkey. *Newsletters on Stratigraphy* 9:49–57.

Antunes, M. T. 1988. The "Proboscidean datum": Evidence from the Miocene of Lisbon. Abstract NATO ARW, *European Neogene Mammal Chronology*, Schloss Reisensburg, Ülm.

———. 1989. The Proboscideans data, age, and paleogeography: Evidence from the Miocene of Lisbon. In *European Neogene Mammal Chronology*, ed. E. H. Lindsay, V. Fahlbusch, and P. Mein, pp. 253–62. New York: Plenum.

Antunes. M. T. and P. Mein. 1986. Petits mammifères du Burdigalien inferiur (Universidade Catolica, Avenida do Uruguay). *Ciencias da Terra (UNL)*, Lisboa 8:123–38.

Arambourg, C. and J. Piveteau. 1929. Les Vertébrés du Pontien de Salonique. *Annales Paléontologiques* 18:59–138.

Azanza, B., E. Menéndez and L. Alcalá. 1989. The middle-upper Turolian and Ruscinian Cervidae in Spain. "Continental faunas at the Miocene/Pliocene boundary," International Workshop, Faenza, March 28–31, 1988. *Bolletino Societá Paleontologia Italia* 28:171–82.

Azzaroli, A., C. de Giuli, G. Ficcarelli, and D. Torre. 1982. Table on the stratigraphic distribution of terrestrial mammal faunas in Italy from the Pliocene to the early middle Pleistocene. *Geofisica Quaternari.* 5:55–58.

———. 1988a. Late Pliocene to early mid-Pleistocene mammals in Eurasia: Faunal successions and dispersal events. *Palaeogeography, Palaeoclimatology, Palaeoecology* 66:77–100.

———. 1988b. Mammal succession of the Plio-Pleistocene of Italy. *Memoire Societé Geologie Italia* 21:213–18.

Bachmayer, F. and R. W. Wilson. 1984. Environmental significance and stratigraphic position of some mammal faunas in the Neogene of eastern Austria. *Sitzungsbericht der Österreichischen Akademie für Wissenschaften, mathemathisch-naturwissenschaftliche Klasse* 193:303–19.

Bandet, Y., B. Donville, and J. Michaux. 1978. Étude géologique et géochronologique du site Villafranchien de Vialette (Puy de Dome). *Bulletin Societé Géologique de France* 20:245–51.

Baubron, J.C., B. Donville, J. Magné, and M.-J. Wallez. 1975. Datation absolué du volcanisme de Beaulieu (Bouches du Rhône, France), Conséquences stratigraphiques. *Bulletin Societé Géologique de France* 17:773–76.

Becker-Platen, J. D., O. Sickenberg, and H. Tobien. 1975. Vertebraten-Lokalfaunen der Türkei und ihre Altersstellung. In *Die Gliederung des höheren Jungtertiärs und Altquartärs in der Türkei nach Vertebraten und ihre Bedeutung für die internationale Neogen-Stratigraphie*, ed. O. Sickenberg, pp. 19–45. Geologisches Jahrbuch 15.

Benda, L. and H. de Bruijn. 1982. Biostratigraphic correlations in the Eastern Mediterranean Neogene. *Newsletters on Stratigraphy* 11:128–35.

Benda, L., K. Heissig, and P. Steffens. 1975. Die Stellung der Vertebraten-Faunengruppen der Türkei innerhalb der chronostratigraphischen Systeme von Tethys und Parathetys. In *Die Gliederung des höheren Jungtertiärs und Altquartärs in der Türkei nach Vertebraten und ihre Bedeutung für die internationale Neo-

gen-Stratigraphie, ed. O. Sickenberg, 110–16. Geologisches Jahrbuch 15.

Benda, L. and J. Meulenkamp. 1979. Biostratigraphic correlations in the Eastern Mediterranean Neogene: Calibration of sporomorph associations, marine microfossil and mammal zones, marine and continental stages, and the radiometric scale. *Annales Géologiques des Pays Helléniques* 1:61–70.

——. 1990. Biostratigraphic correlations in the Eastern Mediterranean Neogene. 9. Integrated biostratigraphic and chronostratigraphic scales. *Newsletters on Stratigraphy* 23:1–10.

Benda, L., J. E. Meulenkamp, and A. van de Weerd. 1977. Biostratigraphic correlations in the Eastern Mediterranean Neogene. 3. Correlation between mammal, sporomorph, and marine microfossil assemblages from the upper Cenozoic of Rhodos, Greece. *Newsletters on Stratigraphy* 6:117–30.

Benson, R. H. and D. Hodell. 1994. Comment on a critical reevaluation of the Miocene/Pliocene boundary as defined in the Mediterranean by F. J. Hilgen and C. J. Langereis. *Earth and Planetary Science Letters* 124:245–50.

Benson, R. H. and K. Rakic-El Bled. 1991. Biodynamics, saline giants, and the late Miocene catastrophism. *Carbonates and Evaporites* 6:127–68.

Berger, J. P. 1983. Charophytes de l'"'Aquitanien" de Suisse occidentale: Essai de taxonomie et biostratigraphie. *Géobios* 16:5–37.

——. 1986. Biozonation préliminaire des charophytes oligocènes de Suisse occidentale. *Eclogae geologica Helvetica* 79:897–912.

——. 1992. Correlative chart of the European Oligocene and Miocene: Application to the Swiss Molasse Basin. *Eclogae geologica Helvetica* 85:573–609.

Berggren, W. A. 1993. Neogene planktonic foraminiferal biostratigraphy of eastern Jamaica. In *Geological Society of America Memoir* 182, ed. R. M. Wright and P. Robinson.

Berggren, W. A., D. V. Kent, and J. A. Van Couvering. 1985a. The Neogene, Part 2: Neogene geochronology and chronostratigraphy. In *The Chronology of the Geological R10*, ed. N. J. Snelling, pp. 211–60. Oxford: Blackwell Scientific Publications.

Berggren, W. A., D. V. Kent, J. F. Flynn, and J. A. Van Couvering. 1985b. Cenozoic geochronology. *Geological Society of America Bulletin* 96:1407–18.

Berggren, W. A., D. V. Kent, C. C. Swisher III, and M.-P. Aubry. 1995. A revised Cenozoic geochronology and chronostratigraphy. In *Geochronology, Time Scales, and Global Stratigraphic Correlations: A Unified Temporal Framework for an Historical Geology*, ed. W. A. Berggren, D. V. Kent, and J. Hardenbol. Society of Economic Paleontologists and Mineralogists Special Publication No. 54: 129–212.

Bernor, R. L. 1985. Systematic and evolutionary relationships of the hipparionine horses from Maragheh, Iran (late Miocene, Turolian age). *Palaeovertébrata* 15:173–269.

——. 1986. Mammalian biostratigraphy, geochronology, and zoogeographic relationships of the late Miocene Maragheh fauna, Iran. *Journal of Vertebrate Paleontology* 6:76–85.

Bernor, R. L., G. D. Koufos, M. O. Woodburne, and M. Fortelius. This volume a. The evolutionary history and biochronology of European and Southwest Asian late Miocene and Pliocene hipparionine horses.

Bernor, R. L., J. Kovar-Eder, D. Lipscomb, F. Rögl, S. Sen, and H. Tobien. 1988. Systematic, stratigraphic, and paleoenvironmental contexts of first-appearing hipparion in the Vienna Basin, Austria. *Journal of Vertebrate Paleontology* 8:427–52.

Bernor, R. L., M. Kretzoi, H.-W. Mittmann, and H. Tobien. 1993a. Preliminary systematic assessment of the Rudabánya hipparions. *Mitteilungen der Bayerischen staatssammlung für Paläontologie und Historische geologie* 33:195–207.

Bernor, R. L., H.-W. Mittmann, and F. Rögl. 1993b. The Götzendorf hipparions. *Annalen des Naturhistorischen Museums, Wien* 95:101–20.

Bernor, R. L., N. Solounias, C. C. Swisher III, and J. A. Van Couvering. This volume b. The correlation of three classical "Pikermian" mammal faunas—Maragheh, Samos, and Pikermi—with the European MN unit system.

Bernor, R. L. and H. Tobien. 1990. The mammalian geochronology and biogeography of Paşalar (middle Miocene, Turkey). *Journal of Human Evolution* 19:551–68.

Blow, W. H. 1969. Late middle Eocene to recent planktonic foraminiferal biostratigraphy. In *Proceedings of the First International Conference on Planktonic Microfossils*, ed. R. Bronnimann and H. H. Renz, pp. 199–421.

Boeuf, O. 1983. *Le site Villafranchien de Chilhac (Haute-Loire), France, Étude paléontologique et biochronologique*. Ph.D. Diss., Université Paris.

Bolli, H. M. and J. B. Saunders. 1985. Oligocene to Holocene low latitude planktic foraminifera Plankton. In *Stratigraphy*, ed. H. M. Bolli, J. B. Saunders, and K. Perch-Nilsen, pp. 155–62. Cambridge: Cambridge University Press.

Boni. 1967. Notizie sul Serravalliano tipo. *Guida alle escursioni del IV Congresso, Committee on Mediterranean Neogene Stratigraphy. International Union of Geological Sciences, 4th International Congress*, ed. R. Selli, pp. 47–63.

Bonis, L. de. 1973. Contribution a l'étude des mammifères de l'Aquitanien de l'Agenais, rongeurs-carnivores-perissodactyles. *Mémoire du Museum Nationale Histoire Naturelle*, Séries C, Sciences de la Terre 2892.

Bonis, L. de, G. Bouvrain, D. Geraads, and K. D. Koufos. 1988. Late Miocene mammal localities of the lower Axios Valley (Macedonia, Greece) and their stratigraphical significance. *Modern Geology* 13:141–47.

——. 1990. New hominid skull material from the late Miocene of Macedonia in Northern Greece. *Nature* 345:712–14.

Bonis, L. de, G. Bouvrain, G. D. Koufos, and J. Melentis. 1974. Première découverte d'un Primate hominoide dans le Miocene supérieur de Macécoine (Grèce). *Compte Rendus Academie des Sciences, Paris* 278:3063–66.

——. 1986. Succession and dating of the late Miocene primates of Macedonia. In *Primate Evolution: Proceedings of the 10th International Congress on Primate Societies*, ed. Else and Lee, pp. 107–14.

Bonis, L. de and G. D. Koufos. 1981. A new hyaenid (Carnivora, Mammalia) from the Vallesian (late Miocene) of Northern Greece. *Science Annales de Faculté de Physique et Mathematique de Université, Thessaloniki* 21:79–94.

Bouvrain, G. 1975. Un nouveau bovidé du Vallésien de Macédoine (Grèce). *Compte Rendus Academie des Sciences, Paris* 280:1357–59.

———. 1982. Révision du genre *Prostrepsiceros* Major. 1891 (Mammalia, Bovidae). *Paläontologische Zeitschrift* 56:113–24.

Bouvrain, G. and L. de Bonis. 1986. *Ouzocerus gracilis* n.g. n. sp., Bovidae (Artiodactyla, Mammalia) du Vallesien (Miocène supérieur) de Macédoine (Grèce). *Géobios* 19:661–67.

Bruijn, H. de. 1984. Remains of the mole-rat *Microspalax odessanus* Topachevski from Karaburung (Greece, Macedonia) and the family Spalacidae. *Proceedings, Koninklijke Nederlandse Akademie van Wetenschappen*, B. 87:417–24.

Bruijn, H. de, R. Daams, G. Daxner-Höck, V. Fahlbusch, L. Ginsburg, P. Mein, and J. Morales. 1992. Report of the RCMNS working group on fossil mammals, Reisensburg 1990. *Newsletters on Stratigraphy* 26:65–118.

Bruijn, H. de, P. Mein, C. Montenat, and A. van der Weerd. 1975. Correlations entre les gisements de rongeurs et les formations marines du Miocène terminal d'Espagne méridionale, I: Provinces d'Alicante et de Murcia. *Proceedings, Koninklijke Nederlandse Akademie van Wetenschappen*, B. 78:1–32.

Bruijn, H. de and A. van der Meulen. 1979. A review of the Neogene rodent succession in Greece. *Annales Geologie Pays Hellenica* 1:207–17.

Bruijn, H. de, M. Sümengen, E. Ünay, G. Sarac, and I. Terlemez. 1988. New Neogene rodent-assemblages from Anatolia (Turkey). Abstract NATO ARW, *European Neogene Mammal Chronology*, Schloss Reisensburg, Ülm.

Bruijn, H. de and W. J. Zachariasse. 1979. The correlation of marine and continental biozones of Kastellios Hill reconsidered. *Annales Geologie Pays Hellenica* 1:219–26.

Burbank, D. W., B. Engesser, A. Matter, and M. Weidmann. 1992. Magnetostratigraphic chronology, mammalian faunas, and stratigraphic evolution of the Lower Freshwater Molasse, Haute-Savoie, France. *Eclogae geologica Helvetica* 85:399–431.

Burchart, J., L. Kasza, and S. Lorenc. 1988. Fission-track zircon dating of tuffitic intercalations (Tonstein) in the brown-coal mine "Belchatów." *Bulletin of the Polish Academy of Sciences, Earth Sciences* 36:281–86.

Campbell, B. G., M. H. Amini, R. L. Bernor, W. Dickenson, R. W. Drake, R. Morris, J. A. Van Couvering, and J. A. H. Van Couvering. 1980. Maragheh: A classical late Miocene vertebrate locality in Northwestern Iran. *Nature* 287:837–41.

Cande, S. C. and D. V. Kent. 1992. A new geomagnetic polarity time scale for the late Cretaceous and Cenozoic. *Journal of Geophysical Research* 97: 13917–51.

———. 1995. Revised calibration of the geomagnetic polarity timescale for the late Cretaceous and Cenozoic. *Journal of Geophysical Research* 100: 6093–95.

Cerdeño, E. 1989. Revisíon dela sistematica de los Rinocerontes del Neógeno de Espangna. *Thesis Universita Complutense de Madrid* 306/89:1–429.

Chumakov, I. S., S. L. Byzova, and S. S. Ganzei. 1988. Kgeokhronologii Meotisa i Ponta vostochnogo Paratetisa. *Doklady Akademii Nauk SSR* 303:178–81.

———. 1992a. Geochronologija i korreljatsija pozdnego Kainozoja Paratetisa. *Rossijskaja Akademie Nauk, Dal'nevost. Otd., Tikhookeanskj Inst. Geografii Nauka, Moskva.*

Chumakov, I. S., S. L. Byzova, S. S. Ganzei, C. Arias, G. Bigazzi, F. P. Bonadonna, J. C. Hadler-Neto, and P. Norelli. 1992b. Interlaboratory fission track dating of volcanic ash levels from Eastern Paratethys. A Mediterranean–Paratethys correlation. *Palaeogeography, Palaeoclimatology, Palaeoecology* 95:287–95.

Chumakov, I. S., S. S. Ganzei, S. L. Byzova, V. Y. Dobrynina, and N. P. Paramonova. 1984. Geokhronologiya Samrata vostochnogo Paratetisa. *Doklady Akademii Nauk SSR* 276:1189–93.

Cicha, I., V Fahlbusch, and O. Fejfar. 1972. Die biostratigraphische Korrelation einiger jungtertiärer Wirbeltierfaunen Mitteleuropas. *Neues Jahrbuch für Geologie und Paläontologie* 140:129–45.

Cita, M. B. 1975. The Miocene/Pliocene boundary: History and definition. *Late Neogene Epoch Boundaries.*

———. 1968. Evolution of the planktonic foraminiferal assemblages in the stratigraphical interval between the type-Langhian and the type-Tortonian and biozonation of the Miocene of the Piedmont. *Giornale Geologia* 35:1–27.

Cita, M. B. and W. H. Blow. 1969. The lithostratigraphy of the Langhian, Serravallian, and Tortonian Stages in the type sections in Italy. *Revista Italiana Paleontologia* 75:549–603.

Cita, M. B. and I. Premoli Silva. 1960. Pelagic forminifera from the type Langhian. *Proceedings of the International Paleontological Union, Norden* 22:39–50.

Clauzon, G. 1982. *Excursionguide: Neogene of Durance, SW France.* Marseilles.

Clauzon, G. and J.-P Aguilar. 1982. Stratigraphy: Geodynamic evolution of the North Provence during the upper and terminal Miocene after the rodents faunas. *Compte Rendus Academie des Sciences, Paris* 294:915–20.

Clauzon, G., J.-P Aguilar, and J. Michaux. 1987. Le bassin Pliocène du Roussillon (Pyrénées-Orientales, France): Exemple d'évolution géodynamique d'une ria méditerranéenne consécutive à la crise de salinité messinienne. *Compte Rendus Academie des Sciences, Paris* 304:585–90.

———. 1987. Mise en évidence d'un diachronisme de 5 Ma au mûr de la Miocène de Valensole (Alpes de Haute Provence, France): Révisions chronostratigraphiques et implications géodynamiques. *Compte Rendus Academie des Sciences, Paris* 305:133–37.

———. 1989. Relation temps-sédimentation dans le Néogène méditerranéen français. *Bulletin Société Géologique de France* (8) V:361–72.

Clauzon, G., J.-P Aguilar, J. Michaux, and J.-P Suc. 1985. "Implications stratigraphiques, géodynamiques et paléogéographiques du nouveau gisement de Rongeurs de Vivès 2: in colloque en hammage à Ch. Dpéret, Perpignan." Résumé.

Clauzon, G., J. Martinell, J.-P. Aguilar, and J.-P. Suc. 1987. *Interim Colloquium RCMNS Livret guide des excursion.*

Colalongo, M. L., A. DiGrande, S. d'Onofrio, L. Gianelli, S. Iaccarino, R. Mazzei, M. Romeo, and G. Salvatorino. 1979. Stratigraphy of late Miocene Italian sections straddling the Tortonian/Messinian boundary. *Bolletino Sociéta Paleontologia Italiana* 18:258–302.

Ctyrocky, P. 1987. Vide in Steininger et al. 1987.

Daxner-Höck, G. 1971. Vertebrata (excl. Pisces) der Eggenburger Schichtengruppe. In *M 1 Eggenburgien–Die Eggenburger Schichtengruppe und ihr Stratotypus*, ed. F. F. Steininger and J. Senes. Bratislava: SAV.

Daxner-Höck, G., H. de Bruijn, and D. Foussekis. 1990. Bericht

1989 über das Projekt "Kleinsäuger" der begleitenden Grundlagenforschung. *Jahrbuch der Geologischen Bundesanstalt, Wien* 133:508–10.

Demarcq, G., P. Mein, R. Ballesio, and J.-P. Romaggi. 1989. Le gisement d'Andance (Coiron, Ardèche, France) dans le Miocène supérieur de la vallée du Rhône: un essai de corrélations marin-continental. *Bulletin Societé Géologique de France* 8:797–806.

Engesser, B. 1980. Insectivora und Chiroptera (Mammalia) aus dem neogen der Türkei. *Schweizer Paläontologische Abhandlungen* 102:47–149.

——. 1989. The late Tertiary small mammals of the Maremma region (Tuscany, Italy), 2d part: Muridae and Cricetidae (Rodentia, Mammalia). *Bolletino della Societá di Paleontologia Italiana* 28:227–52.

Engesser, B., P. Schäfer, J. Schwarz, and H. Tobien. 1993. Paläontologische Bearbeitung des grenzbereiches Obere Cerithienschichten (Corbicula-Schichten) (= Schichten mit Hydrobia inflata) im Steinbruch Rüssingen mit Bemerkungen zur Oligozän/Miozän Grenze im Kalktertiär des Mainzer Beckens. *Mainzer geowissenschaftliche Mitteilungen* 22:247–74.

Fejfar, O. 1988. The Neogene VP sites of Czechoslovakia: A contribution to the Neogene terrestrial biostratigraphy of Europe based on rodents. Abstract NATO ARW, *European Neogene Mammal Chronology*, Schloss Reisensburg, Ülm.

——. 1989. The Neogene VP sites of Czechoslovakia: A contribution to the Neogene terrestrial biostratigraphy of Europe based on rodents. In *European Neogene Mammal Chronology*, ed. E. H. Lindsay, V. Fahlbusch, and P. Mein, pp. 211–36. New York: Plenum.

Fejfar, O. and W. D. Heinrich. 1989. Muroid rodent biochronology of the Neogene, and Quarternary in Europe. In *European Neogene Mammal Chronology*, ed. E. H. Lindsay, V. Fahlbusch, and P. Mein, pp. 91–117. New York: Plenum.

Feru, M., C. Radulesco, and P. Samson. 1980. La faune de micromammifères du Miocène de Comanesti (dèp. d'Arad). *Traveaux de Institute de Speoleologie "Emile Racovitza"* 19:171–90.

Freudenthal, M. and P. Mein. 1989. Description of *Fahlbuschia* (Cricetidae) from various fissûre fillings near la Grive-St. Alban (Isère, France). *Scripta Geologica* 89:1–11.

Gaultier, F., G. Clauzon, J.-P. Suc, J. Cravatte, and D. Violanti. 1994. Age et durée de la crise de salinité messinienne. *Comptes Rendus Academie Sciences Paris*, 318:1103–9.

Geraads, D. 1978. Les Palaeotraginae (Giraffidae, Mammalia) du Miocène supérieur de la region de Thessalonique (Grèce). *Géologie Mediterrané* 5:269–76.

——. 1979. Les Giraffinae (Artiodactyla, Mammalia) du Miocène supérieur de la region de Thessalonique (Grèce). *Bulletin du Museum National d'Histoire Naturelle, Paris* 4C:377–89.

Gianotti, A. 1953. Microfaune della serie tortoniano del Rio Mazzapiedi-Castellania (Tortona-Allesandria). *Rivista Italiana Paleontologia, Memoire* 6:167–308.

Ginsburg, L. 1984. Précisions sur l'age de la série Miocène du bassin de Lisbonne: Volume l'hommage au géologue G. Zbyszewski. *Edition Recherche sûr les Civilisations*, pp. 325–31.

——. 1989. The faunas and stratigraphical subdivisions of the Orleanian in the Loire Basin (France). In *European Neogene*

Mammal Chronology, ed. E. H. Lindsay, V. Fahlbusch, and P. Mein, pp. 157–76. New York: Plenum.

Ginsburg, L. and J. Morales. 1989. Les Ruminants du Miocène inférieur de Laugnac (Lot-et-Garonne). *Bulletin Museum Nationale d'Histoire naturelle 4ième serie* 11, C. 4:201–31.

Ginsburg, L. and S. Sen. 1977. Une faune à micromammifères dans le falun Miocène de Thenay (Loir-et-Cher). *Bulletin Societé Géologique de France* 5–6:223–27.

Giuli, C. de and G. B. Vai. 1988. *Fossil vertebrates in the Lamone Valley, Romagna, Appennines.* Field Trip Guidebook, Université Firenze–Université Bologna. Lithographica Faenza.

Gourinard, Y., J. Magné, and M. J. Wallez. 1987. Présence de la mer burdigalienne dans le centre de l'Aquitane. *Bulletin Societé Histoire Naturelle* 123:147–50.

Grill, R. 1962. *Erläuterungen zur Geologischen Karte der Umgebung von Korneuburg und Stockerau.* Wien: Geologische Bundesanstalt.

Gürbüz, M. 1981. Inönü (KB Ankara) Orta Miysenindeki *Hemicyon sansaniensis* (Ursidae) turnum tanimlanmasi ve stratigrafik yayilimi. *Türkiye Jeologi Kurumu Bülteni* C 24:85–90.

Heintz, E., C. Guérin, R. Martin, and F. Prat. 1974. Principaux gisement Villafranchiens de France: Listes fauniques et biostratigraphie. *Memoire Bureau Regional de Géologique Mediterannée* 78:169–82.

Heissig, K. 1988. The faunal succession in the Bavarian Molasse reconsidered–Correlation of the faunas of the MN 5 and MN 6 Zones. Abstract NATO ARW, *European Neogene Mammal Chronology*, Schloss Reisensburg, Ülm.

Heizmann, E. P. J. and V. Fahlbusch. 1983. Die mittelmiozäne Wirbeltierfauna vom Steinberg (Nördlinger Ries): Eine Übersicht. *Mitteilungen der Bayerischen Staatssammlung für Paläontontologie und historische Geologie* 23:83–93.

Hilgen, F. 1991. Extension of the astronomically calibrated (polarity) time scale to the Miocene/Pliocene boundary. *Earth and Planetary Science Letters* 107:349–68.

Hilgen, F. and C. G. Langereis. 1989. Periodicities of CaCO3 cycles in the Mediterranean Pliocene: Discrepancies with the quasi-periods of the Earth's orbital cycles? *Terra Nova* 1:409–15.

——. 1993. A critical (re)evaluation of the Miocene/Pliocene boundary as defined in the Mediterranean. *Earth and Planetary Science Letters* 118:167–78.

Hochuli, P. 1978. Palynologische Untersuchungen im Oligozän und Untermiozän der Zentralen und Westlichen Paratethys. *Beiträge zur Paläontologie von Österreich* 4:1–132.

Hodell, D. A. and J. P. Kennett. 1986. Late Miocene–early Pliocene stratigraphy and paleoceanography of the South Atlantic and the Southwest Pacific Oceans: A synthesis. *Paleoceanography* 1:285–311.

Hodell, D. A. and F. Woodruff. 1994. Variations in the strontium isotopic ratio of seawater during the Miocene: Stratigraphic and geochemical implications. *Paleoceanography* 9:405–26.

Huegeney, M. and P. Mein. 1965. Lagomorphes et rongeurs du Neogène de Lissieu (Rhône). *Traveaux Laboratoire Geologique Faculté Science Lyon*, n.s., 12:109–23.

Huegeney, M. and M. Ringeade. 1989. Synthesis on the "Aquitanian" Lagomorph and rodent faunas of the Aquitaine Basin (France). In *European Neogene Mammal Chronology*, ed. E. H.

Lindsay, V. Fahlbusch, and P. Mein, pp. 139–56. New York: Plenum.

Huegeney, M. and G. Truc. 1976a. Corrélations stratigraphiques et paléogéographie des formations marines et continentales à la limite Oligo-Miocène dans le SE de la France. *Géobios* 9:363–65.

———. 1976b. Découvertes récentes de Mammifères et de Mollusques dans les formations d'age Oligocène terminal et Aquitanien du SE de la France: Comparaison avec les gisements déjà connus dans la meme région. *Géobios* 9:359–62.

Hünermann, K. A. 1989. Die Nashornskelette (*Aceratherium incisivum* Kaup, 1832) aus dem jungtertiär von Höwenegg im Hegau (Südwestdeutschland). *Andrias* 6:5–116.

Hürzeler, J. and B. Engesser. 1976. Les faunes de mammifèrès neogénes du Bassin de Bacinello (Grosseto, Italie). *Compte Rendus Academie de Sciences, Paris* 283:333–36.

Iaccarino, S. 1985. Mediterranean Miocene and Pliocene planktic foraminifera. In *Plankton-Stratigraphie*, ed. H. M. Bolli, J. B. Saunders, and K. Perch-Nielsen, pp. 284–314. Cambridge: Cambridge University Press.

Jaeger, J.-J., J. Michaux, and B. David. 1973. Biochronologie du Miocène moyen et supérieur continental du Maghreb. *Compte Rendus Academie des Sciences, Paris* 277:2477–80.

Jenkins, D. G., J. B. Saunders, and R. Cifelli. 1981. The relationship of *Globigerinoides bisphenicus* Todd 1954 to *Praeorbulina sicana* (De Stefania) 1952. *Journal of Foraminiferal Research* 11:262–67.

Kappelman, J., S. Sen, M. Fortelius, A. Duncan, B. Alpagut, J. Carbaugh, A. Gentry, J. P. Lunkka, F. McDowell, N. Solounias, S. Viranta, and L. Werdelin. This volume. Chronology and biostratigraphy of the Miocene Sinap Formation of Central Turkey.

Klein Hofmeijer, G. and H. de Bruijn. 1988. The mammals from the Lower Miocene of Aliveri (Island of Evia, Greece), Part 8: The Cricetidae. *Proceedings, Koninklijke Nederlandse Akademie van Wetenschappen* B 91:185–204.

Kollmann, K. 1965. Jungtertiär im Steirischen Becken. *Mitteilungen der Geologischen Gesellschaft, Wien* 57:479–632.

Kondopoulou, D., S. Sen, G. D. Koufos, and L. de Bonis. 1992. Magneto- and biostratigraphy of the late Miocene mammalian locality of Prochoma (Macedonia, Greece). *Paleontologia i Evolucio* 24–25:135–39.

Korotkevich, E. L. 1988. *The History of Hipparion Fauna of Eastern Europe.* Kiew: Naukova Dumka.

Koufos, G. D. 1986. Study of the Vallesian hipparions of the lower Axios valley (Macedonia, Greece). *Géobios* 19:61–79.

———. 1987. Study of the Turolian hipparons of the lower Axios valley (Macedonia, Greece). 2. Locality "Prochoma-1" (PXM). *Paläontologische Zeitschrift* 61:339–58.

———. 1989. The hipparions of the lower Axios valley (Macedonia, Greece): Implications for the Neogene stratigraphy and the evolution of hipparions. In *European Neogene Mammal Chronology*, ed. E. H. Lindsay, V. Fahlbusch, and P. Mein, pp. 321–38. New York: Plenum.

Kowalski, K. 1993a. *Neocometes* SCHAUB and ZAPFE, 1953 (Rodentia, Mammalia) from the Miocene of Belchatów (Poland). *Acta Zoologica Cracoviensis* 36:259–65.

———. 1993b. *Microtocricetus molassicus* (FAHLBUSCH and MAYER, 1975) (Rodentia, Mammalia) from the Miocene of Belchatów (Poland). *Acta Zoologica Cracoviensis* 36:251–58.

Kowalski, K. and H. Kubiak. 1993. *Gomphotherium angustidens* (CUVIER 1806) (Proboscidea, Mammalia) from the Miocene of Belchatów and the Proboscidean Datum in Poland. *Acta Zoologica Cracoviensis* 36:275–80.

Krakhmalnaya, T. 1994. *Hipparion* horses of the Northern coastal areas of the Black Sea. *Neogene and Quaternary Mammals of the Palaecrctic. Conference in Honour of Professor Kazimierz Kowalski*, May, 17–21, 1994. Krakow, Poland.

Krijgsman, W., F. J. Hilgen, C. G. Langenreis, and W. J. Zachariasse. In press a. The age of the Tortonian/Messinian boundary. *Earth and Planetary Science Letters.*

Krijgsman, W., C. G. Langereis, R. Daams, and A. J. van der Meulen. In press b. Magnetostratigraphic dating of the middle Miocene climate change in the continental deposits of the Aragonian type area in the Calatayud-Teruel Basin (Central Spain). *Earth and Planetary Science Letters.*

Krijgsman, W., T. V. Svetlitskaya, and A. L. Chepalyga. 1993. New data on stratigraphy, magnetostratigraphy, and mammal faunas of the late Miocene locality of Novaya Emetovka (Ukrania). *Newsletters on Stratigraphy* 29:77–89.

Langereis, C. G. and M. J. Dekkers. 1992. Paleomagnetism and rock magnetism of the Tortonian-Messinian boundary stratotype at Falconara, Sicily. *Physics of the Earth and Planetary Interiors* 71:100–11.

Langereis, C. G. and G. Hilgen. 1991. The Tosello composite: A Mediterranean and global reference section for the early to early late Pliocene. *Earth and Planetary Science Letters* 104:211–25.

Langereis, C. G., S. Sen, M. Sümengen, and E. Ünay. 1989. Preliminary magnetostratigraphic results of some Neogene localities from Anatolia (Turkey). In *European Neogene Mammal Chronology*, E. H. Lindsay, V. Fahlbusch, and P. Mein, pp. 515–25. New York: Plenum.

Leone, G. 1985. Paleoclimatology of the Casas del Rincon Villafranchian series (Spain) from stable isotope data. *Palaeogeography, Palaeoclimatology, Palaeoecology* 49:61–77.

Lindsay, E. H. 1985. European late Cenozoic biochronology and the magnetic polarity time scale. *National Geographic Society Research Reports* 20:449–56.

Lindsay, E. H., N. D. Opdyke, and N. M. Johnson. 1980. Pliocene dispersal of horse *Equus* and late Cenozoic mammalian dispersal events. *Nature* 287:135–38.

Ly Meng Hour, J. M. Cantagrel, A. De Geer De Herve, and P. M. Vincent. 1983. Revision tephrochronologiques es depôts fossilifères Plio-Pleistocènes des environs de Perrier et Champeix (Puy-de-Dome, France). *Actes Colloque Le villafranchien méditerranéen* 2:407–22.

Magné, J., Y. Gourinard, and M. J. Wallez. 1987. Comparaison des étages du Miocène inférieur définis par stratotypes ou par zones paléontologiques. *Strata, Toulouse* 1:95–107.

Marsini, E. H. and D. Torre. 1990. Review of the Villafranchian arvicolids of Italy. Communicated to International Meeting, Evolution. Phylogeny, and Biostratigraphy of Arvicolids, Reahanov (CSSR), 1987. *Geologica Romana* 26:127–33.

Mauritsch, H. J. and R. Scholger. 1994. *Magnetostratigraphische Untersuchungen im Korneuburger Becken: Bearbeitung der Profile Obergänserndorf und Teiritzberg (Karpat).* Laborbericht

IGCP 329: Paläomagnetiklabor Gams. Austria: Montanuniversität Leoben.

Mein, P. 1975. Résultats du Groupe de Travail des Vertébrés. In *Report on Activity of the RCMNS Working Groups*, ed. J. Senes, pp. 78–81. Bratislava: SAV.

——. 1984. Composition quantitative des faunes de Mammifères du Miocène moyen et supérieur de la région Lyonnaise. (RCMNS Interim-Coll. Mediterranean Neogene continental paleoenvironments and paleoclimatic evolution, Montpellier, 1983). *Paléobiologie Continentale* 14:339–46.

——. 1989a. Updating of MN zones. In *European Neogene Mammal Chronology*, ed. E. H. Lindsay, V. Fahlbusch, and P. Mein, pp. 73–90. New York: Plenum.

——. 1989b. Die Kleinsäugerfauna des Untermiozäns (Eggenburgien) von Maigen, Niederösterreich. *Annalen des Naturhistorischen Museums, Wien* 90:49–58.

Mein, P. and J. Aymar. 1984. Découvertes récentes de Mammifères dans le Pliocène du Roussillon. *Nouvel Archieves Museum Histoire naturelle* 22:69–71.

Mein, P. and J. Michaux. 1970. Un nouveau stade dans l'evolution des rongeurs pliocènes de l'Europe sud-occidentale. *Compte Rendus Academie des Sciences, Paris* (II). 296:1603–10.

Mein, P., E. Moissenet, and R. Adrover. 1983. L'extension et l'age des formations continentales pliocènes fossé de Teruel (Espagne). *Compte Rendus Academie des Sciences, Paris* 296:1603–09.

Mein, P., E. Moissenet, and G. Truc. 1978. Les formations continentales du Néogène supérieur des vallées du Júcar et du Cabriel au NE d'Albacete (Espagne), biostratigraphie et environment. *Document Laboratoire Geologie Faculté de Science, Lyon* 72:99–147.

Mein, P., E. M. Suárez, and J. Agusti. 1993. *Progonomys* Schaub, 1938 and *Huerzelerimys* gen. nov. (Rodentia): Their evolution in Western Europe. *Scripta Geologica* 103:41–64.

Mein, P. and G. Truc. 1990. Faciès et association faunique dans le Miocène supérieur du bassin de Teruel. *Paleontologie Evolucion* 23:121–39.

Meulen, A. J. van der and T. Kolfschoten. 1986. Review of late Turolian to early Biharian mammal faunas from Greece and Turkey. *Memoire Societa Geologia Italiana* 31:201–11.

Michaux, J. 1966. Sur deux faunules de Micromammifères trouvées dans les assises terminales du Pliocène en Languedoc. *Compte Rendus Sommaire, Societé Géologique, France*, pp. 343–44.

——. 1971. Muridae (Rodentia) Neogènes d'Europe Sud-occidentale. Évolution et rapports avec les formes actuelles. *Paléobiologie Continentale* 2:1–67.

Montanari, A., A. Deno, R. Coccioni, V. E. Langenheim, R. Capo, and S. Monechi. 1991. Geochronology, Sr isotope analysis, magnetostratigraphy, and planktonic stratigraphy across the Oligocene-Miocene boundary in the Contessa section (Gubbio, Italy). *Newsletters on Stratigraphy* 23:151–80.

Montanari, A., A. Deno, V. E. Langenheim, and R. Coccioni. 1990. ^{40}Ar/^{39}Ar laser-fusion dating of magnetic polarity reversals and planktonic foraminiferal events across the Aquitanian-Burdigalian boundary at Gubbio, Italy. *Transactions, American Geophysical Union* 71:1295.

Montenat, C., L. Thaler, and J. A. Van Couvering. 1975. La faune de Rongeurs de Librilla: Correlation avec les formations marines du Miocène terminal et les datations radiométriques du volcanisme de Barqueros (Province de Murcia, Espagne méridionale). *Compte Rendus Academie des Sciences, Paris* 281:519–22.

Morales, J. 1984. *Venta del Moro: Su macrofauna de Mamiferos y biostratigraphia continental del Mioceno terminal Mediterráneo.* Universita Complutense Madrid.

Mosna, S. and A. Micheletti. 1968. Microfauna del "Serravalliano," Committee on Mediterranean Neogene Stratigraphy. Proceedings of IVth Session. *Giornale di Geologia* 35:183–89.

Mottl, M. 1957. Bericht über die neuen Menschenaffenfunde aus Österreich, von St. Stefan im Lavanttal, Kärnten. *Carinthia* II:1–67.

——. 1964. *Dorcatherium* aus dem unteren Sarmat von St. Stefan im Lavanttal. *Carinthia* II. 74:22–24.

——. 1970. Die jungtertiären Säugetierfaunen der Steiermark, Südost-Österreich. *Mitteilungen Museum für Bergbau, Geologie und Technik Landesmuseum "Joanneum," Graz* 21:79–168.

——. 1980. Die jungtertiären Säugetierfaunen der Steiermark, Südost-Österreich. *Mitteilungen Museum für Bergbau, Geologie und Technik Landesmuseum "Joanneum," Graz* 31:79–168.

Onofrio, S. d', L. Gianelli, S. Iaccarino, E. Morlotti, M. Romeo, G. Salvatorini, M. Sampo, and R. Sprovieri. 1975. Planktonic foraminifera from some Italian sections and the problem of the lower boundary of the Messinian. *Bolletino Societá Paleontologia Italiana* 14:177–96.

Opdyke, N. D., P. Mein, E. Moissenet, A. Perez-Gonzales, E. H. Lindsay, and M. Petko. 1988. Magnetostratigraphy of upper Neogene mammal bearing sequence of Spain. Abstract NATO ARW, *European Neogene Mammal Chronology*, Schloss Reisensburg, Ülm.

——. 1989. The magnetic stratigraphy of the late Miocene sediments of the Gabriel Basin, Spain. In *European Neogene Mammal Chronology*, ed. E. H.Lindsay, V. Fahlbusch, and P. Mein, pp. 507–14. New York: Plenum.

Ozansoy, F. 1957. Faunes de Mammifères du Tertiare de Turquie et leurs révisions stratigraphiques. *Bulletin of the Mineral Research Exploration Institute, Turkey* (foreign edition) 49:29–48.

——. 1965. Étude des gisements continent aux de Mammifères du Cénozoique de Turquie. *Mémoires Société Géologique de France*, n.s., 44:1–92.

Pevzner, M. A. and E. A. Vangengeim. 1993. Magnetochronological age assignments of middle and late Sarmatian mammalian localities of the Eastern Paratethys. *Newsletters on Stratigraphy* 29:63–75.

Pilbeam, D., M. Morgan, J. C. Barry, and L. Flynn. This volume. European MN units and the Siwalik faunal sequence of Pakistan.

Rabeder, G. 1981. Die Arvicoliden (Rodentia, Mammalia) aus dem Pliozän und dem älteren Pleistozän von Niederösterreich. *Beiträge zur Paläontologie von Österreich* 8.

——. 1985. Die Säugetiere des Pannonien: Chronostratigraphie und Neostratotypen. In *Miozän M6 Pannonien*, ed. A. Papp, A. Jambor, and F. F. Steininger, pp. 440–63. Budapest: Akademii Kiado.

Rabeder, G. and F. F. Steininger. 1975. Die direkten Biostratigraphischen Korrelationsmöglichkeiten von Säugetierfaunen aus dem Oligo/Miozän der Zentralen Parathys. *6. Congress of the Regional Committee of Mediterranean Neogene Stratigraphy, Proceedings* 1:177–83.

Rachl, R. 1983. *Die Chiroptera (Mammalia) aus den Mittelmiozänen Kalken des Nördlinger Rieses (Süddeutschland).* Ph.D. diss., Universität München.

Ringeade, M. 1978. Micromammifères et biostratigraphie des horizons aquitaniens d'Aquitaine. *Bulletin Société Géologique de France* (7) 20:807–13.

Rio, D. and D. Fornaciari. 1994. Remarks on middle to late Miocene chronostratigraphy. *Neogene Newsletter* 1:26–34.

Rio, D., D. Fornaciari, and E. DiStefano. 1994. The Gelasian Stage: A proposal of a new chronostratigraphic unit of the Pliocene series. *Rivista Italiana Paleontologia et Stratigraphia* 100:103–24.

Rio, D. and R. Sprovieri. 1994. Pliocene standard chronostratigraphy: A proposal. *Neogene Newsletter* 1:35–38.

Riveline, J. 1984. Les charophytes du Cenozoique (Danien–Burdigalien) d'Europe occidentale: Implications stratigraphiques. *Memoire Sciences de la Terre Université Pierre and Marie Curie,* 2 vols.

Rögl, F. and G. Daxner-Höck. This volume. Late Miocene Paratethys correlations.

Rögl, F. and F. F. Steininger. 1983. Vom Zerfall der Tethys zur Paratethys. *Annalen des Naturhistorischen Museums, Wien* 85:135–63.

Rögl, F., H. Zapfe, R. L. Bernor, R. Brzobohaty, G. Daxner-Höck, I. Draxler, O. Fejfar, J. Gaudant, P. Herrmann, G. Rabeder, O. Schultz, and R. Zetter. 1993. Die Primatenfundstelle Götzendorf an der Leitha (Obermiozän des Wiener Beckens, Niederösterreich). *Jahrbuch Geologische Bundesanstalt* 136:503–26.

Rook, L., G. Ficcarelli and D. Torre. 1991. Messinian carnivores from Italy. *Bolletino della Societá Paleontologia Italiana Modena* 30:7–22.

Rzebik-Kowalska, B. 1994. Insectivora (Mammalia) from the Miocene of Belchatów in Poland. *Neogene and Quaternary Mammals of the Palaearctic. Conference in honour of Professor Kazimierz Kowalski, May, 17–21, 1994,* pp. 59–60. Kraków, Poland.

Selli, R. 1960. The Mayer-Eymar Messinian. 1867 proposal for a neostratotype. *International Geological Congress, Report 21st session, Norden,* pp. 311–33.

Sen, S. 1986. *Contribution à la magnétostratigraphie et à la paléontologie des formations continentales Néogènes du pourtour Mediterraneen.* Ph.D. diss., Université Paris 6.

——. 1989. *Hipparion* datum and its chronologic evidence in the Mediterranean area. In *European Neogene Mammal Chronology,* ed. E. H. Lindsay, V. Fahlbusch, and P. Mein, pp. 495–505. New York: Plenum.

——. This volume. Present state of magnetostratigraphic studies in the continental Neogene of Europe and Anatolia.

Sen, S., L. de Bonis, N. Dalfes, D. Geraads, and G. D. Koufos. In press. Les gisements de mammifères du Miocène supérieur de Kemiklitepe, Turquie. 1—Stratigraphie et magnétostratigraphie. *Bulletin du Museum National d'Histoire Naturelle,* Paris, 4e séries.

Sen, S., D. Kondopoulou, L. de Bonis, and G. D. Koufos. 1989. Magneto- and biostratigraphy of the late Miocene: Mammalian locality of Prochoma (Macedonia, Greece). *Proceedings of the IXth Congress RCMNS, Barcelona.*

Sen, S. and M. Makinsky. 1983. Nouvelles découvertes de micromammifères dans les faluns miocènes de Thenay (Loire-et-Cher). *Géobios* 16:461–69.

Sen, S. and J.-P. Valet. 1983. A preliminary magnetostratigraphic study of the Neogene of Samos, Greece. *Terra Cognita* 3:1–110.

Sickenberg, O., J. D. Becker-Planten, L. Benda, D. Berg, B. Engesser, A. W. Gaziry, K. Heissig, K. A. Hünermann, P. Y. Sondaar, N. Schmidt-Kittler, K. Staesche, P. Steffens, and H. Tobien. 1975. *Die Gliederung des höheren Jungtertiärs und Altquartärs in der Türkei nach Vertebraten und ihre Bedeutung für die internationale Neogen-Stratigraphie.* Geologisches Jahrbuch 15.

Solounias, N. 1981. The Samos fauna. *Contributions to Vertebrate Paleontology* 6:1–232.

Sovis, W. 1987. *Katalog zur Ausstellung: Projekt "Teiritzberg" Fossilien aus dem Karpat des Korneuburger Beckens, Stockerau.*

Srinivasan, M. S. and J. P. Kennett. 1983. The Oligocene-Miocene boundary in the South Pacific. *Geological Society of America, Bulletin* 94:798–812.

Steffens, P., H. de Bruijn, J. Meulenkamp, and L. Benda. 1979. Field guide to the Neogene of Northern Greece. *Publications of the Department of Geology and Palaeontology, University of Athens* 35:1–44.

Steininger, F. F., A. Albianelli, M.-P. Aubry, M. Biolzi, A. M. Borsetti, F. Cati, R. Corfield, R. Gelati, R. S. Iaccarino, G. Napoleone, F. Ottner, F. Rögl, R. Roetzel, S. Spezzaferri, F. Tateo, G. Villa, and D. Zevenboom. 1994. A *Proposal for the Global Boundary Stratotype Section and Point (GSSP) of the Neogene and the Paleogene/Neogene Boundary.* Vienna.

Steininger, F. F., R. L. Bernor, and V. Fahlbusch. 1989. European Neogene marine/continental chronologic correlations. In *European Neogene Mammal Chronology,* ed. E. H. Lindsay, V. Fahlbusch, and P. Mein, pp. 15–46. New York: Plenum.

Steininger, F. F., F. Rögl, and M. Dermitzakis. 1987. Report on the round-table discussion: "Mediterranean and Parathys Correlations." *Annales Institutes Geologicorum Public: Hungaria* 70:397–421.

Stuchlik, L., A. Szynkiewicz, M. Lancucka-Srodoniowa, and E. Zastawniak. 1990. Results of the hitherto palaeobotanical investigations of the Tertiary brown coal bed "Belchatów" (Central Poland). *Acta palaeobotanica* 30:259–305.

Stworzewicz, E. and A. Szynkiewicz. 1989. Miocenskie slimaki ladowe we wschodniej czesci odkrywki KWB Belchatow. *Kwartalnik Geologiczny* 32:655–62.

Suc, J.-P. 1980. *Contribution à la connaissance du Pliocéne et du Pléistocene inférieur des régions méditerranéennes d'Europe occidentale par l'analyse palynologique des dépots du Languedoc-Rousillon (sûd de la France) et de la Catalogne (nord-est de l'Espagne).* Ph.D. diss., Université de Montpellier.

——. 1982. Palynostratigraphie et paleoclimatologie du Pliocène et du Pleistocène inférieur en Méditerranee nord-occidentale. *Compte Rendus Academie des Sciences, Paris* 294:1003–8.

Suc, J.-P. and W. H. Zagwijn. 1983. Plio-Pleistocene correlations between the North-Western Mediterranean region and North-western Europe according to recent biostratigraphic and palaeoclimatic data. *Boreas* 12:153–66.

Sümengen, M., E. Ünay, G. Sarac, G., H. de Bruijn, I. Terlemez, and M. Gürbüz. 1989. New Neogene rodent assemblages from Anatolia (Turkey). In *European Neogene Mammal Chronology,* ed. E. H. Lindsay, V. Fahlbusch, and P. Mein, pp. 61–72. New York: Plenum.

Swisher, C. C., III. This volume. New ^{40}Ar/^{39}Ar dates and their contribution toward a revised chronology for the late Miocene nonmarine of Europe and West Asia.

Tobien, H. 1968. Palaeontologische Ausgrabungen nach Jungtertiären Wirbeltieren auf der Insel Chios (Griechenland) und bei Maragch (NW Iran). *Jahrbuch der Vereinigung "Freunde der Universität Mainz"* 1968:51–58.

Torre, D. 1987. Pliocene and Pleistocene marine-continental correlations. *Annales Instituti Geologici Publici Hungarici* 70:71–77.

Vai, G. B., I. M. Villa, and M. L. Colalongo. 1993. First direct radiometric datings of the Tortonian/Messinian boundary. *Compte Rendu Academie Science, Paris* 316:1407–14.

Vervloet, C.C. 1966. *Stratigraphical and Micropaleontological Data on the Tertiary of Southern Piemont (Northern Italy)*. Utrecht: Scotanus and Jens.

Weerd, A. van der. 1976. *Rodent Faunas of the Mio–Pliocene Continental Sediments of the Teruel-Alfambra Region, Spain.* Utrecht Micropaleontological Bulletin, Special Publication no. 2.

———. 1979. Early Ruscinian rodents and lagomorphs (Mammalia) from the lignites near Ptolemais (Macedonia, Greece). *Proceedings, Koninklijke Nederlandse Akademie van Wetenschappen,* B. 82:127–80.

Weidmann, M., N. Solounias, R. E. Drake, and G. H. Curtis. 1984. Neogene stratigraphy of the eastern basin, Samos Island, Greece. *Géobios* 17:477–90.

Woodburne, M. O., R. L. Bernor, and C. C. Swisher III. This volume. An appraisal of the stratigraphic and phylogenetic bases for the "Hipparion Datum" in the Old World.

Woodburne, M. O., G. Theobald, R. L. Bernor, C. C. Swisher III, H. König, and H. Tobien. This volume. Advances in the geology and stratigraphy at Höwenegg, southwestern Germany.

Zapfe, H. 1948 (1949). Die Säugetierfauna aus dem Unterpliozän von Gaiselberg bei Zistersdorf in Niederösterreich. *Jahrbuch der geologischen Bundesanstalt, Wien* 93:83–97.

———. 1949. Eine mittelmiozäne Säugetierfauna aus einer Spaltenfüllung von Neudorf a.d. March (CSR). *Anzeiger der Österreichischen Akademie der Wissenschaften, mathematisch-naturwissenschaftliche Klasse* 1949:173–81.

Ziegler, R. 1983. *Odontologische und osteologische Untersuchungen an Galerix exilis (BLAINVILLE) (Mammalia, Erinaceidae) aus den miozänen Ablagerungen von Steinberg und Gildberg im Nördlinger Ries (Süddeutschland).* Ph.D. diss., Universität München.

Zöbelein, H. K. 1988. Die jungtertiären Hoewenegg-Schichten im Hegau (Baden-Württemberg) und ihre Umgebung nach der Literatur. *Mitteilungen der Bayerischen Staatssammlung für Paläontontologie und historische Geologie* 28:173–86.

Appendix 2.1

NEOGENE FAUNA CORRELATION DATA BASE

This data base contains only those mammal localities which: (1) are biostratigraphically well correlated to the Neogene Mammal Faunal Zones (MN zones), and (2) provide correlation tie points to marine biostratigraphies, the geomagnetic polarity time scale, or chronostratigraphic stages, or have isotopic ages. The data base is arranged from the oldest to the youngest Neogene (Miocene and Pliocene) Mammal Faunal Zones (the *MN zones*) followed by the general *Geochronologic Age* of this zone.

Locality: the locality name is given here, as far as possible, arranged after locality, province, and country.

Lithostratigraphic Position: this paragraph provides a general lithologic description of the sequence from which the mammal remains are recovered and cites where the correlation tie points lie in relationship to the faunal horizons. Unfortunately, it is often impossible to extract these data from the literature.

Mammal Correlation: gives the present state of the art of the biostratigraphic unit/zone (MN zone) to which the fauna is correlated.

Correlation Tie Points: lists the different biostratigraphic possibilities for the correlation of the particular mammal fauna.

Paleomagnetic Calibration and *Isotopic Ages:* are given in millions of years (Ma).

Inferred Age: (follows Berggren et al. 1995) for inferring the ages of these Neogene Mammal Faunal Zones, or alternatively Units, with a certain constancy, we use here the most recent compilation of Berggren et al. 1995, since it includes the most recent revision of the geomagnetic time scale by Kent, the most recent radioisotopic determinations by Swisher (this volume) and the most recent correlation of planktonic foraminifera biozonation and magnetobiochronology by Berggren and Aubry.

References: only those references are listed that provide the most recent results.

Remarks: since many of our colleagues have contributed original data on these correlations, these are individually acknowledged within the data base.

The mammal localities within the data base itself are arranged according to their biochronologic position and ages within the specific Mammal Zone, following in part the biochronologic arrangements of mammal localities by de Bruijn et al. (1992: see their comments on pp. 69–71) except in those circumstance where authors to this volume have refined their correlation estimates.

European Mammal Faunal Unit: Agenian

European Mammal Faunal Zone: MN 1

MN 1

Age: early Miocene

Locality: Torrente del Cinca 68, SE Ebro Basin, Spain

Lithostratigraphic Position: an alluvial and lacustrine succession at the base of a fluviatile sequence with channel fill sandstones and flood plain deposits overlain by the carbonate dominated "Mequinenza Unit" (about 500 m) followed by alluvial sequence "Cuesta de Fraga Mudstones" (approximately 35 m) with red and variegated mudstones. This sequence is overlain by the upper lacustrine carbonate unit the "Torrente de Cinca Unit" (approximately 70 m). The locality Torrente de Cinca 68 is situated in a limestone body 45 m below the top of the section (Agusti et al. 1988: fig. 4).

Mammal Correlation: lowermost MN 1 in the *Rhodanomys transiens* Biozone (Agusti et al. 1985, 1987, and 1994; previously in the *Rhodanomys schlosseri* Biozone, Agusti et al. 1988: fig. 5)

Paleomagnetic Calibration: Agusti et al. (1994) report that the geomagnetic reversal pattern recovered from this section can be interpreted in two ways: Option I and Option II. Since Option I is considered the most consistent solution in comparison with the GPTS it is followed here and would place the Torrente del Cinca 68 faunal level within Chron C6Cn.2n.

Inferred Age: (Berggren et al. 1995) Chron C6Cn.2n: base at 23.8 Ma top at 23.67 Ma

References: Agusti et al. 1985, 1987, 1988, 1994, in press

MN 1

Age: early Miocene

Locality: Paulhiac, Bordeaux Basin, Agenais, France

Lithostratigraphic Position: brown marls intercalated in the "Calcaire blanc de l'Agenais" (Hugueney and Ringeade 1989).

Mammal Correlation: MN 1, lower part, according to de Bruijn et al. (1992)

Correlation Tie Points: "Oligocene" by cross correlation, below the "Marnes à Ostrea aginensis" correlated to the Aquitanian transgression into the Bordeaux Basin.

Inferred Age: (Berggren et al. 1995) base of the Aquitanian Stage, 23.8 Ma

References: de Bonis 1973; de Bruijn et al. 1992; Huegeney and Ringeade 1989; Ringeade 1978

MN 1

Age: early Miocene

Locality: La Paillade, France

Mammal Correlation: Mammal Zone A 2 (Aguilar and Michaux, pers. comm., 1990), MN 1; lower MN 1, slightly younger than Paulhiac (de Bruijn et al. 1992)

Correlation Tie Points: foraminifera correlate the section with the Aquitanian Age (Aguilar and Michaux, pers. comm., 1990).

Inferred Age: (Berggren et al. 1995) Aquitanian Stage: 23.8 to 20.5 Ma

References: Aguilar 1974, 1981, 1982a; de Bruijn et al. 1992

Remarks: Aguilar and Michaux (pers. comm., 1990) correlate this locality with Paulhiac.

MN 1

Age: early Miocene

Locality: Les Cevennes, France

Lithostratigraphic Position: the micromammal fauna "Les Cevennes" was discovered in black freshwater marls succeeding the marine blue marls with oysters and bivalves. This freshwater marl is itself succeeded by lignites, which are overlain by a transgressive marine sequence with a basal mollusc lumachelle followed by blue marine marls with foraminifera and selachien teeth.

Mammal Correlation: Mammal Zone A 2 (Aguilar); MN 1 (Aguilar and Michaux, pers. comm., 1990); MN 1 (de Bruijn et al. 1992)

Correlation Tie Points: early Miocene, Blow Zone N 4 and Nannoplankton Zone NN 1, according to Aguilar and Michaux (pers. comm., 1990)

Inferred Age: (Berggren et al. 1995) N 4 (Blow) = M 1 (Berggren et al. 1995), 23.8 Ma to 21.5 Ma

References: Aguilar 1974, 1981, 1982a; de Bruijn et al. 1992

Remarks: according to Aguilar and Michaux (pers. comm., 1990) Les Cévennes correlates with Paulhiac. De Bruijn et al. (1992) position Les Cévennes slightly older than Paulhiac.

MN 1

Age: early Miocene (lowermost)

Locality: Findreuse 3, 4 and 22, 27, 31, 33, Haute-Savoie, France

Lithostratigraphic Position: almost continuous 270 m thick section of sandstones, siltstones, marls, carbonates, and gypsum divided into six lithologic/depositional units of the "Lower Freshwater Molasse" Formation, following unconformably on the lower Cretaceous and transgressively overlain by the "Upper Marine Molasse" Formation. Unit 1 (lacustrine carbonates) yielded from sample numbers 15, 16, 17, 18, 24, and 25 micromammal faunas of Mammal Zone MP-29 (reference locality: Rickenbach); top of unit 2 (braided streams) unit 3 (lacustrine-palustrine carbonates and clastics) and base of unit 4 (playa) yielded from sample numbers 13, 14 (unit 2); 8, 11 (unit 3) and 5, 6, 7 (unit 4) micromammal faunas of the Mammal Zone MP-30 (reference locality: Brochene Fluh 33). Within unit 5 (lacustrine-palustrine carbonates) sample numbers 3, 4 and 22, 27, 31, 33 yielded micromammal faunas of the Mammal Zone MN 1 (reference locality: Fornant 11). The entire section was calibrated paleomagnetically.

Mammal Correlation: faunas from localities Findreuse 3, 4 and 22, 27, 31, 33 according to Burbank and al. (1992) are correlated to Mammal Zone MN 1 (upper part, see Burbank and al. 1992, p. 425, fig. 9, and p. 426, tab. 1).

Correlation Tie Points: the unit 5 belongs to the Charophyte "nitida" and "berdotensis" Zone (Berger 1983, 1986; and Riveline 1984).

Paleomagnetic Calibration: unit 5 (the uppermost lacustrine-palustrine carbonates) containing the micromammal localities belonging without any doubt to the upper part of Mammal Zone MN 1 contains the magnetic polarity Zones N 4–R 5 and N 5. According to the correlation of Burbank and al. (1992, p. 422, fig. 7) with the geomagnetic time scale the magnetic Zone N4 is correlated to Chron C6Cn.1n; R5 to Chron C6Bn.2r; and N5 to Chron C6Bn.2n.

Inferred Age: (Berggren et al. 1995) Chron C6Cn.1n, 23.53 Ma, to the top of Chron C6Bn.2n, 22.80 Ma

References: Berger 1983, 1986; Burbank and al. 1992; Cande and Kent 1994; Riveline 1984

MN 1

Age: early Miocene

Locality: Fornant 13 (11), near the village of Frangy, Haute-Savoie, France

Lithostratigraphic Position: a 370 m-thick section of sandstones, siltstones, marls, carbonates, and gypsum divided into six lithologic/depositional units belonging to the "Lower Freshwater Molasse" Formation, following unconformably on a Lower Cretaceous unit and transgressively overlain by the "Upper Marine Molasse" Formation. Between the 187 m and 265 m levels there is a gap of 80 m. Unit 1 (lacustrine sediments) sample number 7, micromammals indicative of Mammal Zone MP-28; unit 2

(meandering stream sediments) sample number 6 with micromammals indicative of Mammal Zone MP-28; unit 4 (playa
sediments) sample number 13 with micromammals indicative
of Mammal Zone MN 1 (reference locality: Fornant 11) and
above the gap: unit 5 (lacustrine-palustrine sediments) sample
number 11 with micromammals indicative of Mammal Zone
MN 1 (this is the reference locality: Fornant 11).

Mammal Correlation: the micromammal faunas of Fornant samples 11 and 13 are correlated by Burbank and al. (1992) to the
Mammal Zone MN 1 (upper part, see Burbank and al. 1992, p.
425, fig. 9, and p. 426, tab. 1).

Correlation Tie Points: unit 4 belongs to the "notata" Charophyte Zone and unit 5 (above the gap) into the "?nitida" Charophyte Zone (Berger 1983, 1986; and Riveline, 1984).

Paleomagnetic Calibration: the section is paleomagnetically
dated from the base to the gap and unit 4 and correlated at the
top of a longer reversed part (= R2) of the top of the measured
section (from meter 100 until meter 187). The reversed part at
the top of the section containing Fornat sample number 13 (=
MN 1, upper part) is interpreted either as Chron C6Cn.1r or
C6Cn.2r.

Inferred Age: (Berggren et al. 1995) Chron C6Cn.1r, 23.68 to
23.53; Chron C6Cn.2r, 23.99 to 23.80 Ma

References: Berger 1983, 1986; Burbank and al. 1992; Riveline
1984.

Remarks: The paleomagnetic assignment of the upper part of
the section is uncertain.

MN 1

Age: early Miocene

Locality: Weisenau 34b Strasseneinschnitt, near Mainz, Germany

Lithostratigraphic Position: from brackish, calcareous marls,
"Obere Cerithien Schichten"

Mammal Correlation: MN 1 upper part according to Engesser
et al. (1993).

Correlation Tie Points: Nannoplankton Zone NP-25 in samples
38/39 below the mammal bearing sample 34b.

Inferred Age: (Berggren et al. 1995) younger than 23.8 Ma

References: Engesser et al. 1993

European Mammal Faunal Zone: MN 2

MN 2

Age: early Miocene

Locality: Gans, Bordeaux Basin, France

Lithostratigraphic Position: from the "Marnes à Unios du Bazadais."

Mammal Correlation: MN 2a, according to Hugueney and
Ringeade (1989); lowermost MN 2 faunal level (de Bruijn et al.,
1992).

Correlation Tie Points: the "Marnes à Unios de Bazadais" are
correlated with the marine transgression of the Aquitanian (Hugueney and Ringeade 1989).

Inferred Age: (Berggren et al. 1995) marine transgression of
Aquitanian, and therefore above the base of the Aquitanian
Stage and younger than 23.8 Ma (Gourinard et al. 1987; Magné

et al. 1987) using a "grade-dating" methodology: between 22.7
and 21.2 Ma

References: de Bruijn et al. 1992; Gourinard et al. 1987; Hugueney and Ringeade 1989; Magné et al. 1987; Ringeade 1978

MN 2

Age: early Miocene

Locality: Aillas, Bordeaux Basin, France

Lithostratigraphic Position: from the "Marnes à Unios du Bazadais"

Mammal Correlation: MN 2a (Hugueney and Ringeade 1989);
lowermost faunal level of MN 2 (de Bruijn et al. 1992).

Correlation Tie Points: the "Marnes à Unios de Bazadais" are
correlated with the marine transgression of the Aquitanian (Hugueney and Ringeade 1989).

Inferred Age: (Berggren et al. 1995) marine transgression of
Aquitanian and therefore above the base of the Aquitanian Stage
and younger than 23.8 Ma. Inferred age between 22.7 and 21.2
Ma (Gourinard et al. 1987; Magné et al. 1987) using a "grade-
dating" methodology.

References: de Bruijn et al. 1992; Gourinard et al. 1987; Hugueney and Ringeade 1988, 1990; Magné et al. 1987; Ringeade
1978

MN 2

Age: early Miocene

Locality: Quarry Rüssingen, north of Rüssingen, TK 25 Sheet
6314 Kirchheimbolanden R:3434340 H:5498880, south of
Mainz, Germany

Lithostratigraphic Position: section within "Obere Cerithienschichten" and "Corbicula Schichten." Mammals from
sample Rü 005,15 cm of darkgreen to greyish marls with li-
thoclasts, including "Aufarbeitungshorizont" from "Obere Cerithienschichten," RÜ 013: 14 to 18 cm of marls with calcareous
congretions, "Kalkknollenhorizont" from "Untere Corbicula
Schichten"

Mammal Correlation: Lower portion of MN 2a (Engesser et al.
1993).

Correlation Tie Points: Upper part of "Obere Cerithienschichten" and lower part of "Corbicula Schichten" within the Charophyte Zones: *Rantzieniella nitida* (MN 1 in lowermost part to
MN 2a) to *Stephanochora berdotensis* Zone (MN 2a, lower
part).

Inferred Age: (Berggren et al. 1995) the Charophyte Zones: *Rantzieniella nitida* to *Stephanochora berdotensis* Zone correlate to
the Aquitanian stratotype, base of the Aquitanian is estimated to
be 23.8 Ma.

Reference: Engesser et al. 1993

MN 2

Age: early Miocene

Locality: Caunelles, France

Lithostratigraphic Position: rodent fauna out of marine sediments

Mammal Correlation: Mammal Zone A 3 (Aguilar, pers.
comm.); upper part of MN 2a (Steininger et al. 1989); alternatively middle part of MN 2 (de Bruijn et al. 1992)

Correlation Tie Points: foraminifera probably Aquitanian age (Aguilar and Michaux, pers. comm., 1990); Nannoplankton Zone NN 1 (Steininger et al. 1989)

Inferred Age: (Berggren et al. 1995) Aquitanian Stage (lower portion), 23.8 to 20.5 Ma

References: Aguilar 1974, 1981, 1982; de Bruijn et al. 1992; Hugueney and Ringeade 1989; Steininger et al. 1989

MN 2

Age: early Miocene

Locality: Balizac, Bordeaux Basin, Gironde, France

Lithostratigraphic Position: clay level on top of alternating lacustrine and brackish water sediments with a gray limestone on top; overlain by the marine "Gres de Bazas."

Mammal Correlation: MN 2b (Hugueney and Ringeade 1989); medial MN 2 (de Bruijn et al. 1992)

Correlation Tie Points: according to sedimentological interpretations, directly comparable to the Aquitanian stratotypic section.

Inferred Age: (Berggren et al. 1995) A marine transgression of the Aquitanian Stage within the Aquitanian stratotype and therefore younger than 23.8 Ma (lower boundary of the Aquitanian Stage). Inferred age according to Gourinard et al.'s (1987) and Magné et al.'s (1987) "grade-dating" is 19.8 Ma.

References: Alvinerie and Gayet 1971; de Bruijn et al. 1992; Gourinade et al. 1987; Hugueney and Ringeade 1989; Magné et al. 1987; Ringeade 1978.

Remarks: according to Hugueney and Ringeade (1989) Balizac represents a direct correlation point to the marine Aquitanian.

MN 2

Age: early Miocene

Locality: Lespignan, France

Lithostratigraphic Position: rodent fauna from marine deposits.

Mammal Correlation: Mammal Zone A 4 (Aguilar); MN 2b (Steininger et al. 1989); medial MN 2 (de Bruijn et al. 1992)

Correlation Tie Points: Nannoplankton Zone NN 1 (Steininger et al. 1989)

Inferred Age: (Berggren et al. 1995) Aquitanian Stage: 23.8 to 20.5 Ma (lower part of the Aquitanian)

References: Aguilar 1974, 1981, 1982a; de Bruijn et al. 1992; Steininger et al. 1989

MN 2

Age: early Miocene

Locality: La Brete, Bordeaux Basin, Gers, France

Lithostratigraphic Position: in an alternating succession of lacustrine marls and limestones overlying brackish sediments

Mammal Correlation: middle part of MN 2b (Hugueney and Ringeade 1989); middle part of MN 2 (de Bruijn et al. 1992)

Correlation Tie Points: correlated by geological and sedimentological reasons to upper part of the type Aquitanian.

Inferred Age: (Berggren et al. 1995) upper part of the type Aquitanian and therefore well above the base of the Aquitanian Stage and younger than 23.8 Ma. Gourinard et al. (1987) and Magné et al. (1987) give a "grade-dating" estimate of 19.8 Ma.

References: de Bruijn et al. 1992; Gourinade et al. 1987; Hugueney and Ringeade 1988, 1989; Magné et al. 1987; Ringeade 1978

MN 2

Age: early Miocene

Locality: Laugnac, Bordeaux Basin, Lot-et-Garonnes, France

Lithostratigraphic Position: marls intercalated in the "Calcaires gris de l'Agenais."

Mammal Correlation: upper portion of MN 2b (Hugueney and Ringeade 1989); upper portion of MN 2 (de Bruijn et al. 1992).

Correlation Tie Points: correlated by geological and sedimentological criteria to upper part of type Aquitanian.

Inferred Age: (Berggren et al. 1995) upper part of the type Aquitanian and therefore above the base of the Aquitanian Stage and younger than 23.8 Ma. Gourinard et al. (1987) and Magné et al. (1987) estimate an age of 19.8 Ma, using their "grade-dating" methodology.

References: de Bonis 1973; de Bruijn et al. 1992; Ginsburg and Morales 1989; Gourinade et al. 1987; Hugueney and Ringeade 1989; Magné et al. 1987; Ringeade 1978

European Mammal Faunal Unit: Orleanian

European Mammal Faunal Zone: MN 3

MN 3

Age: early Miocene

Locality: Lisboa, Universitá Catolica and Avenue do Uruguay, Portugal

Lithostratigraphic Position: "terrigenous sediments" corresponding to R 1 regressive event (Antunes 1989)

Mammal Correlation: MN 3a (Antunes 1989); lowermost MN 3 (Steininger et al. 1989); lower part MN 3 (de Bruijn et al. 1992); Zone A 5 of Aguilar

Correlation Tie Points: regressive sediments (= R 1) on top of Aquitanian transgression (= C 1) and below the Burdigalian transgression (= C 2). Overlying marine deposits Blow Zone N 5; R 1 (containing local stratigraphic horizons I and II) inferred biostratigraphic correlation: N 4–N 5 Blow Zone.

Inferred Age: (Berggren et al. 1995) younger than the regional Aquitanian transgression but older than the regional Burdigalian transgression. The base of Aquitanian Stage is equal to 23.8 Ma, the base of the Burdigalian Stage is equal to 20.5 Ma; the base of Blow Zone N 5 = M 2 (Berggren et al. 1995) is equal to 21.5 Ma.

References: Antunes 1988, 1989; Antunes and Mein 1986; Ginsburg 1984; Steininger et al. 1989

MN 3

Age: early Miocene

Locality: Estrepouy, Bordeaux Basin, Gers, France

Lithostratigraphic Position: from the "Continental Molasse de l'Armagnac," which overlies the "Calcaires gris de l'Agenais" and the "Marnes à Ostrea aginensis."

Mammal Correlation: MN 3 (Hugueney and Ringeade 1989); lower part of MN 3 (de Bruijn et al. 1992)

Correlation Tie Points: the "Marnes à Ostrea aginensis" belong to sedimentary trangressive cycle of the type Burdigalian of the Bordeaux Basin; i.e., the Agenais. Correlative to the Burdigalian of La Peloua: Nannoplankton Zone NN 2.

Inferred Age: (Berggren et al. 1995) Burdigalian Stage, 20.5 to 16.4 Ma; the Burdigalian transgression in the Bordeaux Basin and Agenais is post 20.5 Ma; Nannoplankton Zone NN 2; if correlative with the base of *Discoaster druggi*, then the age is ca. 23.2 Ma. The top of NN 2 is 18.7 Ma. Inferred "grade dating" age from the "Marne à Ostrea agenensis" yields an age of 19 Ma (re: Gourinard et al. 1987; Magné et al. 1987).

References: de Bruijn et al. 1992; Ginsburg 1974; Gourinade et al. 1987; Hugueney and Ringeade 1988, 1989; Magné et al. 1987; Ringeade 1978

MN 3

Age: early Miocene

Locality: Maigen near Eggenburg, Molasse-Zone, Lower Austria

Lithostratigraphic Position: marine shallow water sands with molluscs, Burgschleinitz Formation (Steininger et al. 1989)

Mammal Correlation: lowermost MN 3 (de Bruijn et al. 1992)

Correlation Tie Points: Nannoplankton Zone NN 2; Blow Zone N 5; Central Paratethys pollen zone NGZ II.

Inferred Age: (Berggren et al. 1995) Blow Zone N 5 = Berggren et al. 1994 Planktonic Foraminifera Zone: M 2: 21.5 to 18.8 Ma; Nannoplankton Zone NN 2. If the base is defined by *Discoaster druggi*, then the age is ca. 23.2 Ma; top of NN 2 is equal to 18.7 Ma.

References: de Bruijn et al. 1992; Mein 1989; Steininger et al. 1989

MN 3

Age: early Miocene

Locality: Eggenburg, Brunnstube-Schindergraben, Molasse-Zone, Lower Austria

Lithostratigraphic Position: coarse marine sands on top of crystalline granitic rocks; Burgschleinitz Formation (Steininger et al. 1989)

Mammal Correlation: MN 3 lower part (de Bruijn et al. 1992)

Correlation Tie Points: Blow Zone N 5; Central Paratethys pollen zone NGZ II

Inferred Age: (Berggren et al. 1995) Blow Zone N 5; Planktonic Foraminifera Zone M 2: 21.5 to 18.8 Ma

References: de Bruijn et al. 1992; Daxner-Höck 1971; Hochuli 1978; Steininger et al. 1989, 1989, 1991

MN 3

Age: early Miocene

Locality: Beaulieu, France

Lithostratigraphic Position: in a volcanic sequence interfingering with sedimentary sequences

Mammal Correlation: Mammal Zone B (Aguilar); MN 3 (Steininger et al. 1989); alternatively upper MN 3 (de Bruijn et al. 1992)

Correlation Tie Points: a contemporaneous sequence of marine sediments containing a foraminifera and ostracod fauna assigned by a few planktonic elements to Blow Zones N6 + N7

Isotopic Age (Ma):   Emplacement of the volcanic sequence is 18.3–17.5 Ma (re: Baubron and al. 1975); alternatively, 18.0 Ma (Aguilar and Michaux, pers. comm., 1990); or 17.5, 18.7 (Steininger et al. 1989).

Inferred Age: (Berggren et al. 1995) Blow Zone N 6 + N 7 (= Berggren et al. 1995); Planktonic Foraminifera Zone M 3 + M 4a and b: 18.8 to 16.0 Ma.

References: Aguilar 1981, 1982a; Aguilar in Steininger et al. 1987; 1989; Aguilar et al. 1986; Baubron et al. 1975; de Bruijn et al. 1992; Clauzon 1982; Clauzon and Aguilar 1982; Steininger et al. 1989

European Mammal Faunal Zone: MN 4

MN 4

Age: early/middle Miocene

Locality: El Casots, Valles Penedes, Spain

Mammal Correlation: *Megacricetodon primitivus* Zone (= MN 4a) of Agusti and Moyá-Solá (1991).

Correlation Tie Points: intercalations of marine horizons with *Globigerinoides bisphaericus* followed by horizons containing *Praeorbulina*.

Inferred Age: (Berggren et al. 1995) approximately 16.0 Ma

References: Agusti and Moyá-Solá 1991

MN 4

Age: early Miocene

Locality: Belchatów coalmine; Belchatow C (Bel-C); Central Poland

Lithostratigraphic Position: According to Stworzewicz and Szynkiewicz (1989), a brown coal environment correlative to the base ?Oligocene/early Miocene. Includes quartz sands (20 to 120 m thickness) with minor coal seams, containing molluscs, covered by a 3 to 5 cm thick volcanic tephra layer (TS-5), succeeded by brown coal layers with TS-4; on top, 40–60 m clays, clays with pebbles, coaliferous sands, and the main coal seam with the faunal horizon of Belchatów C with TS-3 at the very top. The uppermost portion of the section between TS-3 and TS-2 has three coal seams with sand, clay, and lacustrine limestones and the faunal horizon of Belchatów B (freshwater mollusc fauna and mammals) (= Bel-B, Stworzewicz and Szynkiewicz 1989). Fossil pollen has been located from the main seam below TS-3 and the three seams above, between TS-3 and TS-2 (Stuchlik et al. 1990).

Mammal Correlation: Rzebik-Kowalska (1994) report that from the base to the top of the sequence, there are three successive mammal horizons: (1) The lowermost mammal horizon is Belchatow C, and is correlated by the rodents and insectivore assemblage to MN 4. This horizon is found below a "tuffite horizon" = TS-3 dated 18.1 ±1.7 Ma (Burchart et al. 1988). Also found here is the proboscidean *Gomphotherium angustidens* and a chalicothere (Kowalski 1993a; Kowalski and Kubiak 1993). (2) A middle mammal horizon referred here to Belchatow B (= Bel-B) and correlative with MN 5 (Kowasliski 1993a), or MN 5/6 (Rzebik-Kowalska 1994; see below). (3) An upper mammal horizon referred here to Belchatow A, which is tentatively correlated to MN 9 (Kowalski 1993b).

Isotopic Age: Stworzewicz and Szynkiewicz (1989) have reported fission track zircon dates from TS-2 (4 samples) with an average age of 17.05 ±0.69 Ma; TS-3 (3 samples) has an average age of 17.25 ±0.4 Ma. Statistically, the difference between TS-2 and TS-3 is meaningless, with all age values clustering around 17.0 Ma. Rzebik-Kowalska (1994), Kowalski (1993a), and Kowalski and Kubiak (1993) have reported the "lower tuffite" 's (= TS-3) age as being 18.1 ±1.7 Ma.

References: Burchart et al. 1988; Kowalski 1993a, b; Kowalski and Kubiak 1993; Rzebik-Kowalska 1994; Stuchlik et al. 1990; Stworzewicz and Szynkiewicz 1989

MN 4

Age: early Miocene
Locality: Horlak 1a, 1b, Gemerek area, Kayseri-Sivas Basin, Turkey
Lithostratigraphic Position: mammal localities near base of Yenicubuk Formation with sandstone, siltstone, marl, limestone, and lignite. The Horlak 1a, 1b localities are above a prominent gypsum bed of the Burtepe Formation and approximately 120 m below Horlak 2 and Gemerek locality (Sümengen et al. 1989: fig. 2).
Mammal Correlation: local Mammal Zone C 1 correlative to MN 4 according to Sümengen et al. (1989); middle portion of MN 4 according to de Bruijn et al. (1992)
Correlation Tie Points: localities Horlak 1a and 1b are approximately 120 m below Horlak 2 and Gemerek (see below).
Inferred Age: (Berggren et al. 1995) According to the interpretation of geomagnetic results, Langereis et al. (1989) place Horlak 1a and 1b below Chron C5Cn and therefore with an age older than 16.7 Ma.
References: de Bruijn et al. 1988, 1992; Langereis et al. 1989; Sümengen et al. 1989

MN 4

Age: early Miocene
Locality: Gemerek, Horlak 2, Gemerek area, Kayseri-Sivas Basin, Turkey
Lithostratigraphic Position: lower Yenicubuk Formation, sandstone, siltstone, marl, limestone, and lignite. Gemerek and Horak 2 localities approximately 120 m above Horlak 1a, 1b locality (Sümengen et al. 1989: fig. 2).
Mammal Correlation: local Mammal Zone C 2, clearly younger than Horak 1a, 1b (see above), probably MN 4 according to Sümengen et al. (1989).
Paleomagnetic Calibration: paleomagnetic investigations in the Gemerek section demonstrate a small reversed zone (of ca. 20 m), followed by a normal zone (of ca. 70 m), and to the top a reversed zone (of ca. 75 m). After a gap (ca. 80 m) a basalt flow dated at 14.9 ±0.7 Ma. This pattern is considered to represent parts of the upper normal part of Anomaly 5C (Chron 16) (Langereis et al. 1989); = Chron C5C.
Inferred Age: (Berggren et al. 1995) Horlak 2 and Gemerek are

120 m above the Horlak 1a and 1b localities (Langereis et al. 1989) in a normal geomagnetic interval followed by a short reversed portion of Chron C5C here interpreted to be C5Cn.3n and lower part of Chron C5C, interpreted here as being C5Cn.3n, and the lower part of Chron C5Cn.2r: 16.7 to 16.5 Ma.

References: de Bruijn et al. 1988, 1992; Langereis et al. 1989; Sen, this volume); Sümengen et al. 1989

Remarks: de Bruijn (pers. comm., 1992) reports that there is a hiatus above the Yenicubuk Formation, therefore there is no upper age limit other than MN 7 + 8. However, on the basis of its composition, this fauna is closely correlative with Horlak 1a and 1b (see above).

MN 4

Age: early Miocene
Locality; Orechov, Moravia, Czechoslovakia
Lithostratigraphic Position: in littoral marine facies
Mammal Correlation: middle to upper MN 4 (Steininger et al. 1989; de Bruijn et al. 1992)
Correlation Tie Points: below the so-called Oncophora-facies of Central Paratethys; late Ottnangian
Inferred Age: (Berggren et al. 1995) older than 16.5 Ma
References: Cicha et al. 1977; de Bruijn et al. 1992; Fejfar 1988, 1989; Steininger et al. 1989

MN 4

Age: early Miocene
Locality: Aliveri-Kymi, Euboa, Greece
Lithostratigraphic Position: below main coal, underlain with tree roots
Mammal Correlation: lower to middle MN 4 (Steininger et al. 1989); upper MN 4 (de Bruijn et al. 1992)
Correlation Tie Points: Kale-Eskihisar E-Mediterranean pollen zone
References: Benda and de Bruijn 1982; Benda and Meulenkamp 1989; de Bruijn and Van der Meulen 1979; de Bruijn et al. 1992; Klein Hofmijer and de Bruijn 1988

European Mammal Faunal Zone: MN 5

MN 5

Age: early Miocene
Locality: Teiritzberg near Korneuburg, Lower Austria
Lithostratigraphic Position: marine, nearshore environment with clays and silts to sands
Mammal Correlation: middle to upper MN 5, according to Steininger et al. (1989) and de Bruijn et al. (1992).
Correlation Tie Points: upper part of Karpatian Stage based on molluscan fauna.
Paleomagnetic Calibration: Mauritsch and Scholger (1994) lately finished a report on the magnetostratigraphy on this locality and could demonstrate that the entire section including the mammal-bearing part is within a normal magnetozone.
Inferred Age: (Berggren et al. 1995) Central Paratethys Karpatian

Stage with its base approximately at 17.2 Ma and its top at 16.4 Ma. Upper part of the Karpatian Stage within the normal polarity intervals of Chron C5Cn. Because of biostratigraphic reasons (base of Badenian Stage near base of Chron C5n.1n) the measured normal polarity interval can be correlated either to Chron C5Cn.3n, 16.72–16.55 Ma or to Chron C5cn.2n, 16.48 to 16.32 Ma.

References: Daxner-Höck et al. 1989; de Bruijn et al. 1992; Grill 1962; Mauritsch and Scholger 1994; Sovis 1987; Steininger et al. 1989

MN 5

Age: early Miocene
Locality: Eibiswald, Styria, Austria
Mammal Correlation: basal MN 5 (Steininger et al. 1989; de Bruijn et al. 1992)
Correlation Tie Points: Karpatian age
Inferred Age: (Berggren et al. 1995) the Central Paratethys Stage Karpatian; the base is approximately 17.2 Ma, the top is ca. 16.4 Ma.
References: de Bruijn et al. 1992; Kollmann 1965; Mottl 1970; Rabeder and Steininger 1975; Steininger et al. 1989

MN 5

Age: middle Miocene
Locality: Thenay, Loire et Cher, France
Lithostratigraphic Position: Falune de Touraine
Mammal Correlation: MN 5 (de Bruijn et al. 1992)
Correlation Tie Points: Langhian Stage (Ginsburg, pers. comm., 1990)
Inferred Age: (Berggren et al. 1995) base of Langhian Stage: 16.4 and top: 14.8 Ma
References: de Bruijn et al. 1992; Ginsburg and Sen 1977; Sen and Makinsky 1983
Remarks: During the time of the 1992 Reisenburg meeting this locality was retained together with Pontlevoy-Thenay as a reference locality for Mammal Zone MN 5 (de Bruijn et al. 1992).

MN 5

Age: middle Miocene
Locality: Chios, Greece
Lithostratigraphic Position: in sandy clays of the Keremaria Formation, flood plain deposits
Mammal Correlation: upper MN 5 (Steininger et al. 1989; de Bruijn et al. 1992)
Correlation Tie Points: lower Yeni-Eskihisar E-Mediterranean pollen zone
Paleomagnetic Calibration: currently under investigation by Sen.
References: Benda and Meulenkamp 1989; de Bruijn et al. 1992; Steininger et al. 1989; Tobien 1968

MN 5

Age: middle Miocene
Locality: Rimbez, Landes, France

Mammal Correlation: MN 5 (Ginsburg, pers. comm., 1990)
Correlation Tie Points: "Helvetien" with *Pecten subarcuatus*

MN 5

Age: middle Miocene
Locality: Sos, Garonne, France
Lithostratigraphic Position: mammal remains recovered from marine sands
Mammal Correlation: MN 5 (Ginsburg, pers. comm., 1990)
Correlation Tie Points: "Helvetien" with *Pecten subarcuatus*

MN 5

Age: early/middle Miocene
Locality: Pontlevoy-Thenay; France
Lithostratigraphic Position: "Falunes": marine, unsorted near shore deposits with sands to very coarse sands
Mammal Correlation: upper MN 5 (de Bruijn et al. 1992)
Correlation Tie Points: mammal remains intercalated with marine sediments
References: de Bruijn et al. 1992; Ginsburg 1989; Sen and Makinsky 1983
Remarks: During the Reisensburg meeting (1992), this locality, together with the locality of Thenay, was retained as the reference locality for MN 5 (de Bruijn et al. 1992).

MN 5

Age: middle Miocene
Locality: Dumlupinar, Turkey
Mammal Correlation: MN 5 (Benda and Meulenkamp 1989)
Correlation Tie Points: upper Eskihisar E-Mediterranaen pollen zone
Isotopic Age: radiometric date above mammal locality is 14.75 ±0.3 Ma
References: Becker-Platen et al. 1975; Benda and Meulenkamp 1989; Sickenberg et al. 1975; Steininger et al. 1989

MN 5/6

Age: early Miocene
Locality: Belchatów-coalmine; Belchatów B (= Bel-B); Central Poland
Lithostratigraphic Position: Rzebik-Kowalska (1994) reports that from the base to top of the coal mine there are three mammal horizons currently known. The lowermost mammal horizon is referred in this paper to Belchatów C = MN 4 (see above) and is found below a tuffite horizon (TS-3) dated at 18.1 ±1.7 Ma. A middle mammal horizon called Belchatów B correlates with MN 5 based on its rodent assemblage (Kowalski 1993a); however, the insectivore assemblage has led Rzebik-Kowalska (1994) to infer an MN 5/6 age. It is found between the "lower tuffite horizon" (= TS-3) and below a "higher tuffite horizon" (= TS-2) dated at 16.5 ±1.3 Ma. An upper mammal horizon, Belchatow A, is tentatively correlated to MN 9 (Kowalski 1993b) (see Stworzewicz and Szynkiewicz 1989 for the position of tuffite horizons TS-3 and TS-2 see MN 4: Belchatów C).
Mammal Correlation: Kowalski (1993a) correlates the middle

mammal horizon (Belchatów B) with MN 5; Rzebik-Kowalska (1994) correlates the same horizon with MN 5/6, using the insectivore assemblage. These assemblages are found between/above the "lower tuffite horizon" (= TS-3) and below a "higher tuffite horizon" (= TS-2, dated at 16.5 ±1.3 Ma).

Isotopic Age: Rzebik-Kowalska (1994) reports that the "lower tuffite" (= TS-3) is dated 18.1 ±1.7 Ma and the "upper tuffite" (= TS-2) is dated 16.5 ±1.3 Ma (see also Storzewicz and Szynkiewicz 1989).

References: Burchart et al. 1988; Kowalski 1993a, b; Rzebik-Kowalska 1994; Stworzewicz and Szynkiewicz 1989

European Mammal Faunal Unit: Astaracian

European Mammal Faunal Zone: MN 6

MN 6

Age: middle Miocene

Locality: Devinska Nova Ves: fissures 1–3 (Neudorf Spalten 1–3 of Zapfe 1948 1949; Fejfar 1988, 1989); Slovakia

Lithostratigraphic Position: continental fissure fillings in Jurassic limestone surrounded by marine early Badenian deeper water clays with microfauna and partly overlain by marine late Badenian silts with microfauna (see Fejfar 1989:217, fig. 4)

Mammal Correlation: Lowermost MN 6 (Steininger et al. 1989 and de Bruijn et al. 1992; Fejfar, pers. comm., 1994). Bernor believes that since this locality "must predate the Langhian (re: Rögl and Steininger 1983), it must predate MN 6 and is correlative therefore with MN 5."

Correlation Tie Points: The microfauna of the clays surrounding the karstic fissures indicates the base of the Central Paratethys Badenian Stage (Lower Lagenid (Bio-) Zone = M5a of Berggren et al. 1995. The microfauna within the silts directly overlying the fissure filling indicates late Badenian (Bulimina-Bolivina Zone). Since the entire Jurassic limestone hill within which the fissures occur is surrounded by early Badenian sediments, the continental filling must predate the M 5a Zone of Berggren et al. (1994) and predate the Langhian (Rögl and Steininger 1983).

Inferred Age: (Berggren et al. 1995) base of the Central Paratethys Badenian Stage, ca. 16.0 Ma

References: Cicha et al. 1972; de Bruijn et al. 1992; Fejfar 1988, 1989; Rabeder and Steininger 1975; Rögl and Steininger 1983; Steininger et al. 1989; Zapfe 1948, 1949

MN 6

Age: middle Miocene

Locality: Sansan, France

Lithostratigraphic Position: The partly outcropping 40 m of section are composed of sandy marls, silts, sands and sandstones, and calcareous intercalations. The mammal fauna from the upper part of the section is found in facies ranging from sandy marls to silts. This part of the section belongs to the uppermost part of the lower normal and the short reversed part of the magnetostratigraphic sequence (see below).

Mammal Correlation: MN 6 reference locality, medial MN 6 according to de Bruijn et al. (1992). Lower MN 6, according to

Bernor and Tobien (1990), Bernor et al. (this volume a, b) and Sen (this volume).

Geomagnetic Calibration: the lower part of the section contains a long reversed interval, which ends about 32 m above the first reversed magnetic sample. Above this level a shorter normal sequence (about 3 m, containing mammal remains) is followed by a short reversed part (about 1 m, containing mammal remains), followed by a longer normal part (about 5 m); the uppermost 2 to 2.5 m of the section are within a reversed sequence.

Inferred Age: (Berggren et al. 1995) this geomagnetic polarity sequence is correlated pro parte to Chron C5Br to Chron C5ADn.1r; the mammal faunas are correlative with Chron C5Bn.2n and C5Bn.1r: 15.15 to 14.88 Ma.

References: Bernor and Tobien 1990; de Bruijn et al. 1992; Sen, this volume; Sen and Ginsburg (in prep.)

MN 6

Age: middle Miocene

Locality: Devinska Nova Ves, sandhill (Neudorf Sandberg), Slovakia

Lithostratigraphic Position: the mammal remains have been recovered from the transgressive marine nearshore sands.

Mammal Correlation: MN 6 (Steininger et al. 1989 and de Bruijn et al. 1992)

Correlation Tie Points: by micro and macro fauna, middle to upper Badenian

Inferred Age: (Berggren et al. 1995) middle Badenian Stage of Central Paratethys, and therefore younger than 15.2 Ma; (Badenian/Sarmatian boundary is at approximately 13.6 Ma).

References: Cicha et al. 1972; de Bruijn et al. 1992; Fejfar 1988, 1989; Rabeder and Steininger 1975; Steininger et al. 1992

MN 6

Age: middle Miocene

Locality: Luc-sur-Orbieu, France

Lithostratigraphic Position: rodent fauna from marine deposits

Mammal Correlation: Mammal Zone C 3 (Aguilar); MN 6 lower part (de Bruijn et al. 1992)

Correlation Tie Points: benthic and planktonic foraminifera Blow Zone N 9/10

Inferred Age: (Berggren et al. 1995) Blow Zone N 9/10 Planktonic Foraminifera Zone M 6/7: 15.1 to 12.7 Ma

References: Aguilar 1981, 1982; de Bruijn et al. 1992

Remarks: This mammal fauna is correlative with Sansan.

MN 6

Age: middle Miocene

Locality: Veyran, France

Mammal Correlation: Mammal Zone C 3 or C 4 (Aguilar), upper part of MN 6 (re: remarks below)

Correlation Tie Points: Nannoplankton Zone NN 6, according to Aguilar and Michaux (pers. comm., 1990). Nannoplankton Zone NN 5 (Steininger et al. 1989).

Inferred Age: (Rio and Fornacian 1994) base of Nannoplankton

Zone NN 5 (16.0 Ma, top at 13.7 Ma) and top of Nannoplankton Zone NN 6 (11.8 Ma)

References: Aguilar 1981, 1982; Aguilar and Michaux 1984; Rio and Fornacian 1994; Steininger et al. 1989

Remarks: according to Aguilar and Michaux (pers. comm., 1990), the mammal age determination has been revised, and this fauna is younger than Sansan and Luc-sur-Orbieux. On the basis of the *Megacricetodon* its age is correlative to St. Catherine 1 (Aguilar and Michaux 1984).

MN 6

Age: middle Miocene

Locality: Inönü I, localities 24, 24A; Sinap Tepe area south of Sarilar, Central Anatolia, Turkey

Lithostratigraphic Position: localities in upper part of Pazar Formation with a volcanic unit on top

Mammal Correlation: MN 6 by Gürbüz (1981); basal MN 6 here

Inferred Age: (Kappelman et al., this volume) ca. 15.2 ±0.3 Ma

References: Gürbüz 1981; Kappelman et al., this volume

MN 6

Age: middle Miocene

Locality: Paşalar, Turkey

Mammal Correlation: basal MN 6 (Steininger et al. 1989; Bernor and Tobien 1990; de Bruijn et al. 1992)

Correlation Tie Points: Lowermost Eskihisar E-Mediterranean pollen zone

References: Andrews 1989; Benda and Meulenkamp 1989; Benda et al. 1975; Bernor and Tobien 1990; de Bruijn et al. 1992; Engesser 1980; Sickenberg et al. 1975; Steininger et al. 1989

MN 6

Age: middle Miocene

Locality: Steinberg and Goldberg, Franken, Germany

Lithostratigraphic Position: post Ries-event impact lake sediments (travertine)

Mammal Correlation: MN 6 (Steininger et al. 1989 and de Bruijn et al. 1992)

Isotopic Age: younger than Ries Impact Event (dated 14.7 Ma)

References: de Bruijn et al. 1992; Heissig 1988; Heizmann and Fahlbusch 1983; Steininger et al. 1989; Rachl 1983; Ziegler 1983

MN 6

Age: middle Miocene

Locality: Pontigné, Maine et Loire, France

Lithostratigraphic Position: Falun de l'Anjou

Mammal Correlation: MN 6 (Ginsburg, pers. comm., 1990)

Correlation Tie Points: Langhian Stage (Ginsburg, pers. comm., 1990)

Inferred Age: (Berggren et al. 1995) base of the Langhian Stage (16.0)

References: Ginsburg 1989

MN 6

Age: middle Miocene

Locality: Çandir, Turkey

Mammal Correlation: MN 6 upper part, according to Steininger et al. (1989) and de Bruijn et al. (1992)

Correlation Tie Points: Eskihisar E-Mediteranean pollen zone

References: Benda and Meulenkamp 1989; Benda et al. 1975; de Bruijn et al. 1992; Engesser 1980; Sickenberg et al. 1975; Steininger et al. 1989

Remarks: This locality is presently under study by a team directed by E. Gülec (Ankara).

European Mammal Faunal Zone: MN 7 + 8

MN 7

Age: middle Miocene

Locality: Plakia, Greece

Mammal Correlation: lower to middle MN 7 (Steininger et al. 1989)

Correlation Tie Points: base of Yeni-Eskihisar E-Mediterranean pollen zone

References: Benda and Meulenkamp 1990; Steininger et al. 1989

MN 7 + 8

Age: middle Miocene

Locality: La Grenatiere, France

Lithostratigraphic Position: rodent fauna from marine and lacustrine levels

Mammal Correlation: Mammal Zone C 5 (Aguilar); MN 7 lower to middle part (Steininger et al. 1989); lowermost MN 7 + 8 (de Bruijn et al. 1992)

Correlation Tie Points: Nannoplankton Zone NN 6; Blow Zone N 12 (Steininger et al. 1989)

Inferred Age: (Berggren et al. 1995; and Rio and Fornaciari 1994) Blow Zone N 12; Planktonic Foraminiferal Zone M 9, 12.5 to 12.0 Ma; Nannoplankton Zone NN 6, 13.6 to 11.9 Ma.

References: Aguilar 1981, 1982; de Bruijn et al. 1992; Steininger et al. 1989

Remarks: Aguilar and Michaux (pers. comm., 1990) state that based on the stage-of-evolution of the *Megacricetodon*, this fauna is younger than La Grive M (in contrast to de Bruijn et al. 1992: pg. 74, tab. 3).

MN 7 + 8

Age: middle Miocene

Locality: Yeni-Eskihisar 1 (locality No. 2), Turkey

Mammal Correlation: uppermost MN 8 (Steininger et al. 1989); alternatively middle portion of MN 7 + 8 (de Bruijn et al. 1992)

Correlation Tie Points: Yeni-Eskihisar E-Mediterranean pollen zone

Isotopic Age (Ma): Yeni-Eskihisar 1 (locality 1), 13.2 Ma; Yeni-Eskihisar 1 (locality 2), 11.1 Ma. Andrews et al. (1980) have reported that the stratigraphic relationships between the dated tuffs and fossiliferous horizons are unclear. The two tuffaceous levels are separated by only 90 cm of sediments, which cannot justify the age difference claimed by Becker-Platen et al. (1977). For details of this discussion see Andrews et al. 1980.

References: Andrews et al. 1980; Becker-Platen et al. 1975; Benda and Meulenkamp 1990; Benda et al. 1975; de Bruijn et al. 1992; Engesser 1980; Sickenberg et al. 1975; Steininger et al. 1989

MN 7 + 8

Age: middle Miocene
Locality: Santarem, Portugal
Mammal Correlation: MN 7 (Aguilar Zone C 4) (Steininger et al. 1989); MN 7 + 8 (de Bruijn et al. 1992)
Correlation Tie Points: Blow Zone N 12; Nannoplankton Zone NN 6
Inferred Age: (Berggren et al. 1995; and Rio and Fornaciari 1994) Blow Zone N 12 (= Berggren et al. 1995, Planktonic Foraminiferal Zone M 9 12.5 to 12.0 Ma); Nannoplankton Zone NN 6: 13.6 to 11.9 Ma
References: Aguilar 1982a; de Bruijn et al. 1992; Steininger et al. 1989

MN 7 + 8

Age: middle Miocene
Locality: La Grive M, France
Mammal Correlation: middle to upper MN 7 (Aguilar Zone C 4; Steininger et al. 1989); MN 7 + 8 (de Bruijn et al. 1992)
Correlation Tie Points: Blow Zone N 12; Nannoplankton Zone NN 6
Inferred Age: (Berggren et al. 1995; and Rio and Fornaciari 1994) Blow Zone N 12 (= Berggren and al. 1994, Planktonic Foraminiferal Zone M 9, 12.5 to 12.1/12.0 Ma); Nannoplankton Zone NN 6: 13.6 to 11.9 Ma
References: Aguilar 1982; de Bruijn et al. 1992; Freudenthal and Mein 1989; Mein 1984; Steininger et al. 1989

MN 7 + 8

Age: middle Miocene
Locality: Sankt Stefan i.L., Styria, Austria
Mammal Correlation: lower part of MN 8 (Steininger et al. 1989); alternatively, MN 7 + 8 (de Bruijn et al. 1992)
Correlation Tie Points: mammals from lignites intercalated in sediments with lower Sarmatian (= Volhynian) mollusc fauna
Inferred Age: (Berggren et al. 1995) Central Paratethys lower Sarmatian Stage = Eastern Paratethys Volhynian Stage, 13.6 - 12.2 Ma (Rögl and Daxner-Höck, this volume)
References: de Bruijn et al. 1992; Mottl 1964, 1957, 1980; Rabeder and Steininger 1975; Steininger et al. 1989

MN 7 + 8

Age: middle Miocene
Locality: Sofça, Turkey
Mammal Correlation: lower MN 8 (Steininger et al. 1989); upper MN 7 + 8 (de Bruijn et al. 1992)

Correlation Tie Points: Yeni-Eskihisar E-Mediterranean pollen zone
References: Benda and Meulenkamp 1989; Benda et al. 1975; de Bruijn et al. 1992; Engesser 1980; Sickenberg et al. 1975; Steininger et al. 1989

MN 8

Age: middle Miocene
Locality: C. Almirall, Spain
Mammal Correlation: MN 8 (Aguilar Zone C 5), Steininger et al. (1989)
Correlation Tie Points: planktonic foraminifera, Blow Zone N 13/14 boundary; Nannoplankton Zone NN 7
Inferred Age: (Berggren et al. 1995; Rio and Fornaciari 1994) boundary of Blow Zone N 13/14 (= Berggren et al. 1995, Planktonic Foraminiferal Zone boundary M 10/11, 11.7 Ma); Nannoplankton Zone NN 7: 11.9 to 10.8 Ma
References: Aguilar 1982a; Steininger et al. 1989

MN 8

Age: middle Miocene
Locality: Comanesti 1, Romania
Mammal Correlation: MN 8 (Feru et al. 1980)
Correlation Tie Points: in sediments with a lower Sarmatian (= Volhynian) molluscan fauna
Inferred Age: (Berggren et al. 1995) Central Paratethys lower Sarmatian Stage = Eastern Paratethys Volhynian Stage: 13.6 to 12.2 Ma (Rögl and Daxner-Höck, this volume). The Volhynian/Bessarabian boundary has been paleomagnetically recalibrated by Pevzner and Vangengeim (1993) as being 12.4 or 12.18 Ma.
References: Bernor et al. 1988; Feru et al. 1980; Pevzner and Vangengeim 1993; Steininger et al. 1989

MN 8

Age: middle/late Miocene
Locality: Yassören localities 64, 65, Sinape Tepe, Central Anatolia, Turkey
Lithostratigraphic Position: from the fluviatile Sinap Formation, with a total thickness of more than 100 m.
Mammal Correlation: MN 8 (Kappelman et al., this volume)
Inferred Age: (Kappelman et al., this volume): 10.38 Ma for Loc. 64 and a little older for Loc. 65
References: Kappelman et al., this volume; Sen 1989, this volume

European Mammal Faunal Unit: Vallesian

European Mammal Faunal Zone: MN 9

MN 9

Age: late Miocene
Locality: Gaiselberg, Vienna Basin, Lower Austria
Lithostratigraphic Position: fluviatile deposits
Mammal Correlation: basal MN 9 (Bernor et al. 1988; Steininger et al. 1989 1992; de Bruijn et al. 1992)

Correlation Tie Points: Lower Pannonian: local Zone C

Inferred Age: Central Paratethys Stage Pannonian: approximately from 11.5 to 7.1 Ma (Rögl et al. 1993); basal MN 9, 11.2 (Rögl and Daxner-Höck, this volume)

References: Bernor et al. 1988a; 1993a, b; de Bruijn et al. 1992; Rögl et al., 1993; Rögl and Daxner-Höck, this volume; Steininger et al. 1989; Zapfe 1948

MN 9

Age: late Miocene

Locality: Hovorany, Moravia, Czechoslovakia

Mammal Correlation: basal MN 9 (Bernor et al. 1988a); medial MN 9 (de Bruijn et al. 1992)

Correlation Tie Points: Lower Pannonian: local Zone B/C

Inferred Age: Central Paratethys Stage Pannonian: approximately from 11.5 to 7.1 Ma (Rögl et al. 1993); basal MN 9, 11.2 Ma

References: Bernor et al. 1988a, 1993a, b; de Bruijn et al. 1992; Ctyrocky in Steininger et al. 1987; Rögl et al. 1993; Rögl and Daxner-Höck, this volume; Woodburne et al., this volume

MN 9

Age: late Miocene

Locality: Eşme Akçaköy, Turkey

Mammal Correlation: basal MN 9 (de Bruijn et al. 1992).

Isotopic Age: radiometric date, 11.6 ±0.5 Ma

References: de Bruijn et al. 1992; Sen 1989; Sickenberg et al. 1975

MN 9

Age: late Miocene

Locality: Comanesti 2, Romania

Mammal Correlation: Medial MN 9 (Bernor et al. 1988a); alternatively basal MN (Steininger et al. 1989; de Bruijn et al. 1992)

Correlation Tie Points: Lower Pannonian, local Zone C/D

Inferred Age: Central Paratethys Stage Pannonian: from approximately 11.5 to 7.1 Ma (Rögl et al. 1993).

References: Bernor et al. 1988; de Bruijn et al. 1992; Feru et al. 1980; Rögl et al. 1993; Steininger et al. 1989

MN 9

Age: late Miocene

Locality: Vösendorf and Inzersdorf, Vienna, Vienna Basin, Austria

Lithostratigraphic Position: greyish silty clays to sands in near-shore environment of "Pannonian" lake system

Mammal Correlation: basal MN 10 (Steininger et al. 1989; 1992); alternatively medial MN 9 (de Bruijn et al. 1992); medial MN 9, slightly older than Höwenegg (Bernor et al. 1988, 1993a, b; Woodburne et al., this volume)

Correlation Tie Points: middle Pannonian: local Zone E

Paleomagnetic Calibration: the existing outcrop of the Inzersdorf section exhibits a normal polarity, which is correlated to Chron C5n.2n (Lantos and Hodi-Korpas, pers. comm., 1993).

Inferred Age: (Berggren et al. 1995) Chron 5n.2n: 10.95 to 9.92 Ma; Bernor et al.'s correlation with Höwenegg would make this locality slightly older than 10.3 Ma.

References: Bachmayer and Wilson 1984; Bernor et al. 1988, 1993a, b; de Bruijn et al. 1992; Rabeder 1985; Rabeder and Steininger 1975; Rögl et al. 1993; Rögl and Daxner-Höck, this volume; Steininger et al. 1989; Woodburne et al., this volume

MN 9

Age: late Miocene

Locality: Höwenegg, Germany

Lithostratigraphic Position: "Höwenegg-Schichten": whitish, lacustrine marls with tuff-horizons, belonging to the "Obere Süßwasser Molass" Formation

Mammal Correlation: medial MN 9 (Bernor et al. 1993a, b; Woodburne et al., this volume); lower MN 9 (Steininger et al. 1989; de Bruijn et al. 1992)

Paleomagnetic Calibration: entire section within a normal polarity zone correlated to Chron C5n.2n (Woodburne et al., this volume)

Isotopic Age: several radiometric dates published ranging between 12.4 and 9.4 Ma for basalts, tuffs and hornblends. Presently most reliable date: 10.3 Ma (Swisher et al., this volume)

Inferred Age: (Berggren et al. 1995) The most recent age coupled with the section's normal polarity provides a correlation within Chron C5n.2n: 10.95 to 9.92 Ma; the slightly advanced stage-of-evolution of the hipparionine horses from the Höwenegg quarry indicate a medial MN 9 age to Bernor et al. (1993a, b).

References: Bernor et al. 1988, 1993a, b, this volume a; de Bruijn et al. 1992; Hünermann 1989; Swisher, this volume; Woodburne et al., this volume; Zöbelein 1988

MN 9

Age: late Miocene

Locality: Gritsev, Chmielnicki Region, Ukraina

Lithostratigraphic Position: clays with shells and bones filling karstic fissures in limestones of middle Sarmatian (Bessarabian) age (Korotkevich 1988)

Mammal Correlation: medial MN 9 (Kowalski 1993b, Pevzner and Vangengeim 1993, and Krakhmalnaya 1994)

Correlation Tie Points: a Bessarabian mollusc fauna was recovered from the bone bearing clays.

Isotopic Age: Radioisotopically dated 12.4 Ma, correlative with the base of the Bessarabian (Chumakov et al. 1988, 1992 a, b); top of the Bessarabian is calibrated as being 10.2 Ma.

Paleomagnetic Calibration: clays with reversed magnetization (Pevzner and Vangengeim 1993)

Inferred Age: Bessarabian, ca. 12.2–11 Ma (Rögl and Daxner-Höck, this volume). The reversed magnetization is correlated by Pevzner and Vangengeim (1993: fig. 6) to Chron 10, which is equal to Chron C5r.3r, 11.93 to 11.53 Ma or Chron C5r.2r, 11.47 to 11.09 Ma.

References: Chumakov et al. 1988, 1992a, b; Korotkevich 1988; Kowalski 1993b; Krakhmalnaya 1994; Pevzner and Vangengeim 1993; Rögl et al. 1993; Rögl and Daxner-Höck, this volume

MN 9

Age: middle Miocene
Locality: Yassören, locality 94, Sinap Tepe, Central Anatolia, Turkey
Lithostratigraphic Position: from the fluviatile Sinap Formation with a total thickness of more than 100 m
Mammal Correlation: early MN 9, according to Kappelman et al. (this volume) and Sen (this volume)
Inferred Age: (Kappelman et al., this volume) 10.24 Ma
References: Kappelman et al., this volume; Sen 1989, this volume
Remarks: Kappelman et al. (this volume) calculate the MN 8/9 boundary half-way inbetween localities 64 and 94 and estimate an age for this boundary of 10.31 Ma.

MN 9

Age: late Miocene
Locality: Yassören, locality 8a (= Loc. I of Ozansoy 1957; 1965), 87 Sinap Tepe, Central Anatolia, Turkey
Lithostratigraphic Position: from the fluviatile Sinap Formation with a total thickness of more than 100 m. Mammal localities from base to top in stratigraphic order: locality 87 (with one "Hipparion" molar) and locality 8A (with a rich fauna of large and small mammals).
Mammal Correlation: MN 9 (Kappelman et al., this volume; Sen, this volume)
Inferred Age: (Kappelman et al., this volume; Sen, here) 10.31 Ma for locality 87 and 9.77 Ma for locality 8A
References: Kappelman et al., this volume; Ozansoy 1957, 1965; Sen 1989, this volume

MN 9/10

Age: late Miocene
Locality: Bou Hanifia, Horizon BH 1, Algeria
Mammal Correlation: medial to upper MN 9 (Steininger et al. 1989); alternatively, near to the MN 9/10 boundary (Woodburne et al., this volume; Swisher, this volume); BH 5 MN 10 or younger
Correlation Tie Points: beds with planktonic fauna of Blow Zone N 15 about 100 m below mammal horizon BH 1 with "Hipparion." These marine beds intertongue with dated ash beds (see below).
Paleomagnetic Calibration: Horizon BH 1 is situated within a long reversed polarity zone, which is followed by a short normal and again a reversed zone. On top is BH 5 again, in a normal polarity zone (see MN 10 horizon BH 5).
Isotopic Age: radiometric date 12.03 ±0.25 Ma from ash beds at least 100 m below mammal horizon BH 1 with "Hipparion." These ash beds intertongue with marine beds with a planktonic fauna assigned to Blow Zone N 15 (Berggren, pers. comm.).
Inferred Age: (Berggren et al. 1995) Blow Zone N 15 (= Berggren et al. 1995, Planktonic Foraminiferal Zone M 12 and lowermost part of M 13, 11.35 to 10.7 Ma). The reversed part of the section with mammal horizon BH 1 followed by a short normal can either be correlated to Chron C5n.1r and Chron

C5n.1n: 9.92 to 9.74 Ma or to Chron C4Ar.2n and Chron C4Ar.2r: 9.64 to 9.31 Ma.
References: Bernor et al. 1988; Sen 1989; Woodburne et al., this volume

European Mammal Faunal Zone: MN 10

MN 10

Age: late Miocene
Locality: Yassören, locality 84 (Ozansoy 1957; 1965), Sinap Tepe, Central Anatolia, Turkey
Lithostratigraphic Position: from the fluviatile Sinap Formation with a total thickness of more than 100 m.
Mammal Correlation: early MN 10 (Kappelman et al., this volume; Sen, this volume).
Inferred Age: (Kappelman et al., this volume; Sen, this volume) 9.77 Ma
References: Kappelman et al., this volume; Ozansoy 1957, 1965; Sen 1989, this volume

MN 10

Age: late Miocene
Locality: Bou Hanifia, Horizon BH 5, Algeria
Mammal Correlation: BH 5 is MN 10 (or younger, Swisher et al., this volume).
Correlation Tie Points: see explanation for Bou Hanifia I, above.
Paleomagnetic Calibration: see explanation for Bou Hanifia I, above.
Isotopic Age: radiometric date 12.03 ±0.25 Ma from ash beds more than 100 m below mammal horizon BH 1 with "Hipparion." These ash beds intertongue with marine beds with planktonic fauna assigned to Blow Zone N 15 (Berggren, pers. comm.).
Inferred Age: (Berggren et al. 1995) Blow Zone N 15 (= Berggren et al. 1995, Planktonic Foraminiferal Zone M 12 and lowermost part of M 13, 11.35 to 10.7 Ma). The reversed part of the section with mammal horizon BH 1 is succeeded by a short normal interval that can either be correlated with Chron C5n.1r to Chron C5n.1n: 9.88 to 9.74 Ma or with Chron C5n.1r to Chron C4Ar.2r, 9.92 to 9.58 Ma. The beds with BH 5 faunas are in a succeeding normal polarity portion of the sequence. This leads to two possible correlations for the Bou Hanifia faunas: BH 1 is in the reversed part Chron C4Ar.2r and normal part with BH 5 fauna Chron C4Ar.2n: 9.64 to 9.58, or reversed part Chron C4Ar.1r and normal part with BH 5 faunal horizon in Chron C4Ar.1n: 9.30 to 9.23 Ma.
References: Bernor et al. 1988, 1993a, b; Sen 1986, Woodburne et al., this volume

MN 10

Age: late Miocene
Locality: Oued Zra, Morocco
Mammal Correlation: middle to upper part of MN 10 (Steininger et al. 1989); uppermost MN 10 (Sen, pers. comm., 1994)
Isotopic Age: mammal locality younger than: 9.7 ±0.5 Ma
References: Jaeger et al. 1973; Steininger et al. 1989

MN 10

Age: late Miocene

Locality: Kastellios 1, 2, 3, Greece

Mammal Correlation: near the MN 9/10 boundary (Steininger et al. 1989); alternatively MN 10 (Benda and Meulenkamp 1990); alternatively medial MN 10 (de Bruijn et al. 1992); alternatively uppermost MN 10 (KA 1; Mein et al. 1993); see Sen (this volume) for further discussion.

Correlation Tie Points: planktonic foraminifera Blow Zone: N 16; base of Kizilhisar E-Mediterranean pollen zone

Paleomagnetic Calibration: within Chron C4Ar with K 1 in reversed part followed by a normal part, K 2 at the base of the following reversed part followed by a normal part and K 5 in the following reversed part (Opdyke, written comm., 1990).

Inferred Age: Blow Zone N16 = Berggren et al. 1995, Planktonic foraminifera Zone M13a to M13b pro parte: 10.8 to 5.9 Ma. The magnetic pattern is correlated from the base to the top of the section with Chron C4Ar.2r: 9.58 to 9.30 Ma (Kastellios 1) to Chron C4Ar.1r base: 9.23 Ma (Kastellios 2) to Chron C4An: 9.02 Ma (Kastellios 5).

References: Benda and Meulenkamp 1990; de Bruijn and Zachariasse 1979; de Bruijn et al. 1992; Sen et al. 1986; Sen, this volume; Steininger et al. 1989; Woodburne et al., this volume

MN 10

Age: late Miocene

Locality: Ravin de Pluie, Macedonia, Greece

Mammal Correlation: MN 10 (Bonis et al. 1988); alternatively medial MN 10 (de Bruijn et al. 1992); alternatively lower MN 10 (Mein et al. 1993)

Correlation Tie Points: Blow Zone N 16 (Koufos, pers. comm., 1990)

Inferred Age: (Berggren et al. 1995) Blow Zone N 16 (= Berggren et. al. 1994, Planktonic Foraminifera Zone M13a to M13b pro parte): 10.8 to 5.9 Ma

References: Bonis and Koufos 1981; Bonis et al. 1974, 1986, 1988; Bouvrain 1975, 1982; de Bruijn et al. 1992; Koufos 1986, 1989

MN 10

Age: late Miocene

Locality: Lefkon 1, Greece

Mammal Correlation: MN 10 uppermost part, according to Steininger et al. (1989) and de Bruijn et al. (1992)

Correlation Tie Points: lower Kizilhisar E-Mediterranean pollen zone

References: Benda and Meulenkamp 1989, 1990; de Bruijn 1989; de Bruijn et al. 1992; Steininger et al. 1989

MN 10

Age: late Miocene

Locality: Xirochori 1, Macedonia, Greece

Mammal Correlation: MN 10 (Koufos, pers. comm., 1990)

Correlation Tie Points: Blow Zone N 16 (Koufos, pers. comm., 1990)

Inferred Age: (Berggren et al. 1995) Blow Zone N 16 (= Berggren et. al. 1994, Planktonic Foraminifera Zone M13a to M13b pro parte): 10.8 to 5.9 Ma

References: Bonis et al. 1989, 1990

European Mammal Faunal Unit: Turolian

European Mammal Faunal Zone: MN 11

MN 11

Age: late Miocene

Locality: Prochoma 1, Macedonia, Greece

Lithostratigraphic Position: fluviatile deposits

Mammal Correlation: MN 11 (Koufos, pers. comm., 1990); MN 11 (Sen, this volume; Kondopoulou et al. 1992)

Paleomagnetic Calibration: Chron 10 = Chron C4Ar lower part (Sen, pers. comm., 1990). Kondopoulou et al. (1992) report magnetic reinvestigations of 25 m of section that yield the following observations: the basal 5 m have a normal polarity and the remainder of the section has a reversed polarity.

Inferred Age: The current geomagnetic results are correlated tentatively to Chron C4Ar.2n and C4Ar.1r: 9.64 to 9.02 Ma.

References: Bonis et al. 1986, 1988; Geraads 1978, 1979; Kondopoulou et al. 1992; Koufos 1987b, 1989; Sen, this volume

MN 11

Age: late Miocene

Locality: Kayadibi, Turkey

Mammal Correlation: lowermost part of MN 11 (Steininger et al. 1989; de Bruijn et al. 1992)

Correlation Tie Points: lower Kizilhisar E-Mediterranean pollen zone

Isotopic Age: the Bulumya ignimbrite situated below the mammal faunal horizon is dated 9.4 ±0.2 Ma while the Detse ignimbrite above mammal faunal horizon is dated 7.95 ±0.25 Ma.

References: Becker-Platen et al. 1975; Benda and Meulenkamp 1989, 1990; de Bruijn et al. 1992; Sickenberg et al. 1975; Steininger et al. 1989

MN 11

Age: late Miocene

Locality: Kopran, Lower Maragheh, Iran

Lithostratigraphic Position: The Maragheh Formation rests unconformably on top of the "Basal Tuff" dated 10.391 to 10.432 Ma. The Maragheh Formation consists approximately of 300 m of coarsley stratified deposits of brown to tan andesitic volcanic sands and silts. Succeeding an erosional surface follows the "Village Pumice," constituting the top of the Maragheh Formation.

Mammal Correlation: basal MN 11 (Steininger et al. 1989; Bernor et al., this volume b); defines the base of the *Hipparion gettyi* Zone (Bernor et al., this volume b)

Isotopic Age: The lower Maragheh fauna (= faunas in the Kopran and Mirduq sections) have an interpolated age of approximately 8.64 to 8.24 Ma (Swisher, this volume) interpreted by Bernor et al. (this volume b) to be 9 to 8.24 Ma, based on an interpolated date from the Mirduq Tuff to the stratigraphically lower Kopran horizons.

References: Campbell et al. 1980; Bernor 1985, 1986; Bernor et al., this volume b; Steininger et al. 1989; Swisher et al., this volume

MN 11

Age: late Miocene

Locality: Ravin des Zouaves 5, Macedonia, Greece

Mammal Correlation: MN 10/11 (Bonis et al. 1988); MN 11 (de Bruijn et al. 1992)

Correlation Tie Points: Blow Zone N 16/17 (Koufos, pers. comm., 1990)

Inferred Age: Blow Zone N 16/17 (= Berggren et. al., 1995, Planktonic Foraminifera Zone M13a to M13b pro parte and M 14): 10.8 to 5.6 Ma

References: Arambourg and Piveteau 1929; Bonis et al. 1988; Bouvrain 1982; de Bruijn et al. 1992; Geraads 1978, 1979; Koufos 1987, 1989

MN 11

Age: late Miocene

Locality: La Celia (Los Gargantones), Spain

Mammal Correlation: lower MN 11 (de Bruijn et al. 1992)

Isotopic Age: lava beds immediately above the mammal faunal horizon dated 7.2 to 7.6 Ma

References: Agusti 1990, Agusti in Steininger et al. 1987, 1989, de Bruijn et al. 1992

Remarks J. Agusti (writ. comm., 1994) states that a lava bed in the La Celia section is placed immediately above the layer with the typical MN 11 mammal fauna and believes that the base of MN 11 must not be much older than these age determinations. This interpretation is in conflict with the E-Mediterranean calibration of basal MN 11 (ca. 9 Ma; see above).

MN 11

Age: late Miocene

Locality: Crevillente 1 to 3, Spain

Mammal Correlation: Crevillente 2 correlative with basal MN 11, Aguilar Zone D 3 (Steininger et al. 1989); middle part of MN 11 (de Bruijn et al. 1992); lower MN 11 (Mein et al. 1993)

Correlation Tie Points: planktonic foraminifera assigned to Blow Zone N 16

Inferred Age: Blow Zone N 16 (= Berggren et. al., 1995, Planktonic Foraminifera Zone M13a to M13b pro parte): 10.8 to 5.9 Ma

References: Aguilar 1982a; Alcala and Montoya 1991, 1994; de Bruijn et al. 1992; Steininger et al. 1989

MN 11

Age: late Miocene

Locality: Samos: Old Mill Beds, Quarries X, G and 6, Greece

Lithostratigraphic Position: lithological sequence from base to

top: Basal Conglomerate unconformably on Mesozoic basement; Pythagorion Formation: thick-bedded freshwater limestone and paleosoils, lignitic marls, basalts dated 11.2 Ma; Hora Formation: thinly bedded freshwater limestones, with volcanic unit at top dated 9 Ma; Mytilini Formation: floodplain deposits with volcanogenic marls, gravels, soil horizons and rhyolite pumice tuffs and with four members: lowest–Old Mill Beds Member, well-bedded marls and tuffs with paleosoils including Quarries Q X, G and Q 6, best estimated age is 8.33 ±0.05 Ma (Swisher, this volume); Gravel Bed member: channel fills truncating Old Mill Beds; White Beds member: indurated marls and limestones, perhaps quarry Q 4 (7.66 Ma); Main Bone Beds member: volcanoclastic sandy silts, marls, tuffs, local paleosoils, and gravel lenses overlain by a thick limestone conglomerate: quarries: Q 1, Q 2, Q 3, Q 5, S 3, S 4 all late MN 12, approximately 7.3–7.1 Ma. On top of the Mytilini Formation follow with a disconformity, the Marker Tuffs: well-bedded indurated water lain tuffs and marls; quarry L, MN 13 with best estimated age of 6.17 ±0.05 Ma (Swisher, this volume). On top the *Kokkarion Formation:* thick-bedded freshwater algal limestones.

Mammal Correlation: Quarries X, G and 6 from Old Mill Beds: MN 11 (Steininger et al. 1989; Bernor et al., this volume b). Main Bone Beds uppermost MN 12 (Bernor et al., this volume b)

Paleomagnetic Calibration: Old Mill Beds should be in reversed polarity zone.

Isotopic Age: Old Mill Beds 8.33 ±0.05 Ma

Inferred Age: by cross-correlation including radiometric ages and geomagnetic calibration: Old Mill Beds in reversed polarity zone correlated to Chron C4r.1r: 8.22 to 8.07 Ma.

References: Bernor et al., this volume b; de Bruijn et al. 1992; Sen 1986, this volume; Solounias 1981; Steininger at al. 1989; Swisher, this volume; Weidmann et al. 1984

MN 11

Age: late Miocene

Locality: Garkin, Turkey

Mammal Correlation: upper MN 11 (Steininger et al. 1989); alternatively, medial MN 11 (de Bruijn et al. 1992)

Correlation Tie Points: lower Kizilhisar E-Mediterranean pollen zone

Isotopic Age: Radiometric date of 8.6 Ma below faunal horizon

References: Becker-Platen et al. 1975; Benda and Meulenkamp 1989, 1990; de Bruijn et al. 1992; Sickenberg et al. 1975; Steininger et al. 1989

MN 11

Age: late Miocene

Locality: Pertuis, France

Mammal Correlation: Mammal Zone D 3 (Aguilar); medial MN 11 (de Bruijn et al. 1992)

Correlation Tie Points: the rodent-bearing horizon is stratigraphically above marine sediments containing planktonic foraminifera correlative with Blow Zone N 15 or N 16.

Inferred Age: (Berggren et al. 1995) Blow Zone N 15 (= Berggren et al. 1995, Planktonic Foraminifera Zone M 12 and lowermost part of M 13a): 11.35 to 10.5; Blow Zone N 16 (= Berggren et. al. 1994 Planktonic Foraminifera Zone M13a to M13b pro parte): 10.8 to 5.9 Ma

References: Aguilar 1981, 1982a, b; de Bruijn et al. 1992; Clauzon et al. 1987, 1989

Remarks: There are no marine deposits known in southern France younger than N 15 to N 16.

MN 11

Age: late Miocene

Localities: Andance; Le Combier; St. Bauzile; Valréas -CD 56; Lobrieu and Mollon; Ardèche; France

Lithostratigraphic Position: approximately 170 m of section; in the upper part there are approximately 40–45 m of "diatomites principales" separated by lignites from about 10 m of "conglomerats fluviatiles rouges" overlain by "diatomites supérieures" and "argiles rouges." On top the "coulée basaltique du plateau des Coirons" is dated 6.4 Ma.

Mammal Correlation: MN 11 (Demarcq et al. 1989); upper part of MN 11 (de Bruijn et al. (1992); medial MN 11 (Mein et al. 1993)

Correlation Tie Points: diatomites in connection with the marine gulf of the Montelimar Basin in lower Tortonian: Blow Zone N 16.

Inferred Age: (Berggren et al. 1995) Blow Zone N 16 (= Berggren et. al., 1995, Planktonic Foraminifera Zone M13a to M13b pro parte): 10.8 to 5.9 Ma

References: de Bruijn et al. 1992; Demarcq et al. 1989; Michaux 1971

Remarks: Mammal fauna from "diatomite principalé' and "conglomerat supérieur."

MN 11/12

Age: late Miocene

Locality: Pikermi, Greece

Lithostratigraphic Position: Pikermi Formation, pale to dark red clayey silts with limestone conglomerate lenses and layers of hard limey marls

Mammal Correlation: MN 12/13 boundary (Steininger et al. 1989); alternatively MN 13 (Benda and Meulenkamp, 1989); alternatively uppermost MN 12 (de Bruijn et al. 1992); alternatively near the MN 11/12 boundary because of co-occurrence of *Hipparion gettyi* and *Hipparion prostylum* (Bernor et al., this volume b)

Correlation Tie Points: uppermost Kizilhisar E-Mediterranean pollen zone

Inferred Age: 8.3 to 8.2 Ma (Bernor et al., this volume b)

References: de Bruijn 1979; Benda and Meulenkamp 1989, 1990; Steininger et al. 1989; de Bruijn et al. 1992; Bernor et al., this volume b

European Mammal Faunal Zone: MN 12

MN 12

Age: late Miocene

Locality: Middle Maragheh, Iran

Lithostratigraphic Position: on top of a basal tuff (80 m) dated 10.391 to 10.432 Ma follows the Lower Maragheh *Hipparion gettyi* Zone within approximately 300 m, a coarsley stratified unit of brown to tan andesitic volcanic sands and silts. On top of an erosional surface follows the "Village Pumice," which forms the top of the Maragheh Formation.

Mammal Correlation: middle to upper faunas MN 11 (Steininger et al. 1989); the base of the *Hipparion prostylum* Zone is correlative with basal MN 12 for reasons given by Bernor et al. (this volume).

Isotopic Age: Kerjabad mammal fauna with an interpolated age of approximately 8.1 to 7.9 Ma (Swisher, this volume); base of *Hipparion prostylum* Zone in Maragheh section interpolated to be 8.24 Ma (Bernor et al., this volume b).

References: Campbell et al. 1980; Bernor 1985, 1986; Steininger et al. 1989; Bernor et al., this volume b; Swisher, this volume

MN 12

Age: late Miocene

Locality: Samos: Main Bone Beds Member (Quarries: Q 1, Q 2, Q 3, Q 5, S 3 and S 4) (= also Lower and Upper fossiliferous level; Samos 5 [de Bruijn et al. 1992]; Upper fossiliferous level with localities Q 5, Q 1, and A of Sen and Valet 1986), Greece

Lithostratigraphic Position (lithological sequence from base to top): (1) Basal Conglomerate unconformably resting on Mesozoic basement. (2) Pythagorion Formation: thick bedded freshwater limestone and paleosoils, lignitic marls, basalts dated 11.2 Ma. Hora Formation: thin-bedded freshwater limestones, at top volcanic unit dated 9 Ma. (3) The Mytilini Formation consists of floodplain deposits with volcanogenic marls and gravels, soil horizons, and rhyolite pumice tuffs. (4) The Mytilini Formation includes four members. (5) The lowest is the Old Mill Beds member with well-bedded marls and tuffs with paleosoils and including Quarries X, G and 6; best estimated age is 8.33 ±0.05 Ma (Swisher, this volume). (6) The succeeding Gravel Bed member consists of channel fills truncating the Old Mill beds. (7) The succeeding White Beds member consists of indurated marls and limestones and perhaps includes Quarry 4 (date 7.66 Ma; Swisher, this volume). (8) Next is the Main Bone Bed member, which has yielded the majority of the Samos fossil material. It consists of volcanoclastic sandy silts, marls, tuffs, local paleosoils, and gravel lenses overlain by a thick limestone conglomerate; fossil quarries include Q 1, Q 2, Q 3, Q 5, S 3 and S 4, all correlative with late MN 12, between 7.3 and 7.1 Ma. (9) Succeeding the Mytilini Formation is a disconformity and the "Marker Tuffs," with well-bedded indurated water lain tuffs and marls. Solounias (in Bernor et al., this volume b) has identified quarry L from this unit, which has a best estimated age of 6.17 ±0.05 Ma (Swisher, this volume; Bernor et al., this volume b), and is the only portion of the Samos fauna that is MN 13. The top of the sequence includes the Kokkarion Formation, containing thick-bedded freshwater algal limestones.

Mammal Correlation: (from the Mytilini Formation): Bernor et al. (1980) have argued, based on the Samos hipparion lineages, that the bulk of the Samos fauna is ca. 7 Ma; Weidmann et al. (1984) have argued that there is a Lower and Upper fossiliferous level correlative with MN 12; alternatively Steininger et al. (1989) have correlated the bulk of the Samos fossils with basal

MN 13; de Bruijn et al. (1992) correlate the Samos Main Bone Beds member (Quarries: Q 1, Q 2, Q 3, Q 5, S 3 and S 4) with the lower part of MN 13; Bernor et al. (this volume b) correlate the Samos Main Bone Beds with upper MN 12 (ca. 7.66–7.1 Ma).

Paleomagnetic Calibration: The Mytillini Formation, including the Main Bone Bed member, Lower and Upper fossiliferous levels, respectively in a mixed polarity sequence. Samos 5 is in the "Upper level" within a reversed polarity zone.

Isotopic Age: Mytilini Formation, main quarries of the Main Bone Beds bracketed by radioisotopic determinations of 7.276 ± 0.006 Ma and 7.092 ± 0.008 Ma.

Inferred Age: (Berggren et al. 1995) by cross-correlation including radiometric ages and geomagnetic calibration, a lower fauna ca. 8.33 Ma and MN 11 equivalent, and an upper fauna ca. 7.66–7.1Ma and MN 12 equivalent.

References: Bernor et al., this volume b; Sen 1986, this volume; de Bruijn et al. 1992; Sen and Valet 1986; Solounias 1981; Steininger at al. 1989; Swisher et al., this volume; Weidmann et al. 1984

MN 12

Age: late Miocene

Locality: Casa del Acero, Spain

Mammal Correlation: basal MN 12 (Steininger et al. 1989; de Bruijn et al. 1992); medial MN 12 (Mein et al. 1993)

Correlation Tie Points: is included within the third Messinian evaporitic (first evaporite overlies *Globorotalia conomiozea* Zone).

Inferred Age: (according to Gaultier and al. 1994) Messinian evaporitic cycles dated 5.8 to 5.32 Ma

References: Agusti in Steininger 1987; Alcala and Montoya 1991, 1994; in press; de Bruijn et al. 1992; Gaultier et al. 1994; Steininger et al. 1989

Remarks: J. Agusti (May 1994) indicates that the typical MN 12 mammal fauna of Casa del Acero is included within the third evaporitic cycle of the Messinian succession in this basin. Therefore the MN 11/12 boundary must be younger than 7.2 Ma (= base of the Messinian).

MN 12

Age: late Miocene

Locality: Kinik, Turkey

Mammal Correlation: medial MN 12 (Steininger et al. 1989); basal MN 12 (de Bruijn et al. 1992)

Correlation Tie Points: upper Kizilhisar E-Mediterranean pollen zone

References: Benda and Meulenkamp 1989; de Bruijn et al. 1992; Sickenberg et al. 1975

MN 12

Age: late Miocene

Locality: Sholàvand, Upper Maragheh, Iran

Lithostratigraphic Position: see description above for the Lower and Middle Maragheh faunas.

Mammal Correlation: early MN 12 (Steininger et al. 1989 and Swisher et al., this volume); Bernor et al. (this volume b) correlate *Hipparion campelli* Zone with medial MN 12.

Isotopic Age: Sholàvand mammal faunal with an interpolated age of approximately 7.9 to 7.64 Ma (Swisher, this volume), or 8.0 to 7.6 Ma (Bernor et al., this volume b).

References: Bernor 1985, 1986; Bernor et al., this volume b; Campbell et al. 1980; Steininger et al. 1989; Swisher et al., this volume

MN 12

Age: late Miocene

Locality: Fuente Podrida, Cabriel Basin south of Venta del Moro, Spain

Lithostratigraphic Position: locality in Fuente Podrida limestones

Mammal Correlation: MN 12 (Opdyke et al. 1988)

Paleomagnetic Calibration: the Fuente Porida fauna occurs in the first normal magnetozone (N 1) at the base of the section, this magnetozone N 1 is correlated with the lowermost normal of Chron 7 by Opdyke et al. (1989) = Chron C4n.

Inferred Age: 7.2 Ma by Opdyke et al. 1988 = Chron C4n.2n: 8.07 to 7.65 Ma

References: Mein et al. 1978; Opdyke et al. 1988; Sen, this volume

Remarks: Opdyke et al. (1989) argue that the boundary between MN 11 and MN 12 should be only slightly older, ca. 7.3 Ma.

MN 12

Age: late Miocene

Locality: Novaya Emetovka 1, NW Odessa, Ukrania

Lithostratigraphic Position: 5 m of cyclic sedimentation of sandstones, sands, silts, and clays with thin intercalations of tuffaceous limestone together with endemic mollusks. Bone horizon in the lowermost cycle. The sequence follows unconformably on top of upper Sarmatian (ca. 13 m).

Mammal Correlation: Lower portion of MN (Krakhmalnaya et al. 1993)

Correlation Tie Points: early Maeotian by endemic mollusk fauna

Paleomagnetic Calibration: mammal fauna in a reversed polarity zone. See also Novaya Emetovka 2. This geomagnetic pattern is correlated near to base of Chron 6 = obviously thought to correlate within Chron C3Br to C3Bn.

Inferred Age: (Berggren et al., 1995) if we follow this interpretation of the geomagnetic polarity, the reversed zone can be correlated to Chron C3Br: 7.43 to 7.09 Ma; the inferred age of the Maeotian ranges from 9.8 to 7.1 Ma.

References: Krakhmalnaya et al. 1993

MN 12

Age: late Miocene

Locality: Novaya Emetovka 2, NW Odessa, Ukrania

Lithostratigraphic Position: 40 m of cyclic sedimentation of sandstones, sands, silts, and clays with thin intercalations of tuffeceous limestone together with endemic mollusks. The bone horizon is about 10 m above the erosional base. Lowermost cycle of Emetovka missing. The sequence follows unconformably on top of late Sarmatian (ca. 10 m).

Mammal Correlation: lower MN 12 (Krakhmalnaya et al. 1993)

Correlation Tie Points: early Maeotian by endemic molluscan fauna

Paleomagnetic Calibration: the basal part of this Maeotian section is above the erosional surface up to 12 m in a reversed zone (the bone bed is between meters 10 and 11 and equal to R-1, up to meter 26, a normal zone equal to N-2, followed by a short reversed part up to meter 27, 5 equals to R-3 and up to meter 40 at the top of the section, a normal zone equals to N-3). The geomagnetic pattern is correlated to Chron 6 and lower Chron 5. According to the GPTS of Berggren et al. (1995), this geomagnetic pattern can be interpreted as follows: the R-1 is equal to Chron C3Br; N-1 equals to Chron C3Bn; R-2 is equal to Chron C3Ar; N-2, R-3 and N-3 equal to Chron C3An.

Inferred Age: (Berggren et al. 1995) if we follow this interpretation, the bone-bearing horizon would correlate to the reversed base of Chron C3Br: 7.43 to 7.09 Ma.

References: Krakhmalnaya et al. 1993

European Mammal Faunal Zone: MN 13

MN 13

Age: late Miocene

Locality: Molina de Segura, horizon D and 1, Spain

Mammal Correlation: lowermost faunal level of MN 13 (Steininger et al. 1989 and de Bruijn et al. 1992); alternatively, MN 12/13 boundary (Agusti et al. 1987a, b)

Correlation Tie Points: overlies third Messinian evaporitic series

Inferred Age: (to Gaultier et al. 1994) Messinian evaporitic cycles and younger than 5.8 (i.e., 5.8 to 5.32 Ma)

References: Agusti in Steininger et al. 1989; de Bruijn et al. 1992

MN 13

Age: late Miocene

Locality: Samos, Marker Tuffs, Quarry L, Greece

Lithostratigraphic Position: lithological sequence from base to top as described above for Samos

Mammal Correlation: MN 13 correlative (Bernor et al., this volume b)

Isotopic Age: Marker Tuffs dated at 6.2 Ma (6.17 ±0.05 Ma) (Swisher, this volume).

References: Bernor et al. 1980, this volume b; de Bruijn et al. 1992; Solounias 1981; Sen 1986, this volume; Sen and Valet 1986; Steininger et al. 1989; Swisher et al., this volume; Weidmann et al. 1984

MN 13

Age: late Miocene

Locality: Venta del Moro, Spain

Mammal Correlation: lower to medial MN 13 (Steininger et al. 1989); medial MN 13 (de Bruijn et al. 1992)

Paleomagnetic Calibration: within a normal polarity zone correlated with Anomaly 3A = Chron C3An

Inferred Age: (Berggren et al. 1995) Chron C3An: 6.56 to 5.89 Ma. Age inferred by Opdyke et al. (1988) is 5.7 Ma.

References: Azanza et al. 1989; de Bruijn et al. 1992; Cerdeno 1989; Morales 1984; Opdyke et al. 1988; Sen, this volume; Steininger et al. 1989

MN 13

Age: late Miocene

Locality: Crevillente 6, Spain

Mammal Correlation: medial MN 13 (Steininger et al. 1989 and de Bruijn et al. 1992); Aguilar Zone E 2

Correlation Tie Points: *Globorotalia conomiozea* Zone = lower Messinian

Inferred Age: (Gaultier and al. 1994) in Messinian pre-evaporitic series, 7.2 to 5.8 Ma

References: de Bruijn et al. 1975, 1992; Aguilar 1982a; Opdyke et al. 1989; Steininger et al. 1989

MN 13

Age: late Miocene

Locality: Librilla, Spain

Mammal Correlation: medial to upper MN 13 (Aguilar Zone E 2; Steininger et al. 1989); upper MN 13 (de Bruijn et al. 1992)

Isotopic Age: mammal faunal horizon underlain by volcanic unit dated at 7.00, 6.5 and 6.2 Ma

References: Aguilar 1982a; Alberdi et al. 1981; de Bruijn et al. 1975, 1992; Montenat et al. 1975; Steininger et al. 1989

MN 13

Age: late Miocene

Locality: Dytiko 1, 2, 3, Macedonia, Greece

Mammal Correlation: MN 13 (Koufos, pers. comm., 1990); alternatively medial MN 13 (de Bruijn et al. 1992)

Correlation Tie Points: Blow Zone N 17 (Koufos, pers. comm., 1990)

Inferred Age: (Berggren et al. 1995) Blow Zone N 17 (= Berggren et al. 1995, upper Planktonic Foraminifera Zones M 13b and M 14): 7.2 to 5.6 Ma.

References: Arambourg and Piveteaux 1929; Bouvrain 1978; Bonis et al. 1988; Koufos 1987c, 1988b, 1989; de Bruijn et al. 1992

MN 13

Age: late Miocene

Locality: La Alberca, Spain

Mammal Correlation: uppermost faunal horizon of MN 13 (= Aguilar Zone E 3; Steininger et al. 1989 and de Bruijn et al. 1992)

Correlation Tie Points: between marine sediments of Messinian age containing planktonic foraminifera of Zone M 12 (Berggren 1977) and the last occurrence of *Discoaster quinqueramus* indicating a late Messinian age

Inferred Age: (Berggren et al. 1995) slightly older than 5.3 Ma (last occurrence of *Discoaster quinqueramus* is at 5.3 Ma).

References: Aguilar 1982a; de Bruijn et al. 1975, 1992; Mein et al. 1973; Morales 1984; Steininger et al. 1989

MN 13

Age: late Miocene
Locality: Brisighella, Monticino quarry, Faenza, Italy
Lithostratigraphic Position: mammal horizon overlain by marine sediments of lowermost Zanclean age; i.e., deposited after the intra-Messinian tectonic phase (Marabini and Vai 1988)
Mammal Correlation: MN 13 (Guilio et al. 1988; Torre, pers. comm., 1990); alternatively uppermost MN 13 (de Bruijn et al. 1992)
Correlation Tie Points: the overlying marine sediments with Planktonic Foraminifera Zone MPL 1 and Nanno Zone NN 12 (lowermost Zanclean)
Paleomagnetic Calibration: the overlying Zanclean marine beds are in a normal magnetozone and correlated to the Thvera subchron = Chron C3n.4n.
Inferred Age: (Berggren et al. 1995; Rio and Sprovieri 1994) older than base of overlying Planktonic Foraminifera Zone MPL 1 (= Berggren et al. 1995 Planktonic Foraminifera Zone PL 1, 5.6 Ma and older than base of overlying Nannoplankton Zone NN 12, 5.3 Ma). The normal polarity of the marine beds is referred to Chron C3n.4n, mammal-bearing beds older than base of C3n.4n, 5.23 Ma.
References: de Bruijn et al. 1992; de Giulio and Vai 1988
Remarks: Torre (pers. comm., 1990) states that the fossiliferous sediments containing the mammal fauna unconformably overlie the structurally deformed gypsum beds of the intra-Messinian tectonic phase. Italian MN 13 localities are, in biostratigraphic order, Baccinello V 3 and Brisighella above.

MN 13

Age: late Miocene
Locality: Baccinello V 3, Grosseto, Italy
Lithostratigraphic Position: overlain by marine beds
Mammal Correlation: Bacinello V 3 is correlative with medial MN 14 (de Bruijn et al. 1992); alternatively MN 13 (Torre, pers. comm.; see below); alternatively MN 13/14 boundary (Mein et al. 1993).
Correlation Tie Points: overlying marine beds with *Globorotalia margaritae*
Inferred Age: (Torre, pers. comm., 1990) intra Messinian (i.e., between 7.1 to 5.25 Ma; see discussion with the locality of Brisighella)
References: de Bruijn et al. 1992; Engesser 1989; Hürzeler and Engesser 1976; Rook et al. (in press); Torre 1987
Remarks: Torre (pers. comm., 1990) states that in the past this faunal assemblage was correlated with the transition between MN 13 and MN 14 because it included *Dicerorhinus* cf. *megarhinus* and *Tapirus arvernensis*. New elements demonstrate that the assemblage is best correlated with MN 13 and with the pre–intra-Messinian tectonic phase (Rook et al., in press). This correlation is made based upon the structural deformation of the fossiliferous sediments and by the presence of a more primitive *Apodemus* than found from the Cava Monticino (Brisighella; re: Engesser 1989).

MN 13

Age: late Miocene
Locality: La Hornera, Spain
Mammal Correlation: MN 13 (Agusti 1987).
Correlation Tie Points: overlies third Messinian evaporitic series and *Globorotalia conomiozea* Zone
Inferred Age: (Gaultier et al. 1994) Messinian evaporitic series: 5.8 to 5.32 Ma
References: Agusti in Steininger et al. 1989

MN 13

Age: late Miocene
Locality: Rema Marmara, Greece
Mammal Correlation: MN 13 (Benda and Meulenkamp 1989, 1990)
Correlation Tie Points: uppermost Kizilhisar E-Mediterranean pollen zone
References: Benda and Meulenkamp 1989; Steininger et al. 1989

MN 13/14

Age: late Miocene/early Pliocene
Locality: Molina de Segura, Horizon E to 10, Spain
Mammal Correlation: MN 13/14, according to Agusti (1987)
Correlation Tie Points: overlies the third Messinian series in a continous section containing MN 12/13 faunas (see above).
Inferred Age: (Gaultier and al. 1994) Messinian evaporitic series from 5.8 to 5.32 Ma
References: Agusti in Steininger et al. 1989

European Mammal Faunal Unit: Ruscinian

European Mammal Faunal Zone: MN 14

MN 14

Age: early/late Pliocene
Locality: Villalba Alta Rio 1, Teruel graben, Spain
Mammal Correlation: uppermost MN 14 (de Bruijn et al. 1992)
Correlation Tie Points: Lower Alfambrian Stage, AF 1c
Paleomagnetic Calibration: normal polarity interval, interpreted to be the Cochiti Subchron (Mein et al. 1983); = Chron C3n.1n
Inferred Age: (Berggren et al. 1995) C3n.1n: 4.29 to 4.18 Ma
References: Adrover 1987; de Bruijn et al. 1992

MN 14

Age: early Pliocene
Locality: Celleneuve, France
Mammal Correlation: Mammal Zone F 1 (Aguilar); alternatively Zone G1 (Aguilar and Michaux); alternatively MN 14 (Mein 1975, 1989); alternatively basal MN 14 (Steininger et al. 1989 and de Bruijn et al. 1992)
Correlation Tie Points: marine influence proven by shark-teeth

and cysts of hystrichospaerids. NW-Mediterranean pollen zone P II

Paleomagnetic Calibration: reverse polarity in Gilbert (Lindsay 1985), base of Gilbert (Aguilar and Michaux 1984) = base of Chron C3r

Inferred Age: (Berggren et al. 1995) Chron C3r: 5.89 to 5.23 Ma

References: Aguilar and Michaux 1984; Aguilar et al. 1989; de Bruijn et al. 1992; Mein and Michaux 1970; Michaux 1966; Lindsay 1985; Suc and Zagwijn 1983

Remarks: The rodent-fauna indicates an early Pliocene age, not a middle Pliocene age.

MN 14

Age: early Pliocene

Locality: Ptolemais 1, Greece.

Mammal Correlation: medial MN 14 (Steininger et al. 1989; de Bruijn et al. 1992)

Correlation Tie Points: lower Akça E-Mediterranean pollen zone

References: Benda and Meulenkamp 1989, 1990; de Bruijn et al. 1992; Meulen and Kolfschoten 1986; Steininger et al. 1989; Van de Weerd 1979

MN 14

Age: early Pliocene

Locality: Terrats, France

Lithostratigraphic Position: the micromammal fauna was recovered from lignitic marls within a continental sequence.

Mammal Correlation: mammal Zone F 2 (Aguilar); alternatively MN 14 (Mein 1975, 1989), upper MN 14 (de Bruijn et al. 1992)

Correlation Tie Points: NW-Mediterranean pollen zone P II

Paleomagnetic Calibration: reverse and normal polarity in Gilbert (Lindsay 1985); base of Gilbert (Aguilar and Michaux 1984) = Chron C3r to C3n.4n

Inferred Age: (Berggren et al. 1995) Chron C3r to C3n.4n: 5.89 to 5.23 Ma

References: Aguilar and Michaux 1984; de Bruijn et al. 1992; Clauzon et al. 1989; Mein and Michaux 1970; Michaux 1976; Lindsay 1985

Remarks: drillings indicate marine Pliocene below the continental formation.

MN 14

Age: early Pliocene

Locality: Vendargues, France

Lithostratigraphic Position: the mammal fauna from Vendargues was recovered from a continental sequence of about 30 m at the quarry of Pioch Palat discordantly above sandy marls.

Mammal Correlation: Mammal Zone F 2 (Aguilar); alternatively MN 14 (Mein 1975, 1989); alternatively MN 14b (Fejfar and Heinrich 1989); upper MN 14 (de Bruijn et al. 1992)

Correlation Tie Points: NW-Mediterranean pollen zone P II (base)

Paleomagnetic Calibration: reversed polarity of the Gilbert Chron (Lindsay 1985); base of the Gilbert (Aguilar and Michaux 1984; = Chron C3r)

Inferred Age: (Berggren et al. 1995) Chron C3r: 5.89 to 5.32 Ma

References: Aguilar and Michaux 1984; de Bruijn et al. 1992; Lindsay 1985; Mein and Michaux 1970; Suc and Zagwijn 1983

MN 14

Age: early Pliocene

Locality: Spilia 1, Greece

Mammal Correlation: MN 14b according to Fejfar and Heinrich (1989)

Correlation Tie Points: lower Akça E-Mediterranean pollen zone

References: Benda and Meulenkamp 1989, 1990; Fejfar and Heinrich 1989

MN 14

Age: early Pliocene

Locality: Vivès 2, France

Mammal Correlation: Mammal Zone F 1 (Aguilar), MN 14, according to Aguilar and Michaux (pers. comm., 1990).

Paleomagnetic Calibration: reverse event in Gilbert.

Inferred Age: (Berggren et al. 1995) if actually the reversed zone at base of Gilbert, then this would correlate with Chron C3r: 5.89 to 5.23 Ma.

References: Clauzon et al. 1985, 1987; Aguilar et al. 1989

Remarks: the fossiliferous continental deposits follow above marine early Pliocene deposits.

MN 14

Age: early Pliocene

Locality: Elbistan, Turkey

Mammal Correlation: MN 14 (Benda and Meulenkamp 1989, 1990)

Correlation Tie Points: lower Akça E-Mediterranean pollen zone

Isotopic Age: radiometric date of 3.7 Ma

References: Benda and Meulenkamp 1989, 1990; Sickenberg et al. 1975

MN 14

Age: late Pliocene

Locality: Mesas de Asta section, Guadalquivir Basin, Cadiz, Spain

Lithostratigraphic Position: 60 m of section with two lithostratigraphic units. Unit 1 (approximately at 25 m) is a marine off-shore-to-lower shoreface deposit with hardground on top separated by an angular and erosive unconformity from overlying Unit 2 (approximately 35 m). Fluviatile and lacustrine deposits with a micromammal fauna at 40 meter mark of the section.

Mammal Correlation: uppermost MN 14 (Aguirre et al. 1992)

Correlation Tie Points: marine marls in upper part of unit 1 with planktonic foraminifera fauna of P12 Biozone (Berggren et al. 1995).

Inferred Age: (Berggren et al. 1995) Planktonic foraminiferal Biozone P12: 4.2 to 3.55 Ma; mammal horizon younger than 3.55 Ma

References: Aguirre et al. 1992

European Mammal Faunal Zone: MN 15

MN 15

Age: early to late Pliocene
Locality: Villalba Alta Rio 1 and Orrios 1 to Villalba Alta Rio 2 and Orrios 3 section, Spain
Mammal Correlation: upper MN 14 (Orrios 1 and Villalba Alto Rio 1); alternatively lowermost MN 15 (Villalba Alto Rio 2) and medial MN 15 (Orrios 3) (de Bruijn et al. 1992)
Paleomagnetic Calibration: lower Gauss = Chron C2An (Opdyke et al. 1988)
Inferred Age: (Berggren et al. 1995) base Chron C2An, 3.58 Ma
References: Adrover 1978; de Bruijn et al. 1992; Opdyke et al. 1988; Van de Weerd 1976

MN 15

Age: early/late Pliocene
Locality: Villalba Alta Rio 2, Teruel graben, Spain
Mammal Correlation: lowermost MN 15 (de Bruijn et al. 1992)
Paleomagnetic Calibration: in reversed part of the section interpreted as upper Gilbert, above Villalba Alta Rio 1 (Moissenet et al. 1990) = Chron 2Ar
Inferred Age: (Berggren et al. 1995) Chron C2Ar: 4.18 to 3.58 Ma. Inferred age between 3.85 and 3.40 Ma (Moissenet et al. 1990).
References: de Bruijn et al. 1992; Mein et al. 1983; Moissenet et al. 1990
Remarks: Mein (pers. comm., 1992) states that this fauna contains the most archaic *Mimomys* currently known (= *Mimomys vandermeuleni*).

MN 15

Age: late Pliocene
Locality: Dinar-Akçaköy, Turkey
Mammal Correlation: lower MN 14 (Steininger et al. 1989); alternatively lower MN 15 (de Bruijn et al. 1992)
Correlation Tie Points: lower Akça E-Mediterranean pollen zone
References: Benda and Meulenkamp 1989, 1990; de Bruijn et al. 1992; Sickenberg et al. 1975

MN 15

Age: early Pliocene
Locality: Perpignan (= Serrat d'en Vacquer), France
Mammal Correlation: Mammal Zone F3 (Aguilar); MN 15 (Mein 1975, 1989); alternatively medial MN (de Bruijn et al. 1992)
Correlation Tie Points: NW-Mediterranean pollen zone P II
Paleomagnetic Calibration: reverse polarity of Gilbert about 20 m below the fossiliferous level (Lindsay 1985)
References: de Bruijn et al. 1992; Clauzon et al. 1987; Hugueney and Mein 1965; Lindsay 1985; Mein and Aymar 1984; Suc and Zagwijn 1983
Remarks: The mammal locality is 200 m above the boundary between the marine and the continental Pliocene (Mutualité Agricole Drilling). Mein (pers. comm., 1992) states that the presence of *Mimomys* cf. *clavakosi* is a very accurate biostrati-

graphic marker. This suggests that Perpignan is younger than the localities of Ptolemais 3 and Villalba Alta Rio 2, but older than the locality of Sète.

MN 15

Age: late Pliocene
Locality: Escorihuela B, Teruel graben, Spain
Mammal Correlation: MN 15 (Moissenet et al. 1989); alternatively uppermost MN 15 (de Bruijn et al. 1992)
Paleomagnetic Calibration: reversed polarity of the uppermost Gilbert, just before the Gauss lowermost normal (Moissenet et al. 1990), i.e., the upper part of Chron C2Ar/lower part of Chron C2An.
Inferred Age: (Berggren et al. 1995) Chron C2Ar/C2An boundary, ca. 3.58 Ma. Inferred age is 3.4 Ma (Moissenet et al. 1989).
References: de Bruijn et al. 1992; Mein et al. 1983; Moissenet et al. 1989

MN 15

Age: late Pliocene
Locality: Sete, France
Mammal Correlation: uppermost MN 15 (= Aguilar and Michaux Zone G 1) (Steininger at al. 1989; de Bruijn et al. 1992)
Correlation Tie Points: NW-Mediterranean pollen zone P II
References: Aguilar and Michaux 1984; de Bruijn et al. 1992; Steininger et al. 1989; Suc 1980, 1982; Suc and Zagwijn 1983

MN 15

Age: early/late Pliocene
Locality: Villalba Alta Rio 5, Teruel graben, Spain
Mammal Correlation: MN 15 (Moissenet et al. 1989)
Paleomagnetic Calibration: at top of a reversed magnetic zone that is correlative with uppermost Gilbert/Gauss Chron transition (Moissenet et al. 1989); correlative with Chron C2Ar/C2n boundary.
Inferred Age: (Berggren et al. 1995) Chron C2Ar/C2An boundary: 3.58 Ma. Inferred age 3.4 Ma (Moissenet et al. 1989).
References: Mein et al. 1983; Moissenet et al. 1989

MN 15

Age: early Pliocene
Locality: Val di Pugna, Siena, Italy
Lithostratigraphic Position: in sandy marine sediments, macrofaunal assemblage with few biostratigraphic diagnostic elements
Mammal Correlation: MN 15 (Torre, pers. comm., 1990)
Correlation Tie Points: the sandy marine sediments are referred to the *Globorotalia puncticulata* Zone.
Inferred Age: (Berggren et al. 1995) *Globorotalia puncticulata* Zone, 4.6 to 3.6 Ma
References: Azzaroli et al. 1988

MN 15

Age: late Pliocene
Locality: "Karaburun" = Megalo Emvolon, Macedonia, Greece

Mammal Correlation: MN 14 (Benda and Meulenkamp 1989); alternatively MN 15 (Stefens et al. 1979 and de Bruijn 1984)

Correlation Tie Points: lower E-Mediterranean Akça pollen zone; Blow Zone N 19 (Koufos, pers. comm., 1990)

Inferred Age: (Berggren et al. 1995) Blow Zone N 19 (= Berggren et al. 1995 uppermost Planktonic Foraminifera Zone PL 1a and PL 1b): 4.7 to 4.2 Ma

References: Arambourg and Piveteau 1929; Benda and Meulenkamp 1989; de Bruijn 1984

European Mammal Faunal Unit: Villafranchian (Villanyian)

European Mammal Faunal Zone: MN 16

MN 16

Age: late Pliocene

Locality: Escorihuela, Teruel graben, Spain

Mammal Correlation: MN 16 (Moissenet et al. 1989); lowermost MN 16 (de Bruijn et al. 1992)

Paleomagnetic Calibration: normal polarity of lowermost Gauss (Moissenet et al. 1989); correlative with Chron C2An.3n

Inferred Age: (Berggren et al. 1995) Chron C2An.3n: 3.58 to 3.33 Ma. The inferred age is 3.4 Ma (Moissenet et al. 1989).

References: de Bruijn et al. 1992; Moissenet et al. 1989; Van de Weerd 1976

MN 16

Age: late Pliocene

Locality: Triversa sites, Asti, Italy

Mammal Correlation: MN 16a (early Villafranchian; Torre, pers. comm., 1990); alternatively lowermost MN 16 (Steininger et al., 1989 and de Bruijn et al. 1992)

Correlation Tie Points: indirect correlation to the *Globorotalia puncticulata/Globorotalia crassaformis* transition (see remarks).

Paleomagnetic Calibration: reversed-normal-reversed polarity pattern in Fornace RDB quarry assigned to the interval between Kaena and Mammoth of the Gauss Chron (Lindsay et al. 1980) and *not* as indicated in Steininger et al. (1989:36) to the Matuyama.

Inferred Age: (Berggren et al. 1995) The geomagnetic sequence described above can be correlated with either Chron C2Ar/C2An.3n/2r or Chron C2An.2r/2n/1r, with a time interval of 4.18 to 3.0 Ma. Because the *Gt. punticulata/Gt. crassaformis* Zone transition is recognized, the more likely geomagnetic correlation is Chron C2Ar/C2An.3n/2r, 4.18 to 3.22 Ma.

References: Azzaroli et al. 1982; de Bruijn et al. 1992; Giulio et al. 1988; Lindsay et al. 1980; Steininger et al. 1989; Torre 1987

Remarks: (Torre, pers. comm., 1990) The *Mimomys stehlini* from San Giusto (a species also present at Arondelli, Triversa) and a small Villafranchian cervid from Ponte a Elsa were both found in sediments of transitional environments containing pollen that allows their referral to the base of the early Villafranchian. The foraminifera present are commonplace, but a rich molluscan fauna indicates that the sediments can be no younger than the beginning of the Mediterranean's cooling (ca.

3.2 to 3.0 Ma), and therefore are correlative with the upper part of the *Globorotalia puncticulata* Zone (Valleri et al., in press).

MN 16

Age: late Pliocene

Locality: Arcille, Grosseto, Italy

Lithostratigraphic Position: the fossiliferous strata overlie marine sediments of the *Globorotalia puncticulata* Zone.

Mammal Correlation: MN 16a (Torre, pers. comm., 1990); lowermost MN 16 (de Bruijn et al. 1992), below the Triversa locality

Correlation Tie Points: above marine sediments of *Globorotalia puncticulata* Zone

Inferred Age: (Berggren et al. 1995) younger than 3.6 Ma

References: de Bruijn et al. 1992; Hürzeler and Engesser 1976; Torre 1987; Masini and Torre 1989

MN 16

Age: late Pliocene

Locality: Vialette, France

Lithostratigraphic Position: the mammal fauna from Vialette is found in a sequence with white marls at the base, overlain by a volcanic sedimentary sequence containing a volcanic block dated 3.3 Ma, overlain by a basalt flow dated 3.33 ±0.11 Ma (5 samples). The fauna rests on top of this flow. The basalt flow of d'Azanières to the NW is dated 2.18 ±0.15 (2 samples) and 2.60 ±0.20 (1 sample) and is thought to overly prior to its ersosion the faunal horizon of Vialette.

Mammal Correlation: lower part of MN 16 (Steininger et al. 1989; de Bruijn et al. 1992)

Isotopic Age: basalt flow below mammal horizon dated at 3.33 Ma

References: Alberdi and Bonadonna 1987; Azzaroli et al. 1988; Bandet et al. 1978; de Bruijn et al. 1992; Heintz et al. 1974; Ly Meng Hour et al. 1983; Steininger et al. 1989

MN 16

Age: late Pliocene

Locality: Poggio Mirteto, Rome, Italy

Mammal Correlation: lower part of MN 16 (Steininger et al. 1989)

Correlation Tie Points: intertonguing with *Globorotalia aemiliana* (= *crassaformis*) Zone.

Paleomagnetic Calibration: reversed magnetic polarity

Isotopic Age: ash bed dated at 3.32 ±0.3 Ma

References: Alberdi and Bonadonna 1987

MN 16

Age: late Pliocene

Locality: Valdeganga 9, 9b and 16a (like Rincon 2); Valdeganga 4a and 10b (Valdeganga inf.); Júcer Basin, Spain

Mammal Correlation: medial MN 16 (Mein et al. 1990), like Rincon 2; MN 16 (= "Valdeganga inf.") (de Bruijn et al. 1992)

Paleomagnetic Calibration: lowermost part of the reversed polarity interval of the Matuyama (Opdyke et al. 1989), equal to Chron C2r.2r

Inferred Age: (Berggren et al. 1995) Chron C2r.2r: 2.6 to 2.2 Ma
References: Alberdi et al. 1982; de Bruijn et al. 1992; Mein et al. 1989

MN 16

Age: late Pliocene
Locality: Valdeganga 1 and 11, Júcer Basin, Spain
Mammal Correlation: uppermost MN 16 (Mein et al. 1989), together with Rincon 1; MN 16 (= "Valdeganga inf.") (de Bruijn et al. 1992)
Paleomagnetic Calibration: lower, reversed part of Matuyama = Chron C2r.2r
Inferred Age: (Berggren et al. 1995) Chron C2r.2r: 2.6 to 2.2 Ma.
References: Alberdi et al. 1982; de Bruijn et al. 1992; Mein et al. 1990

MN 16

Age: late Pliocene
Locality: Casa del Rincon (= El Rincon 1), Spain
Mammal Correlation: medial MN 16 (Steininger et al. 1989); alternatively upper MN 16 (de Bruijn et al. 1992)
Paleomagnetic Calibration: in a normal magnetic polarity zone
References: Alberdi et al. 1982; Azzaroli 1988; de Bruijn et al. 1992; Leone 1985; Steininger et al. 1989; Torre 1987

MN 16

Age: late Pliocene
Locality: Villaroya, Spain
Mammal Correlation: lowermost MN 17 (Steininger et al. 1989); alternatively medial MN 16 (de Bruijn et al. 1992)
Correlation Tie Points: NW-Mediterranean pollen zone P III
References: Azanza et al. 1989; Aguirre and Morales 1991; de Bruijn et al. 1992; Suc and Zagwijn 1983; Steininger et al. 1989

MN 16

Age: late Pliocene
Locality: El Carasco, Júcer Basin, Spain
Mammal Correlation: MN 16 above the faunal horizons of Valdeganga; alternatively MN 14 and 15 (Mein 1989)
Paleomagnetic Calibration: uppermost part of the Gauss normal polarity the Gauss at the transition to Matuyama, according to Mein et al. (1990) = within Chron C2An.1n
Inferred Age: (Berggren et al. 1995) Chron 2An.1n: 3.0 to 2.6 Ma.
References: Mein 1989; Mein et al. 1990

MN 16

Age: late Pliocene
Locality: Rhodos, Greece
Lithostratigraphic Position: Kritka Formation
Mammal Correlation: MN 16 (Benda and Meulenkamp 1977)
Correlation Tie Points: lower part of *Globorotalia inflata* Zone; upper NN 16 Nannoplankton Zone; upper NN 16; medial Akça E-Mediterranean pollen zone

Inferred Age: (Berggren et al. 1995) approximately 3.0 to 2.4 Ma
References: Benda et al. 1977; Benda and Meulenkamp 1989, 1990

MN 16

Age: late Pliocene
Locality: Montopoli, Pisa, Italy
Lithostratigraphic Position: the fossiliferous mammal horizon is found within marine littoral sands.
Mammal Correlation: MN 16b (Torre pers.comm., 1990); upper MN 16 (de Bruijn et al. 1992)
Correlation Tie Points: the littoral marine sands cap the marine cycle of the Valdarno Inferiore Basin and are still referable to the *Globorotalia crassaformis* Zone.
Paleomagnetic Calibration: the mammal fossils were found above the Gauss/Matuyama palaeomagnetic boundary (Lindsay et al. 1980) = Chron C2An.1n/C2r.2r boundary.
Inferred Age: (Berggren et al. 1995) younger than 2.6 Ma
References: Azzaroli 1988; Bonadonna and Alberdi 1987; de Bruijn et al. 1992; Giulio et al. 1988; Lindsay et al. 1980; Torre 1987

MN 16

Age: late Pliocene
Locality: De Meern drill-site, Netherlands
Mammal Correlation: upper MN 16 (Steininger et al. 1989)
Correlation Tie Points: NW-Mediterranean pollen zone P III/P IV-Pl 1
References: Suc and Zagwijn 1983; Steininger et al. 1989

MN 16

Age: late Pliocene
Locality: Valdeganga 14, 15 (Valdeganga inf.), Júcar Basin, Spain
Lithostratigraphic Position: lacustrine limestones and carbonated silts; Jucar River Formation
Mammal Correlation: lower MN 16; alternatively uppermost MN 16 (= "Valdeganga inf."; de Bruijn et al. 1992)
Paleomagnetic Calibration: normal polarity in uppermost Gauss = Chron C2An.1n
Inferred Age: (Berggren et al. 1995) Chron 2An.1n, 3.0 to 2.6 Ma
References: Alberdi et al. 1982; de Bruijn et al. 1992; Mein et al. 1989

MN 16

Age: late Pliocene
Locality: Les Etouaires, France
Lithostratigraphic Position: the Etouaires local fauna is directly underlain by a sanidine and quartz bearing ash that has been dated.
Mammal Correlation: uppermost MN 16 (Steininger et al. 1989; de Bruijn et al. 1992); alternatively, medial MN 16
Paleomagnetic Calibration: approximately Gauss/Matuyama boundary

Isotopic Age: ash bed below mammal horizon dated at 3.6 and 2.4 Ma by several authors.

References: Alberdi and Bonadonna 1987; Azzaroli 1988; de Bruijn et al. 1992; Heintz et al. 1974; Lindsay et al. 1980; Steininger et al. 1989

MN 16

Age: late Pliocene

Locality: Stranzendorf (sportfield): horizon A and C, Molasse Zone, Lower Austria

Lithostratigraphic Position: Entire sequence about 31 m thick. Lower part to horizon C about 5 m. Horizons A, B and C red loams (= paleo intercalated into the löss sequence).

Mammal Correlation: fauna from Horizon A uncharacteristic for correlation; fauna from Horizon C an upper MN 16 characteristic fauna. Uppermost MN, according to Steininger et al. (1989) and de Bruijn et al. (1992).

Paleomagnetic Calibration: normal from base to Horizon C, equal to the uppermost part of the Gauss (= Chron C2An.1n).

Inferred Age: (Berggren et al. 1995) Chron C2An.1n: 3.04 to 2.58 Ma. Horizon C in uppermost C2An.1n just below boundary to C2r.2r.

References: de Bruijn et al. 1992; Rabeder 1981; Steininger et al. 1989

MN 16

Age: late Pliocene

Locality: Slatina 1, Romania

Mammal Correlation: MN 16 (Andreescu 1981)

Correlation Tie Points: molluscs characteristic for Romanian Stage

Paleomagnetic Calibration: in Lower Matuyama = Chron C2r.2r

Inferred Age: (Berggren et al. 1995) correlative with Chron C2r.2r; 2.6 to 2.2 Ma

References: Andreescu 1981

European Mammal Faunal Zone: MN 17

MN 17

Age: Pliocene

Locality Rocca Neyra (= Roccaneyra), Italy

Mammal Correlation: MN 16 (Azzaroli et al. 1988); alternatively lowermost MN 17 (Steininger et al. 1989 and de Bruijn et al. 1992)

Isotopic Age: radiometric dates are 2.5 to 2.4 Ma

References: Azzaroli et al. 1988; de Bruijn et al. 1992; Heintz et al. 1974; Steininger et al. 1989; Torre 1987

MN 17

Age: late Pliocene

Locality: Gülyazi, Turkey

Mammal Correlation: medial MN 16 (Steininger et al. 1989); MN 16 lower part according to de Bruijn et al. (1992); MN 16b, correlative with Gauss/Matuyama boundary due to co-occurrence of *Equus* and a hipparionine horse ("*Plesiohipparion*" cf. *huangheense*; Bernor and Tobien 1991)

Correlation Tie Points: middle Akça E-Mediterranean pollen zone

References: Benda and Meulenkamp 1989, 1990; Bernor and Tobien 1991; de Bruijn et al. 1992; Van der Meulen and Kolfschoten 1986; Sickenberg et al. 1975; Steininger et al. 1989

MN 17

Age: Pliocene/Pleistocene

Locality: Stranzendorf (sportfield) horizons D, F, G, H, I, L and M; Molasse Zone, Lower Austria

Lithostratigraphic Position: The total thickness of the loess and paleosol sequence is approximately 31 m. From horizon D to top is about 26 m. Horizons D, F, G, H are in brown loams (= paleosols); horizons I, L are in red loams (= paleosols); horizon M is in brown loam (= paleosol).

Mammal Correlation: Horizons D, F, G, and L with characteristic faunas, horizons H and I with smaller faunas. MN 17, according to Rabeder (pers. comm., 1989); MN 17 lowermost faunal level, according to de Bruijn et al. (1992).

Paleomagnetic Calibration: Horizons D, F, G in a lower reversed part, followed by a short normal (this normal is split into two reversed portions). Above this normal a longer reversed portion is found. This reversed portion is followed by a longer normal, which is continuous to the top of the section. Faunal-horizon L begins in the uppermost part of the reversed zone and is continuous to the lowermost part of the following normal. Geomagnetic interpretation: faunal horizon D to G is in the lower reversed part (= Chron C2r.2r). Faunal horizons H and I are within Chron C2r.1n (? Reunion I and II). Faunal horizon L is within uppermost Chron C2r.1r and Chron C2n (= Olduvai Event).

Inferred Age: (Berggren et al. 1995) Faunal horizon D is at the boundary between Chron C2An.1n/C2r.2r, 2.58 Ma. Faunal horizon F is correlative with upper-middle Chron C2r.2r, 2.35 Ma. Faunal horizon G is correlative with uppermost C2r.2r, 2.25 Ma. Faunal horizons H and I are below C2r.1n, 2.15–2.14 Ma. Faunal horizon L is within the boundary of Chron C2r.1r/ C2n, ca. 1.95 Ma.

References: de Bruijn et al. 1992; Rabeder 1981

MN 17

Age: Pliocene/Pleistocene

Locality: Chiliac, France

Mammal Correlation: lower MN 17 (de Bruijn et al. 1992)

Paleomagnetic Calibration: basaltic flow below mammal horizon with reversed magnetization

Isotopic Age: this basalt flow is dated at 1.9 Ma

References: Azzaroli et al. 1988; Boeuf 1983; Alberdi and Bonadonna 1987; de Bruijn et al. 1992

MN 17

Age: Pliocene/Pleistocene

Locality: Slatina 2, Romania

Mammal Correlation: lower MN 17 (Steininger et al. 1989)

Paleomagnetic Calibration: lower part of Matuyama
Inferred Age: (Berggren et al. 1995) Matuyama = Chron C2r, 2.6 to 1.95 Ma
References: Andreescu et al. 1981; Steininger et al. 1989

MN 17

Age: late Pliocene
Locality: Valdeganga 7 and 10, Valdeganga 2, 3 and 6 Valdeganga sup., Júcer Basin, Spain
Mammal Correlation: lower MN 17 (Mein et al. 1989); alternatively medial MN 17 (= "Valdeganga sup."; de Bruijn et al. 1992)
Paleomagnetic Calibration: lower reversed part of Matuyama Chron (Mein et al. 1989), = Chron C2r.2r.
Inferred Age: (Berggren et al. 1995) Chron 2r.2r: 2.6 to 2.25 Ma
References: Alberdi et al. 1982; de Bruijn et al. 1992; Mein et al. 1989

MN 17

Age: Pleistocene
Locality: Tiglian C (= Tegelen), Netherlands
Mammal Correlation: MN 17 uppermost faunal level (Steininger et al. 1989 and de Bruijn et al. 1992)

Correlation Tie Points: NW-Mediterranean pollen zone P IV to Pl 1
References: de Bruijn et al. 1992; Kolfschoten and Van der Meulen 1986; Suc and Zagwijn 1983; Steininger et al. 1989

MN 17

Age: late Pliocene
Locality: Kos, Greece
Mammal Correlation: MN 17 (Benda and Meulenkamp 1990)
Correlation Tie Points: upper part of Akça E-Mediterranean pollen zone
References: Benda and Meulenkamp 1989; 1990

MN 17

Age: early Pleistocene
Locality: Megalopolis, Greece
Mammal Correlation: MN 17 (Benda and Meulenkamp 1990)
Correlation Tie Points: lower part of Megalopolis E-Mediterranean pollen zone
Reference: Benda and Meulenkamp 1989; 1990

3

Late Miocene Paratethys Correlations

F. RÖGL AND G. DAXNER-HÖCK

During the last few decades Neogene Paratethys stratigraphy has been entirely reevaluated using new analytical methods. This stratigraphic research has focused on striking interregional correlations and is conducted under the auspices of IGCP project no. 25 (Tethys-Paratethys Correlation), J. Senes (Bratislava), coordinator. The use of regional stage systems, their inter-basinal correlations, and the paleogeographical development of these regions have been documented in numerous publications. A general overview and compilation of the results was given in the correlation tables provided by Steininger et al. (1985).

The most successful work has been achieved in the area of marine sedimentation. The Paratethys realm extended from western Switzerland eastward to the Lake of Aral in Asia, and has been subdivided into Western, Central, and Eastern biogeographic provinces (with entirely different extensions). This subdivision has been made following observations that these regions had separate sedimentary histories. The failure to recognize these provinces' separate histories has been the cause for great confusion in the statigraphic subdivision and correlation of these basins. From the Oligocene to middle Miocene, Paratethys marine sedimentation has been characterized by connections to open seas and oceans. During the middle Miocene, i.e., at the end of the Central Paratethys' Badenian Stage, biogeographic connection to the marine realm ceased, and broadly distributed endemic aquatic faunas evolved.

With the development of Paratethys endemic faunas problems emerged with regional correlations. While facies development is similar between the Central and Eastern Paratethys, the chronology of their development is asynchronous. The Pannonian Basin (Hungary) and its surroundings became the restricted Central Paratethys Pannonian lake in the late Miocene, whereas east of the Carpathian mountain arc a sea of reduced salinity with Sarmatian facies (but not Sarmatian time) extended as far as the Caspian Basin.

Late Miocene Paratethys Correlation Problems

To understand the problems with late Miocene Paratethys correlations, we must first start with the marine phases of the earlier Miocene. In the early Miocene, basin development is directly related to the compressional phase of the Alpine orogeny, and the subsidence of long west-to-east stretching foredeeps. The Central Paratethys Eggenburgian Sea formed an "aquatic bridge" for connecting the Mediterranean with the Sakaraulian Sea of today's Black Sea–Caspian Sea Basin region (= the Euxinian Basin). This geographically extensive sea had the Indo-Pacific Ocean as its feeding source. This large seaway was interrupted during early Miocene times by regional tectonics associated with the Alpine orogeny, changing the distribution of water masses.

The development of large intra-mountain basins along the inner contour of the Carpathian arc and the establishment of markedly different sedimentary facies in the Eastern Paratethys followed. Drastically changing conditions dominated the middle Miocene marine cycles (comp. Rögl and Steininger 1983). By the middle Miocene, entirely different sedimentary successions developed in the Euxinian Basin. A facies barrier has been recognized along a line bordered in the south by the North Anatolian fault system, continuing to the northwest along the coast of the Black Sea, wherein the embayments of Burgas and Varna Eastern Paratethys facies are still developed.

The reduction in salinity led to the Sarmatian Stage's endemic facies. Only during the early portion of this stage, the Volhynian substage, did the same conditions prevail from the Caspian Sea region to the Vienna Basin. Following correspondence with Barbot de Marny, Suess (1866) concluded that these facies and their correlative relationships required recognition of a Sarmatian Stage. Suess expressed verbally that this stage name is connected with the deposits of the Vienna Basin (Cerithien-Schichten,

Hernalser Tegel). The Sarmatian facies existed further on in the more extended Eastern Paratethys, from the outer side of the Carpathian mountain arc to the Caspian Basin. For the basins within this arc, the facies changed progressively to less salinity until endemic Pannonian facies prevailed.

As the Sarmatian facies conditions continued in the east, the term *Sarmatian sensu lato* was established, and has likewise been consistently used for time equivalents of the Pannonian Stage of the Vienna and Pannonian Basins. The "Sarmatian s.l." has been subdivided in the Eastern Paratethys to include the Volhynian, Bessarabian, and Khersonian substages. These substages are correlative with the late Serravallian-Tortonian Stages of the Mediterranean time scale. The correlation of the Sarmatian sensu stricto and Pannonian boundary has been recently discussed (comp. Papp et al. 1985; Stevanovic 1987; Bernor et al. 1988). It compares roughly with the boundary between the early and the late Bessarabian.

The most important factor for correlating this boundary is the first appearance of the three-toed equid "*Hipparion*," which is considered to be essentially geochronologically instantaneous in time (see Bernor et al., this volume; Swisher, this volume; Woodburne et al., this volume). This horse appeared during the Sarmatian in the Eastern Paratethys, and then later during the Pannonian in the Central Paratethys. Thenius (1959; 1960) has correctly recognized that this apparent asynchroneity was based on the misinterpretation of the "Sarmatian"'s chronologic limits.

The aquatic development of the Pannonian Stage was biogeographically restricted to the intra-Carpathian basins. A succession of molluscan faunas has been used for the stratigraphic subdivision (e.g., Fuchs 1873; Friedl 1932 and Papp 1951 for the Vienna Basin; Neumayr and Paul 1875, Halavats 1903 and 1911, Lörenthey 1905 and 1911, and Stevanovic 1951 for the Pannonian Basin).

All marine organisms became extinct in the Pannonian Basin and Eastern Paratethys by Khersonian time because of the sharp reduction in salinity (to 0.03–0.04%). This environmental shift was followed by a new marine cycle in the Maeotian (Nevesskaya et al. 1984b). Again, strong endemism ensued, but by the end of the Maeotian the large Eastern Paratethys Sea was reduced to a series of smaller isolated basins. At this time a cycle of a nearly freshwater facies followed, with notably similar faunal assemblages in the Pannonian Basin and in the eastern, Euxinian, realm. Recognition of this similarity in Central and Eastern Paratethys faunal assemblages led to the recognition of the Pontian Stage (Barbot de Marny 1869). The Pontian transgression extended southward into the Aegean basins, later giving way to the "Lago Mare" facies of the late Messinian in the Mediterranean. Correlation problems still exist between the Pannonian Basin and the Aegean area during the Pontian (Rögl et al. 1991).

The Pontian in the Vienna Basin

Rögl et al. (1993) have recently revised the Vienna Basin's late Miocene stratigraphy and correlation. Only a brief review is presented here.

According to Fuchs (1873), the Sarmatian Stage is followed by the Congeria beds, which themselves continue into the Paludina beds. Hoernes (1903) correlated the Vienna Basin's Congeria beds with the Pontian Stage, while the succeeding freshwater beds (as found at Eichkogel) were referred to the Levantinian (in comparison with the Paludina beds of Slavonia). Roth von Telegd (1879) erected the Pannonian Stage for the Congeria beds and was followed by Schaffer (1927), who further subdivided this stage into a lower unit, the Pannonian, and an upper unit, the Pontian. More recently, Lörenthey's (1905) conclusion that all late Miocene beds succeeding the Sarmatian are correlative with the Pannonian Stage has been broadly followed for the Vienna Basin (Janoschek 1943; Papp 1948). Papp did not detect any possibilities for a correlation with the Eastern Paratethys's Maeotian and Pontian Stages.

The Pannonian succession was subdivided by Papp (1948; 1951) into a series of letter-stages: "zones" A to H. The boundary between the middle and late Pannonian ("zone" E [= *Congeria subglobosa* Zone] and "zone" F) is marked by the extinction of limnocardiids and most of the Pannonian ostracod taxa. The last congerias occur in Pannonian "zone" F. The upper *Congeria* beds in the Southeastern Pannonian Basin have been correlated with the base of the Dacian Basin's Pontian Stage (Stevanovic 1951): the Radmanest bed or beds with *Congeria ungulacaprae* or *Paradacna abichi*. Despite the fact that these species are not recorded in the Vienna Basin, the Pontian Stage has been correlated there with the late Pannonian "zones" F to H (Papp et al. 1974; Stevanovic 1987). These authors cited the mass occurrence of *Congeria neumayri* and *Congeria zahalkai* in the Vienna Basin's "zone" F, as a facies correlative with the *Congeria ungulacaprae* Zone of the Pannonian Basin. The Pannonian/Pontian boundary would appear to be related to the first emigration of bivalves (limnocardiids, *Pseudoprosodacna*, *Pseudocatillus*) from the Pannonian Basin into the Dacian and Euxinian Basins (Stevanovic et al. 1990).

A more precise biostratigraphic correlation between the Pannonian and Dacian Basins has been achieved by a recent reconstruction of limnocardiid evolution (Müller and Magyar 1992; 1994). Müller and Magyar recognize an evolutionary lineage including: *Lymnocardium decorum ponticum–L. decorum decorum–Prosodacnomya vutskitsi.* The transition from *L. decorum ponticum* to *L. decorum decorum* occurs near the base of the *Congeria balatonica* Zone of the Lake Balaton area. This transition occurs just above the top of the magnetic anomaly 4A (about 8.7

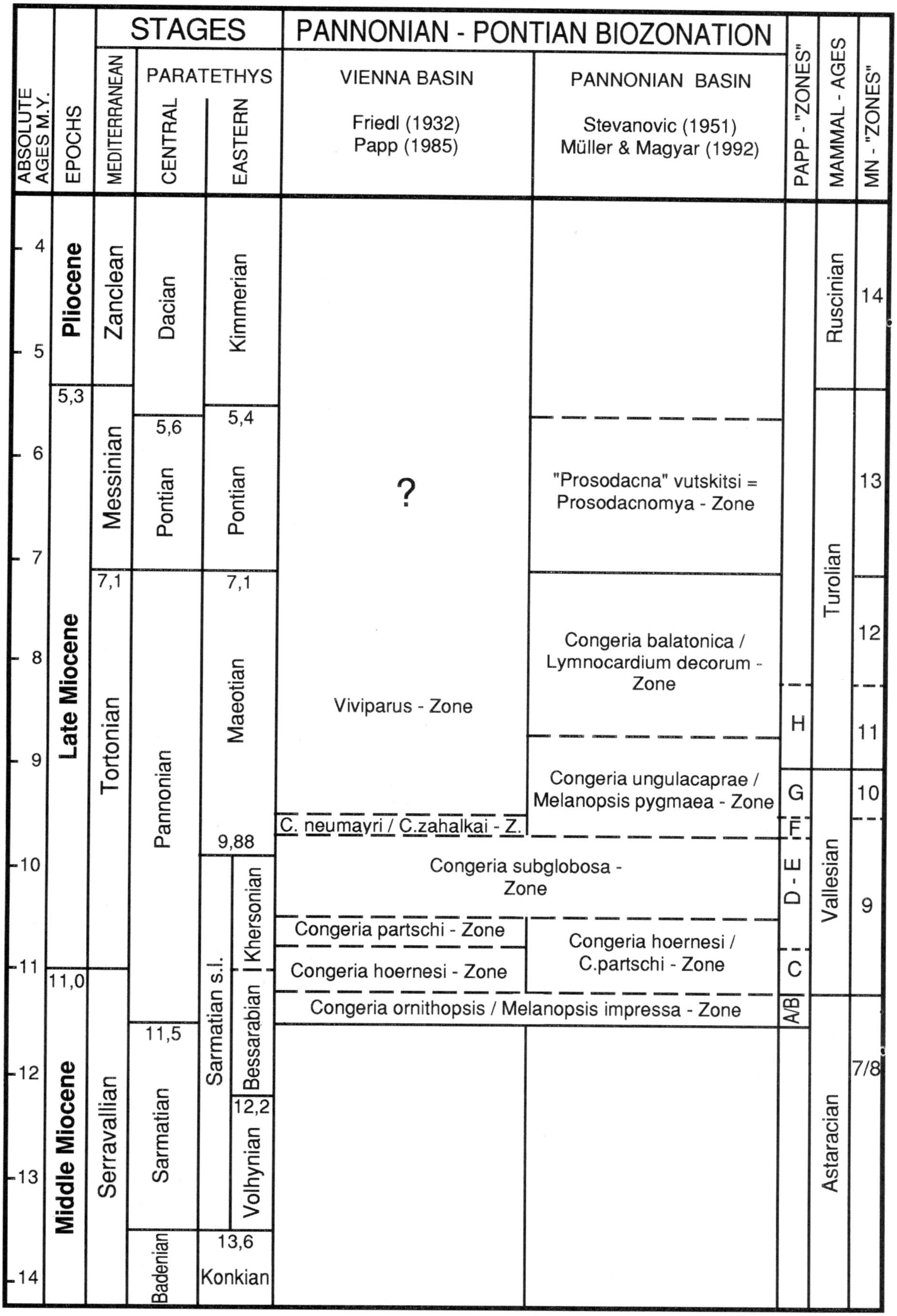

| ABSOLUTE AGES M.Y. | EPOCHS | STAGES | | | PANNONIAN - PONTIAN BIOZONATION | | PAPP - "ZONES" | MAMMAL - AGES | MN - "ZONES" |
		MEDITERRANEAN	PARATETHYS CENTRAL	EASTERN	VIENNA BASIN Friedl (1932) Papp (1985)	PANNONIAN BASIN Stevanovic (1951) Müller & Magyar (1992)			
4	Pliocene	Zanclean	Dacian	Kimmerian				Ruscinian	14
5									
5,3			5,6	5,4					
6		Messinian	Pontian	Pontian	?	"Prosodacna" vutskitsi = Prosodacnomya - Zone			13
7								Turolian	
7,1			7,1	7,1					12
8	Late Miocene	Tortonian	Pannonian	Maeotian	Viviparus - Zone	Congeria balatonica / Lymnocardium decorum - Zone			
9							H		11
						Congeria ungulacaprae / Melanopsis pygmaea - Zone	G	Vallesian	10
				9,88	C. neumayri / C.zahalkai - Z.		Fi		
10				Bessarabian / Khersonian	Congeria subglobosa - Zone		D - E		9
11				Sarmatian s.l.	Congeria partschi - Zone	Congeria hoernesi / C.partschi - Zone	C		
11,0					Congeria hoernesi - Zone				
11,5	Middle Miocene	Serravallian	Sarmatian	Volhynian	Congeria ornithopsis / Melanopsis impressa - Zone		A/B	Astaracian	7/8
12				12,2					
13									
13,6			Badenian	Konkian					
14									

FIGURE 3.1 Correlation chart of the late Miocene in the Vienna and Pannonian Basins (modified from Rögl et al. 1993).

Ma, according to Berggren et al. 1995). A fauna with micromammals currently referred to MN 11 lies stratigraphically above the beds with *L. decorum*. The next evolutionary step, from *Lymnocardium* to *Prosodacnomya*, has been calibrated as being between 7.7 and 7.0 Ma. Furthermore, the *Prosodacnomya vutskitsi* beds have been reported to occur in a drill site at Szarvas, Hungary, that has been calibrated paleomagnetically as being ca. 6.4–6.0 Ma. The mammal locality of Hatvan (MN 13) occurs stratigraphically above beds containing *P. vutskitsi*.

The FAD of the short-lived species *Prosodacnomya dainelli* and *P. sturi* (closely related to *P. vutskitsi*) is recorded in the Dacian Basin's early and middle Pontian horizons (Andreescu 1977; Pavnotescu and Andreescu 1978). According to this correlation, the entire *Congeria ungulacaprae* Zone as well as the *Congeria balatonica* Zone is stratigraphically older, and correlates with the Maeotian. Therefore, indirectly, the beds with *Congeria neumayri/Congeria zahalkai* ("zone" F of the Vienna Basin) are correlative with the Maeotian.

An Absolute Time Scale for the Paratethys

There is a long tradition of radioisotopic dating within the Paratethys Neogene. This has given a good framework for developing the current regional stage system (updated by Vass et al. 1987) for the Central Paratethys. The latest compilation of the middle Miocene to Pleistocene from the Eastern Paratethys has been provided by Chumakov et al. (1992a,b). These values have been corrected according to Berggren et al. (1995) and to the data base of Steininger et al. (this volume).

The early/middle Miocene boundary corresponds to the base of the Badenian and is estimated to be about 16.4 Ma (FAD of *Praeorbulina*). The Badenian/Sarmatian boundary is dated 13.6 Ma, and the Sarmatian s.s./Pannonian boundary is estimated to be approximately 11.5 Ma, somewhat below the middle/late Miocene boundary (base of the Tortonian Stage), ca. 11 Ma. The immigration of "*Hipparion*" correlates closely with the base of Pannonian "zone" C.

The subdivision of the Sarmatian s.l. substages in the Eastern Paratethys is important for correlation with equivalent-aged strata in the Vienna Basin: the Volhynian/Bessarabian boundary is ca. 12.2 Ma (Chumakov et al. 1992a); the Bessarabian/Khersonian boundary is ca. 11 Ma; the Khersonian/Maeotian boundary is 9.88 Ma (according to paleomagnetic recalculations; see Steininger et al., this volume). The first occurrence of "*Hipparion*" is recorded in the late Bessarabian of the Eastern Paratethys.

There has been considerable uncertainty regionally surrounding the age of latest Miocene sediments. Currently it is believed that the Central Paratethys Pannonian/Pontian boundary is somewhere between 8 and 7 Ma. The Eastern Paratethys values have been radioisotopically determined

to be ca. 7 Ma for this same boundary. These latter dates have been based principally on well-preserved glass tuffites found near the base of the Pontian Stage. The top of the Pontian in Azerbaizhan has been dated and is younger than in the Black Sea region (see Nevesskaya et al. 1984a; Chumakov et al. 1992a, top of Babadzhanian ?5.19 Ma). In the Central Paratethys it is estimated to be ca. 5.6 Ma. The Mio–Pliocene boundary is dated 5.32 Ma by Hilgen (1991), which is in agreement with earlier results. The Tortonian/Messinian boundary has recently been correlated by Krijgsman et al. (1994), and with a best estimate of 7.10 Ma, thereby causing the former opinion of a time-equivalent Messinian/Pontian boundary to be reevaluated.

Late Miocene Mammalian Correlations

Suess (1863) was the first to present a mammalian faunal biochronologic sequence in the Vienna Basin. Thenius (1959; 1960) remarked that the Vienna Basin's first appearance of "*Hipparion*" is an isochronous index between the Central and Eastern Paratethys; "*Hipparion*" does not occur in the Vienna Basin's Sarmatian s.s., but rather at the base of Pannonian "zone" C. Bernor et al. (1988; 1989; this volume) and Woodburne et al. (this volume) argue that it is the most primitive and probably oldest-known Old World hipparionine horse. The succeeding Vienna Basin Pannonian Stages C–F are correlative with European MN 9 localities, including the recently redated fauna of Höwenegg (ca. 10.3 Ma; Swisher, this volume; Woodburne et al., this volume).

The Mammal Neogene zonation (MN zones) of Mein (1975; 1979; 1989) are discussed briefly here because they are relevant to the Paratethys mammalian faunas. During the early Vallesian (MN 9), middle Miocene genera dominate the assemblages (*Megacricetodon*, *Democricetodon*, *Eumyarion*, *Eomuscardinus*, *Myoglis*, *Albanensia*, *Gomphotherium*, *Anchitherium*, *Euprox*, *Listriodon*, *Dryopithecus*). "*Hipparion*," *Microtocricetus*, and *Eozapus* are among the first immigrants that accompany this otherwise relictual fauna.

A prominent faunal change occurs in the late Vallesian, MN 10. Murids, including *Progonomys*, *Parapodemus*, and the cricetid *Kowalskia*, become increasingly significant in their proportional occurrence. The more modern character of this zone is marked by the occurrence of further immigrants such as *Epimeriones*, *Pliopetaurista*, and *Graphiurops*. *Progonomys* and a variety of middle Miocene rodents disappear at the end of the Vallesian. According to the most recent data, the duration of MN 10 may be extremely short.

The biofacies transition from middle to late Miocene age faunas is concluded by the beginning of Turolian (MN 11), indicating an extensive number of extinctions. MN 11 and MN 12 are generally uniform. Differences occur

FAUNAL UNITS	MN-"ZONES"	**VERTEBRATE - LOCALITIES** (*Reference faunas)		
		SPAIN - PORTUGAL	WESTERN AND CENTRAL EUROPE	CENTRAL PARATETHYS
RUSCINIAN	15	Layna Gorafe	Sète Wölfersheim *Serrat d`en Vaquer	Csarnota 2 Ivanovce Weze
RUSCINIAN	14	Villalba Alta Rio Orrios Caravaca Alcoy	Baccinello V-3 Celleneuve	Ostramos 1,9,13 *Podlesice
TUROLIAN	13	La Alberca Crevillente 6 *El Arquillo Valdecebro 3	Maramena Brisighella Lissieu Baccinello V-2 Samos 1-5	Polgardi 2 Hatvan Baltavar Polgardi 4
TUROLIAN	12	*Los Mansuetos Concud Casa del Acero	Pikermi Baccinello V-1	Tardosbanya Budapest-Szechenyi
TUROLIAN	11	Tortajada A *Crevillente 1-3 La Celia	Samos X Dorn-Dürkheim Mollon Ambérieu 3	Eichkogel
VALLESIAN	10	Cucalón Can Casablancas *Masia del Barbo Ampudia 3	Ambérieu 1 Soblay, Lefkon Montredon Kastellios	Kohfidisch Suchomasty Neusiedl a.See Götzendorf, Stixneusiedl
VALLESIAN	9	*Can Llobateres Pedregueras 2C Nombrevilla	Höwenegg Hammerschmiede	Rudabanya Vösendorf, Inzersdorf Gaiselberg, Hovorany
ASTARACIAN	7/8	Castell de Barbera San Quirze	Giggenhausen Anwil *La Grive M	Drassburg Nexing

FIGURE 3.2 Late Miocene Mammal Zonation in Europe with the stratigraphic position of the most important mammal localities.

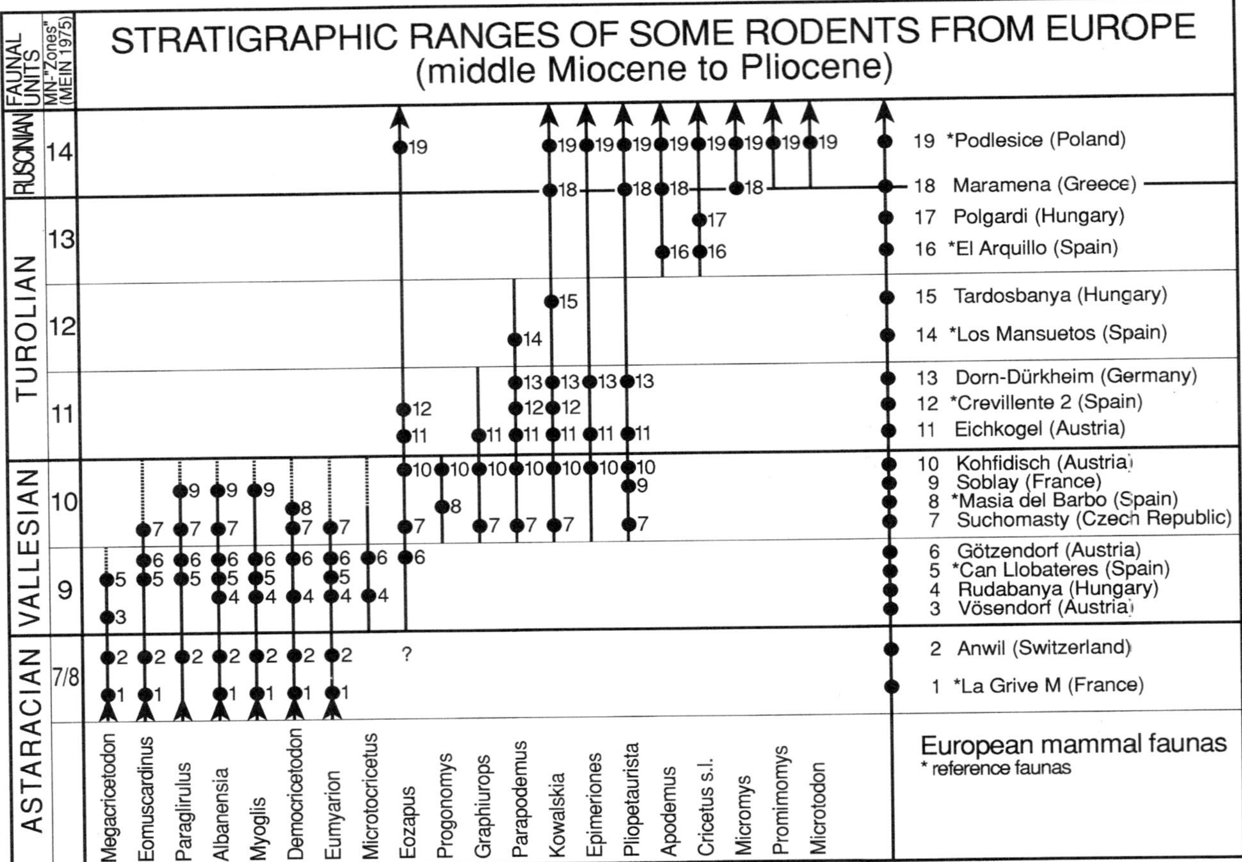

FIGURE 3.3 Stratigraphic range chart of selected rodents in the European late Miocene.

principally in species composition. The early Turolian (MN 11) is characterized by the first occurrence of *Parapodemus lugdunensis*, and the middle Turolian (MN 12) is characterized by the first occurrence of *Parapodemus gaudryi*. In the late Turolian (MN 13), *Parapodemus* is replaced by the advanced murid *Apodemus*, and the modern cricetine *Cricetus* s.l. first appears.

Austrian Late Miocene Mammal Localities

The Austrian middle Miocene (Badenian–Sarmatian) includes a number of localities with fairly species-diverse large mammal faunas; until now, smaller mammals have been generally scarce. We witness a change in these conditions during the late Miocene Pannonian Stage. The oldest Pannonian faunas ("zone" B) occur at Drassburg, Burgenland (Zapfe 1951), with a typical Astaracian (MN 7/8) large mammal fauna, and Bullendorf, in the northern Vienna Basin (Daxner-Höck et al. 1990), with a small mammal fauna. The stratigraphically first-occurring *Hipparion* is recorded in the northern Vienna Basin from the Pannonian C locality of Gaiselberg (Zapfe 1949; basal MN 9). Rich mammal faunas including "Hipparion" occur during "zones" D and E at Vösendorf and Inzersdorf (Bernor et al.

1988; Daxner 1967; Papp and Thenius 1954; Rabeder 1973; Thenius 1950). The last occurrence of *Megacricetodon* at Vösendorf and Inzersdorf is correlative with Pedregueras 2C, Spain, within the early MN 9. *Megacricetodon* is absent from the Hungarian medial-late MN 9 locality of Rudabánya, and this fauna would appear to be intermediate in its age between Vösendorf and the southern Vienna Basin "zone" F locality of Götzendorf (Bachmayer and Wilson 1984; Rögl et al. 1993). The rodents *Democricetodon, Eumyarion, Albanensia, Eomuscardinus, Myoglis,* and *Paraglirulus* survived from the middle Miocene and characterize MN 9 faunas, including Rudabánya and Götzendorf. *Dibolia* and *Lanthanotherium sanmigueli* are stratigraphically younger representatives in these faunas, positioning Götzendorf between MN 9 and MN 10. The modern rodent genus *Eozapus* first occurs at Götzendorf.

The first murids have recently been recorded from Pannonian age fluviatile sands ("zone" G?) at Neusiedl am See, on the western border of the Pannonian Basin (Rögl et al. 1993); the occurrence of *Progonomys cathalai* there suggests a correlation with MN 10. In the famous locality Kohfidisch, in Southern Burgenland (Bachmayer and Wilson 1970; 1978; 1980), *Progonomys woelferi* and *Kowalskia fahlbuschi* are abundant while *Parapodemus lugdunensis,*

Epimeriones, and *Pliopetaurista* have their first occurrences. This assemblage is comparable to the MN 10 faunas of Soblay, France, and Lefkon, Greece.

The Vienna Basin's youngest, species-diverse late Miocene mammal fauna is Eichkogel, correlative with Pannonian "zone" H freshwater beds. The dominance of *Parapodemus lugdunensis* and *Kowalskia skofleki* together with *Pliopetaurista, Epimeriones, Collimys,* and *Eozapus* is characteristic for the early Turolian (MN 11). This fauna correlates with Crevillente 2 (Spain), Dorn-Dürkheim (Germany), and to some extent Csákvár (Hungary).

Later Turolian (MN 12–13) faunas with small mammals have not been found in Austria. The stratigraphic position of some localities with large mammals in the Upper Freshwater Molasse and in the Vienna Basin (e.g., Prottes) could not be made with micromammals, but some indications for a late Turolian age are given by the occurrence of giraffids, hyaenids, and mastodonts typical of the Eastern Mediterranean–Southwest Asian realm. The youngest gravel beds in the northern Vienna Basin (Bernhardsthal) contain the proboscidean *Zygolophodon borsoni*.

Late Miocene Updating

Late Turolian faunas have been reported from the Hungarian Pannonian Basin. There are problems with correlating these faunas with the basin's sedimentary sequence, because the Pannonian mammal faunas occur mostly in karstic facies. One of the most valuable areas for correlation between late Miocene aquatic Eastern Paratethys sediments and the European MN zonation scheme is known from the district of Odessa, on the Ukrainian Black Sea coast. Micromammal faunas have been reported from this area and correlate with the early Turolian locality of Crevillente 4, Spain (Topachevskii and Skorik 1992). This provides a correlation tie-point with the Central Paratethys. The early to middle Turolian and middle to late Maeotian correspond with the late Pannonian of the Vienna Basin, or with the *Congeria balatonica/L. decorum* Zone of the Pannonian Basin (fig. 3.1). The later Turolian is represented in the lectostratotype of the Pontian near Odessa, by the cooccurrence of *Apodemus* and *Cricetus* (Topachevskii et al. 1988). These faunas give a correlation of not older than MN 13 for this Pontian horizon.

In summary, we recognize the following general correlations between the Central-Eastern Paratethys and the Mediterranean stratigraphic sequence:

The Central Paratethys Pannonian Stage is correlative with the Eastern Paratethys late Bessarabian, Khersonian, and Maeotian, which in turn is correlative with the Mediterranean latest Serravallian and the Tortonian.

The Pontian has approximately the same stratigraphic range throughout the Central and Eastern Paratethys aquatic realm, and is correlative with the Mediterranean Messinian Stage. The upper boundary of the Pontian may prove to be time-transgressive from west (older) to east (younger), with the youngest horizons occurring in the Caspian Basin.

In the Vienna Basin, lacustrine sedimentation ceased during the late Pannonian. Pontian age sedimentation at the moment has not been conclusively demonstrated by the fossil record, but may occur in a restricted area of fluviatile sedimentation.

Acknowledgments

We wish to thank all the coworkers of the Götzendorf research project. The results of this working group have enabled us to refine our correlations of the late Miocene Paratethys sequences. Our special gratitude belongs to Ray Bernor, who has discussed with us the different problems of mammalian stratigraphy, critically read the article, and, finally, edited the English. For further discussions of the problems we are grateful to Volker Fahlbusch, Oldrich Fejfar, Gernot Rabeder, Erich Thenius, and Helmut Zapfe. We wish to gratefully acknowledge permission to use the results in print of William A. Berggren and the stratigraphic data base of Fritz F. Steininger. The help of Kriemhild Repp for graphic work is kindly acknowledged. The research has been supported by the Austrian Science Foundation (FWF Proj. P 7525-GEO), the Austrian Geological Survey (Proj. Kleinsäuger), and the "Freunde des Naturhistorischen Museums, Wien."

LITERATURE CITED

Andreescu, I. 1977. Systématique de *Lymnocardiidés prosodacniformes.* Sous-famille Prosodacninae. *Memoires Institut de Géologie et de Géophysique, Bucarest* 26:1–74.

Bachmayer, F. and R. W. Wilson. 1970. Die Fauna der altpliozänen Höhlen- und Spaltenfüllungen bei Kohfidisch, Burgenland (österreich). [Small mammals (Insectivora, Chiroptera, Lagomorpha, Rodentia) from the Kohfidisch fissures of Burgenland, Austria]. *Annalen des Naturhistorischen Museums, Wien* 74:533–87.

———. 1978. A second contribution to the small mammal fauna of Kohfidisch, Austria. *Annalen des Naturhistorischen Museums, Wien* 81:129–61.

———. 1980. A third contribution to the fossil small mammal fauna of Kohfidisch (Burgenland), Austria. *Annalen des Naturhistorischen Museums, Wien* 83:351–86.

———. 1984. Die Kleinsäugerfauna von Götzendorf, Niederösterreich. *Sitzungsberichte österreichische Akademie der Wissenschaften, mathematisch-naturwissenchaftliche Klasse* 193 (6–10): 303–19.

Barbot de Marny, N. 1869. *Geologicheskij ocherk' Khersonskoje Guvernii* [Geological essay of the Khersonian District]. St. Petersburg: B. Demakova.

Berggren, W. A., D. V. Kent, C. C. Swisher III, and M. P. Aubrey. 1995. A revised Cenozoic geochronology and chronostratigraphy. In *Geochronology, Time Scales and Global Stratigraphic*

Correlations: A Unified Temporal Framework for an Historical Geology, ed. W. A. Berggren, D. V. Kent, and J. Hardenbol, pp. 00–00. Society of Economic Paleontologists and Mineralogists Special Publication No. 54: 129–212.

Bernor, R. L., G. D. Koufos, M. O. Woodburne, and M. Fortelius. This volume. The evolutionary history and biochronology of European and Southwest Asian late Miocene and Pliocene hipparionine horses.

Bernor, R. L., J. Kovar-Eder, D. Lipscomb, F. Rögl, S. Sen, and H. Tobien. 1988. Systematic, stratigraphic, and paleoenvironmental context of first-appearing *Hipparion* in the Vienna Basin, Austria. *Journal of Vertebrate Paleontology* 8:427–52.

Bernor, R. L., H. Tobien, and M. O. Woodburne. 1989. Patterns of Old World hipparionine evolutionary diversification and biogeographic extension. In *European Neogene Mammalian Chronology*, ed. E. H. Lindsay, V. Fahlbusch, and P. Mein, pp. 263–319. New York: Plenum.

Chumakov, I. S., S. L. Byzova, and S. S. Ganzey. 1992a. Geochronologija i korreljatsija pozdnego Kainozoja Paratetisa. Moscow: Nauka.

Chumakov, I. S., S. L. Byzova, S. S. Ganzey, C. Arias, G. Bigazzi, F. P. Bonadonna, J. C. Hadler-Neto, and P. Norelli. 1992b. Interlaboratory fission track dating of volcanic ash levels from Eastern Paratethys: A Mediterranean-Paratethys correlation. *Palaeogeography, Palaeoclimatology, Palaeoecology* 95:287–95.

Chumakov, I. S., S. L. Byzova, S. S. Ganzey, V. Dobryina, V. Ya, and N. P. Paramonova. 1984. Geokhronologiya Sarmata vostochnogo Paratetisa. *Doklady, Akademia Nauk SSR* 276:1189–93.

Daxner, G. 1967. Ein neuer Cricetodontide (Rodentia, Mammalia) aus dem Pannon des Wiener Beckens. *Annalen des Naturhistorischen Museums, Wien* 71:27–36.

Daxner-Höck, G., H. de Bruijn, and D. Foussekis. 1990. Bericht 1989 über das Projekt "Kleinsäuger" der begleitenden Grundlagenforschung. *Jahrbuch der Geologischen Bundesanstalt, Wien* 133(3): 508–10.

Friedl, K. 1932. Ueber die Gliederung der pannonischen Sedimente des Wiener Beckens. *Mitteilungen Geologische Gesellschaft in Wien* 24 (1931): 1–27.

Fuchs, Th. 1873. *Erläuterungen zur geologischen Karte der Umgebung Wiens.* Wien: K.k. Geologische Reichsanst.

Halavats, G. H. 1903. A balaton melléki pontusi korú rétegek faunája. A *Balaton tudományos tanúlmányozásának eredményei,* 1, pt. 1, resz. 4(2): 1–74.

——. 1911. *Die Fauna der pontischen Schichten in der Umgebung des Balatonsees.-Resultate der wissenschaftlichen Erforschung des Balatonsees,* 1, Th. 1. Anhang Palaeontologie, v.4 (2). Wien: E. Hölzel.

Hilgen, F. J. 1991. Extension of the astronomically calibrated (polarity) time scale to the Miocene/Pliocene boundary. *Earth and Planetary Science Letters* 107:349–68.

Hoernes, R. 1903. Bau und Bild der Ebenen österreichs. In *Bau und Bild österreichs,* ed. C. Diener, R. Hoernes, F. E. Suess, and V. Uhlig, I-VI, pp. 917–1110. Wien: F. Tempsky.

Janoschek, R. 1943. Das Pannon des Inneralpinen Wiener Beckens. *Mitteilungen des Reichsamtes für Bodenforschung, Zweigstelle Wien* 6:45–61.

Krijgsman, W., F. J. Hilgen, C. G. Langereis, and W. J. Zachari-

asse. 1994. The age of the Tortonian/Messinian boundary. *Earth and Planetary Science Letters* 121:533–47.

Lörenthey, I. 1905. *Adatok a Balaton-melléki pannoniai korú rétegek faunájához és sztratigráfiai helyzetéhez.* A Balaton tudományos tanúlmányozásának eredményei, 4, pt. 3: 1–215. Budapest.

——. 1911. *Beiträge zur Fauna und stratigraphischen Lage der pannonischen Schichten in der Umgebung des Balatonsees.* Resultate der wissenschaftlichen Erforschung des Balatonsees, Bd. 1 (Teil 1), Anhang: Palaeontologie der Umgebung des Balatonsees, 4 (3), 216 p. E. Hölzel, Wien.

Mein, P. 1975. Résultats du Groupe de Travail des Vertébrés. In *Report on Activity of the RCMNS Working Groups (1971–1975),* ed J. Senes, pp. 78–81. Bratislava: SAV.

Mein, P., E. Martin-Suarez, and J. Agusti. 1993. *Progonomys* Schaub 1938 and *Huerzelimys* gen. nov. (Rodentia): Their evolution in Western Europe. *Scripta Geologica* 103:41–64.

Müller, P. and I. Magyar. 1992. Continuous record of the evolution of lacustrine cardiid bivalves in the late Miocene Pannonian lake. *Acta Paleontologica Polonica* 36:353–72.

——. 1994. Stratigraphic significance of Upper Miocene lacustrine cardiid *Prosodacnomya* (Kötcse section, Pannonian Basin, Hungary). *Földtani Közlöny* 122(1992): 1–38.

Neumayr, M. and C. M. Paul. 1875. Die Congerien- und Paludinenschichten Slavoniens und deren Faunen. *Abhandlungen k.k. Geologische Reichsanstalt, Wien* 7(3): IV, 1–106.

Nevesskaya, L. A., I. A. Goncharova, L. B. Ilina, N. P. Paramonova, S. V. Popov, A. K. Bogdanovich, L. K. Gabunia, and M. F. Nosovsky. 1984a. Reginonalnaja stratigraficheskaja shkala neogena Vostochnogo Paratetisa [The regional stratigraphical scale of the Eastern Paratethys]. *Sovetskaya Geologiya* 9:37–49.

Nevesskaya, L. A., A. A. Voronina, I. A. Goncharova, L. Ilina, N. P. Paramonova, S. V. Popov, A. L. Chepalyga, and E. V. Babaak. 1984b. Istoriya Paratetisa. *27th International Geological Congress Moscow, Paleookeanologiya, Kollokvium 03, Doklady* 3:91–101.

Papp, A. 1948. Fauna und Gliederung der Congerienschichten des Pannon im Wiener Becken. *Anzeiger der österreichischen Akademie der Wissenschaften, mathematisch-naturwissenchaftliche Klasse* 85:123–34.

——. 1951. Das Pannon des Wiener Beckens. *Mitteilungen der Geologischen Gesellschaft in Wien* 39–41 (1946–1948): 99–193.

Papp, A., A. Jambor, and F. F. Steininger (eds.). 1985. M_6-Pannonien (Slavonien und Serbien). Serie Chronostratigraphie und Neostratotypen 7:1–636. Budapest: Akademia Kiado.

Papp, A., Fl. Marinescu, and J. Senes. 1974. M_5-Sarmatien (sensu E. Suess 1866). *Die sarmatische Schichtengruppe und ihr Stratotypus.* Serie Chronostratigraphie und Neostratotypen 4:1–707. Bratislava: SAV.

Papp, A. and E. Thenius. 1954. Vösendorf-ein Lebensbild aus dem Pannon des Wiener Beckens. *Mitteilungen der Geologischen Gesellschaft in Wien* 46 (1953): 1–109.

Pavnotescu, V. and I. Andreescu. 1978. Asupra unor Prosodacnine din Pontianul de la Boteni (Jud. Arges). *Studii si Cercetari de Geologie si Geofizica, Seria Geologie* 23:143–55.

Rabeder, G. 1973. *Galerix* und *Lanthanotherium* (Erinaceidae, Insectivora) aus dem Pannon des Wiener Beckens. *Neues Jahrbuch für Geologie und Paläontologie*, Monatshefte. 1973(7): 429–46.

———. 1985. Die Säugetiere des Pannonien. In M_6-*Pannonien (Slavonien und Serbien)*, ed. A. Papp., A. Jambor, and F. F. Steininger, pp. 440–63. Serie Chronostratigrapie und Neostratotypen 7. Budapest: Akademia Kiado.

Rögl, F., R. L. Bernor, M. D. Dermitzakis, C. Müller, and M. Stancheva. 1991. On the Pontian correlation in the Aegean (Aegina Island). *Newsletters in Stratigraphy* 24:137–58.

Rögl, F. and F. F. Steininger. 1983. Vom Zerfall der Tethys zu Mediterran und Paratethys. Die neogene Paläogeographie und Palinspastik des zirkum-mediterranen Raumes. *Annalen des Naturhistorischen Museums, Wien* 85:135–63.

Rögl, F., H. Zapfe, R. L. Bernor, R. Brzobohaty, G. Daxner-Höck, I. Draxler, O. Fejfar, J. Gaudant, P. Herrmann, G. Rabeder, O. Schultz, and R. Zetter. 1993. Die Primatenfundstelle Götzendorf an der Leitha, Niederösterreich (Obermiozän des Wiener Beckens). *Jahrbuch der Geologischen Bundesanstalt, Wien* 136:503–26.

Roth von Telegd, L. 1879. Geologische Skizze des Kroisbach-Ruster Bergzuges und des nördlichen Theiles des Leitha-Gebirges. *Földtani Közlöny* 9:99–114 (Hungarian), 139–50 (German).

Schaffer, F. X. 1927. *Geologische Geschichte und Bau der Umgebung Wiens.* Leipzig-Wien: F. Deuticke.

Steininger, F. F., W. A. Berggren, D. V. Kent, R. L. Bernor, S. Sen, and J. Agusti. This volume. Circum-Mediterranean Neogene (Miocene and Pliocene) marine-continental chronologic correlations of European mammal units.

Steininger, F. F., J. Senes, K. Kleemann, and F. Rögl (eds.). 1985. *Neogene of the Mediterranean Tethys and Paratethys.* Stratigraphic correlation tables and sediment distribution maps. *1:* XIV + 189; 2: XXVI + 536 p. Institute of Paleontology-University Vienna, Wien.

Stevanovic, P. M. 1951. Pontische Stufe im engeren Sinne-Obere Congerienschichten Serbiens und der angrenzenden Gebiete. *Serbische Akademie der Wissenschaften, math.-naturwiss. Klasse,* Serie 2, Sonderausgabe, 187:1–361.

———. 1985. Diskussion der Unterstufen Slavonien und Serbien. In M_6-*Pannonien (Slavonien and Serbien)*, ed. A. Papp, A. Jambor, and F. F. Steininger, pp. 82–85. Serie Chronostratigraphie und Neostratotypen 7. Budapest: Akademia Kiado.

———. 1987. Delimination and correlation of the Pontian and the Messinian Stages on the basis of malacofauna. *Annales Instituti Geologici Publici Hungarici* 70:363–70.

Stevanovic, P. M., L. A. Nevesskaja, Fl. Marinescu, A. Sokac, and A. Jambor. 1990. *Pontien-Pl$_1$ (sensu F. Le Play, N. P. Barbot de Marny, N. I.Andrusov).* Serie Chronostratigraphie und Neostratotypen 8:1–952. Zagreb-Beograd: JAZU and SANU.

Suess, E. 1863. Über die Verschiedenheit und die Aufeinanderfolge der tertiären Landfaunen in der Niederung von Wien. *Sitzungsberichte der k.k. Akademie der Wissenschaften, mathematisch-naturwissenschaftliche* Klasse, Wien 47:306–31.

———. 1866. Untersuchungen über den Charakter der österreichischen Tertiärablagerungen. II. Über die Bedeutung der sogenannten "brackischen Stufe" oder der "Cerithienschichten." *Sitzungsberichte der k.k. Akademie der Wissenschaften, mathematisch-naturwisshaftliche,* Classe, Wien 54:218–357.

Swisher, C. C. III. This volume. New ^{40}Ar/^{39}Ar dates and their contribution toward a revised chronology for the late Miocene nonmarine of Europe and West Asia.

Thenius, E. 1950. Die Säugetierfauna aus den Congerienschichten von Brunn-Vösendorf bei Wien. *Verhandlungen der Geologischen Bundesanstalt, Wien* 1948(7–9): 113–31.

———. 1959. Tertiär. 2. Teil Wirbeltierfaunen. In F. Lotze, *Handbuch der stratigraphischen Geologie* 3(2):1–328. Stuttgart: F.Enke.

———. 1960. Die jungtertären Wirbeltierfaunen und Landfloren des Wiener Beckens und ihre Bedeutung für die Neogenstratigraphie. *Mitteilungen der Geologischen Gesellschaft in Wien* 52 (1959):203–9.

Topachevskii, V. A. and A. F. Skorik. 1992. *Neogenovye i pleistotsenovye nizshie khomyakoobraznye yuga vostochnoi Evropy* [Neogene and lower Pleistocene Cricetidae of Southeastern Europe]. Kiev: Naukova dumka.

Topachevskii, V. A., A. L. Tchepalyga, V. A. Nesin, L. I. Rekoetz, and I. V. Topachevskii. 1988. Mikroteriofauna (Insectivora, Lagomorpha, Rodentia) Lektostratotipu Pontu. *Dopovidi Akademia Nauk Ukrainskoi RSR, Serie B Geologija* 4:73–76.

Vass, D., I. Repcok, K. Balogh, and J. Halmai. 1987. Revised radiometric time-scale for the Central Paratethys Neogene. *Annales Instituti Geologici Publici Hungarici, Budapest* 70:423–34.

Woodburne, M. O., G. Theobald, R. L. Bernor, C. C. Swisher III, and H. König. This volume. Advances in the geology and stratigraphy at Höwenegg, southwestern Germany.

Zapfe, H. 1949. Die Säugetierfauna aus dem Unterpliozän von Gaiselberg bei Zistersdorf in Niederösterreich. *Jahrbuch der Geologischen Bundesanstalt, Wien* 93(1948): 83–97.

———. 1951. *Dinocyon thenardi* aus dem Unterpliozän von Drassburg im Burgenland. *Sitzungsberichte der Österr. Akademie der Wissenschaften, mathematische-naturwissenschaftliche,* Klasse, Abteilung I, Wien, 160:227–41.

4

Present State of Magnetostratigraphic Studies in the Continental Neogene of Europe and Anatolia

S. SEN

During the last two decades it has become evident that magnetostratigraphy is a useful tool for correlating sedimentary deposits that include fossil vertebrates. The most frequent application of magnetostratigraphy has been either with marine sequences in continental exposures or within deep-sea sediments that present a fairly continuous record of sedimentary history and allow for much wider interbasin correlations. Comparatively few studies have focused on the European continental Cenozoic deposits. It is a broadly held belief that terrestrial deposits generally have a discontinuous sedimentation record, occur in isolated basins, and have exposures representing short chronologic intervals. These reasons may partially explain the lack of interest paleomagnetists generally have had in the study of continental deposits. Other circumstances, such as the high sedimentation rate, remagnetization, and the common occurrence of hematite-rich red beds, are practical reasons why paleomagnetists have avoided study of continental sediments.

The purpose of this paper is to review the available magnetostratigraphic data obtained from European and Anatolian middle and late Miocene continental deposits. In the European Neogene mammal chronology, this interval corresponds to units MN 5–MN 13.

Definition and Calibration of Stages and Zones

The stratigraphic units used for European Neogene mammal chronology are stages and zones (re: Fahlbusch 1991 for a review of this problem and his choice of the term *unit* rather than stages and zones). Stages are defined from faunal successions in some continental sedimentary basins (such as Agenian, Vallesian, or Ruscinian), while the MN zones (MN = Mammifères, Néogène) were defined by the first and last occurrences of some key mammalian taxa in Western Europe (Mein 1975, 1989; Fahlbusch 1991; de Bruijn et al. 1992). Each MN zone is characterized by a reference locality.

The definition of a boundary between two stages or two MN zones (= units sensu Fahlbusch 1991), even locally, requires that the mammal localities used as references have a well-established stratigraphic and biostratigraphic context. Unfortunately, this requirement is only satisfied in a few sedimentary basins. The most common circumstance is where mammal localities are found in isolated small outcrops without a physical stratigraphic correlation to any other locality.

The problem of identifying continental stages or MN zone boundaries has recently been discussed at length by Fahlbusch (1991). For him, the problem cannot presently be resolved because of the paucity of available data and the diachronic nature of mammalian events, such as first and last occurrences of zone-defining taxa. However, the stratigraphic code as well as the current practice in mammalian correlations implicitly recognizes such boundaries between stages or MN zones. Pierre Mein, who introduced this zonation, tried himself to calibrate the age of the MN 11/12 boundary (Demarcq et al. 1989; Opdyke et al. 1989).

In this study, the most recently refined magnetic polarity time scale (MPTS) proposed by Cande and Kent (1992) is used to correlate the magnetozones obtained in the studied sedimentary sections. Moreover, the labeling of the polarity zones (= chrons) used here (and also in Berggren et al. 1985 and Cande and Kent 1992) was initially suggested by LaBrecque et al. (1983). Each label indicates a polarity epoch, beginning with a reversed polarity period (e.g., C5r), followed by a normal polarity period (e.g., C5n).

Magnetostratigraphic Studies

There are several Miocene mammal localities spanning the area between Spain and Turkey where magnetostratigraphy has been applied. Some of them yielded magnetic polarity sequences that have been unambiguously correlated to the MPTS. However, in some cases the thickness of the studied sections and/or the low number of recovered magnetozones did not allow the recognition of the magnetic sequence and, therefore, only a rough (and uncertain) correlation to the MPTS was possible using the available biostratigraphic data.

Spain

Opdyke et al. (1989) have studied the Cabriel section, southeast of Valence, which has provided a sequence of 10 magnetozones in a 216-meter section. They correlated this sequence with C3r–C4n1. Two mammal localities (Fuente del Viso and Fuente Podrida) are situated in the sampled section, and two others (Venta del Moro and Balneario) are laterally correlated. The inferred ages by these authors for these localities are 5.6 Ma for Fuente del Viso, 6.0 Ma for Venta del Moro, 7.3 Ma for Fuente Podrida, and about 7.4 Ma for Balneario. Moreover, the first two localities contain late Turolian (MN 13) age faunas. The Fuente Podrida fauna includes *Parapodemus barbarae*, *Occitanomys adroveri*, and *Eliomys truci*, and is correlated with the middle Turolian (MN 12). Using this result, the MN 12/13 boundary should be placed between 7.2 and 5.8 Ma. Thus, Opdyke et al. suggested an age between 7.2 and 6.0 Ma for the MN 12/13 boundary. The locality of Balneario, which yielded *Parapodemus lugdunensis* and characterizes MN 11, is situated about 10 m below the previous locality, and thus is not strictly on the studied section. Its age is deduced from the mean sedimentation rate of the Cabriel section; using this extimated age, Opdyke et al. place the MN 11/12 boundary at 7.3–7.4 Ma. This date does not correlate with the ages recently obtained by Bernor et al. (this volume) in the Samos section in Greece (see also Swisher, this volume).

Another alternative would be to correlate the polarity sequence of the Cabriel section with C3r–C4n2. The occurrence of a 30 m thick normal polarity zone at the bottom of this section reinforces this hypothesis. If this alternative is retained, the ages of the localities Fuente Podrida and Balneario should be calibrated, in correlation with the MPTS of Cande and Kent (1992), as about 7.6 and 7.8 Ma, respectively. The implication of this recalibration is that the age of the MN 11/12 boundary should be about 0.5 Ma older than what Opdyke et al. have proposed. However, even this date (about 7.8 Ma) remains too young compared to the age of 8.24 Ma that is proposed by Bernor

et al. (this volume; re: Swisher, this volume) and Steininger et al. (this volume).

Finally, Can Llobateres, the reference locality for MN 9, has been sampled by a Spanish team (J. Agusti, pers. comm.), but the results are still unpublished.

France

Two localities have thus far been studied in France: Montredon and Sansan. The Montredon section has provided no reliable results because of the very weak magnetic intensities of its sediments (Sen 1988).

Sansan is the reference locality for MN 6 and is correlated with the beginning of this interval (Steininger et al. 1989; Bernor and Tobien 1990). Sansan has been sampled along 40 m of section, together with an additionnal 5-meter outcrop near the bottom of the hill (fig. 4.1). Reliable paleomagnetic data show a long reversed polarity zone followed by two short normal zones in the upper portion of the section. These magnetozones correlate with the polarity of chrons C5B reverse and C5B normal, 15.2 to 15.0 Ma for the Sansan fauna (Sen and Ginsburg, in press).

Former USSR

In Eastern Europe, Pevzner and his collaborators published a number of papers (see Pevzner and Vangengeim 1993) concerned with the magnetic stratigraphy of several mammalian localities in Moldavia, Russia, and Georgia. Most papers are in Russian, and I am unable to discuss these data and their implications at any length.

Central Europe

In Central Europe, the only middle or late Miocene vertebrate locality with associated magnetostratigraphic data is Höwenegg, Germany (Swisher, this volume), which has a normal polarity and is correlative with the long normal of chron 5.

Greece

Thus far, five Greek magnetostratigraphic sections have been made; from oldest to youngest they are: Chios, Biodrak, Kastellios Hill, Prochoma, and Samos.

At Chios a 46-meter section in the Michalou clay pit has been sampled at 32 consecutive levels. This section exhibits reversed polarities except for a 3-meter thick short normal zone near the base. All three horizons that yielded mammalian remains are correlative with MN 5, and are included in a long reversed magnetozone. These polarities cannot be correlated at this time with the MPTS (Kondopoulou et al. 1993). Sampling was pursued during the

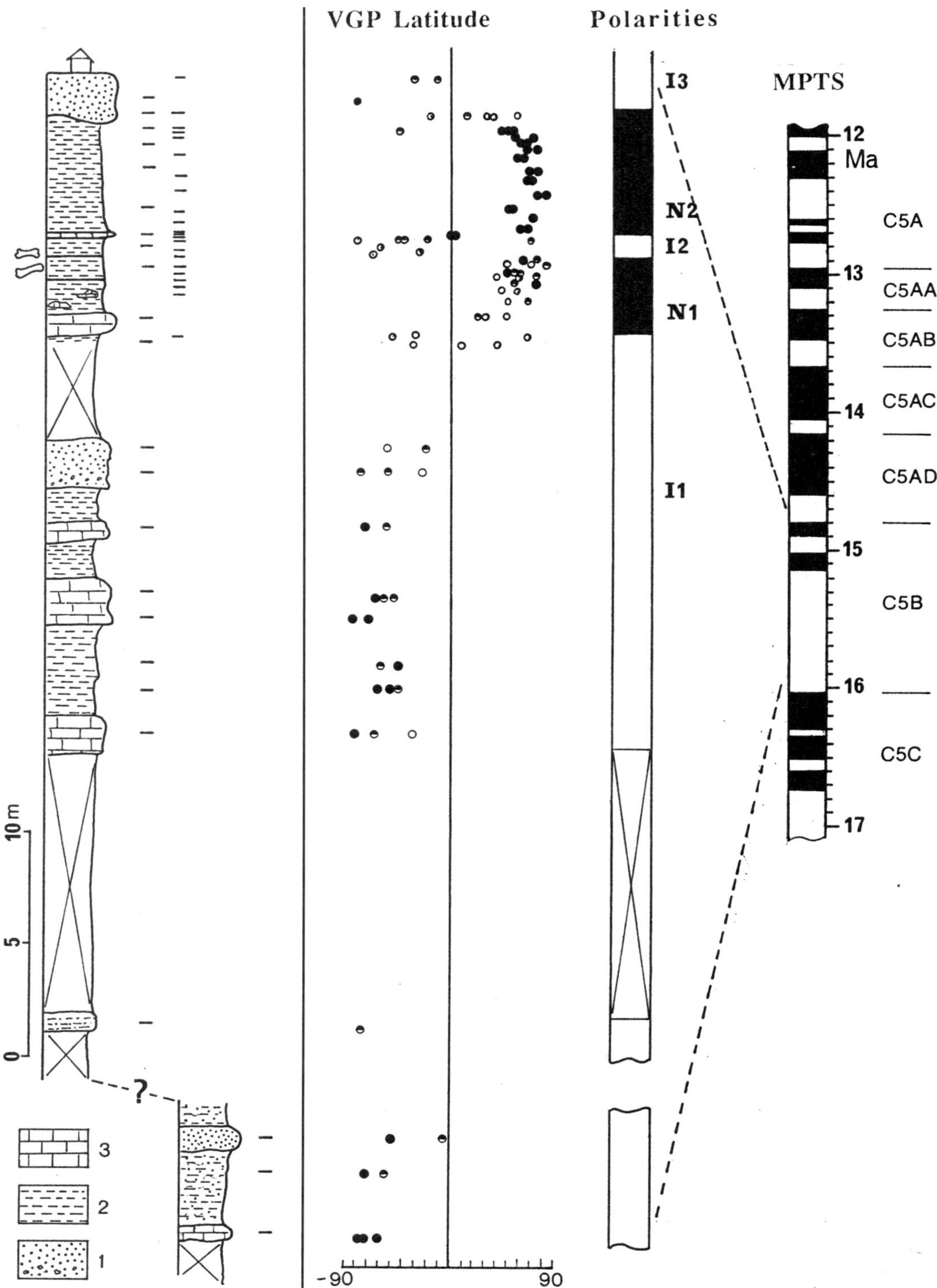

FIGURE 4.1 Magnetic stratigraphy of the Sansan section. The small bars next to the lithostratigraphic column indicate horizons sampled for this study. Virtual geomagnetic pole (VGP) latitudes are indicated for each sample against stratigraphy. The magnetic polarity time scale (MPTS) conforms to Cande and Kent 1992. Lithology: (1) sandstones, (2) clays and marls, (3) limestones.

summer of 1993 to increase the length of the magnetic section. Results are pending laboratory analysis by Sen.

Biodrak is a late Vallesian small mammal locality located approximately 60 km north of Athens. It includes the small mammal taxa *Leptodontomys* sp., *Myomimus multicrestatus*, *Byzantinia nikosi*, *Progonomys cathalai*, and *Schizogalerix* sp. (de Bruijn 1976). A 7-meter thick section that includes seven sampling sites has yielded only reversed polarities (Sen 1986).

Sen et al. (1986) reported on an 88-meter thick section sampled for magnetic stratigraphy at Kastellios Hill (Crete). Kastellios had previously yielded mostly small mammals, but also a few characteristic large mammal taxa from four discrete stratigraphic horizons. From the evidence of the mammalian assemblages, de Bruijn et al. (1971) correlated the Kastellios faunas with the late Vallesian. A later analysis by de Bruijn and Zachariasse (1979) led these authors to conclude that these deposits include the early–late Vallesian boundary. A few meters above the levels with small mammals, some estuarine deposits yielded planktonic foraminifera for which the authors suggested an age nearly correlative with the base of marine planktonic zone N 16 (early Tortonian). Magnetic stratigraphy provided reverse polarities along the entire length of the section (88 m), except for two short normal zones, each of one meter thickness (Sen et al. 1986). These polarities were tentatively correlated with the chron C5r, i.e., between 10.42 and 11.09 Ma in the MPTS of Berggren et al. (1985). However, all biostratigraphic and paleomagnetic correlations relative to this section should be questioned in view of our present understanding about the mammalian correlations.

First, we must consider why the Kastellios Hill mammal faunas were correlated to the mid-Vallesian. The lowest locality, K1, includes *Progonomys woelferi*, *Cricetulodon* cf. *sabadellensis*, *Spermophilinus bredai*, *Muscardinus* cf. *crusafonti*, and *Hipparion* sp. Fifteen meters above this level, locality K2 has yielded only *Progonomys cathalai* associated with *Hipparion* sp. About 12 meters above this, locality K3 contains both of the murids *P. cathalai* and *P. woelferi*.

The biostratigraphic order of these murid species at Kastellios Hill is confusing. In Western Europe *P. cathalai* first appears in the late early Vallesian, while *P. woelferi* first appears in the late Vallesian. De Bruijn and Zachariasse (1979) have concluded from this evidence that *P. woelferi* immigrated into the area before *P. cathalai*. In the same paper, they assigned the association from K1 to MN 9, and the others to MN 10.

Some arguments can be made for a younger age correlation of these faunas. *Progonomys woelferi* is an advanced species probably derived from *P. cathalai*. It was initially described from the Kohfidisch fissure filling in Austria, which is considered to be a latest Vallesian locality (late

MN 10 correlative of Mein 1989). This species has not previously been found elsewhere. Moreover, the predominance of murids in Western European small mammal assemblages is believed to be characteristic of the late Vallesian, since they are very rare in early Vallesian localities (Can Llobateres, Pedragueras 2C, Ampudia 9, Castelnou 1B; see Aguilar et al. 1991). A third argument reinforces this hypothesis. The K1 locality has also yielded *Cricetulodon* cf. *sabadellensis*. In Spain this species first occurs in the late early Vallesian (Can Llobateres), while *C. hartenbergeri* is known to first occur either in the terminal Astaracian or at the beginning of the Vallesian. This last species has also been found at Sinap Tepe in Central Anatolia (Sen 1991), which was previously considered to be an early Vallesian locality. Recent paleomagnetic studies in Anatolia suggest a chron C5N correlation of the long normal magnetozone that includes this locality (Kappelman et al., this volume); this correlation suggests that the fauna has a magnetic age of 9 + Ma (see below). Consequently, the occurrence of a derived species, *C.* cf. *sabadellensis*, at Kastellios Hill, and reversed polarities found in this section, might indicate a younger age. These arguments lead me, in the present biostratigraphic context, to correlate the small mammal associations from Kastellios Hill to the late, if not the latest, Vallesian. This correlation is corroborated by independent information provided by Bernor et al. (this volume).

Some viable arguments do exist for supporting an older age for the Kastellios sequence. First, *Progonomys* first occurs in the Indian subcontinent 11.6 Ma (Barry and Flynn 1989), and shows a rapid evolutionary radiation, which gave rise to several other genera and species. In Europe *Progonomys* is an immigrant form, and one can imagine that it reached the Aegean area somewhat earlier than Western Europe. Such a hypothesis assumes a diachronism in *Progonomys*' invasion and diversification between Eastern and Western Europe. Second, the occurrence of the planktonic foraminiferan *Neogloboquadrina acostaensis* from the K4 level, just above the small mammal localities, has led de Bruijn and Zachariasse (1979) to propose a correlation with the basal part of the zone N 16. In the literature, there is a clear disagreement on the calibration of this biozone, and consequently that of the Serravalian–Tortonian (= middle–late Miocene) boundary in marine sequences. According to Miller et al. (1985) and Berggren et al. (1985), based on data from the DSDP holes 519 in the South Atlantic and 563 in the North Atlantic, the base of this zone is very near the base of the chron C5n, at about 10.2 Ma according to the MPTS of Berggren et al. (1985) or at about 10.7 Ma according to Cande and Kent (1992). This calibration is not compatible with the reversed polarities of the Kastellios Hill section.

These lines of evidence suggest two plausible hypotheses: (1) the planktonic association from Kastellios Hill equates with a younger zone; (2) Berggren et al.'s (1985) calibration of this zone should be older. Barron et al. (1985) argue that the data from the DSDP sites 71, 77, and 289 in the central and western equatorial Pacific provide a correlation of the Serravalian–Tortonian boundary with the base of chron C5r, ca. 11.5 Ma. In their compilation, Steininger et al. (1989) have also retained this calibration. It is not the purpose of this paper to discuss discrepancies between marine and continental Neogene chronologies. However, it is necessary to note that the reversed polarities found at Kastellios Hill conform with this calibration. Otherwise, if Berggren et al.'s calibration is used, the base of the zone N 16 should be younger than 8.9 Ma, and is simply too young to be accepted. In summary, the first argument suggests a younger age for the Kastellios Hill faunas (between 8.9 and 8.3 Ma), while the second line of argument favors an older age (between 11.1 and 10.7 Ma). The present paper retains a younger age correlation for the Kastellios Hill faunas.

The Prochoma locality in Greek Macedonia has a large mammal association correlated with the early Turolian (MN 11). The fluviatile deposits that include this locality have been sampled along a 25-meter thick section. The base of this section has a normal polarity for 5 m, with the remaining 20 m being reverse. An approximate correlation with a part of chron C4Ar has been suggested by Kondopoulou et al. (1992).

Samos Island's Eastern Basin has one of the richest mammalian fossil-bearing deposits in Europe. According to Solounias (1981), Weidmann et al. (1984), Bernor et al. (this volume), and Swisher (this volume), at least fifteen localities yielded fossil mammals, all in the Mytilini Formation. Moreover, the latter authors presented an extensive radiometric analysis of these deposits, bracketing all mammalian localities between 8.5 and 6.1 Ma. The magnetic stratigraphy within a 130-meter thick section provided a succession of seven polarity zones, and they were accurately correlated with chrons C3An2 and C4n1. This correlation assigns an age of about 6.4 Ma to the related mammal localities Q5, Q1, L, and A (Sen and Valet 1986). These localities constitute the named "upper fossiliferous horizon" of the Mytilini Formation (see Bernor et al., this volume, for a further discussion on Samos' chronology and correlation).

Turkey

Five magnetic sections have thus far been studied in Turkey: Kaleköy and Karaözü are both late Vallesian localities in Central Anatolia, about 70 km northeast of Kayseri. Their small mammal associations contain some transitional Vallesian/Turolian species, such as a polymorphic *Progonomys* and a co-occurring primitive *Parapodemus*, indicating a very high level among Vallesian faunas (Sü-

mengen et al. 1989). The sections sampled for magnetostratigraphy include 60 m at Kaleköy and 52 m at Karaözü, and include only reverse polarities, for which Langereis et al. (1989) suggested an approximative correlation within chron C4Ar.

Another locality in the same basin, Gemerek, is represented by a poorly developed middle Miocene small mammal fauna (Sümengen et al. 1989). A 170-meter magnetic section yielded a succession of three polarity zones, which from lowest to highest stratigraphic intervals are: a 15-meter section with reversed polarity; a 70-meter section with normal polarity; an 85-meter section with reversed polarity. A basalt flow 80 m above the sampled section has been dated as 14.9 ± 0.7 Ma. Combining this age with the biochronologic data, Langereis et al. (1989) suggested a correlation with chrons C5Br–C5C, ca. 15.8 Ma.

The fluviatile deposits of the Sinap Tepe area, about 60 km NW of Ankara, are particularly rich in middle and late Miocene large and small mammals. Sen (1991) reported the magnetic stratigraphy of a 55-meter thick section along the Sinap Loc. I, which is known for its Vallesian mammal association, including *Sivapithecus meteai* Ozansoy 1965. The two magnetozones, normal and reverse, found in this section are not sufficiently characteristic to be correlated with the MPTS. Since 1989 an international research team active in the area has obtained new and more detailed magnetostratigraphic results. The fossiliferous early and middle Sinap series have been resampled from a section that is more than 100 m thick. The results of this study is presented in this volume (Kappelman et al., this volume). From the new data, it appears that the normal polarity zone (which includes Loc. I) is in fact longer, covering these series for about 100 m. Kappelman et al. suggest a chron C5n correlation for this magnetozone. Such a correlation implies an age a little older than 9 Ma for the related Sinap Loc. I mammal fauna. We have to emphasize that this locality was considered in previous studies probably to be early Vallesian age. The new correlation implies that this fauna is younger than previously admitted and that the early Vallesian corresponds mainly to chron C5n.

The section studied by Kappelman et al. (this volume) contains several fossiliferous horizons with and without *Hipparion*, allowing a correlation of the locally first-occurring hipparion, and hence the MN 8/9 boundary, ca. 10.3 Ma. These authors have also presented the magnetic stratigraphy of the Kavakdere section, about 10 km west of Sinap Tepe. It contains several rich large mammal faunas. The mammalian systematic determinations from the Kavakdere section are still preliminary and have yet to yield a clear chronology for the fossil mammal sequence.

In Western Anatolia, the 25-meter thick Kemiklitepe section has yielded Turolian large mammal faunas derived from two stratigraphic horizons correlative with MN 11 and MN 12, respectively. Sampling of this section on 17

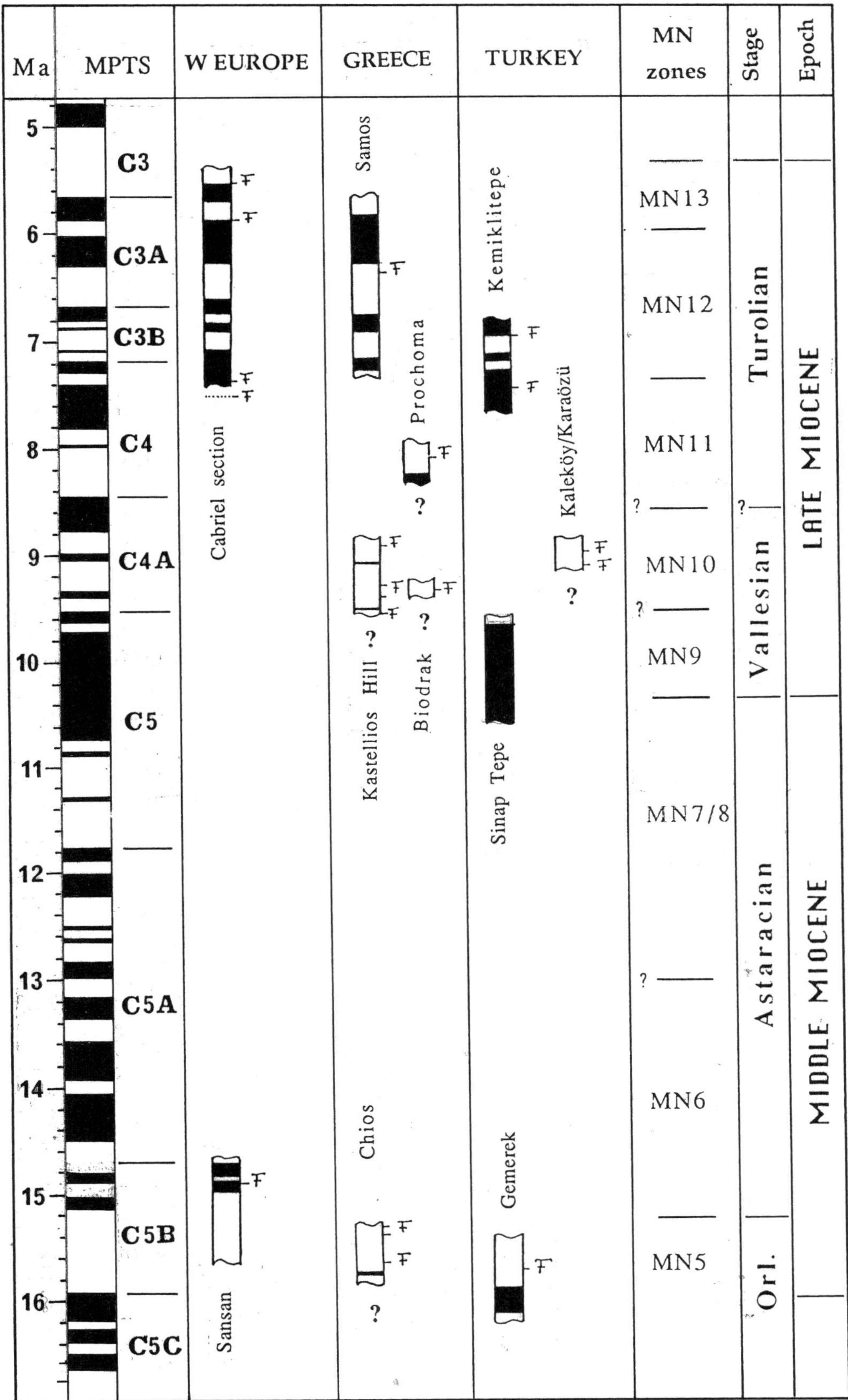

FIGURE 4.2 Correlation chart between the magnetic stratigraphy of sections from Western Europe, Greece, and Turkey and the magnetic polarity time scale (MPTS) of Cande and Kent (1992). Mammal-bearing horizons are indicated at the right side of polarity columns. The sections with a question mark are those presenting an approximate correlation with the MPTS.

levels yielded a succession of five polarity zones. It is strange that fluviatile deposits in such a short section record as this one has so many polarity transitions. Although the Turolian corresponds to a "very busy" magnetic interval in earth history, it was concluded from this study that this section contains sedimentary gaps, thus giving an appearance of low sedimentation rate. This polarity succession was tentatively correlated with the interval of C4n–C3Bn of the MPTS, implicating ages about 7.6 Ma for the lower fossiliferous horizon (loc. KTD) and about 6.9 Ma for the upper fossiliferous horizon (localities KTA and KTB; Sen et al. 1994).

Conclusions

This review of available European and Anatolian middle and late Miocene magnetostratigraphic data reveals that most of the studied sections cover a limited stratigraphic range, and thus yield fragmentary polarity successions insufficient to correlate MN Zones 6–13 securely as a whole with the MPTS. Only results from four sections, Sansan and Gemerek in the middle Miocene, and Cabriel and Samos sections in the late Miocene, provide satisfactory polarity successions, and they have been correlated accurately with the MPTS. In the other sections this correlation is either approximate or nonexistent.

Figure 4.2 summarizes magnetozone correlations from this study area as they correlate with the MPTS sensu Cande and Kent (1992). In each section, the polarity succession and the horizon(s) with mammal fauna(s) are indicated. The sections with a question mark below are those for which only an approximate correlation can be proposed. The number of relevant magnetostratigraphically calibrated sections are still limited and the data is fragmentary, covering only small portions of the MPTS. The situation is such that a secure European Neogene magnetostratigraphic zonation is currently unobtainable.

The principal benefit of these magnetostratigraphic studies is to provide numerical ages for some mammalian faunas and to facilite, to some extent, their correlation. Beyond this, as previously discussed in the context of the "*Hipparion* Datum" (Sen 1989), the fragmentary data allow only a preliminary attempt to calibrate some stages and/or MN zone boundaries. The only boundaries for which a magnetostratigraphic calibration exists is the MN 11/12 boundary in the Cabriel section (Opdyke et al. 1989), and to some extent MN 8/9 boundary in the Sinap Tepe section (Kappelman et al., this volume). The general lack of magnetostratigraphic correlations with well sequenced mammal localities does not presently allow any further precision.

Since this paper was written several relevant studies have appeared. Cande and Kent (1995) published a revised version of their 1992 MPTS. This new calibration provides ages-to-polarity chrons a little older than the previous time scale. The reader is invited to refer to the ages of polarity chrons given in the new version of the MPTS and to convert the correlations and age assignments to mammal localities in the present paper. Krijgsman et al. (1994) published the detailed magnetic stratigraphy of the Armantes and Aragan sections in Central Spain. Their study provides precise correlations with the MPTS for mammal faunas and local biozones assumed as correlative to the interval MN 4–MN 7/8. The ages they propose for MN zone boundaries are notably younger than those of Steininger et al. (this volume).

Acknowledgments

I would like to thank R. Austin, R. Bernor, and Y. Gallet, who critically read this manuscript and suggested improvements to the English.

LITERATURE CITED

Aguilar, J. P., M. Calvet, and J. Michaux. 1991. Présence de *Progonomys* (Muridae, Rodentia, Mammalia) dans une association de rongeurs de la fin du Miocène moyen (Castelnou 1B; Pyrénées Orientales, France). *Géobios* 24:503–508.

Barron, J. A., G. Keller, and D. A. Dunn. 1985. A multiple microfossil biochronology for the Miocene. In *The Miocene Ocean*, pp. 21–35. Memoire, Geological Society of America.

Barry, J. C. and L. J. Flynn. 1989. Key biostratigraphic events in the Siwalik sequence. In *European Neogene Mammal Chronology*, ed. E. H. Lindsay, V. Fahlbusch, and P. Mein, pp. 557–571. New York: Plenum.

Berggren, W. A., D. V. Kent, and J. A. Van Couvering. 1985. The Neogene, Part 2, Neogene geochronology and chronostratigraphy. In *The Chronology of the Geological Record*, ed. N. J. Snelling, pp. 211–260. Memoire of the Geological Society of London 10.

Bernor, R. L., N. Solounias, C. C. Swisher III, and J. A. Van Couvering. This volume. The correlation of three classical "Pikermian" faunas—Maragheh, Samos, and Pikermi—with the European MN unit system.

Bernor, R. L. and H. Tobien. 1990. The mammalian geochronology and biogeography of Paşalar (middle Miocene, Turkey). *Journal of Human Evolution* 19:551–68.

Bruijn, H. de. 1976. Vallesian and Turolian rodents from Biotia, Attica, and Rhodes (Greece). *Proceedings, Koninklijke Nederlandse Akademie van Wetenschappen, B* 79:361–84.

Bruijn, H. de, R. Daams, G. Daxner-Höck, V. Fahlbusch, L. Ginsburg, P. Mein, and J. Morales. 1992. Report of the RCMNS working group on fossil mammals, Reisensburg 1990 *Newsletters on Stratigraphy* 26:65–118.

Bruijn, H. de, P. Y. Sondaar, and W. J. Zachariasse. 1971. Mammalia and foraminifera from the Neogene of Kastelios Hill, (Crete), a correlation of continental and marine biozones. *Proceedings, Koninklijke Nederlandse Akademie van Wetenschappen, B* 74:1–22.

Bruijn, H. de and W. J. Zachariasse. 1979. The correlation of marine and continental biozones of Kastellios Hill reconsidered. *Annales Géologie Pays Hellénica* 1:219–26.

Cande, S. C. and D. V. Kent. 1992. A new geomagnetic polarity time scale for the late Cretaceous and Cenozoic. *Journal of Geophysical Research* 97(B10): 13917–951.

——. 1995. Revised calibration of the geomagnetic polarity time scale for the late Cretaceous and Cenozoic. *Journal of Geophysical Research* 100(B4): 6093–95.

Demarcq, G., P. Mein, P. Ballésio, J. P. Romaggi. 1989. Le gisement d'Andance (Coiron, Ardèche, France) dans le Miocène supérieur de la vallée du Rhône un essai de corrélations marin-continental. *Bulletin Société géologie France* 5: 797–806.

Fahlbusch, V. 1991. The meaning of MN zonation considerations for a subdivision of the European continental Tertiary using mammals. *Newsletters on Stratigraphy* 24:159–173.

Kondopoulou, D., S. Sen, G. D. Koufos, and L. de Bonis. 1993. Paleomagnetic data from the Miocene of the Island of Chios, Greece. *Terra Cogn.*, Abstract 76–77.

——. 1992. Magneto- and biostratigraphy of the late Miocene mammalian locality of Prochoma (Macedonia, Greece). *Paleontologie Evolución.* 24–25:135–139.

Krijgsman, W., C. G. Langereis, R. Daams, and A. van der Meulen. 1994. Magnetostratigraphic dating of the middle Miocene climate change in the continental deposits of the Aragonian type area in the Calatayud-Teruel Basin (Central Spain). *Earth Planetary Science Letters* 128: 513–26.

LaBrecque, J. L., K. J. Hsü, M.F. Carman, A. M. Karpoff, J. A. McKenzie, P. F. Percival, N. P. Petersen, K. A. Pisciotto, E. Schreiber, L. Tauxe, P. Tucker, H. J. Weissert, and R. Wright. 1983. DSDP Leg 73 contributions to Paleogene stratigraphy in nomenclature, chronology, and sedimentation rates. *Palaeogeography, Paleoclimatology, Paleoecology* 42:42–125.

Langereis, C. G., S. Sen, M. Sümengen, and E. Ünay. 1989. Preliminary magnetostratigraphic results of some Neogene localities from Anatolia (Turkey). In *European Neogene Mammal Chronology*, ed. E. H. Lindsay, V. Fahlbusch, and P. Mein, pp. 515–525. New York: Plenum.

Mein, P. 1975. Résultats du Groupe de Travail des Vertébrés. In *Report on Activity of the RCMNS Working Group*, ed. J. Senes, pp. 78–81. Bratislava: SAV.

——. 1989. Updating of MN zones. In *European Neogene Mammal Chronology*, ed. E. H. Lindsay, V. Fahlbusch, and P. Mein, pp. 73–90. New York: Plenum.

Miller, K. G., M.-P. Aubry, M. J. Khan, A. J. Melilo, D. V. Kent, and W. A. Berggren. 1985. Oligocene-Miocene biostratigraphy, magnetostratigraphy, and isotopic stratigraphy of the western North Atlantic. *Geology* 13:257–261.

Opdyke N., P. Mein, E. Moissenet, A. Pérez Gonzalez, E. H. Lindsay, and M. Petko. 1989. The magnetic stratigraphy of the late Miocene sediments of the Cabriel Basin, Spain. In *European Neogene Mammal Chronology*, ed. E. H. Lindsay, V. Fahlbusch, and P. Mein, pp. 507–514. New York: Plenum.

Pevzner, M. A. and E. A. Vangengeim. 1993. Magnetochronological age assignments of middle, and late Sarmatian mammalian localities of the Eastern Paratethys. *Newsletter on Stratigraphy* 29:63–75.

Rögl, F. and G. Daxner-Höck. This volume. Late Miocene Paratethys correlations.

Sen, S. 1986. *Contribution à la magnétostratigraphie et à la paléontologie des formations continentales néogènes du pourtour méditerranéen. Implications biochronologiques et paléobiologiques.* Thèse d'Etat University Paris 6, Mémoires Sciences Terre, 86(19).

——. 1988. Contributions à l'étude du gisement miocène supérieur de Montredon (Hérault) 9-Tentative de magnétostratigraphie. *Palaeovertébra*, Mémoire Extraordinaire, pp. 187–188.

——. 1989. *Hipparion* Datum and its chronologic evidence in the Mediterranean area. In *European Neogene Mammal Chronology*, ed. E. H. Lindsay, V. Fahlbusch, and P. Mein, pp. 495–505. New York: Plenum.

——. 1991. Stratigraphie, faunes de mammifères et magnétostratigraphie du Néogène de Sinap Tepe, province d'Ankara, Turquie. *Bulletin Museum National Histoire Naturelle* 12:243–277.

Sen, S. and J.-P. Valet. 1986. Magnetostratigraphy of late Miocene continental deposits in Samos, Greece. *Earth Planetary Science Letters* 80:167–174.

Sen, S., J.-P. Valet, and C. Ioakim. 1986. Magnetostratigraphy and biostratigraphy of the Neogene deposits of Kastellios Hill (Central Crete), Greece. *Palaeogeography, Paleoclimatology, Paleoecology* 53:321–34.

Sen, S. and L. Ginsburg. In press. *Magnétostratigraphie du Miocène moyen de Sansan (Gers, France)*. Mémoire du Museum Nationale Histoire naturelle, Paris.

Sen, S., L. de Bonis, N. Dalfes, D. Geraads, and G. D. Koufos. 1994. Les gisements de mammifères du Miocène supérieur de Kemiklitepe, Turquie. 1- stratigraphie and magnétostratigraphie. *Bulletin Museum National Histoire Naturelle* 16(1): 5–17.

Solounias, N. 1981. The Turolian fauna from the Island of Samos, Greece. *Contributions in Vertebrate Evolution* 6:1–232.

Steininger, F. F., W. A. Berggren, D. V. Kent, R. L. Bernor, S. Sen, and J. Agusti. This volume. Circum-Mediterranean Neogene (Miocene and Pliocene) marine-continental chronologic correlations of European mammal units.

Steininger, F. F., R. L. Bernor, and V. Fahlbusch. 1989. European Neogene marine/continental chronologic correlations. In *European Neogene Mammal Chronology*, ed. E. H. Lindsay, V. Fahlbusch, and P. Mein, pp. 15–46. New York: Plenum.

Sümengen, M., E. Ünay, G. Sarac, H. de Bruijn, I. Terlemez, and M. Gürbüz. 1989. New Neogene rodent assemblages from Anatolia (Turkey). In *European Neogene Mammal Chronology*, ed. E. H. Lindsay, V. Fahlbusch, and P. Mein, pp. 61–72. New York: Plenum.

Swisher, C. C., III. This volume. New [40]Ar/[39]Ar dates and their contribution toward a revised chronology for the late Miocene nonmarine of Europe and West Asia.

Weidmann M., N. Solounias, R. E. Drake, and G. H. Curtis. 1984. Neogene stratigraphy of the Eastern Basin, Samos Island, Greece. *Géobios* 17:477–490.

5

New ^{40}Ar/^{39}Ar Dates and Their Contribution Toward a Revised Chronology for the Late Miocene of Europe and West Asia

C. C. SWISHER III

Circum-Mediterranean and Southwest Asian late Miocene mammalian history is subdivided into a series of biochronological units that have proven useful for correlation and relative age sequencing of continental nonmarine deposits. The Mammal Neogene (MN) "zones" of Mein (1975; 1979; 1989) are commonly grouped into a series of regional land-mammal "stages" (Fahlbusch 1976; or Neogene mammal faunal units of Steininger et al. 1989; Fahlbusch 1991), and are delimited by first and last occurrences of mammalian genera, first occurrences of immigrant taxa, and commonness of certain temporally restricted taxa. These criteria and the general stage-of-evolution of the fossil faunas have been used successfully to integrate the Neogene nonmarine mammalian history of Europe, North Africa, and West Asia.

Currently, the European–West Asian late Miocene is subdivided into the older Vallesian age (MN 9 and MN 10) and the younger Turolian (MN 11–13; Mein 1975, 1979, 1989; Steininger et al. 1989). The base of the Vallesian (= base of MN 9) is commonly considered coincident with the first appearance of hipparionine horses in the Old World. The rather abrupt appearance of these horses in the stratigraphic record is commonly referred to as the "Hipparion" Datum, often considered to be one of the most important regional correlation indices in the nonmarine strata of Western Europe, and one that affects the entire late Miocene MN unit system. The subsequent evolution and diversification of hipparionines in the late Miocene make them ideal biochronological indices for the correlation and relative age calibration of later Vallesian and Turolian time.

The impetus for this study is an outgrowth of a collaborative project intended to examine the geology, paleontology, and geochronology of the Höwenegg fossil site, Hegau, southern Germany. At the VW Stiftung workshop on middle and late Miocene European and West Asian faunas held at the Schloss Reisensburg, southwest Germany (July 1992), I reported new ^{40}Ar/^{39}Ar dates that bear directly on the age of Höwenegg. When we used ^{40}Ar/^{39}Ar incremental-laser heating techniques, the dates proved to have a precision previously unattainable by conventional K-Ar dating. During discussion, questions arose as to the reliability of the age of a number of key faunas crucial for the calibration of the Vallesian and Turolian mammal ages. It was generally agreed that the MN units and regional land-mammal "stages" (or Neogene mammal faunal units) provide precise regional correlations for the nonmarine circum-Mediterranean region. However, the regional character of terrestrial faunas, as well as associated difficulties in correlating nonmarine and marine sequences, make integration of mammalian biochronologic units into a global framework problematic. The integration of the MN mammal sequence within a global chronology relies heavily on those few representative sites calibrated directly by radioisotopic dates and/or by encompassing magnetostratigraphies that can be unambiguously correlated to the Geomagnetic Polarity Time Scale (GPTS).

The purpose of this study is to provide some new precise dates for key European–West Asian late Miocene vertebrate localities as a basis for giving an independent chronology for MN subdivision. Key to this discussion was the reliability of published K-Ar dates currently in use for the calibration of the Vallesian and Turolian faunas. It was decided to check the accuracy of these K-Ar dates by redating key calibration tie-points at Samos, Greece, and Maragheh, Iran, using similar ^{40}Ar/^{39}Ar dating techniques as those used for redating Höwenegg. Because of the short timeframe for gaining results, it was not possible to collect new samples at the various field sites other than Höwenegg. Samples dated in this study were made available to me by Garniss H. Curtis and Robert E. Drake, who had been involved in the original K-Ar dating project. Further discussion on the correlation of European Vallesian localities can be found in Woodburne et al. (this volume a), and Turolian localities can be found in Bernor et al. (this volume b).

Methodology

Samples for ^{40}Ar/^{39}Ar dating from Höwenegg, Germany, were collected in 1991 and 1992 (also see Woodburne et al., this volume b). Samples from Samos, Greece, and Maragheh, Iran, were made available to us by Drs. G. H. Curtis and R. E. Drake, who published the K-Ar dates for these samples in 1980 and 1984 (Campbell et al. 1980; Weidmann et al. 1984). New mineral separates of sanidine and plagioclase for single crystal ^{40}Ar/^{39}Ar dating were prepared from the same rock samples previously dated.

Samples to be dated were examined under a binocular microscope, then crushed, washed, and sieved to obtain the coarsest minerals. Depending on the sample, either sanidine, plagioclase, and/or hornblende were concentrated using a Frantz magnetic separator, or if coarse enough, hand-picked directly from the matrix. To remove any attached glass or clay, the sanidines and plagioclases were treated with 0.7% hydrofluoric acid in an ultrasonic bath for five minutes followed by a ten-minute ultrasonic rinse in distilled water. The hornblendes were hand-picked from the matrix and treated in an ultrasonic bath with distilled water. The largest, optically freshest appearing crystals were then hand-picked under a binocular microscope.

The minerals were irradiated in two different irradiation packages, both with a centrally located monitor mineral (Fish Canyon Tuff sanidine). The first package included only the Höwenegg samples, and these were irradiated for one hour using the hydraulic rabbit facility of the Omega West research reactor at Los Alamos National Laboratory. The second package included the Samos, Maragheh, and additional Höwenegg samples (6745–01 and 02) and was irradiated for three hours in the Denver Triga Reactor. After irradiation, single crystals of the minerals to be dated were loaded into individual 2-mm diameter wells of a copper sample disk, placed within the sample chamber of the extraction system, and baked out at 200° C for eight hours. Total fusion of the sanidines and plagioclases was accomplished using a 6W Coherent Ar-ion laser with a maximum output during this study of approximately 8 watts.

For the incremental-heating analyses, the laser beam was defocused to heat evenly the 2-mm diameter sample well. For each increment, the sample was heated for 45 seconds by stepwise increase of the output from the Ar-ion laser. The released gases were purified by two Zr-Fe-V getters operated at approximately 150° C, and the argon

was measured in an on-line Mass Analyzer Product 215 noble-gas mass spectrometer operated in the static mode using automated data collection techniques. Laser heating, gas purification, and mass spectrometry were completely automated following computer-programmed schedules. Uncertainty in the fluence-calibration parameter (J) was less than 0.1%. Analytical procedures follow Deino and Potts (1990), Deino et al. (1990), Swisher and Prothero (1990), and Swisher et al. (1992).

Ca and K corrections were determined from laboratory salts: $(^{36}\mathrm{Ar}/^{37}\mathrm{Ar})_{\mathrm{Ca}} = 2.582 \equiv 10^{-4} \pm 4.6\mathrm{x}\ 10^{-6}$, $(^{39}\mathrm{Ar}/^{37}\mathrm{Ar})_{\mathrm{Ca}} = 6.7 \equiv 10^{-4} \pm 2.53 \equiv 10^{-5}$, and $(^{40}\mathrm{Ar}/^{39}\mathrm{Ar})_{\mathrm{K}} = 2.2 \equiv 10^{-2} \pm 3.5 \equiv 10^{-4}$. Mass discrimination throughout this study was monitored by replicate air aliquots delivered from an on-line pipette system. Decay constants are those recommended by Steiger and Jager (1977) and Dalrymple (1979) and are: $\lambda\epsilon + \lambda\epsilon' = 0.581 \equiv 10^{-10}\mathrm{yr}^{-1}$; $\lambda\beta = 4.962 \equiv 10^{-10}\ \mathrm{yr}^{-1}$; $^{40}\mathrm{K}/^{40}\mathrm{K}_{\mathrm{total}} = 1.167 \equiv 10^{-4}$. The uncertainties associated with the individual analyses are one standard deviation (SD), while those that accompany the calculated weighted mean ages of the replicate analyses are standard errors (SE), following Taylor (1982).

The ^{40}Ar/^{39}Ar dates derived from the sanidines were calculated using a J value determined from replicate analyses of individual grains of the co-irradiated monitor mineral Fish Canyon Tuff sanidine (FC) with a reference age of 27.84 Ma. This age is similar to that recommended by Cebula et al. (1986), but slightly modified as a result of in-house intercalibration with MMhb-I with a published age of 520.4 ± 1.7 Ma by Samson and Alexander (1987). For the Samos samples J = 0.0007091 ± 0.0000006; for Maragheh samples and Höwenegg samples 6745–01 and 6745–02, J = 0.0007039 ± 0.0000005; for all other Höwenegg samples J = 0.0007592 ± 0.0000008.

For the incremental-heating analyses, the uncertainties associated with the individual incremental apparent ages are 2σ errors, while those that accompany the calculated weighted mean ages of the plateau increments and weighted means of the replicate analyses are standard errors (SE), following Taylor (1982). The plateau definition used here follows that of Fleck et al. (1977).

The Geomagnetic Polarity Time Scale (GPTS) used in this study is that of Cande and Kent (1992, modified in 1995 and further presented in this volume by Steininger et al.).

Dated Localities

Höwenegg, Germany

The Höwenegg locality (Hegau, southern Germany) is unrivaled for its preservation of complete mammalian skeletons, including 14 skeletons of the morphologically primitive horse *Hippotherium primigenium*. A history of the excavations at Höwenegg, along with the geological, taphonomical, and paleoecological interpretations of the locality, was presented by Tobien (1986), and is further reviewed in Woodburne et al. (this volume, b). The importance of the Höwenegg locality lies not only in its well-preserved mammalian skeletons, including a primitive hipparion (Bernor et al., this volume a; Woodburne et al., this volume a), but also in the fact that it is one of the few Central European vertebrate localities where datable volcanics have been found interbedded within fossiliferous strata.

Dated at 12.5 Ma by K-Ar, the Höwenegg site provided a key calibration for the *"Hipparion"* Datum, a tie-point that affects the relative age assignments of much of the Vallesian. The concept of a *"Hipparion"* Datum (Berggren and Van Couvering 1974) stems from the observation that at a number of sites in Europe, Asia, and North Africa, a species of *"Hipparion"* horse appears abruptly and abundantly at a level stratigraphically above deposits in which that kind of horse is absent (Woodburne et al., this volume a). This abrupt, geographically widespread, first appearance in the stratigraphic record of *"Hipparion"* was considered by Berggren and Van Couvering (1974) and Berggren et al. (1985) to be a consequence of a rapid migration and dispersal of these horses from North America into the Old World, approximately 12.5 Ma, an age derived primarily from early K-Ar dates made on volcanics from the hipparionine-bearing strata at Höwenegg, Germany (Van Couvering and Miller 1971; Berggren and Van Couvering 1974, 1978; and Berggren et al. 1985).

Höwenegg's chronometric dating originates with Lippolt et al's. (1963) work, when they obtained a K-Ar date of 12.4 ±1.1 Ma (recalculated to 12.7 Ma with currently accepted decay constants) on hornblende from an interbedded brownish hornblende-rich "tuffite" (Tobien 1986; Woodburne et al., this volume b). Numerous layers of these so-called tuffites interbedded in the fossil vertebrate–bearing strata at Höwenegg are considered volcanic and volcaniclastic sediments, being laid down primarily as water-lain tuffs. The tuffites appear to have been washed into fluvial sediments consisting of pebble- to cobble-size conglomerates. The conglomerates contain numerous rounded clasts of older volcanic and Mesozoic chalks; however, the matrix itself consists of volcaniclastic sediment with pristine euhedral hornblendes. The lithology of the matrix as well as of the "tuffites" is similar to the underlying "Alterer Tuff" with which it may partially interfinger (Woodburne et al., this volume b).

Although the 12.5 Ma date for Höwenegg is widely quoted in the literature, a number of younger K-Ar dates on rocks from Höwenegg were reported in the middle 1970s that should have brought the 12.5 Ma age for the "Hipparion" Datum into question. Baranyi et al. (1976) reported K-Ar dates on a basaltic cobble extracted from the Höwenegg tuffites. This cobble was originally considered by these authors to be a volcanic "bomb"; however, this interpretation was questioned by Becker-Platen et al. (1977), who considered the cobble to be detrital. Six K-Ar dates on this basalt cobble yielded a mean age of 10.8 ±0.4 Ma, an age approximately 2 Ma younger than the K-Ar date on hornblende from the same deposits (Baranyi et al. 1976). Becker-Platen et al. (1977) considered the cobble to be debris reworked into the Höwenegg "tuffite." Personal observation by me in 1991 and 1992 supports this later interpretation, in that numerous cobbles of basalt and Mesozoic chalk were found in place in the Höwenegg "tuffites." It is equivocal whether these cobbles were reworked into waterlain "tuffites" or whether even the hornblende-bearing matrix is reworked as well, possibly being reworked from or perhaps interfingering with the "Alterer Tuff." The hornblendes from the Höwenegg "tuffites" range in size up to 4 cm in length, and phlogopite, although rare, up to 1 cm. The hornblende's freshness suggests little transport or reworking. There is, however, some remaining ambiguity as to how much time separates the eruption of the volcaniclastic hornblende-bearing matrix and the age of the Höwenegg fauna.

New samples of the "Alterer Tuff," the Höwenegg fossil quarry "tuffite," and a sample of the phlogopite-bearing igneous plug from the adjacent rock quarry were collected for laser incremental-heating ^{40}Ar/^{39}Ar analysis. Hornblende from the "Alterer Tuff" showed definite signs of excess argon. The higher temperature steps, consisting of over 60% of the total percentage of ^{39}Ar released, formed a slightly concave-shaped plateau with a calculated age of 10.36 ±0.19 Ma, whereas the phlogopite from the rock quarry's igneous plug resulted in a well-behaved spectrum, with a plateau age of 9.82 ±0.20 Ma.

Direct dating of the Höwenegg fossil-bearing strata yielded mixed results. Phlogopite from Höwenegg stratigraphic level 33–2 (re: Woodburne et al., this volume) yielded a ^{40}Ar/^{39}Ar plateau age of 343.7 ±2.8 Ma. Similarly, phlogopite separated from a cobble at the same level did not yield a plateau, but yielded a rising spectrum, with the higher temperature steps reaching 350 Ma. The age of these phlogopites clearly indicate the detrital nature of some Höwenegg tuffite minerals, although it is possible that these phlogopites were incorporated in an explosive 10 Ma eruption.

To minimize the potential inclusion of detrital reworked minerals, single crystals of hornblende from the Höwenegg "tuffites" were incrementally heated by the ^{40}Ar/^{39}Ar method. Three laser incremental-heating analyses of hornblende from level 33–2 resulted in plateau ages of 10.58 ±0.06 Ma, 10.86 ±0.04 Ma, 10.63 ±0.05 Ma, and 10.29 ±0.07 Ma, while hornblende from level 33–10 yielded a plateau age of 10.29 ±0.07 Ma. All four spectra

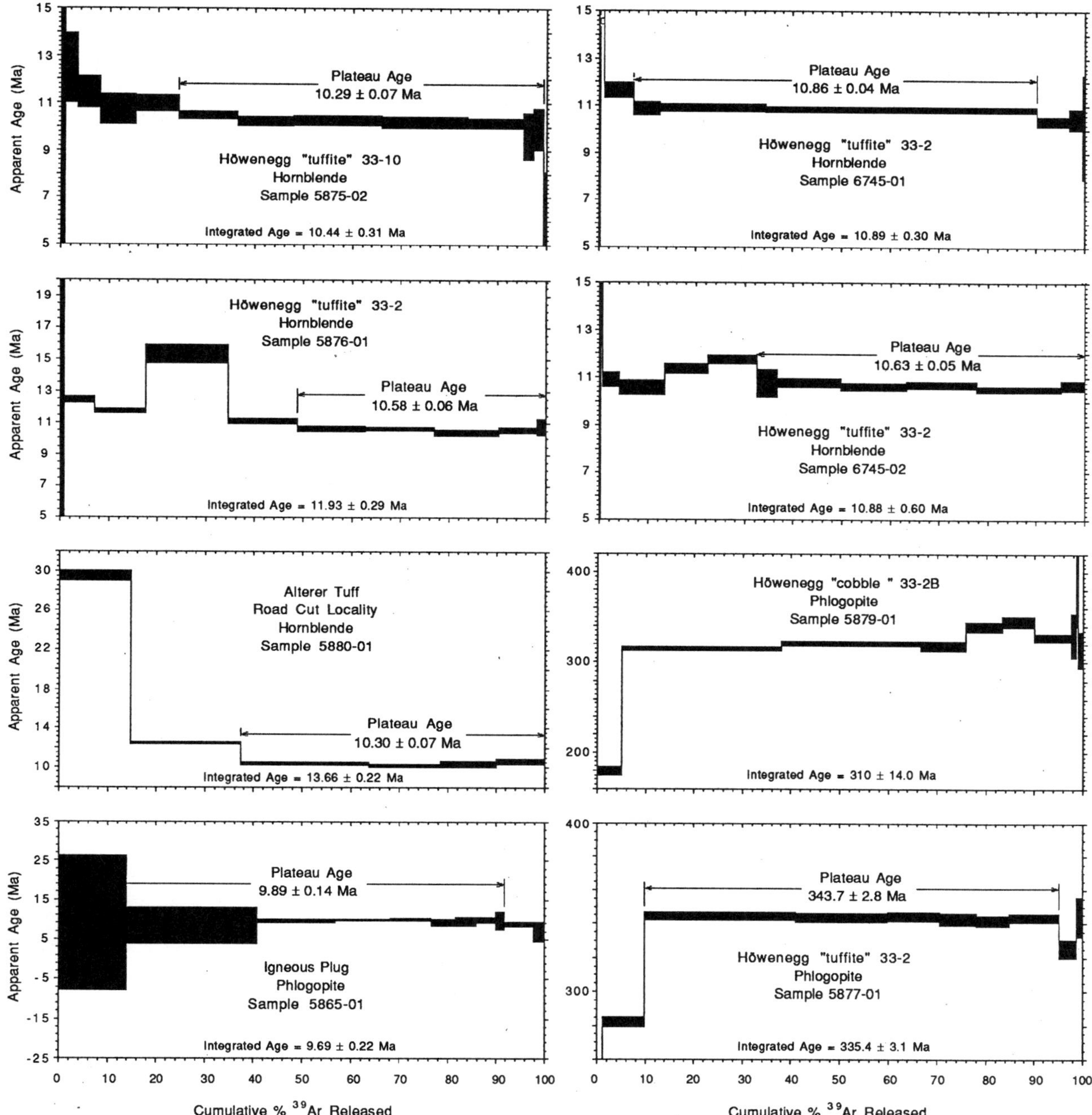

FIGURE 5.1 Integrated ages for various Höwenegg "tuffites."

indicated variable disturbance in the lower temperature steps indicative of excess argon loss (McDougall and Harrison 1988), suggesting that the youngest age is probably the most precise one.

Given the possibility of detrital hornblendes and potentially trapped excess argon in the hornblendes, the ^{40}Ar/^{39}Ar dates obtained here are probably maximum ages, the youngest of these, 10.29 ± 0.07 Ma, being closest to the actual age of the Höwenegg deposits. The ^{40}Ar/^{39}Ar dates obtained on the hornblendes were insufficient to resolve whether the Höwenegg "tuffite" hornblendes were younger than the "Alterer Tuff" hornblende, or whether the two are the same age and actually interfinger one another laterally, or whether the hornblende from the Höwenegg "tuffites" are reworked from the "Alterer Tuff." An age of 10.29 ± 0.07 Ma is, however, consistent with the bracketing ^{40}Ar/^{39}Ar dates of 10.36 ± 0.19 Ma for the "Alterer Tuff" and 9.82 ± 0.20 Ma for the intrusive plug.

The geomagnetic polarity of the Höwenegg deposits is consistent with a 10.3 Ma age. Three hand samples for paleomagnetic analysis were collected from three levels within the Höwenegg fossil quarry, the lowest being below

TABLE 5.1 $^{40}Ar/^{39}Ar$ *Incremental Laser-Heating Analyses of Hornblende from Höwenegg, Hegau, Southern Germany*

Sample	%^{39}Ar	^{37}Ar/^{39}Ar	^{36}Ar/^{39}Ar	^{40}Ar*/^{39}Ar	%^{40}Ar*	Ma	Age ± SD
			Sample 33-10 (Irrad. 77C)				
5875-02A	0.4	1.2047	1.89500	8.6927	1.5	11.867	28.713
5875-02B	0.4	2.4329	0.31066	6.9773	7.1	9.531	4.270
5875-02C	2.6	2.5201	0.23680	9.1404	11.6	12.476	0.744
5875-02D	4.7	2.5443	0.08450	8.4008	25.3	11.470	0.328
5875-02E	7.2	2.5216	0.05176	7.8593	34.2	10.732	0.311
5875-02F	9.0	2.6809	0.02104	8.0502	57.2	10.992	0.172
5875-02G	12.0	2.4995	0.01638	7.6704	62.2	10.475	0.085
5875-02H	11.6	2.5745	0.01273	7.5035	67.7	10.248	0.104
5875-02I	18.0	2.5283	0.01103	7.5206	71.0	10.271	0.102
5875-02J	17.9	2.6936	0.01356	7.4752	66.2	10.210	0.127
5875-02K	11.4	2.6053	0.00940	7.4491	74.2	10.174	0.101
5875-02L	2.2	2.9314	0.00767	7.0387	77.4	9.615	0.496
5875-02M	2.1	2.4595	0.01151	7.2541	69.3	9.908	0.443
5875-02N	0.6	1.7607	0.01961	4.0746	41.8	5.572	1.268
			Sample 33-2 (Irrad. 77C)				
5876-01A	0.9	2.6046	2.33624	28.6479	4.0	38.817	18.471
5876-01B	6.2	2.5573	0.01167	9.1297	73.7	12.461	0.106
5876-01C	10.5	2.3959	0.01120	8.5960	73.3	11.735	0.083
5876-01D	16.9	2.3248	0.00920	11.2353	81.5	15.323	0.264
5876-01E	14.4	2.4777	0.00599	8.1518	83.7	11.131	0.090
5876-01F	13.9	2.5009	0.00676	7.7832	81.1	10.629	0.089
5876-01G	14.4	2.4172	0.00609	7.7884	82.8	10.636	0.053
5876-01H	13.3	2.5753	0.00518	7.5906	85.0	10.367	0.091
5876-01I	7.8	2.5572	0.01105	7.7360	71.5	10.565	0.097
5876-01J	1.8	2.5099	0.00297	7.9024	92.0	10.791	0.256
			Sample 33-2 (Irrad. 94B)				
6745-01B	0.1	4.6810	0.83277	28.9040	10.5	36.336	22.058
6745-01C	0.8	2.882A	0.08110	14.8187	38.4	18.721	0.938
6745-01D	6.1	2.6509	0.01547	9.1991	67.9	11.644	0.162
6745-01E	5.4	2.5681	0.00865	8.6030	78.7	10.892	0.142
6745-01F	21.7	2.4935	0.00853	8.6366	78.9	10.934	0.088
6745-01G	27.5	2.5187	0.00571	8.5692	85.4	10.849	0.064
6745-01H	28.1	2.5727	0.00363	8.5605	91.0	10.838	0.058
6745-01I	6.7	2.6406	0.00254	8.1868	94.0	10.367	0.099
6745-01J	2.8	2.5436	0.00146	8.2427	97.5	10.437	0.221
6745-01K	0.4	2.5778	0.00099	7.9764	99.2	10.101	1.098
6745-02B	0.1	12.6287	0.92489	38.7546	12.4	48.555	39.022
6745-02C	0.4	3.4470	0.23840	14.3289	16.9	18.105	2.899
6745-02D	3.5	2.6976	0.01419	8.6247	68.5	10.919	0.150
6745-02E	9.5	2.5164	0.01173	8.3491	72.0	10.571	0.150
6745-02F	8.7	2.5461	0.00812	8.9819	80.5	11.370	0.103
6745-02G	10.1	2.5533	0.00969	9.2888	77.9	11.757	0.088
6745-02H	4.4	2.5925	0.00563	8.5137	85.5	10.779	0.286
6745-02I	12.9	2.5561	0.00546	8.5163	86.0	10.782	0.093
6745-02J	13.6	2.5392	0.00661	8.3907	82.9	10.624	0.077
6745-02K	14.3	2.6031	0.00427	8.4461	89.1	10.694	0.073
6745-02L	17.4	2.6338	0.00349	8.3039	91.2	10.514	0.059
6745-02M	4.9	2.6004	0.00286	8.4296	93.2	10.673	0.104

$\lambda_\varepsilon + \lambda_\varepsilon' = 0.581 \times 10^{-10}$ yr^{-1}; $\lambda_\beta = 4.962 \times 10^{-10}$ yr^{-1}; $^{40}K/^{40}K_{total} = 1.167 \times 10^{-4}$
* = radiogenic
Irradiation 77C / J = 0.0007592 ± 0.0000008
Irradiation 94B / J = 0.0007039 ± 0.0000005

the dated horizon (re: Woodburne et al., this volume b). Thermal demagnetization of the nine samples to 400 degrees C, indicated that the primary magnetization was residing in magnetite, with a slight overprint residing in goetite. All samples indicate a normal polarity magnetization. The normal polarity of the Höwenegg fossil-bearing layers is consistent with a 10.3 Ma age, indicating a correlation within the long Chron C5n.2n normal (9.92–10.95 Ma). Unfortunately, given the long duration of Chron C5n.2n, this determination does not help in refining the age of the Höwenegg deposits.

Maragheh, Iran

Northwestern Iran's Maragheh Basin contains a thick sequence of late Miocene continental deposits. These deposits are referred to as the Maragheh Formation and consist of approximately 175 meters of vertebrate fossil-bearing volcaniclastic sediments and interbedded tuff beds. The Maragheh Formation unconformably overlies a thick ash flow tuff commonly referred to as the Basal Tuff that locally attains a thickness of over 80 meters, and is in turn capped by the Korde-deh ash flow, although some 300 meters of volcaniclastic sediments were reported by Kamei et al. (1977) to occur above this tuff. The stratigraphy, biostratigraphy, and geochronology of the Maragheh Formation were reported by Kamei et al. (1977), Erdbrink et al. (1976), Campbell et al. (1980), Bernor et al. (1980), and Bernor et al. (this volume).

The main fossil vertebrate-bearing sections are correlated by a series of distinctive tuffs and pumice layers. A distinctive brecciated tuff near the top of the fossil-bearing portion of the section, referred to as the "Loose Chippings Tuff" by Campbell et al. (1980) and Bernor et al. (1980), is used as the 0 m level in their stratigraphic sections, and such usage will be followed here. The lowest volcanic unit is the unconformably underlying Basal Tuff. The marker tuffs of the Maragheh Formation, from lowest to highest stratigraphical level are: the -110m "Ignimbritic Tuff," or "Mordagh or Mordaq Tuff"; the -50m "Gürt Daresah Tuff" = Lower Pumice of Kamei et al. (1977); -20m "Upper Pumice" of Kamei et al. (1977); the 0 m "Loose

TABLE 5.2. $^{40}Ar/^{39}Ar$ *Incremental Laser-Heating Analyses of Biotites and Hornblendes from the Höwenegg Area, Hegau, Southern Germany*

Sample	$\%^{39}Ar$	$^{37}Ar/^{39}Ar$	$^{36}Ar/^{39}Ar$	$^{40}Ar^*/^{39}Ar$	$\%^{40}Ar^*$	Ma	Age ±	SD
			Sample Igneous Plug / phlogopite (Irrad. 77B)					
5865-01A	13.6	0.4786	0.82346	6.3433	2.5	8.675		8.521
5865-01B	28.8	0.3623	0.35625	4.4555	4.1	6.097		2.990
5865-01C	16.1	0.2900	0.05528	6.9906	30.0	9.557		0.406
5865-01D	10.9	0.2731	0.02828	7.3079	46.7	9.990		0.135
5865-01E	8.2	0.5849	0.02700	7.2629	47.8	9.929		0.292
5865-01F	4.7	0.8371	0.02081	7.3963	54.9	10.111		0.417
5865-01G	4.0	3.4492	0.02708	6.6606	46.2	9.107		0.408
5865-01H	4.0	11.6030	0.03371	7.2134	44.1	9.861		0.393
5865-01I	1.8	16.6659	0.02712	6.9905	50.6	9.557		1.090
5865-01J	5.8	50.7316	0.05400	6.4493	34.0	8.819		0.405
5865-01K	2.2	59.8690	0.17530	4.9325	9.1	6.749		1.260
			Sample Alterer Tuff / hornblende (Irrad. 77C)					
5880-01A	14.7	2.2708	0.07936	21.6795	48.2	29.452		0.252
5880-01B	22.4	2.3387	0.00506	9.1147	87.4	12.441		0.066
5880-01C	26.3	2.3259	0.00743	7.5719	78.9	10.341		0.079
5880-01E	15.9	2.2309	0.00265	7.4290	92.4	10.147		0.115
5880-01F	11.6	2.2136	0.00238	7.5485	93.4	10.309		0.132
5880-O1G	9.1	2.4038	0.00181	7.6181	95.6	10.404		0.140
			Sample 33-2B / phlogopite from Höwenegg cobble (Irrad. 77C)					
5879-01A	5.1	0.0648	0.12172	137.5350	79.3	179.176		2.285
5879-01B	32.8	0.0134	0.00711	247.9514	99.2	311.162		1.385
5879-01C	28.7	0.0189	0.00487	256.1099	99.4	320.542		1.321
5879-01D	9.1	0.0516	0.00691	254.1733	99.2	318.320		2.496
5879-01E	7.6	0.1061	0.00602	266.9164	99.3	332.891		2.456
5879-01F	6.6	0.1329	0.00840	277.3120	99.1	344.692		2.869
5879-01G	7.6	0.0839	0.00552	260.6858	99.4	325.781		2.101
5879-01H	1.0	0.2607	0.02434	264.3548	97.4	329.971		11.949
5879-01I	0.5	0.3661	0.01262	324.2392	98.9	397.021		28.347
5879-01J	1.0	0.1046	0.04125	250.2800	95.4	313.844		9.973
			Sample 33-2 / phlogopite from Höwenegg matrix (Irrad. 77C)					
5877-01A2	1.2	0.2841	0.36427	120.7797	52.9	158.276		4.329
5877-01B	8.5	0.0506	0.06924	222.8209	91.6	281.960		1.541
5877-01C	31.1	0.0314	0.00902	276.8300	99.0	344.146		1.334
5877-01D	18.9	0.0358	0.00524	276.3068	99.4	343.554		1.465
5877-01E	10.8	0.0529	0.00481	276.7141	99.5	344.015		1.499
5877-01F	7.4	0.0998	0.00363	273.0939	99.6	339.913		1.874
5877-01G	6.8	0.2382	0.00615	273.9275	99.3	340.858		1.611
5877-01H	10.2	0.2394	0.00431	272.7673	99.5	339.542		1.333
5877-01I	3.8	0.2767	0.00848	259.9269	99.1	324.914		2.647
5877-01J	1.3	0.3161	0.01175	235.5001	98.6	296.752		2.994

$\lambda_\varepsilon + \lambda_\varepsilon' = 0.581 \times 10^{-10}$ yr^{-1}; $\lambda_\beta = 4.962 \times 10^{-10}$ yr^{-1}; $^{40}K/^{40}K_{total} = 1.167 \times 10^{-4}$
* = radiogenic
Irradiation 77B / J = 0.0007599 ± 0.0000010
Irradiation 77C / J = 0.0007592 ± 0.0000008

Chippings Tuff" = "scoria bed" of Kamei et al. (1977), and "trachytic breccia" of Erdbrink et al. (1976); the +7m "Layered Marker Tuff" = pumice falls of Kamei et al. (1977); the +20m Sargizeh Tuff of Kamei et al. (1977), and the +100m "Korde-deh Tuff."

To the east, in the Il'k Chai area, a series of tuffs were projected by Bernor et al. (1980) to occur at levels above the Loose Chippings Tuff, although this tuff did not crop out in this area. These tuffs are the +15m, +60m, and +90m "Lower, Middle, and Upper Tuffs." These tuffs are projected to be above the 0 m Loose Chippings Tuff but below the +100 m Korde-deh Tuff, although no direct stratigraphic correlation was possible as a result of discontinuous outcrops. The Village pumice in the Il'k Chai area was considered anomalous and its stratigraphic level was considered dubious by Bernor et al. (1980), although it may correlate to the west with the +100 m Korde-deh Tuff. No attempt was made by Bernor et al. (1980) to correlate the Il'k Chai tuffs with the Sargizeh or approximately 80 m of air fall tuffs and flows reported between the Loose Chippings Tuff and Korde-deh Tuff in the Kherjabad area by Kamei et al. (1977).

K-Ar and fission-track dates for the major Maragheh Formation tuffs were reported by Erdbrink et al. (1976), Kamei et al. (1977), and Campbell et al. (1980), and subsequently compiled by Bernor et al. (1980) and Bernor (1986). The most detailed work was by Campbell et al. (1980), who reported both K-Ar and fission-track dates for the major Maragheh tuffs. The K-Ar analyses were made by R. Drake and G. Curtis at the University of California, Department of Geology and Geophysics K-Ar Laboratory. Most of the K-Ar dates were made on plagioclase separates that are still on file at the laboratory, now the Berkeley Geochronology Center. The K-Ar dates, although in general yielding fairly consistent results, were based on extremely low radiogenic yields that for the most part except for the basal tuff were less than 15% radiogenic ^{40}Ar. Other samples indicated the inclusion of older detrital material resulting in K-Ar dates that were too old.

Replicate analyses of single plagioclase crystals from the Berkeley separates were analyzed by the ^{40}Ar/^{39}Ar method and resulted in the following weighted means and isochron ages: for the +90 m Upper Pumice (Sample R10), a weighted mean of 11 analyses of 7.42 ±0.107 Ma (SE) and an 11-point isochron of 7.54 ±0.11 Ma (SE); for the +15 m Lower Tuff (Sample R8), a weighted mean of 8 analyses yielded an age of 7.58 ±0.11 Ma (SE) and an 8-point isochron of 7.55 ±0.07 Ma (SE); the +7 m Layered Marker Tuff (Sample R12), a weighted mean of 8 analyses of 7.59 ±0.09 Ma (SE) and an 8-point isochron of 7.64 ±0.03 Ma (SE); the 0 m Loose Chippings Tuff (Sample R11), a weighted mean of 9 analyses of 7.75 ±0.14 Ma (SE) and an 8-point isochron of 7.79 ±0.14 Ma (SE); the -110 m Mordagh Tuff (Sample R3), a weighted mean of 9

analyses of 8.67 ±0.04 Ma (SE) and an 9-point isochron of 8.64 ±0.03 Ma (SE); and the underlying Basal Tuff (Sample R4 and R7), weighted means of 9 analyses each of 10.39 ±0.02 Ma (SE) and 10.43 ±0.02 Ma (SE), because the generally higher than 90% radiogenic yields for the basal tuff isochrons were not calculated.

Paleomagnetic analyses of the Maragheh Formation tuffs have so far produced variable results (Erdbrink et al. 1976 and Kamei et al. 1977). In that no thermal demagnetization of these tuffs was made, the results are considered to be tenuous. According to both Erdbrink et al. (1976) and Kamei et al. (1977), the polarity of the -110 m Ignimbritic Tuff in the Mordaq area appears to be of normal polarity. The ^{40}A/^{39}Ar date of 8.64 ±0.03 Ma for this tuff suggests a possible correlation with Chron 4An of the GPTS. The normal polarity for the 50 m Girt Dareseh Tuff (Erdbrink et al. 1976) suggests a similar correlation, although it has not been reliably dated.

The 0 m Loose Chippings Tuff has a conflicting polarity. Sample II of Erdbrink et al. (1976) and the top of Section 2 of Bernor et al. (1980) indicates a normal polarity for the Loose Chippings Tuff. The ^{40}Ar/^{39}Ar date of 7.75 ±0.14 Ma for the 0 m Loose Chippings Tuff would suggest a correlation with the lower part of Chron 4n.2n. Similarly, the upper part of the section, for example, the +100 m Korde-deh Tuff, also has a normal polarity (Kamei et al. 1977). Although not directly dated, an age for the +100 m Upper Tuff of 7.42 ±0.11 Ma is also consistent with a correlation with Chron 4n.2n.

Inconsistent polarities would include Sample III of Erdbrink et al. (1976). Although no section in this area was made by Bernor et al. (1980), the samples appear to come from the vicinity of Sections 4 and 5 of Bernor et al. (1980) and Bernor et al. (this volume b). Sample III was correlated by Bernor et al. (1980) within the Loose Chippings Tuff; however, it exhibits a reversed polarity, while the Site 2 sample of the Loose Chippings Tuff (Sections 2) is of normal polarity (Erdbrink et al. 1976). Sample VII was also considered by Erdbrink et al. (1976) to have a reversed polarity. This site appears to best correlate with the top of Section 1 of Kamei et al. (1977) and possibly the top of Section 5 of Bernor et al. (1980). This part of the section has not been accurately dated, although Erdbrink et al. (1976) report a K-Ar date of 6.2 ±0.6 Ma. Unfortunately, this sample was not redated by Campbell et al. (1980).

Bernor et al. (1979; 1980) and Bernor (1986) proposed a threefold subdivision of thhe Maragheh Formation based on the stratigraphic distribution of mammalian taxa recorded in seven measured geologic sections. The stratigraphic level of the fossil taxa were reported in meters above or below the 0 m marker tuff.

The Lower Maragheh fauna was derived from rocks primarily in the westernmost Kopran area beginning at the -150 m level, and extending up to the -52 m level,

TABLE 5.3. $^{40}Ar/^{39}Ar$ *Laser Total Fusion Analyses of Plagioclases from Maragheh, Iran.*

Sample	$^{37}Ar/^{39}Ar$	$^{36}Ar/^{39}Ar$	$^{40}Ar^*/^{39}Ar$	$\%^{40}Ar^*$	Age Ma	± SD
Sample R10 / Korde-deh Tuff / +100 m level						
6743-01	9.3184	0.07127	5.8370	22.3	7.397	0.411
6743-02	7.7283	0.07657	5.9594	21.3	7.552	0.455
6743-03	9.9777	0.13748	5.7222	12.5	7.252	0.660
6743-04	10.3505	0.01537	5.9952	62.1	7.597	0.620
6743-05	9.7477	0.11684	5.8542	14.7	7.419	0.389
6743-06	9.7420	0.01270	5.9891	67.3	7.589	0.220
6743-07	10.4495	0.06160	5.6590	24.5	7.172	0.684
6743-08	7.3829	0.03391	5.5428	37.0	7.025	0.286
6743-09	8.9386	0.05922	5.8239	25.7	7.380	0.282
6743-10	12.5029	0.02106	5.8674	53.2	7.436	0.288
6743-11	9.2324	0.01377	6.0763	65.0	7.700	0.421
			Weighted Mean =		**7.420**	**0.107**
			11 pt. Isochron Age (293.7 /0.517) =		**7.536**	**0.111**
Sample R8 / Lower Tuff / +15 m level						
6742-01	9.6999	0.01055	5.9124	72.2	7.492	0.340
6742-02	9.2238	0.01172	5.9069	68.9	7.485	0.240
6742-03	11.7155	0.09894	6.2846	18.1	7.963	0.569
6742-04	10.0388	0.04755	6.0385	31.3	7.652	0.389
6742-05	10.5622	0.04934	6.1452	30.9	7.787	0.366
6742-06	8.4911	0.00779	5.9893	79.3	7.590	0.167
6742-07	9.6106	0.15912	5.9104	11.3	7.490	0.371
6742-09	9.9277	0.01077	5.7563	71.3	7.295	0.696
			Weighted Mean =		**7.579**	**0.106**
			8 pt. Isochron Age (296.0 /0.363) =		**7.550**	**0.071**
Sample R12 / Layered Marker Tuff / +7 m level						
6737-01	10.2078	0.00391	5.9838	95.9	7.583	0.279
6737-02	9.1613	0.00454	6.0827	91.9	7.708	0.246
6737-03	8.1019	0.01652	6.0011	58.9	7.605	0.259
6737-04	9.8976	0.01204	5.9861	68.9	7.586	0.254
6737-05	9.7451	0.01021	5.8159	72.8	7.370	0.271
6737-06	9.0165	0.01938	5.9437	54.5	7.532	0.261
6737-07	9.1854	0.04364	5.9432	32.8	7.531	0.366
6737-09	10.9339	0.00478	6.1003	93.1	7.730	0.237
			Weighted Mean =		**7.592**	**0.094**
			8 pt. Isochron Age (291.6 /1.37) =		**7.642**	**0.033**
Sample R11 / Loose Chippings Tuff / 0 m level						
6741-01	12.3920	0.02877	6.1057	44.9	7.737	0.505
6741-02	12.8220	0.03430	5.9354	39.5	7.522	0.369
6741-03	11.6340	0.00877	6.1069	79.5	7.739	0.829
6741-04	14.1682	0.03342	6.1412	41.3	7.782	0.664
6741-05	16.0864	0.03644	6.1082	39.2	7.740	0.662
6741-06	11.1253	0.01395	6.1827	66.1	7.834	0.287
6741-07	12.1762	0.02482	6.2042	49.5	7.861	0.446
6741-08	13.0245	0.00973	6.1722	78.0	7.821	0.327
6741-09	10.7265	0.01119	6.0654	71.8	7.686	0.315
			Weighted Mean =		**7.748**	**0.136**
			8 pt. Isochron Age (291.5 /0.100) =		**7.787**	**0.139**
Sample R3 / Ignimbritic (Murdaq) Tuff / -110 m level						
6735-01	4.0454	0.01579	6.8297	61.2	8.652	0.083
6735-02	4.1487	0.00725	6.8581	79.4	8.688	0.189
6735-03	4.9355	0.00306	6.7542	93.5	8.557	0.080
6735-04	5.0579	0.00241	6.8428	96.3	8.669	0.106
6735-05	5.3148	0.01716	7.0549	60.4	8.937	0.331
6735-06	5.6756	0.00893	6.7606	75.9	8.565	0.142
6735-07	6.4736	0.00316	6.9222	95.0	8.769	0.144
6735-08	4.7329	0.00327	6.9242	92.6	8.772	0.144
6735-09	3.9722	0.01484	7.0167	63.4	8.889	0.139
			Weighted Mean =		**8.667**	**0.040**
			9 pt. Isochron Age (299.5 /2.61) =		**8.635**	**0.029**
Sample R4 / Basal Tuff						
6736-01	1.4276	0.00092	8.2263	98.2	10.416	0.057
6736-02	2.8414	0.00247	8.1036	94.4	10.262	0.110
6736-03	1.5716	0.00155	8.3301	96.3	10.547	0.073

TABLE 5.3. $^{40}Ar/^{39}Ar$ *Laser Total Fusion Analyses of Plagioclases from Maragheh, Iran. (Continued)*

Sample	$^{37}Ar/^{39}Ar$	$^{36}Ar/^{39}Ar$	$^{40}Ar^*/^{39}Ar$	$\%^{40}Ar^*$	Ma	Age ± SD
6736-04	2.6467	0.00328	8.3120	91.9	10.525	0.124
6736-05	1.2227	0.00195	8.1762	94.6	10.353	0.066
6736-06	1.2468	0.00245	8.1818	93.0	10.360	0.065
6736-07	0.7885	0.00183	8.1549	94.5	10.326	0.058
6736-08	0.9748	0.00240	8.1826	92.9	10.361	0.066
6736-09	2.6938	0.00242	8.3382	94.6	10.558	0.161
				Weighted Mean =	10.391	0.024
			Sample R7 / Basal Tuff			
6739-01	1.0893	0.00062	8.2020	99.0	10.386	0.049
6739-02	0.6162	0.00197	8.1523	93.9	10.323	0.056
6739-03	3.3399	0.00557	8.4014	86.1	10.637	0.123
6739-04	2.7135	0.00575	8.2983	85.0	10.507	0.093
6739-05	0.7690	0.00080	8.1739	98.0	10.350	0.056
6739-06	1.1378	0.00192	8.2533	94.6	10.451	0.045
6739-07	2.8400	0.00302	8.2795	92.8	10.483	0.077
6739-08	2.6589	0.00464	8.4172	88.1	10.657	0.093
6739-09	0.8973	0.00247	8.2613	92.7	10.460	0.044
				Weighted Mean =	10.432	0.020

$\lambda_\varepsilon + \lambda_{\varepsilon'} = 0.581 \times 10^{-10}$ yr^{-1}; $\lambda_\beta = 4.962 \times 10^{-10}$ yr^{-1}; $^{40}K/^{40}K_{total} = 1.167 \times 10^{-4}$
* = radiogenic
Irradiation 94B / J = 0.0007039 ± 0.0000005

where the base of the Middle Maragheh faunas first occurs. The Kopran section lies unconformably on the Basal Tuff and is projected to be -150 m below the Loose Chippings Tuff (Section 1 of Bernor et al. 1980). Unfortunately, the Loose Chippings Tuff does not occur in the Kopran area and its level is projected into the area assuming a consistent regional dip over five kilometers. No dated tuffs have been reported in the Kopran area, and its age is estimated from its projected level below the Loose Chippings Tuff. The upper part of the Lower Maragheh fauna does occur however, between the -110 m Mordagh tuff and the -52 m lower or Gürt Daresah Tuff in Section 2 (Bernor et al. 1980) of the Kherjabad area. Lower Maragheh is considered by Bernor et al. (1979; 1980) to be late Vallesian or early Turolian in age, and given the available K-Ar dates as old as 11 Ma or as young as 8.5 Ma. Mein (1989), Steininger et al. (1989), Bernor et al. (1989), and Bernor et al. (this volume b) correlate Lower Maragheh with MN 11.

Fossil vertebrates that comprise the Middle Maragheh fauna (also referred to as Kerjabad, Ketschawa, Kerdjawa or, Gürt Daresah), occur in rocks within the -52 m to -20 m level (re: sections 2, 3, and 4 of Bernor et al. 1980). The Middle Maragheh fauna has been considered to be medial Turolian age, ca. 8.5–8.0 Ma (Bernor et al. 1979; 1980). Steininger et al. (1989) positioned Middle Maragheh between tuffs ranging in age between 9.3 Ma and 7.75 Ma correlative with C4A and the lower part of C4.

The Upper Maragheh fauna occurs in rocks that range from -20 m to the +7 m Layered Marker Tuff of Bernor et al., (1980) or "pumice falls" of Kamei et al. (1977). An exception would be a site below the Lower Tuff of the Il'k Chai area, projected to occur in the Kherjabad and Shol'avand areas approximately +15 meters above the 0 m Loose Chippings Tuff. The Upper Maragheh fauna is also considered to be medial Turolian (MN 12) age. Steininger et al. (1989) have estimated MN 12 as being ca. 7.5–6.2 Ma and correlative with the upper part of C4 and the lower part of C3A.

Review of the stratigraphy and geomagnetic polarity determinations of some of the Maragheh tuffs, in conjunction with the new $^{40}Ar/^{39}Ar$ dates presented here, lead to a revised age and duration of the Maragheh faunal intervals from those proposed by Steininger et al. (1989) based on the works of Erdbrink et al. (1976), Kamei et al. (1977), Campbell et al. (1980), Bernor et al. (1980), and Bernor (1986). Given the available published data, the proposed correlation between the Kopran section and the central five sections of Bernor et al. (1980) in the Kherjabad and Shol'avand areas based on a regional projection of the 0 m Loose Chippings Tuff is somewhat ambiguous. However, if Van Couvering's (in Campbell et al. 1979; Bernor et al. 1980; Bernor et al., this volume b) observations on the stratigraphic position of the Kopran localities are correct, his 9.0 Ma interpolated age for the lowermost Maragheh fauna would correlate this fauna with the MN 11 reference fauna Crevillente 2 (also ca. 9.0 Ma; Steininger et al. 1989) and is basal MN 11, by definition (re: Fahlbusch 1991).

The new $^{40}Ar/^{39}Ar$ dates for the Maragheh Formation indicate that most of the fauna is bracketed by the +7 m Layered Marker Tuff, with a $^{40}Ar/^{39}Ar$ weighted mean age of 7.59 ±0.09 Ma (SE), and the -110 m Mordaq Tuff, dated at 8.64 ±0.03 Ma (SE), the total duration on the order of approximately 1 Ma (1.4 Ma if accepting the older interpolated date for the western Kopran faunas as being 9.0 Ma) with the following age ranges: Lower Maragheh,

9.0 (interpolated age of the Kopran fauna) to 8.2 Ma; Middle Maragheh, 8.2 Ma (as above) to 8.0 Ma (interpolated age of the upper pumice); Upper Maragheh, 8.0 Ma (as above) to 7.6 Ma (age of the +7 meter Layered Marker Tuff with the inclusion of the 7.79 Ma date for the Loose Chippings Tuff) for the Upper Maragheh fauna (re: Bernor et al., this volume b for a further discussion).

Samos, Greece

The Samos Mytilini Formation is about 220m thick. The basal 60m make up the Old Mill Beds, which have yielded a sparse vertebrate fauna. The major fossil-producing beds are referred to as the Main Bone Beds (MBB) part of the Mytilini Formation The total thickness of the MBB is about 80–90 m, and according to recent fieldwork by Solounias, the known quarries come from a major concentration near the top of this unit and are capped by the "Marker Tuff." K-Ar dates on sanidine separated from tuffs of the Mytilini Formation indicate ages ranging from 8.57 Ma to 5.41 Ma.

Using the best quality data, based on the dated mineral phase, radiogenic yield, and geographic position to fossil localities, the following K-Ar dates are indicated: 8.38 ±0.07 Ma (SK16) and 8.26 ±0.08 Ma (SK3) for the Old Mill Beds (OMB); 7.52 ±0.16 Ma (SK18A), 7.55 ±0.11 Ma (R102, E of the Mytilini Village), 6.89 ±0.06 Ma (R106), 6.98 ±0.15 Ma (SK17), and 7.75 ±0.08 Ma (SK5) for the MBB. The overlying Marker Tuff is dated at four localities: 6.14 ±0.19 Ma (SK1), 6.14 ±0.04 (SK2), 5.41 ±0.17 Ma (SK19), 6.74 ±0.11 Ma (R105), and 6.51 ±0.14 Ma (R105 repeat analysis).

^{40}Ar/^{39}Ar dating of sanidine single crystals from the MBB resulted in the following mean ages: ten analyses of the SK18A sanidine yielded a weighted mean ^{40}Ar/^{39}Ar age of 7.28 ±0.01 (SE), and is in broad agreement with its previous K-Ar date of 7.52 ±0.16 Ma; ten analyses of the SK17 sanidine, a sample yielded a weighted mean ^{40}Ar/^{39}Ar age of 7.09 ±0.01 Ma (SE), an age in agreement with the previous K-Ar date of 6.98 ±0.15 Ma. The SK17 tuff was collected from within Samos Q1, 2 meters above the bonebed; SK18A was collected from somewhat lower in the section. Bernor et al. (this volume b, fig. 2) have demonstrated that the SK17 date is not only in close stratigraphic association with Q1, but also is closely correlative with S3, Q2, and Q5. Of the MBB quarries, only Q4 falls significantly below these stratigraphically. Solounias (in Bernor et al., this volume b) estimates that the maximum age for Q4 is best estimated by sample R102 derived from east of Mytilini Village in the lowermost portion of the MBB sediments. Ten analyses of the R102 sanidine resulted in a weighted mean ^{40}Ar/^{39}Ar age of 7.66 ±0.01 Ma (SE). These results clarify that the bulk of the Samos MBB fossils originate from a rich fossiliferous horizon that is ≥7.1 Ma,

and that only Q4 is significantly older (maximum, 7.7 Ma).

The age of the MBB (including sample R102) is significantly different from the underlying OMB and the overlying Marker Tuffs. The best available K-Ar dates on sanidine from these units indicate a weighted mean age of 8.33 ±0.05 Ma (SE) for the OMB and 6.14 ±0.04 Ma (SE) for the overlying Marker Tuffs.

These findings result in one ambiguous radioisotopic date for the Mytilini Formation. The SK5 sample was reported by Weidmann et al. (1984) as coming from near Q1. Although only one sanidine K-Ar date was reported by these authors, three sanidine analyses were actually made in the Berkeley lab. Two analyses not reported are 8.31 ±0.04 Ma and 8.69 ±0.3 Ma (KA 3901 and 3901R). According to laboratory notes at Berkeley, KA 3901 was further prepared removing some plagioclase from the sanidine separate. The new separate, KA 4039, was dated 7.75 ±0.08 Ma (KA 4039) and is the analysis reported by Weidmann et al. (1984). The explanation of plagioclase removal from the separate is not consistent with the analytical data. Potassium values for KA 3901 averaged 6.03 ±0.18 %K, while KA4039 was 5.98 %K, reversed of what would be expected if the lower-K plagioclase were removed from a sanidine separate. At this point we find no reason why the dates for KA3901 were omitted from Weidmann et al. (1984) other than that their age appears anomalously old.

^{40}Ar/^{39}Ar dating of the SK5 (KA4039) sanidine (10 analyses) resulted in a mean age of 8.58 ±0.01 Ma (SE). This age for the KA4039 SK5 sanidine is consistent with the mean of the unpublished K-Ar dates of the KA3901 SK5 sanidine, and in turn indicates that the K-Ar date for the KA4039 SK5 sample by Weidmann et al. (1984) is anomalously too young.

A date of 8.58 ±0.01 Ma (SE) for SK5 is inconsistent with the other dates in the vicinity of Q1 that indicate an age closer to 7.09 ±0.01 Ma (SE). Again, laboratory notes in the Berkeley files report an inconsistency. The early notation lists SK5 as actually coming from the vicinity of Q5, a locality positioned away from the main quarries and whose stratigraphic position is dubious because of extensive faulting in this area (but see Bernor et al., this volume b). The provenience of the SK5 sample was changed to Q1 by Weidmann et al. (1984). A date of 8.58 ±0.01 Ma (SE) for SK5 is, however, closer in age to the Old Mill Beds, shown here to be about 8.33 ±0.05 Ma (SE). It may be impossible to determine at this point whether SK5 is a detrital sample from the Old Mill beds near Q1, or actually is from Q5, indicating a significantly older age for this site. New samples from Q5 need to be collected to resolve this problem. The stratigraphic provenience of SK5 is dubious and is not useful for calibrating the Mytilini Formation fossil-bearing horizons.

Sen and Valet (1986) reported on the magnetostratigraphy of the MBB and MTs of the Mytilini Formation. They

TABLE 5.4. $^{40}Ar/^{39}Ar$ *Laser Total Fusion Analyses of Sanidines from Samos, Greece.*

Sample	$^{37}Ar/^{39}Ar$	$^{36}Ar/^{39}Ar$	$^{40}Ar^*/^{39}Ar$	$\%^{40}Ar^*$	Age Ma	$\pm$ SD
			Sample SK-17			
6727-01	0.06990	0.00043	5.5695	97.9	7.111	0.017
6727-02	0.04195	0.00017	5.5613	99.1	7.100	0.016
6727-03	0.05946	0.00024	5.5466	98.8	7.082	0.024
6727-04	0.03384	0.00109	5.4445	94.5	6.952	0.045
6727-05	0.05957	0.00033	5.5467	98.4	7.082	0.033
6727-06	0.19326	0.00069	5.5295	96.7	7.060	0.066
6727-07	0.02591	0.00022	5.5495	98.9	7.085	0.020
6727-08	0.07363	0.00021	5.5901	99.0	7.137	0.044
6727-09	0.04978	0.00019	5.5709	99.1	7.112	0.022
6727-10	0.03892	0.00027	5.5332	98.7	7.065	0.024
			Weighted Mean =		**7.092**	**0.008**
			Sample SK-18A			
6724-01	0.03466	0.00021	5.7143	99.0	7.295	0.014
6724-02	0.03856	0.00046	5.6737	97.7	7.244	0.018
6724-03	0.04718	0.00042	5.6814	97.9	7.253	0.015
6724-04	0.03721	0.00041	5.6717	98.0	7.241	0.035
6724-05	0.04929	0.00037	5.6847	98.2	7.258	0.024
6724-06	0.04307	0.00019	5.7115	99.1	7.292	0.018
6724-07	0.04022	0.00037	5.6940	98.2	7.270	0.031
6724-08	0.03189	0.00024	5.7092	98.8	7.289	0.018
6724-09	0.03691	0.00012	5.7190	99.5	7.301	0.018
6724-10	0.04380	0.00026	5.7063	98.7	7.285	0.039
			Weighted Mean =		**7.276**	**0.006**
			Sample R102			
6725-01	0.04739	0.00576	6.0292	78.0	7.697	0.055
6725-02	0.05349	0.00056	6.0408	97.4	7.711	0.046
6725-03	0.04439	0.00056	5.9881	97.4	7.644	0.040
6725-04	0.05941	0.00089	5.9207	95.8	7.558	0.038
6725-05	0.04903	0.00043	5.9516	98.0	7.598	0.045
6725-06	0.04762	0.00494	5.9925	80.5	7.650	0.040
6725-07	0.05395	0.00072	6.0375	96.6	7.707	0.035
6725-08	0.05528	0.00032	6.0276	98.5	7.694	0.032
6725-09	0.04462	0.00239	6.0090	89.5	7.671	0.038
6725-10	0.04228	0.00050	6.0004	97.7	7.660	0.033
			Weighted Mean =		**7.660**	**0.012**
			Sample SK-5			
6723-11	0.06430	0.00043	6.7626	98.2	8.630	0.031
6723-12	0.04208	0.00129	6.6818	94.6	8.528	0.027
6723-13	0.06346	0.00122	6.7178	95.0	8.574	0.033
6723-14	0.03772	0.00061	6.7153	97.4	8.570	0.036
6723-15	0.03767	0.00026	6.7423	98.9	8.605	0.029
6723-16	0.04073	0.00034	6.8007	98.6	8.679	0.035
6723-17	0.04479	0.00073	6.6799	96.9	8.525	0.028
6723-18	0.06354	0.00105	6.7277	95.7	8.586	0.035
6723-19	0.08552	0.00074	6.7081	97.0	8.561	0.031
			Weighted Mean =		**8.580**	**0.010**

$\lambda_\varepsilon + \lambda_\varepsilon' = 0.581$ x 10^{-1}; $\lambda_\beta = 4.962$ x 10^{-10} yr-1; $^{40}K/^{40}K_{total} = 1.167 \times 10^{-4}$
* = radiogenic
Irradiation 94A / J 0.0007091 ± 0.0000006

identified four intervals of reversed polarity with three intervening normal polarity intervals. These intervals were labeled R1/1 through N3/4. It can be ascertained from Weidmann et al. (1984) and Sen and Valet (1986) that the MTs are N3/4, and the MBB R1/1, R2/2, and R3. The upper fossiliferous horizon of the MBB was said to occur in R3.

In their correlation of the Mytilini magnetostratigraphy with the GPTS of Harland et al. (1982), Sen and Valet (1986) indicated incongruent correlations with the determined polarity of the MBB and the GPTS based on avail-able K-Ar dates. Sen and Valet (1986) indicated that the MTs correlated with Chron 5 (Anomaly 3A), while the MBB correlated with the reversed interval of Chron 6. This correlation appeared consistent with the K-Ar dates for the MTs ranging between 5.4 and 6.74, however the K-Ar dates of the MBBs were too old for the suggested correlated age of 6.1–6.4 Ma as indicated for Chron 5 of Berggren et al. (1982). Sen and Valet (1986) suggested a systematic deviation of the K-Ar results for the MBB.

The refined ages presented here, along with a revised GPTS used in this study, lead to a revised correlation of the Mytilini magnetostratigraphy. We propose here that the MTs of primarily normal polarity (N4) with an age of 6.17 ±0.04 Ma (SE) most likely correlates with Chron 3An.2n, estimated from the GPTS to have a duration range of 6.27–6.57 Ma.

The MBB fossil localities date ≥7.1 Ma and have a reversed polarity (R3) that is correlative with Chron 3Br.2r, estimated from the GPTS to have a duration of 7.17 to 7.34 Ma. Quarries of the OMB, such as QX, QG, and Q6, are of unknown polarity; however, their age, 8.33 ±0.05 Ma (SE), is significantly older than the MBB, and should have a reversed polarity correlating with Chron 4r.

Conclusions

Previous age estimates for the Old World *"Hipparion"* Datum range from 9.5 Ma, as formerly estimated from the Siwaliks (Barry et al. 1982), to as old as 12.5 Ma, as determined from K-Ar dates at Höwenegg, Germany. Bernor et al. (1988), Sen (1986; 1989), and Steininger et al. (1989) placed the age of the *"Hipparion"* Datum between 11.5 and 11 Ma. The Höwenegg fauna is considered by most workers typical of early Vallesian/early MN 9 faunas of Central Europe (Mein 1989; Steininger et al. 1989). Steininger et al. (1989) indicate that the Vallesian ranges in age from approximately 11.6 to 9 Ma, coinciding with approximately the range of Chron C5 of La Breque et al. (1983); the early Vallesian, or MN 9, ranges from 11.6 to 10.6 Ma, primarily correlated to Chron C5r, whereas the late Vallesian correlates with C5n. This correlation of the early Vallesian (MN 9) is inconsistent with both the age and magnetic polarity of the Höwenegg site as determined in this paper. Although an age range for the early Vallesian (MN 9) cannot be ascertained here, it is clear that the

TABLE 5.5. *Summary of K-Ar and $^{40}Ar/^{39}Ar$ Dates for Höwenegg, Maragheh and Samos*

Locality/Unit	Sample	Mineral	K-Ar (Ma)	Average $^{40}Ar/^{39}Ar$ (Ma)
Höwenegg	33-10	hb	12.7±1.1[1]	10.29±0.07
		wr	10.8±0.4[2]	
Maragheh				
Ilkhchi				
Upper Pumice (+90m)	R10	p	7.5±0.4	7.42±0.11
Middle Pumice	R5	p	7.7±0.5	
		p	7.2±0.5	
Lower Pumice (+15m)	R8	p	7.9±0.7	7.58±0.11
Shol'avand				
Layered Marker (+7m)	R2	p	8.3±0.9	
	R12	p	6.4±0.6	7.59±0.09
		hb	6.9±0.3	
Loose Chippings (0m)	R11	p	7.4±0.3	7.75±0.14
Murdag				
Mordaq Tuff (-110m)	R3	p	6.4±0.5	8.645±0.029
		p	8.8±0.5	
		p	8.9±0.5	
Ilkhchi				
Basal Tuff	R4	bt	10.1±0.5	
		p	9.3±0.1	10.39±0.02
	R7	bt	11.3±1.0	
		p	9.3±0.1	10.43±0.02
Samos				
Old Mill Beds				
	SK16	s	8.38±0.07	
	SK3	s	8.26±0.08	
Main Bone Beds	SK18A	s	7.52±0.16	7.28±0.01
(lower)	R102	s	7.55±0.11	7.66±0.01
(upper)	R106	s	6.89±0.06	
	SK17	s	6.98±0.15	7.09±0.01
	SK5	s	7.75±0.08	
Marker Tuffs	SK1	s	6.14±0.19	
	SK2	s	6.14±0.04	
	SK19	s	5.41±0.17	
	R105	s	6.74±0.11	
	R105	s	6.51±0.14	

bt = biotite, hb = hornblende, p = plagioclase, s = sanidine, wr = whole rock
[1] Lippolt *et al.* (1963)
[2] Baranyi *et al.* (1976)
[3] Weidmann *et al.* (1984)

Höwenegg site and thus at least part of the early Vallesian (MN 9) correlates with C5n.2n. Mein (1989) and Steininger et al. (1989) considered the Höwenegg fauna to be "basal" MN 9. However, Bernor et al. (1993a,b) and Bernor in Woodburne et al. (this volume a) indicate that the Höwenegg horse has a somewhat derived stage-of-evolution compared to Vienna Basin Pannonian C and D–E hipparions. Based on these observations, and comparisons with other Old World Hipparion LSDs this would suggest a 10.5 Ma age for the Old World *"Hipparion"* Datum (re: Woodburne et al. this volume a).

The Lower Maragheh fauna is correlated here with the early Turolian (MN 11), ca. 9.0 (or minimally, 8.6) to 8.2 Ma (re: Bernor et al., this volume) and is correlative in its upper extent with Samos' Old Mill Beds (8.3 Ma). Although a complete magnetostratigraphic section has not yet been done, the correlation of the Mirduq section of Lower Maragheh, ca. 8.6 to 8.2 Ma, would suggest correlation with Chron 4r.2r. Similarly, the Old Mill Beds of Samos, Greece, comprising fossils collected primarily from localities such QX, Q6, and QG, are correlative with late MN 11 (re: Bernor et al., this volume b). The 8.33 ± 0.05 Ma K/Ar age indicates that this fauna should likewise be derived from strata correlative with Chron 4r.2r.

The Middle and Upper Maragheh faunas are correlative with lower MN 12 (Bernor et al., this volume b). Most of the MBB fossil-bearing quarries are found to be ≥7.1 Ma; only Q4 is older (maximum age = 7.7 Ma). According to the correlations advanced by Rögl and Daxner-Höck and Bernor et al. (this volume b), the bulk of the Samos MBB fossils would be correlative with late MN 12. Sen and Valet (1986) and Steininger et al. (1989), basing their opinion primarily on paleomagnetic correlation of the MBB to the GPTS, have suggested an age of 6.1–6.4 Ma. These findings are inconsistent with the $^{40}Ar/^{39}Ar$ ages presented here. This age is significantly older than that suggested by Sen and Valet (1986) and Steininger et al. (1989), who indicate a correlated age of 6.1–6.4 Ma as indicated for Chron 5 of Berggren et al. (1982). Moreover, the older age proposed here is closer with that indicated by Bernor et al. (1980), based on the correlation of the Samos fossil hipparion horses with those from the Upper Maragheh fauna. However, the $^{40}Ar/^{39}Ar$ ages reported in this study indicate that the Samos MBB fauna is even younger (≥7.1 Ma) than Upper Maragheh's fauna (8.0–7.6 Ma), suggesting little, if any, overlap in time except for the possibility of the Q4 fauna (maximum age = 7.7 Ma).

Acknowledgments

I would like to thank the National Geographic Society for funds (awarded to R. L. Bernor) for fieldwork at Höwenegg and support of the laboratory work that has led to his analysis here. I also thank the VW Stiftung for supporting my participation at the Schloss Reisensburg conference.

LITERATURE CITED

Baranyi, I., H. J. Lippolt, and W. Todt. 1976. Kalium-Argon Alterbestimmungen an tertiären Vulkaniten des Oberrheingraben-Gebietes: II. *Die Alterstraverse vom Hegau nach Lothringen Oberrheinhesse Geologische Abhandlungen* 25:41–62.

Becker-Platen, J. D., L. Benda and P. Steffens. 1977. Litho- und biostratigraphische Deutung radiometrischer Altersbestimmungen aus dem Jungtertiär der Türkei. *Geologische Jahrebuch* 25:139–67.

Berggren, W. A., D. V. Kent, and J. A. Van Couvering. 1985. The Neogene: Part 2, Neogene geocronology and chronostratigraphy; pp. 211–60. *in* N. J. Snelling (ed.) *The Chronology of the Geologic Record.* Geological Society Memoire 10, Blackwell Scientific, Oxford, England.

Berggren, W. A. and J. A. Van Couvering. 1974. The late Neogene-biostratigraphy, geochronology, and paleoclimatology of the last 15 million years in marine and continental sequences. *Palaeogeography, Palaeoclimatology, Palaeoecology* 16:1–216.

——. 1978. Biochronology; pp. 39–55 *in* G. V. Cohee, M. F. Glassner, and H. D. Hedburg (eds.). *Contributions to the Geologic Time Scale.* American Association Petroleum Geologists, Tulsa, Oklahoma.

Bernor, R. L. 1986. The mammalian biostratigraphy, geochronology, and zoogeographic relationships of the late Miocene Maragheh fauna. *Journal of Vertebrate Paleontology* 6:76–95.

Bernor, R. L., G. D. Koufos, M. O. Woodburne, and M. Fortelius. This volume a. The evolutionary history and biochronology of European and Southwest Asian late Miocene and Pliocene hipparionine horses.

Bernor, R. L., J. Kovar-Eder, D. Lipscomb, F. Rögl, S. Sen, and H. Tobien. 1988. Systematic, stratigraphic, and paleoenvironmental contexts of first-appearing Hipparion in the Vienna Basin, Austria. *Journal of Vertebrate Paleontology* 8:427–52.

Bernor, R. L., N. Solounias, C. C. Swisher III, and J. A. Van Couvering. This volume b. The correlation of three classical "Pikermian" mammal faunas—Maragheh, Samos and Pikermi—with the European MN unit system.

Bernor, R. L., H. Tobien, and J. A. Van Couvering. 1979. The mammalian biostratography of Maragheh. *Annales Géologiques des Pays Helléniques* 1979:91–99.

Bernor, R. L., H. Tobien, and M. O. Woodburne. 1989. Patterns of Old World hipparionine evolutionary diversification and biogeographic extension. In *European Neogene Mammal Chronology,* ed. E. H. Lindsay, V. Fahlbusch, and P. Mein, pp. 263–319. New York: Plenum.

Bernor, R. L., M. O. Woodburne, and J. A. Van Couvering. 1980. A contribution to the chronology of some Old World Miocene faunas based on hipparionine horses. *Géobios* 13:705–39.

Campbell, B. G., M. H. Amini, R. L. Bernor, W. Dickenson, R. Drake, R. Morris, J. A. Van Couvering and J. A. Van Couvering. 1980. Maragheh: A classical late Miocene vertebrate locality in Northwestern Iran. *Nature* 28:837–41.

Cande, S. C. and D. V. Kent. 1992. A new geomagnetic polarity timescale for the late Cretaceous and Cenozoic. *Journal of Geopysical Research* 97(B10): 13917–951.

———. 1995. Revised calibration of the geomagnetic polarity time scale for the late Cretaceous and Cenozoic. *Journal of Geophysical Research* 100: 6093–95.

Cebula, G. T., M. J. Kunk, H. H. Mehnert, C. W. Naeser, J. O. Obradovich and J. F. Sutter. 1986. The Fish Canyon Tuff, a potential standard for the ^{40}Ar/^{39}Ar and Fission-Track methods. *Terra Cognita (Abstract)* 6(2): 139–40.

Dalrymple, G. B. 1979. Critical table for the conversion of K-Ar ages from old to new constants. *Geology* 7:558–60.

Deino, A. and R. Potts. 1990. Single-crystal ^{40}Ar/^{39}Ar dating of the Olorgesailie Formation, Southern Kenya Rift. *Journal of Geophysical Research* 95:8453–70.

Deino, A., L. Tauxe, M. Monaghan, and R. Drake. 1990. ^{40}Ar/^{39}Ar age calibration of the litho- and paleomagnetic stratigraphies of the Ngorora Formation, Kenya. *Journal of Geology* 98:567–87.

Erdbrink, D. P., H. N. A. Priem, E. H. Hebeda, C. Cup, P. Dankers, and S. A. P. L. Cloetingh. 1976. The bone-bearing beds near Maragheh in N.W. Iran. *Proceedings, Koninklijke Nederlandse Akademie van Wetenschappen*, B79: 97–113.

Fahlbusch, V. 1976. Report on the international symposium on mammalian stratigraphy of the European Tertiary. *Newsletters on Stratigraphy* 5:160–67.

———. 1991. The meaning of MN zonation: Considerations for a subdivision of the European continental Tertiary using mammals. *Newsletters on Stratigraphy* 41:15–32.

Fleck, R. J., J. F. Sutter and D. H. Elliot. 1977. Interpretation of discordant ^{40}Ar/^{39}Ar age spectra of Mesozoic tholeiites from Antarctica. *Geochimica Cosmochimica Acta* 41:15–32.

Harland, W. B., A. V. Cox, P. G. Llewellyn, C. A. G. Picton, A. G. Smith, and R. Walters. 1982. *A Geologic Time Scale.* New York: Cambridge University Press.

Kamei, T., J. Ikeda, H. Ishida, I. Onishi, H. Partodizar, S. Sasajima and S. Nishomura. 1977. A general reapoort on the geological and paleontological survey in Maragheh area, Northwest Iran. *Memoir Facility of Science, Kyoto University, Geological and Mineralogical Survey* 43:131–64.

Lippolt, H. J., W. Genter, and W. Wimmenauer. 1963. Alterbestimmungen nach der Kalium-Argon-Methode an tertiaren eruptivgesteinen Südwestdeutschlands. *Jahrbuch geologische Landesamt Baden-Württenberg, Freiburg* 6:507–38.

McDougall, I. and T. M. Harrison. 1988. *Geochronology and Thermochronology by the ^{40}Ar/^{39}Ar Method.* Oxford Monographs on Geology and Geophysics no. 9. New York: Oxford University Press.

Mein, P. 1975. Résultats du Groupe de Travail des Vertébrés. In *Report on Activity of the RCMNS Working Group (1971–1975)*, ed. J. Senes, pp. 78–81. Bratislava: SAV.

———. 1979. Rapport d'activité du groupe travail des vertébrés, mise à jour de la biostratigraphie du Néogène basée sur les mammifères. *Annales Géologiques des Pays Helléniques*, hors série, 3:1367–72.

———. 1989. Updating of MN zones. In *European Neogene Mammal Chronology*, ed. E. H. Lindsay, V. Fahlbusch, and P. Mein, pp. 73–90. New York: Plenum.

Rögl, F. and G. Daxner-Höck. This volume. Late Miocene Paratethys correlations.

Samson, S. D. and E. C. Alexander Jr. 1987. Calibration of the interlaboratory ^{40}Ar/^{39}Ar standard, MMhb-I. *Chemical Geology, Isotope Geoscience* 66:27–34.

Sen, S. 1989. *Hipparion* Datum and its chronologic evidence in the Mediterranean area. In *European Neogene Mammal Chronology*, ed. E. H. Lindsay, V. Fahlbusch, and P. Mein, pp. 495–505. New York: Plenum.

Sen, S., J.-P. Valet, and C. Ioakim. 1986. Magnetostratigraphy and biostratigraphy of the Neogene deposits of Kastellios Hill (Central Crete, Greece). *Palaeogeography, Palaeoclimatology, Palaeoecology* 53:321–34.

Steiger, R. H. and E. Jager. 1977. Subcommission on Geochronology: Convention on the use of decay constants in geo- and cosmochronology. *Earth and Planetary Science Letters* 36:359–62.

Steininger, F. F., W. A. Berggren, D. V. Kent, R. L. Bernor, S. Sen, and J. Agusti. This volume. Circum-Mediterranean Neogene (Miocene and Pliocene) marine-continental chronologic correlations of European mammal units.

Steininger, F. F., R. L. Bernor and V. Fahlbusch. 1989. European Neogene marine/continental chronologic correlations. In *European Neogene Mammal Chronology*, ed. E. H. Lindsay, V. Fahlbusch, and P. Mein, pp. 15–46. New York: Plenum.

Swisher, C. C., III, and D. R. Prothero. 1990. Single-crystal ^{40}Ar/^{39}Ar dating of the Eocene-Oligocene transition in North America. *Science* 249:760–62.

Swisher, C. C., III, J. M. Grajales-Nishimura, A. Montanari, E. Cedillo-Pardo, S. V. Margolis, P. Claeys, W. Alvarez, J. Smit, P. Renne, F. J.-M. R. Maurrasse, and G. H. Curtis. 1992. Chicxulub crater melt-rock and K-T boundary tektites from Mexico and Haiti yield coeval ^{40}Ar/^{39}Ar ages of 65 Ma. *Science* 257:954–58.

Taylor, J. R. 1982. *An Introduction to Error Analysis.* New York: Oxford University Press.

Tobien, H. 1986. Die jungtertiare Fosilgrabungsstaette Höwenegg in Hegau (Suedwestdeutschland). Ein Statusbericht. *Carolinea* 44:9–34.

Van Couvering, J. A. and J. A. Miller. 1971. Late Miocene marine and non-marine time scale in Europe. *Nature* 230:559–63.

Weidmann, M., N. Solounias, R. E. Drake, and G. H. Curtis. 1984. Neogene stratigraphy of the Eastern Basin, Samos Island, Greece. *Géobios* 17:477–90.

Woodburne, M. O., R. L. Bernor, and C. C. Swisher III. This volume a. An appraisal of the stratigraphic and phylogenetic bases for the "*Hipparion*" Datum in the Old World.

Woodburne, M. O., G. Theobald, R. L. Bernor, C. C. Swisher III, H. König, and H. Tobien. This volume b. Advances in the geology and stratigraphy at Höwenegg, southwestern Germany.

6

Chronology and Biostratigraphy of the Miocene Sinap Formation of Central Turkey

J. KAPPELMAN, S. SEN, M. FORTELIUS, A. DUNCAN, B. ALPAGUT,
J. CRABAUGH, A. GENTRY, J. P. LUNKKA, F. MCDOWELL, N. SOLOUNIAS,
S. VIRANTA, AND L. WERDELIN

The evolution of large and small mammals in the Neogene of Europe is well documented in numerous collections from across this region. In contrast to its well-documented fauna, the chronology is at best spotty. With only a few exceptions, the chronology that does exist is based on relative faunal correlations having only a limited number of tie-points to the absolute time scale (Lindsay 1989; Steininger et al. 1989). Mein's (1975; 1979; 1989). Mammal Neogene (MN) zonation is the most widely referenced biochronology, and this scheme uses associations between or among various faunal elements and/or the first or last occurrences of key species as the basis for defining its zonations. While the MN zonation has been proven to be useful during the past two decades in establishing relative links between and among fossil localities, the absence of secure anchors to the absolute time scale makes it difficult to address questions concerning the pattern or rates of evolution, or the exact timing of dispersal events. Paleontologists have now reached the point where it is possible to look for larger-scale effects that may influence the patterns of evolution (Barry et al. 1985), but these correlations can only be tested in a convincing way when precise ties between faunal change and large scale regional or global events are made. Establishing a secure chronology for the European Neogene fossil record will permit these questions to be addressed on local, regional, and global scales.

The primary reason for the absence of a secure chronology can be found in the general nature of the European fossil sites themselves: most mammal fossils occur in localized settings having very limited stratigraphic expression. It is usually the case that the fossil localities in one area cannot be linked to fossil occurrences in other areas by direct stratigraphic correlations. In addition, the absence of long stratigraphic sections usually precludes the use of paleomagnetic reversal stratigraphy to formulate chronostratigraphic correlations. Paleomagnetic reversal stratigraphies have been determined for some areas, including the Cabriel section in Spain, with four mammal localities including Venta del Moro (Opdyke et al. 1989; Sen, this volume), and other sites such as Samos (Sen and Valet 1986, this volume; Swisher, this volume; Bernor et al., this volume) and Kastellios Hill (Sen et al. 1986; Sen 1991, this volume; Swisher, this volume; Woodburne et al., this volume). The general absence of volcanic rocks at most European sites further complicates the task since it is not possible to provide absolute dates. The absence of long and continuously fossiliferous outcrops is conspicuous, and the relative dating system of European mammalian evolution stands in sharp contrast to the well-calibrated biochronostratigraphic schemes of Asia (e.g., Barry and Flynn 1989; Barry et al. 1982; Jacobs et al. 1989; Kappelman et al. 1991) and Africa (e.g., Savage and Russell 1983; Pickford 1986). The European record challenges us to produce a precise biochronostratigraphy.

Other areas are sure to offer important contributions to our knowledge of the chronology of European Neogene mammalian evolution. We report here the results of our preliminary investigations of the Sinap Formation of Central Turkey. The fauna from this area has been collected and described since the 1950s, but only recently has attention been directed to the exact stratigraphic relationships of the fossil localities. Turkey is located on the eastern edge of Europe, and most of the research regarding its faunas has shown that many elements are shared with the numerous fossil localities to the west. Perhaps even more important, though, is the fact that Turkey lies at the paleobiogeographical crossroads of Europe, Asia, and Africa. Con-

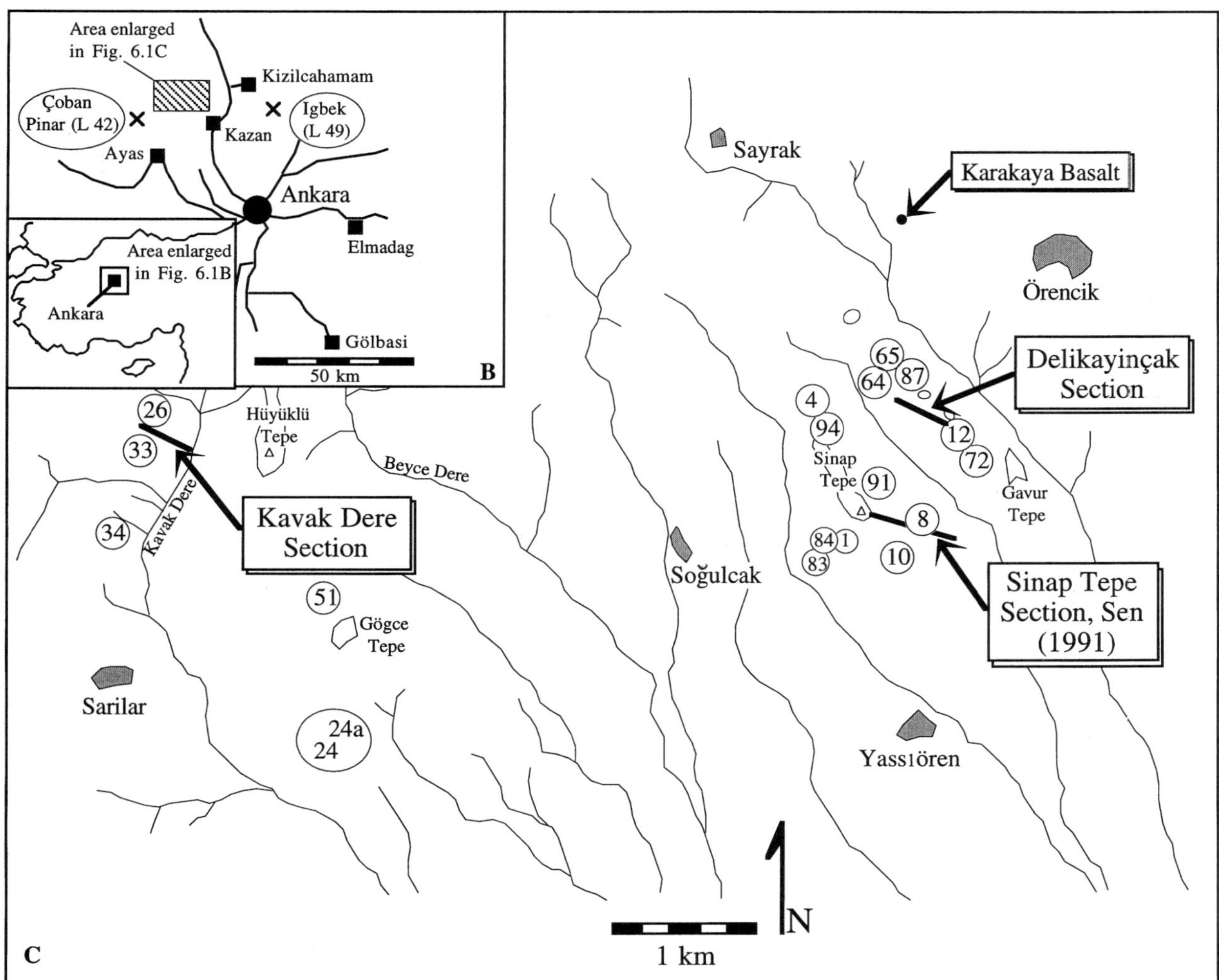

FIGURE 6.1 Map of the Sinap Formation field area showing: a) the location in central Turkey; b) the area north-northwest of Ankara; c) the fossil localities (circled numbers) and geologic sections (dark solid lines) around Sinap Tepe. The names of prominent buttes, hills, and valleys are also listed. Villages are stippled.

sequently, this region witnessed most if not all of the major dispersal events of the Neogene (see Bernor and Tobien 1990). A detailed biochronostratigraphy for Turkey will permit and facilitate intercontinental correlations and provide a broader understanding of Old World Neogene mammalian evolution.

Geological and Paleontological Background of the Sinap Formation

The Sinap Formation outcrops in the area around the village of Yassiören located about 55 km northwest of Ankara in Central Anatolia (fig. 6.1). The formation is named after a prominent butte, Sinap Tepe, that lies to the north-northwest of Yassiören, but other outcrops occur to the west in Kavak Dere as well as to the north at Karakaya. Ozansoy (1957; 1965) named three members of the Sinap

Formation and noted a conformable contact between the lower and middle members, and an erosional unconformity between the middle and upper members.

Later work on the geology of the area around Sinap Tepe was carried out by Öngür (1976), who recognized the validity of the Sinap Formation, named an additional older unit, the Pazar Formation, and further noted that these two formations are separated by a volcanic unit. Ozansoy (1965) recorded a total thickness of 98 m for the entire Sinap Formation, with individual thicknesses of 31 m, 34 m, and 33 m, for the lower, middle, and upper members, respectively. Öngür (1976) generally confirmed these estimates and recorded up to about 80 m for the Pazar Formation. He reported the volcanics to range up to 15 m in thickness.

Little work has been published on the Sinap Formation's sedimentology. Our preliminary observations reveal that the sediments around Sinap Tepe are of fluvial origin and represent small channel, sheet flood, and flood plain deposits, with limited paleosol development. In contrast, the sediments to the west in Kavak Dere appear to document a transition from lacustrine to fluvial conditions as the fault-bounded basin filled with sediments. It is likely that these two areas are part of the same larger depositional system, with the Sinap Tepe sediments representing a flood plain and distal alluvial fan system perhaps closer to the edge of the basin. This pattern of basin evolution is common to other parts of Anatolia as well (see Erol 1981; Inci 1991), and appears to be linked to the extensional neotectonics of Anatolia and the development of the North Anatolian transform fault during the Neogene (Angelier et al. 1981).

Mammal fossils were collected by Ozansoy from all three members of the Sinap Formation at several localities around Sinap Tepe as well as from several other fossil localities in the general vicinity. Ozansoy (1965: tab. 2) characterized the Sinap faunas as follows: the lower Sinap member was reported to be early Pliocene ("Plaisancien"), with the important absence of *Hipparion* from this level; the middle Sinap member fauna as late Pliocene ("Astien") with *Hipparion* abundant; and the upper Sinap member fauna with *Equus* was correlated with the Pleistocene ("Villafranchien"). A recent review of the fauna by Sen (1991) updates these age estimates and suggests that the lower Sinap member predates the Vallesian and is perhaps Astaracian in age while the middle Sinap member is most similar to lower Vallesian sites of Europe. The upper Sinap member is probably early Pleistocene in age (Sen 1991).

Ozansoy noted a number of other fossil localities in the general area around Sinap Tepe (1965: fig. 1) and attempted to tie these into a biozonation that was largely independent of a precise regional stratigraphy. Important fossil occurrences at Kavak Dere to the west of Sinap Tepe were correlated with the lower Sinap member. Other rich sites at Çoban Pinar and Inönü to the west were correlated with the middle Miocene, while the site of Kömürlük Dere to the east was said to predate all of the other known sites (Ozansoy 1965: tab. 2 and fig. 2). Recent comparisons between these collections and the European fossil mammal sequences place the Sinap fossils roughly in the Astaracian to Turolian, but this correlation offers a comparison of relative ages only. It is, however, clear that Ozansoy demonstrated the existence of a Neogene fauna that spans the middle Miocene through Pleistocene.

More recent paleontological work has concentrated on collection and study of the small mammals. Sen (1991) employed screen-washing techniques to recover a small mammal fauna from a middle Sinap member locality at the southeast corner of Sinap Tepe. He recovered a variety of small mammals (e.g., *Schizogalerix sinapensis*, cf. *Ochotona* sp., *Byzantinia dardanellensis*, *Cricetulodon hartenbergeri*, *Myomimus dehmi*, and *Atlantoxerus* sp.) and suggested an age of MN 9. Sen (1991) also provided a preliminary paleomagnetic reversal stratigraphy and demonstrated for the first time that the sediments of the Sinap Formation record a reliable signature of the earth's magnetic field. His paleomagnetic results for about 41 m of section showed one reversal only (normal to reversed polarity) near the upper part of the middle Sinap member. Sen suggested that this reversed interval preceded the long normal polarity of Anomaly 5 (8.98–10.30 Ma; Harland et al. 1982, in Sen 1991).

Chronostratigraphy of the Sinap Formation

Sample Selection and Methods

The area around Sinap Tepe contains a variety of sediments and volcanic rocks that are suitable for dating purposes. Rock samples for paleomagnetic analyses were collected from two stratigraphic sections, one in the Kavak Dere exposures to the north of the village of Sarilar, and the other on the flanks of Delikayinçak Tepe located about 1 km to the west of Sinap Tepe. The two sections were measured with a 1.6 m Jacob's staff and sliding mount Abney level, with the locations given in figure 6.1c. A total of 70 sites were collected through the 184 meters of Kavak Dere, while 45 sites were collected from the 108-meter section at Delikayinçak Tepe. A minimum of three samples per site were collected from unweathered fine-grained sediments (claystones, siltstones, and sandy siltstones) and oriented on the outcrop using a hand rasp and Brunton compass. Sampling avoided obviously bioturbated units and mature paleosols. The mean sampling density of the Kavak Dere section is 2.71 m (SD: 1.69 m, range 0.1–7.5 m), while the Delikayinçak Tepe section is 2.45m (SD: 1.38m, range 0.2–6.0 m).

Sediment samples for paleomagnetic analyses were milled into 2.5 cm^3 cubes using a cut-off saw with masonry blade and stationary vertical platter disc grinder. Measurements were made with a 2G SCT superconducting magnetometer in a magnetically shielded room at the University of Texas.

Numerous volcanic rock samples were collected from several basalt flows and tuffs in the Sinap region for K-Ar and Ar-Ar age determinations as well as paleomagnetic study. Volcanic rock samples were selected from unweathered portions of flows exposed above stream cuts, and weathering rinds were removed from the hand samples using a sledge hammer and other large blocks of the same rock as the anvil at the time the sample was collected. The suitability of each sample for K-Ar analysis was evaluated

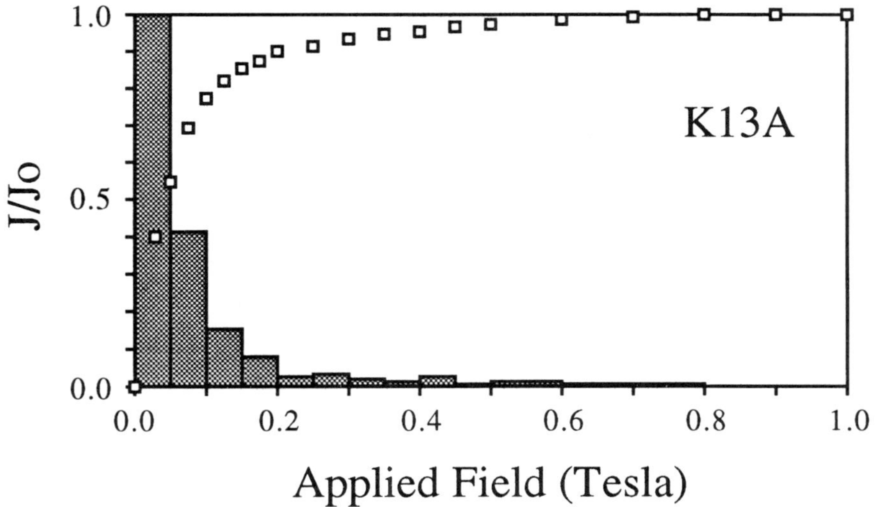

FIGURE 6.2 The results of a study of the acquired isothermal remanent magnetism (IRM) for sample K13a from Kavak Dere. The dominance of low-coercivity grains from 0-.1 Tesla (T) indicates magnetite or maghematite, with a minor contribution of higher coercivity grains (>0.3 T) suggesting pigmentary hematite or goethite. See text for a discussion of the laboratory methodology.

in thin section. At this time only one sample has been dated, but many other samples having broad stratigraphic coverage have been collected and await laboratory analysis. The single age determination of the Karakaya basalt was made in the K-Ar Laboratory at the University of Texas, and the location of this sample site is given in figure 6.1c. Paleomagnetic samples of the volcanic rocks were collected by either taking orientations on large hand samples on the outcrop and later reorienting these in the laboratory in a pan of plaster and then drilling cores, or by drilling and orienting the cores on the outcrop with a Pomeroy Rock Drill and diamond drill bit (ASC Scientific, Cardiff, California).

Rock Magnetic Studies

Studies of the coercivity spectra of several representative sedimentary lithologies were carried out to determine the magnetic mineralogy of the samples. The experimental methodology followed Dunlop (1972) and measured an acquired isothermal remanent magnetism (IRM). Dunlop (1972) showed that the coercivity spectrum can be divided into three mineralogically significant fractions: first, a magnetically soft portion that corresponds to magnetite or maghemite (0–0.1 Tesla [T]); second, an intermediate one that corresponds to specular hematite (0.1–0.3 T); and third, a hard fraction (> 0.3 T) that corresponds to pigmentary hematite with a grain size of < 5 microns. Samples from characteristic lithologies were given an IRM of up to 1 Tesla in increments of 0.025 T from 0–0.2 T, 0.05 T from 0.2–0.5 T, and 0.1 T from 0.5–1.0 T.

The results from a typical IRM acquisition experiment are shown in figure 6.2. Sample K13a is a tan mudstone from Kavak Dere. It demonstrates a dominance of low-coercivity grains from 0–0.1 T indicating magnetite or maghemite, with a minor contribution of higher coercivity grains. This same coercivity pattern is reported by Sen (1991) for his suite of paleomagnetic rock samples from Sinap Tepe, and is typical for the other samples that we have studied.

Thermal Demagnetization

All samples were subjected to stepwise thermal demagnetization using a Schonstedt Thermal Demagnetizer (TSD-1) to remove secondary magnetic components and isolate a stable magnetic vector. Demagnetization protocol generally included 6–10 steps from 0–650°C, and the results for two samples are shown in figure 6.3. Sample G35b (fig. 6.3a) is from the Delikayinçak Tepe section and thermal demagnetization was successful in isolating a stable reversed component at temperatures above 200°C. Sample K31a (fig. 6.3b) is from the Kavak Dere section. Thermal demagnetization produced a uniform decrease in intensity to 650°C. The vector endpoint projection indicates a nearly linear decay to the origin, and this sample is normally magnetized. These thermal demagnetization results are typical for the suite of samples from Delikayinçak Tepe and Kavak Dere and support the inference of magnetite as the primary magnetic carrier as based on the IRM study (fig. 6.2), with some occasional and minor contribution of hematite.

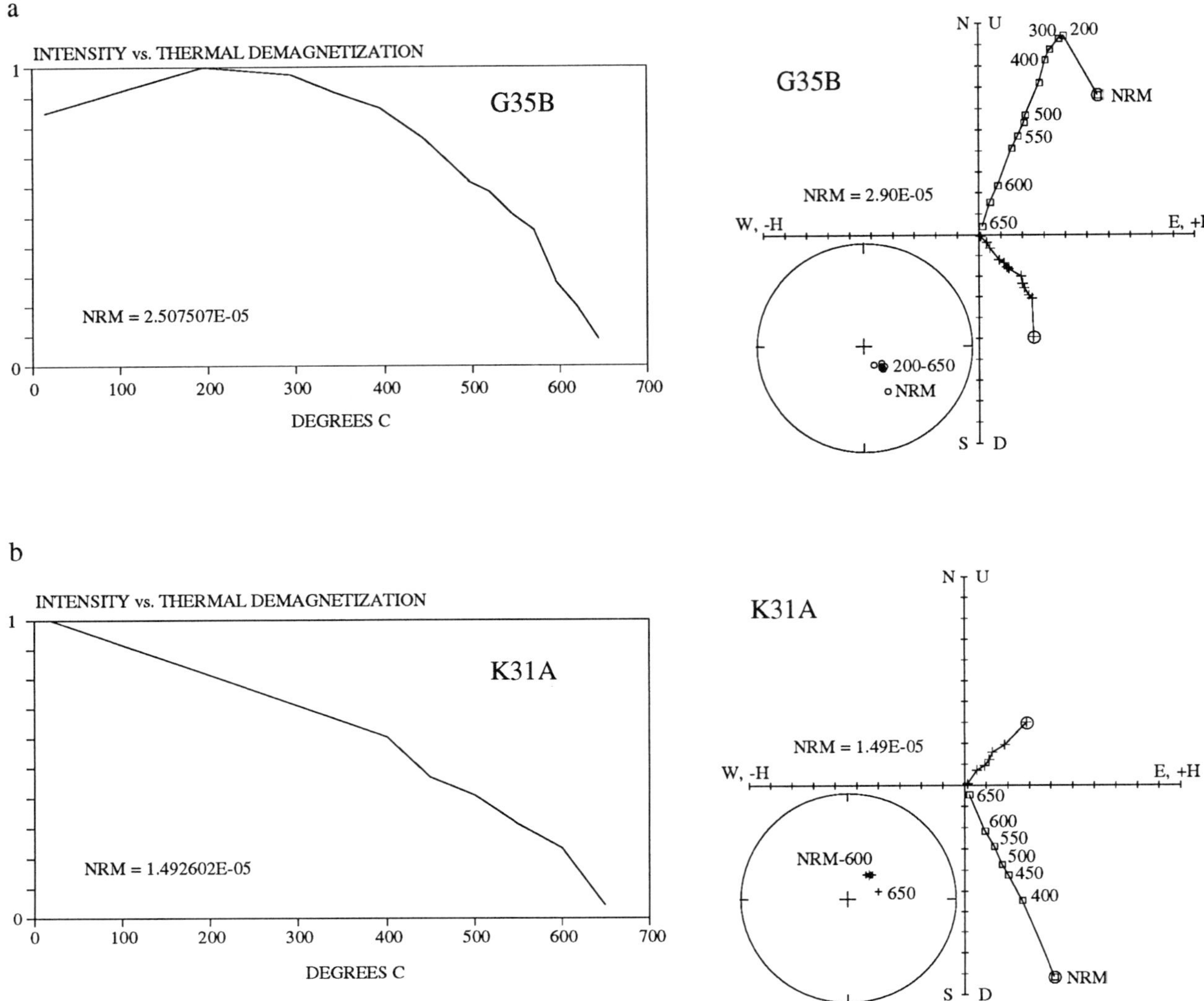

FIGURE 6.3 Results of thermal demagnetization treatments and the vector endpoint diagrams for two representative paleomagnetic sediment samples from the Sinap Formation: a) sample G35b from Delikayinçak Tepe demonstrates a slight increase in intensity with the removal of a normal overprint at around 200°C, with a decrease in intensity and an linear decay to the origin at higher temperatures. This sample is reversely magnetized; b) sample K31a from Kavak Dere shows a decrease in intensity and a linear decay to the origin. Sample K31a is normally magnetized. The steep decrease in intensity from 500 to 600°C indicates the presence of magnetite. Intensities are normalized to the NRM values. Symbols: in stereonets: hollow circle is upper hemisphere, cross is lower hemisphere; in vector endpoint diagrams: hollow square is inclination, cross is declination.

Site Statistics

The principal component analysis (Kirschvink 1980) was used to determine a best fit line for the magnetic vector for each sample from every site, and site means using three or more samples from each site were calculated with Fisher statistics (Fisher 1953). Statistical significance was tested with the Watson criteria (Watson 1956).

A total of 70 sites were collected from Kavak Dere, and of these 51 (73%) satisfied the Watson criteria while 11 (16%) did not. It was not possible to calculate vectors for the samples from the remaining 8 sites (11%). Out of the total of 51 statistically significant sites, 15 exhibited reversed polarity while 36 were normal. The Delikayinçak Tepe section contained 45 sites, 38 (84%) of which satisfied the Watson criteria, while 7 (16%) did not. Only 1 of the 38 statistically significant sites exhibited a reversed polarity. Virtual geomagnetic poles were calculated for each sample for which a vector could be calculated, and for every site that satisfied the Watson criteria.

Stereo plots that compare the declinations and inclinations of the natural remanent magnetism (NRM) measurements with the declinations and inclinations of the statistically significant site means after thermal demagnetization are shown in figure 6.4. The NRM data (figs. 6.4a and 6.4b) show a wide scatter that upon thermal demagnetization divides into normal and reversed populations (figs. 6.4c and 6.4d) although this is much less apparent for Delikayinçak Tepe since there is only 1 reversed site in this section. The present field in the Sinap region (declination = 3.5°, inclination = 57°) falls within the spread of the NRM as well as the demagnetization best fit values from both sections but is more squarely within those from Delikayinçak Tepe. Given that we have found reversely magne-

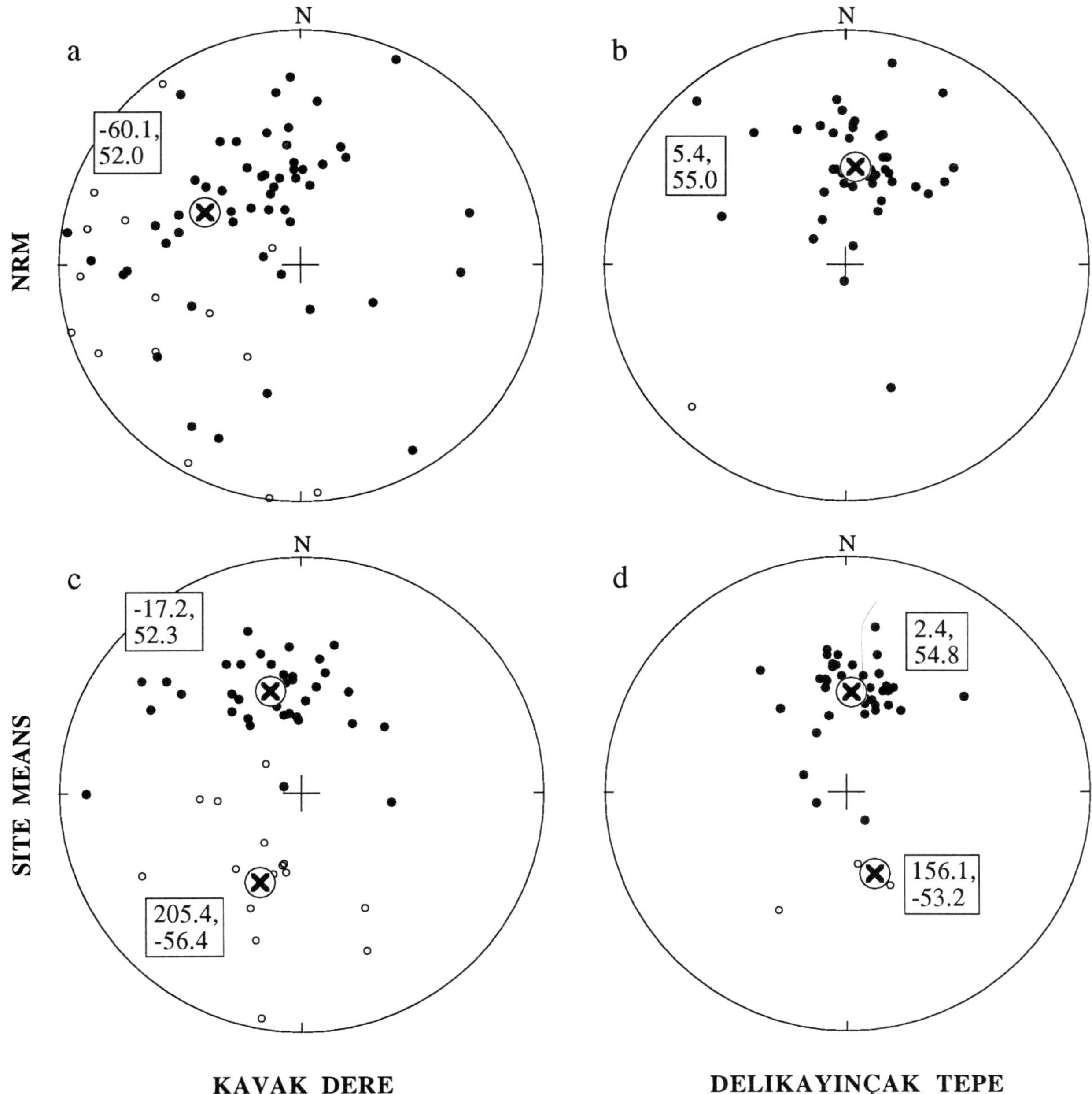

FIGURE 6.4 Stereo plots of Schmidt equal area projections of the natural remanence magnetism (NRM) for one sample from each site (a, b) and for the statistically significant site means (n = 3) calculated from the best fit lines after thermal demagnetization (c, d). Samples from both (a) Kavak Dere and (b) Delikayinçak Tepe show a predominance of normal directions before thermal demagnetization related to a low temperature overprint probably due to pigmentary hematite or goethite. Thermal demagnetization is generally effective in removing the overprint, with resultant normal and reversed populations (c, d) but these means are not exactly antipodal (see text). The two volcanic rock samples that were measured are also illustrated (d: mean declination = 156.1, inclination = -53.2). Symbols: solid circle is lower hemisphere; hollow circle is lower hemisphere; x is population mean.

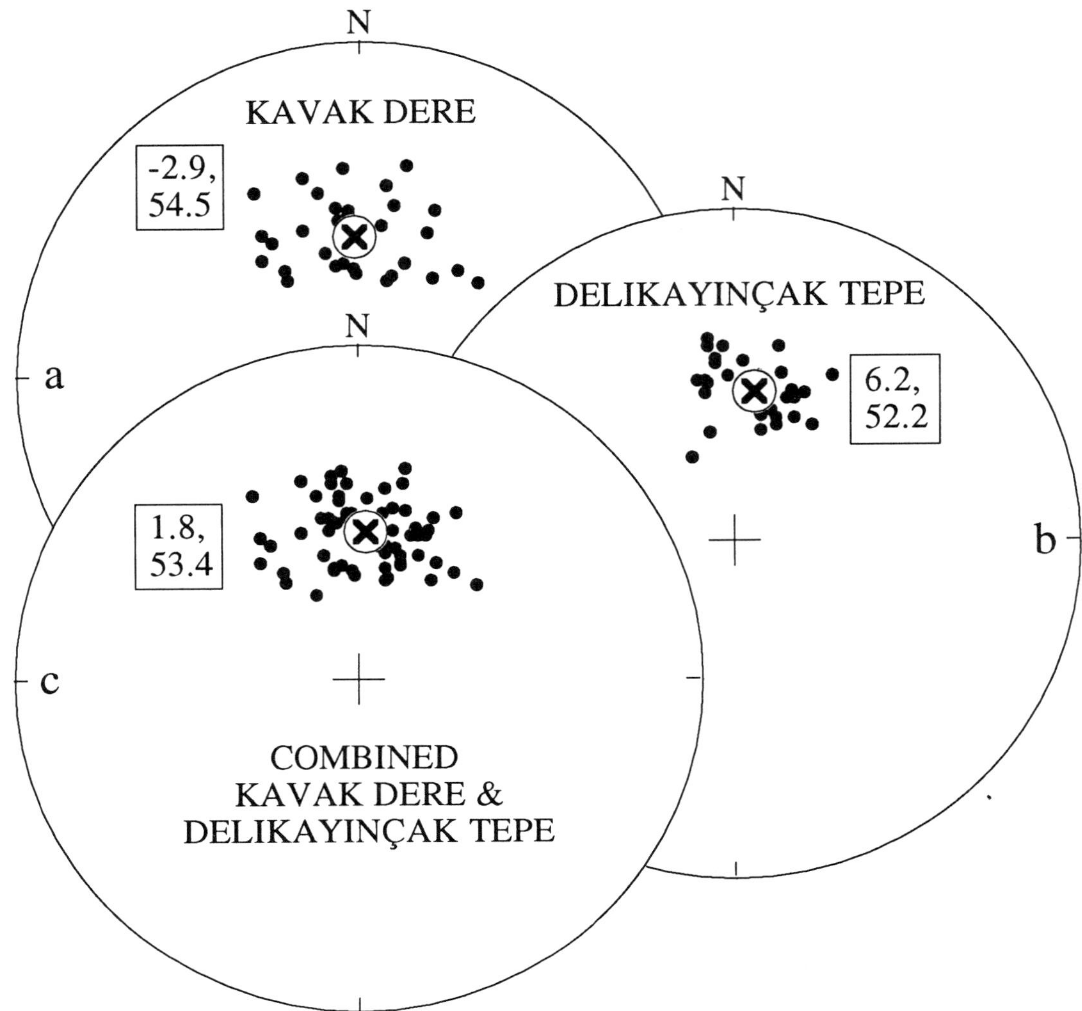

FIGURE 6.5 Stereo plots of Schmidt equal area projections of the statistically significant site means from a test for similarity between the mean site vectors from (a) Kavak Dere and (b) Delikayinçak Tepe. Reversely magnetized sites were inverted, and sites with an angular standard deviation >2 were removed from the calculation. The combined sample is shown in (c). An F-test shows that these means do not differ significantly from one another (see text for discussion). Symbols: solid circle is site mean; x is population mean.

tized samples in each of the two sections, we believe that it is unlikely that the normal directions represent a present field overprint. If so, the remagnetization must have preceded the folding that produced dips of 6°–30° across the region. We feel that it is more likely that little rotation has occurred in this area.

Several statistical tests were conducted to assess the reliability of the demagnetization treatments and the faithfulness of these sediments in recording the ancient field. The first test compared the similarity of the total population from Kavak Dere with that from Delikayinçak Tepe (fig. 6.5). The reversely magnetized samples were inverted, and a test of the Fisher precision estimates excluded those sites having an angular deviation exceeding 2. This test showed the precision estimates to be significantly different from each other at the $0.01 < \rho < 0.05$ level (max k/min k $= X_{k,v}$: $49.6/25.7 = 1.9300_{2,61}$). Although this result exceeds the commonly accepted level of 0.05, these populations are nonetheless quite similar to one another and do not differ at $\rho < 0.01$. A second test compared the mean vectors from each population and showed that there is no difference between the two populations ($F_{2(S-1),\ 2(N-S)} = 2(N-S)/2(S-1) \equiv \Sigma_{Ri}\text{-}R/N\text{-}\Sigma R_i$: $F_{2,122} = 2.7195$, $\rho > 0.05$). This result suggests that the Kavak Dere and Delikayinçak Tepe samples record the same ancient magnetic field signature.

A second statistical test compared the normal and reversed populations from Kavak Dere (see fig. 6.4c). The test of the precision estimates showed no statistical difference between the normal and reversed sample populations (max k/min k $= X_{k,v}$: $10.8/7.8 = 1.3846_{2,49}$; $\rho < 0.05$). However, the mean vectors of the normal and reversed samples were shown to differ from one another at $\rho < 0.05$ ($F_{2(S-1),\ 2(N-S)} = 2(N-S)/2(S-1) \equiv -\Sigma R_i-R/N-\Sigma R_i$: $F_{2,98} = 8.4813$). This result is also apparent from the stereo plot (fig. 6.4c) and suggests that the thermal demagnetization treatment may not have been successful in removing the entire present field normal overprint. However, it is also true that site placement through much of the Kavak Dere section is very dense, and in at least one case the course of a magnetic reversal may be documented. The addition of these statistically significant sites in the above test yields a low precision estimate and a large scatter of data. We have not yet completed the sampling that is necessary to carry out a fold test but plan this for the future across the broad syncline of Kavak Dere.

Volcanic rocks

Several different basalt flows occur in the Sinap region, and we have sampled many of these for K-Ar as well as paleomagnetic studies. The results from one of the paleomagnetic studies is given here and shown in figure 6.6. The dramatic drop in intensity across the 500°C–600°C temperature range (fig. 6.6a) indicates magnetite as the carrier, and this demagnetization pattern is typical for all

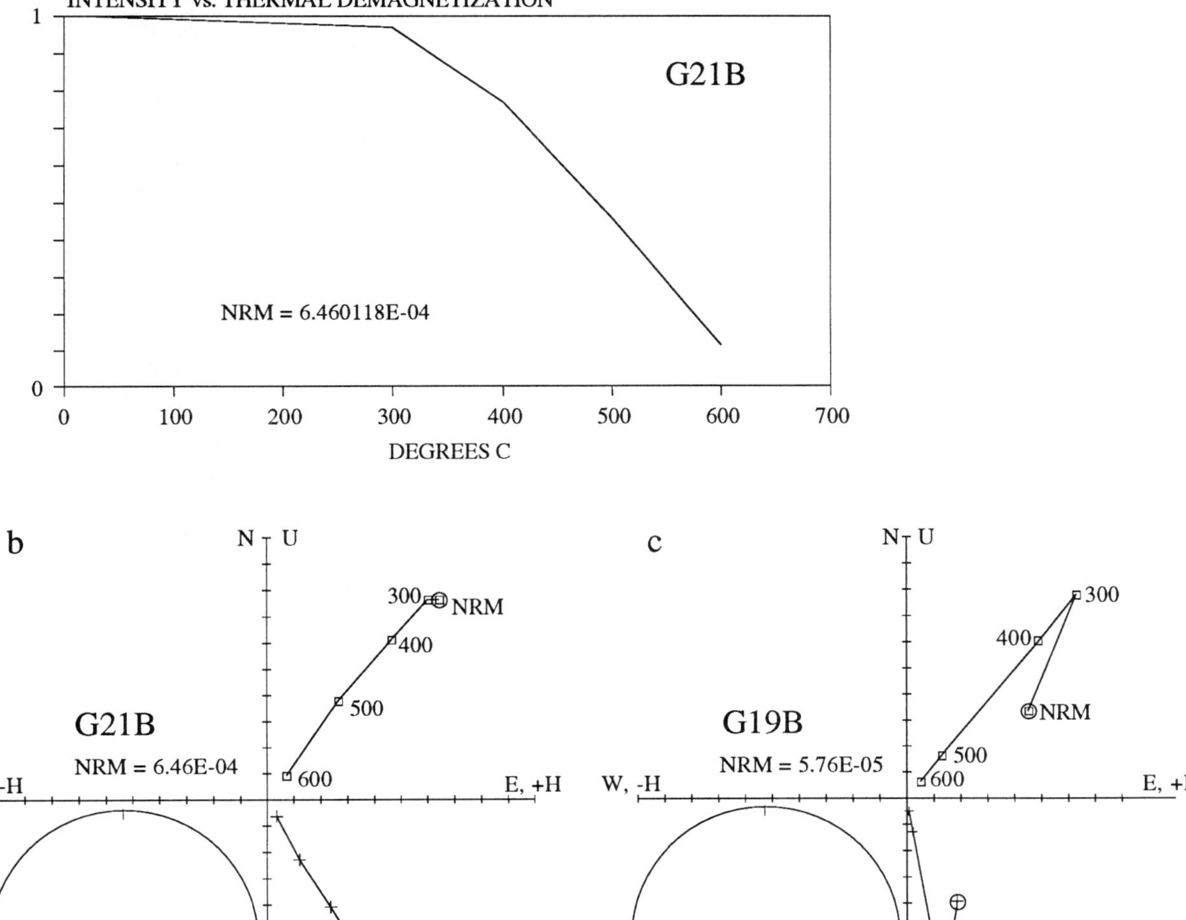

FIGURE 6.6 Results of thermal demagnetization treatments and the vector endpoint diagram for a volcanic sample found near the transition between the older Pazar Formation and the younger Sinap Formation at Karakaya Tepe: a) sample G21b demonstrates a dramatic decrease in intensity at temperatures above 300°C that indicates the presence of magnetite; b) the demagnetization treatments demonstrate the removal of a minor low temperature overprint with a subsequent linear decay toward the origin. This sample is reversely magnetized, and this site mean, along with that of another site from these same volcanics, is shown in figure 6.4d. See table 6.1 for the K-Ar data and the date of this sample. Symbols as in figure 6.3.

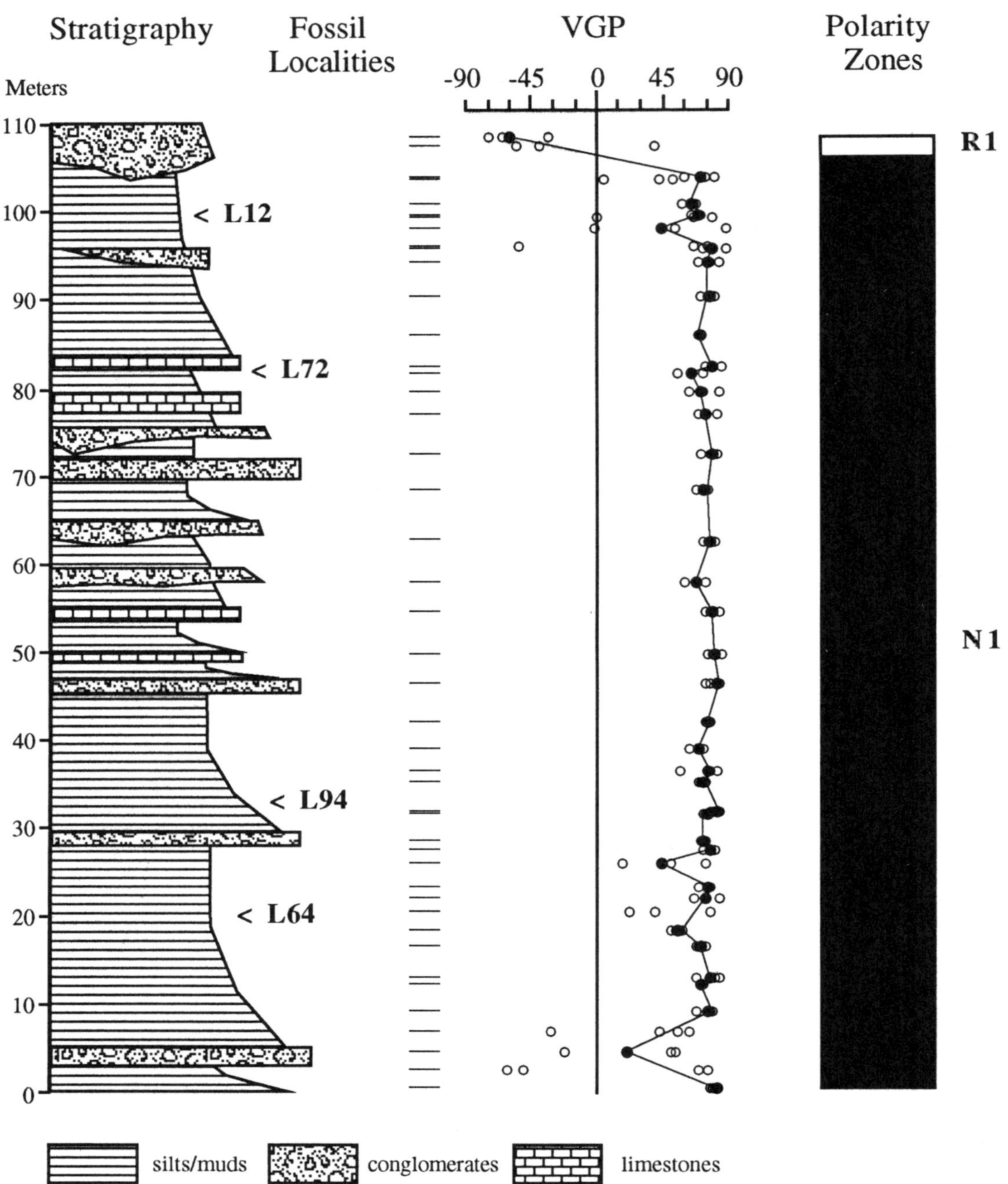

FIGURE 6.7 The stratigraphy and paleomagnetic reversal stratigraphy for Delikayinçak Tepe includes 108 m of fluvial mudstones, sandstones, and conglomerates. Key fossil localities are listed by number to the right of section, and the stratigraphic position of paleomagnetic sample sites are given by the short horizontal bars in the center of the figure. Virtual geomagnetic poles based on the best fit lines are plotted for every sample (hollow circles) and the statistically significant site means (solid circles). The interpretation of the reversal history is indicated by the black and white bars to the far right of the figure. The transition between polarity zones is taken as the midpoint between statistically significant site means of opposite polarity. Black intervals are normal polarity; white intervals are reversed polarity. This section is distinctive in its long interval of normal polarity. See figure 6.9 for the correlation with the geomagnetic reversal time scale (GRTS).

of the basalts that we have measured. The vector endpoint projection (fig. 6.6b) shows a minor low temperature overprint that is removed at temperatures above 300°C with a nearly linear decay to the origin, and the site is reversely magnetized. A calculation that combines the results from this site with other samples from a lower site provides a mean direction of declination = 156.1°, inclination = -53.2° (n = 6, R = 5.965, K = 145, alpha95 = 5.6°). Given the small number of samples obtained from two hand samples, it is unlikely that the effects of secular variation are averaged out.

Sen (1991) sampled a suite of both sediments and volcanics from Karakaya Tepe and reported a mean direction of declination = 172°, inclination = -64° (n = 4, K = 19, alpha95 = 18.5°). Our two sets of measurements both fall within each other's alpha95 cone of confidence limits, and are together roughly antipodal from the mean value from the Kavak Dere and Delikayinçak Tepe sections combined (n = 90, declination = 359°, inclination = 54.7°, K = 11.2, alpha95 = 4.7°).

K-Ar Dating

We have completed a date for the basalt flow that separates the Pazar Formation from the Sinap Formation as mapped by Öngür (1976). This sample was collected in the area to the north-northeast of Sinap Tepe in Karakaya Tepe (see fig. 6.1), and the data are presented in table 6.1. The untreated whole rock material provided a K-Ar date of 15.2 ±0.3 (1 SD) Ma. Many other K-Ar dates have been generated for numerous regions across Turkey (e.g., Becker-Platen et al. 1977; Besang et al. 1977; Benda et al. 1974) but the date for the Karakaya basalt appears to be the first for this particular region. Although this basalt flow is not in direct contact with the Sinap Formation fossil sites, it does provide a maximum age for the fossil-bearing sediments of the Sinap Formation.

Paleomagnetic Reversal Stratigraphy of the Sinap Formation

Our preliminary paleomagnetic reversal stratigraphy for the Delikayinçak Tepe section is presented in figure 6.7. Sen (1991) sampled about 41 m of conformable section through the lower and middle members of Sinap Formation and found that the lower 30 meters were of normal polarity while the upper 11 meters were of reversed polarity. Our mapping has established that there is more than double the thickness of the Sinap Formation section sampled by Sen (1991). The measured section shown in figure 6.7 includes 108 m of fluvial sediments and places several of the Sinap Tepe fossil localities in their superpositional relationships. The section is almost completely normal, with a single reversal found near the very top of the section.

TABLE 6.1. *Whole Rock K-Ar Results for the Karakaya Basalt from NW of Örencik.*

Sample 6-28-1[*]: $\lambda\rho = 4.963 \times 10^{-10}$/yr; $\lambda\epsilon + \epsilon' = 0.581 \times 10^{-10}$/yr; ^{40}K/K = 1.167×10^{-4}.

%K	%^{40}Ar	^{40}Ar × 10^{-6} scc/gm	Age (Ma)	±1 SD
1.131	69	0.6705	15.2	±0.3
1.137	60	0.6782		

[*]This rock was also sampled for paleomagnetic analysis as sample G21; see Figure 6.

We tentatively correlate this reversal event with the reversal found by Sen (1991) in the top of his Sinap Tepe section located about 1 km to the west (fig. 6.8). The correlation is strengthened by the fact that both of the paleomagnetic reversal stratigraphies shown in figure 6.8 contain two shallowing "events" at approximately the same stratigraphic level in the upper part of the normal interval.

We have completed a preliminary paleomagnetic reversal stratigraphy for the Kavak Dere section (fig. 6.9) that samples 184 m of lacustrine and fluvial sediments. Our preliminary paleomagnetic results reveal several polarity intervals of variable but nearly equal thickness. A number

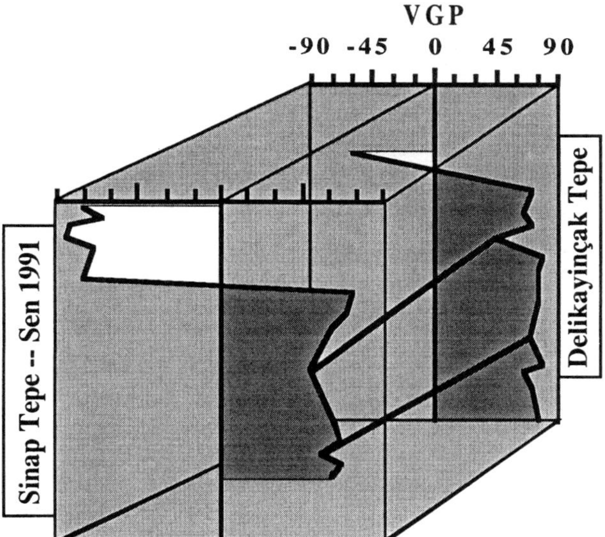

FIGURE 6.8 A correlation between the paleomagnetic reversal stratigraphy of Sinap Tepe by Sen (1991) and that of Delikayinçak Tepe shown as a 3-D diagram (see fig. 6.7). These two sections are located approximately 1 km apart (see fig. 6.1). The figure illustrates the VGP site data for the upper 30 m of normal polarity sediments and the transition to the reversely magnetized interval for each of the two sections. The reversal and two "events" of shallow inclination are found in each of the two sections at approximately the same stratigraphic level (solid dark lines); together these provide support for the correlation between the two sections. Dark stippling indicates normal polarity; white intervals are reversed polarity.

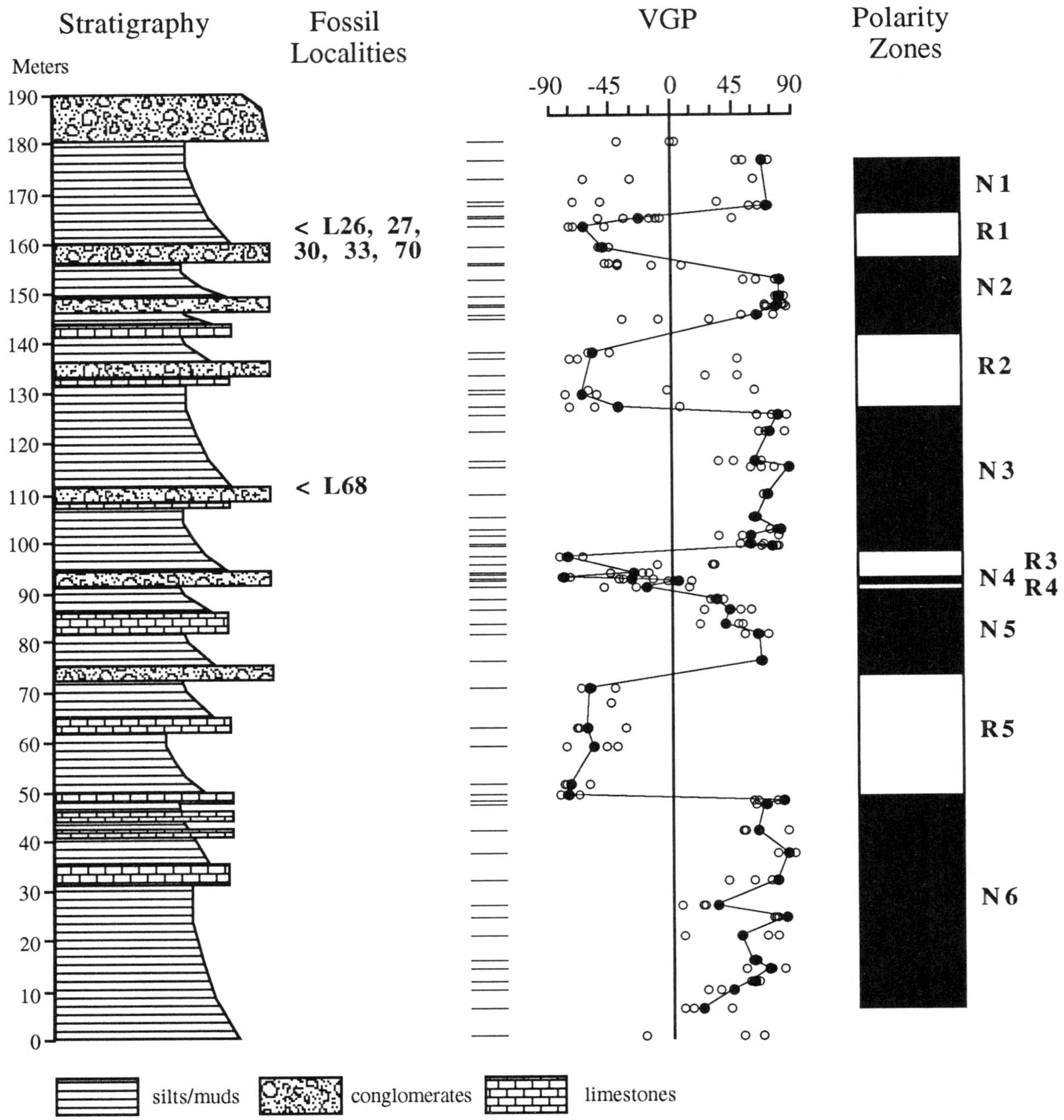

FIGURE 6.9 The stratigraphy and paleomagnetic reversal stratigraphy for Kavak Dere includes 184 m of lacustrine and fluvial mudstones and sandstones in a generally coarsening upward sequence. Symbols are as in figure 6.7. This section demonstrates approximately equal intervals of normal and reversely magnetized sediments, with a finely sampled reversal transition at the 80–90 m level. See figure 6.10 for the correlation with the geomagnetic reversal time scale (GRTS).

of sites collected from apparently bioturbated units do not demonstrate statistical significance.

Correlations with the Geomagnetic Reversal Time Scale

We believe it is likely that the long normal interval within the middle and lower Sinap members at Delikayin-çak Tepe represents the upper portion of Chron C5n (fig. 6.10) with the reversal event at the top of the section marking the Chron C4Ar/C5n boundary. Sen (1991) sampled a much shorter section and correlated his lower 30 m normal interval with one of the brief normal events (about 50 kyr in duration) in Chron C5r that predate the long normal interval of Chron C5n. The increased thickness of the Delikayinçak Tepe section makes this earlier correlation unlikely since it would require a minimum sedimentation rate exceeding 2 m/kyr (107 m/50 kyr). Several brief duration reversal events ("tiny wiggles" of Cande and LaBrecque 1974) are suspected to have occurred within Chron C5n (see Tauxe and Opdyke 1982), and one of the best-documented events is Chron C5n.1r near the younger end of Chron C5n (see fig. 6.10; Steininger et al., this volume). While it is possible that the upper shallowing event in the two Sinap Formation sections represents a

fluctuation in the paleointensity of the earth's magnetic field, it is also possible that it represents an incomplete sampling of this Chron C5n.1r event.

A correlation between the Geomagnetic Reversal Time Scale (GRTS) and the Kavak Dere section is somewhat more problematic. The GRTS contains several distinctive polarity signatures in the late Miocene, but most distinctive of all is the long normal interval of Chron C5n (10.95–9.74 Ma); the approximately 2 m.y. time interval prior to

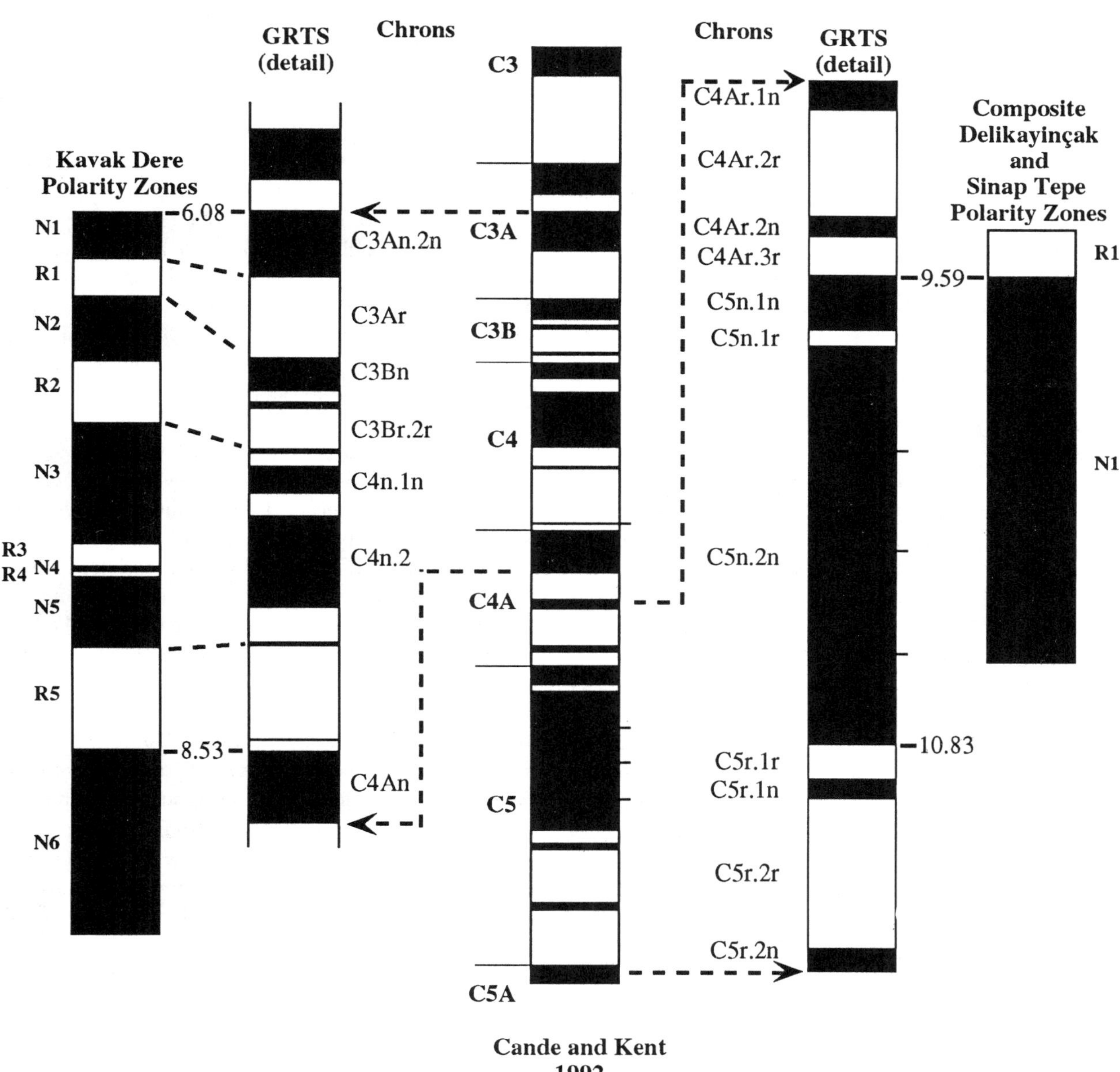

FIGURE 6.10 Correlations between the paleomagnetic reversal stratigraphy of (left) Kavak Dere and (right) the composite Sinap Tepe–Delikayinçak Tepe section and (center) the geomagnetic reversal time scale (GRTS) of Steininger et al. (this volume). The composite Sinap Tepe–Delikayinçak Tepe section combines the results from Sen (1991) with those reported here (see fig. 6.9). This section is distinctive in its long interval of normal polarity, and we correlate this with Chron C5n. The most characteristic aspect of the Kavak Dere section is its approximately equal representation of normal and reversely magnetized sediments. We feel that this section is best correlated with Chrons C3An.2n through C4An, but a confirmation of this will have to await future results. Both sections are figured at the same meter scale.

Chron C5n is of predominantly reversed polarity, while the 4 m.y. interval following Chron C5n is represented by approximately equal durations of normal and reversed polarity. Given the subequal durations of normal and reversed polarity, we provide a preliminary correlation for the Kavak Dere section with Chrons C3An.2n through C4An, for an estimated age range of approximately 9.02–6.27 Ma (fig. 6.10). This correlation should be considered preliminary only, and other correlations with the GRTS are certainly plausible. Additional mapping has revealed conformable underlying exposures of nearly 400 m in Beyçe Dere to the east, and overlying exposures of around 200 m at Çalta to the north, bringing the total Beyçe Dere-Kavak Dere-Çalta section to nearly 800 m. We have begun paleomagnetic sampling in these other sections, and completed sampling on a second 170-m section in Kavak Dere located about 2 km along strike from the section illustrated in figure 6.9. The results of this work, plus radioisotopic dates on a basalt flow and tuffs that occur in the Beyçe Dere section, should provide a much more secure tie between the fossil localities and the time scale.

Discussion

Our preliminary age estimates find support in the faunal correlations as they are presently understood. Recent summaries of the biochronological information are given by Sen (1991) and Alpagut et al. (1992; 1993). We present our updated data set in table 6.2. The oldest presently known faunas of the Sinap Formation are localities 24 and 24A ("Inönü I" of Ozansoy 1965). The fauna was assigned to MN 6 by Gürbüz (1981), and new material recovered since then tends to support this assignment. Although the basalt with the K-Ar date discussed earlier is not in direct association with these localities, both localities appear to be near the upper boundary of the Pazar Formation (as defined by Öngür 1976), which would then give them an age of around 15 Ma. Much younger sites loosely included with the Sinap Formation are found near Çalta (localities 23, 59, and 60) and have a Ruscinian small mammal fauna (Sen 1977), while locality 86 on Sarikol Tepe (= "Sinap Supérieur" of Ozansoy 1957, 1965) has a late Pliocene large mammal assemblage (Sen 1991). It is not yet possible to say exactly how much time from the early middle Miocene to the early Pleistocene is represented by these fossiliferous beds, but it appears that the representation is more complete than has been previously assumed.

So far the only other securely placed middle Miocene sites in addition to localities 24 and 24A are localities 64 and 65 to the northeast of Sinap Tepe (in the "Sinap Inférieur" of Ozansoy 1957, 1965), and these may be closer to the middle/late Miocene boundary. The presence of *Spermophilinus* aff. *bredai*, *Myocricetodon eskihisarensis*, *Peridyromys* sp., and *Democricetodon* sp. at locality 64 sug-

gests a late Astaracian (MN 8) age. Locality 65 has produced a poor fauna of similar aspect, and is stratigraphically below locality 64. No *Hipparion* material has been recovered from these localities, although the large mammal fauna of locality 64 is relatively diverse. If the identification of *Gazella* sp. from locality 64 proves to be valid, it would represent the first pre-Vallesian record of this genus in Anatolia.

The majority of the fossil sites in the area surrounding Sinap Tepe are found at a stratigraphic level above localities 64 and 65. This level is roughly correlative with the middle member of the Sinap Formation as described by Ozansoy (1957; 1965). It should, however, be noted that the sedimentological differences between the "lower" and "middle" members are difficult to distinguish along strike and because of this, Ozansoy's member designations may be of limited utility. Locality 8A at Sinap Tepe ("Locality I" of Ozansoy; see Sen 1991) is stratigraphically higher than localities 64 and 65 and has yielded a small mammal fauna that is referable to the early Vallesian as based on several faunal occurrences. *Byzantinia* is known from locality 8A, which at first glance appears to suggest an older age. Two species of *Byzantinia* are known from Bayrak Tepe I, which Ünay (1981) and Ünay and de Bruijn (1984) have assigned to the late Astaracian, but this age assignment is primarily based on the association of *Spermophilinus*, *Miopetaurista*, *Albanensia* and *Megapedetes*.

Both *Byzantinia* species from locality 8A have some evolved characters that indicate a somewhat younger age than Bayrak Tepe I. Locality 8A has also yielded a new species of *Myocricetodon*, apparently derived from *M. eskihisarensi*, which also indicates a younger age. Perhaps the most important evidence for the age of locality 8A is the cooccurrence there of *Progonomys* cf. *cathalai* and *Cricetulodon hartenbergeri*. The latter species is only recorded from MN 9 localities in Spain. *Progonomys* occurs at Can Llobateres and Pedregueras IIc, and both of these localities are placed in late MN 9 (Agustí 1990; Moyà Solà and Agustí 1990; Alvarez Sierra et al. 1987). A more precise age assignment of locality 8A within the MN 9 zone cannot be made unless we assume an exactly synchronous small mammal turnover from Anatolia to Spain, which appears unlikely.

A very rich small mammal fauna was recently recovered from locality 84, which is near the top of the middle Sinap member at Sinap Tepe and stratigraphically higher than locality 8A. This assemblage has not yet been studied in detail, but preliminary results strongly indicate that it is perhaps significantly younger than locality 8A, with most common taxa being represented by more derived forms.

Dating the first occurrence of *Hipparion* in the Old World is of broad interest. The local first stratigraphic occurrence of *Hipparion* at Sinap is immediately above the locality 64 horizon. The occurrence of a *Hipparion* molar

TABLE 6.2. *Faunal List for Selected Localities of the Sinap Formation*

Localities:	24*	64	87	4	94	72	8B	8A	12	91	83	84	1	51	63	49	26^	42
Schizogalerix sinapensis								×	×	×	×	×						
Schizogalerix sp.		×																
Orycteropus sp.		×	×	×	×	×	×	×	×									
Byzantinia bayraktepensis								×										
Byzantinia ozansoyi								×										
Byzantinia sp. 1 (large)		×																
Byzantinia sp. 2 (small)		×																
Byzantinia sp. 3												×						
Byzantinia sp. 4												×						
Myocricetodon eskihisarensis		×																
Myocricetadon sp. nov.								×										
Democricetodon sp.		×																
Cricetulodon hartenbergeri								×										
Cricetulodon sp.								×	×	×	×	×						
Peridyromys sp.		×																
Myomimus dehmi								×	×	×	×	×						
Progonomys cf. *P. cathalai*								×	×	×	×	×						
Pliospalax sp.												×						
Spermophilinus cf. *S. bredai*		×																
Atlantoxerus sp.								×	×	×	×	×						
Proochotona cf. *P. kalfaense*								×	×	×	×	×						
Ankarapithecus meteai								×	×									
Listriodon cf. *L. latidens*	×																	
Microstonyx sp.																	×	×
Cervidae indet.																×		
Giraffidae indet.		×	×	×	×	×	×	×	×	×	×	×	×	×	×	×	×	
Palaeotragus coelophrys																×		
Turcocerus gracilis	×																	
Hypsodontus sp.	×																	
Caprotragoides stehlini																	×	
Boselaphini spp.											×					×	×	×
Bovidae small sp.		×	×	×	×													
Gazella sp. (*?ancyrensis, capricornis*)						×	×	×	×	×	×	×	×	×	×	×	×	×
Prostrepsiceros cf. *P. rotundicornis*															×			
Prostrepsiceros zitteli																×		
Palaeoreas asiaticus																×		
Nisidorcas planicornis					×													
?Nisidorcas sp.							×	×	×									
Protoryx sp.																	×	
?Criotherium sp.														×	×	×		
Anchitherium sp. (very large)	×																	
Hipparion spp.		×	×	×	×	×	×	×	×	×	×	×	×	×	×	×	×	
Begertherium cf. *B. grimmi*	×																	
Acerorhinus cf. *A. zernowi*																×	×	
Chilotherium cf. *C. samium*														×	×	×		
Chilotherium cf. *C. habereri*																	×	
Stephanorhinus sp. (*pikermiensis*-group)																	×	
Ceratotherium cf. *C. neumayri*																	×	
Rhinocerotidae indet. (large)																×		
Choerolophodon sp.											×							
Amphicyon major	×																	
Ursidae cf. *Indarctos*														×				
Mustelidae indet.											×							
Protictitherium crassum					×													
Protictitherium sp.																×		
Ictitherium intuberculatum						×												
Ictitherium pannonicum																		×
Thalassictis montadai					×													
Percrocutidae indet. (large)																×		
Nimravidae indet.		×																

List includes specimens collected from 1989–1992, with the exception of *Ankarapithecus meteai*.
See Figures 6.1, 6.7, and 6.10 for the geographic location and stratigraphic position of the fossil localities.
Localities 8A, 12, and 91 are in the same stratigraphic interval.
*Includes Inönü localities 24 and 24A.
^ Includes upper Kavakdere localities 26, 27, 33, and 34.

(specimen no. AS.92.859) at locality 87 is important because, although the molar is a surface find, the locality is situated on a local hilltop, and the fossil could not be derived from higher stratigraphic levels. Locality 94 is situated above locality 87 and has yielded a few specimens of *Hipparion. Hipparion* has also been recovered *in situ* in a conglomerate a few meters above locality 94, and also from locality 4, which is about 5 m below locality 94. *Hipparion* seems to be rare at the oldest localities where it occurs. It makes up only 5% of the catalogued specimens at locality 4 (n = 39) and 2% at locality 94 (n = 127); however, it is a dominant form at the next highest localities in the section with values ranging from 31% at locality 12 (n = 113) to 63% at locality 8B (n = 82).

Locality 94 has also yielded good material of the hyaenid *Thalassictis montadai*, which is known from a number of localities from Anatolia and Spain, all attributed to either late MN 8 or early MN 9 (Yeni Eskihisar, Ballestar, Can Mata, Hostalets de Pierola: Werdelin and Solounias 1991). The very small bovid present only at localities 64 and 94 may be an antilopine or a member of the Antilopini-Neotragini basal stock, and would also fit with an age near the middle/late Miocene boundary. It thus appears that the Sinap region's lowest stratigraphic *Hipparion* datum (LSD), which marks the MN 8/9 boundary, occurs near the level of localities 4 and 94. Future discoveries could move this LSD even lower.

Although it is not possible for us to derive an exact age for the lowest stratigraphic datum (LSD) of *Hipparion* and the apparent MN 8 to MN 9 boundary in our section, we are able to provide an estimate. These events appear to be bracketed between localities 64 and 4. Sen (1991) found 11.1 m of reversely magnetized sediments in the top of his section on Sinap Tepe. Our preliminary correlation would link this interval to Chron C4Ar.3r, which has a duration of 0.098 m.y. (see Steininger et al., this volume). Although Sen did not find a normal transition marking the top of this interval, if we assume that the 11.1 m level equals the maximum duration of 0.098 m.y., then locality 64 at approximately 87 m below what we interpret to be the top of Chron C5n has an estimated age of 10.51 Ma (87 m ≡ 0.098 m.y./11.1 m + 9.74 Ma); the estimated age for locality 4 and the LSD of *Hipparion* at about 77 m below the top of Chron C5n is 10.42 Ma (77 m ≡ 0.098 m.y./ 11.1 m + 9.74 Ma). If the actual boundary of the MN 8/9 transition lies halfway between localities 64 and 4, then its age estimate is about 10.46 Ma. Other age estimates are also possible. If we assume that the reversely magnetized interval of 11.1 m represents C5n.1r instead of C4Ar.3r, its older limit of 9.92 Ma would increase the age estimate of the midpoint between localities 64 and 4 to 10.64 Ma. Future work in the Sinap Formation will aim for a finer resolution of this question.

Dates for the *Hipparion* FAD vary widely across the Old World, with some of the variance due to poor dating control (see Sen 1989; Swisher, this volume; Woodburne et al., this volume). It is interesting here to review the age of perhaps one of the best-dated LSDs of *Hipparion*, that reported by Badgley et al. (1986) from the Siwaliks of Pakistan. They recorded the LSD in the Kaulial Kas section (locality Y395) as occurring at the 430 m level within the 940 m long normal interval of Chron 5n, and used the duration (1.44 m.y.) and upper boundary date (8.56 Ma) given by Mankinen and Dalrymple (1979) to estimate an age of 9.22 Ma (430 m/940 m ≡ 1.44 m.y. + 8.56 Ma). When this calculation is repeated using the new duration and boundary date estimates of Steininger et al. (this volume), the age is estimated to be 10.29 Ma (430 m/940 m ≡ 1.21 m.y. + 9.74 Ma). An older estimate is also available. Barry et al. (1982:113) note that the lowest occurrence of *Hipparion* in the Siwaliks is at locality Y550 in the Ratha Kas section, and this locality is 120 m lower than locality Y395 in Kaulial Kas. The age estimate for the *Hipparion* LSD at Y550 is 10.45 Ma (550 m/940 m ≡ 1.21 m.y. + 9.74 Ma). Pilbeam et al. (this volume) provide another estimate for the *Hipparion* LSD based on the Gabhir Kas section located to the south of Kaulial Kas and Ratha Kas, and it has an age of 10.7 Ma. Given the various uncertainties that underlie all of these calculations, perhaps the most important conclusion to draw from this comparison of the Sinap and Siwalik sections is that both *Hipparion* LSDs fall within the middle to lower-middle portion of Chron 5n. While any absolute age estimate will always depend on where the tie points for the time scale are placed, and how these are interpolated to provide boundary estimate for each chron, it seems likely that *Hipparion* entered the Old World within the interval of the middle to lower-middle portion of Chron 5n.

Another local occurrence of interest is that of the late Miocene hominoids. The *Ankarapithecus meteai* lower face specimen (MTA 2125) was recovered from locality 12 at the 105-m level on Delikayinçak Tepe (see fig. 6.7), while a partial mandible reported by Ozansoy (1965) is probably from locality 8A on Sinap Tepe. If our correlation is correct, the estimated age of *Ankarapithecus meteai* is about 9.76 Ma at locality 12 (2 m ≡ 0.098 m.y./11.1 m + 9.74 Ma). This estimate is broadly similar to the age of the bulk of the *Sivapithecus* material from the Nagri Formation of the Siwaliks of Pakistan (Barry 1986), and about 3 Ma younger than the oldest *Sivapithecus* material from the Chinji Formation of the Siwaliks (Kappelman et al. 1991).

Our preliminary results for the Kavak Dere section find fairly good support in the relative age correlations based on the fauna. However, because only preliminary identifications for the ruminants and rhinocerotids are available, the conclusions must be considered extremely tentative. The fossil-poor lower part of the Kavak Dere sequence is especially undiagnostic, but the rich fauna of locality 49 from

I¥bek to the west of the Sinap area may be of equivalent age to Kavak Dere localities 51 and 63 as indicated by the cooccurrence of *Chilotherium* cf. *C. samium* (localities 51 and 49) and *Prostrepsiceros* (localities 63 and 49). The rhinoceroses *Chilotherium* cf. *C. samium* and *Acerorhinus* cf. *A. zernowi* may suggest an age near the Vallesian–Turolian transition for locality 49, and perhaps for lower Kavak Dere (cf. Heissig 1975). The ruminant fauna also supports such an age for locality 49, with the cooccurrence of a small *Gazella* similar to *G. ancyrensis* from the middle Sinap (Tekkaya 1973) with the Turolian forms *Prostrepsiceros zitteli* and *Palaeoreas asiaticus* (cf. Köhler 1987). *Prostrepsiceros* cf. *P. ?rotundicornis* from locality 63 appears to be morphologically more primitive than the Pikermi form and might also fit a transitional Vallesian–Turolian age (re. Bernor et al., this volume).

Ozansoy (1965) originally correlated the entire Kavak Dere fauna with that of the lower Sinap member, and this was first estimated to predate the Vallesian. Later work demonstrated abundant *Hipparion* in the upper Kavak Dere localities; its presence suggests that the fauna is younger and probably early Turolian in age (see Sickenberg et al. 1975). The upper part of the Kavak Dere sequence is represented by a rich large mammal fauna, and an age of MN 10 or MN 11 would be compatible with the identifications so far available (*Chilotherium* cf. *C. habereri*, *Stephanorhinos pikermiensis*–group, *Ceratotherium* cf. *C. neumayri*, *Acerorhinus*, Boselaphini spp., and *Protoryx* sp.), although the occurrence of a good specimen of *Caprotragoides stehlini* from locality 34 is unexpected; this species is known from the middle Miocene in Western Europe (Köhler 1987), but the genus is represented by late Miocene species in Africa and the Siwaliks (Thomas 1981; 1984). The especially well known site of Çoban Pinar (locality 42) is commonly placed in MN 12 (e.g., Mein 1977). Our discovery of *Ictitherium* cf. *pannonicum* from Çoban Pinar confirms the Turolian age placement of this locality. The age estimates for Kavak Dere as based on the preliminary paleomagnetic reversal stratigraphy fall within the middle to younger limits of the Turolian, or MN 11–MN 12, based on the correlation given by Steininger et al. (1989), and it is likely that Çoban Pinar lies at the younger end of this range. Future work will concentrate on providing more secure age estimates for these localities.

Conclusions

Our preliminary work has shown that a remarkably rich, diverse, and well-preserved fossil mammal fauna ranging from the middle Miocene to the Pleistocene is present in the sediments of the Sinap Formation. This formation contains some of the longest conformable stratigraphic sections known in the area considered in this volume. Paleomagnetic reversal stratigraphies for several sections in the Sinap region suggest direct ties between these faunas and the time scale, and radioisotopic dating of the abundant volcanics promises to make these age correlations even more precise. Few if any areas in Western Eurasia outside of the largely endemic Iberian Peninsula offer the same potential for tying the middle of the mammalian Neogene biochronology to an absolute time scale. Anatolia lies at the paleobiogeographical crossroads of Europe, Asia, and Africa, and a well-documented and well-dated fauna will permit more precise intercontinental correlations, and provide a rare opportunity for the fine-scale testing of the effects of global climatic change on faunas of the past.

Acknowledgments

Our work in the Sinap Formation represents a large collaborative effort, and the results presented here reflect the work of many individuals. We would like to especially thank Dr. Lawrence Martin (State University of New York at Stony Brook, U.S.A.), with whom we initiated this project. Many other individuals have participated in and assisted with the fieldwork, laboratory analysis, and faunal identifications. We extend our sincere thanks to P. Andrews, H. Celebi, D. Daegling, A. Ersoy, P. Gibbard, I. Gençtürk, J. Jernvall, B. Richmond, P. Stubblefield, M. Watabe, and R. Wilson, with special thanks to D. Trombatore for bibliographic assistance, and W. Gose for assistance in the paleomagnetic laboratory. J. C. Barry, R. L. Bernor, L. Flynn, and D. Pilbeam reviewed the chapter and offered very useful suggestions and comments. The fieldwork and laboratory work were funded by the following grants: National Science Foundation EAR 9304302 to Kappelman; the L.S.B. Leakey Foundation to Kappelman and Martin; the Academy of Finland to Fortelius; and Ankara University to Alpagut. The research is supported by the General Directorate of Antiquities, T.C. Ministry of Culture and Tourism, and the University of Ankara, Faculty of Language, History, and Geography, Department of Paleoanthropology. To all of these individuals and institutions we extend our very sincere thanks for their generous support and assistance.

LITERATURE CITED

Agustí, J. 1989. The Miocene rodent succession in eastern Spain: A zoogeographical appraisal. In *European Neogene Mammal Chronology*, ed. E. H. Lindsay, V. Fahlbusch, and P. Mein, pp. 375–404. New York: Plenum.

Alpagut, B., M. Fortelius and J. Kappelman. 1992. *Survey Report for the Sinap Formation Project (Ankara, Turkey) 1991*, pp. 315–29. Ankara: T.C. Arastirma Kültür Bakanligi Antitlar ve Müzeler Genel Müdürlügü Sonuclari Toplantisi.

——. 1993. *Survey Report for the Sinap Formation Project (Ankara, Turkey) 1992*, pp. 237–56. Ankara: T.C. Arastirma Kültür Bakanligi Antitlar ve Müzeler Genel Müdürlügü Sonuclari Toplantisi.

Alvarez, Sierra, M. A., E. García Moreno, M. Lopez Martinez, and R. Daams. 1987. Biostratigraphy and palaeoecological interpretation of micromamal faunal successions in the upper Aragonian and Vallesian (middle-upper Miocene) of the Duero Basins (n. Spain). In *Proceedings of the VIIIth RCMNS Congress*, pp. 517–22. Annales Instituti Geologici Publici Hungarici 70.

Angelier, J., J. F. Dumont, H. Karamandersei, A. Poisson, S. Sinsek, and S. Uysal. 1981. Analyses of fault mechanisms and expansion of Southwestern Anatolia since the late Miocene. *Tectonophysics* 75:T1–T9.

Badgley, C. E., L. Tauxe, and F. L. Bookstein. 1986. Estimating error of age interpolation in sedimentary rocks: A bootstrap method. *Nature* 319:139–41.

Barry, J. 1986. A review of the chronology of Siwalik hominoids. In *Primate Evolution*, ed. J. C. Else and P. C. Lee, pp. 93–106. Cambridge: Cambridge University Press.

Barry, J. and L. Flynn. 1989. Key biostratigraphic events in the Siwalik Sequence.In *European Neogene Mammal Chronology*, ed. E. H. Lindsay, V. Fahlbusch, and P. Mein, pp. 557–71. New York: Plenum.

Barry, J., N. M. Johnson, S. M. Raza, and L. L. Jacobs. 1985. Neogene mammalian faunal change in Southern Asia: Correlations with climatic, tectonic, and eustatic events. *Geology* 13:637–40.

Barry, J., E. H. Lindsay, and L. L. Jacobs. 1982. A biostratigraphic zonation of the middle and upper Siwaliks of the Potwar Plateau of Northern Pakistan. *Palaeogeography, Palaeoclimatology, Palaeoecology* 37:95–130.

Becker-Platen, J. D., L. Benda, and P. Steffens. 1977. Litho- und biostratigraphische Deutung radiometrischer Altersbestimmungen aus dem Jungtertiär der Türkei (Känozoikum und Braunkohlen der Türkei, 18). *Geologisches Jahrbuch* B25:139–67.

Benda, L., F. Innocenti, R. Mazzuoli, F. Radicati, and P. Steffens. 1974. Stratigraphic and radiometric data of the Neogene in Northwest Turkey. (Cenozoic and Lignites in Turkey 16.) *Zeitschrift für Deutschen Geologischen Gesellschrift* 125:183–93.

Bernor, R. L., N. Solounias, C. C. Swisher III, and J. A. Van Couvering. This volume. The correlation of three classical "Pikermian age" faunas—Maragheh, Samos, and Pikermi—with the European MN unit system.

Bernor, R. L. and H. Tobien. 1990. The mammalian geochronology and biogeography of Paşalar (middle Miocene, Turkey). *Journal of Human Evolution* 19:551–68.

Besang, C., F.-J. Eckhardt, W. Harre, H. Kreuzer, and P. Müller. 1977. Radiometrische altersbestimmungen an neogenen eruptivgesteinen der Türkei. *Geologisches Jahrbuch* B25:3–36.

Cande, S. C. and D. V. Kent. 1992. A new geomagnetic polarity time scale for the late Cretaceous and Cenozoic. *Journal of Geophysical Research* 97(B10):13917–951.

Cande, S. C. and J. L. LaBrecque. 1974. Behaviour of the Earth's paleomagnetic field from small-scale marine magnetic anomalies. *Nature* 247:26–28.

Dunlop, D. J. 1972. Magnetic mineralogy of unheated and heated red sediments by coercivity spectrum analysis. *Geophysical Journal of the Royal Astronomical Society* 27:37–55.

Erol, O. 1981. Neotectonic and geomorphological evolution of Turkey. *Zeitschrift für Geomorphologie*, Neue Folge, Supplement Band 40:193–211.

Fisher, R. A. 1953. Dispersion on a sphere. *Proceedings of the Royal Society* 217:295–305.

Gürbüz, M. 1981. İnönü (KB Ankara) Orta Miyosenindeki *Hemicyon sansaniensis* (Ursidae) turunum tanimlanmasi ve stratigrafik yayilimi. *Türkiye Jeoloji Kurumu Bülteni* C24 85–90.

Harland, W. B., A. V. Cox, P. G. LLewellyn, C. A. C. Picton, A. G. Smith, and R. Walters. 1982. A *Geologic Time Scale*. Cambridge: Cambridge University Press.

Heissig, K. 1975. Rhinocerotidae aus dem Jungtertiar Anatoliens. *Geologisches Jahrbuch* B15:151–54.

Inci, U. 1991. Miocene alluvial fan-alkaline playa lignite- trona-bearing deposits from an inverted basin in Anatolia: Sedimentology and tectonic controls on deposition. *Sedimentary Geology* 71:73–97.

Jacobs, L. L., L. Flynn, W. Downs, and J. Barry. 1989. *Quo vadis, Antemus?* The Siwalik muroid record. In *European Neogene Mammal Chronology*, ed. E. H. Lindsay, V. Fahlbusch, and P. Mein, pp. 573–86. New York: Plenum.

Kappelman, J., J. Kelley, D. Pilbeam, K. A. Sheikh, S. Ward, M. Anwar, J. C. Barry, B. Brown, P. Hake, N. M. Johnson, S. M. Raza, and S.M. I. Shah. 1991. The earliest occurrence of *Sivapithecus* from the middle Miocene Chinji Formation of Pakistan. *Journal of Human Evolution* 21:61–73.

Kirschvink, J. L. 1980. The least-squares line and plane and the analysis of palaeomagnetic data. *Geophysical Journal of the Royal Astronomical Society* 62:699–718.

Köhler, M. 1987. Boviden des türkischen Miozäns (Känozoicum und Braunkohlen der Türkei 28). *Paleontologia i Evolució* 21:133–246.

Lindsay, E. H. The setting. In *European Neogene Mammal Chronology*, ed. E. H. Lindsay, V. Fahlbusch, and P. Mein, pp. 1–14. New York: Plenum.

Mankinen, E. E. and G. B. Dalyrimple. 1979. Revised geomagnetic polarity time scale for the interval 0–5 m.y. B.P. *Journal of Geophysical Research* 84:615–26.

Mein, P. 1975. Résultats du Groupe de Travail des Vertébrés. In *Report on Activity of the RCMNS Working Group (1971–1975)*, ed. J. Senes, pp. 78–81. Bratislava: SAV.

— —. 1977. Table 1. In *Round Table on Mastostratigraphy of the W. Mediterranean Neogene*, ed. M. T. Alberdi and E. Aguirre. Trabajos Sobre Neogeno Cuaternario 7.

— —. 1989. Updating of MN zones. In *European Neogene Mammal Chronology*, ed. E. H. Lindsay, V. Fahlbusch, and P. Mein, pp. 73–90. New York: Plenum.

Moyà Solà, S. and J. Agustí. 1989. Bioevents and mammal successions in the Spanish Miocene. In *European Neogene Mammal Chronology*, ed. E. H. Lindsay, V. Fahlbusch, and P. Mein, pp. 357–74. New York: Plenum.

Öngür, T. 1976. Kizilcahamam, Camlidere, Celtikci ve Kazan dolayinin jeoloji durumu ve jeotermal enerji olanaklari. Rapp. inédit, MTA, Ankara.

Opdyke, N., P. Mein, E. Moissenet, A. Pérez-González, E. H. Lindsay, and M. Petko. 1989. The magnetic stratigraphy of the late Miocene sediments of the Cabriel Basin, Spain In *European Neogene Mammal Chronology*, ed. E. H. Lindsay, V. Fahlbusch, and P. Mein, pp. 507–14. New York: Plenum.

Ozansoy, F. 1957. Faunes de Mammifères du Tertiaire de Turquie et leurs révisions stratigraphiques. *Bulletin of the Mineral Research Exploration Institute Turkey* (Foreign Edition) 49:29–48.

——. 1965. Étude des gisements continent aux et de Mammifères du Cénozoïque de Turquie. *Mémoires Societe Géologique de France*, N.S., 44:1–92.

Pickford, M. 1986. Cainozoic paleontological sites of Western Kenya. *Munchner Geowissenschaftliche Abhandlungen*, Reihe A. 8:78–81.

Pilbeam, D., M. Morgan, J. C. Barry, and L. Flynn. This volume. European MN units and the Siwalik faunal sequence of Pakistan.

Savage, D. E. and D. E. Russell. 1983. *Mammalian Paleofaunas of the World*. New York: Addison-Wesley Publishers.

Sen, S. 1977. La fauna de rongeurs pliocène Çalta (Ankara, Turquie). *Bulletin du Museum National d'Histoire Naturelle, Paris*, 3ᵉ sér., n° 465, Sciences de la Terre 61:89–172.

——. 1989. *Hipparion* datum and its chronologic evidence in the Mediterranean area. In *European Neogene Mammal Chronology*, ed. E. H. Lindsay, V. Fahlbusch, and P. Mein, pp. 495–505. New York: Plenum.

——. 1991. Stratigraphie, faunes de mammifères et magnétostratigraphie du Néogène de Sinap Tepe, Province d'Ankara, Turquie. *Bulletin du Museum National d'Histoire Naturelle, Paris*, 4e séries, 12:243–77.

——. This volume. Present state of magnetostratigraphic studies in the continental Neogene of Europe and Anatolia.

Sen, S. and J.-P. Valet. 1986. Magnetostratigraphy of late Miocene continental deposits in Samos, Greece. *Earth and Planetary Science Letters* 80:167–74.

Sen, S., J.-P. Valet, and C. Ioakim. 1986. Magnetostratigraphy and biostratigraphy of the Neogene deposits of Kastellios Hill (Central Crete, Greece). *Palaeogeography, Palaeoclimatology, Palaeoecology* 53:321–34.

Sickenberg, F., J. D. Becker-Platen, L. Benda, D. Berg, B. Engesser, W. Gaziry, K. Heissig, K. A. Hünermann, P. Y. Sondaar, N. Schmidt-Kittler, K. Staesche, U. Staesche, P. Steffens, and H. Tobien. 1975. Die Gliederung des höheren Jungtertiärs und Altquartärs in der Türkei nach Vertebraten und ihre Bedeutung für die internationale Neogen-Stratigraphie. *Geologisches Jahrbuch* B15:1–167.

Steininger, F. F., R. L. Bernor, and V. Fahlbusch. 1989. European Neogene marine/continental chronologic correlations. In *European Neogene Mammal Chronology*, ed. E. H. Lindsay, V. Fahlbusch, and P. Mein, pp. 15–46. New York: Plenum.

Swisher, C. C., III. This volume. New ^{40}Ar/^{39}Ar dates and their contribution toward a revised chronology for the late Miocene nonmarine of Europe and West Asia.

Tauxe, L. and N. D. Opdyke. 1982. A time framework based on magnetostratigraphy for the Siwalik sediments of the Khaur area, Northern Pakistan. *Palaeogeography, Palaeoclimatology, Palaeoecology* 37:43–61.

Tekkaya, I. 1973. Une nouvelle espéce de *Gazella* de Sinap moyen. *Bulletin of the Mineral Research and Exploration Institute of Turkey* 80:118–43.

Thomas, H. 1981. Les bovidés miocénes de la formation de Ngorora du Bassin de Baringo (Kenya). *Proceedings, Koninklijke Nederlandse Akademie van Wetenschappen* B84:335–409.

——. 1984. Les Bovidés anté-hipparions des Siwaliks inférieurs (plateau du Potwar, Pakistan). *Mémoires de la Société géologique de France* (N.S.) 145:1–68.

Ünay, E. 1981. Middle and upper Miocene rodents from the Bayraktepe section (Çanakkale, Turkey). *Proceedings, Koninklijke Nederlandse Akademie van Wetenschappen, Amsterdam*, Series B, Physical Sciences. 83:399–418.

Ünay, E. and H. de Bruijn. 1984. On some Neogene rodent assemblages from both sides of the Dardanelles (Turkey). *Newsletters on Stratigraphy* 13:119–32.

Watson, G. S. 1956. A test for randomness. *Monthly Notices of the Royal Astronomical Society*. Geophysics Supplement 7:160–61.

Werdelin L. and N. Solounias. 1991. The Hyaenidae: Taxonomy, systematics, and evolution. *Fossils & Strata* 30:1–104.

Woodburne, M. O., G. Theobald, R. L. Bernor, C. C. Swisher III, H. König, and H. Tobien. This volume. Advances in the geology and stratigraphy at Höwenegg, southwestern Germany.

7

European MN Units and the Siwalik Faunal Sequence of Pakistan

D. PILBEAM, M. MORGAN, J. C. BARRY, AND L. FLYNN

Fossiliferous rocks in Eurasia, especially Western Europe, Southwestern Asia, and Indo-Pakistan record important Neogene faunas within different depositional contexts. In a few instances, including the Siwalik series rocks of the Potwar Plateau in Pakistan, deposition is essentially continuous through almost all of the Neogene. In contrast, in most other areas localities—although frequently more productive and taxonomically rich than those of Pakistan—cannot be placed within secure local or regional stratigraphic frameworks. It is of interest, then, to consider the differences between the two kinds of sequences, contrasting briefly both their strengths and weaknesses, and we do so by examining the European MN faunal units from the perspective of a Siwalik biostratigraphy made possible by a long and relatively continuous depositional sequence.

We first present a general overview of the Siwalik sequence. Then we discuss some of the problems in determining the age of the first appearance of hipparionine horses in the Siwalik sequence, and review whether their appearance there is contemporaneous with their first appearance elsewhere throughout the Old World. We use this well-studied first appearance datum (FAD) as an example of more general problems of correlation and dating, and therefore of the establishment of patterns of faunal change.

The Siwalik Sequence

The fossiliferous Neogene rocks of Northern India and Pakistan are singular in their degree of completeness and represent almost the entire Neogene from about 22 Ma to less than 2 Ma (Keller et al. 1977; Opdyke et al. 1979; Azzaroli and Napoleone 1982; Johnson et al. 1983; Tandon et al. 1984; Johnson et al. 1985; Friedman et al. 1992). This uniquely long faunal sequence records numerous vertebrate taxa and biotic events, and affords a most unusual opportunity to document the pattern of faunal change in the South Asian biogeographic realm over a 20 m.y. interval and to study its dynamics.

The most extensive of the Neogene sediments, the Siwalik formations are widely distributed through Pakistan (fig. 7.1), where they range in age from 18 to 2 Ma (Keller et al. 1977; Opdyke et al. 1979; Johnson et al. 1985). Within Pakistan they are best exposed in the Potwar Plateau. Coeval fossiliferous exposures are found at Dera Bugti, Gaj, Sehwan, Banda Daud Shah, and the Zinda Pir Dome (fig. 7.2). The fossiliferous rocks variously exposed in the Potwar Plateau have a composite thickness of 5 km and have yielded an abundant fauna. Since 1973 we have collected and catalogued 40,000 specimens from over 900 localities on the Plateau. About 50% of these specimens are identified to genus level and 40% to species. More than 80% are mammals.

Siwalik sediments are fluvial and include deposits typical of such environments. The mix of channel sands and finer overbank deposits fluctuated through time in response to regional tectonic and perhaps environmental factors. These fluctuations are used to define the formational boundaries. At least 15 stratigraphic sections over 1,000 m long have been measured by us and together with more than twenty shorter sections have contributed to the development of a comprehensive stratigraphic framework. Many of these sections have been sampled for paleomagnetism and used to build a comprehensive chronostratigraphy for the Siwaliks. Most fossils occur in concentrations that are in or adjacent to small-scale channels and crevasse splays on the floodplains, while occurrences in paleosols are rarer. Most localities appear to be time-averaged attritional accumulations and represent probably only hundreds to thousands of years. Localities can usually be tied to the Geomagnetic Reversal Time Scale (GRTS) with a precision of about 100 ky (Flynn et al. 1990); of course,

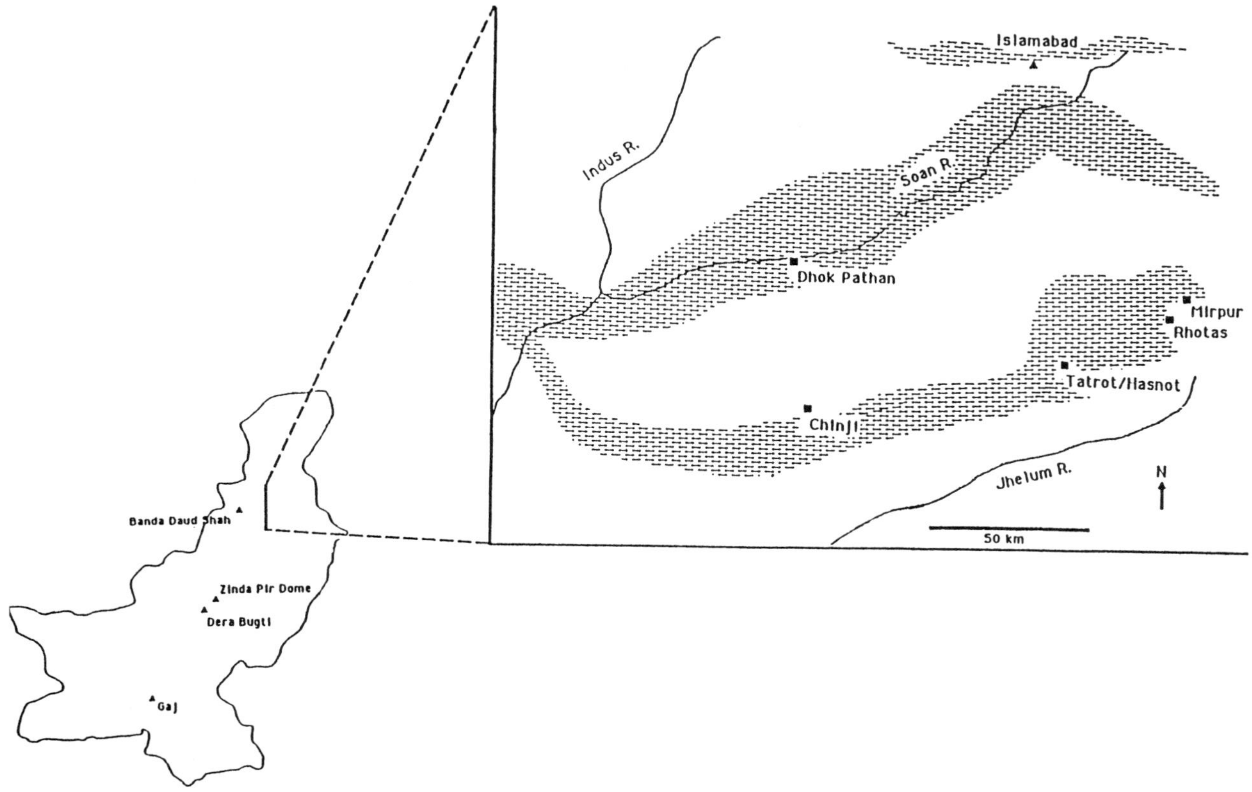

FIGURE 7.1 Map of Pakistan with primary exposures of Siwalik sediments. Inset details Potwar Plateau, with the approximate outcrop area of Miocene sediments indicated by patterning.

locality dates are dependent on the accuracy of the particular time scale, here Berggren et al. 1995, which is based on Cande and Kent (1995).

We have been relatively unsuccessful in recovering well-preserved plant fossils, although detailed lateral studies of paleosols and isotopic analyses of paleosol carbonates permit some reconstruction of vegetation and habitats. Permanent forests and woodlands with some interspersed grasses (mostly C_3) were present before about 9 Ma, after which wooded grasslands became widespread on floodplains (Quade et al. 1989; Morgan et al. 1994; Quade and Cerling, in press). The environmental transition from ca. 9 to ca. 6 Ma in the Potwar, as documented by isotopes, may have been a response to both regional and global climatic shifts. There is no evidence for any significant Miocene environmental patchiness within the Potwar Plateau; indeed, we believe it likely that the plateau is representative of a significantly larger South Asian faunal province.

Faunal Change in the Siwaliks

It is possible to identify a broadly defined "Siwalik fauna," actually a series of very similar successive faunal assemblages, lasting from 18 Ma to ca. 6 Ma, which can be viewed as a precursor of the modern Oriental provincial fauna. The Siwalik fauna is characterized by muroid (cricetid, rhizomyid, murid), sciurid, and ctenodactylid rodents, hipparionines, ruminants (tragulids, bovids, and giraffids), hyaenas, and some archaic taxa such as creodonts and anthracotheres that are rare or absent from contemporaneous faunas elsewhere.

A major goal of any study of a temporally long faunal sequence is to understand the dynamics of biotic change. To that end, we have begun to analyze patterns of faunal change within the Potwar sequence, concentrating on first and last occurrences, species richness, body size, and relative abundances (Barry et al. 1990, 1991; Flynn et al. 1995; Jacobs et al. 1989; Morgan et al. 1995). We have analyzed data in 0.5 m.y. intervals over the period between ca. 18 and 8 Ma. These analyses have primarily concentrated on artiodactyls and rodents, the orders which contribute the majority of Siwalik species and specimens. For each 0.5 m.y. interval we have made an assessment of data quality, with a view to estimating how adequately our record documents the actual species composition of the fauna at successive intervals. Results indicate substantial changes through time in species richness, relative abundances, and body size. For example, we have documented increases and decreases in the relative abundances of bovids and

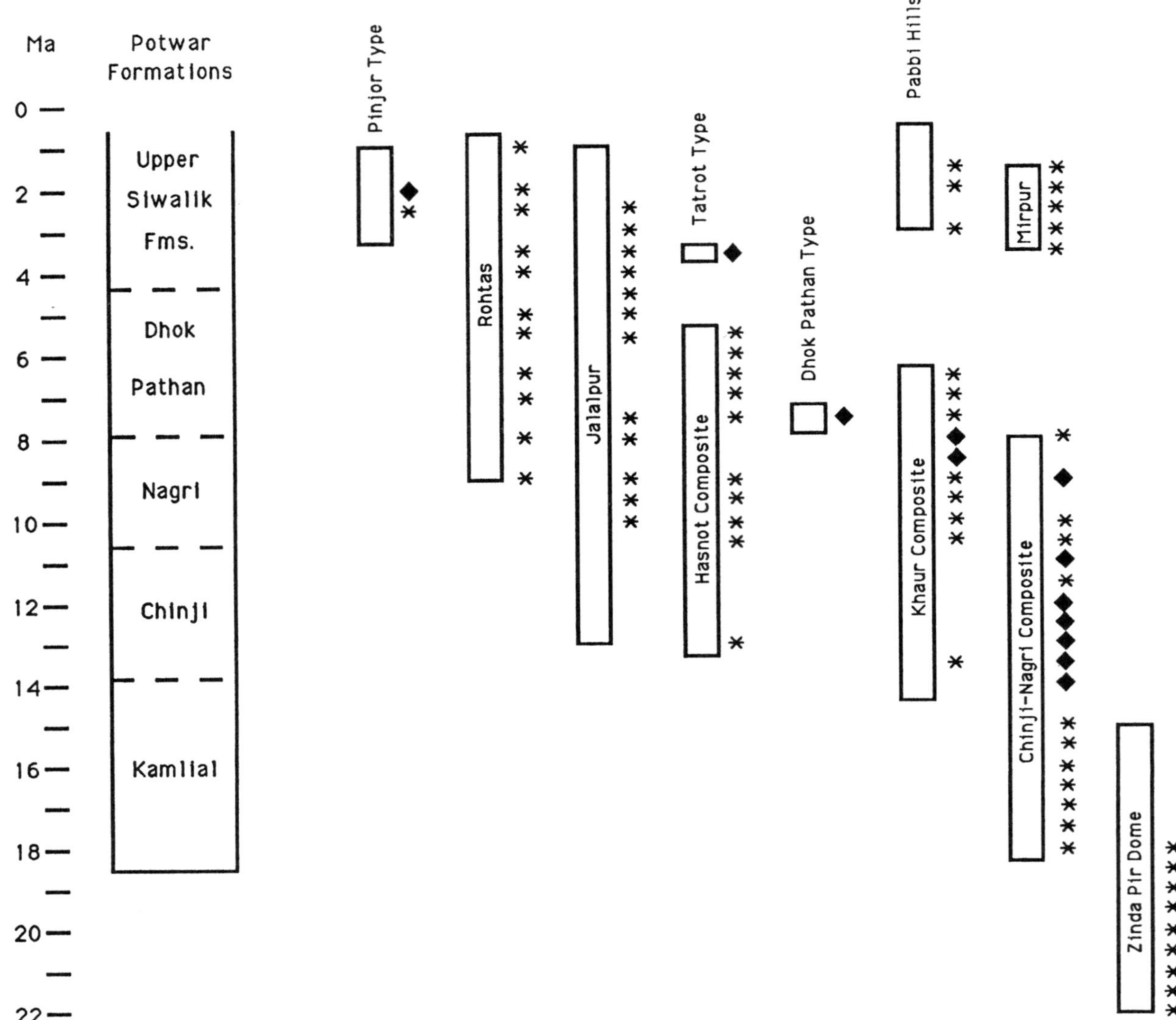

FIGURE 7.2 Stratigraphic sections and occurrences of fossils in Siwalik sediments. Asterisks indicate levels with good, fair, and poor representation; closed diamonds indicate levels with very good representation.

cricetids. There is a general trend toward reduction in species diversity between 13 and 8 Ma, and there is an increase in ruminant and muroid body size.

Our data, which are taken from Flynn et al. (1995) and include species in addition to artiodactyls and rodents, show that both first and last occurrences are found throughout the sequence (fig. 7.3), but are also clustered episodically rather than evenly or randomly distributed through time. First and last occurrence peaks do not generally coincide: a first occurrence maximum with a minimum of last occurrences is centered between 14 and 13.5 Ma, while a last occurrence maximum with a minimum of first occurrences falls between 13 and 12.5 Ma. After 10.5 Ma there are large numbers of first and last occurrences, but they are less clustered in time. Some of the peaks may be associated with regional and/or global environmental

shifts. Figure 7.3 plots first and last occurrences for a number of selected taxa, as noted in the caption, against the GRTS, Siwalik formations, and the European MN units.

Strengths and Weaknesses of the Siwalik Faunal Record

The Neogene record as described here has both strengths and weaknesses. Its strengths can be briefly listed. For a continental record it is relatively complete, and it spans a critical period for mammalian evolution, the Neogene. The faunal record runs from 22 to 2 Ma, and while it is best between 18 and 8 Ma, it is fossiliferous essentially throughout. We are able to generate both a biostratigraphic framework based on superposition and a partial chrono-

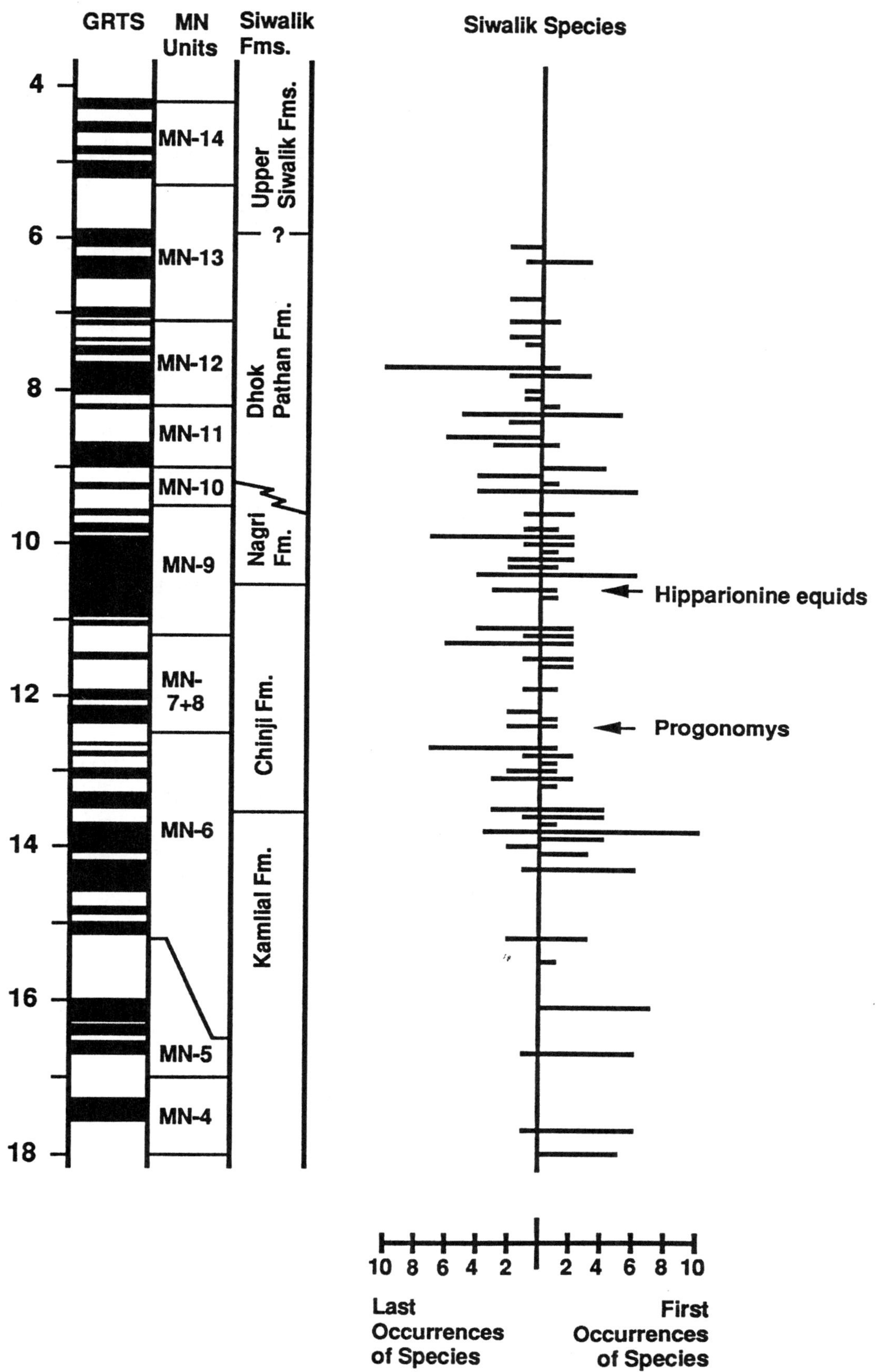

FIGURE 7.3 First and last occurrences of species with well-constrained stratigraphic ranges from the Siwaliks (n = 126, including all rodents, selected artiodactyls, primates, carnivores, creodonts, proboscideans, and perissodactyls) shown with Siwalik Formation boundaries, European MN zones, and the GRTS. Note the general lack of correlation between MN zone boundaries and peaks of faunal change in the Siwaliks.

stratigraphy in the form of faunal units (Barry et al. 1982). An extensive set of paleomagnetic determinations has provided good correlation within and between the various Siwalik sequences, and between the Siwaliks and, through the GRTS, the global record. We currently have a reasonable understanding of local and regional scale environments and the ways in which they change through time. The Potwar Plateau, and indeed the entire regional Neogene sequence as recorded in the Indo-Pakistan subcontinent, is apparently relatively homogeneous and shows little if any large-scale patchiness. Thus broadly contemporaneous faunas are basically similar throughout the region. Finally, we have begun to develop hypotheses concerning possible relationships among patterns of climatic, environmental, and faunal change on regional and global scales.

So much for the strengths. But there are also ways in which the Siwalik record, as well as our interpretations of it, falls short of the ideal. Some of these can be addressed, while others are more stubborn reflectors of the nature of the Siwalik record itself. We have no Höweneggs among our localities. The Siwalik fluvial systems winnowed and scattered the faunal remains, and as a result the sediments contain mostly unassociated and incomplete fossils. Thus our identifications are less certain. Further, as in all fossil sequences, it is difficult to know how closely a known stratigraphic range approximates the true stratigraphic range for all but the most common taxa. In this regard we have conducted intensive biostratigraphic surveys in order to carefully document the first or last occurrences of a few common taxa (for example, hipparionines and hippopotamids). We believe that for these taxa we have established the true first and last appearances, but such work is very labor-intensive, and therefore less likely to be extended to less common taxa.

Our systematic studies also need to be expanded. Currently, we consider that alpha taxonomy of the thryonomyids, muroids, creodonts, hominoids, and other catarrhines is adequate, but considerably more needs to be done before we are able, for example, to provide reasonable estimates of the geographic origins of immigrants and destinations of emigrants. Community reconstruction is still at an early stage of investigation, yet we need to have such a record for a large number of half-million-year time periods in order to make biologically adequate inferences about community change through time. We lack a good plant record, although we continue to search for pollen and other botanical data, and hence our environmental reconstructions are less complete than we would like. Before we can produce the next generation of hypotheses involving possible links between environmental and faunal change we must improve our faunal record, especially in the later Miocene between 8 and 6 Ma. Finally, our localities and taxa are dated almost exclusively through paleomagnetism, so we are dependent on the accuracy of successive global time scales and are hostage to their continuing modification and refinement.

The Western Eurasian Faunal Record

The Western Eurasian Miocene sequence, as reviewed in this volume, also has strengths and weaknesses. It is considerably richer in its taxonomic base than the Siwalik sequence, or indeed any other comparably sized regional sequence, with excellent fossil material as far as completeness, sample size, and systematics are concerned. The paleobotanical and palynological records are also good, and environmental reconstructions adequate. The completeness of animal and plant records is such that the quality of community reconstructions is high. Across Western Eurasia, broadly defined, there is a reasonable understanding of faunal provinciality. Through the use of MN units there is good general knowledge of faunal sequence, correlation, and dating both of key faunas and of particular taxa.

On the other hand, there are correlation problems created by several factors. Within European subregions there are few even moderately long and adequately fossiliferous sequences, so actual superpositional ordering of faunal events is rare. The existence of spacial heterogeneity is also a problem: taxa may appear at different times in different parts of Europe because of ecological differences. There are few good radiometric dates, and few paleomagnetic sequences long enough to permit unambiguous correlation to the GRTS. The MN unit system is an essential biochronological tool for locating sites within a particular time interval, and hence for correlation, but it is province-specific and problems arise therefore when provincial boundaries shift through time. Further problems surround the understandable tendency to see MN unit boundaries as recording clusters of faunal events (first or last occurrence pulses), and with seeing them as reflecting significant regional or global environmental events. Below we discuss this issue in greater detail.

Correlation Problems: GRTS and Radiometric Dates

Correlating and dating particular sites or significant events is a critical challenge for us all, and this is often difficult. As an example of the problem we all face, consider the first occurrence of hipparionines in our records. After intensive targeted field surveys, we are confident that for the Siwaliks we have documented localities that must lie temporally very close to the first appearance of these horses in the region. These sites fall within the long normal Chron C5n, and are stratigraphically 20% above the base of that magnetic interval. Given the ages for the boundaries in the time scale used in this volume (Berggren et al. 1995; Cande and Kent 1995), and assumptions about sediment

TABLE 7.1 *Stratigraphic Position and Large Mammal Content of Gabhir Kas Localities Stratigraphically near the First Occurrence of Hipparionines*

Loc.	Height	Equid.	Rhino.	Tragl.	Bovid.	Giraf.	Suid.	Anthrac.
Y779	1150–1153	X	X	X	X	X	X	X
Y635	1127–1135	X				X		
Y799	1127–1130	X	X				X	
Y646	1097–1102		X					
Y643	1085–1093			X				
Y798	996–998				X	X		
Y636	975–990							
Y804	970–972				X	X	X	

accumulation rates, we now calculate a date of 10.7 Ma for the *"Hipparion"* Datum in the Siwaliks. This new date replaces previous estimates of approximately 9.5 Ma (Barry et al. 1982; Barry and Flynn 1989), and, like those dates, is based on stratigraphically older occurrences of this taxon than those used by Badgley et al. (1986), who interpolated an age of 9.2 Ma. It is of more than historical interest that this "date" has varied with changes in the GRTS. For example, using the same stratigraphic occurrences under the Mankinen and Dalrymple (1979) time scale our date for the *"Hipparion"* Datum in the Siwaliks was 9.5 Ma, while with Berggren et al. (1985) it was 10.1 Ma and with Cande and Kent (1992) it became 10.5 Ma. Clearly, unless a GRTS age estimate has a high probability of being correct, there will be problems with correlating paleomagnetically dated areas (Potwar Plateau of Pakistan, Sinap series of Turkey) with radiometrically dated localities (Höwenegg, Germany, and Ch'orora, Ethiopia), at least at the level to which we all aspire in reconstructing a global continental faunal record.

The oldest records of hipparionines in the Siwaliks are from a section measured by Brian Willis in Gabhir Kas near Sethi Nagri. This section spans the Kamlial through Dhok Pathan Formations and was subjected to intensive paleomagnetic sampling by Johnson et al. (1985). Because of a too-sparse sampling density in the uppermost 450 meters, correlation to the geomagnetic time scale is difficult in the youngest segments of the Gabhir section. But below this level, Johnson et al. (1985) found a long interval of normal polarity that can confidently be correlated to C5n. The top of this normal zone lies between a reversed site at between 1,583 and 1,586 meters and a normal site at between 1,603 and 1,607 meters on Willis's section, and following convention can be placed at the midpoint at 1,595 meters. This transition could represent either the top of C5n.1n or C5n.2n. The poor sampling of the magnetic record above makes it difficult to be certain. The bottom of the Gabhir long normal interval lies between a reversed site at between 991 and 997 meters and a normal site at between 1,037 and 1,040 meters. Following the midpoint convention we place it at 1,016 meters. Since magnetic zones C5r.1n and C5r.2n can be confidently identified

below this transition in the magnetic log of Johnson et al. (1985), we are confident in correlating the base of the Gabhir long normal interval to the bottom of C5n.

Table 7.1 shows the stratigraphic position and large mammal content of localities in this section that are stratigraphically near the first occurrence of hipparionines. The stratigraphically oldest hipparionines are at localities Y635 and Y799, both of which are at approximately 1,130 meters in Willis's section. Localities Y646 and Y643 lie just below this level and do not have hipparionines. Using ages from the Berggren et al. (1995) and Cande and Kent (1995) time scale, and an assumption of constant sediment accumulation rates, we calculate a date of 10.71 Ma for localities Y635 and Y799, if the upper transition of the Gabhir long normal is the top of C5n.1n, and ages of 10.77 and 10.79 Ma for localites Y646 and Y643. If, instead, the upper transition is correlated to the top of C5n.2n (which we regard as less likely), then localities Y635 and Y799 are 10.75 Ma, while localities Y646 and Y643 are 10.80 and 10.84 Ma.

It is a convention among paleomagnetic stratigraphers to place magnetic transitions midway between the actual stratigraphic positions of a pair of reversed and normal sites. If instead of following this convention, we place the magnetic transitions at the highest or lowest possible position so as to get the oldest or youngest possible ages, then we get additional possible ages for the Siwalik hipparionines of 10.67 and 10.76 Ma for the C5n.1n correlation and 10.71 and 10.79 Ma for the C5n.2n correlation. These different ways of determining the age converge on a 10.7 or 10.8 Ma age for the Siwalik *"Hipparion"* Datum. The 10.7 Ma age is our preferred estimate as it results from what we regard as the most likely correlation for the top of the Gabhir long normal interval. It is, however, an estimate for the age of the oldest record of hipparionines in the Siwaliks. It is not necessarily the age of their first appearance, which depends more on evidence relating to their absence. None of the sites listed in Table 7.1 have abundant and diverse fauna, and the absence of equids in some sites could be a result of the problems of paleontological sampling. Localities Y646 and Y643 in particular are poor sites individually, but together might indicate the absence of horses at their level, given the usual ubiquity and high identifiability of equid fossils. Locality Y798 is also not a rich site, but it has bovids and giraffes, and we would expect to find equids at a site with these taxa. In addition, intensive biostratigraphic surveys at the level of Y643 and Y646 and below suggest that equids were truly absent from the Y643/Y646 and older horizons. Thus we think our documented first occurrences must lie temporally very close to the first appearance of these horses in the region.

Let us now turn and look more closely, for example, at the superb hipparionine material from Höwenegg, given a maximum age of 10.3 Ma through a series of high-precision radiometric determinations. The extent to which we can accept that this is older or younger than the Siwalik occurrences (or those from other adequately documented paleomagnetic sequences such as the Turkish Sinap series), or of radiometrically bracketed occurrences like Ch'orora, depends on the accuracy of the radiometric dates of specific localities, the dates used to calibrate particular critical parts of the GRTS ("tie-points"), the quality of the various other records contributing to that time scale, and the assumptions involved in the many interpolations and extrapolations necessary to generate any composite global record.

Many earlier time scales have used as a critical tie-point the age of the base of Chron C5n, because this portion of the GRTS is recognizable as a magnetic signature because of its length. Several sedimentary sequences record this transition and provide both paleomagnetic and radiometric determinations. Two in particular are the Kenyan Baringo Basin and some Icelandic sequences (Deino et al. 1990; McDougall et al. 1984). However, Cande and Kent (1992) use neither in calibrating their timescale, but instead bracket the later Neogene record with tie-points in the Pliocene and at the top of C5Bn, estimating the age of the latter as 14.8 Ma, based on two radiometric dates of 14.6 ±0.4 Ma and 15.0 ±0.3 Ma. This estimated date of 14.8 Ma may be correct, but is misleadingly precise, and any attempt to correlate magnetically dated and radiometrically dated events needs to take this into account. For example, the "*Hipparion*" Datum in the Siwaliks, 10.7 Ma based on Berggren et al. (1995), could fall anywhere between 10.3 and 11.0 Ma if the errors on the two radiometric ages are taken into account. Clearly, a great deal depends on the accuracy of these ages, not just in estimating first occurrence dates for hipparionines in order to evaluate the contemporaneity of inferred first appearances but in providing an important interpolated tie-point for the MN scale (the first appearance of "*Hipparion*" at the base of MN 9).

Cande and Kent (1992) cite Icelandic estimates (McDougall et al. 1984) of 11.07 to 9.64 as compatible with their interpolated ages for lower and upper boundaries of C5n, and disregard a Baringo estimate (Deino et al. 1990) of 10.4 Ma for the base. But there are problems with

both these sequences. In fact, in Baringo (Deino et al. 1990), the youngest radiometrically dated sample in reversed sediments, inferred to be below C5n2 or C5rn1, is 10.51 ±0.03 Ma, and this falls in a part of the column in which paleomagnetic determinations are problematic (Tauxe, pers. comm., 1994). Further, it is closely spaced stratigraphically above a radiometric date of 11.54 Ma. This is not an "unequivocal primary age population; ages represent the youngest components" (Deino et al. 1990:569). Both paleomagnetic and radiometric resampling would be productive. There are also problems with the Icelandic record. First, the Icelandic stratigraphic section is made up of thick lava flows, each of short duration, so geologic time is largely unrepresented. This raises problems with any interpolations in time versus thickness regression analyses. Second, in many cases, superposed lavas do not yield monotonically decreasing ages. Third, the age of a particular magnetic transition, for example the top of C5Ar, assayed in two sections, can differ by almost two hundred thousand years (between 12.52 and 12.70 Ma).

The solution to this problem is to sample sequences where sedimentation is "continuous," or at least where the sediments represent a good fraction of total time, for both radiometric and paleomagnetic data through the reversed to normal magnetic transition at the base of Chron C5N. Both the East African Baringo and the Turkish Sinap sequences would be ideal for providing crucial terrestrial tie-points for the global time scale. It would also be important to sample the Ethiopian site of Ch'orora for paleomagnetism. It records an early hipparionine with bracketing radiometric ages of 10.5 ±0.24 and 10.7 ±0.19 (old decay constants, Tiercelin et al. 1979).

As we become more confident in the age of the base of Chron C5n, we can also become more confident in correlating important paleomagnetically calibrated and radiometrically dated late Miocene localities, and hence in deciding the extent to which the first appearances of hipparionines (or indeed any taxon) are geologically contemporaneous.

"*Hipparion*" and *Progonomys*: A Correlational Test Case

That the contemporaneity of the first appearance of hipparionines is not completely and satisfactorily settled can be illustrated by considering their temporal and stratigraphic relationship to the local first appearances of murid rodents. Throughout much of Europe and Southwest Asia, for example in Spain and Turkey, the first hipparionines precede the first murid rodents (*Progonomys cathalai*) by a considerable stratigraphic interval (Woodburne et al., this volume; Kappelman et al., this volume; Sen 1989). On this basis occurrences of hipparionines with murids are considered to be younger than occurrences without murids, and therefore sites in North Africa (Algeria and Tunisia) where hipparionine and *Progonomys* species occur together are rejected as being potential oldest occurrences, despite some independent counterevidence (Woodburne et al., this volume). However, there is clear evidence that either (or both) the first occurrence of equids or that of murids is time transgressive. In Pakistan *Progonomys*-like species, or indeed true *Progonomys*, occur at 12.3 Ma, well before the first occurrence of hipparionines at 10.7 Ma. It appears that in Turkey first occurrences of hipparionines and *Progonomys* were much closer in time than they were in Spain (Sen 1989). It should be noted that an absence of *Progonomys cathalai* must be evaluated carefully, as it is a rare element when it first appears (Agustí 1989).

Because of intensive sampling in the Siwaliks, we are confident both of the occurrence order and that these first occurrences do document first appearances. But until such intensive searches are conducted elsewhere, we cannot be confident about the local temporal relationships of these two events, the regional geographic contemporaneity of either, nor of their relationship to time. Because of the lack of superposition, in most sequences it will in fact be impossible to recognize and date true first appearances of these (or any other) taxa, and therefore to infer their relationships to each other or to geological time. It is further worth noting that agreement on the species-level taxonomy of any group is of prime importance in making correlations and inferring patterns of faunal change.

Siwalik Faunal Change and European MN Units

Figure 7.3 shows the distribution through time of a selection of species first and last occurrences in the Siwaliks. It can also be seen as a representation of sampling density or data quality, as mentioned earlier. This involves a semiquantitative evaluation of how complete a sampling of original species richness is recorded by our fossil record, and whether there are long gaps between successive fossil horizons. What we have not yet completed is the complementary assessment of species stratigraphic ranges, evaluating how well first and last occurrences document first and last appearances. This is an essential next step in increasing confidence in the extent to which our observed patterns of faunal turnover reflect actual or true patterns of lineage and community change. Of course, without an impossibly high level of record quality, taphonomic factors will always keep the fraction of first or last appearance inferences in which we have high confidence below 100%. Equally, of course, these factors apply to other records. Just as European MN units are revised and refined as the record improves, so too is the Siwalik biostratigraphic sequence.

Accepting the current empirical and analytical levels of both records, what comparative observations can usefully be made? For the Siwalik sequence, the sampling is still less than satisfactory between 18 and 14 Ma, although it documents a clear predominance of first over last occurrences. There is a peak of first occurrences around 14 Ma; this began by at least 14.3 Ma and possibly earlier (there being low fossil productivity before this). There are interesting possible correlations between this peak and a marked global marine isotopic shift in benthic and planktonic $\delta^{18}O$ between 14.5 and 14.1 Ma, which appears to document a major and permanent increase in the Antarctic ice sheet (Flower and Kennett 1993). The last occurrence peak at 13 to 12.5 Ma is clearly younger by about 1 my than the first occurrence peak, and this difference raises interesting issues concerning the patterning of and the dynamics underlying faunal change.

The record is relatively poor between 11 and 10.4 Ma because of the nature of the sedimentary sequence, so the first occurrence peak at 10.4 Ma may be an artifact. Hipparionines appear during this interval, but there is little other faunal activity recorded around this time. After 10.4 Ma the faunal record is good. There are more last than first occurrences, and few significant peaks or correlations of first and last occurrences. There is isotopic evidence from paleosols for a significant environmental shift beginning around 8.2 Ma and lasting for close to 2 my (Quade and Cerling, in press; Behrensmeyer, pers. comm., 1994). Events during this period are dominated by last occurrences, and there may be some patterning to the disappearances.

So the Siwalik record shows a pattern of nonsynchronous first and last occurrences, with some probable relationships to regional and/or global environmental shifts in the middle and late Miocene. There are no obvious parallels between patterns of Siwalik faunal change and the MN zonal record in Europe. This is hardly surprising because the records are not strictly comparable, since true superpositional relationships for localities are rare in Europe. A recent attempt (de Bruijn et al. 1992) to seriate European localities and then to document ranges of selected taxa has the laudable goal of developing a biochronologic framework. The authors note that it is "surprising that the ranges of many genera do not begin or end at the lines separating MN units." In fact, using their data, 50% of rodent first and last occurrences do fall at zonal boundaries, while for artiodactyls the numbers are even higher: 64% for first and 75% for last occurrences. It would be very difficult indeed to subdivide the Siwalik sequence into nine MN-type units in such a way as to retrieve such high percentages of synchronous first and last occurrences.

If we assume that MN unit boundaries reflect in some way important faunal change, and if we further assume that unit boundary dates used here are correct, then we see little correlation of the European and Siwalik faunal sequences. Therefore, global climatic events would not force faunal change on a global scale, or at least there is considerable modulation by regional or local filters. However, it seems likely that the above two assumptions are either partly or entirely incorrect, and that possibly there may be some global effect. We recognize of course that this is a matter of continuing debate among those working on the European faunal sequence.

LITERATURE CITED

Agustí, J. 1989. The Miocene rodent succession in eastern Spain: A zoogeographical appraisal. In *European Neogene Mammal Chronology*, ed. E. H. Lindsay, V. Fahlbusch, and P. Mein, pp. 375–404. New York: Plenum.

Azzaroli, A. and G. Napoleone. 1982. Magnetostratigraphic investigation of the Upper Sivaliks near Pinjor, India. *Rivista Italiana di Paleontologia e stratigrafia* 87:739–62.

Barry, J. C. and L. J. Flynn. 1989. Key biostratigraphic events in the Siwalik sequence. In *European Neogene Mammal Chronology*, ed. E. H. Lindsay, V. Fahlbusch, and P. Mein, pp. 557–71. New York: Plenum.

Barry, J. C., L. J. Flynn, and D. R. Pilbeam. 1990. Faunal diversity and turnover in a Miocene terrestrial sequence. In *Causes of Evolution: A Paleontological Perspective*, ed. R. Ross and W. Allman, pp. 381–421. Chicago: University of Chicago Press.

Barry, J. C., E. H. Lindsay, and L. L. Jacobs. 1982. A biostratigraphic zonation of the middle and upper Siwalks of the Potwar Plateau of Northern Pakistan. *Palaeogeography, Palaeoclimatology, Palaeoecology* 37:95–139.

Barry, J. C., M. E. Morgan, A. J. Winkler, L. J. Flynn, E. H. Lindsay, L. L. Jacobs, and D. Pilbeam. 1991. Faunal interchange and Miocene terrestrial vertebrates of Southern Asia. *Paleobiology* 17(3): 231–45.

Berggren, W. A., D. V. Kent, C. C. Swisher III, and M.-P. Aubry. 1995. A revised Cenozoic geochronology and chronostratigraphy. In *Geochronology, Time Scales, and Stratigraphic Correlation: Framework for an Historical Geology*, ed. W. A. Berggren, D. V. Kent, and J. Hardenbol, pp. 00–00. Society of Economic Mineralogists and Paleontologists Special Publication no. 54: 129–212.

Berggren, W. A., D. V. Kent, L. J. Flynn, and J. A. Van Couvering. 1985. Cenozoic geochronology. *Geological Society of America Bulletin* 96:1407–18.

Bruijn, H. de, R. Daams, G. Daxner-Höck, V. Fahlbusch, L. Ginsburg, P. Mein, and J. Morales. 1992. Report of the RCMNS working group on fossil mammals, Reisensburg 1990. *Newsletters on Stratigraphy* 26:65–118.

Cande, S. C. and D. V. Kent. 1992. A new geomagnetic polarity time scale for the Cretaceous and Cenozoic. *Journal of Geophysical Research* B97:13917–51.

——. 1995. Revised calibration of the geomagnetic polarity time scale for the Cretaceous and Cenozoic. *Journal of Geophysical Research* B100:6093–95.

Deino, A., L. Tauxe, M. Monaghan, and R. Drake. 1990. $^{40}Ar/^{39}Ar$ age calibration of the litho- and paleomagnetic stratigraphies of the Ngorora Formation, Kenya. *Journal of Geology* 98:567–87.

Flower, B. P. and J. P. Kennett. 1993. Relations between Monterey Formation deposition and middle Miocene global cooling: Naples Beach Section, California. *Geology* 21:877–80.

Flynn, L. J., J. C. Barry, M. E. Morgan, D. Pilbeam, L. L. Jacobs, and E. H. Lindsay. 1995. Neogene Siwalik mammalian lineages: Species longevities, rates of change, and modes of speciation. *Palaeogeography, Palaeoclimatology, Palaeoecology* 115:249–64.

Flynn, L. J., D. Pilbeam, L. L. Jacobs, J. C. Barry, A. K. Behrensmeyer, and J. W. Kappelman. 1990. The Siwaliks of Pakistan: Time and faunas in a Miocene terrestrial setting. *Journal of Geology* 98:589–604.

Friedman, R., J. Gee, L. Tauxe, K. Downing, and E. H. Lindsay. 1992. The magnetostratigraphy of the Chitarwata and lower Vihowa Formations of the Dera Ghazi Khan area, Pakistan. *Sedimentary Geology* 81:253–68.

Jacobs, L. L., L. J. Flynn, W. R. Downs, and J. C. Barry. 1989. Quo vadis, *Antemus*? The Siwalik muroid record. In *European Neogene Mammal Chronology*, ed. E. H. Lindsay, V. Fahlbusch, and P. Mein, pp. 573–86. New York: Plenum.

Johnson, G. D., N. Opdyke, S. K. Tandon, and A. C. Nanda. 1983. The magnetic polarity stratigraphy of the Siwalik Group at Haritalyangar (India) and a new last appearance datum for *Ramapithecus* and *Sivapithecus* in Asia. *Palaeogeography, Palaeoclimatology, Palaeoecology* 44:223–49.

Johnson, N. M., J. Stix, L. Tauxe, P. F. Cerveny, and R. A. K. Tahirkheli. 1985. Paleomagnetic chronology, fluvial processes, and tectonic implications of the Siwalik deposits near Chinji Village, Pakistan. *Journal of Geology* 93:27–40.

Kappelman, J., S. Sen, M. Fortelius, A. Duncan, B. Alpagut, J. Crabaugh, A. Gentry, J. P. Lunkka, F. McDowell, N. Solounias, S. Viranta, and L. Werdelin. This volume. Chronology and biostratigraphy of the Miocene Sinap Formation of Central Turkey.

Keller, H. M., R. A. K. Tahirkheli, M. A. Mirza, G. D. Johnson, N. M. Johnson, and N. D. Opdyke. 1977. Magnetic polarity stratigraphy of the Upper Siwalik deposits, Pabbi Hills, Pakistan. *Earth and Planetary Science Letters* 36:187–201.

Mankinen, E. A. and G. D. Dalrymple. 1979. Revised geomagnetic polarity time scale for the interval 0–5 m.y. B.P. *Journal of Geophysical Research* 84(B2):615–26.

McDougall, I., L. Kristjansson, and K. Saemundsson. 1984. Magnetostratigraphy and geochronology of Northwest Iceland. *Journal of Geophysical Research* 89(B8): 7029–60.

Morgan, M. E., C. E. Badgley, G. F. Gunnell, P. D. Gingerich, J. Kappelman, and M. C. Maas. 1995. Comparative paleoecology of Paleogene and Neogene mammalian faunas: Body-size structure. *Palaeogeography, Palaeoclimatology, Palaeoecology* 115:287–317.

Morgan, M. E., J. D. Kingston, and B. D. Marino. 1994. Carbon isotopic evidence for the emergence of C_4 plants in the Neogene from Pakistan and Kenya. *Nature* 267:162–65.

Opdyke, N. D., E. Lindsay, G. D. Johnson, N. Johnson, R. A. K. Tahirkheli, and M. A. Mirza. 1979. Magnetic polarity stratigraphy and vertebrate paleontology of the Upper Siwalik subgroup of Northern Pakistan. *Palaeogeography, Palaeoclimatology, Palaeoecology* 27:1–34.

Quade, J., and T. E. Cerling. In press. Expansion of C_4 grasses in the late Miocene of Northern Pakistan: Evidence from stable isotopes in paleosols. *Palaeogeography, Palaeoclimatology, Palaeoecology.*

Quade, J., T. E. Cerling, and J. R. Bowman. 1989. Development of Asian monsoon revealed by marked ecological shift during the latest Miocene in Northern Pakistan. *Nature* 342:163–66.

Sen, S. 1989. *Hipparion* Datum and its chronologic evidence in the Mediterranean area. In *European Neogene Mammal Chronology*, ed. E. H. Lindsay, V. Fahlbusch, and P. Mein, pp. 495–505. New York: Plenum.

Tandon, S. K., R. Kumar, M. Koyama, and N. Niitsuma. 1984. Magnetic polarity stratigraphy of the Upper Siwalik subgroup, east of Chandigarh, Punjab Sub-Himalaya, India. *Journal of the Geological Society of India* 25:45–55.

Tiercelin, J.-J., J. Michaux, and Y. Bandet. 1979. Le Miocène supérieur du sud de la Dépression de l'Afar, Éthiopie: sédiments, faunes, âges isotopiques. *Société Géologique de France, Bulletin* 21(2): 255–58.

Woodburne, M. O., R. L. Bernor, and C. C. Swisher III. This volume. An appraisal of the stratigraphic and phylogenetic bases for the "*Hipparion*" Datum in the Old World.

8

Advances in the Geology and Stratigraphy at Höwenegg, Southwestern Germany

M. O. WOODBURNE, G. THEOBALD, R. L. BERNOR, C. C. SWISHER III, H. KÖNIG, AND H. TOBIEN

The *"Hipparion"* Datum (see, for example, Berggren and Van Couvering 1978) is recognized as one of the major dispersal events in the late Miocene fossil record of the Old World, and as the basis of the Vallesian mammal age as well (Crusafont-Pairó 1951). Whereas most dispersals throughout Holarctica during the Cenozoic Era transpired from the Old World to the New World, many fewer episodes occurred in the opposite direction (see, for example, Woodburne and Swisher 1995). The Old World *"Hipparion"* Datum is one of these exceptions, and the abrupt appearance of this horse in the stratigraphic record is considered to be so impressive as to be recognized as a singularly important reference for interregional correlation in the late Miocene. During the past four decades, the age of this "event," its systematics, and evidence bearing on correlation have been under review (summarized in Woodburne 1996). Swisher (this volume) and the present work suggest that the concept of the datum is operationally valid as well as geochronologically precise, but also (e.g., Woodburne 1994a) that the nomenclature is incomplete as yet. Hence, in this report, we utilize the term *"Hipparion"* Datum for this concept.

Höwenegg is in the Hegau district of southwestern Germany (fig. 8.1) and has played an important role in the determination of the age of the *"Hipparion"* Datum since 1956 (Tobien 1956). Distinct from most other Old World sites of this age, Höwenegg has long been recognized as having a nearly unique potential of combining radioisotopic data with faunal information important to this topic. Thus all age and systematic refinements of the Höwenegg hipparions are pivotal to the age of the *"Hipparion"* Datum and thus to the age of the Vallesian mammal age in the Old World as well. In this study we combine new information on the radioisotopic age of Höwenegg (Swisher, this volume) with new considerations of its stratigraphy and sedimentology so as to provide an improved basis for correlation to other contexts. Our interpretation of the geologic sequence at Höwenegg differs somewhat from that of prior literature, and bears on the analysis of radioisotopic age and regional correlation.

This project was formulated as a collaboration between S. Rietschel, Director of the Staatliches Museum für Naturkunde, and R. L. Bernor, and was funded by the National Geographic Society to revisit the geological contexts of the Höwenegg site. Paramount in this investigation was to establish greater precision in Höwenegg's chronology and better resolve outstanding questions about its sedimentological context. In this work, Woodburne, Swisher, and König were primarily responsible for stratigraphic details, Theobald for performing sedimentological analyses, Tobien and Bernor for the historical review, and Bernor for the funding and direction of the field portion of this project. In a companion paper, Woodburne, Bernor, and Swisher (this volume) address the relevance of the geochronologic results for the so-called *"Hipparion"* Datum and European intra-Vallesian age correlations.

Abbreviations

Institutional Affiliations

AMNH = Department of Vertebrate Paleontology, American Museum of Natural History, New York

F:AM = Frick: American Mammals, housed in the AMNH collection

HLMD = Hessisches Landesmuseum, Darmstadt

NHMW = Naturhistorisches Museum, Wien

SENK = Senckenberg-Museum, Frankfurt

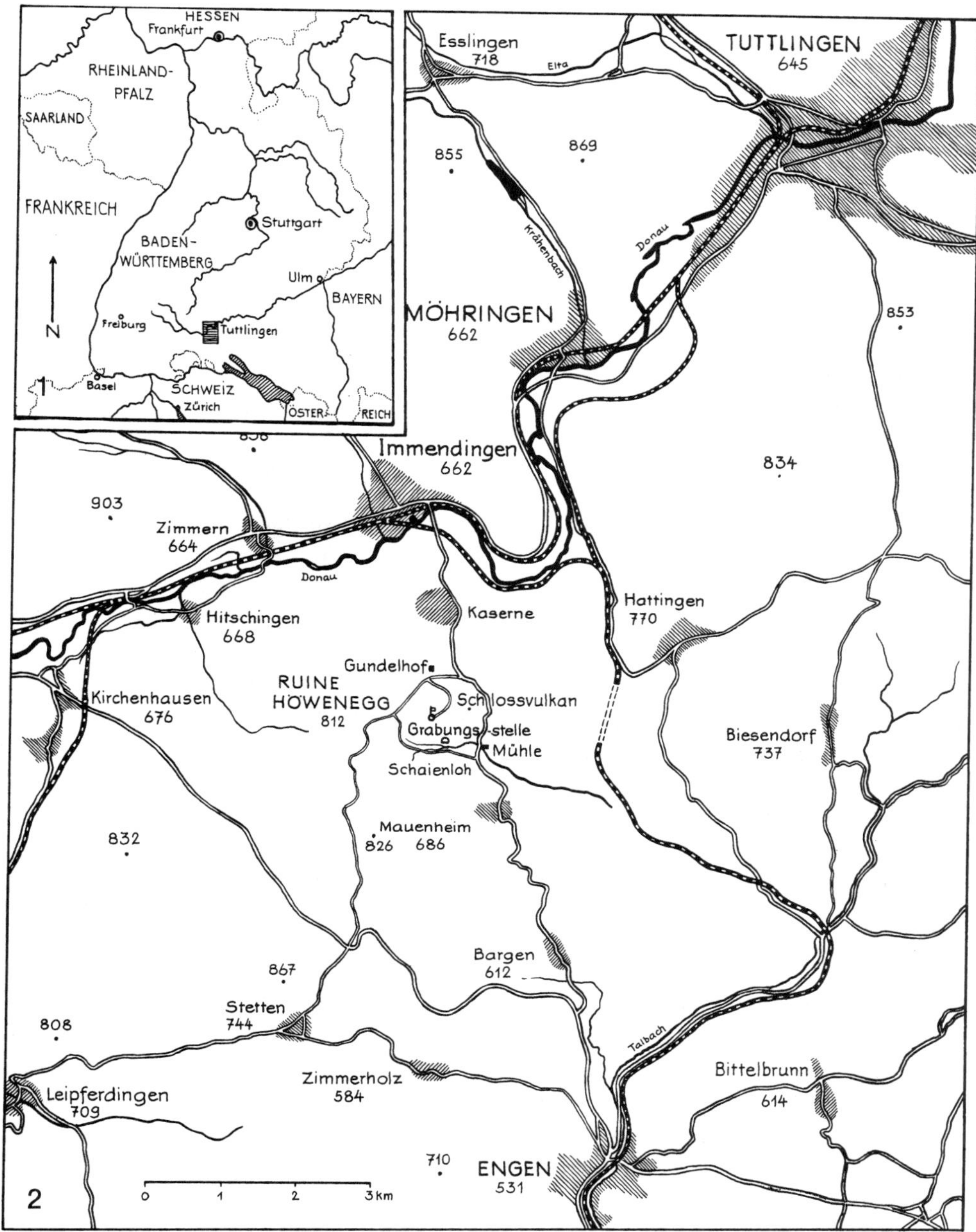

FIGURE 8.1 Map of southwestern Germany, and location of Höwenegg site. After Jörg and Rothausen 1991: fig. 1.

SMNK = Staatliches Museum für Naturkunde, Karlsruhe, Germany

Other Abbreviations

FAD = First Appearance Datum (Berggren and Van Couvering 1978): "changes in the fossil record with extraordinary geographical limits" (Berggren and Van Couvering 1974:ix).

LSD = Lowest stratigraphic datum (Opdyke et al. 1977:324; Lindsay et al. 1984, 1987); lowest stratigraphic appearance of a taxon in local sections.

Ma = megannum in the isotopic time scale (all isotopic ages calculated according to the recommendations of Steiger and Jaeger 1977).

m.y. = an abstract consideration of elapsed time not directly tied to the isotopic time scale.

MN = Mammal Neogene zonation of Mein (1989).

Additional abbreviations are either conventional or are explained in the text.

Methods

Sedimentologic analyses were carried out as described in the following sections.

Carbonate

Carbonate content was determined following procedures in Müller and Gastner 1971, using the "Karbonate-Bombe" (Version: fema, Salzgitter AG). 7.4 g of powdered sample (<0.063 mm) were combined with 4 ml HCl (25%) in an air-tight container. The resultant CO_2 was measured directly with a pressure gauge calibrated in percentage of carbonate. The resultant data were plotted in a histogram (see fig. 8.6 for an example).

Grain-Size Distribution

This was determined by the method described in DIN 18132 (Deutsche Industrie Norm). The weights of the samples obtained are related to the largest grains contained therein, in that the amount of sample material should represent a distinct mean of the investigated sediment. For sample Ho 39, weight was 13 kg; for Höw STP this was 2 kg; and for Ho 39 it was 250 g.

The unlithified sediment in the samples facilitated sieving to obtain grain-size distribution. Mitigation of clustering of grains was accomplished by applying gentle manual pressure during washing with Aquadest (especially in Ho 39).

Sieve sizes utilized were 0.063 mm, 0.125 mm, 0.05 mm, 1 mm, 2 mm, 4 mm, 8 mm, 16 mm, and 31.5 mm. A residue 20% less than 0.125 mm was analyzed by washing out the finer fraction and measuring the remainder with a Bouyoucos-Casagrande Areometer (see also DIN 18132). Decrease in density of particles in suspension in the settling tube was measured in discrete time intervals, and formed the basis for calculating grain sizes (also taking into account the specific weight of the sample and the temperature of the suspension).

The results were plotted in cumulative curves and histograms. Sorting = Q3/Q1 (Trask 1932), and skewness, Sk = Q1 [ep[fy501][cf8];[rs Q3/Md2 (Trask 1932). Sorting values are defined following Sindowski (1961):

<1.20 = very well sorted
1.20–1.50 = well sorted
1.50–2.50 = moderately well sorted
>2.50 = poorly sorted

Skewness shows the tendency of a curve from coarser to finer fraction based on the median grain size (Md). If the skewness is >1, the finer fraction is better sorted; if <1 the coarser fraction is better sorted (Füchtbauer 1988).

The cumulative curves were transformed into log-probability plots for synoptic comparison with the typical log-probability plots for different depositional environments, as portrayed by Visher (1969; e.g., fig. 8.10 herein). The angle of ascent of the different populations (transport by traction, saltation, and suspension), their percent representation in the samples, the position and sharpness of truncation points, and the possible absence of one or two populations form the bases of interpretation and comparison with Visher-plots. This leads to our conclusion that the Höwenegg samples show no evidence of fluviatile deposition.

Geology and Paleontology of the Höwenegg Locality

History of Investigations

The late Miocene volcanic activity at Höwenegg (fig. 8.1) resulted in the nearly unique cooccurrence (for Western Europe) of fossil vertebrates preserved in association with materials amenable to radioisotopic dating. The basic setting is that of a shallow lake formed in a depression in a district of active volcanism (see, for example, Schreiner 1976). Shallow currents transported organic remains entrapped along the lake margins to their final resting places. Due to the periodic occurrence of anoxic, muddy, bottom conditions, the remains were remarkably well preserved.

Höwenegg is thus renowned for its preservation of complete, if variably crushed, mammalian skeletons, including 14 articulated skeletons of *Hippotherium primigenium* (Tobien 1986; Bernor et al., in press). The first fossil found in the Höwenegg district was reported by Deeke (1917). In 1936 a local pharmacist, A. Funk, discovered the actual site of Höwenegg. Wittmann (1937) followed with an initial description of the geology, and Tobien (1938) described some teeth of "*Hipparion*" from the locality, citing the scientific significance of the site. Further investigation of the site was interrupted by the onset of World War II.

A period of intensive excavation resumed between 1950 and 1963. Excavations were initiated by Pfannenstiel and Kircheimer, and financially supported by the Prinz Max zu Fürstenberg. Tobien and Jörg were designated scientific leaders of the excavations, and produced the first detailed descriptions of the work from 1951 to 1965 (Jörg 1951, 1954, 1956, 1957, 1965; Jörg et al. 1955; Tobien 1951, 1952, 1954, 1956, 1957a,b, 1958, 1959a,b, 1962, 1970, 1982; Tobien and Jörg 1959). Other reports relevant to the Höwenegg site included: Rutte (1962), Lutz (1965), Berg (1970), Woodburne and Bernor (1980), Bernor et al. (1980 and 1989), Hünermann (1982), and Forstén (1985). Tobien (1986) gave a detailed summary of this work in German, which we largely excerpt here. The geology of Höwenegg was summarized by Jörg (1954; see fig. 8.2). Geological investigations at Höwenegg also have been pursued by Mäusnest (1976), Schreiner (1963; 1966; 1970; 1976), and Mäusnest and Schreiner (1982), with radioisotopic dating undertaken and discussed by Lippolt et al. (1963), Weiskirchner (1975), Wagner et al. (1975), Baranyi et al. (1976), and Becker-Platen et al. (1977).

A new period of investigation on the Höwenegg geology and paleontology was initiated by Rietschel (SMNU) and Tobien in 1985 to ensure publication of monographic work on the Höwenegg fauna and geology and to give a basis for future work there. Work on the carnivores was presented by Beaumont (1986), turtles by Schleich (1986), the rhinoceros, *Aceratherium*, by Hünermann (1989), and *Chalicotherium grande* by Zapfe (1989). Jörg and Rothausen (1991) summarized the geological campaigns from 1956 to 1963, and discussed the sedimentary environments recorded at the site. Bernor et al. (in press) have presented a monograph on the Höwenegg hipparion, *Hippotherium primigenium*.

The geology of the site will be discussed more fully below. As regards age, Jörg (1951) indicated that the Höwenegg deposits were "upper Miocene," comparable to that of the nearby site of Öhningen, but Tobien (1951) and Jörg (1954) showed that the Öhningen site was older, basing their opinion on the fossil mammals. A correlation to the early Vallesian mammal age of Crusafont-Pairó (1951; which equates to MN 9 in the scheme of Mein 1989 and references cited therein) has been proposed since Berggren and Van Couvering 1974. See also Bernor et al. 1980, 1988, and 1990, Tobien 1986, and Steininger et al. 1989. Swisher (this volume) gives a full discussion of the literature on the radioisotopic geochronology of the Höwenegg site, and this information will not be repeated here.

Höwenegg plant remains are relatively abundant volumetrically, but few in species diversity (Gregor 1982), and considered to be similar to species from other middle Miocene Central European sites.

The invertebrate faunas include principally gastropods and ostracods, and are still in need of a major revision (Lutz 1965:318; Tobien 1986). The whitefish *Leuciscus* is common in the deposits, and is known also from older occurrences in southern Germany. Another Höwenegg fish, *Tinca furcata*, has been found at the late middle Miocene southwestern German locality of Öhningen. The tortoise *Cheirogaster* is known most usually from Paleogene levels of Europe (Schleich 1986). The mammalian fauna includes a mixture of middle Miocene autochthonous forms and earliest late Miocene immigrants (Tobien 1986; Bernor et al. 1988a).

As for the paleoecologic setting, Tobien (1986) has argued that the mammalian assemblage was transported and subsequently deposited in the Höwenegg lake by very low velocity riverine currents. The turtle, catfish, and ostracod fauna indicate shallow water conditions with stable, warm temperatures, while molluscan species indicate that the water was stagnant and oxygen deficient. Based on all

avaliable evidence, Tobien (1986) considered the paleoecology of the Höwenegg site to have been a shallow, warm lake with a periodically oxygen-poor and muddy bottom, the lake being surrounded by a relatively thick deciduous forest developed under the influence of a mild climate. This is concordant with Gregor (1982 [plants]) and Bernor et al. (1988 [plants, mammals, sedimentology]).

Stratigraphic Reevaluation

Here we present information derived from the 1992 excavation of the Höwenegg locality (figs. 8.3, 8.4), and from other observations on the geological setting obtained in recent excavations conducted under the auspices of the Staatliches Museum für Naturkunde, Karlsruhe, in 1986 and 1992 with funding awarded to Bernor by NATO and the National Geographic Society. For reference to previous work, we include one of the major sections described by Jörg and Rothausen (1991), namely Höw 1/50–53 (fig. 8.5, tab 8.1).

Juranagelflüh   The local stratigraphic sequence begins with a unit of largely Jurassic limestone debris, known as the Juranagelflüh (fig. 8.2). This unit ranges in thickness from about 45 to 100 meters, according to Schreiner (1970:239), and was derived from local erosion of the rising alps. The unit is likely of late Miocene age, based on its stratigraphic position below the Höwenegg beds.

Hornblende Tuff   The Juranagelflüh is overlain by a unit of Hornblende Tuff (fig. 8.2) that apparently is 5–15 meters thick (Schreiner 1970:239; also known as "älterer Tuff"— e.g., Tobien 1986; Jörg and Rothausen 1991). According to our data, this is a breccia of basaltic and limestone debris with large (up to 5 cm long) crystals of hornblende and subrounded clusters of hornblende crystals up to ca 4 cm in diameter that might be "bombs" in terms of their origin, based on the pristine nature of the hornblende. According to Schreiner (1970:110) the Hornblende Tuff consists of a basal part of dark gray coarse tuff with occasional relatively large clasts of Jurassic limestone. In the upper part, the tuff is somewhat finer grained, distinctly bedded, and interfingered with the finely bedded white to gray lacustrine calcareous marls of the Höwenegg beds.

Basal outcrops of the Hornblende Tuff are partly covered by scree of the Höwenegg Basalt on the south side of the Höwenegg hill, below the perimeter road (fig. 8.2). Here the Hornblende Tuff forms rounded outcrops of dark brown-gray altered tuff, with only a few hornblende crystals evident. Further down the hillside toward the southwest, an old excavation revealed a deposit of lighter gray to yellow-gray coarse-grained to granule-sized sandstone composed of effectively all components found in the nearby coarse-grained units of the Höwenegg beds: Jurassic lime-

TABLE 8.1 *Explanation of Jörg's Höwenegg Section*

Level	Thickness	Explanation
25	0.18 m	Marls with rhinoceros remains. Colored light gray and violet gray, harder and breaking in small pieces.
26	0.10 m	Tuffite with fist-sized "Auswerflingen"(= bombs)/sedimentary rocks.
27	0.08 m	Marl, light gray, well bedded, with distinct basal contact. Thickness several times less, reaching only 0.04m.
28	0.11 m	Tuffite, like 26, but small "Auswerflingen" dominant.
29	0.30 m max	Marls, light gray, well bedded, with tuffite layers at 0.12 m (0.02 m thick) and 0.05 m (0.02 m thick) above layer 30.
30	0.11 m	Compact marl layer, light gray, breaking in large pieces.
31	0.58 m	Sandy tuffites, olive-green colored, with strongly changing thickness (greatest thickness in the south and in the east). a. 0.14 m Fine-grained tuffite, olive-green colored, rich in snails and containing *Celtis* fruits. Sedimentary "Auswerflingen," maximum nut-sized. b. 0.06 m Sandy marls, hard, gray-green colored. Top and basal contact not distinct. Wood and snail. c. 0.10 m Very fine tuffite, olive-green colored, larger "Auswerflingen" rare. Down to here well-preserved fossiliferous remains, mostly of Rhinoceros. d. 0.28 m Mixed layers of marls and brownish tuffites, with occasional (Eischatringen) olive-green colored, fine-grained tuffites in the top layer (thickness 0.10 m). After a few meters to the north these layers pinch out in the profile. e. 0.04 m Coarse tuffite, brown colored, with larger "Auswerflingen," which come mostly from the Juranagelflüh (= Tertiary disgorgement of Jurassic limestones).
32	0.09 m	Indistinctly bedded marl layer, light-gray colored, breaking into pieces smaller than those in layer 30.
33	0.02 m	Tuffite layer, brightly rust-stained color, with small hazelnut-sized "Auswerflingen."
34	0.36 m	Marls, light gray, breaking into not very sharp pieces, with 2mm-sized layers of tuffite at 0.17 m and 0.32 m below tuffite 33.
35	0.05 m	Sandy tuffite, stain-colored with changing thickness. "Auswerflingen" missing.
36	0.09 m	Indistinctly bedded marls, light gray.
37	0.03 m	Tuffite, like in profile 35.
38	0.51 m	Badly bedded marls, light-gray colored, with thin tuffite layers.
39	0.13 m max	Sandy tuffite, light-brown colored, with some partly head-sized "Auswerfligen" of upper Jurassic limestone. Showing weak classification. Thickness changing strongly, partly only 0.13 m thick.
40	0.10 m	Gray marls, indistinctly bedded, breaking into little pieces.
41	0.03 m max	Tuffite layer, stain-colored, fine-grained, but with maximum egg-sized "Auswerflingen" of Juranagelflüh. The layer partly thins out or doubles by a 0.02 m-thick marl layer.
42	0.45m	Gray marls, in the upper part breaking into smaller and larger pieces. Bedding apparent. In 1953 a badly preserved skeleton of Rhinoceros was found in the upper part.

Translated from the German by H. König.

stone (very abundant and conspicuous) and hornblende crystals. The major difference is that there are no interbedded units of claystone, basalt cobbles were not seen, and the color is not so darkly red-brown. This appears to be the upper part of the Hornblende Tuff unit.

The outcrop appears to be part of a landslide deposit, so it originally was situated farther up slope and, at that point, would be about on strike with, but possibly underneath, that part of the Höwenegg beds exposed in the main trench to the northeast. Both from this and on lithologic grounds, we agree with Schreiner (1970) that the upper part of the Hornblende Tuff interfingers with the Höwenegg beds.

The Hornblende Tuff appears to dip steeply toward the southwest (and is so shown in Schreiner 1970: fig. 26), but within less than a kilometer, in the ridge labeled Schaienloh in Tobien (1986) and in Jörg and Rothausen (1991: fig. 29), and at Daxmühle, both the Hornblende Tuff and overlying Höwenegg beds are effectively horizontal (see also fig. 8.2).

Höwenegg Beds   As indicated above the Höwenegg beds overlie, and apparently partly interfinger with, the Hornblende Tuff. Thickness is at least 12 m. As does the Hornblende Tuff, the Höwenegg beds dip to the southeast (strike about N. 8° W., dip 25°–30° SE). The new measured section (fig. 8.4) is about 4 meters thick, and appears to best correspond to intervals 44–25 of Höw section 1/50–53 in figure 4 of Jörg and Rothausen (1991; based on Jörg 1954; see also fig. 8.5 of this report).

In addition to overall stratigraphic characteristics, figure 8.4 shows the disposition of samples taken for paleomagnetic analysis (Höw Paleomag. 2, etc.), carbonate content

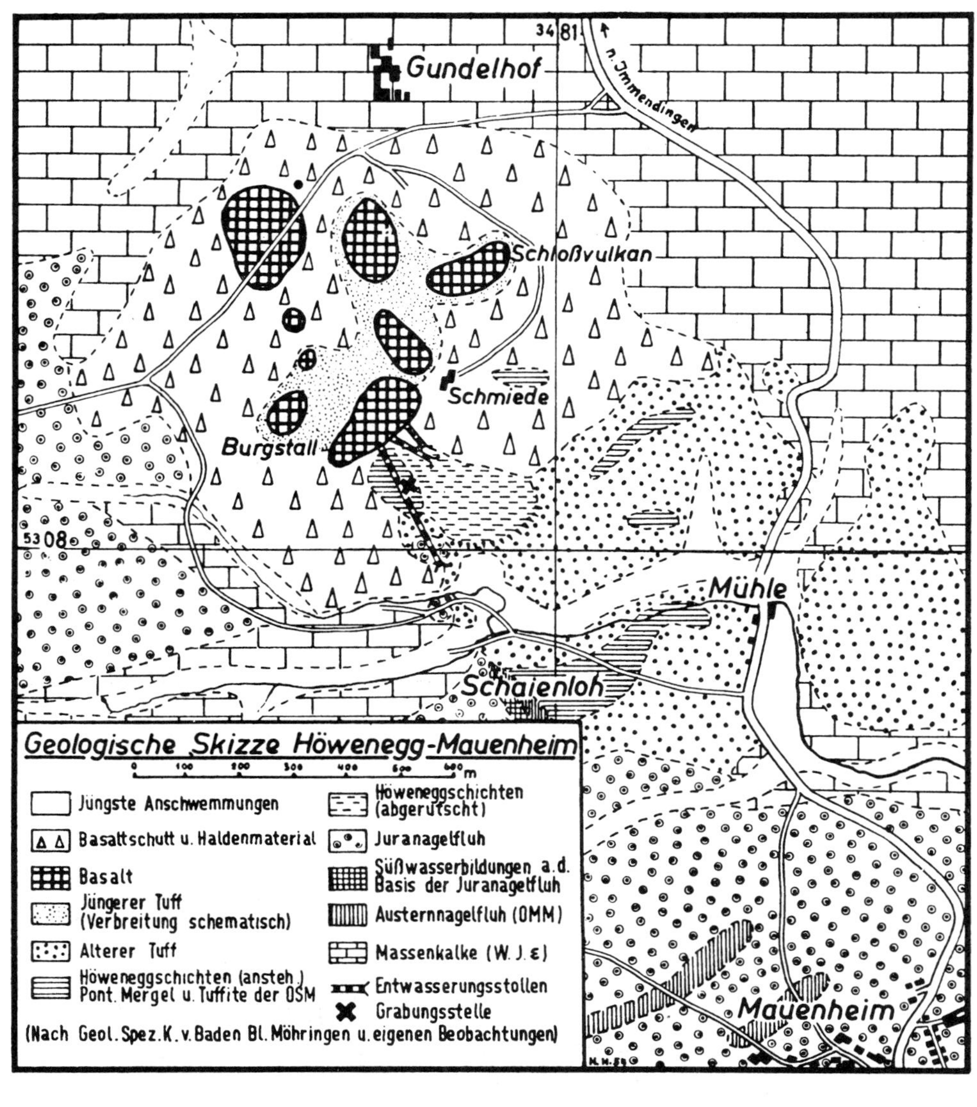

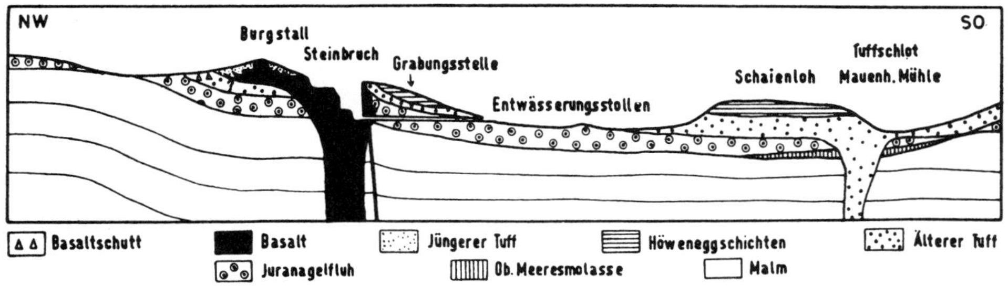

FIGURE 8.2 Geologial map of Höwenegg site, adapted from Jörg 1955. Geological relations modified according to the text.

(KC 2, etc.), pollen (Höp 2, etc.), and fossil mammals (Ho 1–92, etc.). From examining the maps and other data in Jörg and Rothausen (1991) and Tobien (1986), it is clear that the main fossil-producing beds, fish beds, *Dinotherium* beds, etc., are no longer represented in the quarry area, but have been excavated since 1953 and are effectively "in the air." The new Höwenegg section (figs. 8.3 and 8.4) is thus taken at the western edge of the remaining quarry wall after the most recent excavations in 1959. About 25 m to the southwest of the main cut, a test pit (South Test Pit in our terminology) contains another short section of Höwenegg beds. A couplet of gray claystone beds, each ca. 4 cm thick, is interbedded with medium- to coarse-gained red sandstone similar to that described below. The unit above

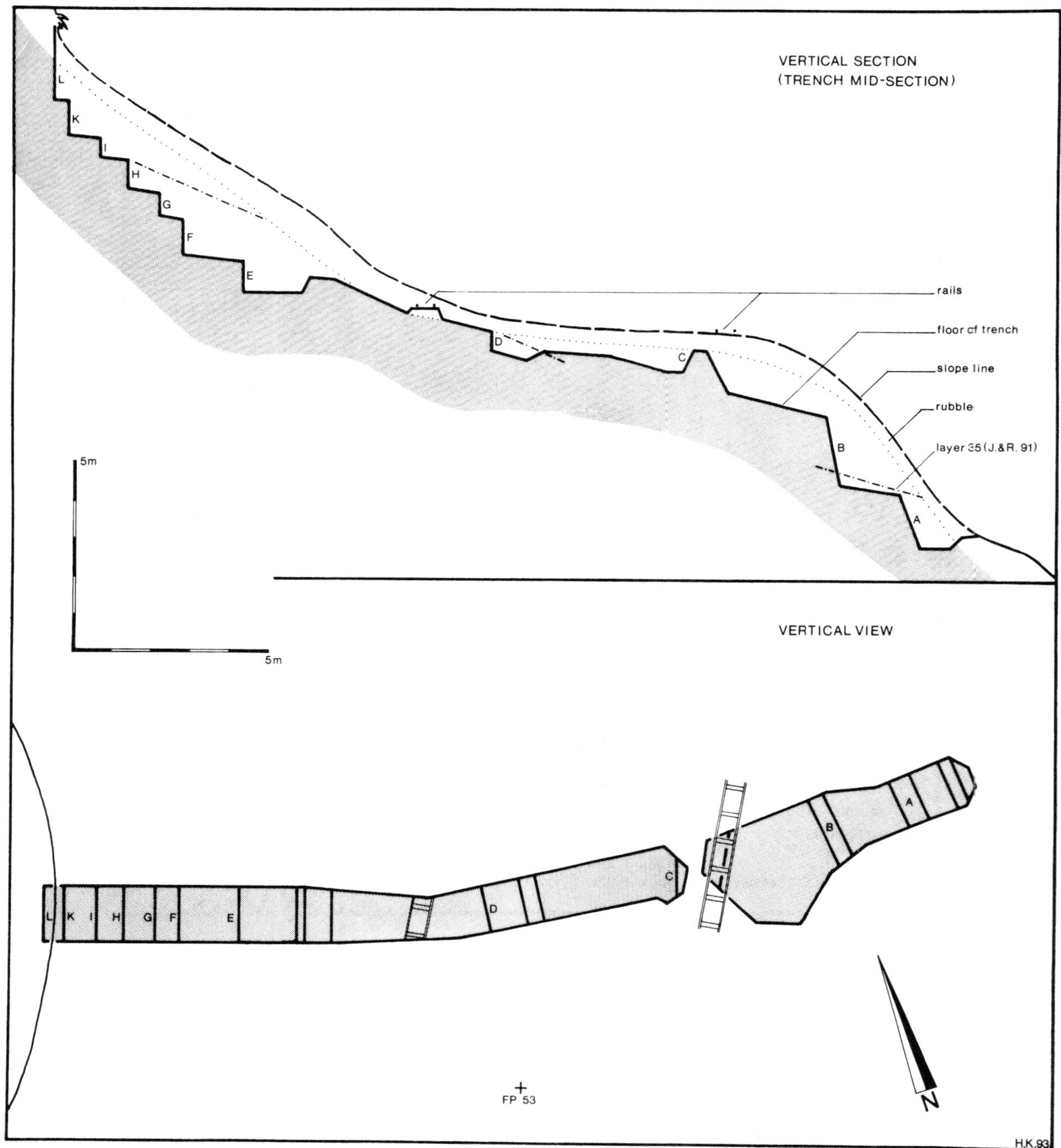

FIGURE 8.3 Profile and plan view of Höwenegg excavations, June/July 1992, by H. König. Rail tracks remain from 1958 excavations. Reference point FP 53 shows the "fixed point" for all measurements made here and in Jörg and Rothausen 1991: fig. 26.

FIGURE 8.4 Stratigraphic section developed for this report, showing proposed correlation to units 25–44 of the Höw section 1/ 50–53 of Jörg and Rothausen 1991. See figure 8.5. Note location of samples for paleomagnetic analysis (Höw Paleomag. 2, etc.), carbonate content (KC 1, etc.), pollen (Höp 1, etc.), and fossil mammals (Ho 11–92, etc.), as well as the disposition of other sites yielding fossil mammals, invertebrates, or woody material. Lateral scale of thickness changes and interfingering of sandstone units are diagrammatic. See text for details. Note apparent concentration of lignitic layers within units 30–32, and conspicuous presence of basaltic cobbles in (but not limited to) unit 39. Otherwise, coarse clasts (cobble to boulder size) are predominantly Jurassic limestone. The samples that yielded Argon $^{39/40}$ radioisotopic dates were obtained from level 33, as indicated.

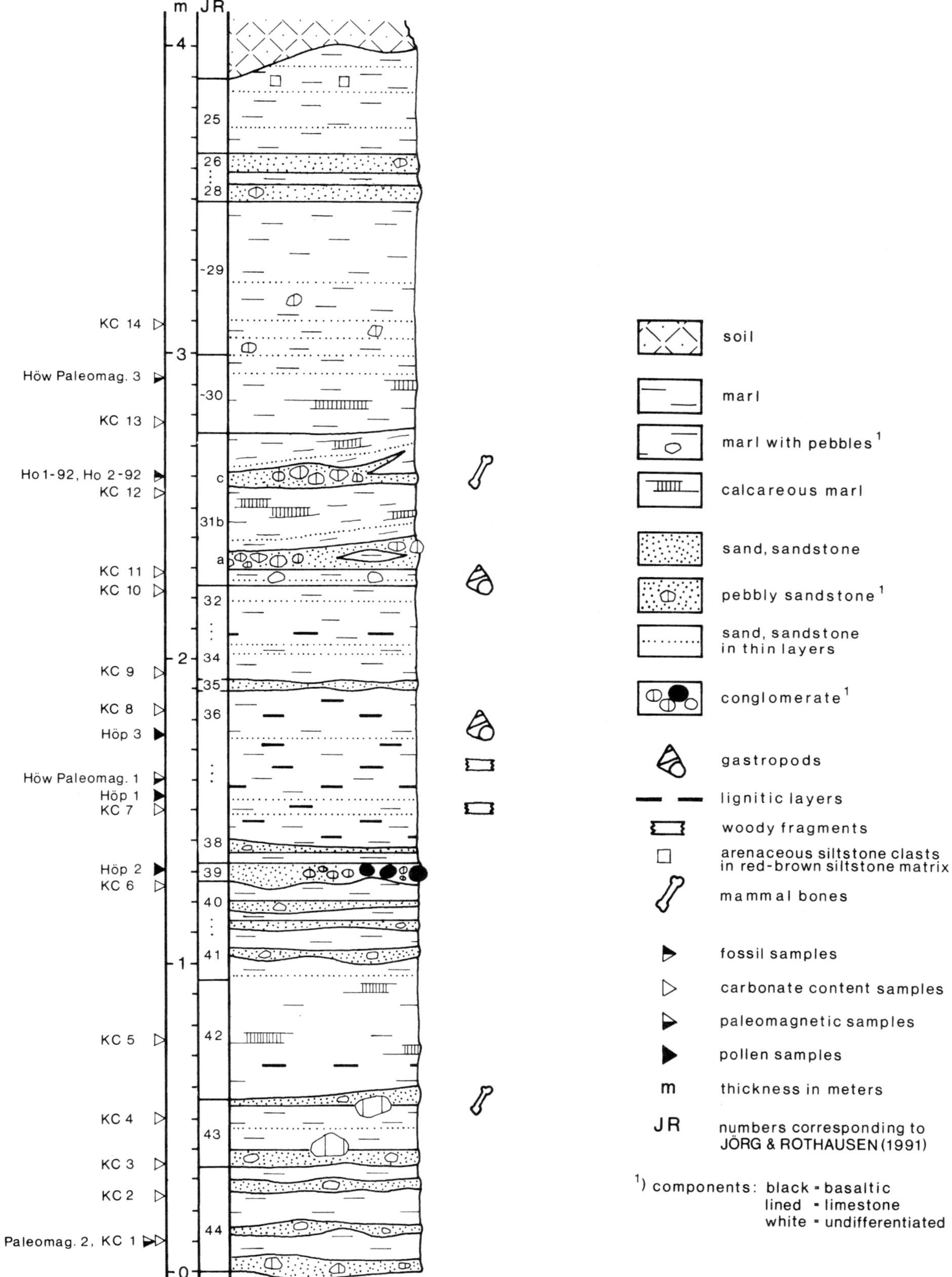

m JR
4
25
26
28
-29
KC 14
3
Höw Paleomag. 3
-30
KC 13
Ho 1-92, Ho 2-92
KC 12
c
31b
KC 11
a
KC 10
32
34
2
KC 9
35
KC 8
36
Höp 3
Höw Paleomag. 1
Höp 1
KC 7
38
Höp 2
39
KC 6
40
41
1
42
KC 5
KC 4
43
KC 3
KC 2
Höw Paleomag. 2, KC 1
44
0
H.K. 93

soil
marl
marl with pebbles [1]
calcareous marl
sand, sandstone
pebbly sandstone [1]
sand, sandstone
in thin layers
conglomerate [1]
gastropods
lignitic layers
woody fragments
arenaceous siltstone clasts
in red-brown siltstone matrix
mammal bones
fossil samples
carbonate content samples
paleomagnetic samples
pollen samples
m   thickness in meters
JR   numbers corresponding to
         JÖRG & ROTHAUSEN (1991)

[1]) components: black = basaltic
                 lined = limestone
                 white = undifferentiated

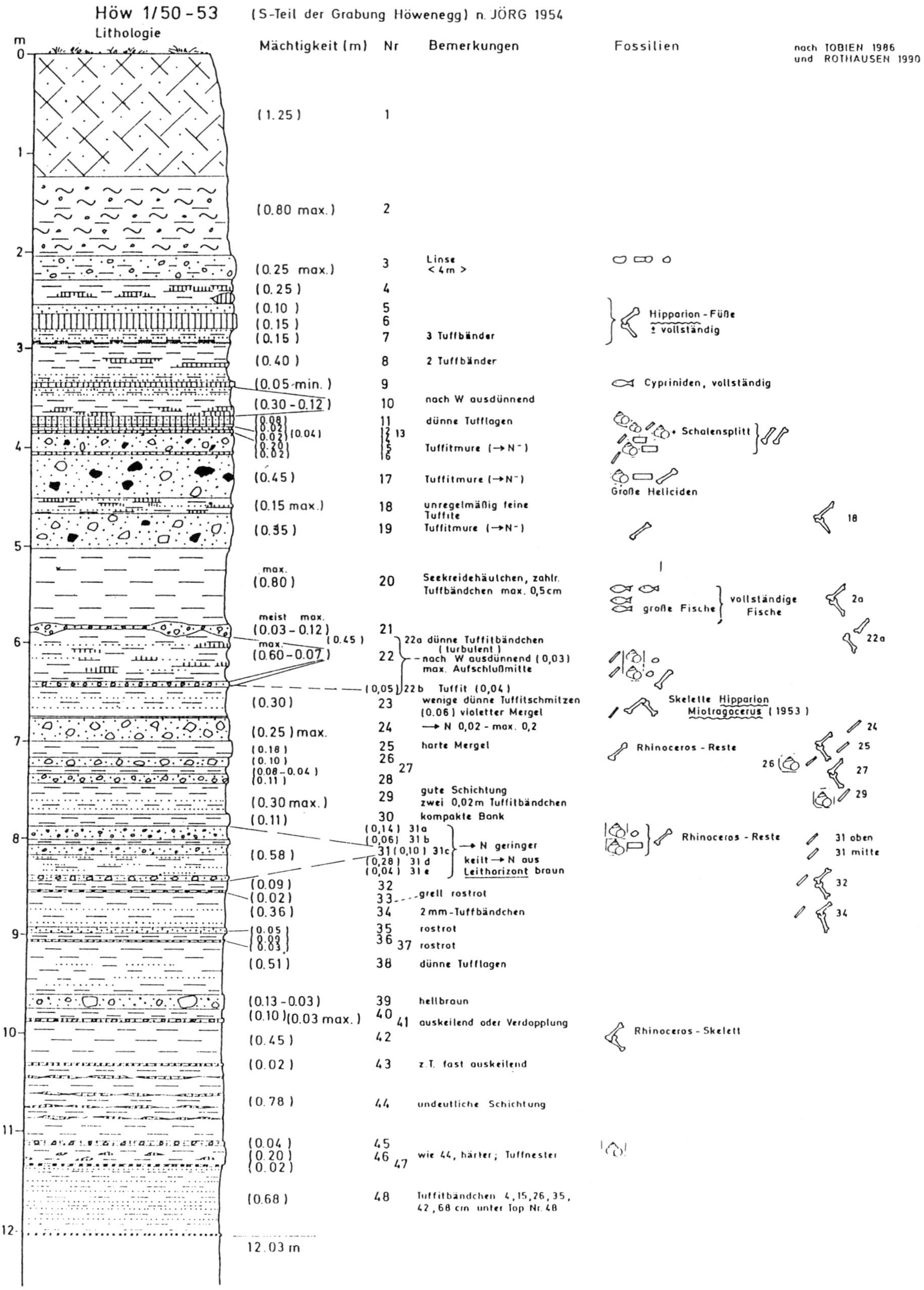

FIGURE 8.5 Simplified version of Profile Höw 1/50–53 (Jörg and Rothausen 1991: fig. 4).

the upper of the two clay beds is distinctly darker than the unit below it, and contains very large crystals of hornblende. The deposit is comparable to, but darker than (and differs in the presence of claystone interbeds from), the outcrop of ?upper Hornblende Tuff described above, which crops out about 20 m to the south (and about 20 m lower in elevation).

Our description (below) is taken from the new Höwenegg section (fig. 8.4) and the South Test Pit. Previous descriptions of these units use the term *tuffit* to describe the sediments (e.g., Jörg 1954; Jörg and Rothausen 1991), a term that implies a volcanogenic origin. While recognizing that crystals of hornblende and boulders of basalt are present in these beds, we prefer to use genetically more ambiguous terms, such as marl and sandstone, and believe that, although well preserved, the volcanic constituents are secondary rather than primary in origin.

Most of the Höwenegg beds are calcareous marl and arenaceous marl, with occasional fine laminations at a scale of parts of millimeters. Often, intervals of marl about 3–5 cm thick are separated by very thin partings of lignite or, less commonly, thin beds (1 cm or thinner) of red-brown fine-grained sandstone (e.g., beds 32–38; see fig. 8.4). Figure 8.6 shows that the carbonate content in marl samples KC1–KC14 is high, ranging from 40% to 60%.

At irregular intervals (ranging from ca. 5 to ca. 50 cm), the marl succession is interbedded with units of fine-grained sandstone that range in thickness from ca. 0.25 cm to 2–3 cm (e.g., unit 35 and others in fig. 8.4). The beds in these units are relatively poorly sorted, but have relatively few particles larger than fine to medium sand size (up to coarse sand or, rarely, granule to small pebble size; figs. 8.7, 8.8). All clastic particles are typically rounded to subrounded, and consist mostly of limestone, apparently derived from the underlying Juranagelflüh. Cement is sparse, and the fabric consists of a grain-supported relatively open matrix at hand lens scale. Bedding is only weakly represented, in part by changes in color density. Presently, it is unclear whether the red color is carried by primary clay constituents or occurs as a film on the clastic particles. Figure 8.6 shows that the carbonate content of the arenaceous intervals (Ho 39, 35, South Test Pit) is much lower (10%–20%) than in the marls. Unit Ho 35 is representative of the relatively thinner fine-grained sandstone units. The grain-size distribution frequencies (figs. 8.7a and 8.8a) are transformed into a log-probability plot (fig. 8.9), which, in comparison with figure 8.10, shows that most of the particles are one millimeter or smaller in diameter and were transported by saltation and in suspension, rather than by traction. The sample is also poorly sorted, which, in combination with the other features, suggests deposition at the distal part of a sand lens.

Also interbedded with the Höwenegg marls are intervals of medium- to coarse-grained sandstone that range in thickness from about 5 to 12 cm. Commonly these units have an irregular, channel-form base, and virtually always a sharp, even upper contact. Preliminary bearings on channel apexes suggest that the depositional azimuth was about N. 30° E. According to Jörg and Rothausen (1991), the other parts of these Höwenegg beds also were deposited by

CARBONATE CONTENT OF HÖWENEGG PROFILE 1992

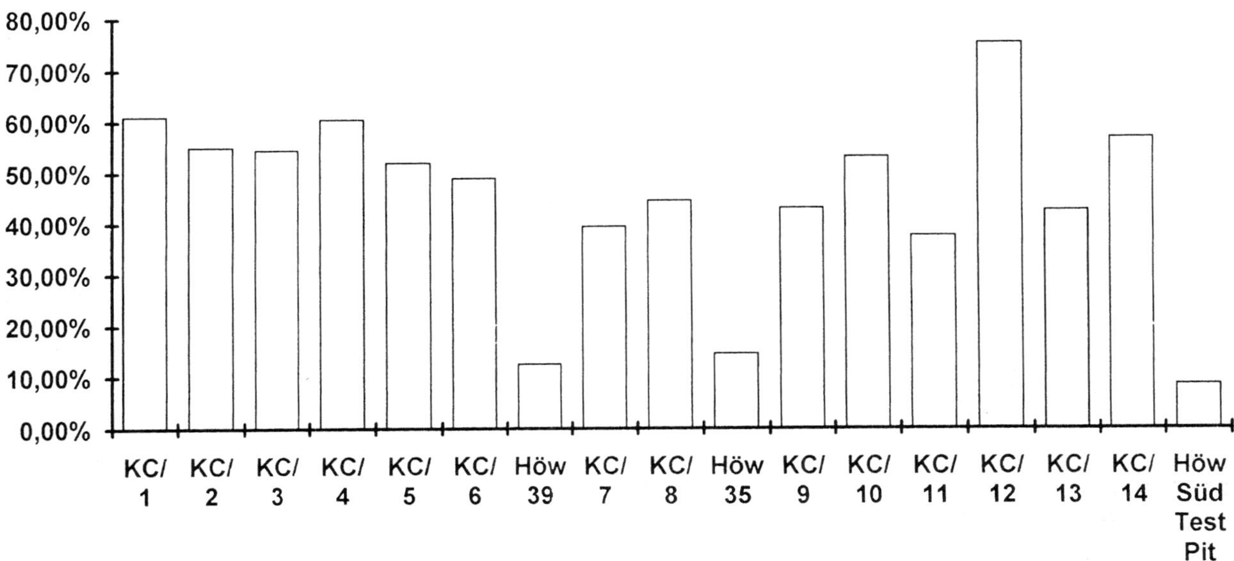

FIGURE 8.6 Histogram of percent carbonate content in samples from new Höwenegg section. KC 1–KC 14 are marl samples; Höw 35, 39 and South (Süd) Test Pit are fine-grained sandstone units. See figure 8.4 for location.

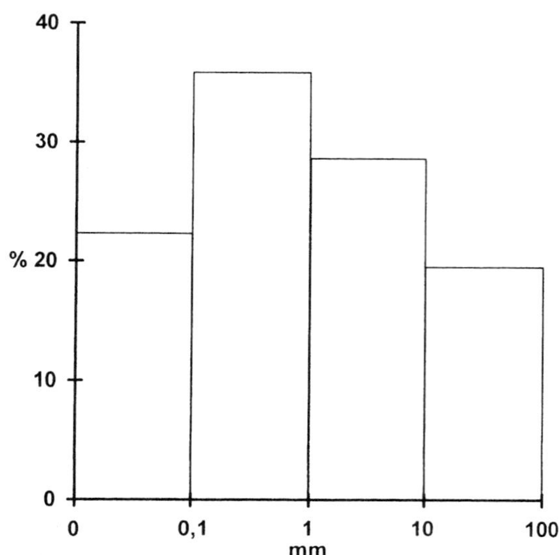

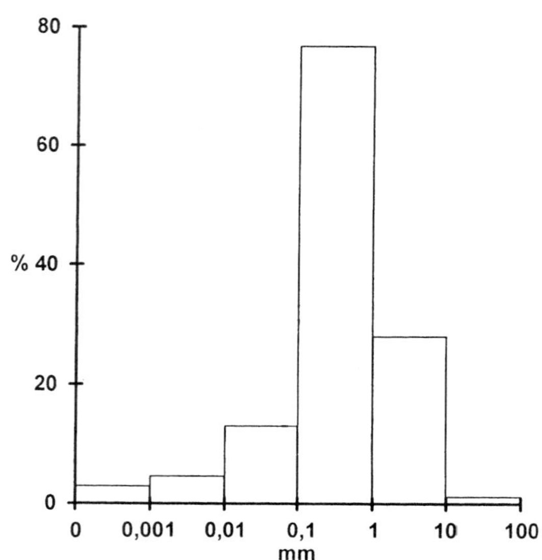

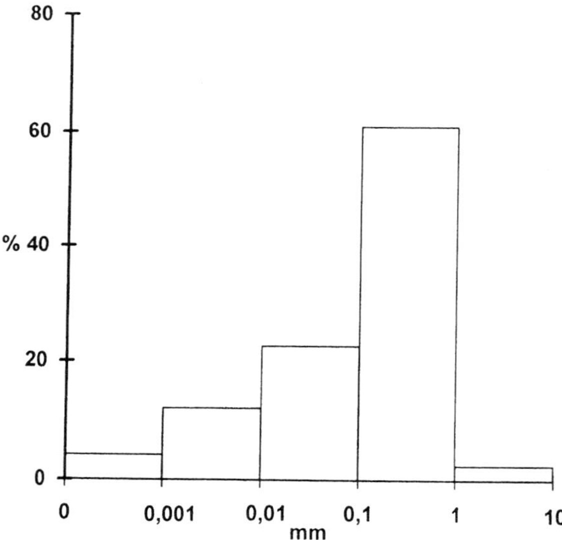

FIGURE 8.7 Histograms of grain-size distribution, fine-grained sandstone units, new Höwenegg trench. (a) Höw 35; (b) Höw 39; (c) Höw STP (South Test Pit). See figure 8.4 for location.

currents that trended northeast (and within 0.5 km to the northeast turned south).

Units Ho 39 and Höw STP are good examples of this type of deposit. Both are poorly sorted. The somewhat greater proportion of traction materials (>2 mm; figs. 8.7b–b, 8.8b–c, 8.9, 8.10) suggest a higher energy (more proximal) regime than Ho 35, consistent with the presence of exotic cobbles in, e.g., unit Ho 39 (see fig. 8.4).

In addition to their being relatively poorly sorted, there seems to be no systematic arrangement of the coarser materials in the coarser-grained units. Grains larger than about granule size are relatively rare; most are subrounded, some are subangular, and most are composed of white Jurassic limestone. These are oriented both semi-parallel to bedding and at sharply discordant angles to bedding. Sparse flakes of biotite and crystals of hornblende also are present but also are not aligned into bedding planes. Centimeter-scale clasts of Höwenegg marl also are contained within

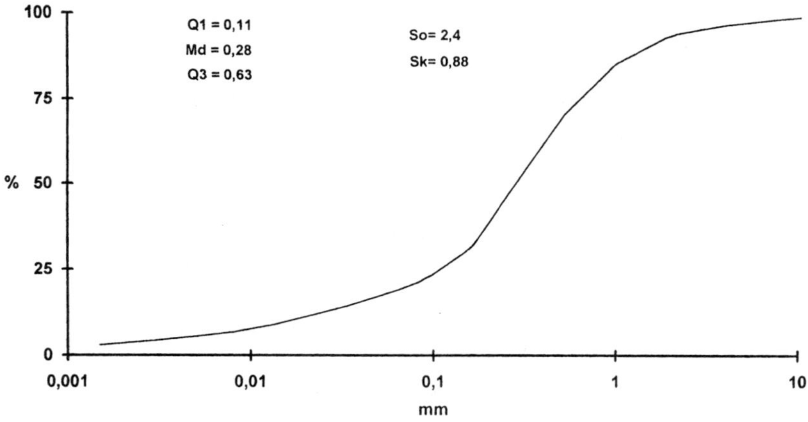

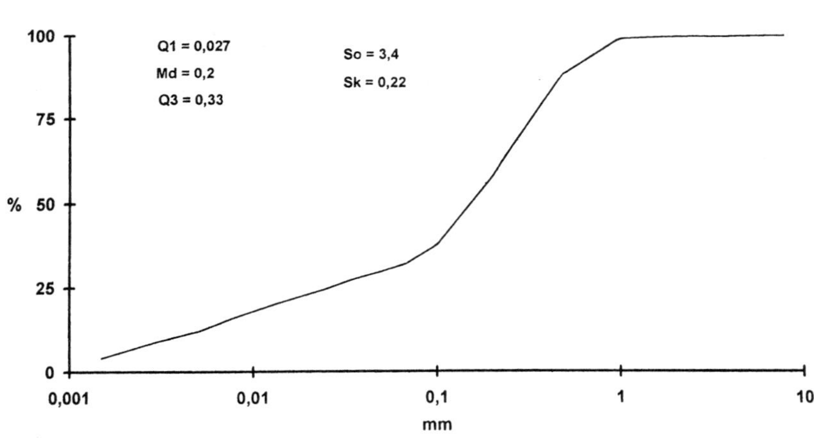

FIGURE 8.8 Grain-size plots, sandstone units, Höwenegg new trench. (a) Höw 35; (b) Höw 39; (c) Höw STP (South Test Pit). See figure 8.4 for locations. As shown, Höw 39 contains a wide distribution of grain sizes.

the sandstone bodies. Some of the grains are composed of quartz. A few subrounded cobbles and even small limestone boulders are present, as well as cobbles and boulders up to 0.5 m long of basalt (these have a distinct weathered rind, but are fresh in the interior). It is important to note that these basalt boulders are unlike the sharply faceted rocks that erode now from the Höwenegg Basalt atop this hill and crop out within ca. 50 meters of the quarry area. It appears unlikely that the Höwenegg Basalt is the source of the basalt boulders in the Höwenegg beds.

The sandstone bodies are characterized as well by having portions that extend laterally from the main unit for about a meter and pinch out within the surrounding marl. Also, some sandstone bodies contain lenses of arenaceous marl within them that exhibit fine scale (less than millimeter) laminations.

The sand bodies are also characterized by the lack of distinct internal bedding, cross-bedding, normal or reversed grading of particles, or lenses or pockets of finer-grained sand of similar composition (as distinct from arenaceous marls). Whereas a fluviatile origin is suggested by the channel-form bases of the units, the lateral (and internal) variations, and pinched-out extensions, bedding features typical of point-bar or certain overbank deposits, seem to be lacking here. The coarser-grained particles (ranging from large pebble to boulder size) appear largely to float in the matrix of mostly medium- to coarse-grained sand.

The sand bodies seem to be better cemented (with calcareous or limonitic cement?) than the layers of fine-grained sandstone described above, but it has yet to be determined whether the clay in the interstices is primary or secondary (the source of the red-brown color of the deposit also is still to be determined).

In summary, the predominance of calcareous marls and

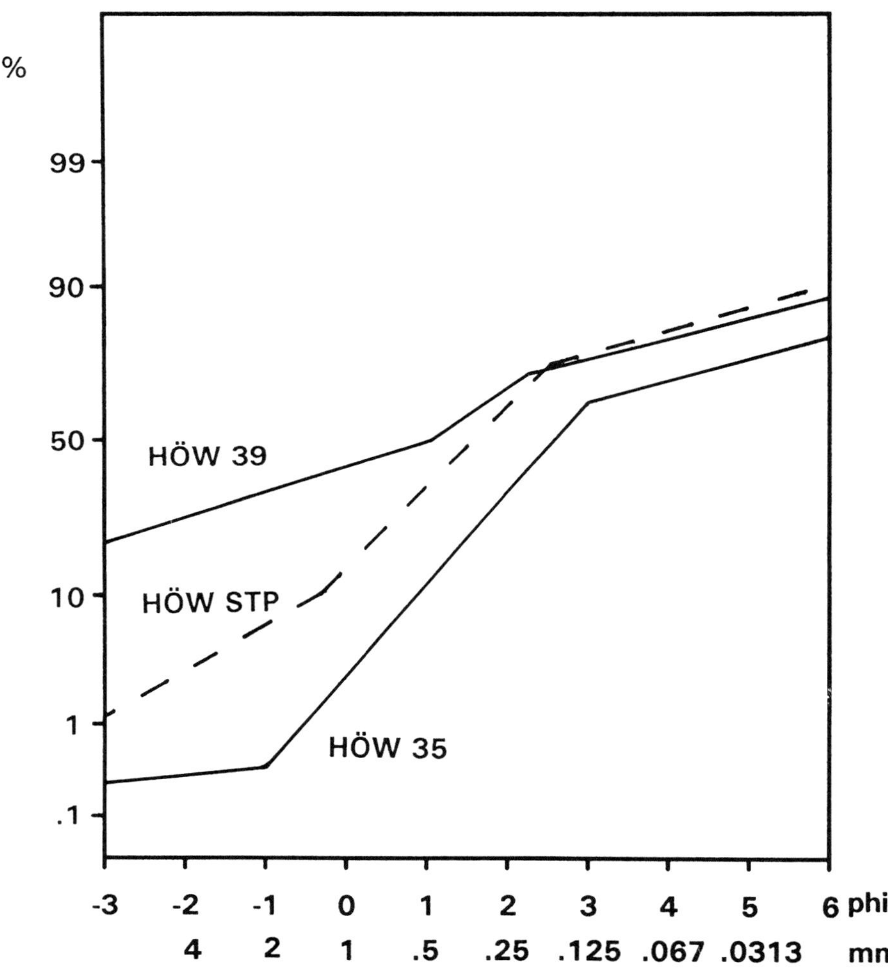

FIGURE 8.9 Grain-size distribution, sandstone units, Höwenegg new trench (see figure 8.8). Cumulative curves transformed into log-probability scale.

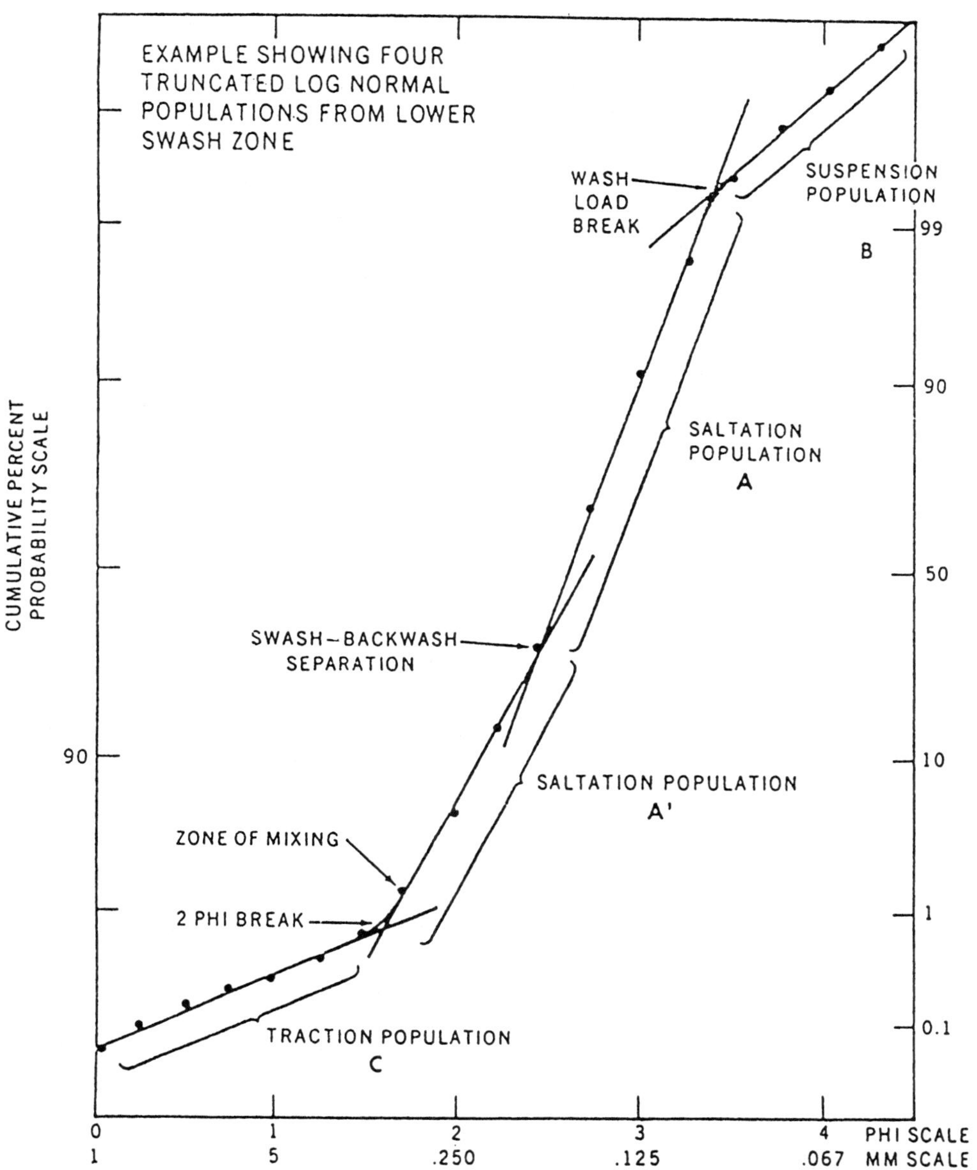

FIGURE 8.10 Relation of sediment transport dynamics to populations and trunctation points in a grain-size distribution (after Visher 1969).

associated fine-scale textural features indicates accumulation in a lacustrine environment with few currents. The poorly sorted and other textural aspects of the sandstone units, lack of bedding, "floating cobbles," etc., suggest deposition as mudflows or lahars rather than by typical fluvial processes.

Sedimentology

The Höwenegg marls were deposited in a low energy, likely lacustrine, environment, with the coarser-grained (sandstone) units reflecting fluvial input from the lake margins. At present the Höwenegg exposures are contained within an area of about one square kilometer. Although lacustrine facies predominate, the fluviatile environment is strongly represented. Jörg and Rothausen (1991) have used the occurrence of plant, fish, and vertebrate remains to postulate a current flowing from about 20° south of west to the northeast in this quarry area, turning southward near Daxmühle less than a kilometer to the east (e.g., fig. 8.2). The lignites suggest intermittent subaerial exposure or at least periodic input from a local shoreline of woody material, and the presence of clasts of Höwenegg beds within the coarser sandstone units indicates that at least the lake margins were subaerially exposed at times. The sequence of events within bed 39 (beginning with a channel-form sandstone body, followed closely stratigraphically by a lignite layer; see fig. 8.4) suggests a degree of depositional periodicity in this part of the section. The well-preserved nature of the plant, fish, and vertebrate remains suggests, also, that the lake waters may have been anoxic in view of the apparently slow sedimentation rate indicated by the fine grain size of the marls. The waters were highly calcareous (as indicated by the calcareous content of the marls), and the preservation of the lignite layers, and the fine-scale laminations of the arenaceous marls, suggests that the beds were not highly bioturbated.

If the main environmental setting was a lake, then the sandstone layers and larger bodies have an external source. Periods of higher energy are suggested by the scoured basal contact of the main sandstone bodies, and the cobble- to boulder-sized clasts they contain. Lateral pinch-out of some layers, and internal lenses of arenaceous claystone, attest to lateral variations in the transport regime, and the lack of distinct bedding in the sandstone units appears to reflect relatively rapid sedimentation without winnowing and sorting by currents operating over a period of time. In that context, the sandstone units appear to represent surge deposits that entered the lake from its margins. The lack of bedding, and the manner in which the coarser clasts float in the matrix at a variety of orientations, leads to a suggestion that the transporting medium may have been viscous, such as a mudflow. The direction of channel axes was approximately parallel to the azimuth suggested for the currents that oriented the fish, wood, and vertebrates. This suggests that the terrigenous input was supplied by currents that trended about parallel to the lake shoreline in this area, rather than at more-or-less right angles to it. The basaltic cobbles and boulders were derived from a pre-Höwenegg source, but not from the Höwenegg Basalt (contra, e.g., Schreiner 1970).

The Geological Sequence

Our stratigraphic interpretation differs somewhat from that previously published (e.g., Schreiner 1970) and bears on the geochronologic data reported on in Swisher (this volume).

The Hornblende Tuff was emplaced (likely as a pyroclastic flow) in a valley developed on the landscape mantled locally by Juranagelflüh (Schreiner 1970). Near the end of Hornblende Tuff deposition, but partly coeval with it, a lake was formed and a 12+-meter thick succession of calcareous marls was deposited in it (Höwenegg beds). The lake filled intermittently, with episodes of input from marginal sources, including lignite layers at a small scale, fine-grained sandstone layers at an intermediate scale, and sandstone bodies with Höwenegg marl clasts at a coarser scale. The presence of coherent marl clasts in the sandstone bodies attests to periods of drying and thus a certain amount of marl lithification prior to their being incorporated into the sandstones. The lake waters were charged

with carbonate because the landscape had been mantled with limestone debris, and this may have led to anoxic conditions that resulted in the fine preservation of the animal remains. It appears that relatively weak currents gave a preferred orientation to plant and fish remains, with perhaps somewhat stronger currents of similar orientation being responsible for orienting carcasses of vertebrate animals. Periodic regimes of relatively high energy apparently resulted in the sandstone bodies, with their large cobbles of limestone and cobbles and boulders composed of basalt. The basalt cobbles were derived either from marginal outcrops of the Hornblende Tuff or from more distant basaltic sources in the general Hegau district. If a mudflow origin of the sandstone units is likely, a local origin for the coarser-grained materials is indicated.

Deposition of the Höwenegg beds is believed to have ceased when the rising Höwenegg Basalt plug tilted the previously horizontal lake beds toward the southeast. Presumably, the marginal basaltic debris that resulted from the eruption (and of the Basaltic Tuff) mantled the landscape relatively widely in this area, but since has been largely eroded away. This sequence differs from that proposed in the previous literature, which states that the Höwenegg Basalt predates the Höwenegg beds and the Hornblende Tuff.

Radioisotopic Chronology and Correlation

Swisher (this volume) has analyzed the Hornblende Tuff, Höwenegg beds basalt boulders, and Höwenegg Basalt using single-crystal argon technology. The first two units give isotopic plateau ages of 10.3 Ma and are analytically indistinguishable from one another. The intrusive plug of Höwenegg Basalt is dated 9.82 ±0.20 Ma, consistent with our interpretation of the geological sequence. Three sites in the Höwenegg beds (Höw Paleomag 1, 2, 3; fig. 8.4) were sampled for paleomagnetic polarity, as described in Swisher (this volume). Each site yielded three samples, and all nine were of normal polarity after standard analytical treatment. The normal magnetozones thus recorded in the Höwenegg succession are consistent with a correlation to the long chron C5 normal interval [= C5n.2n and C5n.1n, with intervening and not always identifiable C5n.1r] in the MPTS of Berggren et al. (1995; Steininger et al., this volume; see also fig. 12), as suggested by the ca. 10.3 Ma radioisotopic age for the Höwenegg sediments.

Conclusions

The fossiliferous deposits at Höwenegg, southwestern Germany, are ca. 10.3 Ma old. The beds are dominantly lacustrine marls, lignites, and clays, deposited in a shallow depression into which carcasses of mammals and remains of other biota were transported under weak currents. Due to the anoxic bottom conditions, the organisms are well preserved, although many are laterally crushed by postmortem lithostatic pressure. The surrounding area was covered by a pre-Höwenegg mantle of limestone and volcanic debris, including the Hornblende Tuff unit that apparently in part interfingered with the lacustrine Höwenegg deposits. Volcanism that resulted in the older tuff is preserved also, in clusters and crystals of hornblende and cobbles of basalt that are secondarily acquired in the Höwenegg deposits. Originally having accumulated at essentially horizontal attitudes, the Höwenegg beds were tilted to the south and east as the plug of Howenegg Basalt was intruded at about 9.8 Ma. The Höwenegg beds thus predate this basalt, rather than the reverse, as traditionally proposed. The association of *Hippotherium primigenium* remains in the Höwenegg beds allows these remains to be calibrated at about 10.3 Ma, one of the few sites of Vallesian age in the Old World where this kind of association can be documented. The Höwenegg site remains as one of the most important cornerstones of geo- and biochronology for mammalian faunas of Vallesian age in the Old World.

Acknowledgments

We would like to thank Professor Dr. S. Rietschel, Director, Staatliches Museum für Naturkunde for kindly granting us permission to undertake this study. W. Munk, E. Biber, and R. Schuppiser provided valuable technical assistance over two field seasons' work at Höwenegg. We wish to give very special thanks to the Burgermeister and people of Immendingen for all the hospitality and help they provided us during the course of our field program. Finally, we gratefully acknowledge the National Geographic Society and the VW Stiftung for providing us the funds to undertake and complete this study. Bernor wishes to thank the Humboldt Stiftung for two one-year fellowships to work on the Höwenegg research project.

LITERATURE CITED

Baranyi, J., H. J. Lippolt, and W. Todt. 1976. Kalium-Argon Altersbestimmungen an tertiären Vulkaniten des Oberrheingrabens. II. Die Alterstraverse vom Hegau nach Lothringen. *Oberrheinische Geologische Abhandlungen* 25:41–62.

Beaumont, G. de. 1986. Les carnivores (Mammifères) du Néogéne de Höwenegg (Hegau, Baden-Würtemberg). *Carolinea* 44:35–43.

Becker-Platen, J. D., L. Benda, and P. Steffens. 1977. Litho- und biostratigraphische Deutung radiometrischer Altersbestimmungen aus dem Jungtertiär der Türkei. *Geologisches Jahrbuch* B25:139–67.

Berg, D. E. 1970. *Die jungtertiäre "Antilopé' Miotragocerus (ein Beispiel für die Problematik paläontologischer Analyse fossiler Tragocerinen/Bovidae, Mammalia)*. Habilitatiowschrift Universität Mainz.

Berggren, W. A., D. V. Kent, C. C. Swisher III, and M.-P. Aubry. 1995. A revised Neogene geochronology and chronostratigraphy. In *Geochronology, Time Scales, and Stratigraphic Correlation: Framework for an Historical Geology*, ed. W. A. Berggren, D. V. Kent, and J. Hardenbol. Society of Economic Mineralogists and Paleontologists Special Publication no. 54: 129–212.

Berggren, W. A. and J. A. Van Couvering. 1974. The late Neogene: Biostratigraphy, geochronology, and paleoclimatology of the last 15 million years in marine and continental sequences. *Palaeogeography, Paleoclimatology, and Paleoecology* 16:1–216.

——. 1978. Biochronology. In *Contribution to the Geologic Time Scale*, ed. G. V. Cohee, M. F. Glaessner, and H. D. Hedberg, pp. 39–55. Amerian Association of Petroleum Geologists Studies in Geology 6.

Bernor, R. L., G. D. Koufos, M. O. Woodburne, and M. Fortelius. This volume. The evolutionary history and biochronology of European and Southwest Asian late Miocene and Pliocene hipparionine horses.

Bernor, R. L., J. Kovar-Eder, D. Lipscomb, F. Rögl, S. Sen, and H. Tobien. 1988. Systematics, stratigraphic, and paleoenvironmental contexts of first-appearing hipparion in the Vienna Basin, Austria. *Journal of Vertebrate Paleontology* 8:427–52.

Bernor, R. L., J. Kovar-Eder, J.-P. Suc, and H. Tobien. 1990. A contribution to the evolutionary history of European late Miocene age hipparionines (Mammalia: Equidae). *Paléobiologie Continentale* 17:291–309.

Bernor, R. L., H. Tobien, L. Hayek, and H.-W. Mittmann. In press. *Hippotherium primigenium* (Equidae, Mammalia) from the late Miocene of Höwenegg (Hegau) Germany. *Andrias* 10.

Bernor, R. L., H. Tobien, and M. O. Woodburne. 1989. Patterns of Old World hipparionine evolutionary diversification and biogeographic extension. In *European Neogene Mammal Chronology*, ed. E. H. Lindsay, V. Fahlbusch, and P. Mein, pp. 263–319. New York: Plenum.

Bernor, R. L., M. O. Woodburne, and J. A. Van Couvering. 1980. A contribution to the chronology of some Old World faunas based on hipparionine horses. *Géobios* 13:35–59.

Crusafont-Pairó, M. 1951. El sistema Miocénico en la depresión española del Vallés Penedés. *International Geological Congress, Report of XVIII Session, Great Britain, 1948* 11:33–43. London.

Daams, R., M. Freudenthal, and A. J. van der Meulen. 1988. Ecostratigraphy of micromammal faunas from the Neogene of Spain. In *Biostratigraphy and Paleoecology of the Neogene Micromammalian Faunas from the Calatayud-Teruel Basin (Spain)*, ed. M. Freudenthal, pp. 287–302. Scripta Geologica Special Issue 1.

Deeke, W. 1917. *Geologie von Baden*. Berlin: Gebrüder Bornträger.

Deutscher Normenausschuss (DNA). 1971. DIN 18123-Baugrund; Untersuchung von Bodenproben, Korngrössenverteilung. In *Deutscher Normenausschuss (DNA)*, pp. 299–307. DIN Taschenbuch 36.

Forstén, A.-M. 1985. *Hipparion primigenium* from Höwenegg/Hegau, FRG. *Annales Zoologici Fennici* 22:417–22.

Füchtbauer, H. 1988. Sandsteine. In *Sedimente und Sedimentgesteine*, ed. H. Füchtbauer. Sediment-Petrologie 11:1–1141.

Gregor, H. J. 1982. *Die jungtertiären Floren Süddeutschlands:*

Paläokarpologie, Phytostratigraphie, Paläoökologie, Paläoklimatologie. Stuttgart: Enke.

Hünermann, K. A. 1982. Rekonstruktion des *Aceratherium* (Mammalia, Perissodactyla, Rhinocerotidae) aus dem Jungtertiär vom Höwenegg (Hegau) (Baden-Württemberg, BRD). *Zeitschrift geologische Wissenschaften Berlin* 7:929–42.

——. 1989. Die Nashornskelette (*Aceratherium incisivum* Kaup 1832) aus dem Jungtertiär vom Höwenegg im Hegau (Südwestdeutschland, Vallesium, Obermiozän). *Andrias* 8:13–64.

Jörg, E. 1951. Über einige Beobachtungen in den öhninger Schichten am Höwenegg. *Mitteilungsblatt der badischen geologisches Landesanstalt* 1950:75–77.

——. 1954. Die Schichtfolge der Fossilfundstelle Höwenegg (Hegau) (pontische Mergel und Tuffite der oberen Süsswassermolasse). *Jahresberichte und Mitteilungen oberrheinischen Geologische Verhandlungen,* n.f., 35:67–87.

——. 1956. Geologische und biostratonomische Beobachtungen an der unterpliozänen Fossilfundstelle Höwenegg/Hegau. *Schriften des Vereins für Geschichte und Naturgeschichte der Baar und der angrenzenden Landesteile in Donaueschingen* 24:198–207.

——. 1957. Tierwelt und Landschaft am Höwenegg (Hegau) zur Unterpliozänzeit. Hegau. *Zeitschrift für Geschichte, Volkskunde und Naturgeschichte* 2:117–25.

——. 1965. *Ophisaurus acuminatus* nov. spec. (Anguidae, Rept.) von der pontischen Wirbeltier-Fundstätte Hewenegg/Hegau. *Beiträge zur naturkundlichen Forschung Südwestdeutschland* 14:21–30.

Jörg, E., H. Rest, and H. Tobien. 1955. Die Ausgrabungen an der jungtertiären Fossilfundstätte Höwenegg (Hegau) 1950–1954: *Beiträge zur naturkundlicher Forschungen Südwest Deutschlands* 14:3–21.

Jörg, E. and K. Rothausen. 1991. Zur Schichtenfolge und Biostratonomie der Wirbeltierfundstelle Höwenegg (Hegau, Südwestdeutschland, Vallesium, Obermiozän). *Andrias* 8:13–64.

Lindsay, E. H., N. D. Opdyke, and N. M. Johnson. 1984. Blancan-Hemphillian land mammal ages and late Cenozoic mammal dispersal events. *Annual Review, Earth and Planetary Science* 12:445–88.

Lindsay, E. H., N. D. Opdyke, N. M. Johnson, and R. F. Butler. 1987. Mammalian chronology and the magnetic polarity time scale. In *Cenozoic Mammals of North America: Geochronology and Biostratigraphy,* ed. M. O. Woodburne, pp. 269–90. Berkeley: University of California Press.

Lippolt, H. J., W. Gentner, and W. Wimmenauer. 1963. Altersbestimmungen nach der Kalium-Argon-Methode an tertiären Eruptivgesteinen Südwestdeutschlands. *Jahreshefte des geologischen Landesamtes Baden-Württemburg* 6: 507–38.

Lutz, A. 1965. Jungtertiäre Süsswasserostracoden aus Süddeutschland. *Geologisches Jahrbuch* 82:271–330.

Mäusnest, O. 1976. Über die Anomalien des erdmagnetischen Feldes im Gebiet einiger Hegauvulkanitvorkommen. *Oberrheinische geologische Abhandlungen* 25:63–74.

Mäusnest, O. and A. Schreiner. 1982. Karte der Vorkommen von Vulkangesteinen im Hegau. *Abhandlungen Geologisches Landesamt Baden-Württemberg* 10:1–48.

Mein, P. 1989. Updating of MN zones. In *European Neogene Mammal Chronology,* E. H. Lindsay, V. Fahlbusch, and P. Mein, pp. 73–90. New York: Plenum.

Müller, G. and M. Gastner. 1971. The "Karbonate-Bombé," a simple device for the carbonate content in sediments, soils, and other materials. *Neues Jahrbuch für Mineralogie Monatsheften* 10:466–69.

Opdyke, N. D., E. H. Lindsay, N. M. Johnson, and T. Downs. 1977. The paleomagnetism and magnetic polarity stratigraphy of the mammal-bearing section of Anza Borrego State Park, California. *Quaternary Research* 7:316–29.

Rutte, E. 1962. Schlundzähne von Süsswasserfischen. *Palaeontographica* A120:165–212.

Schleich, H. H. 1986. Vorläufige Mitteilung zur Bearbeitung der fossilen Schildkröten der Fundstelle Höwenegg. *Carolinea* 44:47–56.

Schreiner, A. 1963. Geologische Untersuchungen am Höwenegg (Hegau). *Jahreshefte des geologischen Landesamtes Baden-Württemberg* 6:395–420.

——. 1966. *Geologische Karte von Baden-Württemberg 1:25000.* Geologisches Landesamt Baden-Würtemberg.

——. 1970. *Erläuterungen zur Geologischen Karte des Landkreises Konstanz mit Umgebung 1:50000.* Geologisches Landesamt Baden-Würtemberg.

——. 1976. *Hegau und westlicher Bodensee.* Sammlungen geologischer Führer 62.

Sindowski, K.-H. 1961. Mineralogische, petrographische und geochemische Untersuchungsmethoden. *Lehrbuch der angewandten Geologie* 1:161–86.

Steiger, R. H. and E. Jaeger. 1977. Subcommission on geochronology, Convention on the use of decay constants in geo- and cosmochronology. *Earth and Planetary Science Letters* 36:359–62.

Steininger, F. F., R. L. Bernor, and V. Fahlbusch. 1989. European Neogene marine/continental chronologic correlations. In *European Neogene Mammal Chronology,* ed. E. H. Lindsay, V. Fahlbusch, and P. Mein, pp. 15–46. New York: Plenum.

Steininger, F. F., J. Senes, K. Kleemann, and F. Rögl. 1985. Neogene of the Mediterranean Tethys and Paratethys, Stratigraphic Correlation Tables and Sediment Distribution Maps, vols. 1 and 2: Institute of Paleontology, University of Vienna Press: XIV + 189; XXV + 536.

Swisher, C.C., III. This volume. New ^{40}Ar/^{39}Ar dates and their contribution toward a revised chronology for the late Miocene of Europe and West Asia.

Tobien, H. 1938. Über *Hipparion*-Reste aus der obermiozánen Süsswassermolasse Südwestdeutschlands. *Zeitschrift der deutschen geologischen Gesellschaft* 90:177–92.

——. 1951. Über die Grabungen in der oberen Süsswassermolasse des Höwenegg (Hegau). *Mitteilungsblatt der badische geologischen Landesanstalt* 1950:72–74.

——. 1952. Über die Funktion der Seitenzehen tridactyler Equiden. *Neues Jahrbuch für Geologie und Paläontologie Abhandlungen* 96:137–72.

——. 1954. Jungtertiäre Wirbeltiere vom Höwenegg/Hegau. *Umschau* 54:559–81.

——. 1956. Zur ökologie der jungtertiären Säugetiere vom Höwenegg (Hegau) und zur Biostratigraphie der europáischen *Hip-*

parion-Fauna. *Schriften des Vereins für Geschichte und Naturgeschichte der Baar* 24:208–23.

———. 1957a. Sobre la Bioestratigrafia de la Fauna de *Hipparion*. *Cursillos Instituto "Lucas Mallada"* 4:121–26.

———. 1957b. Die Bedeutung der unterpliozänen Fossilfundstelle Höwenegg für die Geologie des Hegaus. *Jahreshefte des geologischen Landesamtes Baden-Würtemberg* 2:193–208.

———. 1958. Die Ausgrabungen an der unterpliozänen Fossilfundstätte Höwenegg (Hegau). *Zeitschrift der deutschen geologischen Gesellschaft* 110:617–18.

———. 1959a. Die Ausgrabungen an der pontischen Fossilfundstätte Höwenegg (Hegau). *Cursillos Instituto "Lucas Mallada"* 6:59–62.

———. 1959b. *Hipparion*-Funde aus dem Jungtertiär des Höwenegg (Hegau). *Aus der Heimat* 67:121–40.

———. 1962. Über Carpus und Tarsus von *Deinotherium giganteum* KAUP (Mamm., Proboscidea). *Paläontologische Zeitschrift* 1962:231–38.

———. 1970. Subdivision of Pontian Mammalian Faunas. Committee on Mediterranean Neogene Stratigraphy, Proceedings. Session, Bologna 1967. *Giornale Geologia* 35:1–5.

———. 1982. Osteologische Bemerkungen zum Fussbau von *Hipparion* (Equidae, Mammalia) aus der jungtertiären Wirbeltier-Fundstätte Höwenegg/Hegau. *Zeitschrift der geologischen Wissenschaften* 10:1043–57.

———. 1986. Die jungtertiäre Fossilgrabungsstätte Höwenegg im Hegau (Südwestdeutschland): Ein Statusbericht. *Carolinea* 44:9–34.

Tobien, H. and E. Jörg. 1959. Die Ausgrabungen an der jungtertiären Fossilfundstätte Höwenegg (Hegau), 1955–1959. *Beiträge zur naturkundlichen Forschung Südwestdeutschland* 18:175–91.

Trask, P. D. 1932. *Origin and environment of source sediments of petroleum.* Houston: Houston Gulf Publishing Co.

Visher, G. S. 1969. Grain size distribution and depositional processes. *Journal of Sedimentary Petrology* 21:1074–106.

Wagner, J. J., M. Delaloye, and I. Hedley. 1975. Données géochronométriques et paléomagnétiques sur l'extension du volcanisme du Hegau en Suisse (Ramsen, Schaffhouse). *Comptes Rendus des Séances, SPHN, Genève*, N.S., 10:46–57.

Wagner, W. 1946. Die unterpliozäne Wirbeltierfauna vom Wissberg bei Gau-Weinheim in Rheinhessen. *Wissenschaftliche Veröffentlichungen der Technischen Hochschule Darmstadt* 1:2–11.

———. 1973. Die unterpliozänen Dinotherien-Sande und Ihre Fauna im Gebiet des Blattes 6114 Wörrstadt (Mainzer Becken). *Mainzer geowissenchaftliche Mitteilungen* 2:149–60.

Weiskirchner, W. 1975. Vulkanismus und Magmenentwicklung im Hegau. *Jahresberichte und Mitteilungen des Oberrheinischen Geologischen Vereins*, N.S., 57:117–34.

Wittmann, O. 1937. Deckentuff und Molasse am Höwenegg. Ein Beitrag zur Entwicklungsgeschichte eines Hegauvulkanes. *Jahresberichte und Mitteilungen des Oberheinischen Geologischen Vereins*, N.S., 26:1–32.

Woodburne, M. O. and R. L. Bernor. 1980. On superspecific groups of some Old World hipparionine horses. *Journal of Paleontology* 8:315–27.

Woodburne, M. O., B. J. MacFadden, and M. F. Skinner. 1981. The North American "Hipparion" Datum and implications for the Neogene of the Old World. *Géobios* 14:493–524.

Woodburne, M. O. and C. C. Swisher III. 1995. Land mammal high resolution geochronology, intercontinental overland dispersals, sea-level, climate, and vicariance. In *Geochronology, Time Scales, and Stratigraphic Correlation: Framework for an Historical Geology*, ed. W. A. Berggren, D. V. Kent, and J. Hardenbol, pp. 335-64. Society of Economic Mineralogists and Paleontologists Special Publication 54.

Woodburne, M. O. 1996. Precision and resolution in mammalian chronostratigraphy: Principles, practices, examples. *Journal of Vertebrate Paleontology* 16.

Zapfe, H. 1989. *Chalicotherium goldfussi* Kaup aus dem Vallesium vom Höwenegg im Hegau (Südwestdeutschland). *Andrias* 6:117–28.

9

An Appraisal of the Stratigraphic and Phylogenetic Bases for the "*Hipparion*" Datum in the Old World

M. O. WOODBURNE, R. L. BERNOR, AND C. C. SWISHER III

The "*Hipparion*" Datum (see Berggren and Van Couvering 1978) is recognized as one of the major dispersal events in the late Miocene fossil record of the Old World, and as the basis of the Vallesian mammal age as well (Crusafont-Pairó 1951). Whereas most dispersals throughout Holarctica during the Cenozoic Era transpired from the Old World to the New World, many fewer episodes occurred in the opposite direction (e.g., Woodburne and Swisher, 1995). The Old World "*Hipparion*" Datum is one of these exceptions, and the abrupt appearance of this horse in the stratigraphic record is considered to be so impressive as to be recognized as a singularly important reference for interregional correlation in the late Miocene. During the past four decades, the age of this "event," its systematics, and evidence bearing on correlation have been under review (e.g., Woodburne 1989; Bernor et al. 1989). Woodburne et al. (this volume), Swisher (this volume), and the present work suggest that the concept of the datum is operationally valid as well as geochronologically precise within a span of ca. 0.5 m.y. In the present report, all species important to the Vallesian dispersal event are considered to pertain to the genus *Hippotherium* von Meyer 1829. Accordingly this FAD is herein identified as the *Hippotherium* Datum (e.g., Woodburne 1989).

Thus, the *Hippotherium* Datum represents a "First Appearance Datum" (e.g., Berggren and Van Couvering 1978). A FAD is a probabilistic statement that the first appearance of a taxon was regionally synchronous. Each FAD is composed of a number of LSDs (Lowest Stratigraphic Datum). If these LSDs are closely time-correlative, then a FAD of a given regional scope can be established. The vitality of a FAD is thus dependent on the security with which the LSDs that comprise it can be correlated precisely. In the remainder of this report we thus examine the stratigraphic, geochronologic, and paleomagnetic aspects of the primary sites in Western Europe that bear on the vitality and precision of the *Hippotherium* Datum, and the beginning of the Vallesian mammal age. In addition to citing the stage-of-evolution of the mammalian fauna in each location, we appraise the level of morphologic advancement shown by the hipparion sample, itself, as advocated, e.g., in Woodburne (1987:16L).

Abbreviations

FAD = First Appearance Datum (Berggren and Van Couvering 1978), "changes in the fossil record with extraordinary geographical limits" (Berggren and Van Couvering 1974:ix).

LSD. = Lowest Stratigraphic Datum (Opdyke et al. 1977:324; Lindsay et al. 1984, 1987); lowest stratigraphic appearance of a taxon in local sections.

Ma = megannum in the isotopic time scale (all isotopic ages calculated according to the recommendations of Steiger and Jaeger 1977).

m.y. = an abstract consideration of elapsed time not directly tied to the isotopic time scale.

MN = Mammal Neogene zonation of Mein (1989).

Additional abbreviations are either conventional or explained in the text.

General Overview

The concept of a "*Hipparion*" Datum (Berggren and Van Couvering 1974) stems from the observation that at a number of Old World sites a species of "*Hipparion*" appears abruptly and abundantly at a level stratigraphically above deposits in which this kind of horse is absent, but in which may be found specimens of *Anchitherium* and/or *Leptodontomys* (= *Eomyops*; e.g., Berggren and Van Couvering 1978; Engesser 1979). This abrupt appearance of hipparions is followed stratigraphically in many places in Europe by the rodent *Progonomys cathalai* (see Sen 1989:498 for an illustration of this in Spanish and Turkish deposits).

The first appearance of "*Hipparion*" has been specified

as defining the base of the Vallesian mammal age in Spain (Crusafont-Pairó 1951), and by subsequent workers in recognizing the beginning of the Vallesian in other Old World localities. The *"Hipparion"* FAD is operationally synchronous with the beginning of the Vallesian mammal age. Swisher (this volume) and Bernor et al. (this volume) and we here suggest that the base of the Vallesian is ca. 10.5 Ma (to 11.0 Ma) old, fully 1 to 2 m.y. younger than proposed in previous literature (e.g., Berggren et al. 1985a: fig. 6; 1985b: fig. 2; Steininger et al. 1989), and is contained within Chron C5n.2 (figs. 3 and 4) in the MPTS (Berggren et al., 1995). Localities that are relevant to this interpretation include the Pannonian Stage C localities in the Vienna Basin (Central Paratethys), Austria; Eppelsheim and Höwenegg, Germany; Bou Hanifia, Algeria; the Valles-Penedes, Spain; the Beglia Formation, Tunisia; and Jeltokamenka, C.S.S.R. The Siwalik succession in Pakistan also appears to contain a close correlate of the basal Vallesian. Sen (1989) considers some of the above localities as well as others (Turkey, Crete), and we comment briefly on these also.

Name of the *"Hipparion"* Datum

We utilize the term *Hippotherium Datum*. Bernor et al. (1980; 1989) and Woodburne et al. (1981) have indicated that the species that are considered when the *"Hipparion"* Datum is addressed pertain to a generic level taxon (i.e., it contains a number of species) that is phylogenetically distinct from *"Hipparion"* s.s., and Berggren et al. (1985b:248). Woodburne (1989), and Bernor et al. (1989; this volume) have suggested that this *"Hipparion"* should be known as *Hippotherium* von Meyer 1829 (see also Bernor et al., in press). Other species that are pertinent to a discussion of the *"Hipparion"* Datum (e.g., *"H." koenigswaldi* Sondaar 1961, *"H." catalaunicum* Pirlot 1956, and *"H." africanum* Arambourg 1959) logically should be considered as potentially referable to *Hippotherium*. For present purposes, all species dicussed herein not considered as pertaining to *"Hipparion"* s.s. will be designated as *Hippotherium*. All species referable to *"Hipparion"* s.s. (taxa assigned to cranial Group 3; Woodburne and Bernor 1980) and to *"Hipparion"* s.s. by subsequent workers (see Bernor et al. 1989 for a modern list) are of Turolian and younger age. None appears to have been involved with the early Vallesian-aged dispersal under discussion here.

Source of the Hippotherium Datum

As discussed in Woodburne et al. (1981) and Bernor et al. (1989), a species of the North American genus, *Cormohipparion*, is the most likely source of the taxon represented in the Old World *"Hipparion"* Datum. *Cormohipparion quinni* (late Barstovian: ca. 14–13 Ma; Woodburne, in press) and *Cormohipparion occidentale* (late Barstovian

to late Clarendonian: ca. 13–9 Ma; Woodburne, in press; Bernor et al, this volume) are most commonly cited in this regard, with work still in progress on this question. The topic is discussed more fully below.

Cladistic Analysis of the *"Hipparion"* Datum

In figure 9.1 we advance a cladistic hypothesis for members of the *Hippotherium primigenium* s.s. evolutionary lineage, a lineage largely restricted to the Central Paratethys region after initial prochoresis of the Old World "founding" species. Figure 9.2 summarizes our current phylogenetic hypothesis for this lineage and the closely related regional LSD taxa referred to above. Table 9.1 provides the individual character states used to formulate this hypothesis. Further testing of this hypothesis will be focused principally on filling in missing data by the acquisition of more fossil material, particularly crania. Moreover, this hypothesis requires further rigorous testing by statistical analyses of variation in the continuous variables cited here and used extensively by Bernor et al. (1989). The data used here further build on published work by Bernor et al. (1993a, b) and Bernor and Lipscomb (1991; in press) but should be considered to be provisional pending analysis of many more Central Paratethys *Hippotherium primigenium* populations yet unstudied.

Cormohipparion occidentale has been used for outgroup comparison here, but future work may well synonymise at least this member of *Cormohipparion* with *Hippotherium*. The first step on the tree (fig. 9.1) is Pannonian C *Hippotherium primigenium* characterized by: reduction of the lacrimal foramen (C3B; actually not known until skulls are present in the Pannonian D–E interval); increased plication frequency of pre- and postfossettes (C19A); pli caballins increase in number to become consistently double (C21A); pli caballinid has increased complexity (C40A); metapodials show some lengthening (C50B). The next step, Pannonian D–E and Eppelsheim *Hippotherium primigenium*, is characterized by protocone becoming shortened so that it is lingually flattened and labially rounded (C23E). The next step, Nombrevilla *Hippotherium primigenium*, has protocone losing its lingual flattening and becoming somewhat compressed-ovate (C23F). Autapomorphically, pli caballins become complex (C21C) and metapodials become slightly lengthened (C50C).

The next step includes the Höwenegg *Hippotherium primigenium*, characterized by lower cheek tooth metaconids and metastylids being variably rounded or square (C32–34A/E, C36A/E). The following step, Rudabánya *Hippotherium primigenium*, is characterized by lower cheek tooth metaconids and metastylids being variably square to irregular shaped (C32–34E/D, C36E/D), and metatarsal length (m1) **x** DAW (m10) proportions are elevated outside the 99% confidence interval range of Pannonian C–E, Eppelsheim and Höwenegg *Hippotherium primigenium* (C50D).

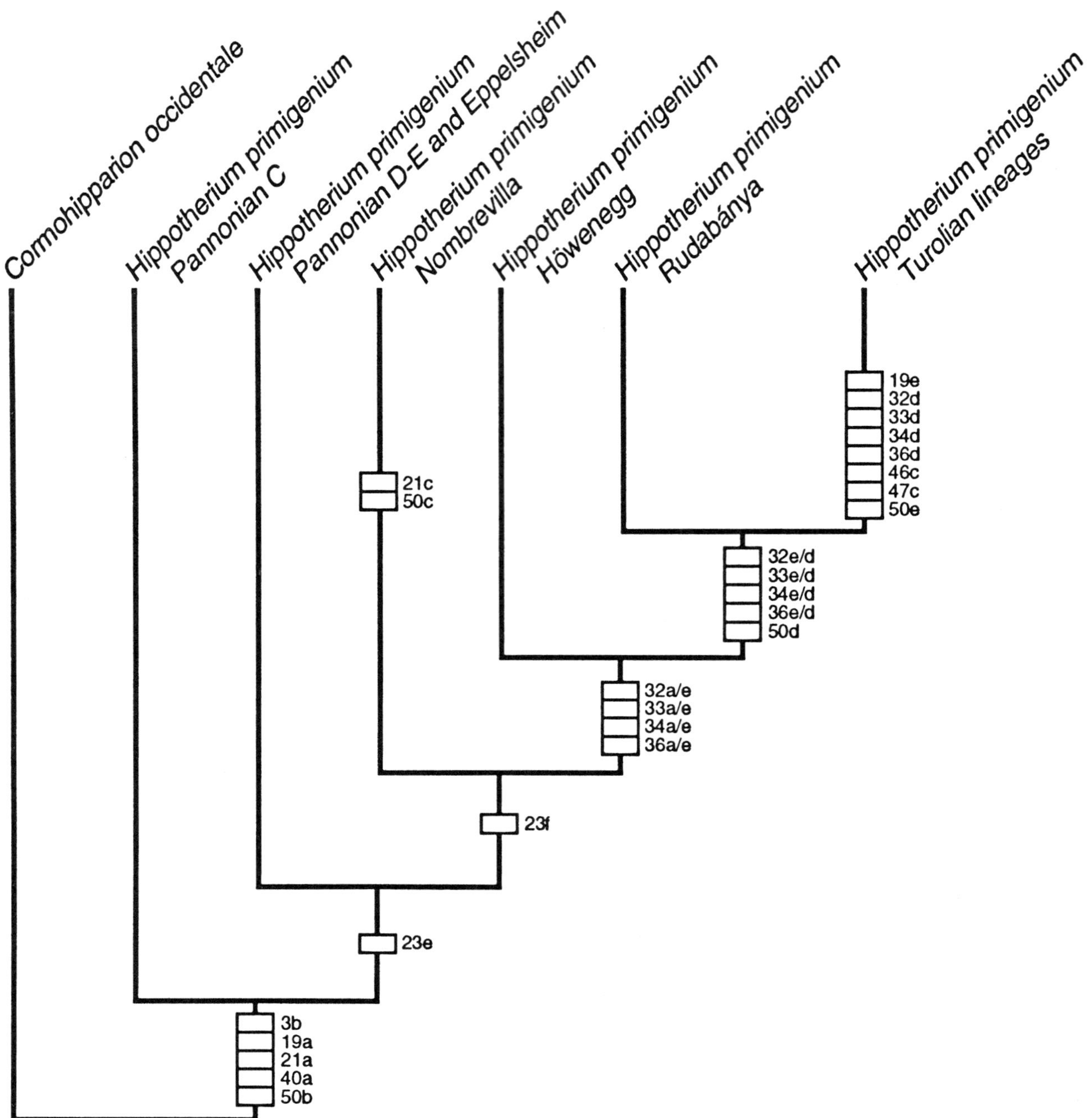

FIGURE 9.1 Cladistic Relationships of the *Hippotherium primigenium*–group.

Hipparionine Character States (following Bernor et al. 1989 and Bernor and Lipscomb 1991):

1) Relationship of lacrimal to the preorbital fossa: A = lacrimal large, rectangularly shaped, invades medial wall and posterior aspect of preorbital fossa; B = lacrimal reduced in size, slightly invades or touches posterior border of preorbital fossa; C = preorbital bar (POB) long with the anterior edge of the lacrimal placed more than half the distance from the anterior orbital rim to the posterior rim of the fossa; D = POB reduced slightly in length but with the anterior edge of the lacrimal placed still more than half the distance from the anterior orbital rim to the posterior rim of the fossa; E = POB vestigial, but lacrimal as in D; F = POB absent; G = POB very long with anterior edge of lacrimal placed less than half the distance from the anterior orbital rim to the posterior rim of the fossa; H = POB becomes vestigial, but lacrimal as in G; I = lacrimal as in G, POB absent.

2) Nasolacrimal fossa: A = POF large, ovoid shape and separated by a distinct medially placed, dorsoventrally oriented ridge, dividing POF into equal anterior (nasomaxillary) and posterior (nasolacrimal) fossae; B = nasomaxillary fossa sharply reduced compared to nasolacrimal fossa; C = nasomaxillary fossa absent (lost), leaving only nasolacrimal portion (when a POF is present).

3) Orbital surface of lacrimal bone: A = with foramen; B = reduced or lacking foramen.

4) Preorbital fossa morphology: A = large, ovoid shape, anteroposteriorly oriented; B = POF truncated anteriorly; C = POF further truncated, dorsoventrally restricted at anterior limit; D = subtriangular shaped and anteroventrally oriented; E = subtriangularly shaped and anteroposteriorly oriented; F = egg-shaped and anteroposteriorly oriented; G = C-shaped and anteroposteriorly oriented; H = vestigial but with a C-shaped or egg-shaped outline; I = vestigial without C-shape outline, or absent; J = elongate, anteroposteriorly oriented; K = small, rounded structure; L = posterior rim straight, with nonoriented medial depression.

5) Fossa posterior pocketing: A = deeply pocketed, greater than 15 mm in deepest place; B = pocketing reduced, moderate to slight depth, less than 15 mm; C = not pocketed but with a posterior rim; D = absent.

6) Fossa medial depth: A = deep, greater than 15 mm. in deepest place; B = moderate depth, 10-15 mm in deepest place; C = shallow depth, less than 10 mm in deepest place; D = absent.

7) Preorbital fossa medial wall morphology: A = without internal pits; B = with internal pits.

8) Fossa peripheral border outline: A = strong, strongly delineated around entire periphery; B = moderately delineated around periphery; C = weakly defined around periphery; D = absent with a remnant depression; E = absent, no remnant depression.

9) Anterior rim morphology: A = present; B = absent.

10) Placement of infraorbital foramen: A = placed distinctly ventral to approximately half the distance between the preorbital fossas anteriormost and posteriormost extent; B = inferior to, or encroaching on, anteroventral border or the preorbital fossa.

11) Confluence of buccinator and canine fossae: A = present; B = absent, buccinator fossa is distinctly delimited.

12) Buccinator fossa: A = unpocketed posteriorly; B = pocketed posteriorly.

13) Caninus (= intermediate) fossa: A = absent; B = present.

14) Malar fossa: A = absent; B = present.

15) Nasal notch position: A = at posterior border of canine or slightly posterior to canine border; B = approximately half the distance between canine and P2; C = at or near the anterior border of P2; D = above P2; E = above P3; F = above P4; G = above M1; H = posterior to M1.

16) Presence of P1: A = persistent and functional; B = lost early.

17) Curvature of maxillary cheek teeth: A = very curved; B = moderately curved; C = slightly curved; D = straight.

18) Maximum cheek tooth crown height: A = <30 mm; B = 30–40 mm; C = 40–60 mm; D = >60 mm maximum crown height

19) Maxillary cheek tooth fossette ornamentation: A = complex, with several deeply amplified plications; B = moderately complex with fewer, more shortly amplified, thinly banded plications; C = simple complexity with few, shortly amplified plications; D = generally no plis; E = very complex.

20) Posterior wall of postfossette: A = may not be distinct; B = always distinct.

21) Pli caballin morphology: A = double; B = single or occasionally poorly defined double; C = complex; D = plis not well formed.

22) Hypoglyph: A = hypocone frequently encircled by hypoglyph; B = deeply incised, infrequently encircled hypocone; C = moderately deeply incised; D = shallowly incised.

23) Protocone shape: A = round q-shape; B = oval q-shape; C = oval; D = elongate-oval; E = lingually flattened, labially rounded; F = compressed or ovate; G = rounded; H = triangular; I = triangular-elongate; J = lenticular; K = triangular with rounded corners.

24) Isolation of protocone: A = connected to protoloph; B = isolated from protoloph.

25) Protoconal spur: A = elongate, strongly present; B = reduced, but usually present; C = very rare to absent.

26) Premolar protocone/hypocone alignment: A = anteroposteriorly aligned; B = protocone more lingually placed.

27) Molar protocone/hypocone alignment: A = anteroposteriorly aligned; B = protocone more lingually placed.

28) P2 anterostyle(26U)/paraconid(26L): A = elongate; B = short and rounded.

29) Mandibular incisor morphology: A = not grooved; B = grooved.

30) Mandibular incisor curvature: A = curved; B = straight.

31) I3 lateral aspect: A = elongate, not transversely constricted; B = very elongate, transversely constricted; C = atrophied.

32) Premolar metaconid: A = rounded; B = elongated; C = angular on distal surface; D = irregular shaped; E = square shaped; F = pointed.

33) Molar metaconid: A = rounded; B = elongated; C = angular on distal surface; D = irregular shaped; E = square shaped; F = pointed.

34) Premolar metastylid: A = rounded; B = elongate; C = angular on proximal surface; D = irregular shaped; E = square shaped; F = pointed.

35) Premolar metastylid spur: A = present; B = absent

36) Molar metastylid: A = rounded; B = elongate; C = angular on proximal surface; D = irregular shaped; E = square shaped; F = pointed.

TABLE 9.1 Character State Distribution of Central European *Hippotherium primigenium* and Related Old World Taxa

	. . . . 1 . . . . 2 . . . . 3 . . . . 4 5
Hpri Pannonian C	ABABDB BAAAAAAAAAB AACBABBBABB
Hpri Pannonian D-E Eppelsheim	CCBBAAAAABBAAAC CCABABEBCBBAAAAAAAAABAB CBABBAABB
Hpri Nombrevilla	ABCBFBCBBAAAAAAAA ABBACBABBBAAC
Hpri Höwenegg	CCBBAAAAABBAAACACCABABFBCBBAAAAAAAAABABAACBABBBAAB EEE E B
Hpri Rudabánya	C BBAAAAABBAAAC CCABABFBCBBAAAAEEEAE AB ACBABBBAAD DDD D
Hpri Hungarian Turolian	CCEBABABCBBA DDDAD AB CCBABCCAAE
Hpri Chorora	ABABEBCBBAAAAAAAAA AB ACBABBBAA
Hnag Nagri (YGSP 330)	C DAAAA CABABEBCBBA

See figure 9.1 for explanation of character states.

The last step is in need of future study, however Bernor et al. (in press) have recognized that putative latest Vallesian and Turolian members of the *Hippotherium primigenium* lineage in Austria and Hungary undergo further transformations of maxillary cheek tooth pre- and postfossette morphology (C19E); mandibular cheek tooth metaconids and metastylids become a more irregular kidney-shape (C32–34D, C36D); mandibular cheek tooth pre- and postflexids become very complex (C46 and 47C); metapodials diversify in their proportions (C50E).

The cladistic relationships of the *Hippotherium primigenium* lineage directly lead to a refined phylogeny (fig. 9.2), and resulting chronologic and biogeographic hypoth-eses. A taxon similar in morphology to Pannonian C hipparions first extended its range into Central Europe, via Northern Asia. There is no evidence of Old World hipparions with the Pannonian C protocone morphology outside of Central Europe, although we assume that they must also have occurred at least in Northern Asia. Postcranial evidence suggests that Pannonian D–E hipparions occurred in Germany and Turkey. Moreover, there is nothing in the morphology of the skeletally limited Nagri (Siwaliks) and Chorora (Ethiopia) hipparions to exclude them from referral to the Pannonian D–E stage-of-evolution. This observation is at least consistent with the geochronologic arguments given here that the Siwalik and East African

37) Molar metastylid spur: A = present; B = absent.
38) Premolar ectoflexid: A = does not separate metaconid and metastylid; B = separates metaconid and metastylid.
39) Molar ectoflexid: A = does not separate metaconid and metastylid; B = separates metaconid and metastylid; C = converges with preflexid and postflexid to abut against metaconid and metastylid.
40) Pli caballinid: A = complex; B = rudimentary or single; C = absent.
41) Protostylid: A = present on occlusal surface; B = absent on occlusal surface, but may be on side of crown buried in cement; C = strong, columnar; D = a loop; E = a small, poorly developed loop; F = a small, pointed projection continuous with the buccal cingulum.
42) Protostylid orientation: A = courses obliquely to anterior surface of tooth; B = less oblique coursing, placed on anterior surface of tooth; C = vertically placed, lies flush with protoconid enamel band; D = vertically placed, lying lateral to protoconid band; E = open loop extending posterolabially.
43) Ectostylids: A = present; B = absent.
44) Premolar linguaflexid: A = shallow; B = deeper, V-shaped; C = shallow U-shaped; D = deep, broad U-shape; E = very broad and deep U-shape.
45) Molar linguaflexid: A = shallow; B = V-shaped; C = shallow U-shaped; D = deep, broad U-shape; E = very broad and deep U-shape.
46) Preflexid morphology: A = simple margins; B = complex margins; C = very complex.
47) Postflexid morphology: A = simple margins; B = complex margins; C = very complex.
48) Postflexid invades metaconid/metastylid by anteriormost portion bending sharply lingually: A = no; B = yes.
49) Protoconid enamel band morphology: A = rounded; B = flattened.
50) Metapodial morphology: A = relatively small, smaller to overlapping in m1 ≡ m10 proportions with Pannonian C-E hpri; B = proportions within m1 ≡ m10 range of Pannonian C–E; C = at uppermost range of Pannonian C–E; D = m1 ≡ m10 proportions elevated beyond 99% confidence interval of Pannonian C–E hipparions; E = diversification of metapodial proportions [probably multiple taxa]. This character is only relevant to figure 9.3.

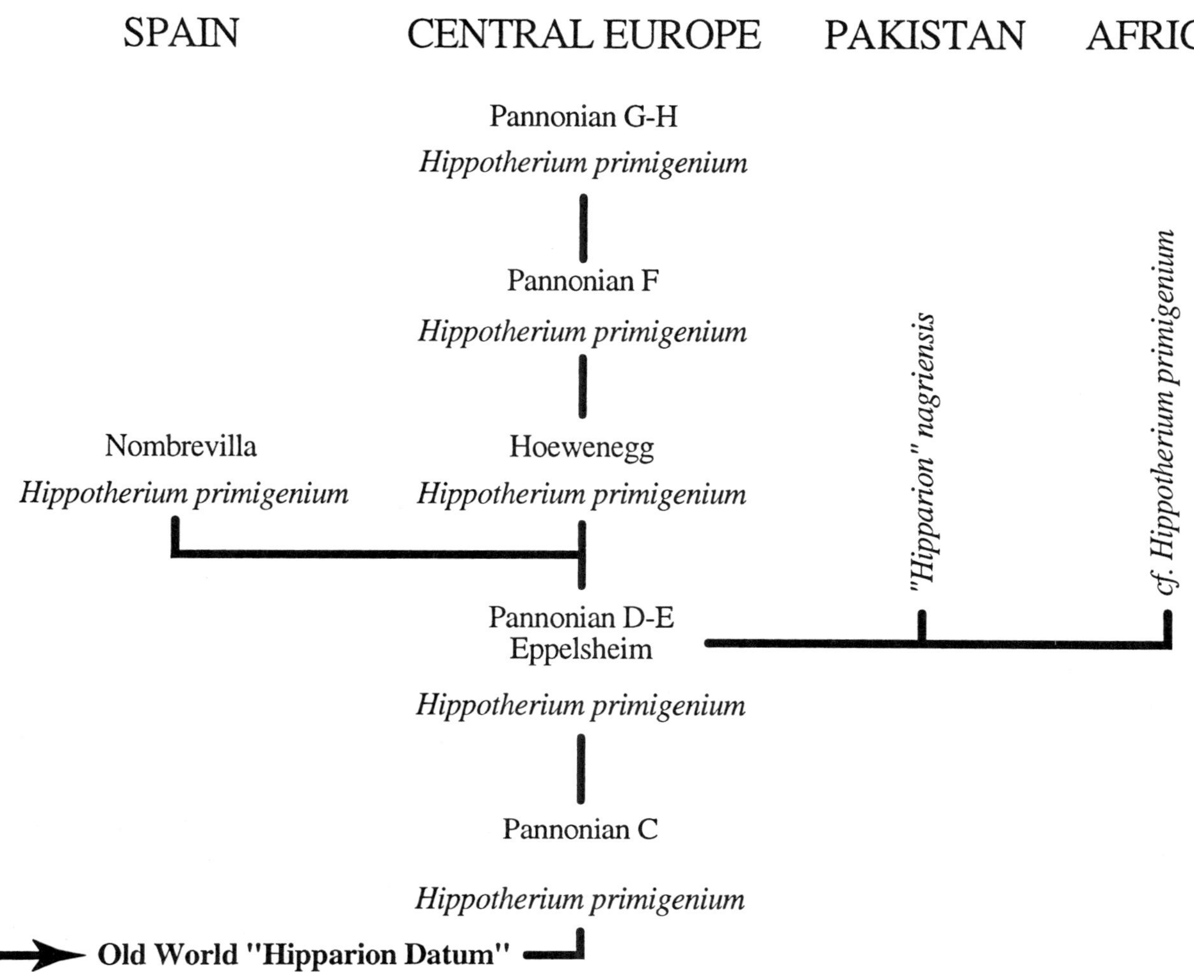

FIGURE 9.2 Phyletic relationships of selected Old World hipparions, based on figure 9.1.

hipparions have an LSD that is closely comparable chronologically with the European hipparion LSD. The LSD of Spanish hipparions, as represented by the Nombrevilla sample, may have occurred slightly later than Pannonian D.

The *Hippotherium primigenium* s.s. clade would then appear to be restricted to Central Europe after plausible deployment into the peri-Mediterranean, Indian subcontinent, and East African regions. The Höwenegg horse is derived compared to Eppelsheim and Pannonian D–E, and therefore hypothesized to be younger. The ca. 10.3 Ma age for the Höwenegg sample (Swisher, this volume; Woodburne et al, this volume) supports an older age for the less derived samples from Eppelsheim and Pannonian D–E (and Pannonian C). Thus the ca. 11.0 Ma age for Pannonian C samples (Rögl and Höck, this volume) is plausible. The Rudabánya hipparions would appear to be yet more derived than those from Höwenegg and are ranked biochronologically younger. Research in progress suggests that the Vienna Basin and Pannonian Basin con-

tained further derived late Vallesian and early Turolian age species of the *Hippotherium primigenium* s.s. lineage. The Spanish *Hippotherium catalaunicum* and North African *Hippotherium africanum* lineage would appear to have been biogeographically isolated from Central European populations soon after their local LSDs. Biochronologically, their first occurrence should be later than *Hippotherium primigenium* if Old World hipparions are truly monophyletic. However, derived members of the Central European *Hippotherium primigenium* evolutionary lineage may well prove to be younger than some of these "Group 1" members.

Stratigraphy and Geochronology of the Hippotherium Datum

In this section we summarize salient data that bear on the vitality of selected *Hippotherium* LSDs in order to evaluate the regional validity and utility of a *Hippotherium* FAD in the Old World (figs. 9.3 and 9.4). Only a few such LSDs are directly calibrated (see also Sen 1989). Most rely on a relatively complex geochronologic framework that utilizes basic lithostratigraphic, biostratigraphic, magnetostratigraphic, or radioisotopic information. The readership should appreciate that change in one aspect of this framework can, and usually does, affect the interrelationships of the others. The status of the various data sets is evolutionary. Thus it is not the intent of the following discussion to arrive at *the* age of the *Hippotherium* Datum. Rather it is an exercise to determine the degree to which various LSDs are mutually consisent and to examine the chronologic range within which the *Hippotherium* Datum likely occurred. In all cases the original evaluation is couched with respect to the geochronological framework presented by Berggren et al. (1995).

Note that in the following discussion Vallesian cooccurrences of *Anchitherium* and *Hippotherium primigenium* are considered likely to be "oldest" *Hippotherium* sites. Cooccurrences of *Hippotherium primigenium* and the rodent *Progonomys cathalai* are considered later Vallesian in age. Sites in which *Hippotherium primigenium* does not occur with either *Anchitherium* or *P. cathalai* in this part of the Vallesian age are likely old, and perhaps, oldest, *Hippotherium* records (e.g., Berggren and Van Couvering 1978).

Central Paratethys

Bernor et al. (1988) presented a modern evaluation of the LSDs of *Hippotherium* in the Vienna Basin and other Paratethys localities, while Steininger et al. (1989; this volume) and Rögl et al. (1993; this volume) have updated various correlations, calibrations, and zonations relevant to the present discussion. Some of the correlations differ from those utilized by Berggren et al. (1985a,b), and we adjust these, as appropriate to the framework presented in Berggren et al. 1995. As background, Steininger et al. (1985) indicated that whereas early Miocene Mediterranean/Paratethys aquatic environments were confluent and enabled their correlation, tectonic fragmentation of the region fostered the development of provincial endemism in marine faunas. By late Miocene time many local stratigraphic/geochronologic sequences were developed that could be correlated only with difficulty to those of other Paratethys districts, much less to those of the Mediterranean realm.

In this general context, sequences bearing *Hippotherium* can be correlated to one another with varying degrees of certainty. For *Hippotherium* LSDs and their eventual contribution to a *Hippotherium* FAD, the biostratigraphic geochronologist is faced with the question: To what extent is one *Hippotherium* LSD securely correlated (i.e., interpreted as being synchronous) with another? Secondly, if the correlation can be established or interpreted to be effectively synchronous, how well can it be calibrated with respect to other geochronologic systems? To what extent

can the Paratethys record contribute to the establishment of a *Hippotherium* FAD beyond the Paratethys region?

As Bernor et al. (1988) have shown, the Central Paratethys Vienna Basin has one of the best medial and late Miocene stratigraphic records, provincially. For present purposes, we follow the correlations presented in Rögl et al. 1993 and this volume. The Pannonian Stage is divided into 8 subdivisions (A–H). In the Vienna Basin (northernmost part), remains of *Hippotherium* first occur in the locality of Horovany (CSSR), correlated "with the top of Pannonian Zone B" (Bernor et al. 1988:435). Bernor et al. (1988:435–36) indicate that *Hippotherium primigenium* extends further, from Pannonian Zone C to Pannonian zones D–E (figs. 9.2, 9.4), or "within the 11–8 Ma time range" (p. 436).

Calibration of the Pannonian succession is indirect and, perhaps understandably, imprecise. Bernor et al. (1988) have suggested that the Vienna Basin "H." *primigenium* sample of Gaiselberg (Pannonian C) has the most primitive upper cheek tooth dental morphology of any Old World hipparion, in that about 35% of the specimens retain a protocone that is elongate to lingually flattened (see also above; and figs. 9.1, 9.2). Lingually flattened elongate upper cheek tooth protocones are commonly found in the proposed sister-taxon to the Old World Vallesian hipparion *Cormohipparion occidentale*, although Woodburne (1996) indicates that this character also occurs in specimens referred to *C. quinni*. In this regard, the Pannonian C sample of *Hippotherium primigenium* is variable and the elongate-oval protocone shape was apparently lost in other early Vallesian populations of Central European *Hippotherium primigenium*. Bernor et al. (1988) thus infer that the elongate-oval shape of the protocone is plesiomorphic for *Hippotherium primigenium* and that populations having lost this feature are chronologically younger than those in which it still is present (e.g., figs. 9.1, 9.2), at least as a working hypothesis.

Whereas the morphology exhibited in the Gaiselberg sample is important for assessing the relative morphologic advancement of other Old World hipparion samples, the relationship between the Gaiselberg and the North American records is less clear. The out-group comparisons prepared by Bernor et al. (1989) were performed with quarry samples of *Cormohipparion occidentale* from the X-Mas/Kat quarries in Nebraska known to be of late Clarendonian age. In terms of the chronology presented in Tedford et al. (1987) and Whistler and Burbank (1992), these quarry samples are on the order of 9–10 m.y. old. Whistler and Burbank (1992), Woodburne (1996), and Bernor et al. (this volume) indicate that the *Cormohipparion occidentale–*group is at least 12.5 Ma old in North America and that the chronology of acquisition of relevant character-states still is under investigation. At the present level of information, the apparent age discrepancy between the Gaiselberg sample and the X-mas/Kat quarries of *C. occidentale* in Nebraska may render comparisons between the two samples irrelevant for closely controlled morphologic/geochronologic analysis.

As regards correlation to the MPTS, the paleomagnetic samples associated with the Gaiselberg hipparion are unknown; but the MPTS signature of the Pannonian D–aged site of Inzersdorf is of normal polarity (Rögl, pers. comm.), consistent at least with it being within magnetochron C5n.2, as utilized in this report. For the moment, the Gaiselberg *Hippotherium primigenium* is taken as being older than the 10.3 m.y. age attributed to the Höwengg sample (Wooburne et al., this volume), but within the time span of magnetochron C5n.2, or approximately 11.0 Ma. At the same time it must be recognized that Rögl and Daxner-Höck (this volume) and Steininger et al. (this volume) suggest that the FAD of *Hippotherium* ("*Hipparion*") is at the base of Pannonian zone C, correlated to an age of 11.5–11.0 m.y., and, until these estimations can be corroborated by recourse to either or both of ^{39}Ar/^{40}Ar radioisotopic calibration or to reconciliation of fission-track ages compiled by, e.g., Chumakov et al. (1992), with either other radioisotopic or magnetic polarity schemes, this issue remains open. The *Hippotherium* FAD could still be older than 11.0 m.y.

Germany

The type locality of *Hippotherium primigenium* is the *Dinotherium* Sands near Eppelsheim, western Germany (Sondaar 1961). The unit consists of a sequence of sand and conglomerate beds up to about 14 meters thick; it is stratigraphically unconformable on *Corbicula*-bearing beds and is overlain by deposits of Quaternary (possibly also late Pliocene) age (Wagner 1964; 1973). The *Corbicula* beds are of Agenian age (Thenius 1959) (ca 20–23.8 Ma; Steininger et al., this volume). The Turolian Dorn-Dürkheim beds (and fauna) differ lithologically from the *Dinotherium* Sands (e.g., Franzen and Storch 1975), but regionally overlie them (Tobien 1988) and are part of the sequence that formed from the ancestral Rhine River (Tobien 1986). The *Dinotherium* Sands cannot be directly correlated with the marine invertebrate, isotopic, or magnetostratigraphic framework. On the basis of biochronology Mein (1989) suggests that the Eppelsheim fauna is older than that from both Nombrevilla (Spain) and from Höwenegg, in ascending order, but still within MN 9. De Bruijn et al. (1992: tab. 4) reiterate the relative chronologic position of Eppelsheim and Höwenegg, but portray the Gaiselberg sample as being younger than both of the others (no discussion provided therein).

The geologic and paleontologic contexts of Höwenegg have been discussed by Woodburne et al. (this volume). Following Bernor et al. (1993a,b) and the scheme of figure 9.2, the Höwenegg sample of *Hippotherium primigenium* is considered to be more derived than those from either Eppelsheim or Gaiselberg. The normal magnetozone associated with the 10.3 Ma calibration at Höwenegg is consistent with these data (fig. 9.3).

In his detailed dental (as well as postcranial) analyses, Sondaar (1961) did not compare the Eppelsheim sample with the one from Höwenegg. On the other hand, Bernor et al. (1989; in press) did make those comparisons and found no significant morphological or metric differences between these populations. Accepting the correlations proposed herein, it appears that the Höwenegg sample is not the geochronologically oldest *Hippotherium* in Western Europe. The local LSD thus should be somewhat older than 10.3 Ma, but in the absence of a detailed morphological comparison, and given that there are no physical means for correlating the two samples, the temporal difference between Höwenegg and Eppelsheim cannot be specified at this time.

North Africa

Arambourg (1959) described "*Hippotherium*" *africanum* from Bou Hanifia, Algeria. Woodburne and Bernor (1980) considered this to be a derived species relative to the Höwenegg taxon, consistent with its occurring stratigraphically 100 m above a tuff dated at 12.03 ± 0.25 Ma (Ameur et al. 1976; Bernor et al. 1988; Sen 1989) and being associated with *Progonomys cathalai* (Jaeger et al. 1973). This is a single level occurrence for *Hippotherium*. Ouda and Ameur (1978:417) indicate that the Bou Hanifia Formation is equivalent to planktonic foraminiferal zone N15. Berggren et al. (1995) provide evidence that zone N15 ranges from 11.4 to 10.9 m.y. From magnetic polarity data, Sen (1989) suggests that *Hippotherium africanum* is older than 10.5 Ma, but younger than 12 Ma. We suggest here that—based on the stratigraphic context, derived status of its hipparion, and geochronologic data—the Bou Hanifia hipparion is more likely to have been about 9.5 Ma old (fig. 9.4).

Spain

Crusafont-Pairó (1951) coined the Vallesian age from the Valles Penedes succession in central Spain, and the presence of "*Hipparion*" in these deposits, in contrast to its absence below, was a clear definition of the new age. Even though the species (proposed here to pertain to *Hippotherium*) occurs abundantly at the oldest Spanish Vallesian type locality (Nombrevilla; Sondaar 1961, MOW personal observation), it is a single-level site in this area and thus the "true" LSD may not be represented. The definition of the Vallesian stands, nevertheless. Subsequent to Crusafont-Pairó (1951), this age concept was extended to other

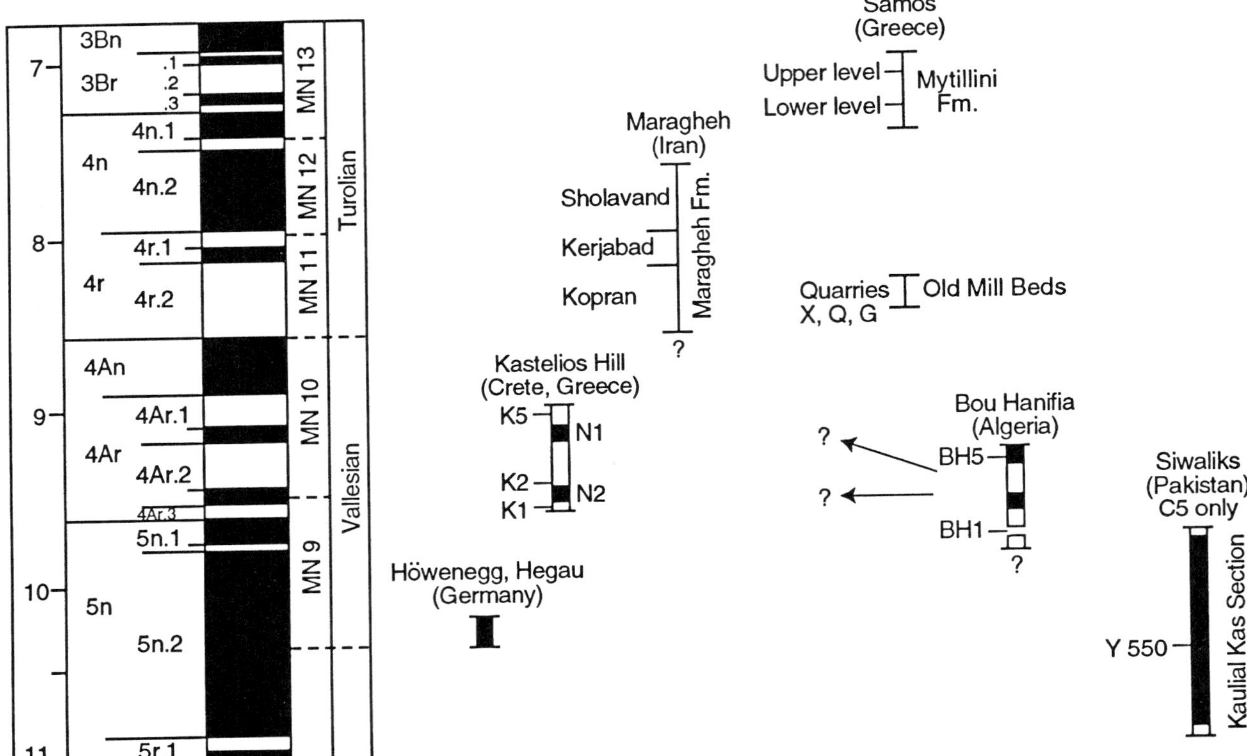

FIGURE 9.3 Chronological relations of selected Old World sites with hipparions, based on new interpretations of chronologic or paleontologic data.

European sites, with a prime correlate being the sequence at Can Mata (Hostalets de Pierola, Spain). Pirlot (1956) described *"Hipparion" catalaunicum* from this site. Additional sites (below) have subsequently been added as (presumptively early) Vallesian correlates, based on a *"Hipparion"* LSD.

Sondaar (1961) named *"Hipparion" koenigswaldi* from the type Vallesian locality of Nombrevilla, and showed (e.g., below) that it differed from *H. primigenium* from the *Dinotherium* Sands, Eppelsheim, Germany, the type locality from which *Hippotherium primigenium* von Meyer 1829 was obtained. The deposits at Nombrevilla, Spain, lack *Progonomys cathalai* and *Hippotherium* (nomenclature used herein) *koenigswaldi* differs in cheek tooth morphology from the Eppelsheim form (below). *Hippotherium koenigswaldi* also occurs at the Spanish locality Pedregueras IA–D, which lacks *Progonomys cathalai* and is considered (Freudenthal 1968) to be stratigraphically below Pedregueras IIC as well as closely correlated with Nombrevilla. Daams et al. (1988:293) also show *Progonomys* as present in Vallesian biozone I (including Pedregueras IIC), but absent in subjacent biozone H (including Nombrevilla). See also table 2 in Rögl and Daxner-Höck, this volume, for Nombrevilla being older than Pedregueras IIC.

"Hipparion" (herein *Hippotherium*) cf. *catalaunicum* is associated with *Progonomys cathalai* at Pedregueras IIC (Freudenthal and Sondaar 1964). This is consistent, at least, with the interpretation that *Hippotherium catalaunicum* is derived relative to, and is chronologically younger than, *Hippotherium koenigswaldi*. In conclusion, *Hippotherium koenigswaldi* is of earliest Vallesian age by definition, and ancillary morphologic and faunal data suggest that *"H." catalaunicum* is younger. In figure 9.1, *Hippotherium koenigswaldi* (= Nombrevilla) is considered to be derived relative to the samples of *Hippotherium primigenium* from Eppelsheim, but more plesiomorphic than that from Höwenegg. No magnetic data are yet known from Nombrevilla, a lapse we hope will soon be rectified. Lacking information to the contrary, we correlate the Nombrevilla sample as shown in figures 9.1 and 9.4.

Tunisia

Some of the oldest samples of *Hippotherium* cited by Berggren et al. (1985b:249; Forstén 1972) are correlated (Wiman 1978:96) approximately to the base of planktonic foraminiferal zone N14 in the Beglia Formation of Tunisia. The base of zone N14 is now correlated by Berggren et al. (1995) at about 11.3 to 11.7 Ma (fig. 9.4). In cooccurring with *Progonomys cathalai*, the Beglia Formation record of

Hippotherium primigenium is not an "oldest" *Hippotherium*. Forstén (1972) reports that *"Hipparion" primigenium* was recovered from thirteen localities that span an interval of about 40–50 meters, but stratigraphic details are not given in the appendix supplied by Robinson and Black. Accordingly, we suggest that the correlation to foraminiferal zone N14 is open to question, and that largely because of its association with *Progonomys*, Beglia Formation hipparion is unlikely to be older than about 9.5 Ma (fig. 9.4).

Ukraine

A geologically "old" *Hippotherium* appears to be represented by a sample from Jeltokamenka (Ukraine), where *Hippotherium* (*"Hipparion"*) and *Anchitherium* cooccur (Gabunia 1981; Berggren et al. 1985b:249) in an early Vallesian fauna. The unit is correlated to about the middle Sarmatian Stage of the Paratethys (Gabunia 1981). That level in the Sarmatian is correlated on invertebrate fossils (NSM 2b; Andreescu 1979) to rocks assigned to the upper half of Magnetic Chron C5A (Berggren et al. 1985b:249). This chron ranges in correlated age from ca. 11.9 to 12.9 Ma (Berggren et al., 1995), with the "upper half" being as old as ca. 12.5 Ma (fig. 9.4). We suggest that the paleomagnetic correlation more likely is to about the middle of Chron C5N.2n and consistent with a correlation older than that of Höwenegg (fig. 9.4).

Siwaliks

Beginning with Falconer and Cautley 1849, hipparionine horses have been known from the important Siwalik sequence of the Potwar Plateau, Pakistan and adjacent India. In fact the richly fossiliferous ca. 5,000-meter-thick nonmarine sequence has yielded the best means of characterizing the Neogene land mammal faunal succession in this part of Asia, and has been the subject of numerous faunal and stratigraphic studies. Barry et al. (1982) report on the *Hippotherium* FAD (the *"Hipparion"* s.l. Interval-Zone) in the Siwalik succession. The discussion is keyed to two reference sections with combined biostratigraphic, lithostratigraphic, and magnetostratigraphic control. The first reference section, Kaulial-Kas, in the Khaur region, is 3,200 m thick, the upper 2,490 m of which are lithologically correlative to the Nagri and Dhok Pathan formations, Nagri being the older. A prominent stratigraphic marker,

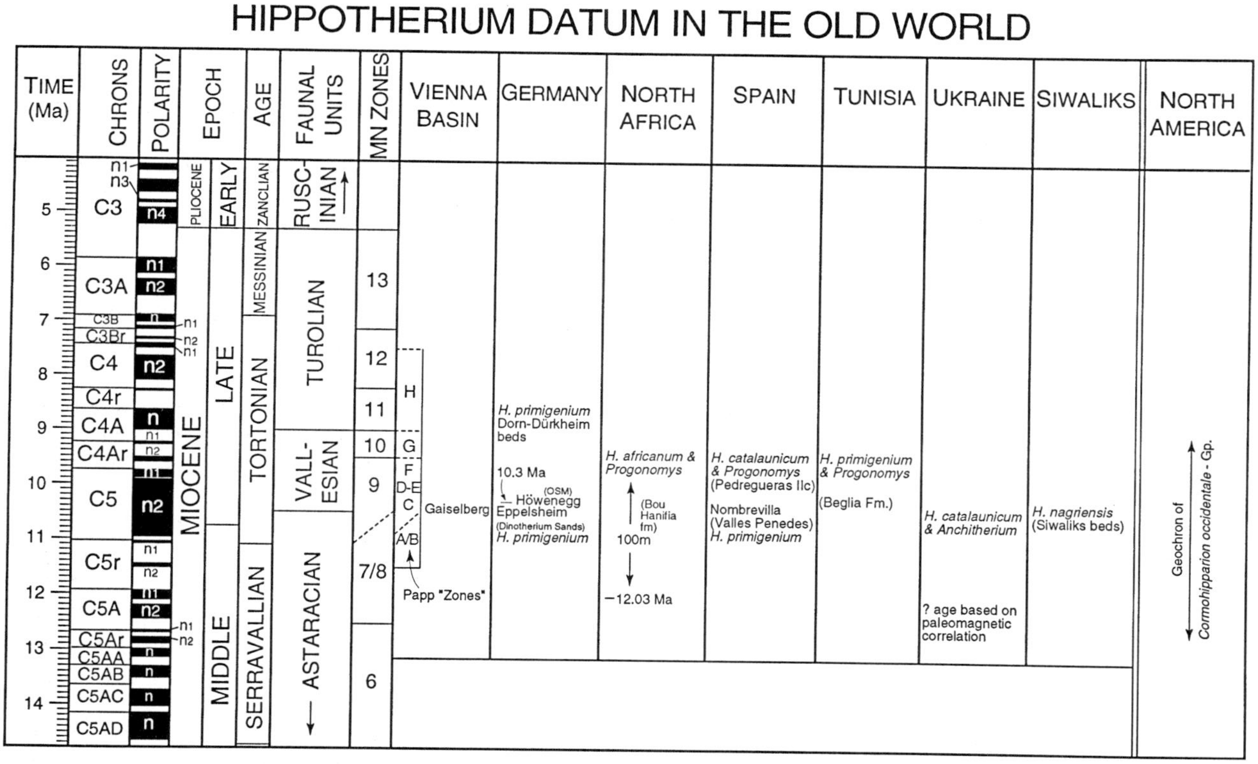

HIPPOTHERIUM DATUM IN THE OLD WORLD

TIME (Ma)	CHRONS	POLARITY	EPOCH	AGE	FAUNAL UNITS	MN ZONES	VIENNA BASIN	GERMANY	NORTH AFRICA	SPAIN	TUNISIA	UKRAINE	SIWALIKS	NORTH AMERICA
5	C3	n1, n3, n4	PLIOCENE / EARLY	ZANCLIAN	RUSCINIAN									Geochron of *Cormohipparion occidentale* - Gp.
6	C3A	n1, n2	MIOCENE / LATE	MESSINIAN	TUROLIAN	13								
7	C3B, C3Br, C4	n1, n2, n1, n2				12	H	*H. primigenium* Dorn-Dürkheim beds						
8	C4, C4r	n2		TORTONIAN		11	H							
9	C4A, C4Ar	n, n1, n2			VALL-ESIAN	10	G, F, D-E, C		10.3 Ma (OSM) Höwenegg Eppelsheim (Dinotherium Sands) *H. primigenium*	*H. africanum & Progonomys* (Bou Hanifia fm) 100m	*H. catalaunicum & Progonomys* (Pedregueras IIc) Nombrevilla (Valles Penedes) *H. primigenium*	*H. primigenium & Progonomys* (Beglia Fm.)		
10	C5	n1, n2				9	C	Gaiselberg					*H. catalaunicum & Anchitherium*	*H. nagriensis* (Siwaliks beds)
11	C5r	n1, n2			ASTARACIAN	7/8	A/B	Papp "Zones"	−12.03 Ma				? age based on paleomagnetic correlation	
12	C5A	n1, n2	MIDDLE	SERRAVALLIAN										
13	C5Ar, C5AA, C5AB	n1, n2, n												
14	C5AC, C5AD	n				6								

FIGURE 9.4 Correlation of all sites discussed in the text, utilizing MPTS data from Berggren et al., 1995. Vienna Basin correlations are after Rögl et al. (1993); southwest Germany is after Swisher (this volume) and Woodburne et al. (this volume). North Africa follows text and Swisher (this volume; B1 and B5 in figure 9.3 are faunal levels adjacent to magnetic polarity profile). Siwaliks follows text. Y 550 (fig. 9.3) is *Hippotherium*-producing locality adjacent to magnetic polarity profile. For all other sites see text. Distribution of *Cormohipparion*–group follows Woodburne (1996).

the "U" sandstone, occurs stratigraphically somewhat above the middle of this 2,490-meter-thick sequence. The lower part of this sequence is comprised of a magnetically normal interval about 940 m thick correlated to chron C5N (Badgley et al. 1986). Based on Berggren et al. (1995), we believe that chron C5N.2 ranges in age from ca. 10.9 to 9.9 Ma and, following the procedures illlustrated in Badgley et al. (1986), that the interpolated age of the Siwaliks *Hippotherium* LSD is about 10.4 Ma (figs. 9.3, 9.4; however, see Pilbeam et al. this volume).

Crete

Kastellios Hill is a locality most likely correlated to MN 9 (fig. 9.3). The predominantly reversed magnetostratigraphy at this site led Sen (1986) to correlate the latest MN 9 K1 site and early MN 10 K2, K3, and K4 with chron C5R, an age now considered older than that at Höwenegg. Along with its later Vallesian age, and thus more likely correlation to some place within chron 4Ar (fig. 9.3), Kastellios Hill is irrelevant to the base of the Vallesian and to the old age of the *Hippotherium* Datum.

East Africa

The earliest Ethiopian hipparionines are recorded in sediments of the Chorora Formation (Sickenberg and Schönfeld 1975). Bernor has studied a small assemblage of these (cited below) and found their morphology to be consistent with Pannonian D and E hipparions (*Hippotherium primigenium* s.s.) from the Central Paratethys. Kunz et al. (1975) reported a K-Ar age of 10.8 ±0.3 Ma (sample M938; corrected for current decay constants from 10.5 Ma) for basalts underlying the formation, and an age of 9.3 ±1.0 Ma (sample M567; corrected from 9.05 Ma) for a trachyite that overlies the Chorora Formation. These ages indicate that the possible age range of the Chorora hipparionines is 10.5 to 9.3 Ma. This is consistent with the advanced morphological similarities shared with samples from Pannonian D–E levels in the Paratethys.

Discussion

Hippotherium LSDs that are directly associated with magnetostratigraphic or radioisotopic data are estimated to be ca. 10.5–11.0 Ma, with the sample from Gaiselberg believed to be the oldest (and first occurring in a biostratigraphically developed stratigraphic sequence), with localities such as Eppelsheim, the Pannonian D–E, and Nombrevilla being somewhat younger. Reanalysis of the relevant Potwar Plateau (Siwaliks) magnetostratigraphic sequence correlates the *Hippotherium* LSD as being ca. 10.4 Ma. Evidence from North Africa is permissive of this corre-

lation, whereas that from East Africa is ambiguous. Until compelling information on direct association of beds with *Hippotherium primigenium* and data to tie the rocks to the marine, radioisotopic, or paleomagnetic time scales are adduced, we set aside the Ukraine records.

Information on the geochronology of potentially relevant samples of *Cormohipparion occidentale* in North America is becoming better known, and we expect that the morphology of these samples will be completely documented in the near future. At that point it may be possible to ascertain a more precise age of the potential ancestor for the Old World *Hippotherium* Datum, and to ascertain the degree to which that compares with the apparent age of that datum as discused here. Within the Old World area discussed here, the Siwaliks section is the only one in Asia well enough developed to be useful in this regard. Unless some kind of provincial factor is operative here, it seems irrational that the Siwalik sequence should be greatly younger than that in Western Europe. More information is required to analyze this situation. For the moment, we consider the prochoresis of Old World *Hippotherium* to have been nominally instantaneous within a span of 0.5 m.y.

LITERATURE CITED

Ameur, R. C., J.-J. Jaeger, and J. Michaux. 1976. Radiometric age of early "*Hipparion*" fauna in Northwest Africa. *Nature* 261:38–39.

Andreescu, I. 1979. Middle-upper Neogene and early Quaternary chronostratigraphy from the Dacic Basin and correlation with neighbouring areas. *VIIth International Congress on Mediterranean Neogene, Athens*, pp. 129–38.

Arambourg, C. 1959. *Vertebrés continentaux du Miocéne superieur de l'Afrique du Nord. Memoire, Publications du Service de la Carte Geologique de l'Algerie*. Paleontologie 4.

Badgley, C., L. Tauxe, and F. Bookstein. 1986. Estimating the error of age interpolation in sedimentary rocks. *Nature* 319:139–41.

Barry, J. C., E. H. Lindsay, and L. L. Jacobs. 1982. A biostratigraphic zonation of the middle and upper Siwaliks of the Potwar Plateau of Northern Pakistan. *Palaeogeography, Palaeoclimatology, and Palaeoecology* 37:95–130.

Berggren, W. A., D. V. Kent, C. C. Swisher III, and M.-P. Aubry. 1995. A revised Cenozoic geochronology and chronostratigraphy. In *Geochronology, Time Scales, and Global Stratigraphic Correlations: A Unified Temporal Framework for an Historical Geology*, ed. W. A. Berggren, D. V. Kent, and J. Hardenbol. Society of Economic Mineralogists and Paleontologists Special Publication No. 54: 129–212.

Berggren, W. A., D. V. Kent, J. J. Flynn, and J. A. Van Couvering. 1985b. Cenozoic geochronology. *Bulletin of the Geological Society of America* 96:1407–18.

Berggren, W. A., D. V. Kent, and J. A. Van Couvering. 1985a. The Neogene; Part 2: Neogene geochronology and chronostratigraphy. In *The Chronology of the Geological Record*, ed. N. J.

Snelling, pp. 211–60. Geological Society of London Memoire 5.

Berggren, W. A. and J. A. Van Couvering. 1974. The late Neogene: Biostratigraphy, geochronology, and paleoclimatology of the last 15 million years in marine and continental sequences. *Palaeogeography, Palaeoclimatology, and Palaeoecology* 16:1–216.

———. 1978. Biochronology. In *Contribution to the Geologic Time Scale*, ed. G. V. Cohee, M. F. Glaessner, and H. D. Hedberg, pp. 39–55. American Association of Petroleum Geologists Studies in Geology 6.

Bernor, R. L., G. D. Koufos, M. O. Woodburne, and M. Fortelius. This volume. The evolutionary history and biochronology of European and Southwest Asian late Miocene and Pliocene hipparionine horses.

Bernor, R. L., J. Kovar-Eder, D. Lipscomb, F. Rögl, S. Sen, and H. Tobien. 1988. Systematics, stratigraphic, and paleoenvironmental contexts of first-appearing hipparion in the Vienna Basin, Austria. *Journal of Vertebrate Paleontology* 8:427–52.

Bernor, R. L., J. Kovar-Eder, J.-P. Suc, and H. Tobien. 1990. A contribution to the evolutionary history of European late Miocene age hipparionines (Mammalia: Equidae). *Paléobiologie Continentale* 17:291–309.

Bernor, R. L., M. Kretzoi, H.-W. Mittmann, and H. Tobien. 1993. Preliminary systematic assessment of the Rudabánya hipparions. *Mitteilungen der Bayerischen Staatssammlung für Paläontologie und historische Geologie* 33: 195–207.

Bernor, R. L. and D. Lipscomb. 1991. The systematic position of *"Plesiohipparion"* aff. *huangheense* (Equidae, Hipparionini) from Gülyazi, Turkey. *Mitteilung der Bayerischen Staatsslammlung für Paläontologie und historische Geologie* 31:107–23.

Bernor, R. L., H.-W. Mittmann, and F. Rögl. 1993b. The Götzendorf hipparions. *Annalen des Naturhistorisches Museum, Wien* 95:101–20.

Bernor, R. L., H. Tobien, L.-A. Hayek, and H.-W Mittmann. In press. *Hippotherium primigenium* (Equidae, Mammalia) from the late Miocene of Höwenegg (Hegau), Germany. *Andrias* 10.

Bernor, R. L., H. Tobien, and M. O. Woodburne. 1989. Patterns of Old World hipparionine evolutionary diversification and biogeographic extension. In *European Neogene Mammal Chronology*, ed. E. H. Lindsay, V. Fahlbusch, and P. Mein, pp. 263–319. New York: Plenum.

Bernor, R. L., M. O. Woodburne, and J. A. Van Couvering. 1980. A contribution to the chronology of some Old World faunas based on hipparionine horses. *Géobios* 13:35–59.

Bruijn, H. de, R. Daams, G. Daxner-Höck, V. Fahlbusch, L. Ginsburg, P. Mein, and J. Morales. 1992. Report of the RCMNS working group on fossil mammals, Reisensburg 1990. *Newsletters on Stratigraphy* 26:65–118.

Chumakov, I. S., S. L. Byzova, S. S. Ganzey, C. Arias, G. Bigazzi, F. P. Bonadonna, J. C. Hadler Neto, and P. Norelli. 1992. Interlaboratory fission track dating of volcanic ash levels from Eastern Paratethys: A Mediterranean-Paratethys correlation. *Palaeogeography, Palaeoclimatology, Palaeoecology* 95:287–95.

Daams, R., M. Freudenthal, and A. J. van der Meulen. 1988. Ecostratigraphy of micromammal faunas from the Neogene of Spain. In *Biostratigraphy and Paleoecology of the Neogene Micromammalian Faunas from the Calatayud-Teruel Basin (Spain)*, ed. M. Freudenthal, pp. 287–302. Scripta Geologica Special Issue 1.

Falconer, H. and P. T. Cautley. 1849. *Equidae, Camelidae, and Sivatherium: Fauna Antiqua Sivalensis*, pt. 9.

Forstén, A.-M. 1972. *Hipparion primigenium* from southern Tunisia. *Extrait de Notes du Service Géologique* 35:7–28.

Franzen, J. L. and G. Storch. 1975. Die unterpliozäne (turolische) wirbeltierfauna von Dorn-Dürkheim, Rheinhessen (SW Deutschland). 1. Entdeckung, Geologie, Mammalia, Carnivora, Proboscidea, Rodentia. Grabungsergebnisse, 1972–1973. *Senckenbergiana lethaea* 56:233–303.

Freudenthal, M. 1968. On the mammalian fauna of the *"Hipparion"*-beds in the Calatayud-Teruel Basin (Prov. Zaragossa, Spain), pt. IV. The genus *Megacricetodon* (Rodentia). *Koninklijke Nederlandese Akademie van Wetenschappen, Proceedings* B71:71–72.

Freudenthal, M. and P. Y. Sondaar. 1964. Les faunes à *"Hipparion"* des environs de Daroca Espagne et leur valeur pour la stratigraphic du Néogene de l'Europe. *Koninklijke Nederlandese Akademie van Wetenchappen, Proceedings* B67:473–90.

Gabunia, L. 1981. Traits esentiels de l'évolution des faunes de Mammifères néogènes de la région mer Noire-Caspienne. *Bulletin Museum National d'Histoire naturelle*, 4th series (3):195–204.

Jaeger, J. J., J. Michaux, and B. David. 1973. Biochronologie du Miocene moyen et supérieur continental du Maghreb. *Comptes Rendus Academie des Sciences, Paris* D277:2477–80.

Kunz, K., H. Kreuzer, and P. Müller. 1975. Potassium-argon age determinations of the Trap basalt of the southeastern part of the Afar Rift. In *Afar Depression of Ethiopia*, ed. A. Pilger and A. Rösler, pp. 370–74. Stuttgart: E. Schweizerbartsche Verlagsbuchhandlung.

Lindsay, E. H., N. D. Opdyke, and N. M. Johnson. 1984. Blancan-Hemphillian land mammal ages and late Cenozoic mammal dispersal events. *Annual Review, Earth and Planetary Science* 12:445–88.

Lindsay, E. H., N. D. Opdyke, N. M. Johnson, and R. F. Butler. 1987. Mammalian chronology and the magnetic polarity time scale. In *Cenozoic Mammals of North America: Geochronology and Biostratigraphy*, ed. M. O. Woodburne, pp. 269–90. Berkeley: University of California Press.

Mein, P. 1989. Updating of MN zones. In *European Neogene Mammal Chronology*, ed. E. H. Lindsay, V. Fahlbusch, and P. Mein, pp. 73–90. New York: Plenum.

Opdyke, N. D., E. H. Lindsay, N. M. Johnson, and T. Downs. 1977. The paleomagnetism and magnetic polarity stratigraphy of the mammal-bearing section of Anza Borrego State Park, California. *Quaternary Research* 7:316–29.

Ouda, K. and R. C. Ameur. 1978. Contribution to the biostratigraphy of the Miocene sediments associated with primitive Hipparion fauna of Bou-Hanifia, Northwest Africa. *Revista Española de micropaleontología* 10:407–20.

Pirlot, P. R. 1956. Les formes européenes du genre *"Hipparion"*. *Memorias y Comunicaciones del Instituto Geologia Barcelona* 14:1–45.

Rögl, F. and G. Daxner-Höck. This volume. Late Miocene Paratethys correlations.

Rögl, F., H. Zapfe, R. L. Bernor, G. Brzobohaty, G. Daxner-Höck, O. Fejfar, J. Gaudant, P. Herrmann, G. Rabeder, O. Schultz, and R. Zetter. 1993. Die Primatenfundstelle Götzendorf and der Leitha (Obermiozän des Wiener Becken, Niederösterreich). *Jahrbuch Geologie B.-A.* 136:503–26.

Sen, S. 1989. "Hipparion" Datum and its chronologic evidence in the Mediterranean area. In *European Neogene Mammal Chronology*, ed. E. H. Lindsay, V. Fahlbusch, and P. Mein, pp. 495–505. New York: Plenum.

Sickenberg, O. and M. Schönfeld. 1975. The Chorora Formation: Lower Pliocene limnical sediments in the southern Afar (Ethiopia). In *Afar Depression of Ethiopia*, ed. A. Pilger and A. Rösler, pp. 277–84. Stuttgart: E. Schweizerbartsche Verlagsbuchhandlung.

Sondaar, P. Y. 1961. Les "Hipparion" de l'Aragon méridional. *Estudios Geologicos Madrid* 17(3–4): 912–16.

Steiger, R. H. and Jaeger, E. 1977. Subcommission on geochronology: Convention on the use of decay constants in geo- and cosmochronology. *Earth and Planetary Science Letters* 36:359–62.

Steininger, F. F., R. L. Bernor, and V. Fahlbusch. 1989. European Neogene marine/continental chronologic correlations. In *European Neogene Mammal Chronology*, ed. E. H. Lindsay, V. Fahlbusch, and P. Mein, pp. 15–46. New York: Plenum.

Steininger, F. F., J. Senes, K. Kleemann, and F. Rögl. 1985. *Neogene of the Mediterranean Tethys and Paratethys, Stratigraphic Correlation Tables and Sediment Distribution Maps*, vols. 1 and 2. Institute of Paleontology, University of Vienna Press.

Steininger, F. F., W. A. Berggren, D. V. Kent, R. L. Bernor, S. Sen, and J. Agusti. This volume. Circum-Mediterranean Neogene (Miocene and Pliocene) marine-continental chronologic correlations of European mammal units.

Swisher, C. C., III. This volume. New ^{40}Ar/^{39}Ar dates and their contribution toward a revised chronology for the late Miocene nonmarine of Europe and West Asia.

Tedford, R. H., T. Galusha, M. F. Skinner, B. E. Taylor, R. W. Fields, J. R. MacDonald, J. M. Rensberger, S. D. Webb, and D. E. Whistler. 1987. Faunal succession and biochronology of the Arikareean through Hemphillian interval (late Oligocene through earliest Pliocene epochs) in North America. In *Cenozoic Mammals of North America: Geochronology and Biostratigraphy*, ed. M. O. Woodburne, pp. 153–210. Berkeley: University of California Press.

Thenius, E. 1959. *Wirbeltierfaunen*. In *Tertiär*, ed. A. Papp and E. Thenius, pp. i–ix, 1–328. Handbuch der Stratigraphischen Geologie. Stuttgart: Ferdinand Enke.

Tobien, H. 1980. A note on the mastodont taxa (Proboscidea, Mammalia) of the "Dinotheriensande" (upper Miocene, Rheinhessen, Federal Republic of Germany). *Mainzer geowissenschaftlicher Mitteilungen* 9:187–201.

——. 1986. Die paläontologische Geschichte der Proboscidier

(Mammalia) im Mainzer Becken. *Mainzer Naturwissenschaftliches Archives* 24:1–4.

——. 1988. The Rhine Graben and Mainz Basin. In *The Northwest Tertiary Basin*, ed. R. Vinken, pp. 395–98. Geologisches Jahrbuch 100.

Wagner, J. J., M. Delaloye, and I. Hedley. 1975. Données géochronométriques et paléomagnétiques sûr l'extension du volcanisme du Hegau en Suisse (Ramsen, Schaffhouse). *Comptes Rendus des Séances, SPHN, Genève*, N.S. 10(1): 46–57.

Wagner, W. 1946. Die unterpliozäne wirbeltierfauna vom Wissberg bei Gau-Weinheim in Rheinhessen. *Wissenschaftliche veröffentlichungen der Technischen Hochschule Darmstadt* 1:2–11.

——. 1973. Die unterpliozänen Dinotherien-Sande und ihre fauna im gebiet des Blattes 6114 Wörrstadt (Mainzer Becken). *Mainzer geowissenchaftlicher Mitteilungen* 2:149–60.

Whistler, D. P. and D. W. Burbank. 1992. Miocene biostratigraphy of the Dove Spring Formation, Mojave Desert, California, and characterization of the Clarendonian mammal age (late Miocene) in California. *Bulletin of the Geological Society of America* 104:644–58.

Wiman, S. K. 1978. Mio-Pliocene foraminiferal biostratigraphy and stratochronology of central and northeastern Tunisia. *Revista Española de Micropaleontología* 10:87–143.

Woodburne, M. O. 1987. Principles, classification, and recommendations. In M. O. Woodburne, ed., *Cenozoic Mammals of North America*, pp. 9–17. Berkeley: University of California Press.

——. 1989. Hipparion horses: A pattern of endemic evolution and intercontinental dispersal. In *The Evolution of Perissodactyls*, ed. D. Prothero and R. M. Schoch, pp. 505–21. Oxford Monographs on Geology and Geophysics 5. New York: Oxford University Press.

——. 1996. Reappraisal of the *Cormohipparion* from the Valentine Formation, Nebraska. *American Museum of Natural History Novitates*.

Woodburne, M. O. and R. L. Bernor, 1980. On superspecific groups of some Old World hipparionine horses. *Journal of Paleontology* 8:315–27.

Woodburne, M. O., G. Theobald, R. L. Bernor, C. C. Swisher, H. König, and H. Tobien. This volume. Advances in the geology and stratigraphy at Höwenegg, southwestern Germany.

Woodburne, M. O., B. J. MacFadden, and M. F. Skinner. 1981. The North American "Hipparion" Datum and implications for the Neogene of the Old World. *Géobios* 14:493–524.

Woodburne, M. O. and C. C. Swisher III. 1995. Land mammal high resolution geochronology, intercontinental overland dispersals, sea-level, climate, and vicariance. In *Geochronology, Time Scales, and Global Stratigraphic Correlations: A Unified Temporal Framework for an Historical Geology*, ed. W. A. Berggren, D. V. Kent, and J. Hardenbol, pp. 338–64. Society of Economic Mineralogists and Paleontologists Special Publication No. 54: 338–69.

10

The Correlation of Three Classical "Pikermian" Mammal Faunas—Maragheh, Samos, and Pikermi—with the European MN Unit System

R. L. BERNOR, N. SOLOUNIAS, C. C. SWISHER III, AND J. A. VAN COUVERING

Superb collections of large mammal fossils from the stratified sites of Maragheh in northwesternmost Iran, Samos in the Eastern Aegean, and Pikermi north of Athens were considered by nineteenth-century paleontologists to represent the climax development of the "savanna faunas," which succeeded more forested, warm-temperate biotopes in Western and Central Eurasia and the Mediterranean Basin during the later Cenozoic (see Bernor et al. 1979; Bernor 1983, 1984). For over a century these faunas, characterized by diverse ungulate species with a marked trend toward hypsodonty, were referred to the middle Pliocene because of the mistaken notion that the first phases of Paratethys regression in Eastern Europe, which coincided with the arrival of *Hipparion*, were correlative with the much later desiccation of the Mediterranean Basin at the end of the Miocene (Van Couvering and Miller 1971). In addition, the Paratethys *Hipparion* faunas occurred in strata with freshwater mollusc faunas that Lyell himself (1856) included within the Pliocene based on survivorship ratios. These faunas are now known to be late Miocene age (re: Rögl et al. 1993; Rögl and Daxner-Höck, this volume).

Application of the term *Pontian* to terrestrial faunas was rejected by Crusafont (1950), who created the Pikermian mammal age for the distinctive "savanna faunas," as typified by the Greek locality of Pikermi. Further work, however, in the richly fossiliferous basins of northern Spain led him to modify this usage in favor of two new "mammal stages," the Vallesian and Turolian (Crusafont 1965). The Turolian, typified at Los Mansuetos, records the full development of the "savanna fauna" in the peri-Mediterranean. Mein (1975) subsequently identified three "Mammal Neogene Zones," MN 11–MN 13, in the Turolian interval, and in later studies confirmed that Los Mansuetos is centrally located in MN 12 (Mein 1989). The material from Maragheh, Samos, and Pikermi was identified with the Turolian

a priori (Crusafont 1965; see Steininger et al. 1989), and refinements of the dating and correlation of these faunas will aid in projecting the detailed history of the Vallesian and Turolian interval throughout Europe and Western Asia.

Historical and Geological Background

The sites at Maragheh, Samos, and Pikermi are notable for abundant, relatively complete, well-fossilized, and sometimes partly articulated bones of large mammals, which occur in dense concentrations. These locations were heavily exploited by vertebrate paleontologists for half a century or more, but interest languished during the Great Depression and was then largely eclipsed by the rapid growth of small mammal paleontology after World War II. The reassignment of the Samos Turolian to late Miocene rather than middle Pliocene by Van Couvering and Miller (1971), which dramatically halved the age of the continental Pliocene as it had been understood everywhere but France, created renewed interest in the dating and correlation of the Turolian faunas of Central Eurasia. In particular, the mammalian biochronology at Samos and Pikermi was reviewed by Solounias (1981a, b) and Weidmann et al. (1984), and at Maragheh by Bernor (1985; 1986).

The Maragheh fauna comes from many sites in the arid, highly seasonal countryside east of the market town of Maragheh in Iranian Azerbaijan, located on the shore of the Urmia (= former Rezaiyeh) salt lake. The fossils, which occur in volcanogenic strata of the dissected outwash plain of the Kuh-e-Sahand stratovolcano, were first collected in 1840 by the Russian explorer Khanikoff. His finds were widely admired and subsequently studied by Abich (1858), Brandt (1870), and Grewingk (1881). Expeditions from Vienna under Pohlig in 1884, and later under

Rodler and Kittl, built up the very fine collection of Maragheh fossils at the Naturhistorisches Museum, Wien. Maragheh fossils collected by Damon for the British Museum of Natural History were the subject of a short note by Lydekker (1886), who recognized similarities between the Maragheh fauna and those of Samos, Pikermi, and the Siwaliks. In 1904 Marcelin Boule undertook the largest and most successful expedition to Maragheh, collecting a magnificent suite of fossils for the Muséum National d'Histoire Naturelle, Paris. Description of this collection occupied Mecquenem for the next twenty years (1905, 1906, 1908, 1911, 1924–1925). A modern review of the systematics and stratigraphic provenance of the classical Austrian and French Maragheh collections has been given by Bernor (1986).

Although the fossil-rich Maragheh localities were (and are today) far from being exhausted, no serious collecting was done for the next fifty years. The first expedition after Boule's was from Tokyo University (Takai,, 1958), and this obtained a small but representative sample. In 1967 Tobien and others (1968) excavated at relatively accessible localities in the middle part of the sequence, stratigraphically as well as geographically speaking, along the Mirduq Chai (Mirduq stream) near Kherjabad. In early 1973 the same area was visited by a Dutch-German group who made a stratigraphically controlled collection of fossils and K-Ar dating samples (Erdbrink et al. 1976); later in the same year a Japanese-Iranian team, also working from Kerjabad, covered a somewhat wider territory in mapping marker beds and sampling for fission-track dating (Kamei et al. 1977). In the summers of 1974, 1975, and 1976, the Iranian-American "Lake Rezaiyeh Expedition," based at Shol'avend, sampled and mapped exposures of fossiliferous and datable Maragheh Formation all along the southern flank of Kuh-e-Sahand, from the town of Maragheh eastward to the village of Ildi Chai (Campbell et al. 1980). Samples were taken for both $^{40}K/^{40}Ar$ and fission track dating. This work extended and amplified the modern studies to include virtually all of the classic sites of Pohlig, Kittl, and Boule throughout a region of more than 300 km². Detailed aspects of Maragheh paleontology, biochronology, and dating have evolved out of all the studies conducted since 1958 and have been summarized by Bernor et al. (1979; 1980) and Bernor (1985; 1986).

The Turolian fauna of Samos is represented in a large, diverse sample from a number of fossiliferous horizons. These deposits, which like Maragheh are composed of volcanogenic sediments with numerous pumice-rich horizons, occupy a downfaulted basin that cuts across eastern Samos (Van Couvering and Miller 1971; Weidmann et al. 1986). They are almost certainly the oldest recognized fossil beds in Western Eurasia and possibly the world, having been known since Homeric times (ca. 1000 *b.c.*) as a place where gigantic bones of "amazons" from a bygone

age could be found encased in rock. Between 1850 and 1924 a succession of well-organized paleontological expeditions recovered more than 30,000 specimens from these deposits.

Solounias's monographs (1981a, b) on the Samos fauna provide a detailed review of paleontology, geology, and history of study. The earliest of four major periods of work at Samos—from 1850 to 1889—was dominated by expeditions of the Barbey-Boissier family conducted by Charles Immanuel Forsyth Major (Major 1888; 1893; 1894). The second period, 1890 to 1920, saw a parade of excavators, among them Bukowski, Borne, Stützel, Henschel, Fraas, Acker, Werner, Kormos, Psilovikos, and Weinberger, who were collecting fossils for sale to German and Austro-Hungarian institutions. From 1921 to 1924 Barnum Brown made a second home on Samos while building the important collection housed at the American Museum of Natural History, New York. In the fourth and last period, from 1963 to 1980, Greek researchers removed representative Samos collections for the universities in Athens and Thessaloniki (see Koufos and Melentis 1984), and Solounias deposited at the University of Colorado (Boulder), and at the Carnegie Museum, collections of small mammals recovered by screening in Samos' Main Bone Beds. In the past few years the stratigraphy and dating of the Samos deposits have undergone extensive analysis (Weidmann et al. 1984; Sen 1986; Sen and Valet 1986).

The Pikermi fauna, perhaps the most celebrated of the three discussed here, comes from a roadside ravine just 21 km from Athens. At Pikermi thousands of fossil vertebrates have been recovered from a number of imbricated bone lenses exposed in the wall of the Megalo Rema stream just north of the village of Pikermi. The red clayey silt matrix, with no discernible volcanics, was apparently derived from weathering of the Penteli Mountain just above the road. The impressively barren marble slopes of the Penteli led Gaudry (1862) to speculate that the piles of fossil bones in the Pikermi beds resulted when rising waters in the Aegean forced the Pikermi animals up onto the rocks, where they starved en masse.

The first professional excavations at Pikermi were opened in 1835 by George Finlay and Anton Lindermayer soon after King Otto I was placed on the throne of newly independent Greece. Their work attracted little attention until 1837, when a furloughed Bavarian soldier attempted to sell a skull of the fossil colobine monkey *Mesopithecus*, filled with calcite "diamonds," to a dealer in Munich and was promptly arrested for grave robbery. Andreas Wagner, who was asked for advice in the matter, wasted little time in arranging with his colleague Johannes Roth to set up excavations at Pikermi, which yielded the basis of the present Munich collection.

In 1853 Hercules Mitsopoulos excavated at Pikermi for the University of Athens, and in 1855 Albert Gaudry began five years of work on a collection for the Muséum National d'Histoire Naturelle (Gaudry 1862), which is now divided among Paris, Lyon, and Dijon. Several other organized excavations followed, including Dames (1882) for the University of Berlin; collectors for the Naturhistorisches Museum, Wien; and Woodward (1901) for the British Museum of Natural History. The last major mission was headed by Othenio Abel in 1921 and 1922 for the Naturhistorisches Museum, Wien (Abel 1922). Between 1970 and 1974 excavations at Chomateri were directed by Professor N. Symeonidis (Symeonidis et al. 1973; Symeonidis and Markopoulou-Diacantoni 1977) for the University of Athens. Chomateri is 2 km east of the Megalo Rema exposures, and the fauna is most likely younger than the classical Pikermi deposits. We do not discuss the Chomateri fauna here, but see the appendix to table 10.1 for a list of species.

Geology and Geochronology

The published Maragheh and Samos ^{40}K/^{40}Ar dates were tested by reprocessing samples in the archives of the Berkeley Geochronology Laboratory. These same samples were originally collected and used for dating at Maragheh by Campbell et al. (1980) and at Samos by Van Couvering and Miller (1971) and Weidmann et al. (1984). New sanidine and plagioclase separates were used for single crystal laser total fusion ^{40}Ar/^{39}Ar analyses. Several of the new

^{40}Ar/^{39}Ar dates (Swisher, this volume) support significant improvements to the stratigraphic interpretations, mammalian systematics, and biochronologic analysis of all three faunas. From this vantage a more precise correlation between the "Pikermian" sites of Maragheh, Samos, and Pikermi and the late Miocene MN zones of Mein (1975; see Fahlbusch 1991) can be proposed.

Maragheh

The base of the Maragheh Formation (Campbell et al. 1980) rests disconformably on the Basal Tuff, a biotite rhyolite ignimbrite up to 80 m in thickness. Our dating confirms that the Basal Tuff is at least 1 m.y. older than the lowest exposed part of the overlying Maragheh Formation; it dips at a slightly greater angle toward Lake Urmia so that Maragheh strata wedge out against it (fig. 10.1). The Maragheh Formation is at least 300 m in stratigraphic thickness but nowhere more than 150 m from base to eroded top. It is a virtually undeformed, coarsely stratified unit composed of brown to tan andesitic volcanic sands and silts, with locally abundant pumice cobbles and pumice-ash beds. It is exposed primarily in steep canyon walls of the streams flowing radially from Kuh-e-Sahand. A thick pyroclastic body, the Village Pumice, is draped over a deeply eroded surface of Maragheh Formation at Ildi Chai and is dated some 2 m.y. younger than the Maragheh tuff beneath it. Kamei et al. (1977) obtained a similarly

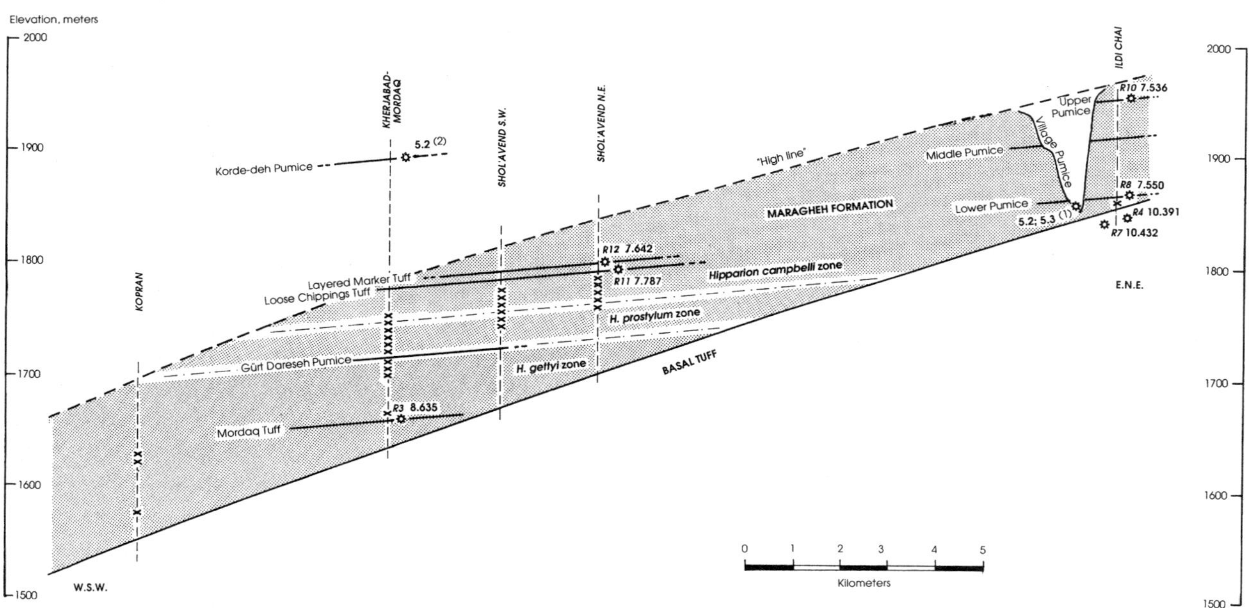

FIGURE 10.1. Lithology and stratigraphy of the Maragheh Formation, adapted from Campbell et al. 1981. The sections, roughly west to east, depict changes in geology along the regional dip in the Maragheh Formation, with elevation at the Basal Tuff contact in meters a.s.l. Sections 2–6 are physically correlated by the Loose Chippings marker bed and other tuff units. Numbers to the left of each column are fossil localities occupied by the Lake Rezaiyeh Expedition (1974–1976). New single-crystal argon/argon ages are given for sampling localities R1, R3, R4, R7, R8, R10, and R12.

young age from the thick Korde-deh Ash in the upper Mirduq Chai. We here designate the unconformable base of the Village Pumice, and (if correlative) the Korde-deh Ash, as the top of the Maragheh Formation. The Village Pumice disconformity is near the top of the section at Ildi Chai but Kamei et al. (1977) reported that another 300 meters of volcaniclastic sediments can be measured above the Korde-deh Ash.

Campbell et al. (1980; see also Bernor et al. 1980) developed an area-wide lithostratigraphic framework for the datable horizons and fossil sites that were located in the 1970s. The sites included some relocated classical sites, and Bernor (1985; 1986: tab. 1, fig. 4) subsequently identified and added others. The stratigraphic framework (fig. 10.1; see also Bernor et al. 1980: pg. 719, fig. 6) is based on seven local sections, the central five of which are physically correlated by the Loose Chippings tuffaceous marker bed, used by Campbell et al. (1980) and Bernor et al. (1980) as their zero level, and a number of other distinctive and datable pyroclastic horizons. In projections of the regional dip from the well-controlled central area, the westernmost section near the village of Kopran is apparently too low in the stratigraphic sequence to include the Loose Chippings unit, and that of Ildi Chai in the east is apparently too high. None of the other stratigraphic markers identified in the central area have been recognized at Kopran or Ildi Chai. Thus the relative position of the westernmost and easternmost fossiliferous sections is based on a defensible stratigraphic reconstruction (fig. 10.1), supported by dating and by paleontology, notably in the succession of *Hipparion* species.

In the central Mirduq-Sholl'avand area, the stratigraphy is controlled by a series of distinctive and datable tuffs and pumice-layers measured by Jacob's staff in meters from Loose Chippings:

+ 25 Sargizeh Tuff of Kamei et al. 1977
+ 7 Layered Marker Tuff (= "pumice falls" of Kamei et al. 1977)
-0- Loose Chippings Pumice Tuff
-25 Upper Pumice of Kamei et al. 1977; trachytic breccia of Erdbrink et al. 1976
-50 Gürt Dareseh Tuff (= Lower Pumice of Kamei et al., 1977)
-110 Mirduq Tuff (= Mordagh Tuff or Ignimbritic Tuff of previous authors)
(unconformity)-Basal Tuff

The Korde-deh Ash in the central area appears to be approximately 100 m above the Loose Chippings level, according to data of Kamei et al. (1977). Ildi Chai, the easternmost and topographically highest of the seven control sections, was correlated to the central area by Campbell et al. (1980) in a simple projection of regional dip (fig. 10.1). In this projection the stratigraphic elevations of three prominent Ildi Chai tuffs above the Loose Chippings datum were estimated as follows:

(unconformity)-Village Pumice
+ 90 Upper Tuff
+ 60 Middle Tuff
+ 15 Lower Tuff
(unconformity)-Basal Tuff

Revised age estimates for Maragheh tuffs are based on samples of andesitic pumice from the Maragheh Formation and biotite rhyolite tuff from the Basal Tuff collected by Van Couvering for $^{40}K/^{40}Ar$ and fisson track age determinations (Campbell et al. 1980). Feldspars from these samples were used for multiple $^{40}Ar/^{39}Ar$ single-crystal analyses (Swisher, this volume), as shown in figure 10.1. Nine analyses of sanidines from two samples of Basal Tuff (samples R4 and R7) gave weighted mean ages of 10.391 ±0.024 Ma. (SE) and 10.432 ±0.020 Ma. (SE), close to the average of earlier K/Ar dates (Campbell et al. 1980). Plagioclase crystals extracted from the Maragheh Formation pumice tuff samples gave stratigraphically consistent isochron ages as follows: 8.635 ±0.029 Ma. (SE) for the -110 m Mirduq Tuff (R3); 7.787 ±0.139 Ma. (SE) for the 0 meter Loose Chippings Tuff (R11); 7.642 ±0.033 Ma. (SE) for the + 7 meter Layered Marker Tuff (R12); 7.550 ±0.071 Ma. (SE) for the + 15 meter Lower Pumice at Ildi Chai (R8); 7.536 ±0.111 Ma for the + 90 meter Upper Pumice at Ildi Chai (R10); (re: Swisher, this volume for more details).

According to these determinations, three reasonably consistent accumulation rates can be calculated for the Maragheh Formation: 130 m/myr from Miqduq Tuff to Loose Chippings, 107 m/myr from Mirduq Tuff to Layered Marker, or 115 m/myr from Mirduq Tuff to the Ildi Chai Lower Pumice. This consistency adds confidence to the interpretative correlation of the Ildi Chai section to the central group (Campbell et al. 1980). The very short duration between lower and upper tuffs at Ildi Chai would, however, suggest a much higher accumulation rate (about 530 m/myr) in the Ildi Chai section itself.

Most of the Maragheh collecting sites fall into the 1.1 m.y. interval between the Mirduq Tuff and the Ildi Chai Lower Pumice. The exceptions are the lower sites at Kopran, which in the stratigraphic reconstruction (fig. 10.1) fall below the projected level of the Mirduq Tuff at 115 m and 150 m below datum. Assuming an average accumulation rate of 120 m/my, the age of the lowest sites (MMTT 41, 43) can be interpolated to be ca. 9 Ma., in agreement with estimates by Bernor et al. (1980), Bernor (1985, 1986), and Steininger et al. (1989). A weathered tuff at Kopran, just above the MMTT 44 locality estimated to be 115 m below datum, was unfortunately not sampled; this should be reinvestigated as a possible exposure of Mirduq Tuff.

The subdivision of the Maragheh fauna into three local biozones is based on the successive stratigraphic ranges of *Hipparion gettyi*, *Hipparion prostylum*, and *Hipparion campbelli*, which Bernor (1985) and Bernor et al. (1989) have argued is a monophyletic lineage. Localities characterized by hipparions with relatively primitive dental and facial morphology of the *gettyi* stage-of-evolution are found in the western, stratigraphically lowest part of the Maragheh Formation from Kopran to the Mirduq Chai. These localities are located in the -150 m to -70 m stratigraphic interval, with interpolated age limits of 9 to 8.2 Ma. The type skull of *Hipparion gettyi* is among material collected from Pohlig's "Kopran I" site, which Bernor (1986) tentatively identified with the MMTT 41 and 43 sites at the lowest fossiliferous level in the formation.

In the Mirduq Chai an unsampled interval separates the highest recognized occurrence of *Hipparion gettyi* at -70 m and the lowest occurrence of *Hipparion prostylum* at -52 m, just below the Gürt Dareseh Tuff. Although distinguishing these species is often difficult, *H. prostylum* is well documented at this level by a suite of skulls, mandibles, and postcrania collected by Tobien and others (1968). Bernor (1985) recognized *Hipparion prostylum* up to the -20 m level, where the range of *Hipparion campbelli* begins. The range of *H. prostylum* at Maragheh is thus quite short, with interpolated age limits of approximately 8.2 to 8.0 Ma in the context of the accumulation rates noted above. *Hipparion campbelli* ranges from the -20 m level at Shol'avand Southwest up to (just below) the Ildi Chai Lower Pumice, with interpolated age limits of 8.0 to 7.6 Ma. In the absence of material from higher in the section this must be considered a minimum upper age limit.

Mein (1989) and Steininger et al. (1989) correlated the "Lower Maragheh" fauna, defined by the presence of *H. gettyi*, to earliest Turolian (earliest MN 11), not younger than 9.3 Ma. They considered the Middle Maragheh fauna with *H. prostylum* to be late early and medial Turolian (late MN 11 and early MN 12), between 9.3 Ma and 7.75 Ma, and the Upper Maragheh fauna with *H. campbelli* to be medial Turolian (later MN 12) between 7.2 and 6.9 Ma. The new age determinations, and the correlations we make with the Western European MN unit system, indicate that the Maragheh faunal seqence is correlative with basal MN 11 to medial MN 12, ca. 9 (interpolated) to 7.6 Ma.

Samos

The fossiliferous sequence at Samos, unlike Maragheh, is marked by extensive normal faulting. In the absence of accurate locality descriptions, the relocation of old quarries and classic collections is often complicated by the presence of beds of quite different ages in close proximity to one another. Schlosser (1904:112) was the first to recognize that Samos' fossils occurred in distinct horizons and that there were important differences in the faunas from different levels. Unfortunately, he published no details despite the fact that his technician, Stützel, excavated at four different locations. In later years more attention was paid to locality provenance, but uncertainties about the stratigraphy continued to plague biostratigraphic analysis. In 1990 a fire swept through the maquis scrub in the Mytilini Valley, exposing the geology to an unprecedented degree. New stratigraphic observations have now clarified some of the remaining questions about locality position that were raised by Meissner (1976), Black et al. (1980), Solounias (1981a, b), Weidmann, et al. (1984), and Sen and Valet (1986).

In Samos' eastern basin, five Miocene formations are recognized, as summarized below, from youngest to oldest:

Kokkarion Formation: thick-bedded freshwater limestone with a central lens of marls and gravels; water-laid Marker Tuffs at the base are a distinct and possibly separate unit.

Mytilini Formation: floodplain deposits of volcanogenic marls and gravels, with soil horizons and rhyolite pumice tuffs; with numerous bone pockets, some of which were developed as quarry excavation (= Q here) by Barnum Brown (American Museum Frick Collection).

Hora Formation: mostly thin-bedded freshwater limestone, with a volcanic unit at the top dated 9 Ma (Weidmann et al. 1984).

Pythagorion Formation: thick-bedded freshwater limestone with paleosols and lignitic marls; basalt within this unit has been dated 11.2 Ma (Weidmann et al. 1984; Ioakim and Solounias 1985), and the pollen suite from lignites below the basalt resembles that of the middle to early late Miocene warm-temperate forests of Central Europe (Kovar-Eder in Bernor et al. 1988).

Basal Conglomerate: conglomerates and marls resting unconformably on Mesozoic marble basement.

The Mytilini Formation is divided into four members (fig. 10.2). The lowest, the Old Mill Beds (informally OMB), consists of well-bedded marls and tuffs with paleosols. Barnum Brown excavated Quarry X in this unit, which appears to have been one of the sites that Forsyth Major and other collectors had also exploited in previous decades. Quarry X represents the stratigraphically lowest large mammal site in the Samos sequence, being situated 10 m above the base of the formation (Solounias 1981a). Also located in the lower part of this member is locality G of the Munich collection near Smakia, while Brown's Q6 is in the upper part of the OMB in the Tholorema cliffs near Kokkari Village. The OMB member is truncated by channels filled by unfossiliferous sandy conglomerates of the Gravel Beds member. Above the Gravel Beds member, the White Beds member (informally termed WB here) is composed of indurated marls and limestones. At the very

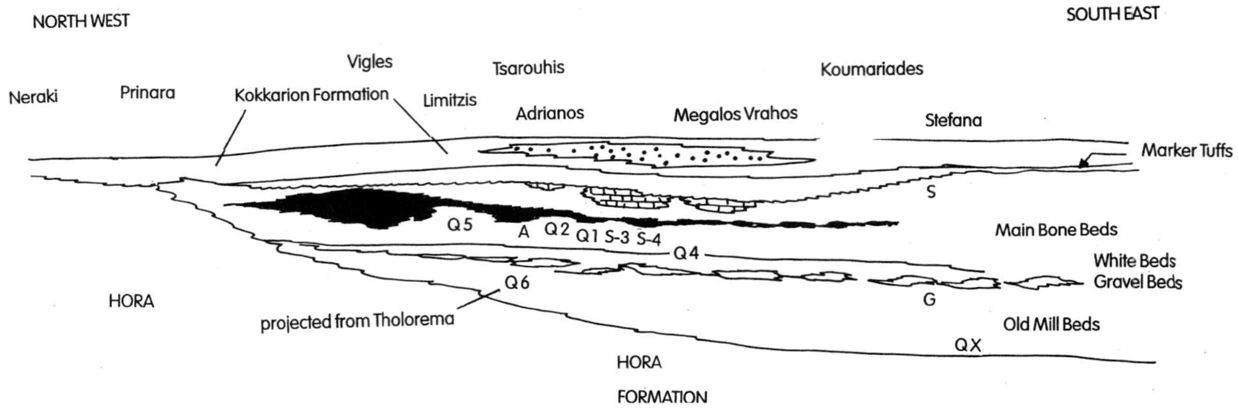

FIGURE 10.2 Stratigraphy and sample localities in the Mytilini Formation. The schematic cross section, viewed from the south, depicts facies relationships and the topographic situation in the fossiliferous region. The Mytilini Formation thickens eastward from the faulted margin of the basin on the left toward the basin center on right. On the skyline the hills from left to right are Prinara, Vigles, Limizis, Andrianos, Tsarouchis, Megalos Vranhos, Koumariades, and Stefanna; the Andrianos Ravine below Tsarouchis is the most fossiliferous location on Samos. The hills are capped by Kokkarion Formation, with Marker Tuff beneath.

top of this unit, but possibly in the lowermost beds of the succeeding Main Bone Beds member in Potamies Ravine, is Brown's Q4.

The Main Bone Beds member (informally MBB) includes volcanogenic sandy silts, marls, and tuffs, with local paleosols and thin lenses of gravel. Brown's quarries Q1, Q2, Q3, Q5, and Solounias's micromammal localities S3 and S4 are located in the lower part of this member. The MBB member is widely exposed along the northwest margin of the Mytilini plain, in ravines feeding the Potamies stream in the Prinara, Vigles, and Platanaki districts, but all of the known bone-bearing pockets in this area have been heavily exploited. Above the lower part of this member a thick, resistant limestone conglomerate is a distinctive stratigraphic marker; it appears to have developed as a talus fan where the Mytilini Formation pinches out against Hora beds at Prinara, and thins abruptly to the southeast.

The position of Brown's Q1, Q2, and Q5 has become clear in the aftermath of the 1990 brush fire (fig. 10.2). It can now be seen that these quarries all lie about 9 m below the marker conglomerate and that Q2 is directly correlative to Solounias' sites S3 and S4. The stratigraphic relationships of Locality A (the Andrianos locality of Forsyth Major) and Q3 are still not precisely known, although both appear also to be in the lower portion of the MBB below the marker conglomerate. Locality A, adjacent to Q1, may be part of the same bone bed displaced upward about 6 m on a fault that runs across the ravine, but it is not possible to see the marker conglomerate above Quarry A because of farm terracing. Q3 was described in Brown's notes as being below the Megalos Vrahos Hill, but its exact location has not been determined. Q4 is located approximately 20 m

stratigraphically below Q1 and Q2 in the contact zone with the White Beds member.

The Mytilini Formation is uncomformably overlain by the Marker Tuffs, a unit of well-bedded, indurated waterlain tuffs and marls that crops out over a wide area, including the center of Kokkari village, with little lateral variation. The base of the Marker Tuff in the Limidzis and Tsarouchis districts is a red marly paleosol with localized fossil bone occurrences (Locality L: fig. 10.2). This marl, not Q5, is the youngest bone-bearing unit on Samos. While Locality L is close enough to Q5 for them to have been confused, Q5 matrix is white and the bone is likewise white and highly indurated, while Locality L matrix is deep red in color and the bone is friable and creamy yellow. All of the Q5 material at the AMNH conforms to the former characteristics and no Locality L material appears to be present. The main part of the Kokkarion Formation is composed of algal limestones, with a lens of "pseudomytilini" marls and paleosols at Tsarouchis that closely resembles the underlying Mytilini Formation.

A major controversy has been resolved with regard to the age of Q5, which for many years was held to be younger than the other Brown quarries but which can now be shown to lie at the same level as Q1 and Q2, about 9 m below the marker conglomerate. The stratigraphic study by Van Couvering and Miller (1971) interpreted Brown's locality map to place Q5 as the highest quarry of the Mytilini sequence. Sondaar (1971) also believed that Q5 was significantly younger than the other Brown quarries based on the grade of evolution observed in the various hipparion and *Orycteropus* species. Gentry (1971) held that Q5 yielded a more advanced species of *Pachytragus, P. crassicornis*, as compared with the *P. laticeps* of the other quarries. Heissig (1975) interpreted the Q5 rhino material as being the most advanced of four stages of evolution in the Samos fauna, in agreement with Schlosser's (1904) concept of four fossiliferous levels. Bernor et al. (1980), on the other hand, argued that the Samos *"Hipparion"* assemblages from all sites in the MMB member represented a single stage-of-evolution in an advanced "Pikermian" fauna of late MN 12 age, and a time range from ca. 7.2 to 7.0 Ma.

This last view is borne out by the new stratigraphic observations and dating reported here. K/Ar ages on sanidine separated from Mytilini Formation tuffs have yielded ages ranging from 8.57 to 5.41 Ma. Swisher (this volume) has assessed which of the previous K/Ar ages have the soundest precision. ^{40}Ar/^{39}Ar dating of single sanidine crystals derived from MBB tuffs resulted in new ages (Swisher, this volume; fig. 10.2 here): SK18A, stratigraphically below localities S3, Q1, Q2 and Q5 and undoubtedly Q3 and locality A = 7.276 ±0.006 Ma; SK17, 2 meters above the Q1 bone bed, within the Q1 quarry area = 7.092 ±0.008 Ma. This gives a best estimate age for all the MBB quarries, except Q4, as being ≥7.1 Ma. Sample R102 derived from east of Mytilini Village and stratigraphically low within the MBB horizons has yielded a ^{40}Ar/^{39}Ar age of 7.660 ±0.012 Ma. Basing his opinion on recent stratigraphic work at Samos, Solounias believes this age gives a good *maximum* age for Q4. Swisher maintains that the Old Mill Beds are currently best estimated by conventional K/Ar as being 8.33 ±0.05 Ma. Also, conventional K/Ar ages on the overlying Marker Tuffs are best estimated as being 6.17 ±0.05 Ma.

Swisher (this volume) has separately reviewed the paleomagnetic correlations in the Samos sequence published by Sen and Valet (1986). We note that the age proposed by these authors for the MBB significantly postdates it because of the major disconformity between the MBB and Marker Tuff, which was not mentioned by earlier workers.

Pikermi

The Pikermi Formation is deposited against the southern flanks of the Penteli Mountain, which has provided most of the fine white marble for the Acropolis. The geology of the region around Athens has never been studied in detail, and even the Pikermi Formation has not been formally characterized or defined (Bachmeyer et al. 1982; Symeonidis et al. 1973; Symeonidis and Markopoulou-Diacantoni 1977). The Neogene stratigraphy in the Athens region is unlike that of the volcano-tectonic basins of Samos and Maragheh in that the Pikermi sediments are derived by weathering in a limestone terrain and contain no pyroclastic or weathered volcanic debris. In the Megalo Rema, the great bulk of the formation is pale to dark red clayey silt, with limestone conglomerate lenses and layers

of hard limey marl. Most bones in the Pikermi sediments, especially skulls, are diagenetically flattened.

The formation dips gently seaward, and is cut by several minor normal faults. The small mammal fauna from Chomateri, about 2 km distant (Symionides and Markopoulo-Diacantoni 1977), is also in the Pikermi Formation but is closer to the sea and is demonstrably younger than the classic Pikermi bone-bearing horizons, ca. late MN 12 (Mein 1989).

Mammalian Faunas and Biochronologic Correlations

Table 10.1 provides a composite faunal list for Maragheh, Samos, and Pikermi, and table 10.2 their known chronologic ranges. We have listed 107 mammalian species belonging to 10 orders: Insectivora (3 species), Primates (2; but see below), Chiroptera (1), Rodentia (7), Carnivora (23), Tubulidentata (1), Proboscidea (4), Hyracoidea (2), Perissodactyla (21), and Artiodactyla (42). Evidence published elsewhere and updated here suggests that these faunas represent several different habitats within a general environment of summer dry, open conditions over a time span on the order of 2 million years. Until now, relatively imprecise correlations have been made between Maragheh, Samos, and Pikermi. Revision of the faunas using the systematic, biochronologic and biogeographic studies presented in this volume, in connection with stratigraphic and chronologic revisions presented here, are the basis of the analysis that follows.

Phylogenetic reconstructions of Maragheh, Samos, and Pikermi mammals include recent systematic studies of hipparionine horses (Bernor 1985; Bernor et al. 1989, this volume b) and hyaenids (Solounias 1981a; Werdelin and Solounias, this volume). The remainder of the fauna is reviewed in this volume from the standpoints of temporal and biogeographic distribution. While we believe that more extensive systematic and biogeographic analyses are still needed for these groups, this general update is useful to more accurately position them chronologically.

The insectivores include *Desmanella dubia, Schizogalerix atticus* and *Schizogalerix moedlingensis.* Engesser and Ziegler (this volume) have reported that *Desmanella* is an old European genus first appearing in the late Oligocene of Middle Europe and subsequently having developed an independent lineage in Turkey during the later Miocene. *Schizogalerix* is largely an Eastern Mediterranean lineage distinct from European *Galerix.* Engesser and Ziegler (op. cit.) refer *S. moedlingensis* to MN 11 (also found at Pikermi) and *S. attica* to MN 12 (also found in Samos MBB).

The primates include two reported species of colobine cercopithecoid, *Mesopithecus pentelici* and *Mesopithecus delsoni.* While we maintain the nomen *Mesopithecus delsoni,* Delson (in Andrews et al., this volume) argues that it

TABLE 10.1 *Mammal Faunas and Their Stratigraphic Position: Maragheh, Samos, and Pikermi*

	Maragheh			Samos			
Taxon	L	M	U	O	W	M	Pikermi
Insectivora							
Desmanella dubia							X
Schizogalerix moedlingensis							X
Schizogalerix atticus						X	
Primates							
Mesopithecus pentelici							X
Mesopithecus delsoni		X					
Chiroptera							
Samonycteris majori						X	
Rodentia							
Spermophilinus cf. *bredai*						X	
Byzantinia hellenicus						X	
Pseudomeriones pythagorasi						X	
Occitanomys cf. *provocator*						X	
?Gerboa sp.			X				
Pliospalax cf. *sotirisi*						X	
Hystrix primigenia						X	
Carnivora							
Simocyon primigenius							X
Ursavus cf. *depereti*						X	
Indarctos atticus						X	X
Indarctos maraghanus		X					
Sinictis pentelici							X
Martes sp.		X					
Martes woodwardi							X
?Plesiogulo sp.							X
Promeles palaeattica		X				X	X
Parataxidea polaki/maraghana		X				X	
Promephitis lartetii						X	X
Enhydriodon laticeps							X
Plioviverrops orbignyi						X	X
Ictitherium viverrinum		X			X	X	X
Hyaenotherium wongii		X	X		X	X	X
Hyaenictis graeca							X
Lycyaena chaeretis						X	X
Belbus beaumonti						X	
Adcrocuta eximia		X	X	X		X	X
Felis attica		X	X			X	X
Metailurus parvulus			X			X	X
Metailurus major						?	X
Machairodus giganteus		X	X			X	X
Tubulidentata							
Orycteropus gaudryi			sp.			X	?
Proboscidea							
Mammut borsoni						X	X
Stegotetrabelodon grandincisivus						X	X
Choerolophodon pentelici	X	X	X			X	X
Deinotherium giganteum			X			X	X
Hyracoidea							
Pliohyrax graecus						X	X
Pliohyrax kruppii						?	
Perissodactyla							
"Hippotherium" brachypus		X					X
"Hippotherium" giganteum					X		
Hipparion gettyi	X		?				X
Hipparion prostylum		X		cf			X
Hipparion campbelli			X				
Hipparion dietrichi						X	
Cremohipparion moldavicum	X	X					
Cremohipparion aff. *matthewi*						?	X
Cremohipparion matthewi	X	X	X			X	
Cremohipaprion nikosi					?		
Cremohipparion mediterraneum				?			X
Cremohipparion proboscideum						X	
Ancylotherium pentelicum		X				X	X
Aceratherium cf. *incisivum*							X
Chilotherium samium				X	X		
Chilotherium persiae	X	X	?				
Chilotherium schlosseri					?		
Chilotherium kowalevskii					?		
"Dicerorhinus" schleiermacheri						X	X
Ceratotherium neumayri		X			X	X	X
Iranotherium morgani		X					
Artiodactyla							
?Microstonyx erymanthius	X	X	X	?	?	X	X
Postpotamochoerus hyotherioides						?	
Dorcatherium naui						?	
Muntiacus sp.						X	
Pliocervus pentelici		X				X	X
Palaeotragus rouneii						X	X
Palaeotragus coelophrys		X	X			X	X
Palaeotragus quadricornis				?			
Samotherium boissieri				X			
Samotherium sp.			?		X	X	
Samotherium neumayri	X	X					
Helladotherium duvernoyi		X				X	X
Helladotherium sp.						?	
Bohlinia speciosa						?	X
Bohlinia attica					?		X
Hoanotherium sp.		X					
Miotragocerus monacensis		X				X	X
Graecoryx valenciennesi						X	X
Tragoportax amalthea	X					X	X
Tragoportax curvicornis						X	
Tragoportax rugosifrons		X			X	X	
Samokeros minotaurus		?				X	
Prostrepsiceros rotundicornis		X				X	X
Prostrepsiceros houtumschindleri	X	X				X	X
Protragelaphus skouzesi		X				X	X
Gazella capricornis/deperdita	X	X	X	X	X	X	X
"Gazella" rodleri		X					
Oioceros rothii		X					X
Oioceros wegneri						X	
Samotragus crassicornis						?	
Samodorcas kuhlmanni						?	
Palaeoreas lindermayeri					X		X
Criotherium argalioides					X	X	
Parurmiatherium rugosifrons						?	
Urmiatherium polaki		X					
Palaeoryx pallasi					?	X	X
Palaeoryx sp.						X	
Sporadotragus parvidens				X	X	?	X
Protoryx carolinae							X
Pachytragus crassicornis					X	X	X
Pachytragus laticeps		X				X	X
Pseudotragus capricornis						X	X

O = Old Mill Beds of Samos
W = White Sands of the Main Bone Beds, Samos
M = Main Bone Beds of Samos

APPENDIX: *Lagamorpha and Rodentia from Chomatari*

	Maragheh			Samos			
	L	M	U	O	W	M	Pikermi
Lagamorpha							
Prolagus cf. *crusafonti*							X
Alilepus sp.							X
Rodentia							
Spermophilinus cf. *bredai*						X	
"Kowalskia" cf. *lavocati*							X
Byzantinia pikermiensis							X
Byzantinia hellenicus						X	
Pseudomeriones pythagorasi						X	
Parapodemus gaudryi							X
Occitanomys cf. *neutrum*							X
Occitanomys cf. *provocator*						X	X
?Gerboa sp.			X				
Pliospalax cf. *sotirisi*						X	
Muscardinus sp.							X
Myomimus cf. *dehmi*							X
Hystrix primigenia						X	X

is the junior synonym of *Mesopithecus pentelicus*. This taxon first occurs in the early Turolian, MN 11, and thereafter through MN 12 and probably into MN 13.

The chiropteran *Samonycteris majori* is derived from the MBB at Samos and has little chronologic value.

The Rodentia are mostly derived from the upper fossiliferous horizons of Samos. De Bruijn and Mein (this volume) cite *Spermophilinus* as being ubiquitous in European faunas during the middle and late Miocene, and further believe it to be an ecological generalist. Its occurrence in Samos' MBB horizons is of little biochronologic consequence.

Cricetodontines achieved their maximum biogeographic extension during MN 6, when they were common from Portugal to Kazakhstan (de Bruijn and Ünay, this volume). *Byzantinia* first occurs in MN 7 + 8 of Asia Minor and belongs to a "type 4" radicle of cricetodontines (de Bruijn and Ünay, this volume) that becomes distinct from the Western European radicle during the middle Miocene. De Bruijn and Ünay correlate *B. hellenicus* with MN 12 and acknowledge a lack of information for MN 11.

Very little is known about the phylogeny and chronology of *Pseudomeriones pythagorasi*. The first reported occurrence of *Occitanomys* is MN 11 of Western and Central Europe (Mein 1989). The relationships of the Western European species *O. sondaari*, *O. adroveri* and *O.* sp. to *Occitanomys* cf. *neutrum* (Pikermi) and *Occitanomys* cf. *provocator* (Pikermi and Samos MBB) are currently unresolved.

?Gerboa sp. is represented by a single mandible from Upper Maragheh (ca. 7.9 Ma, MN 12 correlative). Insufficient analysis has occurred to establish any further correlation.

The spalacid genus *Pliospalax* is first reported from MN 6 in Southwestern Asia (Ünay, this volume). This genus ranges in Southwestern Asia from MN 6 to MN 15 and in Southeastern Europe from late MN 8 to MN 16. Insufficient phylogenetic work as been done to utilize the Samos MBB occurrence of *P.* cf. *sotirisi* for biochronologic correlation.

The oldest greater-European representative of *Hystrix* has been reported from Yeni Esikihisar (MN 8) of Anatolia. All *Hystrix* remains found in the Turolian of Eastern Europe and Anatolia can be attributed to *H. primigenia*, for which Pikermi is the type locality. Its presence at Samos and Pikermi offers little age resolution (Sen, this volume).

The carnivores include a diversity of families: ?Canidae, Ursidae, Mustelidae, Hyaenidae, and Felidae. *Simocyon primigenius* is an enigmatic form belonging to the Simocyoninae, itself variously referred to the Canidae, Amphicyonidae, and Procyonidae (Werdelin this volume). The genus *Simocyon* (*S. diaphorus*) is first reported from Eppelsheim, Germany (MN 9), and later from Montredon (MN 10). *S. primigenius* is well represented by fairly extensive material

at Pikermi. Werdelin (pers. comm.) has recently identified this taxon at Baltavar, Hungary, giving a provincial LAD of MN 13 for this taxon.

The Ursidae are diverse in these faunas. The genus *Ursavus* is a primitive bear first occurring in the early Miocene (MN 3; Mein 1989). The Samos MBB fauna includes *Ursavus* cf. *depereti*, a very late and relatively large surviving member of the genus, which is known elsewhere from Nombrevilla 2, Spain (MN 7; Werdelin, pers. comm.), and Soblay, France (MN 10). Other species of ursids include the Pikermi and Samos MBB species *Indarctos atticus* and the larger Maragheh species *I. maraghanus*. Werdelin (op. cit.) cites the derivation of the pan-Turolian species (MN 11–13) *I. atticus* from the Vallesian species (MN 9 and 10) *I. arctoides*. We recognize the yet larger species *I. maraghanus* from Middle Maragheh (de Mecquenem 1924–1925; Solounias 1981a; Bernor 1986).

The Mustelidae are represented by diverse species: *Sinictis pentelici* (Pikermi), *Martes* sp. (Middle Maragheh), *Martes woodwardi* (Pikermi), *?Plesiogulo* sp. (Pikermi), *Promeles palaeattica* (Upper Maragheh, Samos MBB, and Pikermi), *Parataxidea polaki/maraghana* (Middle Maragheh and upper Samos), *Promephitis lartetii* (upper Samos and Pikermi), *Enhydriodon laticeps* (Pikermi). Werdelin (this volume) has not offered an updated evolutionary and biochronologic reconstruction of the European-West Asian Mustelidae due to their diversity and complicated evolutionary history. He has remarked however that this family's diversity witnessed a marked rise from MN 4 to MN 5 and maintained a plateau until MN 10, when there was a drop in diversity. The greatest diversity among our three faunas is clearly with Pikermi, and likely due to its more forested character. An older age for some Pikermi horizons than the majority of the Maragheh and Samos faunas may also be plausible and should be tested with the phylogenetic relationships of the Mustelidae.

The Hyaenidae are diverse amongst these three faunas. We follow Werdelin (this volume) and Werdelin and Solounias (this volume) in recognizing the distinction of the Percrocutidae from the Hyaenidae, a departure from standard carnivore systematics. *Plioviverrops orbignyi* is the most primitive hyaenid in our assemblage. It belongs to a mongoose-like insectivore/omnivore ecological group distributed in Europe from MN 5 to MN 13. This species's first known occurrence is at Ravin de Pluie (MN 10) and it occurs later at Vathylakkos (MN 11), Pikermi, and Samos MBB.

Ictitherium viverrinum, *Hyaenotherium wongii*, and *Hyaenictis graeca* are characterized by Werdelin and Solounias (op. cit.) as being jackal- and wolf-like meat and bone eaters with a more generalized dentition. *Ictitherium viverrinum* first occurs in Central Europe during early MN 9 (Vösendorf; Pannonian D–E; Rögl et al. 1993; Bernor et al. 1993b; Werdelin and Solounias, op. cit.). It thereafter occurs in Western Europe during MN 10 and continues in Southeastern Europe and Southwestern Asia between MN 11 and MN 12. Upper Maragheh, Samos WS and MBB, and Pikermi all record the occurrence of this taxon. *Hyaenotherium wongii* is first recorded in Central Europe from Höwenegg, Germany (early MN 9, ca. 10.3 Ma; re: Swisher, this volume; Bernor et al. 1993a, b) and continues as a rare element in the Vallesian, becoming more abundant in Southeastern Europe and Southwestern Asia during MN 11–12. This taxon occurs in Middle and Upper Maragheh, Samos WS and MBB, and Pikermi. *Hyaenictis graeca* is recorded only from Pikermi.

Lycyaena chaeretis from Pikermi and Samos MBB is a cursorial meat and bone eater showing a trend toward reduction of the bone-crushing portion of its dentition (Werdelin and Solounias, op. cit.). *Belbus beaumonti* is a transitional bone-cracker occurring solely in upper Samos. *Adcrocuta eximia* is a non-cursorial bone-cracker first occurring during MN 10 in Spain, Austria, and Greece, and continuing its occurrence predominately in Southeastern Europe and Southwestern Asia during the Turolian (MN 11–13). It is present at Middle and Upper Maragheh, Samos WS and MBB, and Pikermi.

The conical-toothed cats are represented by a single taxon, *Felis attica*. This species first occurs during MN 10 and stands at the base of the modern felid evolutionary radiation (Werdelin, this volume). This species is recorded from Middle and Upper Maragheh, Samos MBB, and Pikermi. *Metailurus parvulus* is the smaller of the two species belonging to this genus putatively first occurring at Montredon (MN 10) and continuing its occurrence at Upper Maragheh (Ilkhchi = 7.55 Ma), Samos MBB, and Pikermi. The larger *Metailurus major* is known to occur from Turolian localities (MN 11–13) and is found possibly in Samos MBB and Pikermi. *Machairodus giganteus* is a large machairodont cat first occurring during MN 11 (Crevillente 2), and common through the Turolian in Western Europe, Southeastern Europe, and Southwestern Asia. It occurs in Middle and Upper Maragheh, Samos MBB, and Pikermi through at least MN 12.

The Tubulidentata are represented by a single species, *Orycteropus gaudryi* from Upper Maragheh, Samos MBB, and possibly Pikermi. The genus is first recorded in Eurasia from the basal middle Miocene locality of Paşalar (MN 6; Fortelius 1990). Its occurrence in middle and late Miocene Eurasian localities is rare and insignificant for biochronologic correlations.

The Proboscidea are represented by four taxa: *Mammut borsoni* from Samos MBB and Pikermi; *Stegotetrabelodon grandincisivus* from upper Samos and Pikermi; *Choerolophodon pentelici* from all levels at Maragheh, Samos MBB, and Pikermi; *Deinotherium giganteum* from Upper Maragheh, Samos, and Pikermi. These taxa are sufficiently long-ranging both geographically and chronologically as to offer little insight into relative ages.

The Hyracoidea are represented by two species, *Pliohyrax graecus* from Samos MBB and Pikermi and a possible occurrence of *Pliohyrax kruppii* from upper Samos. These taxa have no discernable value for biochronologic correlations.

The equid remains from Maragheh, Samos, and Pikermi include twelve species belonging to three superspecific lineages (re: Woodburne and Bernor 1980). Bernor et al. (this volume b) have elaborated the evolutionary relationships of these taxa, and provide us with the ability to draw some biochronologic conclusions. *Hippotherium primigenium* is reported from MN 9 of Turkey and as late as MN 10 of Greece (Ravin de Pluie). "*Hippotherium*" *brachypus* is believed to have been derived from "*H.*" *primigenium*, first appearing in MN 10. Pikermi's abundant, and Middle Maragheh's rare, record of "*H.*" *brachypus* supports an earlier Turolian age for Pikermi and Lower/Middle Maragheh. "*Hippotherium*" *giganteum*, occurring in Samos WS, is derived compared to "*H.*" *brachypus* and is reported as first occurring during MN 11.

There are four species of the genus *Hipparion* s.s. in these localities. The first occurring member of the *Hipparion* s.s. lineage is *H. melendezi* (MN 10 of Spain). Bernor et al. (this volume b) argue that the *Hipparion* s.s. species from Maragheh, Samos, and Pikermi represent an evolutionary series: *Hipparion gettyi* (FAD = Lower Maragheh, Pikermi, and probably Samos OMB); *Hipparion prostylum* (FAD = Middle Maragheh, Samos Q6 from the upper portion of the OMB, and Pikermi); *Hipparion campbelli* (FAD = Upper Maragheh, ca. 7.97 Ma); *Hipparion dietrichi* (FAD = Samos MBB, ca. 7.3–7.1 Ma, uppermost MN 12).

There are six species of *Cremohipparion* representing two intrageneric clades. *Cremohipparion moldavicum* (FAD = MN 10 of Moldavia, occurring in Lower and Middle Maragheh) is the most primitive member of this evolutionary radicle and has a morphology from which all other *Cremohipparion* species can ultimately be derived. A dwarf lineage begining with the MN 10 taxon *C. macedoniensis* (Ravin de Pluie) is continued with the more derived taxon *C. matthewi* in the upper portion of Lower Maragheh. A taxon referable to *C.* aff. *matthewi* is rare at Pikermi. *C. matthewi* is known to be abundant in Middle–Upper Maragheh, and specimens from Samos MBB have been referred to this same taxon. *C. nikosi* is represented by the type specimen (Munich; Bernor and Tobien 1989) of unknown provenance (but believed however to come from Samos MBB). Another specimen that compares favorably with *C. nikosi* is AMNH 22908 from Q5. *C. nikosi* is believed to be a plausible ancestor of the more derived MN 13 form from Greece and Spain, *C. periafricanum*.

TABLE 10.2 *Biochronologic Ranges of Maragheh, Pikermi, and Samos Faunas*

Taxon	MN Unit Distribution (6 7 8 9 10 11 12 13)	Maragheh L	Maragheh M	Maragheh U	Samos O	Samos W	Samos M	Pikermi
Insectivora								
Desmanella dubia	<G..........................S..							X
Schizogalerix moedlingensis	S							X
Schizogalerix atticus	S						X	
Primates								
Mesopithecus pentelici	S							X
Mesopithecus delsoni	..S..		X					
Chiroptera								
Samonycteris majori	S						X	
Rodentia								
Spermophilinus cf. *bredai*	<S..........................						X	
Byzantinia hellenicus	G..........................S						X	
Pseudomeriones pythagorasi	?						X	
Occitanomys cf. *provocator*	S?...						X	X
?Gerboa sp.	S?			X				
Pliospalax cf. *sotirisi*	G .... >						X	
Hystrix primigenia	G............S?.....						X	X
Carnivora								
Simocyon primigenius	G......S?.....							X
Ursavus cf. *depereti*	<G.........S........						X	
Indarctos atticus	G......S.....						X	X
Indarctos maraghanus	G........S?		X					
Sinictis pentelici								X
Martes sp.			X					
Martes woodwardi								X
?Plesiogulo sp.								X
Promeles palaeattica				X			X	X
Parataxidea polaki/maraghana			X				X	X
Promephitis lartetii							X	X
Enhydriodon laticeps								X
Plioviverrops orbignyi	<G..............S?.......?						X	X
Ictitherium viverrinum	S.............			X		X	X	X
Hyaenotherium wongii	S.............		X	X		X	X	X
Hyaenictis graeca	S?							X
Lycyaena chaeretis	S?....						X	X
Belbus beaumonti	S						X	
Adcrocuta eximia	S...........?		X	X	X		X	X
Felis attica	S...........?		X	X			X	X
Metailurus parvulus	S...........?			X			X	X
Metailurus major	S.......?						?	X
Machairodus giganteus	S.......?		X	X			X	X
Tubulidentata								
Orycteropus gaudryi	G.................S?......?			sp			X	?
Proboscidea								
Mammut borsoni							X	X
Stegotetrabelodon grandincisivus							X	X
Choerolophodon pentelici	<G	X	X	X			X	X
Deinotherium giganteum	<G			X			X	X
Hyracoidea								
Pliohyrax graecus							X	X
Pliohyrax kruppii							?	
Perissodactyla								
"Hippotherium" brachypus	G...S.....		X				X	
"Hippotherium" giganteum	G......S.....					X		
Hipparion gettyi	G...S.	X			?			X
Hipparion prostylum	G......S	X			?			X
Hipparion campbelli	G........S.			X				
Hipparion dietrichi	G.........S						X	
Cremohipparion moldavicum	GS.....	X	X					
Cremohipparion matthewi	G.....S.....?	?	X				X	
Cremohipparion aff. *matthewi*	G.....S.....						?	X
Cremohipparion nikosi	G.........S.					?		
Cremohipparion mediterraneum	G...S..				?			X
Cremohipparion proboscideum	G.........S.						X	
Ancylotherium pentelicum	S?......?			X			X	X
Aceratherium cf. *incisivum*	G.................S?							X
Chilotherium samium	..S.......					X	X	
Chilotherium schlosseri	S.						?	
Chilotherium kowalevskii	S.						?	

TABLE 10.2 *Biochronologic Ranges of Maragheh, Pikermi, and Samos Faunas* (continued)

Taxon	MN Unit Distribution (6 7 8 9 10 11 12 13)	Maragheh L	Maragheh M	Maragheh U	Samos O	Samos W	Samos M	Pikermi
Chilotherium persiae	S........	X	X	?				
"Dicerorhinus" schleiermacheri	S.............						X	X
Ceratotherium neumayri	S..............?		X			X	X	X
Iranotherium morgani	S?		X					
Artiodactyla								
?Microstonyx erymanthius	G.......S.......?	X	X	X	?	?	X	X
Postpotamochoerus hyotherioides	?						?	
Dorcatherium naui	<G..........S............?						?	
Muntiacus sp.	S						X	
Pliocervus pentelici	S.....		X				X	X
Palaeotragus rouneii	S.............						X	X
Palaeotragus coelophrys	S.............		X	X	X		X	
Palaeotragus quadricornis	S.....					?		
Samotherium boissieri	S.				X			
Samotherium sp.	S.			?		X	X	
Samotherium neumayri	S....	X	X					
Helladotherium duvernoyi	S.....		X				X	X
Helladotherium sp.	S						?	
Bohlinia speciosa	S.....						?	X
Bohlinia attica	S.........					?		X
Hoanotherium sp.	..S		X					
Miotragocerus monacensis	S.........		X				X	X
Graecoryx valenciennesi	...S..						X	X
Tragoportax amalthea	S.........	X					X	X
Tragoportax curvicornis	S.						X	
Tragoportax rugosifrons	S.....		X		X	X		
Samokeros minotaurus	S.....		?				X	
Prostrepsiceros rotundicornis	G...S.....		X				X	X
Prostrepsiceros houtumschindleri	S.....	X	X				X	
Protragelaphus skouzesi	S....		X				X	X
Gazella capricornis/deperdita	S.....	X	X	X	X	X	X	X
"Gazella" rodleri	S		X					
Oioceros rothii	S.		X					X
Oioceros wegneri	S					X		
Samotragus crassicornis	S					?		
Samodorcas kuhlmanni	S.....						?	
Palaeoreas lindermayeri	S.....					X		X
Criotherium argalioides	S.....					X	X	
Parurmiatherium rugosiforns	S.....?						?	
Urmiatherium polaki	S		X					
Palaeoryx pallasi	S..					?	X	X
Palaeoryx sp.	S						X	
Sporadotragus parvidens	S....				X	X	?	X
Protoryx carolinae	S...							X
Pachytragus crassicornis	S			X		X	X	
Pachytragus laticeps	S.......		X			X	X	
Pseudotragus capricornis	S					X	X	

< = earlier FAD than the MN Unit maximum age considered here (MN 6)
> = later than the MN Unit maximum age considered here (MN 13)
O = Old Mill Beds of Samos
W = White Sands of the Main Bone Beds, Samos
M = Main Bone Beds of Samos
G = genus FAD
S = species FAD
T = type species FAD
? = plausible FAD of taxon listed given this biochronologic analysis

APPENDIX: *Lagamorpha and Rodentia from All Sites, Including Chomateri*

Taxon	MN Unit Distribution (6 7 8 9 10 11 12 13)	Maragheh L	Maragheh M	Maragheh U	Samos O	Samos W	Samos M	Pikermi
Lagamorpha								
Prolagus cf. *crusafonti*	...S?..							X
Alilepus sp.	...S?..							X
Rodentia								
Spermophilinus cf. *bredai*	<S........................					X		
"Kowalskia" cf. *lavocati*	G.......S?......T							X

Taxon	MN Unit Distribution								Maragheh			Samos			Pikermi
	6	7	8	9	10	11	12	13	L	M	U	O	W	M	
Byzantinia pikermiensis		G	…	…	S?										X
Byzantinia hellenicus		G	…	…	S								X		
Pseudomeriones pythagorasi							?							X	
Parapodemus gaudryi					G	S?									X
Occitanomys cf. *neutrum*						S?									X
Occitanomys cf. *provocator*						S?…								X	X
?*Gerboa* sp.						S?					X				
Pliospalax cf. *sotirisi*	G	…	…	…	…	>								X	
Muscardinus sp.	G	…	…	…	S?	>									X
Myomimus cf. *dehmi*	G	…	…	…	S?										X
Hystrix primigenia			G	…	S?…									X	X

The other *Cremohipparion* lineage is a group of mixed feeders with pseudoprobosci, *Cremohipparion mediterraneum* (Pikermi and apparently lower Samos) and *Cremohipparion proboscideum* (Samos MBB). Bernor et al. (this volume b) argue that *C. proboscideum* is derived compared to *C. mediterraneum*, and this finds congruence with the biochronologic ranking of the other equid lineages we cite above.

There is a single chalicothere species, *Ancylotherium pentelicum* recorded from Upper Maragheh, Samos MBB, and Pikermi. This is a relatively rare species found in Turolian age horizons from Southeastern Europe and Southwestern Asia.

The Rhinocerotidae are represented by eight species belonging to five genera. The phylogenetic relationships of these taxa are not well understood. They are also sufficiently long-ranging so as not to be particularly useful for biochronologic reconstructions. Mein (1989) reports that *Aceratherium* first occurs during the basal middle Miocene, MN 6. *Aceratherium* cf. *incisivum* is a later-occuring species of this genus reported from Pikermi. The chilotheres are a diverse late Miocene genus-group of aceratherine rhinos that emigrated from Asia (Heissig, this volume). *Chilotherium samium* is first recorded from Anatolia during MN 11, and possibly in late MN 10 (Heissig, op. cit.); it is recorded in our assemblages only from Samos WS and MBB. *Chilotherium persiae* is a close relative of the Chinese species *C. habereri* first occurring during MN 10; *C. persiae* is recorded certainly from Lower and Middle Maragheh, and questionably from Upper Maragheh. *Chilotherium schlosseri* (Samos MBB?) is the largest species of this group, which Heissig argues replaces *C. habereri* in MN12. ?*Chilotherium kowalevskii* is a divergent chilothere questionably occurring in Samos MBB. "*Dicerorhinus*" *schleiermacheri* (Samos MBB) is an MN 9 immigrant into Europe (Mein 1989), continues into MN 10, and is believed to be rarer in the Turolian. This species is present at Pikermi and Samos MBB. *Ceratotherium neumayri* is a late Miocene African immigrant (?MN 9) and savanna grazer; it occurs in Middle Maragheh, Samos WS and MBB, and Pikermi. *Iranotherium morgani* is an enigmatic Asian rhino that occurs in Middle Maragheh and

has been preliminarily reported from the late Miocene of Kenya (Hill, pers. comm.).

The Suidae show a low diversity amongst these three faunas. The genus *Microstonyx* first occurs in Europe during MN 9 (Mein 1989). Fortelius et al. (this volume) report that the derived form, *M. erymanthius*, first occurs in Southwestern Asia during MN 11 and Southeastern Europe during MN 12; it occurs in Lower, Middle, and Upper Maragheh, questionably in Samos OMB and WS, certainly in Samos MBB, and Pikermi. The Pikermi record suggests a possible MN 11 first occurrence in Southeastern Europe. *Postpotamochoerus hyotherioides* is an enigmatic species of uncertain origin that has been reported from Samos MBB.

Dorcatherium naui is a relict tragulid of questionable occurrence in Samos MBB; Gentry and Heizmann (this volume) report its range from MN 9–13. *Muntiacus* sp. is an exotic cervoid questionably occurring in Samos MBB. *Pliocervus pentelici* is a more common cervid recorded from MN 11–13? intervals of Europe (Gentry and Heizmann, this volume); it is recorded from Middle Maragheh, Samos MBB, and Pikermi. The giraffoids are very diverse, including eleven species belonging to no less than five genera. Gentry and Heizmann (this volume), have given detailed biochronologic and biogeographic attributions for these taxa, which we follow here (re: table 10.2). *Palaeotragus rouneii* (FAD = MN 9, Southwestern Asia) is the smallest paleotragine in these assemblages being reported from Samos MBB and Pikermi. *Palaeotragus coelophrys* (FAD = MN 9, Southwestern Asia and Southeastern Europe) is a larger, more ubiquitous form recorded from Middle and Upper Maragheh, and Samos OMB and MBB. *Palaeotragus quadricornis* (FAD = MN 11, Southeastern Europe, Southwestern Asia) is rare, questionably recorded from Middle Maragheh. *Samotherium boissieri* (FAD = MN 11, Southeastern Europe and Southwestern Asia) is a rare form restricted to Samos OMB. *Samotherium neumayri* (FAD = MN 11, Southwestern Asia) is restricted to Lower and Middle Maragheh. *Helladotherium duvernoyi* (FAD = MN 11, Southwestern Asia, Southeastern Europe) is a large grazing sivathere recorded from Middle Maragheh, Samos MBB, and Pikermi. *Bohlinia speciosa* (FAD =

MN 11, Southeastern Europe) is a primitive giraffine recorded from Pikermi and questionably upper Samos. *Bohlinia attica* (FAD = MN 10, Southeastern Europe) is another variant species recorded from Samos WS and Pikermi. *Honanotherium* sp. is yet another lineage of giraffine restricted to Middle Maragheh (FAD = MN 11/12).

The Bovidae are the most diverse family of our assemblage. Gentry and Heizmann (this volume) report *Miotragocerus monacensis* as first occurring in Southeastern Europe and Southwestern Asia from MN 9/10 (also here, Middle Maragheh, Samos MBB, and Pikermi) and *Tragoportax amalthea* as first occurring in Southwestern Asia from MN 10 (also Lower Maragheh, Samos MBB, and Pikermi).

Gentry and Heizmann (this volume) cite a number of Bovidae as first occurring during MN 11: *Tragoportax rugosifrons* (FAD = Southwestern Asia, Southeastern Europe; Middle Maragheh, Samos WS and MBB); *Samokeros minotaurus* (FAD = Southwestern Asia Middle Maragheh and Samos MBB); *Prostrepsiceros rotundicornis* (FAD = Southwestern Asia, Southeastern Europe; Middle Maragheh, Samos MBB, and Pikermi); *Prostrepsiceros houtumschindleri* (FAD = Southwestern Asia, Southeastern Europe; Lower and Middle Maragheh and Samos MBB); *Protragelaphus skouzesi* (FAD = Southwestern Asia, Southeastern Europe; Middle Maragheh, Samos MBB, and Pikermi); *Oioceros rothii* (FAD = Southwestern Asia; Middle Maragheh and Pikermi); *Samodorcas kuhlmanni* (FAD = Southwestern Asia; Samos MBB); *Palaeoreas lindermayeri* (FAD = Southwestern Asia, Southeastern Europe; Samos WS and Pikermi); *Criotherium argalioides* (FAD = Southwestern Asia; Samos WS and MBB); *Parurmiatherium rugosifrons* (FAD = Southwestern Asia; Samos MBB); *Pachytragus laticeps* (Middle Maragheh, Samos WS and MBB); *Pseudotragus capricornis* (FAD = Southwestern Asia; Samos WS and MBB). Gentry and Heizmann (this volume) report a number of taxa as first occurring in MN 12, an assertion that in some cases will need modifying after our analysis (below): *Gazella capricornis/deperdita* is first recorded in Lower Maragheh (MN 11) and is ubiquitous through all levels of our three faunas; *Protoryx carolinae* (Pikermi only); *Pachytragus crassicornis* (upper Maragheh, Samos WS and MBB).

A number of bovids are restricted in their distribution: *Graecoryx valenciennesi* (Samos MBB and Pikermi); *Tragoportax curvicornis* (Samos MBB); *Sporadotragus* (= *Pseudotragus*) *parvidens* (recorded from all levels at Samos and Pikermi); *"Gazella" rodleri* (Middle Maragheh); *Oioceros wegneri* (Samos MBB); *Samotragus crassicornis* (Samos MBB); *Urmiatherium polaki* (Middle Maragheh); *Palaeoryx pallasi* (Samos WS and MBB and Pikermi).

The Maragheh, Samos, and Pikermi faunas demonstrate a complex assembly of open country forms that contrast strikingly with Central European middle and late Miocene faunas. The paleogeographic and paleoenvironmental factors that might have shaped the evolution and dispersal of typical Old World late Miocene "Pikermian" faunas is further addressed by Bernor et al. (this volume a).

Discussion

A central goal of our Schloss Reisensberg workshop was to review and refine the systematic and chronologic bases for later Miocene European MN unit definition and characterization. The earlier view that "Pikermian" megafaunas were referable to the later Turolian MN 12–13 interval (see Mein 1975) has been changing in view of evidence for older age, and several participants in this volume have marshaled evidence that the Pikermian faunas began to develop as early as MN 9 (Bernor et al., this volume; Steininger et al., this volume).

Another element in MN age calibration has been the fact that the time span of the continental late Miocene, including the Vallesian and Turolian ages, has been significantly shortened. A major factor in the change is the recent redating of the early Vallesian site of Höwenegg, now ca. 10.3 Ma (Swisher, this volume), combined with phylogenetic analysis of the provincial Central European *Hippotherium primigenium* s.s. lineage (Bernor et al. 1993a, b, this volume b; Woodburne et al., this volume a). Originally, Berggren and Van Couvering (1974) calibrated the *"Hipparion* Datum" in Western Eurasia at 12.5 Ma, depending heavily on the first K/Ar dating at Höwenegg. Questions about this calibration later arose (see Bernor et al. 1988), and at present Swisher (this volume), Kappelman et al. (this volume), and Woodburne et al. (this volume) agree on a ca. 10.5 Ma for this datum, which defines the base of MN 9.

The upward-revised calibration of late Miocene mammalian biochronology in Europe (Swisher, this volume) is strongly supported by significant "younging" of late Miocene marine microfossil zones in recent years. One key site is Kastellios Hill, Crete, where early MN 10 small mammals occur in beds interdigitated with marine lower Tortonian (Berggren and Van Couvering 1974; Swisher, this volume). Another is at Crevillente, Spain, where Steininger et al. (1989) correlated small mammal faunas at the base of MN 11 (early Turolian) with marine strata close to the planktonic foraminiferal N 16/17 boundary, recently revised to ca. 9 Ma (Steininger et al. 1989). This externally derived boundary age correlates closely with the age we have interpolated for the base of Maragheh's *Hipparion gettyi* Biozone.

The medial Turolian (MN 12) has few calibration points, and the designated reference locality, Los Mansuetos, has only some chronologically long-ranging carnivore taxa common to our three "Pikermian" faunas (i.e., *Adcrocouta eximia, Ictitherium, Plioviverrops, Lycyaena,*

Paramachairodus giganteus, and *Metailurus parvulus;* Mein 1989:87). Steininger et al. (1989) applied the available dates for the base of "Upper Maragheh" *Hipparion campbelli* Zone, 7.2 to 6.9 Ma, as the best estimate for the base of MN 12. The present work places the base of the *campbelli* Zone at ca. 8.0 Ma, and the base of the Middle Maragheh *Hipparion prostylum* Zone at 8.2 Ma, but there is no clear faunal basis for directly correlating either of these levels with Los Mansuetos. On the other hand, Mein (1975, 1989) correlated Mont Lubéron, France, the type locality for *Hipparion prostylum,* with MN 12 based upon its rodent association. Sen (this volume) finds no secure magnetostratigraphic age for the base of MN 12, simply placing it within the middle of Chron 4.

In the absence of an independent and precise geochronology in the relevant Spanish and French sequences, we estimate the base of MN 12 to be correlative with the base of Middle Maragheh's *Hipparion prostylum* Zone based on the species homotaxis shared with Mont Lubéron. Then the Middle Maragheh *Hipparion prostylum* Zone, dated 8.2–8.0 Ma and the Upper Maragheh *Hipparion campbelli* Zone, dated 8.0–7.6 Ma, would be correlative with the earlier portion of MN 12. Samos MBB contains the sister-taxon of *H. campbelli, H. dietrichi,* apparently ranging between 7.7 and 7.1 Ma, but with most of the fossils clearly originating from a restricted stratigraphic horizon calibrated by Swisher (this volume) to be ≥7.1 Ma.

Based on the calibrated Maragheh biostratigraphy, the occurrence of both consecutively occurring *Hipparion* taxa, *H. gettyi* and *H. prostylum,* at Pikermi is suggestive of two possible referrals. If *H. gettyi* and *H. prostylum* actually overlapped in time, then Pikermi would be referable to basal MN 12. If on the other hand Bernor's (1985, 1986) and Bernor et al.'s (1989) observations are valid and *H. gettyi* and *H. prostylum* are indeed time-successive chronospecies, then the classical Pikermi deposits must span a time ranging from later MN 11 across the MN 11/12 boundary. Because of the lack of any data to refute the latter hypothesis, Bernor favors the referral of the Pikermi fauna to late MN 11/early MN 12 with a correlative age of ca. 8.3/8.2 Ma.

Steininger et al. (1989) correlated the base of MN 13 on the basis of some controversial tie points: Molina de Segura (earliest MN 13) is on top of the third evaporitic cycle; Pikermi, MN 12–13, was correlated with the upper Kizilhisar pollen zone; Samos Q5 was correlated with earliest MN 13 because of a spurious correlation with a 6.4–6.1 Ma reversed polarity interval; Venta del Moro (early to middle MN 13) is correlative with Chron 5 (= C3A, 6.9–5.5 Ma). Presently, the base of MN 13 is correlated with the base of the Messinian Stage (Steininger et al., this volume).

The Venta del Moro correlation agrees with Steininger et al.'s (this volume) and Rögl and Daxner-Höck's (this volume) correlation of the base of the Pontian s.s. with the base of the Messinian, ca. 7.1 Ma (Krijgsman et al., in press). The MN 13 reference locality, El Arquillo, Spain, shares no species in common with the classical "Pikermian" faunas reviewed here. Moreover, the only first occurring taxon to appear in Western Europe is the Eastern Mediterranean–Southwest Asian MN 8–13 ranging form, *Hystrix.* The absence of MN 13 first occurring taxa at Maragheh (ca. 9–7.6 Ma), Samos (8.3–7.1 Ma), and Pikermi (ca. ≥8.2 Ma) is congruent with a base MN 13 age = 7.1 Ma. Moreover, we believe it distinctly plausible that the base of the Pontian and Messinian Stages, when the Eastern Paratethys and Mediterranean underwent their final regressions, presented such a strong regional paleoclimatic signal that it provoked a striking increase in seasonality. Moreover, there is evidence of a global pulse and shift from C3 to C4 vegetation ca. 7 Ma (or slightly older) accompanied by increased seasonality globally (Cerling et al. 1989; Quade et al. 1989). European and West Asian late Turolian seasonality was accompanied by the broad dispersion of open country "savanna-like" habitats characteristic of MN 13. We therefore concur in the correlation of the base of MN 13 with the base of the Messinian, ca. 7.1 Ma.

The various lines of evidence advanced here suggest the following MN referrals for Maragheh, Samos, and Pikermi: the Lower Maragheh *Hipparion gettyi* Zone (ca. 9–8.2 Ma) and the Old Mill Beds of Samos (8.3 Ma) are correlative with MN 11; the Middle Maragheh *Hipparion prostylum* Zone (8.2–8.0 Ma), the Upper Maragheh *Hipparion campbelli* Zone (8.0–7.6 Ma) and the bulk of the Samos MBB fossil occurrences (≥7.1 Ma) are correlative with MN 12; Pikermi is correlative with MN 11/12 (ca. 8.3/8.2 Ma). The base of MN 13 is correlative with the Messinian Stage, 7.1 Ma, and includes no fossil-bearing deposits from Maragheh and Samos except possibly Samos' Locality L from the basal marl unit of the Marker Tuffs Member.

Conclusions

Our systematic and chronologic analyses have led to a clearer understanding of the correlations between Maragheh, Samos, and Pikermi and the late Miocene European MN unit system. Moreover, this analysis goes to the heart of this workshop's goals: the apparent existence of pronounced biogeographic provinciality between a Central-Eastern Paratethys province to the north and a Southeast European–Southwest Asian province to the south beginning in the Vallesian. This provinciality would appear to have both paleogeographic and paleoenvironmental bases. The regulation of late Miocene time transgression of "Pikermian" faunas into the Paratethys realm and Western Europe is further explored by Bernor et al. (this volume a).

In conclusion, it is gratifying that new data have become available to support a much clearer understanding of the age and correlations of the Maragheh, Samos, and Pikermi faunas, which have concerned us for so many years, and that with these improvements it is possible for the first time to suggest an outline for a useful biozonation in this incredibly abundant and well-preserved "Pikermian" fauna. In consideration of the goals of this workshop, we are pleased that this study will supply vital clarity in comparing late Miocene history of the well-watered Paratethys province and the region to the south, east, and southwest where the "Pikermian" fauna flourished. The subsequent penetration of "Pikermian" habitats into the Paratethys realm and Western Europe in the latest Miocene is further explored in this volume by Bernor et al. (this volume a).

LITERATURE CITED

Abel, O. 1922. Die Tragödie von Pikermi. *Die Umschau* 26:423–25.

Abich, H. W. Tremblement de terre observé à Tebriz en Septembre 1856, notices physiques et géographiques de M. Khanykof sûr l'Azerbeidjan. *Bulletin de l'Académie (Impériale) des Sciences de Saint Petersbourg. Classe Physico-Mathématique*, pp. 339–352.

Andrews, P., T. Harrison, E. Delson, R. L. Bernor, and L. Martin. This volume. Systematics and biochronology of European and Southwest Asian Neogene catarrhines.

Bachmeyer, V. F., N. Symeonidies, and H. Zapfe. 1982. Die Ausgrabungen in Pikermi-Chomateri bei Athen. *Annalen des Naturhistorischen Museums, Wien* 84:7–12.

Begun, D. 1992. Miocene fossil hominids and the chimp-human clade. *Science* 257:1929–1933.

Berggren, W. A. and J. A. Van Couvering. 1974. The late Neogene: Biostratigraphy, geochronology, and paleoclimatology of the past 15 million years in marine and continental sequences. *Palaeogeography, Palaeoclimatology and Palaeoecology* 16:1–216.

Bernor, R. L. 1983. Geochronology and zoogeographic relationships of Miocene Hominoidea. In *New Interpretations of Ape and Human Ancestry*, ed. R. L. Ciochon and R. S. Corruccini, pp. 21–64. New York: Plenum.

——. 1984. A zoogeographic theater and biochronologic play: The time/biofacies phenomena of Eurasian and African Miocene mammal provinces. *Paléobiologie Continentale* 14:121–42.

——. 1985. Systematic and evolutionary relationships of the hipparionine horses from Maragheh, Iran (late Miocene, Turolian age). *Palaeovertébrata* 15:173–269.

——. 1986. Mammalian biostratigraphy, geochronology, and zoogeographic relationships of the late Miocene Maragheh fauna, Iran. *Journal of Vertebrate Paleontology* 6:76–95.

Bernor, R. L., P. J. Andrews, N. Solounias,, and J. A. H. Van Couvering. 1979. The evolution of "Pontian" mammal faunas: Some zoogeographic, palaeoecologic, and chronostratigraphic considerations. *Annales Géologiques des Pays Helléniques* 1:81–89.

Bernor, R. L., V. Fahlbusch, M. Fortelius, P. Andrews, G. Daxner-Höck, F. Rögl, F. F. Steininger,, and L. Werdelin. This volume a. The evolution of Western Eurasian later Neogene faunas: A chronologic, systematic, biogeographic, and paleoenvironmental synthesis.

Bernor, R. L., G. D. Koufos, M. O. Woodburne, and M. Fortelius. This volume b. The evolutionary history and biochronology of European and Southwest Asian late Miocene and Pliocene hipparionine horses.

Bernor, R. L., J. Kovar-Eder, D. Lipscomb, F. Rögl, S. Sen, and H. Tobien. 1988. Systematics, stratigraphic, and paleoenvironmental contexts of first-appearing hipparion in the Vienna Basin, Austria. *Journal of Vertebrate Paleontology* 8:427–52.

Bernor, R. L., M. Kretzoi, H.-W. Mittmann, and H. Tobien. 1993a. Preliminary systematic assessment of the Rudabánya hipparions. *Mitteilungen der Bayerischen Staatssammlung für Paläontologie und historische Geologie* 33:195–207.

Bernor, R. L., H.-W. Mittmann, and F. Rögl. 1993b. The Götzendorf hipparions. *Annalen des Naturhistorisches Museums, Wien* 95:101–20.

Bernor, R. L. and H. Tobien. 1989. Two small species of *Cremohipparion* (Equidae, Mamm.) from Samos, Greece. *Mitteilungen der Bayerischen Staatssammlung für Paläontologie und historische Geologie* 29:207–26.

Bernor, R. L., H. Tobien, and M. O. Woodburne. 1989. Patterns of Old World hipparionine evolutionary diversification and biogeographic extension. In *Topics on European Mammalian Chronology*, ed. E. H. Lindsay, V. Fahlbusch, and P. Mein, pp. 263–319. New York: Plenum.

Bernor, R. L., M. O. Woodburne, and J. A. Van Couvering. 1980. A contribution to the chronology of some Old World Miocene faunas based on hipparionine horses. *Géobios* 13:25–59.

Black, C. C., L. Krishtalka, and N. Solounias. 1980. Mammalian fossils of Samos and Pikermi. Part 1: The Turolian rodents and insectivores of Samos. *Annals of Carnegie Museum* 49(2): 359–78.

Brandt, J. F. von. 1870. Über die von Herrn Magister Adolph Goebel auf Seiner Persischen Reise bie der Stadt Maragha in der Provinz Aderbeidjan gefundenen Säugethier-Reste. *Denkschrift Naturforscher-Vereins zu Riga* (1870): 1–8.

Bruijn, H. de and P. Mein. This volume. The middle and late Miocene record of the Sciuridae and Petauristidae in France, Central Europe, Southeastern Europe, and Anatolia.

Bruijn, H. de and E. Ünay. This volume. On the evolutionary history of the Cricetodontini from Europe and Asia and its bearing on the reconstruction of migrations and the continental biotope during the Neogene.

Campbell, B. G., M. H. Amini, R. L. Bernor, W. Dickenson, R. Drake, R. Morris, J. A. Van Couvering, and J. A. H. Van Couvering. 1980. Maragheh: A classical late Miocene vertebrate locality in northwestern Iran. *Nature* 287:837–41.

Cerling, T., J. Quade, Y. Wang, and J. R. Bowman. 1989. Carbon isotopes in soils and palaeosols as ecology and palaeoecology indicators. *Nature* 341:138–39.

Crusafont, M.-P. 1950. La cuestion del Ilamado Meotico español. *Arrahona (Savadell)* 1.

——. 1965. Observations á un travail de M. Freudenthal et P. Y. Sondaar sur des nouveaux gisements á *Hipparion* d'Espagne. *Proceedings, Koninklijke Nederlandse Akademie van Wetenschappen*, B 68:121–26.

Dames, W. 1882. Über das Vorkomen fossiler Hirsche im Pliocän von Pikermi. *Sintzunggherichte der Gesellschaft der naturforschenden Freunde, Berlin* 1882:71–72.

Engesser, B. and R. Ziegler. This volume. Didelphids, insectivores, and chiropterans from the later Miocene of France, Central Europe, and Turkey.

Erdbrink, D. P. R., H. N. A. Priem, E. H. Hebeda, C. Cup, P. Dankers, and S. A. P. L. Cloetingh. 1976. The bone-bearing beds near Maragheh in N.W. Iran. *Proceedings, Koninklijke Nederlandse Akademie van Wetenschappen, B* 79:91–113.

Fahlbusch, V. 1991. The meaning of MN zonation: Considerations for a subdivision of the European continental Tertiary using mammals. *Newsletters on Stratigraphy* 24:159–73.

Fortelius, M. 1990. Less common ungulate species from Paşalar, middle Miocene of Anatolia (Turkey). *Journal of Human Evolution* 19:479–87.

Fortelius, M., J. Van der Made, and R. L. Bernor. This volume. Middle and late Miocene Suoidea of Central Europe and the Eastern Mediterranean: Evolution, biogeography, and paleoecology.

Gaudry, A. 1865. *Animaux Fossiles et Géologie de l'Attique.* Paris.

Gentry, A. 1971. The earliest goats and other antelopes from the Samos *Hipparion* fauna. *Bulletin British Museum (Natural History), Geology* 20:234–96.

Gentry, A. and E. Heizmann. This volume. Miocene ruminants of the Central and Eastern Paratethys.

Grewingk, C. 1881. Ueber fossile Saugethiere von Maragha in Persien. *Verhandlungen der K.K. Geologischen Reichsanstalt, Wien* 1881:296.

Heissig, K. This volume. The stratigraphical range of fossil rhinoceroses in the late Neogene of Europe and the Eastern Mediterranean.

Ioakim, Ch. and N. Solounias. 1985. A radiometrically dated pollen flora from the upper Miocene of Samos island, Greece. *Revue de Micropaléontologie* 28:197–204.

Kamei, T., J. Ikeda, H. Ishida, S. Ishida, I. Onishi, H. Partodizar, S. Sasjima, and S. Nishimura. 1977. A general report of the geological and paleontological survey in Maragheh area, North-West Iran. *Memoir Faculty of Science, Kyoto University, Geological and Mineralogical Survey* 43:131–64.

Kappelman, J., S. Sen, M. Fortelius, A. Duncan, B. Alpagut, J. Crabaugh, A. Gentry, J. P. Lunkka, F. McDowell, N. Solounias, and S. Viranta. This volume. Chronology and biostratigraphy of the Miocene Sinap Formation of Central Turkey.

Koufos, G. and J. Melentis. 1984. The late Miocene (Turolian) mammalian fauna of Samos Island (Greece). Study of the collection of Palaeontological Museum of Mytilinii, Samos. 2. Equidae. *Science Annales Faculty Physics and Mathematics, University Thessaloniki* 24:47–78.

Krijgsman, W., F. J. Hilgen, C. G. Langereis, and W. J. Zachariasse. In press. The age of the Tortonian/Messinian boundary. *Earth and Planetary Newsletters.*

Lydekker, R. 1886. On the fossil Mammalia of Maragha in Northwest Persia. *Quarterly Journal of the Geological Society of London* 42:173–76.

Lyell, C. 1856. *A Manual of Elementary Geology,* 6th ed. London: John Murray.

Major, Ch. I. F. 1888. Sur un gisement d'ossements fossiles dans l'île de Samos contemporains de l'âge de Pikermi. *Comptes Rendus Hebd. Sëances Academie des Sciences* 107:1178–81.

———. 1893. Le gisement ossifère de Mytilini et catalogue des ossements fossiles. In *Samos: Étude géologique, paléontologique et botanique,* C. de Stefani, C. I. F. Major, and W. Barbey. Lausanne.

———. 1894. *Le gisement ossifère de Mytilini et catalogue d'ossements fossiles recueills à Mytilini, île de Samos, et déposés au college Gillard, à Lausanne.* Lausanne: Georges Bridel et Cie éditeurs.

Mecquenem, R. de. 1905. Le gisement de vertébrés fossiles de Maragha. *Comptes Rendus Académie des Sciences de Paris* 141:284–86.

———. 1906. Les vertébrés fossiles de Maragha. *La Nature* 24:10.

———. 1908. Contribution à l'étude du gisement des Vertébrés de Maragha et de ses environs. *Annales d'Histoire Naturelle* 1:27–79.

———. 1911. Contribution a l'étude du gisement des Vertébrés de Maragha, deuxième partie. *Annales d'Histoire Naturelle* 1:81–98.

———. 1924–1925. Contribution a l'étude des fossiles de Maragha I. *Annales de Paléontologie* 13:133–60.

Mein, P. 1975. Résultats du Groupe de Travail des Vertébrés. In *Report on Activity of the RCMNS Working Groups (1971–1975),* ed. J. Senes, pp. 78–81. Bratislava: SAV.

———. 1989. European mammal correlations. In *Topics on European Mammalian Geochronology,* ed. E. H. Lindsay, V. Fahlbusch, and P. Mein, pp. 73–90. New York: Plenum.

Meissner, B. 1976. Das Neogen von Ost-Samos. Sedimentationgeschichte und Korrelation. *Neues Jarbuch für Paläontologie Abhandlungen* 152:161–76.

Quade, J., T. E. Cerling, and J. R. Bowman. Development of Asian monsoon revealed by marked ecological shift during the latest Miocene in northern Pakistan. *Nature* 342:163–65.

Rögl, F. and G. Daxner-Höck. This volume. Late Miocene Paratethys correlations.

Rögl, F., H. Zapfe, R. L. Bernor, R. Brzobohaty, G. Daxner-Höck, I. Draxler, O. Fejfar, J. Gaudant, P. Herrmann, G. Rabeder, O. Schultz, R. Zetter. 1993. Primatenfundstelle Götzendorf an der Leitha, Niederösterreich (Obermiozän, Pontien des Wiener Beckens). *Jahrbuch des geologischen Bundesanstalt.*

Schlosser, M. 1904. Die fossilen Cavicornier von Samos. *Beitrage Paläontologie. Österrichten-Ungans* 17:28–118.

Sen, S. 1986. "Contribution á la magnetostratigraphie et á la paléontologie des formations continentales Néogènes du pourtour méditerranéen: Implication biochronologiques et paleobiologiques." Thèse d'Etat, Univ. Paris 6.

———. This volume. Present state of magnetostratigraphic studies in the continental Neogene of Europe and Anatolia.

Sen, S. and J. P. Valet. 1986. Magnetostratigraphy of late Miocene continental deposits in Samos, Greece. *Earth and Planetary Science Letters* 80:167–74.

Solounias, N. 1981a. The Turolian fauna from the Island of Samos, Greece. *Contributions to Vertebrate Evolution* 6:1–232.

———. 1981b. Mammalian fossils of Samos and Pikermi. Part 2: Resurrection of a classic Turolian fauna. *Annals of Carnegie Museum* 50:231–70.

Sondaar, P. Y. 1971. The Samos *Hipparion. Proceedings, Konin-*

klijke Nederlandse Akademie van Wetenschappen, B 74:417–41.

Steininger, F. F., W. A. Berggren, D. V. Kent, R. L. Bernor, S. Sen, and J. Agusti. This volume. Circum-Mediterranean Neogene (Miocene and Pliocene) marine-continental chronologic correlations of European mammal units.

Steininger, F. F., R. L. Bernor, and V. Fahlbusch. 1989. European Neogene marine/continental chronologic correlations. In *Topics on European Mammal Chronology*, ed. E. H. Lindsay, V. Fahlbusch, and P. Mein, pp. 15—46. New York: Plenum.

Swisher, C. C., III. This volume. New ^{40}Ar/^{39}Ar dates and their contribution toward a revised chronology for the late Miocene of Europe and West Asia.

Symeonidis, N. F., F. Bachmeyer, and H. Zapfe. 1973. Ausgrabungen in Pikermi bei Athen, Griechenland. *Annalen des Naturhistorischen Museums, Wien* 77:125–32.

Symeonidis, N. F. and A. Markopolou-Diacantoni. 1977. La faune Pikermienne et le Néogene. *Bulletin Societé Géologique de France* 1977:111–15.

Takai, F. 1958. Vertebrate fossils from Maragha. *Institute of Oriental Culture, University of Tokyo* 26:7–11.

Tobien, H. 1968. Palaeontologische Ausgrabungen nach Jungtertiären Wirbeltieren auf der Insel Chios (Griechenland) und bei Maragheh (N.W. Iran). *Jahrbuch der Vereiningung "Freude der Universität Mainz"* 1968:51–58.

Ünay, E. This volume. On fossil Spalacidae (Rodentia).

Van Couvering, J. A. and J. A. Miller. 1971. Late marine and nonmarine time scale in Europe. *Nature* 230:559–63.

Weidmann, M., N. Solounias, R. E. Drake, and G H. Curtis. 1984. Neogene stratigraphy of the Eastern Basin, Samos Island, Greece. *Géobios* 17:477–90.

Werdelin, L. This volume. Carnivores, exclusive of Hyaenidae, from the later Miocene of Europe and Western Asia.

Werdelin, L. and N. Solounias. This volume. The evolutionary history of hyaenas in Europe and Western Asia during the Miocene.

Woodburne, M. O. and R. L. Bernor. 1980. On superspecific groups of some Old World hipparionine horses. *Journal of Paleontology* 8:315–27.

Woodward, A. S. 1901. On the bone beds of Pikermi, Attika, and on similar deposits in Northern Euboea. *Geologic Magazine* 8:481–86.

II

MAMMALIAN SYSTEMATICS, BIOGEOGRAPHY, AND BIOCHRONOLOGY

11

Didelphids, Insectivores, and Chiropterans from the Later Miocene of France, Central Europe, Greece, and Turkey

B. ENGESSER AND R. ZIEGLER

Didelphids, insectivores, and bats usually occur less frequently than rodents in the European and West Asian Neogene. Moreover, whereas rodent species can be identified using isolated teeth, more complete material is needed for accurate insectivore and bat determinations (i.e., relatively complete jaws). For example, in soricids the antemolar and mandibular condyle are systematically significant. Frequently, bat dental morphology is insufficient for species identification: at the minimum, complete tooth rows are often needed for taxonomic identification. Moreover, many insectivore and bat species show little to no morphological change over long durations. Didelphids are very rare in the European and West Asian Neogene and completely inappropriate for biochronologic correlation. Lastly, the Didelphidae, Insectivora, and Chiroptera have generally been less intensively investigated than other small mammal groups and do not have the comparative background for facilitating interpretations about their distribution.

This article presents range charts of didelphids, insectivores, and bats from France, Central Europe, Greece, and Turkey. The charts span the time from MN 6 to MN 13. Due to insufficient material and the lack of morphological change over a long duration, most of these groups are not suitable for biostratigraphy. However, within the erinaceids, evolutionary lineages and gradual morphological changes can be demonstrated.

Systematics

Family Didelphidae Gray 1821

For the time interval under consideration, didelphids are known only from Central Europe and are represented by a single species, *Amphiperatherium frequens*. Didelphids became extinct by the end of MN 6; late records include Rümikon, Zeglingen, Gallenbach 2b,c, Unterzölling 1a,b, and Gisseltshausen 1a.

Family Plesiosoricidae Winge 1917

Plesiosorex is an archaic genus that ranges back as far as the upper Oligocene (MP 26). The last record of this genus is from Hammerschmiede (MN 9). Within this long stratigraphic range a diversity of species is known but only partly described. The relationships between *P. styriacus*, *P. germanicus*, and *P. schaffneri* are not yet well understood. There is only one record of *Plesiosorex* from Turkey: the locality of Bayraktepe (MN 9, pers. comm. from Dr. Engin Ünay). *Plesiosorex* is conspicuously absent from Western European (France, Spain) localities younger than MN 1 (last record: Chavroches).

Family Soricidae Gray 1821

Almost every Neogene small mammal fauna includes remains of soricids, but only a small portion of these forms have been identified and well described to the species level. Therefore the relationships between the different species are poorly known.

Blarinella dubia first appears during MN 9 (Rudabánya, Hungary). Whereas in Central and most of Western Europe this species occurs only in MN 9 and MN 10, it persists in Spain until MN 13 (El Arquillo 1) and in Italy until MN 13 or 14 (Bacinello V-3).

Family Heterosoricidae Reumer 1987

Heterosorex and *Dinosorex* are difficult to distinguish by isolated teeth. Whereas the sequence *H. neumayrianus-H. subsequens-H. delphinensis* may represent one evolutionary lineage, it is quite probable that within the genus *Dinosorex*

TABLE 11.1 *Temporal Distribution of Taxa*

MN-Zones	6	7/8	9	10	11	12	13
Didelphidae							
Amphiperatherium	5, 6, 9, 30, 50, 51						
A. frequens		†					
Plesiosoricidae							
Plesiosorex							
P. styriacus	30						
P. germanicus	2						
P. schaffneri	30 aff.	32	15 aff.				
P. sp.			14, 18	†			
			55				
Soricidae							
Lartetium							
L. dehmi	4, 5 cf., 24	44					
Miosorex							
M. desnoyersianus	43	44					
M. givensis		44	26 cf.				
Allosorex							
A. gracilidens	24						
A. sp.					47		
Paenelimnoecus							
P. crouzeli	43	44					
P. repenningi			26 cf.	22, 28			
P. sp.		39					

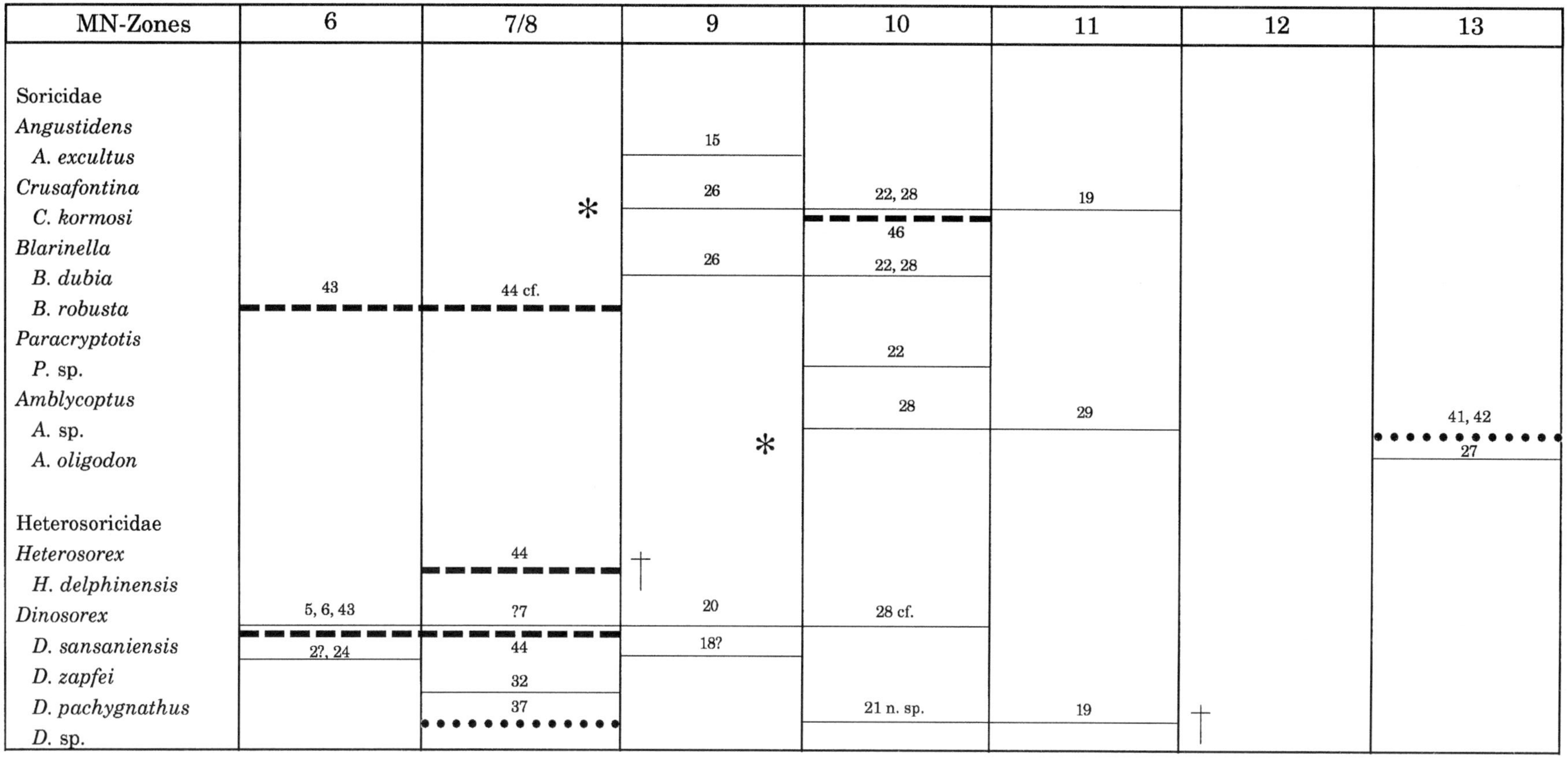

MN-Zones	6	7/8	9	10	11	12	13
Soricidae							
Angustidens							
A. excultus			15				
Crusafontina			26	22, 28	19		
C. kormosi			*	46			
Blarinella			26	22, 28			
B. dubia	43	44 cf.					
B. robusta							
Paracryptotis				22			
P. sp.							
Amblycoptus				28	29		
A. sp.			*				41, 42
A. oligodon							27
Heterosoricidae							
Heterosorex		44					
H. delphinensis			†				
Dinosorex	5, 6, 43	?7	20	28 cf.			
D. sansaniensis	2?, 24	44	18?				
D. zapfei		32					
D. pachygnathus		37		21 n. sp.	19		
D. sp.						†	

TABLE 11.1 *Temporal Distribution of Taxa* (Continued)

MN-Zones	6	7/8	9	10	11	12	13
Talpidae							
Talpa							
T. minuta	24	25, 32	14, 16, 18	21			
T. gilothi	43	44		22	19		
T. vallesensis					19		
Proscapanus							
P. sansaniensis	2, 4, 5, 6, 8, 9 / 43	?7, ?10, 32 / 44	14				
"Scaptonyx*							
"S." edwardsi	24	32 / 44	?17, 18				
Urotrichus							
?U. dolichochir	24	44	26				27
Desmanella							
D. stehlini		32		46 cf.			
cf.D. crusafonti				22			
D. sp.		38	26	20	19		45
D. sickenbergi		37			23,47		
D. cingulata		39					
D. dubia						34	
D. amasyae							41, 42 cf.
"D."quinquecuspidata			15				
Asthenoscapter							
A. meini		44					
Mygalea							
M. antiqua	30 aff., 43						
M. jaegeri	2						
Desmana							
D. sp.				21, 28?			

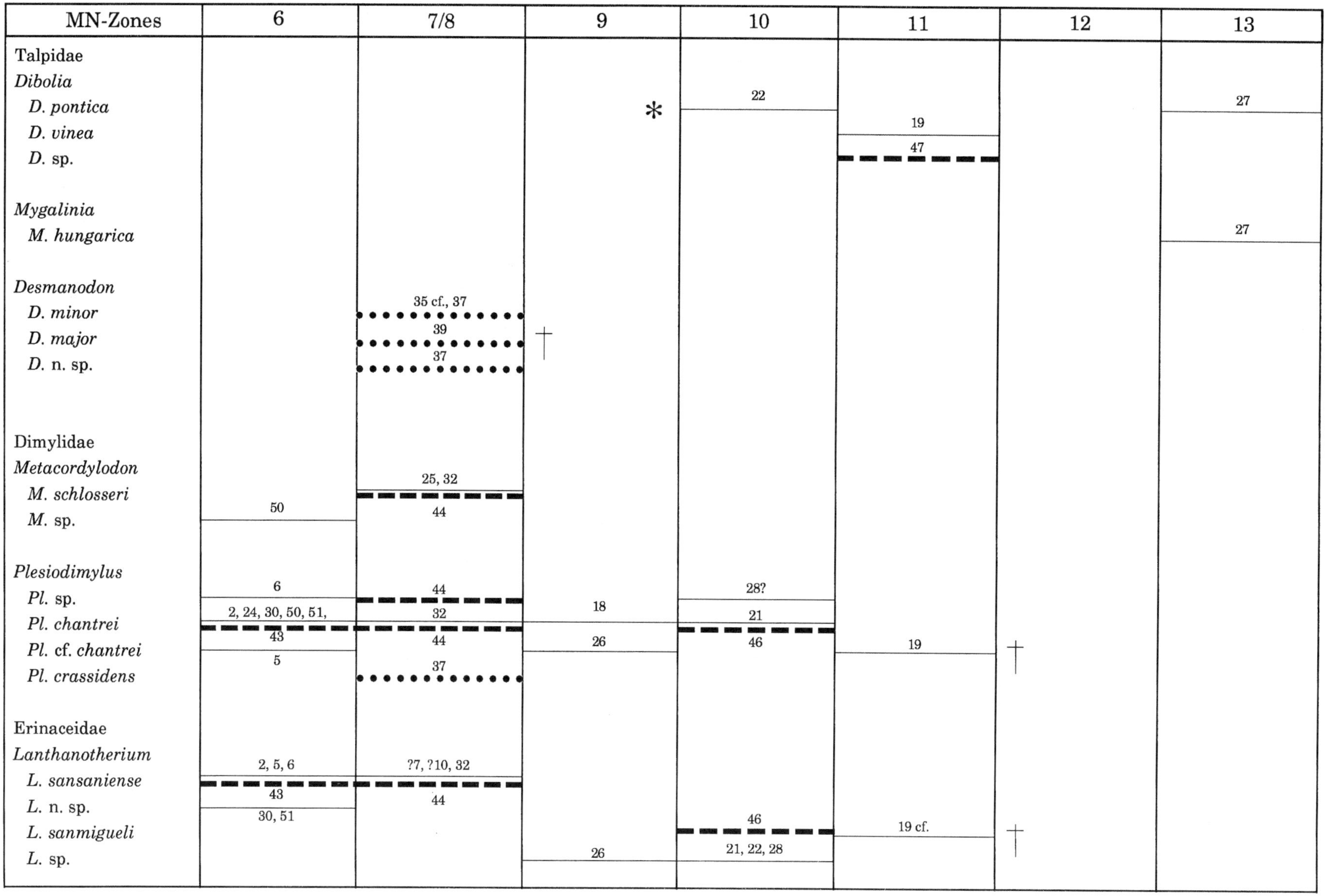

MN-Zones	6	7/8	9	10	11	12	13
Talpidae							
Dibolia							
D. pontica			*	22			27
D. vinea					19		
D. sp.					47		
Mygalinia							
M. hungarica							27
Desmanodon							
D. minor		35 cf., 37					
D. major		39	†				
D. n. sp.		37					
Dimylidae							
Metacordylodon							
M. schlosseri		25, 32					
M. sp.	50	44					
Plesiodimylus							
Pl. sp.	6	44		28?			
Pl. chantrei	2, 24, 30, 50, 51,	32	18	21			
Pl. cf. chantrei	43	44	26	46	19		
Pl. crassidens	5	37					
Erinaceidae							
Lanthanotherium							
L. sansaniense	2, 5, 6	?7, ?10, 32					
L. n. sp.	43	44					
L. sanmigueli	30, 51			46	19 cf.		
L. sp.			26	21, 22, 28			

TABLE 11.1 *Temporal Distribution of Taxa (Continued)*

MN-Zones	6	7/8	9	10	11	12	13
Erinaceidae							
Galerix							
G. exilis	2, 3, 4 / 43						
G. stehlini		44 / 11, 32	14, 18, 26 cf.	22 cf., 28			
G. socialis	30, 51	44		46 aff.	29 / 47		
G. cf. symeonidisi	5, 6, 8, 9, 50	?7, ?10, 31					
G. sp.			20	22	23 / 19		
Schizogalerix							
S. voesendorfensis							
S. moedlingensis							
S. zapfei							
S. sp.							
S. pasalarensis	35	38 aff.					
S. anatolica	36 cf.	38 aff., 39					
S. sinapensis			40				
S. attica						34	
S. n. sp.							41
Mioechinus							
M. sansaniensis	30 cf.	31 aff. / 44					
M. oeningensis		13 / 44					
M. tobieni		39					
M. sp.		38					
Amphechinus							
A. ginsburgi	43						
Postpalerinaceus							
P. intermedius		44					
P. vireti				46			

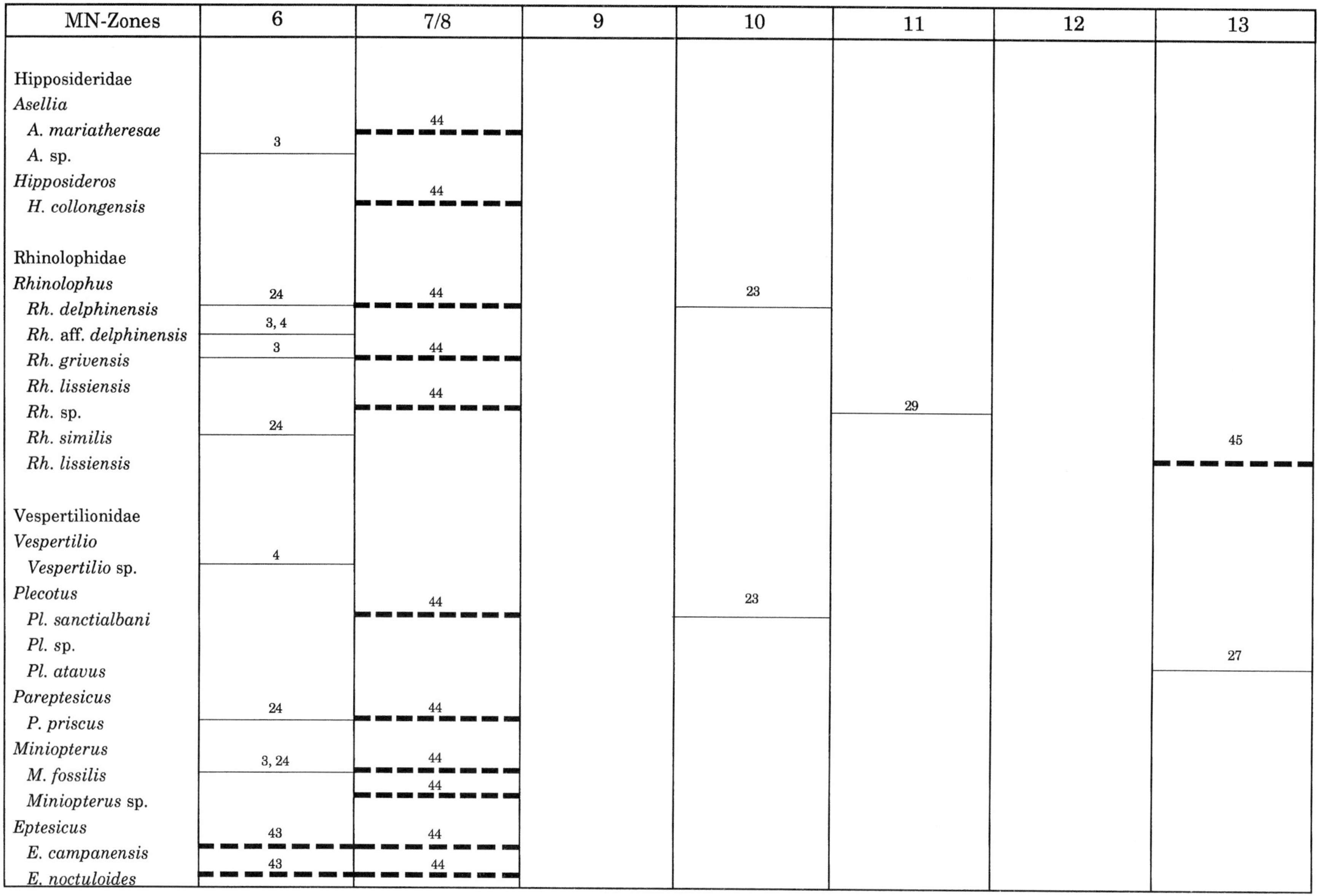

MN-Zones	6	7/8	9	10	11	12	13
Hipposideridae							
Asellia							
A. mariatheresae	3	44					
A. sp.							
Hipposideros							
H. collongensis		44					
Rhinolophidae							
Rhinolophus	24	44		23			
Rh. delphinensis	3, 4						
Rh. aff. delphinensis	3	44					
Rh. grivensis		44					
Rh. lissiensis							
Rh. sp.	24				29		
Rh. similis							
Rh. lissiensis							45
Vespertilionidae							
Vespertilio	4						
Vespertilio sp.							
Plecotus		44		23			
Pl. sanctialbani							
Pl. sp.							
Pl. atavus							27
Pareptesicus	24	44					
P. priscus							
Miniopterus	3, 24	44					
M. fossilis		44					
Miniopterus sp.							
Eptesicus	43	44					
E. campanensis	43	44					
E. noctuloides							

TABLE 11.1 *Temporal Distribution of Taxa (Continued)*

MN-Zones	6	7/8	9	10	11	12	13
Vespertilionidae							
Myotis							
M. antiquus		44					
M. murinoides	43	44					
M. elegans	43						
Myotis sp.		44					
M. boyeri							45
Miostrellus							
M. risgoviensis	3						
Samonycteris							
S. majori						33	
Molossidae							
Mormopterus							
M. helveticus	3, 4	?12, 32					
M. kalorhinus	4						
Tadarida							
T. engesseri	3, 4	32					
T. leptognatha	3						
T. monslapidis	3, 4	32					
T. sp.		44					
Megadermatidae							
Megaderma							
M. lugdunensis	3, 4, 24				29 cf.		
M. vireti		44 cf.		22			45
Miomegaderma							
M. gaillardi	43?	44					

TABLE 11.2 *Listing of Localities and MN Unit Attributions by Country*

No.	locality	MN-zone
Southern Germany		
1	Puttenhausen	5/6
2	Viehhausen	6
3	Goldberg	6
4	Steinberg	6
5	Gallenbach 2b/c	6
6	Gisseltshausen 1a	6
7	Gisseltshausen 1b	?7
8	Unterneul 1a/b	6
9	Unterzolling 1a/b	6
10	Laimering	?7
11	Steinheim	7
12	Böttingen	?7
13	Oeningen	7
14	Aumeister	9
15	Hammerschmiede	9
16	Unterföhring	9
17	Garching	?9
18	Großlappen	9
19	Dorn-Dürkheim	11
Austria		
20	Vösendorf	9
21	Götzendorf	9
22	Kohfidisch	10
23	Eichkogel	11
CSFR		
24	Neudorf	6
Poland		
25	Oppole	7

No.	locality	MN-zone
Hungary		
26	Rudabánya	9
27	Polgardi 2, 4	13
28	Sümeg	10
29	Csákvár	11
Switzerland		
30	Rümikon	6
31	Vermes 2	8
32	Anwil	8
50	Zeglingen	6
51	Schwamendingen	6
Greece		
33	Samos	12
34	Pikermi	12
Turkey		
35	Paşalar	?5/6
36	Çandir	6
37	Çari Cay	7
38	Sofça	7
39	Eskihisar	8
40	Sinap Tepe	9
55	Bayractepe	9
41	Amasya	13
42	Kavurca	13
France		
43	Sansan	6
44	La Grive M, L-7	7/8
46	Montredon	10
47	Mollon	11
45	Lissieu	13

?7	stratigraphy of site uncertain
7?	systematics uncertain
✳	first appearance
†	last occurrence

———————— Central Europe
▬ ▬ ▬ ▬ ▬ France
— — — — — SE Europe (Greece)
• • • • • • • • • Asia Minor (Turkey)

there existed different lineages. *Heterosorex* is reported for the last time from MN 7 (La Grive), while *Dinosorex* is last reported from MN 11 (Dorn Dürkheim).

Family Talpidae Gray 1825

Desmanella is an old genus that is widespread in Europe. It is known already from upper Oligocene and lower Miocene localities (Rottenbuch 1, Eggingen, Ulm-Westtangente). Turkish *Desmanella* evolved independently from its congener in Central Europe, beginning with *D. sickenbergi* from Sariçay (MN 7). *Desmanella* is recorded for the last time from Lissieu (MN 13). The relationships between this genus's species are unclear.

Whereas Central European *Desmanodon* is known only from MN 4 (Rembach), in Turkey it occurs from MN 6 to MN 8. Therefore, it is possible that this genus as well as *Desmanella* dispersed from Central Europe to Turkey and became restricted in the latter area during the middle Miocene due to ecological factors.

Family Dimylidae Schlosser 1887

The genus *Plesiodimylus* has a very long chronologic range: MN 3 to MN 11. Within this time, *Plesiodimylus* shows very little evolutionary change, and in most Neogene zones it is represented by the species *P. chantrei*. On the other hand, Turkish *Plesiodimylus* is reported from only one locality, Sariçay. With its highly specialized dentition the Sariçay *P. crassidens* could be an endemic species.

Family Erinaceidae Bonaparte 1838

Lanthanotherium is a very rare genus. The relationships between this genus's different species are not well understood.

Within the genus *Galerix* we can distinguish two evolutionary lineages. *Galerix* first appears in the Stubersheim 3 fauna (upper MN 3) with the occurrence of *G. aurelianensis*. This species is believed to be ancestral to *G. stehlini* from La Grive. The transition between these two species probably occurred outside Central Europe, because we do not have certain representatives of this lineage within the MN 4 to MN 7/8 time range. *Galerix stehlini* most probably represents a Western European immigrant in the La Grive fauna. The other lineage is first recognized with the occurrence of *G. symeonidisi*, whose earliest record is Petersbuch 2 (MN 4a). This was a very successful species, which replaced *G. aurelianensis* and became a dominant part of the uppermost Orleanian (MN 4b to MN

5) insectivore faunas. The transition of *G. symeonidisi*, which also is sparsely represented in Greece (Aliveri, type locality), to *G. exilis* is well documented in south German MN 5/6 faunas. In MN 6 this lineage is represented solely by a true *G. exilis*. In La Grive and other MN 7/8 faunas this species is replaced by *G. socialis*. This is the most primitive *Galerix* and it could be ancestral to all other European species of *Galerix*. It is obviously a late immigrant from an unknown area of origin.

Schizogalerix first appeared in Turkey and developed independently from *Galerix* after the early middle Miocene. At least three evolutionary lineages are known to be included within this genus. The Turkish species from Paşalar, Çandir, and Yeni Eskihisar may represent a single lineage. The first record of *Schizogalerix* outside Turkey is *S. voesendorfensis*, which can be derived from *S. pasalarensis*. The most northern occurrence is *Schizogalerix* sp. from Dorn-Dürkheim.

The Erinaceinae are less well documented and comparatively poorly known. *Amphechinus* is an archaic genus that flourished in the upper Oligocene and lower Miocene. The relationships between *Amphechinus*, *Mioechinus*, and *Postpalerinaceus* are not clear.

Order Chiroptera Blumenbach 1779

Bats are known to occur at only five Turkish localities: Eşme Akçaköy, Babadat, Eskihisar, Dumlumpinar, and Paşalar. They are not determinable to the genus level.

The localities of Goldberg and Steinberg yielded the best-known and most diverse Central European bat faunas. The La Grive fauna contains sixteen species of bats, but they are not yet published in detail. Less diverse faunas include Neudorf and Sansan.

Since the bats are a very conservative group, most genera also occur in the recent faunas. The stratigraphic record of the genera *Rhinolophus*, *Myotis*, *Hipposideros*, and *Megaderma* extends back into the upper Oligocene. The relationships between the species are poorly known.

Acknowledgments

Thanks are due to Dr. J. Harloff for computer-aided drawing of the tables.

LITERATURE CITED

Andreae, A. 1904. Dritter Beitrag zur Kenntnis des Miozäns von Oppeln i. Schlesien. *Mitteilungen des Roemer Museums* 20:1–22.

Bachmayer, F. and R. W. Wilson. 1970. Small mammals (Insectivora, Chiroptera, Lagomorpha, Rodentia) from the Kohfidisch Fissures of Burgenland, Austria. *Annalen des Naturhistorischen Museums, Wien* 74:533–87.

———. 1984. Die Kleinsäugerfauna von Götzendorf, Niederöster-reich. *Sitzungsberichte der Österreichischen Akademie der Wissenschaften, mathematisch-naturwissenschaftliche Klasse I* 193/6–10:303–19.

Baudelot, S. 1972. *Étude des chiroptères, insectivores, et rongeurs du Miocène de Sansan (Gers)*. Thèse inédite, Université Paul Sabatier Toulouse.

Bruijn, H. de, R. Daams, G. Daxner-Höck, V. Fahlbusch, L. Ginsburg, P. Mein, and J. Morales. 1992. Report on the RCMNS working group on fossil mammals, Reisensburg 1990. *Newsletters on Stratigraphy* 26:65–118.

Butler, P. M. 1948. On the evolution of the skull and the teeth in Erinaceidae with special reference to fossil material in the British Museum. *Proceedings of the Zoological Society of London* 118:446–500.

Crochet, J.-Y. and M. Green. 1982. Contribution à l'étude des micromammifères du gisement miocène supérieur de Montredon (Hérault). 3—Les Insectivores. *Palaeovertébrata* 12:119–31.

Engesser, B. 1972. Die obermiozäne Säugetierfauna von Anwil (Baselland). *Tätigkeitsberichte der naturforschenden Gesellschaft, Baselland* 28:37–363.

———. 1975. Revision der europäischen Heterosoricinae (Insectivora, Mammalia). *Ecologae Geologicae Helvetiae* 68:649–71.

———. 1979. Relationships of some insectivores and rodents from the Miocene of North America and Europe. *Bulletin of the Carnegie Museum of Natural History* 14:1–68.

———. 1980. Insectivora und Chiroptera (Mammalia) aus dem Neogen der Türkei. *Schweizerische Paläontologische Abhandlungen* 102:47–149.

Engesser, B., A. Matter, and M. Weidmann. 1981. Stratigraphie und Säugetierfaunen des mitteleren Miozäns von Vermes (Kt. Jura). *Ecologae Geologicae Helvetiae* 74:893–952.

Fahlbusch, V. and W. Wu. 1981. Puttenhausen: Eine neue Kleinsäuger-Fauna aus der Oberen Süsswasser-Molasse Niederbayerns. *Mitteilungen der Bayerischen Staatssammlung für Paläontologie und historische Geologie* 21:115–19.

Heissig, K. 1989. Neue Ergebnisse zur Stratigraphie der mittleren Serie der Oberen Süsswassermolasse Bayerns. *Geologica Bavarica* 94:239–57.

Heizmann, E. P. J. and V. Fahlbusch. 1983. Die mittelmiozäne Wirbeltierfauna vom Steinberg (Nördlinger Ries): Eine Übersicht. *Mitteilungen der Bayerischen Staatssammlung für Paläontologie und historische Geologie* 23:83–93.

Hürzeler, J. 1939. Säugetierfaunulae aus dem oberen Vindobonien der Nordwest-Schweiz. *Ecologae Geologicae Helvetiae* 32(2): 195–203.

———. 1944. Beiträge zur Kenntnis der Dimylidae. *Schweizerische Paläontologische Abhandlungen* 65:1–44.

Hutchison, J. H. 1974. Notes on type specimens of European Miocene Talpidae and a tentative classification of Old World Tertiary Talpidae (Insectivora: Mammalia). *Géobios* 7:211–56.

Kretzoi, M. 1952. Die Raubtiere der Hipparionfauna von Polgardi. *Annales Instituti Geologici Publici Hungarici* 40:5–42.

———. 1954. Rapport final des fouilles paléontologiques dans la grotte de Csákvár. *Földtani Intézet Evi Jelentesei* 1952:37–68.

———. 1985. A Sümeg-gerinci fauna és faunaszakasz In *Sümeg es Környékének földtani felépitése*, ed. Haas et al. *Geologica Hungarica, Serie geologica* 20:214–20.

Legendre, S. 1985. Molossidés (Mammalia, Chiroptera) cénozoiques de l'Ancien et du Nouveau Monde: Statut systématique; intégration phylogénique des données. *Neues Jahrbuch für Geologie und Paläontologie, Abhandlungen* 170:205–27.

Mayr, H. and V. Fahlbusch. 1975. Eine unterpliozäne Kleinsäugerfauna aus der Oberen Süsswasser-Molasse Bayerns. *Mitteilungen der Bayerischen Staatssammlung für Paläontologie und historische Geologie* 15:91–111.

Rabeder, G. 1973. *Galerix* und *Lanthanotherium* (Erinaceidae, Insectivora) aus dem Pannon des Wiener Beckens. *Neues Jahrbuch für Geologie und Paläontologie, Monatshefte* 1973:429–66.

——. 1973. *Plecotus (Paraplecotus)* aus dem Ober Miozän von Kohfidisch (Burgenland). *Myotis* 11:15–17.

——. 1985. Die Säugetiere des Pannonien. In *Chronostratigraphie und Neostratotypen, Miozän der Zentralen Paratethys. M6, Pannonien*, ed. A. Papp, A. Jambor and F. F. Steininger, pp. 440–63. Budapest: Verlag Ungar. Gbad. Wiss.

Rachl, R. 1983. *Die Chiroptera (Mammalia) aus den mittelmiozänen Kalken des Nördlinger Rieses (Süddeutschland)*. Thesis University München.

Repenning, C. A. 1967. Subfamilies and genera of the Soricidae. *Professional Papers of the United States Geological Survey* 565:1–74.

Reumer, J. W. 1984. Ruscinian and early Pleistocene Soricidae (Insectivora, Mammalia) from the Tegelen (the Netherlands) and Hungary. *Scripta Geologica* 73:1–173.

Rümke, C. G. 1976. Insectivora from Pikermi and Biodrak (Greece). *Proceedings, Koninklijke Nederlandse Akademie van Wetenschappen, B* 79:256–70.

Schötz, M. 1988. Die Erinaceiden (Mammalia, Insectivora) aus Niederaichbach und Maßendorf (Obere Süsswassermolasse Niederbayerns). *Mitteilungen der Bayerischen Staatssammlung für Paläontologie und historische Geologie* 28:65–87.

——. 1989. Die *Plesiosorex*-Funde (Insectivora, Mammalia) aus der Kiesgrube Massendorf (Obere Süsswassermolasse Niederbayerns). *Mitteilungen der Bayerischen Staatssammlung für Paläontologie und historische Geologie* 29:141–57.

Seemann, I. 1938. Die Insektenfresser, Fledermäuse und Nagetiere aus der obermiozänen Braunkohle von Viehhausen bei Regensburg. *Palaeontographica* 89:1–36.

Sen, S. 1990. Stratigraphie, faunes de mammifères et magnétostratigraphie du Néogène de Sinap Tepe, Province d'Ankara, Turquie. *Bulletin du Musée national d'Histoire naturelle de Paris* 4e série. 12(C3–4): 243–77.

Storch, G. 1978. Die turolische Wirbeltierfauna von Dorn-Dürkheim, Rheinhessen (SW-Deutschland). 2. Mammalia: Insectivora. *Senckenbergiana lethaea* 58:421–49.

Storch, G. and Z. Qiu. 1991. Insectivores (Mammalia: Erinaceidae, Soricidae, Talpidae) from the Lufeng hominoid locality, late Miocene of China. *Géobios* 24:601–21.

Stromer, E. 1928. Wirbeltiere aus dem obermiocänen Flinz Münchens. *Abhandlungen der Bayerischen Akademie der Wissenschaften, mathematisch-naturwissenschaftliche Klasse* 32:1–71.

——. 1940. Die jungteriäre Fauna des Flinzes und des SchweissSandes von München: Nachträge und Berichtigungen. *Abhandlungen der Bayerischen Akademie der Wissenschaften, mathematisch-naturwissenschaftliche Klasse* 48:1–102.

Topal, G. 1989. Tertiary and early Quaternary remains of *Corynorhinus* and *Plecotus* from Hungary (Mammalia, Chiroptera). *Vertebrata Hungarica* 23:33–55.

Viret, J. 1951. Catalogue critique de la faune de mammifères de La Grive St.-Alban: Chiroptères, Carnivores, Pholidotes. *Nouvelle Archive du Musée de Lyon* 4:1–197.

Viret, J. and H. Zapfe. 1951. Sur quelques Soricidés miocénes. *Ecologae Geologicae Helvetiae* 44:411–26.

Zapfe, H. 1950. Die Fauna der miozänen Spaltenfüllung von Neudorf an der March (CSR). Chiroptera. *Sitzungsberichte der Österreichischen Akademie der Wissenschaften, mathematischnaturwissenschaftliche Klasse I* 159:51–64.

——. 1951. Die Fauna der miozänen Spaltenfüllung von Neudorf an der March (CSR). Insectivora. *Sitzungsberichte der Österreichischen Akademie der Wissenschaften, mathematisch-naturwissenschaftliche Klasse I* 160:449–80.

Ziegler, R. 1983. *Odontologische und osteologische Untersuchungen an Galerix exilis (Blainville) (Mammalia, Erinaceidae) aus den miozänen Ablagerungen von Steinberg und Goldberg im Nördlinger Ries (Süddeutschland)*. Thesis University München.

——. 1985. Talpiden (Mammalia, Insectivora) aus dem Orleanium und Astaracium Bayerns. *Mitteilungen der Bayerischen Staatssammlung für Paläontologie und historische Geologie* 25:131–75.

——. 1989. Heterosoricidae und Soricidae (Insectivora, Mammalia) aus dem Oberoligozän und Untermiozän Süddeutschlands. *Stuttgarter Beiträge zur Naturkunde, Serie B* 154:1–74.

——. 1990. Didelphidae, Erinaceidae, Metacodontidae und Dimylidae (Mammalia) aus dem Oberoligozän und Untermiozän Süddeutschlands. *Stuttgarter Beiträge zur Naturkunde, Serie B* 158:1–158.

Ziegler, R. and V. Fahlbusch. 1986. Kleinsäuger-Faunen aus der basalen Oberern Süsswasser-Molasse Niederbayerns. *Zitteliana* 14:3–80.

12

Distribution and Biochronology of European and Southwest Asian Miocene Catarrhines

P. ANDREWS, T. HARRISON, E. DELSON, R. L. BERNOR, AND L. MARTIN

European later Neogene primate distribution is reviewed here. Three families of primates are discussed: cercopithecids by Delson, pliopithecids and oreopithecine hominids by Harrison, and other hominids mainly by Andrews and Martin. The chronologic and biogeographic framework for all these groups was initially developed here by Bernor, with subsequent input from the other authors. The final result is a truly collaborative effort in which each author comments on the others' sections. The classification we adopt is described for each group, with notes explaining the rationale behind the taxonomic decisions, but this work is not intended to be a systematic revision of the groups. Accompanying this section is a summary table with a complete listing of the primate-bearing localities in Europe and Southwestern Asia, and the species identifications for each (together with a brief synonymy where relevant), arranged by country and geographic region. Primate distribution patterns are then discussed in the final section and some provisional palaeogeographic conclusions reached.

Systematics

Pliopithecidae

The pliopithecids are a conservative group of catarrhines that had a small to medium body size and that were geographically broadly distributed throughout Eurasia during the Miocene. The family is first recorded during the latest early Miocene (late Orleanean), and it continues well into the late Miocene in both Western and Central Europe (until the late Vallesian) and China (until the latest Miocene). Circumstantial evidence lends support to the claim that the pliopithecids originated in Africa, probably some time during the Oligocene, but no direct antecedents have yet been identified outside Eurasia from the

Paleogene or the early part of the Miocene (Harrison 1987a; Bernor et al. 1988a; Harrison et al. 1991). The pliopithecids are taxonomically diverse, and the family includes at least eleven species, which range geographically from Western Europe to Southern China. Nine of these species are confined to Western and Central Europe, where they comprise at least half of the known catarrhine species from these two biogeographical provinces (tables 12.1 and 12.7).

From comparisons of their dental morphology, pliopithecids appear to have been adapted to a range of dietary behaviors, with taxa inferred to be specialized folivores (i.e., *Anapithecus* and *Laccopithecus*), and others that were apparently more eclectic feeders, capable of exploiting a combination of soft fruits and young leaves (i.e., *Pliopithecus*). Although our knowledge of pliopithecid oro-facial and postcranial anatomy is limited, it is likely, given inferred differences in dietary behavior, their estimated range of body sizes, and their reconstructed ecological associations, that their spectrum of adaptive diversity was broad.

Although abundant and broadly distributed throughout Europe and Asia (being recorded from over forty different localities), pliopithecids are rarely found in association with large hominoids. The localities where they do co-occur include Rudabánya (Hungary), Lufeng (China), Epplesheim (Germany), Castell de Barberà (Spain), and Neudorf-Sandberg (Republic of Slovakia). The observed differences in pliopithecid and large hominoid distribution patterns may possibly reflect inadequate sampling, a conclusion supported to some extent by the fact that two of the sites where a co-occurrence has been established, Rudabánya and Lufeng, are among the most productive primate-bearing localities in Eurasia. However, this explanation seems unlikely in itself, given that pliopithecids and large hominoids co-occur in Europe at only 8.3% of the localities (5 of the 60) from which they are known (see table

12.7), and that some sites with good samples, such as Can Llobateres in Spain and Göriach in Austria, have only yielded the remains of *Dryopithecus* or *Pliopithecus*, respectively. A more likely explanation is that pliopithecids and large hominoids had somewhat different habitat preferences that allowed them only minimal geographical overlap under certain ecological conditions. However, until better data are available on the paleoecology of European Miocene localities, the nature of the ecological partitioning between pliopithecids and large hominoids cannot be resolved.

As mentioned above, the cranio-dental and postcranial anatomy of most pliopithecid species is relatively poorly known. Many of the species are represented by jaw fragments and isolated teeth only. An almost complete cranium of *Pliopithecus vindobonenis* is known from the Republic of Slovakia, and good cranial material of *Laccopithecus robustus* and *Pliopithecus zhanxiangi* has been recovered from sites in China (Zapfe 1958, 1961a; Wu and Pan 1984, 1985; Pan 1988; Harrison et al. 1991). In addition, several partial skeletons of *Pliopithecus vindobonensis* have provided valuable information on postcranial anatomy and its relevance for taxonomic and phylogenetic reconstructions (Zapfe 1958, 1961a; Simons and Fleagle 1973; Fleagle 1983; Harrison 1987a). However, postcranial remains of other species are extremely rare, being limited to a few isolated specimens belonging to *Pliopithecus antiquus*, *Anapithecus hernyaki*, and *Laccopithecus robustus* (Hürzeler 1954; Zapfe and Hürzeler 1957; Kretzoi 1975; Ginsburg and Mein 1980; Begun 1988; Meldrum and Pan 1988). Nevertheless, the available material is adequate to provide a reasonably good assessment of pliopithecid phylogenetic relationships, as well as a sound reevaluation of their alpha-taxonomy.

In the past, authors have argued that pliopithecids bear a close relationship to modern gibbons (Gervais 1849; Hofmann 1893; Hürzeler 1954; Zapfe 1961a; Simons 1972; Simons and Fleagle 1973), and this view has recently received additional support from studies of *Laccopithecus* from China (Wu and Pan 1985; Meldrum and Pan 1988; Fleagle 1988). However, a number of workers have presented a more convincing case that resemblances to hylobatids are due to similarity in overall size and the retention of plesiomorphous features (Remane 1965; Groves 1972, 1974; Delson and Andrews 1975; Ciochon and Corruccini 1977; Szalay and Delson 1979; Ginsburg and Mein 1980; Harrison 1982, 1987a).

It should be further noted that when the family-group name was originally proposed by Zapfe (1961a), the Pliopithecidae also included *Propliopithecus* from the Eo-Oligocene of Egypt. This taxonomic arrangement was widely accepted and expanded on by later workers (Remane 1965; Groves 1972, 1974; Delson and Andrews 1975; Ciochon

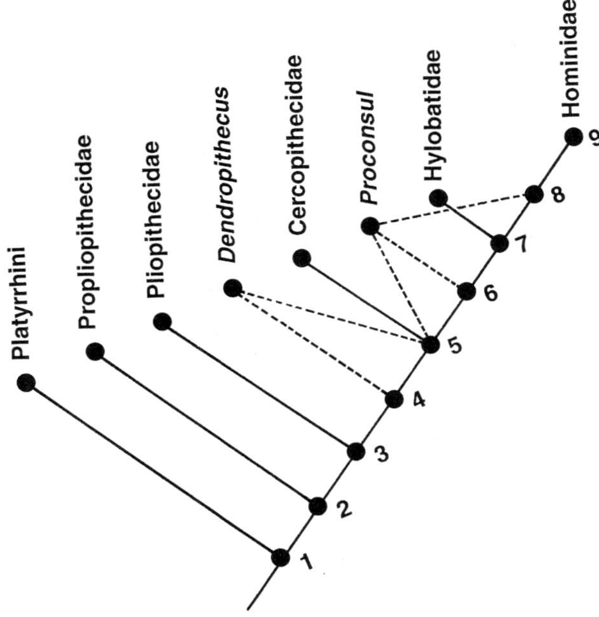

FIGURE 12.1 Cladogram showing the phylogenetic relationships of Pliopithecidae (adapted from Harrison 1987a). Broken lines indicate alternative interpretations of the phylogenetic position of *Dendropithecus* (nodes 4 and 5: Harrison 1982, 1987a) and *Proconsul* (node 5: Harrison 1982, 1987a; node 6: Andrews 1992; node 7: Walker and Teaford 1989).

and Corruccini 1977; Szalay and Delson 1979; Andrews 1980; Fleagle 1986). However, Harrison (1982; 1987a) has demonstrated that the Pliopithecidae, as traditionally conceived, is a paraphyletic group including a number of unrelated, morphologically conservative early catarrhine species. As a result, Harrison includes only Eurasian Miocene genera within the Pliopithecidae, while *Propliopithecus* (including *Aegyptopithecus*) is included in its own family, the Propliopithecidae Straus 1961. This view has received additional support from a number of recent workers (Andrews 1985; Delson 1988; Tattersall et al. 1988; Fleagle 1988), and the nomen Pliopithecidae is used here in this more restrictive sense. The inferred phylogenetic relationships of the Pliopithecidae are shown in figure 12.1.

Cranio-dentally pliopithecid morphology is generally consistent with the ancestral catarrhine morphotype, as inferred by Harrison (1982, 1987a). The partial crania of *Pliopithecus vindobonensis*, *Pliopithecus zhanxiangi*, and *Laccopithecus robustus* show that pliopithecids conform quite closely to the hypothesized ancestral pattern. The pliopithecids share the following key features of the cranium:

1. the face is relatively short and broad
2. the lower face is shallow, with a substantial overlap of

the orbits and nasal aperture in the dorso-ventral plane, and a low subnasal clivus

3. the palate is narrow, especially in the premaxillary region, and the upper toothrows converge anteriorly

4. the region of the incisive canal is represented by a pair of large elliptical openings, located close to the alveolar margin of the incisors, which allow a wide communication between the palate and the nasal fossa

5. the orbits are subcircular, frontally directed, and situated anteriorly over the premolars

6. the orbits have a slightly protruding inferior rim and supraorbital torus (it is uncertain whether this circumorbital rim represents a specialization of the pliopithecids or a feature retained from the primitive catarrhine morphotype)

7. the infraorbital foramen is usually single, and is located close to the inferior margin of the orbit

8. the interorbital region is relatively wide

9. the anterior root of the zygomatic arch originates low on the face, close to the maxillary alveolar margin, just above M1

10. the facial portion of the maxilla bears a distinct canine juga, at least in males, and a shallow canine fossa

11. the maxillary sinus is large, causing inflation of its lateral wall and invading the anterior root of the zygomatic arch, but not so extensively that it penetrates inferiorly between the roots of the cheek teeth

12. the complete postorbital plate retains a wide inferior orbital fissure, but no superior orbital fissure

13. the neurocranium is large in relation to the facial skeleton, and the well-marked temporal lines converge posteriorly, but do not meet to form a sagittal crest, even in males

The structure of the ectotympanic in *Pliopithecus vindobonensis* is critical for determining its phylogenetic status. The ectotympanic forms a short, partially enclosed bony tube, most resembling the condition in some juvenile individuals of extant catarrhine primates. In this respect, *Pliopithecus* is more derived than *Propliopithecus* from the Eo-Oligocene of Egypt, which has a platyrrhine-like annular ectotympanic but is less derived than all extant catarrhines, as well as fossil hominoids such as *Proconsul*, *Sivapithecus*, and *Oreopithecus*, which have a fully enclosed ectotympanic tube (Szalay and Delson 1979; Harrison 1987a). Unfortunately, the ear region of other pliopithecids is completely unknown, so it is uncertain whether this was a characteristic of the entire group. Nevertheless, it is reasonable to assume that *Pliopithecus'* ectotympanic morphology represents an intermediate step in the transformational series between the primitive anthropoid condition and the more derived pattern typical of modern catarrhines (fig. 12.1).

The pliopithecids are characterized by a number of dental specializations that distinguish the family as a monophyletic group. Pliopithecid autapomorphies include:

1. lower incisors are relatively slender and high-crowned (the height of the crown is at least twice that of the mesiodistal length)

2. lower central incisors are waisted toward the base of the crown, giving the tooth a distinctive flask-shaped outline when viewed from the buccal aspect (the lower incisors are broader and more spatulate in *Anapithecus*)

3. p3 is mesiodistally short and high-crowned, with a steeply inclined mesiobuccal honing face for occlusion with the upper canine

4. p4 and lower molars are relatively long and narrow

5. lower molars have a pliopithecine triangle on the talonid basin (except for *Pliopithecus vindobonensis*, which may be secondarily derived in this regard)

6. lower molars increase in size from m1 to m3, with a marked size differential between them

7. upper premolars and molars are relatively short and very broad

8. both M2 and M3 are considerably larger than M1

The postcranium of *Pliopithecus* is generalized and conforms closely to the proposed ancestral catarrhine morphotype (Harrison 1987a). However, in two significant aspects, it is more primitive than all extant catarrhines: the distal humerus is perforated by a large entepicondylar foramen (a feature lacking in all extant and fossil cercopithecids and hominoids, including the early catarrhines from the early Miocene of East Africa, but present in the propliopithecids from the Eo-Oligocene of Egypt); and there is a simple, hinge-like articulation in the carpo-metacarpal joint of the pollex (modern catarrhines have a specialized saddle joint that allows axial rotation of the thumb) (Napier 1961, 1962; Harrison 1987a).

The Pliopithecidae includes two distinct subfamilies, the Pliopithecinae and the Crouzeliinae (see below). The subdivision of the Pliopithecidae into two distinct clades was first proposed by Ginsburg and Mein (1980), who noted that the genus *Pliopithecus* could be distinguished from the other pliopithecids in the morphology of its lower molars. These differences, and additional characteristics used to separate the two subfamilies, are presented in table 12.2 (see also Harrison et al. 1991). The anterior dentition and the facial skeleton are poorly known for most species, especially the crouzeliines, so it is not certain whether consistent morphological differences exist between the subfamilies in these areas. Comparisons of the partial skull of *Laccopithecus* with those of *Pliopithecus* spp., indicate, however, that the cranial morphology of pliopithecids may, in general, be quite uniform. The upper and lower incisors in some of the pliopithecids do exhibit contrasting morphologies, but these appear to be species-specific differences, rather than differences that distinguish the pliopithecids at the subfamilial level. For example, *Laccopithecus robustus* has relatively small incisors, especially the upper central incisors, when compared with those in *Pliopithecus* spp., while *Anapithecus* apparently has slightly broader, more spatulate lower incisors than other pliopithecids.

From comparisons of the upper dentition of *Laccopithecus robustus*, the only crouzeliine in which the upper molars have so far been described, and given the contrasting lower molar morphology in the two subfamilies, it is likely that the upper molars of crouzeliines differ from those of pliopithecines in the following respects: (1) the cusps are higher and more conical in shape, and are connected by sharper occlusal crests; (2) the trigon forms a larger component of the crown; (3) the distal basin is restricted to a narrow, slit-like fissure; and (4) the hypocone is separated by a deep groove from the trigon, rather than connected by a crest to the protocone, as in pliopithecines.

The dental differences between pliopithecines and crouzeliines presumably reflect a fundamental difference in their dietary behavior. Although the pliopithecids as a group were probably adapted to exploit a range of niches that were predominantly folivorous, the more elongated lower molar crowns of crouzeliines, with their more elevated cusps and relatively longer and sharper occlusal crests, imply that the subfamily may have exploited leaves and other soft fibrous plant materials more intensively than did the pliopithecines, which were probably somewhat more frugivorous. The crouzeliines probably occupied niches most comparable to those seen in some of the more frugivorous colobines, such as *Colobus satanas*, or *Alouatta*, which is strikingly similar to *Anapithecus* in molar morphology.

Harrison (1987a) has previously suggested that crouzeliine dentitions appear to be more conservative than those of *Pliopithecus* in retaining several features of the lower molars characteristic of the ancestral anthropoid morphotype. For example, the long, narrow crowns, the high and sharp occlusal crests and cusps, the relatively elongated mesial foveae, the small size of the hypoconulids, the long and obliquely aligned crista obliquae, and the absence (in some species) of true distal foveae in the lower molars of crouzeliines represent a complex of features that are usually interpreted as being more primitive than the morphological pattern typical of pliopithecines. This view is still maintained by Andrews and Martin, but Harrison is now more inclined to accept the alternative explanation that the crouzeliine molar morphology is a derived pattern, associated with more specialized dietary behaviors, that has in some respects secondarily converged on the ancestral anthropoid morphotype.

PLIOPITHECINAE

The subfamily Pliopithecinae includes five species (tab. 12.1). The species are morphologically similar, and all of them can be assigned readily to a single genus, *Pliopithecus*. The species of *Pliopithecus* can be distinguished from each other by a combination of dental morphological

TABLE 12.1 *Classification of the European Pliopithecidae*

Pliopithecoidea
  Pliopithecidae
    Pliopithecinae
      Pliopithecus antiquus (Blainville 1839)
        (including *P. piveteaui* Hürzeler 1954)
      Pliopithecus platyodon Biederman 1863
      Pliopithecus vindobonensis Zapfe and Hürzeler 1957
      Pliopithecus priensis Welcomme et al. 1991
    Crouzeliinae
      Plesiopliopithecus lockeri Zapfe 1961
      Plesiopliopithecus auscitanensis (Ginsburg 1975)
        (= *Crouzelia auscitanensis* Ginsburg 1975)
      Plesiopliopithecus rhodanica (Ginsburg and Mein 1980)
        (= *Crouzelia rhodanica* Ginsburg and Mein 1980)
      Anapithecus hernyaki Kretzoi 1975
       Nov gen. et nov. sp. Moyà-Solà, in prep.
    Indeterminate
      "*Semnopithecus*" *eppelsheimensis* Haupt 1935

differences and overall size. Four of the species, *Pliopithecus antiquus*, *Pliopithecus vindobonensis*, *Pliopithecus platyodon*, and *Pliopithecus priensis*, are known from European sites and will be dealt with in detail below. The fifth species, *Pliopithecus zhanxiangi*, recently described by Harrison et al. (1991), is known from middle Miocene (early Tungurian, correlative with the early Astaracian, MN 6) localities in Northern China. Further discussion of this latter taxon is outside the scope of this present review.

Pliopithecus antiquus   *Pliopithecus antiquus*, the type species of the genus, was the first extinct non-cercopithecid catarrhine to be described (Blainville 1839). The type specimen, from Sansan (MN 6), France, consists of a mandible with complete lower dentition (Blainville 1839; Gervais 1849; Hürzeler 1954; Bergounioux and Crouzel 1965; Simons 1972; Szalay and Delson 1979). The species is known primarily from localities in France, but several isolated specimens referable to this species have been recovered from similar age localities in Switzerland (Stein am Rhein, MN 6) and Germany (Stätzling, Ziemetshausen, and Gallenboch, MN 6; see Heissig 1987). A well-preserved lower jaw fragment with m1–3 from La Grive-Saint-Alban (possibly from the Peyre et Beau quarry) is clearly referable to this species. The specimen was first described by Depéret (1887) as *Pliopithecus antiquus*, race *chantrei*, and later recognized as a distinct subspecies, *Pliopithecus antiquus chantrei*, by Ginsburg (1986), purportedly characterized as having a somewhat reduced m3. The molars, however, are remarkably similar in size and morphology to those of the type specimen from Sansan, differing in a few minor details. The size difference between m2 and m3 is insignificant given the range of variation seen in modern catarrhine species.

The slightly older material from the Faluns de Touraine, Anjou, Pontlevoy-Thenay, and Manthelan in the Loire Valley of France (all correlated with MN 5), was previously attributed to a different species, *Pliopithecus piveteaui* (Hürzeler 1954; Ginsburg 1964, 1975, 1986; Collier 1978, 1979; Szalay and Delson 1979; Ginsburg and Mein 1980). However, the major purported distinction, the smaller relative size of m3 in the Loire Valley sample, is not supported by more recent discoveries. The type specimen, a mandibular fragment of an immature individual with m2 in occlusion and m3 partially erupted, from the locality of Manthelan, is certainly unusual for a pliopithecid in having an m3 smaller in size than m2: the occlusal area of m3 is only 89.4% of that of m2, while in all other pliopithecids it ranges from 101.1% to 127.8%. However, several factors indicate that this difference may be explained as the result of normal pliopithecid intraspecific variation. Firstly, unlike the Göriach sample, which formerly comprised most of the *Pliopithecus antiquus* hypodigm, and in which m3 is substantially larger than m2 (the occlusal area of m3 is 114.8% of m2 on average), the two specimens with m2–3 from Sansan, the type locality for *Pliopithecus antiquus*, have a much smaller discrepancy in size between m2 and m3 (the occlusal area of m3 is 107.8% of m2 in Sansan I and can be reliably estimated to be only 104.3% in Sansan II). Secondly, other species of pliopithecids, including those in which only small samples are available for study, exhibit quite wide ranges of variation in relative m3 size (e.g., *Pliopithecus vindobonensis*, 101.1–127.7%; *Pliopithecus platyodon*, 109.3–127.8%; *Laccopithecus robustus*, 103.2–118.0%). Thirdly, the range of variation in extant catarrhine relative m3 size is great, and more than adequate to encompass the range of variation exhibited by the Sansan sample and the Manthelan specimen. Finally, a number of isolated m3s from the Loire Valley, previously attributed to *Pliopithecus piveteaui*, are as large or even larger than m3 in the type specimen of *Pliopithecus antiquus*, and from this evidence it would seem that m3 reduction in the Manthelan specimen is an isolated case, not typical of the Loire Valley sample as a whole (see figure 4 in Ginsburg and Mein 1980 for a graphic illustration of this point). As a consequence, *Pliopithecus piveteaui* is considered to be insufficiently distinct to merit the recognition of a separate taxon and is included as a junior synonym of *Pliopithecus antiquus* (Harrison 1991a; see also Bergounioux and Crouzel 1965).

A number of additional specimens from France (Meigné-le-Vicomte and Doué-la-Fontaine, MN 9), Spain (Castell de Barberà, MN 9), Switzerland (Kreuzlingen and Rümikon, MN 6), Germany (Diessen am Ammersee, MN 6), and Poland (Opole, MN 7, and Przeworno II, MN 8) can be assigned tentatively to *Pliopithecus antiquus* (Hürzeler 1954; Kowalski and Zapfe 1974; Crusafont-Pairo 1978; Ginsburg 1986, 1989). If these identifications are confirmed, they would greatly extend the species geographic and temporal range. Most of the specimens consist of isolated teeth that are difficult to assign with any degree of taxonomic certainty, although they appear to be morphologically and metrically consistent with *Pliopithecus antiquus*.

The best of the late material referred to this taxon originates from Castell de Barberà, which is considerably younger in age than that from the type locality (Crusafont-Pairo 1978; Crusafont-Pairo and Golpe-Posse 1981). At least two individuals are represented. One individual consists of an associated series of isolated teeth, comprising upper and lower partial toothrows of a subadult female. The second individual is represented by an isolated right p3. The specimens are similar to material from France attributed to *Pliopithecus antiquus*. The minor morphological differences and slightly smaller size of the Castell de Barberà material, when compared with the type specimen from Sansan (a male individual), are easily interpreted as the result of sexual dimorphism. It is worthwhile noting, in this regard, that the lower cheek teeth from Castell de Barberà are closest in size to the dentition in the smaller mandibular fragment from Sansan, which may have belonged to a female individual of *Pliopithecus antiquus*. Nevertheless, the lower molars from Castell de Barberà are relatively narrower than those assigned to *Pliopithecus antiquus* from France, and this feature may later prove to be of some taxonomic significance. However, given the overall similarity in general size and morphology of the material, and the small samples available, it seems best to include the Castell de Barberà specimens in *Pliopithecus antiquus*, and to regard the minor differences as due to intraspecific variation.

Ginsburg (1989) has described a right mandibular fragment preserving a portion of m1 and the well-preserved crowns of m2–3 from Meigné-le-Vicomte, France (MN 9), which he has assigned to *Pliopithecus antiquus*. In addition, a number of *Pliopithecus* isolated teeth have been recovered from Doué-la-Fontaine, a locality of similar age to Meigné-le-Vicomte (Ginsburg 1986, 1989, 1990). These specimens are morphologically and metrically very similar to *Pliopithecus antiquus* from Sansan, and they may thus provide further evidence that the species extended its temporal range from the early Astaracian (MN 5) to the early Vallesian (MN 9).

Pliopithecus vindobonensis   The type specimen of *Pliopithecus vindobonensis* is from Neudorf-Spalte in the Slovakian Republic (MN 6). It is the best-known pliopithecid species, being represented at the type locality by three partial skeletons, in addition to some isolated bones of other individuals (Zapfe and Hürzeler 1957; Zapfe 1958, 1961a). *Pliopithecus vindobonensis* is slightly larger than the type species but comparable in size to *Pliopithecus platyodon*. It is distinguished morphologically from both of these species in the following respects: (1) the lower incisors are relatively higher-crowned; (2) the upper central incisor has a notched lingual cingulum and is relatively broad; (3) the p4 and lower molars tend to be slightly narrower; (4) the pliopithecine triangle on the lower molars is absent or indistinct; (5) the upper molars are slightly broader, with a relatively smaller trigon basin and a less well developed buccal cingulum; and (6) the size differential between the teeth in the upper and/or lower molar series is greater.

In their initial description of *Pliopithecus vindobonensis,* Zapfe and Hürzeler (1957) referred all the material in a new subgenus, *Pliopithecus (Epipliopithecus).* However, Zapfe (1958, 1961a) expressed justified reservations about its level of distinctiveness, and most subsequent authors have tended to abandon the use of a separate subgeneric rank for the taxon (Szalay and Delson 1979; Ginsburg and Mein 1980; Harrison et al. 1991).

Pliopithecus platyodon   The extensive collections of *Pliopithecus* from Göriach, Austria (MN 6) were referred to *Pliopithecus* cf. *antiquus* by Hürzeler (1954). However, these specimens differ from *Pliopithecus antiquus* in their significantly larger size, broader p3, slightly narrower and more rectangular lower molars, and greater size increase from m1 to m3. Although the Göriach specimens are similar in overall size to *Pliopithecus vindobonensis,* they differ in the following characters:

1. upper and lower incisors are lower-crowned
2. I1 has a continuous lingual cingulum without a distinct notch
3. upper premolars are broader
4. p3 is relatively broader with subequal mesial and distal crests
5. the upper molars are somewhat narrower, with better development of the buccal cingulum and a relatively larger trigon
6. P4 has a better-developed lingual cingulum
7. lower molars tend to be slightly broader, with a well-developed pliopithecine triangle.

It is evident from these comparisons that the Göriach sample should be assigned to a separate species of *Pliopithecus.* Sera (1917) referred the Göriach material to a separate species, *Pliopithecus goeriachensis.* However, as noted earlier by Hürzeler (1954), the Göriach sample is very similar in morphology to a crushed palate from Elgg (Switzerland; MN 5), the type specimen of *Pliopithecus platyodon* Biedermann 1863. As a consequence, *Pliopithecus platyodon* has been resurrected as the valid name for the Göriach and Elgg samples, and *Pliopithecus goeriachensis* is considered a junior synonym (Harrison 1991a; Harrison et al. 1991).

Pliopithecus priensis   Welcomme et al. (1991) recently described a new pliopithecid species, *Pliopithecus priensis,* from Priay II, France (upper MN 9) The type specimen is a right mandibular fragment with m1–2. Based on the published illustration of m1, the specimen appears to have its closest affinities with the pliopithecines and is probably referable to the genus *Pliopithecus.* The species is somewhat larger than the other pliopithecines from Europe but is comparable in its size to the early middle Miocene taxon *Pliopithecus zhanxiangi* from Tongxin, China (MN 6). However, it differs from the latter in having relatively more elongated lower molars and a less pronounced size differential between m1 and m2. Additional material will be needed before the precise affinities of *Pliopithecus priensis* can be established with confidence.

Neudorf-Sandberg   An isolated right m3 of a pliopithecid was recovered from Neudorf-Sandberg in the Slovakian Republic (MN 6) at the end of the last century. It has usually been regarded as belonging to *Pliopithecus antiquus* (Glaessner 1931; Pia and Sickenberg 1934; Hürzeler 1954; Zapfe 1961a, 1969; Szalay and Delson 1979). The specimen can be readily identified as belonging to *Pliopithecus,* but its allocation to a particular species is uncertain (see also Hürzeler 1954).

CROUZELIINAE

Compared with the pliopithecines, the crouzeliines are taxonomically more diverse. The European representatives of the subfamily comprise at least six species, belonging to three or four different genera (tab. 12.1); as noted above, the differences between crouzeliines and pliopithecines are summarized in table 12.2. In addition to these species, a late surviving crouzeliine, *Laccopithecus robustus,* is known from the late Miocene of China. The three species of *Plesiopliopithecus* are the earliest recorded crouzeliines in Eurasia, being known from sites in France and Austria correlated with MN 6 and MN 7. *Anapithecus,* a larger and more specialized crouzeliine from Hungary, and possibly also Austria, occurs later in time (late MN 9; ca. 10–9.5 Ma; Bernor et al., this volume). A pliopithecid from Terrassa in Spain (MN 10) probably belongs to an additional, as yet unnamed, genus of crouzeliine.

Plesiopliopithecus lockeri   The holotype of *Plesiopliopithecus lockeri,* a left mandibular fragment with p3–m1, and associated i1 and i2, was recovered from Trimmelkam, Austria (MN 6), in 1959. The block of lignite that originally contained the jaw fragment also preserves the impression of a lower canine, just anterior to p3. From the size of the canine one can be fairly certain that the holotype represents a male individual. This is the only known specimen belonging to the species. The species was first described by Zapfe (1961b) as belonging to the genus *Pliopithecus,* but on the basis of its distinctive premolar and molar morphology he established a new subgenus, *Plesiopliopithecus.* The distinctiveness of the species has

TABLE 12.2 *Differences in the Lower Molars That Distinguish the Pliopithecinae from the Crouzeliinae*

	Pliopithecinae	Crouzeliinae
$M_{1\text{-}2}$	Crown more or less rectangular	Crown narrows mesially in M_1
	No buccolingual waisting	Buccolingual waisting
	Mesial fovea short and broad	Mesial fovea elongated
	Mesial transverse crest slightly oblique	Mesial transverse crest strongly oblique
	Trigonid slightly more elevated than talonid	Trigonid more markedly elevated than talonid
	Cristid obliqua short and more or less mesially directed	Cristid obliqua long and obliquely directed
	Crown moderately long and narrow; breadth-length index of M_1 ($\bar{X}$ = 85.8; Range = 78.3–96.8; N = 23) and M_2 ($\bar{X}$ = 87.9; Range = 77.3–97.1; N = 20)	Crown very long and narrow; breadth-length index of M_1 ($\bar{X}$ = 77.5; Range = 72.3–81.2; N = 8) and M_2 ($\bar{X}$ = 77.3; Range = 72.2–81.4; N = 5)
	Hypoconulid large	Hypoconulid reduced in size
	Hypoconulid situated bucally to the midline of the crown	Hypoconulid situated in the midline or slightly lingually to the midline of the crown
	Distal fovea moderately large and well defined	Distal fovea restricted in size (and poorly defined in *Plesiopliopithecus*)
M_3	Mesial fovea transversely aligned	Mesial fovea slightly obliquely aligned
	Buccal cusps tend to be arranged almost in a line	Hypoconulid more lingually placed than the protoconid and hypoconid
Lower molars	Cusps tend to be low and rounded and voluminous	Cusps tend to be high, conical, and well spaced

been confirmed by subsequent workers who generally prefer to recognize *Plesiopliopithecus* as a separate genus (Ginsburg and Mein 1980; Ginsburg 1986; Harrison 1987a, 1991a; Harrison et al. 1991).

Plesiopliopithecus auscitanensis   A second species of pliopithecid was discovered at Sansan, France (MN 6), during the early 1960s. The type specimen, a mandibular fragment with p4–m2, is the only known specimen assigned to this species. Begounioux and Crouzel (1964, 1965) described the specimen in some detail, and noted a number of distinct differences that separate it from the type specimen of *Pliopithecus antiquus*, including the small size of the teeth, the elongated lower molars with very small hypoconulids, open distal foveae, and a reduced buccal cingula. Bergounioux and Crouzel (1964, 1965) regarded these differences as being of only intraspecific significance, referring the specimen to a new "variety," *Pliopithecus antiquus*, var. *auscitanensis*. Ginsburg (1975) included the specimen in a new genus, *Crouzelia*, and at the same time made the species name *auscitanensis* available (names proposed for varieties or races are unavailable under the provisions of the International Code of Zoological Nomenclature). However, the dentition is very similar in size and morphology to the crouzeliine pliopithecid from Trimmelkamm, and their inclusion in a single genus, *Plesiopliopithecus*, seems justified. With the recovery of further material, they may even prove to be conspecific. The minor differences between the two known specimens provide sufficient justification to recognize two different species. For example, the m1 of *Plesiopliopithecus auscitanensis* can be distinguished from that of *Plesiopliopithecus lockeri* in be-

ing slightly relatively shorter, in having a less well-developed buccal cingulum, and in having a well-defined distal fovea set off from the talonid basin by a distinct crest linking the hypoconulid to the entoconid.

Collier (1978) has described a few isolated teeth from Liet, France (MN 6), which he has referred to *Crouzelia auscitanensis*, as well as to *Pliopithecus antiquus*. Ginsburg and Mein (1980) contend, however, that all of the Liet teeth are referable to *Pliopithecus antiquus*, and this assessment is followed here. Although the placement of this species in *Plesiopliopithecus* results in synonymizing the genus *Crouzelia* within it, the family group taxon based on that genus retains its original name because the type genus was available at the time Crouzeliinae was named by Ginsburg and Mein (1981).

Plesiopliopithecus rhodanica Ginsburg and Mein (1980) named a second species of *Crouzelia, C. rhodanica*, based on an isolated lower molar germ from La Grive-Saint-Alban (Fissure L7), France (MN 7). The presence of a well-defined pliopithecine triangle in the talonid basin and posterior crown narrowing suggests that the tooth is an m2 rather than an m1. It differs from the m2 in *Plesiopliopithecus auscitanensis* in being considerably smaller in size (the *P. auscitanensis* m2 has an occlusal area that is 38.7% larger than that of the La Grive specimen), the crown is relatively narrower (the breadth-length index is 75.0 in the m2 from La Grive, and 79.7 in that from Sansan), the mesial fovea is relatively shorter, and the hypoconulid is reduced to only a vestige. The La Grive specimen is possibly a small individual or an unusual m1 of *Plesiopliopithecus auscitanensis*, but until additional material is available, the La Grive specimen is distinctive enough to be retained as a separate species within the genus *Plesiopliopithecus*.

Anapithecus hernyaki *Anapithecus hernyaki* is represented by an extensive series of cranio-dental specimens and some isolated postcranials. It is the best-known crouzeliine from Europe (Kretzoi 1975; Begun 1988). The species is known principally from Rudabánya, although some specimens recently recovered from Götzendorf, Austria (MN 9/10, ca. 9.5 Ma; Zapfe 1992; Bernor et al. 1993; Rögl et al. 1993), may also prove to be referable to *Anapithecus hernyaki* (Andrews and Bernor, personal observation). Although *Anapithecus* was originally described as a subgenus of *Pliopithecus* by Kretzoi (1975), its relatively large size and distinctive dental features allow the recognition of a separate genus (Ginsburg and Mein 1980; Kretzoi 1984; Harrison 1987a, 1991a; Harrison et al. 1991).

Anapithecus is the largest crouzeliine, having an average lower molar occlusal area 10% larger than that of *Laccopithecus*, 34% larger than that of the Terrassa crouzeliine, and 85%, 89%, and 122% larger than that of *Plesiopliopithecus lockeri*, *Plesiopliopithecus auscitanensis*, and *Plesiopliopithecus rhodanica*, respectively. In fact, *Pliopithecus zhanxiangi* from China is the only pliopithecid larger than *Anapithecus* in dental size (the occlusal areas of its lower molars are on average about 8% larger than those of *Anapithecus*).

Anapithecus hernyaki can be distinguished from the Terrassa crouzeliine in having broader and much more molariform premolars, and relatively broader molars with a more elongated trigonid, a better-defined pliopithecine triangle, a hypoconulid that is more centrally placed, and a smaller, less-distinct distal fovea. It differs from *Laccopithecus* in having a broader p3, with a much more strongly developed metaconid, a broader p4 with a relatively longer talonid basin, and narrower lower molars, with a more distinct buccal cingulum, less pronounced buccolingual waisting of the crown, a more distally positioned hypoconulid, and a correspondingly more obliquely oriented distal fovea. *Anapithecus* can be distinguished from *Plesiopliopithecus* in having a broader p3 with a much more pronounced development of the metaconid, a more molariform p4 with a relatively longer talonid basin and a well-developed pair of distal stylids, and lower molars with a less well-developed buccal cingulum and a hypoconulid located closer to the midline. Table 12.3 summarizes some of the key features that serve to distinguish the different crouzeliine species.

Terrassa Three jaw fragments from "Torrent de Febulines" near Terrassa, northern Spain (MN 10), apparently belonging to a previously undescribed genus and species of crouzeliine, represent the latest occurrence of pliopithecids in Europe (Golpe-Posse 1982; Moyà-Solà, in prep.). The material includes associated mandibular fragments with right p3–m3 and left p3–m2, a right maxillary fragment with P3, and a symphyseal fragment of a subadult female individual, with right and left canines exposed in their crypts and the roots of left i1–2 and right p3–4. The

TABLE 12.3 *Some Key Features of the Lower Dentition That Distinguish the Different Species of the Crouzeliinae*

	Nov. gen. et sp. (Terrassa)	Ples. lockeri (Trimmelkamm)	Ples. auscitan. (Sansan)	Ples. rhodanica (La Grive)	Anapithecus (Rudabánya)	Laccopithecus (Lufeng)
Size of P_3 metaconid	Prominent	Small tubercle	—	—	Prominent	Prominent
Length of P_4 talonid	Talonid much longer than trigonid	Talonid somewhat longer than trigonid	Talonid and trigonid subequal	—	Talonid much longer than trigonid	Talonid somewhat longer than trigonid
Development of distal stylids on P_4	Prominent	Small	Small	—	Prominent	Prominent
Length-breadth proportions of lower molars*	M_1: 81.2 M_2: 72.2 M_3: 62.1	M_1: 78.3 — —	M_1: 78.0 M_2: 79.7 —	— M_2: 75.0 —	M_1: 77.6 M_2: 79.9 M_3: 69.8	M_1: 86.0 M_2: 83.6 M_3: 72.6
Trigonid length of lower molars	Short	Quite short	Long	Long	Long	Long
Development of pliopithecine triangle	Indistinct or absent	Well developed	Well developed	Well developed	Vestigial to well developed	Indistinct or absent
Development of buccal cingulum on lower molars	Moderately well developed	Well developed	Well developed	Well developed	Moderately well developed	Poorly developed
Size and position of hypoconulid	Small, buccally placed	Small, buccally placed	Very small, buccally placed	Vestigial, bucally placed	Small, midline	Small, close to midline
Structure of distal fovea	Small, well defined	Small, well defined	Communicates directly with talonid basin	Communicates directly with talonid basin	Very small, well defined	Very small, well defined

*The indices given for *Anapithecus* and *Laccopithecus* are mean values.

p3 is relatively short, quite high-crowned, with a steep honing face for occlusion with the upper canine. One interesting feature of p3 is that the metaconid is quite prominent, being linked to the protoconid by a sharp, obliquely oriented crest. The p3 is small in relation to p4. The p4 is a long narrow molariform tooth, in which the talonid basin is almost twice as long as the trigonid and the distal margin bears a pair of distinct stylids. The m1 and m2 are long and narrow, and exhibit the following morphological characteristics: a small mesial fovea, with an oblique transverse crest linking the protoconid and metaconid; an elongated talonid basin, with no indication of a pliopithecine triangle (but possibly obliterated by wear); a small, buccally placed hypoconulid; a narrow but distinct distal fovea on m1–2; a narrow ledge-like buccal cingulum mesially and distally. The m3 is much longer than the m2, but slightly narrower. It has broad, almost transversely aligned mesial fovea. The talonid basin is very elongated, and there is a small pit beside the protoconid and hypoconid that may represent a worn trace of the pliopithecine triangle. The hypoconulid is quite small, but distinct and well-developed, and it is situated more or less in line with the protoconid and hypoconid. The distal fovea forms a large heel, which is partially filled by a tuberculum sextum.

The molar morphology confirms that the Terrassa material is referable to the Crouzeliinae, while the size and distinctive characteristics of its teeth provide adequate justification for the recognition of a new species, and probably also a separate genus (Moyà-Solà, pers. comm.; in prep.). In terms of its overall size, the Spanish material is somewhat larger than *Plesiopliopithecus lockeri*, *Plesiopliopithecus auscitanensis*, and *Plesiopliopithecus rhodanica*, but is slightly smaller than *A. hernyaki* and *L. robustus*. Morphologically, it can be distinguished from all other crouzeliines by its unique combination of features of the lower premolar and molar characters (see tab. 12.3). It is interesting that the Terrassa specimens share with *Anapithecus* and *Laccopithecus* the possession of more molariform lower premolars, which includes a combination of a well-developed p3 metaconid, and an elongated talonid with well-developed p4 distal stylids. Because this character complex is not found in the crouzeliine *Plesiopliopithecus*, it could be used as a synapomorphy to support a close relationship between *Anapithecus*, *Laccopithecus*, and the Terrassa species. However, it seems more likely that these shared traits are independently derived, reflecting similar dietary specialization and their larger size. This conclusion is supported by the fact that parallel developments can also be recognized among the pliopithecines.

Can Feliu   In addition to the Terrassa material, an isolated dp4 of a medium-sized crouzeliine has been recovered from Can Feliu, Spain (MN 8). Although the speci-men is older than the Terrassa material, it is concordant in size and form, and may belong to the same species. However, until adequate comparative material is available, it is probably best to refer this specimen to Crouzeliinae gen. and sp. indet.

Pliopithecidae from Eppelsheim and Salmendingen   An isolated pliopithecid upper male canine was collected from Eppelsheim, Germany (MN 9; >10 Ma; Bernor et al. 1993; Swisher, this volume; Woodburne et al., this volume). It was initially identified as being a cercopithecid and named *Semnopithecus eppelsheimensis* (Haupt 1935). It was later identified as a pliopithecid by Hürzeler (1954). Koenigswald (1956) argued that the canine should be included, along with the contemporary small dryopithecine material from Wissberg in Germany, in a new genus *Rhenopithecus*. Szalay and Delson (1979) also argued for its dryopithecine affinities, preferring, however, to assign the specimen to *Dryopithecus brancoi*. More recently, Begun (1989) has once again reaffirmed the pliopithecid affinities of the Eppelsheim canine.

Further comparisons establish that the canine is certainly referable to the Pliopithecidae. However, male pliopithecid upper canines are generally similar in morphology, making any attempt to assess their alpha-taxonomic affinities very difficult. The specimen is somewhat larger than male canines of *Pliopithecus vindobonensis*, and comparable in size to those of *Pliopithecus platyodon*, *Laccopithecus robustus*, and *Anapithecus hernyaki*. Given the close correspondence in age, and the Central European location of both Rudabánya and Eppelsheim, the similarity in size and morphology of the Eppelsheim canine to that of *Anapithecus* may be significant. Pertinent here is the fact that *Rhenopithecus eppelsheimensis* (Haupt 1935; von Koenigswald 1956) has priority over *Anapithecus hernyaki* (Kretzoi 1975).

Begun (1989; 1992b) has further suggested that the femur from Eppelsheim, originally described as *Paidopithex rhenanus* Pohlig 1895, and most commonly attributed to *Dryopithecus*, should also be identified as a pliopithecid. The reasoning behind this claim can be briefly summarized as follows:

1. one of the lower teeth from Salmendingen, Germany (probably MN 11), identified as a dryopithecine dp4 by most previous workers (Branco 1897; Schlosser 1901, 1902; Abel 1902; Simons and Pilbeam 1965), has been misidentified, and is really an m1 of a large-bodied crouzeliine, tentatively identified as an m1 of *Anapithecus* cf. *hernyaki* (Begun 1989)

2. this attribution is supported by the fact that the tooth co-occurs at Salmendingen with a dryopithecine, *Dryopithecus brancoi*, that is very similar to the large hominoid that co-occurs with *Anapithecus* at Rudabánya

3. the supposed crouzeliine molar corresponds in size with the *Paidopithex* femur, which is also derived from a late Miocene site in close geographical proximity to Salmendingen

4. the femur shares a number of morphological and morphometric similarities to the femora of *Pliopithecus vindobonensis* from the Republic of Slovakia

5. the referral of the isolated upper canine from Eppelsheim to the Pliopithecidae, the only other fossil catarrhine from the site, strengthens the case that the *Paidopithex* femur is also a pliopithecid.

Begun's (1989) referral of this material to the Pliopithecidae is intriguing, and requires careful consideration. After all, the correct taxonomic placement of the Eppelsheim femur has implications for determining the appropriate nomenclature for several species of Central European catarrhine. For example, the prior name *Paidopithex rhenanus* Pohlig 1895 could enter into synonymy with either *Anapithecus hernyaki* Kretzoi 1975 or *Dryopithecus brancoi* (Schlosser 1901). However, in Harrison's view (1991b), there are good reasons to accept the earlier suggestion that the Salmendingen tooth is a dp4, rather than an m1, in which case it would fit metrically and morphologically with the dryopithecine material already known from the site (but see below). If this is the case, then Begun's attempt to show that the tooth is consistent in size with the Eppelsheim femur has little relevance for deciding on the pliopithecid affinities of the latter. The tooth is consistent in size with the femur, so they both may belong to a dryopithecine. Moreover, if we exclude the possibility that the Salmendingen tooth belongs to an undescribed species of pliopithecid, which is larger than all other known pliopithecids, then the *Paidopithex* femur is much too large to be attributed to *Anapithecus hernyaki* or to the species represented by the pliopithecid canine from Eppelsheim. On the available evidence, it seems most reasonable to assume that the Eppelsheim femur belongs to a dryopithecine, rather than a pliopithecid. And even if this Salmendingen tooth is an *Anapithecus hernyaki* permanent molar, this says nothing about the identification of the Eppelsheim femur.

There is another problematic specimen from Salmendingen, an isolated m3. It was originally named *Anthropodus brancoi* Schlosser 1901 and then *Neopithecus brancoi* (Abel 1902). It was then referred to *Dryopithecus rhenanus* by Remane (1921) before being transferred to *Dryopithecus* as *D. brancoi* by Abel (1931). It was reassigned to *Pliopithecus* by Hürzeler (1954) and then transferred back to *Dryopithecus* by Szalay and Delson (1979). Despite the fragmentary nature of the type specimen, Szalay and Delson (1979) referred this species to the Spanish MN 8–10 hominoids originally described as *Hispanopithecus laietanus* Villalta and Crusafont 1944 and "*Rahonapithecus sabadellensis*" Crusafont and Hurzeler 1961. Begun (1992c) removed all the Spanish material from *brancoi* and used the name instead for the dryopithecine collection from Rudabánya. New material from Götzendorf and Mariathal assigned to *Dryopithecus brancoi* (Thenius 1982; Zapfe 1989) further complicates the matter because at least some of it is clearly pliopithecid and may come to be grouped with *Anapithecus hernyaki* (Zapfe, pers. comm.). In our view, the Salmendingen m3 cannot be attributed with certainty to either pliopithecids or dryopithecines, and our solution to this is to leave aside the type specimen of *brancoi* as incertae sedis and not apply to it the name for any primate group.

Hominidae

An African origin can be postulated for the Hominidae, which is used here sensu Goodman (1963, 1974) and Andrews (1985, 1992) to include all the great apes and humans (Delson and Andrews [1975] and Szalay and Delson [1979] had included gibbons as well). Early Miocene records for pre-hominid fossils are restricted to Africa, and the earliest known hominoids are from the late Oligocene locality of Lothidok (Eragaleit Beds), Kenya, dated 27.5 to 24.0 Ma (Boschetto et al. 1992). Additionally, the first hominids are also known from Africa with the appearance of the Afropithecini. The earliest hominoid record outside Africa is in Turkey and the Vienna Basin, with the basal MN 6 appearance of the kenyapithecin *Griphopithecus* (ca. 15 Ma; Bernor and Tobien 1990).

The family Hominidae is divided into four subfamilies. The earliest-known members of the family are provisionally included in the Dryopithecinae, which is a paraphyletic group encompassing three tribes (Andrews 1992): the Afropithecini, the Kenyapithecini, and the Dryopithecini. The Afropithecini are restricted to Afro-Arabia. The other subfamilies of the Hominidae are the Ponginae for the orang utan clade, Oreopithecinae for the late Miocene hominoid from Italy, and Homininae for the African ape-human clade. All three of these subfamilies are represented in the European and Southwest Asian Miocene. Their relationships with each other, and with possible ancestral populations of fossil apes, are far from clear at present, and the three subfamilies will therefore be presented here as separate entities (tab. 12.4).

The earliest known tribe of the Dryopithecinae is the Afropithecini. This tribe includes *Afropithecus, Heliopithecus,* and some of the fossils previously attributed to *Kenyapithecus* (for example the samples from Maboko Island and Nachola in Kenya; Andrews 1992). It is possible that the Moroto sample (Pilbeam 1969) and *Otavipithecus* (Conroy et al. 1992) are also referable to this tribe, giving it a geographic distribution from Southwestern Africa (*Otavipithecus*) as far north as the Saudi Arabian plate (*Heliopith-*

TABLE 12.4 *Classification of European and Southwest Asian Hominidae*

Hominidae
  Dryopithecinae
    Kenyapithecini
      Griphopithecus alpani Tekkaya 1974
        (= *Sivapithecus alpani* Tekkaya 1974)
        (= *Sivapithecus darwini* Abel 1902)
      Griphopithecus darwini (Abel 1902)
        (= *Dryopithecus darwini* Abel 1902)
        (= *Austriacopithecus weinfurteri* Ehrenberg 1937)
        (= *Austriacopithecus abeli* Ehrenberg 1937)
        (= *Griphopithecus suessi* Abel 1902)
    Dryopithecini
      Dryopithecus fontani Lartet 1856
      Dryopithecus laietanus (Villalta and Crusafont 1944)
        (= *Hispanopithecus laietanus* Villalta and Crusafont 1944)
        (= *Sivapithecus occidentalis* Villalta and Crusafont 1944)
        (= *Rahonapithecus sabadellensis* Crusafont and Hürzeler
          1961)
        (= *Dryopithecus piveteaui* Crusafont and Hürzeler 1961)
      Dryopithecus carinthiacus Mottl 1957
        (= *Rudapithecus hungaricus* Kretzoi 1975)
        (= *Bodvapithecus altipalatus* Kretzoi 1975)
      Dryopithecus crusafonti Begun 1992
        (= *Dryopithecus fontani* Smith-Woodwood 1914 partim)
  Ponginae
    Sivapithecus meteai (Ozansoy 1957)
      (= *Ankarapithecus meteai* Ozansoy 1957)
  Oreopithecinae
    Oreopithecus bambolii Gervais 1872
  Homininae
    Graecopithecus freybergi von Koenigswald 1972
      (= *Dryopithecus macedoniensis* de Bonis and Melentis
        1974)
      (= *Ouranopithecus macedoniensis* [de Bonis and Melentis
        1974])
  Incertae sedis
    Udabnopithecus garedziensis Burtschak-Abramovich
      & Gabachvili 1950
    Paidopithex rhenanus Pohlig 1895
    Dryopithecus? brancoi Schlosser 1901
      (= *Anthropodus brancoi* Schlosser 1901)
      (= *Neopithecus brancoi* Abel 1902)

ecus). It is beyond the scope of this contribution to discuss this further, but it is important to note that the second dryopithecine tribe, the Kenyapithecini, shares many similarities with the Afropithecini and may indeed be derived from it. The earliest kenyapithecins are only slightly later in time than the afropithecins, and there are thus strong indications that the kenyapithecins originated in Africa. The case of the third tribe, the Dryopithecini, is more problematic. There are substantial morphological differences, as well as a long temporal hiatus, between kenyapithecins and dryopithicins, and it does not appear that they shared an immediate common ancestor.

The second subfamily of the Hominidae to be considered here, Ponginae, is represented by a single species, for which only two specimens are known from Europe and Southwestern Asia. These come from the Sinap Formation in Turkey, from a site dated 9.8 Ma (= latest MN 9; Kappelman et al., this volume), and the species is attributed to the Indo-Pakistan genus *Sivapithecus*. The charac-

ters that are used to assign the fossil species of this genus to the orang utan clade are based on the morphology of the face and palate, and a suite of synapomorphies have been listed (Andrews and Cronin 1982) and subsequently defined in some detail (Ward and Pilbeam 1983; Ward and Kimbel 1983; Ward and Brown 1986). Recent work showing that the postcrania lacked shared great ape characters that would have been expected to be present in an orang utan ancestor (Pilbeam et al. 1990) have been claimed to cast doubt on the affiliation of *Sivapithecus* to the orang utan clade, but in our view the palatal-facial characters are more robust because they encompass several independent functional complexes of the skull, and we maintain the pongine status of the Turkish and Indo-Pakistan specimens. The implication here is that the postcranial characters shared by the orang utan and the African apes, but absent from *Sivapithecus*, are convergent for the orang utan and the African apes; alternatively, there may have been a "reversal" in *Sivapithecus* related to its modified locomotor adaptations.

The third subfamily present in Europe is the Oreopithecinae. It is represented by a single species, *Oreopithecus bambolii*, know from late Miocene sites in Tuscany and Sardinia, Italy (MN 12 and MN 13). This taxon has a uniquely specialized dentition, and previous attempts to elucidate its phylogenetic relationships based on dental morphology alone have produced conflicting results (see Szalay and Delson 1979; Harrison 1987b). However, the postcranial adaptations are clearly very close to those characterizing extant great apes and humans, and several important synapomorphies of the cranium clearly associate *Oreopithecus* with the Hominidae. As the precise relationship of *Oreopithecus* to other hominids is still uncertain, and given the divergent nature of its cranio-dental specializations, we have included it in a separate subfamily.

Homininae is represented by a single genus and species in the Miocene, *Graecopithecus freybergi*, and its place in this subfamily is based on characters such as the African ape pattern of incisive canal and the enlarged supra-orbital tori and glabella (Andrews 1990, 1992; Dean and Delson 1992). Unlike the other subfamilies of Hominidae, which are well represented in other parts of the world, even if rare in Europe, hominine fossils are rare everywhere until the early fossils on the human lineage are known from the Pliocene of Africa.

DRYOPITHECINAE, KENYAPITHECINI

The genus *Kenyapithecus* is known only from Africa, the type species being *Kenyapithecus wickeri* from Fort Ternan, Kenya. The similarities of the Turkish middle Miocene hominoids from Paşalar to *Kenyapithecus* have been recognized since their first description (Andrews and Tobien 1977). They are also similar to the four specimens from middle Miocene deposits of the Vienna Basin at Neudorf Sandberg (MN 6): left and right m3, left M1 and M2 (Abel 1902; Steininger 1967) and to an isolated left m3 from Engelswies (MN 5) in southern Germany (Heizmann 1992). Although originally assigned to two genera and species, we consider that in fact they all belong to the one species, which as first revisers we have identified as *Griphopithecus darwini* (Begun 1987, 1992b; Martin and Andrews 1993). We have attempted a resolution of these similarities by using the genus name available from Neudorf, *Griphopithecus*, for the Paşalar sample, while at the same time grouping this genus with *Kenyapithecus* in the Kenyapithecini (Andrews 1992; Martin and Andrews 1993). Two species of *Griphopithecus* are recognized here, *G. alpani* for the Çandir and Paşalar samples and *G. darwini* for the Neudorf sample. In addition, an unnamed third species is also known, based on a morphologically distinct sample from Paşalar.

Griphopithecus   The type locality of *Griphopithecus alpani* is Çandir (MN 6), and the type specimen is a well-preserved mandible with the crowns of left p3 to m3 and right p4 to m3, described by Tekkaya (1974) as *Sivapithecus alpani*. Andrews and Tekkaya (1976) reassigned the mandible to *Ramapithecus* (now *Kenyapithecus*) *wickeri*, and when material from Paşalar (MN 6; Bernor and Tobien 1990) in Turkey was described, part of the sample was put in the same species (Andrews and Tobien 1977). Later, the Turkish material was recognized as distinct from the African, and, together with much more abundant remains from Paşalar, they were transferred back to *Sivapithecus* (Alpagut et al. 1990). The discovery of more complete material, including several mandibles and maxillae and a portion of premaxilla, now show the presence of a broad incisive canal and a short premaxilla, similar to that seen in early Miocene species of *Proconsul*. The retention of this primitive morphology of the subnasal region, and the absence of the elongated premaxilla and long, narrow incisive canal present in *Sivapithecus* (Ward and Pilbeam 1983), is taken to demonstrate that this taxon does not belong in *Sivapithecus*. The descriptions of the new specimens are in preparation (Andrews, Alpagut, and Martin, in prep.). Discovery of more complete material may show the presence of some of the synapomorphies of the orang utan clade, for example in the orbital region of the skull, but in the absence of such information we see no reason to link the Paşalar hominoids with the orang utan clade.

There is good evidence for the presence of two species at Paşalar. This was originally assessed on the basis of size (Andrews and Tobien 1977), but with increasing collections (Alpagut et al. 1990) it has become apparent that there is extensive overlap in the size distributions of the two species. The overall variability of the Paşalar sample is too great to encompass a single species (Martin and An-

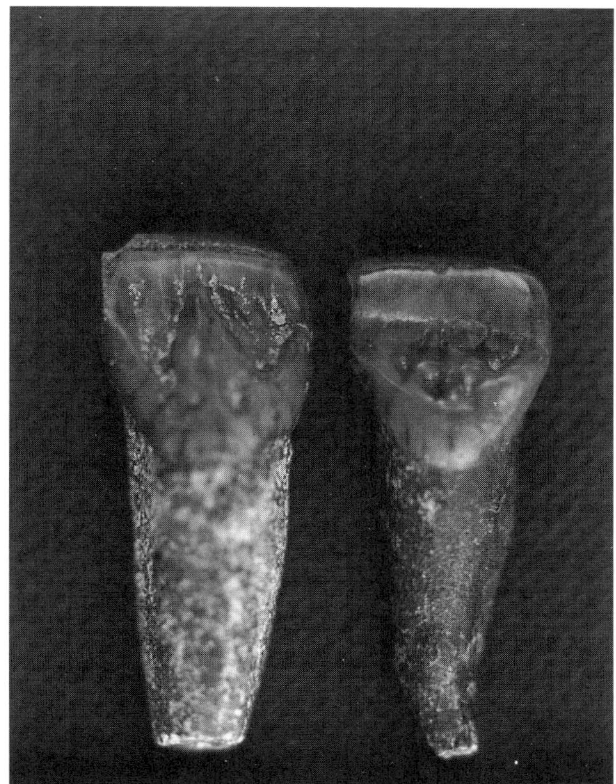

FIGURE 12.2 The two incisor morphs from Paşalar. On the left is one of the largest specimens of incisor with a central lingual pillar, identified here as *Griphopithecus alpani*, and on the right is one of the smallest specimens lacking a pillar. The greater relative breadth of the latter and its lower crown are correlates of this morphology.

drews 1993), and the morphology of teeth like the I1 (fig. 12.2) shows two distinct morphologies, with large and small (male and female) samples of each (in preparation). The common species, about 90% of the sample, is believed to be that recognized from Çandir, and the second species at Paşalar is therefore left without a name. It is possible that it may still be referred to the genus *Sivapithecus*, for the incisor morphology is similar to that described for the Indo-Pakistan specimens (Tattersall and Simons 1969), but the other body parts needed to distinguish this genus from *Griphopithecus* are not known for this species, for which only a few isolated teeth are known. On the other hand, on the basis of its extreme similarity to *alpani*, so great that most teeth cannot be distinguished one from the other, it seems most likely that it should be attributed to the same genus as *alpani*. This is the course we are adopting here.

The similarity of the Paşalar hominoids with the Neudorf Sandberg *Griphopithecus darwini* has also been recognized (Andrews and Tobien 1977). The four Neudorf Sandberg teeth described by Abel (1902) were originally assigned to two genera and species, *Dryopithecus darwini* and *Griphopithecus suessi*. The former species was subsequently transfered to *Sivapithecus darwini* by Lewis (1937; Andrews and Tobien 1977). The attribution to *Sivapithecus* was based on the inferred presence of thick enamel on the molar crowns, and this was also one of the reasons that the Paşalar sample was grouped with this material, combined with the presence of buccal cingula on the lower molars. It is now recognized that these are primitive characters for the hominid clade and do not support the attribution of any part of the Neudorf or Paşalar samples to *Sivapithecus*. An unpublished isolated m3 has been recovered in previous years from Engelswies and may be referable to the same species, although it is believed to come from MN 5 horizons (Heizmann, 1992). If the systematic and age attributions prove to be so, this record would represent the oldest-known entry of hominoids into Europe.

One last point to be considered here is the identification of "*Austriacopithecus weinfurteri*" and *A. abeli* described by Ehrenberg (1937) from Klein Hadersdorf (MN 6) in Austria. Begun (1992b) has shown that these postcranial bones are different from specimens associated with *Dryopithecus* species. The humerus shaft is not as straight and more robust, with a prominent and convex deltoid tuberosity that imparts some degree of anterior convexity to the shaft. On the other hand, the curvature is not as great as in earlier afropithecins, such as the humerus from Maboko Island, and it is also less than that of the two *Sivapithecus* humeri from Pakistan (Pilbeam et al. 1990). Similarly, the olecranon fossa is deeper than that of the Maboko humerus, but not as deep as that from St. Gaudens. It seems very likely, therefore, that these specimens could belong to *Griphopithecus darwini*, as suggested recently by Begun (1992b), on the grounds that the Klein Hadersdorf deposits are similar in age to Neudorf Sandberg and their morphology is different from slightly earlier (e.g., Maboko Island) and slightly later (e.g., St. Gaudens) specimens.

We therefore recommend the following taxonomic changes:

1. synonymize the two taxa from Neudorf, with *suessi* being recognized as the junior subjective synonym of *darwini*
2. include the Klein Hadersdorf taxa *Austriacopithecus weinfurteri* and *A. abeli* in *darwini* (Szalay and Delson 1979; Begun 1992b)
3. retain the genus *Griphopithecus* for the Neudorf and Klein Hadersdorf taxon, hence *Griphopithecus darwini* (Begun 1987, 1992b; Martin and Andrews 1993)
4. transfer the Paşalar and Çandir taxa from *Sivapithecus* to *Griphopithecus* on the basis of their similarity to the Neudorf Sandberg specimens
5. recognize two species of *Griphopithecus* from Paşalar, one referred to the Çandir taxon, now *Griphopithecus alpani*, and the other to an unnamed species of the same genus
6. group the Neudorf and Paşalar/Çandir specimens in the tribe Kenyapithecini in recognition of their close similarity to the East African genus *Kenyapithecus*

DRYOPITHECINAE, DRYOPITHECINI

The species of the Dryopithecini have thinner enamel than the kenyapithecines, comparable to that seen in early Miocene hominoids (Martin 1985; Andrews and Martin 1991). The skull of dryopithecins show some similarities with living great apes, for example in the enlarged glabella (Begun 1992c), and the postcrania also show resemblances to modern apes and humans (Begun 1992b). Begun has interpreted this to indicate sister-group relationship between dryopithecins and the great ape and human clade, but there is a further possibility that they may share characters with the African ape and human clade, what we term here the Homininae (tab. 12.4).

The cranial characters uniting *Dryopithecus* with the hominine clade center on the interpretation of klinorhynchy being present (Shea 1985; Begun 1992c). The klinorhynchous state of basicranial flexion is said to be associated with the presence of a prominent supraorbital torus, prominent glabella, shallow supratoral sulcus, and the development of frontal sinuses (Shea 1985). However, it is still equivocal whether any or all of these characters are present in *Dryopithecus*. Begun (1992c) considers that these characters are present in the common ancestor of all great apes and humans, including the orang utan, which lacks them, but we think it more likely that they were unique to the African ape and human clade (see also Dean and Delson 1992). On the basis of this and other characters of the subnasal region, we suggest below that *Graecopithecus freybergi* belongs to the hominine clade, but it would appear that the cranial characters present in *Dryopithecus* do not support hominine relationship.

Of particular significance here is the interpretation of the polarity of the subnasal morphology of hominines. There is general agreement that the subnasal morphology of hominines is probably convergent on that of the orang utan (Begun 1992c), and while the pongine morphology is known for several *Sivapithecus* species, the hominine morphology is only known for the single species of *Graecopithecus* and is not present in *Dryopithecus*. The condition in *Oreopithecus* also appears to be more similar to the hominine pattern than the condition in *Dryopithecus*, and postcranially it is very much closer. At the very least, therefore, it would appear that both *Oreopithecus* and *Graecopithecus* are more closely related to living African apes and humans than is *Dryopithecus*.

In another significant paper, Begun (1992b) has reviewed the postcranial evidence for hominoid evolution.

He identified two postcranial groups for the European dryopithecines: the Klein Hadersdorf specimens, which he links with earlier, primitive hominoids like *Proconsul* on the basis of symplesiomorphous characters; and *Dryopithecus* from Rudabánya and St. Gaudens (see below), which he claims share advanced characters with the extant great apes, such as straight humeral shafts, flat deltoid planes, and distal humerus modifications for enhanced mobility of the elbow.

This is potentially stronger evidence than the cranial evidence, but two issues need to be addressed. One is that these dryopithecine characters are present in all living great apes, i.e., the Hominidae as a whole, including both pongines and hominines, and they are not necessarily diagnostic of the latter. Many of these characters are present in *Oreopithecus*, and some, although not all, are present in *Sivapithecus* species, but the humeral shaft morphology of *Sivapithecus* lacks the great ape characters present on the *Dryopithecus* humerus (Pilbeam et al. 1990). As a result, the polarity of the postcranial features is difficult to interpret, and it may be questioned whether the changes observed on the *Dryopithecus* postcrania indicate phylogenetic relationship or functional adaptation resulting in convergence.

Dryopithecus fontani   The type species of *Dryopithecus*, this species is known from three mandibles representing male individuals from St. Gaudens, France, MN 8. There are also several isolated teeth from southern Germany and an upper molar (Depéret 1911) and a previously undescribed upper central incisor from La Grive (see below). The Seu d'Urgell (El Firal, Lerida) mandible is included here, but the St. Stephan (Austria) mandible, previously assigned to this species (Simons and Pilbeam 1965), is removed from *D. fontani* (see below).

The Seu d'Urgell mandible has generally been attributed to *D. fontani* since the time of its first description (Smith Woodward 1914; Simons and Pilbeam 1965). Recently, Begun (1992b) has grouped it with the Can Ponsic material in a new species, *Dryopithecus crusafonti*. We agree that these specimens differ from the other Spanish hominoid material that is generally assigned either to *D. laietanus* or to *D. brancoi*, but we do not consider that the case has been made distinguishing them from the type sample of *D. fontani* (see below). This is particularly true of the Seu d'Urgell mandible, which differs in some minor details from *D. fontani*, such as the lower molar cusp proportions and mandibular body proportions. Begun (1992a) argues that these differences justify species separation of the Seu d'Urgell jaw from *fontani*. This view might be justified by additional material, but with existing specimens we do not consider it to be so. Begun (1992a) wisely did not include this specimen in the species hypodigm for *D. crusafonti*, and with only three isolated teeth from the type site (Can Ponsic) with which to compare

FIGURE 12.3 The La Grive upper central incisor, which has a morphology similar to *G. alpani* from Paşalar (see fig. 12.2) and also to the Rudabánya and Can Ponsic specimens.

it, there is little reason for including it in that species.

The La Grive incisor (fig. 12.3) is an isolated left I1 described here for the first time. The crown is intact, but the root is broken buccally from the base of the crown. The root's lingual surface is damaged, but part of the root extends for some distance from the base of the crown. Its dimensions are as follows:

mesiodistal length: 8.6
buccolingual breadth: 7.1
buccal crown height: 11.4
length of root preserved: 10.6

The crown is slender and tall, the breadth/length index being 0.83 and the height/length index 1.3. These values compare closely with those for the Can Ponsic hypodigm of *D. crusafonti* (see tab. 12.5), while the robusticity of the Can Llobateres incisors is greater and the height/length index is lower. There is a prominent lingual pillar arising from a massive lingual swelling at the base of the crown to three quarters of the way along the crown. It is medially placed on the lingual surface, and it is broadly rounded, ending in a rounded apex.

There is reason to believe that the La Grive incisor's

TABLE 12.5 *Comparative Data on Incisor Proportions for Four Samples of* Dryopithecus

	H/L	H/B	B/L
D. crusafonti	1.5–1.6	1.8–1.9	0.82–0.86
D. laietanus*	1.3	1.4	0.93
D. carinthiacus**	1.3–1.4	1.4–1.6	0.88
D. fontani***	1.3	1.6	0.93

H = buccal height
B = buccolingual breadth
L = mesiodistal length
*Can Llobateres sample—*D. brancoi* in Szalay and Delson's (197) usage.
**Rudabánya sample—*D. brancoi* in Begun's (1992c) usage.
***Specimen from La Grive, our data.

morphology is representative of *Dryopithecus fontani*, although in the absence of associated material this cannot be proved. It will be so taken in this paper, and the morphology will form the basis for comparison with other species of *Dryopithecus*. It should be noted in this regard, that the same morphology is seen in the Can Ponsic and Rudabánya samples, but the significance of this will be discussed below.

Dryopithecus carinthiacus   There is good evidence for a second species of *Dryopithecus* in Central Europe, and there are diverse opinions about its proper taxonomic referal. Begun (1992c), following Szalay and Delson (1979), has assigned it to *Dryopithecus brancoi*, which is based on the isolated worn m3 from Salmendingen in southern Germany. Andrews and Martin think that the Salmendingen specimen may be pliopithecid (following Hürzeler 1954, but see above). If it is indeed pliopithecid, the name *brancoi* is unavailable as a species name for *Dryopithecus*. Alternative names available are *Dryopithecus rhenanus*, based on an isolated lower molar from Salmendingen (Schlosser 1901), *Rudapithecus hungaricus*, based on the excellent series of fossils described from Rudabánya, Hungary (Kretzoi 1975; Kordos 1987, 1991), and *Dryopithecus carinthiacus*, based on the mandible from St. Stephan, Austria (MN 8) (Mottl 1957). Kelley and Pilbeam (1986) advocate the retention of the nomen *Rudapithecus hungaricus*, but this has been criticized by Begun and Kordos (1993) on the grounds that the characters identified as being of generic significance by Kelley and Pilbeam are in fact too variable and do not reliably distinguish the Rudabánya sample from other samples of *Dryopithecus*.

Begun and Kordos (1993) have recently redescribed the type specimen of *Dryopithecus brancoi* (from Salmendingen), and argue for differences in its degree of elongation from other species referred to that genus. Within *Dryopithecus* on the other hand, they argue that the Salmendingen molar shares its greatest similarity with the Rudabánya m3 sample and suggest its referral to *D. brancoi*. This interpretation is supported by Delson and Harrison. Andrews, Bernor, and Martin, however, would accept this proposal if indeed the Salmendingen specimen could clearly be attributed to *Dryopithecus*. The nomen *D. brancoi* could be used for the Rudabánya sample, but there remains the possibility that it may be an advanced pliopithecid, as suggested by Hürzeler (1954). For this reason, Andrews, Bernor and Martin are suggesting here that the nomen *Dryopithecus carinthiacus* be applied for all of the Rudabánya *Dryopithecus* material. Finally, Begun and Kordos (1993) have demonstrated that there is no good reason to recognize more than one species of *Dryopithecus* from Rudabánya (contra Andrews and Martin 1987).

The type specimen of *D. carinthiacus* consists of a right mandibular fragment with the crowns of p3 to m1 in place,

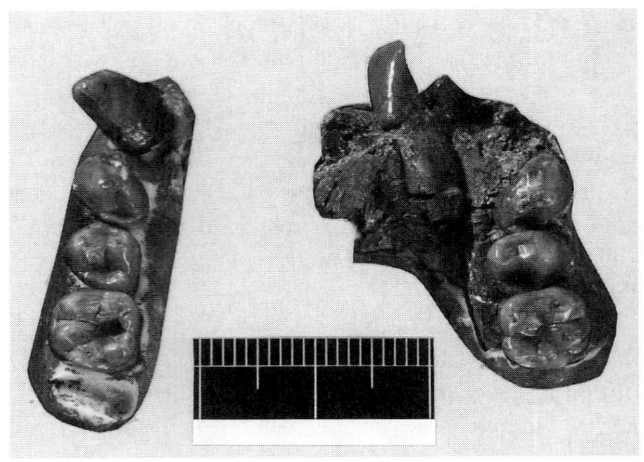

FIGURE 12.4 The mandible from St. Stephan, Austria: *Dryopithecus carinthiacus*. The fragment of mandible on the right has the crowns of p3 to m1 in place, the crown of the lateral incisor displaced, and the left tooth row has c1-m1 with a broken m2 but no part of the mandibular body.

and a displaced right central incisor (fig. 12.4). Associated with this specimen are five isolated teeth, the intact crowns of left c1-m1 and a broken m2 (Mottl 1957). It is similar to the type material of *D. fontani*, differing mainly in its smaller size. In this respect it could be considered a female of this species, as has generally been the case (Simons and Pilbeam 1965). The canine and p3 are small and low crowned in both the type specimen and the Rudabánya material, the former also having a short mesial ridge indicative of female morphology (Kelley and Xu 1991): the *carinthiacus* canine falls in the female range of great ape canines very close to the mean of chimpanzee females. The molars are similar in size and morphology, but the p4 of the *carinthiacus* specimen is narrow and elongated, differing from the corresponding teeth in the Rudabánya sample (and also from the *fontani* type material). This is interpreted for the present as intraspecific variation (but see below).

The large hominoid sample from Rudabánya, Hungary, has been divided in the past into two species (Kretzoi 1975). The majority of the specimens are readily referable to the type specimen of *Rudapithecus hungaricus*, and this group has been in turn referred to *Dryopithecus brancoi* (Szalay and Delson 1979; Begun 1992c), but in view of the possible pliopithecid affinities of the type specimen of *brancoi* (see above), Andrews and Martin reject this assignment. In terms of molar size and morphology it is closer to the *carinthiacus* specimen, and it is recognized as such here. The type specimen of a second species described from Rudabánya, *Bodvapithecus altipalatus*, is distinguished only by its slightly larger size, and it is also synonymized with *carinthiacus*. There appears to be no good reason for the recognition of more than one species.

The size and morphology of the upper central incisor is similar to that of the Can Ponsic material from Spain, and it is also similar to the La Grive incisor, with a broadly rounded but prominent lingual tubercle and similar crown proportions (see below). The distal humerus is similar to the humerus from St. Gaudens with morphology significantly advanced over that of earlier Miocene hominoids (Begun 1992b).

Dryopithecus laietanus   The type site is La Tarumba in Spain (MN 10). We have referred to this species almost all of the Spanish MN 8–10 hominoids, including *Hispanopithecus laietanus* Villalta and Crusafont 1944, "*Rahonapithecus sabadellensis*" Crusafont and Hürzeler 1961, and *Dryopithecus piveteaui* Crusafont and Hürzeler 1961, but we exclude the Can Ponsic material that Begun (1992a) has recently assigned to a new species, *D. crusafonti*. Even with that exclusion, this combination produces a large range of variation for this species.

There is some evidence for more than one species being present at Can Llobateres, particularly when the types of "sabadellensis" and "piveteaui" are compared. There is also some evidence of two morphologies that have similar size distributions. One morphology has upper molars with prominent cusps and ridges, elongated crowns, second and third molars not morphologically reduced, and traces of cingulum on the upper molars. The second morphology differs in having broader crowns with rounded cusps, no cingulum, second and third molars with considerable distal reduction, and the third molar very small. Despite these differences, we do not consider them sufficient to distinguish two species on present evidence, so that we conclude that only one species appears to be represented at Can Llobateres, and the same species is present at La Tarumba, Can Vila, and Sant Quirze, and probably also at Polinya, Can Mata, and Castell de Barberà (Begun 1990; Harrison 1991b).

Another suggestion is that the Can Ponsic material can also be divided into two species, but Begun (1992a) provides good evidence that this is not the case. Begun also shows that the Can Ponsic material belongs to a different species from the Can Llobateres sample (Begun et al. 1990), an observation supported by those of Martin and Andrews when they studied the Spanish collection in 1984.

Dryopithecus crusafonti   This species was described by Begun (1992a) from the type site of Can Ponsic, Spain (MN 9). Begun referred the Seu d'Urgell mandible to this species, but its affinities appear to us to lie with *D. fontani*, as originally described by Smith Woodward (1914; Simons and Pilbeam 1965; see above). Compared with the Can Llobateres and La Tarumba specimens, the Can Ponsic material has I1 high crowned and with a prominent lingual pillar, upper canine crenulated and without a lingual

groove, P3 with less buccal flare, upper molars increasing in size posteriorly, p4 long and narrow, and lower molars with a long metaconid.

Harrison (1991b) has observed that the apparent differences between the Can Ponsic and Can Llobateres samples have been accentuated by differential wear and that they are minor differences that do not justify the recognition of a separate species. This may not be true of the incisor and p4 morphologies, which are the most distinctive features of *D. crusafonti*, and in the view of Martin and Andrews, they are sufficient to justify separation from *D. laietanus*, especially the contrast with the Can Llobateres I1, which is low crowned and lacks a lingual pillar (see fig. 12.5). However, in this respect the Can Ponsic incisors are similar to the isolated incisor from La Grive (see figs. 12.4 and 12.5) which we attribute to *Dryopithecus fontani*. Begun (1992a: fig. 9 and tab. 6) provides good comparative data on the distinctiveness of the central incisor.

The crown dimensions are similar in the La Grive and Rudabánya samples and slightly larger than the Can Ponsic teeth, but the crowns of the Can Ponsic specimens are higher and narrower than in *D. fontani*. They share the presence of a lingual pillar, but on the Can Ponsic specimens the pillar is narrow and pointed, with deeply incised grooves on either side and arising from nearer the base of the crown. On the Rudabánya and La Grive specimens the tip of the pillar is more rounded and the base is broader. The presence of a lingual pillar is probably primitive for hominoids (Andrews 1985) so that the similarity in morphology in the La Grive and Can Ponsic samples is probably not systematically significant.

The narrow p4 is also not as distinct on the Can Ponsic specimen as Begun (1992a) suggests. Apart from the fact that there is only one specimen of this notoriously variable tooth, it is similar in shape and morphology to the p4 on the St. Stephan mandible (see fig. 12.4), which is also relatively elongated and only slightly larger in size (L x B is 7.8 x 7.3 mm for the St. Stephan p4 compared with 7.4 x 6.6 mm for the Can Ponsic specimen). Both specimens are longer than broad, an unusual feature for hominoid p4s and distinct from both the Rudabánya and St. Gaudens samples attributed to *D. fontani*.

Another problem with the *D. crusafonti* hypodigm is the combination of the Seu d'Urgell mandible with the Can Ponsic specimens. There are only two teeth in the latter sample that can be compared with the Seu d'Urgell mandible, and although they are similar in size and morphology (sharing broad buccal cusps, restricted and shallow talonid and trigonid basins, and long post-metaconid cristids), this is not strong evidence for association, particularly because the diagnostic features of the species (i.e., upper central incisor and p4 morphology) are missing from the mandible. Begun (1992a) describes the robustness of

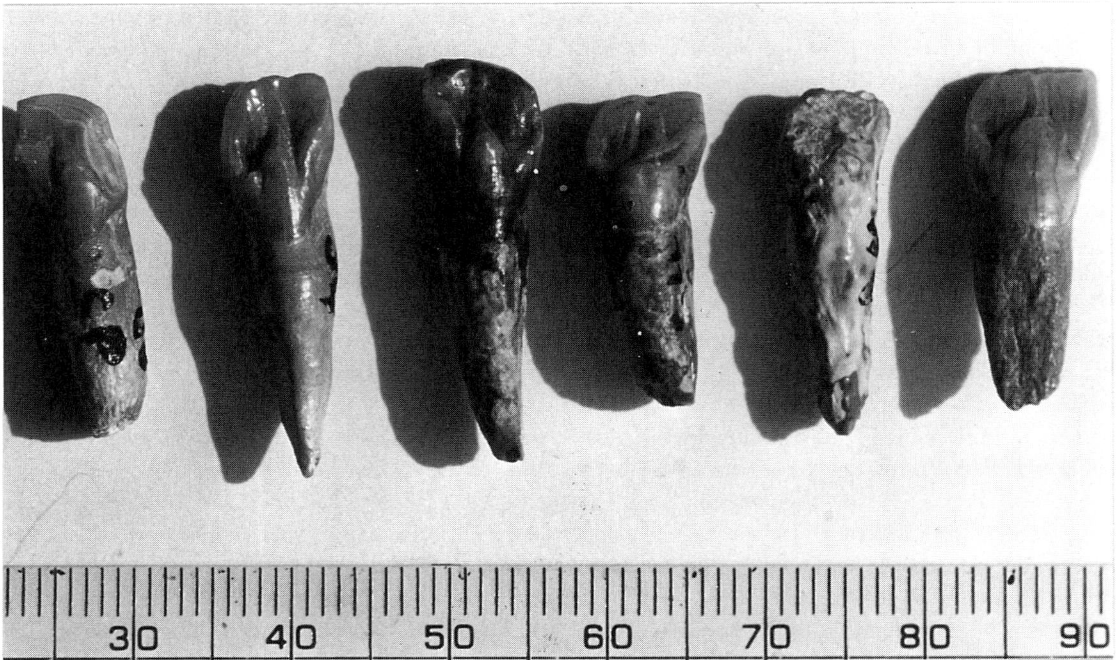

FIGURE 12.5 Incisor morphology from middle Miocene Western European sites. The three specimens on the left are *Dryopithecus laietanus*; next to them are two specimens of *Dryopithecus crusafonti* from Can Ponsic; and on the far right is the La Grive specimen, which we attribute to *Dryopithecus fontani*.

the mandibular ramus by means of the relationship between molar width and ramus width in the Seu d'Urgel mandible, and he concludes that the mandible is remarkably robust (or the molars extremely small relative to mandibular size), so much so as to be distinct from all other species of *Dryopithecus*, but since this feature cannot be measured for the Can Ponsic material it does not necessarily associate the material from the two sites.

To summarize these conclusions for the genus *Dryopithecus*, we make the following attributions, of which we have some confidence in the first three but less in the fourth: (1) *D. fontani* includes the type material from St. Gaudens, as well as La Grive, some isolated teeth from Germany, and the Seu d'Urgell mandible; (2) *D. carinthiacus* includes the St. Stephan mandible and the whole of the Rudabánya hominoid sample; (3) *D. laietanus* includes the samples from most of the Spanish Miocene sites, most notably Can Llobateres, but excluding Can Ponsic and Seu d'Urgell; (4) *D. crusafonti* includes the material from Can Ponsic; this has similarities to, and potentially is synonymous with, *D. laietanus* (Harrison's 1991b view), *D. fontani* (Andrews and Martin's view), or *D. carinthiacus*, in all cases giving the relevant taxon an MN 8–9 range in time. The chronologic range would be ca. 12.5–9.5 Ma (see Rögl and Daxner-Höck, this volume).

Some samples could not be taxonomically assigned, including the *Paidopithex* femur (see discussion at the end of the Pliopithecidae section), and the isolated teeth of

Udabnopithecus. Both of these would appear likely to be *D. fontani*. These apart, it would appear that the four species of *Dryopithecus* recognized here are closely related, so much so that it is difficult to determine the pattern of relationship.

PONGINAE, SIVAPITHECINI

With the removal of the Turkish middle Miocene and Greek thick-enameled hominoids from the orang utan lineage, there remains only a single taxon referred to the Ponginae in Europe and Southwestern Asia: the Turkish species *Sivapithecus meteai*. The type specimen is a fragmentary mandible from the Sinap Formation near Yassiören, Kazan, Turkey; these late MN 9 deposits have a preliminary date of 9.8 Ma (Kappelman et al., this volume). It was originally described as *Ankarapithecus meteai* by Ozansoy (1957), and a maxilla and lower face was described by Andrews and Tekkaya (1980), at which time the species was assigned to the genus *Sivapithecus*. It was suggested later that the Greek material described as *Graecopithecus freybergi* (including "*Ouranopithecus macedoniensis*"—see below) was conspecific with *meteai* (Szalay and Delson 1979; Martin and Andrews 1984), but this view has now to be abandoned because the Turkish face shares characters of the nose and palate with sivapithecins (Andrews and Cronin 1982) and these characters are absent in the Greek sample (de Bonis and Melentis 1978; de Bonis et al. 1990a).

The morphological differences separating *Sivapithecus meteai* from the Indo-Pakistan sivapithecins include characters of the dentition and lower face: (1) the upper central incisors of *S. meteai* are low crowned and extremely broad mesiodistally compared with the robust and high crowned incisors of *S. sivalensis*; (2) the upper lateral incisors of *S. meteai* are more pointed and caniniform; (3) the upper molars are more squared with less constricted occlusal basins; (4) the zygomatic region is broader and has a stronger flare than in *S. sivalensis*. These characters suggest at least a species difference between the Turkish and Indo-Pakistan samples, and it may be that they are generically distinct (in which case the name *Ankarapithecus* is available). With respect to the cranial material, they show a number of similarities that justify their inclusion within the orang utan clade (see addendum).

OREOPITHECINAE

Although usually placed in a separate family of the Hominoidea, we recognize the postcranial similarities shared with the extant great apes and humans, and place *Oreopithecus* in its own subfamily within the family Hominidae. *Oreopithecus bambolii* is a large-bodied arboreal primate adapted for vertical climbing and forelimb suspension (Harrison 1991c). The postcranial similarities shared with the extant hominids, and inferred to be ancestral for the clade, include a suite of characters combined by Harrison into eleven functional complexes, as follows:

1. strongly differentiated usage of fore- and hind-limbs
2. increased potential for raising the forelimb above the head
3. increased potential for full extension and powerful flexion of the forelimb at the elbow
4. greater potential for circumduction at the shoulder and pronation-supination at the elbow and wrist joints
5. increased range of abduction/adduction of the wrist
6. increased potential for powerful grasping with the hands
7. adoption of more orthograde posture
8. increased potential for full extension of the hip and knee joints
9. greater ranges of rotation at the hip and knee, and inversion-eversion at the ankle joint
10. increased potential for body weight to be supported by a single hindlimb
11. increased ability of the foot to grasp and provide powerful push-off from large diameter vertical supports

If this suite of character complexes is shared by oreopithecines and the living great apes and humans, it indicates that it was present also in the common ancestor between them in the middle/late Miocene, and this in turn has implications for the interpretation of the apparently similar characters on the humerus of *Dryopithecus*. It would also support the view that the apparently primitive morphology of *Sivapithecus* is in fact a secondary reversion toward the primitive character state as a result of functional convergence.

Oreopithecus bambolii   The type specimen consists of a mandible from Monte Bamboli, Italy (MN 13). In addition, this species is known from other localities in Maremmia, Tuscany, correlated with MN 12 (Bacinello V1, Casteani, Montemas, and Ribolla), MN 12 or MN 13 (Baccinello *Cardium* horizon), and MN 13 (Baccinello V2) (Szalay and Delson 1979; Azzaroli et al. 1987; Rook et al., 1996). Several isolated teeth of *Oreopithecus bambolii* have also recently been recovered from Fiume Santo in Sardinia (MN 13) (Cordy and Ginesu 1994). An earlier report of *Oreopithecus* from the late Miocene of Russia appears to be unsubstantiated (Laskarev 1909; Hürzeler 1958; Szalay and Delson 1979; Harrison 1986).

Oreopithecus bambolii is one of the best-known fossil catarrhine species, being represented by over fifty individuals (mostly from Baccinello V1), comprising numerous upper and lower jaws, and an almost complete skeleton (Hürzeler 1958; Szalay and Delson 1979; Harrison 1987b, 1991c). Although a number of workers have suggested a possible relationship between *Oreopithecus* and cercopithecids, based on dental evidence (Delson 1979; Szalay and Delson 1979; Rosenberger and Delson 1985), Harrison (1987b, 1991c) and others (Sarmiento 1987; Rose 1988, 1993) have established that postcranially its affinities are with the Hominoidea (see above). Delson (Rosenberger and Delson, in prep.) continues to maintain the possibility that the dental similarities between *Oreopithecus* and the cercopithecid morphotype (such as molar elongation and mirror image symmetry between uppers and lowers, details of cingulum reduction and cusp placement, and development of notches and clefts) reflect unique shared ancestry. In that case, the numerous postcranial similarities noted above would represent convergences due to locomotor adaptation, as has been suggested above for the situation with *Sivapithecus* and *Pongo*. However, the possibility of functional convergence in the postcranium does not take into account the occurrence of derived cranial characters linking *Oreopithecus* with the extant hominids. These include the reduced size of the subarcuate fossa and the morphology of the incisive canal, which is similar to the African ape pattern. These clearly support the relationship of *Oreopithecus* with the Hominidae.

HOMININAE GRAECOPITHECINI

The fourth subfamily of the Hominidae is the Homininae, which is represented in the European Miocene by a single species, *Graecopithecus freybergi* von Koenigswald

1972. This has an unknown relationship with the two extant hominine tribes, Gorillini and Hominini, and it is therefore placed as indeterminate in the Homininae.

Graecopithecus freybergi   The type is a fragmentary mandible from Pyrgos (near Athens, Greece) of uncertain age, probably MN 10 described by Koenigswald (1972). Martin and Andrews (1984), following Szalay and Delson (1979) included in this taxon the more complete material from Ravin de la Pluie, and later collections from Xirochori I (de Bonis et al. 1990) and Nikiti (Koufos 1993, 1994) belong here also (see fig. 12.6). These localities are correlated with MN 10 ca. 9.5–9.0 Ma (Rögl and Daxner-Höck, this volume; Bernor et al., 1993). The northern Greek sample was originally described as *Dryopithecus macedoniensis* (de Bonis et al. 1974) and later placed in a new genus, *Ouranopithecus* de Bonis and Melentis 1977a.

A new partial skull has recently been described (de Bonis et al. 1990a) that confirms the distinction between this taxon and *Sivapithecus*. Schwartz (1990) has described some features linking it with the orang utan lineage, and there is also some evidence linking it with the human and African ape clade (de Bonis et al. 1990a; Dean and Delson 1992). Of particular significance for the latter are the morphology of the subnasal region and the conformation of glabella and the brow ridge development. The premaxilla is elongated and rotated supero-anteriorly, overlapping the maxilla so as to restrict the size of the incisive canal, which may have been the ancestral condition of the great ape and human clade, subsequently retained by the African apes with only minor modification. The further change in the orang utan lineage could therefore have been an additional change from this shared character state (Andrews and Martin 1987). An alternative interpretation, however, is that the unique orang utan morphology developed independently of the African apes from a *Dryopithecus*-like morphology (Begun 1992c), in which case the *Graecopithecus* morphology could be interpreted as an African ape and human synapomorphy.

Similarly, the glabella morphology and the development of the supraorbital torus represent advances over the primitive hominoid condition. The prominent glabella and supraorbital torus in *Graecopithecus* (de Bonis et al. 1990a) are characters shared only with the African apes and hu-

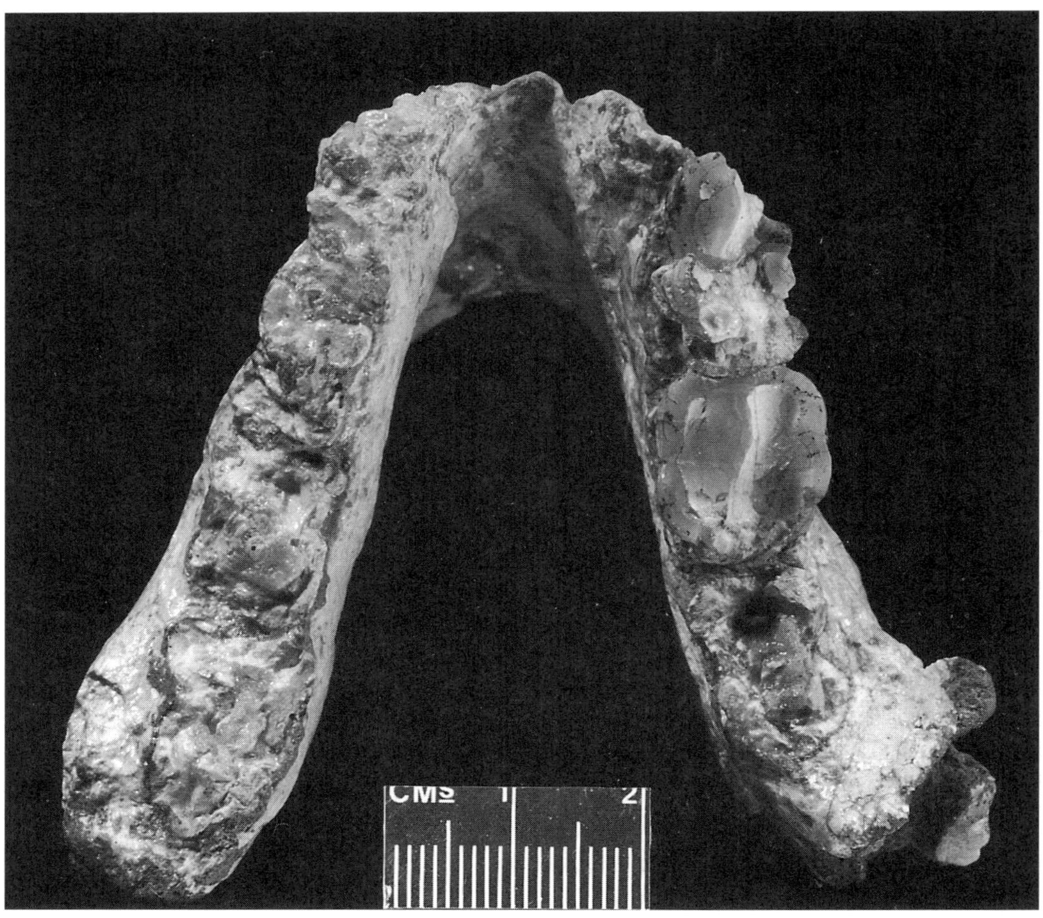

FIGURE 12.6 The type specimen of *Graecopithecus freybergi* from Pyrgos in Greece. The crown of right m2 is intact, with the broken and heavily worn crowns of p4-m1 and roots of m3.

mans in the hominoid clade. These characters have been claimed to be present also in *Dryopithecus* (Begun 1992b, 1992c), whose humeral morphology Begun (1992b) has also interpreted as indicating a link with the African ape and human clade. It is therefore possible that *Dryopithecus* as well as *Graecopithecus* could belong to the African ape and human clade, the Homininae as classified here. Interestingly, *Graecopithecus freybergi* has very thick enamel (Andrews and Martin 1991), which distinguishes it from all other hominoids except for robust australopithecines.

The characters just discussed have been used by de Bonis et al. (1990a) to support the relationship of *Graecopithecus* with the human lineage alone. This view does not appear justified on this evidence (the characters are mainly conservative for all hominids), but they have also put forward the claim for canine reduction that needs to be considered. This claim rests on the identification of the skull XIR-1 as a male. There are several upper canines known from the Greek sites, all of which appear to belong to the one species, and the canine from XIR-1 corresponds to the large end of the range of variation for this tooth. On the other hand, the canine/M1 ratio for this specimen is 94%, which is below the range for males of all extant large hominoids but is within the range for females (Andrews 1990). Two interpretations of these data are possible: either the skull represents a female with a normal sized canine; or it represents a male with reduced canine, as de Bonis has claimed. A new maxilla from Nikiti has been described by Koufos (1994) that is clearly a female and supports the second of these two alternatives. If this is the case, it must then be determined if this is evidence of relationship with the human lineage or is an independent development in this 9.5–9.0 Ma fossil.

Some researchers (e.g., de Bonis et al. 1990a) have suggested that *Ouranopithecus* and *Graecopithecus* should be systematically separated on the grounds that the Pyrgos mandible is too incomplete to use as a type specimen. This runs counter to the Code of Zoological Nomenclature, and moreover the Pyrgos mandible is not that incomplete (see fig. 12.5). Parts of the mandibular body are preserved together with the crown of the right second molar, the damaged crowns of the right p4 and m1, and the roots of the right m3 and left p3–m3 (Martin and Andrews 1984). The size and proportions of both teeth and mandibular body are similar to the female mandibles from Ravin de la Pluie (de Bonis and Melentis 1977b), although the teeth are somewhat larger and the corpus shallower in the Pyrgos specimen. It appears most likely that the latter represents a large female that, perhaps because of its size, appears similar in corpus depth and alveolar planum shape to the Ravin de la Pluie males. Other apparent differences (Begun, pers. comm., 1994) may be the result of deformation of the Pyrgos jaw. In our opinion, there are no distinguishing

features separating these two taxa and therefore no reasonable grounds for recognizing two genera or even species. We have used the same argument for relating the Turkish middle Miocene hominoids to the Vienna Basin taxon *Griphopithecus darwini*, which is based on four isolated teeth, and it might also be claimed that the same argument applies to the use of *brancoi* for the mid-European small dryopithecins, although in the latter case Andrews and Martin consider the family affinities of the type specimen of *brancoi* to be too uncertain for its use as a species of *Dryopithecus*.

Hominid relationships   The relationships of the hominoid higher taxa known from Europe and Southwestern Asia during the middle to late Miocene are depicted here in figure 12.7. Two positive statements are made here: *Graecopithecus freybergi* is recognized as a hominine, related to the African ape and human clade and possibly close to the ancestry of the living species of this group; and *Sivapithecus meteai* is recognized as a member of the pongine clade related to other sivapithecins and the living orang utan. The third subfamily recognized here, Oreopithecinae, contributes to the interpretation of postcranial evolution of the hominids, but these interpretations are still provisional because of the possibility of functional convergence in different hominoid lineages. The fourth subfamily, Dryopithecinae, is even more problematic inasmuch as it is based on the concept of grade rather than clade. The Kenyapithecini has a geographic range from East Africa to Turkey and implicitly the Vienna Basin, and its links are probably with the earlier afropithecins, which are also grouped in Dryopithecinae (Andrews 1992). Dryopithecins appear more advanced, both cranially and postcranially (Begun 1992c), but not sufficiently to warrant grouping with either the pongine or the hominine clade (Begun 1992c; Andrews 1992). They are of similar grade to afropithecins and kenypithecins and for this reason we maintain their grouping in Dryopithecinae.

Cercopithecidae

The cercopithecid monkeys do not appear in the European fossil record until the late Miocene. Only one of the two subfamilies, the Colobinae, is represented until the end of the epoch, but these are widespread across Europe and extend into Southwestern Asia. The earliest record is a single upper premolar from Wissberg, in deposits of probable MN 9 age (Delson 1973; Tobien 1986), but all the other European records are Turolian age (late Miocene, MN 11–13; ca. 9–5.3 Ma; Steininger et al. this volume).

Three species are represented, two of *Mesopithecus* and one referred to *Macaca*. The *Mesopithecus* species are closely similar, and only one is common, *M. pentelicus* (tab. 12.6). *Mesopithecus monspessulanus* is known from

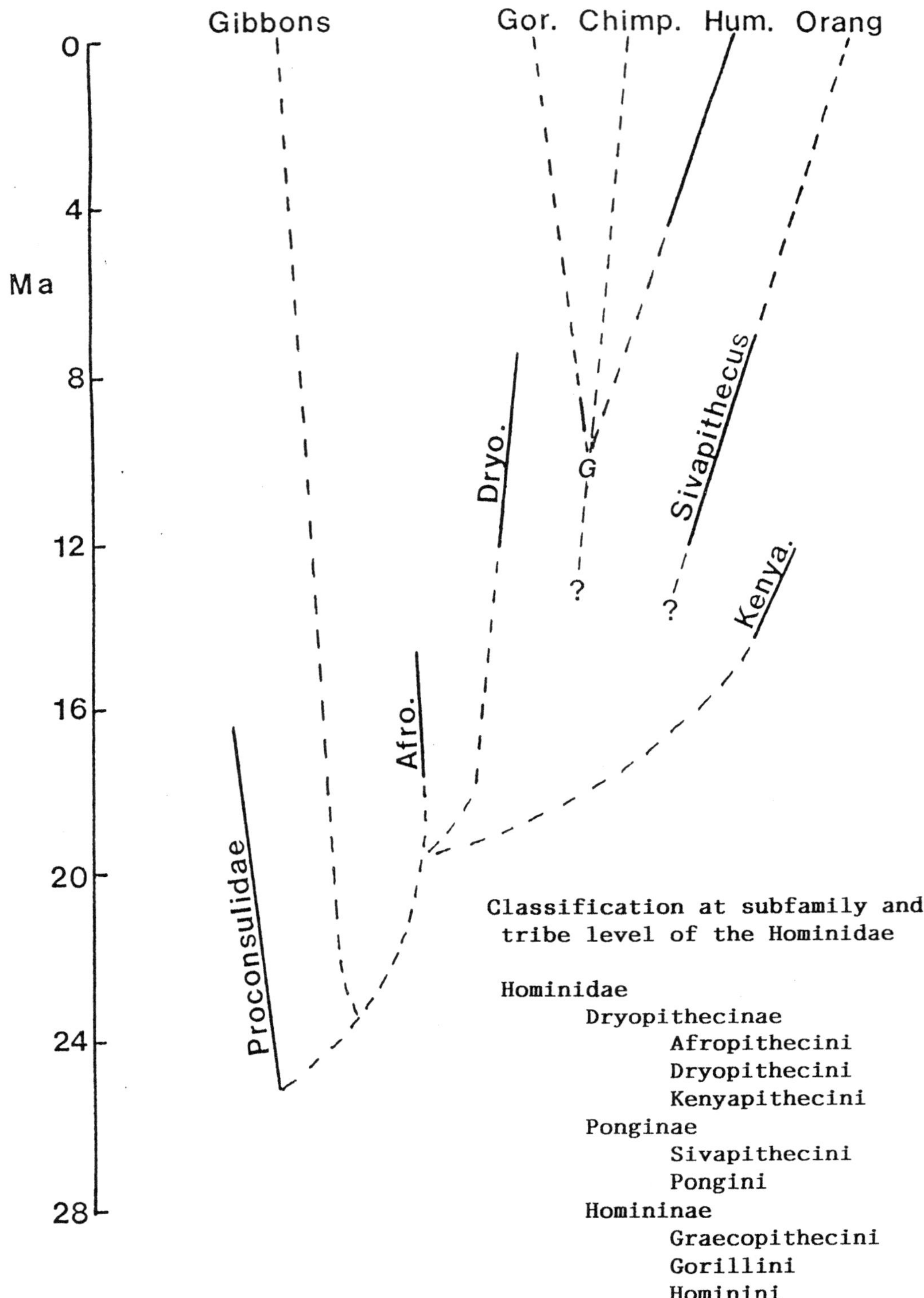

FIGURE 12.7 Phylogeny of the Hominoidea with classification of the Hominidae. The solid lines signify known ranges of fossil taxa, and the dashed lines indicate possible relationships.

TABLE 12.6 *Classification of the European Neogene Cercopithecidae*

Cercopithecidae
  Colobinae
    Colobini
      Subtribe Indet. (??Presbytina)
        Mesopithecus pentelicus Wagner 1839
          (= *M. major* Roth and Wagner 1845)
          (= *M. delsoni* de Bonis et al. 1990b)
          (= *M. p. microdon* Zapfe 1991)
        Mesopithecus monspessulanus (Gervais 1849)
          (= *Semnopithecus monspessulanus* Gervais 1849)
        Dolichopithecus ruscinensis Depéret 1889:
          (*D. cf arvernensis* Depéret 1929: Kretzoi 1954)
  Cercopithecinae
    Papionini
      Macacina
        Macaca sp. indet.

fragmentary material, although it is more common in the Pliocene. Both show some evidence of terrestrial adaptation, the more extreme being found in *M. pentelicus*. In contrast to this, *Macaca* has a more restricted distribution, having been recovered at only one MN 13 locality, but occurring more widely through Europe during the Pliocene (see below). *Dolichopithecus* has been reported from the late Miocene (e.g., by Szalay and Delson 1979), but the locality now appears to be early Pliocene (MN 14?) in age; the material will be briefly discussed to clarify the situation.

COLOBINAE

Mesopithecus The type species of *Mesopithecus* is *M. pentelicus*, known mainly from Southeastern Europe, MN 11–13. The type locality is Pikermi, Greece, which has also yielded the largest sample. Zapfe (1991) designated a new subspecies for a single mandible from the Chomateri locality near Pikermi, distinguished on the basis of relatively small tooth size. Specimens from southern Yugoslavia, Bulgaria, and Northern Greece have generally been referred to the Pikermi species, but recently de Bonis et al. (1990b) have described a small sample from Ravin des Zouaves 5 (RZO, Greece) as *M. delsoni*, which they distinguished by larger m3, longer lower molar and premolar rows, and several mandibular corpus differences. In addition to the specimens from RZO, de Bonis et al. (1990b) reported *Mesopithecus* fossils from the apparently younger Dytiko localities that were tentatively allocated to either *M. pentelicus* or *M. monspessulanus*; these will be discussed under the latter species below.

In order to test the distinctiveness of the putative new species *M. delsoni*, statistical tests were carried out on metrical data obtained from casts of three mandibular specimens kindly provided by L. de Bonis (see comparisons in fig. 12.8). Additional specimens referred to *M. pentelicus* were recovered previously from the nearby localities of Vathylakkos and Ravin X (Arambourg and Piveteau 1929), as well as from the Titov Veles localities in ex-Yugoslavia (now Macedonia; Schlosser 1921; Ciric 1957) that derive from the valley of the same river (named differently across the border). Moreover, Bakalov and Nikolov (1962) described similar material from three sites in Bulgarian Macedonia.

All of these original specimens were measured by Delson, and these data were combined with those obtained from the RZO casts to form a "Macedonian" sample of *Mesopithecus*, with 9–12 individuals for each lower molar measure (mesial width, distal width, and length) and 2–5 individuals for the upper molars; these were compared to a sample of 32–42 individuals per measure for Pikermi. A series of 18 t-tests reveals that although each of these small samples averages slightly larger than the 40–50 specimens from Pikermi measured by Delson, few are statistically different (the 99% confidence level is employed to take into account the large number of t-tests evaluated): m1 length differed at the 98.5% level (even greater distinction was found for the few teeth from RZO and Vathylakkos) and m3 length at the 99.7% level; no other variates approached significance. De Bonis et al. (1990b) reported that both m3 length and [mesial?] breadth differed "highly significantly" (without definition) between the Pikermi and RZO samples, but they included both antimeres of the two large male jaws from RZO, and their measurements were rather larger than those Delson obtained on casts using the same techniques employed for the larger Pikermi sample.

Referring now to the tooth rows, de Bonis et al. (1990b) again found "highly significant" differences between the (bilateral) RZO and Pikermi samples for p3–4 and m1–3 lengths. Molar length was significant at the 99.8% level when both Pikermi sexes were included, but no distinction was found when the two RZO (and Maragheh, see below) males were compared only to seven Pikermi males; the difference in premolar length was not significant even when sexes were combined (Pikermi sample of only four).

Given the few statistically significant differences between the Pikermi and Macedonian samples, a species distinction between *M. delsoni* and *M. pentelicus* is not supported. In terms of the corpus features, the two RZO specimens appear to show plastic deformation leading to an increase in depth, and other characters are not deemed indicative of major taxonomic separation. As all the specimens are of broadly similar age (see below), it is unlikely that even a subspecific distinction is warranted at this time. De Bonis (pers. comm.) rejects this interpretation, arguing that the corpora are not deformed and noting that Delson's measurements on casts differ from his on the originals; in response, the same techniques were used on the whole sample studied here, so that these potential differences in measurement technique cancel out. Delson and de Bonis have agreed to disagree on the status of *M. delsoni* for the

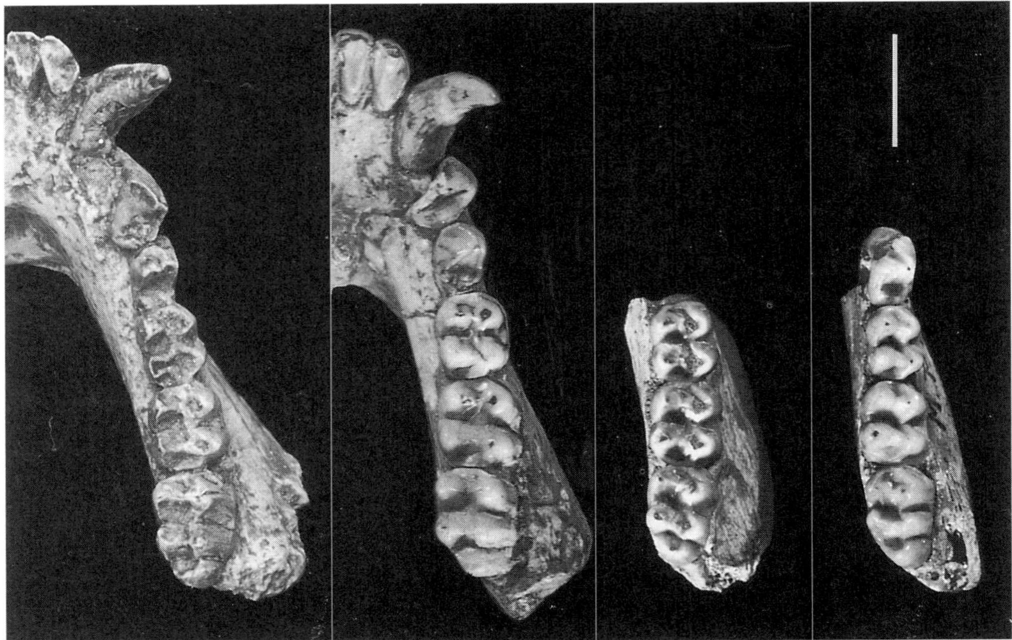

FIGURE 12.8 Right mandibular dentitions of *Mesopithecus* species. Left to right: *M. pentelicus* male from Ravin des Zouaves, i1-m3 (holotype of *M.* "*delsoni*," cast); *M. pentelicus* male from Pikermi, i1-m3; *M.* cf. *pentelicus* (sex unknown) from Baltavar, m1-3; *M. monspessulanus* (sex unknown) from Montpellier, p4-m3. Scale bar = 1 cm.

present. Similarly, the distinction of *M. p. microdon* as a subspecies (from the probably younger Chomateri locality) proposed by Zapfe (1991) based on small teeth in a deep corpus cannot now be supported.

None of these localities has been well dated, and there is disagreement among authors concerning their relative dates. Previous estimates of the age of the main Pikermi locality as early Turolian (MN 11?, e.g., Delson 1973, 1975) were replaced by younger estimates of MN 12/13 (e.g., Mein 1989), based on rodents collected from Chomateri; these do not appear reasonable, given the lack of continuity between the two localities, and the most recent research (Bernor et al., this volume) seems to imply an age equivalent to the lower levels at Maragha and Samos, that is near the MN 11/12 boundary. De Bonis et al. (1987, 1990b) suggested that the Ravin des Zouaves (RZO) locality might be older than Pikermi, while the Dytiko localities would be younger, but local superposition, and unequivocal faunal comparisons outside the Axios Valley are still awaited (but see de Bonis et al. 1992). The ages of the "Saloniki" (Vathylakkos), Bulgarian, and Yugoslav Macedonian localities are even less defined (but see Forstén and Garevski 1989).

Outside this region of Southeastern Europe, Miocene fossil cercopithecids are far less common. One isolated upper premolar from Wissberg (fig. 12.9) was identified by Delson (1973; see also Tobien 1986) as belonging to a colobine of the size of *M. pentelicus*; it is important only in registering the earliest presence (MN 9?) of the family

in Eurasia. Although some workers (e.g., Tobien 1980) have accepted the Wissberg sample as indicative of an early Vallesian age, Bernor among others suggests caution because several of the "Dinotherium-Sands" localities include an admixture of younger faunal elements, of which this tooth could be one. A maxilla from the Meotian (early Turolian) of Grebeniki-1 (Ukraine) has been tentatively termed *M. ukrainicus* by Gremyatskii (1961), but no substantive distinctions were provided, and the specimen is indistinguishable metrically (or morphologically) from any of the above; the same is true for a mandible from Molayan (Afghanistan). On the other hand, the long-known mandible from Maragheh (Iran; middle horizon, basal MN 12, ca. 8.24 Ma, Bernor et al., this volume) was indicated to be possibly distinctive by Delson (1973) and Heintz et al. (1981), and de Bonis et al. (1990b) have referred it to *M. delsoni*. As noted above, it is large but no more likely than the RZO fossils to represent a different species. Its corpus is shallow, rather than deep, as the RZO jaws are said to be. Fragmentary colobine remains from the Indo-Pakistan Siwaliks (see Barry 1987) may be younger but are hard to distinguish from *M. pentelicus*; the possible presence of an earlier late Miocene (MN 11–12) paleoenvironmental/ paleogeographic filter in the Baluchi Hills region between Afghanistan and Pakistan should give rise to caution in synonymizing these taxa (see Bernor 1983, 1984; Brunet et al. 1984; Barry 1987; Delson 1994). De Bonis et al. (1994) have reconsidered the question of this barrier, but we must note that at least for the primates their faunal list is incorrect: *Mesopithecus* is wrongly indicated to occur at Lufeng (perhaps based on an early reference to a cercopithecid, never confirmed) and Taraklia (perhaps based on references such as Simionescu 1930, which confused Taraklia and the nearby Grebeniki); moreover, the presence of *Sivapithecus* at Yassiören does forge a link between the Potwar sequence and Turkish sites.

Several probably latest Miocene (MN 13) colobines are known from Baltavar (Hungary; fig. 12.8), Baccinello V-3, Brisighella, Casino, and Gravitelli (Italy). Only two m3s remain of the Casino sample (see Ristori 1890) and a single dP3 from Baccinello has recently been discussed, but Rook (pers. comm.) indicates that other specimens have been found; all the Gravitelli specimens have been lost. In the past they have usually been referred to *M. monspessulanus* (e.g., Delson 1973; Szalay and Delson 1979), but re-evaluation of tooth size here (see above) suggests that Casino and Baltavar at least are better grouped with *M. pentelicus*; the other Italian fossils must be termed only *Mesopithecus* sp. The exact biochronologic placement of Casino (which cannot be relocated) and Baccinello V3 is questioned, but Gravitelli (now covered by urban development) appears to predate the deposition of gypsum brought on by Mediterranean desiccation, and Baltavar is associated with the "Unio wetzleri" horizon in the terminal Pontian. Two isolated teeth from MN 13 localities in Hungary (Hatvan and Polgardi) were tentatively referred to this species by Delson (1973 et seq.); they are somewhat large for *M. pentelicus*, but much smaller than *D. ruscinensis*, to whose origin they may conceivably relate (see below).

Mesopithecus pentelicus has been described in detail by Gaudry (1862), Delson (1973), Szalay and Delson (1979), and Zapfe (1991). It is a relatively terrestrially adapted form, probably most similar to the living *Semnopithecus entellus*, and recent studies suggest that its Greek habitats were more forested than previously thought (see Solounias and Dawson-Saunders 1988 and Delson 1994; but compare de Bonis et al. 1992).

Mesopithecus monspessulanus's type locality is Montpellier, southern France. It is basal Pliocene age (early MN 14), and most of the species' range lies within the Pliocene (MN 14–17), from France and England through Romania (see fig. 12.8). Several latest Miocene (or earliest Pliocene) localities have yielded fragmentary dental remains previously assigned to this species but transferred above to *M. pentelicus*. However, de Bonis et al. (1990b) indicated that specimens from the Dytiko localities near Ravin des Zouaves included some which are small for *M. pentelicus* and perhaps best referred here.

The reanalysis of dental metrics discussed above demonstrated that lower molars (and the single known M1 and M2) of Pliocene *M. monspessulanus* are smaller than those of *M. pentelicus* at greater than 99.5% probability (except m2 length at 97.5% and m3 length not distinct). A proba-

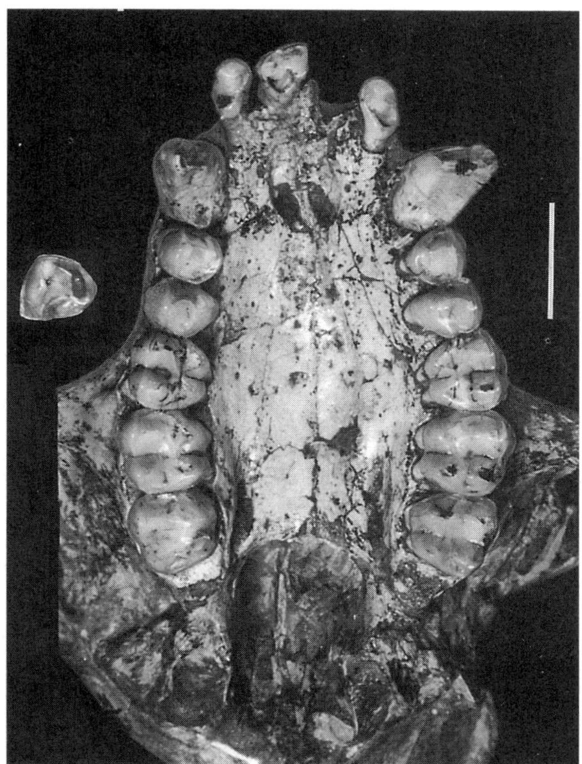

FIGURE 12.9 Upper dentition of *Mesopithecus pentelicus*: palate of male from Pikermi, compared to P4 from Wissberg. Scale bar = 1 cm.

ble female corpus fragment from Dytiko 1 with a measurable m3 differs from *M. pentelicus* in both widths (but not length) at the 99% level but can not be separated from *M. monspessulanus*. A male corpus from Dytiko 2 differs in widths from *M. pentelicus* at greater than 99% and in m3 length at 97%, but is not distinct from *M. monspessulanus*. Both specimens are referred to the latter species. De Bonis et al. (1990b) had suggested that some but not all of the Dytiko 2 specimens be so referred, but attributed most of the other Dytiko remains to *M. pentelicus*. Further study is needed to determine if indeed both species are represented in the Dytiko area, but it is interesting to note the rough (MN 13) contemporaneity of larger individuals identified as *M. pentelicus* in Hungary and Italy and at least some smaller individuals referred to *M. monspessulanus* in Northern Greece. This suggests that the latter may have originated at the end of the Miocene from small population(s) of *M. pentelicus*, which adapted to a more arboreal niche.

Mesopithecus monspessulanus is distinguished by smaller size, especially narrower teeth, and somewhat less terrestrial adaptation of the postcranium (elbow joint). It apparently inhabited mainly more forested environments (in the Pliocene) than did its congeners.

The morphology, evolution, and biogeography of *Mesopithecus* is summarized as follows:

1. *Mesopithecus delsoni* is formally synonymized with *M. pentelicus*, in that it shows few significant size differences, the supposed greater corpus depth may be due to crushing, and other distinctions do not appear to warrant specific (or subspecific) separation.

2. The premolar from Wissberg is potentially the oldest *M. pentelicus*, but the veracity of its MN 9 age has been brought into question by Bernor (here). Specimens from Pikermi, various Macedonian sites, Maragheh, Molayan, and Grebiniki form the homogeneous core of this species. Late MN 13 specimens from Baltavar (Hungary) and Casino (Italy; age less definitive) are transferred to this species, and some material from Dytiko (Greece) may also belong here; MN 13 specimens from Gravitelli and Brisighella and MN 13/14 teeth from Baccinello V3 (Italy) are not determinable to species at present.

3. The Pliocene age *Mesopithecus monspessulanus* is clearly distinct from *M. pentelicus* in its tooth size and elbow morphology. Of the Miocene assemblage, two jaws from Dytiko (Greece) appear best referred to this taxon.

4. Fragmentary evidence suggests that at the end of the Miocene, *M. pentelicus* gave rise (allopatrically and/or vicariantly?) to the smaller *M. monspessulanus* and the larger, more terrestrial forest-dwelling *Dolichopithecus* (see below).

Dolichopithecus ruscinensis   The type locality of this species is at Serrat d'en Vacquer (near Perpignan), Roussillon, southern France, and all specimens come from deposits that are Pliocene in age (MN 14–17?). A single nearly complete ulna from the locality of Pestlörinc (previously Pestszentlörinc, Hungary; Kretzoi 1954, 1969a) can be securely identified as this taxon due to the highly terrestrially adapted nature of its postcranium (fig. 12.10). In previous reviews, Delson (1975; Szalay and Delson 1979) followed Kretzoi (1969a) in dating this locality to early MN 13, but Kordos (pers. comm.; Jasko and Kordos 1990) has now shown that the fauna and stratigraphic position of the site place it in the earlier Pliocene, in MN 14 or 15. Previously, it was suggested that *Dolichopithecus* might have arisen in the more forested region of northern Central Europe at the end of the Turolian, when southern areas saw some diversification of *Mesopithecus*. However, the younger date for Pestlörinc deprives this hypothesis of any direct paleontological support. Similarly, the two large teeth from other Hungarian sites mentioned above (see *M. pentelicus*) might have supported the idea of a large and eventually more terrestrial population of the latter species giving rise to *Dolichopithecus*, but Polgardi (MN 13) only yielded a partial upper canine, and the lower molar from Hatvan is probably comparable in age to Pestlörinc (Kordos, pers. comm.). Further morphological details on the species are

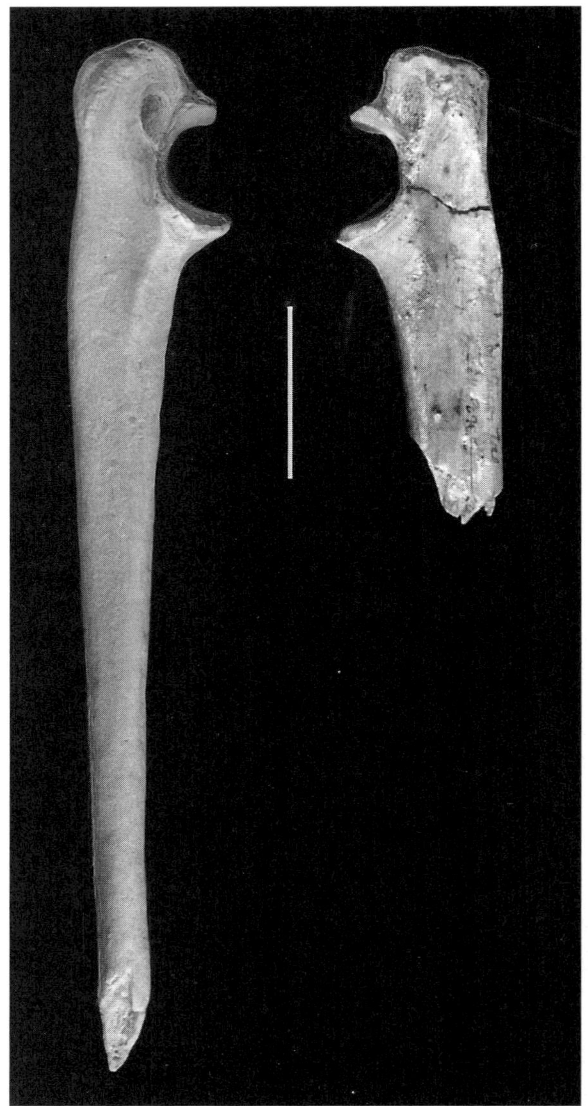

FIGURE 12.10 Medial view of ulnae of ?male *Dolichopithecus ruscinensis*, left to right: Pestszentlôrinc (left), Perpignan (right). Scale bar = 3 cm.

provided by Delson (1973) and Szalay and Delson (1979) as well as Depéret (1890), but at present it is not known from any Miocene site and thus is outside the scope of this chapter.

CERCOPITHECINAE

The most recent addition to the European Miocene primate fauna is the presence of macaque in the karst fissure of Casablanca M, in eastern Spain (*Macaca* sp.; Moyà-Solà et al. 1992). The deposits are correlated with late MN 13 on the basis of rodent and other fauna, but the macaque specimens have not yet been described. *Macaca sylvanus* "subspecies" are known throughout Europe and into Southwestern Asia from early Pliocene through late

Pleistocene, and the species is extant in Northwestern Africa and on Gibraltar.

Chronology and Biogeography

We have reported here twenty-three taxa of Miocene European and Southwest Asian catarrhine primates. Table 12.8 gives our present knowledge of species chronologic ranges by province, while figure 12.11 gives biogeographic ranges by species to identify patterns of biogeographic connections and disconnections. Figures 12.12–12.14 map fossiliferous localities by taxon for three time intervals: middle Miocene, Vallesian, and Turolian. European and West Asian catarrhine primates reveal a complex pattern of biogeographic first occurrences, interprovincial range extension, and vicariance. We attempt to interpret the various lines of evidence for these observations by evolutionary group below.

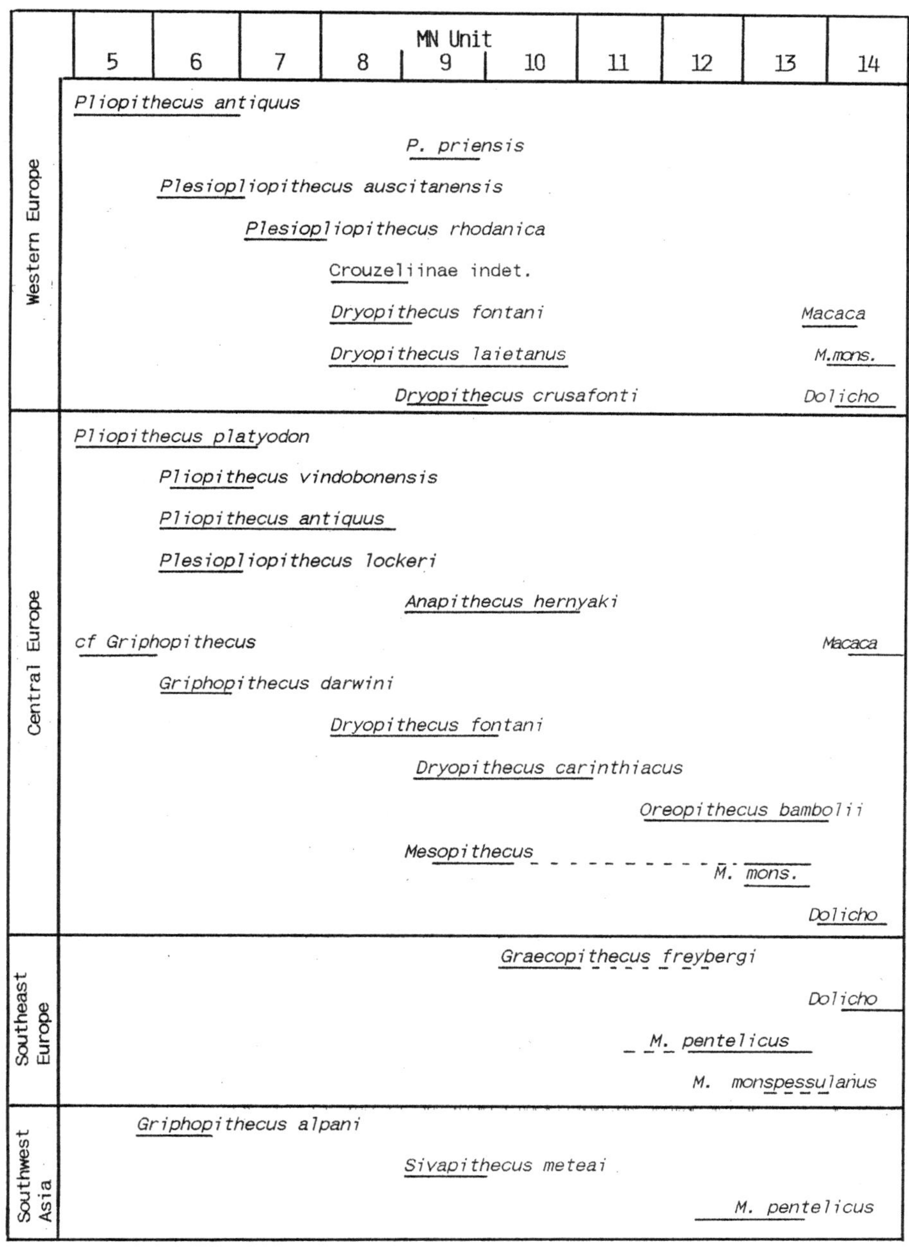

FIGURE 12.11 Distributions of Miocene catarrhine primates in Europe and Southwest Asia. The time scale is given in MN units across the top; four geographic regions are indicated on the left, and the MN ranges of fossil taxa are shown by the solid lines, with uncertain or controversial ranges shown by dashed lines.

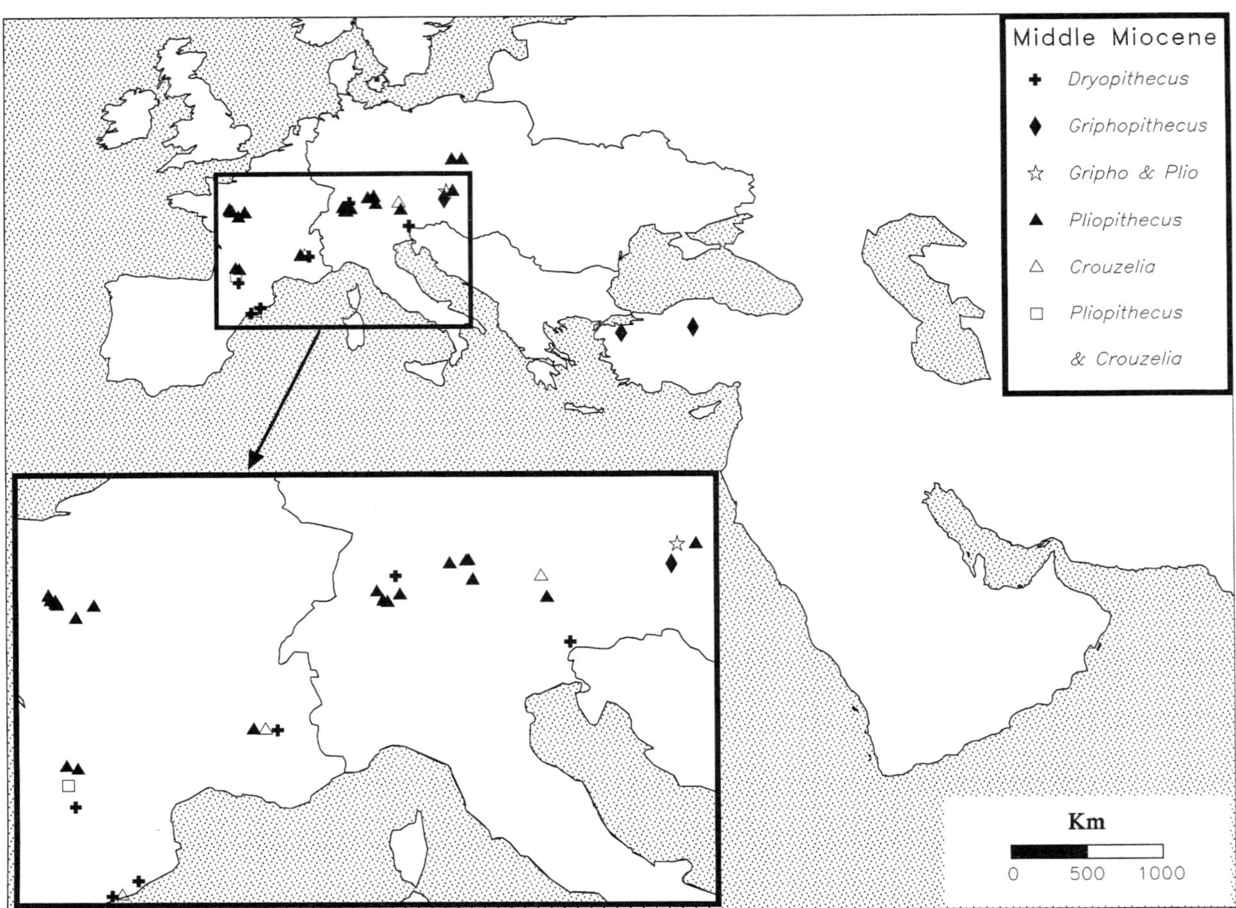

FIGURE 12.12 Distribution of middle Miocene sites yielding catarrhine primates in Europe and Southwest Asia. Central Europe is outlined. The inset enlarges the area boxed on the main map; symbols in the key indicate which taxa are present at each site.

The Pliopithecidae made their first appearance in Europe during MN 5, and became established and geographically widespread by MN 6. The collision of the Afro-Arabian plate with Eurasia during the late Oligocene/early Miocene established a land corridor across which extensive faunal interchange occurred (Adams et al. 1983; Bernor 1983; Whybrow 1984; Thomas 1985; Steininger et al. 1989). The pliopithecids were the earliest catarrhines to migrate into Europe (Thomas 1985; Barry et al. 1985, 1987; Bernor et al. 1988a; Harrison et al. 1991a), and although they almost certainly originated in Africa during the Oligocene, their evolutionary history is unknown until after their arrival in Europe. Pliopithecids are unknown from the Eastern Mediterranean and Southeastern Europe, but they are common and have a wide distribution across Western and Central Europe.

The earliest pliopithecid-bearing localities in Europe are correlated with MN 5. Three species of *Pliopithecus* are represented during the period MN 5–6 interval: *Pliopithecus antiquus* from sites in France and possibly also

Switzerland and Germany; *Pliopithecus vindobonensis* from Neudorf-Spalte, the Republic of Slovakia; and *Pliopithecus platyodon* from Elgg, Switzerland. The earliest crouzeliines, *Plesiopliopithecus lockeri* from Trimmelkam in Austria and *Plesiopliopithecus auscitanensis* from Sansan in France, are probably slightly later, correlated with MN 6, and according to Harrison they were probably derived from the pliopithecines. Andrews's (1980) view that the crouzeliine dental morphology was primitive is receiving support from new analysis of *Anapithecus* material from Rudabánya (Kordos, in prep.).

Harrison et al. (1991a) have recently suggested that the high taxonomic diversity at the time of pliopithecid first appearance may indicate that their initial migration involved multiple species that were derived from a diverse community of pliopithecids that existed prior to their arrival in Europe. If this proves to be the case, it would have important implications for understanding the zoogeographic relationships of Eurasian pliopithecids. On the other hand, there could have been a considerable period

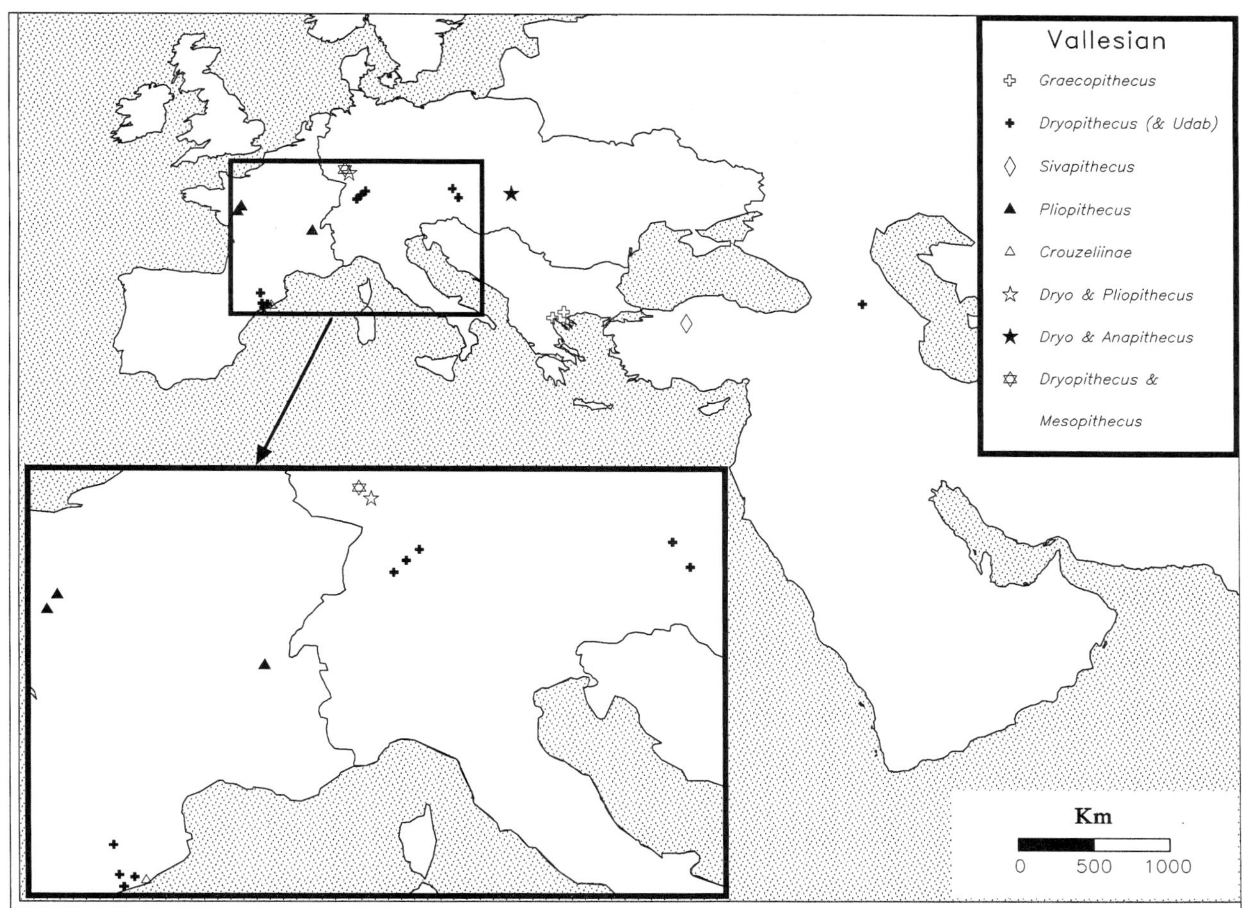

FIGURE 12.13 Distribution of Vallesian sites yielding catarrhine primates in Europe and Southwest Asia. Central Europe is outlined. The inset enlarges the area boxed on the main map; symbols in the key indicate which taxa are present at each site.

of speciation in Southern Europe that is still unknown in the fossil record. Unfortunately, these hypotheses can only be tested with the discovery of ancestral pliopithecids in Europe or Africa.

Although there is a general decline in their relative abundance and species diversity during the middle and late Miocene, pliopithecids remained widespread in Europe throughout the late Astaracian and early Vallesian (MN 7– 9). A number of specimens tentatively referred to *Pliopith-ecus antiquus* are recorded at sites in Poland (Opole, MN 7 and Przeworno II, MN 8), Spain (Castell de Barberà, MN 9) and France (MN 9). If these identifications are confirmed by additional material, they greatly extend the geographic and temporal range of this species, covering most of Europe with a latitudinal range from 42° N to 50° N, and ranging from MN 5 to MN 9. The pliopithecines began to decline in Europe before the crouzeliines, which continue well into the late Miocene of Europe, being represented by abundant remains of *Anapithecus hernyaki* from Rudabánya (MN 9) and an undescribed species from Terrassa, Spain (MN 10). However, the absence of pliopi-

thecids from Turolian sites (MN 11 to 13) strongly suggests that the family had become extinct in Europe by the close of the Vallesian. Nevertheless, a large and specialized crouzeliine, *Laccopithecus robustus*, did survive in China until Turolian correlative time (Wu and Pan 1984; Harrison 1987a; Pan 1988; Pan et al. 1989).

The hominoids made their first appearance in Europe during MN 5 or 6. There is a single lower molar of a large hominoid from Engelswies, which is correlated with MN 5 (Heizmann 1992), and the middle Miocene fauna from Paşalar in Turkey is early MN 6, similar in age to the Neudorf-Sandberg fauna (Steininger et al., this volume). The hominoid recognized from these sites is the genus *Griphopithecus*, which forms a distinct clade with three species, *G. darwini* in Central Europe and *G. alpani* and *G sp* in Southwestern Asia. Andrews and Martin (here; Andrews and Tobien 1977; Andrews 1992) and Begun (1992c) argue for a close evolutionary relationship between *Griphopithecus* and the East African early/middle Miocene form *Kenyapithecus wickeri*. Following Bernor and Tobien (1990) it would appear that their first occurrence corres-

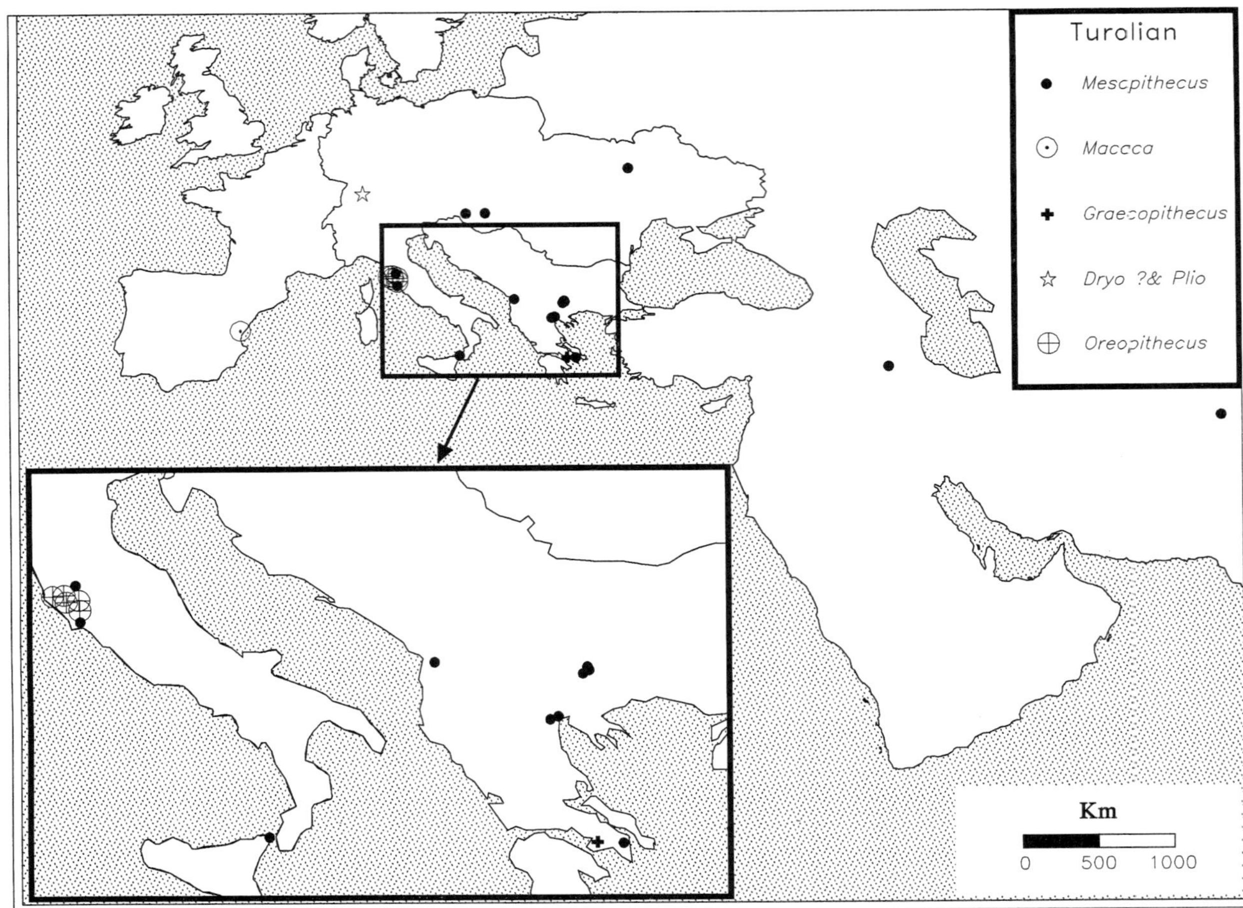

FIGURE 12.14 Distribution of Turolian sites yielding catarrhine primates in Europe and Southwest Asia. Central Europe is outlined. The inset enlarges the area boxed on the main map; symbols in the key indicate which taxa are present at each site.

ponds with the late Langhian regression (= basal MN 6), when a broad continental shelf was exposed between Eurasia and Africa-Arabia.

Begun (1992a, 1992c) has argued that the Kenyapithecini (what Begun refers to as "griphopiths") represent the sister-taxon of all later hominids. We take the opposing view that kenyapithecins are the sister-group to the earlier afropithecins, locating the origin of the former securely in Africa. Moreover, there is a chronologic hiatus between *Griphopithecus* (earlier MN 6) and the first occurrence of *Dryopithecus* (MN 7 + 8) (see: Steininger et al. 1989 and this volume) that apparently witnessed major changes in cranio-dental and postcranial anatomy. For these reasons we consider that the radiations of kenyapithecins and dryopithecins are independent of each other in evolutionary terms, while at the same time we group them together taxonomically on the basis of their common evolutionary grade.

The *Dryopithecus*–group first appears in Western Europe (St. Gaudens, MN 8; Mein 1989; Steininger et al. 1989) and Central Europe (St. Stephan, MN 8, lower Sarmatian s.s. of Vienna-Pannonian Basins, Steininger et al. 1989 and this volume). Whether these first occurring *Dryopithecus* are referable to *D. fontani* alone, or to *D. fontani* (St. Gaudens) and *D. carinthiacus* (St. Stephan), is currently debated, but the latter course is followed here.

There is good evidence for at least two species of western *Dryopithecus* by MN 8, including *D. fontani* from St. Gaudens and *D. laietanus* from Spain. Begun (1992b) has provided evidence for an additional Spanish species, *D. crusafonti*, in MN 8/9, but we are uncertain about the validity of this species (i.e., its distinction from *D. fontani*) and in any case, the specimen from Seu d'Urgell is best identified as *D. fontani*.

The evolutionary pattern of *Dryopithecus* suggests that there was an early MN 8 separation of the Spanish dryopithecines (except for the specimen from Seu d'Urgell, which is on the southern flanks of the Pyrenees) from other Western and Central European species. This suggests a possible MN 7 + 8 dispersal of a primitive member of *Dryopithecus*. *Dryopithecus carinthiacus* is recognized as a clade distinct from Spanish and French taxa, revealing a vicari-

TABLE 12.7. *Miocene Primate Localities in Europe and Southwest Asia*

Primate Localities in France

Locality	MN Zone	Species	References for dates
Pontlevoy-Thenay	MN 5	*Pliopithecus antiquus*	Ginsburg 1986
Manthelan	MN 5	*Pliopithecus antiquus* (= *Pliopithecus piveteaui*)	Ginsburg 1986
La Condoue	MN 5	*Pliopithecus antiquus*	Baudelot and Collier 1978
Faluns of Touraine and Anjou[1]	Upper MN 5	*Pliopithecus antiquus*	Ginsburg 1986, 1989
Liet	Lower MN 6	*Pliopithecus antiquus*	Collier 1978, 1979
Sansan	MN 6	*Pliopithecus antiquus* [type locality] *Plesiopliopithecus auscitanensis* (= *Crouzelia auscitanensis*) [type locality]	Ginsburg 1986
La Grive-Saint-Alban (Peyre and Beau)	MN 6	*Pliopithecus antiquus*	Ginsburg 1986
La Grive–Saint-Alban (L7)	MN 7	*Plesiopliopithecus rhodanica* (= *Crouzelia rhodanica*) [type locality]	Ginsburg 1986
St. Gaudens	MN 8	*Dryopithecus fontani* [type locality]	Mein 1986, 1989
La Grive–Saint-Alban (L3)	MN 8	*Dryopithecus fontani*	Mein 1986, 1989
Doué-la-Fontaine	Lower MN 9	*Pliopithecus antiquus*	Ginsburg 1986, 1989
Meigné-le-Vicomte	MN 9	*Pliopithecus antiquus*	Ginsburg 1989, 1989
Priay II Upper	MN 9	*"Pliopithecus" priensis* [type locality]	Welcomme et al. 1991

[1] Including the following localities: Channay, Denezé, Hommes, Lasse, Noyant-sous-le-Lude, Ponbrault, Pont-Boutard, Pontigné, Rillé, Savigné-sur-Lathan

Primate Localities in Spain

Locality	MN Zone	Species	References for Dates
Sant Quirze	MN 8	*Dryopithecus laietanus*	Mein 1986, 1989
Can Vila	MN 8	*Dryopithecus laietanus* (= *Sivapithecus occidentalis*)	Mein 1986
Can Feliu	MN 8	Crouzeliine indet.	Ginsburg 1986
Seu d'Urgell (El Firal)	Lower MN 9	*Dryopithecus fontani*	Mein 1986
Can Mata 1	MN 9	*Dryopithecus laietanus*	Mein 1986, 1989
Can Ponsic	MN 9	*Dryopithecus crusafonti* [type locality]	Mein 1986, 1989
Castell de Barberà	MN 9	*Dryopithecus laietanus*	Mein 1986
		Pliopithecus cf. *antiquus*	Mein 1986
Can Llobateres	MN 9	*Dryopithecus laietanus* (= *Dryopithecus piveteaui*) (= *Rahonapithcus sabadellensis*)	Mein 1986, 1989
La Tarumba 1	MN 10	*Dryopithecus laietanus* (= *Hispanopithecus laietanus*) [type locality]	Mein 1986, 1989
Polinya II	MN 10	*Dryopithecus laietanus*	Mein 1986
Terrassa	MN 10	Crouzeliine Gen. et sp. nov.	
Casablanca-M	Upper MN 13	*Macaca sp.*	Moyà-Solà et al. 1992

Primate Localities in Austria (Au), Germany (G), and Switzerland (S)

Locality	MN Zone	Species	References for dates
Elgg (S)	MN 5	*Pliopithecus platyodon* [type locality]	Ginsburg 1986
Engelswies (G)	MN 5–6	?*Griphopithecus sp.*	Heizmann 1992
Kreuzlingen (S)	MN 6	*Pliopithecus* cf. *antiquus*	Ginsburg 1986
Rümikon (S)	Lower MN 6	*Pliopithecus* cf. *antiquus*	Engesser et al. 1981
Stein am Rhein (S)	MN 6	*Pliopithecus antiquus*	Ginsburg 1986
Göriach (Au)	Upper MN 6	*Pliopithecus platyodon*	Steininger 1986
Trimmelkam (Au)	MN 6?	*Plesiopliopithecus lockeri* [type locality]	Ginsburg 1986
Diessen am Ammersee (G)	MN 6	*Pliopithecus* cf. *antiquus*	Ginsburg 1986
Stätzling (G)	MN 6	*Pliopithecus antiquus*	Ginsburg 1986
Klein Hadersdorf (Au)	MN 6	*Griphopithecus darwini* (= *Austriacopithecus weinfurteri*) (= *Austriacopthecus abeli*)	Mein 1989
Gallenbach 2(a?) (G)	MN 6	*Pliopithecus antiquus*	Heissig 1989
Ziemetshausen (G)	MN 6	*Pliopithecus antiquus*	Heissig 1989
St. Stephan (Au)	MN 8	*Dryopithecus carinthiacus* [type locality]	Mein 1986, 1989; Mottl 1957

T ABLE 12.7. *Miocene Primate Localities in Austria (Au), Germany (G), and Switzerland (S)(Continued)*

Primate Localities in Austria (Au), Germany (G), and Switzerland (S)			
Locality	MN Zone	Species	References for dates
Mariathal (Au)	MN 9	*Dryopithecus carinthiacus*	Mein 1986, 1990; Thenius 1982
Eppelsheim (G)	MN 9	cf. *Dryopithecus* sp. (= *Paidopithex rhenanus*) Pliopithecid indet. ("*Semnopithecus*" *eppelsheimensis*) [type locality]	Mein 1986, 1989
Wissberg (G)	MN 9	*Dryopithecus* sp. cf. *Mesopithecus pentelicus*	Mein 1986, 1989
Melchingen (G)	MN 9	*Dryopithecus* sp.	Mein 1986, 1989
Götzendorf (Au)	MN 9/10	*Anapithecus hernyaki*	Zapfe 1989
Salmendingen (G)	MN 11	?*Dryopithecus brancoi* (= *Neopithecus brancoi;*) type locality; = pliopithecid or dryopithecine??) ?Pliopithecid (?*Anapithecus hernyaki*)	Mein 1986, 1989
Trochtelfingen (G)	?	*Dryopithecus* sp.	
Ebingen (G)	?	*Dryopithecus* sp.	

Primate Localities in Italy			
Locality	MN Zone	Species	References for Dates
Baccinello V1	MN 12	*Oreopithecus bambolii*	Hürzeler and Engesser 1976); Azzaroli et al. 1986; Harrison and Harrison 1989
Casteani	MN 12	*Oreopithecus bambolii*	Hürzeler and Engesser 1976; Azzaroli et al. 1986; Harrison and Harrison 1989
Montemassi	MN 12	*Oreopithecus bambolii*	Hürzeler and Engesser 1976); Azzaroli et al. 1986; Harrison and Harrison 1989
Ribolla	MN 12	*Oreopithecus bambolii*	Hürzeler and Engesser 1976; Azzaroli et al. 1986; Harrison and Harrison 1989
Baccinello *Cardium* horizon	MN 12/13	*Oreopithecus bambolii*	Rook et al. 1995
Monte Bamboli	MN 13	*Oreopithecus bambolii* [type locality]	Hürzeler and Engesser 1976; Azzaroli et al. 1986; Harrison and Harrison 1989
Baccinello V2	MN 13	*Oreopithecus bambolii*	Hürzeler and Engesser 1976); Azzaroli et al. 1986; Harrison and Harrison 1989
Fiume Santo	MN 13		Cordy and Ginesu 1994; Kotsakis et al. 1995
Gravitelli	MN 13	*Mesopithecus* sp.	Seguenza 1902
Baccinello V3	MN 13/14	*Mesopithecus* sp.	Mein 1989
Casino	MN 13/14	*Mesopithecus* cf. *pentelicus*	Mein 1989

Primate Localities in Bulgaria (B), the Republic of Slovakia (S), Hungary (H), Poland (P), Ukraine (U), and ex-Yugoslavian Macedoni (Y)			
Locality	MN Zone	Species	References for Dates
Neudorf-Spalte (S)	Upper MN 5	*Pliopithecus vindobonensis* [type locality]	Mein 1986; Steininger 1986; Ginsburg and Mein 1980; Ginsburg 1986
Neudorf-Sandberg (S)	MN 6	*Griphopithecus darwini* (= *Dryopithecus darwini*) (= *Griphopithecus suessi*)	Mein 1989
Przeworno II (P)	MN 8	*Pliopithecus* cf. *antiquus*	Kowalski and Zapfe 1974; Ginsburg 1986
Opole (Oppeln) (P)	MN 7	*Pliopithecus* cf. *antiquus*	Steininger 1986; Ginsburg 1986
Rudabánya (H)	Upper MN 9	*Dryopithecus carinthiacus* (= *Rudapithecus hungaricus*) (= *Bodvapithecus altipalatus*) *Anapithecus hernyaki* (=*Rangwapithecus* [*Ataxopithecus*] *sericus*) [type locality]	Mein 1986, 1989; Steininger 1986; Kordos 1987
Grebeniki [1] (U)	MN 11/12	*Mesopithecus pentelicus* (= *M*. "*ukrainicus*")	Mein 1989
Titov Veles (Y)	MN 12/13	*Mesopithecus pentelicus*	Ciric 1957
Kalimanci 2 (B)	MN 12/13	*Mesopithecus pentelicus*	Bakalou and Nikolov 1962
Kromidovo 2 (B)	MN 12/13	*Mesopithecus pentelicus*	Bakalou and Nikolov 1962
Gorna Susica (B)	MN 12/13	*Mesopithecus pentelicus*	Bakalou and Nikolov 1962
Polgardi (H)	MN 13	cf. *Mesopithecus pentelicus*	Mein 1989
Baltavar (H)	MN 13	*Mesopithecus* cf. *pentelicus*	Mein 1989

Primate Localities from Greece (Gr), Iran (Ir), Afganistan (Af), Turkey (T), and Georgia (Ge)

Locality	MN Zone	Species	References for Dates
Paşalar (T)	MN 5–6	*Griphopithecus alpani*	Mein 1989
		Griphopithecus sp.	Andrews 1989
Çandir (T)	MN 6	*Griphopithecus alpani* [type locality]	Mein 1989
Yassiören (T)	MN 9	*Sivapithecus meteai* [type locality];	Mein 1989
Ravin de la Pluie (Gr)	MN 10	*Graecopithecus freybergi* (= *Ouranopithecus macedoniensis*)	Mein 1986
Xirochori 1 (Gr)	MN 10	*Graecopithecus freybergi* (= *Ouranopithecus macedoniensis*)	de Bonis et al. 1990
Udabno (Ge)	MN 10?	*Udabnopithecus garedziensis* [type locality]	Mein 1986 de Bonis et al. 1990
Pyrgos (Tour La Reine) (Gr)	MN 10?	*Graecopithecus freybergi* [type locality]	Mein 1986
Ravin des Zouaves 5 (RZO) (Gr)	MN 11/12	*Mesopithecus pentelicus*	de Bonis et al. 1987, 1989; Mein 1989
Pikermi (Gr)	MN 11/12	*Mesopithecus pentelicus* [type locality]	Mein 1989 Bernor et al. this volume
Vathylakkos (and Ravin X) (Gr)	MN 12/13	*Mesopithecus pentelicus*	de Bonis et al. 1987, 1990; Mein 1989
Pikermi "II"-Chomateri (Gr)	MN 12/13	*Mesopithecus p.* "*microdon*"	Zapfe 1991; Mein 1990
Dytiko locs (1DTK, 2 DIT, 3 DKO) (Gr)	MN 12/13	*Mesopithecus* cf. *pentelicus?* and *Mesopithecus* cf. *monspessulanus*	de Bonis et al. 1987, 1989; Mein 1989
Maraghah (Middle) Beds (Ir)	MN 12/13	*Mesopithecus pentelicus*	
Molayan (Af)	MN 12/13	*Mesopithecus pentelicus*	Heintz et al. 1981

TABLE 12.8 *Biogeographic Ranges of Middle and Late Miocene European and Southwest Asian Catarrhine Primates. Ranges are shown by continuous x's if they cross geographic boundaries. Age ranges are indicated by MN zones, with U signifying upper part of the zone, and L signifying lower part of the zone.*

Taxon	Western Europe	Central Europe	SE Europe	SW Asia	Age Range
Pliopithecus platyodon		x			MN 5–6
Pliopithecus antiquus	xxxxxxxxxxxxxx				MN 5–9
Pliopithecus vindobonensis		x			UMN 6
Pliopithecus priensis	x				UMN 9
Plesiopliopithecus auscitanensis	x				MN 6
Plesiopliopithecus lockeri		x			MN 6
Plesiopliopithecus rhodanica	x				MN 7
Anapithecus hernyaki		x			UMN 9–?11
Pliopithecidae indeterminate		x			MN 9
Oreopithecus bambolii		x			MN 12–13
cf. *Griphopithecus* sp.				x	MN 5–6
Griphopithecus alpani				x	MN 5–6
Griphopithecus darwini		x			MN 6
Sivapithecus meteai				x	MN 9
Graecopithecus freybergi			x		MN 10–?12
Dryopithecus fontani	xxxxxxxxxxxxxx				MN 8–9
Dryopithecus carinthiacus		x			MN 8–9
Dryopithecus laietanus	x				MN 8–10
Dryopithecus crusafonti	x				MN 9
Mesopithecus pentelicus			xxxxxxxxxxxxxxxxxxxxxxxxxxx		MN 11?–13
Mesopithecus monspessulanus	xxxxxxxxxxxxxxxxxxxxxxxxxx				MN 13–15
Dolichopithecus ruscinensis	xxxxxxxxxxxxxxxxxxxxxxxxxxxxxxxx				MN 14–15
Macaca sp.	x				MN 13
Macaca sp.	xxxxxxxxxxxxxxxxxxxxxxxxxxxxxxxxx				MN 13–17

ance of Western and Central European *Dryopithecus* by MN 9. This corresponds to a marked shift in palaeovegetation and climates at the middle/late Miocene boundary (Bernor et al. 1988b, 1990), closely equivalent to the Astaracian/Vallesian boundary (MN 8/9; Swisher, this volume).

Dryopithecus persisted across its range through MN 9 and perhaps into MN 10/11 in Central Europe only. Its extinction coincided with the shift from warm temperate forests to more seasonal and open country biotopes (Bernor 1983; Bernor et al. 1988b).

Coinciding with the temporal range of *Dryopithecus* was the first appearance of a member of the orang utan clade, the Ponginae. The genus *Sivapithecus* is first reported in the Siwaliks as occurring roughly equivalent to the late Astaracian (MN 7 + 8) in the European sequence (Kappelman et al. 1991). The only species with a distribution including the biogeographic regions under consideration in this paper is *Sivapithecus meteai*, from latest MN 9 horizons in Turkey (Kappelman et al., this volume).

A significant morphologic advance is seen in *Graecopithecus*, first known in Southeastern Europe during MN 10. *Graecopithecus* shares a number of characters in common with hominines, although it is uncertain whether they are true synapomorphies, plesiomorphies, or homoplasies. If *Graecopithecus* does prove to share a unique relationship with the African ape-human lineage, then a hitherto undocumented late Miocene biogeographic connection must have existed between Southeastern Europe and East Africa during MN 11 time. It must also be considered that this age is close to the postulated divergence within the African ape and human clade (Andrews 1986) on the basis of molecular clocks (Bailey et al. 1993).

The data presented here are not inconsistent with the hypothesis that both the Homininae and Ponginae may be derived from a European-Southwest Asian early middle Miocene kenyapithecin, close to that known here as the genus *Griphopithecus*. Asian Ponginae would appear to have been phylogenetically distinct by the later half of the middle Miocene. This hypothesis does not necessarily directly support Sen's (1982) argument based on rodent lineages that there was late middle Miocene biogeographic separation between eastern Greece and Western Anatolia.

To summarize, the early middle Miocene East African kenyapithecins would appear to have extended their range into Central Europe and Southwestern Asia during the late Langhian regression, ca. 15.5–15 Ma (Bernor and Tobien 1990). This clade, currently recognized in Eurasia as species of a single genus, *Griphopithecus*, may have evolved into two broadly vicariant clades: the Southeast European hominine species *Graecopithecus freybergi* and the Southwest and South Asian (Siwaliks) clade *Sivapithecus*. If *Graecopithecus* indeed shares synapomorphies with the African great ape-human clade, an early late Miocene biogeographic connection would have to be recognized between the Eastern Mediterranean and East Africa.

Regardless of the competing phylogenetic arguments that would alter the large hominoid paleobiogeography developed here, Eurasian forms are currently best considered to be a natural group, and do show a strong pattern of biogeographic extension followed by vicariance within the middle and late Miocene intervals. This vicariance is plausibly due to the replacement of subtropical and warm temperate forests with increasingly seasonal open country habitats (Bernor 1983).

The hominoid primate *Oreopithecus bambolii* is known from sites in Italy (i.e., Maremmia in Tuscany and Sardinia) tentatively correlated with MN 12 and MN 13 (Hürzeler and Engesser 1976; Harrison and Harrison 1989; Rook 1993; Cordy and Ginesu 1994; Rook et al., 1996). The geographical range of the species, and its association with an endemic vertebrate fauna, indicates that it was an insular form, restricted to a small island chain in the Northern Tethys (Hürzeler and Engesser 1976; Harrison and Harrison 1989). From a biogeographical perspective, much of the fauna (e.g., the rodents, lagomorphs, insectivores, lutrines, and aquatic reptiles) appears to have its closest affinities with taxa from continental Europe that presumably arrived by swimming or rafting across the narrow channel that connected the Ligurian Sea with the Adriatic (Harrison and Harrison 1989). However, the neotragine and alcelaphine bovids probably originated in Africa, and appear to have made their way to Italy via an intermittant trans-Tethyan land connection (Thaler 1973; Hürzeler 1983; Thomas 1984a, 1984b; Harrison 1986; Cordy and Ginesu 1994). Previously, Harrison (1986; Harrison and Harrison 1989) has suggested that *Oreopithecus* may have followed the same route as the bovids, having been derived from an oreopithecine ancestral stock in sub-Saharan Africa. However, an alternative explanation is that *Oreopithecus* was descended from one of the Eurasian hominids, possibly even *Dryopithecus*, and that it survived as a specialized, relictual taxon in an insular setting, long after all other hominids had become extinct in mainland Europe (Harrison and Rook, in prep.). Although provincially quite distinct, these *Oreopithecus*-bearing localities are placed for convenience into the Central European biogeographical province.

Among monkeys, the colobines are a group with low species diversity at any given time in the European and Southwest Asian late Miocene–early Pliocene. The first unequivocal occurrence of colobines is *Mesopithecus pentelicus* reported from Southeastern Europe (Greece) and Southwestern Asia. According to Bernor et al. (this volume), Pikermi would be the oldest certain *Mesopithecus*-bearing locality (MN 11/12, ca. 8.3–8.2 Ma). The age of the single tooth from Wissberg needs clarification, but if it is indeed MN 9 (ca. 10 Ma), it compares with the oldest African colobine *Microcolobus tugenensis* (Benefit and Pickford 1986).

Mesopithecus pentelicus is abundantly represented from MN 11–13 of Central Europe, Southeastern Europe, and Southwestern Asia. It appears to have continued to the end of MN 13 in Italy and Hungary, but was perhaps replaced in Greece by *M. monspessulanus* (known in the Pliocene from Western and Central Europe, MN 14–16/17). A new colobine lineage, *Dolichopithecus ruscinensis*, also was thought to first appear in MN 13 before becoming even more widespread in the Pliocene (Western, Central, and Southeastern Europe, MN 14–16/17), with an apparent congener in Northeastern Asia (see Delson 1994). However, as discussed above, the relevant localities of Pestlörinc and Hatvan are probably earlier Pliocene rather than Miocene in age, vitiating this argument. Although Strasser and Delson (1987) rejected Delson's earlier (1973 et seq.)

hypothesis of the *M. pentelicus* ancestry of *Dolichopithecus* because of apparent polarity reversal of some pedal traits, this hypothesis was resuscitated by Delson (1994) based on the variability of these features in Strasser's modern taxa (Strasser 1988). The widespread occurrence of *D. ruscinensis* in the more "forested" early Pliocene and the possible identification of *M. monspessulanus* at Dytiko suggest that environmental changes at the end of the Miocene could well have led to vicariant character displacement and the allopatric origin of the two mainly Pliocene taxa from *M. pentelicus* (as in Delson 1973). More material from these and other latest Miocene sites would be of great interest in testing this hypothesis, which is rather similar to the pattern observed in the other European catarrhines discussed above.

The presence of a few *Macaca* teeth in the latest MN 13 site of Casablanca-M, eastern Spain, presages the broad geographic occurrence of this genus throughout the Pliocene and Pleistocene as far north as Britain and eastward into the Caucasus, Israel, and Greece. Living species of the genus occur from Pakistan eastward to Japan, in North Africa and Indonesia, but neither fossils nor extant populations occupy the area between the Levant and the Siwaliks. North African late Miocene and early Pliocene cercopithecines are referred to *Macaca*, but definitive diagnosis of the genus on dental and even mandibular morphology is not realistic. The most interesting question, perhaps, is the geographic origin of the Casablanca M fossils.

Various authors have discussed trans-Mediterranean and peri-Gibraltarean dispersal of North African mammals into Europe at the end of the Miocene or early in the Pliocene, often related to Messinian desiccation. Aguilar et al. (1984), for example, found that an assemblage of rodents from the latest Miocene of southern Spain included no North African taxa but only those with closest relatives in Asia or eastern Africa. They thus rejected the possibility of a migration into Spain from North Africa during the Messinian/late Turolian in favor of a route along the northern edge of the Mediterranean.

On the other hand, Benson and Rakic-El Bied (1991; Benson, pers. comm.) discussed in detail the paleoceanography of the Western Mediterranean during the Messinian. They suggested that in the pre-desiccation phase, Mediterranean water flowed out to the Atlantic through the Betic region of southern Spain (north of Gibraltar), while Atlantic water entered the Mediterranean through the Rif passage(s) in Morocco, between the Atlas and the Tangier areas. The region between these two passages probably alternated between being submerged, island arcs or being connected to continental platforms. It is possible that some mammalian and other terrestrial species, including hippopotamus and cercopithecine monkeys, might have traversed this "sweepstakes route" to enter southern Spain and dispersed northward during MN 13. With the refilling

of the Mediterranean by the Zanclean Deluge at the beginning of the Pliocene, whether by dint of a "Gibraltar Waterfall" or less spectacular Moroccan waterway, such transit became impossible, sealing the allopatric isolation of various European vertebrate taxa.

Addendum

Since this article was written, a skull and partial skeleton has been discovered in the Sinap deposits in Turkey (Alpagut et al. in progress). The skull is from a female individual of the same species as the previously described lower face of a male individual from the same site (Andrews and Tekkaya 1980). Originally named *Ankarapithecus meteai*, it was synonymized with *Sivapithecus* as *S. meteai* by Andrews and Tekkaya (1980) as described in this article, but the evidence from the new skull shows it to be lacking several of the key sivapithecin characters linking the group with the orangutan. For example, the orbits are as broad as high, unlike the condition in *Sivapithecus indicus* (Pilbeam 1982), and the brow ridges are relatively well developed, especially for a female individual. The facial profile is only moderately concave and is different from the airorhynch condition in *Sivapithecus*. For this reason we now remove the Turkish species from this genus and resurrect the original name, *Ankarapithecus meteai* (Ozansoy 1957).

Acknowledgments

Andrews, Bernor, and Delson thank the Wenner-Gren Foundation for Anthropological Research for a grant to support the development of studies of catarrhine primate distribution that led to this paper. Harrison received financial support from the Boise Fund and New York University. Delson thanks the PSC-CUNY Faculty Research Award Program for grants 662495, 667370, and 669381, which partially defrayed the cost of his contribution, and Lorraine Meeker for assistance with figures 12.8–12.10. Figures 12.12–12.14 were produced with the Atlas GIS mapping program from Strategic Mapping Inc. Support for Andrews's and Bernor's research was derived in part from grants awarded to Bernor from the National Geographic Society, L. S. B. Leakey Foundation, and Joint American-Hungarian Fund. Finally, we thank David Begun and David Pilbeam for helpful comments on the text. We thank the V.W.-Stiftung for supporting Bernor's and Andrews's participation in this workshop.

LITERATURE CITED

Abel, O. 1902. Zwei neue Menschenaffen aus den Leithakalkbildungen des Weiner Beckens. *Sitzungberichte Akademie der Wissenschaften in Wien, Math-Nat. Klasse* (III), Abt. 1: 1171–208.

———. 1931. *Die Stellung des Menschen im Rahmen der Wirbeltiere*. Jena: G. Fischer.

Adams, C. G., A. W. Gentry, and P. J. Whybrow. 1983. Dating the terminal Tethyan event. In *Reconstruction of Marine Paleoenvironments*, ed. J. E. Meulenkamp, pp. 273–98. Utrecht: Micropaleontology Bulletin

Aguilar, J., L.D. Brandy and L. Thaler. 1984. Les rongeurs de Salobreña (sud de l'Espagne) et le probléme de la migration messinienne. *Paléobiologie Continentale* 14:3–17.

Alpagut, B., P. Andrews, and L. Martin. 1990. New hominoid specimens from the middle Miocene site at Paşalar, Turkey. *Journal of Human Evolution* 19:397–422.

Andrews, P. 1978. A revision of the Miocene Hominoidea of East Africa. *Bulletin of the British Museum (Natural History), Geology Series* 30:85–224.

——. 1980. Ecological adaptations of the smaller fossil apes. *Zeitschrift für Morphologie und Anthropologie* 71:164–73.

——. 1985. Family group systematics and evolution among catarrhine primates. In *Ancestors: The Hard Evidence*, ed. E. Delson, pp. 14–22. New York: Liss.

——. 1990. Lining up the ancestors. *Nature* 345:665.

——. 1992. Evolution and environment in the Hominoidea. *Nature* 360:641–47.

Andrews, P. and J. Cronin. 1982. The relationships of *Sivapithecus* and *Ramapithecus* and the evolution of the orangutan. *Nature* 197:541–46.

Andrews, P. and L. Martin. 1987. Cladistic relationships of extant and fossil hominoids. *Journal of Human Evolution* 16:101–18.

——. 1991. Hominoid dietary evolution. *Philosophical Transactions of the Royal Society* 334:199–209.

Andrews, P. and I. Tekkaya. 1976. *Ramapithecus* from Kenya and Turkey. In *Les plus anciens Hominidés*, ed. P. V. Tobias and Y. Coppens, pp. 7–25. IX International Congress of Prehistoric and Protohististoric Sciences 6.

——. 1980. A revision of the Turkish Miocene hominoid *Sivapithecus meteai*. *Palaeontology* 23:86–95.

Andrews, P. and H. Tobien. 1977. New Miocene locality in Turkey with evidence on the origins of *Ramapithecus* and *Sivapithecus*. *Nature* 268:699–701.

Arambourg, C. and J. Piveteau. 1929. Les vertébrés du Pontien du Salonique. *Annales de Paléontologie* 18:57–139.

Azzaroli, A., N. Boccaletti, E. Delson, G. Moratti, and D. Torre. 1987. Chronological and paleogeographical background to the study of *Oreopithecus bambolii*. *Journal of Human Evolution* 15:533–40.

Bailey, W. J., K. Hayasaka, C. G. Skinner, S. Kehoe, L. C. Sieu, J. L. Slightom, and M. Goodman. 1993. Reexamination of the African hominoid trichotomy with additional sequences from the primate b-globin gene cluster. *Molecular Phylogenetics and Evolution* 1:97–135.

Bakalov, P. and I. Nikolov. 1962. *Les Fossiles de Bulgarie. X. Mammifères Tertiares*. Sofia: Academie des Sciences de Bulgarie.

Barry, J. C, L. L. Jacobs, and J. Kelley. 1987. An early middle Miocene catarrhine from Pakistan with comments on the dispersal of catarrhines into Eurasia. *Journal of Human Evolution* 15:501–8.

Baudelot, S. and A. Collier. 1978. Les faunes miocènes du Haut Armagnac (Gers, France). 1. Les gisements. *Bulletin de la Societe d'Histoire Naturelle de Toulouse* 114:194–206.

Begun, D. R. 1987. *A review of the genus Dryopithecus*. Ph.D. thesis, University of Pennsylvania.

——. 1988. Catarrhine phalanges from the late Miocene (Vallesian) of Rudabánya, Hungary. *Journal of Human Evolution* 17:431–38.

——. 1989. A large pliopithecine molar from Germany and some notes on the Pliopithecinae. *Folia Primatologica* 52:156–66.

——. 1992a. *Dryopithecus crusafonti* sp. nov. a new Miocene hominoid species from Can Ponsic (northeastern Spain). *American Journal of Physical Anthropology* 87:291–309.

——. 1992b. Phyletic diversity and locomotion in primitive European hominoids. *American Journal of Physical Anthropology* 87:311–40.

——. 1992c. Miocene fossil hominids and the chimp-human clade. *Science* 257:1929–33.

——. 1994. Observations on the cranial anatomy of *Ouranopithecus*: Taxonomic and phylogenetic implications. *American Journal of Physical Anthropology* Supplement 18:54.

Begun, D. R. and L. Kordos. 1993. Revision of *Dryopithecus brancoi* Schlosser, 1901, based on the fossil hominid material from Rudabánya. *Journal of Human Evolution* 25:271–85.

Begun, D. R., S. Moyà-Solà, and M. Kohler 1990. New Miocene hominoid specimens from Can Llobateres (Spain) and their geological and paleoecological context. *Journal of Human Evolution* 9:255–68.

Begun, D. R. and L. Kordos. 1993. Revision of *Dryopithecus brancoi* Schlosser, 1901, based on the fossil hominoid material from Rudabánya. *Journal of Human Evolution* 25:271–85.

Benefit, B. and M. Pickford. 1986. Miocene fossil cercopithecoids from Kenya. *American Journal of Physical Anthropology* 69:441–64.

Benson, R. H. and K. Rakic-El Bied. 1991. Biodynamics, saline giants, and late Miocene catastrophism. *Carbonates and Evaporites* 6:127–68.

Bergounioux, F.-M. and F. Crouzel. 1964. Les *Pliopithecus* de Sansan (Gers). *Comptes Rendus de l'Academie des Sciences, Paris* 258:3744–46.

——. 1965. Les Pliopithèques de France. *Annales de Paléontologie* 51:45–65.

Bernor, R. L. 1983. Geochronology and zoogeographic relationships of Miocene Hominoidea. In *New Interpretations of Ape and Human Ancestry*, R. L. Ciochon and R. S. Corruccini, eds., pp. 21–64. New York: Plenum Press.

——. 1986. Mammalian biostratigraphy, geochronology, and zoogeographic relationships of the late Miocene Maragheh fauna, Iran. *Journal of Vertebrate Paleontology* 6:76–95.

Bernor, R. L., L. J. Flynn, T. Harrison, S. T. Hussain, and J. Kelley. 1988a. *Dionysopithecus* from southern Pakistan and the biochronology and biogeography of early Eurasian catarrhines. *Journal of Human Evolution* 17:339–58.

Bernor, R. L., J. Kovar-Eder, D. Lipscomb, F. Rögl, S. Sen, and H. Tobien. 1988b. Systematic, stratigraphic, and paleoenvironmental contexts of first-appearing hipparion in the Vienna Basin, Austria. *Journal of Vertebrate Paleontology* 8:427–52.

Bernor, R. L., J. Kovar-Eder, J.-P. Suc, and H. Tobien. 1990. A contribution to the evolutionary history of European late Miocene age hipparionines (Mammalia: Equidae). *Paléobiologie Continentale* 7:291–309.

Bernor, R. L., H.-W. Mittmann and F. Rögl. 1993. Systematics and Chronology of the Götzendorf *"Hipparion"* (late Miocene, Pannonian F, Vienna Basin). *Annalen des Naturhistorisches Museum, Wien* 95: 101–20.

Bernor, R. L., N. Solounias, C. C. Swisher III, and J. A. Van Couvering. This volume. The correlation of three classical "Pikermian" mammal faunas—Maragheh, Samos, and Pikermi—with the European MN unit system. [contribution 10].

Bernor, R. L. and H. Tobien. 1990. The mammalian geochronology and biogeography of Paşalar (middle Miocene, Turkey). *Journal of Human Evolution* 19:551–68.

Bernor, R. L., H. Tobien, and M. O. Woodburne. 1989. Patterns of Old World hipparionine evolutionary diversification and biogeographic extension. In *European Neogene Mammal Chronology*, ed. E. H. Lindsay et al., pp. 263–319. New York: Plenum.

Biedermann, W. G. A. 1863. *Petrefacten aus der Umgegend von Winterthur. Vol. 2: Die Braunkohlen von Elgg*. Winterthur: Bleuler-Hausheer.

Blainville, H. M. D. de 1839. Ostéographie ou description iconographique Comparée du Squelette et du Systeme Dentaire de mammifères. Baillière, Paris.

Bonis, L. de, G. Bouvrain, D. Geraads, and G. Koufos. 1990a. New hominid skull material from the late Miocene of Macedonia in Northern Greece. *Nature* 345:712–14.

———. 1990b. New remains of *Mesopithecus* (Primates, Cercopithecoidea) from the late Miocene of Macedonia (Greece) with the description of a new species. *Journal of Vertebrate Palaeontology* 10:473–83.

———. 1992. Diversity and paleoecology of Greek late Miocene mammalian faunas. *Palaeogeography, Palaeoclimatology, Palaeoecology* 91:99–121.

Bonis, L. de, G. Bouvrain, D. Geraads, and J. Melentis. 1974. Première découverte d'un primate hominoïde dans le Miocène supérieur de Macédoine (Grèce). *Comptes Rendus de l'Academie des Sciences, Paris* 278D:3063–66.

Bonis, L. de, G. Bouvrain, and G. Koufos. 1987. Late Miocene mammal localities of the lower Axios valley (Macedonia, Greece) and their stratigraphic significance. *Modern Geology* 13:141–47.

Bonis, L. de, M. Brunet, E. Heintz, and S. Sen. 1994. La province gréco-irano-afghane et la répartition des faunes mammaliennes au Miocène supérieur. *Paleontologia i Evolució* 24–25:103–12.

Bonis, L. and J. Melentis. 1977a. Un nouveau genre du primate hominoïde dans le Vallésien (Miocène supérieur) de Macedoine. *Comptes Rendus de l'Academie des Sciences, Paris* 184:1393–96.

———. 1977b. Les primates hominoïdes du Vallésien de Macédoine (Grèce) Étude de la machoire inférieure. *Géobios* 10:849–85.

———. 1978. Les primates hominoïdes du Miocène supérieur de Macédoine. *Annales de Paléontologie* 64:185–202.

Boschetto, H. B., F. H. Brown, and I. McDougall. 1992. Stratigraphy of the Lothidok range, northern Kenya, and K/Ar ages of its Miocene primates. *Journal of Human Evolution* 22:47–71.

Branco, W. 1897. Die menschenähnlichen Zähne aus dem Bohnerz der Schwäbischen Alb. *Jahreshefte des Vereins für vaterländische Naturkunde in Würtemberg, Stuttgart* 54:1–144.

Brunet, M., E. Heintz, Y. Jehenne, and S. Sen. 1982. Der Erste primatenfund im Miozan von Afghanistan. *Zeitschrift für Geologie, Berlin* 10:891–97.

Brunet, M., E. Heintz, and B. Battail. 1984. Molayan Afghanistan and the Khaur Siwaliks of Pakistan: An example of biogeographic isolation of late Miocene mammalian faunas. *Geologie en Mijnbouw* 20:31–38.

Burtschak-Ambramovich, N. and E. Gabachvili. 1950. Découverte d'un anthropoïde fossil en Géorgie. *Priroda, Moscow* 9:70–72.

Ciochon, R. L. and R. S. Corruccini. 1977. The phenetic position of *Pliopithecus* and its phylogenetic relationship to the Hominoidea. *Systematic Zoology* 26:290–99.

Ciric, A. 1957. Die Pikermifauna aus der Umgebung von Titov Veles. *Bulletin du Muséum d'Histoire Naturelle du Pays Serbe, series A.* 8:1–82.

Collier, A. 1978. Découverte de restes de Pliopithèques (Mammalia, Primates) dans les sables fauves de l'Armagnac. *Comptes Rendus de l'Academie des Sciences, Paris* 286:327–30.

———. 1979. Notes sur le Pliopithèque des faluns de Touraine. *Bulletin de la Société Géologique de Touraine* 1:19–23.

Conroy, G. C., M. Pickford, B. Senut, J. Van Couvering, and P. Mein. 1992. *Otavipithecus namibiensis*, first Miocene hominoid from Southern Africa. *Nature* 356:144–48.

Cordy, J.-M. and S. Ginesu. 1994. Fiume Santo (Sassari, Sardiagne, Italie): Un nouveau gisement à Oréopithèque (Oreopithecidae, Primates, Mammalia). *Compte Rendus de l'Academie des Sciences, Paris* 318:697–703.

Crusafont-Pairó, M. 1978. El Gibón fosil (*Pliopithecus*) del Vindoboniense terminal del Vallès. *Paleontologia y Evolución* 13:11–12.

Crusafont-Pairó, M. and Golpe-Posse, J. M. 1981. Estudio de la denticion inferior del primer Pliopithécido hallado en España (Vindobonensis terminal de Castell de Barbera, Cataluña, España). *Bolletin de Informaciones de la Institut de Paleontologia, Sabadell* 13:25–38.

——— and J. Hürzeler. 1961. Les pongidés fossiles d'Espagne. *Comptes Rendus hebdomadaires de l'Academie des Sciences, Paris* 254:582–84.

Dean, D. and E. Delson. 1992. Second gorilla or third chimp? *Nature* 359:676–77.

Delson, E. 1973. *Fossil Colobine Monkeys of the Circum-Mediterranean Region and the Evolutionary History of the Cercopithecidae (Primates, Mammalia)*. Ph.D. diss., Columbia University.

———. 1975. Paleoecology and zoogeography of the Old World monkeys; In *Primate Functional Morphology and Evolution*, ed. R. Tuttle, pp. 37–64. Mouton: The Hague.

———. 1988. Catarrhini. In *Encyclopedia of Human Evolution and Prehistory*, ed. I. Tattersall, E. Delson, and J. Van Couvering, pp. 111–16. New York: Garland.

———. 1994. Evolutionary history of the colobine monkeys in paleoenvironmental perspective. In *Colobine Monkeys: Their Ecology, Behavior and Evolution*, ed. G. Davies and J. F. Oates, pp. 11–43. Cambridge: Cambridge University Press.

Delson, E. and P. Andrews. 1975. Evolution and interrelationships of the catarrhine primates. In *Phylogeny of the Primates: A Multidisciplinary Approach*, ed. W. P. Luckett and F. S. Szalay, pp. 405–46. New York: Plenum.

Depéret, C. 1887. Rechérches sur la succession des faunes de vertébrés miocènes de la Vallée du Rhône. *Archives du Museum d'Histoire Naturelle de Lyon* 4:45–313.

——. 1889. Sur le *Dolichopithecus ruscinensis* nouveau singe fossile du Pliocène du Roussillon. *Comptes Rendus de l'Academie des Sciences* 103:982–83.

——. 1890. Les animaux pliocènes du Roussillon. *Memoires de la societé geologique de France, Paléontologie* 3:1–126.

——. 1911. Sur la découverte d'un grand Singe anthropoide du genre *Dryopithecus* dans le Miocène moyen de La Grive-Saint-Alban (Isère). *Comptes Rendus de l'Academie des Sciences, Paris* 153:32–35.

——. 1929. *Dolichopithecus arvernensis* Depéret: Nouveau singe du pliocène supérieur de Senèze (Haute-Loire). *Travaux du laboratoire de géologie de Lyon* 15:5–12.

Ehrenberg, K. 1937. *Austriacopithecus*, ein neuer menschenaffen-artiger Primate aus dem Miozän von Klein-Hadersdorf bei Poysdorf in Niederösterreich (Nieder-Donau). *Sitzungsberichte der Königliche Bayerischen Akademie der Wissenschaften zu München* 147:71–110.

Engesser, B., A. Matter, and M. Weidman. 1981. Stratigraphie und Säugetierfaunen des mittleren Miozäns von Vermes (Kt. Jura). *Eclogae Geologicae Helvetiae, Basel* 74:893–952.

Fleagle, J. G. 1983. Are there any fossil gibbons? In *The Lesser Apes: Evolutionary and Behavioural Biology*, ed. D. J. Chivers, H. Preuschoft, N. Creel, and W. Brockleman, pp. 437–41. Ediburgh: Edinburgh University Press.

——. 1986. The fossil record of early catarrhine evolution. In *Major Topics in Primate and Human Evolution*, ed. B. Wood, L. Martin, and P. Andrews, pp. 130–49. Cambridge: Cambridge University Press.

——. 1988. *Primate Adaptation and Evolution*. New York: Academic Press.

Forstén, A. and R. Garevski. 1989. Hipparions (Mammalia, Perissodactyla) from Macedonia (Yugoslavia). *Geologica Macedonica* 3:159–206.

Gaudry, A. 1865. *Animaux fossiles et géologie de l'Attique*. Paris: F. Savy.

Gervais, P. 1849. *Zoologie et Paléontologie Française*. Bertrand, Paris.

——. 1872. Sur un singe fossile, d'espèce non encore décrite, qui a été découvert au Monte Bamboli. *Comptes Rendus de l'Academie des Sciences, Paris* 74:1217–23.

Ginsburg, L. 1964. Nouvelle découverte de Pliopthèque dans les faluns Helvétiens de l'Anjou. *Bulletin du Muséum National d'histoire Naturelle, Paris* 36:157–60.

——. 1968. L'évolution des Pliopithèques et l'age de la faune de Sansan. *Comptes Rendus de l'Academie des Sciences, Paris* 266D:1564–66.

——. 1975. Le Pliopithèque des Faluns Helvétiens de la Touraine et de l'Anjou. *Colloque International CNRS* 218:877–86.

——. 1986. Chronology of the European pliopithecids. In *Primate Evolution*, ed. J. G. Else and P. C. Lee, pp. 47–57. Cambridge: Cambridge University Press.

——. 1989. Les Mammifères Vallésians des faluns du synclinal d'Esvres. *Bulletin de la Société d'Études Scientifiques de l'Anjou* 13:35–52.

——. 1990. Les quatre faunes de Mammifères Miocènes des faluns du synclinal d'Esvres (Val-de-Loire, France). *Compte Rendus de l'Academie des Sciences, Paris* 310. Ser. II:89–93.

—— and P. Mein. 1980. *Crouzelia rhodanica*, nouvelle espèce de Primate catarhinien, et essai sur la position systématique des Pliopithecidae. *Bulletin Museum Nationale Histoire Naturelle, Paris* 2:57–85.

Glaessner, M.F. 1931. Neue Zähne von Menschenaffen aus dem Miozän des Wiener Beckens. *Annalen des Naturhistorischen Museums, Wien* 46:15–27.

Golpe-Posse, J. M. 1982. Un pliopitecido persistente en el Vallesiense Medio-Superior de los alrededores de Terrassa (Cuenca del Vallés, España) y problemas de su adaptación. *Boletin Geologico y Minerio* 4:287–96.

Goodman, M. 1963. Man's place in the phylogeny of the primates as reflected in serum proteins. In *Classification and Human Evolution*, ed. S. L. Washburn, pp. 204–34. New York: Aldine.

——. 1974. Biochemical evidence on hominid phylogeny. *Annual Reviews of Anthropology* 3:203–26.

Gremyatskii, M. A. 1961. [The main line of higher primate evolution in the Neogene] (in Russian). *Voprosy Antropoiogii* 7:3–8.

Grine, F. E., D. W. Krause, and L. B. Martin. 1985. The ultrastructure of *Oreopithecus bambolii* tooth enamel: Systematic implications. *American Journal of Physical Anthropology* 66:177–78.

Groves, C. P. 1972. Systematics and phylogeny of gibbons. *Gibbon and Siamang* 1:1–89.

——. 1974. New evidence on the evolution of the apes and monkeys. *Vestnik Ustredniho ustavu geologickeho* 49:53–56.

Harrison, T. 1982. *Small-Bodied Apes from the Miocene of East Africa*. Ph.D. diss., University of London.

——. 1986. New fossil anthropoids from the middle Miocene of East Africa and their bearing on the origin of the Oreopithecidae. *American Journal of Physical Anthropology* 71:265–84.

——. 1987a. The phylogenetic relationships of the early catarrhine primates: A review of the current evidence. *Journal of Human Evolution* 16:41–80.

——. 1987b. A reassessment of the phylogenetic relationships of *Oreopithecus bambolii* Gervais. *Journal of Human Evolution* 15:541–83.

——. 1991a. The taxonomy and phylogenetic relationships of the Pliopithecidae. *American Journal of Physical Anthropology* Supplement 12:89.

——. 1991b. Some observations on the Miocene hominoids from Spain. *Journal of Human Evolution* 20:515–20.

——. 1991c. The implications of *Oreopithecus bambolii* for the origins of bipedalism. In *Origine(s) de la Bipédie chez les Hominidés*, ed. Y. Coppens and B. Senut, pp. 235–44. Paris: Cahiers de Paléoanthropologie, CNRS.

——. 1993. Cladistic concepts and the species problem in hominoid evolution. In *Species, Species Concepts, and Primate Evolution*, ed. W. Kimbel and L. Martin, pp. 345–71. New York: Plenum.

Harrison, T., E. Delson, and J. Guan. 1991. A new species of *Pliopithecus* from the middle Miocene of China and its implications for early catarrhine zoogeography. *Journal of Human Evolution* 21:329–61.

Harrison, T. S. and T. Harrison. 1989. Palynology of the late Miocene *Oreopithecus*-bearing lignite from Baccinello, Italy. *Palaeogeography, Palaeoclimatology, Palaeoecology* 76:45–65.

Haq, B. U., J. Hardenbol, and P. R. Vail. 1987. Chronology of fluctuating sea levels since the Triassic. *Science* 135: 1156–67.

Haupt, O. 1935. Andere Wirbeltiere des Neozoikums. In *Oberrheinischer Fossilkatalog* 4 (9), ed. W. Salomon-Calvi, pp. 1–103. Berlin: Bornträger.

Heintz, E., M. Brunet, and B. Battail. 1981. Cercopithecid primates from the late Miocene of Molayan, Afghanistan, with remarks on *Mesopithecus*. *International Journal of Primatology* 2:273–84.

Heissig, K. 1987. Neue Funde von *Pliopithecus* in Bayern. *Mitteilungen Bayerische Staatssammlung für Paläontologie und historische Geologie* 27:95–103.

———. 1989. Neue Ergebnisse zur Stratigraphie der mittleren Serie der Oberen Süsswassermolasse Bayerns. *Geologica Bavarica* 94:239–57.

Heizmann, E. 1992. Das Tertiär in Südwestdeutschland. *Stuttgarter Beiträge zur Naturkunde, Serie C.* 33:1–90.

Hofmann, A. 1893. Die Fauna von Göriach. *Abhandlungen K-K. Geologie, Reichsanstalt* 6:1–87.

Hürzeler, J. 1954. Contribution à l'odontologie et à la phylogénèse du genre *Pliopithecus* Gervais. *Annales de Paléontologie* 40:1–63.

———. 1958. *Oreopithecus bambolii* Gervais: A preliminary report. *Verhandlungen der Naturforschenden Gesellschaft, Basel* 69:1–49.

———. 1983. Un alcélaphiné aberrant (Bovidé, Mammalia) des "lignites de Grosseto" en Toscane. *Comptes Rendus des l'Academie des Sciences, Paris* 296D:497–503.

Hürzeler, J. and B. Engesser. 1976. Les faunes de mammifères néogènes du Bassin de Baccinello (Grosseto, Italie). *Comptes Rendus des l'Academie Sciences, Paris* 283:333–36.

Jaskó, S. and L. Kordos. 1990. The gravel formations of the area between Budapest, Adony, and Örkény. *Magyar Allami Földtani Intézet, Evi Jelentése* 1988:153–67.

Kappelman, J., J. Kelley, D. Pilbeam, K. Sheikh, S. Ward, M. Anwar, J. Barry, B. Brown, P. Hake, N. Johnson, S. Mahmood Raza, and S. Ibrahim Shah. 1991. The earliest occurrence of *Sivapithecus* from the middle Miocene Chinji Formation of Pakistan. *Journal of Human Evolution* 21:61–73.

Kelley, J. and D. Pilbeam. The dryopithecines: Taxonomy, comparative anatomy, and phylogeny of Miocene large hominoids. In *Comparative Primate Biology*, ed. D. R. Swindler and J. Irwin, pp. 1:361–411. New York: Liss.

Kelley, J. and Q. Xu. 1991. Extreme sexual dimorphism in a Miocene hominoid. *Nature* 352:151–53.

Koenigswald, G. H. R. von. 1956. Gebissreste von Menschenaffen aus dem Unterpliozän Rheinhessens. II. *Verhandelingen der Koninklijke Akademie van Wetenschappen, Natuurkunde* B59:330–34.

———. 1972. Ein unterkiefer eines fossilen hominoiden aus dem Unterpliozän Griechenlands. *Verhandelingen der Koninklijke Akademie van Wetenschappen, Natuurkunde* 39:1000–1009.

Kordos, L. 1987. Description and reconstruction of the skull of *Rudapithecus hungaricus* Kretzoi (Mammalia). *Annales Historico-Naturales Musei Nationalis Hungarici* 79:77–88.

———. 1991. Le *Rudapithecus hungaricus* de Rudabánya (Hongrie). *L'Anthropologie* 95:343–62.

Koufos, G. D. 1993. Mandible of *Ouranopithecus macedoniensis* (Hominidae, Primates) from a new late Miocene locality of Macedonia (Greece). *American Journal of Physical Anthopology* 91:225–34.

Kowalski, K. and H. Zapfe 1974. *Pliopithecus antiquus* (Blainville, 1839) (Primates, Mammalia) from the Miocene of Przeworno in Silesia (Poland). *Acta Zoologica Cracoviensia* 19:19–30.

Kretzoi, M. 1954. Bericht über die Calabrische (Villafranchische) fauna von Kieslang, Kom. Fejer. *Földtani Intézet Evi Jelentése* 1953:239–64.

———. 1969a. Sketch of the late Cenozoic (Pliocene and Quaternary) terrestrial stratigraphy of Hungary. *Földrajzi Kozlemenyek* 176:198–204.

———. 1969b. Geschichte der primaten und der hominisation. *Symposia Biologica Hungarica* 9:23–31.

———. 1975. New ramapithecines and *Pliopithecus* from the lower Pliocene of Rudabánya in north-eastern Hungary. *Nature* 257:578–81.

———. 1984. Uj hominid lelet Rudabányáról. *Anthropologische Közlemenyek* 28:91–96.

Lartet, E. 1856. Note sur un grand singe fossile que se rattache au groupe des singes supérieurs. *Comptes Rendu de l'Academie des Sciences* 43:219–23.

Laskarev, W. 1909. Recherches géologiques dans les environs de Tiraspol. *Mémoires de la Societé des Naturalistes de la Nouvelle Russie* 33:127–48.

Lewis, G. E. 1937. Taxonomic syllabus of Siwalik fossil anthropoids. *American Journal of Science* 34:139–47.

Martin, L. 1985. Significance of enamel thickness in hominoid evolution. *Nature* 314:260–63.

Martin, L. and P. Andrews. 1984. The phyletic position of *Graecopithecus freybergi* von Koenigswald. *Courier Forschungsinstitut, Senckenberg* 69:25–40.

———. 1993. Species recognition in middle Miocene hominoids. In *Species, Species Concepts, and Primate Evolution*, ed. W. H. Kimbel and L. Martin, pp. 393–427. New York: Plenum.

Meldrum, J. and Y. Pan. 1988. Manual proximal phalanx of *Laccopithecus robustus* from the latest Miocene site of Lufeng. *Journal of Human Evolution* 17:719–32.

Mein, P. 1986. Chronological succession of hominoids in the European Neogene. In *Primate Evolution*, ed. J. G. Else and P. C. Lee, pp. 59–79. Cambridge: Cambridge University Press.

———. 1989. Updating of MN zones. In *European Neogene Mammal Chronology*, ed. E. H. Lindsay, V. Fahlbusch, and P. Mein, pp. 73–90. New York: Plenum.

Mein, P., E. Martín-Suárez, and J. Agusti. 1993. *Progonomys* Schaub, 1938 and *Huerzelerimys* gen. nov. (Rodentia): Their evolution in Western Europe. *Scripta Geologica* 103:41–64.

Mottl, M. 1957. Bericht über die neuen Menschenaffenfunde aus Österreich, von Sankt Steven im Levanttal, Kärnten: Carinthica 2. *Mitteilungen des Naturwissenschaftlichen vereins in Karnten.* 67:39–84.

Moyà-Solà, S., J. Pons Moyà, and M. Köhler. 1992. Primates catarrinos (Mammalia) del Neógeno de la península Ibérica. *Paleontologia e Evolució* 23: 41–45.

Napier, J. R. 1961. Prehensility and opposability in the hands of primates. *Symposia of the Zoological Society of London* 5:115–32.

———. 1962. The evolution of the hand. *Scientific American* (December): 56–63.

Ozansoy, F. 1957. Faunes de mammifères du Tertiare de Turquie et leur révisions stratigraphiques. *Bulletin of the Mineral Research and Exploration Institute of Turkey* 49:29–48.

Pan, Y. 1988. Small fossil primates from Lufeng, a latest Miocene site in Yunnan Province, China. *Journal of Human Evolution* 17:359–66.

Pan, Y., D. Waddle, and J. G. Fleagle. 1989. Sexual dimorphism in *Laccopithecus robustus*, a late Miocene hominoid from China. *American Journal of Physical Anthropology* 79:137–58.

Pia, J. and O. Sickenberg. 1934. *Katalog der in den österreichischen sammlugen befindlichen Säugertierreste des jungtertiärs österreichs und der randgebiete.* Leipzig and Vienna: Franz Deuticke.

Pilbeam, D. R. 1969. Tertiary Pongidae of East Africa: Evolutionary relationships and taxonomy. *Bulletin of the Peabody Museum of Natural History* 31:1–185.

Pilbeam, D. R., M. D. Rose, J. C. Barry, and S. M. I. Shah. 1990. New *Sivapithecus* humeri from Pakistan and the relationship of *Sivapithecus* and *Pongo*. *Nature* 348:237–39.

Pohlig, H. 1895. *Paidopithex rhenanus*, le singe anthropomorphe du Pliocène rhénan. *Procès-Verbaux de la Société Belge de géologie* 9:149–51.

Qiu, Z. and J. Guan. 1986. A lower molar of *Pliopithecus* from Tongzin, Ningxia Hui Autonomous Region. *Acta Anthropologica Sinica* 5:201–7.

Remane, A. 1921. Beiträge zur Morphologie des Anthropoidengebisses. *Archive für Naturgeschichte* 87:1–179.

——. 1965. Die Gescichte der Menschenaffen. In ed. G. Heberer, pp. 249–309 *Menschliche Abstammungslehre*. Göttingen: Fischer.

Ristori, G. 1980. Le scimmie fossile *Italiane Studio Paleontogico Bolletino Comitato Geologico d'Italia* 21:178–96, 225–34.

Rögl, F., H. Zapfe, R.L. Bernor, R. Brzobohaty, G. Daxner-Höck, I. Draxler, O. Fejfar, J. Gaudant, P. Herrmann, G. Rabeder, O. Schultz, and R. Zetter. 1993. Die primatenfundstelle Götzendorf an der Leitha, Nierderösterreich. *Jahrbuch Geologie Bundesanstalt* 136:503–26.

Rook, L. 1993. A new find of *Oreopithecus* (Mammalia, Primates) in the Baccinello Basin (Grosseto, Southern Tuscany). *Rivista Italiana di paleontologia e stratigrafia* 99:255–62.

Rook, L., T. Harrison, and B. Engesser. 1996. The taxonomic status and biochronological implications of new finds of *Oreopithecus* from Baccinello (Tuscany, Italy). *Journal of Human Evolution* 30:1–27.

Rose, M. D. 1988. Another look at the anthropoid elbow. *Journal of Human Evolution* 17:193–224.

——. 1989. New postcranial specimens of catarrhines from the middle Miocene Chinji Formation, Pakistan: Descriptions and a discussion of proximal humeral functional morphology in anthropoids. *Journal of Human Evolution* 18:131–62.

——. 1993. Locomotor anatomy of Miocene hominoids. In *Postcranial Adaptations in Nonhuman Primates*, ed. D. L. Gebo, pp. 252–72. DeKalb: Northern Illinois University Press.

Rosenberger, A. L. and E. Delson. 1985. The dentition of *Oreopithecus bambolii*: Systematic and paleobiological implications. *American Journal of Physical Anthropology* 66:222–23.

Roth, J. and A. Wagner. 1845. Die fossilen knochenuberreste von Pikermi in Griechenland. *Abhandlungen der Bayerische Akademie der Wissenschaften* 7:371–88.

Sarmiento, E. E. 1987. The phylogenetic position of *Oreopithecus* and its significance in the origin of the Hominoidea. *American Museum Novitates* 2881:1–44.

Schlosser, M. 1901. Die menschenähnlichen Zähn aus dem Bohnerz der schwäbischen Alb. *Zoologischer Anzeiger* 24:261–71.

——. 1902. Beiträge zur Kenntnis der Säugetierreste aus den süddeutschen Bohnerzen. *Geologische und Palaontologische Abhandlungen* 5:117–258.

——. 1921. Die *Hipparion* fauna von Veles in Mazedonien. *Abhandlungen Bayerischen Akademie der Wissenschaften, Mathematisch-Physisches Klasse, München* 29(4):3–57.

Schwartz, J. H. 1990. *Lufengpithecus* and its potential relationship to an orang-utan clade. *Journal of Human Evolution* 19:591–605.

Seguenza, L. 1902. I vertebrati fossili della Provincia di Messina. *Bolletino della Societa Geologico d'Italia* 21:115–74.

Sen, S. 1982. Biogeographie et biostratigraphie du Neogène continental de la région egéenne: Apports des rongeurs. *Géobios* 6:465–72.

Shea, B. T. 1985. On aspects of skull form in African apes and orangutans, with implications for hominoid evolution. *American Journal of Physical Anthropology* 68:329–42.

Simionescu, I. 1930. Vertebratele Pliocene dela Malusteni (Covurlui). *Academia Romana* 9:83–136.

Simons, E. G. 1972. *Primate Evolution: An Introduction to Man's Place in Nature*. New York: Macmillan.

Simons, E. G. and J. G. Fleagle. 1973. The history of extinct gibbon-like primates. *Gibbon and Siamang* 2:121–48.

Simons, E. G., and D. R. Pilbeam. 1965. Preliminary revision of the Dryopithecinae (Pongidae, Anthropoidea). *Folia Primatologica* 3:81–152.

Smith-Woodward, A. 1914. On the lower jaw of the anthropoid ape (*Dryopithecus*) from the upper Miocene of Lerida, Spain. *Quarterly Journal of the Geological Society* 70:316–20.

Solounias, N. and B. Dawson-Saunders. 1988. Dietary adaptations and paleoecology of the late Miocene ruminants from Pikermi and Samos, Greece. *Palaeogeography, Palaeoclimatology, Palaeoecology* 65:149–72.

Steininger, F. F. 1967. Ein weiterer zahn von *Dryopithecus fontani darwini* Abel 1902 (Mammalia, Pongidae) aus dem Miozan des Wiener Beckens. *Folia Primatologica* 7:243–75.

——. 1986. Dating the Paratethys Miocene hominoid record. In *Primate Evolution*, ed. J. G. Else and P. C. Lee, pp. 71–84. Cambridge: Cambridge University Press.

Steininger, F. F., R. L. Bernor, and V. Fahlbusch. 1989. European Neogene marine/continental chronologic correlations. In *European Neogene Mammal Chronology*, ed. E. H. Lindsay, V. Fahlbusch, and P. Mein, pp. 15–46. New York: Plenum.

Steininger, F. F., W. A. Berggren, D. V. Kent, R. L. Bernor, S. Sen, and J. Agusti. This volume. Circum-Mediterranean Neogene (Miocene and Pliocene) marine-continental chronologic correlations of European mammal units.

Strasser, E. 1988. Pedal evidence for the origin and diversification of cercopithecid clades. *Journal of Human Evolution* 17:225–45.

Strasser, E. and E. Delson. 1987. Cladistic analysis of cercopithecid relationships. *Journal of Human Evolution* 16:81–99.

Szalay, F. S. and E. Delson. 1979. *Evolutionary History of the Primates*. London: Academic Press.

Tattersall, I., E. Delson, and J. Van Couvering, eds. 1988. *Encyclopedia of Human Evolution and Prehistory*. New York: Garland.

Tattersall, I., and E. L. Simons. 1969. Notes on some little-known primate fossils from India. *Folia Primatologica* 10:146–53.

Tekkaya, I. 1974. A new species of Tortonian anthropoid (Primates, Mammalia) from Anatolia. *Bulletin of the Mineralogical Research and Exploration Institute of Turkey* 83:148–65.

Thenius, E. 1982. Ein Menschenaffen (Primates: Pongidae) aus dem Pannon (Jung-Miozän) von Niederösterreich. *Folia Primatologica* 39:187–200.

——. 1983. Zur Paläoklimatologie des Pannon (Jungmiozän) von Niederösterreich. *Neues Jahrbuch der Geologie und Paläontologie* 11:692–704.

Thomas, H. 1984a. Les Bovidae (Artiodactyla: Mammalia) du Miocène du sous-continent Indien, de la peninsule Arabique et de l'Afrique: biostratigraphie, biogéographie et écologie. *Palaeogeography, Palaeoclimatology, and Palaeoecology* 45:251–99.

——. 1984b. Les origines africaines des Bovidae (Artiodactyla, Mammalia) miocènes des lignites de Grosseto (Toscane), Italie. *Bulletin du Muséum d'Histoire naturelle, Paris* 6:81–101.

——. 1985. The early and middle Miocene land connection of the Afro-Arabian plate and Asia: A major event for hominoid dispersal? In *Ancestors: The Hard Evidence*, ed. E. Delson, pp. 42–50. New York: Liss.

Tobien, H. 1980. Taxonomic status of some Cenozoic mammalian local faunas from the Mainz Basin. *Mainzer geowissenschaftlicher Mitteilungen* 9:203–35.

——. 1986. An early upper Miocene (Vallesian) *Mesopithecus* molar from Rheinhessen, FRG. *Primate Report* 14:49.

Villalta, J. F. and M. Crusafont. 1964. Dos neuvos antropomorfos del Mioceno español y su situación dentro de la moderna sistemática de los simidos. *Notas del Comision de investigationes geologicas y mineralogicas* 13:91–139.

Wagner, A. 1839. Fossile ueberreste von einem Affenschádel und anderen Säugethieren aus Griechenland. *Gelehrte Anzeigen der Bayerisches Akademie der Wissenschaften* 38:306–11.

Walker, A. and M. Teaford. 1989. The hunt for *Proconsul*. *Scientific American* (January): 76–82.

Ward, S. C. and B. Brown. 1986. The facial skeleton of *Sivapithecus indicus*. In *Comparative Primate Biology. 1. Systematics, Evolution, Anatomy*, ed. D. R. Swindler and J. Erwin, pp. 413–52. New York: Liss.

Ward, S. C. and W. H. Kimbel. 1983. Subnasal morphology and the systematic position of *Sivapithecus*. *American Journal of Physical Anthropology* 61:157–71.

Ward, S. C. and D. R. Pilbeam. 1983. Maxillo-facial morphology of Miocene hominoids from Africa and Indo-Pakistan. In *New Interpretations of Ape and Human Ancestry*, ed. R. L. Ciochon and R. S. Corruccini, pp. 211–38. New York: Plenum.

Welcomme, J.-L., J.-P. Aguilar, and L. Ginsburg. 1991. Découverte d'un nouveau Pliopithèque (Primates, Mammalia) associé à des rongeurs dans les sables du Miocène supérieur de Priay (Ain, France) et remarques sur la paléogéographie de la Bresse au Vallésien. *Comptes Rendus de l'Academie des Sciences* 313, Sér II:723–29.

Whybrow, P. J. 1984. Geological and faunal evidence from Arabia for faunal migrations between Asia and Africa during the early Miocene. *Courier Forschungsinstitut Senckenberg* 69:189–98.

Wood, B. A. and Q. Xu. 1991. Variation in the Lufeng dental remains. *Journal of Human Evolution* 20:291–311.

Wu, R. and Y. Pan. 1984. A late Miocene gibbon-like primate from Lufeng, Yunnan Province. *Acta Anthropologica Sinica* 3:185–94.

——. 1985. Preliminary observations on the cranium of *Laccopithecus robustus* from Lufeng, Yunnan with reference to its phylogenetic relationship. *Acta Anthropologica Sinica* 4:7–12.

Zapfe, H. 1958. The skeleton of *Pliopithecus* (*Epipliopithecus*) *vindobonensis* Zapfe and Hürzeler. *American Journal of Physical Anthropology* 16:441–58.

——. 1961a. Ein Primatenfund aus der miozänen Molasse von Oberösterreich. *Zeitschrift für Morphologie und Anthropologie* 51:247–67.

——. 1961b. Die Primatenfunde aus der miozänen Spaltenfüllung von Neudorf an der March (Devinská Nová Ves), Tschechoslowakei. *Schweizerische Paläontologische Abhandlungen* 78:4–293.

——. 1969. Primates; pp. 1–16. *in* O. Kühn (ed.), *Catalogus Fossilium Austriae: Ein Systematisches Verzeichnis Alaler auf Österreichischem Gebiet Festgestellten Fossilien*, Springer-Verlag, Vienna.

——. 1989. Pongidenzähne (Primates) aus dem Pontien von Götzendorf, Niederösterreich. *Sitzungsberichte der Österreichen Akademie der Wissenschaften* 197:423–50.

——. 1991. *Mesopithecus pentelicus Wagner aus dem Turolien von Pikermi bei Athen, Odontologie und Osteologie*. Neue Denk schriften des Naturhistorischen Museums in Wien 5.

Zapfe, H. and J. Hürzeler. 1957. Die Fauna der miozänen Spaltenfüllung von Neudorf an der March (CSR). Primates. *Sitzungsberichte der Österreichischen Akademie der Wissenschaften, Abt. 1*, 166:113–23.

13

Eomyids and Zapodids (Rodentia, Mammalia) in the Middle and Upper Miocene of Central and Southeastern Europe and the Eastern Mediterranean

V. FAHLBUSCH AND T. BOLLIGER

Eomyids and zapodids are among the smallest of all rodents, during the Tertiary as well as today. This may be the main reason why they generally have been overlooked in the fossil record. Only with modern collecting and preparation methods (screen washing and acid preparation) have remains of these small mammals (mainly isolated teeth) become known at many localities; in some cases they even exceed other species and are not rare. Eomyids and zapodids represent two different superfamilies of myomorph rodents, each of which is known since the Eocene and which in details have very different evolutionary histories.

Eomyids are of little importance for faunal comparisons between middle and upper Miocene Central Europe and Eastern Mediterranean localities; zapodids are of even less importance. However, in both cases we have to consider that gaps in the distributions of these groups of rodents, both in time and space, may not be due to their actual absence but rather to an incomplete stratigraphic record. For this reason it seems appropriate to summarize our current knowledge about these groups in order to expedite further systematic investigations (see fig. 13.1 and tab. 13.1).

The Eomyidae

Essential phases in the history of eomyids are relatively well known from the Eocene and Oligocene (Uintan through Orellan) in the Western United States and in Europe (MP 25–30) (Fahlbusch 1979; Storer 1987). Recently, three genera of "European" eomyids (*Eomys*, *Eomyodon*, and *Pseudotheridomys*) have been described from China (Wang and Emry 1991). In Europe, eomyid evolutionary history can be followed continuously into the lower Miocene. They again reach a remarkable culmination in Central Europe during MN 3–4 with the genus *Ligerimys* and its relatives (including *"Ligerimys" lophidens* and other successors of *Pseudotheridomys*). Following the extinction of *Ligerimys* at the end of MN 4, eomyids are represented in Central Europe from the late early Miocene (MN 5) through the end of the upper Miocene (MN 13 and later) by two evolutionary lineages. These are represented mainly by the genera *Eomyops* (which may be a synonym of *Leptodontomys*) and *Keramidomys*. However, most species of both these genera are almost entirely documented by isolated teeth.

Both evolutionary lineages exhibit very few changes during this entire time span of ca. 10 m.y., at least as far as can be judged from tooth morphology. Throughout their duration, the species remain small to minute. As far as different species can be distinguished, these are very similar to each other and are best distinguished on the basis of size, less easily by differences in tooth crown morphology. However, for reliable evaluations of size and morphological variability at different sites, a minimum sample size is inevitably necessary; this condition is met at only a few localities. Next to these taxonomic-systematic problems there are large gaps in documentation for certain temporal intervals and areas, which even further hamper our understanding of evolutionary processes in the bioprovinces considered here. Our present knowledge will be summarized briefly in the following text.

The *Eomyops*–Group

In Europe this group is only represented by the genus *Eomyops*. The systematic relationship to the North American genus *Leptodontomys* has not yet been definitively

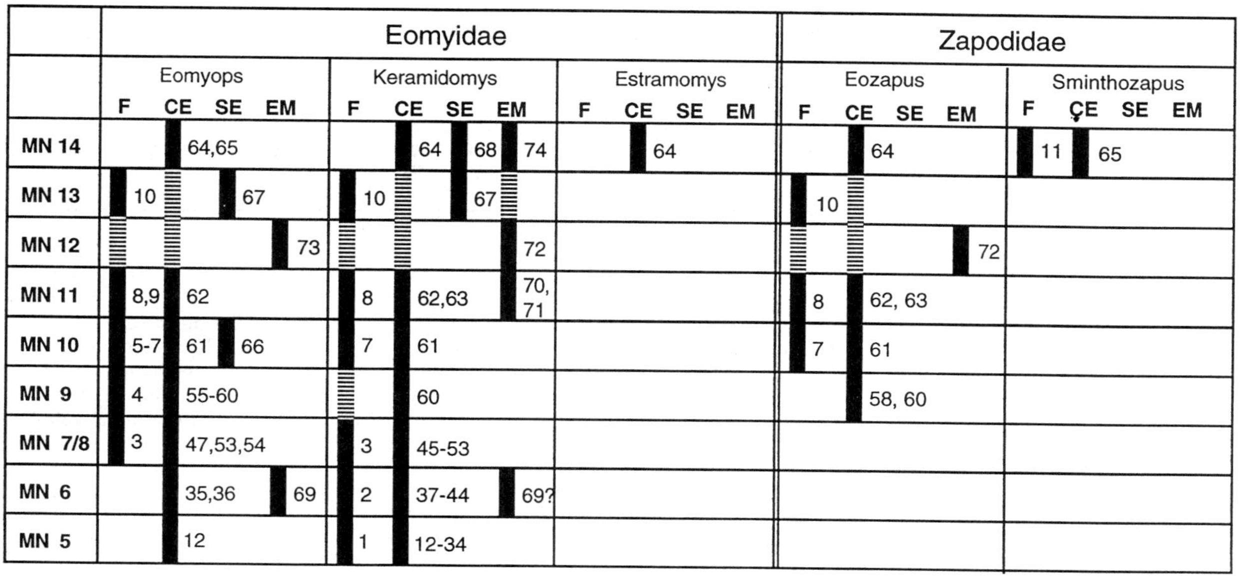

| | Eomyidae | | | | | | | | | | | | Zapodidae | | | | | | | |
| | Eomyops | | | | Keramidomys | | | | Estramomys | | | | Eozapus | | | | Sminthozapus | | | |
	F	CE	SE	EM	F	CE	SE	EM	F	CE	SE	EM	F	CE	SE	EM	F	ÇE	SE	EM
MN 14		64,65				64	68	74		64				64			11	65		
MN 13	10		67		10		67						10							
MN 12				73				72								72				
MN 11	8,9	62			8	62,63		70, 71					8	62, 63						
MN 10	5-7	61	66		7	61							7	61						
MN 9	4	55-60				60								58, 60						
MN 7/8	3	47,53,54			3	45-53														
MN 6		35,36		69	2	37-44		69?												
MN 5		12			1	12-34														

FIGURE 13.1 Distribution of Eomyidae and Zapodidae (numbers correspond with those of table 13.1). Hatched bars indicate a supposed continuity, especially in MN 12 and MN 13.

clarified; the same is true for its origin. New discoveries from the middle and upper Miocene of China (e.g., Fahlbusch et al. 1984; Qiu 1994) foster new doubt as to whether or not *Eomyops* and *Leptodontomys* are independent genera as originally suggested by Engesser (1979). Until this problem is resolved, we provisionally use the nomen *Eomyops* for middle and late Miocene (up to the Pleistocene) European representatives of this small form.

Eomyops can be traced in Central Europe from the upper Orleanian (MN 5) up to the upper Vallesian (MN 10). However, normally there are just a few specimens at just a few localities. For this reason, several details regarding the specific differentiation and phylogenetic relationships are still uncertain. *Eomyops* is somewhat more frequent in its occurrence only at some MN 9 localities, possibly due to ecological factors. Dorn-Dürkheim (Germany; MN 11) and Eichkogel (Austria; MN 11) are the only Turolian age localities in Central Europe that have yielded *Eomyops*. The lack of any MN 12–13 fossil record for this genus is certainly due to sampling error since almost no fossiliferous sediments are known of this age from Central Europe.

In both Southeastern Europe and the Eastern Mediterranean, *Eomyops* is known from only two localities, and even in these places it is very rare and not yet studied in detail. From our present knowledge about the distribution of *Eomyops*, no definite conclusions on biostratigraphy, biogeography, or ecology may yet be drawn. It may be expected that in the future the number of localities providing *Eomyops* will increase in Southeastern Europe as well as in the Eastern Mediterranean. Whether this will provide a clearer picture of this genus's biogeography and ecological tolerances remains, for the time being, unanswered.

The *Keramidomys*–Group

Representatives of this group, including *Keramidomys* and *Estramomys*, are also very small and do not exhibit major changes during their relatively long history. Only *Keramidomys pertesunatoi* and the basal form of the early Pliocene genus *Estramomys* show some major changes in what is otherwise a monotonous tooth morphology record. In comparison with *Eomyops*, the number of localities with *Keramidomys* is somewhat larger during the upper Orleanian (MN 5) and Astaracian (MN 6–8) of Central Europe. For drawing any conclusions about possible causes, however, the present documentation is still too poor. The remarkable decrease of *Keramidomys* beginning with the early Vallesian, and the lack of a Central European MN 12–13 record of the *Eomyops*–group, may be due to the scarcity of fossil localities and/or ecological factors.

In Southeastern Europe and the Eastern Mediterranean, material attributable to *Keramidomys* is so far rather poor and inadequately studied. Currently, it is surprising that, apart from the questionable finding at Çandir, the occurrence in Turkey only begins in the Turolian (MN 11), and even later (MN 13) in Southeastern Europe. However, once again, it should be understood that this observation is based on rather poor data and does not necessarily indicate an absence of *Keramidomys* during the Astaracian and Vallesian.

France (F):

1	Vieux Collonges MN5	2	Sansan MN6	3	La Grive MN7/8	4	Jujurieux MN9
5	Montredon MN10	6	Soblay MN10	7	Ambérieux 1 MN10	8	Ambérieux 2 MN10
9	Ambérieux 3 MN11	10	Lissieu MN13	11	Celleneuve MN14		

Central Europe (CE):

MN5

12	D-Massendorf	13	D-Undorf	14	D-Eitensheim	15	D-Adelschlag
16	D-Langenmoosen	17	D-Niederaichbach	18	D-Pöttmes	19	D-Rosshaupten
20	D-Schönenberg	21	D-Oggenhof	22	D-Gündlkofen	23	D-Gisseltshausen 1b
24	D-Puttenhausen	25	CH-Vermes 1	26	CH-Hüllistein	27	CH-Uerikon-Matt
28	CH-Hombrechtikon-Tobel	29	CH-Wald-Güntisberg	30	CH-Stäfa-Bürgistobel A	31	CH-Stäfa-Schliffitobel B
32	CH-Hombr.-Hotwiel	33	CH-Hombr.-Speerstr.	34	CH-Vermes 2		

MN6

35	D-Affalterbach	36	SO-Neudorf	37	D-Sandelzhausen	38	D-Unterneul 1a
39	SO-Neudorf-Spalte	40	D-Gallenbach 2b	41	D-Gallenbach 3a	42	CH-Schauenberg
43	CH-Chrüzbüel	44	CH-Bichelsee-Gerstel				

MN7/8

45	D-Schwabmünchen	46	D-Kleineisenbach	47	D-Giggenhausen	48	D-Markt Rettenbach
49	PL-Opole (Oppeln)	50	CH-Ergeten	51	CH-Grat	52	CH-Vermes 2
53	CH-Anwil	54	H-Felsötárkány				

MN9

55	CH-Nebelbergweg	56	D-Marktl	57	D-Hammerschmiede	58	A-Götzendorf
59	H-Rudabánya	60	H-Suchomasty				

MN10

61 A-Kohfidisch

MN11

62	D-Dorn Dürkheim	63	A-Eichkogel

MN12 -

MN13 -

MN14

64	PL-Podlesice	65	H-Osztramos 1

South-Eastern Europe (SE):

66	GR-Biodrak MN10	67	GR-Maramena MN13	68	GR-Maritsa MN14

Eastern Mediterranean (EM):

69	TR-Candir MN6	70	TR-Düzyalya 1 MN11	71	TR-Düzyalya 2 MN11	72	TR-Karaözü MN12
73	TR-Bayirköy MN12	74	TR-Igdeli MN14				

A: Austria, CH: Switzerland, SO: Slovakia, D: Germany, GR: Greece, H: Hungary, PL: Poland, TR: Turkey,

The genus *Estramomys* first appears in Central Europe during MN 14 and is of little importance for comparison with the Eastern Mediterranean. Although details of its origin are still unknown, the genus is likely to have originated from *Keramidomys* "somewhere" in the East. Its migration route toward Central and Western Europe was probably from more northerly latitudes; this may also be true for *Keramidomys*.

The Zapodidae

In contrast to the eomyids, which apparently have a Holarctic distribution, the zapodids are first known from the Oligocene of Asia (Wang 1985), and later during the Neogene apparently extended their range into Western Eurasia. It should be qualified that this observation is based on limited Oligocene-Neogene fossil material. Zapodid species are documented to have undergone several migratory extensions from their Asiatic centers of origin to establish new populations in Europe and North America.

Comparison of zapodid evolution between Central Europe and the Eastern Mediterranean is relevant only from the Vallesian onward because after the extinction of the *Plesiosminthus*–group in the European Agenian, zapodids were unknown from MN 3 to MN 7/8 (inclusive) in Europe. From the Vallesian onward zapodids are still very rare. Moreover, it is remarkable that in eastern parts of Central Europe zapodids occur slightly more frequently than in the other areas of this region. Occasionally, they also reached Western Europe, but remain as rare faunal elements there as well. According to recent investigations by Fahlbusch (1992:214) all later Miocene European species belong to the genus *Eozapus*, which is still living in China. Additionally, the genus *Sminthozapus* is known to occur in MN 14, but rarely. It is still conjectural whether *Sminthozapus* is a valid taxon or referable to *Eozapus* (cf. Fahlbusch 1992:215).

No zapodids are known to occur from either Southeastern Europe or the Eastern Mediterranean. While this may be due to gaps in our knowledge, it should be recognized that more than two decades of intensive research, using modern collecting techniques, have failed to uncover evidence of zapodids from Southeastern Europe and Turkey. Again we need to consider whether ecological factors prevented the extention of zapodids into Asia Minor and Southeastern Europe during the upper Miocene. Immigrations to Central and Western Europe consequently must have taken place via more northern latitudes. A detailed ecological explanation is presently not possible.

Conclusions

Analysis of eomyid and zapodid rodents in the middle and upper Miocene of Central and Southeastern Europe and the Eastern Mediterranean is still very limited. An important reason is that these rodents are generally very rare in fossil assemblages, and rich faunas are necessary to retrieve members of these taxa. The extremely rich fauna from Ertemte, Inner Mongolia (Fahlbusch 1992; > 10–12,000 specimens), contained less than 3% zapodid specimens (calculated from the total number of isolated mammal molars). In certain well-documented European micromammal faunas eomyids are represented by less than 0.5%. This demonstrates that many of our micromammal samples treated here are probably much too small to contain any eomyid or zapodid specimens. Another reason is that eomyids and zapodids are very small and may be easily overlooked during screening activities. The current generally practiced collecting methods (screen washing with appropriate screens and concentration of residuals by acidpreparation) assures that this has not been a factor for localities collected during the last thirty years. Yet another reason for zapodid and eomyid spotty records may be the absence or scarcity of fossiliferous sediments for certain time units and geographic areas. This is true in Central Europe for major portions of the Vallesian and Turolian ages.

Given the poor eomyid and zapodid fossil record, any interpretation about their chronologic and biogeographic range must be considered provisional. Eomyid and zapodid distribution may have been influenced by a number of factors beyond their intrinsic scarcity. *Eomyops* and *Keramidomys* are documented in Central Europe for most of the middle and upper Miocene (except in MN 12–13, where no micromammal localities are known at all). Whereas *Keramidomys* is more frequent during MN 5–7/8, and later on decreases, *Eomyops* first occurs in MN 5 but becomes somewhat more frequent in the Vallesian. Both genera are very rare or missing during the Astaracian and Vallesian in Southeastern Europe and the Eastern Mediterranean, and only *Keramidomys* is somewhat better represented after MN 11.

Nothing is known about *Keramidomys*'s center of origin. A broad Eurasian distribution of *Eomyops* is indicated from Europe and the middle and upper Miocene of China. Its poor documentation in Asia Minor (thus far only from two localities), and the slightly more frequent occurrence of *Keramidomys* in that area, may indicate that the species belonging to these genera had very particular, and probably different, ecologic adaptations. This conclusion is apparent in the complete absence of zapodids from the Eastern Mediterranean fossil record. Zapodids, as well as the *Eomyops*–group, appear to have used more northern migratory routes from the East into Central and Western Europe, undoubtedly because of ecological factors. More specific interpretations of eomyid and zapodid biogeographic distribution or ecological requirements is currently not possible with the data at hand.

LITERATURE CITED

Bolliger, T. 1992. Kleinsäugerstratigraphie in der miozänen Hörnlischüttung (Ostschweiz). *Documenta naturae* 75:1–296.

Engesser, B. 1979. Relationships of some insectivores and rodents from the Miocene of North America and Europe. *Bulletin of the Carnegie Museum of Natural History* 14:1–68.

———. 1990. Die Eomyidae (Rodentia, Mammalia) der Molasse der Schweiz und Savoyens. *Schweizerische Paläontologische Abhandlungen* 112:1–144.

Fahlbusch, V. 1979. Eomyidae: Geschichte einer Säugetierfamilie. *Paläontologische Zeitschrift* 53:88–97.

———. 1992. The Neogene mammalian faunas of Ertemte and Harr Obo in Inner Mongolia (Nei Mongol), China. 10. *Eozapus* (Rodentia). *Senckenbergiana lethaea* 72:199–217.

Fahlbusch, V., Z. Qiu, and G. Storch. 1984. Neogene micromammal faunas from Inner Mongolia: Recent investigations on biostratigraphy, ecology, and biogeography. In *The Evolution of the East Asian Environment*, ed. R. O. Whyte, 2:697–707. Hong Kong: University of Hong Kong.

Qui, Z. 1994. Eomyidae in China. *National Science Museum Tokyo Monographs* 8:49–55.

Storer, J. E. 1987. Dental evolution and radiation of Eocene and early Oligocene Eomyidae (Mammalia, Rodentia) of North America, with new material from the Duchesnean of Saskatchewan. *Dakoterra* 3:108–17.

Wang, B. 1985. Zapodidae (Rodentia, Mammalia) from the lower Oligocene of Qujing, Yunnan, China. *Mainzer geowissenschaftliche Mitteilungen* 14:345–67.

Wang, B. and R. J. Emry. 1991. Eomyidae (Rodentia: Mammalia) from the Oligocene of Nei Mongol, China. *Journal of Vertebrate Paleontology* 11:370–77.

14

The Middle and Late Miocene Record of the Sciuridae and Petauristidae in France, Central Europe, Southeastern Europe, and Anatolia

H. DE BRUIJN AND P. MEIN

Sciuroidea are a diverse, possibly polyphyletic, superfamily containing genera that are adapted to very different modes of life. The evolutionary history of the superfamily is insufficiently known to verify whether the three adaptive types—the flying squirrels, the tree squirrels, and the ground squirrels—have originated only once. The main point here is that most specialists agree when it comes to recognizing the adaptive types even on the basis of isolated teeth. In Eurasia the flying squirrels range from the lower Oligocene to Recent, the ground squirrels from the middle Oligocene to Recent, and the tree squirrels from the lower Miocene to Recent. The available evidence suggests that the flying squirrels have descended from an aplodontid stock, the ground squirrels from a paramyid stock, and the tree squirrels from the ground squirrels.

The Fossil Record of the Sciuroidea

Table 14.1 summarizes the ranges of the seven genera of flying squirrels, the five genera of ground squirrels, and the two genera of tree squirrels recognized in Europe and Anatolia during the Neogene. *Tamias* is listed as a tree squirrel, and *Csakvaromys* Kretzoi is considered to be a synonym of *Spermophilinus*. The genera *Palaeosciurus* and *Spermophilinus* are very similar in the shape of the lower jaw as well as in dental morphology. Such differences as the slightly better developed entoconid and the occurrence of a better developed protoconid-anterolophid connection in the lower molars of *Palaeosciurus* are hardly appropriate for generic separation. However, when all the available data are considered in their stratigraphical context, *Palaeosciurus* and *Spermophilinus* seem to represent different groups of squirrels. *Palaeosciurus goti* from the middle Oligocene is small, *P. feignouxi* from the lower Miocene is

large, and *P. fissurae* and *P. sutteri* from the middle Miocene are of medium size. These latter occur in the same assemblages as the much smaller *Spermophilinus* teeth (*S. besana*, ±MN 3–MN 6). In associations later than MN 6 there is again only one sciurid species with *Palaeosciurus/Spermophilinus* morphology per locality. The teeth of the late Neogene species show gradual size increase through time and seem to represent one lineage (*Spermophilinus*).

Since there are reasons to maintain the two genera and since there is no confusion in the literature, we have decided to enter *Palaeosciurus* and *Spermophilinus* as separate genera on the tables. It should be realized however, that generic identification is possible only when the age of the specimens is roughly known.

Incomplete though the fossil record may be, it seems that the essence of the picture will not change by making it more complete. The theoretically expected negative correlation between flying squirrels, indicating forests with tall trees, and ground squirrels, indicating open landscapes, is unlikely to appear other than in single localities. This probably means that there were forests as well as open landscapes in each of the areas recognized throughout the middle and late Miocene.

Table 14.1 gives the ranges of the Eurasian Neogene genera of the Sciuroidea for different geographical areas from MN 6 to MN 14. Zones between subsequent occurrences in the same area, where the genus in question is not (yet) known are indicated with a dot. The areas are: France; Central Europe, encompassing Germany, Switzerland, Italy, Austria, Hungary, Romania, and Poland; Southeastern Europe, containing the Greek mainland and Turkey north of the Dardanelles and Anatolia. The definition of these areas has been a matter of convention and is not based on the supposition that they are bioprovinces. The

TABLE 14.1 *Ranges of the Neogene Eurasian genera of the Sciuroidea in France (Fr.), Central Europe (C.E.), Southeastern Europe (S.E.E.), and Anatolia (An.). The presence of a genus in a specific zone and area is indicated by a line. If a genus has not been found, but is known from earlier and later zones of the same area, the column concerned is indicated with a square.*

(In the tables below a line range is shown as ▬ and a square as ▪; a blank cell is an empty column.)

Petauristidae

	MN	6	7/8	9	10	11	12	13	14
Aliveria	Fr.								
	C.E.								
	S.E.E.	▬							
	An.								
Blackia	Fr.	▬	▬	▬	▬	▬	▪	▪	▬
	C.E.	▬	▬	▬	▪	▬	▪	▪	▬
	S.E.E.								
	An.								
Miopetaurista	Fr.		▬	▪	▪	▪	▪	▪	▬
	C.E.	▬	▬	▬	▪	▬			
	S.E.E.	▪	▪	▪	▪	▪	▪	▬	
	An.		▬						
Albanensia	Fr.	▬	▬						
	C.E.		▬	▬					
	S.E.E.								
	An.		▬						
Forsythia	Fr.		▬						
	C.E.	▪	▬	?					
	S.E.E.								
	An.								
Pliopetaurista	Fr.				▬	▬	▪	▬	
	C.E.				▬	▬	▪	▪	▬
	S.E.E.				▬	▪	▪	▬	
	An.					▬			
Hylopetes	Fr.				▬	▪	▪	▪	▬
	C.E.		▬	▪	▪	▬	▪	▬	
	S.E.E.							▬	
	An.					▬			

Sciuridae

	MN	6	7/8	9	10	11	12	13	14
Atlantoxerus	Fr.								
	C.E.							▬	
	S.E.E.		▬						
	An.		▬	▬	▬	▪	▪	▪	▬
Heteroxerus	Fr.	▬	▬	▬	▬				
	C.E.	▬							
	S.E.E.								
	An.								
Paleosciurus	Fr.								
	C.E.	▬							
	S.E.E.								
	An.	▪?▪							
Spermophilus	Fr.								
	C.E.							▬	
	S.E.E.								
	An.								
Spermophilinus	Fr.	▬	▬	▬	▬	▬	▬	▬	
	C.E.	▬	▬	▬	▬	▬			
	S.E.E.	▬	▬	▬	▪	▬	▪	▬	
	An.	▬	▬	▬	▪	▬	▪	▪	▬
Tamias	Fr.		▬						
	C.E.								▬
	S.E.E.							▬	
	An.	▬	▪	▪	▬	▪	▪	▪	▬
Sciurus	Fr.								
	C.E.								▬
	S.E.E.								
	An.								

localities on which the data presented in table 14.1 are based are given in table 14.2 in order to enable the tracing of particular occurrences.

Conclusions

The erratic ranges of the various genera in the different areas show a confusing pattern. Obviously, many factors determine the absence or presence of a particular genus. The age of the assemblage and the biotope in which it lived do not play a dominating role among these factors because there seems to be no correlation between the presence of, for instance, the Petauristidae and the diversity peaks of rodents (de Bruijn et al., this volume) or the presence of ground squirrels. Some genera appear to have long stratigraphical ranges in all the areas considered, i.e., *Spermophilinus*. This genus is clearly an ubiquist. If we leave *Spermophilinus* and *Tamias* unconsidered, because they probably did not require very specific biotopes, and disregard genera that are either known from only one of the areas considered (*Aliveria*), disappear after MN 6 (*Palaeosciurus*), or appear in MN 14 (*Sciurus*), the picture becomes less complex, but not more easy to understand. The genus *Blackia*, a flying squirrel known exclusively from France and Central Europe during the time interval considered, probably had a much wider geographical range, because it is known from older beds in Southeastern Europe (Aliveri). This example suggests that the fossil record is very incomplete at this moment and therefore deceptive.

TABLE 14.2 *Localities that have yielded the data on which table 14.1 is based.*

France	Central Europe	S.E. Europe	Anatolia	MN
Hautimagne Celleneuve Hauterives	Podlesice Ciuperceni (inf.)	Ptolemais	Iğdeli Kangal Maritsa	14
La Tour	Brisighella Polgardi	Maramena	Amasya	13
Cucuron—stade		Pikermi(chomateri)	Bayirköy	12
Mollon Amberieu 3	Dorn—Dürkheim Eichkogel	Samos 1,4	Düzyayla 1,2	11
Dionay Amberieu 1,2 Soblay Montredon	Kohfidisch	Lefkon	Bayraktepe 2 Dendil, Karaözü	10
Doué—la—Fontaine Priay 2 Jujurieux	Götzendorf	Mahmutköy Kastellios	Yassiören	9
La Grive L7, M, L3	Anwil Opole	Pişmanköy Kalamiş	Yeni Eskihisar Bayraktepe 1	7+8
Sansan Luc—sur—Orbieu	Steinberg Sandelzhausen Ziemetzhausen Ebershausen	Hoşköy Karydia	Çandır 2 Paşalar	6

Sweeping conclusions cannot be drawn until the available information is much more complete. However, the co-occurrence of flying squirrels and ground squirrels in many zones of most areas suggests that the biotope was always diverse. The absence of *Heteroxerus*, which is a very common animal in Southwestern Europe, in Anatolia suggests that it never came so far east. This is peculiar because the closely related xerine *Atlantoxerus* apparently had a circum-Mediterranean dispersal.

The ecological niches occupied by the flying squirrels *Miopetaurista* and *Albanensia* during MN 6–9 seem to be taken by the immigrants *Pliopetaurista* and *Hylopetes* in the upper part of the Neogene.

Tamias, a genus that was considered to be of North American origin until recently, is probably an Asiatic animal that expanded its geographical range during the late Neogene.

Acknowledgments

We thank Dr. G. Daxner-Höck (Vienna), Prof. Dr. V. Fahlbusch (Munich), and Prof. Dr. K. Heissig (Munich) for helping us to assemble the record of the Sciuroidea in Central Europe. We thank Dr. D. F. Mayhew for critically reading the manuscript, Mrs. A. E. de Bruijn-Dudok van Heel for typing it, and Mr. T. van Hinte for making the text figures. The first author (H. de Bruijn) remembers with gratitude the hospitality and friendship of Greek and Turkish institutions and colleagues extended during many field seasons. This study was financially supported by the NATO Scientific Affairs Division, Brussels, Belgium (Grant no. CRG 910750) and by the Faculty of Earth Sciences of the University of Utrecht, the Netherlands.

LITERATURE CITED

Bruijn, H. de, E. Unay, and L. van den Hoek Ostende. This volume. The composition and diversity of small mammal associations from Anatolia through the Miocene.

Bruijn, H. de, A. J. van der Meulen, and G. Katsikatsos. 1980. The mammals from the lower Miocene of Aliveri (island of Evia, Greece). Part I. The Sciuridae. *Proceedings, Koninklijke Nederlandse Akademie van Wetenschappen, B* 83:241–61.

15

Middle and Late Miocene Common Cricetids and Cricetids with Prismatic Teeth

V. FAHLBUSCH

Cricetid rodents are among the most abundant small mammals in early and middle Miocene time, before murids in the upper Miocene and arvicolids in the Pliocene became dominant. But even during the late Miocene there are several generically diverse cricetid groups that characterize certain MN units and geographic areas.

However, a growing problem concerns the taxonomic meaning of terms such as "Cricetidae," "Cricetinae," and "Cricetini" or "cricetid" and "cricetine" rodents. The very content of a family "Cricetidae" and its systematic subdivision is becoming ever more dubious. This is due, at least in part, to the enormous increase of fossil genera, most of which are known only from dentitions or isolated teeth. Further difficulties arise in recognizing parallel evolution between lineages that share a generally simple and unspecialized dental pattern. During the last few years, mainly in paleontological literature, several new "families" and taxa of lower rank have been established on the basis of tooth morphology, sometimes with conflicting meaning. For these reasons a general and easily obtainable definition of these systematic units is no longer possible.

A solution to the taxonomic problems surrounding the "Cricetidae" is far beyond the goal of this brief review. Considering the present amount of material it is also beyond the capability of a single specialist. And it is especially impossible to solve the systematic problems by subdivision of what has once been "Cricetidae" from only a paleontological point of view. Therefore in this review I have attempted to make a very informal grouping of cricetid-like muroid rodents from the time period and geographic area under consideration. In most cases these groups probably do not form natural units. They are used just for easier communication. The groups are the following (fig. 15.1, tab. 15.1):

1. The group of cricetids that are common in the later Orleanian and Astaracian interval from Central and Western Europe, partly ranging up to the Pleistocene, and mostly unspecialized in tooth morphology: *Demo-cricetodon, Megacricetodon, Kowalskia, Eumyarion,* and, provisionally included, *Lartetomys.*

2. The group of relatives, predominantly fossil ones, of the recent Eurasian dwarf hamsters, slightly simplified in dental morphology: *Cricetus* s.l., *Cricetulus, Allocricetus, Mesocricetus, Calomyscus.*

3. Cricetids with a conservative tooth morphology but characterized by increasing crown height: *Collimys, Hypsocricetus.*

4. Gerbillid-like cricetids: *Epimeriones, Pseudomeriones.*

5. Microtoid cricetids and true arvicolids: *Microtocricetus, Pannonicola, Baranomys/Microtodon, Bjornkurtenia, Promimomys.*

The cricetid genera *Cricetodon, Deperetomys, Hispanomys, Ruscinomys,* and *Byzantina,* which may be considered "Cricetodontinae" or "Cricetodontini," are excluded from this review. They have been reviewed by de Bruijn et al. (1993) and de Bruijn and Ünay (this volume).

Middle Miocene to Early Pliocene Common Cricetids in Central and Southeastern Europe and Asia Minor

Among cricetids of group 1, as characterized above, the genus *Eumyarion* displays a rather archaic tooth morphology and resembles some Oligocene cricetids much more than the general Miocene genera. Probably it is descended from those Oligocene (European?) ancestors and appears in Central Europe during the middle Orleanian (MN 4), approximately at the same time as *Democricetodon* and *Megacricetodon. Eumyarion* is present in many faunas up to the end of the Astaracian although only seldomly as frequently as the other two genera. In the Vallesian, *Eumyarion* becomes rather rare and its last occurrence is Suchomasty (MN 10).

Eumyarion is absent in Southeastern Europe and Turkey beginning with the upper Orleanian (MN 5). This may partly be due to poor documentation. On the other hand,

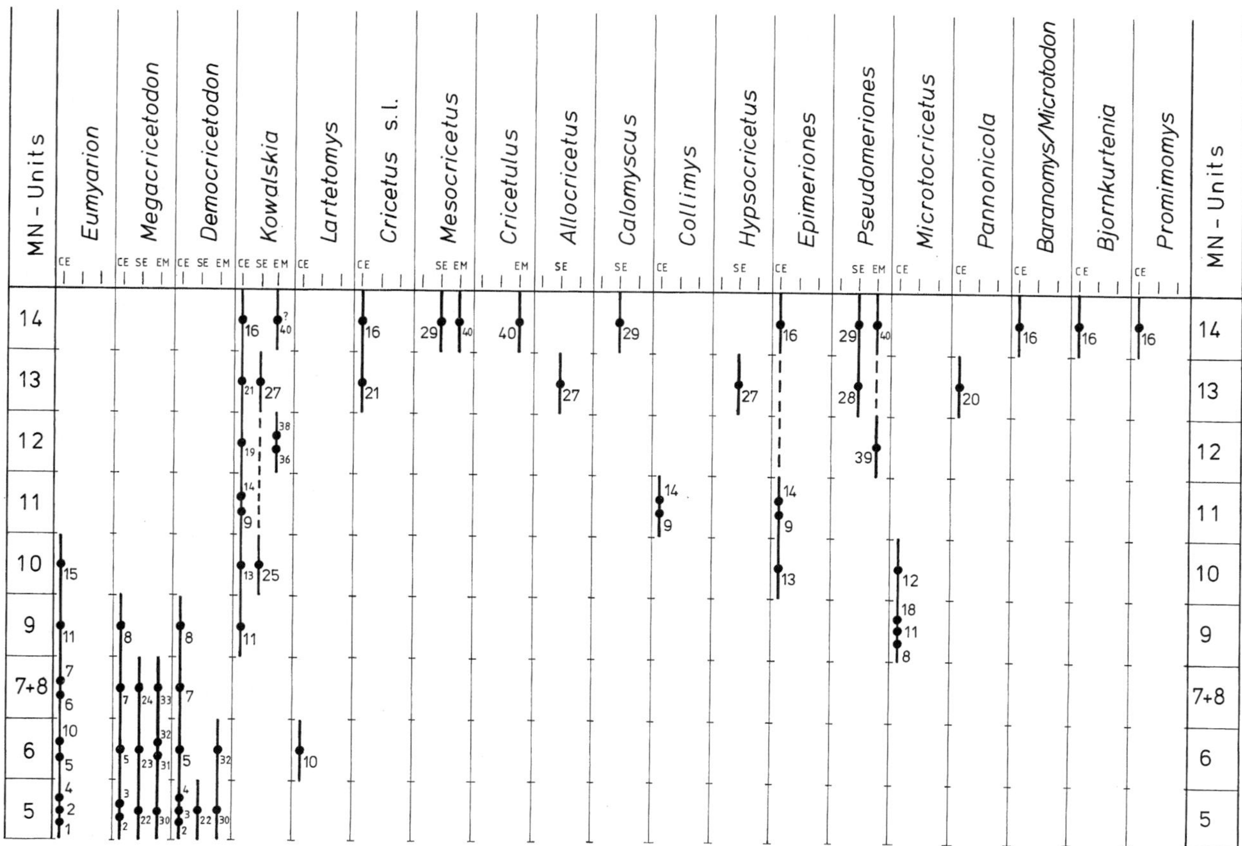

FIGURE 15.1 Stratigraphic and regional distribution of common cricetids and cricetids with prismatic teeth in Central and Southeastern Europe and in Asia Minor. CE = Central Europe, SE = Southeastern Europe, EM = Eastern Mediterranean/ Asia Minor.

the genus is well known in these areas in the early Neogene up to MN 4 (Klein Hofmeijer and de Bruijn 1988; de Bruijn and Sarac 1991). It is quite possible that species of *Eumyarion* migrated north- and northwestward during the Orleanian and later disappeared in Southeastern Europe and Asia Minor. Whether climatic-ecological reasons were responsible may be possible but cannot yet be decided for *Eumyarion*, and not on the basis of just this genus. Considering the large distribution of the genus *Eumyarion* through time and space it seems impossible that all these occurrences should represent the same ecological conditions. This means that the single species of *Eumyarion* must have been ecologically differentiated; this assertion cannot be determined yet with certainty.

Democricetodon and *Megacricetodon* are widespread during the later Orleanian and the entire Astaracian in Central Europe; not until early Vallesian time are they rare before becoming extinct. At many other localities, however, they are the most common micromammals and are relatively well studied. Due to their rapid development in several evolutionary lineages they are useful guide fossils in several areas. Contrarily, the distinction of many species described so far is sometimes rather difficult because the

crown morphology of the cheek teeth does not change much. So in many cases absolute size is a very important feature for species descrimination. We do know, however, about the evolutionary increase and decrease in size. Consequently, the history of these two genera is very complex, and is far from completely known. Comparison of single species from different areas is therefore very tenuous.

The ecological characterization of *Democricetodon* and *Megacricetodon* is similar to *Eumyarion* in that both genera have a broad geographic and chronologic distribution and it is quite certain that all of the many localities where they occur do not represent the same ecological conditions. However, detailed investigations (at least for Central Europe) in which the ecology of these taxa is evaluated have not yet become available. Moreover, it is certain that recent results for distinct species, such as those from Spain, may not be extrapolated for Central European species. For these reasons, it is presently not possible to compare ecological conditions between Central Europe, Southeastern Europe, and Turkey using species of *Democricetodon* and *Megacricetodon*. There are taxonomic and systematic problems in those areas because several new discoveries have not been studied in detail and published. We do

TABLE 15.1 *List of Localities*

Germany/Switzerland
 1. Langenmoosen (MN 5)
 2. Vermes 1 (MN 5)
 3. Puttenhausen (MN 5)
 4. Massendorf (MN 5)
 5. Sandelzhausen (MN 6)
 6. Steinheim (MN 7)
 7. Anwil (MN 8)
 8. Hammerschmiede (MN 9)
 9. Dorn-Dürkheim (MN 11)
Austria
10. Neudorf (Spalte) (MN 6)
11. Götzendorf (MN 9)
12. Neusiedel (MN 10)
13. Kohfidisch (MN 10)
14. Eichkogel (MN 11)
Czechian Republic/Poland
15. Suchomasty (MN 10)
16. Podlesice (MN 14)
Hungary
17. Hasznos (MN 6)
18. Rudabánya (MN 9)
19. Tardosbanya (MN 12)
20. Jaszladany (MN 13)
21. Polgardi (MN 13)
Greece and Turkey West of the Dardanelles
22. Mürefte (MN 5)
23. Pismanköy (MN 6)
24. Kalamis (MN 7/8)
25. Lefkon (MN 10)
26. Pikermi 4 (MN 12)
27. Maramena (MN 13)
28. Monasteri (MN 13)
29. Maritsa (MN 14)
Turkey (Asia Minor)
30. Gemerek (MN 5)
31. Paşalar (MN 6)
32. Çandir (MN 6)
33. Yeni Eskihisar (MN 7/8)
34. Bayraktepe 1 (MN 7/8)
35. Bayraktepe 2 (MN 10)
36. Düsyayla 1, 2 (MN 12)
37. Bayirköy (MN 12)
38. Kaleköy (MN 12)
39. Karaözü (MN 12)
40. Igdeli (MN 14)

not presently know why *Megacricetodon* persists somewhat longer in Central Europe than *Democricetodon* and why *Democricetodon* disappears in the Eastern Mediterranean in MN 6 (or in MN 5 of Southeastern Europe). It is possible that paleontological sampling is not yet sufficient in all areas to understand these patterns of extinction.

The genus *Kowalskia* has been reviewed recently by Daxner-Höck et al. (this volume). This genus's history in the upper Miocene reflects what is seen in the late Orleanian and Astaracian for *Democricetodon* and *Megacricetodon*. With the exception of Maramena (Daxner-Höck 1992), only a few localities in Southeastern Europe and Turkey have provided scanty material of *Kowalskia*, and here determinations are often uncertain and/or detailed studies of the material are lacking. There is as little known about the ecology of *Kowalskia* as for the other genera.

The genus *Lartetomys* is considered here only for the sake of completeness and adds nothing to our understanding of cricetid relationships. *Lartetomys* is known from Central (and Western) Europe only, and in Central Europe only by a few teeth from Neudorf. Its relationship to other cricetids is unknown.

Modernized Cricetids of the Later Neogene

The second group of cricetids comprises genera that are probably much more closely related to each other than the older taxa discussed above. The "modernized" cricetids are all characterized by a secondarily simplified dental morphology. It has been suggested that all these cricetids have an Asian origin, possibly close to some form of *Democricetodon*. However, they are probably not a monophyletic group. Most of these genera are still extant or closely related to extant taxa. "Modernized" cricetids first appear during MN 13, or even later, and are known from only a few localities, often by poor material. Therefore, it is precarious to discuss the regional differentiation within this group even now and, moreover, to seek ecologic explanations for their evolution.

A third group, including the genera *Collimys* and *Hypsocricetus*, displays dental morphologies that may be derived compared to those discussed for groups 1 and 2, but differ from them by a certain tendency to increase crown height. However, their dentitions remain completely rooted and may be termed as being progressive-brachydont forms. The two genera are not related to each other. *Collimys* is known by scanty material from only a few Central European MN 10–11 and MN 14 localities; possibly its history spans back to the Central European middle Miocene. Until now, *Hypsocricetus* is known only from Maramena, Greece (Daxner-Höck 1992). It is similar to some group 2 genera, but it differs from each of them by a somewhat increased crown height.

The dental morphology seen in *Pseudomeriones* and *Epimeriones* corresponds to that of recent gerbillines and may well be derived from the cricetid pattern. Whereas *Pseudomeriones* first occurs in the middle Turolian and is missing in Central Europe, *Epimeriones* is known beginning with late Vallesian, but is restricted to Central Europe. The gaps in geographic and temporal distribution may be due to sampling. Both genera would appear to have an Asian origin.

Microtoid Cricetids and True Arvicolids

The fifth group of Neogene cricetids comprises genera that display certain similarities in cheek tooth morphology to the arvicolids. They were or are still partly assigned to this family. The oldest and only true arvicolid in the area under consideration, *Promimomys*, first occurs in Central Europe during MN 14, as do the other genera. The *Baranomys/Microtodon* and *Bjornkurtenia* lineages are morphologically very similar to arvicolids and may be close to forms from which actual arvicolids descended. On the other hand, they show dental features that are still more reminiscent of common late Miocene cricetids than of arvicolids. But it probably remains a matter of personal taste to which family these genera are attributed.

It has not yet been determined whether the European *Baranomys* is identical with the Asiatic *Microtodon*; there are strong arguments for that on the basis of extensive new material from the type locality of Ertemte. Certainly these are primarily Asiatic faunal elements. The same seems to be the case with *Bjornkurtenia*, which is closely related to *Baranomys/Microtodon* (as well as to *Celadensia* from Spain). All these genera have the following characters in common: molars with pillar-shaped cusps, a distinct trend toward higher-crowned teeth, and retention of the basic cricetid pattern (such as seen in *Democricetodon*).

At first sight, *Microtocricetus* would appear to be a similar genus, and it may be, as for previously discussed genera, of Asiatic origin. Meanwhile, it has been found at a few localities. It is a Central European immigrant characteristic for MN 9–10. *Microtocricetus* displays the same somewhat elevated pillar-shaped main cusps as *Microtodon*, but in a very different arrangement. *Microtocricetus*'s morphology is too apomorphic to be ancestral to later arvicolids. Rather, it represents the result of an early arvicolid-like specialization independently derived from middle Miocene Asian cricetids. It cannot yet be determined whether the genus *Pannonicola* is closely related to *Microtocricetus*; it is poorly known from MN 13 only. Similarly, the latter genus has been considered to be an early arvicolid (Kretzoi 1965), but it may well be a specialized descendant of a cricetid that early displayed a microtoid-like dental morphology.

None of these microtoid cricetids has been found either in Southeastern Europe or Asia Minor. However, it should be kept in mind that most of these taxa are normally very rare faunal elements.

Conclusions

Cricetid rodents are generally considered to be particularly useful for biochronologic correlation because of their relative abundance, wide biogeographic distribution, and partly rapid evolution. These characteristics are exemplified in several regions. However, it becomes increasingly evident that comparisons between cricetid taxa are valid only for restricted biogeographic areas. The example discussed in this contribution, Central Europe in comparison to Southeastern Europe and Asia Minor, permits the following generalizations:

1. Cricetid genera with less differentiated molar patterns are often widely distributed and rather abundant. However, apart from absolute size they exhibit few diagnostic features that are reliable at the species level.

2. Without determination of species and their ecologic differentiation, it is impossible to give evidence about ecologic factors. Genera such as *Democricetodon*, *Megacricetodon*, or *Kowalskia* are useless in this respect.

3. More specialized forms may be useful for chronologic correlation and ecologic specification already at the generic level. Examples therefore may be *Collimys*, *Hypsocricetus*, or *Microtocricetus*. However, they are known from only one or very few localities, and not from different areas.

4. In view of the fairly incomplete state of knowledge, application of cricetid evolutionary history for interregional chronologic or ecologic comparisons is still rather limited.

Acknowledgments

A summary article such as this is the result of collecting data, opinions, arguments, and criticism from colleagues on various occasions, such as meetings, field trips, and symposia, and all this over a long period of time. Not all colleagues can be mentioned here; however, the ones listed here particularly helped to shape my views. Especially appreciated is the manifold support of Dr. H. de Bruijn, Utrecht; Dr. G. Daxner-Höck, Vienna; Dr. B. Engesser, Basel; Professor Dr. O. Fejfar, Prag; Dr. G. Storch, Frankfurt a.M.; Dr. Wu Wenyu, Beijing.

LITERATURE CITED

Bruijn, H. de, V. Fahlbusch, G. Sarac, and E. Ünay. 1993. Early Miocene rodent faunas from the Eastern Mediterranean area. Part III. The genera *Deperetomys* and *Cricetodon*, with a discussion of the evolutionary history of the Cricetodontini. *Proceedings, Koninklijke Nederlandse Akademie van Wetenschappen, B* 96:151–216.

Bruijn, H. de and G. Sarac. 1991. Early Miocene rodent faunas from the Eastern Mediterranean area. Part 1. The genus *Eumyarion*. *Proceedings, Koninklijke Nederlandse Akademie van Wetenschappen B* 94:1–32.

Bruijn, H. de and E. Ünay. This volume. On the evolutionary history of the Cricetodontini from Europe and Asia Minor and its bearing on the reconstruction of migrations and the continental biotope during the Neogene.

Daxner-Höck, G. 1992. Die Cricetinae aus dem Obermiozän von Maramena (Mazedonien, Nordgriechenland). *Paläontologische Zeitschrift* 66:331–367.

Daxner-Höck, G., V. Fahlbusch, L. Kordos, and W. Wu. This volume. The late Neogene cricetid rodent genera *Neocricetodon* and *Kowalskia*.

Klein Hofmeijer, G. and H. de Bruijn. 1988. The mammals of the lower Miocene of Aliveri (island of Evia, Greece). Part 8. The Cricetidae. *Proceedings, Koninklijke Nederlandse Akademie van Wetenschappen, B* 91:185–204.

Kretzoi, M. 1965. *Pannonicola* n.g. n.sp., ein echter Arvicolide aus dem ungarischen Unterpliozän. *Vertebrata Hungarica* 7:131–39.

16

The Late Neogene Cricetid Rodent Genera
Neocricetodon *and* Kowalskia

G. DAXNER-HÖCK, V. FAHLBUSCH, L. KORDOS, AND W. WU

Until the late 1960s there were only a few publications describing an enigmatic group of Pliocene and late Miocene hamster-like rodents that exhibited some dental characters reminiscent of the broadly distributed and commonly occurring early and middle Miocene genera *Democricetodon* and *Megacricetodon*. Subsequently, they were discovered to range geographically from Spain to China and chronologically from the early late Miocene to Pleistocene. These species are principally referred to the genera *Kowalskia* and/or *Neocricetodon*. Knowledge of this group of fossils has been primarily based on analyses of cheek tooth morphology and still leaves many systematic problems unresolved. These problems originate from the frequent lack of sufficient documentation, and in several cases are made more difficult by the increased number of localities. The evolutionary history of these animals is becoming ever more complex, and they may eventually prove to be a polyphyletic group.

Additionally, several nomenclatorial difficulties threaten to hamper our understanding of this group and our communication about it. This is even more regrettable since these genera have the potential for providing long-distance chronologic correlations and biogeographic reconstructions. Therefore, in this treatment we seek to clarify the nomenclatorial problems inherent in using the generic names *Neocricetodon* and *Kowalskia*. We also discuss the specific generic content of what we presently wish to recognize as being *Kowalskia*, and provide a revised generic diagnosis and specific differential diagnoses for all recognized members of this group.

Historical Review and Nomenclatorial Discussion

When listing new fossil occurrences from excavations in the Esterházy cave (Csákvár), Kadic and Kretzoi (1930:49) reported a rodent named "*Neocricetodon schaubi* Kretzoi

(n. gen. n. sp.)." However, the authors provided no description, measurements, figures of specimens, or any indication of this taxon's abundance at the type locality. According to the rules of systematic nomenclature, both the genus and species names were insufficiently defined to allow *Neocricetodon schaubi* to stand as a valid taxon; both names are nomina nuda.

Schaub (1934:26) established a new cricetid genus for what Young (1927:26) had earlier described as "*Cricetulus grangeri*." In 1930 (p. 46) Schaub had already noted that this species did not belong to *Cricetulus*. However, it was in 1934 that Schaub provided more detailed arguments for separating the species from *Cricetulus* and for comparing it to certain Miocene "cricetodontids" (sensu Schaub). His careful redescription and comparison led to his establishing a new monospecific genus, *Neocricetodon*, for which *Neocricetodon grangeri* (Young 1927) automatically became the type species. Schaub did not acknowledge the same generic name, which had been introduced into the literature by Kretzoi (Kadic and Kretzoi 1930). There are good reasons to suppose that Schaub did not know about Kretzoi's naming. There is, however, no doubt about the nomenclatorial validity of the taxon "*Neocricetodon* Schaub 1934," with *Neocricetodon grangeri* (Young 1927) being its type species. Kretzoi (1951:407) cited the presence of "*Neocricetodon schaubi* Kretzoi 1930" at Csákvár. However, Kretzoi now designated a holotype as: a left mandible with i and m1, No. F.I.V. 6004 (concerning the number compare Kordos 1987:73). It was described as differing from *Cricetodon, Plesiodipus* (= *Plesiocricetodon* Schaub), and *Palaeocricetus* by its presence of "non-stephanodont and non-lophodont primitive bunodont molars with strongly developed mesostylid-crest extending to the margin." Kretzoi distinguished this taxon from "*Cricetodon*" *gaillardi* and "*C.*" *breve*, which he attributed to the same genus, by its "more elongated M1 and a stronger frontal cingulum." Kretzoi recognized the genus "*Neocri-*

cetodon" (Schaub 1934) for the Chinese species "*Cricetulus grangeri*" and stated: "In order to replace this preoccupied name I erected the genus *Epicricetodon*. Subsequently, this hardly identifiable group of uncertain relation must be listed as *Epicricetodon grangeri* (YOUNG)—save it belongs to the group established by me."

Kretzoi (1954) again reported the occurrence of "*Neocricetodon schaubi* Kretzoi 1930" at Csákvár. He gave no additional information except that now the upper teeth of this species had also been found.

Fahlbusch (1969) described the genus *Kowalskia*, recognizing *Kowalskia polonica* as the type species and further referring the somewhat larger taxon *K. magna* to the genus. Both taxa were found at Podlesice, Poland (MN 14). He compared these rather "primitive" species to other known Neogene small cricetines, discussing their phylogenetic position. He noted the morphological similarity shared between these species and "*Rotundomys*" *hartenbergeri* and followed Freudenthal (1967:313) in interpreting the Polish taxa as belonging to a lineage independent of *Rotundomys*. Fahlbusch did not compare his *Kowalskia* to "*Neocricetodon* Kretzoi 1930", which he considered to be a nomen nudum, and did not mention it because it was impossible to obtain any information from the existing literature on its morphology. Furthermore, Fahlbusch did not compare it to "*Neocricetodon* Schaub 1934", which was unknown to him at that time.

During the following years many late Miocene and Pliocene localities yielded additional *Kowalskia* material, and several new species were described. These will be dealt with in some detail below. *Kowalskia* is now known to have had a long stratigraphic and broad geographic range. New questions have arisen on this genus's species content as more information has been uncovered on its intraspecific variability and evolutionary patterns. We will address these issues further below.

Kordos (1987) was the first to give a short, reliable description of "*Neocricetodon schaubi* Kretzoi 1930." He considered Kretzoi's 1930 description to be a nomen nudum, but Kretzoi's 1954 (what in fact was in 1951; see above) publication of a "holotype's" catalogue number was considered by Kordos to validate the genus and species "*Neocricetodon schaubi*." Furthermore, Kordos considered *Kowalskia* (at least for *K. fahlbuschi*) to be a synonym of *Neocricetodon*. However, he did not discuss details about the generic identity of these taxa. Kordos did not mention Schaub's *Neocricetodon* of 1934, which undoubtedly is a valid generic name. Neither "*Neocricetodon* Kretzoi 1930" nor "*Neocricetodon* Kretzoi 1951" nor "*Neocricetodon* Kretzoi 1954" is a valid taxon.

Freudenthal and Kordos (1989) compared the type species of *Rotundomys*, *Cricetulodon*, and *Kowalskia*. They concluded that *Kowalskia* is a junior synonym of *Cricetulodon*. This result will be discussed below. From the nomenclatorial point of view, however, it is of further interest that Freudenthal and Kordos (1989: p. 87) came to the conclusion that, referring to Kretzoi 1954 instead of Kretzoi 1951, "*Neocricetodon* is an available name in the sense of the rules for zoological nomenclature" and insofar as "both *Kowalskia* and *Neocricetodon* are to be considered junior synonyms, the correct name to be used is *Neocricetodon*." The authors, as Kordos in 1987, did not take into consideration "*Neocricetodon* Schaub 1934."

Freudenthal et al. (1991: p. 25) continued to use the genus "*Neocricetodon* Kretzoi 1930", applying it in a very broad morphological sense and including a dozen species within it. Whether all of them belong to a single genus will be discussed below. They certainly cannot be attributed to the nomen nudum "*Neocricetodon* Kretzoi." Whether they belong to "*Neocricetodon* Schaub 1934" will also be treated below.

When Wu Wenyu (1991) discussed possible relationships of Chinese and European *Kowalskia* species she included in her comparison casts of the Csákvár species "*Neocricetodon schaubi*" and stated (p. 284): "It seems probable that "*Neocricetodon*" *schaubi* is closely related to *K.* cf. *fahlbuschi* from Eichkogel, perhaps corresponding to *K.* sp. from Vösendorf."

Daxner-Höck (1992) discussed the relationships between *Kowalskia* Fahlbusch and *Neocricetodon* Kretzoi and clearly stated (p. 344): "Nach unserem heutigen Wissensstand sind *Neocricetodon* und *Kowalskia* eine generische Einheit." In contrast to Kordos (1987:73) and Freudenthal and Kordos (1989:87), and because of the insufficient description of *Neocricetodon* by Kretzoi in 1930 or 1951/1954 respectively, she decided to further apply the better-known taxon *Kowalskia*. She did not consider *Neocricetodon* Schaub 1934 in her treatment.

We remain convinced that "*Neocricetodon* Kretzoi 1930" is a nomen nudum, and that *Neocricetodon* Schaub 1934 is a valid taxon. We maintain that "*Neocricetodon* Kretzoi 1951" and "*Neocricetodon* Kretzoi 1954" are invalid taxa. Quite another issue is whether the species referred to *Kowalskia* Fahlbusch 1969 and *Neocricetodon grangeri* (Young 1927) belong to the same genus; i.e., whether *Kowalskia* Fahlbusch 1969 is a junior synonym of *Neocricetodon* Schaub 1934. This will be discussed further below.

Redescription of the Type Material of *Neocricetodon grangeri* (Young 1927) (fig. 16.1, tab. 16.1)

This species's type material is housed by the Universitet Paleontologiska Museet, Uppsala, Sweden. Wu (here) was recently been provided the opportunity to study the original type material.

Holotype   A pair of lower jaws, a fragmentary skull, and a few postcranial skeletal remains from Chia Yu Tsun, Yûshe

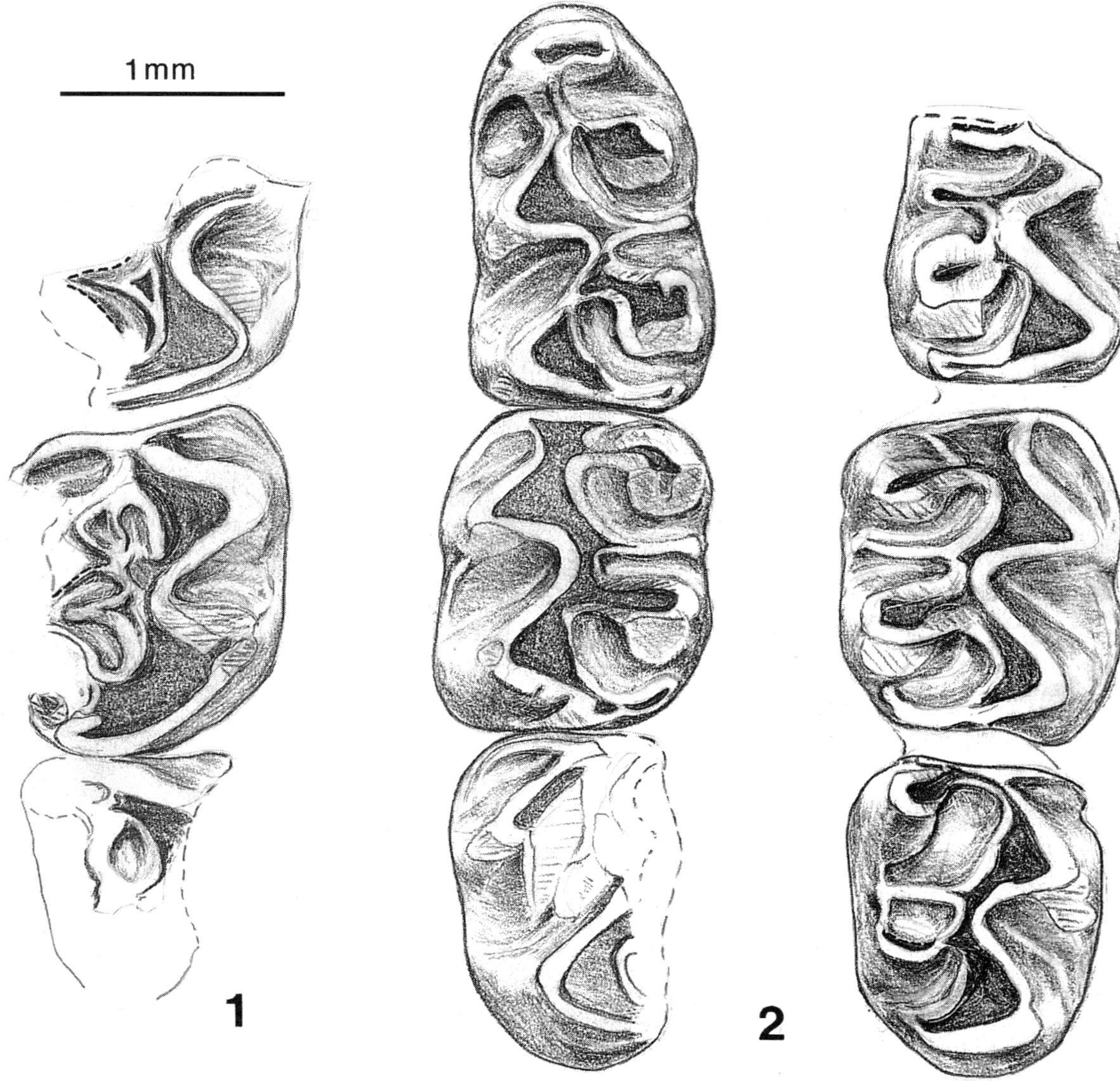

FIGURE 16.1 *Neocricetodon grangeri* (Young 1927). Dentitions of the holotype. Uppsala Universitet Paleontologiska Museet, Sweden. Drawings by Wenyu Wu.

County, Shanxi Province, China. All the remains undoubtedly belong to the same individual. The original descriptions by both Young (1927) and Schaub (1934) were based on insufficiently cleaned specimens.

Description

After a recent cleaning of the dentitions, Wu has been able to make the following new morphological observations: m1 of the right jaw and m3 of the left jaw are partly broken; they are preserved, however, in the opposite ramus (fig. 16.1). The teeth are low-crowned and moderately worn. m1 is rather narrow anteriorly. The anteroconid is slightly divided from behind; its front wall is smooth; its labial side is connected posteriorly to the anterior arm of the protoconid through an anterolophulid that is very low

and nearly completely worn. The metalophid and hypolophid extend slightly obliquely toward the anterolophulid and ectolophid, respectively. The mesolophid is low, reaching the lingual border of the tooth and ending in a mesostylid. Low labial walls weakly dam the protosinusid and sinusid. No ectomesolophid is present in the sinusid. The posterolophid is low; it extends close to the posterior slope of the entoconid and thus closes a narrow posterosinusid. m2 and m3 are similiar to the posterior aspect of m1. A labial anterolophid is present; it extends downward to the labial border of the protoconid base, closing the protosinusid; the lingual anterolophid is about half as long as the labial one. m3 entoconid is slightly reduced and well developed.

The skull's right dentition is incompletely preserved and its left dentition is missing. M1 exhibits a short meso-

TABLE 16.1 *Measurements on the Type Specimen of* Neocricetodon grangeri

left m1: 1.80 × 1.10	right m1: — × 1.05
left m2: 1.45 × 1.25	right m2: 1.50 × 1.23
left m3: 1.40 × —	right m3: 1.40 × 1.15
	right M2: 1.50 × ca. 1.25

loph that extends halfway along the crown's length. Both lingual and labial anterolophs are present in M2, reaching the lingual or labial border respectively; the lingual cingulum, however, stands somewhat lower than the labial one. The protolophule I and II are developed on both M2 and M3. A funnel-shaped feature is present between the protolophules and protocone. The mesoloph extends halfway and is close to the front slope of the metacone.

This type material is very poor, and the M1, which is very important for species determination, is not complete, it is presently not possible to decide whether *Neocricetodon* and *Kowalskia* should stand as two different genera. In some morphological features *Neocricetodon grangeri* is similar to *Kowalskia*, in others it is not. As long as we lack convincing arguments for the individuality of these two genera, we should continue to further use the better-defined *Kowalskia*. Additional cricetine material was collected in 1987, 1988, and 1991 by the joint Chinese-American team working in the Yûshe Basin. This material compares closely in its morphology with the holotype and will be described later by the participants of that research project. If this material proves to be identical to *Kowalskia*, this name will fall in the synonymy of *Neocricetodon*, which has priority over *Kowalskia*.

Emended Diagnosis of *Kowalskia* and Its Differential Diagnosis

Type species   *Kowalskia polonica* Fahlbusch 1969; Podlesice (Poland), MN 14.

Diagnosis   A small- to medium-sized Cricetinae with low-crowned molars.

Lower molars   m1 anteroconid is elongate and wide, anteriorly rounded, split from behind into two cusps (in some cases, three to four); the anterolophulid is simple and connected to the labial part of the anteroconid or forked (2–3 branches), sending a branch to each one of the anteroconid cusps and ending at their base. The extension of the metalophulid and the hypolophulid is transverse or slightly oblique. The mesolophid is long and reaches the molar border, or it is of medium length; very rarely it is short; an ectomesolophid is sometimes present as well as an extra ridge in the protosinusid. The m2 and m3 are similar to the posterior part of m1. The m3 entoconid is

well developed; the m3 mesolophid may be shorter than in m1 or m2.

Upper molars   M1 anteroconus is wide and deeply split from behind, anteriorly rounded and smooth, sometimes slightly dent. The anterolophule is either a single crest connected to the lingual anterocone cusp, or it is forked and connected to both anterocone cusps at their base, or it is forked, sending a free labial transverse spur into the anterosinus. The paracone and metacone are coniform and symmetrical and situated opposite the protocone and hypocone; protolophule and metalophule are frequently double, sometimes simple; the posterior protolophule is mainly stronger than the anterior one; the mesoloph is generally long, frequently reaching the molar border and ending in a mesostyle; sometimes it is of medium length, ending free in the mesosinus, sporadically bending backward and touching the metacone. A labial and lingual anteroloph (cingulum) is present in M2 and M3. M1 has 3–4 roots, M2 has 4 roots, M3 has 3 roots.

Differential diagnosis

The Mio-, Plio-, and Pleistocene cricetine genera differ from *Kowalskia* in a number of characters, as listed below.

Democricetodon Fahlbusch 1964: *Fahlbuschia* Mein and Freudenthal 1971, and *Pseudofahlbuschia* Daams and Freudenthal 1988 anterocone and anteroconid short, simple, or slightly split; lingual M1 anteroloph descends from anteroconus and closes the protosinus; paracone and metacone inclined posteriorly; M3/m3 more reduced; protolophule and metalophule frequently simple; anterior protolophule always absent in *Pseudofahlbuschia*.

Megacricetodon Fahlbusch 1964:   m1 anteroconid long, simple, or deeply split anteriorly; M1 anterocone long as well and generally split from the anterior aspect; protolophule and metalophule single, transverse or projecting posteriorly; M1 and M2 paracone each exhibit a paracone spur; M3/m3 moderately reduced.

Cricetulodon Hartenberger 1966:   Sinuses/sinusids are wide; slender crests joined by high cusps; anterocone is short, asymmetrical, and split slightly posteriorly; the labial one is the principal cusp; anterolophule is simple and strongly built, connecting the lingual portion of the anterocone with the protocone; a strong lingual anteroloph closes the protosinus; the protolophule and metalophule are principally simple (posterior), if double, the posterior one dominates; Ms are 3-rooted; anterolophulid is connected to the dominant lingual anteroconid cusp; metalophid and hypolophid are obliquely oriented; mesolophid is absent or short; m3 entoconid is reduced.

Rotundomys Mein 1965:   The molar pattern is generally similiar to *Cricetulodon*, but more advanced; anterocone/anteroconid wide; mesoloph/mesolophid absent; anterior protolophule absent; posterosinus absent.

Hypsocricetus Daxner-Höck 1992:   Tooth crown distinctly higher, long, and narrow; opposing upper molar labial and lingual main cusps encircle well-developed funnels; presence of thick and high crests; anterocone and anteroconid deeply split anteriorly; the two anterocone cusps and the forked anterolophule encircle a well-defined funnel; the mesoloph is not free (it is short or of medium length and turned into an anterior metalophule); the posterior metalophule is absent; M1 has 4 roots; M3 lacks metacone and hypocone, but the posterior funnel is enclosed by a strong crest; anterolophulid and ectolophid are strongly developed, with high crests; m1 mesolophid is absent, in m2 it is absent or short, in m3 it is short or of medium length.

Cricetulus Milne Edwards 1867: *Allocricetus* Schaub 1930, and *Cricetus* Leske 1779   Anterocone/anteroconid superficially or deeply split; mesolophid nearly or completely reduced; metalophid and hypolophid oblique; funnels are well developed between the anterocone cusps, the opposing labial, and principal lingual cusps; mesoloph is not free — it is short or of medium length and turns into an anterior metalophule; metalophule is absent; posterosinus is absent; M1–2 have four roots.

Mesocricetus Nehring 1898:   Molars are narrow; mesoloph/mesolophid absent, short or of medium length; anterocone is deeply split; funnels are well developed between the upper molars' labial and lingual cusps, including the anterocone cusps; M2–3 labial anteroloph is very strongly developed and has a free end; lingual anteroloph is absent; M1–2 have four roots; m3 is long, and posteriorly less reduced.

Cricetinus Zdansky 1928:   Mesolophid is absent or varies from being short to long; mesoloph is short or of medium length, not free but terminating as an anterior metalophule.

Sinocricetus Schaub 1930:   Tooth crowns are slightly higher-crowned, more robust, long, and narrow; strong crests are present between the cusps and the sinuses/sinusids; anterocone/anteroconid are distinctly bicuspid and deeply split posteriorly; mesoloph is of short or medium length; m1-m2 mesolophid is generally short or absent; m3 is long and extends to the molar border; metalophulid and hypolophulid are obliquely oriented.

Nannocricetus Schaub 1934:   Tooth crown is long and slender; anterocone is deeply split anteriorly; protolophule

is simple in M1 (no anterior protolophule), double in M2; mesoloph is short and turned into an anterior metalophule; anteroconid is slender, simple, or slightly split; anterolophulid is simple and connected to the labial anteroconid cusp; mesolophid is absent.

Pseudocricetus Topachevski and Scorik 1992: *Stylocricetus* Topachevski and Scorik 1992, and *Odessamys* Topachevski and Scorik 1992   Mesoloph/id is nearly completely missing; anterocone/id frequently to predominantly deeply split anteriorly; metalophid and hypolophid strongly antero-obliquely oriented.

What Belongs and What Does Not Belong to *Kowalskia*?

In the previous section, we restricted the content of the genus *Kowalskia*. In this section we primarily follow the original generic diagnosis and description, taking into account recent improved information about its temporal and geographic distribution as well as intraspecific variation. We include the following species in *Kowalskia*:

K. polonica Fahlbusch 1969; Podlesice (Poland) [type species]; MN 14

K. magna Fahlbusch 1969; Podlesice (Poland); MN 14

K. schaubi (Kretzoi 1951); Csákvár (Hungary); MN 11

K. ? lavocati (Hugueney and Mein 1965); Lissieu (France); MN 13

K. fahlbuschi Bachmayer and Wilson 1970; Kohfidisch (Austria); MN 10

K. intermedia Fejfar 1970; Ivanovce (Slowakia); MN 15

K. moldavica Lungu 1981; Moldavia; MN 10?

K. occidentalis (Aguilar 1982); Crevillente 2 (Spain); MN 11

K. gansunica Zheng and Li 1982; Tiansu (China); MN 11?

K. yinanensis Zheng 1984; Yinan (China); MN 15

K. skofleki (Kordos 1987); Tardosbánya (Hungary); MN 12

K. nestori Engesser 1989; Baccinello (Italy); MN 13/14

K. polgardiensis (Freudenthal and Kordos 1989); Polgardi 4 (Hungary); MN 13

K. neimengensis Wu 1991; Ertemte 2 (China); MN 13?

K. similis Wu 1991; Ertemte 2 (China); MN 13?

K. browni Daxner-Höck 1992; Maramena (Greece); MN 13/14

K. progressa Topachevski and Scorik 1992; MN 9

Most of these species were attributed to the genus *Kowalskia* at the time of their original descriptions. Exceptions include:

K. schaubi (Kretzoi 1951); original determination: *Neocricetodon schaubi* Kretzoi 1930. Remarks: This taxon falls morphologically within the range of variation of *Kowalskia*. However, the morphologic and metric variation of this species remains unknown due to the current paucity of material (Kordos 1987:73).

K. occidentalis (Aguilar 1982): original determination: *Cricetulodon occidentalis*, Aguilar 1982. Remarks: On the strength of the anterocone/id morphology, the considerable length of the mesoloph/id, and the relatively low connecting lophs, this species corresponds most closely to the diagnosis of *Kowalskia*. In *Cricetulodon* (cf. differential diagnosis), the connecting lophs/ids are relatively high and more massively built, the mesoloph/id is often short or missing, the synclines are relatively wider. Based on morphological characters, *occidentalis* is closer to *Kowalskia* than it is to *Cricetulodon*. However, the material from Crevillente 2 (type locality of *occidentalis*) includes many teeth in which the mesoloph/id is rather long, but in its distal part fairly low and indistinct. The m3s from this locality are marked by a rather reduced entoconid. On one hand *K. occidentalis* displays, in comparison to the considerably older *C. sabadellensis*, rather symplesiomorphic characters that we take as being indicative of its referral to the *Kowalskia*–group. On the other hand, given the presence of advanced characters, it may well be ancestral to the Spanish species *"Neocricetodon" plinii*, *"Neocricetodon" lucentensis* and *"Kowalskia" meini*, which, however, are too advanced to be attributed to the genus *Kowalskia*. When more material of *K. schaubi* is available we will be able to determine whether or not *K. occidentalis* is a species synonym.

K. skofleki (Kordos 1987): original determination: *Karstocricetus skofleki* Kordos 1987. Remarks: According to the generic diagnosis for *Karstocricetus* Kordos 1987 (p. 75), there are no significant differences between this genus and *Kowalskia* (cf. Daxner-Höck 1992:344).

K. polgardiensis (Freudenthal and Kordos 1989): original determination: *Cricetus polgardiensis* Freudenthal and Kordos 1989. Remarks: This taxon exhibits a tendency toward shortening of the mesoloph/mesolophid and a remarkably high range of variation in certain morphological features (based on very rich fossil material). This species compares well in its variation to the genus *Kowalskia* as defined here.

K. ?lavocati (Hugueney and Mein 1965): original determination: *Cricetulus lavocati* Hugueney and Mein 1965. Remarks: The known material is very poor and a definite generic attribution is not currently possible. Otherwise, Pradel (1988) notes the genus *Cricetulus* only since the late Pleistocene.

Species not included, but (some) possibly close to *Kowalskia* include: *"K." meini* Agusti 1986, Casa del Acero (Spain); *"N." plinii* Freudenthal et al. 1991, Crevillente 15 (Spain); *"N." lucentensis* Freudenthal et al. 1991, Crevillente 17 (Spain); *Cricetus kormosi* Schaub 1930, Polgardi 2 (Hungary), attributed to *Pseudocricetus* Topachevski and Scorik 1992 (p. 103); *K. yananica* Zheng et al. 1985, Yanan (China); *K. ? dalinica* Wang 1988, Dali (China); *K.*

complicidens Topachevski and Scorik 1992; Franzovka 2, Ukraine.

The seven species listed above do not belong to *Kowalskia* for one or more of the following reasons: mesoloph/id is distinctly shortened or lost; metaloph is progressively lost; instead of an anterior metaloph crest there is a short or medium-length mesoloph turned posteriorly and connected to the metacone; there is a tendency toward development of more or less closed funnels between anterocone cusps as well as between anterior and posterior opposing main cusps in M1; the oblique position of metalophulid and hypolophulid; in *"K." meini* and *"N." lucentensis* there is a tendency for the anterolophid to connect with the lingual anteroconid; the ectoloph or "pseudectoloph" in M1–2 of *"K." complicidens*.

These forms belong either to one or the other modernized Plio-Pleistocene to Recent cricetines (e.g., *Allocricetus*, *Cricetulus*, *Tscherskia*), or they represent still unrecognized new genera within this highly complex chapter of late Tertiary and Quaternary cricetids.

Conclusions

A critical review of all relevant literature since 1930 clearly indicates that *Neocricetodon* is a valid genus of cricetine rodents and that it was correctly established by Schaub in 1934, with the type species being the late Miocene Chinese form *Neocricetodon grangeri* (Young 1927). The nomen *"Neocricetodon"* first applied by Kretzoi in 1930 for the late Miocene (Csákvár, Hungary) taxon *"N." schaubi* is a nomen nudum. Whether the genus *Kowalskia* Fahlbusch 1969 is a junior synonym of *Neocricetodon* Schaub 1934 is questionable and cannot be decided for the time being because the type material (*"Neocricetodon grangeri"*) is too poor. Therefore, the genus *Kowalskia* is retained until more material of the genus *Neocricetodon* becomes known.

The species referred to *Kowalskia*, as originally described, have to be restricted in number (presently 15–16), and include members ranging from early late Miocene to Pleistocene in age. The *Kowalskia*–group is unified by a combination of primitive and advanced characters. We summarize these here in an emended diagnosis, and compare them by differential diagnoses to closely related genera.

All species with distinctly divided anterocones/conids, strongly reduced or missing mesolophs/lophids, and well-developed funnels ("Rautentrichter") between opposed inner and outer main cusps are excluded from *Kowalskia*. Even in this restricted sense, *Kowalskia* displays a very long and complex evolutionary history that is far from being completely understood and probably comprises more than one evolutionary lineage. Restricted so far to the investigation of almost entirely isolated teeth, it needs much more effort to reconstruct its actual phylogeny.

Acknowledgments

The subject of this article has been discussed for a rather long time with several colleagues. We are especially grateful to Dr. T. Bolliger, Zürich; Dr. H. de Bruijn, Utrecht; Dr. B. Engesser, Basel; and Professor Dr. O. Fejfar, Prague. We also thank these colleagues for giving us access to valuable material under their care for comparison. NSF grant BSR 9020065 to L. J. Flynn funded Wu's research trip to Uppsala.

LITERATURE CITED

Aguilar, J. P. 1982. Contributions á l'étude des micromammifères du gisement miocène supérieur de Montredon (Herault). *Palaeovertébrata* 12:75–118.

Agusti, G. 1986. Nouvelles espèses des Cricetidés vicariantes dans le Turolien moyen de Fortuna (Prov. Murcia, Espagne). *Géobios* 19:5–11.

Bachmayer, F. and R. W. Wilson. 1970. Die Fauna der altpliozänen Höhlen- und Spaltenfüllungen bei Kohfidisch, Burgenland (Österreich). *Annalen des Naturhistorischen Museums, Wien* 74:533–87.

Daxner-Höck, G. 1972. Cricetinae aus dem Alt-Pliozän vom Eichkogel bei Mödling (Niederösterreich) und Vösendorf bei Wien. *Paläontologische Zeitschrift* 46:133–50.

———. 1980. Rodentia (Mammalia) des Eichkogels bei Mödling (Niederösterreich). 1. Spalacinae und Castoridae. 2. Übersicht über die gesamte Nagetierfauna. *Annalen des Naturhistorischen Museums, Wien* 83:135–52.

———. 1992. Die Cricetinae aus dem Obermiozän von Maramena (Mazedonien, Nordgriechenland). *Paläontologische Zeitschrift* 66:331–67.

Engesser, B. 1989. The late Tertiary small mammals of the Maremma region (Tuscany, Italy). 2nd part: Muridae and Cricetidae (Rodentia, Mammalia). *Bollettina della Societa Paleontologica Italiana* 28:227–52.

Fahlbusch, V. 1969. Pliozäne und Pleistozäne Cricetinae (Rodentia, Mammalia) aus Polen. *Acta Zoologica Cracoviensia* 14:99–137.

Fejfar, O. 1970. Die plio-pleistozänen Wirbeltierfaunen von Haynácka und Ivanovce (Slowakei, CSSR). VI. Cricetidae (Rodentia, Mammalia). *Mitteilungen der Bayerischen Staatssammlung für Paläontologie und historische Geologie* 10:277–96.

Freudenthal, M. 1967. On the Mammalian Fauna of the *Hipparion*-Beds in the Calatayud-Teruel Basin. Part III. *Democricetodon* and *Rotundomys* (Rodentia). *Proceedings, Koninklijke Nederlandse Akademie van Wetenschappen, B* 70:298–315.

Freudenthal, M. and L. Kordos. 1989. *Cricetus polgardiensis* sp. nov. and *Cricetus kormosi* Schaub 1930 from the late Miocene Polgardi localities (Hungary). *Scripta Geologica* 89:71–100.

Freudenthal, M., J. I. Lacomba, and E. Martin-Suarez. 1991. The Cricetidae (Mammalia, Rodentia) from the late Miocene of Crevillente (prov. Alicante, Spain). *Scripta Geologica* 96:9–46.

Hugueney, M. and P. Mein. 1965. Lagomorphes et Rongeurs du Néogène de Lissieu (Rhône). *Traveaux des Laboratoires de Géologie de la Faculté des Sciences de Lyon*, n.s., 12:109–23.

Kadic, O. and M. Kretzoi. 1930. Ergebnisse der weiteren Grabungen in der Esterházyhöhle (Csákvárer Höhlung). *Mitteilungen über Höhlen- und Karstforschung, Zeitschrift des Hauptverbandes deutscher Höhlenforscher* 2:45–49.

Kordos, L. 1987. *Karstocricetus skofleki* gen. n., sp. n. and the evolution of the late Neogene Cricetidae in the Carpathian Basin. *Fragmenta Mineralogica et Palaeontologica* 13:65–88.

Kretzoi, M. 1951. The Hipparion-Fauna from Csákvár. *Földtani Közlöny* 81:384–417.

———. 1954. Rapport final des fouilles paléontologiques dans la grotte de Csákvár. *Földtani Intézet Evi Jelentesei* 1952:37–68.

Lungu, A. N. 1981. Hipparionovaja fauna Sarmata Moldavii (nasekomojadnye, zaiceobraznye i gryzuny). [Hipparion-Fauna of the middle Sarmatian from Moldavia (Insectivora, Lagomorpha, Rodentia)]. *Ztinca*: 1–118.

Schaub, S. 1930. Quartäre und jungtertiäre Hamster. *Abhandlungen der Schweizerischen Palaeontologischen Gesellschaft* 49:1–49.

———. (1934): Über einige fossile Simplicidentaten aus China und der Mongolei. *Abhandlungen der Schweizerischen Palaeontologischen Gesellschaft* 54:1–40.

Topachevski, V. A. and A. F. Scorik. 1992. Neogene and Pleistocene Cricetidae of early evolutionary stage of South Eastern Europe. *Academy of Sciences of Ukraine Institute of Zoology*: 1–242.

Wang Hong. 1988. An early Pleistocene Mammalian Fauna from Dali, Shaanxi. *Vertebrata Palasiatica* 26:59–72.

Wu Wenyu. 1991. The Neogene mammalian faunas of Ertemte and Harr Obo in Inner Mongolia (Nei Mongol), China.-9. Hamsters: Cricetinae (Rodentia). *Senckenbergiana lethaea* 71:257–305.

Young, C.C. 1927. Fossile Nagetiere aus Nord-China. *Palaeontologica Sinica, C* 5:1–82.

Zheng, S. 1984. A new species of Kowalskia (Rodentia, Mammalia) of Yinan, Shandong. *Vertebrata Palasiatica* 22:251–60.

Zheng, S. and Y. Li. 1982. Some Pliocene Lagomorphs and Rodents from Loc. 1 of Songshan, Tianzu Xian, Ganseu Province. *Vertebrata Palasiatica* 20:35–44.

17

On the Evolutionary History of the Cricetodontini from Europe and Asia Minor and Its Bearing on the Reconstruction of Migrations and the Continental Biotope During the Neogene

H. DE BRUIJN AND E. ÜNAY

The taxonomical data that form the basis of this review have appeared in the *Proceedings of the Koninklijke Neder-landse Akademie van Wetenschappen* (de Bruijn et al. 1993). The reason for presenting some of the results of that study in this paper is that they are relevant for the reconstruction of the continental biotope during the Neogene. The Cricetodontini are understood as comprising *Cricetodon* (inclusive *Turkomys*), *Hispanomys*, *Ruscinomys* (inclusive *Pseudoruscinomys*), and *Byzantinia* (inclusive *Pararuscinomys*).

The best records of the Cricetodontini are from Southwestern Europe, where they range from MN 5 to MN 15, and from Asia Minor, where they range from MN 1 to MN 12. The available information from the two intermediate areas recognized—Central Europe consisting of Switzerland, south Germany, and Hungary, and Southeastern Europe consisting of Greece and Turkey north of the Dardanelles—is unfortunately less complete (figure 17.1).

Since the Cricetodontini appear in Asia Minor long before they are known from elsewhere, we shall first briefly review their evolutionary history in that area. The stages-of-evolution observed in Anatolia will then be compared with the European record.

The (composite) dentitions figured on figures 17.2–5 are all **x** 17.5 and from the left side.

The Cricetodontini from Anatolia

The oldest occurrence of *Cricetodon* in Anatolia is from the locality Inkonak (Sivas area), where the genus is represented by a few isolated teeth only. Other occurrences assigned to MN 1 come from the section of Kilçak (Çankiri area). In the lower part of MN 1 the genus is represented by one species per locality, but in the upper part of MN 1 two species may be present. The teeth of these *Cricetodon*

species are small and low-crowned, the M1 and M2 have three roots, the protoloph and the metaloph of the M2 are directed forward, the posterior spur of the paracone is always present, and the m1 has a double metalophulid (metalophulid 2 seems to be homologous with the mesolophid). This is what we call Cricetodontini type 1 (figures 17.2, 17.4).

The only rodent associations from Anatolia assigned to MN 2 are Harami 1, 2, and 3. Unfortunately none of these associations contains *Cricetodon*.

The assemblage from Keseköy (Kizilcahamam area) assigned to MN 3 contains some *Cricetodon* remains. These teeth are much larger than those of type 1 but smaller than all those from Western Europe. The M1 have three roots and the M2 four. The protoloph and metaloph of the M2 are directed posteriorly, the anterocone of the M1 is rather narrow and only indistinctly divided into two cusps, the posterior spur of the paracone is absent, and the m1 has a posteriorly directed metalophulid as well as a short mesolophid. We call this stage-of-evolution Cricetodontini type 2 (figures 17.2, 17.4).

The Anatolian assemblages assigned to MN 4, Horlak 1 and 2 (Sivas area), each contain a different species of *Cricetodon*. The M1 is still three rooted and the M2 four rooted, the anterocone of the M1 is rather narrow and not distinctly divided into two cusps, the posterior spur of the paracone is rather well developed but does not reach the metacone, and the m1 has a posteriorly directed metalophulid as well as a short mesolophid. These assemblages are morphologically intermediate between type 2 and the *Cricetodon* species from MN 5 and 6 assigned to type 3.

The only assemblage from Anatolia assigned to MN 5 contains only one M2 of *Cricetodon*. This specimen is very similar in overall morphology to the material from Paşalar and Çandir. These latter associations, both assigned to MN

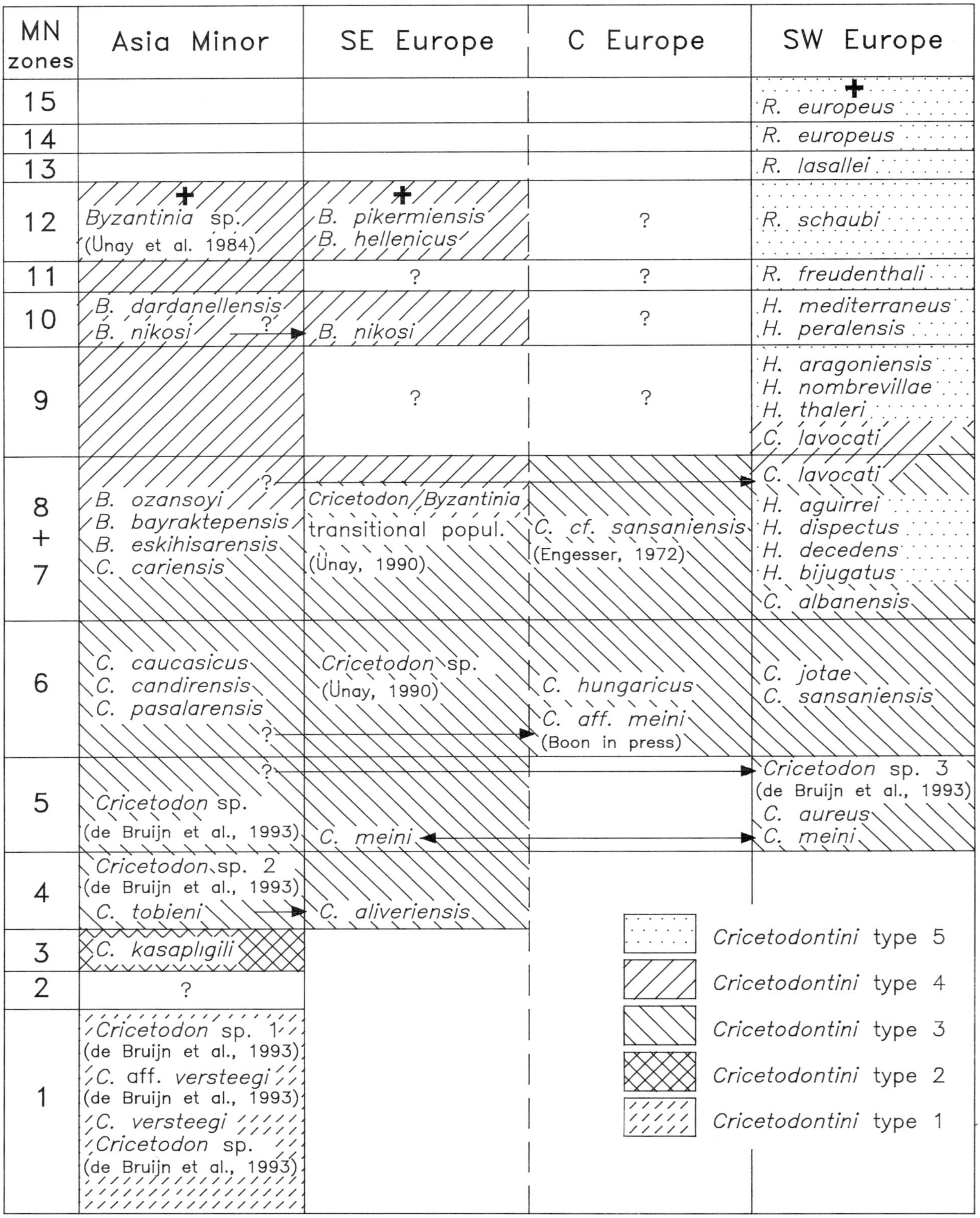

MN zones	Asia Minor	SE Europe	C Europe	SW Europe
15				*R. europeus* ✝
14				*R. europeus*
13				*R. lasallei*
12	*Byzantinia* sp. ✝ (Ünay et al. 1984)	*B. pikermiensis* ✝ *B. hellenicus*	?	*R. schaubi*
11		?	?	*R. freudenthali*
10	*B. dardanellensis* *B. nikosi* ? →	*B. nikosi*	?	*H. mediterraneus* *H. peralensis*
9		?	?	*H. aragoniensis* *H. nombrevillae* *H. thaleri* *C. lavocati*
8 + 7	*B. ozansoyi* *B. bayraktepensis* *B. eskihisarensis* *C. cariensis*	*Cricetodon/Byzantinia* transitional popul. (Ünay, 1990)	*C.* cf. *sansaniensis* (Engesser, 1972)	*C. lavocati* *H. aguirrei* *H. dispectus* *H. decedens* *H. bijugatus* *C. albanensis*
6	*C. caucasicus* *C. candirensis* *C. pasalarensis*	*Cricetodon* sp. (Ünay, 1990)	*C. hungaricus* *C. aff. meini* (Boon in press)	*C. jotae* *C. sansaniensis*
5	*Cricetodon* sp. (de Bruijn et al., 1993)	*C. meini*		*Cricetodon* sp. 3 (de Bruijn et al., 1993) *C. aureus* *C. meini*
4	*Cricetodon* sp. 2 (de Bruijn et al., 1993) *C. tobieni* →	*C. aliveriensis*		
3	*C. kasaplıgili*			
2	?			
1	*Cricetodon* sp. 1 (de Bruijn et al., 1993) *C.* aff. *versteegi* (de Bruijn et al., 1993) *C. versteegi* *Cricetodon* sp. (de Bruijn et al., 1993)			

Legend:
- *Cricetodontini* type 5
- *Cricetodontini* type 4
- *Cricetodontini* type 3
- *Cricetodontini* type 2
- *Cricetodontini* type 1

FIGURE 17.1 The Geographic and Stratigraphic Ranges of the Various Eurasian Cricetodontini and Their Allocation to the Different Categories

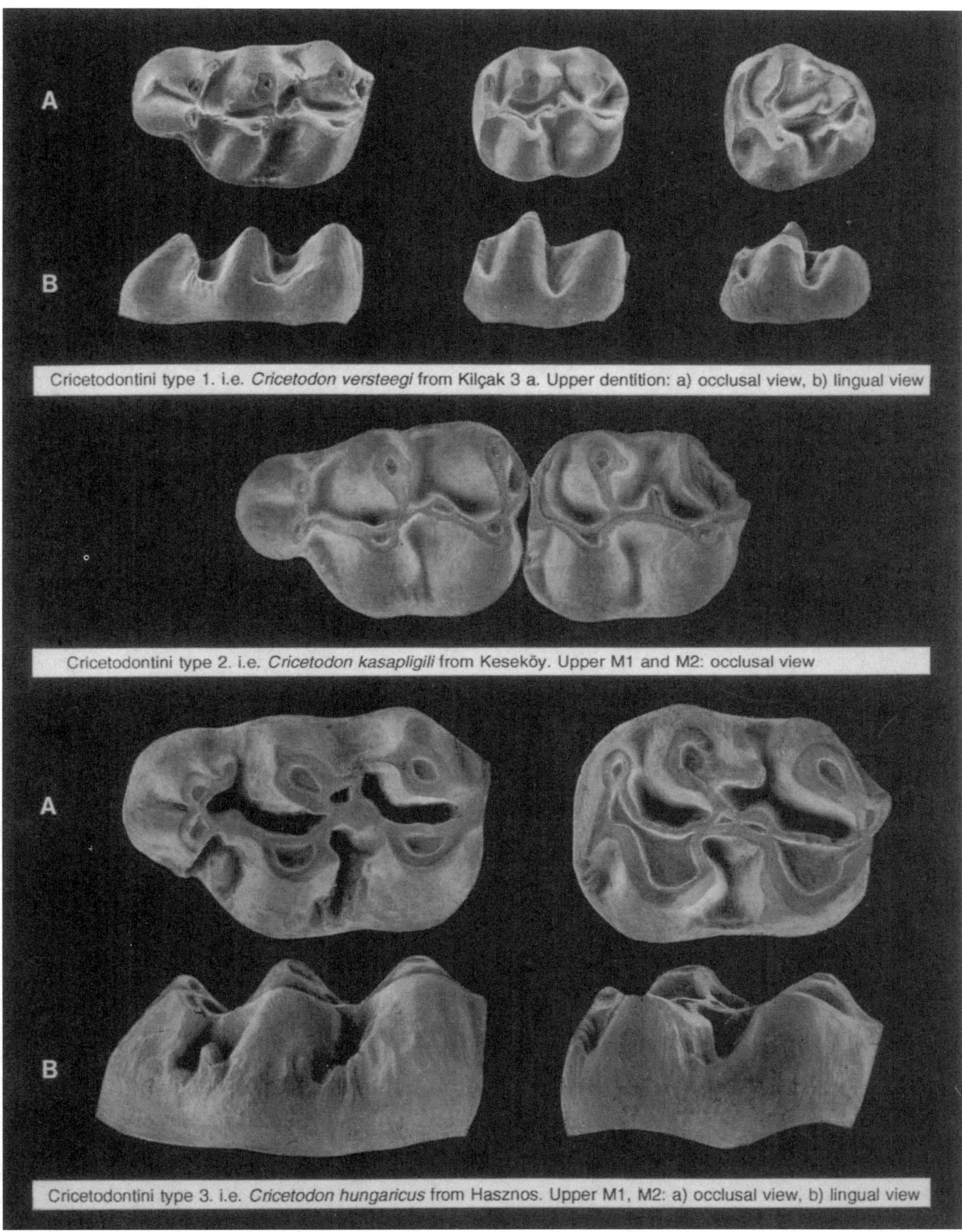

Cricetodontini type 1. i.e. *Cricetodon versteegi* from Kilçak 3 a. Upper dentition: a) occlusal view, b) lingual view

Cricetodontini type 2. i.e. *Cricetodon kasapligili* from Keseköy. Upper M1 and M2: occlusal view

Cricetodontini type 3. i.e. *Cricetodon hungaricus* from Hasznos. Upper M1, M2: a) occlusal view, b) lingual view

FIGURE 17.2

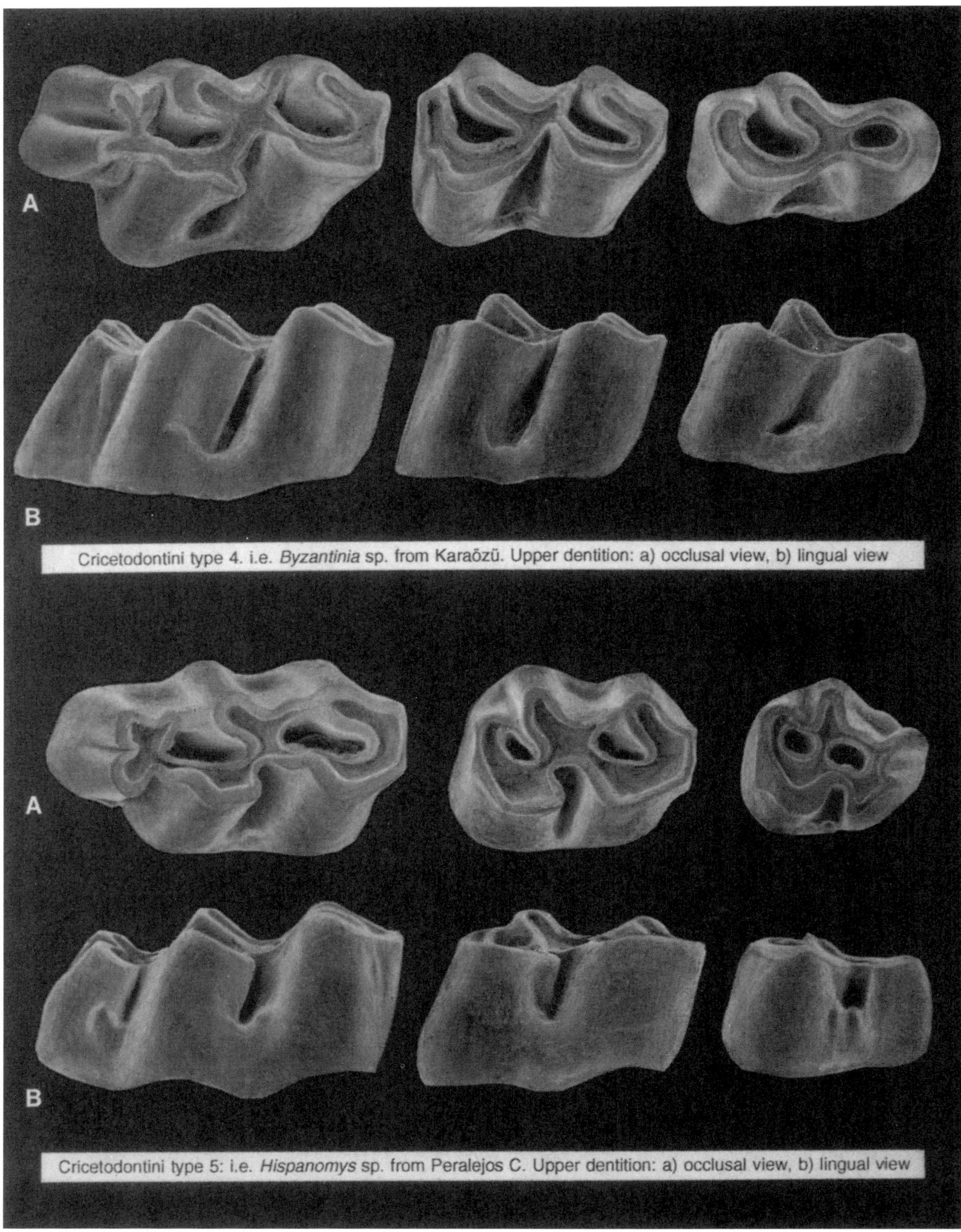

A
B
Cricetodontini type 4. i.e. *Byzantinia* sp. from Karaözü. Upper dentition: a) occlusal view, b) lingual view
A
B
Cricetodontini type 5: i.e. *Hispanomys* sp. from Peralejos C. Upper dentition: a) occlusal view, b) lingual view

FIGURE 17.3

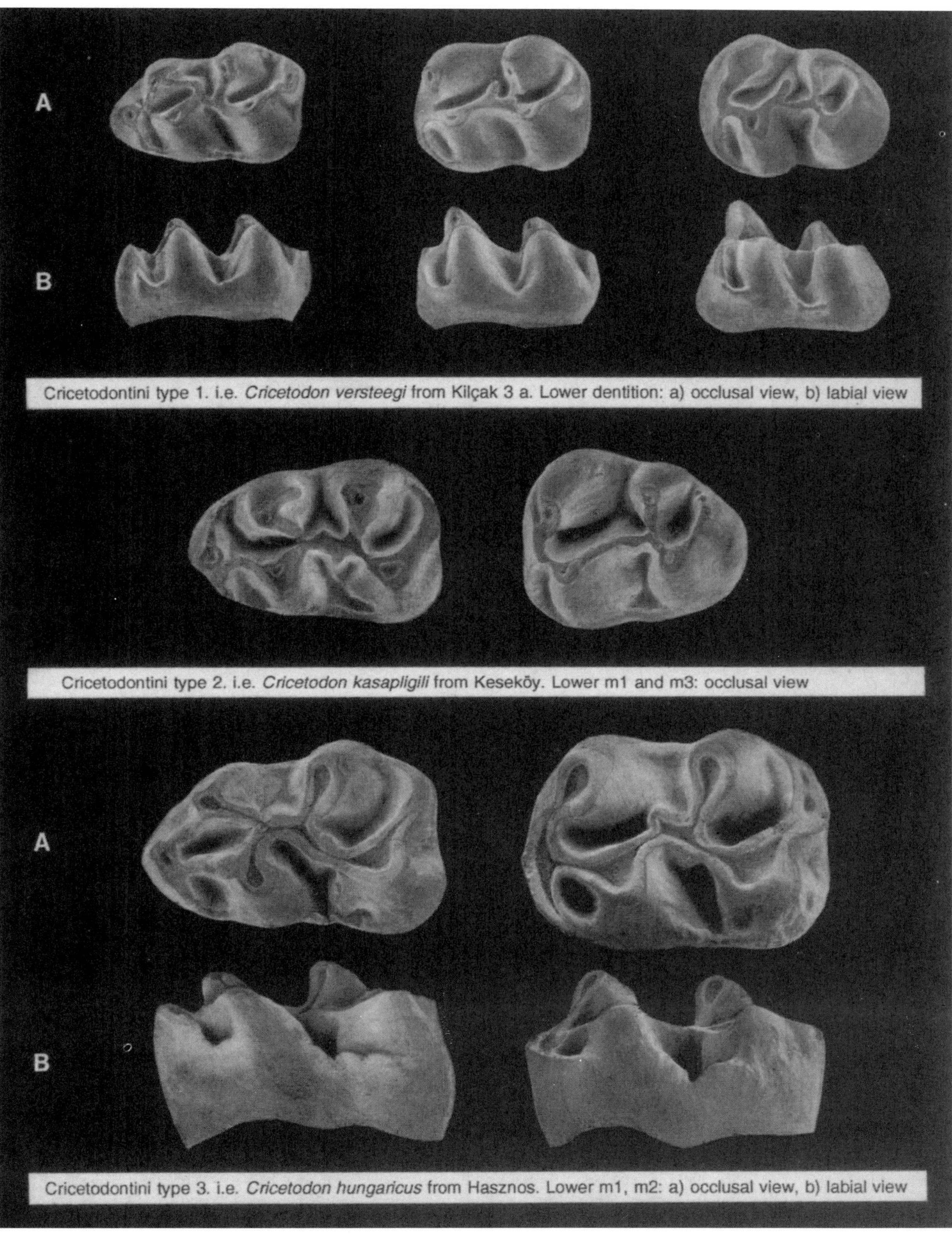
A
B
Cricetodontini type 1. i.e. *Cricetodon versteegi* from Kilçak 3 a. Lower dentition: a) occlusal view, b) labial view
Cricetodontini type 2. i.e. *Cricetodon kasapligili* from Keseköy. Lower m1 and m3: occlusal view
A
B
Cricetodontini type 3. i.e. *Cricetodon hungaricus* from Hasznos. Lower m1, m2: a) occlusal view, b) labial view

FIGURE 17.4

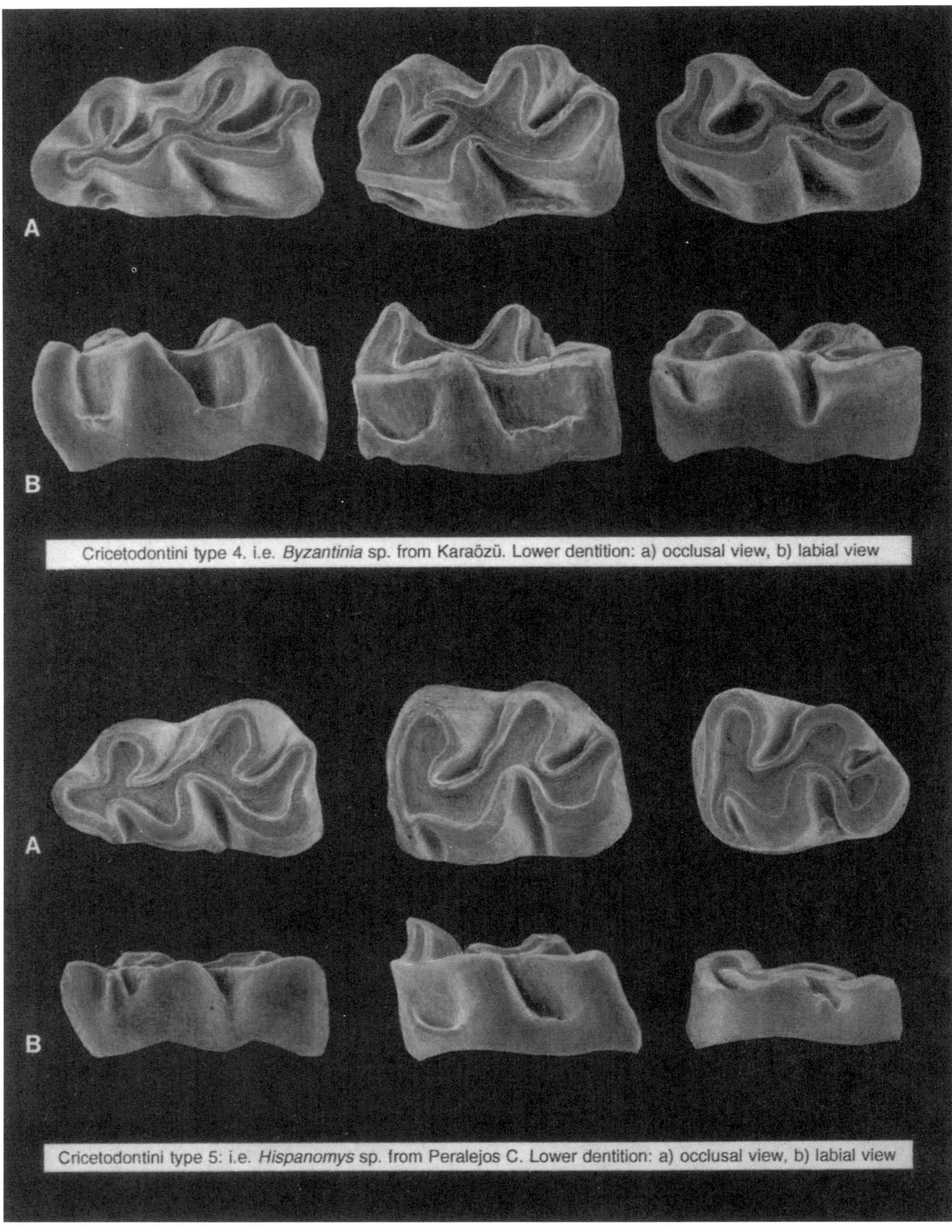

Cricetodontini type 4. i.e. *Byzantinia* sp. from Karaözü. Lower dentition: a) occlusal view, b) labial view

Cricetodontini type 5: i.e. *Hispanomys* sp. from Peralejos C. Lower dentition: a) occlusal view, b) labial view

FIGURE 17.5

6, have teeth that are larger than those of type 2, the M1 as well as the M2 have four roots, the anterocone of the M1 is still rather narrow, but its division into two equal parts is distinct. The posterior spur of the paracone varies in length and may reach the metacone. The posteroloph of the M2 does not continue beyond the point where the metaloph reaches it, and the m1 usually has a double metalophulid as well as a small mesolophid. We call this stage-of-evolution, which is characteristic for *Cricetodon* proper, Cricetodontini type 3 (figures 17.2, 17.4).

The Cricetodontini from the Anatolian and Greek associations assigned to MN 7/8–12 are a rather diverse group of species showing appreciable differences in size and crown height. Intraspecific variability seems to be greater than in the older Cricetodontini. The assemblages from MN units 7/8–12 from Anatolia and Greece share a number of characteristics: the M1 and the M2 have four roots as in type 3; the anterocone of the M1 is rather high and deeply split; the cusps are incorporated in the lophs, which are directed forward in the lower molars and backward in the upper molars. The posterior spur of the anterocone of the M1 and of the paracone of the M1, M2, and M3 are usually long and high, forming a continuous ectoloph. At the same time this selenodont type of dentition develops, the lower as well as the upper teeth acquire a longitudinally "stretched" appearance. Most peculiar is that the semi-hypsodont teeth of this type are not worn as in most other cricetids, but develop concave wear surfaces. We call this stage-of-evolution Cricetodontini type 4 (figures 17.3, 17.5). The only Southwestern European species that possibly belongs in this group is *"Cricetodon" lavocati*.

We conclude that the evolutionary history of the Anatolian Cricetodontini, complex as it is, allows the recognition of four distinct evolutionary grades. The Cricetodontini from Southwestern Europe assigned to the genera *Hispanomys* and *Ruscinomys*, ranging from MN 7/8 to MN 15, comprise a large number of species that share dental characteristics that differentiate them from the type 4 Cricetodontini. These characteristics are: the M1 has often five roots; their cheek teeth, in particular the M3 and m3, are not longitudinally "stretched" and wear flat. We call this stage-of-evolution, which is exclusively known from Southwestern Europe, Cricetodontini type 5 (figures 17.3, 17.5).

Discussion

The type 1 and type 2 Cricetodontini comprise a limited number of species that have been found exclusively in Anatolia so far (table 17.1). The diversification of the tribe starting in MN 4 is accompanied by migration westward, first into Southeastern Europe and later (MN 5) into Southwestern Europe. During MN 6 the tribe reaches its maximum geographic range, from Kazahkstan to Portugal (with its first occurrence in Central Europe), but all the species show the stage-of-evolution characterizing type 3. In assemblages assigned to MN 7/8 the homogeneity in dental characteristics among the Cricetodontini disappears and a Southwest European group of species begins to differentiate from a Southeast European–Asiatic group. This differentiation becomes complete in assemblages assigned to MN 10. The Cricetodontini in the eastern province show a rapid decline in number of specimens as well as species after MN 10 (Ünay and de Bruijn 1984) and become extinct around the MN 12/13 boundary interval.

The Cricetodontini of the western province also show a reduced species diversity after MN 10 and seem to have retracted into a smaller geographical area. Single species, however, continued to exist until the MN 15/16 boundary interval.

Conclusions

The Cricetodontini seem to have appeared as immigrants into Asia Minor around the Oligo-Miocene boundary interval, because there are no potential ancestors known from the area from older levels than MN 1. During the time interval covered by MN 1–3 the tribe remains modest in numbers of both specimens and species. Local evolution may well account for the evolutionary change from type 1 to type 2.

The diversification and geographical expansion of the tribe after the MN 3/4 boundary interval coincides with the drop in diversity of the insectivore faunas in Anatolia. This phenomenon has been interpreted as indicating that the climate became drier, which would mean that Cricetodontini preferred drier biotopes (de Bruijn et al. 1993). The success of the tribe lasted in Anatolia until MN 11, which seems to have been a period of expansion of tropical rain forests in the area (de Bruijn et al. 1993). This change of biotope may eventually have been the cause of their local extinction by the end of MN 12.

In Southwestern Europe the Mid-Vallesian Crisis, supposedly a change toward a drier climate, does not have much effect on the abundance and diversity of the Cricetodontini. In this area where the MN 11 wet period does not seem to have occurred (de Bruijn et al. 1993) the tribe survived until MN 15 although their diversity diminished. Possibly the warm wet interval of MN 14 that is tentatively correlated with the widespread early Pliocene transgression in the Mediterranean caused their extinction in Southwestern Europe in MN 15.

The prolonged difference (MN 7/8–12) between an Asiatic/Eastern European group of species, type 4, classified in *Byzantinia* and a Southwest European group of species, type 5, classified in *Hispanomys* and *Ruscinomys* suggests that there must have been an effective barrier between these two areas during the late Miocene. In the absence of a fossil record of the Cricetodontini from Central Europe later than MN 7/8 nothing can be said about the nature and position of this barrier. Could it be that the Central European climate during the Vallesian and early Turolian was too wet for the Cricetodontini?

Acknowledgments

We gratefully acknowledge the support received from the General Directorate of the M.T.A. (Mineral Research and Exploration Institute of Turkey). This research was financially supported by the NATO Scientific Affairs Division (Grant no. CRG 910750) and the Faculty of Earth Sciences, Utrecht, the Netherlands. We are grateful to Dr. D. Mayhew for critically reading the manuscript, to Mrs. A. E. de Bruijn-Dudok van Heel for typing the manuscript, to Mr. W. den Hartog for making the S.E.M. photographs, to Mr. J. Luteijn for making the plates, and to Mr. T. van Hinte for drawing the text figures.

LITERATURE CITED

Boon, E. 1991. *Die Cricetiden und Sciuriden der oberen Süsswasser-Molasse von Bayerisch-Schwaben und ihre stratigraphische Bedeutung.* Thesis München.

Bruijn, H. de, V. Fahlbusch, G. Saraç, and E. Ünay. 1993. Early Miocene rodent faunas from the Eastern Mediterranean area. Part III. The genera *Deperetomys* and *Cricetodon*, with a discussion of the evolutionary history of the Cricetodontini. *Proceedings, Koninklijke Nederlandse Akademie van Wetenschappen,* B 96:151–216.

Ünay, E. 1990. *Turkomys pasalarensis* Tobien: Its range of variation in the type locality Paşalar, Turkey. *Journal of Human Evolution* 19:437–43.

Ünay, E. and H. de Bruijn. 1984. On some Neogene rodent assemblages from both sides of the Dardanelles, Turkey. *Newsletters on Stratigraphy* 13:119–32.

18

A Current Understanding About the Anomalomyidae (Rodentia): Reflections on Stratigraphy, Paleobiogeography, and Evolution

T. BOLLIGER

The first published account on *Anomalomys* (*A. gaudryi,* La Grive, France) was given by Gaillard (1900). Additional species have subsequently been discovered and described by a number of authors (Viret and Schaub 1946; Fejfar 1972; Daxner-Höck 1980; Klein Hofmeijer and de Bruijn 1985; Kordos 1989; Bolliger 1992). Schaub (1925:66) and Stehlin and Schaub (1951:327) first mentioned the similarity of *Eumyarion* (*Cricetodon helveticum*) to *Anomalomys*. Fejfar (1972:180) further noted the striking morphological resemblance between *Anomalomys* and European *Cotimus* (now referred to *Eumyarion*), but argued that the two forms were not evolutionarily related. Klein Hofmeijer and de Bruijn (1985:197) questioned Stehlin and Schaub's (1951:367) assertion that *Argyromys* was related to *Anomalomys,* and alternatively suggested that *Anomalomys'* ancestor could be found somewhere within Southern Russian small Oligocene age cricetids. Later, de Bruijn and Sarac (1991:1, 13) argued for a *Eumyarion–Anomalomys–Prospalax* lineage, proposing that *Anomalomys* most probably descended from a small Anatolian *Eumyarion*–species morphologically similar to MN 1–MN 3 *Eumyarion microps* and *Eumyarion intercentralis.*

The genus *Prospalax* was recognized by Méhely (1908) based on the species *P. priscus*. *P. priscus* was previously described by Nehring (1897) under the nomen *Spalax priscus*. Both of these authors believed that recent *Spalax* was a direct descendant of *Prospalax*. Stehlin (1923:259) believed that *Prospalax* represented an evolutionary sidebranch of the *Rhizospalax–Spalax* lineage. Kretzoi (1971:111, 114) argued that *Rhizospalax* was morphologically basal to *Spalax,* but concluded that the two genera developed separately from *Prospalax* and *Anomalomys.* Fejfar (1972:188) followed Kretzoi in advocating the distinction of *Prospalax* from *Spalax,* but further argued that *Anomalomys* could be a successor of *Argyromys aralensis,* as previously suggested by Stehlin and Schaub (1951:367). Janossy (1972:43), like Kretzoi (1971), suggested that *Prospalax* clearly belonged to the Anomalomyidae. De Bruijn (1984:420) has argued that the families Spalacidae and

Anomalomyidae have been distinct since the early Miocene and that their phylogenetic relationship is not a close one.

Some anomalomyid species have been referred by Kordos (1985b) to a new genus, *Anomalospalax* (including the species *A. viretschaubi* and *A. tardosi*). Kordos (1985b:30) further argued that *Prospalax petteri* is a possible descendant of *Anomalomys rudabanyensis* and that *Anomalospalax tardosi* is a descendant of an intermediate stage-of-evolution taxon *Anomalomys gaudryi-gaillardi.* Both conclusions were later questioned by Kordos (1989:306) when he argued that *Anomalospalax* was an immigrant and that *Prospalax petteri* descended from an intermediate stage of the *Anomalomys gaudryi–Anomalomys gaillardi* lineage. He argued that *Prospalax petteri* should be included in a separate genus, *Allospalax* Kretzoi 1971, but did not specify how *Prospalax priscus* (and *Prospalax kretzoi*) could be related to the Anomalomyidae.

Following these various observations, Kordos (1989:305) proposed a new anomalomyid phylogeny, creating several new subgenera. Figure 18.1 summarizes some of the competing phylogenetic hypotheses of the Anomalomyidae. The confusion surrounding intergeneric relationships of the younger Anomalomyidae will not be resolved here, but some new hypotheses will be advanced. Furthermore, I follow the proposals set forth by de Bruijn et al. (1992) concerning the referral of localities to MN units, one result being to combine the MN 7 and MN 8 intervals into a single MN 7/8 unit.

A New Interpretation of the Origin and Evolutionary Relationships of the Anomalomyidae

It is currently impossible to resolve the phylogeny of the Anomalomyidae. Enough information is in hand however to show where the problems are and to pose some cogent hypotheses about anomalomyid relationships. Data are not yet sufficient to formulate evolutionary "lineages," however

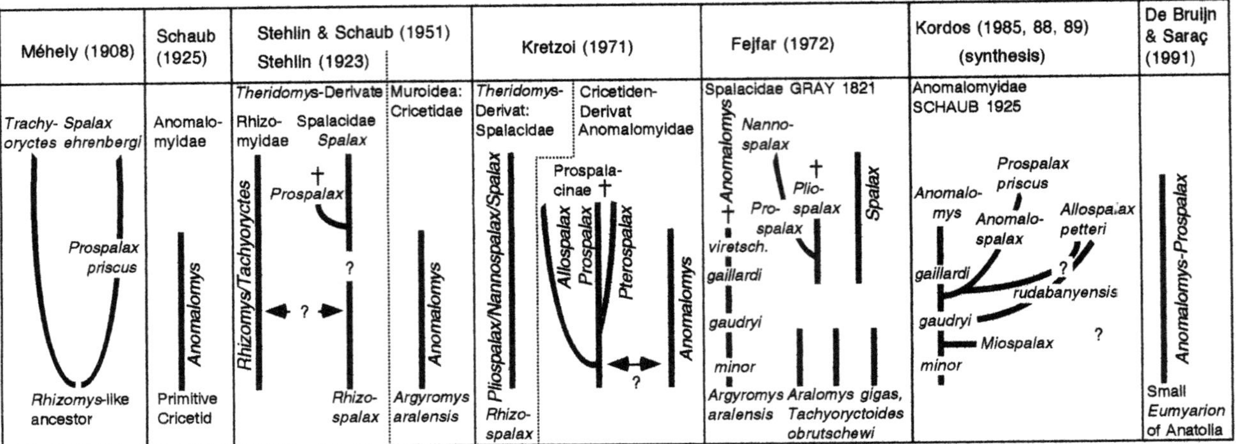

FIGURE 18.1 Summary of phylogenetic hypothesis by various authors.

I do suggest plausible evolutionary relationships based on tooth morphology. Despite the much increased data base, the family's evolutionary complexity is such that we can only pose probable phylogenetic relationships based on a few morphological characters. This is a feature of the record often common to paleontology and one that should be kept in mind when referring taxa to zoological genera and species.

I have reviewed all the available literature and material on Miocene Anomalomyidae. Figure 18.2 gives the tooth-morphological species relationships. Figure 18.3 gives the geographic distribution of species. Figures 18.4 and 18.5 provide information on species size variation and tooth morphology/distribution of sites. Possible migrations due perhaps to climatic changes will be discussed further below.

As previously mentioned, there has been a great deal of confusion surrounding anomalomyid and spalacid relationships because of their convergent development of hypselodont teeth and their morphological analogies. Morphological features common to these groups most probably developed independently, perhaps due to similar ecological preferences. A similar dental specialization can be observed in Oligocene *Rhizospalax* (Fejfar 1972), the middle Miocene *Pliospalax*–group (Ünay 1981), Mio–Pliocene Rhizomyidae (Flynn 1982), the Pleistocene to Recent *Prospalax*–group, and the *Spalax*–group (Fejfar 1972). I basically agree with de Bruijn and Sarac (1991) that *Eumyarion* is closely related to *Anomalomys*, but I am more skeptical about the likelihood that *Anomalomys* is derived from *Eumyarion*. I believe that they share a common ancestor of an Oligocene Asian origin.

The earliest representatives of both European *Anomalomys* immigration events are known from Greece and Turkey. Fejfar (1972) believed that a simple size increase occurred within the Anomalomyidae. Daxner-Höck (1980) and Bolliger (1992) have demonstrated that a much more complex evolutionary history occurred. The development of occlusal morphology is complex, showing a pervasive trend toward hypselenodont molars whose occlusal pattern simplifies during wear. It may be argued that there is no direct descent within Europe of the later medium- to large-sized forms of *Anomalomys* from the earlier small forms (figs. 18.2 and 18.3). This is why I postulate a second European anomalomyid immigration in MN 6 (FAD at Sariçay and Neudorf Sandberg) from somewhere eastward.

The medium- to large-sized *Anomalomys* diversified into new lineages: *Anomalospalax tardosi* was probably derived from a form similar to *Anomalomys gaillardi*, while *Prospalax petteri* probably evolved from a form like *Anomalomys gaudryi*. The details of these proposed relationships are currently difficult to demonstrate because there seems to be a large number of local species partially representing transitional evolutionary stages. Moreover, many localities lack sufficient material to make secure species identifications. Kordos (1989:306) believes that *Anomalospalax* was a recent immigrant. It may be that *Prospalax petteri* and *Prospalax priscus* were also new immigrants, but it is currently difficult to demonstrate these assertions with the data at hand.

Significance of Anomalomyidae in Small Mammal Assemblages

Anomalomyids are generally rare in faunal assemblages, occasionally representing as much as 20% of a fossil sample. They do not appear to be especially predated upon by particular raptors (Andrews 1990), and their possible ecological biases are poorly understood. Only geologically younger species from the Carpathian region seem to occur in relatively great abundance. The Anomalomyidae are belived to have lived underground something like recent *Spalax* (Fejfar 1972:189; Görner and Hackethal 1989:247). Anomalomyid migration has been believed to be related to regional environmental change. Moreover, the Anomalomyidae are believed to have some value for approximate

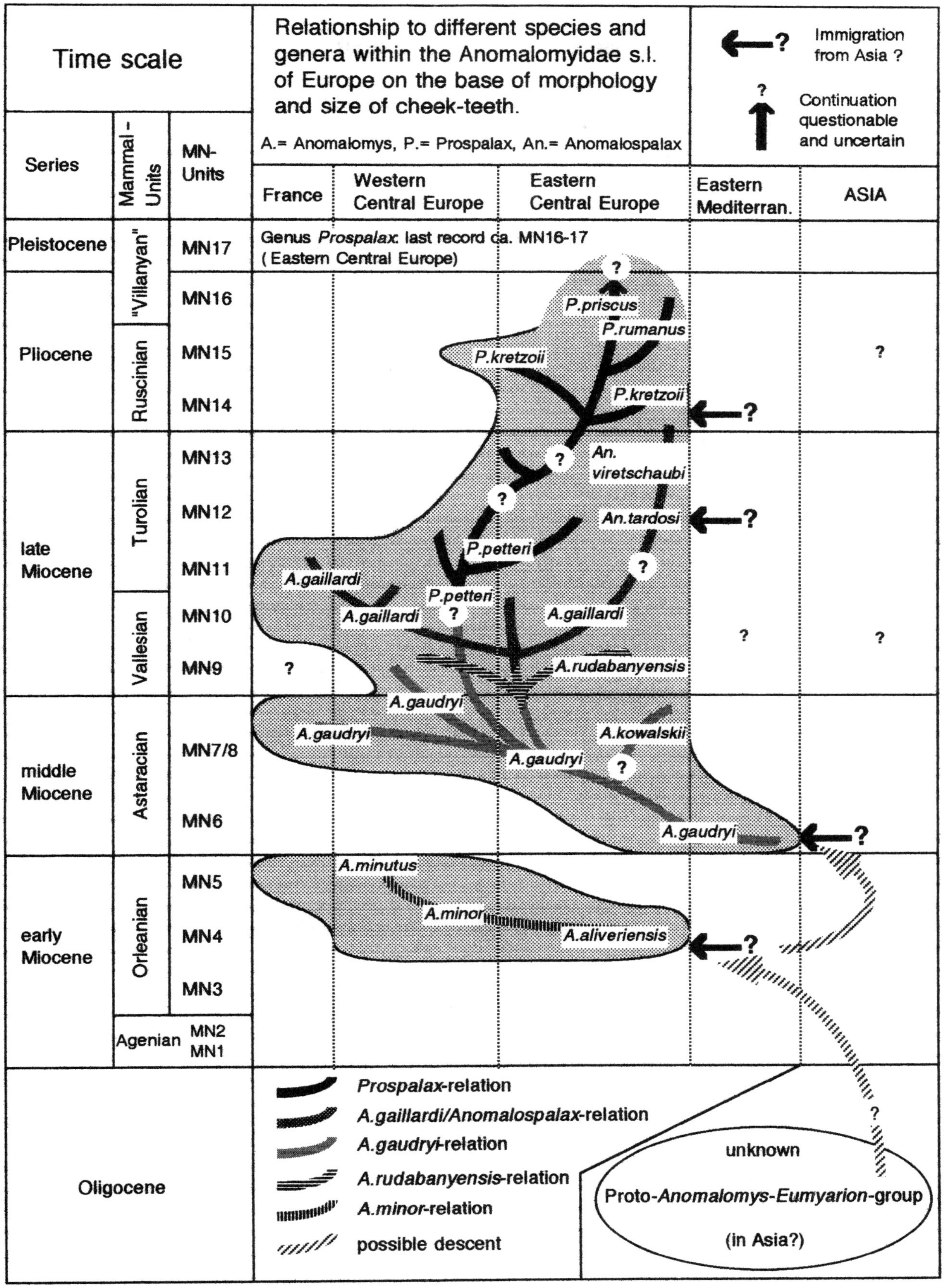

FIGURE 18.2 Relationship of Anomalomyidae. There is a broad diversification of taxa during the late Miocene, making it difficult to reconstruct more precise hypotheses of relationship.

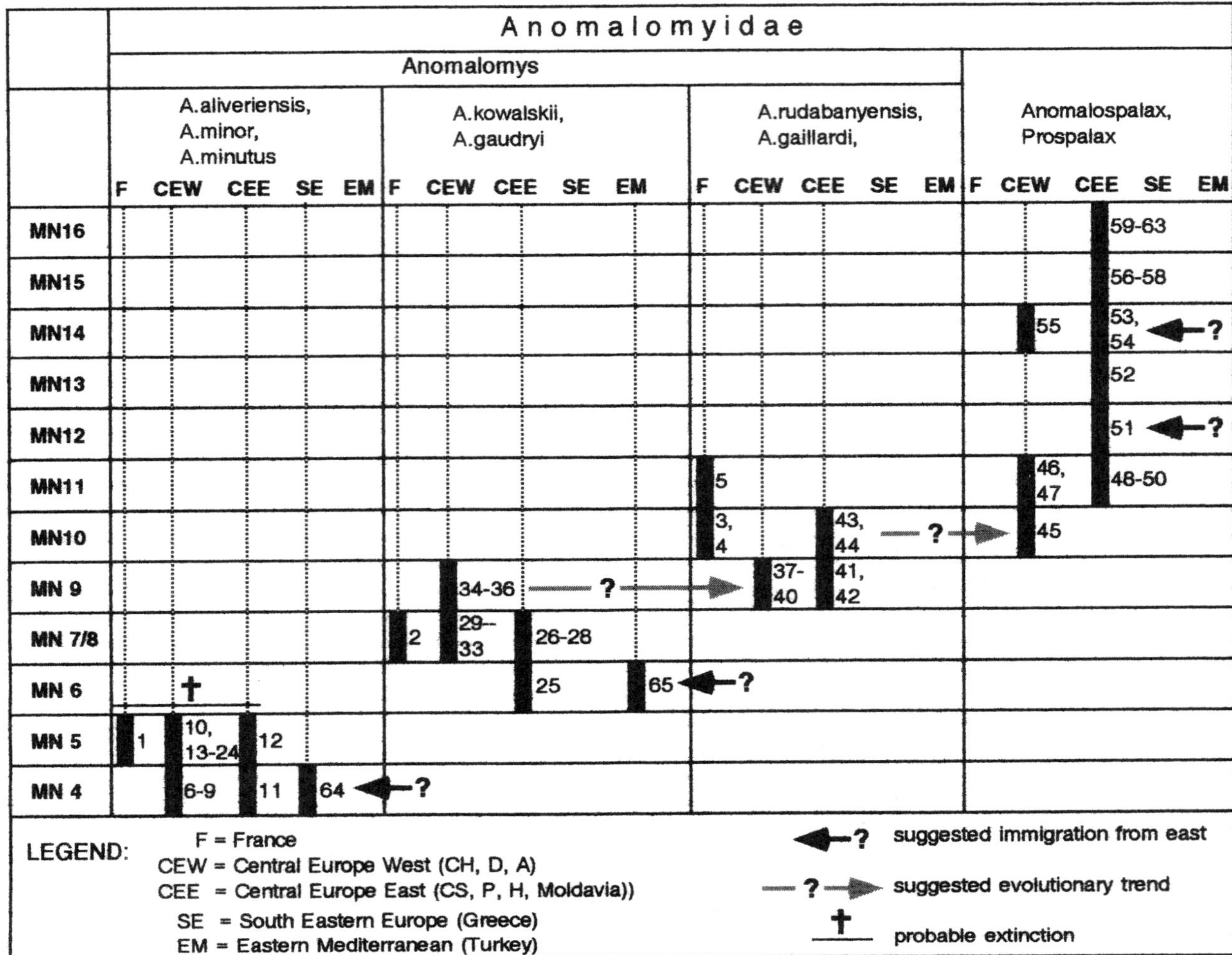

FIGURE 18.3 Stratigraphic and geographic distribution of Anomalomyidae from MN 4 to MN 16/17. Numbers of localities refer to those of table 18.1.

biochronologic correlations in Central Europe s.l. because of evolutionary trends, including the development of crown size, crown height, and enamel pattern from the early to late Miocene (see below on stratigraphic distribution of Anomalomyidae).

Taxonomy of the Anomalomyidae

Provided below is a short overview of the taxonomy of the Anomalomyidae. Detailed diagnoses are not included.

VALID NAMES:

Anomalomys Gaillard 1900

ANOMALOMYS MINOR–GROUP

Anomalomys aliveriensis Klein Hofmeijer and de Bruijn 1985   MN 4. This is a small species known only from Greece. It is closely related to *A. minor*.

Anomalomys minor Fejfar 1972   MN 4–MN 5. This is the first described of three known small *Anomalomys* species. It is broadly distributed throughout Europe.

Anomalomys minutus Bolliger 1992   MN 5 (youngest). This is the smallest species of *Anomalomys*. It is only known from the north Alpine Molasse Basin.

ANOMALOMYS GAUDRYI–GROUP

Anomalomys gaudryi Gaillard 1900   MN 6–MN 9. This is the genotype species, known to occur in Central Europe and France. It is distinctly larger than the three previously mentioned species, showing a slightly simpler molar enamel pattern.

Anomalomys kowalskii Kordos 1989   Valid (?), MN 7/8. This species has been described based on material from Opole 2 and Szentendre. It shows only a few differences with *Anomalomys gaudryi* and may prove to be a junior synonym of that species.

Anomalomys gaillardi–Group

Anomalomys gaillardi Viret and Schaub 1946   MN 9–MN 11. This is the largest known species of the genus, with quite hypsodont molars and an apparently massive crown. It has a recorded range of France to Moldavia, but the true identity is partly uncertain because of poor material.

Anomalomys rudabanyensis Kordos 1989   MN 9. This has been described from Rudabánya and found to be intermediate in size between *A. gaudryi* and *A. gaillardi*. m1 and m2 anteroconids are more separated in *Anomalomys rudabanyensis*. The upper molars' fold system is more simplified than in *Anomalomys gaudryi*. The *Anomalomys*-species from Hillenlohe therefore is very closely related to it. The Götzendorf species is morphologically similar, but distinctly larger, and has thus been provisionally named *Anomalomys* cf. *gaillardi* (fig. 18.5). The poor material from Nebelbergweg is also supposed to be more closely related to *Anomalomys gaillardi*.

Prospalax Méhely 1908

Prospalax petteri Bachmayer and Wilson 1970   MN 10–MN 11. This species is widely distributed over the eastern portion of Central Europe. It may have evolved from *Anomalomys gaillardi*, but shows a further simplification of the enamel pattern, finally leading to that of *Prospalax priscus*. Within a transitional stage of development from *Anomalomys* to *Prospalax*, it may be difficult to find good arguments for separation. Kordos (1989) referred the species *P. petteri* to the formerly synonymous genus *Allospalax*, deciding that it was neither *Anomalomys* nor *Prospalax*; this is as yet uncertain, and I have elected to leave it within *Prospalax* for the time being.

Prospalax rumanus Simionescu 1930   MN 16? *P. rumanus* is only known from Malusteni, Romania, and its inclusion within the genus *Prospalax* is uncertain.

Prospalax priscus (Nehring 1897)   MN 15–MN 16/17. This is the genotype species. It is large and only known in later Pliocene aged eastern Central European horizons.

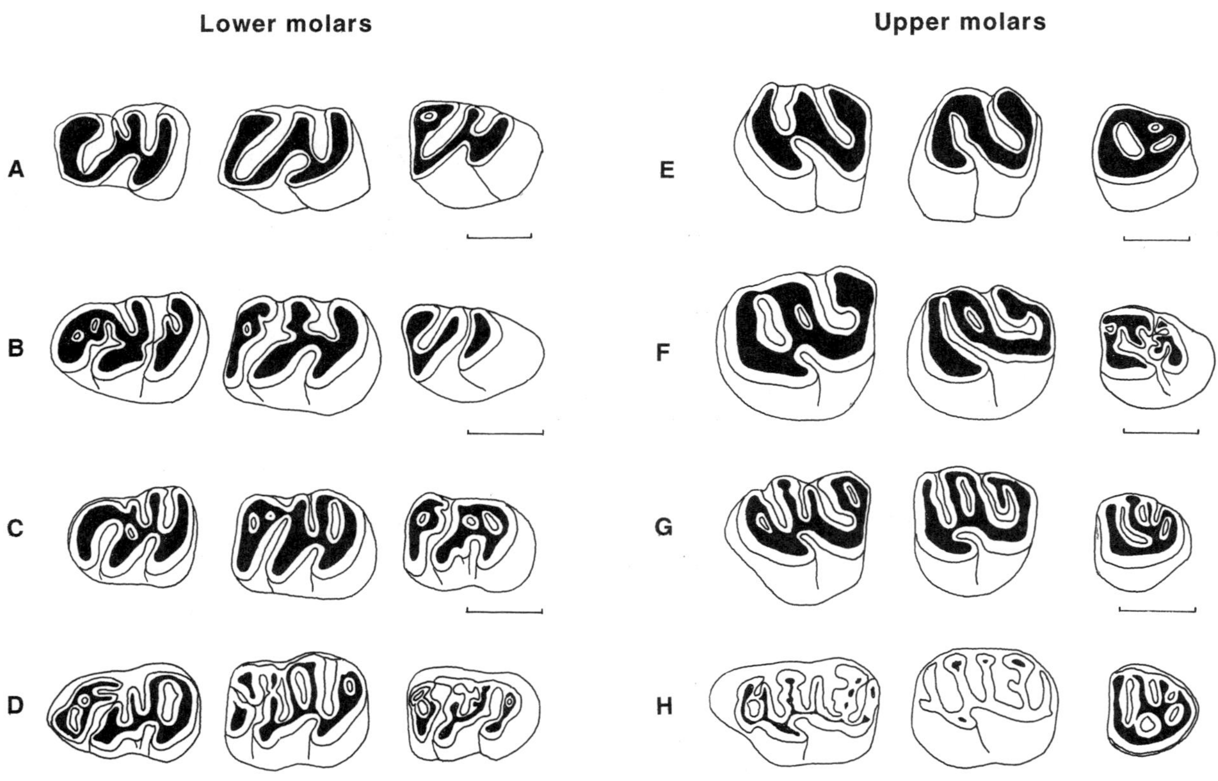

FIGURE 18.4 Dentition of some Anomalomyidae, showing the tendency toward a simplified enamel-pattern. A,E: *Anomalospalax tardosi* from Tardosbánya, MN 12, after Kordos 1985. B,F: *Anomalomys gaillardi* from Montredon, MN 10, after Viret and Schaub 1946 and Fejfar 1972. C,G: *Anomalomys gaudryi* from La Grive, MN 7/8, after Viret and Schaub 1946. D,H: *Anomalomys aliveriensis* from Aliveri, MN 3/4, after de Bruijn and Sarac 1991. Note that teeth are enlarged to fit all drawings at approximately equal size; bars are equal to 1 mm.

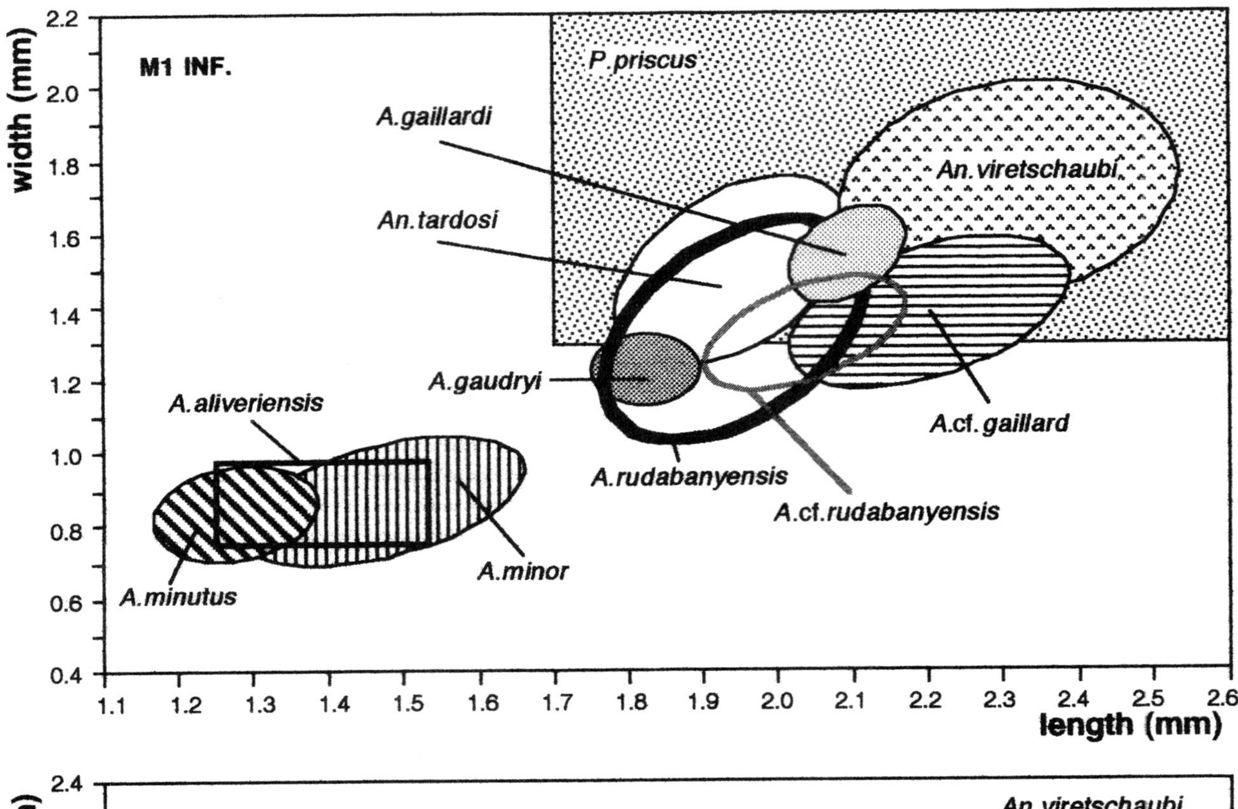

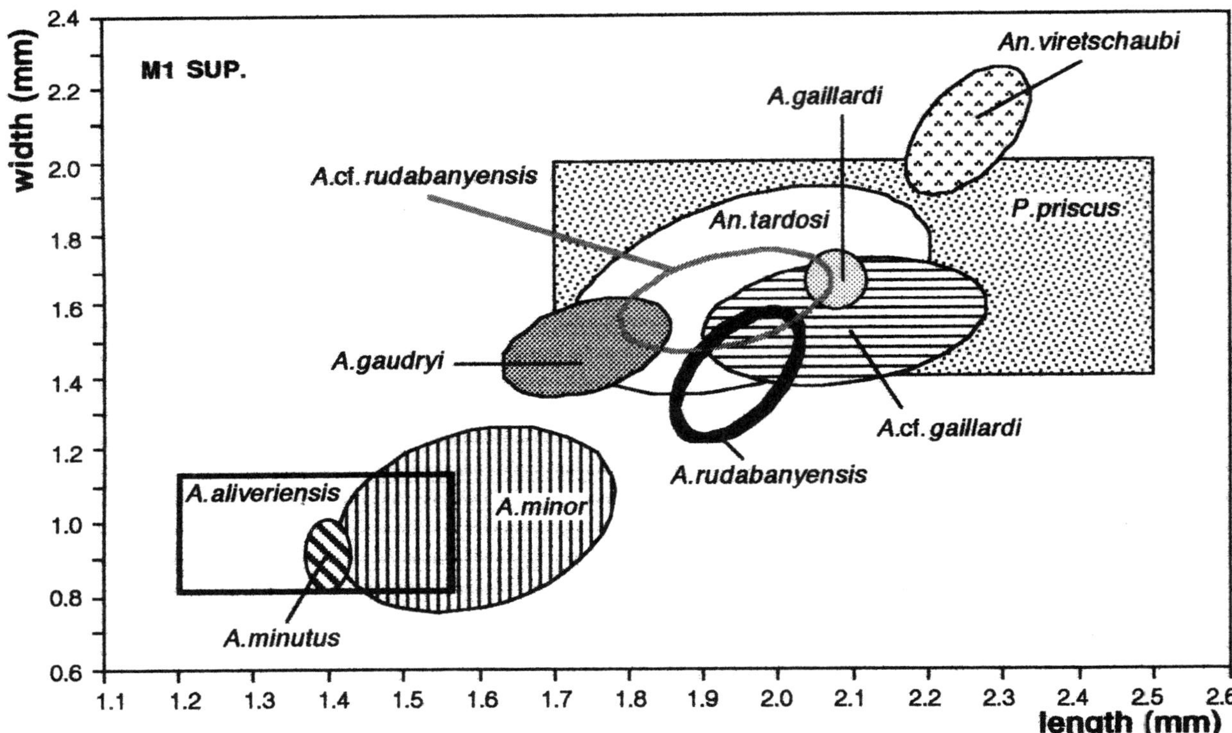

FIGURE 18.5 Size variation of some important and characteristic Anomalomyidae populations.

Legend with sources of data:

Anomalomys aliveriensis:	MN 4	Aliveri, Greece	Klein Hofmeijer and de Bruijn 1985:194
A. minor:	MN 4	Jona, Switzerland	Bolliger 1992:288
A. minutus:	MN 5	Hombrechtikon, Switzerland	Bolliger 1992:28
A. gaudryi:	MN 7/8	La Grive, France	Material NHM, Wien
A. rudabanyensis:	MN 9	Rudabánya, Hungary	Kordos 1989:300
A. cf. rudabanyensis:	MN 9	Hillenlohe, Germany	Material PIUM. München
A. cf. gaillardi:	MN 9	Götzendorf, Austria	Material NHM, Wien
A. gaillardi:	MN 10	Montredon, France	Material NHM, Wien (casts)
Anomalospalax tardosi:	MN 12	Tardosbánya, Hungary	Kordos 1985b:41
Anomalospalax viretschaubi:	MN 13	Polgárdi, Hungary	Kordos 1985b:42
Prospalax priscus:	MN 16/17	Weze 1, Poland	Sulimski 1964:218

Anomalospalax Kordos 1985b

Anomalospalax tardosi Kordos 1985b   MN 12. The genus is very closely related to *Anomalomys* and *Prospalax*. However, the molars are generally broader than in these genera. The enamel folds are always connected.

Anomalospalax viretschaubi (Kretzoi 1971)   MN 13. This is one of the largest known anomalomyid species, having a size more or less equal to *Prospalax priscus*. It is known only from the late Turolian locality of Polgardi (MN 13). Unfortunately, no recent drawings or photos have been published.

SYNONYMS

Anomalomys gernoti Daxner-Höck 1980   Synonym of *Prospalax petteri* Bachmayer and Wilson 1970. The holotype is a slightly damaged tooth, described as an m2. However three teeth figured by Daxner-Höck (1980: figs. 1c, g, h) are very probably all m3s of *Prospalax petteri*. The tooth fragment mentioned by Daxner-Höck is an m2, not an m1. These teeth fall within the range of *Prospalax petteri*, both metrically and morphologically. Only the smaller teeth (Daxner-Höck 1980: fig. 1a,b,i) could be of another species (*Anomalomys gaudryi*–related), but having seen the size-variation of the *Anomalomys*–species from Götzendorf, it seems possible that they only represent the smallest teeth of a single population. It is currently difficult to decide whether the Eichkogel and Kohfidisch samples of *Prospalax petteri* represent a single species in that the Kohfidisch teeth show a slightly simpler enamel fold pattern. Based on the Eichkogel sample, it seems plausible that *Prospalax petteri* evolved from a form closely related to *Anomalomys gaudryi*.

Anomalomys viretschaubi Kretzoi 1971   Synonym of *Anomalospalax viretschaubi* (Kretzoi 1971). Unfortunately, Kordos (1985b) has provided no figures or plates of his new material, but has reported that this species size-variation suggests a relationship to *Anomalospalax tardosi*.

Allospalax plenus Kretzoi 1971   Synonym of *Prospalax petteri* Bachmayer and Wilson 1970. This synonymy has previously been advanced by Kordos (1989:305). However, Kordos believes that *Prospalax petteri* is referable to a genus other than *Prospalax priscus* and therefore has nominated the nomen *Allospalax*. As argued above, I prefer maintaining the nomen *Prospalax petteri* and synonymizing *Allospalax* within it.

Miospalax monacensis Stromer 1928   Synonym of *Anomalomys gaudryi* Gaillard 1900. The synonymy has already been partly explained by Stromer (1940:31).

Pterospalax rumanus (Simionescu 1930)   Synonym of *Prospalax rumanus* (Simionescu 1930). Kretzoi (1971) mentions the close relationship between "*Pterospalax*" and *Prospalax*, and argues for generic separation based on differences in mandibular features. I doubt that the purported morphological differences are enough to separate this form from *Prospalax*.

DOUBTFUL NAMES

Prospalax kretzoii Janossy 1972   The problem of this taxon's generic membership has been discussed by Kordos (1985:30).

Prospalax macoveii Simionescu 1930   This taxon may belong to the Spalacidae.

Stratigraphic Distribution of the Anomalomyidae

The list given in table 18.1 does not give a complete inventory of all known European localities with *Anomalomys*, but rather it records all of the available published localities. Spanish occurrences, as well as findings from MN 4 and MN 14–17 localities, have been included for the sake of completeness. The tabulation provided here clearly demonstrates that most occurrences are from Central Europe, which is probably in part due to longer and more intensive collecting. This is especially the case in areas such as Greece and Turkey where collections have been ongoing for only about twenty years. It turns out that most *Anomalomys* species are valuable for correlation, provided that enough material is collected. Even a poor *Anomalomys* fauna may be used for a first rough biochronologic correlation. A biochronologic ranking using anomalomyid species is given below.

Thus far we can distinguish seven species belonging to the genus *Anomalomys*. The small-sized *Anomalomys aliveriensis* Klein Hofmeijer and de Bruijn 1985, *Anomalomys minor* Fejfar 1972, and *Anomalomys minutus* Bolliger 1992 can be placed in an *Anomalomys minor*–group. The medium-sized *Anomalomys gaudryi* Gaillard 1900 and *Anomalomys kowalskii* Kordos 1989 species are referable to an *Anomalomys gaudryi*–group. The rather large and high-crowned *Anomalomys gaillardi* Viret and Schaub 1946 is referred to the *Anomalomys gaillardi*–group. *Anomalomys rudabanyensis* Kordos 1989 also is referable to this group, but only in a broad systematic sense. The various groups' biochronologic distributions are interpreted as follows: *Anomalomys minor*–group (small size), MN 4–5; *Anomalomys gaudryi*–group (medium size), MN 6–9; *Anomalomys gaillardi*–group (medium to large size) MN 9–11.

Nr.	Locality	Country	MN	mentioned in ("this work"= newly [re-]considered)	Species
(Numbers to the left refer to the distribution-sketch of fig.1)					
France (F):					
1	Vieux-Collonges	F	MN5	Mein (1958); Fejfar (1972)	A. sp.
2	La Grive	F	MN7/8	Viret and Schaub (1946)	A. gaudryi
3	Montredon	F	MN10	Viret and Schaub (1946)	A. gaillardi
4	Soblay	F	MN10	Fejfar (1972)	A. (cf.) gaillardi
5	Mollon	F	MN11	Viret and Schaub (1946)	A. cf. gaillardi
Central Europe (CE):					
6	Jona-Tägernaustr.	CH	MN4	Bolliger (1992)	A. minor
7	Rauscheröd	D	MN4	Ziegler and Fahlbusch (1986)	A. minor
8	Rembach	D	MN4	Ziegler and Fahlbusch (1986)	A. minor
9	Forsthart	D	MN4	Ziegler and Fahlbusch (1986)	A. minor
10	TU2, B.*, 34m	D	MN4-5	Eberhard (1986)	A. minor
11	Yáralja 21, B.*, 265m	H	MN4	Kordos (1985a, 1987, 1989)	A. minor
12	Franzensbad	CS	MN5	Fejfar (1972)	A. minor
13	Rothenstein 1/13	D	MN5	Rummel (1992), this work	A. minor
14	Massendorf	D	MN5	Schötz (1980)	A. minor
15	Niederaichbach	D	MN5	Schötz (1980)	A. minor
16	Bellenberg 2	D	MN5	Boon (1991)	A. minor
17	Puttenhausen	D	MN5	Wu (1982)	A. minor
18	Langenmoosen	D	MN5	Fejfar (1972)	A. minor
19	Undorf	D	MN5	Fejfar (1972)	A. minor
20	Goldern	D	MN5	Mayr et al. (1988); Spitzlberger (1989)	A. minor
21	Arth	D	MN5	Fiest under work; this work	A. minor
22	Ponholz	D	MN5	(Fahlbusch 1985); this work	A. cf. minutus
23	Gisseltshausen 1a	D	MN5	(Heissig 1989); this work	A. cf. minutus
24	Tobel-Hombrechtikon	CH	MN5	Bolliger (1992)	A. minutus
25	Neudorf 2 (Spalte)	CS	MN6	Fejfar (1972)	A. gaudryi
26	Opole (Oppeln) 2	P	MN7/8	(Kowalski 1967; Fejfar 1972); Kordos (1989)	A. kowalskii
27	Szentendre	H	MN7/8	Kordos (1987, 1989)	A. kowalski
28	Hasznos	H	MN7/8	Kordos (1987, 1989)	A. gaudryi
29	Steinheim	D	MN7/8	Schaub (1925)	A. gaudryi
30	Petersbuch 7	D	MN7/8	Rummel under work; this work	A. gaudryi
31	Le Locle s.I.S.	CH	MN7/8	Kälin in prep.; this work	A. cf. gaudryi
32	Chlihörnli 1000m	CH	MN7/8	Bolliger and Eberhard (1989); Bolliger (1992)	A. gaudryi
33	Anwil	CH	MN7/8	Engesser (1972)	A. gaudryi
34	Grosslappen	D	MN8-9	(Stromer 1928; Fejfar 1972); this work	A. gaudryi
35	Aumeister	D	MN8-9	(Stromer 1940; Fejfar 1972); this work	A. gaudryi
36	Hammerschmiede	D	MN8	Mayr and Fahlbusch (1975)	A. gaudryi
37	Hillenlohe	D	MN8	(Jung and Mayr 1980); this work	A. cf. rudabanyensis
38	Nebelbergweg	CH	MN8-9	Kälin in prep.; this work	A. cf. gaillardi
39	Götzendorf 1	A	MN8	This work	A. cf. gaillardi
40	Vösendorf	A	MN8	Rabeder (1985)	? A. sp.
41	Rudabanya 1/2	H	MN8	(Kretzoi et al. 1976); Kordos (1989)	A. rudabanyensis
42	Suchomasty	CS	MN9	Fejfar (1989)	A. gaillardi
43	Bujori (Kotovskovo)	M**	MN10?	Lungu (1981), Kordos (1989)	A. gaillardi
44	Csákvár	H	MN10-11	Kordos (1987, 1989)	A. gaillardi
45	Kohfidisch	A	MN10	Daxner-Höck (1980)	P. petteri
46	Eichkogel	A	MN11	Daxner-Höck (1980)	P. petteri, (p.p. "A. gernoti")
47	Dorn-Dürkheim	D	MN11	Franzen & Storch (1975); Daxner-Höck (1980)	P. petteri
48	Sümeg	H	MN11	(Kretzoi 1971; Kordos 1989); this work	P. petteri
49	Tihany	H	MN11	(Kordos 1989); this work	P. petteri
50	Nyárád	H	MN11	Kordos (1989)	P. petteri
51	Tardosbánya	H	MN12	(Kordos 1985b, 1988)	An. tardosi
52	Polgárdi	H	MN13	(Kretzoi 1971; Kordos 1985b, 1988)	An. viretschaubi
53	Podlesice	P	MN14	(Mein 1989)	? A. sp.
54	Osztramos 1	H	MN14	Jánossy (1972)	P. kretzoi
55	Gundersheim-Findling	G	MN15	Storch and Fejfar (1989)	P. kretzoi
56	Weze	P	MN15	Kretzoi (1962); Sulimski (1964); Kordos (1988)	P. priscus
57	Csarnóta 1, 2, B.*	CS	MN15	Kordos (1988)	P. priscus
58	Ivanovce	CS	MN15	Kordos (1988)	P. priscus
59	Hajnácka	CS	MN16	Kordos (1988)	P. priscus
60	Rebielice-Królewskie	CS	MN16	Kretzoi (1962)	P. priscus
61	Malusteni	Rom	MN16?	Simionescu (1930); Kormos (1932)	P. rumanus
62	Beremend 5,11, (Villany)	H	MN16-17	Méhely (1908); Kordos (1988)	P. priscus
63	Nagyharsány 1, (Villany)	H	MN16-17	(Nehring 1897; Méhely 1908; Kordos 1988)	P. priscus
B.*=Bohrung, M**=Moldavia, ex-UssR, Rom= Romania					
South-Eastern Europe (SE):					
64	Aliveri	G	MN3/4	Hofmeijer and de Bruijn (1985)	A. aliveriensis
Eastern Mediterranean (EM):					
65	Sariçay	T	MN6-7?	Ünay (1978); Kordos (1989)	A. gaudryi
***Anomalomys* of Spain:**					
	San Quirico	E	MN7/8?	Viret and Schaub (1946)	A. gaudryi
	Can Llobateres	E	MN9	Viret and Schaub (1946)	A. cf. gaillardi

The MN-unit referrals conform to those given by referenced authors and/or the list given by Mein (1989), and/or more recent proposals of the 1992 Reisensburg Workshop sponsored by the VW Stiftung.

Regional Distribution and Migration Patterns

Although information on the origin and evolution of the Anomalomyidae has increased significantly over the last few years, we still have much to learn about the group. In order to provide a clearer differentiation of Central European localities yielding *Anomalomys*, an eastern region (Czechoslovakia [CS], Poland [P], Austria [A], Hungary [H], and Moldavia [M]) has been distinguished from a western region (Switzerland [CH], Germany [D], and France [F]) in figures 18.2 and 18.3.

Aliveri, Greece, has yielded the oldest known *Anomalomys*, suggesting an eastern origin for the family. The family then first occurs in Central Europe during MN 4, and by MN 5 it is recorded from France. By the end of MN 5, the *Anomalomys minor*–group is nearly extinct, having its last occurrence recorded by the smallest *Anomalomys* species, *A. minutus* (north Alpine Molasse Basin). Without an apparent intermediate stage-of-evolution, *Anomalomys gaudryi* and *Anomalomys kowalskii* first occur in MN 6 of Turkey (Sariçay) and Central Europe (Opole, Neudorf) apparently as the result of immigration from the east. To the west these species first occur in MN 7/8 faunas. From MN 9 onward a discontinuous development occurs, leading to generally larger species such as *Anomalomys rudabanyensis*, *Anomalomys gaillardi*, the two members of *Anomalospalax* and *Prospalax petteri*. The genus *Anomalospalax* has so far only been found in eastern Central Europe. Whether the later forms of *Prospalax* (*Prospalax kretzoi*, *Prospalax rumanus*, *Prospalax priscus*) descend directly from *Prospalax petteri* is uncertain.

To summarize, it would appear that members of the *Anomalomys*–group (sensu lato) were probably introduced from Asia at least two times, once in MN 4 and again in MN 6–7. The evolution of the *Anomalomys gaudryi*–group (and *Anomalomys kowalskii*) is not directly derived from the European *Anomalomys minor*–group. The second European immigration of *Anomalomys* led to a significant diversification of the group, with a large concentration of species being centered in eastern Central Europe. This diversification center was undoubtedly regulated by ecological factors particular to this area.

Although *Anomalomys* is not considered to be closely related to the spalacids, the Anomalomyidae obviously inhabited a similar habitat. It is believed that this habitat included steppe-like or at least open park-like areas. That *Anomalomys* was successfully introduced all over Europe during MN 4/5 and again during MN 7/8 suggests that climatic conditions changed significantly during those times. Kordos (1989) has suggested a third introduction, with the first occurrence of *Anomalospalax*, during MN 12. However, this "event" is not certain, since *Anomalospalax* may as well have developed from an endemic population of *Anomalomys gaillardi*.

Paleobiogeographic Aspects of Anomalomyidae: A Summary

As figure 18.2 depicts, the Anomalomyidae and their relatives occur principally in Central Europe s.l. However, the first known *Anomalomys*, *A. aliveriensis*, is known from Greece, supporting the hypothesis of an Asian origin. The first *Anomalomys gaudryi*–forms occur in eastern Central Europe and Asia Minor (Sariçay). There is no intermediate form of *Anomalomys minor*–*Anomalomys gaudryi*, so again, migration from the east during MN 6 seems the only plausible explanation. The later *Anomalomys* descendants (from MN 12 onward) occur only in eastern Central Europe and seem to be quite abundant there. Intrestingly, all extant spalacid species (*Microspalax* and *Spalax*) occur from the Balkans eastward and south to Africa (Görner and Hacketal 1988:247), suggesting a similar preference in their habitat. It is distinctly possible that the oldest small anomalomyids had quite different behaviors and ecological preferences than the later-derived species. In that our understanding of anomalomyid evolution is still poor, future research should concentrate on behavior as reflected in the postcranial skeleton and the paleobiological associations of their localities.

Since this article was written and submitted for publication more data concerning the Anomalomyidae has become available. Kowalski (1994) published an article concerning *Anomalomys* in which he refers to *A. kowalskii* as a younger synonym of *A. gaudryi*. The following recent findings (Kowalski 1994) should be noted:

A. minor: Belchatow B and C (Poland); Hirschtal (Switzerland)

A. gaudryi: Belchatow A (Poland); Hostalets de Pierola, Sant Quirize (Spain)

A. gaillardi: Trinxera Sud Autopista, Can Llobateres, La Bastida, Can Petit, Can Jofresa, Ballester (Spain)

Bolliger and Rummel (1994) have published findings of *Anomalomys gaudryi* from Petersbuch 6, 10, 14, 17, and 18. Finally, information about the following (as yet) unpublished findings is listed here according to the data base in process in Helsinki:

A. gaudryi: La Gloria (Spain)

A. cf. rudabanyensis: Dytiko (Hungary)

A. gaillardi: Terrassa (Spain)

Acknowledgments

The author is grateful to the following colleagues for lending original material for study and providing useful discussions on the evolution of anomalomyids: Dr. V. Fahlbusch, Dr. G. Daxner-Höck, Dr. H. de Bruijn, W. Fiest, Dr. K. Heissig, Dr. D. Kälin, Dr. L. Kordos, Dr. H. Mayr, M.

244   T. Bollinger

Rummel, and Dr. W. Wu. Thanks go to R. Bernor for corrections to the grammatical style of this contribution.

LITERATURE CITED

Andrews, P. 1990. *Owls, Caves, and Fossils: Predation, Preservation, and Accumulation of Small Mammal Bones in Caves, with an Analysis of the Pleistocene Cave Fauna from Westerbury-Sub-Mendip, Somerset, UK.* London: British Museum (Natural History).

Bachmayer, F. and R. W. Wilson. 1970. Die Fauna der altpliozänen Höhlen- und Spaltenfüllungen bei Kohfidisch, Burgenland (Österreich). *Annalen des Naturhistorischen Museums, Wien* 74:533–87.

Bolliger, T. 1992. Kleinsäugerstratigraphie in der miozänen Hörnlischüttung (Ostschweiz). *Documenta naturae* 75:1–296.

Bolliger, T. and M. Eberhard. 1989. Neue Faunen- und Florenfunde aus der Oberen Süsswassermolasse des Hörnligebietes (Ostschweiz). *Vierteljahrsschrift der Naturforschenden Gesellschaft in Zürich* 134:109–38.

Bolliger, T. and M. Rummel. 1994. Säugetierfunde aus Karstspalten-Die komplexe Genese am Beispiel eines Steinbruches bei Petersbuch, südliche Frankenalb (Bayern). *Mitteilungen der Bayerischen Staatssammlung für Paläontologie und historische Geologie* 34:239–64.

Boon, E. 1991. *Die Cricetiden und Sciuriden der Oberen Süsswassermolasse von Bayerisch-Schwaben und ihre stratigraphische Bedeutung.* Ph.D. diss., University of München.

Bruijn, H. de. 1984. Remains of the mole-rat *Microspalax odessanus* Topachevsky from Karaburun (Greece, Macedonia) and the Family Spalacidae. *Proceedings, Koninklijke Nederlandse Akademie van Wetenschappen, B* 87:417–25.

Bruijn, H. de, R. Daams, G. Daxner-Höck, V. Fahlbusch, L. Ginsburg, P. Mein, and J. Morales. 1992. Report of the RCMNS working group on fossil mammals, Reisensburg 1990. *Newsletters on Stratigraphy* 26:65–118.

Bruijn, H. de and G. Sarac. 1991. Early Miocene rodent faunas from the Eastern Mediterranean area. Part I. The Genus *Eumyarion*. *Proceedings, Koninklijke Nederlandse Akademie van Wetenschappen, B* 94:1–36.

Daxner-Höck, G. 1980. Rodentia (Mammalia) des Eichkogels bei Mödling (Niederösterreich) 1. Spalacinae und Castoridae, 2. Übersicht über die gesamte Nagetierfauna. *Annalen des Naturhistorischen Museums, Wien* 83:135–52.

Eberhard, M. 1986. Litho- und Biostratigraphie im Oberen Süsswassermolasse Fächer der Adelegg (Südbayern). *Jahrbuch der Geologischen Bundesanstalt* 129:5–39.

Engesser, B. 1972. Die obermiozäne Säugetierfauna von Anwil (Baselland). *Tätigkeitsberichte der Naturforschenden Gesellschaft Baselland* 28:35–363.

Fahlbusch, V. 1985. Säugetierreste (*Dorcatherium, Steneofiber*) aus der miozänen Braunkohle von Wackersdorf/Oberpfalz. *Mitteilungen der Bayerischen Staatssammlung für Paläontologie und historische Geologie* 25:81–94.

Fejfar, O. 1972. Ein neuer Vertreter der Gattung *Anomalomys* Gaillard 1900 (Rodentia, Mammalia) aus dem europäischen Miozän (Karpat). *Neues Jahrbuch für Geologie und Paläontologie, Abhandlungen* 141:168–93.

——. 1989. The Neogene VP sites of Czechoslovakia: A contribution to the Neogene terrestric biostratigraphy of Europe based on rodents. In *European Neogene Mammal Chronology*, ed. E. H. Lindsay, V. Fahlbusch, and P. Mein, pp. 211–36. New York: Plenum.

Flynn, L. J. 1982. Systematic revision of Siwalik Rhizomyidae (Rodentia). *Géobios* 15:327–89.

Franzen, J. L. and G. Storch. 1975. Die unterpliozäne (turolische) Wirbeltierfauna von Dorn-Dürkheim, Rheinhessen (SW-Deutschland). 1. Entdeckung, Geologie, Mammalia: Carnivora, Proboscidea, Rodentia. Grabungsergebnisse 1972–73. *Senckenbergiana lethaea* 65:233–303.

Gaillard, C. 1900. Sur un nouveau rongeur miocène. *Compte rendue de l'académie des sciences* 130:1–2.

Görner, M. and H. Hackethal. 1988. *Säugetiere Europas* Stuttgart: Ferdinand Enke Verlag.

Heissig, K. 1989. Neue Ergebnisse zur Stratigraphie der mittleren Serie der Oberen Süsswassermolasse Bayerns. *Geologica Bavarica* 94:239–57.

Janossy, D. 1972. Middle Pliocene Microvertebrate Fauna from the Osztramos Loc. 1. (Northern Hungary). *Annales historico-naturales musei nationalis hungarici* 64:27–52.

Jung, W. and H. Mayr. 1980. Neuere Befunde zur Biostratigraphie der Oberen Süsswassermolasse Süddeutschlands und ihre palökologische Deutung. *Mitteilungen der Bayerischen Staatssammlung für Paläontologie und historische Geologie* 20:159–73.

Klein Hofmeijer, G. and H. de Bruijn. 1985. The mammals from the lower Miocene of Aliveri (Island of Evia, Greece). Part 4: The Spalacidae and Anomalomyidae. *Proceedings, Koninklijke Nederlandse Akademie van Wetenschappen, B* 88:185–98.

Kordos, L. 1985a. A Magyarorsz gi Eggenburgi-Szarmata képzödnyek sz razföldi gerinces maradv nyai, biozon cíja és rétegtani korrel ciója. A *Magyar llami Földtani intézet évi jelentése az 1983 évröl:* 157–65.

——. 1985b. Lower Turolian (Neogene) *Anomalospalax* gen. n. from Hungary and its phylogenetic position. *Fragmenta Mineralogica et Palaeontologica* 12:27–42.

——. 1987. Neogene Vertebrate Biostratigraphy in Hungary. *Annales Instituti Geologici Publici Hungarici* 70:393–96.

——. 1988. The appearence in Europe of the Genus *Spalax* (Rodentia) and the problem of the Plio–Pleistocene boundary. A *Magyar llami Földtani intézet évi jelentése az 1986 évröl:* 469–91.

——. 1989. Anomalomyidae (Mammalia, Rodentia) remains from the Neogene of Hungary. A *Magyar llami Földtani intézet évi jelentése az 1987 évröl:* 293–311.

Kormos, T. 1932. Neue pliocäne Nagetiere aus der Moldau. *Paläontologische Zeitschrift* 14:193–200.

Kowalski, K. 1967. Rodents from the Miocene of Opole. *Acta Zoologica Cracoviensia* 12:1–33.

——. 1994. Evolution of *Anomalomys* GAILLARD 1900 (Rodentia, Mammalia) in the Miocene of Poland. *Acta zoologica cracoviense* 37/1:163–76.

Kretzoi, M. 1962. A Csarnótai Fauna és Faunaszint. A *Magyar llami Földtani intézet évi jelentése az 1959 évröl:* 297–395.

——. 1971. Bemerkungen zur Spalaciden-Phylogenie. *Vertebrata Hungarica* 12:111–21.

Kretzoi, M., E. Krolopp, H. Lörincz, and I. Plfalvy. 1976. Flora, Fauna und stratigraphische Lage der unterpannonischen Prähominiden-Fundstelle von Rudabánya (NO-Ungarn). *A Magyar llami Földtani intézet évi jelentése az 1974 ëvröl:* 365–94.

Lungu, A. N. 1981. *Hipparionovaja fauna Srednevo Sarmata Maoldavii (nasekomojadnye, zaiceobraznye i gryzuny)* [Hipparion-Fauna from middle-Sarmatian of Moldavia (Insectivora, Lagomorpha, Rodentia)]. Kishinev: Ztinca. (Booklet written in Russian)

Mayr, H. and V. Fahlbusch. 1975. Eine unterpliozäne Kleinsäugerfauna aus der Oberen Süsswasser-Molasse Bayerns. *Mitteilungen der Bayerischen Staatssammlung für Paläontologie und historische Geologie* 15:91–111.

Mayr, H., E. Rieber, and G. Spitzlberger. 1988. Die Fossilfundstelle Goldern bei Landshut (Untermiozän, Karpat). *Sitzungsberichte der Bayerischen Akademie der Wissenschaften, Mathematisch-Naturwissenschaftliche Klasse* 4:63–78.

Méhely, L. 1908. *Prospalax priscus* (Nehring), die pliozäne Stammform der heutigen *Spalax*-Arten. *Annales historico-naturales musei nationalis hungarici* 6:305–16.

Mein, P. 1979 Les mammifères de la faune sidérolithique de Vieux-Collonges. *Nouvelles Archives du Muséum d'histoire naturelle de Lyon* 5:1–122.

———. 1989. Updating of MN zones. In *European Neogene Mammal Chronology*, E. H. Lindsay, V. Fahlbusch, and P. Mein, pp. 73–90. New York: Plenum.

Nehring, A. 1897. Über mehrere neue Spalax-Arten. *Sitzungsberichte der Gesellschaft der naturforschenden Freunde zu Berlin* 10:163–83.

Rabeder, G. 1985. Die Säugetiere des Pannonien. In *Chronostratigraphie und Neostratotypen*, ed. A. Papp, A. Jambor, and F. F. Steininger, pp. 440–43. Miozän der zentralen Paratethys, Budapest, Band 7.

Rummel, M. 1992. *Neue tertiäre fossilführende Spaltenfüllungen der südlichen Frankenalb.* Unpublished diploma-thesis, Erlangen.

Schaub, S. 1925. Die hamsterartigen Nagetiere des Tertiärs und ihre lebenden Verwandten. *Abhandlungen der Schweizerischen Paläontologischen Gesellschaft* 45:1–112.

Schötz, M. 1980. *Anomalomys minor* Fejfar 1972 (Rodentia, Mammalia) aus zwei jungtertiären Fundstellen Niederbayerns. *Mitteilungen der Bayerischen Staatssammlung für Paläontologie und historische Geologie* 20:119–32.

Simionescu, I. 1930. Vertebratele Pliocene dela Malusteni (Covurlui). (Les Vertébrés pliocènes de Malusteni). *Academia Romania, Publicatiunile Fondului Vasile Adamachi* Tomul 9/49: 83–151 (résumé en francais, pp. 137–46).

Spitzlberger, G. 1989. Die Miozänfundstelle Goldern bei Landshut. *Geologica Bavarica* 94:371–407.

Stehlin, H. G. 1923. *Rhizospalax poirrieri* Miller and Gidley und die Gebissformel der Spalaciden. *Verhandlungen der Naturforschenden Gesellschaft in Basel* 34:233–63.

Stehlin, H. G. and S. Schaub. 1951. Die Trigonodontie der simplicidenten Nager. *Schweizerische Paläontologische Abhandlungen* 67:1–385.

Storch, G. and O. Fejfar. 1989. Gundersheim-Findling, a ruscinian rodent fauna of Asian affinities from Germany. In *European Neogene Mammal Chronology*, ed. E. H. Lindsay, V. Fahlbusch, and P. Mein, pp. 405–12. New York: Plenum.

Stromer, E. 1928. Wirbeltiere im obermiozänen Flinz Münchens. *Abhandlungen der bayerischen Akademie der Wissenschaften, Mathematisch-naturwissenschaftliche Abteilung* 32:1–71.

———. 1940. Die jungtertiäre Fauna des Flinzes und des Schweiss-Sandes von München: Nachträge und Berichtigungen. *Abhandlungen der bayerischen Akademie der Wissenschaften, Mathematisch-naturwissenschaftliche Abteilung* N.F. 48:1–71.

Sulimski, A. 1964. Pliocene Lagomorpha and Rodentia from Weze 1 (Poland). *Acta Palaeontologica Polonica* 9:149–261.

Ünay, E. 1978. *Pliospalax primitivus* n. sp. (Rodentia, Mammalia) and *Anomalomys gaudryi* Gaillard from the *Anchitherium* Fauna of Sariçay (Turkey). *Türkiye Jeoloji Kurumu Bülteni* (Bulletin of the Geological Society of Turkey) 21:121–28.

———. 1981. Middle and upper Miocene Rodents from the Bayraktepe Section (Canakkale, Turkey). *Proceedings, Koninklijke Nederlandse Akademie van Wetenschappen, B* 84:217–38.

Wu, W. 1982. Die Cricetiden (Mammalia, Rodentia) aus der Oberen Süsswasser-Molasse von Puttenhausen (Niederbayern). *Zitteliana* 9:37–80.

Viret, J. and S. Schaub. 1946. Le genre *Anomalomys*, rongeur néogène et sa répartition stratigraphique. *Eclogae Geologicae Helvetiae* 39:342–52.

Ziegler, R. and V. Fahlbusch. 1986. Kleinsäuger-Faunen aus der basalen Oberen Süsswasser-Molasse Niederbayerns. *Zitteliana* 14:3–80.

19

On Fossil Spalacidae (Rodentia)

E. ÜNAY

The spalacids are a Southeast European, West Asian, and African group of muroid rodents that are adapted to a strictly fossorial mode of life. The systematics and phylogeny of the spalacids and of those muroid families whose morphology shows a similar burrowing adaptation, the Anomalomyidae, Rhizomyidae, and Tachyoryctoididae, have been extensively discussed by Petter (1961), Topachevski (1969), de Bruijn et al. (1970), and de Bruijn et al. (1981). These authors have suggested that the anomalomyids are ancestral to the spalacids. However, Fejfar (1972) argued that the spalacids descended from the tachyoryctoidids. Black (1972) and Chaline, Mein, and Petter (1977) considered rhizomyids and tachyoryctids to share a common ancestor. More recent studies (Ünay 1978; de Bruijn 1984; Klein Hofmeijer and de Bruijn 1985) have shown that the Spalacidae, Anomalomyidae, Rhizomyidae, and Tachyoryctoididae are discrete families of burrowing muroids with long independent evolutionary histories. While the early species belonging to these groups are strikingly dissimilar to one another morphologically, they later show convergent evolution in a number of characters.

The Spalacidae are a monophyletic group and can be defined as follows: Muroidea with semi-hypsodont, rooted, and massive teeth; primitively, cheek teeth are low crowned and have a complex occlusal pattern; in more advanced forms the teeth become higher crowned and have a simpler occlusal pattern; geologically younger (and more advanced) species have reduced mesoloph(id)'s with lophs and cusps fused resulting in a Z- or an S-shaped pattern; cusps are confluent with short, thick, high ridges in later representatives, inclined to be so in older ones; m1 and m2 lingual sinusids narrow and posteriorly directed; m1 lacks a well-developed metalophid except in *Heramys*; older (and more primitive) species have anterocone and anteroconid retracted, incorporating the protocone and the anteriorly situated, enlarged metaconid. The family Spalacidae includes the genera *Spalax* Güldenstaedt 1770, *Pliospalax* Kormos 1932, *Heramys* Klein Hofmeijer, and de Bruijn 1985, and *Debruijnia* n. gen.

Taxonomy

Spalacidae Gray 1821

GENUS *DEBRUIJNIA* N. GEN.

Type Species   *Debruijnia arpati* n. sp.

Holotype   Fragment of a left mandible with m1 and m2, KE. 5001, (fig. 19.1), measurements in table 19.1.

Derivatio nominis   The genus is named after Dr. Hans de Bruijn in recognition of his work on spalacids; the species is named after the geologist Esen Arpat in recognition of his support of my work as a small mammal paleontologist.

Type Locality   Keseköy

Type Level   Orleanian (MN 3)

Diagnosis   Cheek teeth semi-hypsodont and bunolophodont; cusps and high lophs dominate the occlusal surface, narrow valleys are of secondary importance; m1 and m2 have strongly developed protoconid and hypoconid posterior arms; M1 has a retracted anterocone that may be divided into two interior cusps; M3 has a lingually closed sinus; m1 and m2 have posteriorly directed sinusids; mandibular diastema is short and wide, and the foramen mentale is situated below the anterior root of the m1 as in all true spalacids.

Differential Diagnosis   *Debruijnia* differs from *Heramys* in its larger size, presence of a larger M3 with respect to the M1, a larger length/width ratio of M1, and more primitive dental characters such as the prominent anterocone(id), the stronger mesoloph(id), the longer posterior arm of protoconid and hypoconid, and the more complicated M3. The lophs are less confluent with cusps in *Debruijnia*

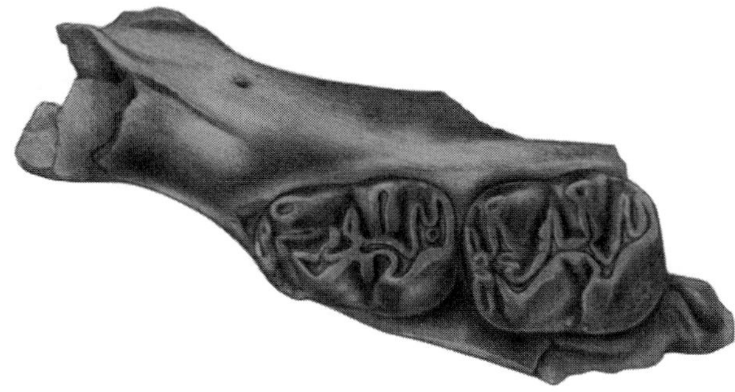

FIGURE 19.1 Holotype of *Debruijnia arpati* from Keseköy (Bolu area, Anatolia). Fragment of a left mandible with m1 and m2. (x 14).

TABLE 19.1 *Measurements on* Debruijnia arpati *n. sp.*

	Length			Width	
	Range	Mean	N	Mean	Range
M1	25.6–29.4	27.1	4	19.9	19.0–21.0
M2	20.0–23.6	21.8	9	20.5	18.6–22.8
M3	17.0–20.0	18.1	10	18.2	16.8–20.0
m1	23.2–26.2	24.6	6	19.3	17.8–20.5
m2	23.0–25.8	24.6	8	20.7	19.6–22.0
m3	21.6–22.0	21.7	3	17.8	16.6–18.4

than in *Pliospalax*. In addition, *Debruijnia* differs from *Pliospalax* in having less hypsodont cheek teeth, the presence of an M1 anterocone, more frequent presence of a protocone anterior arm, well-developed protoconid and hypoconid posterior arms, and an open anterolabial sinusid in the lower molars. *Debruijnia* differs from *Spalax* in being less hypsodont and retaining primitive "cricetid" characters.

Description (fig. 2 and fig. 3)   The M1 has three roots and is longer than wide. The large anterocone is retracted. The anterior wall of the anterocone is smooth in all specimens. A valley on the interior wall divides the anterocone into two cusps in three of six specimens; in the other three specimens this division is not so clear. Irregular ridges descending from the anterocone may be present in the antero-labial valley. The anterocone is separated from the paracone, but connected by a high ridge to the protocone. The protocone anterior arm is very strong in one specimen and short, but distinct, in the others. The transverse or anteriorly directed protoloph is connected to the endoloph behind the protocone. The paracone bears a posteriorly directed spur. The mesoloph is long. The metaloph is transverse and connected to the hypocone in unworn specimens and to the posteroloph in worn ones. The posteroloph is separated from the metacone by a notch. The sinus is oblique anteriorly.

The M2 is sub-square in its outline and has three roots.

The anteroloph labial branch is long and connected to the paracone. The lingual branch is absent. The protoloph, which is usually constricted lingually, is directed slightly forward. A strong protocone anterior arm is present in one specimen, while in this specimen the M2 protoloph is incomplete. There may occur irregular ridges descending from the lingual end of the protoloph into the antero-labial valley. The mesoloph is mostly very long, reaching the labial border, where it may connect to the posterior spur of the paracone. The transverse metaloph is connected to the hypocone. The posteroloph is separated from the metacone by a notch. The sinus is directed forward.

The M3 has a rounded outline and three roots. The anteroloph labial branch is strong and may either be connected to the paracone or separated from it by a notch. The lingual branch is absent. The long transverse protoloph is connected to the anterior part of the protocone. The mesoloph is long in nine out of eleven specimens and reaches the labial border. In four out of eleven specimens, there is an additional ridge (protolophule) descending from the paracone and connecting to the endoloph. The metaloph, which may be interrupted, is directed posteriorly. The posteroloph is short. The sinus is closed lingually by the fusion of the closely situated protocone and hypocone, but it is more or less distinct as a fold on the lingual wall in eight out of eleven specimens.

The m1 is much longer than it is wide and has two roots. The anterior part of the occlusal surface is blunt. The anteroconid, which is situated on the longitudinal axis of the crown, is distinct as a cusp in six out of seven specimens. The lingual branch of the anterolophid is connected to the metaconid in five specimens, in two others it is separated from that cusp by a notch. The labial branch of the anterolophid delimits an anterosinusid. The metaconid is situated anteriorly. The metalophid is connected to the protoconid in two specimens, absent or interrupted in four others. The posterior arm of the protoconid is long and reaches the lingual border in one specimen. The transverse or anteriorly directed mesolophid is very long except for one specimen. It reaches the lingual border in two specimens. The transverse hypolophid is connected to the ectolophid in front of the hypoconid. The posterior arm of the hypoconid is very long. One specimen has a long ectomesolophid. In the others the ectomesolophid is short or absent. The strong posterolophid is separated from the entoconid by a notch in all but one m1. The lingual sinus is directed posteriorly.

The m2 is longer than it is wide and has 2 roots. The lingual and labial branches of the anterolophid are equally well developed and usually extend to the base of the metaconid and protoconid. The transverse metalophid is connected to the anterolophulid. The transverse ridges in the central basin are difficult to homologize: in every specimen there is one very strong ridge (mesolophid?) and there

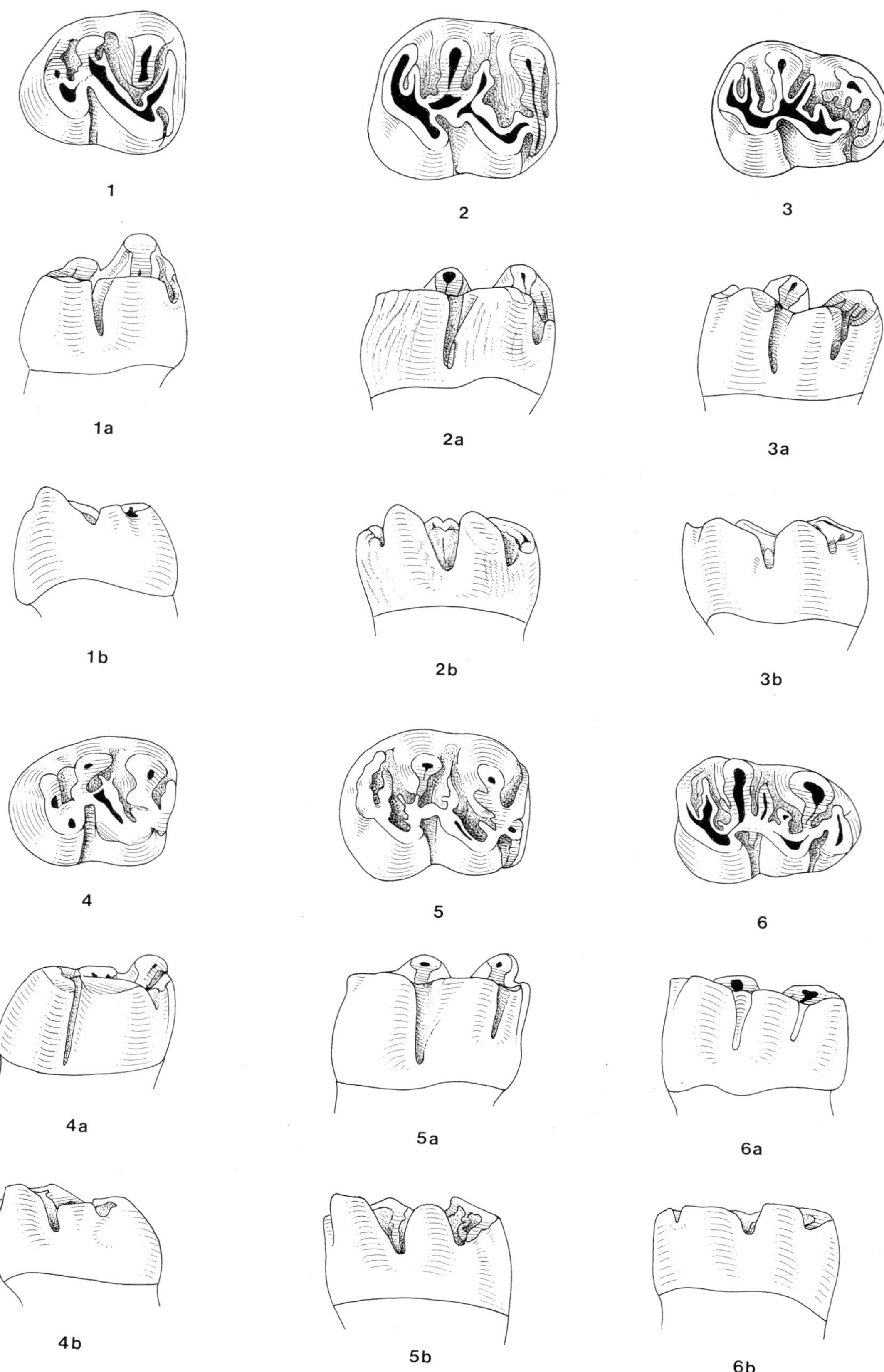

FIGURE 19.2 *Debruijnia arpati* from Keseköy. 1: m3 dext. (occlusal view); 1a: same specimen (labial view); 1b: same specimen (lingual view). 2: m2 dext. (occlusal view); 2a. same specimen (labial view); 2b: same specimen (lingual view). 3: m1 dext. (occlusal view); 3a: same specimen (labial view); 3b: same specimen (lingual view). 4: m3 dext. (occlusal view); 4a: same specimen (labial view); 4b: same specimen (lingual view). 5: m2 dext. (occlusal view); 5a: same specimen (labial view); 5b: same specimen (lingual view). 6: m1 dext. (occlusal view); 6a: same specimen (labial view); 6b: same specimen (lingual view). All figures are approximately x14; 1, 2, and 6 are reversed.

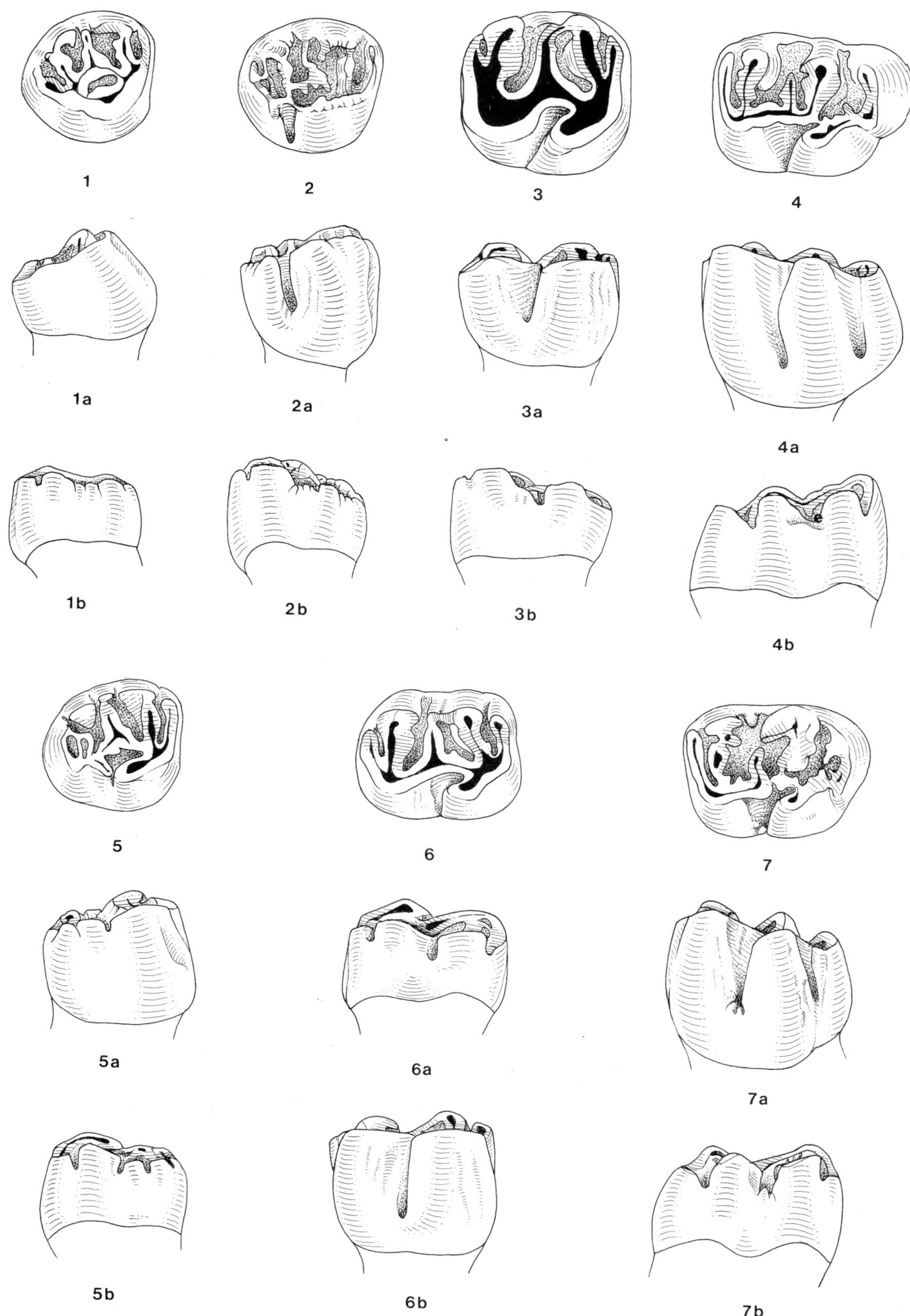

FIGURE 19.3 *Debruijnia arpati* from Keseköy. 1: M3 dext. (occlusal view); 1a: same specimen (lingual view); 1b: same specimen (labial view). 2: M3 dext. (occlusal view); 2a: same specimen (lingual view); 2b: same specimen (labial view). 3: M2 dext. (occlusal view); 3a: same specimen (lingual view); 3b: same specimen (labial view). 4: M1 dext. (occlusal view); 4a: same specimen (lingual view); 4b: same specimen (labial view). 5: M3 dext. (occlusal view); 5a: same specimen (lingual view); 5b: same specimen (labial view). 6: M2 dext. (occlusal view); 6a: same specimen (lingual view); 6b: same specimen (labial view). 7: M1 dext. (occlusal view); 7a: same specimen (lingual view); 7b: same specimen (labial view). All figures are approximately ≡14; 1 and 5 are in reverse.

may be a weak ridge either in front (posterior arm of the protoconid?), or behind it. The strong hypolophid is transversely or anteriorly directed connecting to the ectolophid anterior to the hypoconid. The posterior arm of the hypoconid is strong. One specimen has a weak ectomesolophid, in the others this ridge is absent. The posterolophid is separated from the entoconid. The sinus is directed posteriorly.

The m3 is longer than it is wide and has 2 roots. The symmetrically developed lingual and labial branches of the anterolophid reach the base of the metaconid and protoconid in one specimen, in two others these ridges are weak. The transverse metalophid curves slightly forward, where it connects to the anterolophid. The posterior arm of the protoconid is absent. The mesolophid is short in one specimen but reaches the lingual border in three others. In these it is fused with the entoconid to enclose a funnel. The short transverse hypolophid is connected to the ectolophid anterior to the hypoconid. The posterior arm of the hypoconid is very long in two specimens, short in a third one, and absent in the fourth specimen. The posterolophid may reach the base of the entoconid or may be separated from it by a notch. The sinus is transverse or directed slightly posteriorly.

Phylogeny, Stratigraphic and Geographic Distribution

The geographic range of the fossil Spalacidae approximately corresponds to the distribution of the family's extant species. In spite of a number of new finds made during the last decade, the spalacid fossil record is still limited and does not permit detailed phylogenetic reconstruction. The best record of the Spalacidae is from Southwestern Asia, where they range from MN 3 to Recent. In Southeastern Europe they range from MN 4 to Recent. In Central Europe they are first known to appear much later in time (MN 15?).

The oldest spalacid is *Debruijnia* (MN 3), from Keseköy, Central Anatolia. *Debruijnia* has more characters in common with cricetids than spalacids. However, a number of characters mark its affinity with the spalacids: its semihypsodont cheek teeth and short high ridges that tend to be confluent with massive cusps; the retracted loph-like anterocone/id; the tendancy of the large anteriorly situated metaconid to be confluent with the anterolophid; the antero-labial lower molar sinusid, which is similar to *Heramys* and *Pliospalax*, and the deep narrow posteriorly directed m1 and m2 sinusid, which are derived characters.

The combination of these primitive dental characters and great geologic antiquity suggests that *Debruijnia* is ancestral to other spalacids, that the spalacids originated in Anatolia, and that their divergence from cricetids probably took place in the middle or late Oligocene. That the Asia

Minor Oligocene rodent assemblages have not yet yielded a potential ancestor of *Debruijnia* is probably because of the sparseness of these assemblages as a whole. The spalacids are known to have dispersed from Anatolia into the Balkans, Russia, the Near East, and Africa at several different times.

At Aliveri (Southeastern Europe), the Spalacidae are represented by the genus *Heramys*, in which the M1 anteroloph is still prominent and somewhat separated from the protocone, and the antero-labial sinusid of the m2 and m3 is labially open. This genus has an unusually small M3, which precludes it from being placed at the base of spalacid evolution. However, its morphology and age allow it to be derived from *Debruijnia*. *Heramys* is represented by one species, *Heramys eviensis*.

The genus *Pliospalax* is first recorded in Southwestern Asia during MN 6. *Pliospalax* is characterized by the presence of two lingual and two labial reentrant folds on the m1, by an M1 anterocone that is entirely incorporated into the protoloph, and by an antero-labial sinusid developed as a lake even in unworn teeth. Spalacids with dental patterns intermediate between *Debruijnia* and *Pliospalax* have not yet been found. The known age range of *Pliospalax* in Southwestern Asia is MN 6–MN 15, and in Southeastern Europe it is late MN 8—MN 16, with gaps in MN 9 and MN 11 and 12.

The fragmentary new finds from Anatolia suggest that the evolution of *Pliospalax* is complex and includes many more lineages than formerly recognized. *P. marmarensis* from Paşalar is the oldest known species of this genus. The poorly documented *Pliospalax* from Çandir is of the same size as *P. marmarensis*, but the M1's lingual sinus is transverse and the mesoloph is longer. In *P. marmarensis* the M1's lingual sinus is directed anteriorly and the mesoloph is short. Moreover, the two Çandir M3's are distinctly larger than those of *P. marmarensis* from Paşalar, so there seems to have been at least two MN 6 species in Central Anatolia. *P. primitivus* from Sarıçay and *P. canakkalensis* from Bayraktepe 1 (MN 7/8) are both larger than the *Pliospalax* n. sp. 1 from Çandir and *P. marmarensis*. Their M1 have an anteriorly directed sinus. Ünay (1990) concluded that *P. primitivus* is derived from *P. marmarensis*, and it now seems that the smaller lower-crowned *Pliospalax* n. sp. 1 and the larger higher-crowned *P. canakkalensis* belong to another lineage: the Çandir species' M3 has exactly the same morphology (lingual and labial sinuses confluent), as seen in *P. canakkalensis*.

In Pişmanköy (MN 8, Southeastern Europe), *Pliospalax* is documented (Ünay and de Bruijn 1984) by only five and one-half teeth. These specimens are about the same size as the largest teeth of *P. canakkalensis*. The distinctly longer mesolophid of the two available m1s suggest that they represent another species, *Pliospalax* n. sp. 2.

The poor MN 9–10 *Pliospalax* remains from Southwestern Asia cannot be identified to the species rank. In Southeastern Europe, *Pliospalax compositodontus* is present in two localities assigned to MN 10, Andreevka and Berezan (Ukraine). Two more *Pliospalax* species (*Pliospalax* n. sp. 3 and 4) are documented from Düzyayla 1 (MN 11, Anatolia). Although the material is not sufficient to define new species, it is obvious that both of them are new. One is documented by only one M1 (44.5 x 43.0), which is much larger than any known *Pliospalax*. The other, represented by one broken M2, three M3s, one and one-half m2s, and seven m3s, is larger than the geologically younger *P. macoveii* and *P. tourkobouniensis*, and with the exception of the M3 as large as the smallest teeth of older species *P. primitivus* and *P. canakkalensis*. The M3 is much smaller, and the m3 is extremely reduced posteriorly. *Pliospalax macovei* is known from MN 13 in Southeastern Europe, from MN 15 in Southwestern Asia (Çalta), and MN 15 in Central Europe (Malusteni and Beresti), distinguishing it as having had the longest stratigraphic and broadest geographic distribution of all *Pliospalax* species. One M3 and one m3 from Igdeli in Southwestern Asia (MN 14) complicate the picture even more, because these two teeth are larger than those in contemporaneus species *P. macovei*. These teeth are as large as the older *P. primitivus*, and the M3 shows a trace of the lingual sinus, which is known to be absent in *P. macovei* and *P. primitivus*. This new species is entered as *Pliospalax* n. sp. 5. The geologically youngest *Pliospalax* is *P. tourkobouniensis* from Tourkobounia 1 (MN 16, Southeastern Europe).

The classification of the extant spalacids is also complex. Méhely (1913) recognized one genus of *Spalax* with four subgenera: *Nannospalax*, *Mesospalax*, *Macrospalax*, and *Microspalax*. Corbett (1978) recognized one genus *Spalax* including three species. Topachevski (1969) maintained two genera, *Microspalax* and *Spalax*, while Gromova and Baranova (1981) also recognized two genera, *Nannospalax* and *Spalax*.

Savic and Nevo (1990) have chosen to use the generic nomen *Spalax* to avoid confusion until a taxonomic revision based on chromosomal and genetic data can be used to resolve evolutionary relationships. To my knowledge there is no work dealing with the distinctive dental characteristics of *Microspalax*, *Macrospalax*, *Mesospalax*, *Nannospalax*, and *Spalax*. I therefore follow Savic and Nevo (1990) and use the generic nomen *Spalax* for all the younger spalacids that are usually characterized by two lingual and one labial re-entrant fold on the m1.

Some species of *Spalax*, such as *S. microphatalmus*, preserve the antero-labial fold in some m1s. *Spalax odessanus* (MN 15) is the geologically oldest species referred to this genus. It appears in Southeastern Europe (Karaburun and Odessa). The stratigraphic ranges of *Spalax* and *Pliospalax* overlap because the last record of *Pliospalax* is from the late Villanyian of Tourkobounia. *Spalax minor* and *S.*

ehrenbergi appear in the early Biharian of Southeastern Europe (Nogaisk, Ukraine) and Southwestern Asia (Ubeidiya). *Spalax advenus* is a Central European late Biharian species (Zyrany, Villany 8, Betfia 2, etc.). *Spalax* migrated into Africa during the late Pleistocene (Lay and Nadler 1972).

Acknowledgments

I thank Mr. G. Saraç (Ankara) and Dr. H. de Bruijn (Utrecht) for collecting a larger part of the material discussed above. The project was financially supported by the NATO Scientific Affairs Division (CRG 910750) and the faculty of Earth Sciences, Utrecht, the Netherlands. The figures have been made by Mr. J. Luteijn (Utrecht). Dr. H. de Bruijn critically read the manuscript, and Mrs. A. E. de Bruijn-Dudok van Heel helped to edit the manuscript despite incompatible computer programs.

LITERATURE CITED

Black, C. C. 1972. Revision of fossil rodents from the Neogene Siwalik Beds of India and Pakistan. *Paleontology* 15:238–66.

Bruijn, H. de. 1984. Remains of the mole-rat *Microspalax odessanus* Topachevski, from Karaburun (Greece, Macedonia) and the family Spalacidae. *Proceedings, Koninklijke Nederlandse Akademie van Wetenschappen, B* 87:417–25.

Bruijn, H. de, M. R. Dawson, and P. Mein. 1970. Upper Pliocene Rodentia, Lagomorpha, and Insectivora (Mammalia) from the Isle of Rhodes (Greece). I, II, III. *Proceedings, Koninklijke Nederlandse Akademie van Wetenschappen, B* 73:535–84.

Bruijn, H. de, S. T. Hussain, and J. J. M. Leinders. 1981. Fossil rodents from the Murree Formation near Banda Daud Shah, Kohat, Pakistan. *Proceedings, Koninklijke Nederlandse Akademie van Wetenschappen, B* 84:71–99.

Chaline, J., P. Mein, and F. Petter. 1977. Les grandes lignes d'une classification evolutive des Muroidea. *Mammalia* 41:245–52.

Corbett, G. B. 1978. The mammals of the Palearctic region: A taxonomic review. *British Museum (Natural History)* 3:5.

Fejfar, O. 1972. Ein neuer Vertreter der Gattung *Anomalomys* Gaillard, 1900 (Rodentia, Mammalia) aus dem Europäischen Miozän (Karpat). *Neue Jahrbuch für Geologie und Paläontologie, Abhandlungen* 141:168–93.

Gromova, I. M. and G. L. Baranova. 1981. Catalogue of mammals in USSR. Nauka. (in Russian)

Klein Hofmeijer, G. and H. de Bruijn. 1985. The mammals from the lower Miocene of Aliveri (Island of Evia, Greece). The Spalacidae and Anomalomyidae. *Proceedings, Koninklijke Nederlandse Akademie van Wetenschappen, B* 88:185–98.

Kordos, L. 1988. The appearance in Europe of the genus *Spalax* (Rodentia) and the problem of the Plio–Pleistocene boundary. *Magyar Allami Földtani Intézet* 1988:369–491.

Lay, D. M. and C. F. Nadler. 1972. Cytogenetics and origin of North African *Spalax* (Rodentia, Spalacidae). *Cytogenetics* 11:279–85.

Méhely, L. 1913. Species generis *Spalax*. Die Arten der Blindmäuse in systematischer und phylogenetischer Beziehung. *Ma-*

thematische und Naturwissenschaftliche Berichte aus Ungarn 28:1–385.

Petter, F. 1961. Affinités des genres *Spalax* et *Brachyuromys* (Rongeurs, Cricetidae). *Mammalia* 25:485–97.

Savic, I. R. and E. Nevo. 1990. The Spalacidae: Evolutionary history, speciation, and population biology. In *Evolution of Subterranean Mammals at the Organismal and Molecular Levels*, ed. E. Nevo and O. A. Reig, pp. 129–53. New York: Wiley-Liss.

Topachevski, V. O. 1969. Fauna of the USSR: Mammals, Molerats, Spalacidae. *Akademie Nauk USSR*, n.s., 99:1–29 [translated into English, *Amerind Publications Co. Pvt. Ltd.* 1976(3):119–32).

Ünay, E. 1978. *Pliospalax primitivus* and *Anomalomys gaudryi* Gaillard from the *Anchitherium* fauna of Sariçay. *Bulletin Geologie Societe, Turkey* 21:121–28.

——. 1981. Middle and upper Miocene rodents from the Bayraktepe Section (Canakkale, Turkey). *Proceedings, Koninklijke Nederlandse Akademie van Wetenschappen, B* 84:217–38.

——. 1990. A new species of *Pliospalax* (Rodentia, Mammalia) from the middle Miocene of Paşalar, Turkey. *Journal of Human Evolution* 19:445–53.

Ünay, E. and H. de Bruijn. 1984. On some Neogene rodent assemblages from both sides of the Dardanelles, Turkey. *Newsletters on Stratigraphy* 13:119–32.

20

The Genera of the Murinae, Endemic Insular Forms Excepted, of Europe and Anatolia During the Late Miocene and Early Pliocene

H. DE BRUIJN, J. A. VAN DAM, G. DAXNER-HÖCK, V. FAHLBUSCH, AND G. STORCH

Morphoclines leading from a primitive cricetid pattern of the M1 to the Murinae, the Myocricetodontinae, and the Dendromurinae pattern will be demonstrated on the basis of specimens in the collections of the Institute of Earth Sciences, Utrecht. The Murinae became the most diverse group of rodents in Europe and Western Asia shortly after their arrival in the middle Vallesian (for the Mid-Vallesian crisis, see Fortelius et al., this volume). The (sub)family is represented by at least twelve genera in Europe and Anatolia during the late Miocene and early Pliocene. The stratigraphical and geographical ranges of these genera will be discussed.

The contents of this paper are not the same as in the oral presentation by P. Mein during the 1992 Reisensburg meeting, but we have tried to cover the same subjects and maintain the same style.

The Morphology of the First Upper Molar in Some Selected Cricetinae, Murinae, Myocricetodontinae, and Dendromurinae (figure 20.1)

In many biological characteristics the Murinae cluster with other Muroidea as well as with the Zapodidae, Heteromyidae, Geomyidae, Gliridae, Sciuridae, and Aplodontidae (George 1985). This group of families is considered to have been derived from the Ischyromyoidea branch of the Rodentia (Hartenberger 1980) and not from the Ctenodactyloidea branch as suggested by Flynn et al. (1985). The oldest record of an unquestionable muroid still seems to be *Eucricetodon schaubi* (Zdansky 1930) from the upper Eocene of the He-Ti Formation of Shansi, China (Li and Ting 1984), which suggests that the Ischyromyoidea underwent an extensive radiation in Asia during the Eocene. This is an unexpected conclusion because the Eocene faunas from that area are dominated by Ctenodactyloidea.

The dawn of the Murinae is better documented by

fossils than that of Muroidea, but the phylogenetic relationships between candidates that may be at the basis of this subfamily are controversial. The views expressed by Flynn et al. (1985), Lindsay (1988), and Tong and Jaeger (1993) on the phylogenetic reconstruction of the Megacricetodontinae (sensu Lindsay 1988), the Myocricetodontinae, the Dendromurinae, and the Murinae are not compatible. The fossil record seems too incomplete to allow reconstruction of the phylogenetic context. A relationship that seems reasonably well established is that the Gerbillinae are derived from the Myocricetodontinae (Jaeger 1977; Tong 1989). Even the contents of (sub)families in terms of genera is for most groups still controversial when fossils are considered. The development of the postero-labial sinus of the M1 and M2—long and straight (= primitive) in the Dendromurinae and Cricetomyinae and short or absent (= derived) in the Myocricetodontinae, Murinae, and Gerbillinae—seems to provide a good criterion to separate these groups.

The hypothetical stages in the transformation of the (myo)cricetid dental pattern to the murid pattern have been shown by Jacobs (1977) and Butler (1985), but it is interesting to see how this transformation is documented by fossils (figure 20.1). The first upper molars on figure 20.1 are not figured to scale and not arranged in stratigraphical order, but derived patterns are at the top and primitive patterns at the bottom. The *Spanocricetodon* configuration (figure 20.1:12), with slightly forward directed protoloph and metaloph, symmetrically developed free-ending arms of the protocone and hypocone, strong longitudinal ridge, and long postero-labial sinus, is interpreted as primitive. Peculiar in this tooth is that the lingual cingulum is nevertheless well developed, forming a ledge in front of the anterocone and around the protocone into the lingual sinus and that the anterocone is weakly divided into two cusps.

The teeth in the next row (figure 20.1:9,10,11) have in

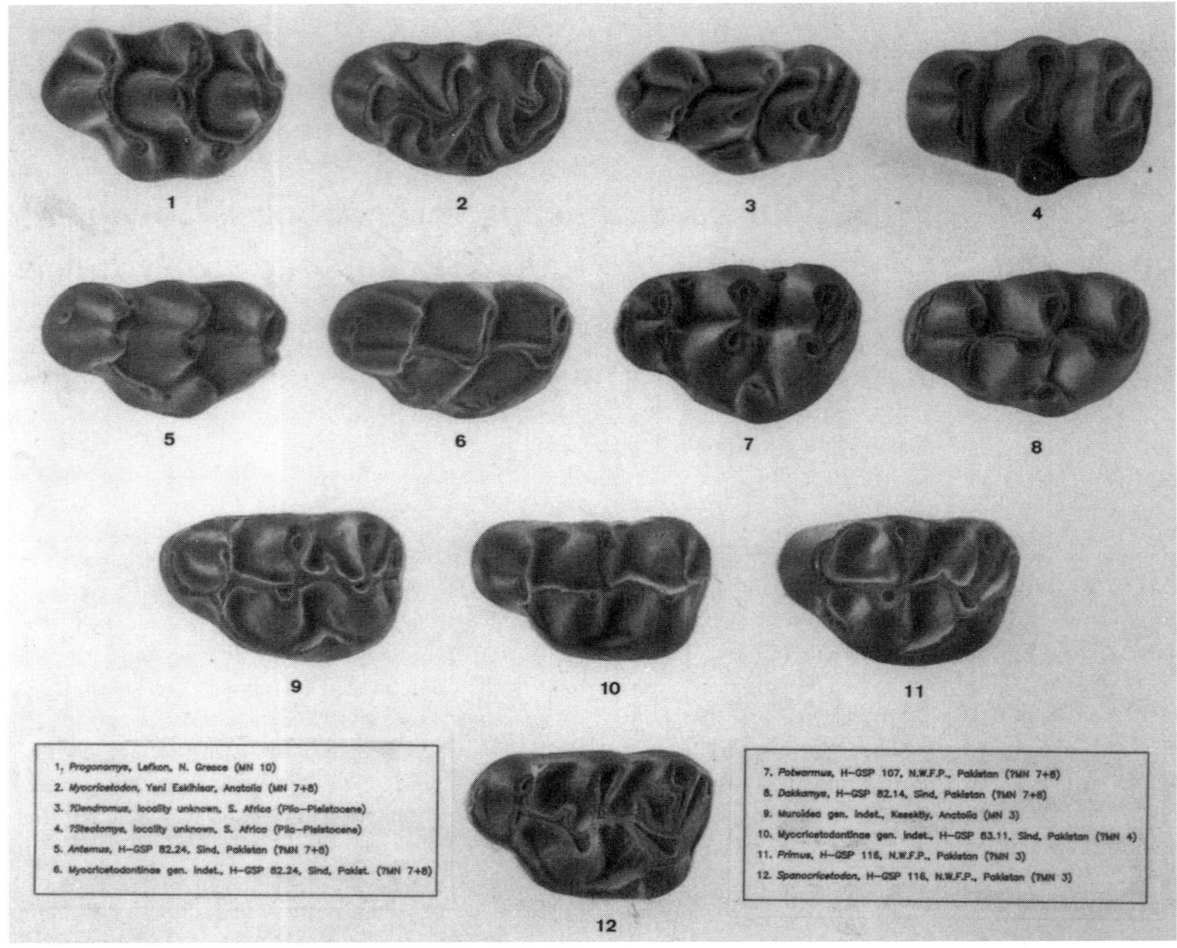

FIGURE 20.1 The M1 on figure 20.1 are all figured as if they are from the left side. The length of these teeth has been made the same in order to facilitate morphological comparison and to draw attention to differences in shape. The figures are not to scale and not arranged in stratigraphical order. The figures are meant to show morphoclines as available in the fossil record and not phylogeny.

different respects a slightly more modern pattern. The tooth of 9 shows a protoloph and metaloph that are directed somewhat posteriorly, the anterior arm of the protocone is absent, the longitudinal ridge is strong, and the posterior labial sinus is long. This tooth is unusual in that the lingual cingulum and the anterior cingulum are much stronger than in the *Spanocricetodon* tooth (figure 20.1:12). Consequently, the protocone and the hypocone have a more labial position.

The tooth of 10 differs from *Spanocricetodon* in having the protoloph and metaloph directed posteriorly and having lost the anterior arms of the protocone as well as the hypocone. The lingual cingulum is weak and does not extend in front of the split anterocone. The configuration of this tooth with incipient reduction of the postero-labial sinus makes it a perfect structural ancestor for the Myocricetodontinae. The tooth of *Primus* (figure 20.1:11) is in many respects as primitive as the M1 of *Spanocricetodon*

from the same locality. The protoloph and metaloph are directed forward, the anterior arm of the protocone is absent, and the postero-labial sinus is long. The anterocone of the M1 of *Primus* is not divided, and the lingual cingulum is weak and stops at the protocone. However, a different cingulum, which seems to initiate from the hypocone, is developed. This configuration is common in the otherwise very different Cricetodontini.

The teeth in the third row (fig. 20.1:5–8) all are more derived than the ones in the second row (fig. 20.1:9–11), but in different ways. Item 5 shows *Antemus*, a genus that is known from the Indian subcontinent and from Libya (Fejfar, pers. comm.). In this tooth, the short protoloph and metaloph still have the original direction. However, the anterior arms of the protocone and hypocone are absent, and so is the longitudinal ridge. The postero-labial sinus is short, and the lingual cingulum has expanded but has an interruption at the level of the protocone. The two cusps of the lingual cingulum (t1 and t4?) are situated much nearer to one another than in later Murinae. The anterior cingulum is "reduced" to a weak cusp that is situated antero-lingually of the labial cusp of the anterocone.

The anterocone, protoloph, and metaloph of the tooth of 6 are developed as parallel blades. The longitudinal ridge is weak, the anterior arms of the protocone and hypocone are absent, and the postero-labial sinus is short. Lingually there is not much of a cingulum, but there are two weak cusps in a similar position as the t1 and t4 of the Murinae. The anterior part of the longitudinal ridge of this M1 is situated more lingually than the posterior part, which seems to foreshadow the configuration seen in later Myocricetodontinae.

Fig. 20.1:7 shows the M1 of *Potwarmus primitivus*. The short protoloph and metaloph are still directed forward. The anterior arms of the protocone and hypocone are absent, the longitudinal ridge is very weak, and the postero-labial sinus is rather long. The anterior part of the lingual cingulum is absent, and the enterocone is large as in extant Dendromurinae. In general habitus this tooth is, however, not very different from that of *Antemus* (5).

Fig. 20.1:8 shows the M1 of *Dakkamys*. The short protoloph and metaloph are directed forward, the anterior arms of the protocone and metacone are absent, and the longitudinal ridge is strong. The postero-labial sinus is of medium length. The anterior cingulum and enterocone are isolated structures.

Fig. 20.1:1 shows the situation in *Progonomys*. The short protoloph is directed forward, but the metaloph is directed backward. The longitudinal ridge is absent, and so are the anterior arms of the protocone and hypocone. The postero-labial sinus is very short and the central row of cusps is inflated. The anterior part of the lingual cingulum (t1) is longer than the labial cusp of the anterocone (t3), and the enterocone (t4) is as large as the paracone and connected to the hypocone (t8).

In the M1 of *Myocricetodon* (20.1:2) the anterior part of the longitudinal ridge is situated much more lingually than the posterior part. The anterior arms of the protocone and hypocone are absent, and the protoloph and metaloph have disappeared because the triangular protocone and paracone and the hypocone and metacone touch. The postero-labial sinus is absent. Lingually there is just the enterocone, which is connected to the posterior corner of the protocone by a ridge. The incipiently triangular cusps and the zigzag course of the longitudinal ridge are reminiscent of the configuration seen in the Arvicolidae, a group that—for other reasons–seems to have nothing to do with the Myocricetodontinae.

Fig. 20.1:3 shows the M1 of *Dendromus*. The longitudinal ridge and the anterior arms of the protocone and hypocone are absent and the postero-labial sinus is long.

The protoloph and metaloph are directed forward, and the paracone and metacone are connected. The anterior cingulum is developed as a cusp and the large enterocone is situated off the protocone.

The M1 of *Steatomys* (20.1:4) lacks both longitudinal connections and the anterior arms of the protocone and metacone. The short protoloph and metaloph are almost transverse, and the wide anterocone, consisting of two cusps, does not bear an anterior cingulum. The enterocone is large and isolated and causes a bulge on the lingual side. The long postero-labial sinus is preserved, but this valley is closed labially by the posteroloph that is connected to the metacone.

The teeth figured in figure 20.1 have been selected to show the large range in the morphology of the M1 from a primitive *Spanocricetodon* (20.1:12) to the well-established patterns of the Murinae (1), the Myocricetodontinae (2 and 6), and the Dendromurinae (20.1:3 and 4) and not to suggest phylogeny. Assignment to subfamily of the teeth in the two middle rows (20.1:5–11) remains more or less a matter of taste at this stage. Construction of cladograms tends to oversimplify the situation because these leave no leeway for intermediate character states. Using that technique makes for example *Antemus* (20.1:5) a murid and *Potwarmus* (20:1:7) a dendromurid on the basis of presence, respectively absence of the t1. How should we allocate the tooth of item 6? The conclusion seems to be that we are unable to allocate species unequivocally to subfamily on the basis of isolated teeth and that potential ancestors of the Murinae, the Dendromurinae, and the Myocricetodontinae were present in Southern Asia and Africa since the early Miocene (MN 3). The M1 of *Notocricetodon* from Songor (Lavocat 1973) is after all not essentially different from that of *Spanocricetodon*.

The two main questions that concern us here are: (1) What is the ancestor of the true Murinae? and (2) Was it some unknown ancestor of the Murinae that migrated through Asia and Africa or else well-defined Murinae that managed to enlarge their geographical range? We think the available information rather suggests migration in several stages. Whatever the case may have been, it seems well established that the first Murinae to arrive in Europe and Anatolia were at the *Progonomys* stage-of-evolution.

The Vallesian, Turolian, and Lower Ruscinian Genera of the Murinae from Southwestern Europe, Central Europe, Southeastern Europe, and Anatolia

Each line in the column of a particular genus on figure 20.2 indicates the range of that genus—from left to right—in Southwestern Europe, Central Europe, Southeastern Europe, and Anatolia. These geographical areas are defined and interpreted as by de Bruijn and Ünay (this vol-

FIGURE 20.2 The Morphology of the First Upper Molar of the Twelve Murine "Genera" of Europe and Anatolia and Their Stratigraphical Ranges

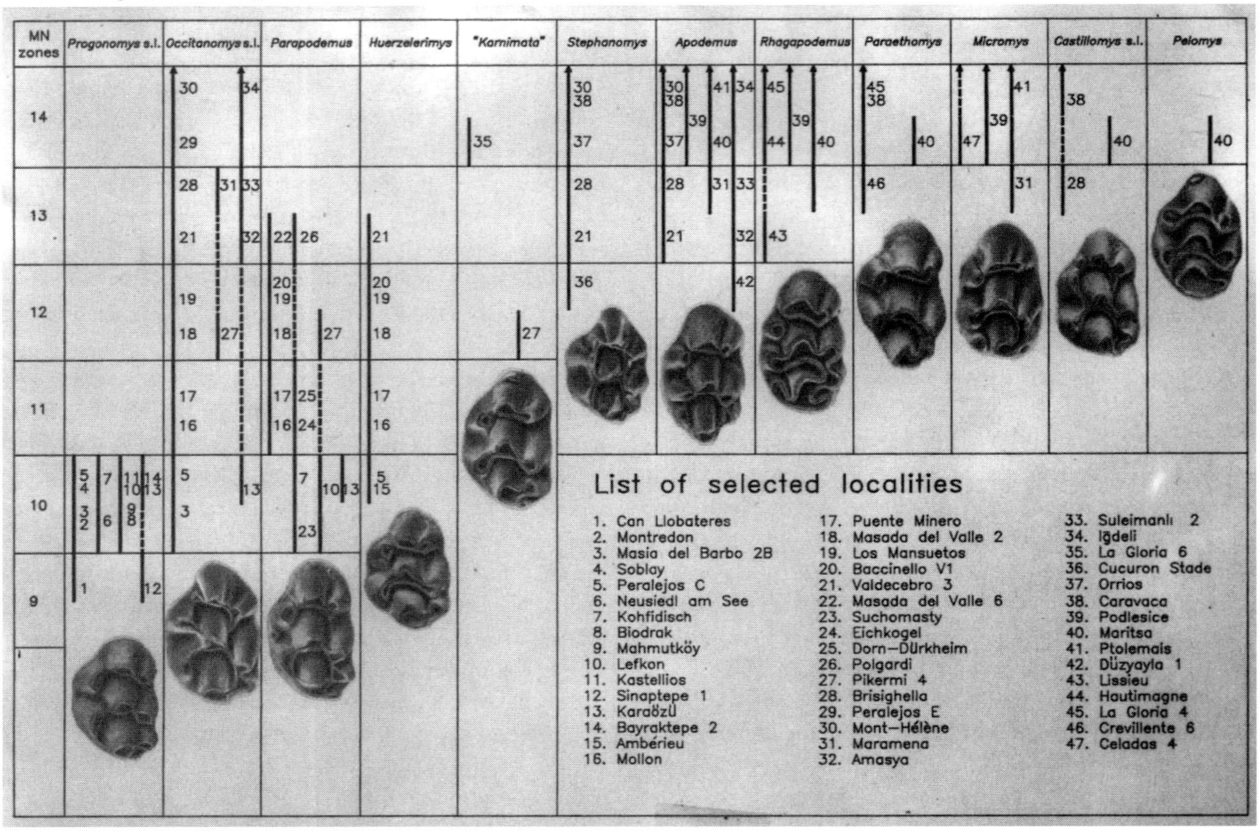

The vertical lines in each column indicate the successive ranges of a particular genus of the Murinae in four geographic areas. These are, from left to right, Southwestern Europe, Central Europe, southeastern Europe, and Anatolia. For the definition of parts of Europe we refer to de Bruijn and Ünay (this volume). The numbers refer to the selected localities listed. The figures of the left M1 at the bottom of each colomn are not to scale.

ume). If the record of a genus is discontinuous in a specific geographical area, the known occurrences are connected by a broken line. The numbers on the right side of each line giving the range of a genus refer to selected localities from which that genus is known.

The first column is indicated as *Progonomys* s.l. because it includes the specimens from Can Llobateres that are considered to belong to a lineage that is different from that including the type species of the genus (Mein et al. 1993).

Occitanomys is given as *Occitanomys* sensu lato because it includes the subgenera *Rhodomys* Martin Suárez and Mein 1991 and *Hansdebruijnia* Storch 1995. Strictly speaking it is not required to use s.l. here, but since fossil subgenera tend to have short lives in the literature because they are either upgraded or abolished, we decided to anticipate the qualification. *Castillomys* is given as *Castillomys* s. l. because it includes the genus *Centralomys* de Giuli 1989. Finally, the column *Karnimata* is given as (quote) "*Karnimata*" because the type species of this genus has been synonymized with *Progonomys woelferi* (Mein et al. 1993). Unfortunately these authors do not introduce a new name to house the other species that have been assigned to *Karnimata* and are not *Progonomys*.

Discussion

Progonomys appears as an immigrant in Southern Europe and Anatolia during the upper part of MN 9. Its absence in Central Europe in that zone is remarkable because it does not seem to be due to lack of appropriate fossil localities. The extinction of *Progonomys* in the upper part of MN 10 is probably a pseudo-extinction, because the dental morphology of *Occitanomys sondaari* is intermediate between *Progonomys* and *Occitanomys*.

Occitanomys first occurs in Southwestern Europe—that is, if *Progonomys sondaari* (van de Weerd 1976) is transferred to *Occitanomys* as suggested by Mein et al. (1993). This action almost automatically results in seeing *Occitanomys* s.s. as a descendant of the Southwest European population of *Progonomys*. In Southwestern Europe as well as in Anatolia *Occitanomys* ranges into the upper Ruscinian, and there seems to be no reason to suppose that this is not the case in Central Europe and Southeastern Europe, since the number of Turolian faunas from these areas is limited. However, the Turolian faunas of Dorn-Dürkheim and Eichkogel include hundreds of *Parapodemus lugdunensis* specimens, but not a single *Occitanomys* tooth.

TABLE 20.1 *Recent References for Localities Listed in Fig. 20.2*

1.	Can Llobateres	MN 9	Mein et al. 1993
2.	Montredon	MN 10	Aguilar 1982
3.	Masia del Barbo 2B	MN 10	van de Weerd 1976
4.	Soblay	MN 10	Mein 1984
5.	Peralejos C	MN 10	van de Weerd 1976
6.	Neusiedl am See	MN 10	Rögl et al. 1993
7.	Kohfidisch	MN 10	Bachmayer et al. 1980
8.	Biodrak	MN 10	de Bruijn 1976
9.	Mahmutköy	MN 10	Ünay et al. 1984
10.	Lefkon	MN 10	de Bruijn, 1989
11.	Kastellios	MN 10	Mein, P. et al. 1993
12.	Sinaptepe 1	MN 9	Kappelman et al. 1995
13.	Karaözü	MN 10	Sümengen et al. 1990
14.	Bayraktepe 2	MN 10	Ünay 1984
15.	Ambérieu	MN 10	Mein, P. et al. 1993
16.	Mollon	MN 11	Michaux 1971
17.	Puente Minero	MN 11	Alcalá et al. 1991
18.	Masada del Valle 2	MN 12	van de Weerd 1976
19.	Los Mansueto	MN 12	van de Weerd 1976
20.	Baccinello V-1	MN 12	Engesser 1989
21.	Valdecebro 3	MN 13	van de Weerd 1976
22.	Masada del Valle 6	MN 13	van de Weerd 1976
23.	Suchomasty	MN 10	Fejfar 1989
24.	Eichkogel	MN 11	Daxner-Höck 1977
25.	Dorn-Dürkheim	MN 11	Franzen, J. et al.
26.	Polgardi	MN 13	Schaub 1938
27.	Pikermi 4	MN 12	de Bruijn 1976
28.	Brisighella	MN 13	de Giuli 1989
29.	Peralejos E	MN 14	Adrover et al. 1988
30.	Mont-Hélène	MN 14	Aguilar et al. 1976
31.	Maramena	MN 13	de Bruijn 1989
32.	Amasya	MN 13	Becker-Platen et al. 1975
33.	Suleimanli 2	MN 13	unpubl. material, collection MTA Ankara
34.	Igdeli	MN 14	Sümengen et al. 1990
35.	La Gloria 6	MN 14	Mein et al. 1990
36.	Cucuron Stade	MN 12	Mein et al. 1979
37.	Orrios 1	MN 14	de Bruijn et al. 1992
38.	Caravaca	MN 14	de Bruijn et al. 1975
39.	Podlesice	MN 14	Nadachowski, A. 1989
40.	Maritsa	MN 14	de Bruijn et al. 1970
41.	Ptolemais	MN 14	van de Weerd 1979
42.	Düzyayla 1	MN 12	unpubl. material, collection MTA Ankara
43.	Lissieu	MN 13	Hugueney and Mein 1965
44.	Hautimagne	MN 14	Mein and Michaux 1970
45.	La Gloria 4	MN 14	Mein et al. 1990
46.	Crevillente 6	MN 13	de Bruijn et al. 1975
47.	Celades 4	MN 14	Mein et al. 1983

Parapodemus appears first in Central Europe, but its absence in the records from Southeastern Europe and Anatolia is probably attributable to lack of information. We think that *Parapodemus* is an Eastern European or Western Asiatic descendant of *Progonomys* that appears as an immigrant in Southwestern Europe during MN 11 and subsequently becomes extinct in the lower part of MN 13. In Southeastern Europe and Anatolia populations that have *Parapodemus* and *Apodemus*, morphotypes are common, and *Parapodemus* grades gradually into *Apodemus*. This results in the relatively early pseudo-extinction of *Parapo-*

demus and the early appearance of *Apodemus* in that area.

Huerzelerimys has not been found east of northern Italy and seems therefore to have descended from a Southwestern European *Progonomys* population (Mein et al. 1993). The last occurrence of this genus, which is characterized by size increase through time, is in the lower part of MN 13 in Spain, where it seems to become extinct without leaving descendants.

"Karnimata" occurs in two European localities only. If these specimens really belong to the same genus as the material from Pakistan described by Jacobs as has been suggested by Mein et al. (1993), the European occurrences are to be interpreted as immigrants from Asia. However, we have not compared these specimens and have therefore no opinion on this subject.

Stephanomys is exclusively known from Southwestern Europe, where it is a common constituent of most late Turolian and early Ruscinian assemblages. Although the strongly developed stephanodonty of its later representatives is reminiscent of *Occitanomys*, the geologically older representatives (upper part of MN 12, lower part of MN 13) are more primitive than contemporaneous *Occitanomys*. Therefore it seems more likely that *Stephanomys* has developed independently from *Occitanomys* and descended from a Southwest European population of *Parapodemus*.

Apodemus undoubtedly developed more than once from different Southeast European–Western Asiatic populations of *Parapodemus*. The undescribed association from Düzyayla 1 (Central Anatolia) contains two species that both have M1, with the *Parapodemus* and the *Apodemus* morphology as intraspecific variants. The larger of these two species has a t7 in 71% of the M1 (N = 55) and is consequently assigned to *Apodemus*; the smaller species has a t7 in 32% of the M1 (N = 53) and is assigned to *Parapodemus*. This assemblage is of interest because it shows that *Apodemus* is polyphyletic. This polyphyly is also suggested by the Western European record of this genus. In that area populations with a mixture of *Parapodemus* and *Apodemus* morphotypes are only known to occur in larger species, while the smaller species seems to appear as an immigrant during the late Turolian. The extant species *sylvaticus* and *flavicollis* resemble the smaller fossil species, while the extant species *mystacinus* resembles the larger fossil species. Electrophoretic analysis indeed suggests a rather early divergence of *mystacinus* from the other species of the subgenus *Sylvaemus* (Filippucci 1992).

Rhagapodemus can hardly be anything else than a descendant of *Apodemus*. The presence of highly derived specimens in the fauna from Lissieu (lower part of MN 13) seems an anachronism, because these specimens are more evolved than *Rhagapodemus hautimagnensis* from MN 14, a species that is known to have a very large geographical range (France, Sardinia, Greece). Since the fauna from

Lissieu comes from a karst infill, it cannot be excluded that it contains a mixture of fauna elements of different ages. *Rhagapodemus* is common in Ruscinian faunas from Central Europe and Southeastern Europe but has not (yet) been found in Anatolia.

Paraethomys arrives as an immigrant in Southwestern Europe (upper part of MN 13) and is known to have had a circum-Mediterranean range soon after its arrival. Its origin is not known, but since Coiffait et al. (1985) have shown that there has been fauna exchange between Southwestern Europe and North Africa during the late Miocene, it may well have come from Africa. Judging by the dental pattern, *Paraethomys* may have descended from "*Karnimata.*"

Micromys appears in the upper part of MN 13 of N. Greece (Storch and Dahlmann, in prep.). The *Micromys* population from Maramena is considered to be primitive in resembling *Apodemus*, from which it presumably descended. *Micromys* seems to have developed in Southeastern Europe, where it became quite diverse during the Ruscinian (van de Weerd 1979). In Southwestern Europe and Central Europe *Micromys* appears as an immigrant. The absence of *Micromys* in the Anatolian record is probably artificial and due to lack of information.

Castillomys s.l. includes the genus *Centralomys* (de Giuli 1989) with the species *benericetti* (de Giuli 1989) from Brisighella (Tuscany) and *magnus* (Sen 1977) from Çalta (Anatolia) and Maritsa (Rhodes). A problem is that the difference between *Occitanomys* and *Castillomys* is not clear-cut when species from outside Spain and France are considered.

Castillomys s.s. seems to appear as an immigrant in Spain in the upper part of MN 14. The genus is a common constituent of Southwestern European faunas of late Ruscinian and Villanyian age. The ancestor of *Castillomys* s.s. is not known, and we have no idea of the origin of the first representative, *C. gracilis* from Caravaca (Spain).

Pelomys is within the area and interval studied known from Maritsa (island of Rhodes) only. Similar small rats have, however, become known from strata of roughly the same or slightly older levels in Pakistan (*Parapelomys robertsi* from Dhok Pathan), Algeria (*Pelomys* cf. *europeus* from Amama 3), and Abu Dhabi (some undescribed teeth of a small rat from the upper Turolian Baynunah Formation).

Acknowledgments

H. de Bruijn is grateful to Mrs. Wilma Wessels and Mr. Constantin Theocharopoulos for discussions of the phylogenetic relationships of the "Megacricetodontinae" (sensu Lindsay 1988), the Murinae, the Myocricetodontinae, and the Dendromurinae. The material that has become available through financial support of the Smithsonian Institution grant 708710000-13 and U.S. National Science Foundation grant DEB. 8003601 awarded to S. T. Hussain and of the North Atlantic Treaty Organisation grant C.R.G. 910750 awarded to H. de Bruijn has substantially influenced our perspective of the various groups of Muroidea discussed below.

The photographs and reproductions in figure 20.1 and table 20.1 were made by Mr. W. den Hartog and retouched by Mr. J. Luteijn. Table 20.1 has been made by Mr. T. van Hinte. We thank Mrs. A. E. de Bruijn-Dudok van Heel for typing and editing the text and Dr. D. F. Mayhew for suggesting grammatical improvements.

H. de Bruijn gratefully acknowledges the hospitality received in Pakistan through the G.S.P. (Geological Survey of Pakistan) and in Turkey through the M.T.A. (Institute for Mineral Research and Exploration of Turkey).

LITERATURE CITED

Adrover, R., P. Mein, and E. Moissenet. 1988. Contribución al conocimiento de la fauna de roedores del Plioceno de la Región de Teruel. *Teruel* 79:91–151.

Aguilar, J.-P. 1982. Contributions à l'étude des micromammifères du gisement Miocène supérieur de Montredon (Hérault). 2. Les rongeurs. *Palaeovertebrata* 12:81–117.

Aguilar, J.-P., M. Calvet, and J. Michaux. 1976. Descriptions des rongeurs Pliocènes de la faune du Mont-Hélène (Pyrénées-Orientales, France), nouveau jalon entre les faunes de Perpignan (Serrat-d'en-Vacquer) et de Sète. *Palaeovértebrata* 16:127–44.

Alcalá, L., C. Sesé, E. Herráez, and R. Adrover. 1991. Mamíferos del Turoliense inferior de Puente Minero (Teruel, España). *Boletín de la Real Sociedad Española de Historia Natural (Sección Geológica)* 86:205–51.

Bachmayer, F. and R. W. Wilson. 1980. A third contribution to the fossil small mammal fauna of Kohfidisch (Burgenland), Austria. *Annalen des Naturhistorischen Museums, Wien* 83:351–86.

Becker-Platen, J. D., O. Sickenberg, and H. Tobien. 1975. Die Gliederung der känozoischen Sedimente der Türkei nach Vertebraten-Faunengruppen. *Geologisches Jahrbuch, B* 15:19–46.

Bruijn, H. de. 1976. Vallesian and Turolian Rodents from Biotia, Attica and Rhodes (Greece). I. *Proceedings of the Koninklijke Nederlandse Akademie van Wetenschappen, B* 79:361–84.

———. 1989. Smaller mammals from the upper Miocene and lower Pliocene of the Strimon Basin. Part 1. Rodentia and Lagomorpha. *Bollettino della Società Paleontologica Italiana* 28:189–95.

Bruijn, H. de, R. Daams, G. Daxner-Höck, V. Fahlbusch, L. Ginsburg, P. Mein, and J. Morales. 1992. Report of the RCMNS working group on fossil mammals, Reisensburg 1990. *Newsletters on Stratigraphy* 26:56–118.

Bruijn, H. de, M. R. Dawson, and P. Mein. 1970. Upper Pliocene Rodentia, Lagomorpha, and Insectivora (Mammalia) from the Isle of Rhodes (Greece). I, II, and III. *Proceedings of the Koninklijke Nederlandse Akademie van Wetenschappen, B* 73:535–84.

Bruijn, H. de, C. Montenat, and A. van de Weerd. 1975. Correlations entre les gisements des rongeurs et les formations marines du Miocène terminal d'Espagne meridional (Prov. d'Alicante et de Murcia). *Proceedings of the Koninklijke Nederlandse Akademie van Wetenschappen, B* 78:282–313.

Bruijn, H. de and E. Ünay. This volume. On the evolutionary

history of the Cricetodontini from Europe and Asia Minor and its bearing on the reconstruction of migrations and the continental biotope during the Neogene.

Bruijn, H. de and W. J. Zachiarasse. 1979. The correlation of marine and continental biozones of Kastellios Hill reconsidered. *Annales Géologiques des Pays Helléniques*, hors série, 1:219–26.

Butler, P. M. 1985. Homologies of molar cusps and crests, and their bearing on assessments of rodent phylogeny. In *Evolutionary Relationships Among Rodents*, ed. W. P. Luckett and J.-L. Hartenberger. New York: Plenum. NATO ASI series A 92:381–401.

Coiffait, B. and J. J. Jaeger. 1985. Découverte en Afrique du Nord des genres *Stephanomys* et *Castillomys* (Muridae) dans un nouveau gisement de microvertébrés néogènes d'Algérie Orientale: Argoub Kemellal. *Proceedings of the Koninklijke Nederlandse Akademie van Wetenschappen, B* 88:167–83.

Daxner-Höck, G. 1977. Muridae, Zapodidae und Eomyidae (Rodentia, Mammalia) des Eichkogels bei Mödling (Niederösterreich). *Paläontologische Zeitschrift* 51:19–31.

Engesser, B. 1989. The late Tertiary small mammals of the Maremma region (Tuscany, Italy). 2nd part: Muridae and Cricetidae (Rodentia, Mammalia). *Bollettino della Società Paleontologica Italiana* 28:227–52.

Farjanel, G. and P. Mein. 1984. Une association de mammifères et de pollens dans la formation continentale des "Marnes de Bresse" d'âge Miocène supérieur à Ambérieu (Ain). *Géologie de la France* 1–2:131–48.

Fejfar, O. 1989. The Neogene VP Sites of Czechoslovakia: A contribution to the Neogene terrestric biostratigraphy of Europe based on Rodents. In *European Neogene Mammal Chronology*, E.H. Lindsay, V. Fahlbusch and P. Mein eds., pp. 211–36. New York: Plenum Press.

Filippucci, M. G. 1992. Allozyme variation and divergence among European, Middle Eastern, and North African species of the genus *Apodemus* (Rodentia, Muridae). *Israel Journal of Zoology* 38:193–218.

Flynn, L. J., L. L. Jacobs, and E. H. Lindsay. 1985. Problems in Muroid phylogeny: Relationship to other rodents and origin of major groups. In *Evolutionary Relationships Among Rodents*, ed. W. P. Luckett and J. L. Hartenberger. NATO ASI Series A 92:589–616.

Fortelius, M., L. Werdelin, P. Andrews, R. L. Bernor, A. Gentry, L. Humphrey, H.-W. Mittmann, and S. Viranta. This volume. Provinciality, diversity, turnover, and paleoecology in land mammal faunas of the later Miocene of Western Eurasia.

Franzen, J. and G. Storch. 1975. Die unterpliozäne (turolische) Wirbeltierfauna von Dorn-Dürkheim, Rheinhessen (SW Deutschland). 1. Entdeckung, Geologie, Mammalia: Carnivora, Proboscidea, Rodentia. Grabungsergebnisse 1972–1973. *Senckenbergiana lethaea* 56:233–303.

George, W. 1985. Reproductive and chromosomal characters of ctenodactylids as a key to their evolutionary relationships. In *Evolutionary Relationships Among Rodents*, ed. W. P. Luckett and J. L. Hartenberger. NATO ASI Series A 92:453–74.

Giuli, C. de. 1989. The rodents of the Brisighella latest Miocene fauna. *Bollettino della Società Paleontologica Italiana* 28:197–212.

Hartenberger, J. L. 1980. Données et hypothèses sur la radiation initiale des rongeurs. *Palaeovértebrata, Mémoire Jubilaire R. Lavocat*: 285–301.

Hugueney, M. and P. Mein 1965. Lagomorphes et Rongeurs du Néogène de Lissieu (Rhône). *Travaux Laboratoire de Géologie Faculté des Sciences Lyon*, n.s. 12:109–23.

Jacobs, L. L. 1977. A new genus of murid rodent from the Miocene of Pakistan and comments on the origin of Muridae. *Paléobios* 25:1–11.

Lavocat, R. 1973. Les rongeurs du Miocène d'Afrique orientale. I. Miocène inférieur. *Mémoires et traveaux de l'institut de Montpellier* 1:1–279.

Li, C. K. and S. Y. Ting. 1984. The Paleogene mammals of China. *Bulletin of the Carnegie Museum of Natural History* 21:1–98.

Lindsay, E. H. 1988. Cricetid rodents from Siwalik deposits near Chinji village. Part 1: Megacricetodontinae, Myocricetodontinae, and Dendromurinae. *Palaeovértebrata* 18:95–154.

Martin Suárez, E. 1991. Revision of the genus *Castillomys* (Muridae, Rodentia). *Scripta Geologica* 96:47–80.

Mein, P. 1979. Une faune de petits mammifères d'âge Turolien Moyen (Miocène Superieur) à Cucuron (Vaucluse): Données nouvelles sur le genre *Stephanomys* (Rodentia) et consequences stratigrafiques. *Géobios* 12:481–85.

——. 1984. Composition quantitative des faunes de mammifères du Miocène moyen et supérieure de la région lyonnaise. *Paléobiologie Continentale* 14:239–46.

Mein, P., E. Martin Suárez, and J. Agusti. 1993. *Progonomys* Schaub, 1938 and *Huerzelerimys* gen. nov. (Rodentia): Their evolution in Western Europe. *Scripta Geologica* 103:41–64.

Mein, P. and J. Michaux. 1970. Un nouveau stade dans l'évolution des rongeurs pliocènes de l'Europe sud-occidentale. *Comptes Rendues Académie des Sciences, Paris* 270:2780–83.

Mein, P., E. Moissenet and R. Adrover. 1983. L'extension et l'âge des formations continentales pliocènes du fossé de Teruel (Espagne). *Comptes Rendues Academie des Sciences, Paris* 296:1603–10.

——. 1990. Biostratigraphie du Néogène Supérieur du bassin de Teruel. *Paleontologia i Evolució* 23:121–39.

Michaux, J. 1971. Muridae (Rodentia) Néogènes d'Europe sud-occidentale. Évolution et rapports avec les formes actuelles. *Paléobiologie Continentale* 2:1–67.

Nadachowski, A. 1989. Gryzonie-Rodentia. In *History and Evolution of the Terrestrial Fauna of Poland*, ed. K. Kowalski. Folia Quaternaria 59–60:151–76.

Rögl, F., H. Zapfe, R. L. Bernor, R. L. Brzobohaty, G. Daxner-Höck, I. Draxler, O. Fejfar, P. Herrmann, G. Rabeder, O. Schultz, and R. Zetter. 1993. Die Primatenfundstelle Götzendorf an der Leitha (Obermiozän des Wiener Beckens, Niederösterreich). *Jahrbuch der Geologischen Bundesanstalt* 136:503–26.

Schaub, S. 1938. Tertiäre und quartäre Murinae. *Abhandlungen der Schweizerischen Paläontologischen Gesellschaft* 61:1–38.

Sen, S. 1977. La faune de rongeurs Pliocènes de Çalta (Ankara, Turquie). *Bulletin du Muséum National d'Histoire Naturelle* 465:89–171.

Sümengen, M., E. Ünay, H. de Bruijn, I. Terlemez, and M. Gürbüz. 1989. New Neogene rodent assemblages from Anatolia (Turkey). In *European Neogene Mammal Chronology*, E. H. Lindsay, V. Fahlbusch, and P. Mein, eds., pp. 61–72. New York: Plenum Press.

Storch, G. and Dahlmann. 1995. The vertebrate locality of Maramena (Macedonia, Greece) at the Turolian–Ruscinian boundary. *Münchner Geowissenschaftliche Abhandlungen* 28(A): 121–32.

Tong, H. 1989. Origine et évolution des Gerbillidae (Mammalia, Rodentia) en Afrique du Nord. *Mémoires de la Societé Géologiques de France* 155:1–120.

Tong, H. and J.-J. Jaeger. 1993. Muroid rodents from the middle Miocene Fort Ternan locality (Kenya) and their contribution to the phylogeny of muroids. *Palaeontographica,* A 229:51–73.

Ünay, E. and H. de Bruijn. 1984. On some rodent assemblages from both sides of the Dardanelles (Turkey). *Newsletters on Stratigraphy* 13:118–32.

Weerd, A. van de. 1976. Rodent faunas of the Mio-Pliocene continental sediments of the Teruel-Alfambra region, Spain. *Utrecht Micropaleontological Bulletins,* Special Publication 2:1–217.

——. 1979. Early Ruscinian rodents and lagomorphs (Mammalia) from the lignites near Ptolemais (Macedonia, Greece). *Proceedings of the Koninklijke Nederlandse Akademie van Wetenschappen,* B 82:127–70.

Zdansky, O. 1930. Die alttertiären Säugetiere Chinas nebst stratigraphischen Bemerkungen. *Palaeontologica Sinica,* C IV (2): 1–87.

21

Middle and Late Miocene Gliridae of Western, Central, and Southeastern Europe

G. DAXNER-HOCK

Almost 75% of the middle and late Miocene glirid genera can be evolutionarily traced to their Oligocene or early Miocene origin. A most remarkable change in glirid associations occurred during the Vallesian, when many old elements became extinct and several newcomers invaded Central Europe. The diversity of glirid genera and species decreased rapidly during the Turolian to the level seen today.

Figure 21.1 gives the stratigraphic ranges of Western, Central, and Southeast European Gliridae during the middle–late Miocene interval (MN 6–13, ca. 15–5.3 Ma). For each MN unit only a few representative glirid faunas (1–31) have been entered. Table 21.1 lists the glirid species occurring in some selected faunas from Western and Central Europe. Numbers 1–20 indicate the glirid faunas as referenced in figure 21.1.

Discussion

Unfortunately, the Gliridae have a poorer fossil record in Greece and Turkey than in Western and Central Europe. Therefore, my discussion is necessarily limited to these latter areas. *Bransatoglis*, *Paraglirulus*, and *Glirulus* are known from older levels in the Eastern Mediterranean, and their absence in the middle and late Miocene is almost certainly the result of the poor record. *Myoglis*, *Prodryomys*, and *Graphiurops* are absent from the Eastern Mediterranean, and they may never have extended their range into this area. Glirid temporal and geographic distribution give the impression that they evolved along similiar lines across all of Central and Western Europe.

Glirid genus and species diversity is great during the early Miocene. Their diversity and abundance decreased during the middle and late Miocene. All of the middle Miocene (MN 6–7/8) genera are also known from the early Miocene, but some of these (*Prodryomys*, *Bransatoglis*,

Glirudinus) seem to have disappeared during the middle Miocene; others (*Miodyromys*, *Myoglis*, *Eomuscardinus*, and *Paraglirulus*) continued and occur as rare elements during the Vallesian. The last record of *Microdyromys* and *Vasseuromys* is from the early Turolian (MN 11). The genera *Glis* and *Glirulus* are extant, but the geographical distribution of the latter has become restricted, and today is limited to Japan. In contrast with the large group of archaic genera, only a few new taxa appeared during the Vallesian (MN 9–10). One of these, *Graphiurops*, disappeared during the early Turolian, while others, *Muscardinus*, *Eliomys*, and *Myomimus*, still have living representatives.

Acknowledgments

I would like to thank Mrs. K. Repp for the graphics and my colleagues E. Ünay (Ankara), H. de Bruijn (Utrecht), and V. Fahlbusch (München) for contributions and discussion. I wish to thank H. de Bruijn (Utrecht) for critically reviewing the manuscript.

LITERATURE CITED

Agusti, J. 1981. *Glis vallesiensis* n. sp., nouveau gliridé (Rodentia, Mammalia) du Neogene de seu d'Urgell (Catalogne, Espagne). *Géobios* 14:543–47.

Bachmayer, F. and R. W. Wilson. 1980. A third contribution to the fossil small mammal fauna of Kohfidisch (Burgenland). *Annalen des Naturhistorischen Museums, Wien* 83:351–86.

Baudelot, S. 1966. Complément à l'etude de la faune des Rongeurs de Sansan: Les Glirides. *Bulletin de la Société géologique de France*, (5) 7:758–64.

Baudelot, S. and L. de Bonis. 1966. Nouveaux Glirdés (Rodentia) de l'Aquitanien du bassin d'Aquitaine. *Comptes Rendus Sommaire des Séances Société Géologique de France* 9:341–42.

Bruijn, H. de. 1966. Some new Miocene Gliridae (Rodentia, Mammalia) from the Calatayud Area (Prov. Zaragoza, Spain). I.

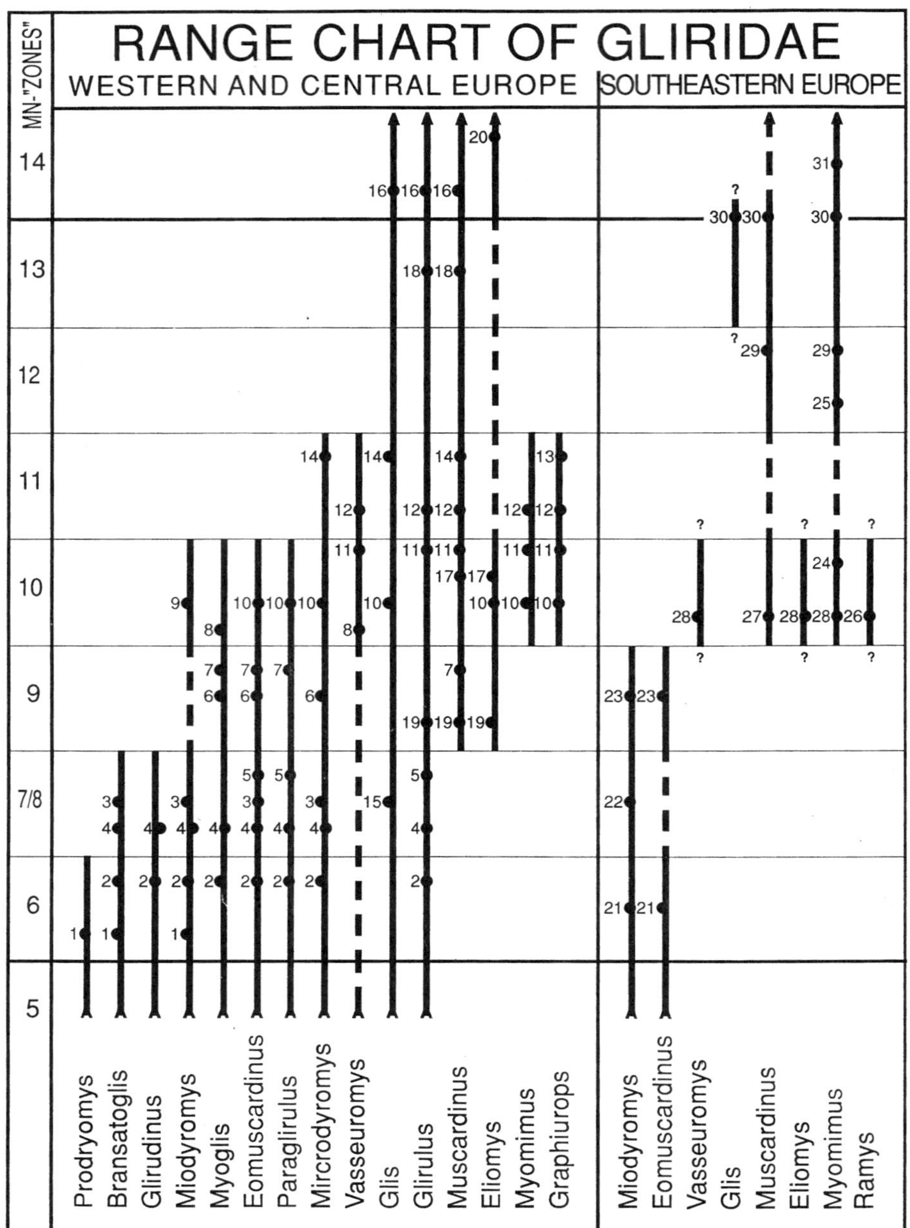

FIGURE 21.1 Stratigraphic range chart of Western, Central, and Southeast European Gliridae.

Table 21.1 *Species of* Gliridae *from Selected Faunas (1–20) of the Middle and Late Miocene in Western and Central Europe*

Prodryomys Mayr 1979
  satus (1)
Bransatoglis Hugueney 1967
  cadeoti (1)
  astaracensis (2, 3, 4)
Glirudinus de Bruijn 1966
  undosus (2, 4)
Miodyromys Kretzoi 1943
  aegercii (1, 2, 3, 4)
  hamadryas (3, 4)
  alter (9)
Myoglis Baudelot 1966
  meini (2, 4, 6, 7, 8)
Eomuscardinus Hartenberger 1966
  sansaniensis (2, 3, 4, 5)
  vallesiensis (6, 7, 10)
Paraglirulus Engesser 1972
  werenfelsi (2, 4, 5, 7, 10)
Microdyromys de Bruijn 1966
  koenigswaldi (3, 4)
  complicatus (2, 3, 6, 10), sp. (14)
Vasseuromys Baudelot and Bonis 1966
  pannonica (8, 11, 12)
Glis Brisson 1762
  vallesiensis (15)
  minor (10, 14, 16)
Glirulus Thomas 1906
  diremptus-conjunctus (2, 4)
  conjunctus (5, 19)
  lissiensis (11, 12, 18)
Muscardinus Kaup 1829
  hispanicus (19, 17)
  austriacus (11)
  heintzi (17)
  vireti (14, 18)
  davidi (18)
  pliocaenicus (12, 16)
Eliomys Wagner 1843
  reductus (19)
  assimilis (10, 19)
  truci (10, 17, 20)
Myomimus Ognev 1924
  dehmi (10, 11, 12)
Graphiurops Bachmeyer and Wilson 1980
  austriacus (10, 11, 12, 13)

Proceedings, Koninklijke Nederlandse Akademie van Wetenschappen, B 69:1–21.

——. 1976. Vallesian and Turolian Rodents from Boiotia, Attica, and Rhodes (Greece). I. *Proceedings, Koninklijke Nederlandse Akademie van Wetenschappen, B* 79:361–84.

——. 1989. Smaller mammals from the upper Miocene and lower Pliocene of the Strimon Basin, Greece. Part 1. Rodentia and Lagomorpha. *Bollettino della Società Paleontologica Italiana* 28:189–95.

Bruijn, H. de, R. Daams, G. Daxner-Höck, V. Fahlbusch, L. Ginsburg, P. Mein, and J. Morales. 1992. Report on the RCMNS working group on fossil mammals, Reisensburg 1990. *Newsletters on Stratigraphy* 26:65–118.

Daams, R. 1981. The dental pattern of the dormice *Dryomys, Myomimus, Microdyromys,* and *Peridyromys. Utrecht Micropaleontological Bulletins, Special Publication* 3:1–115.

Daxner-Höck, G. and H. de Bruijn. 1981. Gliridae (Rodentia, Mammalia) des Eichkogels bei Mödling (Niederösterreich). *Paläontologische Zeitschrift* 55:157–72.

Engesser, B. 1972. Die obermiozäne Säugetierfauna von Anwil (Baselland). *Tätigkeitsberichte der Naturforschenden Gesellschaft Baselland* 28:37–363.

Hartenberger, J. L. 1966. Les rongeurs du Vallésien (Miocène supérieur) de Can Llobateres (Sabadell, Espagne): Gliridae et Eomyidae. *Bulletin de la Société géologique de France*, (7) 8:596–604.

Hugueney, M. 1967. Les Gliridés (Mammalia, Rodentia) de l'Oligocène supérieur de Coderet-Branssat (Allier). *Comptes Rendus Académie des Sciences de Paris* 3:91–92.

Hugueney, M. and P. Mein. 1965. Lagomorphes et Rongeurs du néogène de Lissieu (Rhône). *Travaux des Laboratoires de Géologie de la Faculté des Sciences de Lyon*, n.s., 12:109–23.

Kowalski, K. 1963. The Pliocene and Pleistocene Gliridae (Mammalia, Rodentia) from Poland. *Acta Zoologica Cracoviensia* 8:533–67.

Kretzoi, M. 1943. Ein neuer Muscardinide aus dem Ungarischen Miozän. *Földtany Közlöny* 73:271–74.

Mayr, H. 1979. *Gebissmorphologische Untersuchungen an miozänen Gliriden (Mammalia, Rodentia) Süddeutschlands.* Ph.D. diss. München.

Mein, P. 1984. Composition quantitative des faunes de Mammifères du Miocene moyen de la region Lyonnaise. *Paléobiologie Continentale* 14:339–46.

Mein, P. and J. Michaux. 1970. Un nouveau stade dans l'évolution des Rongeurs pliocènes de l'Europe sud-occidentale. *Comptes Rendus Académie des Sciences,* D. 270:2780–83.

Nadachowski, A. 1989. Gryzonie—Rodentia. In *History and Evolution of the Terrestrial Fauna of Poland,* ed. K. Kowalski, pp. 151–75. *Folia Quaternaria* 59–60:151–76.

Ünay, E. and H. de Bruijn. 1984. On some Neogene rodent assemblages from both sides of the Dardanelles, Turkey. *Newsletters on Stratigraphy* 13:119–32.

Wu, W. 1990. Die Gliriden (Mammalia, Rodentia) aus der Oberen Süsswasser-Molasse von Puttenhausen (Niederbayern). *Mitteilungen der Bayerischen Staatssammlung für Paläontologie und historische Geologie* 30:65–105.

22

Late Miocene Hystricidae in Europe and Anatolia

S. SEN

Compared to other rodent families, the Hystricidae are relatively rare in late Miocene European and Anatolian localities. Only one genus, *Hystrix*, is thus far known from Europe and Anatolia. When present, fossil representation is usually quite poor and often includes only fragmentary jaw and postcranial material.

Occurrences of *Hystrix*

In the geographic area under consideration in this volume, the oldest *Hystrix* has been reported from Yeni Eskihisar, Turkey (Sickenberg et al. 1975), but has not yet been described. *Hystrix* is presently not know from Vallesian age localities, but is quite frequent in Turolian age ones.

The Hystricidae contrast with other rodent families in their restricted distribution to only a few European and Anatolian late Miocene age localities. Figure 22.1 gives the spatial distribution of Turolian *Hystrix*. In late Pliocene and Pleistocene localities, the occurrence of *Hystrix* is more frequent, but often includes fragmentary remains, except in Italy (*H. etruscus* Bosco 1898) and Israel (*H. angressi* Frenkel 1970).

All *Hystrix* remains found in the Eastern European and Anatolian Turolian can be referred to *H. primigenia* (the type locality being Pikermi, Greece). This species further occurs in some European Plio–Pleistocene localities including Weze (Poland), Serrat d'en Vacquer (France), Layna, and Villaroya (Spain). It has not been established whether this species truly survived the entire Pliocene or whether speciation was involved. There is a lack of adequate fossil material to resolve this issue.

Remains of a species from Häder, Germany, referred by Schlosser (1984) to "*Hystrix suevica*," are in fact referable to the castorid, *Anchitheriomys* (van Weers 1993). However, *Hystrix* sp. does occur in the Central European fissure fillings of Salmendingen (MN 10?) and Kohfidisch (MN 11).

Kretzoi (1951) briefly described *Miohystrix parvae* genus and species nov. from Csákvar (Hungary), but has neither figured nor illustrated the referred material. Kretzoi (1952) further reported the occurrence of *H. primigenia* from Polgardi (MN 13) in an updated faunal list. Because of the lack of description or illustration of this Hungarian material, these localities are not included in figure 22.1.

The type material of *H. primigenia* (Wagner 1848) comes from Pikermi, Greece. It is found in other Greek Turolian localities such as Alifakas and Dytiko 3 (Macedonia), Halmyropothamus of Evia Island, and Samos (see de Bonis et al. 1992). It also occurs in Kalimanci level IV (Bulgaria; Sen and Kovatchev 1987) and at Umen Dol (Yugoslavian Macedonia; Garevski 1956). Further dental material referred to *Hystrix* sp. was discovered from Bayirköy and Çoban Pinar, Turkey (Ünay and de Bruijn 1984; Ozansoy 1965). More recently, some fragments of upper and lower jaws have been found at Kemiklitepe, Western Anatolia (Sen 1994). *H. bessarabica* Riabinin 1929 from Taraklia, Moldavia, was synonymized with *H. primigenia* by Sulimski (1960). The Moldavian locality of Chimislia has further yielded a lower incisor similar in size to that of *H. primigenia*. In summary, except for the few fragmentary specimens from Central Europe (Salmendingen and Kohfidisch), all Eastern European and Anatolian Turolian age *Hystrix* are referable to *H. primigenia*.

The evolutionary trends of European *Hystrix* are difficult to define because of poor fossil representation. However, Sen and Kovatchev (1987) and Koliadimou and Koufos (1991) have observed an increase in size and hypsodonty and a concomitant decrease in robusticity index within Eastern European and Anatolian *Hystrix*. These indicies independently predict that the localities of Pikermi (MN 11/12 of Bernor et al., this volume) and Kemiklitepe upper horizon (MN 12) are older than Dytiko and Kalimanci IV (both referable to MN 13).

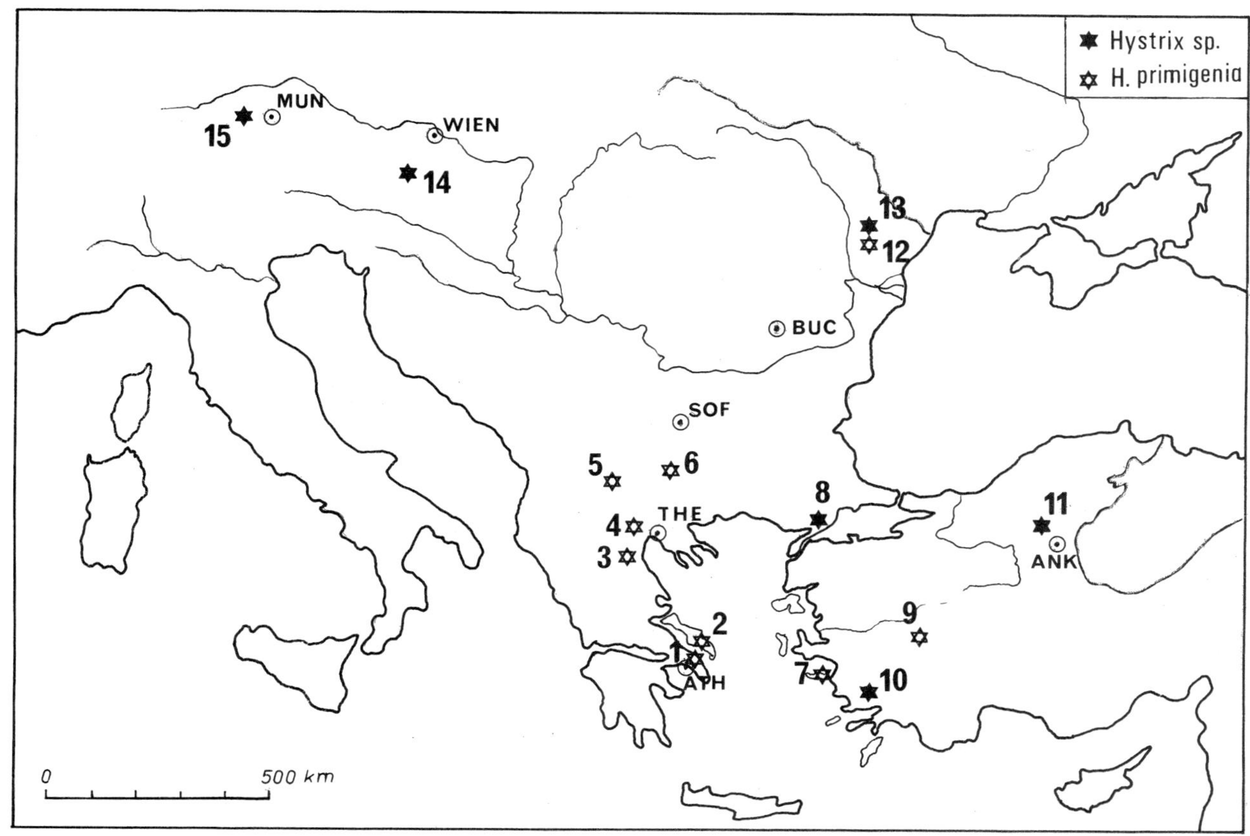

FIGURE 22.1 Spatial distribution of the Turolian *Hystrix* in Central and Eastern Europe and Anatolia: (1) Pikermi; (2) Halmyropotamus; (3) Alifakas; (4) Dytiko 3; (5) Umen Dol; (6) Kalimanci IV; (7) Samos; (8) Bayirköy; (9) Kemiklitepe upper level; (10) Yeni Eskihisar; (11) Çoban Pinar; (12) Taraklia; (13) Chimislia; (14) Kohfidisch; (15) Salmendingen.

LITERATURE CITED

Bernor, R. L., N. Solounias, C. C. Swisher III, and J. A. Van Couvering. This volume. The correlation of three classical "Pikermian" mammal faunas—Maragheh, Samos, and Pikermi—with the European MN unit system.

Bonis, L. de, G. Bouvrain, D. Geraads, and G. D. Koufos. 1992. A skull of *Hystrix primigenia* from the late Miocene of Macedonia (Greece). *Neue Jahrbuch für Geologie und Palaöntologie Monatschefte* 1992:75–87.

Garevski, R. 1956. Neue Fundstellen der Pikermifauna im Mazedonien. *Acta Musei Macedonici Scientificarium Naturalium* 4:69–96.

Koliadimou, K. and G. D. Koufos. 1991. The Hystricidae from the Pleistocene of Macedonia (Greece) and a review of the European representatives of the family. *Bulletin of the Geological Society of Greece* 25:453–71.

Kretzoi, M. 1951. The *Hipparion* fauna from Csákvar. *Földtani Közlöny* 81:384–417.

——. 1952. Die Raubtiere der Hipparionfauna von Polgardi. *Magyar allami Földtani Intézet Evkönyve* 40:1–35.

Ozansoy, F. 1965. Étude des gisements continentaux et des mammiféres du Cénozöique de Turquie. *Mémoire Societé géologie France, Paris*, n.s., 102:1–92.

Riabinin, A. N. 1929. Faune de mammiféres de Taraklia. I. Carnivora vera, Rodentia, Subungulata. *Trudy Geologicheskij Muzej Akademija Nauk SSSR* 5:75–115 and 127–134 (in Russian).

Sen, S. 1994. Les gisements de mammiféres du Miocéne supérieur de Kemiklitepe, Turquie 5. Rongeurs, Tubulidentés et Chalicothéres. *Bulletin Museum Nationale Histoire Naturelle*, 4° série, section C, 16:97–111.

Sen, S. and D. Kovatchev. 1987. The porcupine *Hystrix primigenia* (Wagner) from the late Miocene of Bulgaria. *Proceedings, Koninklijke Nederlandse Akademie van Wetenschappen, B* 90:317–323.

Sickenberg, O., J. D. Becker-Platen, L. Benda, D. Berg, B. Engesser, W. Gaziry, K. Heissig, K. A. Hünermann, P. Y. Sondaar, N. Schmidt-Kittler, K. Staesche, U. Staesche, P. Steffens, and H. Tobien. 1975. Die Gliederung des höheren Jungtertiärs und Altquartärs in der Turkei nach Vertebraten und ihre Bedeutung für die internationale Neogen-Stratigraphie. *Geologische Jahrbuch, Hannover*, Series B. 15:1–167.

Sulimski, A. 1960. *Hystrix primigenia* (Wagner) in the Pliocene fauna from Wéze. *Acta Palaeontologica Polonica* 5:319–35.

Ünay E. and H. de Bruijn. 1984. On some Neogene rodent assemblages from both sides of the Dardanelles, Turkey. *Newsletters on Stratigraphy* 13:119–32.

Weers, D. J. van. 1993. Teeth morphology and taxonomy of the Miocene rodent *Anchitheriomys suevicus* (Schlosser 1884), with notes on the family Hystricidae. *Proceedings, Koninklijke Nederlandse Akademie van Wetenschappen, B* 96:81–89.

23

The Composition and Diversity of Small Mammal Associations from Anatolia Through the Miocene

H. DE BRUIJN, E. ÜNAY, AND L. V. D. HOEK OSTENDE

Shifts in relative abundance of groups and changes in diversity within groups as observed in successive associations of fossil small mammals probably are the result of many, often interacting, phenomena. Physical causes—such as changes in climate, sea level, tectonics, latitude, altitude, ocean currents, and geography—and biotic causes, such as changes in vegetation, population dynamics, niche diversity, migration, and evolution—may all influence the composition and diversity of associations to a different degree. But there are more reasons to be cautious in interpreting thanatocoenoses of small mammals: selection by predators and/or streams prior to burial and by collecting techniques as well as by differences in suitability for preservation as fossils influence collections drastically. If we want to compare associations from the same time slice but from different areas, we also have a correlation problem.

Reconstruction of the evolution of Neogene continental biotopes on the basis of associations of fossils, the goal of this symposium, requires that the effect of the different processes that influence the composition of a particular association are recognizable. In order to achieve this we would need many well-studied associations from each time slice and evenly dispersed sampling in time and space. Obviously, this situation is not, and will never be achieved, so our source of information is, and will remain, inadequate. However, much would be gained if the information on existing collections of small mammals were made available. It is with this issue in mind that we present our data from Anatolia.

The Data Set

Figures 23.1 and 23.2 give the relative abundance of the Rodentia, Lagomorpha, and Insectivora as well as the diversity of these orders on the generic level for, respectively, the intervals MN 1–3 and MN 4–14 of Anatolia. The relative abundance of the Rodentia and Insectivora has been calculated on the basis of the sum of the M1, M2, m1, and m2 remains available. The percentage of Lagomorpha has been calculated on the basis of the sum of the P3 and p3 times two. In both figures the number in the left column gives the total number of the specimens available from a particular locality. First appearances of rodent genera that are supposedly immigrants into Anatolia are entered in the column to the far right.

The last occurrences of *Meteamys* and the Cricetodontini are also given in this column. Among the rodents the Muroidea are the dominating group throughout the Neogene.

Comparison of the figures 23.1 and 23.2 shows that the relative abundance of the Rodentia, Lagomorpha, and Insectivora fluctuates more during the early Miocene than during the middle Miocene to early Pliocene and that the Insectivora are more abundant as well as more diverse in the early Miocene assemblages. The sudden drop of the percentage of the insectivores from about 25% to 5–10% in combination with a drop in their diversity from about 7 genera to about 3 genera at the MN 3/4 boundary interval suggests that the order suffered some crisis. Though the nature of this crisis cannot be identified, we adopt the working hypothesis that the climate became drier during at least part of the year. The earlier suggestion (de Bruijn and van der Meulen 1979) that this change in climate may be due to the closure of the connection between the Mediterranean Sea and the Indian Ocean still seems a realistic possibility. This same climatic change is thought to have induced the migration of the so-called Miocene muroids *Democricetodon*, *Megacricetodon*, and *Cricetodon* from Western Asia into Southwestern Europe. Figures 23.1 and 23.2 also show that first occurrences through immigration are rather evenly dispersed over the various MN zones.

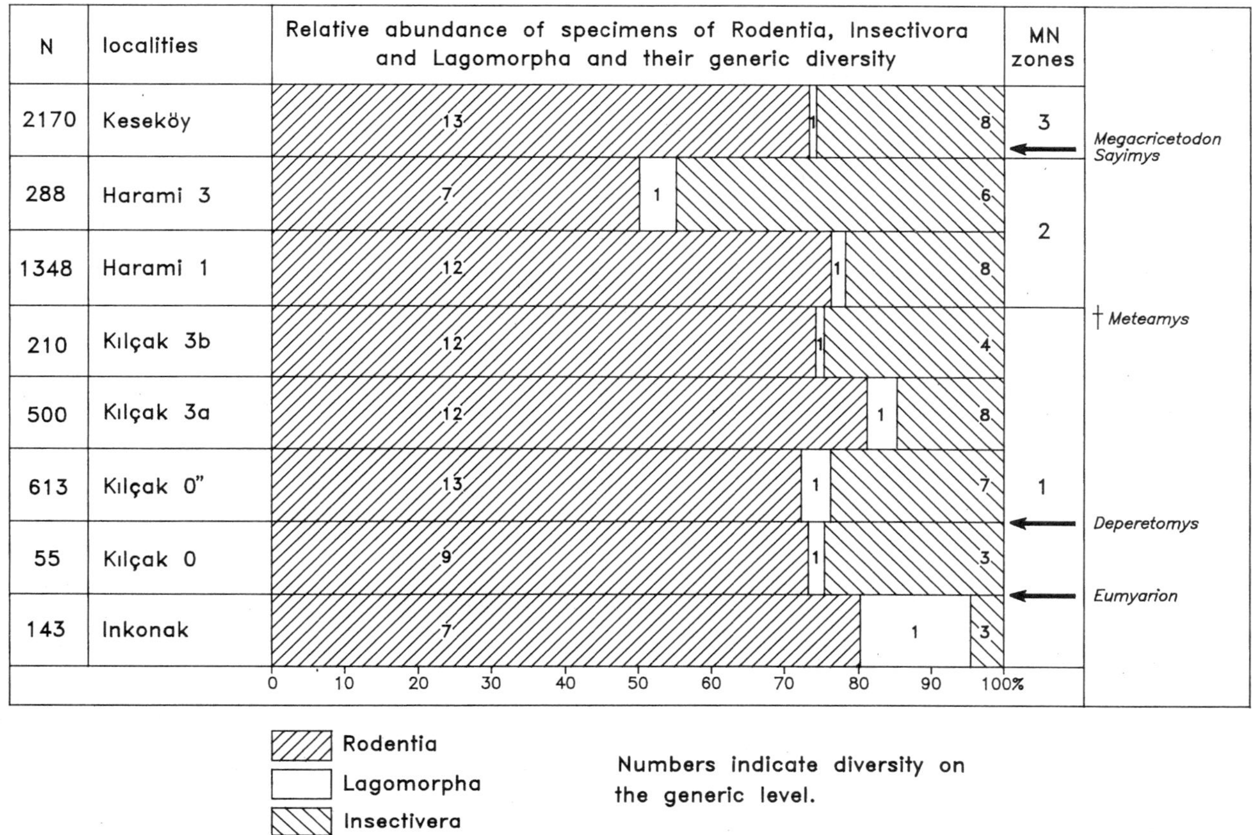

FIGURE 23.1 Relative abundance of the Rodentia, Lagomorpha, and Insectivora in eight early Miocene assemblages from Central Anatolia. N refers to the total number of M1, M2, m1, m2 of the Rodentia and the Insectivora plus the number of P3 and p3 of the Lagomorpha times two per locality. The number of genera per order recognized in each assemblage is entered in the relevant column. First appearances of some selected rodent genera are indicated by an arrow.

This, however, may be an artifact, because these zones certainly do not represent equal time intervals. Moreover, the quality of our documentation is obviously very different for different time slices.

Figure 23.3 summarizes the generic diversity of the rodents per zone in Anatolia from MN 4 to MN 14 (left column). The average number of rodent genera in our sequence appears to be 8.2. When the number of genera in a zone is below the average, the interval between the observed number and the average is hatched. When the observed number of genera is above the average the interval is given in black. The pattern obtained is unfortunately strongly influenced by the differences in documentation between zones. For six of the ten zones only one association is available, and the total number of specimens in some of these zones is rather limited (fig. 23.2). This means that the pattern of this graph will probably change appreciably when more and better faunas become known. Similar graphs for Aragon (van de Weerd and Daams 1978) and the reference localities of the zones (de Bruijn et al. 1992) are given in the second and third columns for comparison.

It is of interest to note that the average number of rodent genera per zone is somewhat higher in Anatolia, where documentation is poorer, than in Aragon. This is due to the high diversity peaks in MN 6, MN 11, and MN 14. Since the dips in diversity in MN 4–5, MN 9, and MN 12 in Anatolia are at least partly the result of poor documentation during these intervals we conclude that the real diversity of the rodents was higher in Anatolia than in Aragon during the time interval considered. Comparison of the data from Anatolia and Aragon to the reference localities is of interest because, although the latter were mainly selected for their high diversity and their richness in specimens, the fluctuation in diversity in this artificial sequence is remarkably high. The graphs giving the generic diversity in Aragon and in the sequence of the reference localities show a high degree of correlation. The very high peaks in the MN 7/8 (La Grive) and MN 14 (Podlesice) are considered to be the expression of the choice of reference faunas from fissure fillings that may contain fauna elements that never lived in the same biotope.

The correlation of the diversity pattern in the Anatolian and the Aragonian sequence is poor, but if we take the

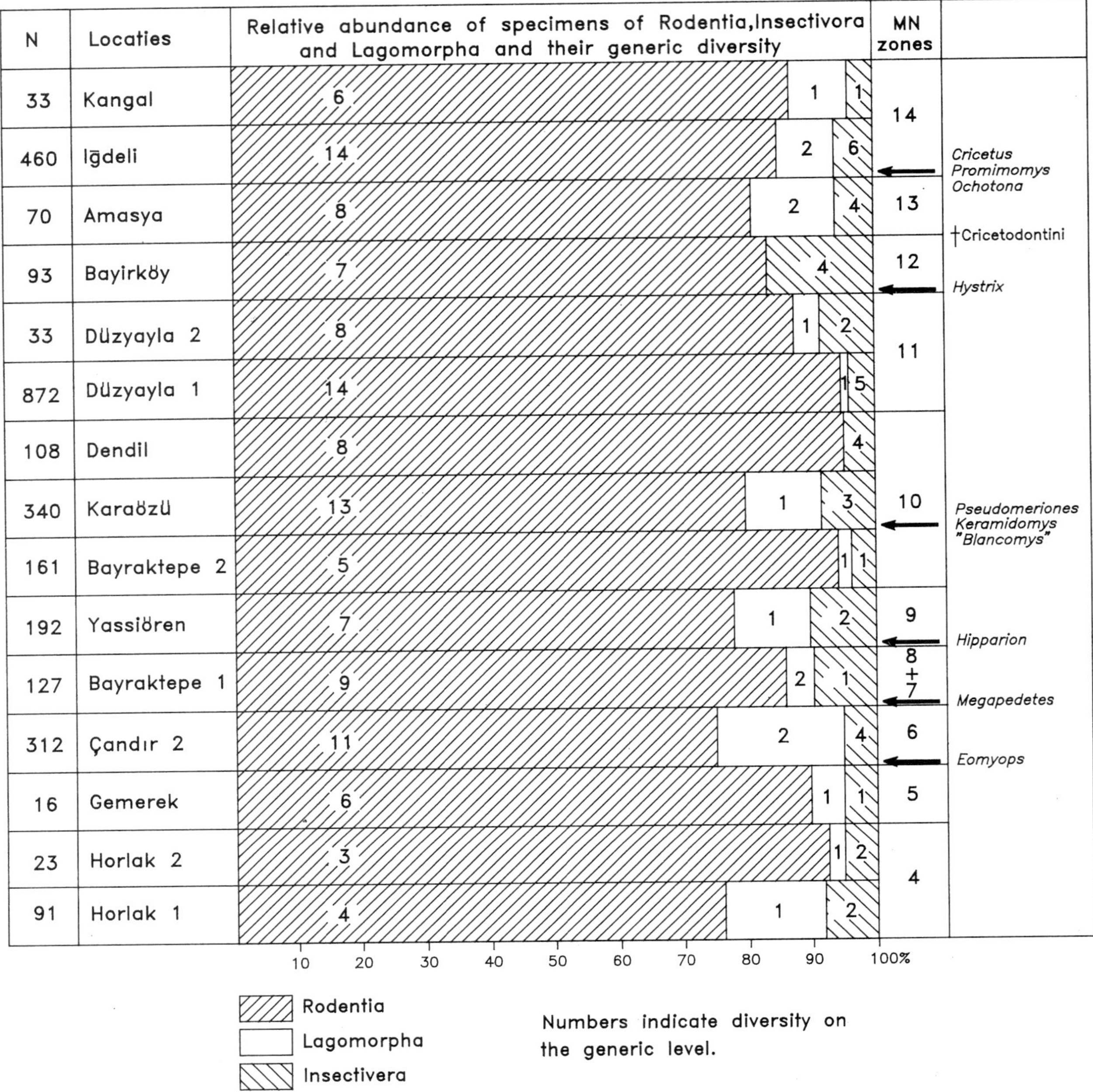

FIGURE 23.2 Relative abundance of the Rodentia, Lagomorpha, and Insectivora in fifteen assemblages from Anatolia ranging in age from the middle Miocene to the early Pliocene. The total number of specimens and the relative abundance of the three orders have been calculated on the basis of the M1 + M2 + m1 + m2 for the Rodentia and the Insectivora, and the number of the P3 and p3 times two for the Lagomorpha. The generic diversity per order is indicated by a number in the relevant column. First appearances of some selected genera of mammals are indicated by an arrow.

often limited documentation into account only one essential difference exists between the graphs for Anatolia and Aragon: the mid-Vallesian crisis, a reality in the Southwestern European sequence, is missing in Anatolia. The peak in MN 11 in Anatolia, based on the Düzyayla assemblages containing forest as well as open country indicators, correlates with rodent faunas of low generic diversity in Southwestern Europe even when we allow for mistakes in correlation. Apart from this discrepancy during the late Miocene, the faunal developments in Anatolia and Spain were probably not essentially different, showing peaks in diversity during MN 6 and MN 14. This similarity gains interest when we realize that many of the taxa involved are not the same in both areas. The high diversity in MN 6 is probably partly due to the long duration of this zone and partly the result of high niche diversity. The high diversity in MN 14

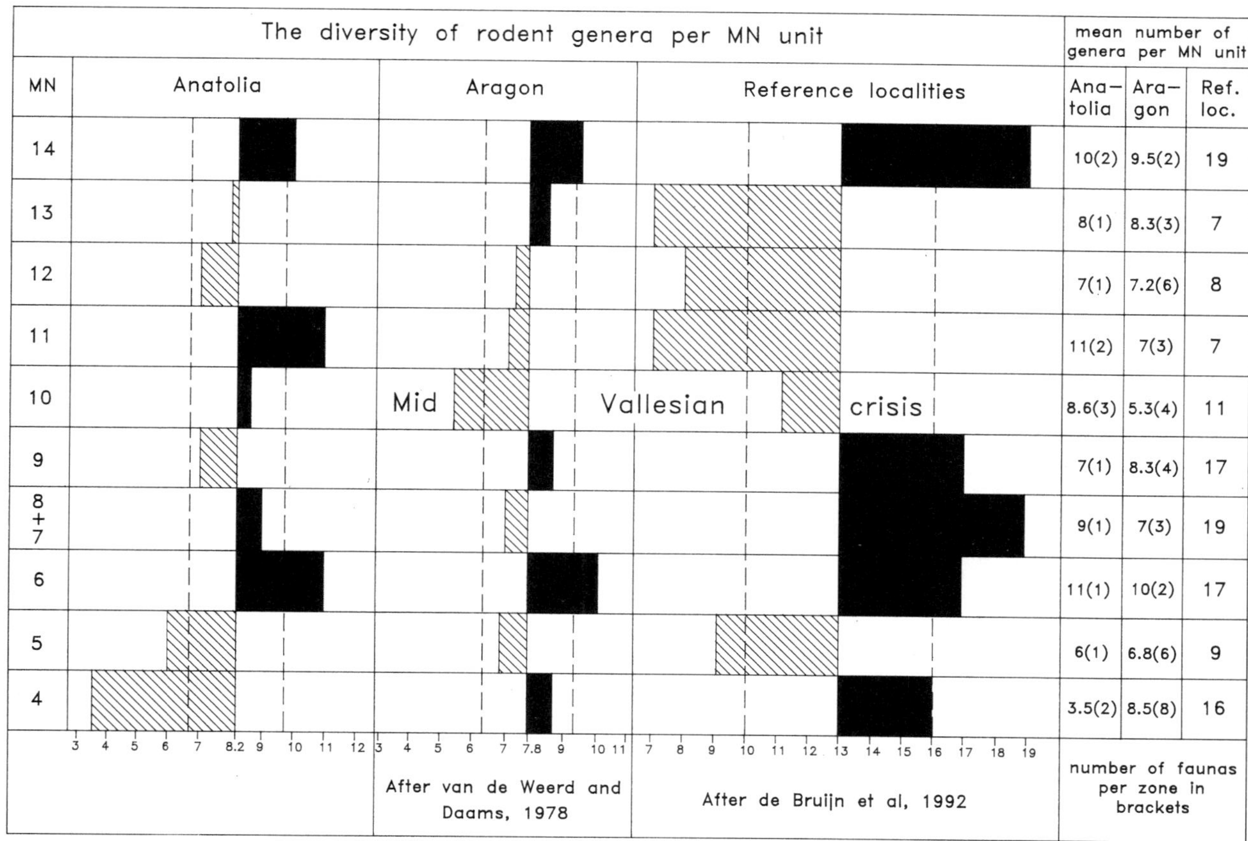

MN	mean number of genera per MN unit		
	Ana-tolia	Ara-gon	Ref. loc.
14	10(2)	9.5(2)	19
13	8(1)	8.3(3)	7
12	7(1)	7.2(6)	8
11	11(2)	7(3)	7
10	8.6(3)	5.3(4)	11
9	7(1)	8.3(4)	17
8+7	9(1)	7(3)	19
6	11(1)	10(2)	17
5	6(1)	6.8(6)	9
4	3.5(2)	8.5(8)	16

FIGURE 23.3 The (average) number of rodent genera per MN unit for Anatolia (see above), Aragon (van de Weerd and Daams 1978), and the reference localities of the MN units (de Bruijn et al. 1992). If the number of genera in a particular zone is larger than the average for the sequence concerned, this is indicated in black. Numbers smaller than the average are hatched. The number of assemblages per MN unit and per area is given in brackets.

seems to correlate with the broad Mediterranean transgression during the early Pliocene.

Conclusions

The data presented, unsophisticated though they are, allow some general statements on the evolution of the continental biotope in Anatolia and Southwestern Europe during the Miocene if one assumes that the diversity observed in associations of small mammals reflects niche diversity. We also assume that the highest niche diversity is found where tropical forests and open country meet, i.e., gallery forests in a warm climate.

The relatively high diversity of the rodents and insectivores in Anatolia during MN 1–3 suggests that the climate was warm as well as wet during that interval. The decline of the insectivores during the MN 3/4 boundary interval suggests that the climate became drastically drier during at least part of the year. This event may reflect the closure of the connection between the Mediterranean Sea and the Indian Ocean. The high diversity of the rodents in MN 6 all over the Mediterranean suggests that gallery forests be-

came wide spread again during that time interval. The mid-Vallesian crisis, restricted to the Southwestern Mediterranean area, indicates a local deterioration of the climate. Perhaps there was reduced rainfall during at least part of the year? This phenomenon somehow did not affect Anatolia. The Messinian crisis, correlated with MN 13, does not appear to be reflected in the diversity pattern of small mammals of the Western Mediterranean or Anatolia. During the early Pliocene forests became more widespread again all over the Mediterranean and the climate was probably warm and wet. This event may correlate with the early Pliocene transgression.

Acknowledgments

We thank Mr. T. van Hinte for drawing the tables, Dr. D. F. Mayhew for critically reading the manuscript, and Mrs. A. E. de Bruijn-Dudok van Heel for typing the manuscript. The research would not have been possible without the financial support of the faculty of Earth Sciences, Utrecht, the Netherlands, and the NATO Scientific Affairs Division (Grant no. CRG 910750), Brussels, Belgium.

LITERATURE CITED

Bruijn, H. de, and A. J. van der Meulen. 1979. A review of the Neogene rodent succession in Greece. Congress on Mediterranean Neogene, Sept. 1979. *Annales Géologiques des Pays Helléniques*, hors série, I:207–17.

Bruijn, H. de, R. Daams, G. Daxner-Höck, V. Fahlbusch, L. Ginsburg, P. Mein, and J. Morales. 1992. Report of the RCMNS working group on fossil mammals, Reisensburg 1990. *Newsletters on Stratigraphy* 26:65–118.

Weerd, A. van de, and R. Daams. 1978. Quantitative composition of rodent faunas in the Spanish Neogene and paleoecological implications. *Proceedings, Koninklijke Nederlandse Akademie van Wetenschappen*, B 81:448–73.

24

Carnivores, Exclusive of Hyaenidae, from the Later Miocene of Europe and Western Asia

L. WERDELIN

This paper represents a compilation of data on the late early Miocene (MN 4)–late Miocene (MN 13) Carnivora (exclusive of the Hyaenidae, see Werdelin and Solounias, this volume) of Europe and Western Asia. It is in part based on original work (Percrocutidae, Felidae, and Canidae) and partly on a literature survey (remaining families). Because of the varying states of systematic refinement of the different families, the data are somewhat uneven in quality. However, the purpose of this paper has simply been to compile and display some general data at a gross level to allow for comparisons with other taxa discussed in this volume and elsewhere.

As is the case with other compilations of this type, the results depend on the quality of the available data. Because of the aforementioned uncertainties regarding the taxonomy and systematics within certain groups (especially the Mustelidae), no attempt has been made to discuss trends within family subgroups. The geographic coverage is also uneven in many groups, much more so than in the case of hyaenas (Werdelin and Solounias 1991; this volume). This is due largely to a problem intrinsic to this compilation: the relative dominance in some groups of a very few taxon-rich localities (Vieux Collonges, Sansan, La Grive St. Alban, Can Llobateres). This problem is discussed under the heading Mustelidae below, since this is the group most affected.

One general problem is that I have no information on the subdivision of La Grive St. Alban: which taxa are present in level M (MN 7) and which are present in level L3 (MN 8; re: Mein 1989). Unless specific information is available, all taxa are therefore considered to be present in both MN 7 and MN 8.

This results in artificially increased diversity levels for these MN units (although this effect may be minimal), and also (and more important) reduced numbers of exits in MN 7 and entries in MN 8. This means that when summarized in a diagram the data on entries and exits may show

artifactual "dips" in these places. I refer to this as the "La Grive effect." Because of the uneven geographic coverage no attempt has been made in this analysis to study and compare trends in subareas of Europe and Western Asia, such as was possible with the Hyaenidae (Werdelin and Solounias, this volume).

Materials and Methods

This synthesis covers the area from Spain and France in the west to Kazakhstan in the east. As in the case of the Hyaenidae (Werdelin and Solounias, this volume), the reason for extending the geographic range to Spain and Kazakhstan is that their carnivore faunas have been found to be closely related to the rest of Europe and West Asia.

Chronologically, this treatment covers the periods MN 4 through MN 13. MN 4 is included simply in order to obtain data on entries and turnover for MN 5, although it should be noted that MN 4 is a very important unit for carnivore evolution, its taxa representing a transition from an earlier (late Oligocene to early Miocene) archaic carnivore fauna to the "typical" Miocene fauna. Very few carnivore taxa survive the MN 3/4 interval into MN 5.

It should be noted that the data do not represent all records of carnivores, but only those from localities that have a reasonably stable MN referral. While there may be some serious omissions in this review, this procedure ensures that the best data are included.

The data have been quantified for several categories. The first is simply the number of species present in an MN unit. The second consists of the number of entries in (= at the beginning of, in terms of the present resolution) each MN unit and the number of exits in (= at the end of) each unit. For the individual families these are given as raw counts and for the entire family as percentages (i.e., normalized against the number of taxa present). The third quantified category is relative turnover. This is here defined as the number of events (entries + exits) in an MN unit, divided by the number of species present in that unit. This will give a number between 0 and 2, where 0 means no events occurred (no species were present), 1 represents an average of one event (an entry or exit) per species, and 2 represents the fact that all species present in a unit both entered and exited in that unit. Many other quantifications could have been undertaken, but these will suffice to show the very general patterns sought herein.

Systematic Framework

Percrocutidae

The percrocutids have traditionally been considered as being members of the Hyaenidae, whom they most resemble in the morphology of their permanent dentition. However, recent work by Schmidt-Kittler and associates (Chen Guanfang and Schmidt-Kittler 1983; Schmidt-Kittler 1976) on the deciduous dentition has demonstrated significant differences between the dp4s of percrocutids and hyaenas. In these features, percrocutids more closely resemble viverrids and felids than hyaenids, suggesting that a relationship with the latter is doubtful. This impression is reinforced by the discovery of a percrocutid skull referred to *Dinocrocuta gigantea* from Hezheng, Gansu, China (Qiu Zhanxiang et al. 1988). Although it is difficult to be certain from the available illustrations, this form shows some curious basicranial characters that appear to differ from those of contemporary and younger hyaenas. Moreover, characters of the deciduous dentition would seem to offer two possibilities: either the condition in percrocutids is apomorphic, in which case they would be allied with viverrids and felids, or the condition seen in percrocutids is plesiomorphic, in which case the hyaena condition is derived and a synapomorphy for this family exclusive of percrocutids. In either case, the percrocutids cannot be included in the Hyaenidae as presently defined (Werdelin and Solounias 1991; this volume), and rather merit placement in their own family, the Percrocutidae.

The taxonomic history of the percrocutids is rather complicated. The earliest works describing percrocutids (Andrews 1916; Colbert 1939; Pilgrim 1913, 1932; Schlosser 1903) generally referred them to new species of the extant genera *Hyaena* or *Crocuta*. Kretzoi (1938) introduced the generic names *Percrocuta*, *Adcrocuta*, and *Allohyaena* for some taxa that had previously and have subsequently been considered to belong to the percrocutids. However, Kretzoi's work was largely ignored until Kurtén (1957) established the similarity of a number of the previously described forms to each other and their differences from the extant species. Thereafter, the genus nomen *Percrocuta* was used intermittently, until Ficcarelli and Torre (1970) reintroduced the issue by establishing the distinc-

tiveness of *Percrocuta* and *Adcrocuta*. This distinction was reaffirmed by Schmidt-Kittler (1976) in what amounts to a revolutionary conception for the percrocutids. He showed here and subsequently (Chen Guanfang and Schmidt-Kittler 1983) that, while *Percrocuta* differs from hyaenas in deciduous dental characters, *Adcrocuta* is a normal hyaena (later shown by Werdelin and Solounias 1990, 1991 to be the sister taxon to *Crocuta*). He also erected the subgenus *Dinocrocuta* for some larger percrocutids. Most recently, Howell and Petter (1985) revised the group, showing the distinctiveness of *Percrocuta* and *Dinocrocuta* and placing the genus *Allohyaena* among the percrocutids, which none of the earlier authors had done. The discussion of the European and Western Asian percrocutids that follows is for the most part based on the revision by Howell and Petter. Points of difference will be noted in the text.

"Crocuta" abessalomi   This species is known from a skull, two mandibles with teeth, and two isolated teeth, all from the Belomechetskaja locality in Georgia (Gabunia 1958, 1973; Howell and Petter 1985), which is broadly correlative with Sansan (MN 6). This is the best known of all the percrocutids discussed herein, which gives an indication of the group's rarity. This taxon is referred by Howell and Petter (1985) to their small percrocutid group (genus *Percrocuta sensu stricto*). Since there is a skull, albeit crushed and distorted, available in the material, it is possible that renewed study of this specimen's basicranium will help resolve the question of percrocutid affinities.

Percrocuta miocenica   This species is known from three mandibles and a mandible fragment from Prebreza in the former Yugoslavia, and a maxillary fragment and isolated teeth from Paşalar, Turkey (Pavlovi and Thenius 1965; Schmidt-Kittler 1976). Both of these localities are referred to MN 6. This taxon is also referred by Howell and Petter (1985) to the small percrocutid group. These authors also discount the suggested synonymy between this taxon and *P. abessalomi*, mainly on the basis of P4 morphology, which is derived in *P. abessalomi* and lower dentition morphology, which is more derived in *P. miocenica*.

Percrocuta sp. 1   This taxon is represented by a series of isolated teeth from Çandir, Turkey (MN 6; Schmidt-Kittler 1976; Bernor and Tobien 1990). This tooth series appears to be more evolutionarily derived than that of *P. miocenica*, and almost certainly represents a different species. Schmidt-Kittler (1976) suggests affinities with the Mongolian *P. tungurensis*, but as noted by Howell and Petter (1985), the morphology of P4 would seem to contradict this interpretation.

Percrocuta sp. 2   This species is represented by an isolated P2 from La Grive St. Alban. It was referred by Viret

(1951) to *Hyaena* sp., and later (most probably correctly) to *Percrocuta* by Howell and Petter (1985). This form's age is probably MN 7 (La Grive M). Unfortunately, P2 is not sufficiently diagnostic to provide any information regarding specific affinities. It is nevertheless the oldest record of a percrocutid in Western Europe.

Dinocrocuta minor   This taxon is represented by a mandible from the early Vallesian (MN 9) locality of Yassiören, Turkey (Ozansoy 1965). Howell and Petter (1985) also refer a fragmentary P4 from Yeni Eskihisar (Turkey) to this taxon, but without explicit justification. Since this tooth is not present in the type material, the referral cannot be evaluated and I leave it as being questionable but possible. Soria (1980) referred a number of isolated teeth from Concud to *D. minor*, but Howell and Petter (1985) have presented several reasons why this attribution is unlikely. Given Concud's MN 12 age, a referral to *Adcrocuta* would seem possible, and is perhaps corroborated by the presence in both taxa of p1, which is absent in all undoubted percrocutids except for the Sahabi specimen of *D. senyureki*. On the other hand, the available measurements tend to align the Concud material with *Dinocrocuta*. No resolution to this dilemma appears imminent, but it should be noted that Alcala et al.'s (1992) recent review of Spanish Miocene carnivores omits any record of *Percrocuta* from Concud.

Dinocrocuta senyureki   This taxon is represented by a skull fragment, three partial mandibles, and isolated teeth from Yassiören, Turkey, an isolated P4 from Eşme Akçaköy, Turkey, isolated teeth from Kayadibi and Inönü, Turkey, and a lower jaw from Sahabi, Libya. Both Yassiören and Eşme Akçaköy are early Vallesian (MN 9) while Kayadibi is early Turolian (MN 11). Inönü is late Vallesian or early Turolian (MN 10–11), and Sahabi member U lies close to the Turolian-Ruscinian boundary, but is probably MN 13. The species has been synonymized with the Asian *D. gigantea* by Soria (1980), but, as previously noted by Howell and Petter (1985), this is surely not the case.

Dinocrocuta salonicae   This taxon is known from a maxillary fragment (Andrews 1916; Beaumont 1979). The locality is not certainly known, but the specimen has been attributed to the MN 11 locality of Diavata by Howell and Petter (1985) and Koufos (1989). This species has also been synonymized with *D. gigantea* by Soria (1980), and this synonymy remains a distinct possibility, although the P4 protocone is much more developed in *D. salonicae*.

Dinocrocuta sp.   This form was referred by Soria (1980) and Alcala et al. (1992) to *D. gigantea* for some large teeth and other fragments collected from the MN 12 locality of Los Aljezares, Spain. However, these specimens are significantly smaller than Asian material referred to this species, and the attribution must be discarded. On the other hand, it seems evident that a species of *Dinocrocuta* is present at Los Aljezares, and this is therefore the youngest occurrence of the Percrocutidae in Europe.

Allohyaena kadici   This form is represented by two mandibles and numerous isolated teeth reported from Csákvár, Hungary (Kretzoi 1938, 1951, 1952a). *A. kadici* is evidently synonymous with *Xenohyaena csakvarensis* (Howell and Petter 1985; Schmidt-Kittler 1976). However, opinions regarding the affinities of this species differ. Howell and Petter (1985) consider that it belongs to the Percrocutidae and that it is closely related to *D. minor*, while Schmidt-Kittler (1976) specifically discounts this relationship. It is clear from figures and Howell and Petter's discussion (1985) that *A. kadici* generally resembles *D. minor*. However, the specific characters that they use to relate *Allohyaena* and *Dinocrocuta* would seem to be plesiomorphic ones: the presence of a third root on P3 and the large P4 protocone. Furthermore, *A. kadici* exhibits yet another plesiomorphic character, apparently overlooked by Howell and Petter (1985), that sets it apart from all percrocutids: the presence of an m2, the alveolus of which is particularly clearly seen in the specimen MAFI Ob. 3740, and which is also apparent in the holotype Ob. 3736 (B. Kurtén, pers. comm.; Werdelin and Kurtén ms.).

This observation is incompatible with the view forwarded by Howell and Petter (1985) and would place *A. kadici* outside the monophyletic group composed of the remaining percrocutids (which do lack m2), even if it were related to them. On the other hand, this only compounds the systematic dilemma, since there is no other taxon to which *A. kadici* would seem closely related. This leaves us Schmidt-Kittler's (1976) view that *Allohyaena* is a distinct lineage without a close relationship to the percrocutids. For the present I have elected to questionably maintain it within the Percrocutidae. However, I do not see how the synonymy of *Allohyaena* within *Dinocrocuta* can be maintained (Howell and Petter 1985), and I have restricted the latter genus to the nomen *Allohyaena kadici*.

Discussion   Judging from their known fossil record, percrocutids appear to have been mainly an Asian group, with a distribution that extended into the southeastern part of Europe, but only intermittently west of the Adriatic (two records, from France and Spain). They had a low but steady diversity, with from one to three species present in Europe and Western Asia during the MN 6–12 interval (figs. 24.1, 24.2).

The ecological relationship between percrocutids and true hyaenids is obvious, but perhaps not as simple as might be thought. The earliest percrocutids, such as *P. abessalomi* and *P. miocenica*, were medium-sized forms, with a dentition that was only moderately adapted to bone-

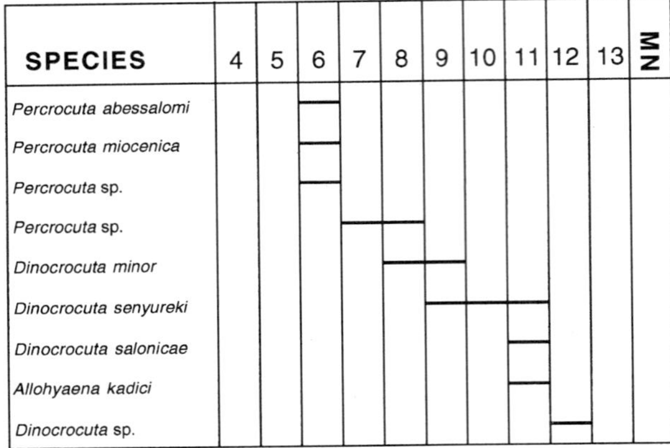

FIGURE 24.1 Biochronologic range chart for Percrocutidae of Europe and Western Asia during the later Miocene.

cracking. In overall morphology these forms, up to and including *Dinocrocuta minor*, match true hyenids such as *Ikelohyaena abronia* and *Leecyaena lycyaenoides*, forms that Werdelin and Solounias (this volume) have categorized as transitional bone-crackers. Therefore, it is not strictly true to say that the percrocutids occupied a hyaena-like, or bone-cracking, niche before hyaenas did. It may be more to the point to say that they were fairly generalized scavengers, with a tendency toward enlargement of the premolars and a general increase in body size. It is not until the occurrence of very large forms such as *Dinocrocuta senyureki* and *D. salonicae* that there is clear evidence for adaptations to bone-cracking, and even so the adaptations in this direction are limited compared to later hyaenids such as *Pachycrocuta* or *Crocuta*. In *D. senyureki*, for example, the premolars are not particularly bulbous, as would be expected of a bone-cracker, but seem merely to be extremely large. The only percrocutid exhibiting adaptations to bone-cracking matching those seen in derived hyaenids is *Dinocrocuta algeriensis* from North Africa.

About half of the European and West Asian percrocutid finds are associated with hyaenid remains. At Paşalar, *P. miocenica* is associated with the small, civetlike *Protictitherium gaillardi* and *P. intermedium* (Werdelin and Solounias, this volume). At Çandir, *P.* sp. 1 is associated with *P. intermedium*, while at La Grive St. Alban, *P.* sp. 2 is associated with *P. gaillardi* and *P. crassum*, as well as *Plioviverrops gaudryi* and the larger *Thalassictis certa* (although the internal stratigraphy of these finds at La Grive is not clear). Of the *Dinocrocuta* species, *D. minor* is associated with *Protictitherium cingulatum* and *T. montadai* at Yeni Eskihisar, while at Yassiören it is associated with *P. crassum* and a jackal-sized form, *Ictitherium intuberculatum*, as well as with *D. senyureki*. This latter species is associated with *P. crassum* at Eşme Akçaköy.

Howell and Petter (1985) state that *Dinocrocuta* re-

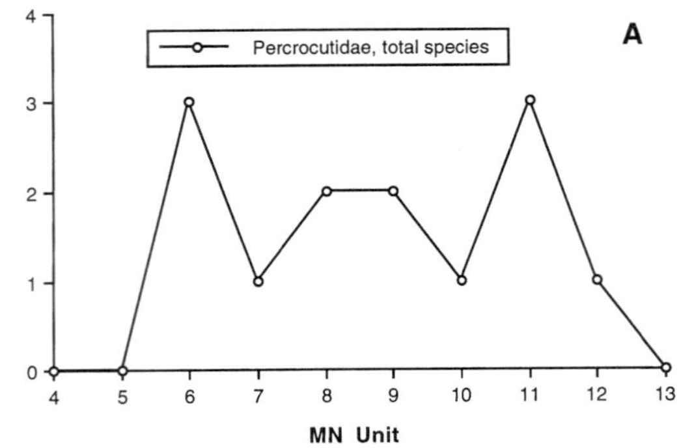

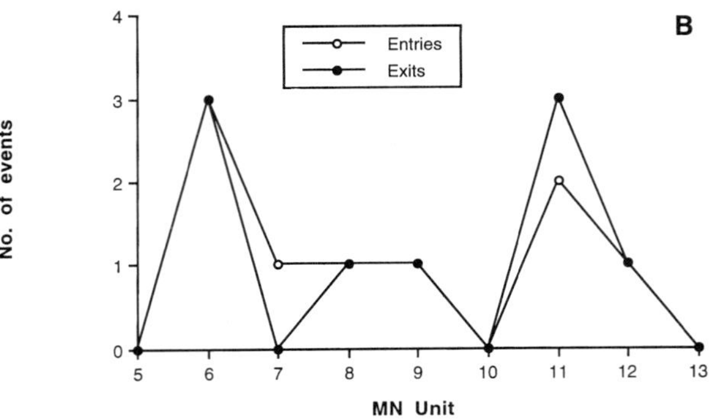

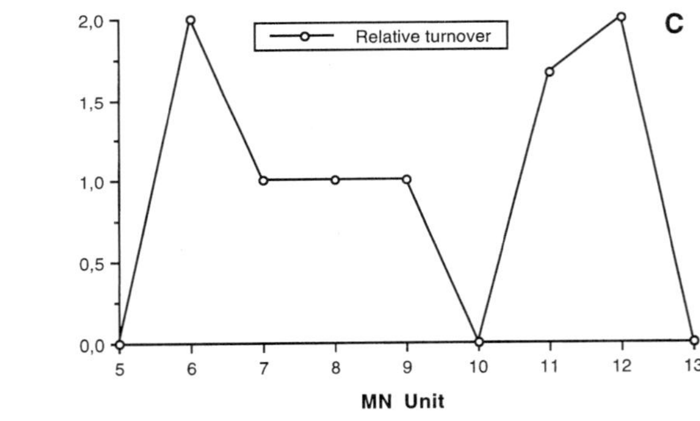

FIGURE 24.2 Fossil record of Percrocutidae of Europe and Western Asia quantified: (a) total number of species per MN unit; (b) number of entries and exits per MN unit; (c) relative turnover per MN unit.

mains are rarely associated with *Adcrocuta* in Vallesian-Turolian localities, and that this is probably not a coincidence. However, this is not necessarily true, since of the five possible localities (Inönü, Kayadibi, Diavata, Los Aljezares, and, if *A. kadici* is included, Csákvár), *Adcrocuta eximia* is known from two, Los Aljezares and Csákvár (A. advena, Kretzoi, pers. comm. to B. Kurtén). Given the rarity of *Dinocrocuta* remains, two out of five localities is a considerable number, suggesting that these taxa may not have excluded each other to the degree implied by Howell and Petter (1985). On the other hand, it is clear that percrocutids were absent from the typical Pontian localities (Turolian correlatives) such as Samos and Pikermi, and in China Baode and the Yûshe Basin (Flynn et al. 1991; Tedford et al. 1991).

Precisely why the percrocutids became extinct is not clear. They were never common, and given their presence in Sahabi (MN 13), it is not inconceivable that they became extinct along with the majority of the upper Miocene carnivores at the end of the Turolian in conjunction with the Messinian salinity crisis. The dispersion of derived hyaenas may have assisted in this extinction. However, this seems unlikely given that percrocutids and *Adcrocuta* coexisted from the late Vallesian onward; eliciting competitive exclusion as a cause for percrocutid extinction would not appear to be a viable explanation. One last aspect of percrocutid evolution should be considered here. Unlike the Miocene bone-cracking hyaenid *Adcrocuta eximia*, percrocutids seem to have had a propensity for speciation and, generally speaking, relatively restricted geographic ranges. This suggests that there is some sort of underlying ecological difference shaping the evolution of these two groups. This difference extends to living hyaenas, which are few in number and have broad geographic ranges (Werdelin and Solounias 1991). However, in the absence of any known percrocutid postcranial remains, we cannot know what this difference may have been.

Canidae

The Canidae were for most of their history a Nearctic group, and had an extensive radiation there (Martin 1989). It was not until the late Miocene that this family established itself in Eurasia (fig. 24.3). Unfortunately, there have been some chronologic problems raised by these early Eurasian canid records, as will be expounded upon briefly below. Late Miocene Eurasian canid records are very few, and the group apparently did not really begin radiating there until the Pliocene (Kurtén 1974).

Canis cipio   This taxon has been reported from Concud, Spain, where it is represented by a right maxillary fragment with P3-M2 (Crusafont Pairó 1950). It has also been somewhat doubtfully identified from an m1 from Los Mansuetos, Spain. Both of these localities are of medial Turo-

SPECIES	4	5	6	7	8	9	10	11	12	13	MN
Canis cipio									—		
'*Canis*' sp.										—	
Nyctereutes donnezani										—	

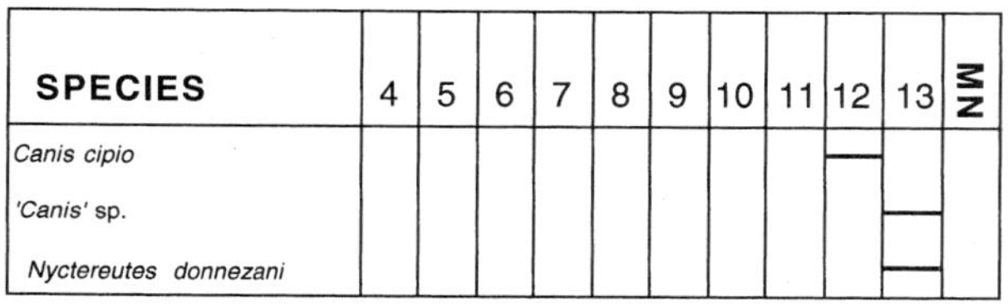

FIGURE 24.3 Biochronologic range chart for Canidae of Europe and Western Asia during the later Miocene.

lian age (MN 12). The material exhibits fairly generalized *Canis* characteristics, reminiscent of the North American Hemphillian age form *Canis davisi*, but is clearly distinguishable from that species (Rook et al. 1991). It is clear that the genus *Canis* as presently conceived is paraphyletic and that these species belong to a lineage separate from *Canis* sensu stricto. Pending a revision of the group, it would be unwise to change its taxonomic referral here.

Canis sp.   This form is represented by material of the upper and lower dentition, lower jaw, and parts of the postcranial skeleton of a small species from Brisighella, Italy (MN 13; Rook et al. 1991). This species is smaller than the Concud and Los Mansuetos form *Canis cipio*, being more similar to *Canis davisi* in both morphology and size. It also compares closely with the primitive early Pliocene *Canis* from Chang Pa Kou and Chao Chuang, China (Zdansky 1927 and pers. obs.). This and similar forms may be close to the common ancestry of *Canis* and *Vulpes*.

Vulpes sp.   A P4 and metatarsal from Venta del Moro, Spain (MN 13) were referred to *Canis* sp. by Morales and Aguirre (1976). Alcala et al. (1992) have, alternatively, referred this specimen to *Vulpes* sp. The tooth is smaller than that of *Canis cipio*, and closely comparable to *Canis* sp. from Brisighella and Chao Chuang in length, but is noticeably narrower, which is presumably the rationale behind Alcala et al.'s (1992) referral.

Nyctereutes cf. *donnezani*   This taxon is represented by a small assemblage of maxillary fragments probably belonging to a single individual, as well as a mandibular fragment from Venta del Moro (Morales and Aguirre 1976). These specimens represent a primitive *Nyctereutes*-like form, slightly larger than *N. donnezani* from Serrat d'en Vacquer and Layna. This taxon is distinguished by its narrower P4. The specific identity of these forms has not been well established, and the Venta del Moro form may represent a new species. The material is too limited for a systematic revision.

Spanish and Italian Turolian canids are biogeographically and/or biochronologically enigmatic, inasmuch as they are the oldest representatives of the family in Europe and that they must have arrived in Eurasia from North America via China. Yet there is no trace of canids in the abundant late Miocene carnivore materials from China. The Miocene canid occurrences reported by Zdansky (1924, 1927) all now appear to be derived from early Pliocene age localities. The same is true of the canid remains from other Chinese localities, such as Chao Chuang, at which *Chasmaporthetes lunensis*, a Ruscinian-Villafranchian species (Kurtén and Werdelin 1988; Qiu 1987) is present. In the Yûshe Basin, canids appear in the upper part of the Gaozhuang Formation (Nanzhuanggou and Culiugou Members), some 100–150 m above the Miocene–Pliocene boundary and after the first appearance of *Pliocrocuta* sp. (Flynn et al. 1991; Tedford et al. 1991). Hence the available data do not support the pre-Ruscinian occurrence of canids in China, and cast new doubt about their late Miocene first occurrence in Europe. No answer to this question is forthcoming at present, and it is especially difficult in view of the fact that the earliest occurrences are from Spain, the geographic area furthest removed from the postulated evolutionary source (North America).

Viverridae

This group has historically been a wastebasket category for small Feloidea, and there has generally been very little good evidence that Miocene fossil taxa actually belong here rather than in one of the other feloid families. A natural viverrid group has recently emerged with the recognition that many small Eurasian feloids thought to belong to the Viverridae are actually referable to the Hyaenidae (Beaumont 1969; Beaumont and Mein 1972; Schmidt-Kittler 1976; Werdelin and Solounias 1991; but see Semenov 1989). A second step was taken when Hunt (1991) discovered that the viverrid auditory bulla pattern is present in early Miocene fossil material. What still remains to be resolved is a clear phylogenetic subdivision of the Miocene taxa into groups assignable to the Viverridae and Herpestidae. Unfortunately, this subdivision requires well-preserved material, especially from the basicranial region, and therefore most taxa cannot certainly be assigned to either of these families. I have ignored this issue entirely here, and the taxa assigned to either the Viverridae or Herpestidae by previous authors are all placed in the Viverridae here.

Semigenetta elegans   This taxon occurs in the early Miocene localities (MN 4 or younger) of Erkertshofen 2 (Roth 1989) and Artenay (Ginsburg 1989). The species is also well known from Wintershof-West (Dehm 1950). It differs from the following *Semigenetta* species mainly in its size (Heizmann 1973). Otherwise, it is not clear if any real specific distinction can be made between them.

Semigenetta sansaniensis   This species has been described under at least two other nomina: *S. mutata* from La Grive St. Alban (Viret 1951) and *S. repelini* from Vieux Collonges (Mein 1958). However, Heizmann (1973) has shown that all of this material belongs in a single species, for which the name *S. sansaniensis* has priority.

Semigenetta ripolli   This species was described by Petter (1976) from Can Llobateres, Spain (MN 9). It is distinguished from *S. sansanensis* by few characters other than its significantly smaller size. *S. ripolli* is the size of *S.*

elegans, but I agree with Petter (1976) that the chronologic difference between these taxa seems to indicate that they are specifically distinct.

Viverrictis modica   This species was described from La Grive St.Alban by Gaillard (1899) as *Viverra modica*. It was later reported from Vieux Collonges (Mein 1958) and Sansan (Ginsburg 1961). The material from the two former localities was revised by Beaumont (1973), who created the subgenus *Viverrictis* for the material. It seems very unlikely indeed that this middle Miocene material should belong to the extant genus *Viverra*, and I am here using *Viverrictis* as the generic name. Beaumont (1973) notes that this species is common at Vieux Collonges, but rare at La Grive (and Sansan), while the reverse is true of *Leptoplesictis aurelianensis* (see below). What this means paleoecologically is unclear.

"Viverra" oeningensis   This species is represented by a complete skeleton, first described by Mantell (in Murchison 1835), who believed that it was a fox, and later by Owen (1846), who showed its viverrid affinities. Unfortunately, the skull and dentition of this specimen are not well preserved, and it is at present not clear whether it belongs to any of the described Miocene viverrid genera or species.

Viverra sp.   This taxon is represented by a viverrid lower jaw fragment with c and m1 from Baccinello V3, Italy (Rook et al. 1991). It compares most closely with *Viverra*, but neither the generic nor specific relationships can be clarified without additional material. As it stands this specimen is the only record of the Viverridae from the Turolian of Europe and Western Asia.

Leptoplesictis aurelianensis   This species is considered by most authors to belong to the extant genus *Herpestes*. Presently, comparisons to *Herpestes* are limited to plesiomorphic characters, and I prefer to keep this species distinct from all living taxa at the generic level. The nomen *Leptoplesictis*, in itself doubtfully monophyletic, is available. *L. aurelianensis* is known from slightly more localities than *V. modica*, and spans a slightly greater stratigraphic range (from MN 4 to MN 8). Beaumont (1973) notes that, unlike *V. modica*, this species is rare at Vieux Collonges but more common at La Grive.

?Leptoplesictis cadeoti   This species was originally described as *Semigenetta cadeoti* by Roman and Viret (1934). It has not been recorded from any Miocene locality since, but Roth (1989) suggests that it might belong to *Herpestes* (= *Leptoplesictis* in the present usage), and I have followed this suggestion here.

"Herpestes" dissimilis   This species was reported from Hostalets de Pierola inferior by Villalta Comella and Cru-

safont Pairó (1943). Unfortunately, I have not had the opportunity to examine this publication, and cannot comment further on this taxon.

Jourdanictis grivensis   This species was described by Viret (1951) on the basis of some few remains from La Grive St. Alban. Subsequently, this material has been revised by Beaumont and Mein (1972) and Beaumont (1973), with the result that it is now based on only two pieces of lower jaw. These are quite distinct from other middle Miocene viverrid species. Its affinities are not clear, but Beaumont (1973) suggests that it is close to *Viverra*.

Lophocyon carpathicus   This enigmatic species was described by Fejfar and Schmidt-Kittler (1987) on the basis of some skull and mandibular fragments with teeth. Its basicranium reveals that it is a member of the Viverridae, but with advanced adaptations to herbivory in its molarized premolars and high-crowned cheek teeth. Fejfar and Schmidt-Kittler (1987) ally this genus with *Sivanasua* and *Euboictis* into a new subfamily of Viverridae, Lophiocyoninae. The material is from Kosice-Bankov, Slovakia, and is of probable Astaracian age.

Discussion   It is evident that the European and Western Asian viverrid record consists of two entirely separate data sets (fig. 24.4). The first is their modest early and middle Miocene evolutionary radiation, beginning morphologically with a species such as *Herpestides antiquus* and reaching its acme in MN 7 + 8 with as many as five species. These forms disappear at the end of the Astaracian, with their last record being the Spanish form *Semigenetta ripolli*. There is no good evidence that any of these species are closely related to recent viverrid genera or species.

The second set is the single record of *Viverra* sp. from Italy. This form appears to be more closely related to living taxa and almost certainly represents an immigration event from Africa or Asia. However, the Viverridae again became extinct, as the modern European viverrids are the product of yet another immigration event from Africa.

It should be noted that the viverrid record undoubtedly suffers from taphonomic biases. These will be further considered in the section on the Mustelidae below.

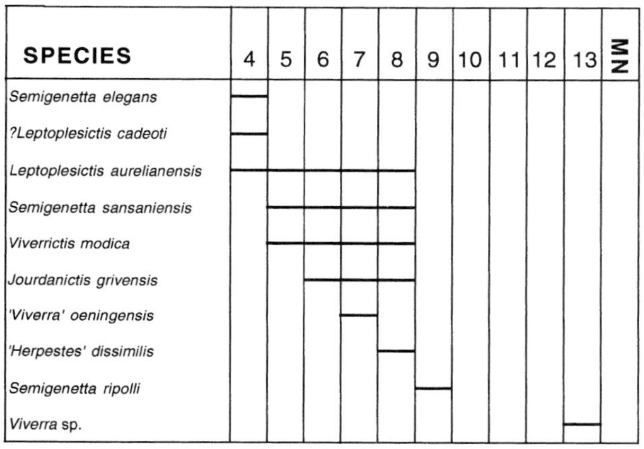

Amphicyonidae

The Amphicyonidae and Nimravidae are unique in that they are remnants of earlier evolutionary radiations, and they become extinct in the Miocene. All other families discussed here either first appear in the Miocene (Percrocutidae, Hyaenidae, Felidae, Viverridae) or, if not, continue through the epoch until today (Mustelidae, Ursidae, Canidae).

The European Paleogene radiation of the Amphicyonidae has recently been revised by Springhorn (1977). A corresponding work for the Miocene taxa has been made by Kuss (1965). This older work is currently being revised (Viranta, in prep.). In the interim I have tried to formulate a consensus position regarding the taxonomy and nomenclature of the various amphicyonid species. It should be noted that this is based entirely on a literature survey, just as in the case of the Ursidae and Mustelidae, below.

Amphicyon major   This is the most common Miocene amphicyonid, found at many of the most important European localities from MN 4 through MN 9. The Sansan sample is particularly important, because it includies most of the skeleton (Ginsburg 1961). With the exception of a possible occurrence at Paşalar (Schmidt-Kittler 1976), it is known only from Western and Central Europe, as is true of all the other species of amphicyonid discussed here except *A. caucasicus*.

Amphicyon giganteus   This is the largest European amphicyonid, known only from a few sites ranging from MN 4 to MN 6.

Amphicyon olisiponensis   Alcala et al. (1992) report this taxon as only occurring at Buñol. No other information is currently available.

Amphicyon castellanus   This very late form has been recorded at a few MN 7–9 Spanish localities.

Amphicyon steinheimensis   This form is smaller than, but otherwise similar to, *A. major*, which it appears to replace at Steinheim (Heizmann 1973) and Göriach (Thenius 1949).

Amphicyon caucasicus   This species, which is recorded only from Belomechetskaja (Gabunia 1973), clearly represents the genus *Amphicyon*, but the status of the species is unclear because of the limited material.

SPECIES	4	5	6	7	8	9	10	11	12	13	MN
Semigenetta elegans	—										
?Leptoplesictis cadeoti	—										
Leptoplesictis aurelianensis	——	——	——	——	——						
Semigenetta sansaniensis		——	——	——							
Viverrictis modica		——	——	——	——						
Jourdanictis grivensis			——	——							
'Viverra' oeningensis				——	——						
'Herpestes' dissimilis					——						
Semigenetta ripolli						——					
Viverra sp.										——	

FIGURE 24.4 Biochronologic range chart for Viverridae of Europe and Western Asia during the later Miocene.

278   L. Werdelin

Simamphicyon batalleri This species is recorded in the Can Llobateres faunal list published by Crusafont Pairó and Kurtén (1976), but is omitted from the list given by Alcala et al. (1992). I have no further data available regarding its status.

Amphicyonopsis serus This is another extremely rare form, known from Steinheim and possibly from La Grive St. Alban (Heizmann 1973).

Pseudocyon sansaniensis This poorly known genus has been the subject of much taxonomic confusion, as discussed by Kuss (1965). However, it is known from a few MN 4–6 localities (as well as earlier ones), Sansan being the last record of the species.

Pseudarctos bavaricus As in the case of several of the other species, this form is represented by scanty material from a number of localities between MN 5 and MN 8 in age.

Thaumastocyon bourgeoisi This aberrant form, often considered to be an ursid, is known from a few localities of early middle Miocene age, such as Pontlevoy and the Synclinal d'Esvres (Ginsburg 1989; 1990).

Thaumastocyon dirus This species is known from Los Valles do Fuentidueña by three teeth (Ginsburg et al. 1981). Despite the limited material and the chronologic gap separating this form from *T. bourgeoisi*, the attribution seems well supported. The species is somewhat larger than *T. bourgeoisi*, and more derived in several features of the dental morphology.

Agnotherium grivense This is another poorly represented form, best known from La Grive St. Alban (Viret 1951).

Agnotherium antiquus This species, which was described from Eppelsheim by Kaup (1832–1839; 1861), is still only known from a very few MN 9 localities (Pedreguerez, possibly Can Llobateres). It is interesting to note that Kurtén (1976) recorded the presence of this species at Bled Douarah, which is also generally placed in MN 9.

Cynelos helbingi The genus *Cynelos* is basically an Oligocene–early Miocene holdover, and all the Miocene forms are poorly represented. This species's type locality is Wintershof-West (Dehm 1950), and for the interval considered in this treatment, it only occurs at Baigneaux, Buñol, and Erkertshofen 2 (all MN 4).

Cynelos bohemicus This is the youngest *Cynelos* known, being represented from Esvres and Pontlevoy (Ginsburg 1989; 1990).

Cynelos schlosseri This species is similar to, but smaller than, *C. helbingi*. It is represented at Wintershof-West, Bézian, and Artenay (the latter two localities are MN 4).

Ysengrinia valentiana Like *Cynelos*, this genus is an Oligocene–early Miocene holdover. The only recorded occurrence for this time period is at Buñol (Alcala et al. 1992).

Discussion As can be seen from figure 24.6a, there is a slow but steady decline (excepting two minor peaks in MN 7 and 9) in the total number of Amphicyonid species from Europe and Western Asia during the MN 4–12 interval, when the last species becomes extinct (fig. 24.5). Figure 24.6b shows that this pattern is due to the fact that, although entries and exits generally balance each other, when they do not, exits tend to dominate, which in the long term leads to a reduced number of species. Relative to the number of species present, amphicyonid turnover is quite constant and low until MN 10, when there is an increase (fig. 24.6c).

There is good reason to believe that the pattern seen in the Amphicyonidae is a reflection of the replacement during the middle and early late Miocene of subtropical forest by temperate forest and woodland (Bernor 1983, 1984), and that amphicyonids may have been replaced by species that were either more catholic in their dietary requirements (e.g., Ursinae) or more cursorially adapted (Hyaenidae and eventually Canidae).

Ursidae

The Ursidae are another carnivore group that is in need of revision and for which I have relied on a compilation of records from many literature sources. The Miocene is an important period for ursid evolution because of the decline and extinction of Hemicyoninae and the divergence of the Agriotheriinae and Ursinae, the latter of course being especially important for the subsequent evolution of the family. In the absence of an adequate revision of the group, I have made no attempt here to analyze these subfamilies separately, although this is an important work for the future.

Hemicyon sansaniensis This species is known from a number of MN 5–6 localities, such as Sansan (Ginsburg 1961) and Paşalar (Schmidt-Kittler 1976).

Hemicyon goeriachensis This species differs from *H. sansaniensis* in some details of the dentition, as cited by Heizmann (1973) and Schmidt-Kittler (1976). However, these taxa are apparently not found together at any site, and the distinction between them must be considered somewhat doubtful. The species is recorded from MN 6 to MN 8.

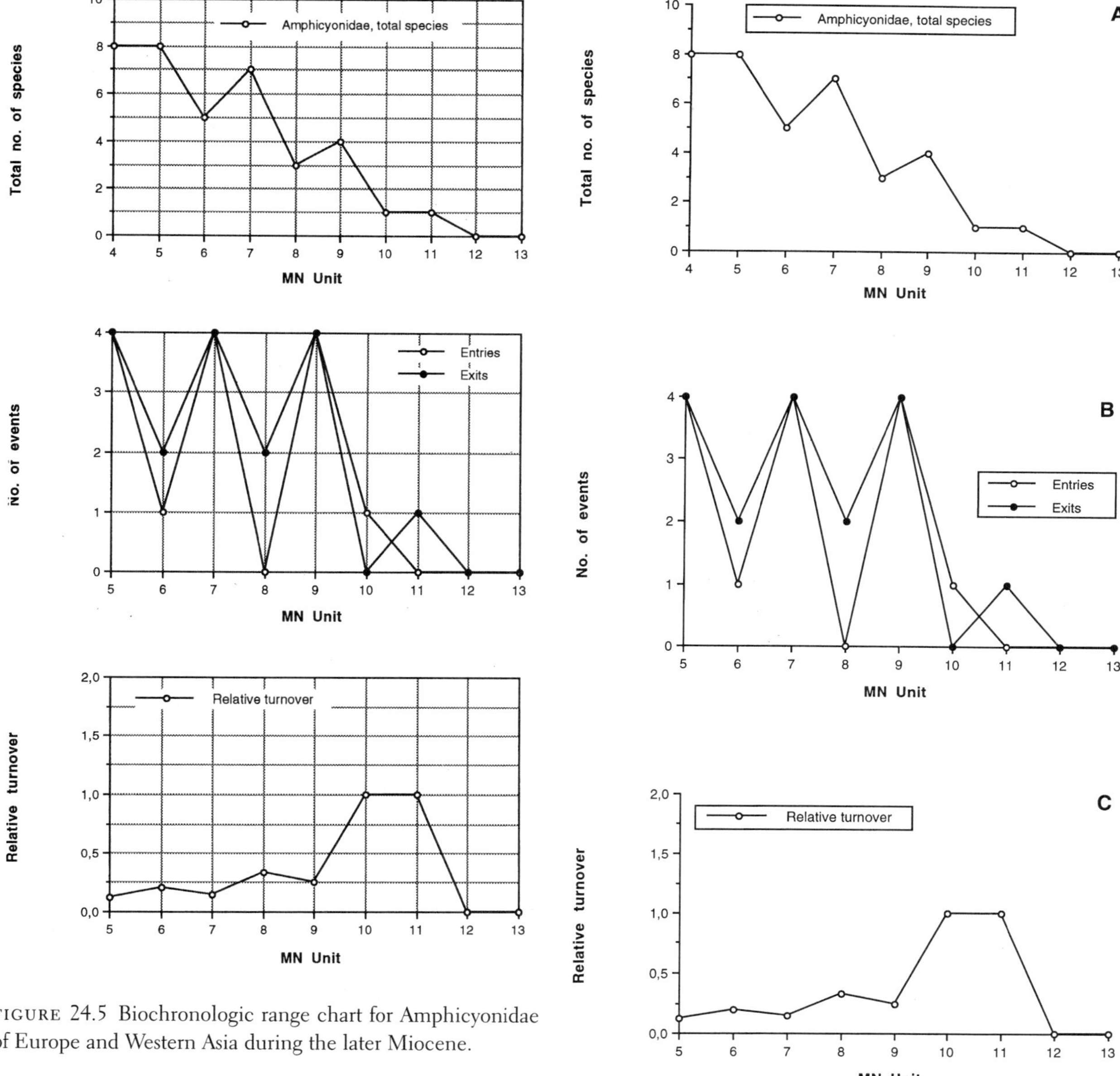

FIGURE 24.5 Biochronologic range chart for Amphicyonidae of Europe and Western Asia during the later Miocene.

FIGURE 24.6 Fossil record of Amphicyonidae of Europe and Western Asia during the later Miocene quantified: (a) total number of species per MN unit; (b) number of entries and exits per MN unit; (c) relative turnover per MN unit.

Plithocyon stehlini   This is an early species known from a few localities, MN 5 or older, notably Bézian, Pontlevoy (Ginsburg 1989; Ginsburg and Bulot 1982) and Vieux Collonges (Mein 1958; Heizmann 1973). This species's referral to *Plithocyon* was suggested by Heizmann (1973).

Plithocyon bruneti   This species is more derived than *P. stehlini*, but more primitive and smaller than *P. armagnacensis*. It is known from a few MN 5 localities and has been suggested to be the ancestor of *P. armagnacensis* (Ginsburg 1980).

Plithocyon armagnacensis   This species is well represented at Sansan (Ginsburg 1961), but is otherwise poorly known. Its last recorded appearance is at Hostalets de Pierola (Alcala et al. 1992).

Dinocyon thenardi   This extremely rare species is known only from a few specimens at La Grive St. Alban (Viret 1951), and a single tooth from Poysbrunn, Austria (Thenius 1947a), a locality of somewhat uncertain age.

Ursavus brevirhinus   This is the smallest of the four species known from the European and Western Asian middle and late Miocene. The Turolian form *U. ehrenbergi* is known only from the island of Euboea (Thenius 1947b). This fauna has not been specifically placed in any MN unit, and it is not treated further here. *Ursavus brevirhinus* is known from a number of localities between MN 4 and MN 11, although the last record, at Csákvár, is poorly documented (pers. comm. M. Kretzoi to B. Kurtén). The last well-documented appearance of this species is at Can Llobateres (MN 9; Crusafont Pairó and Kurtén 1976).

Ursavus primaevus   This species is larger and more derived than *U. brevirhinus*, but is otherwise similar to Thenius (1949) has synonymized them, but I believe that they are valid species. *U. primaevus* is the less common of the two, being known from MN 6–9, where the last occurrence is from Can Llobateres and Can Ponsich (Crusafont Pairó and Kurtén 1976).

Ursavus depereti   This is the largest *Ursavus* and not well known, being recorded from only a handful of localities of generally late Vallesian age. The only well dated of these is Soblay (MN 10). It is also known from Luzinay and Melchingen. A record from Nombrevilla 2 (MN 7; Alcala et al. 1992) must be considered to be doubtful. The specimen might conceivably come from Nombrevilla 1 (MN 9).

Agriotherium roblesi   A few teeth are known of this large species, which is only certainly known to occur at Venta del Moro (MN 13; Morales and Aguirre 1976). There are other questionable occurrences of a similar taxon, *Agriotherium* sp., from Alcoy (Gervais 1852) and Csákvár (Kretzoi 1952a).

Indarctos vireti   The validity of this species is the subject of some controversy (Crusafont Pairó and Kurtén 1976; Petter and Thomas 1986), which cannot be resolved here. Nevertheless, this form is limited to the Valles-Penedes and the localities of Can Llobateres and Can Ponsich (MN 9).

Indarctos arctoides   This species is known from a few Vallesian age localities (MN 9–10). It is apparently the ancestor of *I. atticus*.

Indarctos atticus   This species is larger than its presumed ancestor and known from many more localities ranging from MN 11 to MN 13. It is the *Indarctos* species typical of classic "Pikermian" age sites such as Pikermi, Samos, and Maragheh (Solounias 1981; Bernor et al., this volume).

Discussion   The total number of European and Western Asian ursid species (fig. 24.7) increases from two in MN 4, to six or seven in MN 6, after which there is a brief plateau and a subsequent drop back to the original number by MN 10 (fig. 24.8a). Since these shifts are relatively minor, the same is true of the patterns of entries and exits (fig.

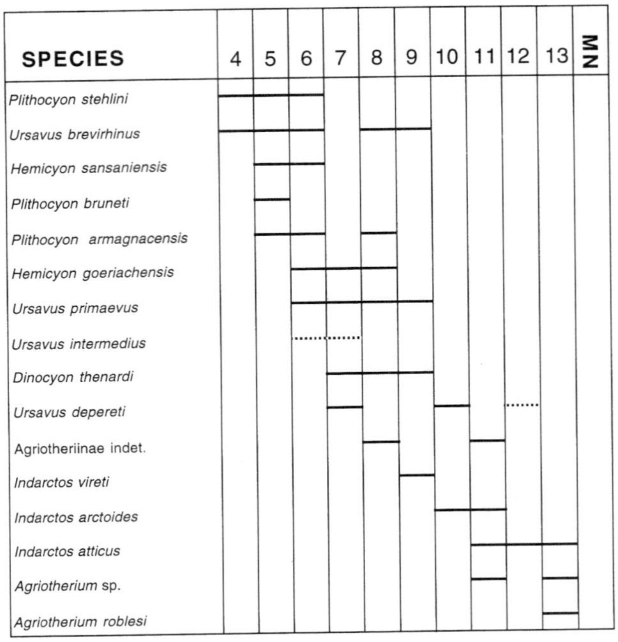

FIGURE 24.7 Biochronologic range chart for Ursidae of Europe and Western Asia during the later Miocene.

24.8b). The relative turnover seen in this group remains at a fairly high level (fig. 24.8c). This level is higher than seen in the Hyaenidae (Werdelin and Solounias, this volume), which might indicate one of two things: either this is a real phenomenon, and ursid turnover is higher than in most other carnivore groups for the time period and area under consideration here, or the family is taxonomically oversplit and there is a larger proportion of pseudoextinctions and pseudooriginations than seen in many other groups. Given the morphological diversity seen in this group, I find it likely that the evolutionary phenomenon cited here is real.

Nimravidae

Like the Amphicyonidae, this group is a remnant of an earlier, more extensive evolutionary radiation. This radiation is well documented both taxonomically and morphologically (re: Hunt 1987; Neff 1983; Scott and Jepsen 1936) in North America, but the European record is less extensive and more poorly documented. It consists of two phases: an early to middle Oligocene radiation mainly documented in the Quercy region (Ginsburg 1979), and then the three Miocene species discussed here. In the late Oligocene, nimravids are apparently absent from Europe and Western Asia.

Prosansanosmilus peregrinus   This species is the most primitive European Miocene nimravid species. It has only

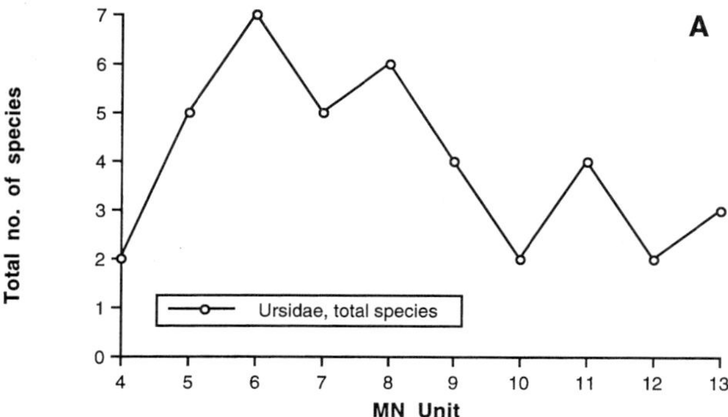

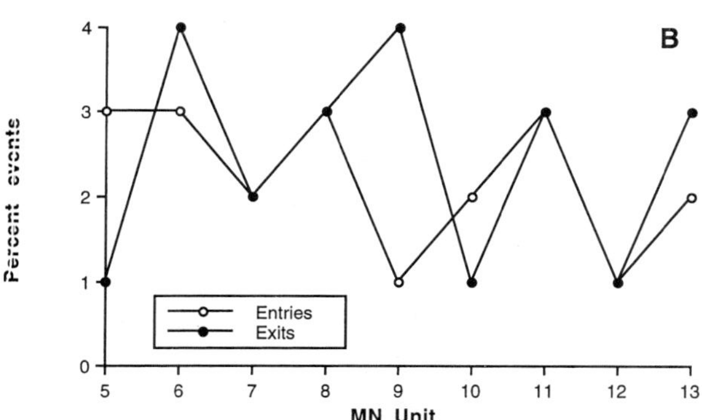

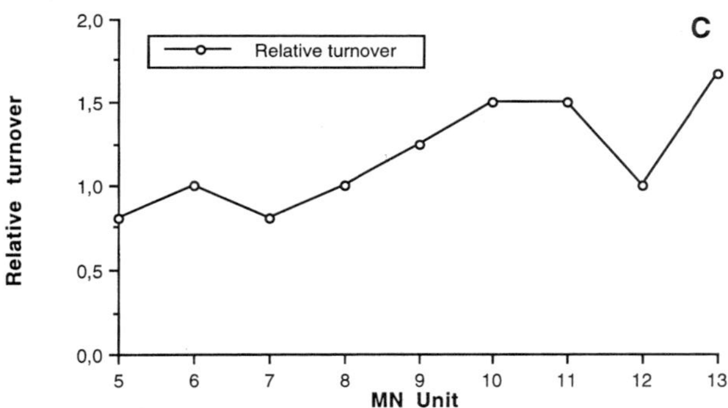

FIGURE 24.8 Fossil record of Ursidae of Europe and Western Asia during the later Miocene quantified: (a) total number of species per MN unit; (b) number of entries and exits per MN unit; (c) relative turnover per MN unit.

recently been recognized (Heizmann et al. 1980), and is known from only a few MN 4 localities.

Sansanosmilus palmidens   Morphologically, this species is intermediate between *P. peregrinus* and *S. jourdani*. It is only well known from Sansan (MN 6) and may also be present at a few other localities of similar age (Ginsburg 1961, 1989, 1990).

SPECIES	4	5	6	7	8	9	10	11	12	13	MN
Prosansanosmilus peregrinus											
Sansanosmilus palmidens											
Sansanosmilus jourdani											

FIGURE 24.9 Biochronologic range chart for Nimravidae of Europe and Western Asia during the later Miocene.

Sansanosmilus jourdani   This, the largest of the three nimravid species, is also the best known of them. It is present at a number of localities dating between MN 6 and MN 9, and is best known from a series of localities in the Valles-Penedes Basin (Alcala et al. 1992; Villalta Comella and Crusafont Pairó 1943).

Discussion   There is little to add to the outline of species presence shown here (fig. 24.9). At first glance it might appear that nimravids finally become extinct with the rise of sabertoothed felids around MN 8–9. However, Miocene nimravids were always rare, were apparently not sympatric, and did continue on for some time after the appearance of such pivotal felid taxa as *Machairodus aphanistus* (see below). Furthermore, *M. aphanistus* and *Sansanosmilus jourdani* co-occur at several localities (Los Valles do Fuentidueña, Can Ponsich, Can Llobateres, Santiga, and Höwenegg—all MN 9 localities), suggesting that they may have been able to coexist in some environmental settings. One might instead speculate that if felids with similar adaptational specializations as nimravids had anything to do with the demise of the latter, it was through an intrinsically greater rate of speciation. This is indicated by the rapid radiation of machairodonts after their initial appearance as contrasted with the low diversity of nimravids throughout the middle Miocene.

Felidae

The Felidae represent a diverse group that presents some intricate taxonomic and phylogenetic problems. The family probably originated in the late Oligocene with *Proailurus* and subsequently radiated rapidly in the middle Miocene, with the machairodont sabertooths showing the greatest number of species. Partly because of the rapidity of this radiation, and partly because of the dearth of useful systematic characters in most felid dentitions, the taxonomy and systematics of many Miocene taxa is obscure, as are the relationships of taxa such as *Felis attica* and *F. antediluviana* to the extant Felidae.

This overview is based on an assessment, partly original, partly from the literature, of which taxa may be valid. No attempt has been made to adjust the generic groupings to reflect evolutionary relationships. The genera used here are simply those which are commonly cited for the species indicated.

Pseudaelurus quadridentatus   This is the largest European species of *Pseudaelurus*. It has been recorded from MN 4 (Artesilla and Buñol; Alcala et al. 1992) to MN 9 (Los Valles do Fuentidueña; Ginsburg et al. 1981).

Pseudaelurus turnauensis   This is the only one of the four species of *Pseudaelurus* to appear before MN 4. This species is also known from a number of MN 3 localities (e.g., Wintershof-West; Dehm 1950). It is the smallest and putatively most primitive of the four species, but also the last to become extinct, being reported from Dorn-Dürkheim (MN 11; Franzen and Storch 1975).

Pseudaelurus romieviensis   This species is intermediate in size between *P. lorteti* and *P. quadridentatus*. It is known from comparatively few localities between MN 4 (Bézian, La Romieu, Baigneaux; Ginsburg 1989; Ginsburg and Bulot 1982; Roman and Viret 1934) and MN 7/8 (Steinheim, Häder; Heizmann 1973).

Pseudaelurus lorteti   This species is slightly larger than *P. turnauensis*. It is known from a large series of MN 4 localities (Artenay, Moratines; Alcala et al. 1992; Ginsburg 1989) to MN 8 (La Grive St. Alban, Hostalets de Pierola inferior; Alcala et al. 1992; Viret 1951).

Stenailurus teilhardi   This species is very close to *Metailurus*, as noted in the original publication (Crusafont Pairó and Aguirre 1972). There remains some doubt as to its generic distinction from *Metailurus*, which will remain until better *Stenailurus* material becomes available. However, the current evidence, particularly P4 morphology, indicates that *Stenailurus* is distinct from *Metailurus*.

Felis attica   This is a typical small conical-toothed felid known from Turolian age horizons (Beaumont 1961). This species has a position that is pivotal in the phylogeny of the conical-toothed cats, standing at the base of the modern felid radiation. Its first occurrence is at Karain (MN 10; Schmidt-Kittler 1976).

Metailurus parvulus   This is the smaller of the two *Metailurus* species. A number of species were synonymized under this nomen by Thenius (1951), including *Felis leiodon* Weithofer, and *Metailurus minor* Zdansky. It is particularly well known from the late Miocene of China (Zdansky 1924; Werdelin, in prep.). It is further well known from MN 11 to MN 13, with a not entirely certain record from MN 10 (Montredon; Beaumont 1988).

Metailurus major   This is the larger of the two *Metailurus* species. It is known at a number of Turolian age localities (MN 11–13).

Felis antediluviana   This species is known only from Eppelsheim, where it is represented by a mandibular ramus (Kaup 1832–1839). The identity and relationships of this form are entirely unknown.

Miomachairodus pseudailuroides   This is the earliest and most primitive machairodont cat. It is known from only two localities in Turkey, Yeni Eskihisar (MN 8) and Eşme Akçaköy (MN 9; Schmidt-Kittler 1976).

Paramachairodus orientalis   This is a smaller species than *M. aphanistus* and is known from a few Turolian localities (e.g., Polgardi; MN 13; Kretzoi 1952b). Pilgrim (1931) suggested that *P. orientalis* and *Machairodus maximiliani* from China (Zdansky 1924) might be conspecific. In view of this I am referring the record of the latter species from Venta del Moro (Morales 1984) to *P. orientalis*.

Machairodus romeri   This species, described from Kücükyözgat by Senyürek (1957), resembles *M. aphanistus* but differs in its more primitive carnassial and other dental characteristics (Schmidt-Kittler 1976).

Machairodus kurteni   This is a large *Machairodus* with reduced premolars described from Kalmakpai by Sotnikova (1991).

Machairodus laskarevi   This species was described from Kalfa by Lungu (1978). According to Sotnikova (1991), this form shows the characters of *M. aphanistus*, but retains a p2, which is a primitive characteristic. A specific differentiation on this basis is questionable, but since the data available to me are limited, I prefer to retain the distinction between these species here.

Machairodus aphanistus   This is the common Vallesian and Turolian age machairodont, ranging from MN 9 (e.g., Eppelsheim, Can Llobateres) to (possibly) MN 11 (Mahmutgazi; Schmidt-Kittler 1976).

Machairodus giganteus   Beaumont (1975) referred the typical Turolian *Machairodus* to this species, which ranges from MN 11 (Lower Maragheh, Crevillente 2) to MN 13 (e.g., Baccinello, Venta del Moro).

Machairodus copei   This is a species of *Machairodus* known only from Grebeniki (Pavlow 1914). It is considered a synonym of *M. aphanistus* by Beaumont (1975), but Sotnikova (1991) suggests that there are indeed characters

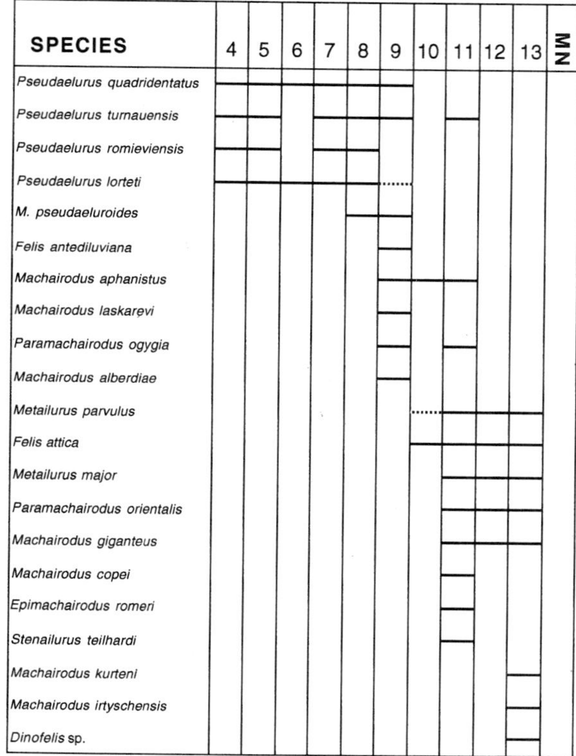

SPECIES	4	5	6	7	8	9	10	11	12	13	MN
Pseudaelurus quadridentatus											
Pseudaelurus turnauensis											
Pseudaelurus romieviensis											
Pseudaelurus lorteti											
M. pseudaeluroides											
Felis antediluviana											
Machairodus aphanistus											
Machairodus laskarevi											
Paramachairodus ogygia											
Machairodus alberdiae											
Metailurus parvulus											
Felis attica											
Metailurus major											
Paramachairodus orientalis											
Machairodus giganteus											
Machairodus copei											
Epimachairodus romeri											
Stenailurus teilhardi											
Machairodus kurteni											
Machairodus irtyschensis											
Dinofelis sp.											

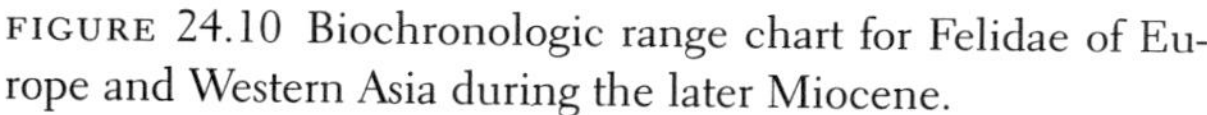

FIGURE 24.10 Biochronologic range chart for Felidae of Europe and Western Asia during the later Miocene.

to differentiate them. In view of the documented species variability within the genus *Machairodus*, it is difficult to make a firm decision on this issue without examining the material, and I will therefore keep these species separate.

Paramachairodus ogygia   This species is related to *P. orientalis*, but is smaller. Its only certain occurrence is at Eppelsheim (MN 9), but Alcala et al. (1992) have referred specimens from Crevillente 2 and Puente Minero (MN 11) to this species.

Machairodus alberdiae   This *Machairodus* is in some respects intermediate between *M. aphanistus* and *M. giganteus*, and is known only from Los Valles do Fuenti-dueña (Ginsburg et al. 1981).

Machairodus irtyschensis   This species (MN 13; Orlov 1936) differs from *M. aphanistus* and *M. giganteus* in its reduced p3. A relationship with African Pliocene *Homotherium* would appear possible.

Dinofelis sp.   Only one (somewhat doubtful) record of *Dinofelis* is known from the Miocene of Western Eurasia, Venta del Moro (MN 13; Alcala et al. 1992; Morales and Aguirre 1976). This is the earliest record of the genus in

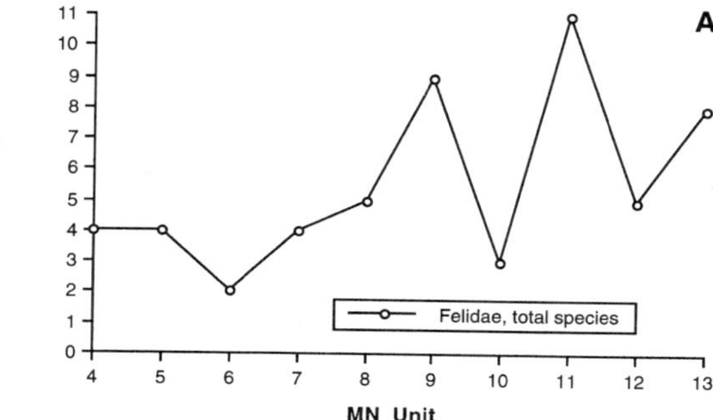

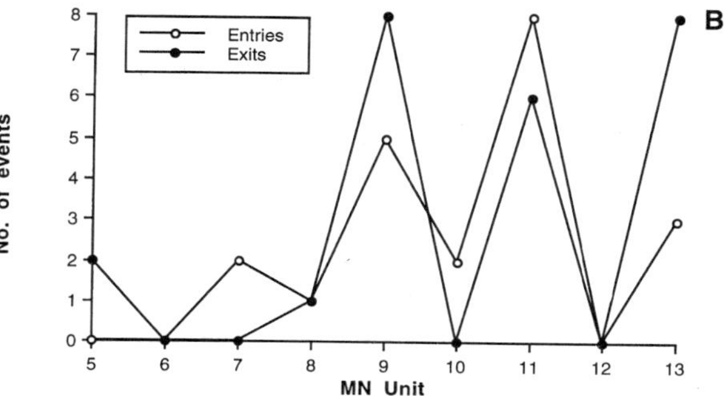

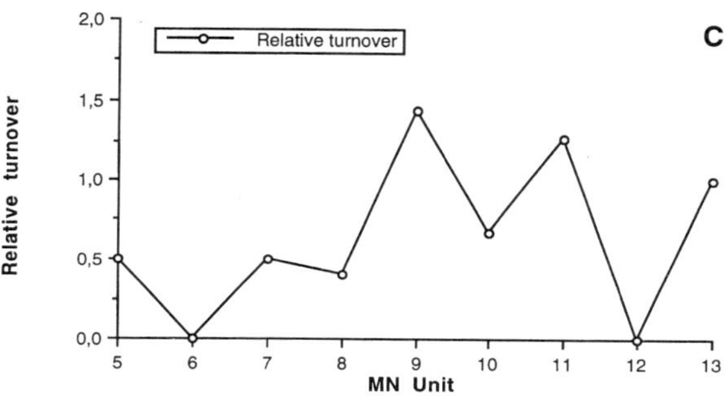

FIGURE 24.11 Fossil record of Felidae of Europe and Western Asia during the later Miocene quantified: (a) total number of species per MN unit; (b) number of entries and exits per MN unit; (c) relative turnover per MN unit.

Eurasia, since the material described by Zdansky (1924) comes from a Ruscinian locality (Loc. B).

Discussion   Felid diversity (fig. 24.10) is constant from MN 4 to MN 7 (the drop to two species in MN 6 is surely an artifact, since the two missing species occur both before and after this time period). In MN 8 an increase begins with the first appearance of machairodont cats, *Mioma-chairodus pseudaeluroides* (Schmidt-Kittler 1976). The subsequent pattern is one of rises and falls in diversity (fig. 24.11a), although it is hard to say to what extent the MN

10 drop is influenced by the availability of suitable sites of that age. The pattern of entries and exits shows the same ups and downs (fig. 24.11b). Relative felid turnover (fig. 24.11c) is obviously low until MN 9, after which we again see an up and down pattern, with peaks at MN 9 and MN 11. The former of these peaks can be correlated with the global regressive phase (Serravallian regression), which particularly affected the Mediterranean-Paratethys region (Rögl and Steininger 1983). This peak correlates with events in other groups as well (Werdelin and Solounias, this volume; Fortelius et al., this volume). The MN 11 peak has no obvious climatic or tectonic correlates, but marks a strong and broad extension of typical "Pikermian" age faunas (Bernor et al., this volume).

Mustelidae

This carnivore family is the one most in need of revision. A great number of species have been named, but few of these have been critically reappraised in the light of all the material now available. Many species have been referred to the genus *Martes*, but the relationships between the Miocene forms and extant species is obscure. The Miocene species are often referred to this genus on the basis of plesiomorphic characters. I have been able to critically evaluate only a very few mustelid species for this paper. Out of necessity, then, what follows here must include a number of pseudotaxa—i.e., species resulting from phyletic change rather than taxonomic splitting events.

To compound these difficulties, the Mustelidae are subject to a greater taphonomic bias than the other carnivore families discussed here, insofar as they include a number of small forms found only at very specific sites. These sites, which have an inordinate influence on mustelid diversity throughout the time period under investigation, include Vieux Collognes (Mein 1958; 12 species out of 19 present in MN 5), La Grive St. Alban (Viret 1951; all 12 species from MN 7, all 13 species from MN 8) and Can Llobateres (Petter 1976; 13 of 17 species from MN 9). These sites therefore tend to bias the mustelid data. However, we must work with what we have, and hope that the above mentioned localities are representative of their time periods.

The mustelid material is too extensive to allow the consideration of each species separately, and since there is no firm evidence to show that the genera used in the literature represent natural groups, these cannot be substituted for the species in this survey. Therefore, what follows is simply a list of the species used in this study. Some of these may undoubtedly be synonyms, while others have been overlooked. It may be further noted that this list includes species and/or genera referred by Schmidt-Kittler (1981) to the Procyonidae.

The mustelid species recognized here include: *Ischyr-ictis bezianensis, I. mustelinus, I. zibethoides, I. anatolicus, I. petteri, I. helbingi; Hoplictis florancei, H. noueli; Mionictis artenensis, M. dubia; Potamotherium miocenicum, P. valetoni; Martes munki, M. sainjoni, M. filholi, M. delphinensis, M. cadeoti, M. collongensis, M. burdigalensis, M. sansaniensis, M. transitoria, M. mellibulla, M. andersoni, M. basilii, M. woodwardi; Trocharion albanense; Plesiogale postfelina; Paralutra jaegeri; Plesiomeles pusilla, P. cajali, P. meini; Proputorius pusillus, P. sansaniensis; Taxodon sansaniensis; Heterictis oppolensis; Trochotherium cyamoides; Melidellavus leptorhynchus; Anatolictis laevicaninus; Sabadellictis crusafonti; Hadrictis fricki; Promephitis pristinidens, P. maeotica, P. lartetii, P. brevirostris; Mesomephitis medius; Sivaonyx hessicus, S. lehmani; Limnonyx pontica, L. sinizeri; Lutra hessica; Trochictis narcisoi; Eomellivora hungarica, E. orlovi; Circamustela dechaseauxi; Marcetia santigae; Plesiogulo crassa, P. brachygnathus, P. monspessulanus; Perunium ursogulo; Paraenhydriodon csakvarensis; Parataxidea maraghana, P. polaki, P. crassa; Promeles palaeattica; Baranogale adroveri, Enhydriodon lluecai, E. latipes; Mustela leporinum; Sinictis pentelici; Nannomephitis crassidens; Mellivora benfieldi; Paramartes pococki; Polgardia pannonica.*

Discussion   Mustelid diversity records a marked rise from MN 4 to MN 5 (fig. 24.12a), part of which is the effect of Vieux Collonges. After this there is a plateau of high diversity until MN 10, when there is a notable drop in diversity. Part of this drop is due to the absence in MN 10 of good mustelid localities, but part may be real, since the diversity never again reaches the level seen in MN 9.

From the diagram showing relative turnover we can see that mustelid turnover is generally high throughout the middle and late Miocene in Europe and Western Asia (fig. 24.12c). Part of this high turnover must be ascribed to pseudoextinction and pseudoorigination, but it is clear that turnover is higher than in most groups; as high as seen in the Ursidae (fig. 24.8c) and higher than seen in the Felidae (fig. 24.11c) and Hyaenidae (Werdelin and Solounias, this volume). However, within this high turnover pattern there is some fluctuation. The lowest turnover is seen in MN 7. Part of this is due to the "La Grive effect" (see introductory remarks above), but it is not entirely expected, since it also reflects relatively low entry numbers (fig. 24.12b) despite the difference in mustelid faunas between Sansan (MN 6) and La Grive St. Alban (MN 7 + 8). The highest turnover, on the other hand, is in MN 9, which correlates with high turnover events in other groups (see Werdelin and Solounias, this volume; Fortelius et al., this volume).

Family indet. *Simocyoninae*

This group (the genera *Alopecocyon* and *Simocyon*) has variously been referred to the Canidae, Amphicyonidae,

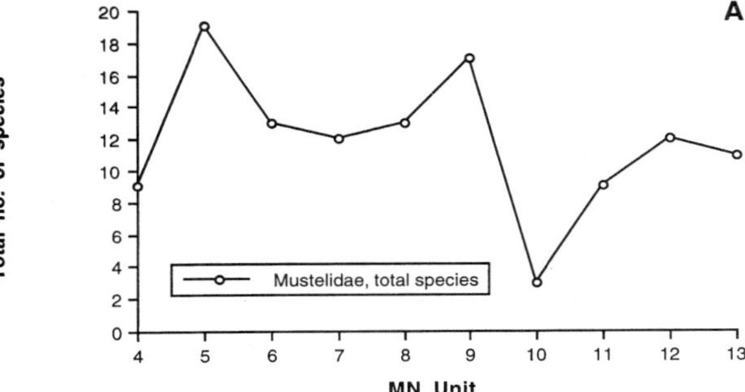

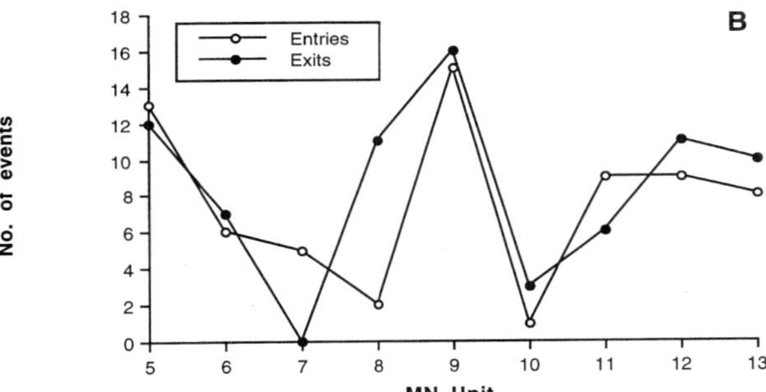

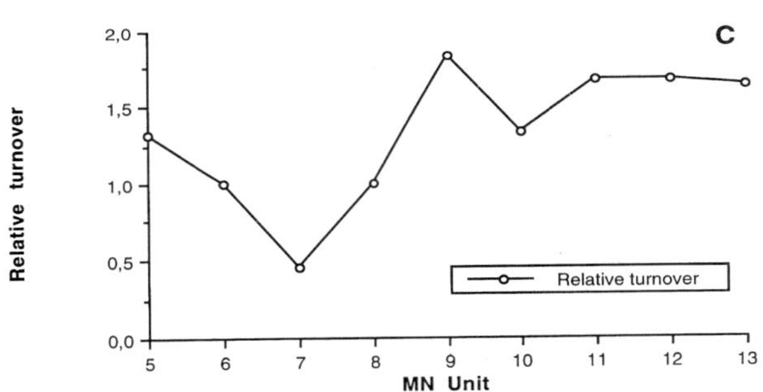

FIGURE 24.12 Fossil record of Mustelidae of Europe and Western Asia during the later Miocene quantified: (a) total number of species per MN unit; (b) number of entries and exits per MN unit; (c) relative turnover per MN unit.

and Procyonidae. No consensus regarding their evolutionary relationships seems to have been achieved, and I prefer to leave the question entirely open here.

Alopecocyon leptorhynchus   This small form is known from a number of localities ranging between MN 5 and MN 8 in age. However, it is never abundant, being generally represented by only a small number of specimens. At Sansan it is represented by only a single partial lower jaw (Ginsburg 1961).

Alopecocyon getti   This species is known from Vieux Collognes by only three isolated teeth (Mein 1958). It exhibits

some minor differences from *A. leptorhynchus*, especially in M1. However, the taxonomic value of these characters is questionable.

Simocyon diaphorus   This species is certainly known only from Eppelsheim (MN 9; Kaup 1832–1839). The type specimen has been lost. It is often referred to *Metarctos* (Pilgrim 1931). As indicated by Beaumont (1988), the differences between *Metarctos* and *Simocyon* are slight, and both have seemingly evolved from *Alopecocyon*-like ancestors. I have chosen here to refer the Eppelsheim form to *Simocyon*. The *Simocyon* from Montredon has on occasion been referred to this species, but this seems not to be proven (see Beaumont 1988 for a discussion).

Simocyon hungaricus   This species is known only from Csákvar. Kretzoi (1951) argues that it is intermediate in its stage-of-evolution between *S. diaphorus* and *S. primigenius* in its reduction of the premolars. This observation is congruent with the chronologic position of Csákvar (MN 11) between Eppelsheim (MN 9) and occurrences of *S. primigenius* (MN 12–13).

Simocyon primigenius   This relatively rare species is known from a few MN 12–13 localities. The Pikermi material is relatively extensive and includes complete skulls. A renewed investigation of this material may help determine this group's phylogenetic relationships (Schmidt-Kittler 1981).

Others

There are some other taxa of unclear phylogenetic relationship. These include the following:

Schlossericyon viverroides   This form has generally been referred to the Procyonidae (Ginsburg 1961; Pilgrim 1931; Viret 1951), but Schmidt-Kittler (1981) has argued that it in fact belongs to the Viverridae. It is known from Sansan and La Grive St. Alban.

Parailurus anglicus   This form is probably a relative of *Ailurus* (Boyd-Dawkins 1888; Pilgrim 1931), but current evidence suggests that its Miocene age is highly questionable, *pace* Savage and Russell (1983).

The Carnivora as a Whole

This final discussion will pool all data on carnivores, including data given here and data given on the Hyaenidae by Werdelin and Solounias (this volume). It should be clear from the discussion and figures above that some carnivore families show quite similar diversity and turnover patterns, while others differ. Pooling these data should therefore show a more diffuse pattern, and this is also the case. Even so, some fluctuations are notable. From MN 5 to MN 9 there is a steady, high carnivore diversity in Europe and Western Asia (fig. 24.13a). This diversity peaks in MN 9, with 49 known species. From MN 9 to MN 10 there is a severe drop in diversity. For reasons noted above (lack of suitable localities in MN 10) at least part of this drop is an artifact. However, even if most of this drop is an artifact, the subsequent pattern reflects a decrease in carnivore diversity through the Turolian from 45 in MN 11, to 36 in MN 12, to 34 in MN 13. The last difference is

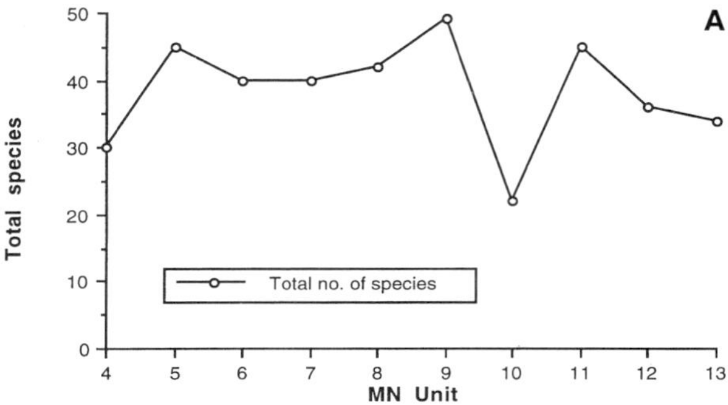

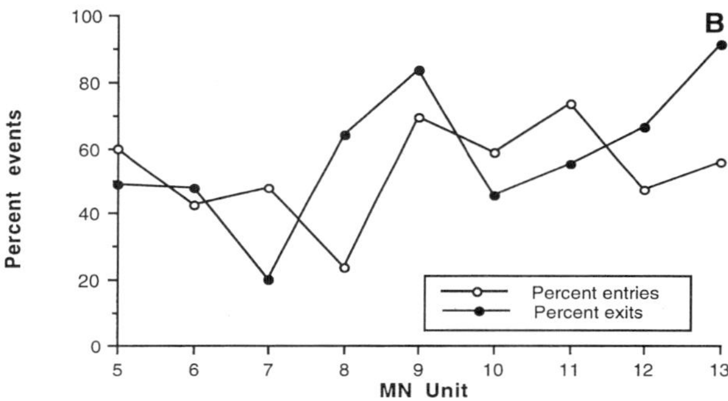

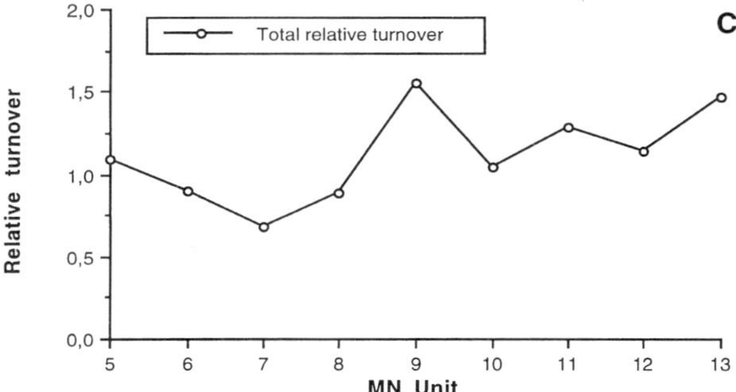

FIGURE 24.13 Fossil record of Carnivora of Europe and Western Asia during the later Miocene quantified: (a) total number of species per MN unit; (b) number of entries and exits per MN unit; (c) relative turnover per MN unit.

hardly significant, but the difference between MN 11 and MN 12—13 probably is. At the end of the Turolian there is a mass extinction, involving at least 31 of the 34 species.

The pattern of the percent of species present in an MN unit that exit in that unit (fig. 24.13b) is particularly interesting for large-scale carnivore diversity and turnover dynamics. Aside from the drop in MN 7 exits (due to the "La Grive effect," see above), it consists of two phases. The first runs from MN 5 to MN 9, showing a relatively steady rise in exits (given that MN 7 would fit the pattern under normal circumstances), culminating in MN 9, at which time 84% of the species present in that unit exit at the end of it. After MN 9 there is a sharp drop in relative exits, followed by another steady rise, this time culminating in MN 13, where, as noted, 31 out of 34 species (92%) exited.

The fluctuations in entries are on the other hand much less marked. If the drop during MN 8 is discounted (the "La Grive effect" again), the pattern seen is one of a slight drop from MN 5 to MN 6, a low plateau to MN 9, where there is a rise, followed by another plateau and a minor drop to MN 12–13.

Viewing this as total relative turnover (fig. 24.13c), we again see a two-phase pattern. The first is from MN 5 to MN 8 and is a period of relatively low turnover (average for this period: 0.89, i.e., less than one event per species per unit). A second phase is from MN 10 to MN 13, which is a period of relatively high turnover (average for this time period: 1.24, i.e., more than one event per species per MN unit). The difference between these two phases is in fact statistically significant (Mann-Whitney U-test: $Z = 2.021$, $p = 0.043$). Between these phases lies MN 9, which shows an elevated turnover (1.54).

There is no certain reason that can be given for this two-phase pattern. Given the heuristic hypothesis that turnover patterns are mainly driven by features of the external environment (see Vrba 1985 for a discussion), one might suggest that increased carnivore turnover reflects increased environmental fragmentation. This fragmentation may have been the result of the marked shift from tropical forest to temperate forest and woodland ecosystems during the course of the middle and late Miocene.

Acknowledgments

My work on carnivores is financed by the Swedish Natural Science Research Council (NFR).

LITERATURE CITED

Alcala, L., J. Morales, and D. Soria. 1992. El registro fósil neógeno de los carnívoros (Creodonta y Carnivora, Mammalia) de España. *Paleontologia i Evolució* 3:55–66.

Andrews, C. W. 1916. Note on some fossil mammals from Salonica and Imbros. *Geological Magazine* 6:540–43.

Beaumont, G. de. 1961. Recherches sur *Felis attica* Wagner du

Pontien Eurasiatique avec quelques observations sur les genres *Pseudaelurus* Gervais et *Proailurus* Filhol. *Nouvelles Archives du Muséum d'Histoire Naturelle de Lyon* 6:17–45.

——. 1969. Brèves remarques sur *Plioviverrops* Kretzoi (Carnivora). *Bulletin de la Société, Vaudoise des Sciences Naturelles* 70:1–7.

——. 1973. Contribution a l'étude des viverridés (Carnivora) du Miocène d'Europe. *Archives des Sciences, Genève* 26:285–96.

——. 1975. Recherches sur les Félidés (Mammifères, Carnivores) du Pliocène inférieur des sables à *Dinotherium* des environs d'Eppelsheim (Rheinhessen). *Archives des Sciences, Genève* 28:369–400.

——. 1979. Qu'est-ce que la "*Hyaena salonicae* Andrews" (Mammifère Carnivore). *Archives des Sciences, Genève* 32:247–50.

——. 1986. Les Carnivores (Mammifères) du Néogene de Höwenegg/Hegau, Baden-Württemberg. *Carolinea* 44:35–45.

——. 1988. Contributions a l'étude du gisement Miocène supérieur de Montredon (Herault). Les grands mammifères. 2-les carnivores. *Paleovértebrata* Mémoire extraordinaire: 15–42.

Beaumont, G. de and P. Mein. 1972. Recherches sur le genre *Plioviverrops* Kretzoi (Carnivora, ?Hyaenidae). *Archives des Sciences* 25:383–94.

Bernor, R. L. 1983. Geochronology and zoogeographic relationships of Miocene Hominoidea. In *New Interpretations of Ape and Human Ancestry*, ed. R. L. Ciochon and R. S. Corruccini, pp. 21–64. New York: Plenum.

——. 1984. A zoogeographic theater and biochronologic play: Time/biofacies phenomena of Eurasian and African Miocene mammal province. *Paléobiologie Continentale* 14:121–42.

Bernor, R. L., N. Solounias, C. C. Swisher III, and M. O. Woodburne. This volume. The correlation of three classical "Pikermian" faunas—Maragheh, Samos, and Pikermi—with the European MN unit system.

Bernor, R. L. and H. Tobien. 1990. The mammalian geochronology and biogeography of Paşalar (middle Miocene, Turkey). *Journal of Human Evolution* 19:551–68.

Boyd-Dawkins, W. 1888. On *Ailurus anglicus*, a new carnivore from the Red Crag. *Quarterly Journal of the Geological Society, London* 44:228–31.

Chen Guanfang, and N. Schmidt-Kittler. 1983. The deciduous dentition of *Percrocuta* Kretzoi and the diphyletic origin of hyaenas (Carnivora, Mammalia). *Paläontologische Zeitschrift* 57:159–69.

Colbert, E. H. 1939. Carnivora of the Tung Gur Formation of Mongolia. *Bulletin of the American Museum of Natural History* 76:47–81.

Crusafont Pairó, M. 1950. El primer representante del genero *Canis* en el Pontiense eurasiático (*Canis cipio* nova sp.). *Boletin de la Real Sociedad Espana de Historia Natural, Sección Geológica* 48:43–51.

Crusafont Pairó, M. and E. Aguirre. 1972. *Stenailurus*, félidé nouveau, du Turolien d'Espagne. *Annales de Paléontologie (Vertébrés)* 58:211–23.

Crusafont Pairó, M. and B. Kurtén. 1976. Bears and bear-dogs from the Vallesian of the Vallés Pénédes Basin, Spain. *Acta Zoologica Fennica* 144:1–29.

Dehm, R. 1950. Die Raubtiere aus dem Mittel-Miocän (Burdigalium) von Wintershof-West bei Eichstätt in Bayern. *Abhandlungen der Bayerischen Akademie der Wissenschaften, Mathematisch-naturwissenschaftliche Klasse* Neue Folge 58:1–141.

Fejfar, O. and N. Schmidt-Kittler 1987. *Lophocyon carpathicus* n. gen. n. sp. aus dem Jungtertiär der Ostslowakei und eine neue Unterfamilie der Schleichkatzen (Viverridae). *Palaeontographica* 199: 1–22.

Ficcarelli, G. and D. Torre. 1970. Remarks on the taxonomy of hyaenids. *Paleontographia Italica* 36:13–33.

Flynn, L. J., R. H. Tedford, and Q. Zhanxiang. 1991. Enrichment and stability in the Pliocene mammalian fauna of North China. *Paleobiology* 17:246–65.

Fortelius, M., J. Van der Made, and R. L. Bernor. This volume. Middle and late Miocene Suoidea of Central Europe and the Eastern Mediterranean: Evolution, biogeography, and paleoecology.

Franzen, J. L. and G. Storch. 1975. Die unterpliozäne (turolische) Wirbeltierfauna von Dorn-Dürkheim, Rheinhessen (SW-Deutschland). 1. Entdeckung, Geologie, Mammalia: Carnivora, Proboscidea, Rodentia. Grabungsergebnisse 1972–1973. *Senckenbergiana lethaea* 56:233–303.

Gabunia, L. 1958. On fossil carnivore remains from Belomechetskaya locality (northern Caucasus). *Vertebrata Palasiatica* 2:249–52. [in Russian]

——. 1973. Fossil vertebrates of the fauna of Bjelometschkaya. Akademia Nauk, Georgian SSR, Tbilisi, 138 pp. [in Russian]

Gaillard, C. 1899. Mammifères miocènes nouveaux ou peu connus de La Grive St. Alban. *Archives du Museum d'Histoire Naturelle* de Lyon 7:1–79.

Gervais, P. 1852. Description des ossements fossiles de Mammifères rapportés d'Espagne par M.M. Verneuil. *Bulletin de la Société, Géologique de France* 10:147–67.

Ginsburg, L. 1961. La faune des carnivores Miocènes de Sansan (Gers). *Mémoires du Muséum National d'Histoire Naturelle, Paris*, n.s., Series C. 9:1–190.

——. 1979. Revision taxonomique des Nimravini (Carnivora, Felidae) de l'Oligocène des Phosphorites du Quercy. *Bulletin du Muséum National d'Histoire Naturelle, Paris*, Series 4, 1:35–49.

——. 1980. *Plithocyon bruneti* nov. sp., Hemicyoninae (Ursidae, Carnivora, Mammalia) du Miocène de France. *Compte Rendu Sommaire des Séances de la Société Géologique de France* 22:232–35.

——. 1989. The faunas and stratigraphical subdivisions of the Orleanian in the Loire Basin (France); pp. 157–76 *in* E.H. Lindsay, V. Fahlbusch and P. Mein (eds.), *European Neogene Mammal Chronology*, Plenum, New York.

——. 1990. Les quatre faunes de Mammifères miocène des faluns du synclinal d'Esvres (Val-de Loire, France). *Comptes Rendus de l'Academie des Sciences, Paris* 310:89–93.

Ginsburg, L. and C. Bulot. 1982. Les carnivores du Miocène de Bézian près de Romieu (Gers, France). *Proceedings, Koninklijke Nederlandse Akademie van Wetenschappen* B, 85:53–76.

Ginsburg, L., J. Morales and D. Soria. 1981. Nuevos datos sobre los carnivoros de los Valles do Fuentidueña (Segovia). *Estudios geológicos* 37:383–415.

Heizmann, E. P. J. 1973. Die Carnivoren des Steinheimer Beckens. B. Ursidae, Felidae, Viverridae sowie Ergänzungen und Nachtrichten zu den Mustelidae. *Palaeontographica*, Supplement-band 8:1–95.

Heizmann, E. P. J., L. Ginsburg and C. Bulot. 1980. *Prosansanosmilus peregrinus,* ein neuer machairodonter Felide aus dem Miocän Deutschlands und Frankreichs. *Stuttgarter Beitrage zu Naturkunde,* Series B. 58:1–27.

Howell, F. C. and G. Petter. 1985. Comparative observations on some middle and upper Miocene hyaenids. Genera: *Percrocuta* Kretzoi, *Allohyaena* Kretzoi, *Adcrocuta* Kretzoi (Mammalia, Carnivora, Hyaenidae). *Géobios* 18:419–76.

Hunt, R. M., Jr. 1987. Evolution of the aeluroid Carnivora: Significance of the auditory structure in the nimravid cat *Dinictis. American Museum Novitates* 2886:1–74.

——. 1991. Evolution of the aeluroid Carnivora: Viverrid affinities of the Miocene carnivoran *Herpestides. American Museum Novitates* 3023:1–34.

Kaup, J. J. 1832–1839. *Déscription d'ossements fossiles de mammifères inconnus jusqu'a présent qui se trouvent au muséum grand-ducal de Darmstadt.* Darmstadt.

——. 1861. *Beiträge zur näheren Kenntnis der urweltlichen Säugethiere 5.* Darmstadt-Leipzig.

Koufos, G. D. 1989. The hipparions of the lower Axios Valley (Macedonia, Greece): Implications for the Neogene stratigraphy and the evolution of hipparions. In *European Neogene Mammal Chronology,* ed. E. H. Lindsay, V. Fahlbusch, and P. Mein, pp. 321–38. New York: Plenum.

Kretzoi, M. 1938. Die Raubtiere von Gombaszög nebst einer übersicht der Gesamtfauna. *Annales Museum Nationale Hungaricum* 31:89–157.

——. 1951. A Csákvári *Hipparion*-fauna. *Földtani Közlön* 81:384–417.

——. 1952a. Befejezó jelentés a Csákvári barlang ósléngtani feltárásáról. *Földtani Intezset* 40:37–69.

——. 1952b. Die Raubtiere der Hipparionfauna von Polgárdi. *Annales Instituti Geologici Publici Hungarici* 40:1–42.

Kurtén, B. 1957. *Percrocuta* Kretzoi (Mammalia, Carnivora), a group of Neogene hyaenas. *Acta Zoologica Cracoviensia* 2:375–404.

——. 1974. A history of coyote-like dogs. *Acta Zoologica Fennica* 140:1–38.

——. 1976. Fossil Carnivora from the late Tertiary of Bled Douarah and Cherichira, Tunisia. *Notes du Service Géologique, Tunis* 42:177–214.

Kurtén, B. and L. Werdelin. 1988. A review of the genus *Chasmaporthetes* Hay 1921 (Carnivora, Hyaenidae). *Journal of Vertebrate Paleontology* 8:46–66.

Kuss, S. E. 1965. Revision der europäischen Amphicyoninae (Canidae, Carnivora, Mamm.) ausschiesslich der voroberstampischen Formen. *Sitzungsberichte der Heidelberger Akademie der Wissenschaften, Mathematisch-Naturwissenschaftlige Klasse* 1965:5–168.

Lungu, A. N. 1978. The Hipparion fauna of the middle Sarmatian of Moldavia. Shtiintsa, Kishinev. [in Russian]

Martin, L. D. 1989. Fossil history of the terrestrial Carnivora. In *Carnivore Behavior, Ecology, and Evolution,* ed. J. L. Gittleman, pp. 536–68. London: Chapman and Hall.

Mein, P. 1958. Les mammifères de la faune sidérolithique de Vieux-Collonges. *Nouvelles Archives du Muséum d'Histoire Naturelle, Lyon* 5:1–122.

——. 1989. Updating of MN zones. In *European Neogene Mammal Chronology,* ed. E. H. Lindsay, V. Fahlbusch, and P. Mein, pp. 73–90. New York: Plenum.

Morales, J. 1984. *Venta del Moro: Su macrofauna de Mamiferos y bioestratigrafia continental del Mioceno terminal mediterraneo.* Ph.D. diss., Universidad Complutense, Madrid.

Morales, J. and E. Aguirre. 1976. Carnivoros de Venta del Moro. *Trabajos Sobre Neogeno-Cuaternario* 5:31–81.

Murchison, R. I. 1835. On a fossil fox found at Oeningen near Constance, with an account of the deposit in which it was imbedded. *Transactions of the Geological Society of London* (2) 3:277–92.

Neff, N. A. 1983. *The Basicranial Anatomy of the Nimravidae (Mammalia: Carnivora): Character Analyses and Phylogenetic Inferences.* Ph.D. diss., City University, New York.

Orlov, Y. A. 1936. Tertiäre Raubtiere des Westlichen Sibiriens. I. Machairodontinae. *Trudy, Paleontologicheskij Instituta, Akademia Nauk, SSR* 5:111–52.

Owen, R. 1846. On the extinct fossil viverrine fox of Öhningen showing its specific characters and affinities to the family Viverridae. *The Quarterly Journal of the Geological Society of London* 3:55–60.

Ozansoy, F. 1965. Étude des gisements continentaux et des mammiférès du Cénozoique de Turquie. *Mémoires de la Société Géologique de France* 102:1–92.

Pavlovi, M. and E. Thenius. 1965. Eine neue Hyäne aus dem Miozän Jugoslaviens und ihre phylogenetische Stellung. *Anzeiger der österreichischen Akademie der Wissenschaften, mathematisch-naturwissenschaftlige Klasse* 177–85.

Pavlow, M. 1914. Mammiféres Tertiaires de la la nouvelle Russie. *Nouvelles Mémoires de la Société, des Naturalistes de Moscou* 17:6–52.

Petter, G. 1976. Étude d'un nouvel ensemble de petits carnivores du Miocène d'Espagne. *Géologie médterranéene* 3:135–54.

Petter, G. and H. Thomas. 1986. Les Agriotheriinae (Mammalia, Carnivora) Néogènes de l'ancien monde. Presence du genre *Indarctos* dans la faune de Menacer (ex-Marceau), Algérie. *Géobios* 19:573–86.

Pilgrim, G. E. 1913. Correlation of the Siwaliks with mammal horizons of Europe. *Records of the Geological Survey of India* 43:264–326.

——. 1931. *Catalogue of the Pontian Carnivora of Europe in the Department of Geology.* London: British Museum (Natural History), London.

——. 1932. The fossil Carnivora of India. *Paleontologica Indica,* n.s. 18:1–232.

Qiu Zhanxiang. 1987. Die Hyaeniden aus dem Ruscinium und Villafranchium Chinas. *Münchner Geowissenschaftliche Abhandlungen* A9:1–108.

Qiu Zhanxiang, Xie Junyi, and Yan Defa. 1988. Discovery of the skull of *Dinocrocuta gigantea. Vertebrata Palasiatica* 26:128–38.

Roman, F. and J. Viret. 1934. La faune de mammifères du Burdigalien de La Romieu. *Mémoires de la Société Géologique de France* Nouvelle Serie, 21:1–67.

Rook, L., G. Ficcarelli, and D. Torre. 1991. Messinian carnivores from Italy. *Bollettino della Societcà Paleontologica Italiana* 30:7–22.

Rögl, F. and F. F. Steininger. 1983. Vom Zerfall der Tethys zu Mediterran und Paratethys-Die neogene Paläeographie und Palinspastik des zirkum-mediterranen Raumes. *Annalen des Natur-historisches Museums, Wien* 85:135–63.

Roth, C. 1989. Die Raubtierfauna (Carnivora, Mammalia) der untermiozänen Spaltenfüllung von Erkertshofen 2 bei Eichstätt/Bayern. *Mitteilungen der Bayerische Staatssammlung für Paläontologie und historische Geologie* 29:163–205.

Savage, D. E. and D. Russell. 1983. *Mammalian Paleofaunas of the World.* New York: Addison-Wesley.

Schlosser, M. 1903. Die fossilen Säugetiere Chinas nebst einer Odontographie der rezenten Antilopen. *Abhandlungen der Bayerischen Akademie der Wissenschaften* 22:1–221.

Schmidt-Kittler, N. 1976. Raubtiere aus dem Jungtertiär Kleinasiens. *Palaeontographica, A.* 155:1–131.

———. 1981. Zur Stammesgeschichte des marderverwandten Raubtiergruppen (Musteloidea, Carnivora). *Eclogae Geologicae Helvetiae* 74:753–801.

Scott, W. B. and G. L. Jepsen. 1936. The mammalian fauna of the White River Oligocene. Pt I. Insectivora and Carnivora. *Transactions of the American Philosophical Society* 28:1–153.

Semenov, Y. A. 1989. Ictitheres and morphologically related hyaenas of the Neogene of the USSR. Naukova Dumka, Kiev. [in Russian]

Senyürek, M. 1957. A new species of *Epimachairodus* from Kücükyozgat. *Belleten Türkei Tarih Kurumu.* 21:1–60.

Solounias, N. 1981. The Turolian fauna from the island of Samos, Greece. *Contributions to Vertebrate Evolution* 6:1–232.

Soria, D. 1980. *"Percrocuta"* y *"Adcrocuta"* (Hyaenidae, Mammalia) en el mioceno superior del area de Teruel. *Estudios Geológicos* 36:143–61.

Sotnikova, M. V. 1991. A new species of *Machairodus* from the late Miocene Kalmakpai locality in eastern Kazakhstan (USSR). *Annales Zoologici Fennici* 28:361–69.

Springhorn, R. 1977. Revision der alttertiären europäischen Amphicyonidae (Carnivora, Mammalia). *Palaeontographica,* Series B, 158:26–113.

Tedford, R. H., L. J. Flynn, Q. Zhanxiang, N. D. Opdyke, and W. R. Downs. 1991. Yûshe Basin, China: Paleomagnetically calibrated mammalian biostratigraphic standard from the late

Neogene of Eastern Asia. *Journal of Vertebrate Paleontology* 11:519–26.

Thenius, E. 1947a. *Dinocyon thenardi* aus dem Miozän österreichs. *Sitzungsberichte der österreichisches Akademie von Wissenschaften, mathematisch-naturwissenschaftliche Klasse* 14:209–24.

———. 1947b. *Ursavus ehrenbergi* aus dem Pont von Euböa (Griechenland). *Sitzungsberichte der Akademie von Wissenschaften, Wien* 156:225–49.

———. 1949. Die Carnivoren von Göriach (Steiermark). Beiträge zur Kenntnis der Säugetierreste des steirischen Tertiärs IV. *Sitzungsberichten der österreichische Akademie der Wissenschaften, Mathematisch-Maturwissenschaftlige Klasse,* Abteilung I. 158:695–762.

———. 1951. Zur odontologischen Characteristik von *"Felis"* leiodon aus dem Pont von Pikermi. *Neues Jahrbuch für Geologie und Paläontologie, Monatshefte* 1951:88–96.

Villalta Comella, J. F. de and M. Crusafont Pairó. 1943. Los vertebrados del Mioceno continental de la cuenca del Vallés-Pénédes (Prov. de Barcelona). I. Insectívoros. II. Carnívaros. *Boletin del Instituto Geológico y Minero de Espana* 56:147–336.

Viret, J. 1951. Catalogue critique de la faune des mammiféres Miocènes de La Grive Saint-Alban (Isère). *Nouvelles Archives du Muséum d'Histoire Naturelle, Lyon* 3:1–104.

Vrba, E. S. 1985. Environment and evolution: Alternative causes of the temporal distribution of evolutionary events. *South African Journal of Science* 81:229–36.

Werdelin, L. and N. Solounias. 1990. Studies of fossil hyaenids: The genus *Adcrocuta* and the interrelationships of some hyaenid taxa. *Zoological Journal of the Linnean Society* 98:363–86.

———. 1991. The Hyaenidae: Taxonomy, systematics, and evolution. *Fossils & Strata* 30:1–104.

———. This volume. The evolutionary history of hyaenas in Europe and Western Asia during the Miocene.

Zdansky, O. 1924. Jungtertiäre Carnivoren Chinas. *Paleontologia Sinica,* Series C 2(1): 1–149.

———. 1927. Weitere Bemerkungen für fossile Carnivoren aus China. *Palaeontologia Sinica,* Series C, 4(4): 1–30.

25

The Evolutionary History of Hyaenas in Europe and Western Asia During the Miocene

L. WERDELIN AND N. SOLOUNIAS

Hyaenas are among the most common later Miocene carnivores. Their taxonomy, systematics, and evolution have recently been reviewed by Werdelin and Solounias (1991). Although a small group today, with only four monotypic genera, hyaenas are both common and diverse in the fossil record, with at least 69 known species (not all of them named as yet). Of these, at least 35 are known from the middle and late Miocene of Europe and Western Asia (see appendix). Thus the material studied in this paper represents about half the known hyaenid diversity. This taxonomic diversity is accompanied by a similarly greater diversity in adaptations than seen in living hyaenas, which show two extreme dietary adaptations: insectivory on social insects (*Proteles cristatus*) and scavenging/bone-cracking (the other three species). The spotted hyaena is also an active pack-hunting carnivore (Kruuk 1972; Mills 1990). In the Miocene, however, locomotor adaptations ranged from being semi-arboreal (*Protictitherium*), to generalized cursorial (e.g., *Hyaenotherium*, some *Ictitherium*, *Hyaenictitherium*), to specialized cursorial (e.g., *Lycyaena*, some *Ictitherium*). Similarly, the Miocene hyaenas show a broad range of dietary adaptations, from insectivory, to combined meat and bone eaters, to specialized carnivores. These ecomorphs will be described further below.

Since the completion of our previous work (Werdelin and Solounias 1991), some new information has come to light and new analyses have been made. This presentation is a refinement of our earlier work, and serves to address the issues particular to this workshop volume. A cladogram of hyaenid phylogeny is presented in figure 25.1.

The earliest hyaenid record recognized by Werdelin and Solounias (1991) was that of *Plioviverrops gervaisi* from Vieux Collonges, France. At that time our information indicated that Vieux Collonges should be placed in MN 4b (Mein 1979). However, more recent information (Mein 1989) now places it in MN 5. In addition, we have verified the occurrence of *Protictitherium gaillardi* at Vieux Col-

longes (Mein 1958), previously omitted from our record. However, we have also noted the presence of this same species at Bézian, France (Ginsburg and Bulot 1982). This locality is approximately coeval with La Romieu, the reference locality for MN 4.

Thus we still take MN 4 as the earliest record of the Hyaenidae. As was discussed extensively elsewhere (Werdelin and Solounias 1991), hyaenids subsequently radiated until the end of the Miocene, at which time the majority of existing hyaenid species became extinct and replaced by hyaenas of modern type; i.e., large, bone-cracking, and scavenging forms. This radiation follows three phylogenetic phases. The first of these involves a general increase in size, a reduction of the posterior molars, and a suite of morphological changes in the carnassials (reduction of m1 talonid, trend toward equality in size of m1 paraconid and protoconid, shifting of P4 protocone from a position anterior to the paraconid to a position level with the paraconid). These trends occur in the genera *Ictitherium*, *Thalassictis*, *Hyaenotherium*, *Miohyaenotherium*, and *Hyaenictitherium*.

Subsequent evolution follows two quite distinct directions. One includes the genera *Lycyaena*, *Hyaenictis*, and *Chasmaporthetes* and involves an evolution toward more cursorial and meat-eating types. The other passes through a series of genera, including the genera *Hyaena*, *Parahyaena*, and *Crocuta*, and involves a trend toward bone-cracking and, with the exception of *Crocuta*, less cursorial forms. This evolution is discussed further below and also in Werdelin and Solounias 1991 and Werdelin et al. 1994.

Materials and Methods

This review of European and Western Asian hyaenids includes the area from Spain and France in the west to Kazakhstan in the east. The easternmost localities included are Pavlodar, Kalmakpai, and Bota Mojnak, all in Kazakh-

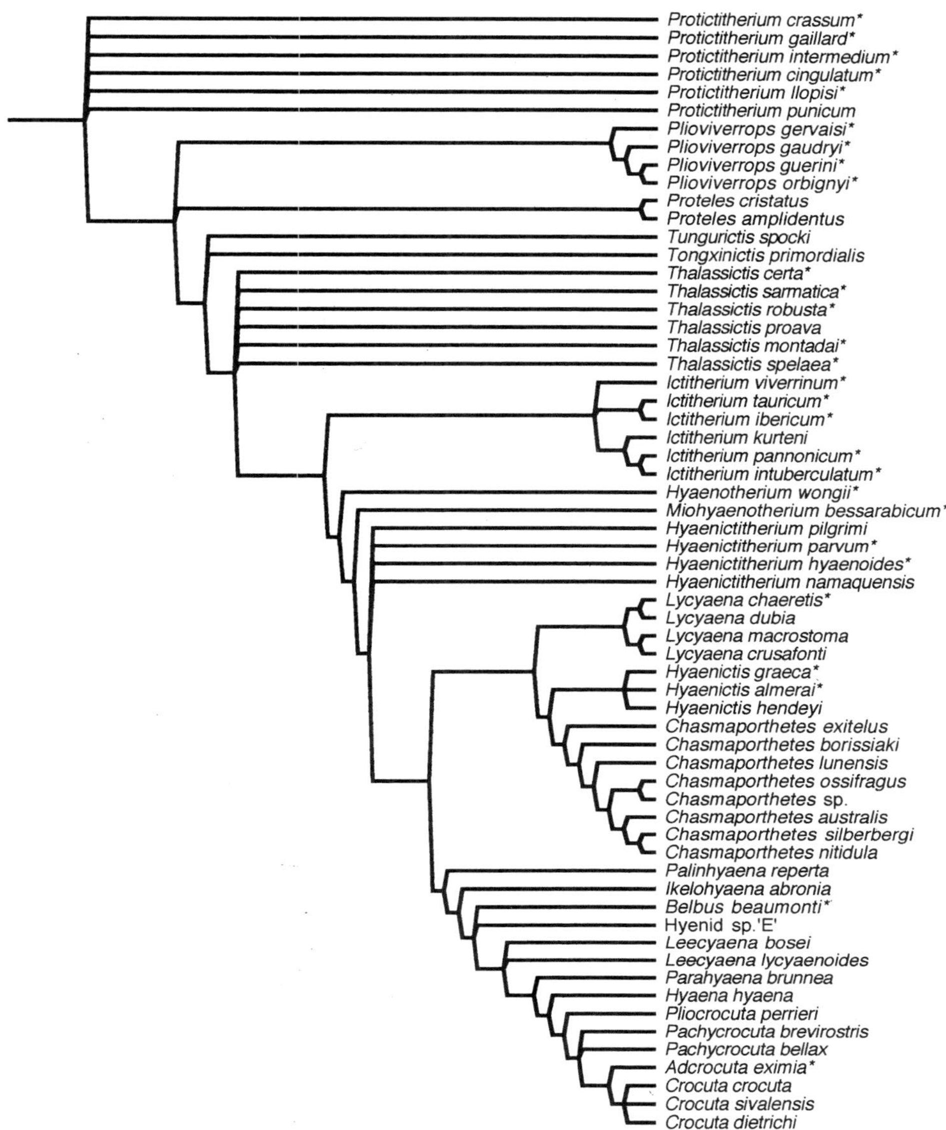

FIGURE 25.1 Phylogeny of the Hyaenidae. Modified after Werdelin and Solounias 1991. Starred taxa are those present in Europe and Western Asia between MN 4 and 13.

stan. This is a somewhat enlarged area, relative to those considered in other articles in this volume. The reason for extending the study area further to the west is that Spain (as does France), has a diverse early and middle Miocene hyaenid fauna that is largely absent from the rest of Europe and Western Asia (except Turkey). In the east, there is no clear demarcation between the late Miocene hyaenid faunas of Greece and Iran and those of Kazakhstan (Werdelin 1988a,b; Werdelin and Solounias 1990, 1991). However, the Chinese faunas include species such as *Palinhyaena reperta* and *Lycyaena dubia*, which are not known from more westerly localities. The easternmost localities expressing the West Asian hyaenid fauna is Kazakhstan, and therefore it has been included here as part of that region.

Chronologically, the period surveyed extends from MN 4 (see above) to MN 13, which is about 18–5 Ma, according to the latest chronologic correlations (Steininger et al. 1989; this volume). As noted above, this period covers the time from the appearance of hyaenids to the acme of their diversification in the late Miocene. It should be emphasized that only one of the species discussed here has a possible stratigraphic range that spans the Mio–Pliocene boundary in this area (see below), although there are indications that some species, e.g., *Hyaenotherium wongii*, do extend into the early Pliocene of the Yûshe Basin, China (Flynn et al. 1991; Tedford et al. 1991).

Hyaenid species occurrences have been tabulated in appendix 25.1. The great majority of ages assigned to local-

APPENDIX 25.1 *List of Occurrences of Hyenas in Europe and Western Asia by Species and MN Unit.*

Species	Type	MN Unit				
		4	5	6	7	8
Protictitherium gaillardi	1	Bézian	Pontlevoy, Vieux Collonges, Castelnau-d'Arbieu, Esvres	Arroyo del Val, Paracuellos, ?Pasalar	La Grive	La Grive, Castell de Barbera, Hostalets de Pierola
Protictitherium intermedium	1			Çandir, Pasalar		
Protictitherium crassum	1		Esvres		La Grive	La Grive, Sofça Yeni Eskihisar
Protictitherium cingulatum	1					
Protictitherium llopisi	1					
Protictitherium sumegense	1					
Protictitherium csakvarense	1					
Protictitherium sp.	1			?Przeworno 2		
Plioviverrops gervaisi	2		Vieux Collonges, Calatayud			
Plioviverrops gaudryi	2				La Grive	La Grive
Plioviverrops orbignyi	2					
Plioviverrops guerini	2					
Plioviverrops faventinus	2					
Plioviverrops sp.	2				Los Andurriales	
Ictitherium tauricum	3					
Ictitherium viverrinum	3					
Ictitherium ibericum	3					
Ictitherium intuberculatum	3					
Ictitherium pannonicum	3					
Ictitherium sp.	3					
Thalassictis certa	3				La Grive	La Grive
Thalassictis montadai	3					Yeni Eskihisar, Hostalets de Pierola
Thalassictis spelaea	3					
Thalassictis robusta	3					
Thalassictis sarmatica	3					
Hyaenotherium wongii	3					
Miohyaenotherium bessarabicum	3					
Hyaenictitherium parvum	3					
Hyaenictitherium hyaenoides	3					
Hyaenictitherium sp.	3					
Lycyaena chaeretis	4					
Lycyaena sp.	4					
Hyaenictis graeca	4					
Hyaenictis almerai	4					
Belbus beaumonti	5					
Adcrocuta eximia	6					

Species	Type	MN Unit				
		9	10	11	12	13
Protictitherium gaillardi	1	Can Llobateres, Can Ponsich, Hostalets de Pierola, Santiga	Ravin de la Pluie			
Protictitherium intermedium	1					
Protictitherium crassum	1	Can Llobateres, Los Valles do Fuentidueña, Esme Akçaköy, Yassiören, Kalfa, Eppelsheim, ?Estevar, ?Nombrevilla, Santiga	Montredon, Sevastopol, ?Pentalophos	Mahmutgazi, Küçükyozgat		
Protictitherium cingulatum	1					
Protictitherium llopisi	1		Can Bayona	Can Bayona		
Protictitherium sumegense	1			Sümeg		
Protictitherium csakvarense	1			Czákvár		
Protictitherium sp.	1					Dytiko
Plioviverrops gervaisi	2					
Plioviverrops gaudryi	2					
Plioviverrops orbignyi	2		Ravin de la Pluie	Vathylakkos 2, Prochoma	Pikermi, Samos	
Plioviverrops guerini	2			Piera, Crevillente 2, Puente Minero	Concud, Los Mansuetos, Los Aljezares	
Plioviverrops faventinus	2					Brisighella
Plioviverrops sp.	2	Los Valles do Fuentidueña				
Ictitherium tauricum	3		Sevastopol			
Ictitherium viverrinum	3		Montredon, Vösendorf	Ravin des Zouaves 5, Vathylakkos, Grebeniki,	Pikermi, Chobruchi, Titov Veles, Belka, Novoelisavetovka, Samos	
Ictitherium ibericum	3					Bazalethi
Ictitherium intuberculatum	3	Yassiören				
Ictitherium pannonicum	3			Valdecebro 5	Chobruchi, Cherevichnoe, Novaja Emetovka	Polgardi
Ictitherium sp.	3		Kohfidisch, Akin, Ravin des Zouaves 1	Küçükyozgat, Sümeg	Los Mansuetos	Pavlodar, El Arquillo, Venta del Moro
Thalassictis certa	3					
Thalassictis montadai	3	Can Mata, Hostalets de Pierola, Ballestar, Can Ponsich	Can Barra			
Thalassictis spelaea	3	Gritsev				
Thalassictis robusta	3	Kishinev	Höwenegg			
Thalassictis sarmatica	3	Kishinev				
Hyaenotherium wongii	3			Vathylakkos 2–3, Ravin des Zouaves 5, Maragheh, ?Grebeniki	Pikermi, Cherevichnoe, Bota Mojnak, Samos	
Miohyaenotherium bessarabicum	3			Udabno	Belka, Çimislia	
Hyaenictitherium parvum	3			Grossulovo	Belka, Novaja Emetovka 2, ?Novoelisavetovka, Taraklia, Tudorovo	Pavlodar

APPENDIX 25.1 *List of Occurrences of Hyenas in Europe and Western Asia by Species and MN Unit.* (*Continued*)

Species	Type	MN Unit				
		9	10	11	12	13
Hyaenictitherium hyaenoides	3			Maragheh, Grossulovo	Kalmakpai	
Hyaenictitherium sp.	3	Los Valles do Fuentidueña		Czákvár		
Lycyaena chaeretis	4				Pikermi, Samos	
Lycyaena sp.	4			Valdecebro 5	Los Mansuetos	Brisighella, El Arquillo
Hyaenictis graeca	4				Pikermi	
Hyaenictis almerai	4				Sant Miquell de Taudell	
Belbus beaumonti	5				Samos	
Adcrocuta eximia	6		Ravin de la Pluie, Ravin des Zouaves 1, Masio del Barbo, Karain, Viladecavalls, Kohfidisch, Xirochori	Prochoma, Ravin des Zouaves 5, Maragheh, Piera, Mahmutgazi, Grebeniki, Novoukrainka, Dorn-Dürkheim, Kuçukyözgat, Czákvár	Kalimantsi, Mt. Leberon, Pikermi, Samos, Çimislia, Los Aljezares, Los Mansuetos, Kinik, Belka, Chobruchi, Cherevichnoe, Novaja Emetovka, Novoelisavetovka, Titov Veles, Çoban Pinar, Halmyropotamos, Pena del Macho, Taraklia, Kuyutarla, Kavak Dere, Kalmakpai	Dytiko, Baltavar, Polgardi, El Arquillo, Bazalethi, Amasya, Pavlodar

ities in this table are taken from Mein 1989, with additional data obtained from Agusti 1989, Semenov 1989, and Kappelman et al., this volume. The main source of data for this table and the following analyses has been Werdelin and Solounias 1991, updated with a number of new sources and previous omissions (Alcala et al. 1992; Bulot et al. 1992; Bonis and Koufos 1991; Bonis et al. 1992; Ginsburg and Bulot 1982).

Ecomorphs

The morphological characteristics of hyaenid species, along with the phylogeny shown in figure 25.1, make it possible to subdivide the species occurring in Europe and Western Asia during the Miocene into six categories or ecomorphs. These ecomorphs are as follows:

1. Civetlike insectivore/omnivore morph (confined to *Protictitherium* spp.). These forms have a very generalized civetlike dentition, with a full set of premolars and molars and a lower first molar characterized by a low paraconid, a high protoconid, a substantial metaconid, and a large, tricuspid talonid. The postcranial skeleton has not yet lost its adaptation for an arboreal mode of life and retractile claws were present (Semenov 1989). This suite of characters would indicate at least a semi-arboreal existence and a diet consisting of small mammals, birds, and insects. Type 1 occurs from MN 4 to MN 13.

2. Mongooselike insectivore/omnivore type (confined to *Plioviverrops* spp.). In this group the sectorial portion of the dentition has been reduced, increasing the number of high, puncture-crushing cusps on the molars and premolars. The paraconid, protoconid, and metaconid become progressively subequal in height through a relative reduction of the protoconid and enlargement of the metaconid. This suite of changes may be the result of an adaptation to increased insectivory. The claws are not retractile, while the skeleton was apparently more adapted to a terrestrial than an arboreal mode of life. Type 2 is known from MN units 5 to 13.

3. Jackal- and wolflike meat and bone eaters (several genera, notably *Ictitherium, Thalassictis, Hyaenotherium* and *Hyaenictitherium*). These forms are generally characterized by an unspecialized carnivorous dentition much resembling that of canids, but with a somewhat greater emphasis on bone-eating. The carnassial has a reduced metaconid and talonid relative to Types 1 and 2. Within this group we see some changes, such as increased size, reduction of the m1 talonid and the posterior molars, and a relative increase in size of the m1 paraconid, making paraconid and protoconid subequal in height. The postcranial skeleton is adapted to terrestrial locomotion, but in the majority of cases does

not show any significant adaptations to cursoriality. An exception to this is found in undescribed material from Lothagam, Kenya, which does show clear signs of increased cursorial abilities. Compared to Types 1 and 2, this group would appear to be adapted more toward open environments; it is presumably not coincidental that the diversification of Type 3 species follows the increase in middle and late Miocene seasonal woodlands (Bernor 1983). Type 3 is known from MN 7 to MN 14.

4. Cursorial meat and bone eaters (genera *Lycyaena, Hyaenictis,* and *Chasmaporthetes*). These forms have been called catlike elsewhere (Kurtén and Werdelin 1988), but catlike is really a misnomer in that these forms only show a trend toward a reduction of the bone-crushing portion of their dentition, while at the same time developing an extended sectorial dentition. The posterior molars (m2, M1–2) are reduced or entirely lost. The p4 cingular cusp is progressively lost in Type 4 (see Werdelin et al. 1994). Within this group, morphological changes include: an increase in size of the anterior premolar accessory cusps, loss (in some *Chasmaporthetes*) of P1, and loss of the m1 metaconid. The skeleton, where known, shows clear indications of increased cursorial abilities and this would seem to indicate a further adaptation to open grassland environments. Type 4 is known from MN 9 into the early Pleistocene.

5. Transitional bone-crackers (genera *Palinhyaena, Ikelohyaena, Belbus,* and *Leecyaena*). These are the first forms, phylogenetically speaking, to show adaptations toward bone-cracking of the type seen in modern hyaenas. The premolars increase in size (especially width), and show a tendency toward the development of asymmetrical main cusps. This is particularly apparent in *B. beaumonti* (Werdelin and Solounias 1991: fig. 16). Unlike in Type 4, the p4 cingulum cusp is emphasized. The postcranial skeleton is not certainly known in any of these forms. These forms appear in some of the same localities as late Type 3 species, and therefore presumably occupied similar environments. However, their morphology indicates a substantial shift in ecological role (niche) toward that occupied by modern hyaenas, especially *Hyaena hyaena* and *Parahyaena brunnea*. Type 5 is known from MN 12 to the early Pleistocene.

6. Bone-crackers (genera from *Hyaena* in fig. 25.1). These forms show advanced, sometimes extreme, adaptations to bone-cracking, with enlarged bone-cracking premolars. Some were cursorial (such as *Crocuta*), while others were not (*Adcrocuta*). It has been claimed by Orlov (1939) that *Adcrocuta* had long slender limbs. However, examination of the Pikermi *Adcrocuta* postcranial material indicates that this genus instead had

short stocky limbs (Pilgrim 1931; Werdelin, in prep.). Type 6 is known from MN 10 to Recent.

Some known forms have been excluded from this classification. *Proteles*, the aardwolf, is a specialized termite eater. It is possibly related to *Plioviverrops*, but the relationship is uncertain. The genera *Tongxinictis* and *Tungurictis* do not fit neatly into any of the groups. They may be closely related, and combine some derived hyaenid characteristics of the auditory bulla and dentition with an apparent evolution in the direction of civets. The dentition of *Tungurictis*, for example, is extremely civetlike in its adaptation to meat-eating and lack of bone-cracking adaptations.

Diversity and Turnover

The study area has been separated into five regions: Western Europe (localities from Spain, France, and Italy), Central Europe (Germany, Austria, and Poland), Southeastern Europe (the former Yugoslavia, Bulgaria, and Greece, except Samos), Western Asia (Samos, Anatolia, Iran, Georgia, and Kazakhstan), and the Hungary–Black Sea area (Hungary, Romania, Moldavia, and the Ukraine). This subdivision of the study area follows the biogeographic distribution of hyaenas. We consider Samos to be a part of Western Asia because of its proximity to the Turkish coast. As we will see below, there are clear distribution differences between the Greek Turolian hyaenas on the one hand and those of Romania, Hungary, and Moldavia on the other. It is hoped that the regions thus formed will correspond to true biogeographic units rather than being simply arbitrary divisions on the map. Hyaenid species biochronologic ranges are given in figure 25.2.

In terms of absolute species diversity (fig. 25.3), the study area can be more or less confidently separated into three units. The first comprises Western Europe only, where hyaenid diversity rises rapidly from its beginning in MN 4 to a plateau with 5 species in MN 8. This diversity is then maintained until a moderate drop after MN 12. The second unit comprises Central Europe as defined herein, where hyaenas are never diverse, reaching a peak of 4 species in MN 10 and disappearing after MN 11. The third unit, comprising the remaining three regions, shows a generally increasing diversity trend from the Astaracian to the Turolian. Toward the end of the Turolian (MN 12–13) there is a drop in diversity in all three regions.

Total species diversity over the entire study area reveals an almost continuous rise from 1 species in MN 4, to 16 in MN 11 (fig. 25.4). Subsequently, there is a minor drop in diversity to 14 in MN 12 and 10 in MN 13. This drop is mostly due to the drop in diversity in the Hungary–Black Sea region. As noted above, only one of the species discussed here (*Plioviverrops faventinus*) survives into the

lower Ruscinian (MN 14), and hyaenas are then replaced by a taxonomically different set of species belonging to genera *Pliocrocuta*, *Pachycrocuta*, and *Chasmaporthetes*. The origin of these taxa would appear to be Asia.

Species turnover (total number of entries and exits per MN unit; fig. 25.5) shows basically the same pattern, although the difference between Western Europe and the three easterly regions is less. However, because of the obvious high correlation between species diversity and turnover (fig. 25.4; table 25.1), we decided instead to plot total turnover per species present in an MN unit (i.e., turnover divided by species diversity). This ratio can vary between 2 (all species present have a range of only one MN unit) and 0 (no entries or exits in a unit). It is only defined if there are species present in a unit. When this is not the case, the value has been set to zero. The data from the individual regions are of reduced value because of the limited amount of data, so most of this discussion is based on data from the entire study area, as shown in figure 25.6.

In figure 25.6 there are two clear peaks in turnover per species present: one at MN 5–6 and one at MN 9–10. Due to the low number of species (two and three), it is not clear that the first peak, at MN 5–6, represents anything other than a random fluctuation, although it is suggestive that it coincides with a peak in turnover and loss of diversity of suoids in this area (Fortelius et al., this volume). The MN 9 peak also corresponds to an episode of increased suoid turnover. The extinction of the hyaenid guild in Europe at the end of the Miocene is clearly seen in this diagram, but is dispersed over MN 12–13, rather than being concentrated only in the terminal Miocene. It is believed that this extinction was complete. However, Alcala et al. (1992) have recently reported the presence of one species, *Plioviverrops faventinus*, from the early Ruscinian (MN 14) deposit of La Gloria 4. This material has yet to be described, and therefore cannot be evaluated at this time.

The pattern for the entire study area can now be compared with the patterns for the individual regions (fig. 25.7). These show that the peak at MN 5–6 is primarily due to turnover in Central Europe and Western Asia (although there is a limited amount of data this is readily deducible from the pooled data). The peak at MN 9 is due mainly to turnover in Central Europe and the Hungary–Black Sea region.

The question naturally arises whether these turnover fluctuations are driven by entries (originations), or exits (extinctions), or both. There are two ways of looking at this problem. Either we can consider the entire time span as a whole or we can look at the individual turnover peaks and try to interpret what lies behind them. For the entire chronologic range as a unit we can look at the correlation between entries and exits on the one hand, and diversity and turnover on the other (table 25.1). Not unexpectedly, both the number of entries and the number

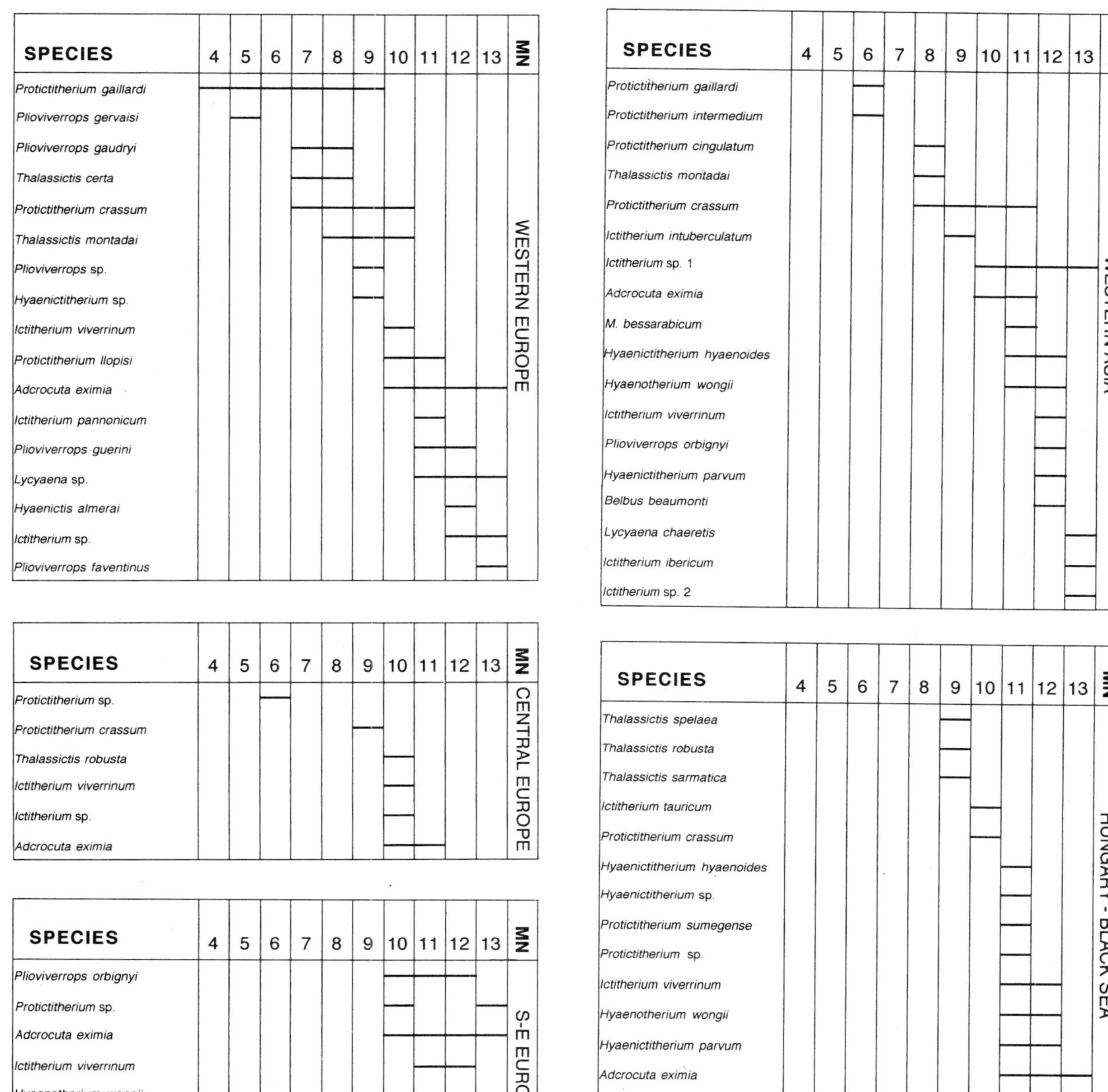

FIGURE 25.2 Biochronologic range charts for the taxa and geographic areas studied in this paper.

of exits show high correlations with turnover and diversity. The correlation between total species diversity and entries is higher than between total diversity and exits, which is to be expected given the general increase in species number throughout the period (fig. 25.4). On the other hand, the correlation between exits and turnover is higher than between entries and turnover, but the difference is hardly significant. Thus, these data do not answer the question whether entries or exits control turnover patterns. (Note that the low correlation between

entries and exits simply indicates that the average existence of a species in an area spans more than one MN unit.)

To answer our question, we must therefore look at patterns of entries and exits throughout the entire interval. These are given as raw data in figure 25.8. Here we can see that the curve for entries rises fairly steadily until MN 11, after which it drops considerably. This is again a correlate of (or rather an important reason for) the increasing number of species throughout the period.

We have normalized entries and exits by the number of

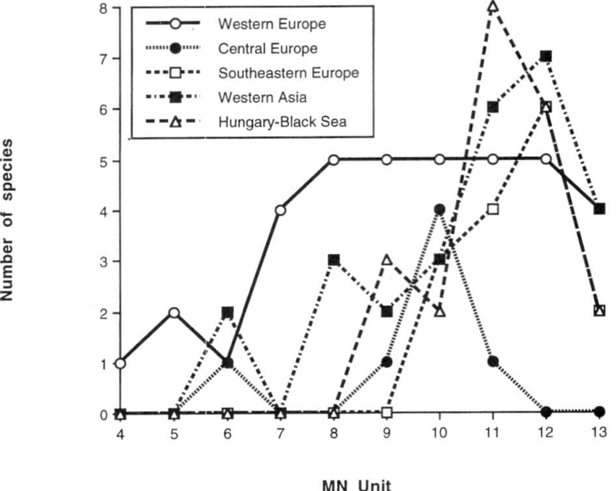

FIGURE 25.3 Hyaenid diversity plotted as number of species present per MN unit and geographic area.

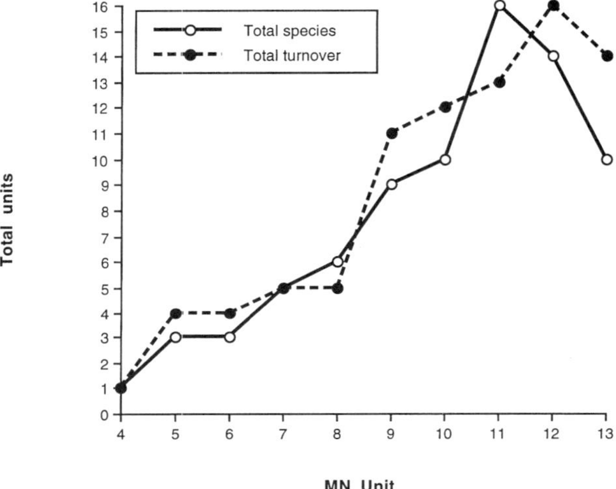

FIGURE 25.4 Hyaenid diversity plotted in terms of number of species present per MN unit and turnover plotted in terms of total number of entries and exits per MN unit over all of Europe and Western Asia.

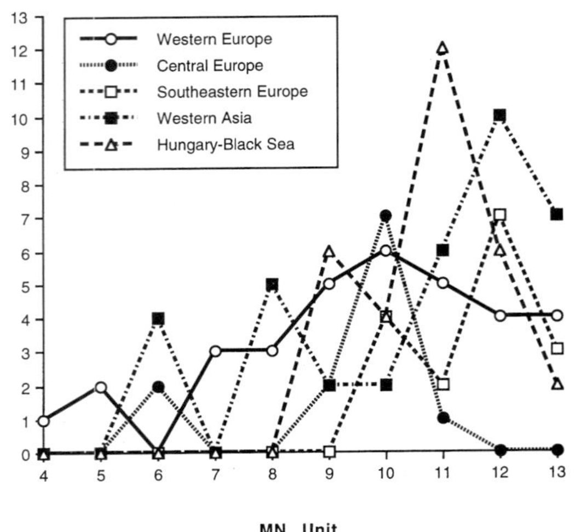

FIGURE 25.5 Turnover among hyaenas plotted as total number of entries and exits per MN unit and geographic area.

TABLE 25.1

	Total Species	Total Turnover	Entries	Exits
Total species	1			
Total turnover	.967	1		
Entries	.830	.833	1	
Exits	.790	.852	.423	1

species present in an MN unit and transformed this number into a percentage. The vertical axis in figure 25.9 shows the percentage of species that either entered or exited a given MN unit. For example, the single species present in MN 4 entered in this unit, giving a percentage of 100, while all but one species present in MN 13 exited at the end of that unit, giving a percentage of 90. There are two important things to note in this figure. One is that the peaks in percent exits generally correspond to the turnover peaks (MN 5–6 and 9), suggesting that turnover in this group, time, and geographic area is mainly driven by exits (local or global extinctions). The other thing to note is that the entry peaks tend to lag behind the exit peaks by about one unit. This pattern is not incontrovertibly clear, since the third entry peak spans three units (9–11). However, if the graph had been extended into the Pliocene and MN units 14 and 15, we would again have seen a peak in entries following the exit peak in MN 13. In summary, the picture we see is one of exits, presumably driven by external factors, followed by entries of species to fill up the voids thus created.

The peaks in relative turnover are broadly synchronous with events in the physical environment. The MN 6 peak corresponds to the beginning of the Paratethys Badenian salinity crisis (Rögl and Steininger 1983), a global sea-lowering event that influenced continental Old World mammalian biogeographic patterns (Bernor and Tobien 1990), and several other global environmental changes (cf. Meulen and Daams 1992). The second peak, at MN 9, also occurs during a global regressive phase (Serravallian regression) that particularly affected the Mediterranean-Paratethys (Rögl and Steininger 1983). However, the fact that the peak correlates with events in other groups as well (Fortelius et al., this volume) indicates that whatever the driving force, it was not limited to hyaenas. The exit peak in MN 13 may be related to the Messinian salinity crisis, but it should be noted that this is part of a global extinction of Miocene hyaenid groups (see below) and may therefore depend more on global environmental changes than on changes affecting only Europe and Western Asia. That the

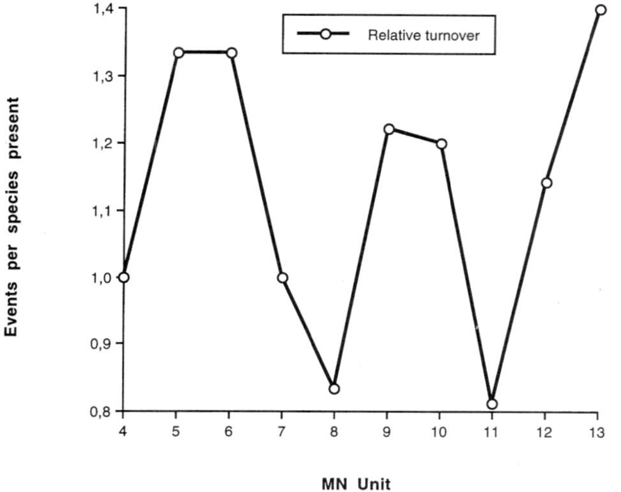

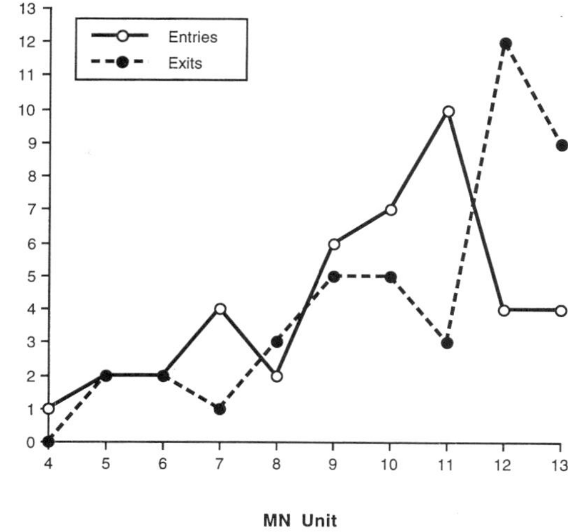

FIGURE 25.6 Relative turnover among hyaenas in Europe and Western Asia plotted as number of entries + exits per MN unit, divided by the number of species present in each unit.

FIGURE 25.8 Number of entries and exits of hyaenas per MN unit for Europe and Western Asia.

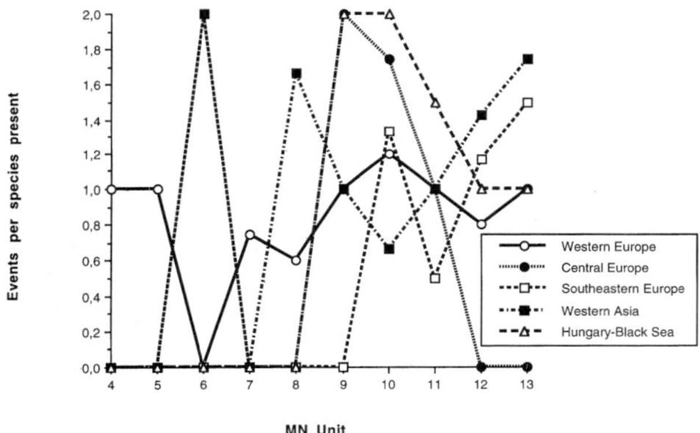

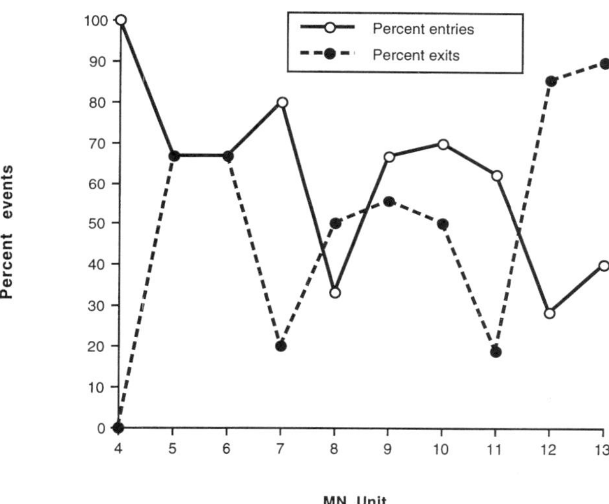

FIGURE 25.7 Relative turnover among hyaenas plotted in the same manner as figure 25.6, but for each of the five study areas separately.

FIGURE 25.9 Entries and exits plotted as percentage of species present in each MN unit for Europe and Western Asia.

extinction event may have been promoted by the deforestation of Europe and installation of a temperate biota at the base of the Ruscinian (Bernor 1983, 1984) cannot be ruled out, since it apparently occurs slightly later in both China (Flynn et al. 1991; Tedford et al. 1991) and Africa (Werdelin, in prep.).

Paleoecology

Paleoecological Analysis of European and Western Asian Hyaenid Ecomorphology

In this section we analyze the Miocene European and Western Asian fossil hyaenid record from an ecomorphological standpoint. In the list of ecomorphs, indications of which taxa belong to which type were presented. This list

can be compared with the phylogeny given in figure 25.1. From this comparison it can be seen that the first three types are successively evolved parts of the Hyaenidae stem group. The fourth is a monophyletic group, the fifth is the stem group of modern bone-cracking hyaenas, and the sixth is again a monophyletic group. Since, strictly speaking, only sister taxa can be compared in an evolutionary sense inasmuch as they are the only taxonomic units of precisely the same age (Vrba 1980, 1984), the six ecomorphs are not equivalent evolutionary units. However, we are convinced that our classification by ecomorph is adequate in making them equivalent in eco-functional terms, and thus in our opinion subject to valid comparisons.

The known biochronologic ranges of the various types

are indicated above and shown in figure 25.10. Since the phylogeny of the Hyaenidae has been advanced, it is possible to go beyond the fossil record and look at ghost lineages (Norell 1992). Type 1 cladistically has less advanced characters than Type 2, implying that the latter cannot have an older first occurrence. Also, Type 4 and Types 5 + 6 are sister taxa, which implies that they must have closely equivalent actual first occurrences. Finally, Type 5 is cladistically more archaic than Type 6 and cannot have an actual chronologically younger first occurrence. These ghost lineages are shown in dashed lines in figure 25.10. This shows the sort of additional stratigraphic knowledge that can be anticipated from a more complete fossil record. In the following discussion we will, however, consider only the known data.

Figure 25.10 concerns hyaenas globally, but there are very few modifications that need to be made when the field of inquiry is limited to Europe and Western Asia. In fact, the only substantial difference is that Type 4 does not appear in this region until MN 11. The oldest Type 4 species, *Lycyaena crusafonti*, occurs in North Africa (Tunisia). This could either indicate a migration of Type 4 forms from Eurasia to North Africa, or it could be the result of a migration of Type 3 forms to North Africa with subsequent

evolution of Type 4 in Africa. The postcranial characters seen in the Lothagam hyaenid may indicate that the latter is the correct option, as may our current understanding of migration patterns between North Africa and Europe in the Miocene (Tchernov 1992).

Hyaenids first appear in Western Europe. This is true both of Type 1 and Type 2 hyaenas, and might be taken at face value to indicate that this is the center of origin for hyaenas. It is apparent, however, from the presence of *Tongxinictis* in China in a fauna interpreted to be broadly correlative with MN 6 (Qiu 1989; Qiu et al. 1988) that there is much that we do not known about the early evolution of the family, and that we should therefore be slightly apprehensive about making statements regarding centers of origin.

What does seem clear from the record, however, is that Type 1 and Type 2 hyaenas are unique to our study area, with one exception, *Protictitherium punicum*, from Bled Douarah, Tunisia. No members of this group have as yet been found in the extensive Miocene material from China, indicating that, in fact, they were not present there. Instead, forms such as *Tongxinictis* and *Tungurictis* may have been the ecological equivalents of these two types in China. The presence of *P. punicum* in Tunisia in MN 9 indicates the presence of a biogeographic connection between Western Eurasia and North Africa before or during that time. Studies by Bernor (1983, 1984), Tchernov (1992), and Thomas et al. (1982) indicate that there was considerable faunal exchange between North Africa and Eurasia through the Levant during the Astaracian, while the Vallesian was a time of biogeographic isolation between these regions.

Type 3 hyaenas are essentially ubiquitous throughout Europe and Western Asia. As currently recognized, hyaenid taxonomy reveals a compositional difference between the Turolian (MN 11–13) faunas of Greece and Western Asia (Pikermi, Samos, Vathylakkos, Ravin des Zouaves 5, and Maragheh) and those of the area from Hungary to the Black Sea. In the former we see the presence of such common Eurasian forms as *Ictitherium viverrinum*, *Hyaenotherium wongii*, and others. We also see the presence of Type 4 and 5 genera such as *Lycyaena*, *Hyaenictis*, and *Belbus* in one or many of these localities. In the Hungary–Black Sea area these taxa are rarely present and less abundant when they are, while other taxa, such as *Ictitherium pannonicum*, *I. ibericum*, *Miohyaenotherium bessarabicum*, and *Hyaenictitherium parvum* take their place. No Type 4 or 5 taxa are present in this region before the Ruscinian.

The differences in the Turolian hyaenid faunas from Greece and the Black Sea areas suggest environmental provincialism of these areas possibly beginning in MN 9. A similar pattern is seen in hipparionine horses (Bernor et al. 1989; this volume). These areas were certainly biogeographically contiguous during the later Turolian (Rögl and

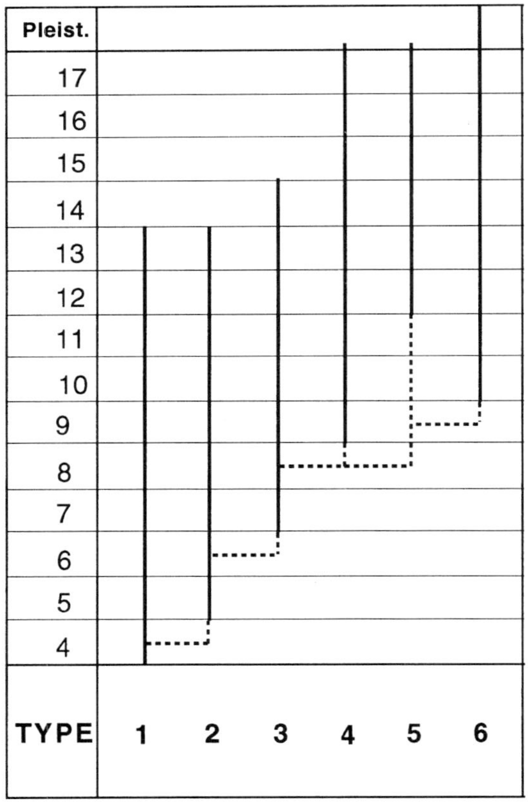

FIGURE 25.10 Biochronologic range charts for the six ecomorphs discussed in the text. Ghost lineages are shown as dashed lines.

Steininger 1983) and some hyaenids (*H. wongii, I. viverrinum, A. eximia*) have been found in both areas.

On a smaller scale, there are some interesting differences between localities within the same region. Since the number of specimens is usually very small, it is for the most part difficult to make anything out of these differences, but in the case of Samos and Pikermi there is enough material to make some interesting comparisons. These two localities are generally similar in their hyaenid faunas, including common species such as *I. viverrinum, H. wongii*, and *Adcrocuta eximia*. However, the less common forms at the two localities are different. At Pikermi we find *Hyaenictis graeca*, while at Samos we find *Belbus beaumonti*. This difference may be due to differences in the age of the localities (re: Bernor et al., this volume). However, it does not seem unlikely that some ecological differences are also involved. This observation is reinforced by the fact that *H. wongii* is common at Samos but rare at Pikermi, while *I. viverrinum* is common at Pikermi but rare at Samos. In the Chinese Miocene localities studied by Werdelin (1988a, b), *H. wongii* is generally the commonest of the two species. At one locality, however, Yan Mu Kou (Zdansky's 1924, Loc. 49), *I. viverrinum* is the commoner of the two species, matching the pattern seen at Pikermi. This similarity between Pikermi and Loc. 49 is reinforced by the common presence of the rhinoceros *Stephanorhinus pikermiensis* at Loc. 49. This species is otherwise rare in the Chinese Turolian localities (Fortelius, pers. comm., 1993).

Another interesting point regarding Type 3 hyaenas is the absence of the common Chinese species *Hyaenictitherium hyaenoides* from areas west of the Black Sea. It is present at Maragheh, and at Chobruchi, Grebeniki, and Novoelisavetovka, but it is absent from Samos, Pikermi, and all other European localities.

As far as Type 4 and Type 5 hyaenas are concerned there is less to be said, for the simple reason that there are very few of these species present in the region and time span involved here. With one exception, *Hyaenictis almerai* from Spain, these forms are not known west of Greece, indicating a biogeographic distinction between Southeastern Europe and Western Asia on the one hand, and Western Europe on the other. Type 6 hyaenas are represented in the Miocene by *Adcrocuta eximia*, which is more or less ubiquitous in the Turolian. However, none of these three types presents any significant patterns in the Miocene, since, with the exception of *Lycyaena* and the rare *Hyaenictis*, their radiation does not really occur until the Ruscinian, at which time species such as *Chasmaporthetes lunensis, Pliocrocuta perrieri*, and *Crocuta crocuta* first appear.

A comparison between Chinese and Southeast European and West Asian late Miocene hyaenid faunas shows a striking similarity in taxic distributions (Werdelin, in prep.). A major difference between Chinese and Euro-pean/Asian hyaenid faunas is that there are no Type 1 or 2 hyaenas present in China. The Type 3 hyaenas are generally the same, i.e., *I. viverrinum* and *H. wongii. H. hyaenoides* is, as noted, present in China but absent from Greece. The Type 4 hyaenas are similar, but appear to be taxonomically different, with *L. chaeretis* present in Greece and *L. dubia* present in China. These may thus be considered vicariants. *Hyaenictis graeca*, present in Southeastern Europe (Pikermi only), and *Chasmaporthetes exitelus* from China may also be seen as ecological equivalents despite belonging to different genera. This pattern of vicariant species is particularly clear for the Type 5 hyaenas, where the Chinese *P. reperta* is replaced in Western Asia (Samos only) by the very similar but clearly more derived *B. beaumonti*. The Type 6 hyaena is *A. eximia* in all faunas.

A final issue with regard to the six types of hyaenas concerns their turnover patterns, and whether these are entirely synchronous. We know from the analysis presented earlier here that there are two clear peaks of relative turnover, in MN 5–6 and 9. Figure 25.11 is a diagram constructed in the same fashion as figure 25.6, but with the data separated into the different types of hyaenas. Types 4, 5 and 6 are united in this diagram, since all three are small groups, but together represent derived hyaenas relative to groups 1–3. The diagram shows a number of peaks, of which the first, that of Type 2 in MN 5, can be discounted, since it is based on only a single record. We can see that the turnover peak for MN 6 is due entirely to turnover in Type 1 hyaenas for the simple reason that there were no other hyaenas present at that time.

The MN 9 turnover peak is due to turnover in both Type 2 and Type 3 hyaenids, but there is again only a single record of the first type. It can perhaps be considered significant that there is no indication of increased turnover among Type 1 hyaenas in MN 9. Instead, a peak for this type appears in MN 11, but this is not accompanied by

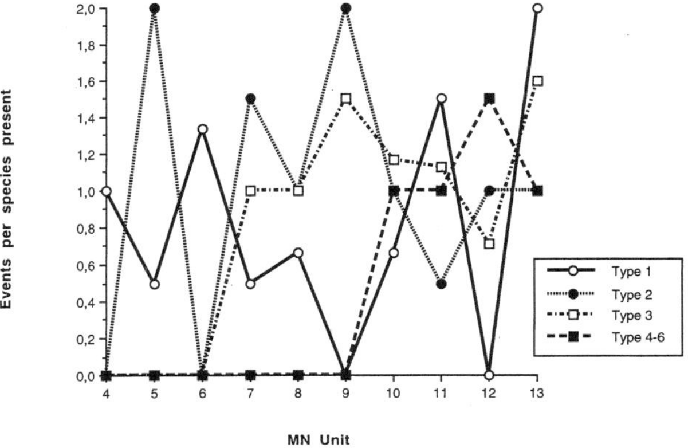

FIGURE 25.11 Relative turnover among hyaenid ecomorphs in the five study areas, plotted as number of entries + exits per MN unit divided by number of species present.

turnover among Type 3 hyaenas, and therefore does not register in figure 25.6.

Regional Differentiation

One question which we must ask and which has been only briefly touched upon so far is the regional differentiation of hyaenid faunas across the study area. Differences in taxonomic composition can be studied by means of the Simpson coefficient (100 times the number of shared taxa, divided by the number of taxa in the smaller fauna), extensively used in analyses of this kind in the past (Flynn 1987; Simpson 1947). Unfortunately, the number of hyaenid taxa in this investigation is inadequate for individual MN comparisons. However, we have made some comparisons between all geographic areas for the Vallesian (MN units 9–10) and the Turolian (MN Units 11–13). The results are given in table 25.2. Numbers in parentheses indicate that there is only one taxon in the smaller fauna.

It must be kept in mind that the numbers presented in table 25.2 are based on a very limited taxonomic sample. However, to the extent that anything meaningful can be gleaned from these data it is that there appears to be a pattern of increased differentiation from the Vallesian to the Turolian between Western Europe on the one hand and Southeastern Europe and Western Asia on the other. At the same time there appears to be a reduced differentiation between the Black Sea–Hungary region and Southeastern Europe and Western Asia. It should be noted that the only meaningful comparison between pre-Vallesian faunas is between Western Europe and Western Asia. This comparison gives a coefficient of 60, supporting our interpretation of regional provincial differentiation between these two areas.

We can also study regional differentiation in terms of the six ecomorphs that were defined above. Figure 25.12 shows biochronologic range charts for these six types in the five regions under study, as well as giving an indication of their regional relative abundances. Of course, this type of analysis is as much subject to the vagaries of localities and preservation as any of the preceding ones. However, a pattern does seem to emerge from figure 25.12. Hyaenid types 1, 2, 3, and 4 all have their earliest records in Western Europe, which could be seen as an indication that their origin lies in this region, although, as noted earlier, there is an earlier occurrence of Type 4 hyaenas outside the present study area. Support for this observation is found in our phylogenetic reconstruction (fig. 25.1). Types 5 and 6 hyaenas, on the other hand, have their first occurrences in Southeastern Europe and Western Asia. Again, this is fully compatible with the phylogeny, which shows forms either endemic to, or present in, China to be the sister taxa to the single Type 5 taxon (*Belbus beaumonti*, plesiomorphic sister taxon *Palinhyaena reperta*), and the single Type 6 taxon (*Adcrocuta eximia*, plesiomorphic sister taxon *Pachycrocuta brevirostris*).

In terms of relative abundances, there is a difference between Western and Southeastern Europe on the one hand, and Western Asia and the Hungary–Black Sea region on the other. The former regions have their hyaenid diversity fairly evenly distributed over the six types, with no type highly dominant in terms of taxonomic diversity. The two latter regions, on the other hand, have fewer types present, and Type 3 hyaenas are considerably more diverse in terms of the number of species than the other types.

Synthesis

With the relevant comparisons made above in mind, it now becomes possible to synthesize the known information on hyaenids from Europe and Western Asia during the later Miocene. Greece and Turkey, although closely adja-

TABLE 25.2 *Simpson Coefficients of Faunal Similarity Between the Study Areas for the Two Time Periods Shown*

	Vallesian				
	W. E.	C. E.	S-E. E.	W. A.	B. S.
W. E.	—				
C. E.	60	—			
S-E. E.	33	33	—		
W. A.	50	50	33	—	
B. S.	20	40	0	25	—

	Turolian				
	W. E.	C. E.	S-E. E.	W. A.	B. S.
W. E.	—				
C. E.	(100)	—			
S-E. E.	14	(100)	—		
W. A.	13	(100)	71	—	
B. S.	25	(100)	43	60	—

See main text for discussion.

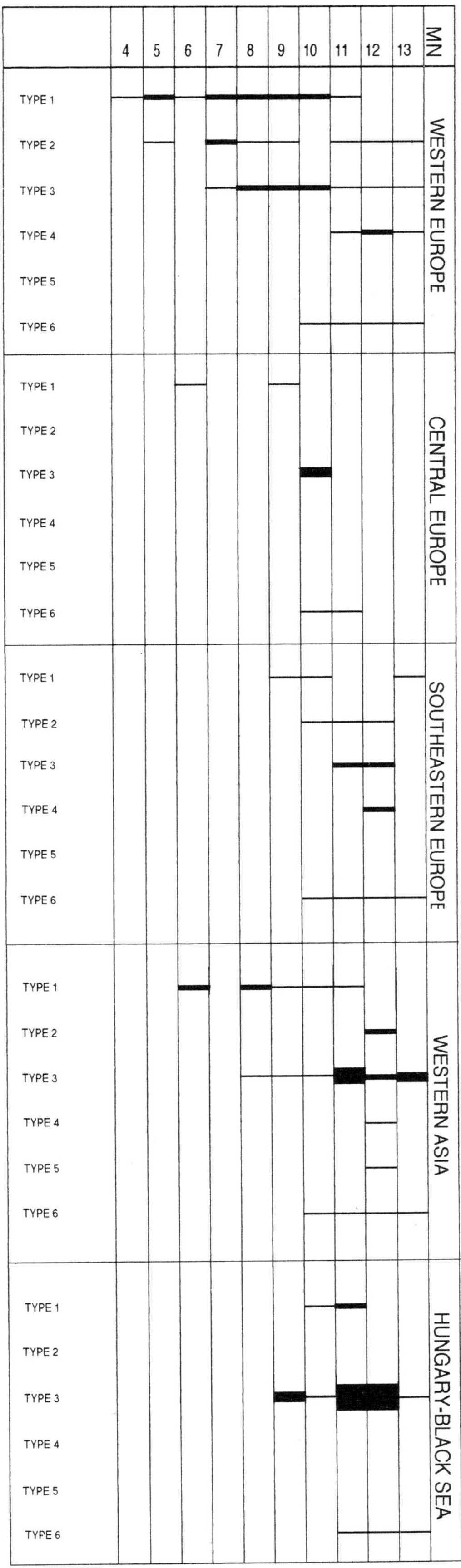

cent to one another, have different hyaenid faunas. However, the difference is mostly due to the fact that as far as sites with hyaenid remains are concerned, the majority of the Greek ones are younger than most Turkish ones. The Turkish hyaenid faunas, on the other hand, are structurally similar to the contemporary ones of Western Europe, including Spain, indicating that if there were Greek faunas of the same age available, they would most likely include the same type of hyaenids; hyaenids such as *Protictitherium*, now unknown from Greece, with the exception of the youngest record of the genus, material from Dytiko attributed to *P. crassum* by Koufos (1980). This generic allocation seems clearly correct, but the specific identification is questionable.

With this in mind, what we see in the fossil record is a gradual development of a highly structured carnivore guild, apparently extending their resource base during the course of the Miocene, while at the same time partitioning it more finely. The earliest hyaenas were small, partly arboreal, omnivore/insectivore forms (Types 1 and 2) that apparently originated in Western Europe. The diversity of Types 1 and 2 hyaenas at any one time is low, although this may be partly due to sampling effects. The limited data suggest a minor (in terms of absolute numbers) turnover peak in MN 5–6, which was more or less coincident with the Parathys Badenian Stage Salinity Crisis (Rögl and Steininger 1983). This time marks the first appearances of hyaenas in Central Europe and Western Asia (Anatolia; re: Appendix 25.1).

By MN 7, Type 3 hyaenas had originated, probably from some less arboreal Type 1 hyaena. This group of hyaenas may also have its origin in Western Europe, but dispersed rapidly into other parts of Eurasia. Evolution of Type 3 hyaenas represents the "core" of the family's radiation as a whole. The first Type 3 hyaenas belong to the genus *Thalassictis* and are characterized by a reduction of the posterior molars and increased body size. Even at the earliest stages a pattern emerges in which there is one small and one large Type 3 hyaena in the faunas. In the earliest Western European faunas these are *T. certa* and *T. montadai*; in the Black Sea area, *T. robusta* and *T. spelaea*. This circumstance persists until the MN 9 turnover peak. At this time *Ictitherium* evolves, either from older Type 3 hyaenas or independently from Type 1 forms. It is characterized by an emphasis on the premolars but less reduction of the posterior molars than *Thalassictis*, and the carnassials are less sectorial than in the latter. *Ictitherium* spp. seems to have replaced *Thalassictis* spp. as the smaller Type 3 hyaenas in faunas after MN 9. Slightly later we see the evolution of *Hyaenotherium wongii*, an extremely successful intermediate form, with the reduced posterior

FIGURE 25.12 Chart of biochronologic range and abundance of the six adaptational types in the five study areas.

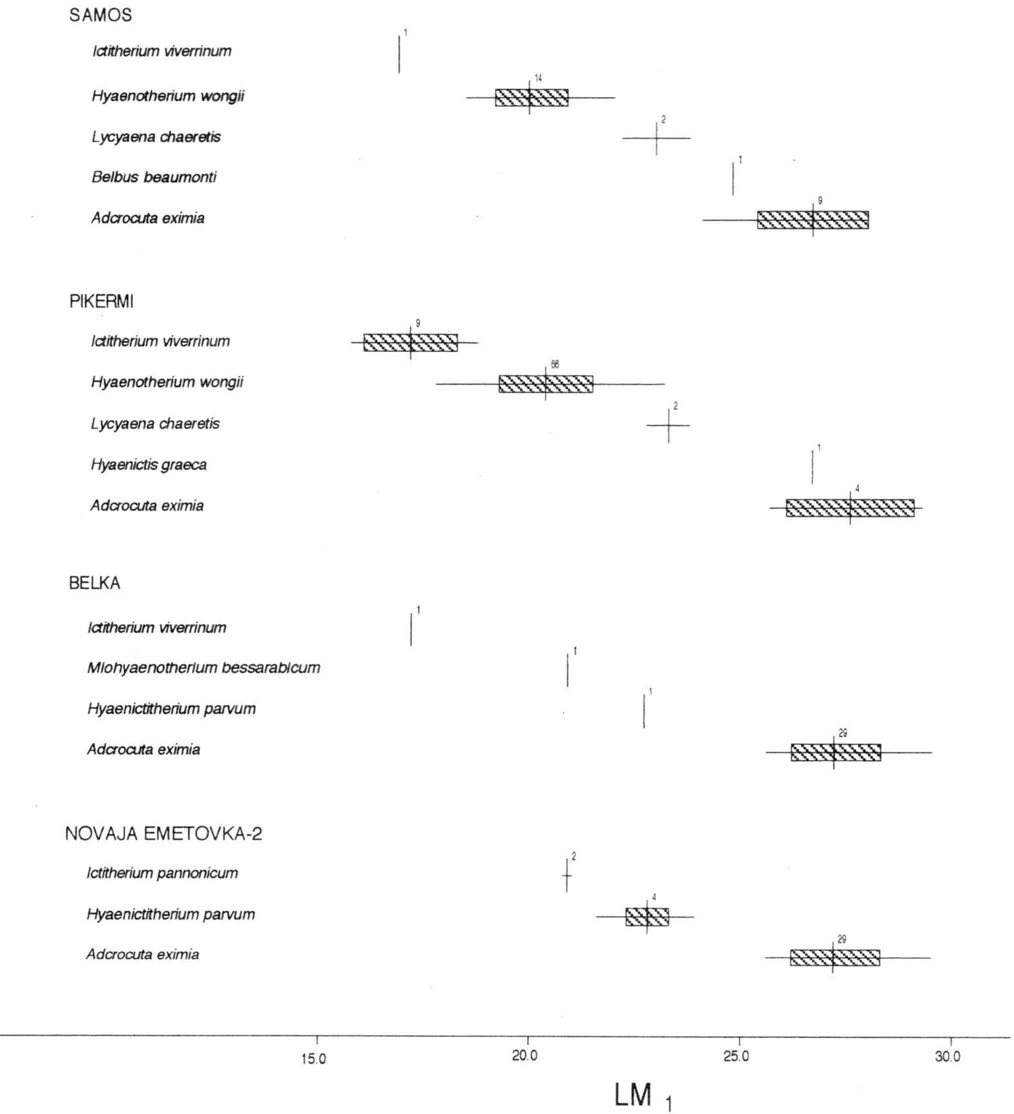

FIGURE 25.13 Diagram showing lower carnassial length for hyaenid species from four localities: Samos (Western Asia), Pikermi (Southeastern Europe), and Belka and Novaja Emetovka-2 (Hungary–Black Sea area). Vertical lines indicate means, the boxes cover one standard deviation to either side of the mean, and the horizontal lines indicate the observed range. The numbers represent the size of the sample. In the case of Belka and Novaja Emetovka-2 no data on *Adcrocuta eximia* were available, wherefore data from a combined Chinese sample were used (Werdelin and Solunias 1990). Note the apparent regularity in size steps despite the fact that different taxa fill the slots in the different faunas.

molars of *Thalassictis* and the less sectorial dentition of *Ictitherium*. At the same time, larger forms began to evolve, such as *Miohyaenotherium bessarabicum*, *Hyaenictitherium parvum* (these two are in fact only doubtfully distinct), and *H. hyaenoides*. These later large forms seem for the most part to have an Asian origin, as evidenced both by their earliest appearances and by their geographic distribution, which is clearly biased toward the east.

In the Turolian hyaenid faunas we see an evolved and more complex, but highly stable, pattern emerging. We have one small Type 3 form, an intermediate one, and a large one (the latter may in some areas be replaced by a small Type 4 form, or both may be present). Thus, in Greece we have *I. viverrinum* (small Type 3 form), *H. wongii* (medium Type 3 form), and *L. chaeretis* (small Type 4 form). In Iran we have *I. viverrinum*, *H. wongii*, and *H. hyaenoides* (large Type 3 form). In China we find these same species along with *L. dubia* (small Type 4 form). In the Black Sea area we find *I. viverrinum*, *T. wongii*, and *H. parvum* or *M. bessarabicum*.

So far the taxa mentioned have basically been ordered by size. We see this as an indication of some degree of competitive displacement due to competition for similar resources, much as has been documented for some extant carnivore guilds (Dayan et al. 1989, 1990; see Pimm and Gittleman [1990] for a review). This seems, in fact to be borne out by the limited data so far investigated (fig. 25.13) and will be discussed in detail elsewhere (Werdelin, in prep.). An additional complication to the pattern arises with the evolution of Type 5 hyaenas. These forms seem to have a different resource utilization pattern, possibly a more scavenging mode of life, and they are similar in size to some Type 3 hyaenas in the same faunas. In China the Type 5 hyaena known is, as noted before, *P. reperta*, while in Greece it is *B. beaumonti*. Finally, all of the Turolian faunas include *A. eximia*, the first type 6 hyaena. This form is much larger than any of the others and was not likely to be in competition with them to any great extent, although in some cases *A. eximia* does fit the size differentiation pattern developed by the Type 3 forms (fig. 25.13). That *A. eximia* would not compete with cursorial or semi-cursorial forms is indicated by the fact that, unlike extant *C. crocuta*, it does not show any cursorial adaptations at all; it rather had a very stocky build with short, robust limbs (Werdelin, in prep.). It may just have lumbered up to carcasses and taken what it liked by virtue of its size. The available evidence again argues for an eastern origin for both Type 5 and Type 6 hyaenas.

For all of its shortcomings, the scenario given here would seem a plausible one, and represents a pattern that was stable throughout the later Miocene. At the end of the Miocene, it was disrupted for some possibly climatic reason, and never recovered. At Langebaanweg, South Africa (possibly of an age equivalent to the beginning of MN 14), we see the last remnants of this, although even there it is considerably altered from that seen in the late Miocene (Werdelin et al. 1994; Werdelin, in prep.). Later faunas have entirely lost this pattern and show only Type 4 and 6 hyaenas. Instead, canids invaded the former areas of Type 3 hyaenas and now show the same kinds of ecomorphs as these forms did in the Miocene.

Acknowledgments

This work is an outgrowth of our previous work on hyaenas, and we thank all those people (too numerous to mention) who helped us in various ways with that work. We would, however, like to express our gratitude to R. Bernor and M. Fortelius for critical reading of the manuscript. The work of LW has been financed by grants from the Swedish Natural Science Research Council and the Swedish Council for Planning and Coordination of Research.

LITERATURE CITED

Agusti, J. 1989. The Miocene rodent succession in eastern Spain: A zoogeographical approach. In *European Neogene Mammal Chronology*, ed. E. H. Lindsay, V. Fahlbusch, and P. Mein, pp. 375–404. New York: Plenum Press.

Alcala, L., J. Morales, and D. Soria. 1992. El registro fósil neógeno de los carnívoros (Creodonta y Carnivora, Mammalia) de España. *Paleontologia i Evolucio* 3:55–66.

Bernor, R. L. 1983. Geochronology and zoogeographic relationships of Miocene Hominoidea. In *New Interpretations of Ape and Human Ancestry*, ed. R. L. Ciochon and R. S. Corruccini, pp. 21–64. New York: Plenum Press.

——. 1984. A zoogeographic theater and biochronologic play: Time/biofacies phenomena of Eurasian and African Miocene mammal province. *Paléobiologie Continentale* 14:121–42.

Bernor, R. L., N. Solounias, C. C. Swisher III, and J. A. Van Couvering. This volume. The correlation of three classical "Pikermian" faunas—Maragheh, Samos, and Pikermi—with the European MN unit system.

Bernor, R. L. and H. Tobien. 1989. The mammalian geochronology and biogeography of Paşalar (middle Miocene, Turkey). *Journal of Human Evolution* 19:551–68.

Bernor, R. L., H. Tobien, and M. O. Woodburne. 1989. Patterns of Old World hipparionine evolutionary diversification and biogeographic extension. In *European Neogene Mammal Chronology*, ed. E. H. Lindsay, V. Fahlbusch, and P. Mein, pp. 263–319. New York: Plenum Press.

Bonis, L. de, G. Bouvrain, D. Geraads, and G. Koufos. 1992. Diversity and paleoecology of Greek late Miocene mammalian faunas. *Palaeogeography, Palaeoclimatology, Palaeoecology* 91:99–121.

Bonis, L. de and G. Koufos. 1991. The late Miocene small carnivores of the lower Axios Valley (Macedonia, Greece). *Géobios* 24:361–79.

Bulot, C., L. Ginsburg, and P. Tassy. 1992. Le gisement à mammifères miocènes de Castelnau-d'Arbieu (Gers). Données nouvelles et implications biostratigraphiques. *Comptes rendus des séances de l'Academie des Sciences, Paris* 314:533–37.

Dayan, T., D. Simberloff, E. Tchernov, and Y. Yom-Tov. 1989. Inter- and intraspecific character displacement in mustelids. *Ecology* 70:1526–39.

——. 1990. Feline canines: Community-wide character displacement among the small cats of Israel. *American Naturalist* 136:39–60.

Flynn, J. J. 1987. Faunal provinces and the Simpson coefficient. *Contributions to Geology, University of Wyoming, Special Paper* 3:317–38.

Flynn, L. J., R. H. Tedford, and Q. Zhanxiang. 1991. Enrichment and stability in the Pliocene mammalian fauna of North China. *Paleobiology* 17:246–65.

Fortelius, M., J. Van der Made, and R. L. Bernor. This volume. Middle and late Miocene Suoidea of Central Europe and the Eastern Mediterranean: Evolution, biogeography, and paleoecology.

Ginsburg, L. and C. Bulot. 1982. Les carnivores du miocène de Bézian près de Romieu (Gers, France). *Proceedings, Koninklijke Nederlandse Akademie van Wetenschappen, B* 85:53–76.

Kappelman, J., S. Sen, M. Fortelius, A. Duncan, B. Alpagut, J. Crabaugh, A. Gentry, J. P. Lunkka, F. McDowell, N. Solounias, S. Viranta, and L. Werdelin. This volume. Chronology and biostratigraphy of the Miocene Sinap Formation of Central Turkey.

Koufos, G. 1980. Palaeontological and stratigraphical study of the Neogene continental deposits of the Axios River basin. *Scientific Annals of the Faculty of Physics and Mathematics, University of Thessaloniki* 19:1–322.

Kruuk, H. 1972. *The Spotted Hyaena: A Study of Predation and Social Behavior.* Chicago: University of Chicago Press.

Kurtén, B. and L. Werdelin. 1988. A review of the genus *Chasmaporthetes* Hay 1921 (Carnivora, Hyaenidae). *Journal of Vertebrate Paleontology* 8:46–66.

Mein, P. 1958. Les mammifères de la faune sidérolithique de Vieux-Collonges. *Nouvelles Archives du Muséum d'Histoire Naturelle, Lyon* 5:1–122.

——. 1979. Rapport d'activité du groupe de travail vertébrés mise à jour de la biostratigraphie du Néogène basée sur les mammifères. *Annales Géologiques des Pays Helléniques* 1979:1367–72.

——. 1989. Updating of MN zones. In *European Neogene Mammal Chronology*, ed. E. H. Lindsay, V. Fahlbusch, and P. Mein, pp. 73–90. New York: Plenum Press, New York.

Meulen, A. J. v. d. and R. Daams. 1992. Evolution of early–middle Miocene rodent faunas in relation to long-term palaeoenvironmental changes. *Palaeogeography, Palaeoclimatology, Palaeoecology* 93:227–53.

Mills, M. G. L. 1990. *Kalahari Hyaenas: The comparative behavioural ecology of two species.* London: Unwin Hyman.

Norell, M. A. 1992. Taxic origin and temporal diversity: The effect of phylogeny. In *Extinction and Phylogeny*, ed. M. J. Novacek and Q. D. Wheeler, pp. 89–118. New York: Columbia University Press.

Orlov, J. A. 1939. On the structure of the extremities of *Crocuta eximia* Roth et Wagner. *Comptes Rendus de l'Académie des Sciences de l'URSS* 2233:533–35.

Pilgrim, G. E. 1931. *Catalogue of the Pontian Carnivora of Europe in the Department of Geology.* London: British Museum (Natural History).

Pimm, S. L. and J. L. Gittleman. 1990. Carnivores and ecologists on the road to Damascus. *Trends in Ecology and Evolution* 5:70–73.

Qiu, Z. 1989. The Chinese Neogene mammalian biochronology: Its correlation with the European Neogene mammalian zonation. In *European Neogene Mammal Chronology*, ed. E. H. Lindsay, V. Fahlbusch and P. Mein, pp. 527–56. New York: Plenum Press.

Qui, Z., Ye Jie, and C. Jingxuan. 1988. A new species of *Percrocuta* from Tongxin, Ningxia. *Vertebrata Palasiatica* 26:116–27.

Rögl, F. and F. F. Steininger. 1983. Vom Zerfall der Tethys zu Mediterran und Paratethys-Die neogene Paläographie und Palinspastik des zirkum-mediterranen Raumes. *Annalen des Naturhistorisches Museums, Wien* 85A:135–63.

Semenov, Y. A. 1989. Ictitheres and morphologically related hyaenas from the Neogene of the USSR. Kiev: Naukova Dumka. [in Russian]

Simpson, G. G. 1947. Evolution, interchange, and resemblance of the North American and Eurasian Cenozoic mammalian faunas. *Evolution* 1:218–20.

Steininger, F. F., W. A. Berggren, D. V. Kent, R. L. Bernor, S. Sen, and J. Agusti. This volume. Circum Mediterranean Neogene (Miocene and Pliocene) Marine-Continental Chronologic Correlations of European Mammal Units.

Steininger, F. F., R. L. Bernor, and V. Fahlbusch. 1989. European Neogene marine/continental chronologic correlations. In *European Neogene Mammal Chronology*, ed. E. H. Lindsay, V. Fahlbusch, and P. Mein, pp. 15–46. New York: Plenum Press.

Tchernov, E. 1992. Eurasian-African biotic exchanges through the Levantine corridor during the Neogene and Quaternary. In *Mammalian Migration and Dispersal Events in the European Quaternary*, ed. W. von Koenigswald and L. Werdelin, pp. 103–23. Courier Forschungsinstitut Senckenberg 153.

Tedford, R. H., L. J. Flynn, Q. Zhanxiang, N. D. Opdyke, and W. R. Downs. 1991. Yûshe Basin, China: Paleomagnetically calibrated mammalian biostratigraphic standard from the late Neogene of Eastern Asia. *Journal of Vertebrate Paleontology* 11:519–26.

Thomas, H., R. L. Bernor, and J. J. Jaeger. 1982. Origines du peuplement mammalien en Afrique du Nord. *Géobios* 15:283–97.

Vrba, E. S. 1980. Evolution, species, and fossils: How does life evolve? *South African Journal of Science* 76:61–84.

——. 1984. Evolutionary pattern and process in the sister-group Alcelaphini-Aepycerotini (Mammalia: Bovidae). In *Living Fossils*, ed. N. Eldredge and S. M. Stanley, pp. 62–79. New York: Springer-Verlag.

Werdelin, L. 1988a. Studies of fossil hyaenids: The genera *Ictitherium* Roth & Wagner and *Sinictitherium* Kretzoi, and a new species of Ictitherium. *Zoological Journal of the Linnean Society* 93:93–105.

——. 1988b. Studies of fossil hyaenids: The genera *Thalassictis* Gervais ex Nordmann, *Palhyaena* Gervais, *Hyaenictitherium* Kretzoi, *Lycyaena* Hensel, and *Palinhyaena* Qiu, Huang & Guo. *Zoological Journal of the Linnean Society* 92:211–65.

Werdelin, L. and N. Solounias. 1990. Studies of fossil hyaenids: The genus *Adcrocuta* and the interrelationships of some hyaenid taxa. *Zoological Journal of the Linnean Society* 98:363–86.

——. 1991. The Hyaenidae: Taxonomy, systematics and evolution. *Fossils & Strata* 30:1–104.

Werdelin, L., A. Turner, and N. Solounias. 1994. Studies of fossil hyaenids: The genera *Hyaenictis* Gaudry and *Chasmaporthetes* Hay, with a reconsideration of the Hyaenidae of Langebannweg, South Africa. *Zoological Journal of the Linnean Society* 111:197–217.

Zdansky, O. 1924. Jungtertiäre Carnivoren Chinas. *Paleontologia Sinica*, series C, 2(1): 1–149.

26

The Evolutionary History and Biochronology of European and Southwest Asian Late Miocene and Pliocene Hipparionine Horses

R. L. BERNOR, G. D. KOUFOS, M. O. WOODBURNE, AND M. FORTELIUS

Old World hipparionine horses have been studied for more than 150 years. As would be expected from such a protracted period of study, there is an extensive systematic literature that has been difficult to organize for an evolutionary reconstruction. However, a good taxonomic foundation for Eurasian taxa was developed through a series of monographic studies (Gromova 1952; Gabunja 1959; Forstén 1968). Woodburne and Bernor (1980) reviewed the Eurasian and North African record and assembled four provisional superspecific groups. This systematic arrangement sharply departed from the practice of including all hipparionine species into a single genus, "*Hipparion*," still maintained by some workers.

Woodburne and Bernor's (1980) taxonomic scheme was found to have biochronologic use, and in many cases led to further interprovincial biochronologic refinements (Bernor et al. 1980). A series of papers has followed, revising hipparions from local faunas as well as old regional assemblages including Iran (Bernor 1985), Indo-Pakistan (Mac-Fadden and Woodburne 1982; Bernor and Hussain 1985), Greece (Koufos 1986; 1987a,b,c; 1988a, b; 1990), Spain (Alberdi 1974a), North Africa (Bernor et al. 1987b), China (Qiu et al. 1987; Bernor et al. 1990b), and the Vienna Basin (Bernor et al. 1988). These and previous regional reports have been extensively reviewed and reevaluated by Alberdi (1989), Bernor et al. (1989), and Woodburne (1989).

Our current understanding supports the view that Old World hipparionines diversified into several lineages, some monospecific and others multispecific (re: Bernor et al. 1989). Some of these lineages have interprovincial to intercontinental ranges, and as they become more highly resolved systematically, they become informative for correlation and biogeographic reconstructions. This contribution is aimed at identifying those hipparionine taxa that we currently recognize from Europe and Southwestern Asia and their evolutionary relationships to other Holarctic and Ethiopian taxa. We will use this reconstruction to refine MN correlations, reconstruct moments of biogeographic connection, and evaluate potential relationships between evolutionary patterns and regional-to-global abiotic events.

Abbreviations and Definitions

Institutions

AMNH = American Museum of Natural History

ANSP = Academy of Natural Sciences, Philadelphia

BSP = Bayerische Staatssammlung für Paläontologie und historische Geologie, München, Germany

F:AM = Frick: American Mammals, in the collections of the AMNH

GIU = Geologische Institut, Münster

GIUB = Geologica Instituta Universita Barcelona

HLMD = Hessiches Landesmuseum, Darmstadt

HGS = Hungarian Geological Survey, Budapest

IPPS = Instituto Provincial de la Ciudad de Sabadell, Barcelona

LGPUT = Laboratory of Geology and Palaeontology, University of Thessaloniki, Greece

LSNK = Staatliches Museum für Naturkunde, Karlsruhe

MNHN = Museum Nationale d'Histoire Naturelle, Paris

NHMW = Naturhistorisches Museum, Wien

SENK = Senckenberg Museum, Frankfurt

UNSM = University of Nebraska State Museum, Lincoln, Nebraska

Technical Terms

IOF = infraorbital foramen

M1MSTHT = M1 mesostyle height

POB = preorbital bar
POF = preorbital fossa
TRL = tooth row length

Chronological Terms

FAD = First Appearance Datum, a regionally significant change in the fossil record (e.g., Berggren and Van Couvering 1974)

Ma = Megannum in the geochronologic time scale

MPTS = Magnetic polarity time scale; we specifically implement the updated time scale given in Steininger et al. 1989; this volume.

m.y. = million years in duration or interval not tied directly to the radioisotopic time scale

Stratigraphic and Mammal Age Terms

Fauna and Local Fauna follow Tedford (1970).

HSD = Highest Stratigraphic Datum, a local stratigraphic last appearance (e.g., Lindsay et al. 1984)

LSD = Lowest Stratigraphic Datum, a local stratigraphic last appearance (e.g., Lindsay et al. 1984)

North America = Hemingfordian, Barstovian, Clarendonian, Hemphillian, Blancan; intervals of the North American land mammal age sequence (e.g., Wood et al. 1941; Woodburne 1987), based on characteristic associations of mammalian taxa.

Europe = Vallesian, Turolian, and Ruscinian; intervals of the European land mammal age sequence, commonly termed units (sensu Fahlbusch 1991; Steininger et al., this volume)

MN (Mammal Neogene) zones = represent refinements in the European mammal biochronologic system discussed in several articles in this volume.

Geographic Abbreviations

NA = North America
Eu = Europe
WEu = Western Europe
CEu = Central Europe
SEE = Southeastern Europe
SWA = Southwestern Asia (includes Turkish localities)
SAs = Southern Asia
EAs = Eastern Asia
EAf = East Africa

Character State and Continuous Variable Abbreviations

C = accompanied by number, refers to characters given in the text here

DAW = metapodial III distal articular width

MC III = metacarpal III
MT III = metatarsal III
mm = millimeters (all measurements after AMNH recommendations as presented by Eisenmann et al. 1988 and rounded to tenth's of mm)
var = measured variable

Nomenclature and Taxonomy

Nomenclature for dental position and coronal morphology is standard. Fossette plication count procedure follows Eisenmann et al. (1988).

"Hipparionine" or "hipparion" designates horses with an isolated protocone on maxillary premolar and molar teeth and, as far as known, tridactyl feet, including species of the following genera: *Cormohipparion, Neohipparion, Nannippus, Pseudohipparion, Hippotherium, Hipparion, Cremohipparion, "Sivalhippus," Eurygnathohippus* (= senior synonym of *"Stylohipparion"*), *"Plesiohipparion," Proboscidipparion*. Recent characterizations of these taxa can be found in MacFadden 1984, Bernor and Hussain 1985, Webb and Hulbert 1986, Hulbert 1987a, Qiu et al. 1987, Bernor et al. 1988 and 1989, Woodburne 1989, Hulbert and MacFadden 1991, and Bernor et al. 1993a and 1993b.

Hipparion s.s. = Following MacFadden (1980; 1984:53), Woodburne and Bernor (1980:1329), Woodburne et al. (1981:496), MacFadden and Woodburne (1982:187), Bernor and Hussain (1985:134), Bernor (1985:180), Bernor et al. (1989:298), and Woodburne (1989:205) in restricting this nomen to a specific lineage of horses that has a facial fossa positioned high on the face. The posterior pocket becomes reduced and eventually lost, and confluent with the adjacent facial surface (includes Group 3 of Woodburne and Bernor 1980:1329). We differ with some of the previous authors in the specific referrals of *Hipparion* s.s.; this issue has most recently been discussed in detail by Bernor et al. (1989:298–302).

"Hipparion": Following Woodburne and Bernor (1980:1328), MacFadden and Woodburne (1982:187), Bernor and Hussain (1985:34), Bernor (1985:180), Bernor et al. (1988:428), and Bernor et al. (1989), these are Old World hipparion horses with facial morphologies that differ from *Hipparion* s.s. (includes superspecific taxa listed above) and belong to distinct, separate lineages.

Methods for Quantification and Interpretation of Biogeographic Patterns

Available sample sizes are small, and the data points are scattered in time and space. Some taxa are represented in areas and MN intervals by large samples across several localities, but many are based on few or even single isolated specimens. The taphonomic background for specimens is mostly unknown, and the entire material is inherently biased, because only localities with hipparionine horses were included in this study. Statistically significant differences would be impossible to demonstrate for most comparisons no matter what technique was used, and we have based most of our analysis on simple graphic representations of range charts and time series plots. We have used two simple statistics to support our interpretations: Van Valen's Probability of Extinction and Simpson's Index of Faunal Resemblance, as described below.

For each region and interval, the Probability of Extinction is simply calculated as the number of extinctions in an MN unit divided by the taxa at risk of the same interval (Van Valen 1984). We estimate the number of taxa at risk simply as the species diversity of the unit. In his original study, Van Valen used a somewhat more complicated formula to estimate this parameter (the number of taxa present at the beginning of the interval, minus half the extinctions, plus half the originations in the interval), but suggested that modifications should be made as appropriate. We have used the simple version because our data points are very few, because we do not recognize a chronology within units, and because we believe that extinctions are most likely concentrated toward the ends of MN units. The simple formula also has the advantage of allowing a quick reference back to the primary data. Using the formula given by Van Valen produces a virtually identical pattern, however (with the curious twist that the values range between 0 and 2 rather than 0 and 1, as in the simplified version).

For any pair of taxonomic lists, Simpson's Index of Faunal Resemblance is calculated as:

$$100 \, C \, / \, N_1$$

whereby the number of shared taxa (C) is divided by the number of taxa in the smaller of the two faunas (N_1), and the result expressed as a percentage (Simpson 1947). Simpson's index has recently been shown to be inferior to statistics that include the number of taxa missing from both faunas (Archer and Maples 1987, 1989; Maples and Archer 1988), but this variable cannot be reliably estimated for the present material. Simpson's index is used here as a convenient (quasi-quantitative) descriptive aid.

All taxa were used to calculate faunal resemblance, but lineages were treated as single entities for diversity and turnover.

Systematics

Cormohipparion Complex

As named, this is a wholly North American superspecific taxon of hipparionine horses. It ranges in age from late Barstovian to late Hemphillian (or from about 16 to 5 Ma) and is the most likely group from which Old World

hipparions are derived (re: Woodburne et al., this volume a). Throughout the geochronologic range of the genus, its species display increased size, hypsodonty, and dental complexity, although not in a linear fashion. The POF is well developed in Barstovian and Clarendonian taxa but is effectively lost by the late Hemphillian, comparable to the situation documented in a number of other hipparionine lineages (Bernor et al. 1989; Woodburne 1989). In fact, the reduction in POF pocketing presages the reduction of the fossa itself, and this is under way by the late Barstovian (e.g., *C. quinni* contra *C. goorisi*, earlier late Barstovian).

Below, the named species are briefly cited along with their provenience and age. This is followed by a discussion on the most pertinent specimens included within the nomina, *Cormohipparion quinni* and *C. occidentale*—taxa most consistently identified as being directly related to the origin of the Old World *Hippotherium primigenium* Complex, particularly those of early Vallesian age (Berggren and Van Couvering 1974; Woodburne 1989, 1994a; Bernor et al. 1988, 1989). We wish to note that Hulbert (1989; 1993) and Hulbert and MacFadden (1991) advocate removing *C. goorisi* from *Cormohipparion*. That suggestion is not followed here, pending further review.

Cormohipparion Skinner and MacFadden 1977

Type species   *Neohipparion occidentale* Leidy 1856

Included species   *Cormohipparion goorisi* MacFadden and Skinner 1981 (late Barstovian, Texas); *C. quinni* Woodburne 1994b (later Barstovian, western North America); *C.* subgenus *Notiocradohipparion* Hulbert 1988. Species of the subgenus, *Notiocradohipparion: C. (N.) plicatile* (MacFadden 1984; Hulbert 1988; late Clarendonian, Texas); *C. (N.) ingenuum* (MacFadden 1984; Hulbert 1988; late Clarendonian, Texas); *C. (N.) emsliei* Hulbert 1987a (late Hemphillan, Florida), are not considered further.

Type locality and age   Little White River (Skinner and Taylor 1967); Minnechaduza Fauna, late Clarendonian (e.g., Webb 1969; Tedford et al. 1987).

Definition of Cormohipparion   Note that this definition does not depend on a reappraisal of the number of OTUs actually represented by the nomen *C.* "*occidentale*." The definition offered here is taken from MacFadden 1984:146, with **boldface changes** added, based on Woodburne's observations of MacFadden's broad hypodigm: "Medium to large **size,** and mesodont to hypsodont hipparion. Mean TRL ranges from 116.83 to 138.00 mm. Unworn or little worn M1MSTHT *ca.* **34–60** mm. The POF is prominent with a relatively well-developed and usually continuous anterior rim **with IOF located above P3 and consistently**

very close to the anteroventral tip of the POF. Posteriorly this fossa also has a well-developed and pocketed rim. In general, the fossa's outline is oval or teardrop in shape and situated far anteriorly on the face resulting in a very wide POB. The anterior margin of the lacrimal bone **enters into the rear of the POF (primitive species) or does not reach the rear of the fossa (advanced species). The fossa may be lost or severely reduced in most advanced species.** Protocones are rounded with anterior spur, oval, or elongate (with angular margins). Fossette borders are moderately to very plicated. Protostylids are usually present and moderately developed. **Premolar ectoflexids do not penetrate isthmuses except in late wear.** Pli caballinids are moderate to absent."

Note that this definition includes *C. goorisi*. Note also that the dimensions of the lacrimal shown in MacFadden and Skinner 1981 and subsequently appear to be wrong. The frontal "shingles over" the lacrimal in the type specimen, and the drawing of the juvenile specimen does not match what Woodburne (1994b) interprets to be present in the actual specimen (F:AM 73947). Woodburne (1994b) interprets the lacrimal in *C. goorisi* to have actually reached, and entered, the rear of the POF for a short distance, and the badly cracked skull F:AM 73942 is consistent with that kind of morphology. A prime factor in the new interpretation of the shape and extent of the lacrimal in this species is aided by F:AM 99387, a partial cranium with muzzle, alveoli for incisors, right and left canines present, as are dP1, P2-M3 in early occlusal wear, from F:AM locality McMurry Pit 1. This specimen shows a "normal" lacrimal penetrating the rear of the POF with the naso-lacrimal suture being about 9 mm above the maxillo-lacrimal as they enter the POF. The tip of the lacrimal is eroded by the rear of the POF such that about 3–5 mm of its normally pointed tip is missing.

Cormohipparion goorisi MacFadden and Skinner 1981

This taxon is based on F:AM 73490 from Trinity River Pit 1, Fleming Formation, San Jacinto County, Texas Gulf Coastal Plain (MacFadden and Skinner 1981:220). The type specimen and hypodigm are late Barstovian age and are geographically restricted to the Gulf Coastal Plain of Texas and Florida (ca. 15 Ma; Hulbert and MacFadden 1991:36), which includes material from the Sweetwater Branch site, Arcadia Formation, Polk County, Florida, not mentioned by MacFadden and Skinner (1981).

Woodburne (in press a) has appraised the relationship between *C. goorisi* and its nominal sister-taxon, *Merychippus insignis*. That review shows that *Cormohipparion goorisi* can be characterized on the basis of its hypodigm from Trinity River Pit 1: being of relatively small size; having a short muzzle; nasal notch retracted to a position about midway between C1 and P2; a POB that is relatively wide

(ca. 24 mm); the POF is distinctly expressed with strongly defined anterior, dorsal, posterior, and ventral borders; the POF is strongly pocketed posteriorly (virtually to a position opposite the anterior edge of the orbit), but its facial expression is relatively short (ca. 48 mm) in comparison to the cheek tooth row length (ca. 116 mm); the IOF is located above the P3/P4 boundary, at or very close to the anteroventral rim of the POF; the lacrimal is dorsoventrally narrow and reaches the rear of the POF; dP1 is persistently large (ca. 10 mm. long, but shorter than in *M. insignis*), two rooted, and present into old age; upper cheek teeth are mesodont (but taller than in *M. insignis*, ca. 26–34 tall at the mesostyle in the unworn condition); the protocone is isolated from the protoloph in P2 except in late wear, as in the other cheek teeth; fossette borders are relatively complex in the upper half of the cheek tooth crown, with as many as six but usually four plications on the opposing faces of the pre and post-fossettes; the pli prefossette loop is usually conspicuously large and bifid posterolingually; protocones are virtually uniformly not spurred; those of P2 are nearly circular, whereas those of the other premolars and molars are more elongate; the hypoconal groove commonly lacks plications; dp1 is not preserved in the lower cheek tooth dentiton; metaconids and metastylids are subequal in size and mostly have rounded outlines and are separated by shallow but U-shaped linguaflexids; premolar ectoflexids virtually never penetrate the metaconid/metastylid isthmuses and maintain a position at or labial to the pre- or postflexid throughout wear; protostylids are developed on p3-m1 in later wear and are usually little more than an angular bend in the enamel at the pertinent part of the tooth; mandibular rami have short symphyses and become markedly shallower in depth from the anterior extent of m3 to p2.

In comparison to its most plausible sister-taxon, *Merychippus insignis*, *Cormohipparion goorisi* differs, and appears to be more derived, in all of the above-mentioned characters. Although a number of intervening steps are indicated by Hulbert and MacFadden (1991), the changes involved in comparing *M. insignis* and *C. goorisi* appear to involve modifications that result in the latter evolving a somewhat shorter basicranium, overall narrower skull, shorter POF, as well as being proportionately larger in cheek tooth length, preorbital region of face, muzzle length, POB length, and a more deeply pocketed POF. In addition, *C. goorisi* has a much more hypsodont and complex cheek tooth dentition, a lower jaw that is more attenuated anteriorly, and a lower cheek tooth dentition in which the premolar ectoflexids penetrate the metaconid/metastylid isthmus, and develop protostylids on P3-M3 only in later wear.

Based on the foregoing description, *C. goorisi* exhibits a number of differences from *Merychippus insignis* in cranial and dental characters. If each species can be established as standing at the base of their respective clades, then clearly the generic-rank taxa that they each represent are phyletically separate, and neither was the ancestor of the other.

CORMOHIPPARION QUINNI WOODBURNE, 1996b
CORMOHIPPARION SPHENODUS WOODBURNE ET AL. 1981 (IN PART): 503
MERYCHIPPUS SPHENODUS OSBORN 1918 (IN PART): 112

Cormohipparion quinni typically occurs in the Valentine Formation, Nebraska, below the uppermost Burge Member, and in correlative deposits (Sand Canyon beds) of northeast Colorado. The fauna from these deposits is of late, but not latest, Barstovian age, or about 15–13 Ma.

This species has been considered (under the taxon *C. sphenodus*) to be a plausible ancestor of first occurring Old World hipparions (Woodburne et al. 1981; Bernor et al. 1989). It may be a sister-taxon of the *Cormohipparion occidentale*–group, in that these two appear to be more similar to one another than to *C. goorisi*.

Cormohipparion quinni differs from *C. goorisi* in being larger; posterior pocket of POF extends posteriorly about 1/2 the length of the lacrimal (instead of extending medial to the anterior border of the orbit); upper cheek teeth are higher crowned (ca. 35–40 mm tall at the mesostyle, rather than 28–34 mm tall); protocone is isolated nearly to the base of the crown, and lacks an anterior spur except in very early stages of wear; protocone is more elongate; enamel pattern of upper cheek teeth is more complex; P1 is shorter as compared to P2 (length P1 is ca 40% length of P2 instead of nearly 50%, as in *C. goorisi*); lower cheek teeth are higher crowned and with stronger protostylids, longer isthmuses on p3-p4, and better separated metaconid-metastylid on p2; cranium has POF only slightly pocketed posteriorly.

Contrary to Woodburne et al. (1981) and MacFadden (1984), the lacrimal in *C. goorisi* resembles that of *C. quinni* in extending to, and being eroded by, the rear of the POF.

Cormohipparion quinni is best represented by specimens from the Crookston and Devil's Gulch members of the Valentine Formation, with only a few specimens (dentition) represented in the AMNH collections in the Cornell Dam Member. Voorhies (1990:A185–A188) allocates a juvenile cranium from the Norden Bridge Quarry (UNSM 352554) to *Neohipparion republicanus*. This skull has been considered to be a representative of *C. quinni* by Woodburne (1996b).

AMNH specimens from the Cornell Dam Member have mostly worn dentitions, but even so virtually all cheek teeth are as tall as, or taller than, unworn cheek teeth in the same position in the dentitions of *C. goorisi*, and have a dental morphology compatible with that of specimens in the Crookston Bridge and Devil's Gulch members. Thus, based both on dental and cranial morphology, the *Cormohipparion* seen in the Cornell Dam Member on one hand differs from *C. goorisi*, and on the other, is similar to *C. quinni* from the Crookston Bridge and Devil's Gulch members of the Valentine Formation.

Remarks   In comparison to *Cormohipparion goorisi*, the Valentine Formation species is larger and higher crowned and has a more complex dental morphology. Unworn upper cheek tooth heights range from 35 to 45 mm in this species, versus 26–34 mm in *C. goorisi*. Upper cheek tooth protocones are more elongate and overall lose the anterior spur (if even present) in very early wear. Both relatively and actually dP2 appears to be reduced in the Valentine species.

In its cranium, the POF is pocketed only about 15 mm posteriorly, or about 1/2 the distance from the rear of the fossa and the anterior border of the orbit. This is in contrast to the condition exhibited in *C. goorisi* in which the pocket passes posteriorly well medial to the anterior edge of the orbit and is both actually and relatively deeper than in the Valentine form. In both species, the IOF is located superior [dorsal] to P3, and the anterior rim of the POF varies from relatively faint (Cornell Dam and Crookston Bridge members) to more sharply defined (Devil's Gulch Member). The POB is not as long anteroposteriorly as in *C. occidentale*, so that the lacrimal still touches (or enters into) the rear of the POF.

Based on mandibles definitely associated with crania in the Valentine Formation, the lower dentition of this species appears to differ from that of *C. goorisi* in having a better separated metaconid/metastylid on p2, more prominent protostylids on the teeth, and more strongly developed isthmuses on p3 and p4.

CORMOHIPPARION "OCCIDENTALE" (LEIDY) 1856

The type series of *Cormohipparion occidentale* is derived from probable late Clarendonian age deposits along the Little White River in South Dakota (Skinner and Taylor 1967). Skinner and MacFadden's (1977) hypodigm of this species includes a series of specimens derived from late Barstovian to early Hemphilian age localities in California, early to late Clarendonian age in Texas, ?late Barstovian to possibly early Hemphillian in Nebraska, late Clarendonian in South Dakota and New Mexico (modified from MacFadden 1984; see below).

Revised diagnosis   Entries in **boldface** show modifications from MacFadden's (1984:162) species diagnosis. **In comparison to C. quinni: Larger** and **more** hypsodont. Mean TRL 138.00 mm; unworn or little M1MSTHT ca. **60** mm (see also Hulbert 1988: 242). Protocones oval and **elongate.** Fossette borders more complex, **but not as complicated as in C. emsliei. POF with moderately developed (ca. 5 mm or less) posterior pocket. Lacrimal well exposed but does not penetrate the rear of POF. POB**

wider. In lower molars protostylids **better** developed, and sometimes isolated. p2 metaconid and metastylid more strongly separated.

Comment on the allocation of the type material   As shown in Osborn (1918; fig. 140 [ANSP 11287]), the type of *C. occidentale* consists of five upper cheek teeth, RP2, RP3, RM2, another right upper molar (?M1) and LP3. The four teeth from the right side are reversed as illustrated in Skinner and MacFadden (1977; text- fig. 4). Leidy (1869:281, in Osborn 1918:176) notes that the teeth are "between a third and a half worn away." As figured by Osborn (1918; fig. 140), the RP2 is about 31 mm tall at the mesostyle; RP3 about 38 mm tall, and RM2 about 42 mm tall. Based on Leidy's (1969:281) statement, these teeth would have been approximately 60 mm tall in the unworn condition, and this is borne out by study of other specimens referred to this species. The pattern in the type dentition is well developed, and shows that the protocone is an elongate oval, with a slightly convex lingual border in P2 and P3; the pli caballin is trifid (P2) to bifid (P3) and single in M2 and ?M1. The enamel pattern is complex, with plication formulae being 4.6.4.1 (RP2), 4.5.3.1 (RP3), 5.7.4.1 (LP3), and 4.6.5.1 (M2). Both M2 and ?M1 show an enlarged pli protoconule. The protocone length/width ratios are 0.43 (RP2); 0.31 (RP3); 0.29 (LP3); and 0.40 (RM2). The hypocone projects posteriorly in the premolars, more posterolingually in the molars. These specimens are in advanced stages of wear, but in actual mesostyle height are comparable to virtually unworn teeth at the same tooth position in *C. quinni*. The enamel complexity is significantly greater, and protocone length/width ratios significantly lower, than those aspects of the upper cheek tooth dental morphology of *C. quinni*.

Discussion   An early result of the 1980 "*Hipparion* Conference" (Eisenmann et al. 1988) was the recognition that the late Clarendonian cranial sample of *C. "occidentale"* from Hans Johnson Quarry, Nebraska, actually represents two morphotypes (Eisenmann et al. 1987). In parallel with Eisenmann et al.'s (1987) observations, Woodburne began compiling cranial and dental characteristics of late Barstovian and Clarendonian age quarry samples from Texas to Nebraska. Some of the results of these studies were briefly summarized in Bernor et al. (1989). Woodburne's current work reveals that of the two morphologies found in Hans Johnson, X-Mas, Kat, and correlative quarries in Nebraska; morph 1 of Eisenmann et al. (1987) is the most common (about 90% of the cranial specimens), the largest, and dentally most complex of the two morphologies contained in those quarries. Accordingly, the morphology exhibited by the upper cheek tooth type series of *C. occidentale* is believed by Woodburne (in progress) to be referable to morph 1 of Eisenmann et al. (1987). Woodburne (in prog-

ress) has identified yet a third morph of *C. "occidentale"* represented by specimens from MacAdams Quarry (early Clarendonian), Texas.

Whistler and Burbank (1992) have recently identified the *Cormohipparion "occidentale"* species-group in the Dove Springs Formation, California. They correlate this group's local first/last occurrence with magnetic chrons C5Ar.1n (chron interval = 12.708–12.678 Ma) and C4Ar.1r (chron interval = ca. 9.230–9.025 Ma; magnetic correlations follow revisions presented in Steininger et al., this volume), respectively. The HSD of *C. occidentale* in the Dove Springs Formation occurs just above the base of the *Paronychomys/O. diabloensis* Assemblage Zone considered by Whistler and Burbank (1992) to be correlative with the base of the Hemphillian.

Trends in late Barstovian and Clarendonian samples of Cormohipparion, *and the Origin of the Old World Vallesian taxon,* Hippotherium primigenium   Presently, both *Cormohipparion quinni* and *Cormohipparion "occidentale"* would appear to be more closely related to Old World hipparions than is *C. goorisi*. A pooled *C. quinni-C. "occidentale"* species-group differs from *C. goorisi* in its larger size, more reduced POF, more hypsodont and complex upper cheek tooth dentition, more strongly separated p2 metaconid/metastylid. At present, data do not falsify Woodburne et al.'s (1981) hypothesis that *C. quinni* (then under the name *C. sphenodus*) might be the ultimate source of the "*Hipparion*" Datum on morphological grounds. This hypothesis is not currently corroborated by the species' known chronologic range: Valentine age, 14–13 Ma.

The *Cormohipparion "occidentale"* species-group first occurs about 12.7 Ma (if not earlier, within the Devil's Gulch Member of the Valentine Formation; Woodburne, 1996b), and at least by the early Clarendonian began to diversify; the MacAdams Quarry material may prove to represent the initial diversification of this species-group (Woodburne, in progress). Although *C. "occidentale"* species-group is a likely source for the Old World "*Hipparion*" Datum, it appears that later Clarendonian aged populations are morphologically too derived and too late in time to contain the actual sister-taxon of first occurring Old World hipparions. Given that the MacAdams Quarry sample is closely equivalent in its age to estimates for first occurring Old World Hipparions (Swisher, this volume; Woodburne et al., this volume), it might inevitibly prove to be the most plausible source for the Old World Hipparion Datum.

Hippotherium primigenium Complex

The *Hippotherium primigenium* Complex includes a large suite of taxa that evolved closely following the first,

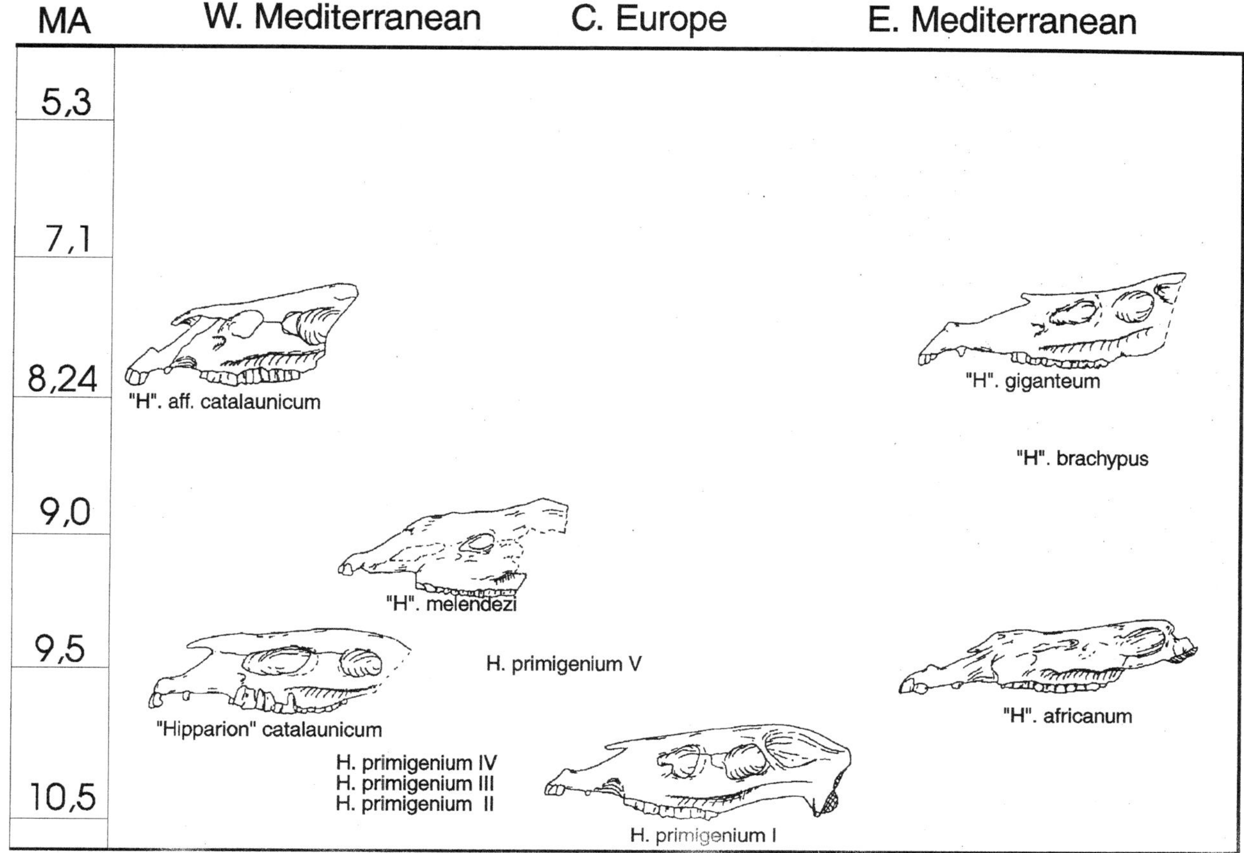

FIGURE 26.1 The *Hippotherium primigenium*–Complex

and perhaps only, immigration of North American hipparions (fig. 26.1). These taxa are broadly referable to Group 1 of Woodburne and Bernor (1980) and Bernor et al. (1980). Bernor et al. (1988) presented a cladistic analysis of this complex that demonstrated that their original hypothesis of relationship had been based on symplesiomorphic characters. An expanded analysis of this group using forty-nine craniodental and postcranial characters by Bernor and Lipscomb (1995) confirmed this earlier interpretation and found further refinements in the branching pattern. The taxa relevant to the geographic area under study include: *Hippotherium primigenium* I, *Hippotherium primigenium* II (*Hippotherium primigenium* s.s. von Meyer 1829), *Hippotherium primigenium* III, *Hippotherium primigenium* IV, *Hippotherium primigenium* V, "*Hippotherium*" *koenigswaldi*, "*Hippotherium*" *brachypus*, "*Hippotherium*" *giganteum*, "*Hippotherium*" *catalaunicum*, "*Hippotherium*" aff. *catalaunicum*, and "*Hippotherium*" *depereti*. Other taxa referable to this group found outside the study area include: the North African taxa "*Hippotherium*" *africanum* and "*Hippotherium*" *sitifense*; the Chinese taxa "*Hippotherium*" *weihoense* (Qiu et al. 1988) and "*Hippotherium*" *dermatorhinum* (Bernor et al. 1989; Bernor et al. 1990b); the Siwaliks taxon "*Hippotherium*" *nagriensis* (Hussain 1974;

MacFadden and Woodburne 1982; Bernor and Lipscomb, 1995); *Hippotherium primigenium* s.l. from Chorora, Ethiopia (Bernor, under study).

HIPPOTHERIUM PRIMIGENIUM S.S. VON MEYER 1829

The *Hippotherium primigenium* type series originates from the Dinotherium Sands locality of Eppelsheim, Germany (MN 9). Forstén (1968) has claimed that the specimens originally figured by Hermann von Meyer (1833: Taf. 30 and 31, figs. 1–30; Taf. 32, figs. 31–33) were lost from the HLMD during World War II. Recently, Tobien discovered these specimens wholly, or nearly wholly, deposited in the SENK. Bernor (in progress) has made a survey of this collection and cross-correlated most of the line-drawing illustrations with von Meyer's publication. Von Meyer never designated a type. However, M1421 (Taf. 30, fig. 17; Taf. 31, fig. 18), a mandibular fragment with p2–4, was incorrectly assigned the nomen *Equus caballus primigenius*, and is certainly a member of *Hippotherium primigenium* s.s., is one of the more complete specimens, and has page priority over other specimens. These various attributes lead us to nominate SENK M1421 as the lectotype for *Hippotherium primigenium*. The type series includes at least twelve specimens that can be cross-correlated with figures published by von Meyer (1833).

The lectotype specimen is characterized by: p2 with pointed anterostyle; premolars with short ectoflexids that do not separate metaconid-metastylid; pli caballinid usually single in p2–4; premolars without clearly expressed protostylids on the occlusal surfaces (probably due to early stage-of-wear); linguaflexid is shallow on p2–3, and "V"-shaped on p4; metaconids-metastylids are rounded on all premolars; pre- and postflexids are elongate with complex margins on all premolars. These characters are consistent with those seen in Pannonian D-E hipparions from the Vienna Basin, referred by Bernor et al. (1988) to *Hippotherium primigenium* (s.s.)

The most salient characters defining *Hippotherium primigenium* are: orbital surface of the lacrimal bone with a sharply reduced or absent foramen; nasal notch retracted well anterior to P2: POF subtriangular shaped, anteroventrally oriented, pocketed deeply posteriorly, medially deep, with a strongly delineated peripheral border outline, and a distinctly defined anterior rim; when present, dP1 large and dp1 vestigial to absent; maxillary cheek teeth moderately curved-to-straight, unworn specimens with a maximum crown height of approximately 50 mm, pre- and post-fossette enamel plications complex and maintain several deeply amplified plis until well beyond middle stage-of-wear, posterior wall of postfossette is always distinct, pli caballins mostly double, hypoglyph deeply incised well beyond the middle stage-of-wear, protocone commonly lingually flattened and labially rounded in middle adult stage-of-wear and clearly isolated from the protoloph until very late stage-of-wear, protocone spur very rare to absent, protocone more lingually placed than the hypocone, P2 anterostyle/paraconid elongate; mandibular premolars and molars with rounded metaconids and metastylids, premolar ectoflexids do not separate metaconid and metastylid, although they do so in the molars, lower cheek teeth frequently with complex pli caballinids, protostylids commonly present in middle adult stage-of-wear, and expressed as enclosed circular structures but never strong and columnar, ectostylids virtually always absent, linguaflexids usually shallow to slightly V-shaped on the premolars, more V-shaped on the molars, pre- and postflexids with complex margins; protoconid enamel band rounded; metapodials robustly built but lengthened compared to *C. occidentale*.

Bernor et al. (1988; 1989) reported that *Hippotherium primigenium* s.s. had not at that time been conclusively demonstrated to have a geographic range extending outside of Central Europe. Their concept of a *Hippotherium primigenium* evolutionary lineage was restricted to *Hippotherium primigenium* stages I–V by Bernor et al. (1993b; Woodburne et al., this volume a). The Eppelsheim hipparion is morphologically indistinguishable from the Pannon-

ian D-E hipparions cited by Bernor et al. (1988), and is referable to *Hippotherium primigenium* stage II of Bernor et al. (1993b).

Hippotherium primigenium stage II's first occurrence is regarded here to be somewhat older than the Höwenegg hipparion (*Hippotherium primigenium* stage III), and slightly younger than the Pannonian C horse (*Hippotherium primigenium* stage I). Also, Bernor provides evidence (see Woodburne et al., this volume a) that this morphology is closely approximated by "*Hippotherium*" *nagriensis* from the Siwaliks and *Hippotherium primigenium* from Chorora (Ethiopia). Therefore, we hypothesize that this could be the stage-of-evolution from which other known Old World hipparions were derived.

Hippotherium primigenium Stage-of-Evolution I

Bernor et al. (1988) identified the Pannonian C assemblage from Gaiselberg, Austria, as being more primitive than Pannonian D-E hipparions in the presence of elongate-oval protocones among 35% of the maxillary cheek tooth sample. They identified this character as a symplesiomorphy with North American *Cormohipparion occidentale*. Subsequently, Bernor (pers. obs.) has noted this same morphology in the Vienna Basin Pannonian C assemblage from Marienthal, demonstrating this character's morphological consistency and biochronologic value. Presently, this character would appear to precede the apparent shortening and tapering of protocone's length, which gives the lingually flattened and labially rounded morphology that typifies *Hippotherium primigenium* II, and is generally distributed among many Old World hipparionine lineages.

Hippotherium primigenium Stage-of-Evolution II

As per diagnosis above.

Hippotherium primigenium Stage-of-Evolution III

This stage-of-evolution is well represented by the articulated skeletons from Höwenegg, Germany (Tobien 1959; Bernor et al. 1993c). Since this sample is so complete and comes from a restricted stratigraphic/chronologic range, it is critical for understanding the skeletal morphology of the *Hippotherium primigenium* s.s. evolutionary group. Bernor et al. (1993b and c) have reported that the Höwenegg hipparion is derived from *Hippotherium primigenium* stages I and II in its rounded-to-squared mandibular cheek tooth metaconid and metastylid morphology. Koufos (1986) has recognized *Hippotherium primigenium* in the late Vallesian locality of "Ravin de la Pluie," RPl (Macedonia, Greece). The locality is referred to MN 10 (Bonis et al. 1988). The available material contains a mandible and few dental and postcranial remains retained by the LGPUT. Recently Koufos et al. (1991) reported a form similar to *Hippotherium primigenium* from the locality of

Nikiti (NKT), Macedonia, Greece, as *H.* cf. *primigenium*; the locality is referred to MN 10/11. The mandible from RPl has a relatively short snout, which is narrow across the diastema but has a broad, arcuate incisor tooth row. The upper cheek teeth have richly plicated enamel. The protocone is short and free from the protoloph even in the advanced wear teeth. The pli caballin is large and simple-to-double. The lower cheek teeth have a well-developed protostylid and a rounded-to-squared metaconid and metastylid. This feature is similar to the Höwenegg hipparion. The lower premolars are very wide in comparison with the molars, a feature which is also present in the Höwenegg sample. The metapodials from RPl are short and robust. In comparison with those from Eppelsheim the RPl metapodials seem to be more slender, indicating a younger late Vallesian (MN 10) age (Koufos 1986). With the lack of cranial material, Bernor believes that the characters uniting the Greek specimens with Central European *Hippotherium primigenium* s.s. may be symplesiomorphous.

The RPl *H. primigenium* does indeed have some morphological similarities with that from Höwenegg (*Hippotherium primigenium* III). This view is corroborated by the observation that the Rudabánya hipparion exhibits square-to-irregular shaped metaconids and metastylids (seen in stage *Hippotherium primigenium* IV; Bernor et al. 1993b), reinforcing the older correlation with the Höwenegg stage (III)-of-evolution. However, if *Hippotherium primigenium* existed in Southeastern Europe later than MN 10, then the Central European *Hippotherium primigenium* s.s. lineage may have been disjunct from the Southeast European *Hippotherium primigenium* s.l. lineage. More cranial and postcranial material is needed from Southeastern Europe to resolve this outstanding issue.

Swisher (this volume) has calibrated Höwenegg as having a maximum age of 10.3 Ma, based on determinations of multiple volcanic samples stratigraphically within, below, and above the fossil-bearing horizons (see also Sen, this volume).

HIPPOTHERIUM PRIMIGENIUM STAGE-OF-EVOLUTION IV

Bernor et al. (1993a, b), Rögl et al. (1993), and Woodburne et al. (this volume a) have identified this stage-of-evolution from Rudabánya, Hungary. The Rudabánya hipparion exhibits mandibular cheek teeth with variably square-to-irregular shaped metaconids and metastylids. Also, Bernor et al. (1993a) reported that a well-preserved metatarsal had a statistically significantly more elongate maximum length dimension than all other Central European *Hippotherium primigenium* s.s. (stages I–III) yet studied. These characters suggest that Rudabánya is younger in age than Höwenegg, but is older than Götzendorf (= Pannonian F, uppermost MN 9; Rögl et al. 1993; Bernor et al. 1993b). As discussed further below, the MN 9/10

boundary would appear to be circa 9.5 Ma (Swisher et al., this volume). That would necessitate a biochronologic correlation of Rudabánya between 10 and 9.5 Ma.

HIPPOTHERIUM PRIMIGENIUM STAGE-OF-EVOLUTION V

Bernor et al. (1993b) have recognized continued transformation of mandibular cheek tooth morphology in this advanced stage-of-evolution. In *Hippotherium primigenium* V metaconids and metastylids are kidney-shaped or a highly irregular crenulated morphology through adult middle stage of wear. Assemblages with this morphology include the MN 9/10 locality of Götzendorf (kidney-shaped), MN 10 locality of Sümeg and MN 11 locality of Csákvar (irregular-shaped). The Götzendorf assemblage is very small, being limited to a few fragmentary cheek teeth and postcranials. The Sümeg and Csákvar assemblages are also fragmentary, but represented by many more specimens. An initial study of these materials by Bernor et al. (1993b) suggests that there is more species diversity among these three referred localities than currently recognized. Indeed, the kidney-shape of the Götzendorf hipparions metastylid may precede the irregular pattern found in late Vallesian and Turolian members of this Central European lineage.

"HIPPOTHERIUM" KOENIGSWALDI SONDAAR 1961

"Hippotherium" koenigswaldi was named by Sondaar (1961) for a large assemblage of hipparion teeth and postcranial remains from Nombrevilla (lower MN 9), Spain (Aragon). Woodburne (1996 a) has cited a number of characters that he believes to be primitive for Old World hipparions including: (1) more strongly developed lower cheek tooth protostylid; (2) protostylid connecting to the protoconid in later stage of wear; (3) shallower premolar ectoflexids; (4) larger size; (5) less complex upper cheek tooth dentition; (6) upper cheek tooth protocones more ovate. These characters, coupled with stratigraphic associations, suggest that Nombrevilla is the locally first occurring hipparion in Spain (Freudenthal and Sondaar 1964; Freudenthal 1968), and perhaps the most primitive and oldest European hipparion yet reported.

We present an alternative hypothesis here (Woodburne et al., this volume a) based on a character state analysis of the "*Hippotherium* Complex." Bernor argues that the Nombrevilla horse's protostylid and ectoflexid morphologies are similar (albeit somewhat variable) to Central European members of the *Hippotherium primigenium* s.s. clade; its ovate protocones would appear to be derived relative to Eppelsheim and Pannonian C-E horses; the occurrence of highly ornamented pli caballins (not less complex enamel ornamentation) is a derived character (and potentially autapomorphic for this clade) compared to *Hippotherium primigenium* stages I–V. Moreover, metatarsal ML x DAW dimensions are found at the uppermost range for Central

European Pannonian C-E hipparions and the Höwenegg hipparion (Sondaar 1961: tab. 21).

Bernor concludes that despite the Nombrevilla hipparion sample's low stratigraphic occurrence, it may very well be that this single site does not record the European *"Hipparion"* Datum. The number of autapomorphies found compared to *Hippotherium primigenium* stages I and II (at least) suggest that some vicariance occurred after extension of *Hippotherium primigenium* into Spain.

"HIPPOTHERIUM" CATALAUNICUM (PIRLOT 1956)

Pirlot (1956) recognized this taxon from Hostalets de Pierola (Supérieur), Spain. Freudenthal and Sondaar (1964) reported this species from stratigraphically higher horizons (Pedregueras IIC) than the Nombrevilla assemblage (Pedregueras I). *Hippotherium catalaunicum* is derived compared to *Hippotherium primigenium* s.s. in its very elongate and anteroposteriorly oriented POF (Woodburne and Bernor 1980; Bernor et al. 1980 and Bernor et al. 1989). Bernor et al. (1989) have further noted that a Turolian hipparion reported by Alberdi (1974a) from Piera, *"Hipparion mediterraneum"* (*"Hippotherium"* aff. *catalaunicum* here) retains a faintly delimited POF with much the same length and orientation characteristics of *"Hippotherium" catalaunicum* and is probably related to it.

The POF morphology exhibited in *"Hippotherium" catalaunicum* is shared by the North African taxon *"Hippotherium" africanum* (Bernor et al. 1989). Bernor et al. (1987a) have suggested that North African *"Hipparion" sitifense* s.s. (= *"Hippotherium" sitifense*) may be an endemic derivative of *"Hippotherium" africanum*. If these interpretations prove to be correct, *"Hippotherium" catalaunicum*-*"Hippotherium" africanum* are a Vallesian clade that shared a common biogeographic range. It is most plausible that this clade became vicariant after MN 9, with endemic forms subsequently evolving in Spain (*"Hippotherium"* aff. *catalaunicum*) and North Africa (*"Hippotherium" sitifense*).

"?HIPPOTHERIUM" SITIFENSE (POMEL 1897)

As noted by Boné and Singer (1965) the nomen *Hipparion sitifense* was originally erected by Pomel for two upper cheek teeth from St. Arnaud Cemetery (Algeria). Forstén (1968) noted that Arambourg (1956) included all the known material from St. Arnand, Ain el Hagj Baba, St. Donand, and Mascara (Algeria) into his hypodigm of *Hipparion sitifense*. This material was restudied by Boné and Singer (1965), and again referred to *"Hippotherium" sitifense*. Forstén (1968) includes all of the small-sized hipparions from the Iberian peninsula in this species. This is not accepted by others who have studied the Spanish material (Alberdi 1974a). An upper permanent cheek tooth from St. Arnaud, Algeria, has been assigned as the type specimen

for *Hipparion sitifense* (Forstén 1968). Bernor et al. (1987a) referred dental and postcranial material of a small early Pliocene hipparion from Sahabi (considered by Howell 1987 to be late Miocene age) to *"Hipparion" sitifense*.

"?Hippotherium" sitifense is characterized by its small size, flattened protocone, simple enamel plication, absence of the ectostylid, and well-developed lateral digits (Boné and Singer 1965). The very limited skeletal material, as well as uncertainties of the type material's actual provenience, makes *"?Hippotherium" sitifense*'s characterization and recognition difficult.

"?HIPPOTHERIUM" DEPERETI SONDAAR 1974

This taxon was erected by Sondaar (1974) for some isolated teeth and postcranials recovered from Montredon and Soblay (France). The Montredon material was revised with the addition of some new material (lacking cranial remains) by Eisenmann (1988). Both localities are referred to MN 10 (Mein 1989). Eisenmann (1988) refers the maxilla from the locality of Diavata (Macedonia, Greece), described by Koufos (1985) as *Hipparion* sp., to *"H."* depereti.

"Hippotherium" depereti is a medium-sized species with very massive limb bones, moderate enamel plication, oval protocone, deep ectoflexid, and moderate hypsodonty (Sondaar 1974; Eisenmann 1988). *"Hippotherium" depereti* has primitive characters common to the *Hippotherium* Complex, including: rich enamel plication and short and robust metapodials. At Montredon *"?Hippotherium" depereti* is associated with a small-sized hipparion, which Eisenmann (1988) compares closely to *Cremohipparion macedonicum*. Koufos (1986) notes that *Cremohipparion macedonicum* is associated with *"Hippotherium"* at *Cremohipparion macedonicum*'s type locality (MN 10). According to Koufos, *"Hippotherium" depereti* may mark an early appearance of medium-sized hipparions in the late Vallesian (along with *Hipparion melendezi* and *Hipparion gettyi*). On the other hand, *"Hippotherium" depereti*'s short and robust metapodials and the rich cheek tooth enamel plication are characters shared among members of the *Hippotherium primigenium* evolutionary lineage and *"Hippotherium" brachypus*.

"HIPPOTHERIUM" BRACHYPUS HENSEL 1862

Hensel (1862) reported a suite of short and robust metapodials from Pikermi that he referred to *Hipparion brachypus*. Gaudry (1862–1867) cited the presence at Pikermi of an hipparion with short and robust metapodials together with a slender metapodial form. However, Gaudry referred both taxa to *Hipparion gracile*. Later, Gromova (1952), Pirlot (1956), Forstén (1968), and Dermitzakis (1976) studied the Pikermi hipparions and distinguished two taxa based on postcranial morphology. Forstén (1968) referred the Pikermi morph with short and robust metapodials to *Hipparion primigenius*. Koufos (1987a) revised the material from Pikermi, referring the large-sized hipparion with short robust metapodials to *Hipparion brachypus*.

The *"Hippotherium" brachypus* type material included some metapodials described by Hensel (1862). However, this material has been lost for some time. There is a cast of a Pikermi hipparion manus figured by Abel (1927). If the original specimen were located, it could possibly be considered to be a suitable lectotype of *"Hippotherium" brachypus* (Heissig, pers. comm. to Koufos). In the event that Hensel's original specimens are not found (including proposed lectotype and syntypes), Koufos (1987a) has proposed that the metacarpal, BMNH M11240 (pl. VII.2), could be used as a suitable neotype, and the metatarsal BMNH M11265 (pl. VIII.2) as well as a series of metapodials of the Gaudry collection at the BMNH (Pik 42, 46, 48, 52, 54, 59, and 104) could be potentially suitable topotypes.

Woodburne and Bernor (1980), Bernor et al. (1980), and Bernor et al. (1989) have recognized three hipparion species at Pikermi: *Hipparion* sp. Group 1 (= *"Hippotherium" brachypus* of Koufos 1987), *H. mediterraneum* (= same taxon of Koufos 1987) and *H.* cf. *prostylum*. This last taxon, *Hipparion prostylum*, has recently been restudied by Bernor at the BMNH. Half exposed in its original matrix, and surrounded by a breccia of other mammalian bone, is a complete skull, BMNH M42603. This skull has a facial and snout morphology somewhat advanced compared to *Hipparion gettyi*, similar to the Lectotype specimen (BMNH M33603) from Mont Lubéron (re: Bernor 1985:198, 203), and less advanced than materials referred to *Hipparion prostylum* from Middle Maragheh (BMNH collection). This would, for Bernor, make Pikermi (8.3–8.2 Ma) somewhat older than the classic MNHN localities at Maragheh (ca. ≤8.2 Ma; Bernor et al., this volume; Swisher, this volume).

The classic Pikermi collections stored in the Museum of Geology and Palaeontology at the University of Athens are currently under study by Dr. G. Theodorou. As with Koufos's work, Theodorou has not yet been able to identify a third Pikermi hipparion species. This may be either because *Hipparion prostylum* is rare at Pikermi or because its metapodial proportions overlap too closely with *Cremohipparion mediterraneum* to be distinguished. In every case, the BMNH skull differs greatly from *Cremohipparion mediterraneum* in having a longer preorbital bar and in having a narrower, more slightly built snout (Bernor, pers. obs.). Further statistical analysis of the Pikermi collection, including all of the known material, is necessary to test the various hypotheses cited here.

"Hippotherium" brachypus as characterized and defined by Koufos (here) is characterized by large size, elongated skull, wide muzzle, simple and oval POF, elongate POB, rich enamel plication in the upper cheek teeth, elliptical protocone, dP1 present, crenulated or plicated flexids in the lower cheek teeth, and short-robust metapodials. More morphological details are given in Koufos (1987a).

Tobien collected a fine assemblage of hipparionine horse materials from the Middle Maragheh horizons, Iran. Included in this assemblage is an unpublished single short and robust metapodial potentially referable to *"Hippotherium" brachypus* (Bernor, pers. obs.). The co-occurrence of this species with *Hipparion prostylum* in Middle Maragheh further supports the correlation between this Maragheh horizon and Pikermi (Bernor et al. 1980; Bernor et al., this volume).

Forstén (1978) has cited the occurrence of hipparions with short and robust metapodials from the Bulgarian localities of Kalimanci and Kromidovo (Bulgaria), and referred them to *H. primigenium*. Kojumdgieva et al. (1982) have reported the occurrence of several fossiliferous horizons from Kalimanci and two from Kromidovo. All are referred to Turolian age except for the lowermost, Kalimanci 1, dated to the late Vallesian. The association of *"Hippotherium" brachypus* and *Cremohipparion mediterraneum* in these localities (Forstén 1978) and the opinion of Forstén (1968) that *"Hippotherium" brachypus* from Pikermi is similar to *Hippotherium primigenium* support the idea that these occurrences of *Hippotherium primigenium* may actually prove to be referable to *"Hippotherium" brachypus*. Moreover, Forstén (1978:34) refers the Kalimanci hipparions to *"Hipparion" mediterraneum* (sensu Nikolov 1973), which differ from the Pikermi form of *"Hipparion" mediterraneum* and rather correspond to *"Hippotherium" brachypus* (sensu Koufos, here).

"Hippotherium" giganteum Gromova, 1952

Gromova (1952) recognized *"Hipparion" giganteum* craniodental and mandibulodental material from the Meotian (= Vallesian/Turolian) age locality of Grebeniki (MN 11; Mein 1989), the Ukraine. Woodburne and Bernor (1980) further recognized material from Samos (AMNH 20628 and material in the NHMW and SENK) that is referable to this taxon (Bernor et al. 1989). The Samos *"Hippotherium" giganteum* material is known from the main bone-bearing beds (= MN 12; Bernor et al., this volume).

"Hippotherium" giganteum is the largest known member of the *Hippotherium* Complex. It is characterized by: its lengthened snout; POF deep medially and posteriorly and shortened anteroposteriorly and dorsoventrally, anterior rim absent; maxillary and mandibular cheek teeth as in *Hippotherium primigenium* except for possible greater enamel complexity and somewhat reduced maxillary cheek tooth hypoglyph incision. *"Hippotherium" gigant-

eum's increased size and snout length would appear to be derived from the condition exhibited by Pikermi horses referred here to *"Hippotherium" brachypus*. Further study is needed to test this hypothesis.

Hipparion s.s.–Group

The *Hipparion* s.s.–group includes a number of West European, Southeast European, Southwest Asian, and South Asian taxa (fig. 26.2). These include for Bernor: *Hipparion melendezi, Hipparion gettyi, Hipparion prostylum, Hipparion dietrichi,* and *Hipparion campbelli,* and are derived from an Old World form belonging to the *Hippotherium* Complex. Woodburne does not include *Hipparion melendezi* and *Hipparion gettyi* in *Hipparion* s.s., and believes that the ancestor of the group is North American in origin. Bernor recognizes no North American *Hipparion* s.s. We refer the reader to Bernor et al. (1989) for a review of these competing hypotheses.

The salient features that characterize Old World *Hipparion* s.s. include: medium size; POB length reduced compared to *Hippotherium*, but with the anterior edge of the lacrimal placed still more than 1/2 the distance from the anterior orbital rim to the posterior POF rim; nasal notch consistently at or near the anterior border of P2; POF progressively reduced in dorsoventral and anteroposterior dimensions, posterior pocketing, medial depth and peripheral rim expression; infraorbital foramen usually encroaches on the anteroventral border of the POF; dP1s absent in adults; maxillary cheek teeth with a maximum crown height of about 50 mm, pre- and postfossettes moderately complex, often having many plis, but of markedly shortened amplitude, posterior wall of postfossette distinct, pli caballins variably double or single, hypoglyphs deeply to moderately deeply incised, protocones oval-shaped to round, isolated from protoloph until very late stage of wear, protocone spurs very rare to absent; mandibular dentition with elongate P2/p2 anterostyle/paraconid, metaconids and metastylids rounded, protoconids reduced to absent and most often covered by cement, ectostylids absent, premolar and molar linguaflexids variably V- to shallow U-shaped; metapodials elongate and slender.

Hipparion melendezi Alberdi 1974a

Alberdi (1974a) named this species for a horse from the late Vallesian horizons (MN 10) of Castilla la Vieja, Spain. Bernor et al. (1989) have characterized *Hipparion melendezi* as being: a medium-sized horse with reduced POF

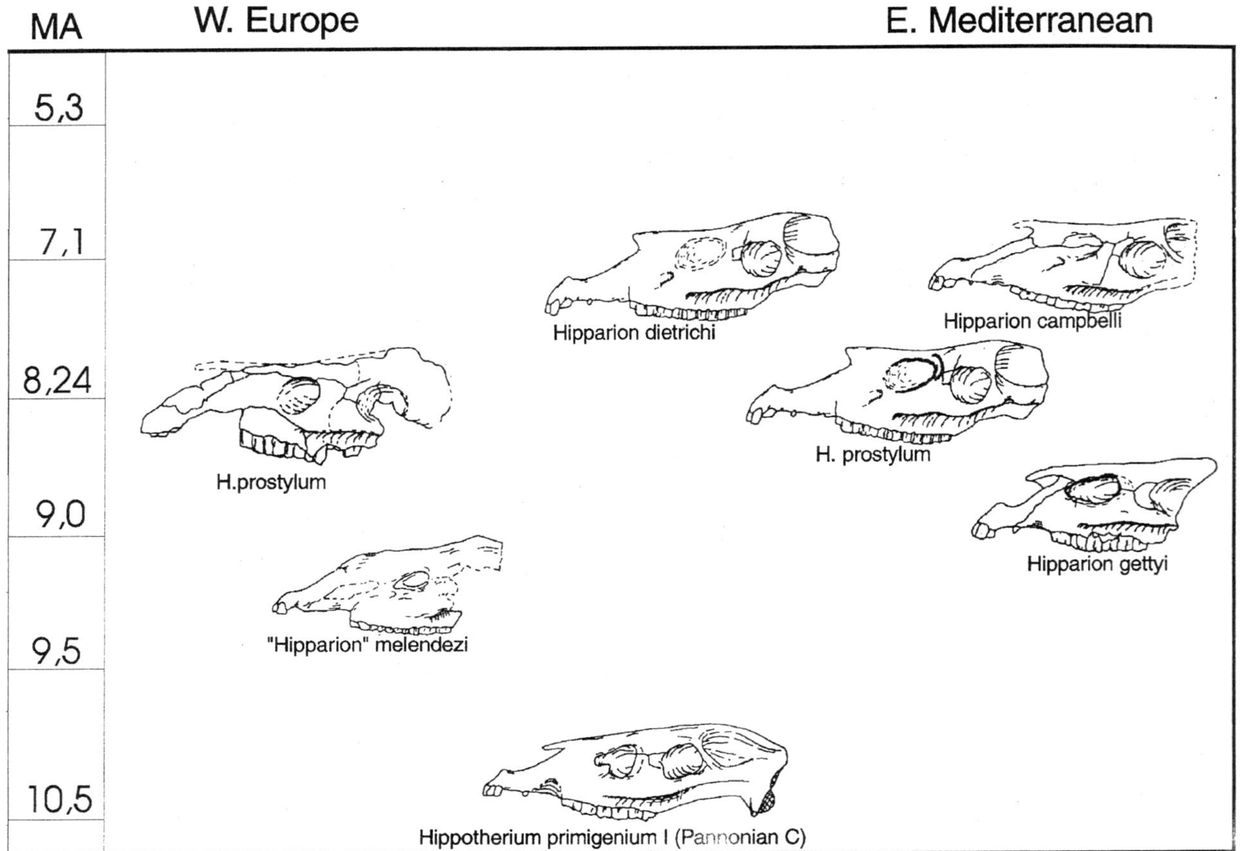

FIGURE 26.2 The Old World *Hipparion* s.s.–group.

complex (sharply reduced length and dorsoventral dimension, reduced peripheral rim, medial depth and posterior pocketing, and absent anterior rim); cheek teeth with oval to rounded protocone and reduced hypoglyph incision; metapodials elongate and slender (re: Alberdi 1974). All of these characters are congruent with the inclusion of *Hipparion melendezi* in *Hipparion* s.s.

While Bernor et al. (1989) recognized its derived craniodental and postcranial morphology, they preferred *Hipparion melendezi*'s referral to a monospecific clade within the paraphyletic Group 1 Complex. Bernor presently prefers to recognize these characters as being synapomorphies with other members of *Hipparion* s.s.

HIPPARION GETTYI BERNOR 1985

Hipparion gettyi was recognized by Bernor (1985, 1986) for a skull derived from the lower biostratigraphic interval at Maragheh, ca. 9 Ma (basal MN 11 equivalent as currently understood; Bernor et al., here; Steininger et al., this volume). Recently, Bernor has discovered a well-preserved Samos specimen of *Hipparion gettyi* in a repository of the HGS, and believes that it may be derived from the lowest stratigraphic horizons at Samos. A slightly advanced form of *Hipparion gettyi* is also known to occur at Pikermi. *Hipparion gettyi* is derived compared to *Hippotherium primigenium* in the following characters: reduced (medium) size; POF rotated dorsally to be anteroposteriorly directed, medial depth and peripheral border outline reduced, anterior rim lacking; maxillary cheek teeth with less complex plications of the fossettes, hypoglyph somewhat less deeply incised, protocone with an oval shape. Whereas Bernor takes these characters as indicating a phylogenetic relationship with *Hipparion* s.s., Woodburne remains skeptical of its inclusion within this clade. Clearly more skeletal material is needed to resolve this issue.

HIPPARION PROSTYLUM GERVAIS 1849

The type material for *Hipparion prostylum* is derived from the Turolian locality of Mont Lubéron (= Mont Leberon or Cucuron), France. Material referable to *Hipparion prostylum* is also found from Middle Maragheh, Pikermi, Samos (recent determination of HGS collections by Bernor, pers. obs.), and arguably Thessaloniki. Koufos (pers. obs.) has not recognized this taxon in material derived from the Axios Valley (Macedonia, Greece), while Woodburne and Bernor (1980) and Bernor et al. (1980) referred Arambourg's collection from Saloniki to *Hipparion* cf. *prostylum*. All of these localities are MN 11/12 in age (see further explanation of this age attribution below).

Hipparion prostylum is characterized by: medium size closely similar to *Hipparion gettyi*; reduction of POF dorsoventrally, anteroposteriorly, in medial depth, posterior pocketing and strength of peripheral rim outline; maxillary cheek teeth with reduced plication frequency, pli caballins variably double or single, hypoglyphs variably deep to moderately deeply incised, protocones modally oval-shaped; mandibular cheek teeth primitively retaining a rounded morphology; postcranium with elongate-slender metapodials.

HIPPARION DIETRICHI WEHRLI 1941

The type specimen of *Hipparion dietrichi* originates from Samos, Greece (GIM I/7; Wehrli 1941, Taf. 17, fig. 4; Taf. 20; re: Sondaar 1971). There is an extensive series of hipparion from this fauna distributed through literally dozens of European and American museums. *Hipparion dietrichi* is one of the most abundant of the Samos taxa, but due to the failure of early collectors to keep accurate stratigraphic records, the age range of this and other Samos taxa cannot always be certainly determined.

Sondaar (1971) referred a large series of AMNH skulls originating from Samos quarries Q1 and Q4 to *Hipparion dietrichi*; the localities are referred to the medial Turolian (MN 12; Bernor et al., here). Koufos and Melentis (1984) have reported *Hipparion dietrichi* from Andrianos Ravin, which they believe may be the same as AMNH Q1. Koufos (1987b, c, 1988b) further identified a large series of cranial and postcranial remains from "Ravin des Zouaves 5," "Prochoma 1," "Vathylakkos 1,2,3" (all probably MN 12, Macedonia, Greece) as being referable to *Hipparion dietrichi*.

Hipparion dietrichi's salient characters include: maintenance of medium size; nasals remain conservative in their retraction to anteriormost P2; POF reduced to a flattened, open outline with a thickened posterior rim; cheek tooth plication frequency has become simplified, protocones oval to rounded, hypoglyphs shallow. Direct postcranial associations from Samos are unknown and impossible to ascertain because of the great diversity of hipparion species. However, the Axios Valley metapodials referred to *Hipparion dietrichi* are elongate and slender (Koufos 1990).

HIPPARION CAMPBELLI BERNOR 1985

The nomen *Hipparion campbelli* was assigned for a complete adult female skull found from Upper Maragheh, ca. 7.6 Ma (Bernor et al., this volume). It is not clearly known from any other locality. This material is chronologically referable to the medial Turolian (MN 12).

Hipparion campbelli is morphologically and phylogenetically closely related to *Hipparion dietrichi*. *Hipparion campbelli* is characterized by: medium size; nasal incision derived, being retracted to P2 mesostyle; POF highly reduced, retaining an egg-shaped outline and only a very small, sharply delimited segment of the posterior fossa rim; simplified enamel plications, oval to rounded proto-cones, shallow hypoglyphs. An associated lower jaw has: a primitive incisor morphology (as in *Hippotherium primigenium*), lacking grooving, curved, and with I3 not transversely constricted; premolar and molar metaconids and metastylids rounded; premolar ectoflexids not separating metaconid and metastylid, whereas in the molars they do, pli caballinids and ectostylids absent; protostylids rarely present, linguaflexids shallow on premolars and deeper on molars. Associated metapodial material is elongate and slender.

Hipparion s.s. has also been reported from Asia. Bernor et al. (1989) report that the Siwalik species *Hipparion antelopinum* and the Chinese species *Hipparion hippidiodus* both belong to this clade. These occurrences suggest a middle Turolian dispersal of the *Hipparion* s.s. clade across all of Eurasia (Bernor et al. 1989). There is no reported occurrence of *Hipparion* s.s. from Africa.

HIPPARION CONCUDENSE PIRLOT 1956

Pirlot (1956) described the hipparion assemblage from Concud, Spain, recognizing a new subspecies, *Hipparion mediterraneum concudense*. This taxon is distinguished from *"Hipparion" mediterraneum mediterraneum* by its smaller size and more plicated upper cheek tooth enamel. Later, Sondaar (1961) and Alberdi (1974b) recognized *Hipparion concudense* as a distinct species. Concud has been referred to the medial Turolian MN 12 by Mein (1989). Sondaar (1961) nominated a series of upper cheek teeth (P40-1960) retained by the IPPS as the species lectotype. A skull with the mandible of *Hipparion concudense* has been described from the locality of Las Hoyeles (Teruel, Spain) by Alberdi (1974b). The species is unreported outside of Spain.

The most remarkable characteristics of the species as given by Sondaar (1961) and Alberdi (1974a, b) are: medium size, small POF, elongate POB (37 mm), moderate enamel plication in the upper teeth, well-developed protostylid, and slender metapodials. These characters suggest a close comparison to *Hipparion dietrichi*.

Sondaar (1961) and later Alberdi (1974) further described a small subspecies of *Hipparion concudense*, *Hipparion concudense aguirrei*, from Los Mansuetos (MN 12). The type specimen is a right upper toothrow with P3-M2 (P162-1958) stored in IPPS (Sondaar 1961). It also is known from Spain only.

Cremohipparion–Group

Qiu et al. (1988) recognized *Cremohipparion* as a subgenus of *Hipparion* (*Hipparion* [*Cremohipparion*]). Bernor and Tobien (1989) raised its rank to the genus level because of clear synapomorphies distinguishing this species

diverse group. Species considered to belong to the *Cremohipparion* lineage are: *Cremohipparion moldavicum*, *Cremohipparion macedonicum*, *Cremohipparion matthewi*, *Cremohipparion nikosi*, *?Cremohipparion periafricanum*, *Cremohipparion mediterraneum*, *Cremohipparion proboscideum*, *Cremohipparion forstenae*, and *Cremohipparion licenti* (fig. 26.3).

Salient features of the *Cremohipparion*–group include: mostly medium sized; lacrimal foramen reduced to absent; nasal notch primitively incised to the anterior border of P2, becoming retracted to strongly retracted in some derived species; POB short with lacrimal invading posterior rim of POF; POF subtriangular shaped and anteroventrally to anteroposteriorly oriented, posterior pocketing sharply reduced, deep to shallow medially, peripheral border outline strong to weak, anterior rim distinct to faded; infraorbital foramen placed inferior to and encroaching upon anteroventral border of the POF; buccinator fossa primitively distinct and well separated from POF; primitively no canine fossa or malar fossa, but acquiring these fossae in some derived species; lacking persistent and functional dP1; maxillary cheek teeth moderately curved to straight, and maximum crown height is between 45 and 50 mm, fossette ornamentation primitively moderately complex becoming simplified in some derived species, postfossette posterior wall always distinct, pli caballins primitively double becoming single in some derived species, hypoglyph primitively deeply incised becoming moderately to shallowly incised in some derived species, protocone always isolated from protoloph, lacking protoconal spur and lingually placed relative to hypocone, primitively lingually flattened and labially rounded becoming rounded in some derived species; P2 anterostyle/parastylid primitively elongate; mandibular incisors never grooved, always curved and I3s never transversely constricted, premolar and molar metaconids and metastylids always rounded, ectoflexids do not separate metaconids and metastylids on the premolars but do so on the molars, pli caballinids generally absent, protostylids rare to lost, ectostylids absent, linguaflexids shallow to shallow V-shaped, molars have linguaflexids somewhat deeper and V-shaped.

CREMOHIPPARION MOLDAVICUM GROMOVA 1952

The nomen *Cremohipparion moldavicum* was constructed by Gromova (1952) for a series of cranial and postcranial material from the Meotian locality of Taraklia, Republic of the Ukraine. Mein (1989) refers this locality to lowermost MN 12. Gromova (1952) also reports this species from Novo-Elizavetovka, the Ukraine, and Bernor (1985) and Bernor et al. (1989) have recorded a similar form, *Cremohipparion* aff. *moldavicum*, from Lower and Middle Maragheh (ca. 9–8 Ma). The likely age range of these localities is therefore within the 9–7 Ma range, and

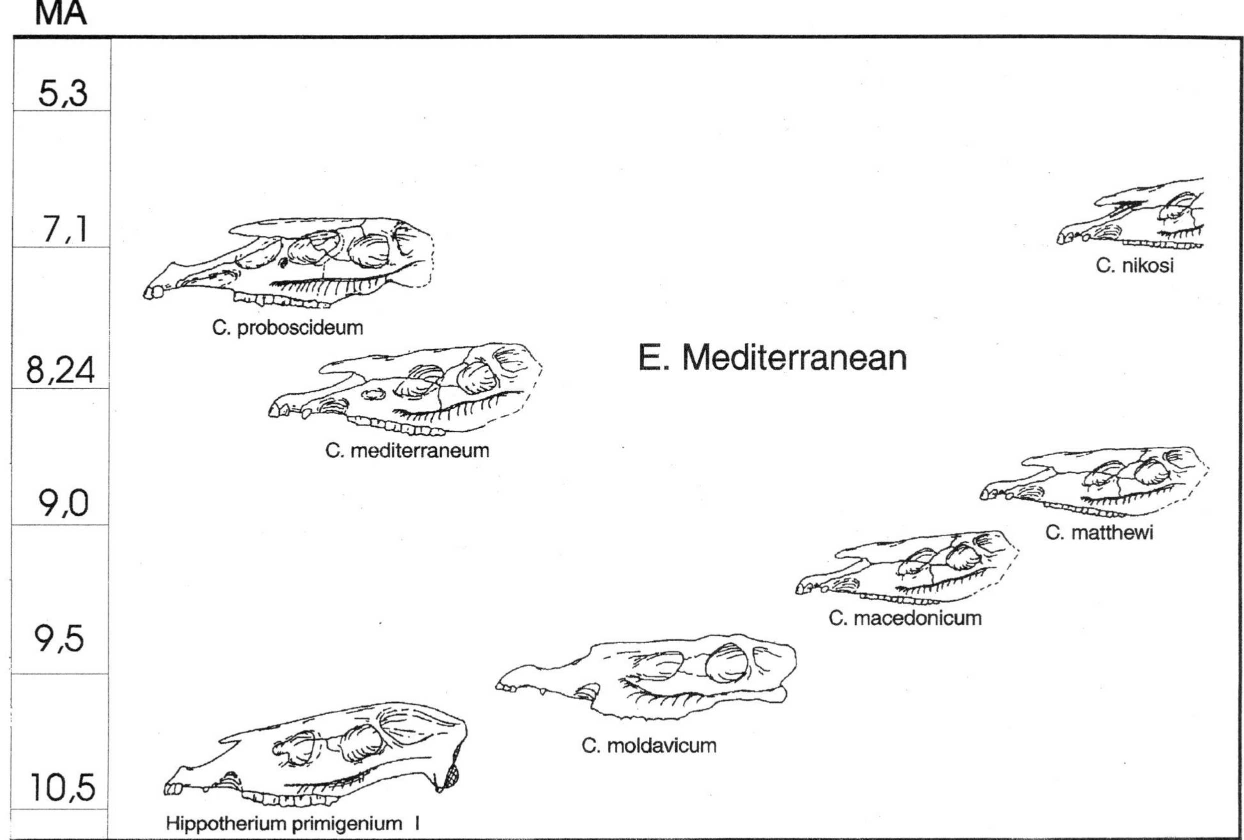

FIGURE 26.3 The *Cremohipparion*–group.

according to new correlations given here (Bernor et al., this volume; Steininger et al., this volume), would span the early–middle Turolian.

Cremohipparion moldavicum is the most primitive member of this group. Its morphological characterization is as given above (primitive features). The most remarkable characteristics for *Cremohipparion moldavicum* are: sharp reduction of POB length by loss of the maxillary bony shelf covering the posteriormost aspect of the POF (giving the deep pocket), and subsequent invasion of the posterior POF rim by the lacrimal bone. Despite these strong autapomorphies, *Cremohipparion moldavicum*'s POF morphology and level of nasal notch incision is the same as Pannonian D–E and Höwenegg members of *Hippotherium primigenium*. Moreover, *Cremohipparion moldavicum* exhibits the following salient characters: POF has an anteroposterior orientation and sharply reduced posterior pocketing; maxillary cheek teeth exhibit some reduction of plication amplitude and hypoglyph incision; lower cheek teeth reduce complexity of enamel ornamentation and expression of protostylid, while retaining primitive rounding of metaconids and metastylids. In many, but certainly not all, characteristics (i.e., reduced plication frequency), *Cremohipparion moldavicum* represents a near

basal morphology of the *Cremohipparion* evolutionary radiation.

CREMOHIPPARION MACEDONICUM KOUFOS 1984

The species was originally based on a small suite of maxillae, mandibles, and postcranial remains from the late Vallesian (MN 10) locality of "Ravin de la Pluie," RPl (Macedonia, Greece) (Koufos 1984; 1986). Latter it was recognized in the early Turolian localities of "Ravin des Zouaves 5" (RZO), "Prochoma 1" (PXM), and "Vathylakkos 2, 3" (VTK, VAT) by cranials and postcranials (Koufos 1987 b, c, 1988b). It is also known from the Macedonian (Greece) locality of "Nikiti 1" (NKT) by one mandible and some postcranials; the locality is referred to MN 10/11 (Koufos et al. 1991). Koufos (pers. obs.) recognizes a form similar to *Cremohipparion macedonicum* from the locality of N. Triglia (Macedonia, Greece). This taxon's holotype is a mandible with both toothrows, stored in LGPUT (coll. no RPl-21). Bernor believes it plausible that the small hipparion from Lower/Middle Maragheh is closely related to this taxon; presently too little material is known to substantiate this possibility.

Its morphological characters include: small size, relatively long and wide muzzle, single, oval, deep and well-defined POF, short POB (28 mm). The posterior extent of the nasal cavity is situated above the anterostyle of P2. The upper cheek teeth have rich-to-moderate enamel plication, elliptical protocone, small, single-to-double pli caballin. The lower cheek teeth have elliptical-to-rounded metaconids and metastylids, slightly plicated or crenulated enamel in the flexid's borders, and a small pli caballinid only in the little worn premolars. The metapodials are elongate and slender. This small-sized hipparion preserves some primitive features, such as the rich enamel plication, rounded metaconids and metastylids, short POB and plicated or crenulated flexid borders that are similar to the *H. primigenium*–group and *Cremohipparion moldavicum*. On the other hand, its small size, the shape of the protocone, and the elongated and slender metapodials suggest a relationship with the *C. matthewi-C. nikosi* lineage.

Cremohipparion matthewi (Abel 1926)

The species has been found in Samos and referred to the nomen *Hipparion minus* by Forsyth Major (1894). The type specimen is a complete skull and associated mandible from an unknown locality of Samos, Greece. It is stored in the Hungarian Geological Survey, Budapest (collection number OK/557). *Cremohipparion matthewi* is common at Samos Q5 (Sondaar 1971; Woodburne and Bernor 1980). It is represented by cranial and postcranial remains from Dytiko (Macedonia, Greece; LGPUT collections; Koufos 1980, 1988a); from Middle and Upper Maragheh, Iran (MN 12; Bernor 1985; Bernor et al., this volume; Swisher et al., this volume); from the medial-late Turolian localities of Kemiklitepe A-B (Turkey; Koufos and Kostopoulos, in review); from southern Yugoslavia (Forstén and Garevsici 1989); and from Bulgaria (Forstén 1978).

There has been an extensive "generalized" hypodigm cast for this taxon including virtually all Southeast European, Southwest Asian, and Ukranian small Turolian age horses (Forstén 1968). In more recent years, however, morphological differences between small horses have suggested a greater diversity in this extended hypodigm than previously appreciated.

Woodburne and Bernor (1980) and Bernor et al. (1980) noted a marked diversity in POF morphologies among Samos small-sized hipparionines. Koufos (1984 and this volume) recognized an early member of this clade from Thessaloniki, *Cremohipparion macedonicum*. Bernor and Tobien (1989) recognized a derived member of this lineage from Samos, *Cremohipparion nikosi*. Small-sized hipparions appearing to belong to potentially different lineages include: "*Hipparion*" *coelophyes* from China (Bernor et al. 1990) and "*Hipparion sitifense*" from East Africa (Hooijer and Maglio 1974; Bernor, pers. obs.).

Cremohipparion matthewi is small. It has a short and narrow muzzle, short nasal notch with its posterior extent situated above P2, short POB, single-oval and deep POF, simple enamel plication, small and simple pli caballin, moderate parastylid, simple enamel in the flexid's borders, pli caballinid absent and elongated, and slender metapodials.

Cremohipparion matthewi is quite similar morphologically to *Cremohipparion moldavicum*. It is derived compared to *Cremohipparion moldavicum* in its: further reduced size; POF's reduced medial depth, posterior pocketing, peripheral border outline, and anterior rim morphologies. *Cremohipparion matthewi* is more primitive than *Cremohipparion moldavicum* in its retention of an anteroventrally oriented POF and more consistent retention of double pli caballins (Bernor et al. 1989). The occurrence of *Cremohipparion macedonicum* in Greek late Vallesian horizons, and retention of primitive "Group 1" characters in the *Cremohipparion macedonicum-Cremohipparion matthewi* lineage not found in *Cremohipparion moldavicum*, suggest an early divergence of this small horse clade from other members of the *Cremohipparion* evolutionary group.

Cremohipparion nikosi Bernor and Tobien 1989

Bernor and Tobien (1989) recognized this taxon from Samos based on a cranial fragment. This species is virtually identical to *Cremohipparion matthewi* in size and cranial morphology except that the nasal incision is not placed at the anterior border of P2, but rather retracted to P4 mesostyle. The stratigraphic provenience of this specimen is unknown, but since Bernor and Tobien (1989) cite evidence for progressive steps of nasal retraction within the Samos small *Cremohipparion* series (*Cremohipparion macedonicum–Cremohipparion matthewi–Cremohipparion nikosi*), they suggest that this specimen must have come from a higher fossil-bearing unit.

?Cremohipparion periafricanum Villalta and Crusafont 1957

The species was nominated for dental and postcranial remains derived from the Turolian locality of Valdecebro II (Villalta and Crusafont 1957). Sondaar (1961) and Alberdi (1974) concurred on the validity of this species, and Alberdi (1974a) referred Valdecebro II to uppermost MN 13. Villalta and Crusafont (1957) did not assign a holotype. However, Forstén (1968) referred an upper and lower cheek tooth series from the GIUB to the type *Hipparion periafricanum*. Alberdi (1974a) reported that the holotype referred to by Forstén cannot be found in Barcelona, and responded by nominating a right maxillary cheek tooth row with P2-M3 to it as the holotype (?lectotype) of *Hipparion periafricanum* (stored at IPPS). Koufos (1980; 1988a; 1990) has referred a series of dental remains from "Dytiko 1"

DTK (MN 13; Macedonia, Greece) to *Hipparion periafricanum*.

Sondaar (1961) characterized *?Cremohipparion periafricanum* as being small size (smaller than *H. matthewi*), having very simple maxillary cheek tooth enamel plication, absence of pli caballin in the molars, early connection of the protocone with the protoloph, absence of the pli protoloph and pli hypostyle in the upper cheek teeth, and elongate, slender metapodials. The referred skeletal material from Spain and Greece is scant. However, Koufos (1988a) has reported a maxilla of *?Cremohipparion periafricanum* from Thessaloniki and believes that this specimen shows a number of similarities with both the Spanish material and Samos *Cremohipparion nikosi* of Bernor and Tobien (1989). *?Cremohipparion periafricanum* possibly represents the most evolved species of the *Cremohipparion macedonicum–Cremohipparion matthewi–Cremohipparion nikosi–?Cremohipparion periafricanum* lineage. Its more certain inclusion within the *Cremohipparion*–group is dependent on currently unknown cranial material.

CREMOHIPPARION MEDITERRANEUM ROTH AND WAGNER 1855

Wagner (1848) originally studied Pikermi hipparion material, referring it to *Equus primigenium*. Later, Roth and Wagner (1855) referred the Pikermi hipparion to *Hippotherium gracile* var. *mediterraneum*. Hensel (1860) followed, including all the known Pikermi material in a single species, *H. mediterraneum*. Later, Hensel (1862) distinguished two species of hipparion from Pikermi: one with elongate and slender metapodials (*Cremohipparion mediterraneum*, here) and another with short and robust metapodials ("*Hippotherium*" *brachypus*, here). Forstén (1968) suggested that the skull and associated mandible figured by Wagner (1848: Taf. 1, fig. 1) should be nominated as the lectotype of "*Hipparion*" *mediterraneum*. As reported by Bernor (1985:218), the missing specimen to which Forstén refers is believed to have been lost during allied bombing of the BSP (München). In lieu of Wagner's (1848) specimen, Bernor (1985:218) has proposed that a very fine skull preserved by the MNHN, Pik. 259, be recognized as the Neotype of "*Hipparion*" *mediterraneum* (= *Cremohipparion mediterraneum* of Bernor and Tobien 1989 and us here).

Besides Pikermi, which Bernor correlates with the Lower/Middle Maragheh boundary (MN 11/12 equivalent; ca. 8.3–8.2 Ma), *Cremohipparion mediterraneum* has been reported from the late Turolian (MN 13) localities of Dytiko (Macedonia, Greece), the medial Turolian (MN 12) localities of Kemiklitepe A-B (Turkey), and the possibly early Turolian (MN 11) locality of Kemiklitepe D (Koufos 1980, 1988a; Koufos and Kostopoulos, in review). Forstén (1978) also reports *Cremohipparion mediterraneum* from the Turolian localities of Kalimanci and Kromidovo (Bulgaria).

Cremohipparion mediterraneum is characterized by: medium size; presence of three distinct facial fossae (preorbital, caninus, and buccinator); short preorbital bar; lacrimal invading POF; nasal notch above P2 mesostyle; skull relatively short; snout short and narrow; complex to moderate maxillary cheek tooth enamel plication; double to single and small pli caballin; protostylids small and rare; simple enamel complexity of the pre- and postflexids; pli caballinid rare; elongate slender metapodials.

Cremohipparion mediterraneum is similar in size to *Cremohipparion moldavicum*. It is derived compared to *Cremohipparion moldavicum* in: the presence of a distinct canine fossa (likely for the origin of a large caninus muscle to assist in retraction of the lips) and modest retraction of the nasals to P2 mesostyle. *Cremohipparion mediterraneum* is primitive relative to *Cremohipparion moldavicum* in its retention of: anteroventrally oriented POF; maxillary cheek tooth fossette plication complex (at least in the Lectotype; Koufos has observed moderate enamel complexity); pli caballins consistently double, hypoglyphs deeply incised and protocone shape often lingually flattened and labially rounded. The retention of primitive *Hippotherium*-like characters in *Cremohipparion mediterraneum*, which are apparently lost in *Cremohipparion moldavicum*, and the presence of distinct synapomorhpies shared with *Cremohipparion proboscideum* (below) suggest that the *Cremohipparion moldavicum/Cremohipparion mediterraneum–Cremohipparion proboscideum* lineages were disjunct by the earlier Turolian.

CREMOHIPPARION PROBOSCIDEUM STUDER 1911

Cremohipparion proboscideum was recognized by Studer (1911) for cranial material from Samos distinguished by apomorphies of the facial region. Sondaar (1971) also recognized the presence of this taxon from Samos AMNH Q1 and Q5. Koufos (1987b) has reported this species from the early Turolian (MN 11) locality of "Ravin des Zouaves 5" (RZO) (Macedonia, Greece), and from Andrianos Ravin, Samos (Koufos and Melentis 1984). *Cremohipparion proboscideum* has also been reported from Southern Yugoslavia (Forstén and Gareversuski 1989) and Romania (Forstén 1980).

Cremohipparion proboscideum is a medium- to large-sized hipparion remarkable for its elongate snout, hypertrophied preorbital and caninus fossae, more strongly retracted nasals (to mesostyle of P3), and longer snout than *Cremohipparion mediterraneum*. Bernor et al. (1989) argue that *Cremohipparion proboscideum* is derived directly from *Cremohipparion mediterraneum*. This suggests that Pikermi is older than those localities that have *Cremohipparion proboscideum*. Koufos's (1987b) report of *Cremohipparion proboscideum* from the earliest Turolian (lowermost MN 11) of the Axios Valley may suggest a non–ancestral-descendant relationship of *Cremohipparion mediterraneum–Cremohipparion proboscideum*. However, Bernor believes that the MN 11 correlation for *Cremohipparion proboscideum*–bearing localities must be further examined; he prefers a later MN 12/13 referral based on the relative stage-of-evolution of *Cremohipparion mediterraneum* and *Cremohipparion proboscideum*.

Bernor et al. (1989) have reported two species of *Cremohipparion* from China, *Cremohipparion forstenae* and *Cremohipparion licenti*. *Cremohipparion forstenae* is a common late Miocene medium- to large-size species with three distinct facial fossae typical for the *Cremohipparion*–group. *Cremohipparion licenti* is a rare and somewhat smaller early Pliocene species distinguished by sharply retracted nasals (to mesostyle of M1), presence of four facial fossae (buccinator, caninus, preorbital and malar), caninus fossa posteriorly pocketed, extreme simplicity of enamel ornamentation, and round protocones. *Cremohipparion* evidently made a Turolian biogeographic extension into China and subsequently evolved in isolation from the Southeast European theater of *Cremohipparion mediterraneum–Cremohipparion proboscideum* evolution.

The *"Plesiohipparion"*–Group

The *"Plesiohipparion"*—Group is a clade within the *"Sivalhippus"* Complex (sensu Bernor et al. 1989, Bernor and Lipscomb 1991; 1995). The *"Sivalhippus* Complex" includes a number of late Miocene, Pliocene and Pleistocene horses ranging throughout Eurasia and Africa. Current systematic work is aimed at better appreciating this late radiating group's diversity, and resolving the taxonomic and phylogenetic problems that currently exist (fig. 26.4).

The genus *"Plesiohipparion"* was recognized as a subgenus of *Hipparion* by Qiu et al. (1987), and included two species, *Hipparion (Plesiohipparion) houfenense* and *Hipparion (Plesiohipparion) huangheense*. Qiu et al. (1987) further reported that hipparions referable to this clade had been found in Mongolia, Baikal, Kazakhstan, the Caucasus and Spain. In Europe, they reported that this clade was represented by the species *Hipparion rocinantis*. The authors also commented that *Hipparion turkanense* and most African Pliocene and Quaternary hipparion species share characters of the lower cheek tooth metaconids and metastylids with *Hipparion (Plesiohipparion)*.

Bernor and Lipscomb (1991) provisionally raised *Hipparion (Plesiohipparion)* to the rank *"Plesiohipparion,"* rec-

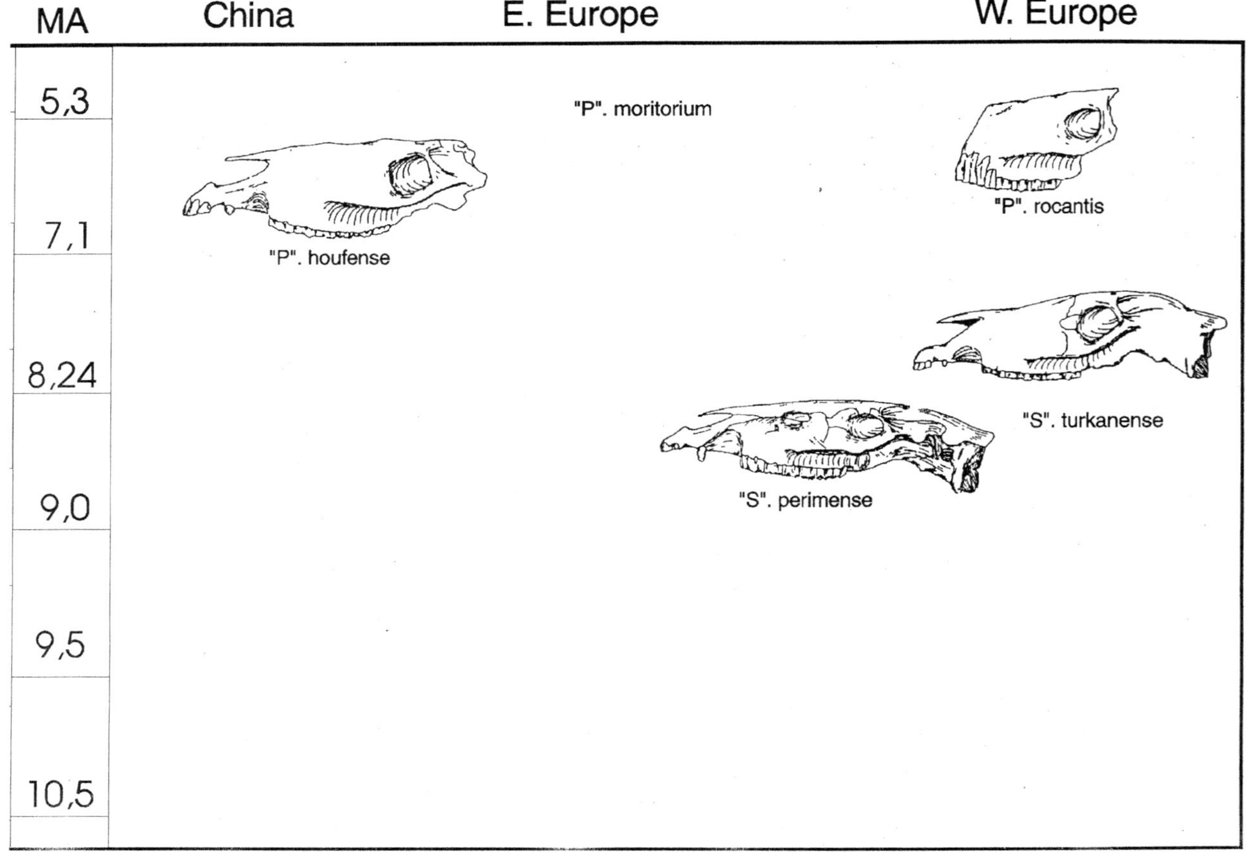

FIGURE 26.4 The *"Sivalhippus"*–Complex.

ognizing the Chinese species (Qiu et al. 1987) *"Plesiohipparion" houfenense* (latest Miocene), *"Plesiohipparion"* aff. *houfenense* (n. sp., late Pliocene), and *"Plesiohipparion" huangheense* (late Pliocene), the Spanish species *"Plesiohipparion" rocinantis crusafonti* (early Pliocene), and the Turkish (late Pliocene) species *"Plesiohipparion"* aff. *huangheense*. We further refer the Hungarian (late Pliocene) species *"Plesiohipparion" moritorum* to this group, and provisionally, *"?Plesiohipparion" crassum* and *"?Plesiohipparion" longipes*.

The *"Sivalhippus"* Complex is recognized by 8 million years in Asia and East Africa by a number of cooccurring evolutionary trends: sharply increased size; progressive loss of POF; increased absolute crown height; lower cheek teeth with metaconids and metastylids that acquire strongly angular facing borders, separated by broad and deep U-shaped linguaflexids; metapodials that become longer and more massive, indicating increased body size. Early members of this clade such as *"Sivalhippus" perimense* and *"Sivalhippus" theobaldi* retain primitive *Hippotherium* characters including: complex enamel ornamentation, double or multiple pli caballins, protocones lingually flattened and labially rounded, hypoglyphs remain deeply incised; lower cheek teeth with columnar protostylids.

Members of the *"Plesiohipparion"* clade are distinguished by the loss of the POF, acquisition of grooved incisors and reduction of maxillary and mandibular fossette complexity (Bernor and Lipscomb 1991; 1995).

"Plesiohipparion" rocinantis Hernandez-Pacheco 1921

Alberdi (1972: figs. 96–99; 1974) described specimens of two Spanish hipparions that she refers to *Hipparion rocinantis rocinantis* (early Pliocene) and *Hipparion rocinantis crusafonti* (late Pliocene). She has argued that they are time-successive chronosubspecies. Zhegallo (1978), Qiu et al. (1987) and Bernor et al. (1989) closely agree that *"Plesiohipparion" rocinantis* shares evolutionary relationships with the late Miocene Chinese species *"Plesiohipparion" houfenense* in cranial, dental, and postcranial anatomy (also = morphogroup 6 of Alberdi 1989). Cranial and postcranial characters are common with those described for the *"Sivalhippus"* Complex earlier, and derived characters of POF, incisors, and fossette complexity are as described for the *"Plesiohipparion"*–group.

"Plesiohipparion" moritorum (Kretzoi 1954)

Kretzoi (1954) named this species for a limited suite of dental and postcranial material from Kislang, Hungary. Since this material cooccurs with *Equus* (Feijfar, pers. comm.), its age is regarded to be late Pliocene (re: MN 17, ca. 2.6 Ma: Lindsay et al. 1980; Bernor and Lipscomb 1991; Steininger et al., this volume). A single maxillary cheek tooth studied by Bernor has complex fossette plications and a derived small, rounded protocone. A single lower cheek tooth has a broad U-shaped linguaflexid. A complete metatarsal is large and robustly built (ML = 276.8 mm; DAW = 43.1 mm), very similar to the dimensions given by Qiu et al. (1987) and Bernor et al. (1989: fig. 10d, p. 313) for Chinese *"Plesiohipparion" houfenense*. Koenigswald (1970) reported the presence of *"Plesiohipparion" houfenense* from Roumania (Malusteni, Beresti) and from England (Red Crag of Suffolk).

"Plesiohipparion" aff. *huangheense*

Bernor and Lipscomb (1991) recognized this taxon from Gülyazi, Turkey. The Gülyazi hipparion is late Pliocene age (MN 17; ca. 2.5 Ma) and cooccurs with *Equus*. Skeletal material limited to a left M1 and a left mandibular fragment with p3–m3 is virtually identical for similarly aged material from China, on which the type is based (Qiu et al. 1987).

This species is distinguished from other members of the *"Plesiohipparion"*—group by: presence of pointed metaconids and metastylids, molar ectoflexid converging with preflexid and postflexid to abut against metaconid and metastylid, pli caballinid rudimentary to single, protostylid present on the occlusal surface as a posterolabially extending open loop. The Turkish *"Plesiohipparion"* aff. *huangheense* differs from Chinese *"Plesiohipparion" huangheense* in its less deeply incised hypoglyph and broader premolar and molar linguaflexids. Bernor and Lipscomb (1991; 1995) have hypothesized a distinct late Pliocene extension of *"Plesiohipparion" huangheense* into Europe. If this hypothesis is accepted, the consequence would be that *"Plesiohipparion"* aff. *rocinantis* and *"Plesiohipparion" huangheense* are separate clades that should be distinguished at some superspecific taxonomic rank. Presently, we prefer to leave this possible taxonomic revision to a time when more fossil evidence and analysis can be had.

"?Plesiohipparion" crassum Gervais 1859

This taxon was first described by Depéret (1890) from Perpignan correlated by Mein (1989) to MN 15. Subsequently, skeletal material was reported from Gödollo, Hungary, and referred to this species by Mottl (Gromova 1952). The species has also been recognized from the locality of Alcoy, Spain (Alberdi 1974a), and referred by Alberdi (1986) as being MN 13/14. Mein (1989) refers Alcoy to MN 14. *"?Plesiohipparion" crassum* has also been identified from the lignite deposits of Ptolemais (Macedonia, Greece); the level (upper parts of the lignitic sequence) in which *H. crassum* was found is considered to be MN 15 by Koufos (1982). The lower part of the lignitic sequence has been correlated by the micromammalian fauna to MN 14 (Van de Weerd 1979). Some teeth similar to *H. crassum* have also been identified from the locality of Apollakia (Rhodos, Greece; MN 15, Benda et al. 1977). A skull from Perpignan (MN 13) described by Depéret (1890) is considered to be the species Lectotype (Forstén 1968; Alberdi 1974a).

"*?Plesiohipparion*" *crassum* is poorly represented skeletally. Because the skull was crushed, Depéret (1890) only figured the upper dentition. The palate is large and has long cheek tooth rows (length P2–M3 = 170 mm). The POF is unknown due to crushing. The upper cheek teeth have highly plicated enamel and the metapodials are very short and robust. Short metapodials have been reported by Qiu et al. (1987) and Bernor et al. (1989) for "*Sivalhippus*" *platyodus*. The known morphology of this taxon, coupled with its early Pliocene age, suggests that its referral to the "*Plesiohipparion*"–group is plausible.

"*?PLESIOHIPPARION*" *LONGIPES* GROMOVA 1952

This taxon was first recognized at the locality of Pavlodar, Kazakhstan, and assigned to a late Miocene or early Pliocene age (Gromova 1952). This material included only isolated teeth and postcranial remains. Later, "*?Plesiohipparion*" *longipes* was identified from the early Pliocene Turkish locality, Çalta (Heintz et al. 1975). Çalta was subsequently referred to MN 15 by Sen et al. (1978). The species has also been recognized from the MN 15 (Koufos et al. 1991) locality of Megalo Emvolon (Macedonia, Greece) by Steffens et al. (1979). There is also a metatarsal fragment maintained by the MNHN (collection of Arambourg) from Megalo Emvolon that appears to belong to this species (Koufos, pers. obs.).

Gromova (1952) has characterized "*Hipparion*" *longipes* as having: large size, moderate enamel plication in the upper cheek teeth, short and wide protocone, very long and slender extemities, and long metapodials. The very large size and the elongate and robust metapodials are characteristic of most species belonging to the "*Plesiohipparion*"–group.

"*HIPPARION*" *FISSURAE* CRUSAFONT AND SONDAAR 1971

The species is only known from the locality of Layna (Spain) (Crusafont and Sondaar 1971; Alberdi 1974a). It is referred to uppermost MN 15 by Mein (1989). The holotype is a left MT III, housed by the IPPS, while the rest of the known material includes isolated teeth and few postcranial remains (Crusafont and Sondaar 1971). *H. fissurae* is medium-sized with more or less hypsodont teeth and elongate slender metapodials. There is too little skeletal material to make a superspecific referral of this taxon.

Evolutionary Biogeography

Figure 26.5 presents the time of first appearances, chronologic ranges, and biogeographic distribution of the taxa under consideration here. Figure 26.6 gives the chronologic ranges and, where known, evolutionary relationships of hipparionine taxa within the different provinces. We summarize these relationships here.

The first-occuring Old World hipparion is *Hippotherium primigenium* I. This taxon is stratigraphically first recorded in the Vienna Basin from the Pannonian Stage C localities of Gaiselberg and Marienthal. This population of *Hippotherium primigenium* is apparently short-lived, and unknown in its morphology outside this time and restricted biogeographic area. The next evolutionary step in Central Europe, *Hippotherium primigenium* II, is more broadly distributed in Vienna Basin Pannonian Stage D and E strata and the German Dinotherium Sands localities. This stage hipparion is believed by Bernor (in Woodburne et al., this volume a) to be the form that extended its biogeographic range into Western Europe, Southeastern Europe, Southwestern Asia, South Asia, and East Africa during the so-called "Hipparion" Datum event.

In Central Europe, the *Hippotherium primigenium* s.s. lineage persists, evolving through later MN 9 into stages III (Höwenegg, ca. 10.3 Ma) and IV (Rudabánya and Götzendorf; ca. 10–9.5 Ma). The *Hippotherium primigenium* lineage would then appear to differentiate into a stage V, which may or may not represent more than one taxon, during the Turolian (Sümeg, MN 10; Csakvár, MN 11). In a recent study of the Dorn Dürkheim, Germany, hipparions, Bernor and Franzen (in press) suggested that there may have been a late Vallesian cladogenetic event separating the Austro-Hungarian population of *Hippotherium primigenium* from the German one, which is evolutionarily more conservative.

Western Europe does not have a member of the *Hippotherium primigenium* s.s. lineage. "*Hippotherium*" *koenigswaldi* is a closely related species exhibiting some minor autapomorphies in the dentition and postcranial skeleton; separation from *Hippotherium primigenium* s.s. II is slight. Differentiation of Western European hipparions begins in MN 10 with the first occurrences of "*Hippotherium*" *catalaunicum*, "*Hippotherium*" *depereti*, and *Hipparion melendezi*. The sister taxon of each Western European MN 10 hipparion species cannot be determined with the material at hand. However, it would appear that "*Hippotherium*" *catalaunicum* is related to MN 12 taxon "*Hippotherium*" aff. *catalaunicum*. Also, *Hipparion melendezi* would appear to be a member of the *Hipparion* s.s. lineage and related, although not directly, to *Hipparion prostylum* which first occurs in MN 11. The MN 12 taxon *H. concudense* may be another member of the *Hipparion* s.s. clade, but there is no cranial material critical for establishing this relationship. The MN 13 species *Cremohipparion periafricanum* is a very small horse that Koufos (here, and elsewhere) has argued is a late, terminal member of the *Cremohipparion* dwarf lineage including: *Cremohipparion macedonicum–Cremohipparion matthewi–Cremohipparion nikosi–Cremohipparion*

FIGURE 26.5 Biogeographic Ranges of Several European, North African, West Asian Hipparionine Lineages

Taxon	C. Europe	N. Africa	W. Europe	SE Europe	SW Asia	FAD	MN
Hippotherium Complex							
H. primigenium I	−−X−−					<10.5	BMN9
H. primigenium II	−−X−−					>10.3	LMN9
H. primigenium III	−−X−−			−−−−		10.0	MMN9–10
H. primigenium IV	−−X−−					<10.0	UMN9
H. primigenium V	−−X−−					9.5	UMN9–11
H. "koenigswaldi"			−−X−−			<10.5	LMN9
"H". brachypus				−−−−	−−−−	8.5	MN10–12
"H". giganteum				−−−−	−−−−	8.5	MN11–13
"H". africanum		−−X−−				9.5	MMN9
"H". sitifense		−−X−−				7.0	MN13
"H". catalaunicum			−−X−−			9.0	MN9–10
"H". aff. catalaunicum			−−X−−			8.5	MN12
"H". depereti			−−−−			<9.0	MN10
Hipparion s.s. group							
H. melendezi			−−X−−			9.0	MN10
H. gettyi				−−−−	−−−−	9.0	MN11
H. prostylum			−−−−−	−−−−	−−−−	8.5	MN11
H. dietrichi				−−X−−		<7.6	MN13
H. campbelli					−−X−−	7.6	MN12
H. concudense			−−X−−			<8.0	MN12
Cremohipparion group							
C. moldavicum				−−X−−	−−−−	9.0	MN10–12
C. macedonicum				−−X−−	−−−−	9.0	MN10–11
C. matthewi				−−−−	−−−−	8.5	MN12–13
C. nikosi					−−X−−	?<7	MN13
C. mediterraneum				−−X−−		8.5	MN11–13
C. proboscideum				−−X−−	−−−−	?<7	MN11–13
"C". periafricanum			−−X−−	−−−−		<7	MN13
"Plesiohipparion" group							
"P". rocinantis			−−−−			<5	MN14
"P". aff. huangheense					−−−−	2.5	MN17
"P". moritorum	−−−−					2.5	MN17
"P". crassum			−−−−	−−−−		<5	MN14–15
"P$_i$. longipes			−−−−		−−−−		MN15
Hipparionine group indet.							
"H". fissurae			−−X−−			<5	MN15

periafricanum ranging from MN 10 through MN 13. There are three taxa reported from Western Europe during Pliocene intervals MN 14 and 15: "*Plesiohipparion*" *rocinantis*, "*Plesiohipparion*" *crassum*, and "*Hipparion*" *fissurae*. The "*Plesiohipparion*" taxa are of Asian origin, and "*Hipparion*" *fissurae* is apparently an endemic Western European form.

Western Europe then would appear to have exhibited an early MN 9 biogeographic connection with Central Europe with the *Hippotherium primigenium* II s.s. lineage. Western Europe also exhibited a biogeographic connection with Southeastern Europe/Southwestern Asia during MN 10/11 with the *Hipparion* s.s. lineage, and again in MN 13 with the dwarf radicle of the *Cremohipparion* lineage. During the early Pliocene there was a migration of Eastern Asian horses across Southwestern Asia, Southeastern Europe, and Western Europe, represented by the "*Plesiohipparion*" taxa cited here ("*Plesiohipparion*" *crassum* and "*Plesiohipparion*" aff. *rocinantis*). These observations lead us to interpret that Western Europe exhibited episodic biogeographic connections through the late Miocene and Pliocene accompanied by periods of isolation and endemic evolution. This pattern

is exhibited across no less than four superspecific lineages: *Hippotherium primigenium* s.s., *Hipparion* s.s., *Cremohipparion*, and the *Sivalhippus* Complex.

Southeastern Europe contains no bonafide MN 9 occurrence of *Hippotherium primigenium* s.s., but we can see no reason why it should not have had one, since both Turkey and Middle Europe do. Koufos (here, and elsewhere) has identified the *Hippotherium primigenium* morphology in MN 10 horizons from Macedonia. "*Hippotherium*" *giganteum* was a large Turolian species apparently derived from the *Hippotherium primigenium* s.s. lineage.

MN 10 also witnessed the first occurrence of the short-limbed form "*Hippotherium*" *brachypus*, the early *Hipparion* s.s. form *Hipparion melendezi*, and two early members of the *Cremohipparion* lineage, *Cremohipparion moldavicum* and *Cremohipparion macedonicum*. The Turolian interval continues with the evolution of many of these lineages. *Hipparion gettyi* (MN 11) evolves through a series of stages: *Hipparion prostylum* (lower MN 12), *Hipparion campbelli*, and *Hipparion dietrichi* (upper MN 12). *Cremohipparion moldavicum* persists as a monospecific lineage

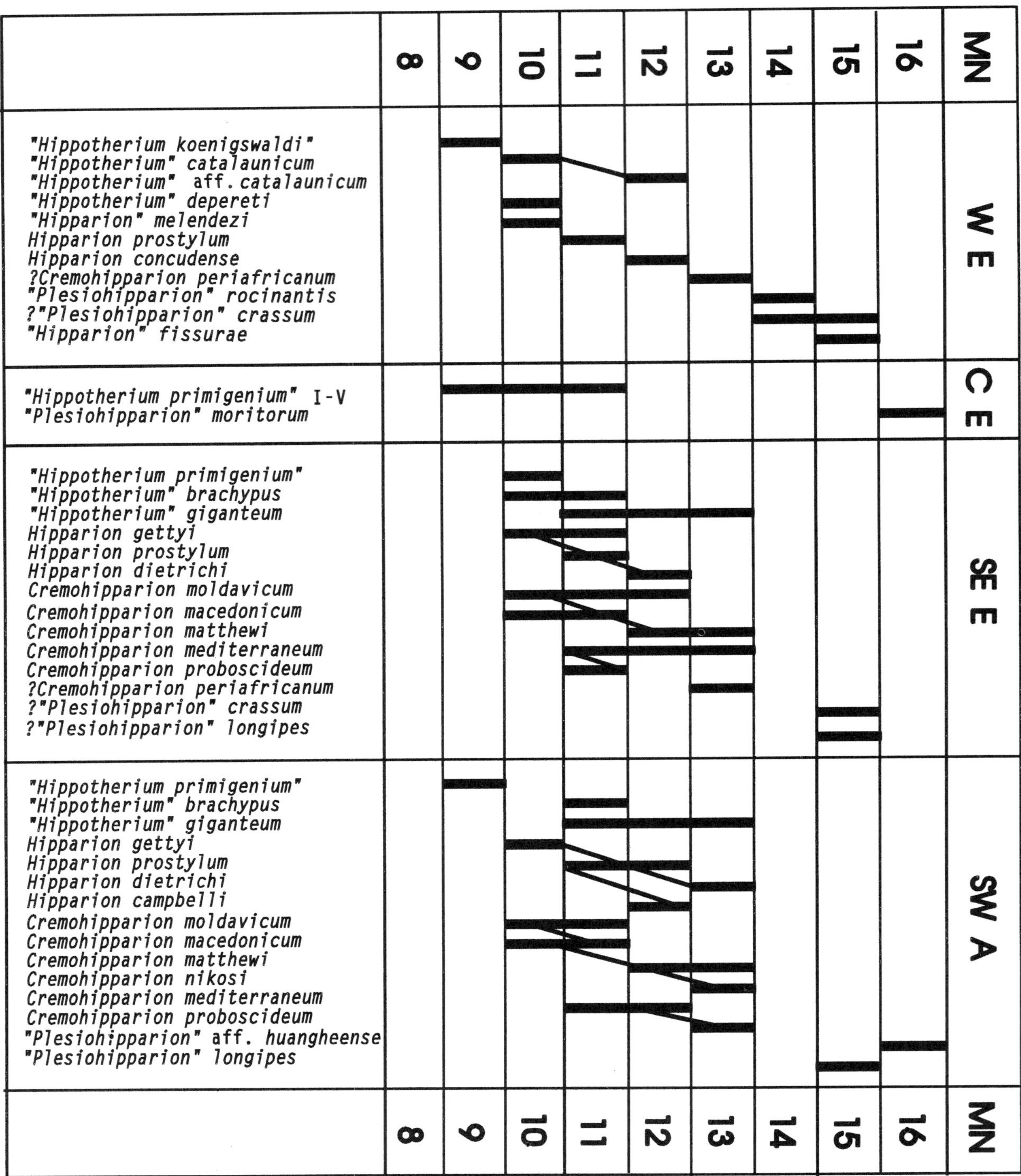

FIGURE 26.6 Chronologic ranges and phylogenetic relationships of European and Southwest Asian Hipparionine horses.

through MN 11 and 12, and is the likely population from which *Cremohipparion mediterraneum* (MN 11–13) and *Cremohipparion proboscideum* (?MN 11–12) are derived. *Cremohipparion macedonicum* would appear to be related to the interprovincial species *Cremohipparion matthewi* (MN 12–13) and *Cremohipparion periafricanum* (MN 13).

The three *"Plesiohipparion"* taxa, *"Plesiohipparion" moritorum*, *"?Plesiohipparion" crassum* and *"?Plesiohipparion longipes*, are early Pliocene members of the *"Sivalhippus"* Complex.

Southwestern Asia shows close affinities with Southeastern Europe throughout this interval. Indeed, events cited here between these two areas differentially are very likely related to events in the former southern U.S.S.R. (Ukraine, Moldavia, and Crimea). Southwestern Asia records the MN 9 first occurrence of the *Hippotherium primigenium* s.s. lineage, and its later derivatives *"Hippotherium" brachypus* and *"Hippotherium" giganteum*. The *Hipparion* s.s. lineage is represented here with the first two steps found in Southeastern Europe, *Hipparion gettyi* (MN 11) and *Hipparion prostylum* (MN 12). The third step, *Hipparion campbelli*, is related cladogenetically to *Hipparion dietrichi*. The Southwest Asian *Cremohipparion* lineage is likewise represented here with the occurrence of *Cremohipparion moldavicum* (MN 10/11), *Cremohipparion macedonicum* (MN 10/11)–*Cremohipparion matthewi* (MN 12/13)–*Cremohipparion nikosi* (MN 13). Of these, *Cremohipparion nikosi* is the only one without certain evolutionary ties outside Southwestern Asia; however, it may be that *Cremohipparion nikosi* is the sister taxon of the transprovincial form *Cremohipparion periafricanum*. The *Cremohipparion mediterraneum* (MN 11/12)–*Cremohipparion proboscideum* (MN 13) clade is likewise represented from the westernmost portion of this range. The Pliocene members of the *Sivalhippus* Complex found in this province include *"Plesiohipparion"* aff. *huangheense* and *"?Plesiohipparion" longipes*. The biogeographic connections between Southwestern Asia and Southeastern Europe are extensive during this period. Indeed, the only evidence of disjunction between these two provinces is found in the terminal evolutionary stages of the *Hipparion* s.s.–group (SWAs = *Hipparion campbelli*; SEE = *Hipparion dietrichi*) during MN 12 and the occurrence of *"Plesiohipparion"* aff. *huangheense* in Southwestern Asia (MN 17) and *"?Plesiohipparion" crassum* in Southeastern Europe (MN 15).

The Southwest Asian and Southeast European provinces would appear to have been connected during the late Vallesian and Turolian as an hipparion "species-factory" that periodically exported taxa to Western Europe, but not to Central Europe (except possibly during the late Turolian, MN 13). At the end of the Miocene, these species became entirely extinct and were succeeded by larger, higher-crowned horses of the *"Sivalhippus"* Complex derived from East and South Asia.

Faunal Similarity, Diversity, and Turnover Patterns

Simpson's index, especially when applied to "faunas" with a small number of taxa, gives an extremely rough and incomplete quantification of the relationships. Similarity is greatly emphasized over difference: a fauna with a single taxon will be 100% similar to a fauna with 15 taxa, if that single taxon is shared between the faunas. Nonetheless, it does provide as good a quantitative estimation of faunal similarity as any alternative method currently available (fig. 26.7; also see Fortelius et al., this volume, for further discussion).

The indices obtained are given in table 26.1. The principal result is clear enough and robust: Central Europe is very different from the other regions, lacking the diversified hipparion assemblage that is present in all the other regions. Southeastern Europe and Southwestern Asia are extremely similar to each other throughout the interval (index for MN 9–16 = 86%). Western Europe, although diverse, is not so similar to the eastern provinces (indices for WE versus SEE and SWA 27% and 14%, respectively, for the same period). If the regions are lumped into a western block (WE + CE) and an eastern block (SEE + SWA), the index for MN 9–16 is 31%. With 15 taxa in

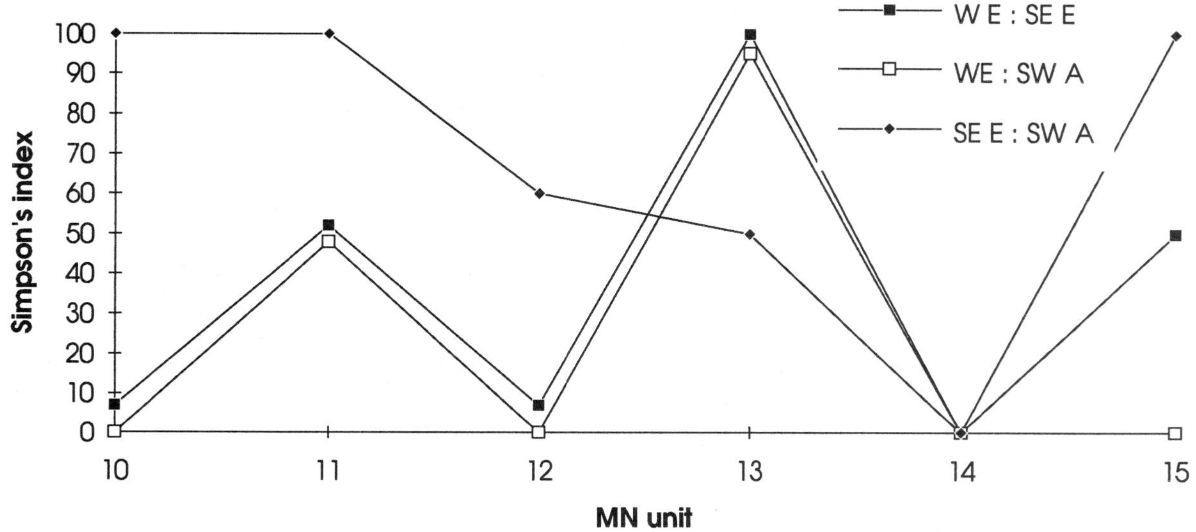

FIGURE 26.7 Simpson's Faunal Resemblance Indices for European and Southwest Asian Hipparionine horses.

TABLE 26.1 *Regional Faunal Resemblance Expressed as Simpson's Index*

Region	WE	CE	SEE	SWA	Interval
		0	27	9	MN 9–16
		0	—	0	MN 9
		0	0	0	MN 10
		0	50	50	MN 11
WE		—	0	0	MN 12
		—	100	100	MN 13
		—	—	—	MN 14
		—	50	0	MN 15
		—	—	—	MN 16
	0		50	50	MN 9–16
	0		—	100	MN 9
	0		100	0	MN 10
	0		0	0	MN 11
CE	—		—	—	MN 12
	—		—	—	MN 13
	—		—	—	MN 14
	—		—	—	MN 15
	—		—	0	MN 16
	27	50		86	MN 9–16
	—	—		—	MN 9
	0	100		100	MN 10
	50	0		100	MN 11
SEE	0	—		60	MN 12
	100	—		50	MN 13
	—	—		—	MN 14
	50	—		100	MN 15
	—	—		—	MN 16
	9	50	86		MN 9–16
	0	100	—		MN 9
	0	0	100		MN 10
	50	0	100		MN 11
SWA	0	—	60		MN 12
	100	—	50		MN 13
	—	—	—		MN 14
	0	—	100		MN 15
	—	—	—		MN 16

Shared taxa divided by taxa in the smaller fauna times one hundred.

one supraprovincial "fauna" and 14 in the other, this is a reasonable estimate of similarity. The pattern is quite stable throughout the study period, except that MN 11 and especially MN 13 represent periods of increased similarity between east and west. MN 12–13 also appears to be a period of increased differentiation between the two eastern provinces (MN 14 lacks either horses or data, but resemblance is restored by MN 15). This pattern is similar to some extent to that seen for the suids, with the greatest regional uniformity in MN 11.

Diversity

Diversity is low everywhere in MN 9 (fig. 26.8; no data for Southeastern Europe), but in MN 10 there is a marked increase in all regions except Central Europe, where the *H. primigenium* lineage remains the sole taxon until its extinction in MN 10/11. The maximum diversity for Western Europe (three taxa) is reached in MN 10, and diversity there subsequently fluctuates between one and two taxa. In Southeastern Europe and Southwestern Asia the maximum (eight and six taxa, respectively) is reached in MN 11, and the following decline is slow (one taxon per MN unit) until MN 14, when it drops abruptly to zero for both regions. After this diversity is again very low everywhere, between zero and two taxa per MN unit.

Turnover

Absolute turnover (fig. 26.9) reflects changes in diversity, except that MN 12 stands out as a minimum for both the eastern regions. There is a weak suggestion of antiphasic timing of changes between Western Europe and the eastern provinces (eastern diversity maxima in MN 11 and MN 13, western in MN 10 and MN 12), but this may be spurious. Relative turnover shows high values for MN 9–10, MN 13, and MN 15, except for CE. The relative diversity curves (fig. 26.10) for the eastern regions are again strikingly similar, while Western Europe shows generally higher turnover and a minimum at MN 11 rather than MN 12.

Probability of Extinction

Probability of extinction clearly shows the rising trend toward MN 12/13 clearly (fig. 26.11). The dominant peak

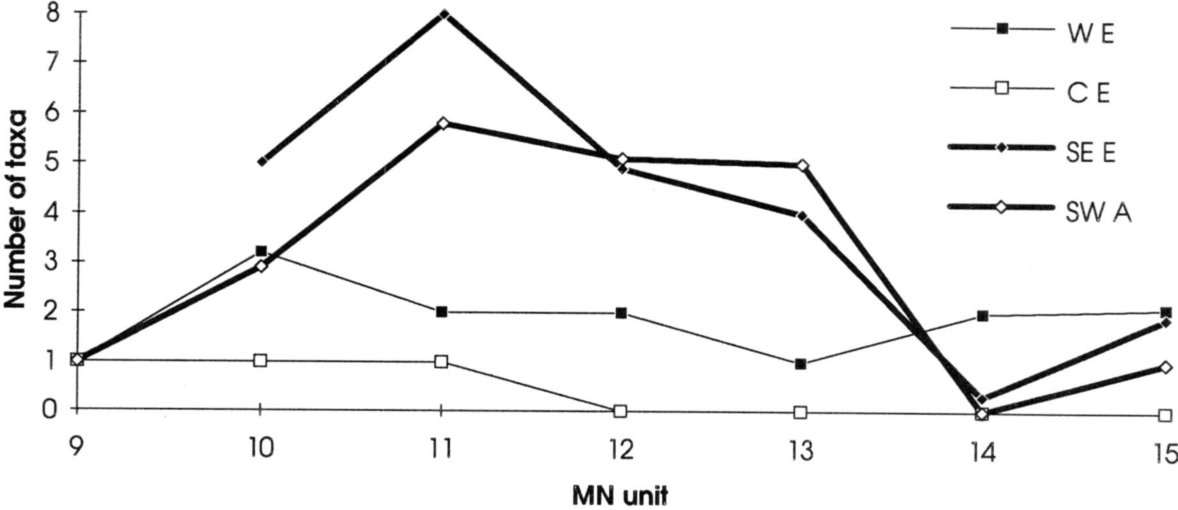

FIGURE 26.8 Diversity of European and Southwest Asian Hipparionine Horses.

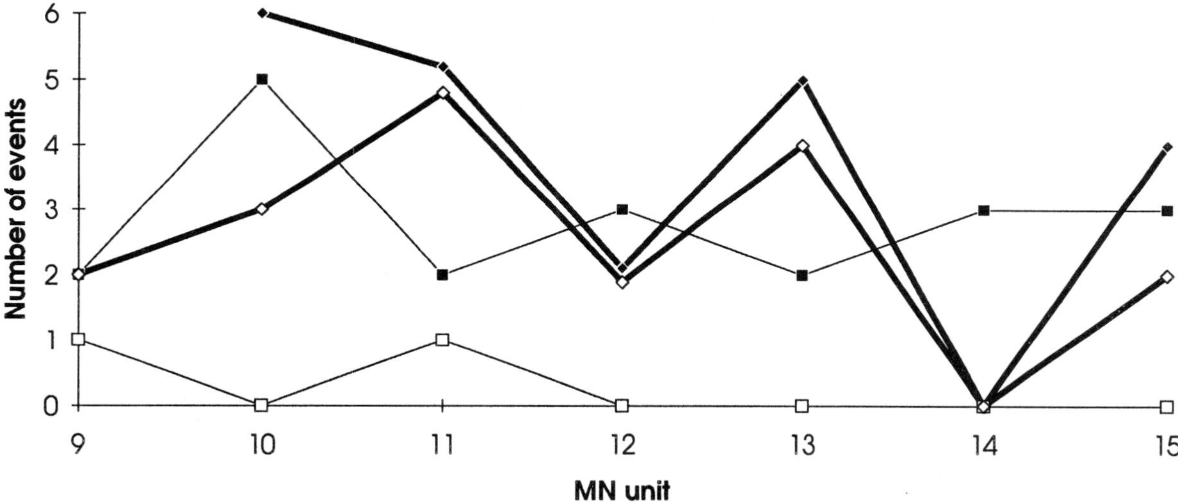

FIGURE 26.9 Absolute turnover of European and Southwest Asian Hipparionine horses.

for the eastern region is that of MN 13, while Central Europe peaks with the extinction of the single lineage in MN 11. Western Europe and Southwestern Asia also have high values at MN 15, after the low (or lack of data) in MN 14. In the Vallesian and early Turolian extinction was a less dominant component of turnover, with origination playing a correspondingly greater role.

Paleoecology

The paleoecology of European and West Asian hipparions has been only briefly addressed elsewhere. Bernor et al. (1988) discussed the paleoecological and paleoenvironmental contexts of Central European first-occurring (early Vallesian) hipparions and concluded that those *Hippotherium primigenium* populations inhabited warm temperate woodlands with low seasonality. This was quite in contrast to traditional dogma that the first occurrence of Old World hipparion heralded a dramatic and rapid shift from forested to grassland environments. Moreover, early Vallesian Central European environments contrasted strikingly with those that the likely ancestral group, the *Cormohipparion "occidentale"* Complex inhabited: seasonal, open country grassland environments (Webb 1983).

Bernor et al. (1990a) contrasted Central European and Perimediterranean/Southwest Asian paleovegetation regimes and found that the former harbored forested environments through most of the late Miocene and concomitant low hipparion species diversity, while the latter harbored more seasonal open country environments and relatively high hipparion species diversity. Our results here are congruent with these earlier interpretations.

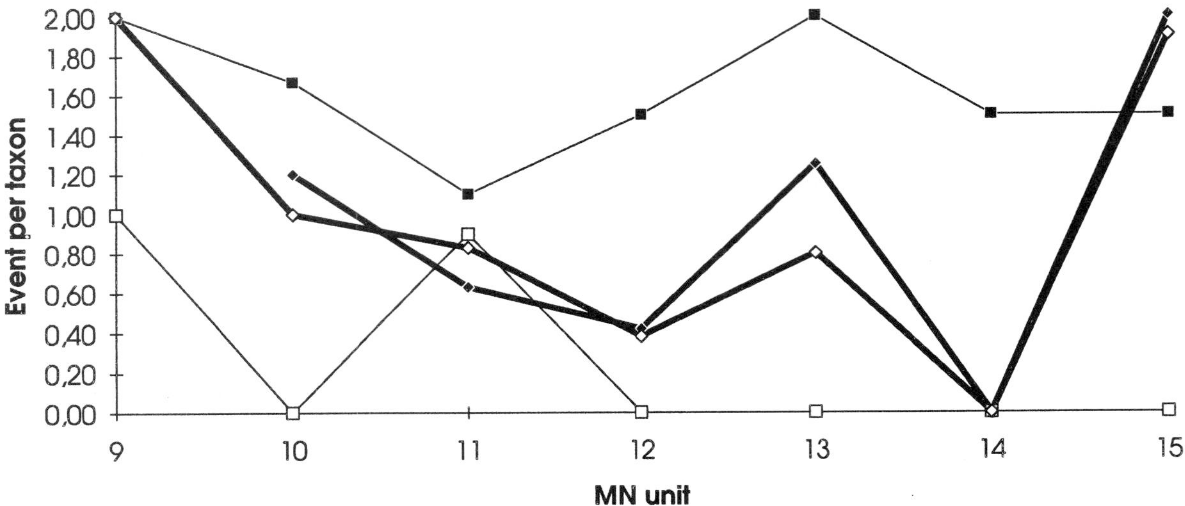

FIGURE 26.10 Relative Turnover of European and Southwest Asian Hipparionine horses.

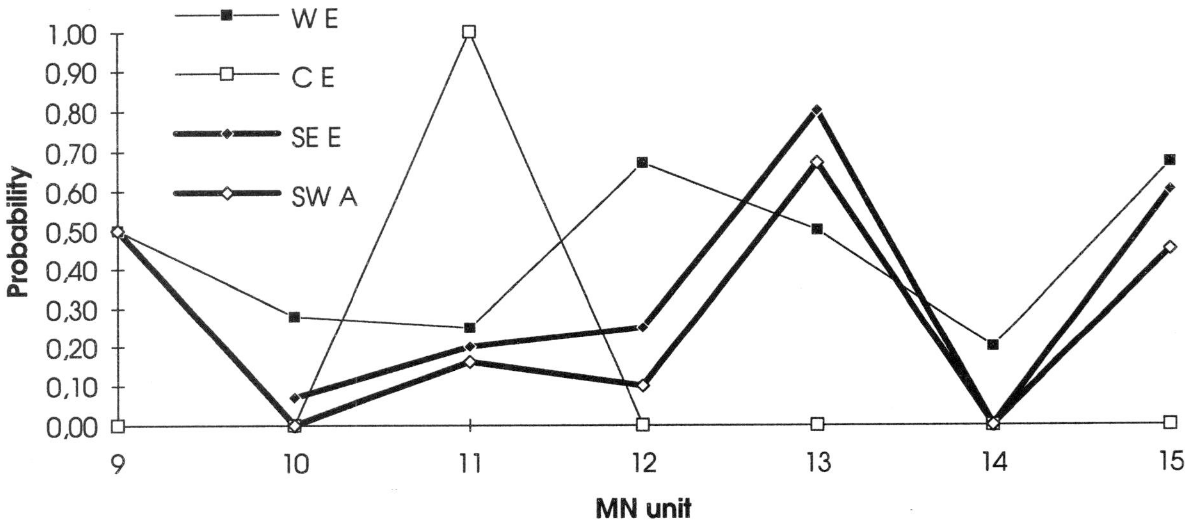

FIGURE 26.11 Probability of Extinction of European and Southwest Asian Hipparionine horses.

Hayek et al. (1991) made a preliminary analysis of dental microwear on a suite of North American middle and late Miocene hipparions and their sister taxa, and the Old World hipparion radiation. The North American sister taxon to all hipparions, *Merychippus insignis*, was characterized as having been adapted to a mixed browsing/grazing regime. The *Cormohipparion* clade showed a marked shift toward grazing, with *Cormohipparion goorisi* and *Cormohipparion "occidentale"* being categorized as grazers, while the intermediate form *Coromohipparion sphenodus* (= *quini* here) was characterized as being a mixed feeder. The Old World hipparions sampled were relatively few. However, a consistent pattern uncovered is that those taxa which retained or elaborated facial fossae structures and

retracted nasals (to support a mobile upper lip for more selective feeding), had a browsing component to their diet, whereas those taxa that did not retract nasals and lost facial fossae structures were categorized as grazers.

If we take these various observations as a guide, we can hypothesize the following series of testable paleoecological hypotheses:

1. Central European hipparions of the *Hippotherium* Complex retain well-developed preorbital fossae, highly complex enamel plications, deep medio-lateral grooving of the cheek teeth, and robust metapodial proportions, all indicative of a browse component to the diet and adaptation to warm temperate, highly equable environ-

ments. These environments were retained at least through the Vallesian age, and probably well into the Turolian of Central Europe.

2. Large horses of the *Hippotherium* Complex, including *"Hippotherium" brachypus* and *"Hippotherium" giganteum*, had short metapodials (*"Hippotherium" brachypus* only) complex enamel ornamentation, mediolateral grooving of the cheek teeth, and prominent POF's. These horses are believed to have been very likely adapted to more forested habitats adjacent to or within grassland biomes of Southeast European and Southwest Asian provinces. Other large horses such as the Spanish *"Hippotherium" koenigswaldi* and *"Hippotherium" catalaunicum* probably retained the same warm temperate woodland preferences.

3. Members of the *Hipparion* s.s.–group show a progressive loss of the POF, have little nasal retraction and elongation of the metapodials. Moreover, they are generally associated with seasonal, open country faunas (Bernor 1983; Bernor et al. 1989). Within this clade, the one species studied for microwear, *Hipparion dietrichi*, showed a clear grazing adaptation (Hayek et al. 1991). Species belonging to the *Hipparion* s.s. clade (as defined here) would appear to have been largely grazers.

4. The *Cremohipparion*–group showed two evolutionary/paleodietary pathways. *Cremohipparion moldavicum* is the most primitive member of the group and may well have been a mixed browse/graze eater. The dwarf lineage *Cremohipparion macedonicum–Cremohipparion matthewi–Cremohipparion nikosi–Cremohipparion periafricanum*, progressively decreased its size, and, where known, the postcrania became progressively more elongate and slender and the POF disappeared. All available lines of evidence suggest that this was yet another lineage adapted to open seasonal grassy woodland environments and eating graze. The larger Chinese form *Cremohipparion forstenae* (Bernor et al. 1991) lost its facial fossae and was shown by Hayek et al. (1991) to have grazed. The *Cremohipparion mediterraneum–Cremohipparion proboscideum* clade elaborated the facial fossae to a more prominent POF, and intermediate and well-developed buccinator fossae. Microwear studies revealed a mixed feeding adaptation with a significant browse component. The Chinese species *Cremohipparion licenti* more strongly retracted nasals (to the level of M1 mesostyle), further developed the facial fossae so that a malar fossa was added and the buccinator fossa had a posterior pocket as further attachment area for the muscle origin. All of these facial structures probably served to support and move a highly mobile snout (tapiroidlike pseudo-probosis). The dietary microwear studies revealed a strong browsing dietary adaptation.

No paleodietary studies have been made on members of the *Sivalhippus* Complex. However, all European Pliocene members of this clade have, where known, no POF, have quite high-crowned cheek teeth and elongate, heavily built limbs. We would suspect that these forms mimicked *Equus* somewhat and had a significant proportion of graze in their diets.

Later Neogene European and Southwest Asian hipparions are evolutionarily diverse and were undoubtedly adapted to a variety of ecological niches. Convergent evolution in facial, dental, and postcranial anatomy is significant across the lineages we have reported here. Our review of this group has impressed us that the bioprovincial patterns we have unraveled are quite real, and that there are underlying regional tectonic, and very likely global, processes that shaped their adaptation and evolution. We believe that further work on this group would do well to target continued detailed morphologic and systematic study, refinement of the chronologic framework, and study of the relationships between lineage evolutionary change and paleoecological context.

Summary and Conclusions

We have cited the presence of thirty-two late Miocene to medial Pliocene hipparionine taxa from Central Europe, Western Europe, North Africa, Southeastern Europe, and Southwestern Asia. These include a minimum of four superspecific evolutionary complexes: the *Hippotherium* Complex (13 taxa), the *Hipparion* s.s.–group (6 taxa), the *Cremohipparion*–group (7 taxa), and the *"Plesiohipparion"*–group (6 taxa) belonging to the *Sivalhippus* Complex.

Hippotherium primigenium stage I from Vienna Basin Pannonian C strata is hypothesized to be the most primitive and first-occurring Old World hipparionine. It is believed to be derived from a group of closely related North American taxa referred by MacFadden (1984) to a single taxon, *"Cormohipparion" occidentale*. *Hippotherium primigenium* stage II has a dental and postcranial morphology that is common to, or can be shown to precede, the morphology of common first occurring hipparions in Western Europe, the Siwaliks, China, North Africa, and East Africa. Succeeding *Hippotherium primigenium* stages (III–V) were confined mostly, or entirely, to Central Europe (see Woodburne et al., this volume a, for further details). Other members of the *Hippotherium* Complex are cited here from all provinces under consideration and are interpreted to be vicariant derivatives of *Hippotherium primigenium* s.s. Most members of this group retained well-developed facial fossae, complex enamel ornamentation on the cheek teeth, and robust metapodials (all primitive characters). These characters, and for those circumstances where the environmental settings are known (Bernor et al. 1988; Bernor et al. 1990a), indicate that members of the *Hippotherium* Complex were semi-cursorial denizens of warm temperate forest environments.

The *Hipparion* s.s. clade has been cited here as being derived from a member of the *Hippotherium* Complex (but for alternative hypotheses see Bernor et al. 1989 and Woodburne 1989). Species belonging to this group are recorded from Western Europe, Southeastern Europe, Southwestern Asia, Southern Asia (Bernor and Hussain 1985), and China (Bernor et al. 1990b). *Hipparion* s.s. species are all medium sized, and show trends toward loss of the preorbital fossa, simplification of cheek tooth ornamentation, and elongation of metapodials. Their environmental context, and where known their dental microwear, suggests that members of this group were adapted to more seasonal open country woodlands and a more grazing habitus.

The *Cremohipparion*–group is also believed to have been derived directly from the *Hippotherium* Complex. Species belonging to this group are concentrated in the former Southern USSR (Moldavia, the Ukraine, Crimea areas), Southeastern Europe, and Southwestern Asia. There are closely related derivatives also in China (Qiu et al. 1988; Bernor et al. 1987b; 1990b). There are two fundamental evolutionary pathways exhibited in this group: (1) a trend toward extreme dwarfism accompanied by loss of the facial fossa and elongation of metapodials (*Cremohipparion macedonicum*, *Cremohipparion matthewi*, *Cremohipparion nikosi*, *Cremohipparion periafricanum*); (2) a trend toward deep incision of the nasal notch and hypertrophy and multiplication of facial fossae (*Cremohipparion moldavicum*, *Cremohipparion mediterraneum*, *Cremohipparion proboscideum*, *Cremohipparion licenti*). The dwarf lineage would appear to be predominately a grazer adapted to late Miocene open country environments.

European and Southwest Asian species belonging to the "*Plesiohipparion*"–group represent immigrants of an Asian radicle of the Eurasian-African "*Sivalhippus*" Complex. The "*Sivalhippus*" Complex represents an extensive secondary late Miocene–middle Pleistocene hipparionine radiation (Bernor et al. 1989). These were nearly entirely large horses mostly having highly reduced or absent facial fossae, high-crowned cheek teeth, and robustly built (variously short to very long) metapodials. There may have been an African dwarf lineage derived from a member of this group (Bernor, in progress). Their environmental preference was probably highly varied in that they were so morphologically diverse and biogeographically broadly distributed.

Hipparionine horses show a strong provincial distribution. Central Europe would appear to have a provincially distinct hipparion fauna by the later half of MN 9 (*Hippotherium primigenium* stage III or IV); species diversity remained low there. Southeastern Europe and Southwestern Asia were an hipparionine superprovince and "species factory" that episodically exported taxa to Western Europe, Asia, and Africa. Western Europe exhibited a combination of its own modest endemic evolutionary and immigration events. Hipparionine faunal similarity increased modestly between Western and Eastern Europe during MN 11 and MN 13. Diversity maxima are recorded in Western Europe during MN 10 and in Southeastern Europe and Southwestern Asia during MN 11; Central Europe retained a low diversity throughout the late Miocene and Pliocene. Species turnovers are highest across our study area during MN 10, MN 13, and MN 15. Probability of extinction was highest during MN 12/13 and MN 15. There was a massive extinction of European and Southwest Asian hipparionines just prior to the lower Pliocene boundary.

The evolutionary and biogeographic patterns given here would appear to be regulated by a combination of global and regional abiotic events. The first occurrence of Old World hipparions and the massive late Miocene/early Pliocene extinctions witnessed across Eurasia are transprovincial and suggest a global climatic influence (Bernor et al. 1989). However, the various late Miocene and Pliocene transprovincial immigration events that were succeeded by vicariant evolution were most likely regulated by regional tectonics and provincial paleoenvironmental differentiation. A particularly sharp contrast existed between late Miocene environments that bordered the periphery of the Central and Eastern Paratethys and supported more forested environments and those Perimediterranean areas that supported more seasonal open country habitats. The 1:1 relationship between global climatic events and hipparionine evolutionary events cannot be supported by this record.

Acknowledgments

Bernor and Woodburne would like to thank the National Science Foundation and the National Geographic Society for funding their research on hipparionine systematics and geochronology. Bernor would further like to thank the Humboldt Stiftung for supporting this research over a year's sabbatical in Karlsruhe, Germany. All authors would like to gratefully acknowledge funding from the VW Stiftung for supporting our attendence at this workshop, the Staatliches Museum für Naturkunde and its Director, Professor Dr. S. Rietschel, for so much of the logistical support, and the town of Immendingen and the Schloss Reisensburg for its gracious hospitality during the workshop.

LITERATURE CITED

Abel, O. 1926. *Amerikafahrt Eindrücke, Beobachtungen und Studien eines Naturforschers auf einer Reise nach Nordamerika und Westindien.* Jena: Fischer Verlag.

———. 1927. *Lebensbilder aus der Tierwelt der Vorzeit*, 2d ed. Jena: Fischer Verlag.

Alberdi, M.-T. 1974a. El genero *Hipparion* en España. *Trabajo sobre Neogeno-Cuaterno* 1:1–146.

———. 1974b. Descripción del primer craneo de *Hipparion* procedente del Neogeno de Teruel (España). *Boletin Geologico y Minero, Revista bimestral* 72:5–14.

———. 1989. A review of Old World hipparionine horses. In *The Evolution of Perissodactyls*, ed. D. R. Prothero and R. M. Schoch, pp. 234–61. Oxford: Oxford University Press.

Archer, A. W. and C. G. Maples. 1987. Monte Carlo simulation of selected binomial similarity coefficients. 1. Effect of number of variables. *Palaios* 2:609–17.

Benda, L., J. Meulenkamp, and A. Van de Weerd. 1977. Biostratigraphic correlations in the Eastern Mediterranean Neogene: Correlation between mammal, sporomorph, and microfossil assemblages from the upper Cenozoic of Rhodos, Greece. *Newsletters on Stratigraphy* 6:117–30.

Berggren, W. A. and J. A. Van Couvering. 1974. The late Neogene: Biostratigraphy, geochronology, and paleoclimatology of the past 15 million years in marine and continental sequences. *Palaeogeography, Palaeoclimatology, and Palaeoecology* 16:1–216.

Bernor, R. L. 1978. *The Mammalian Systematics, Biostratigraphy, and Biochronology of Maragheh and Its Importance for Understanding Late Miocene Hominoid Zoogeography and Evolution*. Ph.D. Diss., University of California, Los Angeles.

———. 1983. Geochronology and zoogeographic relationships of Miocene Hominoidea. In *New Interpretations of Ape and Human Ancestry*, ed. R. L. Ciochon, and R. S. Corruccini, pp. 21–64. New York: Plenum.

———. 1984. A zoogeographic theater and biochronologic play: The time/biofacies phenomena of Eurasian and African Miocene mammal provinces. *Paléobiologie Continentale* 14:121–42.

———. 1985. Systematic and evolutionary relationships of the hipparionine horses from Maragheh, Iran (late Miocene, Turolian age). *Palaeovertébrata* 15:173–269.

Bernor, R. L. and J. Franzen. In press. The late Miocene (early Turolian) mammalian fauna from Dorn Dürkheim, Reissenhessen (SW-Deutschland) Mammalia: Perissodactyla, Equidae. *Courier Forschungsinstitut Senckenberg*.

Bernor, R. L., K. Heissig, and H. Tobien. 1987. Early Pliocene Perissodactyla from Sahabi, Libya. In *Neogene Paleontology and Geology of Sahabi*, ed. N. T. Boaz, A. El-Arnauti, A. W. Gaziry, J. de Heinzelin, and D. D. Boaz, pp. 233–54. New York: Liss.

Bernor, R. L. and S. T. Hussain. 1985. An assessment of the systematic, phylogenetic, and biogeographic relationships of Siwalik hipparionine horses. *Journal of Vertebrate Paleontology* 5:32–87.

Bernor, R. L., J. Kovar-Eder, D. Lipscomb, F. Rögl, S. Sen, and H. Tobien. 1988. Systematics, stratigraphic, and paleoenvironmental contexts of first-appearing Hipparion in the Vienna Basin, Austria. *Journal of Vertebrate Paleontology* 8:427–52.

Bernor, R. L., J. Kovar-Eder, J.-P. Suc, and H. Tobien. 1990. A contribution to the evolutionary history of European late Miocene age hipparionines (Mammalia, Equidae). *Paléobiologie Continentale* 17:291–309.

Bernor, R. L., M. Kretzoi, H.-W. Mittmann, and H. Tobien. 1993a. A preliminary systematic assessment of the Rudabánya hipparions (Equidae, Mammalia). *Mitteilungen der Bayerischen Staatssammlung für Paläontologie und historische Geologie* 33:195–207.

Bernor, R. L. and D. Lipscomb. 1991. The systematic position of "*Plesiohipparion*" aff. *huangheense* (Equidae, Hipparionini) from Gülyazi, Turkey. *Mitteilung der Bayerischen Staatssammlung für Paläontologie und historische Geologie* 31:107–23.

———. 1995. A consideration of Old World hipparionine horse phylogeny and global abiotic processes. In *Paleoclimate and Evolution, with Emphasis on Human Origins*, ed. E. Vrba et al., pp. 163–177. New Haven: Yale University Press.

Bernor, R. L., H.-W. Mittmann, and F. Rögl. 1993b. Systematics and chronology of the Götzendorf "*Hipparion*" (late Miocene, Pannonian F, Vienna Basin). *Annalen des Naturhistorisches Museum, Wien* 95(A):101–20.

Bernor, R. L., Z. Qiu, and L.-A. Hayek. 1990b. Systematic revision of Chinese hipparion species described by Sefve 1927. *American Museum of Natural History, Novitates* 2984:1–60.

Bernor, R. L., Z. Qiu, and H. Tobien. 1987. Phylogenetic and biogeographic bases for an Old World hipparionine horse geochronology. *Annales Instituti Geologici Publici Hungarici* 70:43–53.

Bernor, R. L., N. Solounias, C. C. Swisher III, and J. A. Van Couvering. This volume. The correlation of three classical "Pikermian" faunas—Maragheh, Samos, and Pikermi—with the European MN unit system.

Bernor, R. L. and H. Tobien. 1989. Two small species of *Cremohipparion* (Equidae, Mammalia) from Samos, Greece. *Mitteilungen der Bayerischen Staatssammlung für Paläontologie und historische Geologie* 29:207–26.

———. 1990. The mammalian geochronology and biogeography of Paşalar (middle Miocene, Turkey). *Journal of Human Evolution* 19:551–68.

Bernor, R. L., H. Tobien, L.-A. Hayek, and H.-W.Mittmann. In press. *Hippotherium primigenium* (Equidae, Mammalia) from the late Miocene of Höwenegg (Hegau), Germany. *Andrias* 10:1–00.

Bernor, R. L., H. Tobien, and M. O. Woodburne. 1989. Patterns of Old World hipparionine evolutionary diversification and biogeographic extension. In *Topics on European Mammalian Chronology*, ed. E. H. Lindsay, V. Fahlbusch, and P. Mein, pp. 263–319. New York: Plenum.

Bernor, R. L., M. O. Woodburne, and J. A. Van Couvering. 1980. A contribution to the chronology of some Old World Miocene faunas based on hipparionine horses. *Géobios* 13:25–59.

Boellstorf, J. and M. F. Skinner. 1977. A fission-track date from post-Rosebud, early Valentine rocks. *Proceedings of the Nebraska Academy of Science, Lincoln, 87th Annual Meeting*, pp. 39–40.

Boné, E. L. and R. Singer. 1965. *Hipparion* from Langebaanweg, Cape Province and a revision of the genus in Africa. *Annals of the South African Museum* 48:273–397.

Bonis, L. de, G. Bouvrain, and G. D. Koufos. 1988. Late Miocene mammal localities of the lower Axios valley (Macedonia, Greece) and their stratigraphical significance. *Modern Geology* 13:141–47.

Cande, O. and D. J. Kent. 1992. Revised geomagnetic polarity time scale. *Journal of Geophysical Research* 97 (B10): 13917–51.

Cope, E. D. 1873. Report of the stratigraphy and Pliocene vertebrate paleontology of northern Colorado. *Bulletin of the U.S. Geological Survey of the Territories* 1:12.

——. 1874. Report on the vertebrate paleontology of Colorado. *Annual Report of the U.S. Geological and Geographic Survey of the Territories* 1874:522.

——. 1889. A review of the North American species of *Hippotherium*. *Proceedings of the American Philosophical Society* 26:447–58.

Crusafont Pairo, M. and P. Y. Sondaar. 1971. Une nouvelle espece d'*Hipparion* du Pliocéne terminal d'Espagne. *Palaeovertébrata* 4:59–66.

Dalrymple, G. B. 1979. Critical tables for conversion of K-Ar ages from old to new constants. *Geology* 7:558–60.

Depéret, C. 1890. Les animaux pliocénes du Rousillon. *Memoire Societé Géologie France* 3:1–64.

Dermitzakis, M. 1976. Observations on the metapodials of *Hipparion* from Pikermi (Attica, Greece). *Proceedings, Koninklijke Nederlandse Akademie van Wetenschappen, B* 79:18–28.

Eisenmann, V. 1988. Contributions a l'étude du gisement Miocene supérieur de Montredon (Herault): Les grands mammifères. 5—Le Perisssodactyles Equidae. *Palaeovertébrata*, Memoire Extraordinaire: 65–96.

Eisenmann, V., M. T. Alberdi, C. de Giuli, and U. Staesche. 1988. *Studying Fossil Horses*. Leiden: Brill.

Eisenmann, V., P. Y. Sondaar, M.-T. Alberdi, and C. de Giuli. 1987. Is horse phylogeny becoming a playfield in the game of theoretical evolution? *Journal of Vertebrate Paleontology* 7:224–29.

Evernden, J. F., D. E. Savage, G. H. Curtis, and G. T. James. 1964. Potassium-Argon dates and the Cenozoic mammalian geochronology of North America. *American Journal of Science* 262:145–98.

Fahlbusch, V. 1991. The meaning of MN zonation: Considerations for a subdivision of the European continental Tertiary using mammals. *Newsletters on Stratigraphy* 24:159–73.

Forstén, A. 1968. Revision of the Palearctic *Hipparion*. *Acta Zoologica Fennici* 119:1–134.

——. 1978. A review of the Bulgarian *Hipparion* (Mammalia, Perissodactyla). *Géobios* 11:31–41.

Forstén, A. and R. Garevski. 1989. Hipparions (Mammalia, Perissodactyla) from Macedonia, Yugoslavia. *Geologia Macedonia* 3:159–206.

Freudenthal, M. 1968. On the mammalian fauna of the *Hipparion*-beds in the Calatayud-Teruel Basin (Prov. Zaragossa, Spain), part IV. The genus *Megacricetodon* (Rodentia). *Proceedings, Koninklijke Nederlandse Akademie van Wetenschappen* B74:417–41.

Freudenthal, M. and P. Y. Sondaar. 1964. Les faunes à *Hipparion* des environs de Daroca Espagne et leur valeur pour la stratigraphie du Neogène de l'Europe. *Proceedings, Koninklijke Nederlandse Akademie van Wetenschappen* B, 67:473–90.

Gabunja, L. 1959. *Histoire du genre Hipparion*. Moscow: Academie Science Géorgie Institut Paléobiologie.

Gaudry, A. 1862–1867. *Animaux fossiles et géologie de l'Attique*. Paris.

Gromova, V. 1952. Le genre *Hipparion*. *Bureau Reserche géologie et mineralogie, CEDP* 12:1–288.

Hayek, L.-A., R. L. Bernor, N. Solounias, and P. Steigerwald. 1991. Preliminary studies of hipparionine horse diet as measured by tooth wear. In *Bjorn Kurtén: A Memorial Volume*, ed. A. Forstén, M. Fortelius, and L. Werdelin. *Annales Zoologici Fennici* 28: 187–200.

Heintz, E., L. Ginsburg and S. Sen. 1975. *Hipparion longipes* Gromova du Pliocene de Çalta (Ankara, Turkey): Le plus dolichopodial des hipparions. *Proceedings, Koninklijke Nederlandse Akademie van Wetenschappen, B* 78:77–82.

Hensel, R. 1860. Uber *Hipparion mediterraneum*. *Abhandlungen der Akademie der Wissenschaften, Berlin* 1860:356–63.

——. 1862. Uber die Reste einiger Säugetierarten von Pikermi in der Münchener Sammlung. *Monatsberichte der K. p. Akademie Wissenschaften, Berlin* 27:560–69.

Hernandez-Pacheco, E. 1921. La Ilanura manchega y sus mamíferos. *Comisión de Investigaciones paleontológicas y prehistóricas, Memorias, Madrid* 28:1–41.

Hooijer, D. A. and V. J. Maglio. 1974. Hipparions from the late Miocene and Pliocene of northwestern Kenya. *Zoologische Verhandelingen* 134:1–34.

Howell, F. C. 1987. Preliminary observations on Carnivora from the Sahabi Formation (Libya). In *Neogene Paleontology and Geology of Sahabi*, ed. N. T. Boaz, A. El-Arnauti, A. W. Gaziry, J. de Heinzelin, and D. D. Boaz, pp. 152–81. New York: Liss.

Hulbert, R. C. 1987a. A new *Cormohipparion* (Mammalia, Equidae) from the Pliocene (latest Hemphillian and Blancan) of Florida. *Journal of Vertebrate Paleontology* 7:451–68.

——. 1987b. *Cormohipparon* and *Hipparion* (Mammalia, Perissodactyla, Equidae) from the late Neogene of Florida. *Florida State Museum, Bulletin of Biological Sciences* 33:229–338.

——. 1988. *Cormohipparon* and *Hipparion* (Mammalia, Perissodactyla, Equidae) from the late Neogene of Florida. Florida State Museum, Bulletin of Biological Sciences. 33:229–338.

——. 1989. Phylogenetic interrelationships and evolution of North American late Neogene Equinae. In *The Evolution of Perissodactyls*, ed. D. R. Prothero and R. Schoch, 176–196. New York: Oxford University Press.

Hulbert, R. C. and B. J. MacFadden. 1991. Morphological transformation and cladogenesis at the base of the adaptive radiation of Miocene hypsodont horses. *American Museum of Natural History, Novitates* 3000:1–61.

Koenigswald, G. H. R. von. 1970. *Hipparion* from the Pleistocene of Europe, especially from the Red Crag of East Anglia. *Palaeogeography, Palaeoclimatology, Palaeoecology* 8:261–64.

Kojumdgieva, W., I. Nikolov, P. Nedjakov, and A. Busev. 1982. Stratigraphy of the Neogene in Sandinski Graben. *Geologica Balcanica* 12:69–81.

Koufos, G. D. 1980. *Palaeontological and Stratigraphical Study of the Neogene Continental Deposits of the Axios Basin (Macedonia, Greece)*. Ph.D. diss. *Science Annals of the Faculty of Physics and Mathematics, University of Thessaloniki* 19(11):1–322.

——. 1982. *Hipparion crassum* Gervais 1859, from the lignites of Ptolemais (Macedonia-Greece). *Proceedings, Koninklijke Nederlandse Akademie van Wetenschappen, B* 85:229–39.

——. 1984. A new hipparion (Mammalia, Perissodactyla) from the Vallesian (late Miocene) of Greece. *Paläontologie Zeitschrift* 58:307–17.

——. 1985. *Hipparion* sp. (Equidae, Perissodactyla) from Diavata (Thessaloniki, northern Greece). *Bulletin of the British Museum of Natural History (Geology)* 38:335–45.

——. 1986. Study of the Vallesian hipparions of the lower Axios Valley (Macedonia, Greece). *Géobios* 19:61–79.

——. 1987a. Study of the Pikermi hipparions. Part I: Generalities and taxonomy. *Bulletin Museum Nationale Histoire Naturelle Paris*, 4e séries, 9, section C, no. 2: 197–252. Part II: Comparisons and odontograms. Bulletin Museum Nationale Histoire Naturelle Paris, 4e séries, 9, section C, no. 2: 327–63.

——. 1987b. Study of the Turolian hipparions of the lower Axios Valley (Macedonia, Greece). 1. Locality "Ravin des Zouaves-5" (RZO). *Géobios* 20:293–312.

——. 1987c. Study of the Turolian hipparions of the lower Axios Valley (Macedonia, Greece). 2. Locality "Prochoma-1" (PXM). *Paläontologie Zeitschrift* 61:339–58.

——. 1988a. Study of the Turolian hipparions of the lower Axios Valley (Macedonia, Greece). 3. Localities of Vathylakkos. *Paleontologie i Evolucion* 22:15–39.

——. 1988b. Study of the Turolian hipparions of the lower Axios Valley (Macedonia, Greece). 4. Localities of Dytiko. *Palaeovertébrata* 18:187–239.

——. 1989. The hipparions of the lower Axios Valley (Macedonia, Greece): Implications for the Neogene stratigraphy and the evolution of hipparions. In *European Neogene Mammal Chronology*, ed. E. Lindsay, V. Fahlbusch, and P. Mein, pp. 321–38. New York: Plenum.

Koufos, G. D. and J. Melentis. 1984. The late Miocene (Turolian) mammalian fauna of Samos Island (Greece). Study of the collection of Palaeontological Museum of Mytilinii, Samos. 2. Equidae. *Science Annals of the Faculty of Physics and Mathematics, University of Thessaloniki* 24:47–78.

Koufos, G. D., G. E. Syrides, K. K. Koliadimou. 1991. A Pliocene primate from Macedonia (Greece). *Journal of Human Evolution* 2:283–94.

Koufos, G. D., G. E. Syrides, K. K. Koliadimou, and D. S. Kostopoulos. 1991. Un nouveau gisement de Vértèbres avec hominoide dans le Miocene supérieur de Macédoine (Grèce). *Compte Rendu Academie Science Paris*, ser. II, 313:691–96.

Kretzoi, M. 1954. Bericht über die Calabrische (Villafranchische) Fauna von Kisláng, Kom. Fejér. *Földtani Intézet. Evkönvve, Betürendes Mutató* 1953:239–65.

Leidy, J. 1856. Notices of some remains of extinct Mammalia recently discovered by Dr. F. V. Hayden in the badlands of Nebraska. *Proceedings of the Academy of Natural Sciences, Philadelphia* 8:59.

Lindsay, E. H. 1972. Small mammal fossils of the Barstow Formation, Calfiornia. *California University Publications in Geological Sciences* 93:1–104.

Lindsay, E. H., N. D. Opdyke, and N. M. Johnson. 1984. Blancan-Hemphillian land mammal ages and late Cenozoic mammal dispersal events. *Annual Review Earth and Planetary Sciences* 12:445–48.

MacFadden, B. J. 1980. The Miocene horse *Hipparion* from North America and from the type locality in Southern France. *Palaeontology* 23(3): 617–35.

——. 1984. Systematics and phylogeny of *Hipparion, Neohipparion, Nannippus*, and *Cormohipparion* (Mammalia, Equidae) from the Miocene and Pliocene of the New World. *American Museum of Natural History Bulletin* 179:1–195.

——. 1987. Systematics, phylogeny, and evolution of fossil horses: A rational alternative to Eisenmann et al. (1987). *Journal of Vertebrate Paleontology* 7:230–35.

MacFadden, B. J. and M. F. Skinner. 1981. Earliest Holarctic hipparion, *Cormohipparion goorisi* n. sp., (Mammalia, Equidae) from the Barstovian (medial Miocene) Texas Gulf Coastal Plain. *Journal of Paleontology* 55:619–27.

MacFadden, B. J. and M. O. Woodburne. 1982. Systematics of the Neogene Siwalik hipparions (Mammalia, Equidae) based on cranial and dental morphology. *Journal of Vertebrate Paleontology* 2(2):185–218.

Mein, P. 1989. Updating MN zones. In *European Neogene Mammal Chronology*, ed. E. Lindsay, V. Fahlbusch, and P. Mein, pp. 73–93. New York: Plenum.

Meyer, H. von. 1829. Taschenbuch für die gesammte Mineralogie. *Zeitschrift für Minerale*, n.s., 23:150–53.

——. 1833. Beiträge zur Petrefactenkunde, Fossile Säugethiere. *Nova acta Academie caesareae Leopoldino-Carolinae germanicae naturae curiosorum, Halle* 16:423–516.

Nikolov, I. 1973. Über einige Adaptivprozesse bei den Hipparionen. *Bulletin Geolgie, Institut Sofia, Series, Paleontology* 22:71–80.

Osborn, H. F. 1918. Equidae of the Oligocene, Miocene, and Pliocene of North America: Iconographic type revision. *American Museum of Natural History Memoirs* 2:1–326.

Pirlot, P. L. 1956. Les formes européennes du genre *Hipparion*. *Memoire Instituto Geólogico Diputación Provincial, Barcelona* 14:1–122.

Qiu, Z., H. Weilong, and G. Zhihui. 1987. The Chinese hipparionine fossils. *Palaeontologica Sinica*, series C, 175:1–250.

Radinsky, L. D. 1984. Ontogeny and phylogeny in horse skull evolution. *Evolution* 38:1–15.

Rögl, F., H. Zapfe, R. L. Bernor, R. Brzobohaty, G. Daxner-Höck, I. Draxler, O. Fejfar, J. Gaudant, P. Herrmann, G. Rabeder, O. Schultz, and R. Zetter. Die Primatenfundstelle Götzendorf an der Leitha, Niederösterreich (Obermiozän des Wiener Beckens). *Jahrbuch der Geologischen Bundesanstalt, Wien* 136:503–26.

Roth, J. and A. Wagner. 1855. Die fossilen knochenüberreste von Pikermi in Griechenland. *Abhandlen bayerische Akademie Wissenschaften, München* 7:371–464.

Sen, S., P. Y. Sondaar, and U. Staesche. 1978. The biostratigraphical applications of the genus *Hipparion* with special references to the Turkish representatives. *Proceedings, Koninklijke Nederlandse Akademie van Wetenschappen, B* 81:370–85.

Simpson, G. G. 1947. Evolution, interchange, and resemblance of North American and Eurasian Cenozoic mammalian faunas. *Evolution* 1:218–20.

Skinner, B. F. and B. E. Taylor. 1967. A revision of the geology and paleontology of the Bijou Hills, South Dakota. *American Museum of Natural History, Novitates* 2300:1–53.

Skinner, M. F. and B. J. MacFadden. 1977. *Cormohipparion* n. gen. (Mammalia, Equidae) from the North American Miocene (Barstovian-Clarendonian). *Journal of Paleontology* 51:912–26.

Sondaar, P. Y. 1961. Les *Hipparion* de l'Aragon méridional. *Estudios geologicos Intituto "Lucas Mallada"* 17: 209–305.

——. 1971. The Samos *Hipparion*. *Proceedings, Koninklijke Nederlandse Akademie van Wetenschappen, B* 74:417–41.

Steffens, P., H. de Bruijn, J. Meulenkamp, and L. Benda. 1979. Field guide to the Neogene of northern Greece (Thessaloniki area and Strimon Basin). *Publication of the Department of Geology and Palaeontology, University of Athens*, ser. A, 35:1–14.

Steininger, F. F., W. A. Berggren, D. V. Kent, R. L. Bernor, S. Sen, and J. Agusti. This volume. Circum-Mediterranean Neogene (Miocene and Pliocene) marine-continental chronologic correlations of European mammal units.

Steininger, F. F., R. L. Bernor, and V. Fahlbusch. 1989. European Neogene marine/continental chronologic correlations. In *Topics in European Mammalian Geochronology*, ed. E. H. Lindsay, V. Fahlbusch, and P. Mein, pp. 15–46. New York: Plenum.

Studer. 1911. Eine neue Equidenform aus dem Obermiocän von Samos. *Verhandlungen der Deutschen zoologischen Gesellschaft* 20–21:192–200.

Swisher, C. C., III. This volume. New ^{40}Ar/^{39}Ar dates and their contribution toward a revised chronology for the late Miocene of Europe and West Asia.

Tedford, R. A. 1970. Principles and practices of mammalian geochronology in North America. *North American Paleontological Convention, C, Proceedings F*, pp. 666–703.

Tedford, R. A., T. Galusha, M. F. Skinner, B. E. Taylor, R. W. Fields, J. R. Macdonald, J. M. Rensberger, S. D. Webb, and D. P. Whistler. 1987. Faunal succession and biochronology of the Arikareean through Hemphillian interval (late Oligocene through earliest Pliocene Epochs), North America. In *Cenozoic Mammals of North America: Geochronology and Biostratigraphy*, ed. M. O. Woodburne, pp. 153–210. Berkeley: University of California Press.

Tobien, H. 1959. *Hipparion*-Funde aus dem Jungtertiär des Höwenegg (Hegau). *Aus der Heimat* 67:121–40.

Van Valen, L. 1984. A resetting of Phanerozoic community evolution. *Nature* 307:50–52.

Villalta, J. and M. Crusafont Pairo. 1967. Dos nuevas especies de *Hipparion* del Pikermiense Espanol. *Cursos y Conferencio Instituto Lucas Mallada* 4:65–69.

Wagner, A. 1848. Urweltliche Säugethierreste aus Griechenland. *Abhandlen bayerische Akademie Wissenschaften, München* 5:333–78.

Webb, S. D. 1969. The Burge and Minnechaduza Clarendonian mammalian faunas of north-central Nebraska. *University of California Publications in Geological Sciences* 78:1–191.

——. 1983. The rise and fall of the late Miocene ungulate fauna in North America. In *Coevolution*, ed. M. H. Nitecki, pp. 267–306. Chicago: University of Chicago Press.

Webb, S. D. and R. C. Hulbert. 1986. Systematics and evolution of *Pseudohipparion* (Mammalia, Equidae) from the late Neogene of the Gulf Coastal Plain and the Great Plains. *Contribution to Geology, University of Wyoming Special Papers* 3:237–72.

Weerd, A van der. 1979. Early Ruscinian rodents and lagomorphs (Mammalia) from the lignites near Ptolemais (Macedonia, Greece). *Proceedings, Koninklijke Nederlandse Akademie van Wetenschappen, B* 52:127–70.

Wehrli, H. 1941. Beitrag zur Kenntnis der Hipparionen von Samos. *Paläeontologie Zeitschrift* 22:321–86.

Whistler, D. P. and D. W. Burbank. 1992. Miocene biostratigraphy of the Dove Spring Formation, Mojave Desert, California, and characterization of the Clarendonian mammal age (late Miocene) in California. *Geological Society of America Bulletin* 104:644–58.

Wood, H. E., H. R. W. Chaney, J. Clark, E. H. Colbert, G. L. Jepsen, J. B. Reeside, and C. Stock. 1941. Nomenclature and correlations of the North American continental Tertiary. *Bulletin, Geological Society of America* 52:1–48.

Woodburne, M. O. 1987. *Cenozoic Mammals of North America*. Berkeley: University of California Press.

——. 1989. *Hipparion* horses: A pattern of endemic evolution and intercontinental dispersal. In *The Evolution of Perissodactyls*, ed. D. R. Prothero and R. M. Schoch, pp. 197–233. New York: Oxford University Press.

——. 1996a. Precision and resolution in mammalian chronostratigraphy: Principles, practices, examples. *Journal of Vertebrate Paleontology* 16.

——. 1996b. Reappraisal of the *Cormohipparion* from the Valentine Formation, Nebraska. *American Museum of Natural History, Novitates.*

——. In press a. Craniodental morphology of *Merychippus insignis* in comparison with the base of the *Cormohipparion* radiation in North America. *Journal of Vertebrate Paleontology.*

Woodburne, M. O. and R. L. Bernor. 1980. On superspecific groups of some Old World hipparionine horses. *Journal of Paleontology* 8(4): 315–27.

Woodburne, M. O., R. L. Bernor and C. C. Swisher III. This volume a. The stratigraphic and phylogenetic bases for the "*Hipparion*" Datum in the Old World.

Woodburne, M. O., B. J. MacFadden, and M. F. Skinner. 1981. The North American "*Hipparion*" datum and implications for the Neogene of the Old World. *Géobios* 14:493–524.

Woodburne, M. O., G. Theobald, R. L. Bernor, H. König, C. C. Swisher III, and H. Tobien. This volume b. Advances in the geology and stratigraphy at Höwenegg, southwestern Germany.

Zhegello, V. I. 1978. The Hipparions of Central Asia. *Trudy Soviet Mongolian Palaeontological Expedition* 7:1–152.

27

The Stratigraphical Range of Fossil Rhinoceroses in the Late Neogene of Europe and the Eastern Mediterranean

K. HEISSIG

Fossil rhinoceroses are common faunal elements in Europe and Western Asia during the late Neogene. They are a rather conservative group, not exhibiting many distinctive intrafamily characters, and rather complete skeletal material is needed to make an accurate specific determination. So the considerable later Neogene rhino radiation is largely masked by insufficient materials, making a step-by-step evolutionary reconstruction impossible. Often the most common and completely preserved materials belong to species having a vast chronologic and geographic distribution and providing little valuable information for correlation. However, other rhinoceros lineages are distinctive, and their migration and extinction events are often useful for correlation, albeit a rather coarse one. Only a single multispecific lineage can be tracked across the late Miocene–Pliocene boundary.

Despite the limitations seen for geochronologic correlations (de Bruijn et al. 1992), most rhinoceros species are good ecological indicators. The composition of rhino faunas varies even within the same temporal interval, and over short geographic distances. Unfortunately, rhino paleoecology is not yet well understood, leaving an unrealized treasure of information yet to be developed.

Time-Space Distribution of Middle Miocene–Pliocene European and West Asian Rhinoceroses

The Middle Miocene

During the middle Miocene the Aceratherini and Teleoceratini are the dominant rhinoceroses in both geographic areas considered here. *Brachypotherium brachypus*, *Alicornops simorrensis*, and *Hoploaceratherium*, with its successive species *H. tetradactylum* and *H. bavaricum*, occur throughout the entire region. In Europe the early Miocene holdovers, *Prosantorhinus germanicus*, a small

teleoceratine, and *Plesiaceratherium fahlbuschi*, an Aceratherini, become extinct in MN 6, contracting in their geographic distribution to Central Europe. The single occurrence of *P. mirallesi* in the isolated Central European locality of Georgensgmünd is also its last appearance. Of all these forms, only *Hoploaceratherium* was not already present in the late (MN 5) early Miocene. The successive species of *Hoploaceratherium* can be distinguished only by cranial characters.

The two Rhinocerotini species *Lartetotherium sansaniense* and "*Dicerorhinus*" *steinheimensis* first occurred in Europe during the late early Miocene, and there is no record of them in the Eastern Mediterranean. In the Eastern Mediterranean the influence of Asian faunas is seen by the occurrence of the elasmotherine genus *Begertherium* (Fortelius 1990) with its two successive species, co-occurring with the common genera *Alicornops*, *Hoploaceratherium*, and *Brachypotherium*. The genus *Diceros*, an African immigrant, is found very rarely in Eastern Mediterranean middle Miocene age horizons.

Unfortunately, the middle Miocene is represented by only a few French rhinoceros-bearing localities yielding the most common species. There is no record of rare Western European elasmotherine rhinos after MN 5. Rhinoceroses are not known during this time from Southern Europe, and only from two Southeast European localities, with the only determinable species being *Lartetotherium sansaniense*.

Some of these species have apparently been restricted to very peculiar habitats; their occurrence is not correlated with age or any particular sedimentary environment. Other species occurred over vast geographic areas and for long chronological durations. There is little known about the actual ecology of fossil browsing rhinos. All of the middle Miocene European forms would appear to have been browsers. Thenius (1951) and Heissig (1972) have remarked that *Brachypotherium* did not prefer swampy wood-

FIGURE 27.1 (stratigraphic range and localities — left column group)

age	Prosantorhinus germanicus	Brachypotherium brachypus			Brachypotherium goldfussi	Plesiaceratherium fahlbuschi	Plesiaceratherium mirallesi	Hoploaceratherium tetradactylum		Hoploaceratherium bavaricum	Alicornops simorrensis		
	CE	F	CE	EM	CE	CE	CE	F	EM	CE	F	CE	EM
MN 13													
MN 12-13													
MN 12													
MN 11-12													
MN 11													
MN 10-11													
MN 10													
MN 9-10					39								
MN 9			42		33,34					40-42			
MN 7+8-9										32			
MN 7+8		22; 23	25; 26	27; 28					29		22-24	25	27,28; 30
MN 6-7+8	10		14,18; 19	21									
MN 6	12,14	4	5-7; 9-12	15; 16			9-12; 14	8	4	16		4	13
MN 5-6			1; 2										1

FIGURE 27.1 (stratigraphic range and localities — right column group)

age	Alicornops alfambrensis	Aceratherium incisivum			Chilotherium Subc. intermedium	Chilotherium Acerorhinus zernowi		Chilotherium kiliasi	Chilotherium samium	Chilotherium habereri	Chilotherium kowalevskii		Chilotherium schlosseri
	F	F	CE	SEE	EM	SEE	EM	SEE	EM	EM	SEE	EM	EM
MN 13													78
MN 12-13													77
MN 12													72; 75
MN 11-12													70
MN 11								67		66; 67	62	65,67; 68	
MN 10-11			52					61				60	
MN 10		48	48					60	54	55-58			
MN 9-10			43				46						
MN 9			33,35; 38,39	44	47	45	47						
MN 7+8-9													
MN 7+8													
MN 6-7+8													
MN 6													
MN 5-6													

FIGURE 27.1 Stratigraphic range and localities bearing Aceratheriinae species in France (F), Central Europe (CE), Southeastern Europe (SEE), and the Eastern Mediterranean (EM).

lands. The content of the shrub's wooden parts in the diet of the Aceratherini was probably higher than that preferred by *Lartetotherium sansaniense*. In contrast to the European woodlands, the Eastern Mediterranean semi-hypsodont Elasmotherini clearly inhabited dry areas and were adapted to abrasive food that had relatively poor nutritive quality. The presence of some European browsing species is indicative of the occurrence of restricted bush or woodland environments. The lack of Rhinocerotini corresponds with their preference for very soft plants.

The Late Miocene

Rhinoceros geographic distribution and ecological preferences change dramatically in the late Miocene, most probably because of regional climatic change. Sedimentary environments probably were also affected, because rhino-bearing localities were now numerous in the Eastern Mediterranean, whereas Central European localities younger than the earliest late Miocene (MN 9) are rare.

The change is more radical in the Eastern Mediterranean than in Europe, and the dissimilarities in both regions increase considerably. Whereas no rhino lineage continues across the middle/late Miocene boundary in the Eastern Mediterranean region, the appearance of new elements in Central and Western Europe is followed by a gradual decline in the occurrence of survivors during MN 9 and MN 10. During this time we find the following surviving species: *Brachypotherium goldfussi*, which differs in no essential details from the earlier species *B. brachypus*; *Lartetotherium sansaniense*; *Hoploaceratherium bavaricum*; and *Alicornops simorrensis*. In Western Europe the short-limbed species *Alicornops alfambrensis* Cerdeño 1989 occurs, apparently derived from *A. simorrensis*. There are

The figure is a stratigraphic range chart (genus and species × MN age). Its columns are transcribed below, split by taxonomic group; each sub-table repeats the age column. Region codes: France (F), Central Europe (CE), Southern Europe (SE), Southeastern Europe (SEE), Eastern Mediterranean (EM).

age	Begertherium tekkayai (EM)	Begertherium grimai (EM)	Diceros sp. (EM)	primaevus (EM)	Ceratotherium neumayri (CE)	Ceratotherium neumayri (SEE)	Ceratotherium neumayri (EM)
MN 17							
MN 16-17							
MN 16							
MN 15							
MN 14-15							
MN 14							
MN 13							
MN 12-13						76	78 77
MN 12						74	75
MN 11-12							70-72
MN 11					69		68 65
MN 10-11				61			60
MN 10						54	55-57
MN 9-10							
MN 9						45	47
MN 7+8-9							
MN 7+8		27-31					
MN 6-7+8							
MN 6	15	16	16				
MN 5-6							

age	"Dicerorhinus" steinheimensis (F)	"Dicerorhinus" steinheimensis (CE)	Lartetotherium sansaniensis (F)	Lartetotherium sansaniensis (CE)	Lartetotherium sansaniensis (SEE)	"Dicerorhinus" schleiermacheri (F)	"Dicerorhinus" schleiermacheri (CE)	"Dicerorhinus" schleiermacheri (SEE)
MN 17								
MN 16-17								
MN 16								
MN 15								
MN 14-15								
MN 14								
MN 13								
MN 12-13								
MN 12						73		
MN 11-12								
MN 11								63
MN 10-11							59	
MN 10						53 48-51		
MN 9-10		42					39	
MN 9		33					36,37 33,34	
MN 7+8-9								
MN 7+8	23	25	23	26 25				
MN 6-7+8		17			20			
MN 6		12 10	4	11				
MN 5-6		3		2 1				

age	"Dicerorhinus" megarhinus (F)	"Dicerorhinus" megarhinus (CE)	"Dicerorhinus" megarhinus (SE)	"Dicerorhinus" megarhinus (SEE)	"Dicerorhinus" jeanvireti (F)	"Dicerorhinus" jeanvireti (CE)	"Dicerorhinus" jeanvireti (SE)	"Dicerorhinus" miguelcrusafonti (F)
MN 17								
MN 16-17								
MN 16					101-103	108	104-107	
MN 15	98	100 99						98
MN 14-15	84-89	90	91-96	97				
MN 14	80-82		83					
MN 13								
MN 12-13								
MN 12								
MN 11-12								
MN 11								
MN 10-11								
MN 10								
MN 9-10								
MN 9								
MN 7+8-9								
MN 7+8								
MN 6-7+8								
MN 6								
MN 5-6								

age	Stephanorhinus pikermiensis (SEE)	Stephanorhinus pikermiensis (EM)	Stephanorhinus etruscus (F)	Stephanorhinus etruscus (SE)
MN 17			112 113	
MN 16-17			109-111	105
MN 16			102	106
MN 15				
MN 14-15				
MN 14				
MN 13				
MN 12-13	76			
MN 12				
MN 11-12				
MN 11				
MN 10-11				
MN 10				
MN 9-10				
MN 9				
MN 7+8-9				
MN 7+8				
MN 6-7+8				
MN 6				
MN 5-6				

FIGURE 27.2 Stratigraphic range and localities bearing Rhinocerotinae species in France (F), Central Europe (CE), Southern Europe (SE), Southeastern Europe (SEE), and the Eastern Mediterranean (EM).

only two immigrants: *Aceratherium incisivum* and *"Dicerorhinus" schleiermacheri*, which continue together with the latest survivor, *Brachypotherium*, into MN 10 and later. The last appearance of these forms is masked by their sparse fossil record. Only *"Dicerorhinus" schleiermacheri* transgresses the late Miocene/early Pliocene boundary, continuing in the early Pliocene as *"D." megarhinus*.

The Elasmotherini and other middle Miocene lineages became extinct in the Eastern Mediterranean at this time and were replaced by an Asian-African rhinoceros fauna.

The Central Asian lineage *Chilotherium* s.l. and the African lineage *Ceratotherium neumayri* invade Southeastern Europe and the eastern margin of Central Europe at this time.

There are several occurrences of *Chilotherium* reported from Italy, France, Greece, and Spain, but currently the evidence can only sufficiently support its occurrence in Greece. The French and Spanish remains have been assigned by Cerdeño (1989) to *Alicornops alfambrensis*. The chronologically long-ranging genus *Stephanorhinus pikermiensis* first occurs in terminal late Miocene horizons of Southeastern Europe and portions of the Eastern Mediterranean. The single record of a true *Diceros* from Kayadibi suggests some ecological differences within the Eastern Mediterranean. *Ceratotherium neumayri* shows some size increase, together with a slight tendency of progressive hypsodonty during the late Miocene interval.

Three subgenera of *Chilotherium* arise from the second Asian Aceratheriinae radiation. Two subgenera there can be traced back to the middle Miocene. Their arrival in the Eastern Mediterranean region begins with *Chilotherium (Subchilotherium) intermedium*, recorded from only one Anatolian locality. The few undetermined remains from France and Spain may probably belong to this species. *Chilotherium (Acerorhinus)*, also known from Eastern Asian middle Miocene age horizons, lacks the specializations typical of *Chilotherium* s.s. It occurs at several Anatolian localities with the species *zernowi*, known also from Eastern Europe. The Anatolian record seems to be confined to MN 9 and MN 10, possibly extending into MN 11. Its Eastern Asian relatives are more advanced.

The evolutionary relationships of species included within *Chilotherium* s.s. are not yet well understood. The most primitive form dentally, *Chilotherium samium*, first occurs in Anatolia possibly as early as late MN 10, but certainly by MN 11. The holotype *Ch. samium* comes from an unknown horizon at Samos. *Chilotherium kiliasi* (Geraads and Koufos 1989) occurs at Pentalophos (MN 10). This species was originally described as being referable to *Aceratherium*, but indeed it is typical of a primitive *Chilotherium* and may be closely related to *Chilotherium samium*, the only species with female mandibles not broadened anteriorly. It has to be added that the larger form of this species, reported in Heissig (1975) from Çorak Yerler, is possibly another species belonging to the same lineage.

Chilotherium kowalevskii is the only species of this time period exhibiting characters diverging from other members of this group. *Chilotherium habereri*, also known from China, first occurs a little earlier in time (MN 10) and is possibly closely related to the Maragheh species *Ch. persiae* (for its stratigraphic and chronologic record at Maragheh, see Bernor et al., this volume). It is replaced in MN 12 by the latest and largest *Chilotherium* species, *Ch. schlosseri*, which may very well be derived from *Ch. habe-*

reri; however, intermediate forms are known only from Eastern Asia. All of these species have near relatives in Eastern and Central Asia, but the lineages are difficult to disentangle. *Chilotherium habereri* seems to be representative of the central stock, from which at least the larger and more hypsodont middle and late Turolian species *Ch. anderssoni* (the Chinese genotype species) and *Ch. schlosseri* are derived. The Maragheh species *Ch. persiae* is also a close relative.

Despite our poor understanding of European rhinoceros ecology, rhino faunal development is consistent with the evolution of a continuous woodland environment. The extinction of most rhino species during the late Miocene corresponds to progressive cooling and development of more temperate woodlands.

To the south and southeast we find a gradual change in landscape to increasingly open habitats and, at the same time, the extension of this more Mediterranean type of climate to the north. Nevertheless, we may conclude fundamental differences between the late and middle Miocene rhino faunas by the distinct faunal break found in the Eastern Mediterranean region. The dominant form, *Ceratotherium neumayri*, was at least partly a grass eater, whereas the genus *Chilotherium* most likely occupied diverse ecological niches, indicated by the number of species. Certainly there were a number of non-gramimean herbs that formed their diet. The most brachydont *Chilotherium (Acerorhinus)* may have been the only widespread browser in this region. Only one locality has yielded the African browsing rhinoceros genus *Diceros*, indicative of more densely vegetated areas.

The Pliocene

The basal Pliocene marks yet another turnover in rhinoceros faunas. Very few late Miocene rhinoceros species persist during the Pliocene in the Eastern Mediterranean. Also, Central and Southeastern Europe have very few rhinoceros-bearing localities. Quite to the contrary, Western and Southern Europe have numerous Pliocene rhinoceros localities.

The Mio–Pliocene boundary includes the principal extinction event for European rhinoceroses. All Aceratherini and Teleoceratini disappear from Europe and the Eastern Mediterranean. From MN 14 onward, *Brachypotherium*, *Diceros*, and *Ceratotherium* are restricted to Africa. In Europe only one major lineage has survived the Mio–Pliocene boundary: *"Dicerorhinus" schleiermacheri*. This lineage evolves through the successive Pliocene species: *"D." schleiermacheri—"D." megarhinus—D. jeanvireti*.

"Dicerorhinus" megarhinus is the only known basal Pliocene rhinoceros. *"Dicerorhinus" miguelcrusafonti*, a species of unknown origin, first appears during MN 15 and has a geographic distribution restricted to the Iberian Peninsula

and southern France. "*Diceros*" *megarhinus* is replaced at the beginning of the late Pliocene by its descendant, "*D.*" *jeanvireti*. "*D.*" *jeanvireti* coexists for a short time with the smaller form *Stephanorhinus etruscus*, which itself is probably derived from the Turolian form *St. pikermiensis*. *Stephanorhinus etruscus* is the only rhinoceros species that survives into the Pleistocene.

We have very little knowledge of the Eastern Mediterranean rhino fauna during this temporal interval. Most specimens are insufficient for determination. Some bones from the Turkish MN 17 locality Gülyazi (Gauss/Matuyama boundary, ca. 2.6 Ma; Bernor and Lipscomb 1991; Steininger et al. this volume) conform in their size to *St. etruscus*, but the morphological characters of this form are atypical for the species. Two lower molars of undetermined age from Kobasi, provisionally referred by Heissig (1973) to *Stephanorhinus ringstroemi*, may actually be referable to either "*D.*" *schleiermacheri* or "*D.*" *megarhinus*.

Nearly all of the European and Western Asian Pliocene rhinos had brachyodont cheek teeth and were browsers. The only rhino with somewhat higher crowns was the Western European species "*Dicerorhinus*" *miguelcrusafonti*, suggesting an adaptation to a drier Mediterranean climate in Spain and southern France during MN 15. Knowledge of Eastern Mediterranean Pliocene rhino faunas is insufficient to make meaningful comparisons with the rest of Europe.

LITERATURE CITED

Alpagut, B., P. Andrews, and L. Martin. 1989. Miocene paleoecology of Paşalar, Turkey. In *European Neogene Mammal Chronology*, ed. E. H. Lindsay, V. Fahlbusch, and P. Mein, pp. 443–59. New York: Plenum.

Arambourg, C. and J. Pivetau. 1929. Les vertébrés du Pontien de Salonique. *Annales de Paléontologie* 18:59–82.

Bakalov, P. and I. Nikolov. 1962. Mammifères Tertiaires (Bulg., frz. Resumé). *Fossilite na Bulgarija* 10:1–162.

Bascieri, F. and A. G. Segre. 1957. Notizie sul ritrovamento di fauna a Rinoceronte etrusco e macairodo all'Argentario (Prov. di Grosseto). *Quaternaria* 4:195–97.

Bernor, R. L. and D. Lipscomb. 1991. The systematic position of "*Plesiohipparion*" aff. *huangheense* (Equidae, Hipparionini) from Gülyazi, Turkey. *Mitteilungem der Bayerischen Staatssammlung für Paläontologie u. historische Geologie* 31:107–23.

Bernor, R. L., N. Solounias, C. S. Swisher III, and J. A. Van Couvering. This volume. The correlation of three classical "Pikermian" faunas—Maragheh, Samos, and Pikermi—with the European MN unit system.

Cerdeño, E. 1989. Revision de la sistematica de los Rinocerontes del Neogeno de España. Thesis Doctoral 306/89: 1–429, Madrid.

Bonis, L. de, G. Bouvrain, D. Geraads, and G. Koufos. 1991. Composition and species diversity in late Miocene faunal assemblages of Northern Greece. *Bulletin Geologie Société, Greece* 25:395–404.

Bruijn, H. de, R. Daams, G. Daxner-Höck, V. Fahlbusch, L. Ginsburg, P. Mein, and J. Morales. 1992. Report on the RCMNS working group on fossil mammals, Reisensburg 1990. *Newsletters on Stratigraphy* 26:65–118.

Fejfar, O. 1964. The lower Villafranchian vertebrates from Hajnacka near Filakovo in Southern Slovakia. *Rozpravy Ustredniho Ustavu Geologickeho* 30:1–115.

Fortelius, M. 1990. Less common ungulate species from Paşalar, middle Miocene of Anatolia (Turkey). *Journal of Human Evolution* 19:489–508.

Fortelius, M. and K. Heissig. 1989. The phylogenetic relationships of the Elasmotherini (Rhinocerotidae, Mamm.). *Mitteilungen der Bayerischen Staatssammlung für Paläontologie und historische Geologie* 29:227–33.

Geraads, D. 1988. Révision des Rhinocerotinae (Mammalia) du Turolien de Pikermi; Comparaison avec les formes voisines. *Annales de Paléontologie (Vertébrés)* 74:13–41.

Geraads, D. and G. Koufos. 1989. Upper Miocene Rhinocerotidae from Pentalophos-1, Macedonia, Greece. *Palaeontographica* A 210:151–68.

Ginsburg, L. 1974. Les rhinocérotidés du Miocène de Sansan (Gers). *Comptes Rendus Académie Sciences, Paris* 278:597–600.

Guerin, C. 1980. Les Rhinocéros (Mammalia, Perissodactyla) du Miocène terminal au Pleistocène supérieur en Europe occidentale. *Documents Laboratoire Geologie, Lyon* 79:1–1182.

Heissig, K. 1972. Geologische und paläontologische Untersuchungen im Tertiär von Pakistan 5: Rhinocerotidae aus den unteren und mittleren Siwalik Schichten. *Abhandlungen Bayerische Akademie Wissenschaften Mathematisch-naturwissenschaft, Kl. München* 152:1–112.

——. 1975. Rhinocerotidae aus dem Jungetertiär Anatoliens. In *Die Gliederung des höheren Jungtertiärs und Altquartärs in der Türkei nach Vertebraten und ihre Bedeutung für die internationale Neogen-Stratigraphie*, ed. O. Sickenberg et al., pp. 145–51. *Geologisches Jahrbuch, B, Hannover* 15:3–167.

——. 1976. Rhinocerotidae (Mammalia) aus der Anchitherium-Fauna Anatoliens. *Geologisches Jahrbuch, B, Hannover* 19:3–121.

——. 1984. Nashornverwandte (Rhinocerotidae) aus der Oberen Susswassermolasse und ihre Bedeutung für deren Lokalstratigraphie. *Inzburger Heimathefte* 2 (August Wetzler Memorial Volume): 62–74.

——. 1989. Neue Ergebnisse zur Stratigraphie der Mittleren Serie der Oberen Süsswassermolasse Bayerns. *Geologica Bavarica* 94:239–57.

Heizmann, E. 1973. *Die Carnivoren des Steinheimer Beckens*. Palaeontographica Supplement, B. 8:1–95.

Hünermann, K. A. 1989. Die Nashornskelette (*Aceratherium incisivum* KAUP 1832) aus dem Jungtertiär vom Höwenegg im Hegau (Südwestdeutschland). *Andrias* 6:5–116.

Kowalski, K. 1990. Stratigraphy of Neogene Mammals of Poland. In *European Neogene Mammal Chronology*, ed. E. H. Lindsay, V. Fahlbusch, and P. Mein, pp. 193–209. New York: Plenum.

Osborn, H. F. 1900. Phylogeny of the Rhinoceroses of Europe. *Bulletin of the American Museum of Natural History* 13:229–67.

Pavlović, M. B. 1963. *Dicerorhinus* aff. *sansaniensis* (LARTET) aus dem Jungmiozän Serbiens (Serbian, German summary). *Annales Geologie Peninsule Balkanique* 30:61–75.

Schlosser, M. 1902. Beiträge zur Kenntnis der Säugethierreste aus den Süddeutschen Bohnerzen. *Geologie, Palaontologische Abhandlungen N.F.* 5:117–258.

Steininger, F. F., W. A. Berggren, D. V. Kent, R. L. Bernor, S. Sen, and J. Agusti. This volume. Circum Mediterranean Neogene (Miocene and Pliocene) Marine-Continental Chronologic Correlations of European Mammal Units.

Thenius, E. 1951. Die Rhinocerotiden des Wiener Jungtertiärs. Anzeige mathematische u. naturwissenschaften Kl. *Österreich Akademie Wissenschaften, 1951, Wien* 13:343–47.

Name	Author	Locality	References	Seen	Frequency
		Aceratheriinae—Teleoceratini			
Prosantorhinus germanicus	Wang 1929	Georgensgmünd	Heissig 1972	y	c
Prosantorhinus germanicus	Wang 1929	Thannhausen	Heissig 1984	y	c
Prosantorhinus germanicus	Wang 1929	Derching	Heissig 1984	y	c
Prosantorhinus germanicus	Wang 1929	Stätzling	Heissig 1984	y	c
Prosantorhinus germanicus	Wang 1929	Friedberg	Heissig 1984	y	c
Brachypotherium brachypus	Lartet 1837	Simorre	Guérin 1980	y	d
Brachypotherium brachypus	Lartet 1837	Sansan	Guérin 1980	n	s
Brachypotherium brachypus	Lartet 1837	La Grive	Guérin, 1980	n	?
Brachypotherium brachypus	Lartet 1837	Thannhausen	Heissig 1984	y	s
Brachypotherium brachypus	Lartet 1837	Derching	Heissig 1984	y	d
Brachypotherium brachypus	Lartet 1837	Stätzling	Heissig 1984	y	c
Brachypotherium brachypus	Lartet 1837	Friedberg	Heissig 1984	y	c
Brachypotherium brachypus	Lartet 1837	Unterzolling	Heissig 1984	y	c
Brachypotherium brachypus	Lartet 1837	Enghausen	Heissig 1989	y	c
Brachypotherium brachypus	Lartet 1837	Ziemetshausen ib	Heissig 1984	y	c
Brachypotherium brachypus	Lartet 1837	Prambach	Heissig 1984	y	s
Brachypotherium brachypus	Lartet 1837	Langenneufnach	Heissig 1984	y	s
Brachypotherium brachypus	Lartet 1837	Auloh b. Schönbrunn	Heissig 1984	y	s
Brachypotherium brachypus	Lartet 1837	Steinheim a. Albuch	Guérin 1980	y	c
Brachypotherium brachypus	Lartet 1837	Massenhausen	Heissig 1984	y	d
Brachypotherium brachypus	Lartet 1837	Wartenberg	Heissig 1984	y	s
Brachypotherium brachypus	Lartet 1837	Meßkirch	Schlosser 1902	y	c
Brachypotherium brachypus	Lartet 1837	Heudorf	Schlosser 1902	n	s
Brachypotherium brachypus	Lartet 1837	Przeworno	Kowalski 1990	y	s
Brachypotherium brachypus	Lartet 1837	Opole	Kowalski 1990	n	s
Brachypotherium brachypus	Lartet 1837	Pasalar	Alpagut, Andrews & Martin 1990	n	s
Brachypotherium brachypus	Lartet 1837	Candir	Heissig 1976	y	s
Brachypotherium brachypus	Lartet 1837	Tüney	Heissig 1976	y	s
Brachypotherium brachypus	Lartet 1837	Sofca	Heissig 1976	y	s
Brachypotherium brachypus	Lartet 1837	Catakbagyaka	Heissig 1976	y	c
Brachypotherium goldfussi	Kaup 1834	Eppelsheim	Guérin 1980	y	s
Brachypotherium goldfussi	Kaup 1834	Gauweinheim	Guérin 1980	y	c
Brachypotherium goldfussi	Kaup 1834	Melchingen	Schlosser 1902	n	s
		Aceratheriinae—Aceratherini			
Plesiaceratherium fahlbuschi	Heissig 1972	Thannhausen	Heissig 1984	y	c
Plesiaceratherium fahlbuschi	Heissig 1972	Stätzling	Heissig 1984	y	c
Plesiaceratherium fahlbuschi	Heissig 1972	Friedberg	Heissig 1984	y	c
Plesiaceratherium fahlbuschi	Heissig 1972	Mainburg	Heissig 1984	y	c
Plesiaceratherium fahlbuschi	Heissig 1972	Unterzolling	Heissig 1984	y	c
Plesiaceratherium mirallesi	Crusafont et al. 1955	Georgensgmünd	Yan & Heissig 1986	y	c
Hoploaceratherium tetradactylum	Lartet 1837	Sansan	Guérin 1980	y	d
Hoploaceratherium aff. tetradactylum	Lartet 1837	Candir	Heissig 1976	y	c
Hoploaceratherium aff. tetradactylum	Lartet 1837	Yeni Eskihisar	Heissig 1976	y	s
Hoploaceratherium bavaricum	Stromer 1902	Niedernkirchen	Heissig 1984	y	s
Hoploaceratherium bavaricum	Stromer 1902	Wartenberg	Heissig 1984	y	s
Hoploaceratherium bavaricum	Stromer 1902	München	Heissig 1984	y	c
Hoploaceratherium bavaricum	Stromer 1902	Oberföhring	Heissig 1984	y	s
Alicornops simorrensis	Lartet 1851	Simorre	Guérin 1980	y	c
Alicornops simorrensis	Lartet 1851	Villefranche d. Astarac	Guérin 1980	y	c
Alicornops simorrensis	Lartet 1851	Sansan	Guérin 1980	y	s
Alicornops simorrensis	Lartet 1851	Tutzing	Heissig 1984	y	s
Alicornops simorrensis	Lartet 1851	Steinheim a. Albuch	Heizmann 1973	y	c
Alicornops simorrensis	Lartet 1851	Meßkirch	Schlosser 1902	n	s
Alicornops simorrensis	Lartet 1851	La Grive	Guérin 1980	y	s
Alicornops simorrensis	Lartet 1851	Sofca	Heissig 1976	y	c
Alicornops simorrensis	Lartet 1851	Catakbagyaka	Heissig 1976	y	c
Alicornops simorrensis	Lartet 1851	Yaylacilar	Heissig 1976	y	c
Alicornops alfambrensis	Cerdeño 1989	Montredon	Cerdeño 1989	n	c
Aceratherium incisivum	Kaup 1832	Eppelsheim	Guérin 1980	y	c
Aceratherium incisivum	Kaup 1832	Höwenegg	Hünermann 1989	y	d
Aceratherium incisivum	Kaup 1832	Wien, Belvedere	Guérin 1980	y	s
Aceratherium incisivum	Kaup 1832	Oppenheim	Guérin 1980	y	c
Aceratherium incisivum	Kaup 1832	Melchingen	Schlosser 1902	n	s
Aceratherium incisivum	Kaup 1832	Montredon	Guérin 1980	y	c
Aceratherium incisivum	Kaup 1832	Priay	Guérin 1980	y	c
Aceratherium incisivum	Kaup 1832	Merischleri	Bakalov & Nikolov 1962	n	s
Chilotherium (Subchil.) intermedium	Lydekker 1884	Esme-Akcaköy	Heissig 1975	y	s
Chilotherium (Chil.) samium	Weber 1905	Kayadibi	Heissig 1975	y	c
Chilotherium (Chil.) samium	Weber 1905	Corak Yerler	Heissig 1975	y	s

Name	Author	Locality	References	Seen	Frequency
Chilotherium (Chil.) habereri	Schlosser 1903	Kücükcekmece	Heissig 1975	y	s
Chilotherium (Chil.) habereri	Schlosser 1903	Gülpinar	Heissig 1975	y	s
Chilotherium (Chil.) habereri	Schlosser 1903	Basbereket	Heissig 1975	y	s
Chilotherium (Chil.) habereri	Schlosser 1903	Kavakdere	Heissig 1975	y	c
Chilotherium (Chil.) habereri	Schlosser 1903	Selcik	Heissig 1975	y	s
Chilotherium (Chil.) habereri	Schlosser 1903	Samos	Heissig 1975	y	c
Chilotherium (Chil.) kowalevskii	Pavlov 1913	Corak Yerler	Heissig 1975	y	c
Chilotherium (Chil.) kowalevskii	Pavlov 1913	Garkin	Heissig 1975	y	c
Chilotherium (Chil.) kowalevskii	Pavlov 1913	Karacahasan	Heissig 1975	y	s
Chilotherium (Chil.) kowalevskii	Pavlov 1913	Losengrad	Bakalov & Nikolov 1962	n	s
Chilotherium (Chil.) schlosseri	Weber 1905	Mahmutgazi	Heissig 1975	y	s
Chilotherium (Chil.) schlosseri	Weber 1905	Kinik	Heissig 1975	y	c
Chilotherium (Chil.) schlosseri	Weber 1905	Eski Bayirköy	Heissig 1975	y	s
Chilotherium (Chil.) schlosseri	Weber 1905	Amasya	Heissig 1975	y	s

Rhinocerotinae—Elasmotherini

Name	Author	Locality	References	Seen	Frequency
Begertherium tekkayai	Heissig 1974	Pasalar	Fortelius & Heissig 1989	y	d
Begertherium grimmi	Heissig 1974	Candir	Fortelius & Heissig 1989	y	c
Begertherium grimmi	Heissig 1974	Sofca	Heissig 1976	y	c
Begertherium grimmi	Heissig 1974	Yaylacilar	Heissig 1976	y	d
Begertherium grimmi	Heissig 1974	Catakbagyaka	Heissig 1976	y	c
Begertherium grimmi	Heissig 1974	Zivra	Heissig 1976	y	s
Begertherium grimmi	Heissig 1974	Yeni Eskihisar	Heissig 1976	y	s

Rhinocerotinae—Rhinocerotini

Name	Author	Locality	References	Seen	Frequency
Diceros primaevus	Arambourg 1959	Kayadibi	Heissig 1976	y	s
Ceratotherium neumayri	Osborn 1900	Hauskirchen	Thenius 1956	y	s
Ceratotherium neumayri	Osborn 1900	Pikermi	Geraads 1989	y	c
Ceratotherium neumayri	Osborn 1900	Samos	Geraads 1989	y	d
Ceratotherium neumayri	Osborn 1900	Thessaloniki	Arambourg & Pivetau 1929	n	c
Ceratotherium neumayri	Osborn 1900	Pentalophos	Geraads & Koufos 1990	n	c
Ceratotherium neumayri	Osborn 1900	Kalimanzi	Bakalov & Nikolov 1962	y	c
Ceratotherium neumayri	Osborn 1900	Esme Akcaköy	Heissig 1976	y	d
Ceratotherium neumayri	Osborn 1900	Kavakdere	Heissig 1976	y	c
Ceratotherium neumayri	Osborn 1900	Gülpinar	Heissig 1976	y	c
Ceratotherium neumayri	Osborn 1900	Basbereket	Heissig 1976	y	s
Ceratotherium neumayri	Osborn 1900	Balciklidere	Heissig 1976	y	s
Ceratotherium neumayri	Osborn 1900	Corak Yerler	Heissig 1976	y	c
Ceratotherium neumayri	Osborn 1900	Garkin	Heissig 1976	y	c
Ceratotherium neumayri	Osborn 1900	Karacahasan	Heissig 1976	y	s
Ceratotherium neumayri	Osborn 1900	Mahmutgazi	Heissig 1976	y	d
Ceratotherium neumayri	Osborn 1900	Kinik	Heissig 1976	y	c
Ceratotherium neumayri	Osborn 1900	Eski Bayirköy	Heissig 1976	y	c
Ceratotherium neumayri	Osborn 1900	Amasya	Heissig 1976	y	s
"Dicerorhinus" steinheimensis	Jäger 1839	Steinheim a. Albuch	Guérin 1980	y	s
"Dicerorhinus" steinheimensis	Jäger 1839	Thalhausen	Heissig 1984	y	s
"Dicerorhinus" steinheimensis	Jäger 1839	Derndlmühle	Heissig 1984	y	s
"Dicerorhinus" steinheimensis	Jäger 1839	Eibiswald	Thenius 1951	y	c
"Dicerorhinus" steinheimensis	Jäger 1839	Thannhausen	Heissig 1984	y	s
"Dicerorhinus" steinheimensis	Jäger 1839	Derching	Heissig 1984	y	s
"Dicerorhinus" steinheimensis	Jäger 1839	Wartenberg	Heissig 1984	y	s
"Dicerorhinus" steinheimensis	Jäger 1839	Eppelsheim	Osborn 1900	y	s
"Dicerorhinus" steinheimensis	Jäger 1839	La Grive	Guérin 1980	n	s
Lartetotherium sansaniense	Lartet 1851	Sansan	Ginsburg 1974	y	c
Lartetotherium sansaniense	Lartet 1851	La Grive	Guérin 1980	y	s
Lartetotherium sansaniense	Lartet 1851	Stätzling	Heissig 1984	y	s
Lartetotherium sansaniense	Lartet 1851	Massenhausen	Heissig 1984	y	c
Lartetotherium sansaniense	Lartet 1851	Hirblingen	Heissig 1984	y	s
Lartetotherium sansaniense	Lartet 1851	Lurtz	Heissig 1984	y	s
Lartetotherium sansaniense	Lartet 1851	Meßkirch	Schlosser 1902	n	c
Lartetotherium sansaniense	Lartet 1851	Heudorf	Schlosser 1902	n	s
Lartetotherium sansaniense	Lartet 1851	Pozlata	Pavlovic 1963	n	s
"Dicerorhinus" schleiermacheri	Kaup 1832	Eppelsheim	Guérin 1980	y	c
"Dicerorhinus" schleiermacheri	Kaup 1832	Gauweinheim	Guérin 1980	y	c
"Dicerorhinus" schleiermacheri	Kaup 1832	Esselborn	Guérin 1980	y	c
"Dicerorhinus" schleiermacheri	Kaup 1832	Dorn Dürkheim	—	y	s
"Dicerorhinus" schleiermacheri	Kaup 1832	Melchingen	Schlosser 1902	n	s
"Dicerorhinus" schleiermacheri	Kaup 1832	Salmendingen	Schlosser 1902	n	s
"Dicerorhinus" schleiermacheri	Kaup 1832	Soblay	Guérin 1980	n	d
"Dicerorhinus" schleiermacheri	Kaup 1832	Aubignas	Guérin 1980	n	c
"Dicerorhinus" schleiermacheri	Kaup 1832	Charmouille	Guérin 1980	n	c
"Dicerorhinus" schleiermacheri	Kaup 1832	Luberon	Guérin 1980	y	c
"Dicerorhinus" schleiermacheri	Kaup 1832	Lyon-Croix-Rousse	Guérin 1980	n	s
"Dicerorhinus" schleiermacheri	Kaup 1832	Montredon	Guérin 1980	y	c
"Dicerorhinus" schleiermacheri	Kaup 1832	Vienne	Guérin 1980	y	c
"Dicerorhinus" schleiermacheri	Kaup 1832	Dobrisch	Bakalov & Nikolov 1962	n	c

Name	Author	Locality	References	Seen	Frequency
"Dicerorhinus" megarhinus	de Christol 1834	Montpellier	Guérin 1980	y	d
"Dicerorhinus" megarhinus	de Christol 1834	St. Laurent-des-Arbres	Guérin 1980	n	d
"Dicerorhinus" megarhinus	de Christol 1834	Celleneuve	Guérin 1980	n	s
"Dicerorhinus" megarhinus	de Christol 1834	Ille-sur-Têt	Guérin 1980	n	s
"Dicerorhinus" megarhinus	de Christol 1834	Lens-Lestang	Guérin 1980	n	s
"Dicerorhinus" megarhinus	de Christol 1834	Meyrargues	Guérin 1980	n	c
"Dicerorhinus" megarhinus	de Christol 1834	Millas	Guérin 1980	n	c
"Dicerorhinus" megarhinus	de Christol 1834	Perpignan	Guérin 1980	y	d
"Dicerorhinus" megarhinus	de Christol 1834	Pézenas	Guérin 1980	n	s
"Dicerorhinus" megarhinus	de Christol 1834	Vendargues	Guérin 1980	n	s
"Dicerorhinus" megarhinus	de Christol 1834	Antwerpen	Guérin 1980	n	c
"Dicerorhinus" megarhinus	de Christol 1834	Baccinello	Guérin 1980	n	s
"Dicerorhinus" megarhinus	de Christol 1834	Dusino/San Paolo	Guérin 1980	n	s
"Dicerorhinus" megarhinus	de Christol 1834	Fango Nero	Guérin 1980	n	s
"Dicerorhinus" megarhinus	de Christol 1834	Imola	Guérin 1980	n	s
"Dicerorhinus" megarhinus	de Christol 1834	Monte Biancano	Guérin 1980	n	s
"Dicerorhinus" megarhinus	de Christol 1834	Monte Giogo	Guérin 1980	n	c
"Dicerorhinus" megarhinus	de Christol 1834	Palaia	Guérin 1980	n	s
"Dicerorhinus" megarhinus	de Christol 1834	Wölfersheim	Guérin 1980	n	s
"Dicerorhinus" megarhinus	de Christol 1834	Chrabarsko	Bakalov & Nikolov 1962	y	c
"Dicerorhinus" jeanvireti	Guérin 1972	Vialette	Guérin 1980	y	d
"Dicerorhinus" jeanvireti	Guérin 1972	Desnes	Guérin 1980	n	c
"Dicerorhinus" jeanvireti	Guérin 1972	Capannoli	Guérin 1980	n	s
"Dicerorhinus" jeanvireti	Guérin 1972	Perrier-Etouaires	Guérin 1980	n	c
"Dicerorhinus" jeanvireti	Guérin 1972	Vincent	Gúerin 1980	n	s
"Dicerorhinus" jeanvireti	Guérin 1972	Dusino	Guérin 1980	n	c
"Dicerorhinus" jeanvireti	Guérin 1972	Montopoli	Guérin 1980	n	c
"Dicerorhinus" jeanvireti	Guérin 1972	Villafranca d'Asti	Guérin 1980	n	c
"Dicerorhinus" jeanvireti	Guérin 1972	Hajnacka	Fejfar 1964	n	c
"Dicerorhinus" miguelcrusafonti	Guérin & Santafé 1978	Perpignan	Cerdeño 1989	n	c
Stephanorhinus pikermiensis	Toula 1906	Pikermi	Geraads 1988	y	c
Stephanorhinus pikermiensis	Toula 1906	Samos	Geraads 1988	y	s
Stephanorhinus ringstroemi	Arambourg 1959	Köprübasi	Heissig 1975	y	s
Stephanorhinus etruscus	Falconer 1859	Valdarno sup.	Guérin 1980	n	c
Stephanorhinus etruscus	Falconer 1859	Argentario	Bascieri & Segre 1957	n	?
Stephanorhinus etruscus	Falconer 1859	Barberino d.Mugello	Guérin 1980	n	c
Stephanorhinus etruscus	Falconer 1859	Borgo San Lorenzo	Guérin 1980	n	s
Stephanorhinus etruscus	Falconer 1859	Castelnuovo Berardenga	Guérin 1980	n	s
Stephanorhinus etruscus	Falconer 1859	Ferrere d'Asti	Guérin 1980	n	s
Stephanorhinus etruscus	Falconer 1859	Incisa Belbo	Guérin 1980	n	s
Stephanorhinus etruscus	Falconer 1859	Montopoli	Guérin 1980	n	s
Stephanorhinus etruscus	Falconer 1859	Olivola	Guérin 1980	n	d
Stephanorhinus etruscus	Falconer 1859	San Clemente a Signano	Guérin 1980	n	s
Stephanorhinus etruscus	Falconer 1859	Villafranca d'Asti	Guérin 1980	n	s
Stephanorhinus etruscus	Falconer 1859	Chagny	Guérin 1980	n	s
Stephanorhinus etruscus	Falconer 1859	Perrier-Etouaires	Guérin 1980	n	c
Stephanorhinus etruscus	Falconer 1859	Puimoisson	Guérin 1980	n	s
Stephanorhinus etruscus	Falconer 1859	Puy-en-Velay	Guérin 1980	n	c
Stephanorhinus etruscus	Falconer 1859	Saint-Vallier	Guérin 1980	n	d
Stephanorhinus etruscus	Falconer 1859	Saint-Vidal	Guérin 1980	n	s

28

Middle and Late Miocene Suoidea of Central Europe and the Eastern Mediterranean: Evolution, Biogeography, and Paleoecology

M. FORTELIUS, J. VAN DER MADE, AND R. L. BERNOR

The good stratigraphic and paleogeographic record of the Paratethys realm (Thenius 1959; Rögl and Steininger 1983, 1984; Bernor et al. 1988; Steininger et al. 1989; Rögl et al., in press; Bernor et al. 1993) and the rich terrestrial fossil record found around its margins provide an opportunity to study fossil land mammal and paleofloral assemblages in relationship to a background of changing land-sea configurations. The Eastern Mediterranean–Southwest Asian area is even richer in diverse continental vertebrates that are often associated with substantial lithostratigraphic sequences and volcanic debris suitable for radioisotopic analysis. The intensity of recent study on later Neogene paleontology, stratigraphy, and geochronology has created a rich data base for us to utilize in our effort to synthesize the evolutionary history of middle and late Miocene terrestrial mammals.

The Suoidea constitute a relatively common and diverse group, ecologically as well as taxonomically, that underwent moments of rapid turnover during the middle and late Miocene. Suoids are also sufficiently well known to allow judicious use of the published literature in the few places where one of us has not been able to study original specimens. We summarize data on the geographic and temporal distribution of this group and reconstruct patterns of change in species diversity, turnover, and extinction.

The area canvassed in this review extends from Germany and Poland in the northwest to Iran in the southeast. The temporal range is MN 6 to MN 13 (Mein 1975; 1990), circa 15.5 to 5 Ma (Steininger et al. 1989). During most of the time considered here the Paratethys paleoenvironments changed dramatically. At its geographic maximum, the Paratethys extended from the Rhône Valley north and eastward through southern Germany, and eastward along the north flank of the Alps to the Caucasus and beyond as far as the Aral Sea. Its progressive areal contraction was provoked by the Alpine Orogeny and was presumably a major cause of the attendant episodic shifts in conti-

nental climates and associated ecosystems. These events had direct consequences for individual marine and terrestrial species (Rögl and Steininger 1983, 1984; Bernor 1983, 1984; Bernor et al. 1988, 1990; Potts and Behrensmeyer et al. 1992; Rögl et al. 1993). An important aspect of the later Miocene Paratethys is that relatively minor global sea-level changes had a major effect on the distribution of the Paratethyan water masses (Rögl and Steininger 1984), and therefore may have affected the dispersal potential of land mammals in the region.

The base of the Langhian Stage, correlative with the late Orleanian (= MN 5; Bernor, in Steininger et al. this volume), records the final Indomalaysian marine transgression (Rögl and Steininger 1983) into the Vienna Basin. This marine incursion nourished a diverse subtropical biome that included the latest-occurring crocodiles and palm trees. Rupture of this marine connection led to the Serravallian (= Badenian-Sarmatian of the Paratethys) regression and development of progressive brackish water conditions in the Paratethys realm. The succeeding late Miocene interval witnessed marked paleoenvironmental shifts. First, in the Pannonian Stage, brackish water conditions were replaced by freshwater fluviatile and lake ecosystems that included paleogeographic connections between the Vienna Basin, southward to the Pannonian Basin, and then southeastward through the "Iron Gates" into the very large Eastern Paratethys (Rögl and Steininger 1983, 1984; Bernor et al. 1988, 1993; Rögl et al., in press). The latest Miocene Pontian Stage marks an extreme regression and desiccation of the Paratethys system and is correlative with the Mediterranean realm's Messinian Salinity Crisis.

Being largely adapted to subtropical and warm temperate forests and woodlands, the later Neogene Suoidea are sensitive indicators of environmental change, which would appear to have been progressive, albeit episodic. Roughly speaking, the time considered here includes two desiccation episodes in Europe, the Paratethys middle Badenian

(= Serravallian Stage; middle Miocene) Salinity Crisis and the paleo-Mediterranean/Paratethyan Messinian (latest Miocene) Salinity Crisis. The former corresponds with the origin of the Eurasian *"Listriodon* mammal fauna" with numerous African and Western Asian elements (Thenius 1979; Bernor 1983; de Bruijn and Hussain 1984). The latter marks the maximum extension of so-called Pontian open country woodlands, and directly precedes the basal Pliocene's sharp shift to cooler temperate forests (Bernor et al. 1979; Bernor 1983, 1984). The interval considered here also marks some major shifts in environments, including those of the early Astaracian (MN 6), early Vallesian (MN 9), the mid-Vallesian crisis of Western Europe (MN 9/10), and the medial Turolian (MN 12).

The most broadly applicable chronological system in Europe is the one framed as biochronologic units, the Mammal Neogene correlation system. This biochronologic framework has been developed and progressively refined by Mein (1975; 1979; 1989). Beyond Mein's monumental contribution to unifying European, West Asian, and North African Neogene mammal faunas within this system, there have been recent proposals to standardize methods for European MN correlations (Bernor 1983, 1984; Lindsay 1989; Lindsay and Tedford 1989; Fahlbusch 1991) as well as to tie it to a global geochronologic system (Steininger et al. 1989; several articles in this volume). A drawback of all evolutionary studies using a biochronological time scale is that it is inherently difficult to address questions of differential timing of turnover events in different areas, or questions relating to the undoubtedly variable tempo of change. Until more ties to the global time scale become available there is very little that can be done about this problem, but as it turns out several patterns do emerge quite clearly, even under this limited temporal control and resolution.

In order to use the suoid material for this investigation it was necessary to revise the group's systematics and integrate this result with the MN system. To minimize the effects of purely taxonomic turnover we have tried to distinguish anagenetic evolution from range extension/replacement wherever it seems justifiable. We wish to emphasize that the systematic aspects of this ad hoc overview are preliminary and tailored for the special aims of our analysis: recognition of biogeographic patterns within the MN correlation framework. Fortelius and Bernor furthermore wish to emphasize that Van der Made's contribution to this synthesis has been crucial.

Basing far-reaching general interpretations on a single group is not ideal, especially when freshly updated data for the major faunal groups is available in the same volume. However, at this writing the data is not yet at hand, and we wish to take this "blind" opportunity to pose hypotheses that can be tested by comparison with other groups in the synthetic biogeographic chapters that follow here.

Abbreviations

BMNH = The Natural History Museum, London
BSPHGM = Bayerische Staatssammlung für Paläontologie und historische Geologie, München
FISF = Forschungs-Institut Senckenberg, Frankfurt
HLD = Hessisches Landesmuseum, Darmstadt
IGF = Istituto de Geologia, Firenze
IVAU = Instituut voor Aardwetenschappen Utrecht, Faculteit Geologie en Geofysica, Rijksuniversiteit te Utrecht
MGL = Museum Guimet, Lyon
MNHN = Museum national d'Histoire naturelle, Paris
MTA = Maden Tetkik ve Arama Enstitüsü, Ankara
NMB = Naturhistorisches Museum, Basel
NMM = Naturhistorisches Museum, Mainz
NSSW = Naturwissenschaftliche Sammlungen der Stadt Winterthur
PIMUZ = Paläontologisches Institut und Museum der Universität, Zürich
PDTFAU = Paleoantropoloji, Dil ve Tarih Cografya Facültesi, Ankara Üniversitesi
SLJG = Steiermärkisches Landesmuseum Joanneum, Graz
SMNK = Staatliches Museum für Naturkunde, Karlsruhe
SMNS = Staatliches Museum für Naturkunde, Stuttgart
TMH = Teylers Museum, Haarlem

Conventions

Age Correlations

The age of the localities (usually in MN units) is taken from the literature and updated where needed as a result of correlation studies provided in this volume. The locality references and their MN attributions are given in appendix 28.1.1. A reference in brackets means that the age is our interpretation of information given by that source. Apart from discussing correlation with physical changes, we have not attempted to analyze the data in terms of absolute time, leaving that for later geochronologic chapters, and we have treated all MN units as having no internal chronology.

For consistency we have adopted the following policies regarding age: (1) an age must not be based exclusively on suoids; (2) uncertain ages are not allowed to extend the range of a taxon but are accepted as support of more secure data; (3) an age such as MN 9/10 is not used to extend a range up from MN 9 or down from MN 10, but if it is the only occurrence of a taxon in an area it is counted present in both, with entry in MN 9 and exit in MN 10; (4) a taxon is counted as present if it is reported from the preceeding and succeeding MN unit of the area. We have consciously departed from our age attribution policies once, by placing Veltheim in MN 5/6 based on the occurrence of *Bunolistriodon latidens* (no other chronologically useful information appears to be available). For the analyses of diversity, turnover, and faunal similarity we have lumped MN 7 and 8 and MN 12 and 13. The main effect of keeping these units separate is the addition of seemingly major features based on evidence that is tenuous indeed.

Systematics

We have made one pragmatic decision concerning systematics: to accept paraphyletic taxa such as *Bunolistriodon*. This practice promotes nomenclatural stability, is useful for paleoecological analysis, and still allows one to informally recognize species-level lineages. We have also decided to recognize diversity rather than synonymy where we believe it is warranted, the rationale being that when some real difference is known or strongly suspected, it should not be suppressed as a taxonomic convenience. Phylogenetic analysis of Neogene Suoidea will certainly lead to the recognition of several paraphyletic groups and precipitate taxonomic reorganization. However, this is work for the future. We have omitted authors for all supra-generic taxa because we have discovered several commonly perpetuated errors but have not yet been able to clarify the issues to our satisfaction.

Materials and Methods

This study is based on 24 suoid taxa from over 100 localities, resulting in over 180 species-locality combinations. Only localities with suoid fossils were included in this investigation. appendix 28.1.1 lists the localities and the collections studied for each species. If no collection is indicated, the locality is listed for that species on the basis of a published description only (note that our identification is sometimes different from the original one). All taxa known from the study area between MN 6 and MN 13 were included, with a few taxa outside this range added for clarity.

Body Size Estimates

Body mass and M2 length for the recent suoids *Sus scrofa*, *Potamochoerus porcus*, *Hylochoerus meinertzhageni*, *Phacochoerus aethiopicus*, and *Tayassu tajacu* were taken from the data base used by Fortelius (1985, 1990) and used to calculate a least squares regression equation with body mass as the dependent variable (the second molar was judged to be least affected by differences in relative molar size between taxa). The equation used was: **body mass** = 1.08 × (LM2) $^{2.99}$. Data for LM2 of the fossil taxa were taken from the private notes of the authors, and from the

literature (see table 28.2). The resulting body mass estimates were rounded to two significant digits or the nearest whole kg, and are obviously rough approximations. They were used to construct four ad hoc size classes: 1–20 kg, 21–80 kg, 81–200 kg and 201–1000 kg, and the analysis is couched in terms of these classes. Especially for classes one and two the estimates were clumped in distinct clusters, suggesting that they may have some real (i.e., biological) meaning. For two taxa length of m2 was used to provide a rough approximation (as indicated in the table). We emphasize that all the estimates are very imprecise, and discourage far-reaching conclusions, especially based on absolute values.

Quantification of Patterns

Available sample sizes are small, and the data points scattered in time and space. Some taxa are represented in an area and interval by large collections from many localities, but many are based on few or even single specimens. The taphonomic background is mostly unknown, and the whole material is inherently biased, because only localities with pigs were included. Statistically significant differences would be impossible to demonstrate for most comparisons no matter what technique was used, and we have based most of our analysis on simple graphic representations of range charts and time series plots. We discuss sampling error in relation to the general nature of the fossil material in our assessment of specific features found. We have used two simple statistics to support our interpretations: Probability of Extinction and Simpson's Index of Faunal Resemblance, as described below.

For each region and interval, the Probability of Extinction is simply calculated as the number of extinctions in an MN unit divided by the taxa at risk in the same interval (Van Valen 1984). We estimate the number of taxa at risk simply as the species diversity of the unit. In his original study Van Valen used a somewhat more complicated formula to estimate this parameter (the number of taxa present at the beginning of the interval, minus half the extinctions, plus half the originations in the interval), but suggested that modifications should be made as appropriate. We have used the simple version because our data points are very few, because we do not recognize a chronology within units, and because we believe that extinctions are concentrated toward the ends of MN units. The simple formula also has the advantage of allowing simple reference back to the primary data. Using the formula given by Van Valen produces a virtually identical pattern, however (with the curious twist that the values range between 0 and 2 rather than 0 and 1, as in the simplified version).

For any pair of taxonomic lists, Simpson's Index of Faunal Resemblance is calculated as:

$$100 \ C \ / \ N_1$$

whereby the number of shared taxa (C) is divided by the number of taxa in the smaller of the two faunas (N_1), and the result is expressed as a percentage (Simpson 1947). Simpson's index has recently been shown to be inferior to statistics that include the number of taxa missing from both faunas (Archer and Maples 1987, 1989; Maples and Archer 1988), but this variable cannot be reliably estimated for the present material. Since the comparisons are between ecologically and taphonomically heterogeneous composite faunas, conclusions can only be general and tentative. In this preliminary survey Simpson's index is only used as a convenient (quasi-quantitative) descriptive aid.

All taxa were used to calculate faunal resemblance, but lineages were treated as single entities for diversity and turnover.

Systematic Overview

In this section we present a taxonomic revision and a brief evolutionary documentation of the Suoidea relevant to our study area. Our purpose is mainly to clarify muddled issues and give credence to the various systematic decisions we have made. The biogeographic aspects are separately summarized at the beginning of the following section.

Tayassuidae

The middle and late Miocene Tayassuidae (if, indeed, that is what these primitive suoids are) of Western Eurasia appear to belong to two different groups. The first comprises the *Taucanamo-Schizochoerus* combination, with a probable common ancestry (Pickford 1978). The second is *Albanohyus*, which is quite different and may well have had a long independent history.

A curious taxonomic problem concerning the smaller European Tayassuidae must be dealt with here. *Barberahyus castellensis* would appear to be synonymous with *Albanohyus pygmaeus*, while most of the material usually referred to *Albanohyus pygmaeus* or *Taucanamo pygmaeum* seems to belong in *Taucanamo grandaevum*. The history of this problem began when Depéret (1892) based the species *"Choeromorus" pygmaeus* on three specimens from La Grive and included *Colobus grandaevus* Fraas 1870 (for the Steinheim form) in the synonymy. No later author seems to have applied the nomen proposed by Fraas (see also Chen 1984). Simpson (1945) introduced *Taucanamo* for *"Choerotherium" sansaniense*, the type species, and for *"Choeromorus" pygmaeus*. Later, Ginsburg (1974) erected the genus *Albanohyus* for *"Taucanamo" pygmaeum*. According to Ginsburg and Bulot (1987), *Albanohyus pygmaeus* occurs only at La Grive (MN 7) (type material, contra Fortelius and Bernor 1990) and Artenay (MN 4). Morphologically, the mandible from Artenay is close to *T. sansaniense*, but smaller, and we refer it

to *Taucanamo*. The Steinheim material, usually assigned to *A. pygmaeus*, was referred to *T. sansaniense* by Ginsburg and Bulot (1987), but the specimens are much smaller than those from Sansan (Chen 1984; Fortelius and Bernor 1990), and have relatively shorter premolars and last molars (fig. 28.2). We have chosen to treat the Steinheim form as a distinct species under the valid name *Taucanamo*

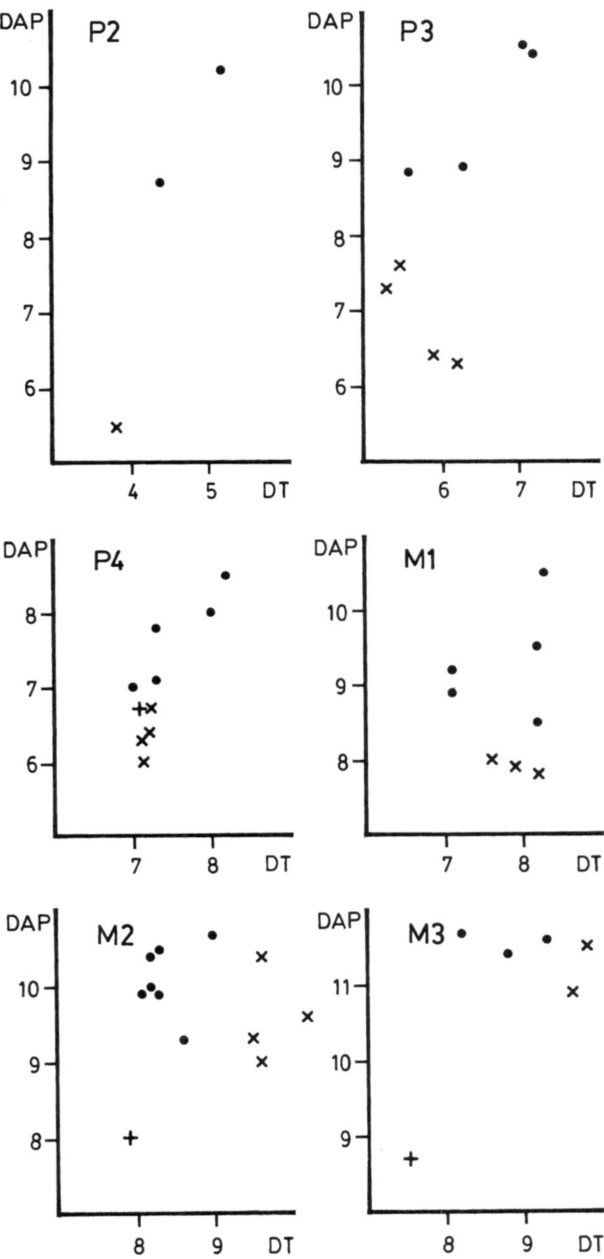

FIGURE 28.1 Bivariate plot of upper cheek tooth dimensions (in mm) of *Albanohyus* and *Taucanamo*, to show differences in size and proportion discussed in the text. DAP = maximum length, DT = basal width. Symbols: plus sign = type of *A. pygmaeus* from La Grive; oblique crosses = *A. pygmaeus* ("*Barberahyus castellensis*") from Castell de Barberà; dots = *T. grandaevum* from Steinheim.

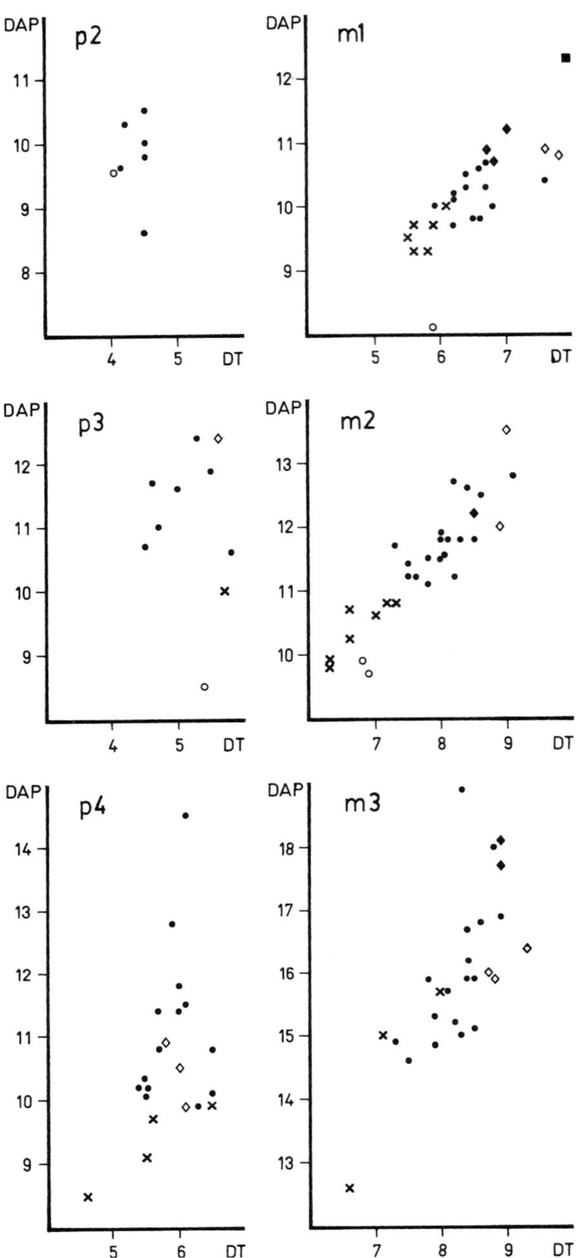

FIGURE 28.2 Bivariate plot of lower cheek tooth dimensions (in mm) of *Taucanamo*, to show size differences discussed in the text. DAP = maximum length, DT = basal width. Symbols: circles = *Taucanamo* sp. from Petersbuch 2 (MN 4); dots = *T. sansaniense* from Sansan; black diamonds = *T. inonuensis* from Paşalar; open diamonds = *T. inonuensis* from Münzenberg; black square = *T. inonuensis* from Sandelzhausen; oblique crosses = *T. grandaevum* from Steinheim.

grandaevum. Chen (1984) and Van der Made (1992b) pointed out that the material from Przeworno 2 seems to belong in *Barberahyus castellensis* Golpe Posse. It has subsequently appeared that *Barberahyus* is morphologically identical to the lectotype of *Albanohyus pygmaeus*, and we

have synonymized them here. The type of *A. pygmaeus* (La Grive, MN 7) is smaller than that of *B. castellensis* (Castell de Barberà, MN 8), which might reflect either size increase or variability. If there is a real size increase, the nomen *A. castellensis* might be used for the larger, younger form.

The main differences between *Taucanamo* and *Albanohyus* should be listed here. The most obvious concerns the premolars, small and short in *Albanohyus*, large and elongated in *Taucanamo* (fig. 28.1). The molars of *Taucanamo* show a slight tendency toward lophodonty, whereas in *Albanohyus* the cusps are well separated from each other. *Albanohyus* also has the paraconule isolated from the protocone and fused to the cingulum, while in *Taucanamo* the paraconule is fused to the protocone. *Taucanamo's* molars are also relatively narrower than those of *Albanohyus* (fig. 28.1). *Albanohyus'* metapodials are morphologically similar to those of the Suidae, in that the distal articulation has a median crest that extends to the dorsal surface, whereas the metapodials of *Taucanamo* have the median crest limited to the plantar and distal surfaces. *Taucanamo sansaniense* is a larger species than *T. grandaevum*, with relatively elongated premolars and last molar (fig. 28.2). The even larger *Taucanamo* from Central Europe (fig. 28.2) could be referred to *T. sansaniense*, but may prove to compare more closely with the large Anatolian form *T. inonuensis*. If so, the latter species has a geographical range extending from Central Anatolia to Germany. Nothing is known about the genus from Southeastern Europe before MN 5 or Southwestern Asia before MN 6, but *Taucanamo* from Wintershof Ost and Petersbuch 2 (MN 4) is smaller than *T. sansaniense* (fig. 28.2) and has the size that is common for Western European *Taucanamo* of that period.

The earliest known *Taucanamo* is from Artenay (MN 4). It is smaller and has a relatively shorter last molar than the Sansan (MN 6) population. *Taucanamo inonuensis* (MN 5–6) is larger than both, and m3 in the Paşalar population has a tendency (like in the Sansan population but more pronounced) to increase the size of its third lobe. This would suggest that *T. inonuensis* is a more derived species than *Taucanamo sansaniense*. However, since the differences are small, it might be argued that *T. inonuensis* is not sufficiently different from *T. sansaniense* for separation at the species level. We have made the separation here to emphasize that regional differentiation (vicariance) is seen in this genus. The relatively short premolars and last molars of *Taucanamo grandaevum* (MN 6–8) seem to set it apart from the other two species, but whether it is more primitive or more derived is difficult to assess.

Schizochoerus has lophodont molars and short, broad premolars. The genus is known only from Spain, Eastern Europe, and Turkey. There are two species: the large *S. vallesiensis* and a smaller, unnamed species. Pickford

(1978) reported "*Schizochoerus*" from a locality near Ganda Kas in the Siwaliks. This material contrasts with European *Schizochoerus* in having elongated premolars, as in *Taucanamo* and a peccary from Lufeng, China (Van der Made and Han 1994). We suspect that further revision of "*Schizochoerus*" sensu Pickford (1978) will reveal multiple lineages.

A smaller species of *Schizochoerus* was described as "*Schizochoerus* cf. *gandakasensis*" by Pickford and Ertürk (1979) and Pickford (1978). This species is represented at Sinap by a skull fragment with short and wide premolars and contrasts with *?Taucanamo gandakasensis* in being a true *Schizochoerus*. It is about 30% smaller (linear) than *S. vallesiensis*, which is also found at Sinap (Pickford and Ertürk 1979). This is somewhat more than the 15% size difference commonly seen in sympatric suoids (Van der Made 1990a). Whether sympatry or an evolving lineage fits the evidence better would depend on the precise provenance of the Sinap specimens, which is unknown. The sediments around Sinap Tepe are currently known to span at least the interval of MN 8 to MN 9, with most fossiliferous localities in MN 9 (see Kappelman et al., this volume). Pickford's (1978) suggestion that *Schizochoerus* replaced *Taucanamo* by evolution would support the interpretation that the Sinap material represents a lineage that progressively increased in size. As a conservative solution, we have decided to record *Schizochoerus* from Sinap as one lineage (for diversity and turnover) with two taxa (for geographic range and size), for the MN 8–9 interval.

Material from Çandir, described by Pickford and Ertürk (1979) as "Genus indet. cf. *Taucanamo*," is difficult to interpret. There is a p4 with two main cusps and the m3 is large, both characters that occur in *Schizochoerus*. The molars are also more lophodont than suggested by Pickford and Ertürk's (1979: fig. 97) line drawing, and this form is distinctly smaller than Sinap's small species of *Schizochoerus*. We treat the Çandir species as a separate taxon here, recognizing that it might represent an early stage of the *Schizochoerus* lineage. *Bunolistriodon michali* from Chios was mistakenly referred to *Schizochoerus* by Nikolov and Thenius (1967). This probably explains why Hünermann (1975) referred the sublophodont listriodont from Paşalar to *Schizochoerus*. The Chios specimen probably belongs to the *Bunolistriodon latidens* lineage (see below).

Suidae

Listriodontinae There is a long-standing controversy over the validity of Arambourg's genus *Bunolistriodon* (Leinders 1975). *Bunolistriodon*, as presently recognized, is most probably a paraphyletic group united by plesiomorphic characters, in particular the absence of fully lopho-

dont teeth or "horns." It is likely that other members of the Listriodontinae, including *Listriodon*, *Lopholistriodon*, and *Kubanochoerus* evolved from different "*Bunolistriodon*" species.

Bunolistriodon lockharti is unambiguously known only from MN 4–5 in Western and Central Europe, and would thus be excluded from the present review, except that it seems important to discuss its relationships with other sublophodont forms. Sublophodont teeth that are smaller, relatively narrower, and have a less angular cusp morphology than typical for *B. lockharti* occur in MN 6 at Prebreza, Paşalar, and Çandir. In these localities the incisors are extremely wide (fig. 28.4; see also Fortelius and Bernor 1990; Fortelius et al. 1996) and the male upper tusks are very large, even more than in *Bunolistriodon latidens* (Biedermann 1873) from Veltheim. *Bunolistriodon latidens* has usually been synonymized with *B. lockharti*, but it seems useful to resurrect it for forms that share the enlarged anterior dentition and small, narrow postcanines. The extreme development of the anterior dentition in the Paşalar and Çandir populations (fig. 28.3) would justify the recognition of a new species for these later forms (Fortelius and Bernor 1990). The material from Inönü I compares closely with *B. latidens* s.s., and appears less derived than the Paşalar form. *Bunolistriodon michali* (Paraskevaidis 1940) from Chios is based on a single indeterminate worn M3, and should probably be restricted to the type specimen in the interest of stability. It is highly probable that it belongs somewhere in the *B. latidens* lineage.

The molars and especially the premolars of *Bunolistriodon latidens* appear to be considerably more primitive (similar to *Kubanochoerus*, for example), than those of *B. lockharti*. The two species also overlap in time, and it thus appears quite unlikely that *B. lockharti* and *B. latidens* share a direct ancestor-descendant relationship (Fortelius et al. 1996). It is odd that the type specimen of *Bunolistriodon latidens* seems to represent the only unequivocal occurrence of the lineage in Central Europe. Lack of incisors and canines precludes secure determination of sublophodont material from such Central European MN 6 localities as Georgensgmünd or Stätzling (Mein 1989; de Bruijn et al. 1992), but it is plausible that they represent members of the *Bunolistriodon latidens* evolutionary lineage.

An evolutionary transition from any European or Southwest Asian *Bunolistriodon* to *Listriodon* is improbable to say the least. *Listriodon* undoubtedly evolved from sublophodont ancestors, but the transition is not certainly recorded in the material known at present. *Listriodon* appeared simultaneously (within MN 6) across West Asia and Europe. A morphological gradient is apparent from the Indian subcontinent (primitive?), westward to Anatolia (intermediate?) and Europe (derived?). Anatolian *Listriodon* material has so far universally been referred to *L. splendens* (or *L.* cf. *splendens*, by Fortelius and Bernor 1990). In most

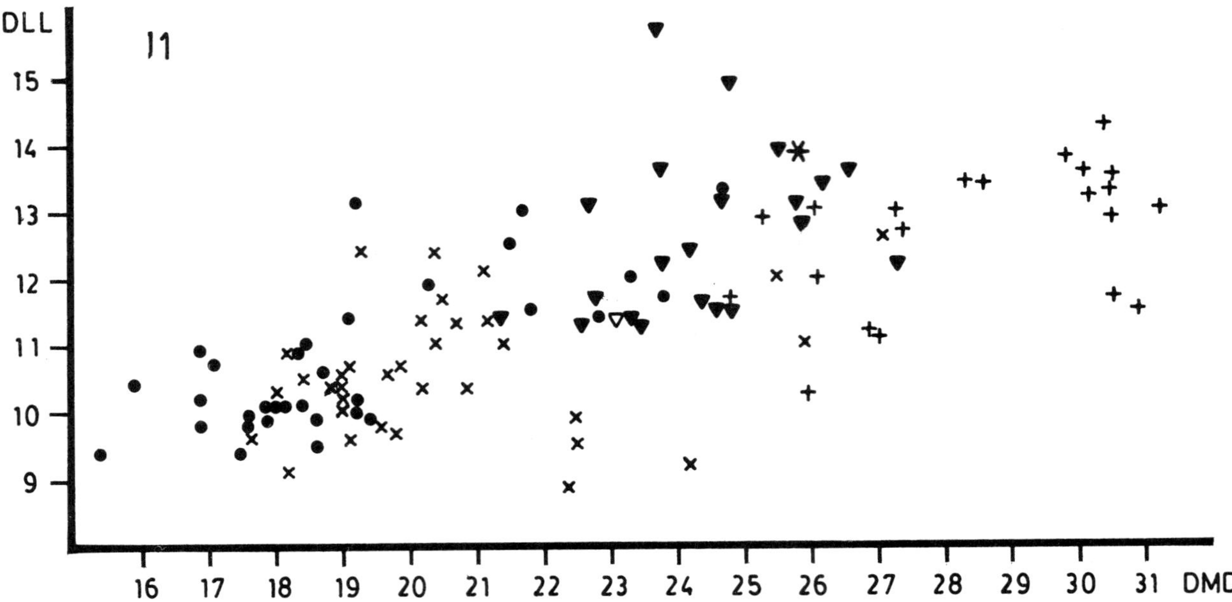

FIGURE 28.3 Bivariate plot of upper central incisor dimensions of selected Listriodontinae. Symbol: dots = European *Bunolistriodon*; plus signs = European *Listriodon*; oblique crosses = *Listriodon* from the Indian subcontinent; black triangles = *Listriodon* from Paşalar (Anatolia, MN 6); open triangle = *Listriodon* from Sarıçay (Anatolia, MN 7); asterisk = *Listriodon* from Simorre.

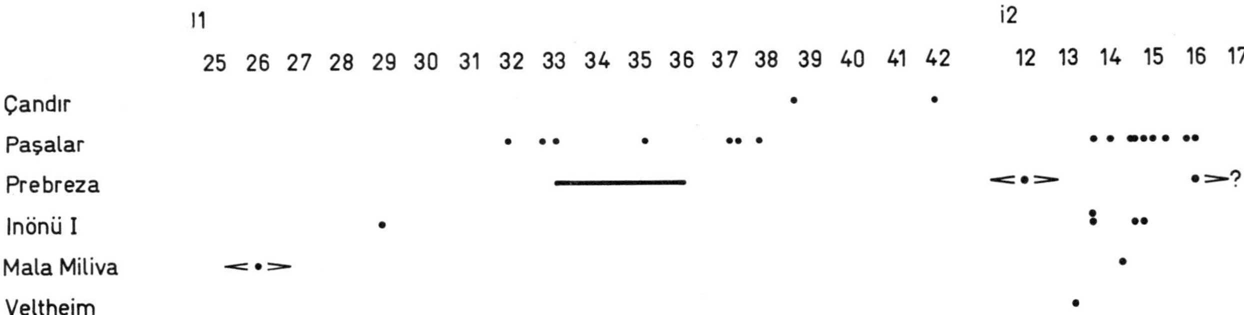

FIGURE 28.4 Increase of the maximum mesio-distal dimension ("width") of upper and lower central incisors in the *Bunolistriodon latidens* lineage (bottom to top). The relative positions of Mala Miliva, Paşalar, and Çandir are based on independent evidence, and Prebreza is known to be younger than Mala Miliva (appendix 28.1; Mein 1989). The stratigraphic positions of Inönü I and Veltheim are essentially based on the trend figured. Data are original except for Mala Miliva (Petronijević 1967) and Prebreza (measured off figures in Pavlović 1969; the bar indicates range of values obtained from different views).

features, such as in the relative width of the molars and the degree of P4 lophodonty, it is morphologically intermediate between *L. splendens* and the somewhat smaller Indian species *L. pentapotamiae* (fig. 28.3 and plentiful unpublished data). The Anatolian form shares at least one nonmetric character with *L. pentapotamiae*, an upturned m3 talonid. *Listriodon* last occurs in Anatolia during MN 8, but remained a common Central and Western European element throughout MN 9.

Ginsburg (1980) recognized two subspecies of European *Listriodon* distinguished by size: the Astaracian *Listriodon splendens splendens* and the Vallesian *Listriodon splendens major*. Detailed study has failed to support this size distinction as such, but there is good evidence that the incisors became mesio-distally broader (fig. 28.5), the canines larger, the premolars more molariform, and the posterior molars relatively smaller during the species's history (Van der Made 1992a,b). The size pattern appears more complex. Most of the early *Listriodon splendens* material seems to be small (e.g., Sansan), but the material from Simorre (MN 7?) is even smaller, and the largest material on record is actually from MN 6 of Spain (Arroyo del Val I). Future recognition of multiple vicariant and/or sympatric taxa is possible, perhaps even probable, but at

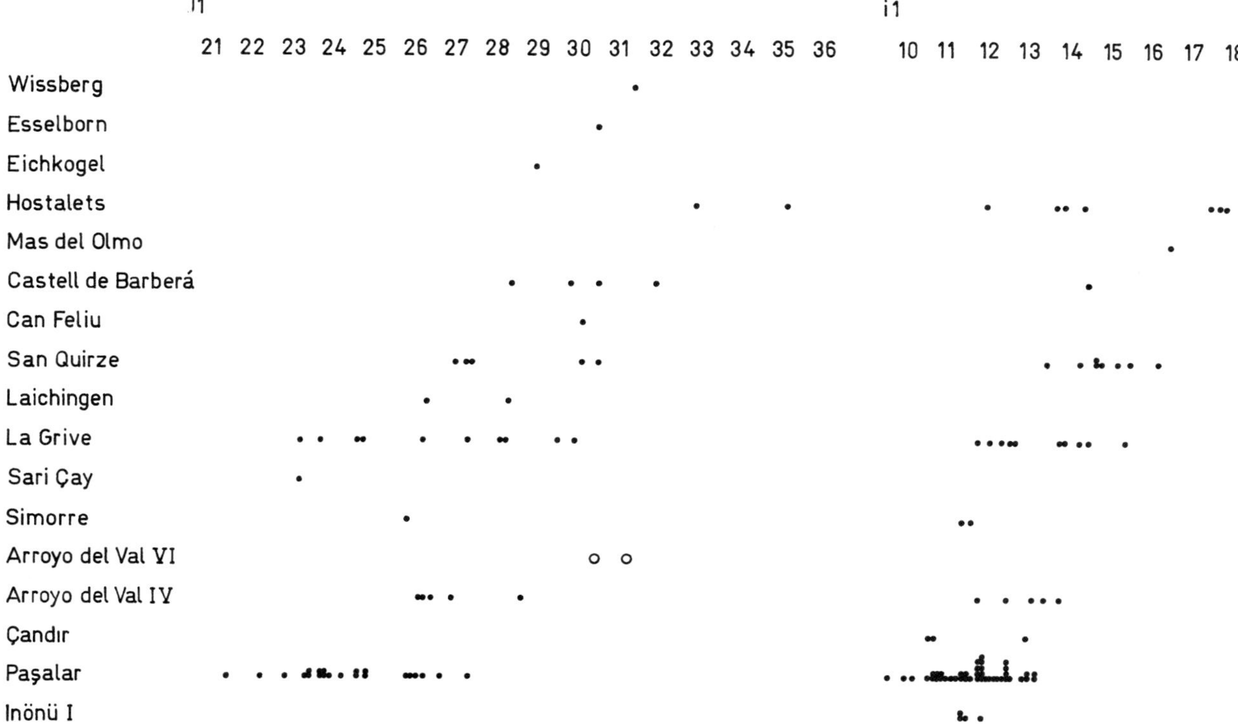

FIGURE 28.5 Increase of the maximum mesio-distal dimension ("width") of upper and lowers central incisors in the *Listriodon splendens* lineage (bottom to top). The relative positions of the localities are based on literature cited in the summary table; for the West European localities Mein 1989, Azanza and Menendez 1989/1900, and Golpe-Posse 1972 have been referred to. The circles indicate incisors from Arroyo de Val VI, a locality that has produced extremely large cheek teeth of *Listriodon* (Arroyo del Val IV and Simorre have normal or small cheek teeth). Arrows indicate approximate measurements.

present it does not seem feasible to assign material to any such taxon with much confidence.

Two MN 6 species of *Kubanochoerus* Gabunia 1955 are known from the eastern portion of our study area: *K. robustus* from Belometchetskaia and *K. khinzikebirus* from Inönü I. The latter species was placed in a separate genus, *Libycochoerus* Arambourg 1961, by Pickford and Ertürk (1979), but the main (apparent) difference between the genera seems to be due to sexual dimorphism in the development of the "horns" (Guan Jian and Van der Made 1994), and we have synonymized them here.

Hyotheriinae *Hyotherium soemmeringi* is the latest surviving member of the European Hyotheriinae (exit in MN 6), and was particularly abundant in Central Europe (Schmidt-Kittler 1971). The species has been listed from Petrovac in Serbia (Laskarev 1937), but from the description it appears more likely that the material represents a tetraconodont.

Tetraconodontinae Thenius (1952) recognized two morphs ("Rassen") of European *Conohyus*: *C. simorrensis simorrensis*, from Simorre, Neudorf Sandberg, Göriach, and a few additional minor localities, and *C. simorrensis*

steinheimensis, from Steinheim and Klein Hadersdorf. He distinguished these forms on premolar size and shape, and regarded them as ecologically distinct. Chen (1984) considered them to be separate lineages, but also suggested that *C. steinheimensis* might be a "fairly direct descendant" of *C. simorrensis*. Chen also quantified the differences in premolar size and shape between the taxa, and indicated that if *C. steinheimensis* descended from *C. simorrensis*, then the Asian trend toward enlarged premolars and reduced molars was reversed in Europe. Alternatively, *C. steinheimensis* could represent a more primitive state (short last molars, small premolars, and no labial bulge on p4). At any rate, there is no sign in Europe of the extreme hypertrophy of the premolars seen even in early Asian tetraconodonts, so the European taxa must represent one or more conservative derivatives of the group. Their tetraconodont status is clearly shown by the uniquely specialized premolar constellation, however.

Van der Made (1989) proposed a lineage of *Conohyus* spanning the interval MN 5–9 from *C. simorrensis* to *C. ebroensis* Azanza Asensio 1986. The lineage is characterized by the increase in molar size and decrease or stasis in premolar size, with development of a diastema between the canines and premolars in *C. ebroensis*, trends visible in

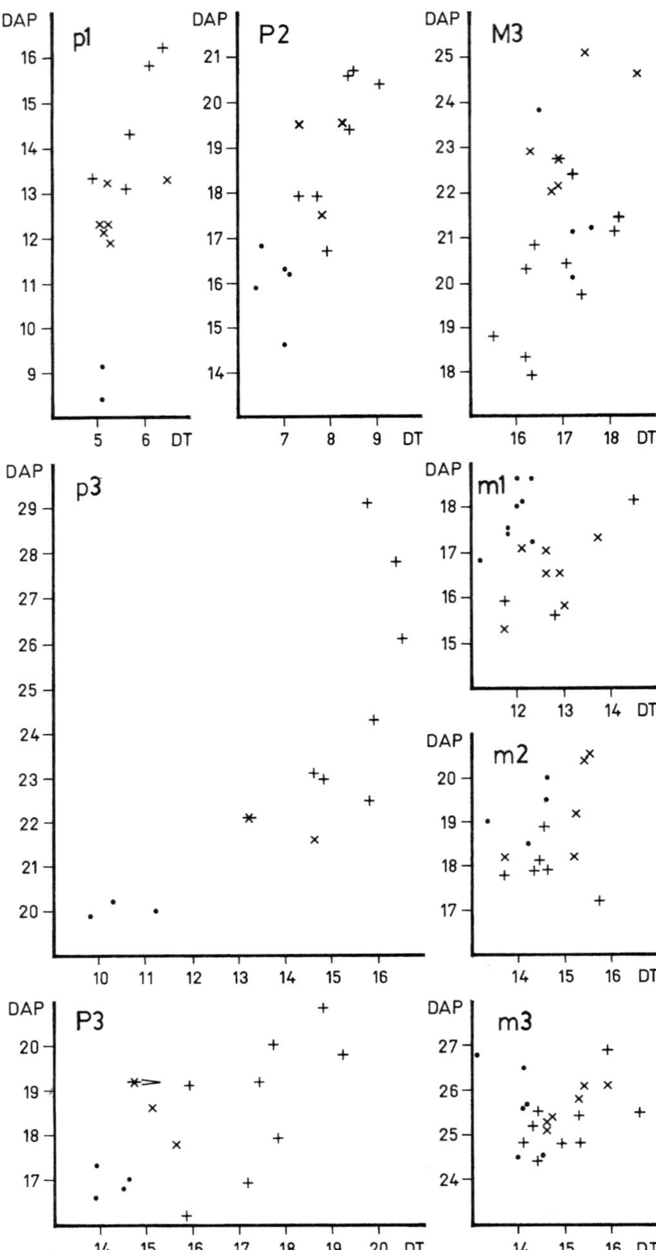

FIGURE 28.6 Bivariate plot of relative lower cheek tooth dimensions of *Conohyus* and *Parachleuastochoerus*, to show differences discussed in the text. Symbols: plus signs = *Conohyus simorrensis* from Göriach; oblique crosses = *Conohyus simorrensis* from Paşalar; asterisk = *Conohyus simorrensis* from Kleineisenbach; dots = *Parachleuastochoerus steinheimensis* from Steinheim.

the Central European material as well (fig. 28.6). This lineage might possibly be derived from *Conohyus sindiensis*, but only material attributable to *C. simorrensis* is found in the study area.

Heissig (1989), following an observation by Hünermann (1968), described a new small species of *Conohyus* from the lower part of the Bavarian Obere Süsswassermolasse, *C. huenermanni*. According to Heissig, this taxon is characterized by its broad and brachydont p4 and three-rooted p3 (as in *C. simorrensis*), and also occurs in the Deinotherien-

sande. Heissig suggests a possible derivation of *C. huenermanni* from *C. simorrensis* by size decrease. However, *C. steinheimensis* and *C. huenermanni* both resemble *Parachleuastochoerus crusafonti* Golpe Posse 1972 more than either resembles *C. simorrensis* (Van der Made 1992a), and we have accordingly transferred "*C.*" *huenermanni* to *Parachleuastochoerus* Golpe Posse 1972. We recognize *P. steinheimensis* in MN 6–9 of Western Europe and MN 7–9 of Central Europe.

Parachleuastochoerus huenermanni is known from MN

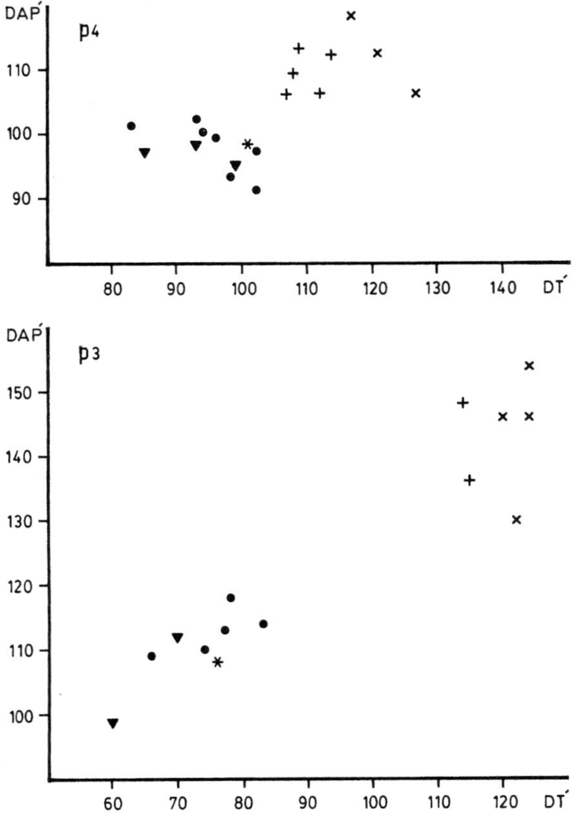

FIGURE 28.7 Bivariate plot of relative lower third and fourth premolar size in *Conohyus* and *Parachleuastochoerus*, to show the difference between the genera. The points are based on individual associated dentitions. DAP' = maximum length of the tooth as a percentage of m1 length; DT' = basal width of the tooth as a percentage of m1 width. Symbols: dot = *P. steinheimensis* (various localities); triangle = *P. huenermanni* (Breitenbrunn); inverted triangle = *P. crusafonti* (Can Llobateres); plus sign = *C. simorrense* (various localities); oblique cross = *C. sindiense* (various Indian localities); asterisk = *C. ebroensis* (Fonte do Pinheiro). *C. ebroensis* plots with *Parachleuastochoerus* for this character, but is part of a different lineage with increasing molar size (Van der Made 1989).

9 of Central Europe, while the *Parachleuastochoerus huenermanni–crusafonti* lineage is recorded in MN 8–10 of Western Europe (Spain). *Parachleuastochoerus* does not have a p4 with a labial bulge, the cheek teeth are narrower than in *Conohyus*, and premolars tend to be smaller relative to the molars (fig. 28.7). Large collections of *P. steinheimensis* show that there is variability in the degree of fusion of the p3's posterior roots, and the character thus appears unreliable for single specimens. *Parachleuastochoerus steinheimensis* shows size increase with time, whereas in Spain the small *P. huenermanni* is replaced by the still smaller *P. crusafonti*, which has not been found in the study area.

There are probable lineages within the *Nyanzachoerus*

(? = *Sivachoerus*) and *Notochoerus*–groups, but no representatives of these lineages have been reported from this study area. However, *Conohyus ebroensis* from the Iberian Peninsula is very similar to *Nyanzachoerus devauxi*, and a circum-Mediterranean distribution for this group is possible, as is seen in the "*Hipparion*" catalaunicum–"*Hipparion*" africanum lineage (Bernor et al. 1989; this volume). It has become duly apparent that the *Conohyus, Sivachoerus, Nyanzachoerus, Notochoerus* lineages are a potentially monophyletic group with a complex evolutionary history that spanned Eurasia and Africa.

Suinae   Ginsburg (1980) and Van der Made and Moyà-Solà (1989) included *Propotamochoerus provincialis* in *Korynochoerus*. However, the type species of *Propotamochoerus* is *P. hysudricus* rather than the "strange" *P. salinus* (Pilgrim 1925 contra Pilgrim 1926). Since the genotype of *Korynochoerus, K. palaeochoerus,* is very similar to *P. hysudricus,* one could argue instead that *Korynochoerus* should be synonymized with *Propotamochoerus,* as we have tentatively done here. Alternatively, one could retain *Korynochoerus* for *P. palaeochoerus,* since it is more conservative in I1 morphology than the other known *Propotamochoerus* (lower, less linguolabially flattened crown), but with a distal cusp (derived character). Similarly, its skull has a parietal crest (primitive) and an elevated occiput (derived). The combination of these characters makes it unlikely that *P. palaeochoerus* is the ancestor of any younger species of *Propotamochoerus.*

Propotamochoerus palaeochoerus has been reported to have ranged from MN 8–13/14, but this may be a mistaken interpretation. Material from localities of the Baccinello V3 level (MN 13) has the same size as *P. palaeochoerus,* but differs in I1 morphology (flatter and higher crown) and in having much narrower I2 and I3. It seems more likely that the Baccinello V3 *Propotamochoerus* sp. is an immigrant rather than a direct descendant from *Propotamochoerus palaeochoerus* (and there is, suggestively, a remarkable morphological resemblance between the Baccinello species and "*Propotamochoerus*" hyotherioides from Lufeng). *Propotamochoerus palaeochoerus* is well known and abundant in MN 8–9 in Western and MN 9–10 in Central Europe. It is not known to which species the scarce material from MN 10/11–13 belongs.

Propotamochoerus palaeochoerus may conceivably extend its range in Western Europe to MN 12 or even MN 13, or the same "*Propotamochoerus*" sp. that occurs at Baccinello may be present there. For the analyses we have treated *P. palaeochoerus* as exiting Western Europe with the last unambiguous records in MN 9, a conservative decision, as we show below. *Propotamochoerus provincialis* differs from *P. palaeochoerus* in skull structure and incisor morphology, in which characters it resembles most other *Propotamochoerus.* Derivation of *Microstonyx* (? = *Hippo-*

358 Fortelius, van der Made, and Bernor

potamodon) *major* from *Hippopotamodon antiquus* seems probable (Thenius 1972; Ginsburg 1980; Van der Made and Moyà-Solà 1989), and we have treated the two species as a single lineage, even though *Hippopotamodon sivalense* (which could be synonymous with *H. antiquus*) persisted on the Indian subcontinent after the transition had occurred in the west (Van der Made and Hussain 1989; Van der Made 1990b; however, work currently in progress by Fortelius, Bernor, and Fessaha has cast some doubt on the reality of the *Hippopotamodon–Microstonyx* transition).

Microstonyx erymanthius has generally been synonymized with *M. major* (Trofimov 1954; Hünermann 1968; Thenius 1972). The two taxa were considered to be valid subspecies by Van der Made and Moyà-Solà (1989) and Van der Made et al. (1992). Van der Made (in prep.) regards them as valid species, mainly because of a rapid elongation of the I2 and I3 seen in *M. erymanthius* during MN 11 and MN 12. We have treated them as separate taxa in this contribution, but it should be made clear that the recognition of a

Hippopotamodon–Microstonyx lineage was considered important mainly from the point of view of the analysis, and that the question of whether *M. major* or the dentally more derived *M. erymanthius* is more directly related to *H. antiquus* cannot be definitely settled here.

The skull and mandible from Samos, described by Thenius (1950) as *Potamochoerus* (*Postpotamochoerus* nov. subgen.) *hyotherioides*, differs from *Potamochoerus porcus* and all *Propotamochoerus* (Van der Made and Moyà-Solà 1989), including the Lufeng "*Propotamochoerus*" *hyotherioides*. At present the affinities of this species are not clear and should be restudied.

A mandible from Karain (Anatolia) belongs to the Suinae, as indicated by its premolars, but cannot be referred to any of the species known from elsewhere because of its greatly elongated m3. It appears to be related to *Kolpochoerus*.

Patterns of Biogeography, Diversity, Turnover, and Paleoecology

For comparative purposes the area studied is considered as belonging to three "provinces" (more or less arbitrarily delimited regions): (1) Central Europe (Germany, Switzerland, Austria, Poland, Czechoslovakia, and Hungary), (2) Southeastern Europe (the former Yugoslavia, Greece, and Turkish Thrace, Bulgaria, Romania, Moldavia, the Ukraine, and the areas of the former Soviet Union between the Black and Caspian Seas, southward to the Caucasus), and (3) Southwestern Asia (Anatolia, including islands close to the current mainland [= Samos], and Iran). We have referred two Georgian localities to Southwestern Asia because placing Belometchetskaia in Southeastern Europe would affect the biogeographic pattern in a misleading

way. Reference is also made to the excellent record from Western Europe (France, Spain, and Portugal).

The data presented in the previous section comprises 24 taxa and over 180 species-locality occurrences (see appendix 28.1), averaging about 8 localities per taxon. The distribution is strongly skewed, however, with a few common taxa accounting for the majority of occurrences. Most taxa range over more than one MN unit, and the number of data points per area and time slice is low, except for the most common species (it is common for the presence of a taxon in an MN unit to be based on a single locality, sometimes a single specimen). Sampling error, mostly unknown taphonomic biases, and several other factors contribute to make the data unsuitable for quantitative treatment in any strict sense, but quantification is nevertheless heuristically useful.

Southeastern Europe has an especially poor record, with complete gaps in MN 7/8 and MN 13. Appendix 28.1 is summarized graphically in charts and plots (figs. 28.8– 28.13), with data for Western Europe taken (with minor changes) from Van der Made 1992a,b. We begin this analysis with a biogeographic overview of the taxa and their history in a somewhat broader perspective than that of the study area. This is followed by simple quasi-quantitative analyses of the general patterns seen within the study area

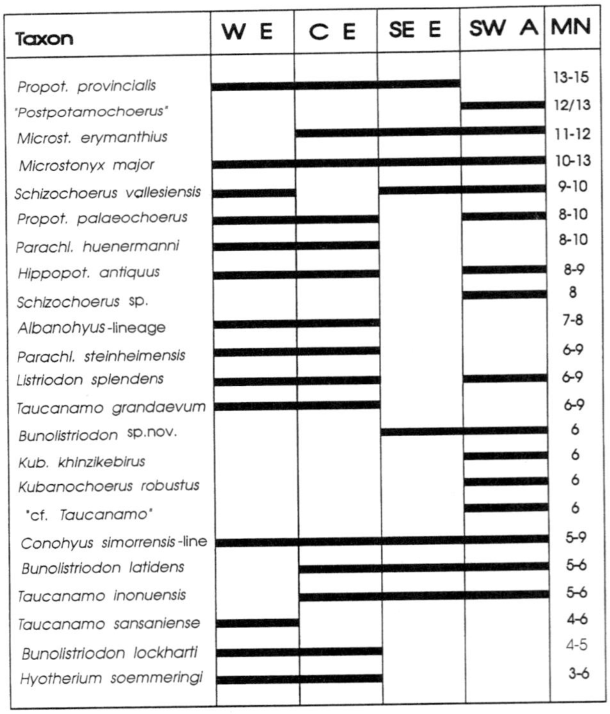

Taxon	W E	C E	SE E	SW A	MN
Propot. provincialis					13-15
'*Postpotamochoerus*'					12/13
Microst. erymanthius					11-12
Microstonyx major					10-13
Schizochoerus vallesiensis					9-10
Propot. palaeochoerus					8-10
Parachl. huenermanni					8-10
Hippopot. antiquus					8-9
Schizochoerus sp.					8
Albanohyus -lineage					7-8
Parachl. steinheimensis					6-9
Listriodon splendens					6-9
Taucanamo grandaevum					6-9
Bunolistriodon sp.nov.					6
Kub. khinzikebirus					6
Kubanochoerus robustus					6
'*cf. Taucanamo*'					6
Conohyus simorrensis -line					5-9
Bunolistriodon latidens					5-6
Taucanamo inonuensis					5-6
Taucanamo sansaniense					4-6
Bunolistriodon lockharti					4-5
Hyotherium soemmeringi					3-6

FIGURE 28.8 Biochronologic range chart of the taxa included in this chapter. WE = Western Europe, CE = Central Europe, SE E = Southeast Europe, SW A = Southwest Asia. A star indicates a lineage of more than one taxon.

of regional differentiation, diversity, and turnover. We conclude with a review of relevant paleoecological information.

Biogeographic Reconstruction

The peccaries have a relatively sparse evolutionary and biogeographic record. *Taucanamo* first occurs in our study area during MN 4. *T. sansaniensis* is known from Western Europe during MN 4–6. A larger and somewhat more derived form, *T. inonuensis*, is likely derived from *T. sansaniensis* and occurs in MN 5–6 strata of Central Europe, Southeastern Europe, and Southwestern Asia. Both *T. sansaniensis* and *T. inonuensis* are subsequently replaced in Western and Central Europe by the smaller form *T. grandaevum* (MN 6–8). *Schizochoerus* (*S.* sp.) is first reported from MN 8 of Southwestern Asia, plausibly being derived from *Taucanamo*. During MN 9 *Schizochoerus vallesiensis* appeared in Southwestern Asia and Southeastern Europe but seems to have reached Western Europe (Spain) only in MN 10, suggesting an east-west migratory extension. *Albanohyus pygmaeus* is probably unrelated to *Taucanamo* and *Schizochoerus*. It occurs in Central and Western Europe during MN 7 + 8, and may be related to African forms from Fort Ternan and Langebaanweg.

The Suidae are far better represented and more diverse than the Tayassuidae in our study area. The listriodontine (sensu lato) pigs are particularly well represented in middle Miocene age horizons. *Bunolistriodon lockharti* is known only from MN 4–5 of Western and Central Europe. Evidence reviewed above suggests that the "*Bunolistriodon*" *latidens* evolutionary lineage is distinct from *B. lockharti*, occurring in MN 5 and MN 6 horizons of Central Europe, Southeastern Europe, and Southwestern Asia.

Listriodon would appear to be South Asian (MN 5 correlative) in origin. *Listriodon splendens* sensu lato seemingly extended its range along an east-west gradient with the MN 6 Anatolian morph being more primitive than Central and Western European morphs. Bernor, Fortelius, Hussain, and Van der Made (in prep.) have recognized a "*Bunolistriodon*" grade form from the Lower Manchars (Sind, Pakistan; ca. 16 Ma; = MN 5 equivalent) that could be a sister taxon to *Listriodon*. If so, this would support the hypothesis that *Listriodon* originated in Asia. European *L. splendens* was established in Europe between MN 6 and MN 9 and seems to show poorly understood geographic and/or ecological diversification as well as a general trend of increasing emphasis on the anterior dentition and molarization of the premolars.

Kubanochoerus could conceivably be derived from an African "*Libycochoerus*"-like form. It extended its range only into the eastern portion of our study area and into China (Qiu 1990; Mammal unit III, Astaracian correla-

tive). *Kubanochoerus* is represented by two species in MN 6 of Southwestern Asia: *K. robustus* and *K. khinzikebirus*.

The Hyotheriinae established a cosmopolitan Old World distribution in the earliest Miocene. They were particularly abundant in Central Europe, but also occurred in Western Europe during MN 5–6.

The Tetraconodontinae have a complicated evolutionary history. *Conohyus* (*C. sindiensis*) is first known from horizons in South Asia correlated as being 16 Ma (Bernor et al. 1988; MN 5 correlative) and similarly aged horizons in East Asia (Pope and Bernor 1990). Earliest members of *Conohyus sindiensis* already show marked hypertrophy of the premolars (labial bulging) that is either reversed in some European lineages, or, more probably, never occurred there, implying polyphyly for *Conohyus*, as it is currently applied. It would appear that two or three "*Conohyus*"-like lineages evolved in Europe. First, the *C. simorrensis–C. ebroensis* lineage, plausibly derived from South Asian *C. sindiensis*, is recognized as ranging during MN 5–9 in Western Europe, MN 6–8 in Central Europe, MN 5 in Southeastern Europe, and MN 6 in Southwestern Asia. Second, the *Parachleuastochoerus steinheimensis* lineage, with small premolars and generally narrow cheek teeth, occurred in Western Europe during MN 6–9 and Central Europe during MN 7–9. A group of smaller species, the *P. huenermanni–P. crusafonti* lineage, is found during MN 9 in Central Europe and MN 8–10 in Spain. Future study should focus more closely on the evolutionary relationships of the *Conohyus–Parachleuastochoerus* lineages and the question of Eurasian *Conohyus* monophyly.

The Suinae first occur in Europe at the end of the middle Miocene. *Propotamochoerus* (= *Korynochoerus*) *palaeochoerus* definitely occurs in Western Europe during MN 8–9 and Central Europe during MN 9–10, but in both areas fragmentary material may extend the range up to MN 13. Several Turolian and Pliocene suines occur, but their relationships are unclear.

Propotamochoerus provincialis occurs during MN 13–15 in Western, Central, and Southeastern Europe; *Propotamochoerus* sp. (or spp.?) at Baccinello V3 (MN 13) and a few MN 14/15 Central European localities; and "*Postpotamochoerus hyotheroides*" during MN 12/13 in Southwestern Asia (Samos). For much of the smaller suine material from MN 10 to MN 14 taxonomic attribution remains problematic, and the geographic as well as the temporal distributions of these taxa are therefore very incompletely documented. The relationships of these late European *Propotamochoerus* (s.l.) to the African *?Potamochoerus* (= *Kolpochoerus*) *afarensis* and the Chinese *Propotamochoerus hyotheroides* should also be studied (Van der Made and Han 1994). According to Pickford (1988) the *Propotamochoerus*-group arose in South Asia from a hyotherine pig around 11 Ma, and subsequently extended its range westward into Southwestern Asia and Europe. The regional chronology and correlations remain unresolved, however.

P. palaeochoerus appears in MN 8 in Western Europe, while new chronostratigraphic data record *Propotamochoerus* as first appearing in the Potwar Plateau as late as ca. 9.5–9.1 Ma (Barry, pers. comm.). It is, however, quite likely that *Propotamochoerus* also occurs in the Chinji Formation, but has previously been misidentified as *Hyotherium* (Van der Made, pers. obs.; Barry, pers. comm.). The chronology and origin of *Propotamochoerus* cannot be settled until the separation of *Hyotherium* and *Propotamochoerus* (in the Siwaliks material as well as in European collections) is reexamined critically.

The *Hippopotamodon-Microstonyx*–group is an apparently monophyletic clade of large suines. *Hippopotamodon* (= *Microstonyx*) apparently first extended its range from Asia into Europe during the late Miocene. *Hippopotamodon antiquus* is recorded from MN 9 of Western and Central Europe, and has been suggested to be closely related to or identical with *H. sivalense*, which persisted until about 7 Ma in the Siwaliks (Pickford 1988). *Microstonyx major* diverged in the late Vallesian and extended its range throughout the study area between MN 10 and MN 13. *Microstonyx erymanthius* was yet a third lineage of this group known from MN 11–12 of Central and Southeastern Europe and Southwestern Asia. At Maragheh, Iran, it is known to last occur near the base of Upper Maragheh, ca. 7.4 Ma, upper MN 12 (Bernor 1986; Bernor et al., this volume; Swisher, this volume).

Faunal Similarity

Species distributions reveal three geographic groupings (fig. 28.9): Western, Eastern, and "cosmopolitan" (we use these terms only with reference to the study area and interval). Of the 15 taxa found in Central Europe during MN 6–13, 12 are also found in Western Europe, but only 8 are known from Southwestern Asia. Conversely, of the 15 taxa known from Southwestern Asia during the entire interval, only 8 are known from Central and 5 from Western Europe. It would also seem that the "cosmopolitan" group (including *Schizochoerus*, *Listriodon*, *Conohyus*, *Propotamochoerus*, and *Hippopotamodon-Microstonyx*) has an Asiatic origin, while the western group (*Taucanamo grandaevum*, *Hyotherium* and *Parachleuastochoerus*) seems autochthonous within the study area. Within our spatio-temporal limits *Albanohyus* also behaves like a western taxon, but it may be related to African forms and its plausible broader geographic distribution is an interesting aspect that should be further investigated.

The regional faunal similarity can be roughly quantified (although not tested statistically, for a large number of reasons) using Simpson's Index of Faunal Resemblance

(table 28.1). The overall diversity is very low, and the index consequently gives a very coarse indication of the patterns, but it definitely does provide a convenient descriptive handle. The main results as they appear in figure 28.10 are easily summarized (we discuss their reliability together for all these analyses): (1) Southeastern Europe shows complete (100%) similarity to Southwestern Asia (only in terms of Simpson's index) for all the MN units that it can be calculated; (2) Western and Central Europe are very similar throughout the interval (index 80–100%), except for a somewhat dubious drop to 50% in MN 10 caused by the appearance in Western Europe of *Schizochoerus* and the (uncertain) disappearance there of *Propotamochoerus palaeochoerus*; (3) Western and Central Europe behave as a block toward Southwestern Asia throughout the interval. The index ranges from 33 to 67% except for an episode of "total resemblance" (index 100%) in MN 11, caused by the presence of *Microstonyx major* over the entire area; (4) The general trend from MN 6 to MN 11 is one of increasing similarity between Europe and Southwestern Asia, but the trend is reversed at the end because of the appearance in Western and Central Europe of *Propotamochoerus provincialis* and of "*Postpotamochoerus hyotheroides*" at Samos (here treated as part of Anatolia).

Diversity and Turnover

Diversity (i.e., species richness) shows the same basic pattern for all areas: high in the early Astaracian (MN 6), declining to a low in the late Vallesian (MN 10) and early Turolian (MN 11) (figs. 28.8 and 28.11). The diversity curves for Western and Central Europe are quite similar (fig. 28.11a), even though the individual taxa are partly different. The main difference in pattern is that there is little change from MN 6 to MN 9 in Central Europe, whereas an increase to an all-time high in MN 7 + 8 and MN 9 is seen in Western Europe. In both regions there is a substantial decrease from MN 9 to MN 10, after which diversity stays low everywhere. For Southwestern Asia the most distinct change is the drop from MN 6 to MN 7 + 8, while the drop from MN 9 to MN 10 is virtually absent there.

The total number of entries and exits per MN unit ("Absolute Turnover") shows distinct peaks at MN 6 and MN 9, the former for Asia and the latter for Europe (fig. 28.11b). The high value for Southwestern Asia in MN 6 is partly artificial, due to the absence of data for MN 5, which results in all taxa present being counted as entries. From MN 10 onward the pattern is roughly similar between Western Europe, Southeastern Europe, and Southwestern Asia, with a minimum at MN 11. Central Europe appears to be out of phase with the other regions, with a rising trend from MN 10 to MN 12/13. Closer scrutiny

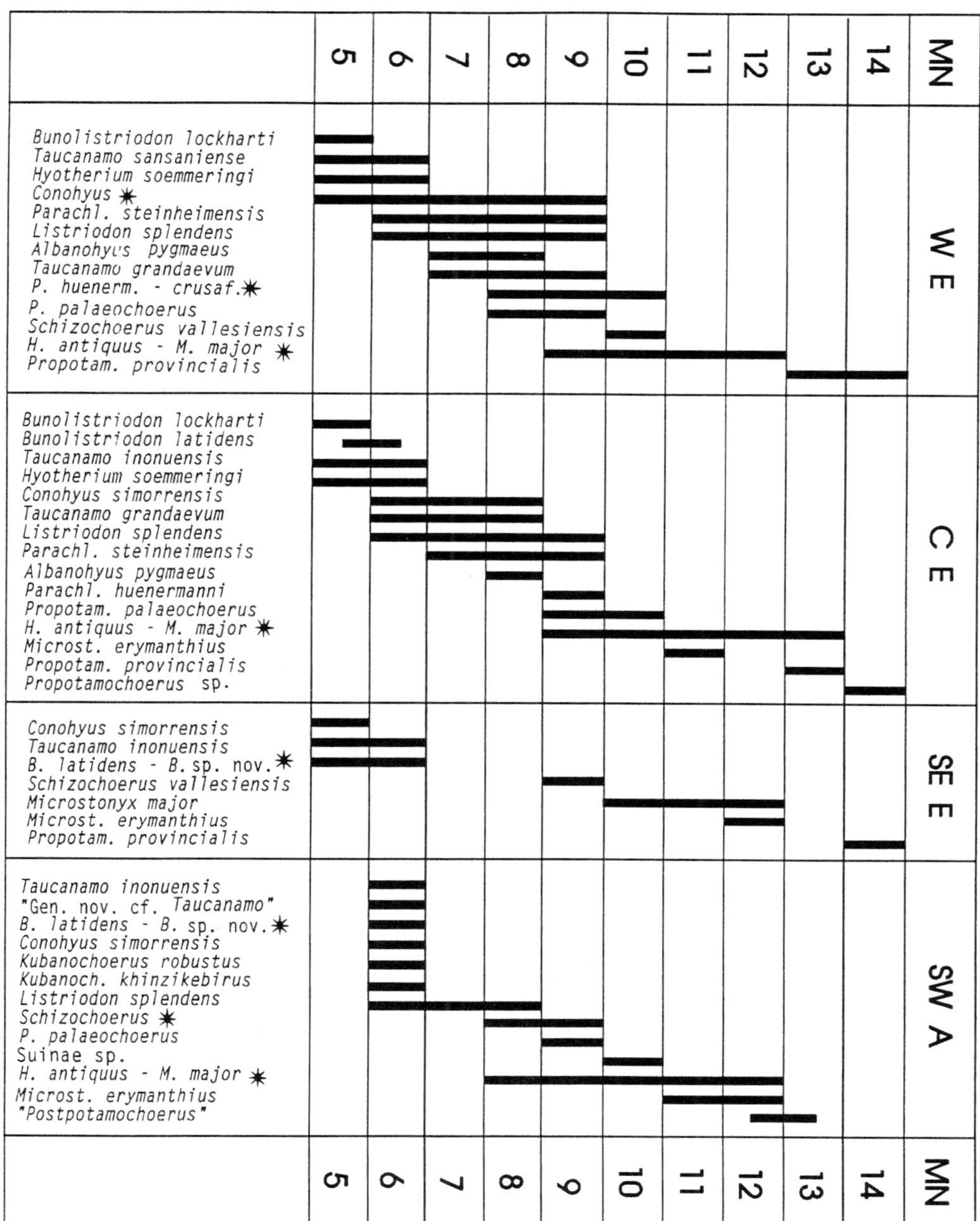

FIGURE 28.9 Biogeographic range chart for the taxa included in this chapter. WE = Western Europe, CE = Central Europe, SE E = Southeast Europe, SW A = Southwest Asia.

reveals that the synchronous pattern is mainly due to extinctions, the pattern of entries being similar but more noisy (see also Werdelin and Solounias, this volume). A comparison with Simpson's index (fig. 28.10, table 28.1) shows that similarity between Southwestern Asia and Western-Central Europe decreased after each turnover event (MN 6, MN 9, and MN 12).

Since the number of turnover events is generally highly correlated with diversity (in this case, r=0.76; P<0.001), two relative parameters were also calculated: "Relative Turnover" (absolute turnover divided by diversity) and "Probability of Extinction" (the number of extinctions divided by the number of taxa at risk, the latter estimated in this case simply as the number of taxa present). Relative Turnover (fig. 28.12a) shows a sinking trend from MN 6 to MN 11, with a marked rise to MN 12/13. The MN 9 peak is weakly indicated for Western and Central Europe and not at all for Southwestern Asia, and Central Europe again appears out of phase with the other regions, with a minimum at MN 10 rather than MN 11. The very high values

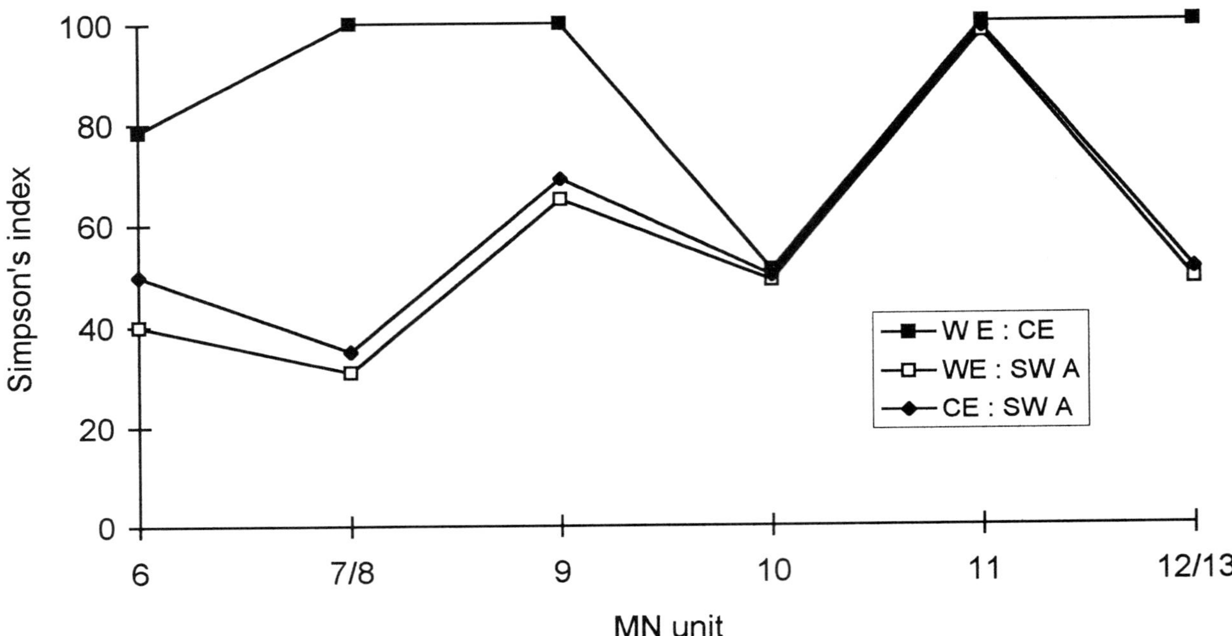

FIGURE 28.10 Graphic representation of Simpson's Index of Faunal Resemblance, showing high resemblance between Central and Western Europe as well as the highly similar resemblance trends of these regions with respect to Southwest Asia. See text for discussion. WE = Western Europe, CE = Central Europe, SE E = Southeast Europe, SW A = Southwest Asia.

TABLE 28.1. *Regional faunal resemblance expressed as Simpson's Index (shared taxa divided by taxa in the smaller fauna times one hundred).*

Region	W E	C E	SE E	SW A	Interval
		79	60	40	MN 6–13
		80	50	40	MN 6
Western		100	—	33	MN 7/8
Europe		100	—	67	MN 9
		50	100	50	MN 10
		100	100	100	MN 11
		100	50	50	MN 12/13
	79		100	50	MN 6–13
	80		50	50	MN 6
Central	100		—	33	MN 7/8
Europe	100		—	67	MN 9
	50		100	50	MN 10
	100		100	100	MN 11
	100		50	50	MN 12/13
	60	100		100	MN 6–13
	50	50		100	MN 6
Southeast	—	—		—	MN 7/8
Europe	—	—		—	MN 9
	100	100		100	MN 10
	100	100		100	MN 11
	50	50		100	MN 12/13
	40	50	100		MN 6–13
	40	50	100		MN 6
Southwest	33	33	—		MN 7/8
Asia	67	67	—		MN 9
	50	50	100		MN 10
	100	100	100		MN 11
	50	50	100		MN 12/13

for Southeastern Europe in MN 9 and MN 12/13 may be due to lack of material. The MN 9 value simply represents the entry and exit in that interval of the only taxon present (*Schizochoerus*), while the MN 12/13 value represents the dubious entry of one and exit of both species of *Microstonyx* in MN 12.

Probability of Extinction (fig. 28.12b) also peaks at MN 6 and MN 9, but the highest values are now seen in MN 12/13. The curve for Central Europe is now strikingly different from the others, lacking a peak at MN 9 and being instead uniformly high from MN 7 + 8 to MN 11. Although the diversity is too low for any great confidence, this could indicate that different controls operated on extinction in Central Europe than in the Tethys-Paratethys realm.

How reliably these patterns reflect reality is difficult to assess. They are certainly affected by various factors reflecting the incomplete nature of the data that could be referred to collectively as sampling error, and the sample sizes are well below the minima needed for statistical significance (e.g., Koch 1987). Sampling error is primarily present at two levels: (1) within the composite faunal lists based on several localities from a given region and time interval; and (2) among these composite faunal lists. The effects are compounded at the second level, which unfortunately is the most relevant to our analyses. In practice, however, the problems boil down to relatively simple specific issues that lend themselves to critical discussion.

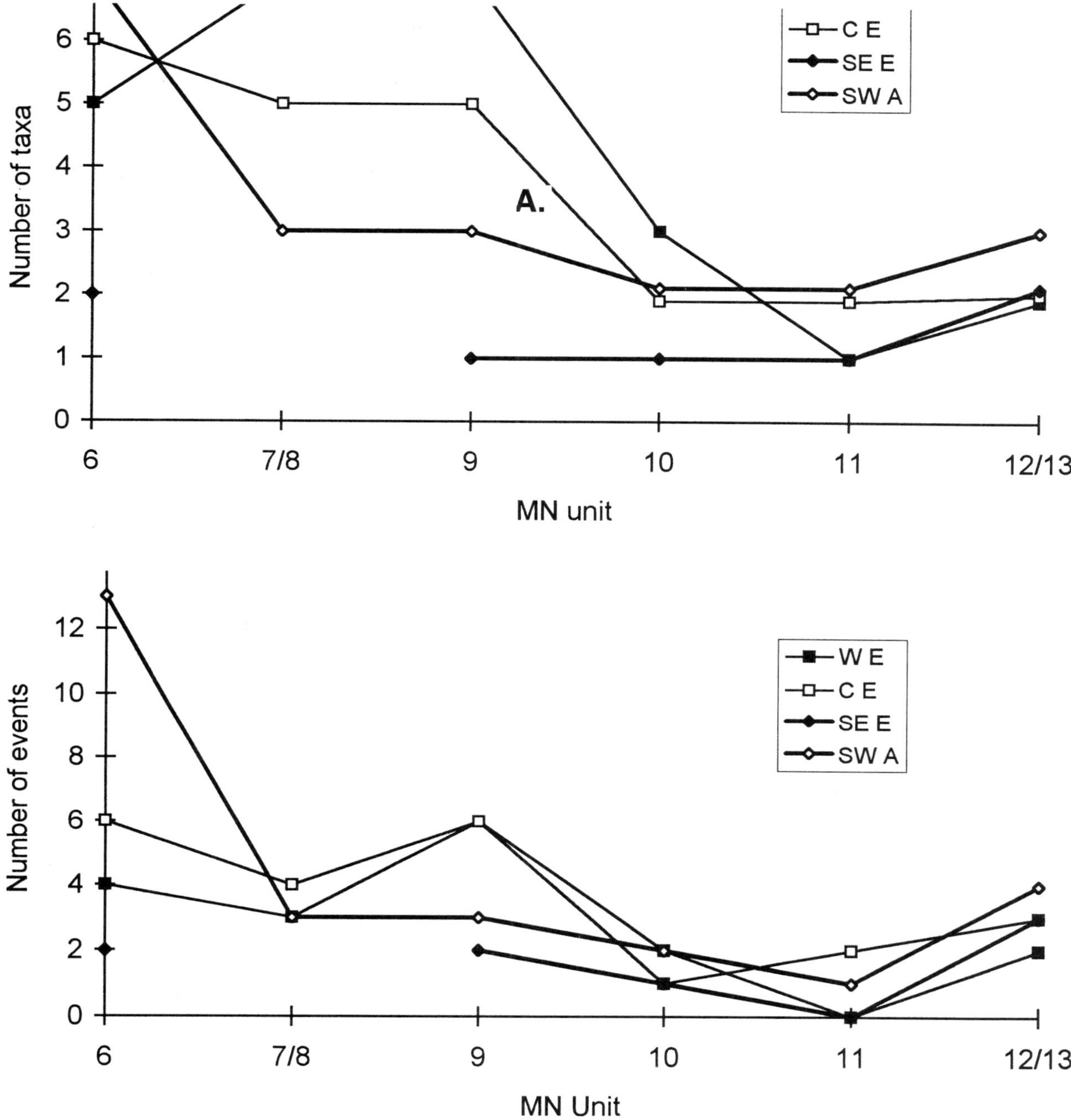

FIGURE 28.11 28.11A. Species diversity and absolute turnover between MN 6 and MN 12/13 for the different regions. See text for discussion. Diversity is the number of species or lineages recognized for that MN unit and region. 28.11B. Absolute turnover is the total number of entries and exits of species or lineages for the MN unit and region. WE = Western Europe, CE = Central Europe, SE E = Southeast Europe, SW A = Southwest Asia.

For example, the MN 6 turnover peak for Southwestern Asia might be inflated by poor sampling of MN 7 + 8 as well as by the artificial "entries" due to lack of data for MN 5. *Conohyus* is known from later deposits in Europe and the Siwaliks, and might have been present in between as well, while "Gen. nov. cf. *Taucanamo*" may have given rise to *Schizochoerus*, in which case its exit is an artifact of taxonomy. The exits of *Taucanamo*, the *Bunolistriodon latidens* lineage, and *Kubanochoerus* (2 taxa) are most probably genuine, however, as is the entry of *Listriodon*,

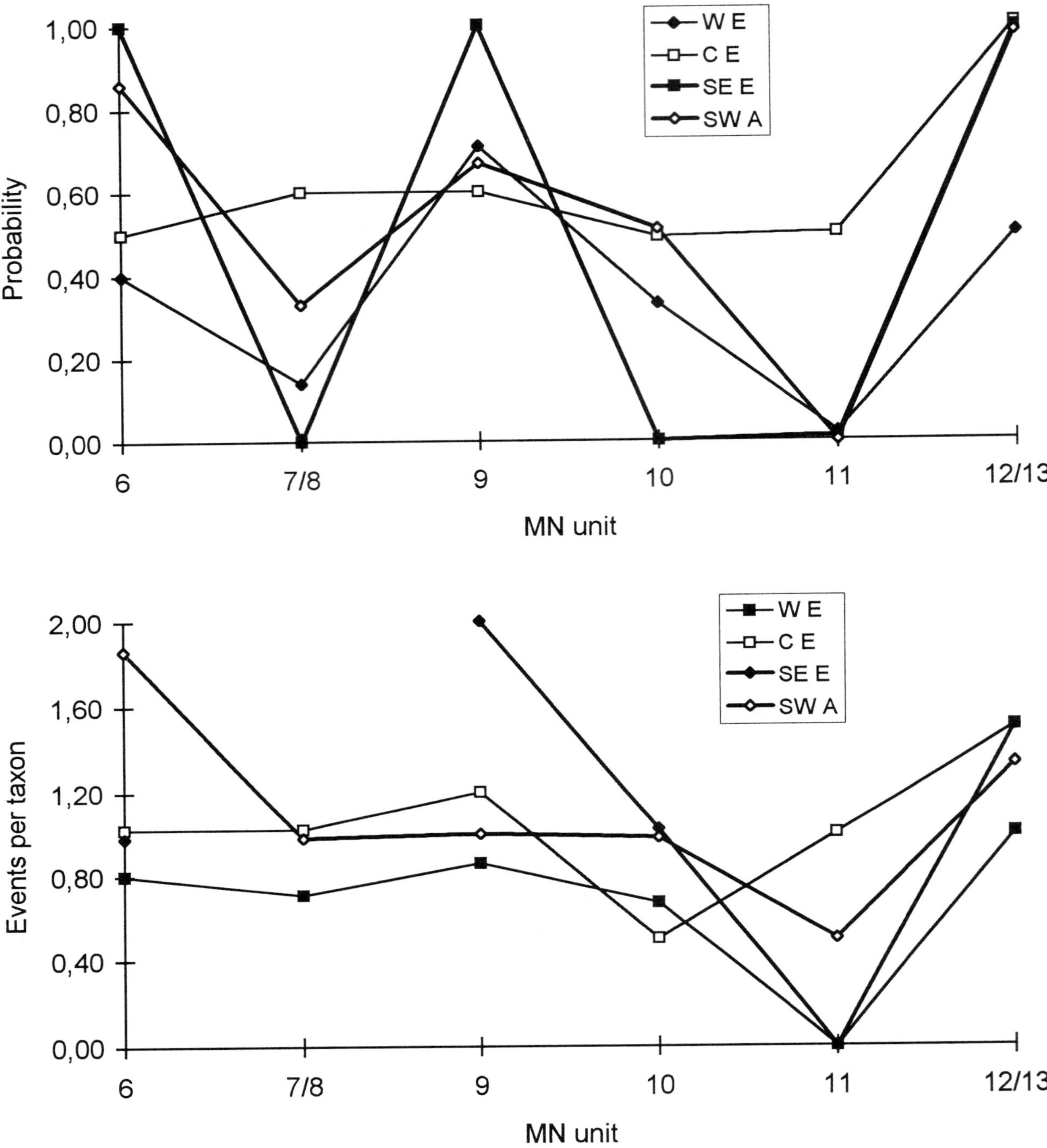

FIGURE 28.12 28.12A. Relative turnover and probability of extinction between MN 6 and MN 13 for the different regions. See text for discussion. Relative turnover is absolute turnover divided by diversity. 28.12B. Probability of extinction is the number of extinctions divided by an estimate of the number of taxa at risk (estimated here as the number of taxa present). WE = Western Europe, CE = Central Europe, SE E = Southeast Europe, SW A = Southwest Asia.

since they concern the entire area. The fact that all three Anatolian MN 6 localities have a rich suoid fauna while all four Anatolian MN 7 + 8 localities have yielded material only of *Listriodon* but not of any other taxon present in MN 6 also indicates that sampling error is not the main explanation of this pattern. Even a conservative interpreta-

tion of the evidence therefore suggests that this event was significantly stronger in Southwestern Asia than in Europe.

The somewhat arbitrary decision to terminate the range of *Propotamochoerus palaeochoerus* with MN 9 in Western Europe emphasizes the Western European MN 9 turnover peak, but it is only one of six changes recorded so the effect is relatively small (extending its range would strengthen the pattern of diachronous east-to-west change suggested by our analysis, an effect we wished to avoid). The difference in turnover patterns (especially probability of extinction) between Central Europe and the other regions is affected by many samples and also appears unlikely to be a pure artifact, especially given the highly congruent synchroneity exhibited by all the other regions, including the poorly sampled Southeastern Europe.

Probably the most difficult pattern to interpret is the Turolian one. The suoid material is extremely limited for the end of the interval, and unlikely to give a reliable picture of the distribution of the taxa that have been recorded. Taken literally, the evidence suggests that regional differentiation increased greatly in MN 12/13 (fig. 28.10), following the (apparent?) regional extinction of *Microstonyx* outside Central Europe, and the appearance of *Propotamochoerus provincialis* in Western and Central Europe. However, as discussed above, *P. provincialis* appears closely related to Chinese and African taxa, and its absence from the eastern part of the study area may well be a sampling artifact. On the other hand it seems likely that *Sus arvernensis* entered Sardinia in MN 13, even though the oldest finds are from MN 14 in Spain, so the late Turolian diversity was probably greater than recorded by our data (sampling error operating in the opposite direction). The change in MN 12/13 may appear more radical than it probably really was, but there is no denying the contrast to MN 11, when only *Microstonyx* is recorded in relatively rich material from the entire study area.

It may therefore be concluded that although some details are undoubtedly spurious, the main pattern appears quite robustly founded: three intervals of marked turnover (MN 6, MN 9, and MN 12/13) are seen against a background of more gradual, diachronous change proceeding from east to west. This impression is further supported by the paleoecological review that follows here.

Suoid Paleoecology

Suoids are the archetypal omnivores, and most living species have strikingly broad dietary and environmental adaptations. This may explain why suoid paleoecology has received comparatively little study: little can be concluded from living analogs. Accepting this limitation, we provide a brief review here based mainly on body mass and dental morphology.

The taxa included in this analysis range in estimated body mass from less than 10 kg to 850 kg, which we arbitrarily divide into four size classes: 1–20 kg; 21–80 kg; 81–200 kg; 201–1000 kg (table 28.2). Only in MN 6 of Southwestern Asia and MN 9 of Western Europe are all four size classes present simultaneously (fig. 28.13), indicating that these were maxima of ecological as well as

taxonomic diversity. Most probably the reason is the diversity of habitats available, either locally as mosaics in Anatolia (Andrews 1990) or regionally as in Western Europe (Agustí et al. 1984). The general body mass trend for all areas is that of increasing mean size accompanying decreasing size diversity, until only Class 4 (*Microstonyx*) is found in MN 11–12. After the extinction of *Microstonyx* in MN 12/13 all taxa are Class 3. There is a striking gradient in this pattern from east to west, with changes occurring first in Anatolia and last in Spain. For example, the last Class 1 taxa are found in MN 6 of Anatolia, in MN 8 of Central, and MN 9 of Western Europe (Spain). Similarly, Class 2 disappears from Anatolia during MN 8, from Central Europe during MN 9 and from Spain during MN 10. From MN 6 to MN 9 the dominance of Classes 1 and 2 is strong in Central and Western Europe, whereas the distribution of size is more evenly distributed in Anatolia during this interval, with Classes 3 and 4 beginning to dominate there from MN 9 onward.

The autecological significance of suoid body size is not clear. The smallest living suoid, *Sus salvanius* (6–10 kg), inhabits tall grasslands of the Himalayan foothills, and *Tayassu tajacu* (17–25 kg) ranges from wet and dry tropical forests to arid open habitats (Macdonald 1984), providing evidence that the smallest suoids can function in quite diverse environments. However, all of the fossil Class 1 taxa under consideration have a dental morphology with high pointed cusps and variable loph development. Functionally, the teeth would appear to be analogous with certain cercopithecoid primates and tragulids, and it seems probable that they were forest forms.

Class 2 taxa are dentally more diverse, although the molars of *Hyotherium* and the tetraconodonts are still relatively close to the primitive suoid pattern. The combination of thick dental enamel and conical premolars with hyaenalike macrowear might indicate that cracking hard food items such as seeds was an important component of tetraconodont species feeding behavior. Thenius (1952: footnote p. 121) postulated, on faunal and sedimentological criteria, that *Conohyus simorrensis* was a denizen of marshy forests ("Sumpfwaldbewohner") while *Parachleuastochoerus steinheimensis* was adapted to drier conditions ("Trockenstandorte"). We are unable to elucidate this matter further, but the later history of the African tetraconodonts amply demonstrates that the group was capable of invading open habitats. There is one lophodont Class 2 form, the small *Schizochoerus* from Sinap.

Of the recent suoid species, the two larger peccaries and the bush pig are in Class 2, and although they occur in a wide range of habitats, all are at least partly forest animals. Class 2 spans the body mass range between 35 and 80 kg, where foraging in open habitats becomes increasingly feasible physiologically as size increases (Wheeler 1992). The fact that size did not increase signifi-

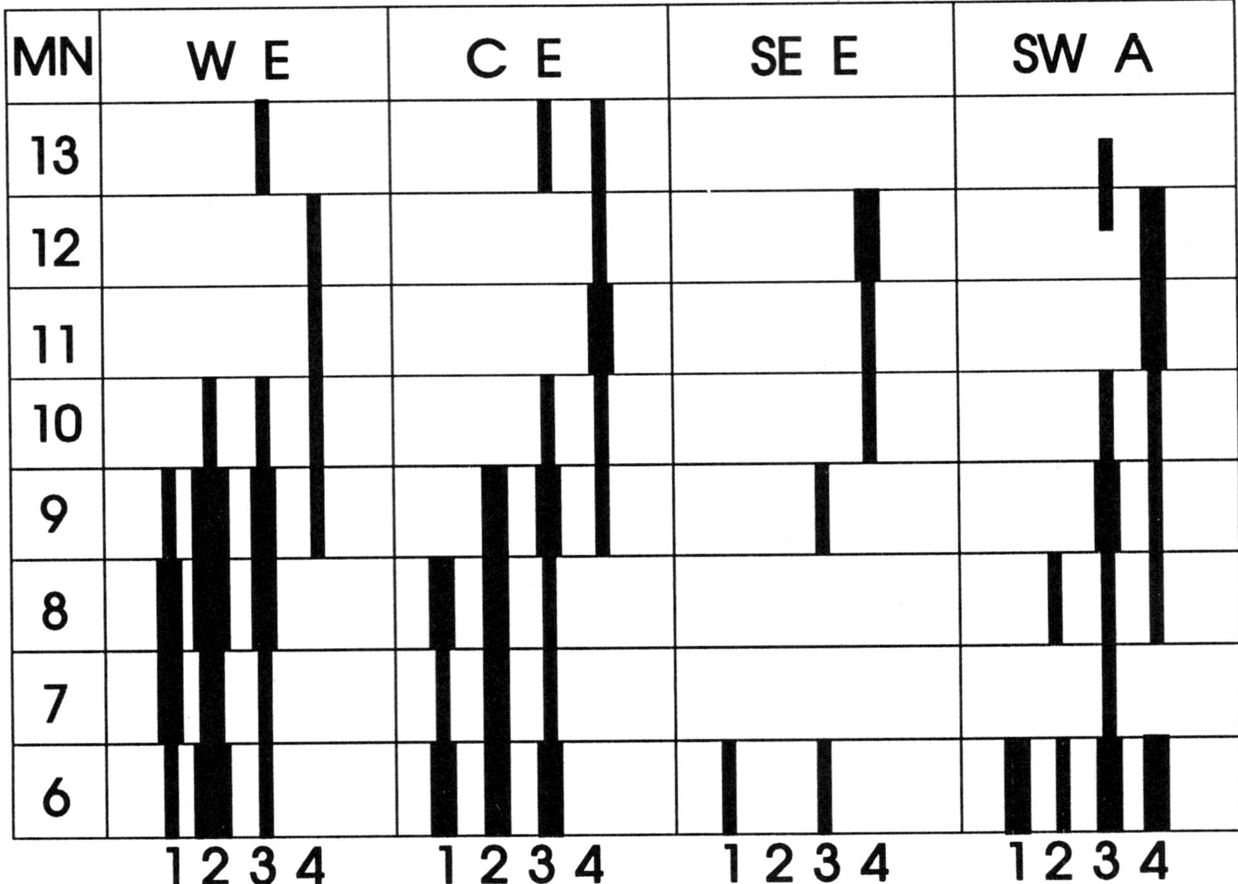

FIGURE 28.13 Distribution of size classes 1–4 (less than 20 kg, 21–80 kg, 81–200 kg, and 201–1,000 kg). The width of the column indicates the number of taxa in that size class for the MN unit for region. Note increase in mean size accompanied by decrease in diversity, proceeding time-transgressively from east to west. WE = Western Europe, CE = Central Europe, SE E = Southeast Europe, SW A = Southwest Asia.

cantly during the history of these taxa might perhaps argue against predominantly open habitat associations. The only exception is *Schizochoerus* sp., which probably gave rise to *S. vallesiensis* of Class 3. It occurred at Sinap in MN 8, together with the Class 3 *Listriodon splendens*, in what appears to have been an open and at least partly arid habitat (Alpagut et al. 1994).

Class 3 suoids have the greatest dental diversity. *Bunolistriodon* retains a *Hyotherium*-like molar morphology in combination with the derived listriodont incisal cropping device. *Listriodon* and *Schizochoerus* are both fully lophodont. *Propotamochoerus* and a few other suines also belong to Class 3. Thenius (1950) regarded *P. palaeochoerus* as a forest form and "*Postpotamochoerus hyotheroides*" as a "savanna" form, mainly on the basis of faunal associations. It is difficult, however, to separate the temporal from the ecological element in this argument. Of the recent suoids, *Babyrousa* and *Phacochoerus* belong in Class 3, the latter being the only living suoid that relies on grass for bulk food and is essentially confined to open habitats (Kingdon 1979). Medium-sized *Sus scrofa* also belongs to this class,

but the various subspecies span Classes 2–4. Suines are among the many examples in mammalian evolution of the "bear syndrome," or proliferation of minor molar cusps concomitant with the loss of individual cusp identity. Such a dental morphology is unlikely to contribute much beyond the label "omnivore" and, indeed, there may be little else to conclude. Suoids in the 100-kg range undoubtedly had the basic physiological capability to live in open habitats.

Basing his opinion on faunal associations, Thenius (1970) suggested that *Listriodon splendens* occupied open habitats. Late Miocene–Pliocene African *Nyanzachoerus* exhibited a similar locomotor adaptation by general lengthening the distal limb elements (McCrossin 1987). Recently, Hunter and Fortelius (1994) compared the sympatric listriodont species from Paşalar (*Bunolistriodon* sp. nov. and *Listriodon* cf. *splendens*), and found significant differences in microwear between them, indicating a more uniform diet with less inclusion of minerogenic hard particles in the *Listriodon*. This conclusion supports the hypothesis that *Listriodon* had diverged from the traditional omnivo-

TABLE 28.2 *Body Mass Estimates Based on Second Upper Molar Length*

Taxon/locality	Ref	N	LM2 (mm)	Mass (kg)
Class 1: 1–20 kg				
Albanohyus pygmaeus (LaGrive)		4	9.2	9
Taucanamo grandaevum (Steinheim)		4	9.9	11
Taucanamo sansaniense (Sansan)		8	10.4	13
Taucanamo inonuensis (Inönü I)		1	11.0	16
"Gen. indet. cf. *Taucanamo*" (Çandir) (est. fr. m2)	1	1	(11)	(16)
Class 2: 21–80 kg				
Schizochoerus sp. (Lower? Sinap)	1	1	14.7	37
Parachleuastochoerus huenermanni (various locs)		4	14.9	39
Hyotherium soemmeringi (various localities)	2	9	17.5	63
Conohyus simorrensis (various localities)		3	18.1	69
Parachleuastochoerus steinheimensis (Steinheim)		2	18.4	73
Class 3: 81–200 kg				
Bunolistriodon sp. nov. (Paşalar)		2	19.6	88
Listriodon cf. *splendens* (Paşalar)		21	19.8	91
Bunolistriodon latidens (Inönü)		2	21.4	110
Schizochoerus vallesiensis (Middle Sinap)		1	21.7	120
Propotamochoerus palaeochoerus (various localities)	3	21	22.0	120
"*Postpotamochoerus*" (Samos)	4	1	23.0	140
Propotamochoerus provincialis (various localities)			23.9	160
"*Suinae* sp." (Karain) (estimated from m2)		1	(25)	(180)
Class 4: 201–1,000 kg				
Microstonyx erymanthius (Pikermi)	5	13	28.6	270
Microstonyx major (various localities)	6	12	30.5	330
Kubanochoerus robustus (Belometchetskaia)	7	1	34.7	480
Hippopotamodon sivalense (for *H. antiquus*)	8	6	35.3	510
Kubanochoerus khinzikebirus (Inönü I)		1	41.8	850

Data from 1 = Pickford & Ertürk 1979; 2 = Schmidt-Kittler 1971; 3 = Hünermann 1968; 4 = Thenius 1950; 5 = Pearson 1928; 6 = Van der Made et al. 1992; 7 = Gabunia 1973; 8 = Pickford 1988; otherwise original.

rous suoid feeding behavior, and that it probably did not engage in rooting.

Schizochoerus seems to always have been rare. It is possible, however, that a lineage from the unnamed Çandir tayassuid in MN 6 to *Schizochoerus* sp. in MN 8 and *S. vallesiensis* in MN 9 occurs in Anatolia. If that is the case it would imply that *Schizochoerus* became large only toward the end of its history, perhaps after the local extinction of *Listriodon*. *Schizochoerus vallesiensis* first occurs in Spain during MN 10, coincident with the disappearance of *Listriodon*. Pickford (1978) has speculated that this large, lophodont tayassuid filled a niche vacated by *Listriodon*, a suggestion that would fit its biogeographic history well.

Class 4 shows a similarly mixed ecological content, with two groups that probably differed significantly from each other in habitat use. The lifestyle of *Kubanochoerus* remains as much of an enigma as does the significance of its "horns." The only true Class 4 taxon among living suoids is *Hylochoerus meinertzhageni*, which appears to require a "yearlong and plentiful supply of green fodder and dense cover" (Kingdon 1979). *Hylochoerus meinertzhageni* shows strong and continuous sexual competition among males. The sexually dimorphic "horns" of *Kubanochoerus* (Guan Jian and Van der Made 1994) might indicate a similar social lifestyle. A sexually dimorphic, large herbivore with male "horns" might also fit the extended version (Janis 1982) of Jarman's (1974) bovid category B of polygamous species with territorial males. These ungulates have brachy-

dont or mesodont molars and select strongly for nutritious individual food items, often with minor seasonal variability in the diet, and they occur in habitats ranging from forest to woodland-grassland mosaics. Both occurrences of *Kubanochoerus* are from MN 6 faunas in Southwestern Asia (Belometchetskaia and Inönü I), in association with both brachydont and hypsodont ungulates (Gabunia 1973; Gürbüz 1981; see also Andrews 1990).

The *Hippopotamodon*—*Microstonyx* lineage seems to have more definitely open habitat affiliations. From cranial morphology and faunal associations, Thenius (1972) argued that *Microstonyx major* inhabited open country environments. This interpretation is supported by a series of later studies on Turolian mammalian megafaunas (Bernor et al. 1979; Bernor 1983, 1984; Solounias 1981). Today, *Sus scrofa* is successful in the steppe, and demonstrates that large suines do have the potential to function in open habitats. It is significant, however, that *Microstonyx major* became extinct in the circum-Mediterranean area during the MN 12/13 climax of open country savannalike faunas. No hypsodont, grazing suoids are known to have occurred in Western Eurasia, in contrast to the Indian subcontinent and especially East Africa.

To summarize, the fossil Class 1 taxa were probably forest forms. Class 2 contains a mixture of forms but would seem to be predominantly composed of species that did not inhabit open environments. Class 3 is an ecologically heterogeneous group with the majority of taxa exhibiting open habitat associations. Class 4 contains forms from woodland-grassland mosaic habitats and (later on) forms more definitely associated with open woodlands. The pattern seen in figure 28.13 thus strongly suggests that the increase in mean size and concomitant loss of taxonomic and ecological diversity was associated with a loss of environmental diversity, involving especially the reduction of forest habitats.

Discussion

The results presented here generally correspond to a pattern already recognized from other sources (see Potts and Behrensmeyer et al. 1992 for a recent review). The high diversity seen in MN 6 is undoubtedly related to the immigration of land mammals from Africa and West Asia (Thenius 1979; Bernor 1983; de Bruijn and Hussain 1984). The subsequent trend of declining diversity would follow from the general reduction of warm subtropical woodlands between 15.5 and 9 Ma succeeded by the explosive expansion of open country habitats between 9 and 7 Ma, to a maximum equilibrated extension between 7 and 5 Ma (Bernor 1983; Bernor et al. 1988, 1990; recent correlations by Steininger et al. 1989; Rögl et al. 1993; Bernor et al. 1993; several articles in this volume). Whereas diversity of most ungulate groups increased as the forest habitats became more diversified in the late Astaracian and early Vallesian, this seems to apply to the suoids only in Central and Western Europe. Anatolian suoid diversity dropped precipitously both taxonomically and ecologically following the MN 6 high. Indeed, the most compelling pattern to emerge from our analysis is the time-transgressive east-to-west gradient of change, with extinction of small forms of predominantly forested habitats (size Classes 1 and 2), first from Anatolia and last from Spain, followed by their replacement with larger taxa of eastern origin and with more uniform, open habitat associations (size Classes 3 and 4; figs. 28.8 and 28.13).

The timing of evolution and extinction events is suggestively synchronous with physical changes. The initial sharp drop in suoid diversity (and extinction peak) after MN 6 occurs closely coincident with the Serravallian regression (= Paratethyan "Badenian Salinity Crisis," sensu Rögl and Steininger 1984), and directly following the last extension of Indomalaysian marine currents into Central Europe. This regressive phase further corresponds to active Alpine orogeny and a series of closely spaced global sea lowering events at 16.5, 15.5 and 13.5 Ma (Haq et al. 1987; Bernor and Tobien 1990) and the rapid shift in the oceanic oxygen isotope record, indicating an episode of rapid Antarctic glaciation (Raymo and Ruddiman 1992; Van der Meulen and Daams 1992; Flower and Kennett 1993).

Scattered data from other vertebrate groups reinforce the pattern seen in the suoid fauna. In MN 6 the cold-sensitive crocodiles still occured in Central Europe (Scherer 1981) while in France their last record is in MN

5 (Antunes and Ginsburg 1989). Flamingoes are abundant in MN 6, but in MN 7 they become rare as ducks become more abundant (Hesse, pers. comm.). Tragulid diversity decreased in Central Europe from 4 species in MN 5–6 to 1 species in MN 7–12, but stayed high in warmer regions (Van der Made 1992c). Van der Meulen and Daams (1992) interpreted the change in Spanish rodent faunal composition as evidence for decreasing temperature during MN 6, with particularly sharp temperature declines documented at the beginning and end of the interval. We conclude that there is good evidence for an early Astaracian cooling in Europe.

The second extinction peak, during MN 9, has less obvious physical correlates but might relate to a significant global environmental event. There is a global cooling spike recorded in the oceanic oxygen record at 11.5 Ma, and the culmination of the European Serravallian regression occurs at 10.5 Ma (depending on the calibration of the MN units). The "*Hipparion* Datum" now appears to correlate with this latter event (Swisher, this volume; Woodburne et al., this volume). It may be that the global cooling spike, closely followed by the terminal Serravallian event, and aggravated by continued Alpine orogeny, had the effect of driving terrestrial environments progressively further toward more open conditions, perhaps with increased seasonality (Bernor 1983; Bernor et al. 1988, 1990). Fossil floras from the Eastern Mediterranean and Spain both indicate a decrease in temperature at this time, with regional fluctuations in humidity (Benda 1971; Benda and Meulenkamp 1990; Baltuille et al. 1992).

The final extinction peak occurred during MN 12/13, and may well correspond to the terminal Miocene Paratethys and Mediterranean regression (ca. 7–5 Ma; Rögl et al. 1993). If so, it may be significant that Central Europe, displaced geographically from the Mediterranean, and continuing to support a viable lake system, is the only region where *Microstonyx* is definitely known not to have become locally extinct until MN 13.

The provinciality of the study area would appear to be strongly affected by the Central and Eastern Paratethys system. During the Astaracian, when diversity is high, Simpson's Index of Faunal Resemblance reveals that suoid species are segregated into an Eastern and a Western province (fig. 28.10). A bottleneck passage through the alpine foredeep (Austria-Germany) may have been episodically operational, but clearly did not result in increased faunal similarity, since the differentiation was greatest in MN 7 + 8. It is interesting in this context that the extinction pattern found for Central Europe differs from that of the three other regions, but the species-level reasons for the difference do not lend themselves to simple explanation, and in any case a handful of species is a poor data base for deciphering the timing of distant biogeographic connections. It is certainly probable that Central Europe differed climatically from the more southern regions of the Tethys-

Paratethys realm, and different environmental controls may have operated there, even if the composition of the fauna was quite similar to that of adjacent regions.

Taxonomic and ecologic diversity decreased, and the uniformity between regions increased from MN 6 until MN 11. The main mechanism behind the faunal unification was replacement of the taxonomically and ecologically diverse suoid fauna by considerably less diverse eastern immigrants (*Schizochoerus*, *Hippopotamodon-Microstonyx*, and *Propotamochoerus*). This replacement occurred during the Astaracian in Southwestern Asia and the Vallesian in Europe, where the last native taxa are found in Spain. In Southwestern Asia, there appears to be a gap between the local MN 6 extinction and the appearance of *Propotamochoerus* and *Hippopotamodon* in MN 8. If this is not a sampling artifact, it would strengthen the impression that no direct interaction was involved between natives and immigrants. The extinction peaks at MN 6 and MN 9 both decreased faunal uniformity, and the (less rhythmic) immigration pattern seems to have been mainly responsible for increased uniformity in most of the area during most of the time. We correlate the early Turolian uniformity of Western Eurasian suoid faunas with the expansion of open country woodlands reported by Bernor (1983) and Bernor et al. (1990). This phase came to an end with the extinctions in MN 12/13, and was followed by a less uniform, though still extremely sparse suoid fauna.

A mirror image of the suoid pattern is seen for the hipparionine horses (Bernor et al., this volume), while the pattern for primates is more similar to that of the suoids (Andrews et al., this volume). It is entirely plausible, of course, that taxa from more continental regions should be ecologically preadapted to the increasingly dry and open habitats that developed in Europe during the Vallesian and Turolian. It is unlikely that much competition was involved, since there is little ecological overlap between the natives and the invaders. Suoid succession would appear to be related to rapid shifts in the originally diverse local habitats, which precipitated the extinction of native forms and their replacement by immigrants from more uniform, open environments.

Acknowledgments

We wish to thank the following individuals for their permission to study material under their care, and for the generous help offered during our work: B. Alpagut, S. Alpaslan, O. Baysal, T. Bolliger, H. de Bruijn, B. Engesser, A. Gentry, D. Goujet, W. Gräf, Ç. Ertürk, V. Fahlbusch, J. L. Franzen, L. Ginsburg, Z. Gözler, K. H., E. P. J. Heizmann, A. Hesse, M. Hugueney, K. A. Hünermann, J. Hürzeler, W. von Koenigswald, L. Kordos, H. Lelièvre, H. Lutz, P. Mein, S. Moyà-Solà, R. Niederl, Ü. Özdemir, M. Philippe, G. Plodowski, S. Rietschel, G. Saraç, H.-K. Schmutz, F. Schrenk, S. Sen, P. Y. Sondaar, I. Tekkaya, H. Tobien, L. Trunko, E. Ünay, and J.. de Vos. We thank J. Barry, J. Damuth, A. Forstén, L. Van Valen, and L. Werdelin for discussion and valuable comments on the manuscript. M. Fortelius is grateful to the Academy of Finland, J. Van der Made to the Faculty of Geology of the Rijksuniversiteit Utrecht, and R. L. Bernor to the Humboldt Stiftung, the National Science Foundation, the National Geographic Society, the L. S. B. Leakey Foundation, and Howard University for support of the research leading to this study. We all thank the VW Stiftung for support during the Schloss Reisensburg workshop.

LITERATURE CITED

Abusch-Siewert, S. 1983. Gebissmorphologische Untersuchungen an eurasiatischen Anchitherien (Equidae, Mammalia) unter besonderer Berücksichtigung der Fundstelle Sandelzhausen. *Courier Forschungsinstitut Senckenberg* 62:1–401.

Agustí, J., S. Moyà-Solà, and J. Gibert. 1984. Mammal distribution dynamics in the eastern margin of the Iberian Peninsula during the Miocene. *Paléobiologie Continentale* 14:33–46.

Alpagut et al. 1994. *Survey report for the Sinap Formation project (Ankara, Turkey)* 1992. T.C. Kültür Bakanligi Anitlar ve Müzeler Genel Müdürlügü. Ankara: Arastirma Sonuçlari Toplantisi.

Andrews, P. 1990. Palaeoecology of the Miocene fauna from Paşalar, Turkey. *Journal of Human Evolution* 19:569–82.

Andrews, P., T. Harrison, E. Delson, R. L. Bernor, and L. Martin. This volume. Distribution and biochronology of European and Southwest Asian Miocene catarrhines.

Andrews, P., P. J. Whybrow, and C. B. Stringer. 1980. Stratigraphy and palaeontology of Miocene deposits at Yeni Eskihisar, Turkey. *Newsletters on Stratigraphy* 9:49–57.

Antunes, M. T. and L. Ginsburg. 1989. Les Crocodiliens des faluns miocènes de l'Anjou. *Bulletin du Muséum national d'Histoire naturelle*, 4e serie, 11, section C, 2:79–99.

Archer, A. W. and C. G. Maples. 1987. Monte Carlo simulation of selected binomial similarity coefficients (I): Effect of number of variables. *Palaios* 2:609–17.

———. 1989. Response of selected binomial coefficients to varying degrees of matrix sparseness and to matrices with known data interrrelationships. *Mathematical Geology* 21:741–53.

Azanza, B. and E. Menendez. 1992. Los Ciervos fósiles del néogeno español. *Palontologia i Evolució* 23 (1989/1990): 75–82.

Bach, F. 1908. *Listriodon splendens* H. v. M. aus Steiermark. *Verhandlungen der kaiserlich königlichen geologischen Reichsanstalt* 5/6:117–18.

Bakalov, P. and I. Nikolov. 1962. [Les fossiles de Bulgarie 10. Mammifères tertiaires] (Bulgarian, French summary).

Baltuille, J. M., J. D. Becker-Platen, L. Benda, and Y. I. Calzaga. 1992. A contribution to the subdivision of the Neogene in Spain using palynology. *Newsletters on Stratigraphy* 27:41–57.

Becker-Platen, J. D., O. Sickenberg, and H. Tobien. 1975. Die Gliederung der känozoischen Sedimente der Türkei nach Vertebraten-Faunengruppen. *Geologisches Jahrbuch* (B) 15:19–45.

Benda, L. 1971. Principles of the palynologic subdivision of the Turkish Neogene. (Känozoikum und Braunkohlen der Türkei 3). *Newsletters on Stratigraphy* 1:23–26.

Benda, L. and J. E. Meulenkamp. 1990. Biostratigraphic correlations in the Eastern Mediterranean Neogene. 9. Sporomorph associations and event stratigraphy of the Eastern Mediterranean. *Newsletters on Stratigraphy* 23:1–10.

Bernor, R. L. 1983. Geochronology and zoogeographic relationships of Miocene Hominoidea. In *New Interpretations of Ape and Human Ancestry*, ed. R. L. Ciochon and R. S. Coruccini, pp. 21–64. New York: Plenum Press.

———. 1984. A zoogeographic theater and a biochronologic play: The time/biofacies phenomena of Eurasian and African Miocene mammal provinces. *Paléobiologie Continentale* 14:121–42.

———. 1986. Mammalian biostratigraphy, geochronology, and zoogeographic relationships of the late Miocene Maragheh fauna, Iran. *Journal of Vertebrate Paleontology* 6:76–95.

Bernor, R. L., P. J. Andrews, N. Solounias, and J. A. H. Van Couvering. 1979. The evolution of "Pontian" mammal faunas: Some zoogeographic, palaeoecologic, and chronostratigraphic considerations. *Annales Géologiques des Pays Helléniques*, hors série, 1:81–89.

Bernor, R. L., J. Kovar-Eder, D. Lipscomb, F. Rögl, S. Sen, and H. Tobien. 1988. Systematic, stratigraphic, and paleoenvironmental contexts of first-appearing *Hipparion* in the Vienna Basin, Austria. *Journal of Vertebrate Paleontology* 8:427–52.

Bernor, R. L., J. Kovar-Eder, J.-P. Suc, and H. Tobien. 1990. A contribution to the evolutionary history of European late Miocene age hipparionines. *Paléobiologie Continentale* 17:291–309.

Bernor, R. L., H.-W. Mittmann, and F. Rögl. 1993. Systematics and Chronology of the Götzendorf "*Hipparion*" (late Miocene, Pannonian F, Vienna Basin). *Annalen des Naturhistorisches Museum, Wien* 95:101–20.

Bernor, R. L., N. Solounias, C. C. Swisher III, and J. A. Van Couvering. This volume. The correlation of three classical "Pikermian" faunas—Maragheh, Samos, and Pikermi—with the European MN unit system.

Bernor, R. L. and H. Tobien. 1990. The mammalian geochronology and biogeography of Paşalar (middle Miocene, Turkey). *Journal of Human Evolution* 19:551–68.

Bernor, R. L., H. Tobien, and M. O. Woodburne. 1989. Patterns of Old World hipparionine evolutionary diversification. In *European Neogene Mammal Chronology*, ed. E. H. Lindsay, V. Fahlbusch, and P. Mein, pp. 263–319. New York: Plenum Press.

Biedermann, W. G. A. 1873. *Petrefacten aus der Umgegend von Winterthur 4. Reste aus Veltheim.* Winterthur: J. Westfehling.

Blainville, H. M. D. de. 1847. *Ostéographie ou description iconographique comparée du squelette et du système dentaire des mammifères récents et fossiles pour servir de base a la Zoologie et a la Geologie. Vol 4. Sur les Hippopotames et les Cochons.* Paris: J. B. Baillière et Fils.

Bruijn, H. de, R. Daams, G. Daxner-Höck, V. Fahlbusch, L. Ginsburg, P. Mein, and J. Morales. 1992. Report of the RCMNS working group on fossil mammals, Reisensburg 1990. *Newsletters on Stratigraphy* 26:65–118.

Bruijn, H. de and S. T. Hussain. 1984. The succession of rodent faunas from the Lower Manchar Formation (Southern Pakistan) and its relevance for the biostratigraphy of the Mediterranean Miocene. *Paléobiologie Continentale* 14:191–204.

Chen Guanfang. 1984. Suidae and Tayassuidae (Artiodactyla, Mammalia) from the Miocene of Steinheim a. A. (Germany). *Palaeontographica* 184:79–83.

Dehm, R. 1934. *Listriodon* im Südbayerischen Flinz (Obermiozän). *Zentralblatt für Mineralogie, Geologie und Paläontologie,* Abteilung B: 513–28.

———. 1937. Neue tertiäre Spaltenfüllungen im südlichen Frankischen Jura. *Zentralblatt für Mineralogie, Geologie und Paläontologie,* Abteilung B: 349–69.

———. 1980. Über ein neues *Hyotherium* (Suidae, Schweineverwandte) aus der oberen Süsswassermolasse Südbayerns. *Annalen des Naturhistorischen Museums, Wien* 83:49–57.

Engesser, B. 1972. Die obermiozäne Säugetierfauna von Anwil (Baselland). *Tatigkeitsberichte der naturforschenden Gesellschaft Basellands* 28:37–363.

Fahlbusch, V. 1991. The meaning of MN zonation: Considerations for a subdivision of the European continental Tertiary using mammals. *Newsletters on Stratigraphy* 24:159–73.

Flower, B. P. and J. P. Kennett. 1993. Relations between Monterey Formation deposition and middle Miocene global cooling: Naples Beach section, California. *Geology* 21:877–80.

Fortelius, M. 1985. Ungulate cheek teeth: Developmental, functional, and evolutionary interrelations. *Acta Zoologica Fennica* 180:1–76.

———. 1990. Problems with using fossil teeth to estimate body sizes of extinct mammals. In *Body Size in Mammalian Paleobiology,* ed. J. Damuth and B. J. MacFadden, pp. 207–28. Cambridge: Cambridge University Press.

Fortelius, M. and R. L. Bernor. 1990. A provisional systematic assessment of the Miocene Suoidea from Paşalar, Turkey. *Journal of Human Evolution* 19:509–28.

Fortelius, M., J. Van der Made, and R. L. Bernor. 1996. A new listriodont suid, *Bunolistriodon meidamon* sp. nov., from the middle Miocene of Anatolia. *Journal of Vertebrate Paleontology* 16.

Fraas, O. 1870. Die Fauna von Steinheim, mit Rücksicht auf die miocenen Säugetier- und Vogelreste des Steinheimer Beckens. *Jahreshefte des Vereins für vaterländische Naturkunde in Württemberg, Stuttgart* 26:145–306.

Gaal, I. von 1943. Unterpliozäne Säugetierreste aus Hatvan in Ungarn. *Geologica Hungarica, Series Palaeontologica* 20:1–120.

Gabunia, L. K. 1960. Kubanochoerinae, nouvelle sous-famille de porcs du miocene moyen du Caucase. *Vertebrata Palasiatica* 4:87–98.

———. 1973. Fossil vertebrate fauna of Belomechetskaya. Tblisi: Metsniereba. [in Russian]

———. 1981. Traits essentiels de l'évolution des faunes de Mammifères néogènes de la région mer Noire-Caspienne. *Bulletin du Muséum national d'Histoire naturelle,* C 2., Paris 4, serie 3, section C, 2:195–204.

Gaudry, A. 1862–1867. *Animaux Fossiles et Geologie de l'Attique, d'après les recherches faites en 1855, 1856 et 1860.* Paris: F. Savy.

Ginsburg, L. 1974. Les Tayassuides des phosphorites du Quercy. *Palaeovertébrata* 6:55–85.

———. 1980. *Xenohyus venitor,* suide nouveau (Mammalia, Artiodactyla) du Miocène Inférieur de France. *Géobios* 13:861–77.

Ginsburg, L. and Ch. Bulot. 1987. Les Suiformes (Artiodactyla, Mammalia) du Miocène de Bézian (Gers). *Bulletin du Museum national d'Histoire naturelle,* 4e série 9, section C, 4:455–69.

Glazek, J., J. Oberc, and A. Sulimsky. 1971. Miocene vertebrate faunas from Przeworno (Lower Silezia) and their geological setting. *Acta Geologica Polonica* 21:473–515.

Golpe-Posse, J. M. 1972. Suiformes del Terciario Espanol y sus yacimientos. *Palaeontología y Evolución* 2:1–197.

Guan Jian and J. Van der Made. 1994. Fossil Suidae from Dingjiaergou near Tongxin, China. *Memoirs of Beijing Natural History Museum* 53:151–99.

Gürbüz, M. 1981. Inönü (KB Ankara) Orta Miyosenindeki *Hemicyon sansaniensis* (Ursidae) türünüm tanimlanmasi ve stratigrafik yayilimi. *Türkiye Jeoloji Kurumu Bülteni* C24:85–90.

Haq, B. U., J. Hardenbol, and P. R. Vail. 1987. Chronology of fluctuating sea levels since the Triassic. *Science* 235:1156–67.

Heissig, K. 1989. *Conohyus huenermanni* n. sp., eine kleine Schweineart aus der Oberen Süsswassermolasse Bayerns. *Mitteilungen der Bayerischen Staatssammlung für Paläontologie und historische Geologie* 29:235–40.

Hofmann, A. 1888. Beiträge zur Säugetierfauna der Braunkohle des Labitschberges bei Gamlitz in Steiermark. *Jahrbuch der kaiserlich königlichen geologischen Reichsanstalt* 38:545–62.

———. 1893. Die Fauna von Göriach. *Abhandlungen der kaiserlich königlichen geologischen Reichsanstalt* 15:1–87.

Hofmann, A. and A. Zdarsky. 1904. Beitrag zur Säugetierfauna von Leoben. *Jahrbuch der kaiserlich königlichen geologischen Reichsanstalt* 54:577–94.

Hünermann, K. A. 1968. Die Suidae (Mammalia, Artiodactyla) aus den Dinotheriensanden (Unterpliozän + Pont) Rheinhessens (Südwestdeutschland). *Schweizerische Paläontologische Abhandlungen* 86:1–96.

———. 1975. Die Suidae aus dem Türkischen Neogen. *Geologisches Jahrbuch*, Reihe B 15:153–56.

Hunter, J. P and M. Fortelius. 1994. Comparative dental occlusal morphology, facet development, and microwear in two sympatric species of *Listriodon* (Mammalia, Suidae) from the middle Miocene of Western Anatolia (Turkey). *Journal of Vertebrate Paleontology* 14:105–6.

Janis, C. M. 1982. Evolution of horns in ungulates: Ecology and palaeoecology. *Biological Reviews (Cambridge)* 57:261–318.

Jarman, P. 1974. The social organisation of antelope in relation to their ecology. *Behaviour* 48:213–67.

Kappelman, J., S. Sen, M. Fortelius, A. Duncan, B. Alpagut, J. Crabaugh, A. Gentry, J. P. Lunkka, F. McDowell, N. Solounias, S. Viranta, and L. Werdelin. This volume. Chronology and biostratigraphy of the Miocene Sinap Formation of Central Turkey.

Kaup, J. J. 1833. *Description d'ossements fossiles de mammifères.* Heft 2: 1–31. Darmstadt.

———. 1859. *Beiträge zur näheren Kenntnis der urweltlichen Säugethiere.* Heft 4: 1–16. Darmstadt.

Kingdon, J. 1979. *East African mammals. III B. Large Mammals.* London: Academic Press.

Kittl, E. 1889. Reste von *Listriodon* aus dem Miozän Nieder Österreichs. *Beiträge zur Paläontologie und Geologie Österreich Ungarns und des Orients, Wien* 7:233–49.

Klähn, H. 1925. Die Säuger des badischen Miozäns. *Palaeontographica* 66:163–244.

Koch, C. F. 1987. Prediction of sample size effects on the measured temporal and geographic distribution patterns of species. *Paleobiology* 13:100–107.

Kowalski, K. 1990. Stratigraphy of Neogene mammals of Poland. In *European Neogene Mammal Chronology*, ed. E. H. Lindsay, V. Fahlbusch, and P. Mein, pp. 193–204. New York: Plenum Press.

Kubiak, H. 1981. Suidae and Tayassuidae (Artiodactyla, Mammalia) from the Miocene of Przeworno in Lower Silesia. *Acta Geologica Polonica* 31:59–70.

Laskarev, V. 1937. Über *Hyotherium soemmeringi* var. *media* H. v. Meyer. *Annales géologiques de la Péninsule Balkanique (Beograd)* 14:87–96.

Leinders, J. 1975. Sur les affinités des Listriodontinae bunodontes de l'Europe et de l'Afrique. *Bulletin du Muséum national d'Histoire naturelle* 341:197–204.

Leinders, J. and J. Meulenkamp. 1978. A *Microstonyx* tooth from eastern Crete, palaeogeografical implications of Cretan Tortonian mammal associations. *Proceedings, Koninklijke Nederlandse Akademie van Wetenschappen, B* 81:416–24.

Lindsay, E. H. 1989. The setting. In *European Neogene Mammal Chronology*, ed. E. H. Lindsay, V. Fahlbusch, and P. Mein. New York: Plenum Press.

Lindsay, E. H. and R. Tedford. 1989. Development and application of land mammal ages in North America and Europe: A comparison. In *European Neogene Mammal Chronology*, ed. E. H. Lindsay, V. Fahlbusch, and P. Mein, pp. 601–24. New York: Plenum Press.

Macdonald, D. 1984. *The Encyclopedia of Mammals.* London: Unwin.

Made, J. Van der. 1989. A *Conohyus*-lineage (Suidae, Artiodactyla) from the Miocene of Europe. *Revista Española de Paleontología* 4:19–28.

———. 1990a. Masticatory adaptations, size, and niches of related sympatric suids. *VI Jornadas de Paleontologia, Abstracts*, p. 37.

———. 1990b. Paleobiogeography of *Hippopotamodon* and *Microstonyx* in relation to climate. *IXth RCMNS Congress, Barcelona 1990, Abstracts*, p. 223.

———. 1992a. Iberian Suoidea. *Paleontología y Evolución* 23 (1989/1990): 83–97.

———. 1992b. A range-chart for European Suidae and Tayassuidae. *Paleontología y Evolución* 23:99–104.

———. 1992c. Migrations and climate. In *Mammalian Migration and Dispersal Events in the European Quaternary*, ed. W. von Koenigswald and L. Werdelin. Senckenberg: Courier Forschungsinstitut. 153:27–37.

———. In preparation. *The fossil pig from the upper Miocene of Dorn-Dürkheim in Germany.* Senckenberg: Courier Forschungsinstitut.

Made, J. Van der and Han Defen. 1994. Suoidea from the upper Miocene hominoid locality of Lufeng, Yunnan Province, China. *Proceedings, Koninklijke Nederlandse Akademie van Wetenschappen, B* 97:27–82.

Made, J. Van der and S. T. Hussain. 1989. "*Microstonyx*" *major* (Suidae, Artiodactyla) from Nagri. *Estudios Geologicos* 45:409–16.

Made, J. Van der, P. Montoya, and L. Alacalà. 1992. *Microstonyx* (Suidae, Mammalia) from the upper Miocene of Spain. *Géobios* 25(3): 395–413.

Made, J. Van der and S. Moyà-Solà. 1989. European Suinae (Artiodactyla) from the late Miocene onwards. *Bolletino della Società Paleontologica Italiana* 28:329–39.

Maples, C. G. and A. W. Archer. 1988. Monte Carlo simulation of selected binomial similarity coefficients (II): Eeffect of sparse data. *Palaios* 3:95–103.

Mayr, H. 1979. *Gebissmorphologische Untersuchungen an miozänen Gliriden (Mammalia, Rodentia) Süddeutschlands.* München.

McCrossin, M. L. 1987. Postcranial remains of fossil Suidae from the Sahabi Formation, Libya. In *Neogene Paleontology and Ge-*

ology of Sahabi, ed. N. T.Boaz, A. El-Arnauti, A. W. Gaziry, J. de Heinzelin, and D. D. Boaz, pp. 267–86. New York: Liss.

Mecquenem, R. de 1911. Contribution a l'étude du gisement des Vertebrés de Maragha, deuxième partie. *Annales d'histoire naturelle* 1:81–98.

——. 1924. Contribution a l'étude des fossiles de Maragha I. *Annales de Paléontologie* 13:133–60.

Mein, P. 1975. Proposition de biozonation du Néogène méditerranéen a partir des mammifères. *Trabajos Sobre Neogeno Cuaternario* 4:112.

——. 1977. Table 1. In M. T. Alberdi and E. Aguirre, Round table on mastostratigraphy of the W. Mediterranean Neogene. *Trabajos Sobre Neogeno Cuaternario* 7.

——. 1979. Rapport d'activité du groupe travail des vertébrés, mise à jour de la biostratigraphie du Néogène basée sur les mammifères. *Annales Géologiques des Pays Helléniques*, hors série, 3:1367–72.

——. 1989. Updating of MN zones.In *European Neogene Mammal Chronology*, ed. E. H. Lindsay, V. Fahlbusch, and P. Mein, pp. 73–90. New York: Plenum Press.

Meulen, A. J. van der and R. Daams. 1992. Evolution of early–middle Miocene rodent faunas in relation to long-term palaeo-environmental changes. *Palaeogeography, Paleoclimatology, Palaeoecology* 93:227–53.

Meyer, H. von. 1829. Letter to Dr. Bronn. *Zeitschrift für Mineralogie, Taschenbuch* 23:150–52.

——. 1834. *Die fossilen Zähne und Knochen von Georgensgmünd in Bayern*. Frankfurt a. M: J. D. Sauerlander.

——. 1846. Mitteilungen an Prof. Bronn gerichtet (Brief). *Neues Jahrbuch für Mineralogie, Geologie und Palaeontologie* 1846:462–76.

Mottl, M. 1939. Die mittelpliozäne Säugetierfauna von Gödöllö bei Budapest. *Mittheilungen aus dem Jahrbuch der K. Ungarischen geologischen Anstalt (Budapest)* 32:257–350.

——. 1955. *Hyotherium palaeochoerus*, ein neuer Suide aus dem Unterplioän der Steiermark. *Mitteilungen des Museums für Bergbau, Geologie und Technik am Landesmuseum Joanneum (Graz)* 15:69–76.

——. 1957. Bericht über die neuen Menschenaffenfunde aus Österreich, von St. Stefan im Lavanttal, Kärnten. *Carinthia II (Klagenfurt)* 67:39–84.

——. 1966. Neue Säugetierfunde aus dem Jungtertiär der Steiermark VII. Ein vollständiger *Hyotherium palaeochoerus* Schädel aus dem Altpliozän (Pannon) Südösterreichs. *Mitteilungen des Museums für Bergbau, Geologie und Technik am Landesmuseum Joanneum (Graz)* 28:3–31.

——. 1970. Die jungtertiären Säugetierfaunen der Steiermark, Südösterreichs. *Mitteilungen des Museums für Bergbau, Geologie und Technik am Landesmuseum Joanneum (Graz)* 31:1–92.

Nikolov, I. and E. Thenius. 1967. *Schizochoerus* (Suidae, Mammalia) aus dem Pliozän von Bulgarien. *Annalen des kaiserliche-königlichen naturhistorischen Museum (Wien)* 71:329–40.

Ozansoy, F. 1965. Étude des gisements continentaux et de mammifères du Cénozoique de Turquie. *Mémoires de la Société géologique de France* (N.S.) 44:1–92.

Paraskevaidis, I. 1940. Eine obermiocäne Fauna von Chios. *Neues Jahrbuch für Mineralogie, Geologie und Paläontologie* 83:363–442.

Pavlović, M. B. 1969. [Miozän-Säugetiere des Toplica-Beckens–

paläontologisch-stratigraphische Studie.] (Serbocroatian, German summary). *Annales géologiques de la Péninsule Balkanique (Beograd)* 34:269–394.

Pavlow, M. 1913. Mammifères tertiaires de la Nouvelle Russie. Avec un article géologique du Prof. A.P. Pavlow. 1-re Partie. Artiodactyla, Perissodactyla (*Aceratherium kowalevskii* n.s.). *Nouvelles Mémoires de la Societé impériale des naturalistes de Moscou* 17:1–68.

Pearson, H. S. 1928. Chinese fossil Suidae. *Palaeontologia Sinica* (C) 5:1–75.

Peters, K. F. 1869. Zur Kenntniss der Wirbeltiere aus den Miozänschichten von Eibiswald in Steiermark. II *Amphycyon-Hyotherium. K. Akademie der Wissenschaften (Vienna), Mathematisch-naturwissenschaftlichen Klasse; Denkschriften* 29:189–214.

Petronijević, Z. M. 1967. [Die mittelmiozäne und untersarmatische ("Steirisché') Säugetierfauna Serbiens] (Serbocroatian, German summary). *Paleontologia Jugoslavica* 7:1–160.

Pickford, M. 1978. The taxonomic status and distribution of *Schizochoerus* (Mammalia, Tayassuidae). *Tertiary Research* 2:29–38.

——. 1988. Revision of the Miocene Suidae of the Indian Subcontinent. *Münchner Geowissenschaftliche Abhandlungen A* 12:1–90.

Pickford, M. and C. Ertürk. 1979. Suidae and Tayassuidae from Turkey. *Bulletin of the Geological Society of Turkey* 22:141–54.

Pilgrim, G. E. 1925. Presidential address to the geological section of the 12th Indian Science Congress. *Proceedings 12th Indian Scientific Congress*, pp. 200–218.

——. 1926. The Fossil Suidae of India. *Memoirs of the Geological Survey of India*, New Series, 8(4).

Pope, G. G. and R. L. Bernor. 1990. A new early Miocene fauna from northern Thailand. *Journal of Human Evolution* 19:811–15.

Potts, R., and A. K. Behrensmeyer (rapporteurs), with R. E. Taggart, W. G. Spaulding, J. A. Harris, B. Van Valkenburgh, L. D. Martin, J. D. Damuth, and R. Foley. 1992. Late Cenozoic terrestrial ecosystems. In *Terrestrial Ecosystems Through Time*, ed. A. K. Behrensmeyer, J. Damuth, W. DiMichele, R. Potts, H.-D. Sues, and S. L. Wing, pp. 419–514. Chicago: Chicago University Press.

Qiu, Z. 1990. The Chinese Neogene mammalian biochronology: Its correlation with the European Neogene mammalian zonation. In *European Neogene Mammal Chronology*, ed. E. H. Lindsay, V. Fahlbusch, and P. Mein, pp. 527–56. New York: Plenum Press.

Qui, Z., J. Ye, and F. Huo. 1988. Description of a *Kubanochoerus* skull from Tongxin, Ningxia. *Vertebrata Palasiatica* 26:1–19.

Raymo, M. E. and W. F. Ruddiman. 1992. Tectonic forcing of late Cenozoic climate. *Nature* 359:117–22.

Rinnert, P. 1956. Die Huftiere aus dem Braunkohlen miozän der Oberpfalz. *Palaeontographica*, Abteilung A. 107:1–65.

Roger, O. 1900. Wirbeltierreste aus dem Dinotheriensand III. *Berichte des Naturwissenschaftlichen Vereins für Schwaben und Neuburg (e.v.) (Augsburg)* 34:53–70.

Rögl, F. and F. F. Steininger. 1983. Vom Zerfall der Tethys zu Mediterran und Paratethys. *Annalen des Naturhistorisches Museum, Wien* 85:135–63.

——. 1984. Neogene Paratethys, Mediterranean, and Indo-Pacific seaways: Implications for the paleobiogeography of marine and terrestrial biotas. In *Fossils and Climate*, ed. P. Brenchley, pp. 171–200. Chichester: Wiley.

Rögl, F., H. Zapfe, R. L. Bernor, R. Brzobohaty, G. Daxner-Höck, I. Draxler, O. Fejfar, J. Gaudant, P. Herrmann, G. Rabeder, O. Schultz, and R. Zetter. 1993. Primatenfundstelle Götzendorf an der Leitha, Niederösterreich (Obermiozän, Pontien des Wiener Beckens). *Jahrbuch des geologischen Bundesanstalt* 136:503–26.

Scherer, E. 1981. Die mittelmiozäne Fossil-Lagerstätte Sandelzhausen 12. Crocodylia (abschliessender Bericht). *Mitteilungen der Bayerischen Staatssammlung für Paläontologie und historische Geologie* 21:81–87.

Schlosser, M. 1902. Beiträge zur Kenntnis des Säugetierreste aus den süddeutschen Bohnertzen. *Geologische un paläontologische Abhandlungen (neue Serie) (Berlin)* 5:117–58.

Schmidt-Kittler, N. 1971. Die obermiozäne Fossillagerstätte Sandelzhausen 3. Suidae, Artiodactyla, Mammalia. *Mitteilungen der Bayerischen Staatssammlung für Paläontologie und historische Geologie* 11:129–70.

Senyürek, M. S. 1952. A study of the Pontian Fauna of Gökdere (Elmadag), South-East of Ankara. *Türk Tarih Kurumu Belleten*, XVI 64:451–92.

Sickenberg, O. (ed.). 1975. Die Gliederung des höheren Jungtertiärs und Altquartärs in der Türkei nach Vertebraten und ihre Bedeutung für die internationale Neogen Stratigraphie. *Geologisches Jahrbuch* B. 15:1–167.

Simpson, G. G. 1945. The principles of classification and a classification of mammals. *Bulletin of the American Museum of Natural History* 85:1–350.

——. 1947. Evolution, interchange, and resemblance of the North American and Eurasian Cenozoic mammalian faunas. *Evolution* 1:218–20.

Solounias, N. 1981. Mammalian fossils of Samos and Pikermi. Part 2. Resurrection of a classic Turolian fauna. *Annals of the Carnegie Museum* 50:231–70.

Stehlin, H. G. 1899/1900. Geschichte des Suiden-Gebisses. *Abhandlungen der schweizerischen paläontologischen Gesellschaft* 26/27:1–335.

——. 1914. Übersicht über die Säugetiere der schweizerischen Molasse-formation, ihre Fundorte und ihre stratigraphische Verbreitung. *Verhandlungen der naturforschenden Gesellschaft, Basel* 25:179–202.

Steininger, F. F., R. L. Bernor, and V. Fahlbusch. 1989. European Neogene marine/continental chronologic correlations. In *European Neogene Mammal Chronology*, ed. E. H. Lindsay, V. Fahlbusch, and P. Mein, pp. 15–46. New York: Plenum Press.

Stromer, E. 1928. Wirberltiere im obermiozänen Flinz Münchens. *Abhandlungen der bayerischen Akademie der Wissenschaften (München), Mathematisch-naturwissenschaftliche* Abteilung 32 (1).

Swisher, C. C., III. This volume. New ^{40}Ar/^{39}Ar dates and their contribution toward a revised chronology for the late Miocene nonmarine of Europe and West Asia.

Thenius, E. 1949. Über die Säugetierfauna aus dem Unterpliozan von Ilhan bei Ankara (Türkei). *Sitzungsberichte der Österreichische Akademie von Wissenschaften, mathematisch-naturwissenschaftliche Klasse I* 158. (9–10): 656–61.

——. 1950. *Postpotamochoerus* n. subg. *hyotherioides* aus dem Unterpliozän von Samos (Griechenland) und die Herkunft der Potamochoeren. *Sitzungsberichte der Österreichische Akademie von Wissenschaften, mathematisch-naturwissenschaftliche* Klasse I 159 (1–5): 25–36.

——. 1952. Die Säugetierfauna aus dem Torton von Neudorf an der March (CSR). *Neues Jahrbuch für Geologie und Paläontologie*, Abhandlungen 96:27–136.

——. 1955. *Sus antiquus* aus Ligniten von Sophades (Thessalien) und die Alterstellung der Fundschichten. *Annales Géologiques des Pays Helléniques* 6:199–206.

——. 1956. Die Suiden und Tayasuiden des steirischen Tertiärs. *Sitzungsberichte der Österreichische Akademie von Wissenschaften, mathematisch-naturwissenschaftliche* Klasse I 165 (4–5): 337–82.

——. 1959. Tertiär. 2 Teil. Wirbeltierfaunen. In F. Lotze, *Handbuch der Stratigraphischen Geologie, vol. 3 (2)*. Stuttgart: F. Enke.

——. 1970. Zur Evolution und Verbreitungsgeschichte der Suidae (Artiodactyla, Mammalia). *Zeitschrift für Säugetierkunde* 35:321–42.

——. 1972. *Microstonyx antiquus* aus dem Alt-Pleistozän Mittel-Europas: Zur Taxonomie und Evolution der Suidae (Mammalia). *Annalen des Naturhistorischen Museum, Wien* 76:539–86.

——. 1979. Afrikanische Elemente in der miozänen Säugetierfauna Europas. *Annales Géologiques des Pays Helléniques*, hors série, 3:1201–8.

Tobien, H. 1951. Die Aufzeichnungen H. G. Stehlin's über die pliozänen Säugetierreste von Herbholzheim bei Freiburg i. Br. *Mitteilungsblatt der Badischen Geologischen Landesanstalt* 1950:78–84.

Trofimov, B. A. 1954. [The fossil suids of the genus *Microstonyx*. In: *Tertiary Mammals, part 2. On the mammalia of the southern SSSR and Mongolia*] (Russian). *Dokladi Akademia Nauk SSSR* 47:61–99.

Vacek, M. 1900. Ueber Säugethierreste der Pikermifauna vom Eichkogel bei Mödling. *Jahrbuch der kaiserlich königlichen geologischen Reichsanstalt* 50:169–86.

Van Valen, L. M. 1984. A resetting of Phanerozoic community evolution. *Nature* 307:50–52.

Vekua, A. K. and V. M. Trubikhin 1988. [About a new location of fossil mammals in eastern Georgia] (Russian, English Summary). *Bulletin of the Academy of Sciences of the Georgian SSR* 132. (1):197–200.

Werdelin, L. and N. Solounias. This volume. The evolutionary history of hyaenas in Europe and Western Asia during the Miocene.

Wheeler, P. E. 1992. The thermoregulatory advantages of large body size for hominids foraging in savanna environments. *Journal of Human Evolution* 23:351–62.

Woodburne, M. O., R. L. Bernor, and C. C. Swisher III. This volume. An appraisal of the stratigraphic and phylogenetic bases for the "*Hipparion* Datum" in the Old World.

Zittel, K. A. von. 1891/1893. *Handbuch der Palaeontologie*, Band IV. Munich.

Appendix 28.1 *Later Neogene Suoidea of Central Europe and the Eastern Mediterranean: Summary Table*

Locality	Collection	Description	Age	Age Reference
Taucanamo grandaevum (Fraas, 1870)				
?Neudorf Sandberg		Thenius 1952	MN 6	Mein 1990
Steinheim	SMNS,NMB, SMNK	Fraas 1870, Stehlin 1899/1900, Chen 1984	MN 7	Mein 1990
Anwil	NMB	Engesser 1972	MN 8	Mein 1990
Taucanamo aff. *T. sansaniense* (Lartet, 1851)				
Petersbuch 2	BSPHGM		MN 4	Mein 1990
Winterhof Ost		Dehm 1937		
Taucanamo inonuensis (Pickford and Ertürk, 1979)				
Münzenberg (Leoben)	SLJG	Thenius 1956	MN 5	(Mein 1990)
Mala Miliva		Petronijevic' 1967	MN 5	Mein 1990
Göriach	SLJG	Thenius 1956 Hofmann 1893	MN 6	Mein 1990, V. d. Made 1989
Inönü I	MTA	Pickford & Ertürk 1979	MN 6	Gürbüz 1981
Pasalar	PDTFAU, MTA	Fortelius & Bernor 1990	MN 6	Mein 1990
Prebreza		Pavlovic' 1969	MN 6	Mein 1990
Sandelzhausen	BSPHGM		MN 6	Mein 1990
Schizochoerus sp.				
?Lower Sinap	MTA	Pickford & Ertürk 1979	MN 8	Kappelman et al., this vol.
Schizochoerus vallesiensis (Crusafont and Lavocat 1954)				
Middle Sinap	MNHN	Ozansoy 1965	MN 9	Kappelman et al., this vol.
Nsebar	BSPHGM (cast)	Nikolov & Thenius 1967	MN 9	Bernor et al. 1988
Kalfa		Pickford 1978	MN 9?	Gabunia 1981
"Genus nov. cf. *Taucanamo*"				
Çandir	MTA	Pickford & Ertürk 1979	MN 6	Mein 1990
Albanohyus pygmaeus (Deperet 1892)				
Przeworno 2	SMNS	Kubiak 1981	MN 8	Mein 1990
Bunolistriodon lockharti (Pomel, 1848)				
Langenau 1	SMNS		MN 4	Abusch-Siewert 1983
Langenau 2	BSPHGM		MN 4	Abusch-Siewert 1983
Ravensburg	SMNS		MN 4	Abusch-Siewert 1983
Engelswies	NMB	Stehlin 1899/1900 Klähn 1925	MN 5	Abusch-Siewert 1983
Gerlenhofen	SMNS (casts)			
Grimmelfingen	SMNS	Stehlin 1899/1900		
?Langenenslingen	NMB (casts)	Stehlin 1899/1900		
Bunolistriodon latidens (Biedermann, 1873)				
Mala Miliva		Petronijevic 1967	MN 5	Mein 1990
Veltheim	NSSW	Biedermann 1873, Stehlin 1899/1900	MN 5/6	
Inönü I	MTA	Pickford & Ertürk 1979	MN 6	Gürbüz 1981
?Georgensgmünd	NMB	Stehlin 1899/1900	MN 6	Abusch-Siewert 1983
?Stätzling		Stehlin 1899/1900	MN 6	Abusch-Siewert 1983
Bunolistriodon sp. nov.				
Pasalar	PDTFAU, MTA, PIMUZ	Fortelius & Bernor 1990	MN 6	Mein 1990
Çandir	MTA, PIMUZ	Pickford & Ertürk 1979	MN 6	Mein 1990
Prebreza		Pavlovic 1969	MN 6	Mein 1990
Listriodon splendens (Meyer, 1846)				
Çandir	MTA, PIMUZ	Pickford & Ertürk 1979	MN 6	Mein 1990
Inönü I	MTA	Pickford & Ertürk 1979	MN 6	Gürbüz 1981
Neudorf Sandberg		Thenius 1952	MN 6	Mein 1990
Pasalar	PDTFAU, MTA, PIMUZ	Fortelius & Bernor 1990	MN 6	Mein 1990
Nussdorf		Kittl 1889, Zittel 1891– 1893, Stehlin 1914	MN 6/7	
Korethi		Gabunia 1981	MN 7	Mein 1990
Laichingen	SMNS	Hünermann 1968	MN 7	Abusch-Siewert 1983

Locality	Collection	Description	Age	Age Reference
Sariçay	MTA		MN 7	Mein 1990
Steinheim	SMNS, NMB	Stehlin 1899/1900, Chen 1984	MN 7	Mein 1990
Yaylacilar	PIMUZ		MN 7	Steininger et al. 1990
La Chaux-de-Fonds		Blainville 1847, Stehlin 1899/1900	MN 7/8	(De Bruijn et al. 1992)
Anwil	NMB		MN 8	Mein 1990
Friedberg	BSPHGM		MN 8?	Abusch-Siewert 1983
Kleineisenbach	BSPHGM		MN 8	Mein 1990
Lower Sinap	MNHN		MN 8	Kappelman et al. this vol.
St. Stefan		Mottl 1957	MN 8	Mein 1990
Löffelbach		Bach 1908	MN 8/9	(Bach 1908)
Massenhausen	BSPHGM		MN 8/9	Abusch-Siewert 1983
Wartenberg	IVAU, BSPHGM		MN 8/9	Abusch-Siewert 1983
Aumeister (Flinz)		Stromer 1928	MN 9	Mein 1990
Eichkogel		Vacek 1900	MN 9	(Vacek 1900)
Esselborn	HLD	Hünermann 1968	MN 9	Abusch-Siewert 1983
Grosslappen (Flinz)		Stromer 1928	MN 9	Mein 1990
Hammerschmiede	BSPHGM		MN 9	Mayr 1979
Markt Rettenbach	BSPHGM	Dehm 1934	MN 9	
Wissberg	NMM	Hünermann 1968	MN 9	Abusch-Siewert 1983
Atzgersdorf		Kittl 1889		
Haselbach		Thenius 1956		
Kaisersteinbruch		Kittl 1889		
Locle		Stehlin 1899/1900, 1914	isol. finds	
Loreto		Kittl 1889		
Mannersdorf		Kittl 1889		
Mauer		Kittl 1889		
Mering	NMB, BSPHGM	Roger 1900		
Morsingen	NMB			
Mösskirch		Schlosser 1902	mixed	Klähn 1925
Prittlbach	BSPHGM			
Soóskut		Kittl 1889		
St. Georgen	BSPHGM			
St. Margarethen		Kittl 1889		

Kubanochoerus robustus (Gabunia, 1955)

Locality	Collection	Description	Age	Age Reference
Belometchetskaia		Gabunia 1960, 1973	MN 6	Mein 1990

Kubanochoerus khinzikebirus (Wilkinson, 1976)

Locality	Collection	Description	Age	Age Reference
Inönü I	MTA	Pickford & Ertürk 1979	MN 6	Gürbüz 1981

Hyotherium soemmeringi (Meyer, 1829)

Locality	Collection	Description	Age	Age Reference
Eibiswald	SLJG	Peters 1869, Thenius 1956	MN 5	Mein 1990
Engelswies	NMB	Klähn 1925	MN 5	Abusch-Siewert 1983
Münzenberg (Leoben)	SLJG	Thenius 1956	MN 5	(Mein 1990)
Seegraben (Leoben)	SLJG	Hofmann & Zdarsky 1904, Thenius 1956	MN 5	(Mein 1990)
Buchenthal	PIMUZ	Stehlin 1899/1900	MN 5/6	(Stehlin 1914)
Georgensgmünd	FISF, NMB	Meyer 1834, Stehlin 1899/1900	MN 6	Mein 1990
Göriach	SLJG	Thenius 1956 fig.20 c,d	MN 6	Mein 1990
Rümikon	NMB		MN 6	Bolliger p. comm. 1992
Sandelzhausen	BSPHGM	Schmidt-Kittler 1971	MN 6	Mein 1990
Stätzling	BSPHGM		MN 6	De Bruijn et al. 1992
Thannhausen		Dehm 1980	MN 6	Abusch-Siewert 1983
Viehhausen		Rinnert 1956	MN 6	Abusch-Siewert 1983
Feisternitz	SLJG	Thenius 1956		
Fohnsdorf	SLJG			
Hüllistein	PIMUZ			
Labitschberg (Gamlitz)	SLJG	Hofmann 1888		
Michelsberg	BSPHGM			
Pettrachmühle	BSPHGM			
Rothenstein 6	BSPHGM			

| Schönegg | SLJG | | | |

Locality	Collection	Description	Age	Age Reference
Voitsberg		Thenius 1956		
Vordersdorf		Thenius 1956		
Walda	BSPHGM			
Wissberg	NMM		reworked	
Zangtal	SLJG			

Conohyus simorrensis (Lartet, 1851)

Locality	Collection	Description	Age	Age Reference
Mala Miliva		Petronijevic 1967	MN 5	Mein 1990
Elgg	PIMUZ, NMB	Kaup 1859, Stehlin 1899/1900	MN 6	Bollinger p. comm. 1992
Göriach	SLJG, NMB	Hofmann 1893, Stehlin 1899/1900, Thenius 1956, V. d. Made 1989	MN 6	Mein 1990
Pasalar	PDTFAU, MTA, PIMUZ	Fortelius & Bernor 1990	MN 6	Mein 1990
Kleineisenbach	BSPHGM		MN 8	Mein 1990
Au	SLJG			
Bâlâ		Pickford & Ertürk 1979		isolated find
Pichelsberg	BSPHGM			
Rosenthal	SLJG			
St. Oswald	SLJG			
?St. Georgen	BSPHGM	Stehlin 1899/1900		
Tutzing	NMB	Stehlin 1899/1900		
Urlau	NMB, SMNS	Stehlin 1899/1900		

Parachleuastochoerus steinheimensis (Fraas, 1870)

Locality	Collection	Description	Age	Age Reference
Steinheim	SMNS, NMB	Fraas 1870, Chen 1984	MN 7	Mein 1990
Przeworno 1		Kubiak 1981, Glazek, Oberc & Sulimski 1971	MN 8	(Kowalski 1990, Mein 1990)
Przeworno 2		Kubiak 1981, Glazek, Oberc & Sulimski 1971	MN 8	Mein 1990
Anwil	NMB	Engesser 1972	MN 8	Mein 1990
Wartenberg	IVAU, BSPHGM		MN 8/9	Abusch-Siewert 1983
Eppelsheim	HLD	Kaup 1833, 1859, Hünermann 1968	MN 9	Mein 1990
Esselborn	HLD	Hünermann 1968	MN 9	Abusch-Siewert 1983
Hammerschmiede	BSPHGM cast		MN 9	Mayr 1979
Wissberg	HLD, NMM	Chen 1984	MN 9	Abusch-Siewert 1983
Hinterauerbach	BSPHGM			

Parachleuastochoerus huenermanni (Heissig, 1989)

Locality	Collection	Description	Age	Age Reference
Breitenbrunn		Heissig 1989	MN 8/9	Heissig p. comm. 1992
Charmoille	HLD cast		MN 9	Mein 1990
Esselborn	FISF, HLD	Hünermann 1968	MN 9	Abusch-Siewert 1983
Wissberg	HLD, NMM	Hünermann 1968	MN 9	Abusch-Siewert 1983

Propotamochoerus palaeochoerus (Kaup, 1833)

Locality	Collection	Description	Age	Age Reference
Eppelsheim	HLD, FISF	Hünermann 1968	MN 9	Mein 1990
Esselborn	HLD	Hünermann 1968	MN 9	Abusch-Siewert 1983
Wissberg	HLD, SMNK, IVAU	Hünermann 1968	MN 9	Abusch-Siewert 1983
Isarbett	BSPHGM	Stehlin 1899/1900, Schmidt-Kittler 1971, Stromer 1928	MN 9?	
Vösendorf		Mottl 1955	MN 10	Steininger et al. 1990
Johnsdorf	SLJG	Mottl 1966, Schmidt-Kittler 1971, Thenius 1972		
München U-Bahn Tunnel	BSPHGM			
?Middle Sinap	MNHN		MN 9	Kappelman et al. this vol.
?Dintesheim	HLD	Hünermann 1968		
?Lassnitzhöhe		Mottl 1955		
?Pyhra		Mottl 1955		
?Wolfsheim	HLD	Hünermann 1968		

Propotamochoerus provincialis (Gervais, 1859)

Locality	Collection	Description	Age	Age Reference
? Hatvan		Gaal 1943	MN 13	Mein 1990
Kardia	IVAU		MN 14	Mein 1990

Locality	Collection	Description	Age	Age Reference
		Propotamochoerus sp.		
?Gödöllö		Mottl 1939	MN 14	Mein 1990
Herbholzheim	NMB	Tobien 1951	MN 14/15	(Tobien 1951)
		Hippopotamodon antiquus (Kaup, 1833)		
Yeni Eskihisar 2	MTA		MN 8	Steininger et al. 1990
Eppelsheim	HLD	Stehlin 1899/1900, Hünermann 1968	MN 9	Mein 1990
Esme Akçaköy	PIMUZ	Hünermann 1975	MN 9	Mein 1990
Wissberg	HLD	Hünermann 1968	MN 9	Abusch-Siewert 1983
Middle Sinap	MNHN	Ozansoy 1965	MN 9	Kappelman et al. this vol.
Karasian	PIMUZ			
		Microstonyx major (Gervais, 1848–1852)		
Berislav		Trofimov 1954	MN 10	Mein 1990
Eldar		Trofimov 1954	MN 10	Mein 1990
Grossulovo		Trofimov 1954	MN 10	Mein 1990
Stratzing		Thenius 1972	MN 10?	
Çorak Yerler	PIMUZ		MN 11	Mein 1990
Grebeniki		Pavlow 1913, Trofimov 1954	MN 11	Mein 1990
Kayadibi	PIMUZ		MN 11	Mein 1990
Çobanpinar	MTA, PDTFAU		MN 12	Mein 1977
Taraklia		Trofimov 1954	MN 12/13	Mein 1990
Polgardi	NMB		MN 13	Mein 1990
Dzedzvtakhevi		Vekua & Trubhikhin 1988	Meotian	Vekua & Trubhikhin 1988
Gökdere (Elma Dag)		Senyürek 1952		
?Salmendingen	NMB		mixed fauna?	(Abusch-Siewert 1983)
		Microstonyx erymanthius (Roth & Wagner, 1854)		
Çevril	PIMUZ		MN 10/11	Sickenberg et al. 1975
Dorn Dürkheim	FISF	V. d. Made (in prep.)	MN 11	Mein 1990
?Kavakdere	PIMUZ, PFTFAU		MN 11	Kappelman et al. this vol.
Mahmutgazi	SMNK		MN 11	Mein 1990
Samos	NMB, HLD		MN 11	Mein 1990
Kinik	PIMUZ		MN 12	Mein 1990
Titov Veles	IVAU		MN 12	Mein 1977
Kerassia	IVAU		MN 12/13	V.d.Made & Moyà-S. 1989
Pikermi	BMNH, MGL, TMH	Gaudry 1862, Stehlin 1899/1900, Pearson 1928	MN 12/13	Mein 1990
Tudurovo		Pavlow 1913, Trofimov 1954	MN 12/13	Mein 1990
Sophades		Thenius 1955		
		Microstonyx major/erymanthius		
Csávákar	NMB		MN 11	Mein 1990
Garkin	PIMUZ		MN 11	Mein 1990
Kayadibi Sarisik Inleri	PIMUZ		MN 11	Mein 1990 (Becker-Platen et al. 1975)
Lower Maragheh	IVAU, NMB, HLD	De Mecquenem 1911, 1924	MN 11-12	Mein 1990 (Bernor, 1986)
Ilhan		Thenius 1949	MN 12	Mein 1977
Crete		Leinders & Meulenkamp 1978		
Kalimanci		Bakalov & Nikolov 1962		
?Sungurlu	PIMUZ			
		"*Potamochoerus (Postpotamochoerus) hyotheroides*" (Thenius, 1950)		
Samos		Thenius 1950	MN 12/13	Mein 1990
		Suinae sp.		
Karain	PIMUZ		MN 10	Mein 1990

29

Miocene Ruminants of the Central and Eastern Tethys and Paratethys

A. W. GENTRY AND E. P. J. HEIZMANN

In the Old World the suborder Ruminantia at the present day comprises the family Tragulidae (chevrotains) plus the infraorder Pecora (Cervidae, Giraffidae, and Bovidae, to which some would add Moschidae for the musk deer of Asia). The suborder also includes extinct families and the New World Antilocapridae.

Early Miocene pecorans were not yet differentiated into the later families. At that period *Dremotherium* and various other genera were found in Europe and *Walangania* in Africa. Cervoids and giraffoids could have had predominantly vicariant origins within the Pecora, giraffoids within Africa, Arabia and/or India, and cervoids further north in Eurasia. Undoubted cervids then appeared from among cervoids in the early Miocene of Europe and further east. Later again, cervids migrated to the Siwaliks and Northern Africa, but this postdated the Miocene. Meanwhile giraffids dispersed to Eurasia (beyond India) in the middle Miocene, and by the late Miocene were successful in the region under consideration in this volume.

Bovids may have had more complicated initial intercontinental dispersals. *Eotragus* appeared in Western Europe late in the early Miocene at the same time as proboscideans and could have been an immigrant with them. By the middle Miocene it had become more widespread in the Old World and was now accompanied by *Tethytragus* and *Hypsodontus*. Boselaphines also appeared before the end of the middle Miocene of Europe, Africa and the Siwaliks, but not in China. Gazelles seem to have evolved from an Afro-Arabian stock containing *Homoiodorcas* and only dispersed to other continents in the late Miocene.

We have surveyed the occurrences of ruminants in 161 localities attributed to the middle and late Miocene (MN 6–MN 13) of Central and Southeastern Europe and Southwestern Asia (tab. 29.1). Central Europe is taken as Germany, Switzerland, Czechoslovakia, Hungary, Poland, and Austria. Southeastern Europe comprises Yugoslavia, Romania, Bulgaria, Greece, and the regions north and east of the Black Sea in Moldavia, Ukraine, Russia, and Georgia. Southwestern Asia is the Sub-Paratethyan Province of Be-

rnor (1983), comprising Asiatic Turkey across to Iran. This classification entails Pikermi being in Southeastern Europe and Samos in Southwestern Asia.

Table 29.2 shows the numerical distribution of localities. It can be noted that over half of them are in Central Europe and less than 15% in Southwestern Asia. Of middle Miocene localities very few occur in Southeastern Europe or Southwestern Asia. Among late Miocene localities, Vallesian ones predominate in Central Europe and Turolian ones in Southeastern Europe and Southwestern Asia. We have also referred in this chapter to localities in Italy, Arabia, and North Africa and to other localities at present of uncertain or unknown age.

The famous mammals of the Mytilini Formation, Samos, occur in the Old Mill Beds and in the Main Bone Beds (Solounias 1981a, b). Such signs of evolutionary change as have been detected among the Samos mammals are out of alignment with stratigraphical interpretations (Solounias 1981b:261). The Old Mill Beds and the Main Bone Beds have previously been interpreted as being ca. 8.26 and 7.75 Ma, respectively (Weidmann et al. 1984). The upper fauna of the Main Bone Beds was thought to date from around 6.9 Ma, but Sen and Valet (1986) preferred an age of 6.1–6.4 Ma in a period of reversed magnetic polarity. Swisher (this volume) and Bernor et al. (this volume b) provide evidence that the age of the Old Mill Beds is 8.3 Ma and the bulk of the Main Bone Beds is between 7.3 and 7.1 Ma. Bernor et al. (this volume b) refer the Old Mill Beds fauna to uppermost MN 11 and the remainder of the Main Bone Bed faunas to late MN 12, with the bulk of this latter sample being correlative with uppermost MN 12. For a review of the stratigraphic and chronologic positions of the various Samos quarries, please refer to Bernor et al. (this volume b).

Giraffoids

The middle Miocene giraffid *Giraffokeryx* must be an immigrant to Chios, Paşalar, Prebreza, and Belometsch-

TABLE 29.1 *Distribution of Better Substantiated Giraffid and Bovid Species in Central and Eastern Paratethys*

	5	6	7	8	9	10	11	12	13
Giraffokeryx cf. *punjabiensis*	E	AE							
Palaeotragus coelophrys					AE	E	A	A	
Palaeotragus roueni					A		AE	AE	
Palaeotragus asiaticus									E
Samotherium boissieri							AE	AE	
Decennatherium macedoniae						E			
Helladotherium duvernoyi							AE	AE	C
Bohlinia attica						E	AE	E	E
Eotragus clavata	C	CE	C						
Miotragocerus sp. or spp.		E		A	C		C		
Miotragocerus monacensis/pannoniae					C	C			
Tragoportax leskewitschi					E	E	E		
Tragoportax gaudryi							A	AE	E
Tragoportax amalthea						E	AE	AE	E
Tragoportax rugosifrons							AE	AE	
Protragocerus chantrei			C		C				
Austroportax latifrons			C						
Samokeros minotaurus							A	A	
Tyrrhenotragus gracillimus								I	
Gazella ancyrensis					A				
Gazella capricornis								E	
Gazella deperdita							A	A	X
Gazella schlosseri						E	E	E	
Prostrepsiceros vallesiensis						E			
Prostrepsiceros zitteli							AE	A	
Prostrepsiceros houtumschindleri							AE		
Prostrepsiceros rotundicornis							AE	E	
Prostrepsiceros fraasi					A		A	A	
Ouzocerus gracilis						E			
Protragelaphus skouzesi							AE	AE	
Protragelaphus theodori									E
Palaeoreas elegans					A	X			
Palaeoreas lindermayeri							AE	AE	E
Palaeoreas asiaticus							A		
Palaeoreas zouavei							E		
Nisidorcas planicornis						A	AE		
Hispanodorcas rodleri (= *orientalis*)							A	E	
Oioceros (= *Samotragus*) *praecursor*						E			
Oioceros rothi (incl. *Samotragus crassicornis*)							A	AE	
Oioceros wegneri							A	A	
Oioceros atropatenes							A	A	
Oioceros (= *Samotragus*) *occidentalis*									I
Samodorcas kuhlmanni							A	A	
Hypsodontus serbicus	E	E	X						
Hypsodontus pronaticornis		A							
Turcocerus gracilis		A							
Tethytragus spp.		AC	C			A			
Mesembriacerus melentisi						E			
Plesiaddax inundatus							A		
Urmiatherium polaki								A	
Parurmiatherium rugosifrons							X	A	X
Criotherium argalioides							A	A	
Palaeoryx pallasi							AE	AE	E
Protoryx enanus			A						
Protoryx solignaci				A	A				
Protoryx carolinae								E	
Pachytragus laticeps							A	A	
Pachytragus crassicornis							A	AE	
Pseudotragus capricornis							A	AE	
Pseudotragus parvidens							A	AE	
Procobus melania							E	E	
Moldoredunca amalthea							?		
Maremmia hauptii								I	I
Maremmia lorenzi								I	I
?Redunca sp.									A

Top row of numbers = MN zones. A = SW Asia, C = Central Europe, E = SE Europe, I = Italy, X = area unknown.

TABLE 29.2 *Numerical Distribution of 161 Localities for Fossil Ruminants by Geographical Region and Biochronology*

	Central Europe		Southeastern Europe		Southwestern Asia	
Turolian (MN 11–13)	6	4%	28	17%	13	8%
Vallesian (MN 9–10)	16	10%	15	9%	2	1%
Other Upper Miocene	17	11%	11	7%	-	-
Total Upper Miocene	39	25%	54	33%	15	9%
Middle Miocene (MN 6–8)	44	27%	3	2%	6	4%

Note: The readings are also given as percentages of the total number of localities.

eskaya, since giraffids had only been present previously in Africa and Arabia and at Bugti. *Injanatherium* of Arabia is very similar and survived into the late Miocene of Iraq (Brunet and Heintz 1983:288).

The best-known fossil giraffes are of Turolian age. *Palaeotragus roueni*, type species of its genus, is small for a giraffid, with ossicones upright and more or less parallel in anterior view (unlike middle Miocene forms), molarized lower premolars, and lengthened limbs. We follow Geraads (1986:473) in accepting a larger species, *P. coelophrys*, including as junior synonyms *Achtiaria expectans* Borissiak 1914 from Sebastopol, *A. borissiaki* Alexejeva 1930 from Eldar, *Palaeotragus quadricornis* Bohlin 1926 from Samos, and *P. hoffstetteri* Ozansoy 1965 from Yassiören. To these we add *P. berislavicus* Korotkevich 1970 from Berislava (first published in 1957 but in Ukrainian and without illustrations). The lower premolars of this species are less advanced than in *P. roueni*. It is very probable that *P. coelophrys* is a temporal predecessor, if not an ancestor, of *P. roueni* and *Samotherium boissieri*. *Samotherium boissieri* is much larger than *P. roueni* but has ossicones rather like that species. No very noticeable limb lengthening has occurred, and the lower premolars become differently specialized from *P. roueni*. The cheek teeth are higher, and the premolar row may be shorter than in other giraffids. Some *Samotherium* from later localities such as Andrianos on Samos may be from a larger and more advanced form, *S. b. majori* Bohlin 1926.

Bohlinia attica has very long legs, perhaps more so than in *P. roueni*, and must be closely related to *Giraffa* despite some differences in its limb bone characters (Geraads 1979:380). The sivathere *Helladotherium duvernoyi* is very large. Limb bone proportions are most similar to those in *Samotherium boissieri* except that individual bones are more robust. In later sivatheres the metapodials became notably short.

It is very likely that the Vallesian *Decennatherium pachecoi* of Spain and *D? macedoniae* of our region are also sivatheres according to the morphology of the skull discussed by Morales (1985). *Helladotherium* replaces *Decennatherium* in the Turolian and occurs as far west as

Hungary. It is interesting that the lower P3s and possibly the lower P4s of *D. pachecoi* and *D? macedoniae* are somewhat more advanced than those of *Helladotherium* in the attenuation of the transverse ridge of the entoconid (Morales and Soria 1981: fig. 7; Geraads 1989a: fig. 1). Limb bones of *D. pachecoi*, however, were longer and more gracile, and hence more primitive for a sivathere than in *Helladotherium*, according to the measurements and illustrations of Morales and Soria (1981). *Decennatherium* in Europe was probably similar to the Siwaliks Miocene sivatheriines *Bramatherium* and *Hydaspitherium* in showing an enlarged anterior pair of ossicones (or one single fused anterior ossicone) and a less prominent posterior pair. *Birgerbohlinia* of the Spanish Turolian, on the other hand, had a large posterior pair of ossicones and a smaller anterior pair (Montoya and Morales 1991), a pattern more reminiscent of later *Sivatherium*. It will be interesting in the future to discover the state of the ossicones in *Helladotherium*.

?Bovoidea

The little-known pecoran *Amphimoschus* survives into MN 5 and perhaps MN 6 in Central Europe (Rinnert 1956). Its remains are customarily assigned to *A. artenensis*, but in Western Europe this species occurs only in MN 3–4, preceding the larger type species *A. ponteleviensis* of MN 5 (Ginsburg 1989:172). We have linked *Amphimoschus* with *Hispanomeryx* as questionable members of the superfamily Bovoidea. *Hispanomeryx* has its type species in the Spanish Vallesian and has been recorded in pre-Vallesian deposits at Çandir, Chios ("*Cervus lunatus*") and possibly Paşalar (Moyà-Solà 1986; Paraskevaidis 1940). It was accepted into the Moschidae by Janis and Scott (1987:66), apparently in the belief that Morales, Moyà-Solà, and Soria (1981) had described the anterior distal gully on the metatarsal as closed. However, Moyà-Solà (1986:269) confirmed that the gully is open. He regarded the family Hispanomerycidae as the sister group to the Bovidae.

Early Bovids

Eotragus, type species *E. clavata* (Gervais 1850) [= *E. sansaniensis* Lartet 1851; see Gentry 1994:129] from Sansan, is a notably primitive bovid of Western and Central Europe, and it occurs in these regions earlier than other bovids (from MN 4 onward as *E. artenensis*). It may belong to the Boselaphini, but there is no unanimity on this point. Its final occurrence was in MN 7. *Eotragus* has been cited many times outside of Europe, e.g., Negev, Israel, in China, and in the Siwaliks, where it may predate its European appearance (Tchernov et al. 1987; Ye 1989; Thomas 1984b:57). There is often some doubt about the claimed identifications, e.g., Gentry (1970:261) for Fort Ternan, Kenya. *Tethytragus* (Azanza and Morales 1994) is a middle Miocene genus that is definitely not a boselaphine. It extends into Central and Western Europe but has not yet been found in Russia. It is allied with *Caprotragoides* and *Gentrytragus* may be on or close to the ancestry of Caprini via species like *Protoryx enanus.*

Hypsodontus is very different from Boselaphini, *Eotragus*, or *Tethytragus* and indicates an early dichotomy in the Bovidae, or possibly a pecoran parallelism with them. It is very specialized in its early acquisition of hypsodont cheek teeth and torsion of its horn cores. It and some related genera, if validly separate from one another, achieved a broad Old World distribution in the middle Miocene: China, Russia, India, Turkey, Southwestern Asia, Southeastern Europe, and Africa. In Arabia and China it seems to occur prior to the middle Miocene and perhaps as early as *Eotragus* does in Pakistan and Europe (see Gentry 1990a:544; Azanza and Morales 1994). *Hypsodontus serbicus* of Yugoslavia, the westernmost occurrence of the genus (Pavlović 1969), is probably more specialized than the type species *H. miocenicus* from Belometscheskaya. *H. pronaticornis* from Turkey is larger and even more specialized. *Turcocerus* Köhler 1987 was founded for smaller species related to *Hypsodontus.*

Many previous authors have noticed the putative link of this group of bovids with the late Miocene *Oioceros*, suggested by the existence and direction of torsion of the spiraled horn cores, and this has led to the attribution of *Hypsodontus* to Caprini or Caprinae (e.g., Gentry 1970). This could well be a mistaken conclusion, about both the tribal affiliation of *Oioceros* and the relationship of *Hypsodontus* to *Oioceros*. It is possible, however, that *Hypsodontus* could be related to the late Miocene ovibovine genera *Mesembriacerus, Urmiatherium, Criotherium,* and *Tsaidamotherium,* mentioned below.

The Neotragini are a probably paraphyletic African tribe (Gentry 1992). The middle Miocene *?Homoiodorcas* sp. from Al Jadidah, Saudi Arabia (Thomas 1983), is more primitive and less obviously neotragine than the much later type species *H. tugenicum* in Kenya. Its horn cores and dentition are very slightly smaller than Pikermi *Gazella*, and the metatarsal appreciably smaller. Perhaps it is related to later *Gazella* but not yet sufficiently evolved to be counted as an antilopine.

Boselaphine Bovids

Definite boselaphines are known from the middle and late Miocene. They are difficult to work with, and there are many problems with the generic distinctions, with horned/ hornless females, and with horn cores of incompletely mature males. Moyà-Solà's (1983:198, figs. 59, 60) division

of the Boselaphini is crucial for an understanding of the evolution of the tribe. He linked the middle Miocene *Austroportax latifrons* and some allied forms with Bovini and separated them from the more primitive *Protragocerus*, the more advanced *Miotragocerus*, and *Tragoportax*. *Tragoportax* was extremely successful and widespread in the late Miocene, but it and its group became extinct at the end of that period or in the earliest Pliocene.

At the beginning of the history of this last group of Boselaphini, it is hard to draw generic distinctions between the species of *Protragocerus, Miotragocerus,* and *Tragoportax*. *Protragocerus* is known from late in the middle Miocene and late Miocene, the other two genera nearly entirely from the late Miocene. The range of morphology would allow two rather than three generic names. The type specimen of the type species of *Miotragocerus, M. monacensis,* is distinguishable (more mediolateral compression of horn cores, tips not reapproaching) from the type specimen of *Protragocerus chantrei,* type species of its genus, but similar to *M. pannoniae* and *Tragoportax gaudryi.* Indeed, Solounias (1981a) applied the name *M. monacensis* to many late Miocene fossils of this group at Samos and Pikermi. However, both Moyà-Solà (1983) and Bouvrain (1988) have accepted a specialized concavity at the top of the metatarsal lateral surface as characterizing *Miotragocerus* in Central Europe and use *Tragoportax* for the remaining late Miocene forms. We follow them here.

In general, the more primitive species of this group occur earlier and overlap with the more advanced ones. The species *M. monacensis, M. pannoniae, M. gradiens,* and earlier examples of *T. leskewitschi* are more primitive than *T. amalthea, T. salmontanus, T. rugosifrons, T. browni,* and later *T. gaudryi.* However, identifications at localities other than the type locality for any particular named species can be doubtful. The record of *M. pannoniae* at its type locality (Sopron) is one of the rare occurrences of *Miotragocerus* in Central or Eastern Europe before the advent of the "*Hipparion*" faunas. Another possible middle Miocene *Miotragocerus* is present at Belometscheskaya under the name *Paratragocerus caucasicus.* Boselaphines are not known from the Vallesian of Greece, and in Turkey only one occurrence at Eşme Akçaköy is recorded by Köhler (1987:139), but in the Turolian they increased and diversified, alongside *Gazella*, as the most widespread of all bovids. Moyà-Solà's (1983) identification of the two species, *T. amalthea* and *T. gaudryi,* at Pikermi can be agreed on and welcomed. A more obvious Turolian innovation is *T. rugosifrons* (Schlosser 1904), of which Bouvrain and Bonis (1984a) give an early record in the early Turolian and which lasts into later MN 12 (Solounias 1981a: fig. 30H). Korotkevich (1980) also refers to the presence of multiple contemporaneous tragocerine lineages in regions north of the Black Sea.

Many generic and specific names have been founded in the Russian literature for boselaphine fossils. It looks as if Vallesian *T. leskewitschi* gradually evolved more advanced characters, e.g., a stronger transverse ridge between the horn bases that made it more like *T. gaudryi*. Records of *T. leskewitschi* overlap *T. gaudryi* from the Turolian onward. Early in the Turolian, *T. frolovi* (Pavlow 1913) appeared. Bouvrain (1988:14) placed *T. frolovi* in *T. rugosifrons*, but Pavlow's Tchobroutchi fossil looks more like a *T. amalthea*. If it is indeed *T. amalthea* or *amalthea*-like, then *T. rugosifrons* could be unknown in Russia, although Godina and David (1973) cited it for Taraklia. *T. gaudryi* and *T. frolovi* survive into the latest Turolian, MN 13, but *T. rugosifrons* has yet to be found at that age.

The Samos *Samokeros minotaurus* Solounias 1981a is perhaps a member of the Bovini, but no relationships to later genera have yet been detected. The origin of the bovine *Bubalus* and its relatives appears to have occurred on the Indian subcontinent with the appearance of forms like the Dhok Pathan (Turolian equivalent) *Pachyportax* and the Tatrot (Pliocene) *Proamphibos*. The more northerly *Bos* and the African *Syncerus* have less apparent origins.

Gazelles

Gazella ancyrensis Tekkaya 1973 from Yassiören, of likely MN 9 age, is small and more like an ancestor than a member species of *Gazella*. Vallesian *Gazella* is also known in the Ukraine and Algeria and in MN 11 becomes widespread. The classification of Miocene gazelles was discussed by Gentry (1970:295–300). Kurtén (1952) found in the Chinese "*Hipparion*" faunas that a more hypsodont gazelle can often be distinguished from a less hypsodont one. This is also true for *Gazella deperdita* (= *G. pilgrimi*) at Samos versus *G. capricornis* at Pikermi, but elsewhere Turolian gazelles are difficult to classify into species. The gazelles of the regions north of the Black Sea give the impression (Pavlow 1913: pl. 2, figs. 1 and 15) that the dorsal orbital rims are perhaps wider than in the European species. We therefore attribute them to *G. schlosseri* Pavlow 1913. Korotkevich (1968) detected two groups in these faunas, one of which foreshadows Asiatic gazelles of the so-called subgenus *Procapra*. This is a very plausible idea since the wide ranging extant *G. subgutturosa* must be related to subgenus *Procapra*, and it is highly unlikely that this group would have evolved in Africa or Western Europe. Most other living gazelles can be arranged in three groups: a group of three large-sized African species (so-called subgenus *Nanger*), two smaller Sub-Saharan species (*G. thomsoni, rufifrons*), and lastly a mass of southern Palaearctic species centred on *G. dorcas* of North Africa and extending to peninsular India as *G. bennetti*. It is the ancestry of this third group that could be expected to occur among those late Miocene gazelles of our region not showing trends toward subgenus *Procapra*.

Spiral-Horned Antelopes with Anticlockwise Torsion

Knowledge of the late Miocene spiral-horned antelopes with anticlockwise torsion on the right side has been much extended in the last decade or so. They are especially characteristic of our region and are known from MN 9 onward. The most prominent genus of this group is *Prostrepsiceros*. Bouvrain (1982) altered the arrangement of *Prostrepsiceros* species proposed by Gentry (1971), and we now follow her in recognizing *P. houtumschindleri, P. zitteli, P. rotundicornis,* and *P. fraasi* as separate species.

Prostrepsiceros houtumschindleri has rather large horn cores with a strong posterior and no anterior keel. Most *P. zitteli* have an anterior keel and a much weaker posterior keel, but Bouvrain and Thomas (1992) allot the small *Prostrepsiceros* at Gebel Hamrin to *P. zitteli*, and here there is a postero-external keel and a feebler anterior one, more akin to the larger *P. houtumschindleri*. Both species could derive from *P. vallesiensis*, in which anterior and posterior keels are present. *P. rotundicornis* and *fraasi* have only traces of a keel that looks vestigial on an otherwise smoothly rounded surface. This keel descends to a medial or even a posteromedial insertion and raises questions about its homology and the possible rotation of horn core insertions on the skull surface. *P. fraasi*, especially the Samos specimen, is larger than *P. rotundicornis* and has a more open spiral of its horn cores. *P. fraasi* cannot be accurately referred to an MN zone, but its possible Yassiören record could make it as early as MN 9. Pavlow's frontlets of "*Protragelaphus skouzesi*" from Grebeniki are too small and supraorbital pits too large for that species, yet the horn cores do have a strong posterior keel. Bouvrain referred them to *Prostrepsiceros zitteli* despite discrepancies with keels; we prefer to allocate them to *P. houtumschindleri*. The two sites of Grebeniki and Maragheh give *P. houtumschindleri* the least extensive geographical range within its genus although apparently on opposite sides of the Eastern Paratethys. It is also the only species of this group to occur in the Ukraine. Maragheh *P. houtumschindleri* is the species among *Prostrepsiceros* that is most like *Protragelaphus*.

Ouzocerus gracilis differs from later *Protragelaphus skouzesi* by its less divergent and more uprightly inserted horn cores. It would be interesting to know if it shows any vestiges of an anterior keel on the horn cores, such as occasionally occur in the later species (Mecquenem 1924–1925: pl. 6, fig. 6). *P. theodori* is only a slightly later species than *P. skouzesi*, but it does have horn insertions closer together and a shorter premolar row. *Palaeoreas lindermayeri* was first described from Pikermi, where it has large horn cores with a strong posterior and a weaker anterior keel. The rare Samos *P. lindermayeri* probably has a late MN 11 age (Lausanne collection from Stephana) and has horn cores less tightly spiraled and less divergent. *P. asiaticus*, a Turkish species from MN 11, is doubtfully separate from *P. lindermayeri*. The Vallesian to Turolian *P. elegans* has an anterior keel, but the posterior keel is weaker and there is more mediolateral compression. This species is very like the lectotype *Prostrepsiceros zitteli* and threatens nomenclatural disorder in the future.

Nisidorcas planicornis (see Köhler 1987) is a Turolian species with only a weak development of torsion in its horns. Its type locality, interestingly, is Piram (formerly Perim) Island, India. *Nisidorcas*-like horn cores have recently come to light in Vallesian levels of the Sinap Formation (Kappelman et al., this volume). Bouvrain (1992) does not accept all these genera as Antilopini; she aligns *Protragelaphus* with Caprinae and leaves *Ouzocerus* and *Palaeoreas* in an unresolved position.

Spiral-Horned Antelopes with Clockwise Torsion

Samotragus was placed in the Antilopini by Bouvrain and de Bonis (1985), and the type species, *S. crassicornis*, comes from Samos. Apart from its horn cores it is not large-sized. *Samotragus* is supposed to differ from the long-known *Oioceros rothi* by shorter horn cores, single deep longitudinal groove on the horn core, weak antero-posterior instead of medio-lateral compression, stronger torsion, pronounced diminution of horn cores distally, no postcornual fossa, larger supraorbital pits, and shorter premolar rows. We suggest that the horn cores of *O. rothi* could be females, so that *S. crassicornis* is the junior synonym and male of *Oioceros rothi*. Several bovid lineages seem to have acquired horned females in the late Miocene (Gentry 1990b:208, quoting and extending observations of Thomas).

Oioceros (including *Samotragus*) had a long late Miocene history. *O. praecursor* comes from the Vallesian of Ravin de la Pluie and *O. occidentalis* from the later Turolian of Italy (Bouvrain and de Bonis 1985; Masini and Thomas 1989). *O. praecursor* still had hornless females (Bouvrain and de Bonis 1985: fig. 2), and males had smaller horn cores than in *O. rothi*. A closely related species to *O. rothi*, or a variety within it, is *Oioceros wegneri*, known from Samos, some Turkish localities, and also, probably, as *Paraoioceros improvisus* from the Caucasus (Köhler 1987; Meladze 1985). The distinctive multiple grooves on its horn cores are concentrated in the same region as the large groove of males of *O. rothi* (= *Samotragus crassicornis*). The insertions may be less upright in profile. *Hispanodorcas* has torsion of its horn cores in the same direction as *Oioceros*, but weaker. *H. orientalis* differs slightly from the Spanish type species, *H. torrubiae* (MN 12). Females are perhaps hornless. *Hispanodorcas* could represent a likely horn core morphology for an *Oioceros* ancestor. The apparently steep cranial roof of the Grebeniki "*Gazella deperdita*" (Pavlow 1913: pl. 2, fig. 14) suggests *?Hispanodorcas rodleri* (Pilgrim and Hopwood 1928) as a possible identification.

Ovibovini

The first discovered among a group of late Miocene ovibovines with intense specializations of their horn cores and basioccipital/atlas articulations was *Urmiatherium polaki* Rodler 1889 of Maragheh, later determined to be from Upper Maragheh (Bernor 1986). The much smaller *Parurmiatherium rugosifrons* Sickenberg 1933 from Samos is extremely similar. *Plesiaddax* is also very similar but more primitive, its various species differing from *Urmiatherium* by their outwardly directed horn cores, nonunited frontals' boss anterior to horn cores, and smaller facets on the back of the basioccipital. All these forms could really be regarded as one genus, and insofar as any torsion can be detected in their horn cores, it is clockwise on the right. *Criotherium argalioides* of Samos, however, is more strikingly different in the spiraling of its horn cores, this being anticlockwise on the right. Hence, this group too, like Antilopini, contains species with opposing directions of torsion.

Urmiatherium and its allies are found in Southeastern Europe and Southwestern Asia but are not so far documented in the northern Black Sea region or the Caucasus. It was pointed out above that *Urmiatherium, Criotherium, Mesembriacerus* Bouvrain 1975, and the Asiatic *Tsaidamotherium* might be in the same tribe as the middle Miocene *Hypsodontus*, very long separate from all other bovids or even diphyletic with them. This idea, as yet not investigated, would leave a taxonomic problem with Plio–Pleistocene and extant Ovibovini. Perhaps *Ovibos* itself could descend from near *Mesembriacerus* (Bouvrain and de Bonis 1984b), while *Budorcas* is related to Caprini via a form like *Palaeoryx* (Gentry 1971; 1992). In this case *Budorcas* would be in a different tribe from Ovibovini. Post-Miocene ovibovines or relatives of *Budorcas* have been found in Sub-Saharan Africa, China, and as late immigrants to North America. It was only during the Pleistocene that they became a relict group.

Caprinae Other Than Ovibovines

A prominent group among late Miocene bovids of our region is that embraced by the generic names *Palaeoryx, Protoryx,* and *Pachytragus*. They have been most fully described from Pikermi and Samos. *Pachytragus* is the most amenable of these genera for classification. The lectotype of *P. crassicornis* (Schlosser 1904: pl. 11, fig. 11), type species of its genus, appears to be conspecific with other Samos fossils (Gentry 1971: pl. 4) that have teeth definitely more advanced than *Protoryx carolinae* from Pikermi or *Palaeoryx pallasi* from Pikermi and Samos. These advanced teeth are central to our concept of the genus *Pachytragus*. *Pachytragus laticeps* (Andree 1926), another Samos species, has slightly less advanced teeth, leaving the larger and rather rare *Protoryx carolinae* to stay as type species of *Protoryx*. *Pachytragus crassicornis* and *P. laticeps* appear to overlap temporally at Samos, although the latter species is morphologically suited to be ancestral to the former. The lectotype of *P. crassicornis* is likely to be from the Old Mill Beds of the Mytilini Formation (Solounias 1981b:261) and therefore uppermost MN 11, while a Lausanne skull, No. 22, and the American Museum of Natural History fossils from Quarry 1 of *P. laticeps* are late MN 12 in age (Bernor et al., this volume b).

The small *Protoryx enanus* of MN 7 is important as the earliest occurrence of its genus and in our view of Caprinae, too. The species could descend from *Tethytragus*, although it and related genera appear to have survived into the late Miocene of Turkey, the Siwaliks, and Africa. The slightly later (MN 8–9) Tunisian and Turkish species "*Pachytragus*" *solignaci* has primitive teeth and is also best placed in *Protoryx*, as already done by Köhler (1987:172).

Arambourg (1959:98) noted that the *Palaeoryx pallasi* lectotype horn appeared giraffidlike in the original illustration, but it is in fact a bovid. *Palaeoryx* may include *Prodamaliscus* Schlosser 1904. A probable bovid the size of *Palaeoryx pallasi* occurs at Baccinello (Del Campana 1918: pl. 4, fig. 10). The tribal and subfamily attribution of *Palaeoryx*, *Protoryx*, and *Pachytragus* is still not agreed. Traditionally all three were regarded as Hippotragini or in an extinct grouping, the Pseudotraginae. Gentry (1971) proposed that they were Caprinae, and the last two Caprini. Erdbrink (1988:151–54) preferred to retain them as Hippotragini, while Bouvrain and de Bonis (1984b) did not accept that *Palaeoryx* was an ovibovine (but were also doubtful about *Budorcas* as an ovibovine). The advanced teeth of *Pachytragus crassicornis* (reduced basal pillars, prominent mesostyle, and very flat or even concave labial wall of metacone on upper molars) are very caprinelike. *P. crassicornis* is associated with *Capra aegagrus* on the cladogram of Gentry (1992: fig. 6). This group is most abundantly known from Southeastern Europe and Southwestern Asia and also in some northern Black Sea faunas.

Antelopes of African Affinity

Thomas (1980; 1984c) showed that distinctively African bovids became recognizable very late in the Miocene at Mpesida and Lukeino in Kenya: e.g., reduncines and tragelaphines. Bovids of apparent African origin do occur at some southern sites within our area. Thomas et al. (1982:289) mention a possible reduncine at Gravitelli, Sicily. The horn core in question is shown in Seguenza (1902:154: pl. 6, figs. 23–25). There is also a possible late Turolian reduncine in Turkey (Köhler 1987:214). These could be part of a Messinian spread from Africa and be linked with the appearance of reduncines in the Siwaliks (top of Dhok Pathan or shortly afterward–slightly later than 7.0 Ma). However, *Redunca eremopolitana* Erdbrink (1982) is based on wrongly oriented female horn cores and on teeth, both belonging to a *Pachytragus* or some later caprine.

Maremmia hauptii in the Tuscan Turolian (Baccinello and other localities) is a dentally precocious offshoot close to the ancestry of African Alcelaphini (Thomas 1984a: fig. 4). Teeth are more advanced than in early Pliocene Alcelaphini of Langebaanweg, South Africa. *Tyrrhenotragus gracillimus* is apparently a European neotragine, and occurs with *Maremmia*. They seem to represent a temporary (perhaps 2.0 Ma long) irruption of African forms in a restricted part of Southern Europe. Neither alcelaphines nor neotragines would be likely inhabitants of swamps at Baccinello. Present-day neotragines tend to occur in thickets in rather dry country and in bushland. Alcelaphines are not noted for drought endurance, but neither are they so dependent on habitats in the vicinity of water as are most reduncines.

Changes Within the Miocene in European Ruminant Faunas

In our own region in MN 5 or 6, *Palaeomeryx kaupii* evolves into the more advanced *P. eminens*, while *Procervulus* and larger *Lagomeryx* species (*L. praestans*, *L. rutimeyeri*) are no longer present. *Stephanocemas*, *Dicrocerus*, *Heteroprox*, and *Micromeryx* all appear. *Euprox* also appears and seems to survive longer or is commoner later than are *Dicrocerus* and *Heteroprox*. The bovid *Eotragus* comes to cooccur with *Giraffokeryx*, *Tethytragus*, and *Hypsodontus* around MN 5/6, although *Hypsodontus* had been present earlier in China and perhaps Arabia. Boselaphines do not appear until MN 7, and are not known in Turkey at that time, although they were present far to the west in Spain in MN 5. All these changes do not show much clustering around any particular temporal interval, and there is a possible hiatus between the changes in cervoids and those in the bovids and giraffids.

In the late Miocene ruminants become more differentiated regionally and more remote from their evolutionary origins than were middle Miocene ruminants. An initiation of new forms, especially bovids, takes place in the Vallesian, but their expansion and diversification seem to be delayed until the Turolian. To some extent the patchiness of the fossil record would make for nonoccurrences in

TABLE 29.3 *Contrasts in Middle and Upper Miocene Pecoran Faunas in Central Europe and Southwestern Asia*

	No. of species in Central Europe	No. of species in Southwestern Asia
Late Miocene		
Bovidae (Boselaphini)	2–3	2–4
Bovidae (other tribes)	0	many
Cervidae	2–3	1–2
Giraffidae	0	4–5
Middle Miocene		
Bovidae (Boselaphini)	2–3	0
Bovidae (other tribes)	1	2–3
Cervidae	2–3	2–3
Giraffidae	0	1

Note: The Middle Miocene giraffid extends westwards to Yugoslavia.

certain areas, and there is a definite shortfall of Vallesian as against Turolian sites in our lists for Southeastern Europe and Southwestern Asia (tab. 29.2).

Palaeomeryx disappears early in the Vallesian. Somewhat later more species of deer appear, like *Procapreolus loczyi* or *Pliocervus pentelici*, in which the first fork or brow tine is higher above the basal burr and in which there is at least one further fork distally on the main beam. Among giraffids, *Giraffokeryx* is succeeded by *Palaeotragus coelophrys* and a little later again by *P. roueni* and *Samotherium* (although the *Giraffokeryx*-like *Injanatherium* continues in Arabia). *Bohlinia* first occurs in the later Vallesian (MN 10). Representation of the sivathere group changes from *Decennatherium* to *Helladotherium*. Among Vallesian bovids there are representatives of Boselaphini (Russia, Turkey), *Prostrepsiceros*, *Ouzoceros*, *Nisidorcas*, and *Palaeoreas* (Greece, Turkey), *Oioceros* (Greece), *Mesembriacerus* (Greece), and *Protoryx* (Turkey). *Hypsodontus* has disappeared. The Sub-Paratethyan Province (Bernor 1983) comes to be inhabited by a notably large number of ruminant species in the Turolian (MN 11–12). The pecoran list for Samos, even after some lumping, totals 23 species, most of them bovids. This contrasts with Central Europe, which in all the late Miocene contains only Boselaphini, cervids, and tragulids, albeit advanced on middle Miocene forms (see tab. 29.3—Western Europe also contains *Gazella*). The regions north of the Black Sea have a similar fauna to Central Europe, with the addition of *Palaeotragus*, *Samotherium* and sometimes *Helladotherium*, *Prostrepsiceros*, *Oioceros*, or *Palaeoryx*. Most of this rich Turolian ruminant fauna disappears in MN 13 (Köhler 1993).

European Changes in Relation to Those Elsewhere

Chinese middle and late Miocene ruminants are more like European ones than are those of either Africa or the Siwal-

iks. At a period thought to be equivalent to Orleanian or Astaracian one finds *Lagomeryx*, *Stephanocemas*, *Dicrocerus*, and *Palaeomeryx*, although there appears to be no record of *Heteroprox* or *Euprox*. At Tung Gur the pre-*Hipparion* giraffid *Giraffokeryx tungurensis* is notable for its short premolar row. Among middle Miocene bovids neither boselaphines nor *Tethytragus* are known. One feature of major interest in China is the claimed early occurrence of *Hypsodontus* (Chen 1988; Li and Qiu 1980; Azanza and Morales 1994), at least as early as *Eotragus* in Europe. Once the *Hipparion* faunas are in place, the ruminant changes look very similar to Europe, with more advanced deer and more diverse giraffids. Boselaphines and gazelles are present, also *Urmiatherium* and members or relatives of the *Palaeoryx-Protoryx-Pachytragus*–group. However, there are no spiral-horned Antilopini until the advent of *Antilospira* during the Pliocene (Tedford et al. 1991:522).

In the Siwaliks there are no deer in pre-*Hipparion* levels; *Giraffokeryx* is the main giraffid; there is possibly an *Eotragus*; *Protragocerus* and other boselaphines, *Sivoreas*, *Hypsodontus*, and rare *Caprotragoides* are present. *Sivoreas eremita* resembles *Prostrepsiceros vallesiensis*, the earliest species of *Prostrepsiceros*, in the two keels and the compression of its horn cores, and this could affect the question of whether the species is a boselaphine or an antilopine (Thomas 1984b:40). Thomas (1984b:53) accepts previous claims (Pilgrim 1937:801; Gentry 1970:299) for *Gazella* in the Chinji zone—probably earlier than in Europe.

Once "*Hipparion*" arrives in the Siwaliks, one also finds larger and more diverse boselaphines (*Tragoportax*, *Selenoportax*, *Pachyportax*), the tiny *Elachistoceras*, continuing *Gazella*, and persisting *Caprotragoides*. Among the poorly understood giraffes *Giraffokeryx* disappears, perhaps around 10.0 Ma, the sivathere *Bramatherium* appears, and there is also present *Giraffa punjabiensis* Pilgrim 1911, which may indeed be a *Giraffa* if it is not an *Injanatherium* (? = surviving *Giraffokeryx*) or akin to *Palaeotragus coelophrys*.

Earlier African faunas, equivalent to the early Miocene Songhor-Rusinga faunal interval in Kenya, contain *Dorcatherium* and *Propalaeoryx*, the latter being a pecoran not known to have carried horns or antlers. The later Maboko, Zelten, and other faunas show a middle Miocene aspect, but their correlation to the MN scale of Europe is poorly known and their ruminants are different. The antlered giraffoid *Climacoceras* has succeeded *Propalaeoryx*, from which it may be descended (see Hendey 1978:27), the giraffid *Canthumeryx* has appeared, and also *Prolibytherium*, with palmated antlers and teeth without any cervidlike characters. Incompletely known bovids are also present, including a possible *Eotragus*, a *Gazella* at Majiwa (Thomas 1984d:83), and possibly an early relative of the Ngorora *Homoiodorcas*. Somewhat later faunas of Africa, e.g., Fort Ternan and Beni Mellal, contain a more advanced *Climacoceras*, the giraffid *Giraffokeryx* (with very long legs at Fort Ternan) superseding *Canthumeryx*, and bovids such as *Protragocerus* and the nonboselaphines *Hypsodontus*, *Gentrytragus*, and *Benicerus*. These faunas have a somewhat Astaracian aspect (MN 6–8) in European terms, but there are discrepancies in that the *Hypsodontus* is less hypsodont and the *Gentrytragus* more hypsodont than their Eurasian counterparts at, say, Paşalar. Also present on the evidence from Ngorora (Thomas 1981) is a *Sivoreas*. In the Baringo deposits "palaeotragines" were present until about 10 Ma (Hill et al. 1985:771, fig. 4), which is slightly older than the last supposed *Giraffokeryx* in the Siwaliks.

This two-step change, first to faunas of the Maboko-Zelten level, then to the Fort Ternan–Beni Mellal level, suggests a drawn-out evolution, but the component of immigration may not be absent. Thomas (1985:43) drew attention to the second step of the change occurring in Arabia between the faunas of the Dam Formation or its terrestrial equivalents and the fauna of the Hofuf Formation.

Details of the *Hipparion* faunas are less clear in Africa than in Europe, with limited information on the late Miocene available from Menacer, Bou Hanifia, Sahabi, Wadi Natrun, Mpesida, Lukeino, Lothagam, and the Wembere-Manonga Formation. No deer are known on the continent. *Palaeotragus germaini* at Bou Hanifia already had longer legs than did the similarly sized *Palaeotragus coelophrys* and *Samotherium* of Eurasia. Fossils accepted as Giraffinae appear at Lukeino at around 6.0 Ma. The oldest African sivatheriine is from Douaria, Tunisia (Geraads 1985:317; 1989b:786), or the Wembere-Manonga Formation, Tanzania (Gentry, in press) very late in the Miocene. By the time of Langebaanweg, African giraffids can be satisfactorily referred to *Sivatherium* and *Giraffa* (Harris 1976). Boselaphines may persist, as suggested at Menacer (Thomas and Petter 1986), but large-sized *Miotragocerus* or *Tragoportax* at the end of the late Miocene may have immigrated only shortly before their extinction. Neotragines are present, e.g., *?Raphicerus* sp. at Menacer (Thomas and Petter 1986:368). *Gazella* is present at Bou Hanifia. *Prostrepsiceros libycus* at Sahabi is as large as *P. fraasi* at Samos and does not have the same keels as in *P. vallesiensis*, *P. houtumschindleri*, and *P. zitteli*. *Oioceros* has yet to be found in the late Miocene of North Africa, although the Pliocene *Parantidorcas* Arambourg 1979 of Ain Boucherit looks as if it could be a descendant of Turolian *Oioceros* or *Hispanodorcas*. The North African Vallesian *Damalavus boroccoi* may be related to *Palaeoryx pallasi*.

Modern tribes of African bovids appear in Sub-Saharan Africa during or toward the end of the late Miocene— Tragelaphini, Reduncini, Hippotragini, Alcelaphini, and *Aepyceros*. Similar changes are also hinted at in North Africa, e.g., Reduncini and Alcelaphini at Wadi Natrun (Gentry 1980: 254, 296) and the *Redunca* aff. *darti* of Lehmann and Thomas (1987) at Sahabi.

Paleoecology

There must have been a pronounced difference between the late Miocene palaeoenvironments of Central Europe and those of the Sub-Paratethyan Province. The Turolian faunas of Pikermi and Samos were once likened to African savanna faunas, and similarities were seen in the presence of horses, rhinoceroses, giraffids, and multitudes of bovids, and in the near absence of cervids. In the past decade it has become apparent that this is an inadequately analyzed comparison. Most of the Turolian giraffids and bovids were not taxonomically akin to present-day African ones. The *Palaeotragus-Samotherium*–group of giraffids was different from *Giraffa* and *Okapia*, and even *Helladotherium* may have been a different sivathere stock from *Sivatherium*. The principal bovid groups in the Greek-Turkish Turolian were Boselaphini, spiral-horned Antilopini, *Urmiatherium* and its relatives, and Caprinae; those in Kenya (to take an East African example) in the recent past have been Tragelaphini, Cephalophini, Neotragini, Antilopini without spiral horns, Reduncini, Hippotragini, and Alcelaphini (tab. 29.4).

Characteristics of the late Miocene habitats in the region under consideration here remain conjectural in some details. Gabunia and Chochieva (1982) thought that woodland persisted in the west through the Vallesian, while more open habitats appeared in the east and southeast. Köhler (1993) judged from a comprehensive study of ruminant morphological adaptations that habitats in the Sub-Paratethyan Province in MN 6–9 and in MN 11 were drier and more open than in Western Europe. Solounias and Dawson-Saunders (1988) originally assessed many of the ruminants of Samos and Pikermi as browsers, and thought that the predominant biotope could not be a savanna, since *Parurmiatherium* was the only definite grazing ruminant in the faunas. This seemed to mean that the predominant

TABLE 29.4 *Numbers of Pecoran Species in the Greek-Turkish Turolian and in Present-Day East Africa*

	Greek-Turkish Turolian	Kenya Recent
Cervidae	1	—
Giraffidae	5 (1)	1
Boselaphini-Bovini	5 (2)	1 (1)
Tragelaphini	—	6
Cephalophini	—	6
Neotragini and Antilopini	16	8 (2)
Reduncini and Hippotragini	1	7 (7)
Alcelaphini	—	5 (5)
Urmiatherium and relatives	4 (1 or 2)	—
Caprinae	6	—

Note: Numbers in brackets are grazers within the preceding total.

biotope could not have been a savanna with abundant grasses, but it could still have been a savanna in the sense of a biotope with restricted tree cover. Moreover, it is also very likely that the first ruminants venturing into open habitats would have spent only a short season here each year, with limited effects on their morphology. Later studies of *Samotherium boissieri* and of two species of *Tragoportax* at Samos showed some grazing by these animals as well (Solounias et al. 1988; Solounias and Hayek 1993). The grazers in the Greek-Turkish Turolian thus constitute 12% of the pecoran fauna, those in Kenya 44% (tab. 29.4). Evidently there was not an endless dry grass steppe nor a dense, closed forest in the Sub-Paratethyan Province. Some grasses were present, there must have been extensive openings in the tree cover, and the dryness of the climate could not have been severe. Köhler (1993: diagrams 28, 31, 33 in comparison with 15, 17, 19) shows that certain postcranial traits of Miocene pecorans were less advanced than modern pecoran inhabitants of open/dry habitats. Bonis et al. (1992) concluded that there were generally open conditions in the late Miocene of Macedonia but that the late site of Dytiko was probably more forested. The ecological interaction between cervids and boselaphines in Eurasia would make an interesting investigation.

Small-sized pecorans are rare in the late Miocene faunas, whereas in the Kenya list all five neotragines and all but one of the cephalophines (29% of the total in tab. 29.4) are smaller than *Gazella thomsoni*. Such small species probably were present in the late Miocene, but hitherto unrepresented for reasons of taphonomy or collecting bias. They have come to light in the Sinap succession (Kappelman et al., this volume) and *"Lagomeryx celer"* Kretzoi 1954 from the Hungarian Turolian appears to be the teeth of a tiny bovid.

To what extent are late Miocene faunas an evolutionary stage in the development of modern faunas, and to what extent are they a set of faunas integrated into their own environments and then disruptively displaced by later faunas? *Gazella* and *Pachytragus* probably left descendants in later faunas. A *Prostrepsiceros* species somewhere in Eurasia must have been ancestral to the extant Indian blackbuck, *Antilope cervicapra*, while *Protragelaphus* may be related to later *Gazellospira* and *Spirocerus*. It is also possible to see resemblances in the horn cores of *Prostrepsiceros fraasi* and *libycus* to the African *Aepyceros* (in which case *Aepyceros* might be an antilopine and not in or close to Alcelaphini [cf. Gentry 1990b]). *Samotherium, Palaeotragus, Tragoportax, Urmiatherium,* and *Criotherium,* on the other hand, are all forms without modern descendants, as also probably are *Palaeoreas, Oioceros, Palaeoryx,* and others. Probably any fauna tends to adapt to its circumstances, and the longer it has for this process, the less flexible it becomes for evolving as a whole at a subsequent period of fast environmental change. Köhler (1993) believed that

very mature ecological systems containing ruminants were especially vulnerable at times of faunal change, and this seems to have been the fate of the Sub-Paratethyan ruminants.

Faunal Movements and Dispersals

A number of phases of mammal evolution have been detected during the Miocene. Important events discussed by various authors include the following:

1. The arrival of Proboscidea and numbers of other exotic mammals in the European record. This event can best be placed later in MN 3 (Bulot and Ginsburg 1993), probably corresponding to the later Burdigalian and dated around 18 Ma. Tassy (1989) points out that different proboscidean species did not all immigrate at the same time.
2. In East Africa a change from early Miocene faunas (Songhor, Rusinga, etc.) to those like the middle Miocene fauna of Fort Ternan, Kenya. This change took place over a period of time from about 16.0 to 14.0 Ma (Pickford 1981). Thomas (1985:43) postulated the same change in Arabia between the faunas of the Dam and Hofuf Formations.
3. The arrival of hipparionine horses in Eurasia at around 10.5 Ma (Swisher, this volume; Woodburne et al., this volume). Other mammals also showed changes around this time, which is close to the middle/late Miocene boundary.
4. Some movements of certain species late in the late Miocene after the Turolian native pecorans had been much reduced, probably arising from the Messinian drying of the Mediterranean. This produced some sporadic records or claims for African antelopes in our region, as noted above.

The proboscidean event is too early to concern this volume, but the others are not. However, the arrangement of fossil sites into the MN unit intervals, whatever its other advantages, does not easily permit an analysis. In areas where bio- and chronostratigraphic studies are possible, such as Spain (Moyà-Solà and Agusti 1989), appearances of new mammals in the Miocene do not cluster into as few as four time levels.

The following summary of observations is what looks likely for giraffids and bovids at the time of writing.

1. *Eotragus* is known earlier in Europe than in China (Halamagai Formation) or North Africa (Zelten), but there is a possibility that it is even older in the Siwaliks—more than 18.0 Ma according to Barry and Flynn (1989:565). If *Eotragus* did come to Europe via an immigration event, the Siwaliks or some other southern region is the most likely source. The early Miocene

Songhor-Rusinga faunas of East Africa and those of Namibia however do not contain bovids. In China and Arabia *Hypsodontus* seems to occur at least as early as *Eotragus* does in Europe.

2. *Homoiodorcas* does not occur in northern continents and could have had an ancestry in Arabo-Africa going back beyond its Al Jadidah, Zelten, and Maboko possible occurrences. *Gazella* and the Antilopini could have originated in Africa or the Siwaliks from Neotragini. Thomas (1984b:53; 1984d:83) notes early *Gazella* in the Chinji zone of the Siwaliks and from Majiwa, Kenya.

3. Among the scrappily known bovids of the middle Miocene one can distinguish definite boselaphines. Records from MN 5 in Spain could be earlier than anywhere else, and boselaphines become widespread in Astaracian and equivalent faunas in Europe, Africa, and the Siwaliks but not in China. *Tethytragus* is a similarly widespread nonboselaphine member of Astaracian-level faunas, and has the same continental distribution pattern as boselaphines. In Turkey, the Siwaliks and Africa it seems to survive into the late Miocene.

4. *Hypsodontus* and *Giraffokeryx* also became widespread near the start of the middle Miocene, *Giraffokeryx* from Africa and *Hypsodontus* probably from Eastern Asia.

5. There was an initiation of new pecorans, especially bovids, in the Vallesian, and their main diversification is manifest in the Turolian. Caprines had evolved from *Tethytragus* and now appeared in force, and there also appeared *Urmiatherium* and its relatives and spiral-horned Antilopini. Gazelles also dispersed at this time from their possible African origins. The boselaphines entered China, and in the Siwaliks they and giraffids alone diversified. Evidence is lacking from Africa for such massive changes at this time.

6. Toward the end of the late Miocene Bovini or bovine-like bovids appear in Europe and Africa, and perhaps a little later in the Siwaliks. Also toward the end of the late Miocene *Aepyceros*, Tragelaphini and Reduncini appear in Africa. Reduncini appeared in the Siwaliks at this time and perhaps elsewhere, and neotragines and apparent alcelaphines in Tuscany.

We thank the organizers of the VW Workshop at Immendingen and Schloss Reisensburg for inviting us to participate in such a stimulating meeting.

LITERATURE CITED

Alexejeva, A. K. 1930. Die obersarmatische Säugetierfauna von Eldar. *Travaux du Musée géologique Académie Sciences de l'USSR* 7:167–204.

Andree, J., 1926. Neue Cavicornier aus dem Pliocän von Samos. *Palaeontographica, Stuttgart* 67:135–75.

Arambourg, C. 1959. Vertébrés continentaux du Miocène supérieur de l'Afrique du Nord. *Publications du Service de la Carte Géologique de l'Algérie (Nouvelle Série) Paléontologie, Mémoires* 4:1–159.

———. 1979. *Vertébrés Villafranchiens d'Afrique du Nord (Artiodactyles, Carnivores, Primates, Reptiles, Oiseaux)*. Paris: Singer-Polignac.

Azanza, B. and J. Morales. 1994. *Tethytragus* nov. gen. et *Gentrytragus* novv. gen. Deux nouveaux Bovidés (Artiodactyla, Mammalia) du Miocène moyen. *Proceedings, Koninklijke Nederlandse Akademie van Wetenschappen* B 97:249–82.

Barry, J. C. and L. Flynn. 1989. Key biostratigraphic events in the Siwalik sequence. In *European Neogene Mammal Chronology*, ed. E. H. Lindsay, V. Fahlbusch, and P. Mein, pp. 557–71. New York: Plenum.

Bernor, R. L. 1983. Geochronology and zoogeographic relationships of Miocene Hominoidea. In *New Interpretations of Ape and Human Ancestry*, ed. R. L. Ciochon and R. S. Corruccini, pp. 21–64. New York: Plenum.

———. 1986. Mammalian biostratigraphy, geochronology, and zoogeographic relationships of the late Miocene Maragheh fauna, Iran. *Journal of Vertebrate Paleontology* 6:76–95.

Bernor, R. L., G. D. Koufos, M. O. Woodburne, and M. Fortelius. This volume a. The evolutionary history and biochronology of European and Southwest Asian late Miocene and Pliocene hipparionine horses.

Bernor, R. L., N. Solounias, C. C. Swisher III, and J. A. Van Couvering. This volume b. The correlation of three classical "Pikermian" mammal faunas—Maragheh, Samos, and Pikermi—with the European MN unit system.

Bohlin, B. 1926. Die Familie Giraffidae. *Palaeontologica Sinica* C41:1–178.

Bonis, L. de, G. Bouvrain, D. Geraads, and G. Koufos. 1992. Diversity and paleoecology of Greek late Miocene mammalian faunas. *Palaeogeography, Palaeoclimatology, Palaeoecology* 91:99–121.

Borissiak, A. 1914. Mammifères fossiles de Sebastopol. *Mémoires Commemoratif géologie St. Petersburg*, n.s., 87:1–154.

Bouvrain, G. 1975. Un nouveau bovidé du Vallésian de Macédoine (Grèce). *Comptes Rendus de l'Académie des Sciences, Paris* D280:1357–59.

———. 1982. Révision du genre *Prostrepsiceros* Major 1891. *Paläontologische Zeitschrift* 56:113–24.

———. 1988. Les *Tragoportax* (Bovidae, Mammalia) des gisements du Miocène supérieur de Ditiko (Macédoine, Grèce). *Annales de Paléontologie* 74:43–63.

———. 1992. Antilopes à chevilles spiralées du Miocène supérieur de la province Gréco-Iranienne: Nouvelles diagnoses. *Annales de Paléontologie* 78:49–65.

Bouvrain, G. and L. de Bonis. 1984a. Étude d'un miotragocère du Miocène supérieur de Macédoine (Grèce). In *Actes du symposium paléontologique G. Cuvier*, ed. E. Buffetaut, J. M. Mazin, and E. Salmon, pp. 35–49. Montbéliard.

———. 1984b. Le genre *Mesembriacerus* (Bovidae, Artiodactyla, Mammalia): Un ovibociné primitif du Vallésian (Miocène supérieur) de Macédoine (Grèce). *Palaeovertébrata* 14:201–23.

———. 1985. Le genre *Samotragus* (Artiodactyla, Bovidae), une antilope du Miocène supérieur de Grèce. *Annales de Paléontologie* 71:257–99.

Bouvrain, G. and H. Thomas. 1992. Une antilope à chevilles spiralées: *Prostrepsiceros zitteli* (Bovidae). Miocène supérieur du Jebel Hamrin en Irak. *Géobios* 25:525–33.

Brunet, M. and E. Heintz. 1983. Interprétation paléoecologique et relations biogéographiques de la faune de vertébrés du Miocène supérieur d'Injana, Irak. *Palaeogeography, Palaeoclimatology, Palaeoecology* 44:283–93.

Bulot, C. and L. Ginsburg. 1993. Gisements à mammifères miocènes du Haut-Armagnac et âge des plus anciens proboscideens d'Europe occidentale. *Comptes Rendus de l'Académie des Sciences, Paris* 316:1011–16.

Chen, G. F. 1988. Remarks on the *Oioceros* species (Bovidae, Artiodactyla, Mammalia) from the Neogene of China. *Vertebrata Palasiatica* 26:169–72.

Del Campana, D. 1918. Considerazioni sulle Antilopi terziarie della Toscana. *Palaeontographica Italica*, Modena, 24:147–233.

Erdbrink, D. P. B. 1982. A fossil reduncine antelope from the locality K2 east of Maragheh, N.W. Iran. *Mitteilungen der Bayerischen Staatssammlung für Paläontologie und historische Geologie* 22:103–12.

——. 1988. *Protoryx* from three localities east of Maragheh, N.W. Iran. *Proceedings, Koninklijke Nederlandse Akademie van Wetenschappen* B91:101–59.

Gabunia, L. K. and K. I. Chochieva. 1982. Co-evolution of the *Hipparion* fauna and vegetation in the Paratethys region. *Evolutionary Theory* 6:1–13.

Gentry, A. W. 1970. The Bovidae (Mammalia) of the Fort Ternan fossil fauna. In *Fossil Vertebrates of Africa*, ed. L. S. B. Leakey and R. J. G. Savage, 2:243–324. London: Academic Press.

——. 1971. The earliest goats and other antelopes from the Samos Hipparion fauna. *Bulletin, British Museum (Natural History), Geology* 20:229–96.

——. 1980. Fossil Bovidae (Mammalia) from Langebaanweg, South Africa. *Annals South African Museum* 79:213–337.

——. 1990a. Ruminant artiodactyls of Paşalar, Turkey. *Journal of Human Evolution* 19:529–50.

——. 1990b. Evolution and dispersal of African Bovidae. In *Horns, Pronghorns, and Antlers*, ed. G. A. Bubenik and A. B. Bubenik, pp. 195–227. New York: Springer-Verlag, New York.

——. 1992. The subfamilies and tribes of Bovidae. *Mammal Review* 22:1–32.

——. 1994. The Miocene differentiation of Old World Pecora (Mammalia). *Historical Biology* 7:115–58.

——. In press. Fossil ruminants (Mammalia) from the Manonga Valley, Tanzania. In *Neogene Paleontology of the Manonga Valley, Tanzania*, ed. T. Harrison. Plenum: New York.

Geraads, D. 1979. Les Giraffinae (Artiodactyla, Mammalia) du Miocène supérieur de la région de Thessalonique (Grèce). *Bulletin du Muséum Nationale d'Histoire Naturelle* C4:377–89.

——. 1985. *Sivatherium maurusium* POMEL (Giraffidae, Mammalia) du Pléistocène de la République de Djibouti. *Palaeontologische Zeitschrift* 59:311–21.

——. 1986. Remarques sur la systématique et la phylogénie des Giraffidae (Artiodactyla, Mammalia). *Géobios* 19: 465–77.

——. 1989a. Un nouveau Giraffidé du Miocène supérieur de Macédoine (Grèce). *Bulletin du Muséum Nationale d'Histoire Naturelle* C4:189–199.

——. 1989b. Vertébrés fossiles du Miocène supérieur du Djebel Krechem el Artsouma (Tunisie centrale). Comparaisons biostratigraphiques. *Géobios* 22:777–801.

Ginsburg, L. 1989. The faunas and stratigraphical subdivisions of the Orleanian in the Loire Basin (France). In *European Neogene Mammal Chronology*, ed. E. H. Lindsay, V. Fahlbusch, and P. Mein, pp. 157–76. New York: Plenum.

Godina, A. Y. and A. I. David. 1973. *Neogene Occurrences of Vertebrates on the Territory of the Moldavian SSR*. Kishinev: Shtiintza.

Harris, J. M. 1976. Pliocene Giraffoidea (Mammalia, Artiodactyla) from the Cape Province. *Annals South African Museum* 69:325–53.

Hendey, Q. B. 1978. Preliminary report on the Miocene vertebrates from Arrisdrift, South West Africa. *Annals South African Museum* 76:1–41.

Hill, A., R. Drake, L. Tauxe, M. Monaghan, J. C. Barry, A. K. Behrensmeyer, G. Curtis, B. F. Jacobs, L. Jacobs, N. Johnson, and D. Pilbeam. 1985. Neogene palaeontology and geochronology of the Baringo Basin, Kenya. *Journal of Human Evolution* 14:759–73.

Janis, C. M. and K. M. Scott. 1987. The interrelationships of higher ruminant families with special emphasis on the members of the Cervoidea. *American Museum Novitates* 2893:1–85.

Kappelman, J., S. Sen, M. Fortelius, A. Duncan, B. Alpagut, J. Crabaugh, A. Gentry, J.-P. Lunkka, F. McDowell, N. Solounias, S. Viranta, and L. Werdelin. This volume. Chronology and biostratigraphy of the Miocene Sinap Formation of Central Turkey.

Köhler, M. 1987. Boviden des turkischen Miozäns (Kanozoikum und Braunkohlen der Turkei 28). *Paleontologia Evolucio* 21:133–246.

——. 1993. Skeleton and habitat of Recent and fossil ruminants. *Münchner Geowissenschaftliche*, Abhandlungen A, 25:1–88.

Korotkevich, E. L. 1968. On the problem of taxonomic position of Miocene gazelles in the south of the USSR. *Vestnik Zoologii Kiev* 4:42–50.

——. 1970. [Mammals from late Sarmatian *Hipparion* fauna of Berislava.] *Prirodnaya Obstanovka: Fauny Proshlogo Kiev* 5:24–121.

——. 1980. Tragocerinae of Eastern Europe. *Vestnik zoologie* 4:6–10.

Kurtén, B. 1952. The Chinese *Hipparion* fauna. *Commentationes Biologicae Helsinki* 13:1–82.

Lartet, E. 1851. Notice sur la colline de Sansan. *L'Annuaire du Département du Gers* 3–45.

Lehmann, U. and H. Thomas. 1987. Fossil Bovidae from the Mio–Pliocene of Sahabi, (Libya). In *Neogene Paleontology and Geology of Sahabi*, ed. N. T. Boaz, A. El-Arnauti, A. W. Gaziry, J. de Heinzelin, and D. D. Boaz, pp. 323–35. New York: Liss.

Li, C. K. and Z. D. Qiu. 1980. Early Miocene mammalian fossils of the Xining Basin, Qinhai. *Vertebrata Palasiatica* 18:210–14.

Masini, F. and H. Thomas. 1989. *Samotragus occidentalis*, a new bovid from the late Messinian of Italy. *Bollettino della Società Paleontologica Italiana* 28:307–16.

Mecquenem, R. 1924–1925. Contribution à l'étude des fossiles de Maragha. *Annales de Paléontologie* 13:135–60; 14:1–36.

Meladze, G. K. 1985. *Review of the* Hipparion *Faunas of the Caucasus*. Tbilisi: Akademiya Nauk Gruzinskoi SSR.

Montoya, P. and J. Morales. 1991. *Birgerbohlinia schaubi* Crusafont 1952 (Giraffidae, Mammalia) del Turoliense inferior de Crevillente-2 (Alicante, España): Filogenia e historia biogeografica de la subfamilia Sivatheriinae. *Bulletin du Muséum Nationale d'Histoire Naturelle* C4:177–200.

Morales, J. 1985. Nuevos datos sobre *Decennatherium pachecoi* (Crusafont, 1952) (Giraffidae, Mammalia): Descripción del craneo de Matillas. *COLPA* [= Coloquios de la Catedra de Paleontologia] 40:51–58.

Morales, J., S. Moyà-Solà, and D. Soria. 1981. Presencia de la familia Moschidae (Artiodactyla, Mammalia) en el Vallesiense de España: *Hispanomeryx duriensis* novo gen. nova sp. *Estudios geologicos Instituto Geologicas, Madrid* 37:467–75.

Morales, J. and D. Soria. 1981. Los artiodactilos de Los Valles de Fuentidueña (Segovia). *Estudios geologicas Instituto Geologicas, Madrid* 37:477–501.

Moyà-Solà, S. 1983. Los Boselaphini (Bovidae Mammalia) del Neogeno de la peninsula Iberica. *Publicaciones de Geologia, Universitat Autonoma de Barcelona* 18:1–236.

———. 1986. El genero *Hispanomeryx* Morales et al. (1981): Posición filogenética y systematica. Su contribución al conocimiento de la evolución de los Pecora (Artiodactyla, Mammalia). *Paleontologia i Evolucío* 20:267–87.

Moyà-Solà, S. and J. Agusti. 1989. Bioevents and mammal successions in the Spanish Miocene. In *European Neogene Mammal Chronology*, ed. E. H. Lindsay, V. Fahlbusch, and P. Mein, pp. 357–73. New York: Plenum.

Ozansoy, F. 1965. Étude des gisements continentaux et des mammifères du Cénozoique de Turquie. *Mémoire Société Géologie France, Paris*, n.s, 102:1–92.

Paraskevaidis, I. 1940. Eine obermiocäne Fauna von Chios. *Neues Jahrbuch für Mineralogie Geologie und Paläontologie Beilägebände, B* 83:363–442.

Pavlović, M. B. 1969. Miozän-Säugetiere des Toplica-Beckens. *Geoloski Anali Balkanskoga Poluostrva* 34:269–394.

Pavlow, M. 1913. Mammifères Tertiaires de la Nouvelle Russie 1, Artiodactyles, Perissodactyles. *Nouveaux Mémoires de la Société impériale des Naturalistes, Moscou* 17:1–67.

Pickford, M. 1981. Preliminary Miocene mammalian biostratigraphy for western Kenya. *Journal of Human Evolution* 10:73–97.

Pilgrim, G. E. 1911. The fossil Giraffidae of India. Memoirs Geological Survey, India. *Palaeontologia Indica*, n.s., 4, 1:1–29.

———. 1937. Siwalik antelopes and oxen in the American Museum of Natural History. *Bulletin of the American Museum of Natural History* 72:729–874.

Rinnert, P. 1956. Die Huftiere aus dem Braunkohlenmiozän der Oberpfalz. *Palaeontographica, A.* 107:1–65.

Rodler, A. 1889. Uber *Urmiatherium polaki* n.g., n.sp. *Denkschrift Akademie Wissenschaften, Wien* 56:315–22.

Rodler, A. and K. A. Weithofer. 1890. Die Wiederkäuer der Fauna von Maragha. *Denkschrift Akademie Wissenschaften, Wien* 57:753–72.

Schlosser, M. 1904. Die fossilen Cavicornier von Samos. *Beitrage Paläontologie Geologie Ost-Ungarn* 17:28–118.

Seguenza, L. 1902. I vertebrati fossili della Provincia di Messina, 2, Mammiferi e geologia del piano Pontico. *Bollettino della Società Geologica Italiana* 21:115–75.

Sen, S. and J.-P, Valet. 1986. Magnetostratigraphy of late Miocene continental deposits in Samos, Greece. *Earth and Planetary Science Letters* 80:167–74.

Sickenberg, O. 1933. *Parurmiatherium rugosifrons*, ein neuer Bovide aus dem unterpliozän von Samos. *Palaeobiologica Wien* 5:81–102.

Solounias, N. 1981a. The Turolian fauna from the island of Samos, Greece. *Contributions to Vertebrate Evolution* 6:1–232.

———. 1981b. Mammalian fossils of Samos and Pikermi. Part 2. Resurrection of a classic Turolian fauna. *Annals of the Carnegie Museum* 50:231–70.

Solounias, N. and B. Dawson-Saunders. 1988. Dietary adaptations and paleoecology of the late Miocene ruminants from Pikermi and Samos in Greece. *Palaeogeography, Palaeoclimatology, Palaeoecology* 65:149–72.

Solounias, N. and L.-A. C. Hayek. 1993. New methods of tooth microwear analysis and application to dietary determination of two extinct antelopes. *Journal of Zoology* 229:421–45.

Solounias, N., M. Teaford, and A. Walker. 1988. Interpreting the diet of extinct ruminants: The case of a non-browsing giraffid. *Paleobiology* 14:287–300.

Swisher, C. C., III. This volume. New ^{40}Ar/^{39}Ar dates and their contribution toward a revised chronology for the late Miocene of Europe and West Asia.

Tassy, P. 1989. The "Proboscidean Datum Event": How many proboscideans and how many events? In *European Neogene Mammal Chronology*, ed. E. H. Lindsay, V. Fahlbusch, and P. Mein, pp. 237–52. New York: Plenum.

Tchernov, E., L. Ginsburg, P. Tassy, and N. F. Goldsmith. 1987. Miocene mammals of the Negev (Israel). *Journal of Vertebrate Paleontology* 7:284–310.

Tedford, R. H., L. J. Flynn, Z. Qiu, N. D. Opdyke, and W. R. Downs. 1991. Yûshe Basin, China: Paleomagnetically calibrated mammalian biostratigraphic standard from the late Neogene of Eastern Asia. *Journal of Vertebrate Paleontology* 11:519–26.

Tekkaya, I. 1973. Une nouvelle espèce de *Gazella* de Sinap moyen. *Bulletin Mineral Research and Exploration Institute of Turkey* 80:118–43.

Thomas, H. 1980. Les bovidés du Miocène supérieur des couches de Mpesida et de la formation de Lukeino (district de Baringo, Kenya). In *Proceedings of the 8th Pan-African Congress of Prehistory, Nairobi 1977*, ed. R. E. F. Leakey and B. A. Ogot, pp. 82–91. Nairobi.

———. Les Bovidés miocènes de la formation de Ngorora du Bassin de Baringo (Kenya). *Proceedings, Koninklijke Nederlandse Akademie van Wetenschappen* B84:335–409.

———. 1983. Les Bovidae (Artiodactyla, Mammalia) du Miocène moyen de la Formation Hofuf (Province du Hasa, Arabie Saoudite). *Palaeovertébrata* 13:157–206.

———. 1984a. Les origines africaines des Bovidae miocènes des lignites de Grosseto (Toscane, Italie). *Bulletin du Muséum Nationale d'Histoire Naturelle* C4:81–101.

———. 1984b. Les Bovidés anté-hipparions des Siwaliks inférieurs (plateau du Potwar, Pakistan). *Mémoire Société Géologique de France*, n.s., 145:1–68.

———. 1984c. Les Bovidae du Miocène du sous-continent Indien, de la peninsule Arabique et de l'Afrique. *Palaeogeography, Palaeoclimatology, Palaeoecology* 45:251–99.

———. 1984d. Les Giraffoidea et les Bovidae miocènes de la formation Nyakach (Rift Nyanza, Kenya). *Palaeontographica, Stuttgart,* A 183:64–89.

——. 1985. The early and middle Miocene land connection of the Afro-Arabian plate and Asia: A major event for hominoid dispersal. In *Ancestors: The Hard Evidence,* ed. E. Delson and F. S. Szalay, pp. 42–50. New York: Liss.

Thomas, H., R. L. Bernor, and J. J. Jaeger. 1982. Origine du peuplement mammalien en Afrique du Nord durant le Miocène terminal. *Géobios* 15:283–97.

Thomas, H. and G. Petter. 1986. Révision de la faune de mammifères du Miocène supérieur de Menacer (ex-Marceau), Algérie: Discussion sur l'âge du gisement. *Géobios* 19:357–73.

Weidmann, M., N. Solounias, R. E. Drake, and G. H. Curtis. 1984. Neogene stratigraphy of the eastern basin, Samos island, Greece. *Géobios* 17:477–90.

Woodburne, M. O., R. L. Bernor, and C. C. Swisher III. This volume. An appraisal of the stratigraphic and phylogenetic bases for the "*Hipparion* Datum" in the Old World.

Ye, Jie. 1989. Middle Miocene artiodactyls from the northern Junggar Basin. *Vertebrata Palasiatica* 27:37–52.

III

PALEOBOTANY, PALEOBIOGEOGRAPHY, AND PALEOECOLOGY

30

Floristic Trends in the Vegetation of the Paratethys Surrounding Areas During Neogene Time

J. KOVAR-EDER, Z. KVACEK, E. ZASTAWNIAK, R. GIVULESCU, L. HABLY, D. MIHAJLOVIC, J. TESLENKO, AND H. WALTHER

Continental fossil floras often lack sufficient possibilities for reliable age determination. Because of this deficiency, age determinations have often been based merely on floristic comparisons. This method followed the notion that the higher the degree of floristic correspondence, the closer the age assignment must be.

A correlation system has been established here for the Paratethys realm and surrounding areas. It is based primarily on those floras that have been dated by independent means, including radioisotopic, magnetostratigraphic, mammalian biochronologic, and marine/continental correlations. Under these circumstances a date has been derived directly from plant-bearing sediments (underlined localities here in the tables), but not by the plant remains (pollen, leaves, fruits) themselves. Also, those fossil plant localities dated indirectly by independent means (in the above sense) of stratigraphically sub- or superjacent horizons are included (indicated by broken lines). Furthermore, we include those localities that are dated by well-based regional stratigraphic correlations (localities not underlined, table 30.1, table 30.2, figure 30.1).

A first step in this direction was made by Kovar-Eder (1988b) for a more restricted area (the Molasse zone north of the Alps and the Pannonian area). Other localities are referred to in the text only when their stratigraphic evidence is still less reliable and they are situated in a floristically relevant area, or adjacent to the surrounding Paratethys region. In future investigations, fossil plant assemblages that lack even this kind of information will be included.

The data presented here are part of a developing data base project. This data base includes information on the local basin and depositional facies, lithologic context of the plant-bearing sediment, the method of age determination, and, of course, the floristic content (species lists). These data are used to determine long-term changes in paleovegetation communities and gain a greater understanding of the regional paleogeographic and climatic trends that regulate them. Since these trends are of long chronologic duration, our treatment extends backward in time to the late Oligocene.

Key Taxa to Paleogeographical Distribution in Time

Paleotropic Elements

Tetraclinis salicornioides (Unger) Kvacek (pl. 30.1:1; tab. 30.3)   Twig fragments of a conifer (family Cupressaceae), well flattened, usually less than 1 cm in length. They are composed of cupressoid (closely attached) small leaves fused in verticles of four. Only rarely are there preserved remains of branches containing more than one or two verticles.

This species is well known under various synonyms: *Libocedrus salicornioides*, *Libocedrites salicornioides*, *Hellia salicornioides*. The existence of such an extensive synonymy is indicative of the debate surrounding this taxon's affinity to the extant genus. The discovery of attached and associated cones finally corroborated the assignment to *Tetraclinis*. Today, there is only a single extant species, *Tetraclinis articulata*, which occurs in the Southern Mediterranean (Atlas Mountains, Morocco) and is now adapted to the Mediterranean type of climate.

The branch fragments of *Tetraclinis salicornioides* are resistant to water transport and are therefore often found in fluviatile deposits (e.g., drift sands, associated with fruit- and seed taphocoenoses). This plant is also common in volcanogenic and lacustrine sediments. According to the associated plant remains, *T. salicornioides* is considered to

TABLE 30.1 *Sites Arranged in Stratigraphical Order*

Table gives information on the region, age, method of dating, depositional facies, and reference. The numbers on the left correspond with those in figure 30.1.

Nr.	Name of flora	Country/Region	Age	Determined by	Facies	Source
1	Eger Wind brickyard	Hungary	Egerian, lower NP 25	nannoplankton	marine	Kvacek & Hably 1991
2	Linz	Austria	Egerian, NP 25	nannoplankton	marine	Kovar 1982
3	Eferding	Austria	Egerian, NP 25	nannoplankton, foraminifera,	marine	Kovar-Eder & Berger 1987
4	Krumvir	CSFR	Egerian	foraminifera	marine	Knobloch 1975
5	Pomaz	Hungary	Egerian, NP 25	nannoplankton, molluscs	marine	Hably 1992a
6	Vertesszölös	Hungary	Egerian, NP 25	nannoplankton	marine	Hably 1990
7	Corus	Rumänien	Egerien	regional geology	brackish	Ticleanu & Givulescu 1978
8	Velka Causa	Slovakia	Eggenburgian	foraminifera, molluscs	marine	Sitar & Kvacek in press
9	Nesuchyne	Bohemia	MN 3a	mammals in overlying strata	fluviatile	Buzek & Kvacek 1988
10	Bilina	Bohemia	MN 3a	mammals of underlying strata	fluviatile	Buzek & al. 1992
11	Brestany	Bohemia	MN 3a	mammals of underlying strata	limnic	Buzek & al. 1992
12	Cheb	Bohemia	MN 4	mammals	continental	Buzek & al. in press
13	Lipovany	Moravia	Eggenburgian uppermost	regional geology	continental, volcanic	Nemejc & Knobloch 1973
14	Ipolytarnoc	Hungary	Ottnangian	regional geology	continental, volcanic	Hably 1985
15	Köflach	Austria	Ottnangian/Karpatian, MN 4/5	mammals in upper complex	continental, lignite facies	Meller 1992
16	Engelswies	Germany	MN 5	mammals in overlying strata	limnic	Schweigert 1992
17	Massendorf	Germany	MN 5	mammals in plant-bearing strata	fluviatile/limnic	Gregor 1983b
18	Mydlovary	Bohemia	MN 5, Mydlovary Formation	mammals in the Mydlovary Form.	limnic	Knobloch & Kvacek in press
19	Olesnik	Bohemia	MN 5, Mydlovary Formation	mammals in the Mydlovary Form.	limnic	Knobloch & Kvacek in press
20	Kameny Ujezd	Bohemia	MN 5, Mydlovary Formation	mammals in the Mydlovary Form.	fluviatile	Knobloch & Kvacek in press
21	Ledenice	Bohemia	MN 5, Mydlovary Formation	mammals in the Mydlovary Form.	limnic	Knobloch & Kvacek in press
22	Magyaregregy	Hungary	Badenian	regional geology	continental, limnic	Palfalvy 1981, Hably, pers. comm.
23	Nogradszakal Bertec.v.	Hungary	Badenian, lower	radiometric dating	continental	Andreanszky 1959, Kordos-Szakaly 1984
24	Weingraben	Austria	Badenian	regional geology	limnic	Berger 1953, Bachmayer & al. 1991
25	Smolin	Moravia	Badenian	foraminifera	marine	Knobloch 1969
26	Belchatow level TS2	Poland	16,5 ± 1,3 m.a.	radiometric date	continental, lacustic	Stuchlik & al. 1990
27	Swoszowice	Poland	Badenian	regional geology	marine, brackish	Il'inskaya 1964
28	Czernica	Poland	Badenian	regional geology	marine, brackish	Raniecka-Bobrowska 1957
29	Wieliczka	Poland	Badenian, middle	regional geology	marine	Kolasa 1982
30	Tuzla	Bosnia-H.	Badenian, NN 5, NN6	nannoplankton, foraminifera	marine	Pantic, Mihajlovic & Vrabac 1988
31	Devinska Nova Ves	Slovakia	Badenian, upper	foraminifera, mammals reg.geol.	marine	Berger 1951
32	Pistynka	Ukraine	Badenian, upper	foraminifera, molluscs	marine, brackish	Shvareva 1983
33	Kosov	Ukraine	Badenian, upper	molluscs of underlying strata	brackish	Shvareva 1983
34	Zalestsy	Ukraine	Badenian, upper	molluscs, foraminifera	brackish	Shvareva 1989
35	Krepkaya river	Russia	Badenian, upper	molluscs	marine	Teslenko 1975, 1977

No.	Locality	Country	Stage	Evidence	Environment	Reference
36	Krynka river	Russia	Badenian, upper	molluscs, foraminifera	brackish	Krishtofovich & Baykov-skaya 1965
37	Zagubica	Serbia	Badenian, upper	regional geology	continental	Milovanovic & Mihajlovic 1984
38	Steinheim	Germany	MN 7	regional geology	limnic	Gregor 1983a
39	Türkenschanze	Austria	Sarmatian, lower	molluscs	brackish	Berger & Zabusch 1953
40	Stawiany	Poland	Sarmatian, lower	foraminifera	brackish	Zastawniak 1980
41	Mlyny	Poland	Sarmatian, lower	molluscs, foraminifera	brackish	Zastawniak 1980
42	Amvrosievka	Ukraine	Sarmatian, lower	molluscs	brackish	Pimenova 1954
43	Zaporozhe	Ukraine	Sarmatian, lower	molluscs	brackish	Teslenko 1975
44	Racsa	Romania	Sarmatian, lower	molluscs	brackish	Ticleanu & Givulescu 1982
45	Borod	Romania	Sarmatian, lower	molluscs	brackish	Suraru & al. 1978
46	Daia-Sacadate	Romania	Sarmatian, lower	molluscs	brackish	Givulescu 1975b
47	Feleac	Romania	Sarmatian, lower	molluscs	brackish	Givulescu 1957b
48	Deva	Romania	Sarmatian, lower	molluscs	brackish	Ticleanu & Artin 1982
49	Radoboj	Croatia	Sarmatian, lower	molluscs	brackish	Unger 1869
50	Beograd, Pancevo br.	Serbia	Sarmatian, lower	molluscs, foraminifera	brackish	Pantic & Mihajlovic 1977
51	Negotin	Serbia	Sarmatian, lower	molluscs	brackish	Ercegovac 1963
52	Bozdarevac	Serbia	Sarmatian, lower	molluscs	brackish	Stevanovic & Pantic 1954
53	Saranovo	Serbia	Sarmatian, lower	molluscs	brackish	Pantic 1956
54	Comanesti	Romania	Sarmatian, middle	regional geology	brackish	Givulescu 1968
55	Hernals	Austria	Sarmatian	molluscs	brackish	Berger 1953
56	Erdöbenye	Hungary	Sarmatian	regional geology	continental, limnic	Kovats 1856, Andreanszky 1959
57	Balaton	Hungary	Sarmatian	regional geology	continental, volcanic	Andreanszky 1959
58	Tallya	Hungary	Sarmatian	regional geology	continental, limnic	Andreanszky 1959
59	Saly	Hungary	Sarmatian	regional geology	continental	Andreanszky 1959
60	Stare Gliwice	Poland	Sarmatian	foraminif., molluscs in underly.str.	brackish	Szafer 1961
61	Kokoszyce	Poland	Sarmatian	reg.geol., foraminifera nearby	brackish	Reichenbach 1919, Dyjor (pers.comm.)
62	Mukatchevo	Ukraine	Sarmatian	regional geology	marine	Staub 1890
63	Rudabánya	Hungary	Pannonian, lower	mammals ???	brackish	Kretzoi & al. 1975, Pal-falvy 1981
64	Dubona I	Serbia	Pannonian, lower	molluscs	brackish	Mihajlovic 1977
65	Sremska Kamenica	Serbia	Pannonian, lower	molluscs	brackish	Pantic & Mihajlovic 1976
66	Gheghie	Romania	Pannonian, lower	regional geology	brackish	Givulescu 1960
67	Valea Crisului	Romania	Pannonian B/C	regional geology	brackish	Givulescu 1962, 1975c
68	Cornitel	Romania	Pannonian B/C	regional geology	brackish	Givulescu 1957a
69	Großenreith	Austria	Pannonian	mammals	continental, fluviatile	Kovar-Eder 1988a
70	Schneegattern	Austria	Pannonian	mammals	continental, fluviatile	Kovar-Eder 1988a
71	Lohnsburg	Austria	Pannonian	mammals	continental, fluviatile	Kovar-Eder 1988a
72	Laaerberg	Austria	Pannonian E	molluscs, regional geology	brackish	Berger 1955b
73	Vösendorf	Austria	Pannonian E, MN 9	molluscs, mammals	brackish	Berger 1955a
74	Delureni	Romania	Pannonian E	bivalves, ostracods	brackish	Givulescu 1975a
75	Dubona II	Serbia	Pannonian E	molluscs	brackish	Mihajlovic 1977
76	Sprendlingen	Germany	MN 9/10	mammals	continental, fluviatile	Meller 1989
77	Postorna	Moravia	Pannonian F	molluscs	brackish	Knobloch 1969
78	Moravska Nova Ves	CSFR	Pannonian F	molluscs	brackish	Knobloch 1969
79	Tihany	Hungary	Pannonian, MN 11/12	molluscs, mammals	continental, fluviatile	Hably in Müller & al. 1989

TABLE 30.1 *Sites Arranged in Stratigraphical Order (Continued)*

Table gives information on the region, age, method of dating, depositional facies, and reference. The numbers on the left correspond with those in figure 30.1.

Nr.	Name of flora	Country/ Region	Age	Determined by	Facies	Source
80	Odesti	Romania	Pannonian[x]	molluscs (45 m above)	brackish, lignite facies	Givulescu & al. 1986
81	Sinersig	Romania	Pannonian[x]	regional geology	brackish, lignite facies	Givulescu & Florei 1960
82	Iharosbereny	Hungary	Pontian	molluscs, ostracods	brackish, lignite facies	Hably 1992b
83	Racsa	Romania	Pontian[y]	regional geology	brackish	Givulescu & Edelstein 1981
84	Turt	Romania	Pontian[y]	molluscs in underlying strata	brackish	Givulescu & al. 1975
85	Chiuzbaia	Romania	Pontian[y]	regional geology	continental	Givulescu 1990
86	Pincina	Slovakia	- 4,9 - 7,14 m.a.	radiometric date	volcanic	Sitar, Kvacek & Buzek 1989
87	Grocka	Serbia	Pontian, upper[y]	molluscs	brackish	Pantic 1956
88	Osojna	Serbia	Pontian, upper[y]	molluscs	brackish	Pantic 1956
89	Gabbro	Italy	Messinian	regional geology	marine/limnic	Berger 1957
90	Berezinka	Ukraine	Dacian	molluscs,ostrac.underly.str.	continental,limnic-fluv., volcanic	Il'inskaya 1968
91	Il'nica	Ukraine	Dacian	molluscs in underlying strata	continental,limnic-fluv., volcanic	Il'inskaya 1968
92	Velikiy Rakovets	Ukraine	Dacian	molluscs in underlying strata	continental,limnic-fluv., volcanic	Il'inskaya 1968
93	Hajnacka	Slovakia	MN 16	mammals of underlying strata	volcanic	Sitar, Kvacek & Buzek 1989

[x] former lower Pontian, now regarded as Pannonian
[y] Pontian in the sense of (Stevanovic et al., 1989)

TABLE 30.2 *Stratigraphic and Geographic Distribution of Plant Taphocoenoses*

In this and the following tables, strata have been dated by independent means (i.e., not by the fossil plants themselves). Such data include nannoplankton, foraminifera, and isotopes for radiometric dating. Gaps correspond either to a deficiency in the fossil record or to deficiency of floras dated by independent means (e.g., late Oligocene/lower Miocene of Southeastern Europe).

The stratigraphic table relies on Steininger et al. 1989. According to Rögl et al. (1993) and Rögl and Daxner-Höck (this volume), the range of the Pannonian and Pontian is modified. The Pannonian now includes the MN 9—12, the Pontian being very short, comprising the uppermost MN 12 and the main part of MN 13.

Neither underlined nor broken-underlined are those sites for which the stratigraphic position has been determined by reliable regional geology.

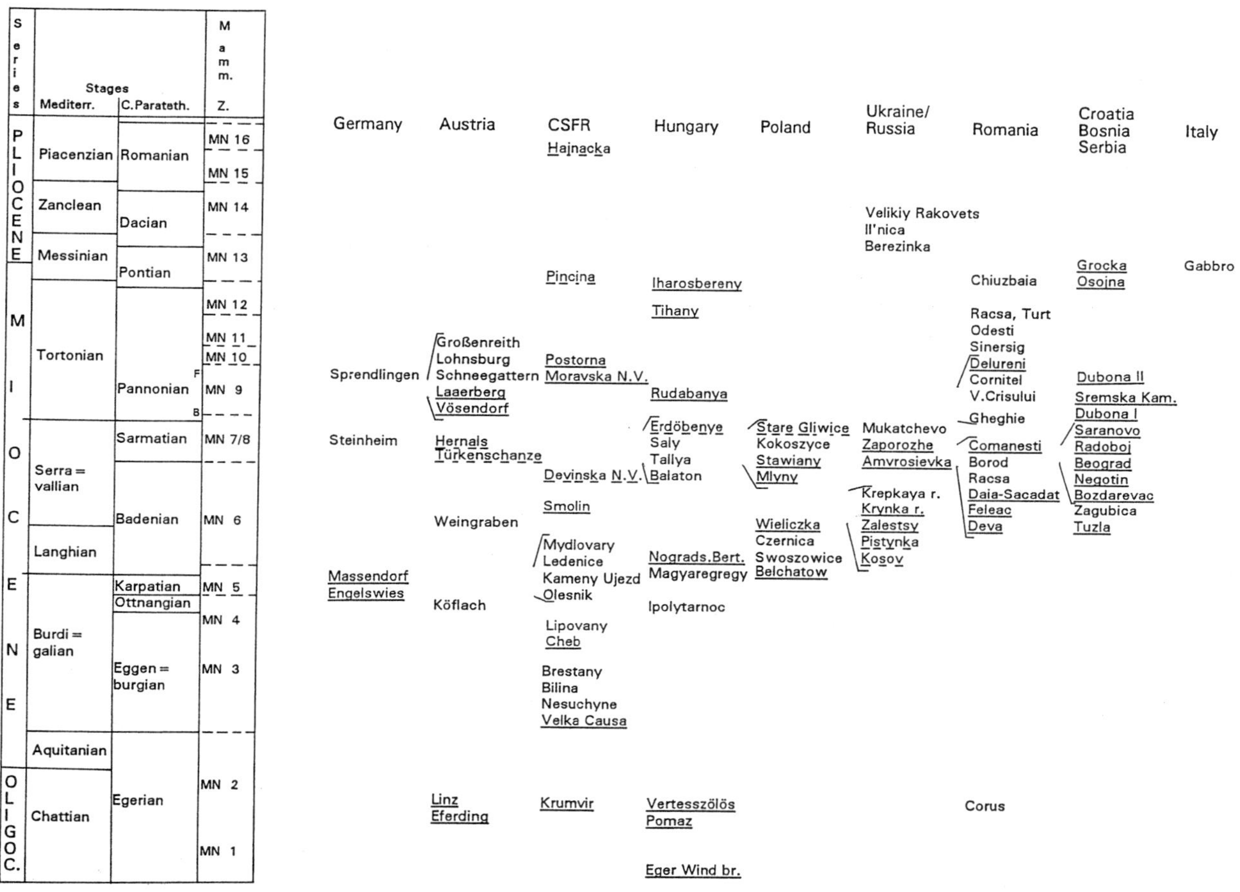

Series	Stages — Mediterr.	Stages — C.Parateth.	Mamm. Z.
PLIOCENE	Piacenzian	Romanian	MN 16 / MN 15
PLIOCENE	Zanclean	Dacian	MN 14
PLIOCENE	Messinian	Pontian	MN 13
MIOCENE	Tortonian	Pannonian	MN 12 / MN 11 / MN 10 / MN 9 (F / B)
MIOCENE	Serravallian	Sarmatian	MN 7/8
MIOCENE	Serravallian	Badenian	MN 6
MIOCENE	Langhian	Badenian	MN 6
MIOCENE	Burdigalian	Karpatian	MN 5
MIOCENE	Burdigalian	Ottnangian	MN 4
MIOCENE	Burdigalian	Eggenburgian	MN 3
MIOCENE	Aquitanian	Egerian	MN 2
OLIGOC.	Chattian	Egerian	MN 1

have been a thermophilous element that constitued a portion of the undergrowth of mixed evergreen and deciduous forests.

The earliest records of *T. salicornioides* are derived from the late Eocene in Saxony (Weisselster Basin; Mai and Walther 1985) and from England (Reid and Chandler 1926). In the Oligocene it is common throughout Europe as far southeastward as the Balkan Peninsula. In the early Miocene *T. salicornioides* is common until the Karpatian Stage (Kvacek 1989). On the Balkan Peninsula, the dating of early Miocene plant assemblages is less exact: Slanci, Ravna Reka, Vrdnik, and Jancici in Serbia are some examples (Pantic 1956; Mihajlovic 1978, 1980). It is also known from Central European plant assemblages correlated with the middle Miocene (Badenian and Sarmatian Stages). Many localities are dated by stratigraphically less valuable means, i.e., Lerch, Bavaria (Jung 1968), and Handlova, Slovakia (Takac 1974). Well-documented younger (Pannonian/Pontian) records are confined to Southern and Southeastern Europe. Late Miocene Central European records are either not convincing (Moravska Nova Ves, Moravia) or primarily or at least partly dated by paleobotanical criteria (Sosnica [former Schossnitz: Sadowska 1991], Gozdnica [Zastawniak 1992], Poland). In Western Europe *T. salicornioides* occurs sporadically in early Pliocene deposits (Frankfurt, Germany; Mädler 1939). The last refugium in latest Miocene and early Pliocene time is Southern/Southeastern Europe and the Caucasian region,

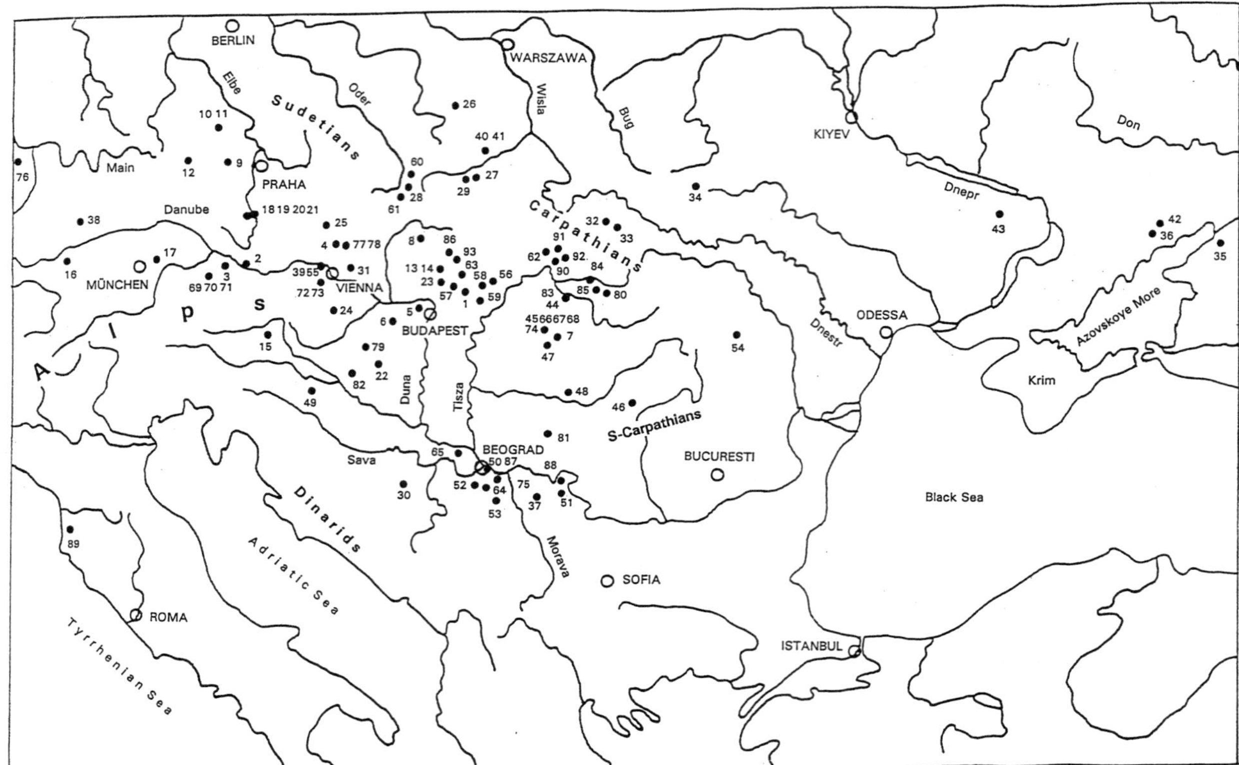

FIGURE 30.1 Geographic map of the investigated region, Paratethys surrounding areas, and position of the fossil sites. The numbers correspond to those in table 30.1.

including: Kodor (Kolakovskiy et al. 1970), Duab (Kolakovskiy 1958), Pitsunda (Kolakovskiy 1962), Gul'ripsh (Kolakovskiy and Shakryl 1978), and Chochkhati (Purtseladze and Tsagareli 1974).

Platanus neptuni (Ettingshausen) Buzek, Holy, and Kvacek (pl. 30.1:2; tab. 30.4)   Simple elongated leaves quite variable in length. The leaf-margin is finely and regularly bluntly dentate except at the entire-margined base. The texture is firm.

The systematic position within the plane trees (Platanaceae) has been determined by anatomical features and associated inflorescences. The only living counterpart, *Platanus kerrii*, occurs in streamside environments of Laos and Vietnam.

The fossil record suggests that *Platanus neptuni* occurs principally in predominantely fluviatile and partly volcanic deposits and is believed to have preferred habitats similar to the extant species *Platanus kerrii*.

The ancestors of *P. neptuni* that have composed leaves are known to occur in the Central European late Eocene horizons (Saxonia, Bohemia). *P. neptuni* is common throughout Central and West European Oligocene plant assemblages (France, Germany, Bohemia, Moravia, Austria, Hungary). The existence of this species in Balkan Peninsula Oligocene age horizons has been documented

at the locality of Divljana, Serbia (Mihajlovic 1985). There are no records of this taxon in the numerous East European and Asian plant assemblages thus far described.

In Central Europe we can trace the presence of *P. neptuni* into the Ottnangian Stage. Thus far we have been unable to find a satisfactory fossil record of early Miocene plant assemblages from the Balkan Peninsula. Despite the fact that we lack cuticles of this species there, we believe on other morphological criteria that this tree occurs in the early Sarmatian of Croatia and Serbia. Even younger (Pannonian) are the Romanian data partly based on morphological and cuticular evidence. After the Pannonian, *P. neptuni* becomes extinct in Europe. We remain uncertain about possible records from the late Miocene of Kodor (Kolakovskiy 1964), where *Arbutus elegans* has been reported. This material has unfortunately been destroyed since its initial discovery and description.

Daphnogene sp. (pl. 30.1:3; tab. 30.5)   Simple entire-margined coriaceous leaves with characteristic cinnamoid venation. They are extremely variable in shape, from very narrow to lanceolate/ ovate (most common forms) to round. Synonyms include *Cinnamomum* and *Cinnamomophyllum* sp. div. The number of fossil species referable to this genus is presently unclear.

The genus *Daphnogene* is referable to the Lauraceae.

PLATE 30.1 Paleotropical elements: (1) *Tetraclinis salicornioides*, Linz, Austria, Egerian, ×1; (2) *Platanus neptuni* Ipolytarnoc, Hungary, Ottnangian, ×1; (3) *Daphnogene* sp., Eferding, Upper Austria, Austria, Egerian, ×1;(4) *Engelhardia orsbergensis*, Tuzla, Bosnia-Herzegowina, Badenian, ×1; (5) *Engelhardia macroptera*, fruit, Cheb, Bohemia, MN 4, ×2.

Arctotertiary elements: (6) *Quercus pseudocastanea*, Lohnsburg, Austria, Pannonian, ×1; (7) *Alnus ducalis*, Laaerberg, Vienna, Austria, Pannonian E, ×1; (8) *Acer vindobonense*, Laaerberg, Vienna, Austria, Pannonian E, ×1; (9) *Byttneriophyllum tiliifolium*, Postorna, Moravia, Pannonian F, ×1.

TABLE 30.3 *Stratigraphic and Geographic Distribution of* Tetraclinis salicornioides

Series	Mediterr.	C.Parateth.	Mamm. Z.	Germany	Austria	CSFR	Hungary	Poland	Ukraine/Russia	Romania	Croatia Bosnia Serbia	Italy
PLIOCENE	Piacenzian	Romanian	MN 16									
	Piacenzian	Romanian	MN 15									
	Zanclean	Dacian	MN 14									Gabbro
	Messinian	Pontian	MN 13							Chiuzbaia		
MIOCENE	Tortonian	Pannonian	MN 12									
	Tortonian	Pannonian	MN 11									
	Tortonian	Pannonian	MN 10			? Moravska N.V.				Delureni / Cornitel V.Crisului Gheghie	Dubona II[x)]	
		Pannonian	MN 9									
	Serravallian	Sarmatian	MN 7/8				Erdöbenye	Stare Gliwice Kokoszyce Mlyny		Deva	Radoboj Beograd Negotin, Zagubica	
	Langhian	Badenian	MN 6		Weingraben			Wieliczka				
	Burdigalian	Karpatian	MN 5	Massendorf	Köflach							
	Burdigalian	Ottnangian	MN 4			Cheb	Ipolytarnoc					
	Burdigalian	Eggenburgian	MN 3			Velka Causa Bilina Nesuchyne						
	Aquitanian	Egerian	MN 2	Linz Eferding			Vertesszölös[x)]			Corus		
OLIGOC.	Chattian	Egerian	MN 1									

Despite different far-ranging investigations, its exact generic position within this family has not yet been decided. The members of the Lauraceae characteristically range from tropical to warm temperate climatic zones. Most of them require an equable, humid climate. Therefore, the fossils are considered to be thermophilous elements.

Daphnogene is common in European leaf assemblages beginning with the late Eocene and extending throughout the entire Oligocene and early Miocene. It constitutes the best-recognizable and most common paleotropic element. The Badenian/Sarmatian records are by far less numerous, and *Daphnogene* constitutes by then a mere accessory element in deciduous leaf taphocoenoses. Although numerous Sarmatian floras have been described from Poland and the Ukraine, only one report of *Daphnogene* has been made there from the locality of Dzierzyslaw (former Dirschel: Kräusel 1921; Sarmatian according to S. Dyjor, pers. comm.). Southeastward on the Balkan Peninsula, the Sarmatian, and to a lesser extent the Pannonian, plant assemblages retain abundant fossil remains of *Daphnogene*. Central European Pannonian occurrences are rare (in Moravska Nova Ves, Moravia, 3 specimens out of more than 700; in Neuhaus, Austria, Kovar-Eder et al. 1995) and are only known from a few sites. *Daphnogene* survives until the Pliocene in Southern and Southeastern Europe, including: Pichegu, France (Roiron 1979), and Duab and Pitsunda in the Caucasus (Kolakovskiy 1958; 1962).

Engelhardia orsbergensis (Wessel and Weber) Jähnichen, Mai, and Walther; *Engelhardia macroptera* (Brongniart) Unger (pl. 30.1:4 and 5; tab. 30.6)   The leaflets of compound leaves are recognizable by slight asymmetrical leaf-base and lamina, almost no petiole (subsessile), and widely spaced, sharp, outwardly curved denticles of the leaf-margin. The winged fruits, *Engelhardia macroptera*, are commonly associated in leaf assemblages. The involucrum (wing) is deeply trisected and is additionally characterized by a small basal lobe covering the fruit.

Of the numerous synonyms, we list here only the most common ones: *Palaeocarya orsbergensis, Myrica lignitum* pro parte, *Engelhardia brongniartii, Rhus juglandogene,* and *Engelhardia detecta.*

Engelhardia orsbergensis is an extinct member of the Juglandaceae family, series Engelhardieae. It shows close affinities to the East Asiatic *Engelhardia roxburghiana* but also to the Central American genus *Oreomunnea.* Both

TABLE 30.4 *Stratigraphic and Geographic Distribution of* Platanus neptuni

Series	Stages Mediterr.	Stages C.Parateth.	Mamm. Z.	Germany	Austria	CSFR	Hungary	Poland	Ukraine/Russia	Romania	Croatia Bosnia Serbia	Italy
PLIOCENE	Piacenzian	Romanian	MN 16									
PLIOCENE	Piacenzian	Romanian	MN 15									
PLIOCENE	Zanclean	Dacian	MN 14									
PLIOCENE	Messinian	Pontian	MN 13									
MIOCENE	Tortonian	Pontian	MN 12									
MIOCENE	Tortonian	Pannonian	MN 11									
MIOCENE	Tortonian	Pannonian	MN 10									
MIOCENE	Tortonian	Pannonian	MN 9							/Delureni Cornitel \V.Crisului		
MIOCENE	Serravallian	Sarmatian	MN 7/8								/Radoboj Beograd \Negotin	
MIOCENE	Serravallian / Langhian	Badenian	MN 6									
MIOCENE	Burdigalian	Karpatian / Ottnangian	MN 5									
MIOCENE	Burdigalian	Ottnangian / Eggenburgian	MN 4			Lipovany Cheb	Ipolytarnoc					
MIOCENE	Burdigalian	Eggenburgian	MN 3			Velka Causa Brestany Bilina						
MIOCENE / OLIGOC.	Aquitanian / Chattian	Egerian	MN 2		Linz Eferding	Krumvir	Vertesszölös Pomaz					
OLIGOC.	Chattian	Egerian	MN 1				Eger-Wind br.					

x) unfigured records

TABLE 30.5 *Stratigraphic and Geographic Distribution of* Daphnogene *sp.*

Series	Stages Mediterr.	Stages C.Parateth.	Mamm. Z.	Germany	Austria	CSFR	Hungary	Poland	Ukraine/Russia	Romania	Croatia Bosnia Serbia	Italy
PLIOCENE	Piacenzian	Romanian	MN 16									
PLIOCENE	Piacenzian	Romanian	MN 15									
PLIOCENE	Zanclean	Dacian	MN 14									
PLIOCENE	Messinian	Pontian	MN 13							Chiuzbaia	Grocka	Gabbro
MIOCENE	Tortonian	Pontian	MN 12									
MIOCENE	Tortonian	Pannonian	MN 11									
MIOCENE	Tortonian	Pannonian	MN 10									
MIOCENE	Tortonian	Pannonian	MN 9	Sprendlingen[v]		Moravsk.N.V.[v]				/Delureni Cornitel V.Crisului \Gheghie	Dubona II Dubona I	
MIOCENE	Serravallian	Sarmatian	MN 7/8		Hernals Türkenschanze		Saly Tallya			/Borod Deva \Feleac	/Saranovo /Radoboj Beograd \Negotin \Bozdarevac Zagubica Tuzla	
MIOCENE	Serravallian / Langhian	Badenian	MN 6			Smolin		Swoszowice	Pistynka Kosov			
MIOCENE	Burdigalian	Karpatian / Ottnangian	MN 5	Massendorf Engelswies	Köflach	/Mydlovary Ledenice Kameny Ujezd \Olesnik	Nograds.Bert. Magyaregregy					
MIOCENE	Burdigalian	Ottnangian / Eggenburgian	MN 4			Lipovany Cheb	Ipolytarnoc					
MIOCENE	Burdigalian	Eggenburgian	MN 3			Velka Causa Brestany Bilina						
MIOCENE / OLIGOC.	Aquitanian / Chattian	Egerian	MN 2		Linz Eferding	Krumvir	Vertesszölös Pomaz			Corus		
OLIGOC.	Chattian	Egerian	MN 1									

v) in Sprendlingen one specimen, in Moravska Nova Ves 3 specimens out of many hundred

TABLE 30.6 *Stratigraphic and Geographic Distribution of* Engelhardia orsbergensis–E. macroptera

Series	Stages Mediterr.	Stages C.Parateth.	Mamm. Z.	Germany	Austria	CSFR	Hungary	Poland	Ukraine/ Russia	Romania	Croatia Bosnia Serbia	Italy
PLIOCENE	Piacenzian	Romanian	MN 16									
PLIOCENE	Piacenzian	Romanian	MN 15									
PLIOCENE	Zanclean	Dacian	MN 14									
PLIOCENE	Messinian	Pontian	MN 13							Chiuzbaia		Gabbro
MIOCENE	Tortonian	Pannonian	MN 12									
MIOCENE	Tortonian	Pannonian	MN 11									
MIOCENE	Tortonian	Pannonian	MN 10									
MIOCENE	Tortonian	Pannonian	MN 9				Rudabanya			Delureni[x], Cornitel, V.Crisului	Sremska Kam.	
MIOCENE	Serravallian	Sarmatian	MN 7/8					Stare Gliwice		Borod, Daia-Sacadat, Feleac	Radoboj, Negotin	
MIOCENE	Serravallian	Badenian	MN 6		Weingraben			Wieliczka			Tuzla	
MIOCENE	Langhian											
MIOCENE	Burdigalian	Karpatian	MN 5			Mydlovary, Ledenice, Kameny Ujezd	Magyaregregy					
MIOCENE	Burdigalian	Ottnangian	MN 4			Lipovany	Ipolytarnoc					
MIOCENE	Burdigalian	Eggenburgian	MN 3			Velka Causa, Brestany, Bilina						
OLIGOC.	Aquitanian / Chattian	Egerian	MN 2	Linz			Pomaz[x]			Corus		
OLIGOC.	Chattian	Egerian	MN 1				Eger-Wind br.					

x) unfigured records
y) could not be verified

taxa live today in mesophytic, mixed evergreen/deciduous forests. Similar ecological requirements can be assumed from the fossil assemblages. Its ancestors are known from the middle Eocene of Messel, Germany (Wilde 1989).

Beginning with the latest Eocene locality of Bembridge (Reid and Chandler 1926), and following in the Oligocene with the species *Engelhardia orsbergensis* and *E. macroptera*, this group is widely distributed over almost all of Europe and the Caucasian region (Jähnichen et al. 1977; 1984). It also commonly occurs in early Miocene Central European leaf assemblages. Badenian to Sarmatian age records are still numerous, but those of *E. macroptera* from Moldavia, Russia, and the Ukraine (Jähnichen et al. 1977) have not yet been verified. Records of *E. orsbergensis* are not available. From the Caucasus, *Engelhardia* is confidently reported from Medzhuda (Avakov 1979, revised by Jähnichen et al. 1984).

In Central Europe, the *E. orsbergensis–E. macroptera* assemblages only very rarely cross the Sarmatian/Pannonian boundary (Rudabánya and Rozsaszentmarton, Hungary; Kretzoi et al. 1975; Palfalvy 1981). In Romania and Serbia further Pannonian records are known. *Engelhardia* is rarely reported from Pliocene age strata, based almost entirely on incomplete fruit remains (Berga, Germany, Mai and Walther 1988; Rochessauve, Grangeon 1958). Only in Gabbro (Italy) is there no doubt about its presence (involucres).

Arctotertiary Elements

Quercus pseudocastanea Goeppert emend. Walther and Zastawniak—and roburoid oaks (pl. 30.1:6; tab. 30.7) Oak-leaves that are characterized by a coarsely lobed lamina like the extant *Quercus robur*.

Leaf-size is extremely variable, ranging from a few cm up to more than 15 cm in dimension. The specific determination of these fossils is problematic because various extreme leaf-forms occur together that may represent more than one natural species. Fossil leaves belonging to this taxon have been described under a variety of nomina. The most common are *Quercus pseudocastanea*, *Q. pseudorobur*, and *Q. roburoides*. They are assigned to the sections Cerris and/or Robur within the family Fagaceae.

This group is common in fluviatile and limnic facies. It is not found in lignitic facies and, therefore, not character-

TABLE 30.7 *Stratigraphic and Geographic Distribution of* Quercus pseudocastanea *and roburoid oaks*

Stratigraphic framework

Series	Stages (Mediterr.)	Stages (C.Parateth.)	Mamm. Z.
PLIOCENE	Piacenzian	Romanian	MN 16 / MN 15
PLIOCENE	Zanclean	Dacian	MN 14
PLIOCENE	Messinian	Pontian	MN 13
MIOCENE	Tortonian	Pontian	MN 12 / MN 11 / MN 10
MIOCENE	Tortonian	Pannonian	MN 9
MIOCENE	Serravallian	Sarmatian	MN 7/8
MIOCENE	Serravallian	Badenian	MN 6
MIOCENE	Langhian	Badenian	MN 6
MIOCENE	Burdigalian	Karpatian	MN 5
MIOCENE	Burdigalian	Ottnangian	MN 4
MIOCENE	Burdigalian	Eggenburgian	MN 4 / MN 3
MIOCENE	Aquitanian	Egerian	MN 2
OLIGOC.	Chattian	Egerian	MN 2 / MN 1

***Quercus pseudocastanea* and roburoid oaks**

Stratigraphic level	Germany	Austria	CSFR	Hungary	Poland	Ukraine/Russia	Romania	Croatia Bosnia Serbia	Italy
Romanian/Pliocene			Hajnacka						
Pontian (MN 13)			Pincina				Chiuzbaia	Grocka Osojna[x]	Gabbro
Pannonian	Sprendlingen	Großenreith Lohnsburg Vösendorf	Moravska N.V.				V.Crisului		
Sarmatian				Erdöbenye Balaton	Kokoszyce Mlyny[z]	Mukatchevo[x] Amvrosievka Krynka r. Zalestsy	Racsa		
Badenian (MN 6)					Czernica				

x) unfigured record
z) doubtful record, very fragmentary

istic of swamp forests. It is believed that these oaks preferred riparian habitats and mesophytic forests.

Probable ancestors of the European Miocene/Pliocene roburoid oaks have been reported from the late Oligocene of northeast Kazakhstan (Altai Mountains), from Bukhtarma (Rayushkina 1979), and westward from the North Aralian region at Altyn-Shokysy (Zhilin and Andrejev 1984).

In the Paratethys realm, the first evidence of this group has been documented from the Badenian of the Krynka River, South Russia, Zalestsy, and Czernica (Carpathian Foredeep). During the Sarmatian the roburoid oaks were common in the Eastern Paratethys area: Bursuk, Naslavtsy in Moldavia (Shtefyrtsa 1974, Yakubovskaya 1955), and as far southeastward as Ukraine (Amvrosievka, Donec region; Mukatchevo in the Transcarpathians). In Hungary (Erdöbenye, Balaton) and Slovakia (Turiec-Basin, Mociar, Nitra-Basin; Sitar 1969, 1973; Takac 1974), the age of these plant taphocoenoses has been determined on regional geological grounds. Reliable Sarmatian records of this group are lacking further to the west (Bohemia, Austria, and southern Germany), and in the south, although many fossil plant assemblages of this age do exist. The exact age of Auben-

ham and Achldorf, Bavaria (Knobloch 1986; 1988), is still the subject of much debate. During the Pannonian Stage these oaks are well represented over almost all the area cited above, with the exception of the more southerly parts (western Balkan Peninsula).

This group first appears in the Balkan Peninsula (Grocka, Osojna in Serbia; Kurilo near Sofia, Bulgaria, Stojanoff and Stefanoff 1929) and Italy (Gabbro) during the late Miocene/Pliocene. During the Pliocene roburoid oaks are present at almost every site in Central and Western Europe (Hajnacka, Slovakia [Sitar et al. 1989]; Willershausen and Berga, Germany [Knobloch 1990; Mai and Walther 1988]; Gerce, Hungary [Fischer and Hably 1991]; Domanski Wierch, Poland [Zastawniak 1972]).

Alnus ducalis (Gaudin) Knobloch (pl. 30.1:7; tab. 30.8)   Simple long-petiolate leaves with a sharply simple serrate leaf-margin. Most characteristic is the distinctly emarginate leaf apex.

The binomina *Alnus hoernesi* and *A. stenophylla* are synonyms of *Alnus ducalis*. Based on leaf shape, the most similar extant species is *Alnus matsumurae*, which lives in Japanese montane forests. *A. ducalis* preferred alluvial

TABLE 30.8 *Stratigraphic and Geographic Distribution of* Alnus ducalis

Series	Stages Mediterr.	Stages C.Parateth.	Mamm. Z.	Germany	Austria	CSFR	Hungary	Poland	Ukraine/Russia	Romania	Croatia Bosnia Serbia	Italy
PLIOCENE	Piacenzian	Romanian	MN 16									
	Piacenzian	Romanian	MN 15									
	Zanclean	Dacian	MN 14									
	Messinian	Pontian	MN 13							Chiuzbaia		Gabbro
MIOCENE	Tortonian	Pannonian	MN 12				Tihany x)					
	Tortonian	Pannonian	MN 11									
	Tortonian	Pannonian	MN 10									
	Tortonian	Pannonian	MN 9		Schneegattern Laaerberg	Moravska N.V.						
	Serravallian	Sarmatian	MN 7/8		Hernals		Saly					
	Serravallian / Langhian	Badenian	MN 6						Krepkaya r. x)			
	Langhian / Burdigalian	Karpatian / Ottnangian	MN 5									
	Burdigalian	Ottnangian	MN 4									
	Burdigalian	Eggenburgian	MN 3									
	Aquitanian	Egerian	MN 2									
OLIGOC.	Chattian	Egerian	MN 1									

x) unfigured records
The determinations of this species from Mlyny and Stawiany (Zastawniak, 1980) have been revised by Zastawniak as Alnus sp.. Alnus hoernesi
(Pimenova, 1954: 43) from Amvrosievka (Ukraine) has been corrected by Knobloch (1969: 72). All lack the characteristic feature of the emarginate leaf-apex

soils, most probably regularly flooded riparian environments, but apparently avoided swampy facies.

Generally speaking, there are only a few well-documented records of this species. Of the Badenian/Sarmatian only the ones from Saly (Hungary) and Hernals (Vienna Basin) are unambiguous. The first rich occurrences are documented in the Pannonian of the Vienna Basin (Laaerberg, Moravska Nova Ves) and the Molasse zone north of the Alps (Schneegattern). Remarkable are the numerous latest Miocene/Pliocene discoveries in the Mediterranean (Skoura and Likudi in Greece [Knobloch and Velitzelos 1986; Velitzelos and Knobloch 1986]; Gabbro, Montaine, and Sarzanello, in Italy [Berger 1957; Gaudin and Strozzi 1858]; Théziers and St. Marcel in the Rhône Valley, Vaucluse, France [Depape 1922; Ballesio et al. 1979]). In the Caucasian region, A. ducalis is reported from the Pliocene localities of Kodor (Kolakovskiy 1964) and Gul'ripsh (Kolakovskiy and Shakryl 1978), Malye Shiraki (Kolakovskiy and Ratiani 1967), and Duab (Kolakovskiy 1958), and the Pleistocene locality of Gumista (Ratiani 1970). The stratigraphic and biogeographic utility of these leaf fossils has proven to be great.

Acer vindobonense (Ettingshausen) Berger (pl. 30.1:8; tab. 30.9)   Maple leaves usually having 5 to 7 slender and regular lobes, coarsely and irregularely dentate.

They have been described under the following synonyms: *Acer sanctae-crucis, Acer polymorphum pliocenicum,* and *Acer nordenskjoeldii* pro parte. The latter binomen was originally based on East Asiatic material and subsequently applied to European findings. Although the morphological similarity is striking, we hesitate to regard the East Asiatic occurrences as conspecific with the European ones. The leaves of *Acer vindobonense* obviously belong to the section Palmata Pax of the Aceraceae family.

Like *Quercus pseudocastanea* and the roburoid oaks, this maple tree occurs in riparian and mesophytic forests. The extant members of the section Palmata inhabit deciduous and mixed-mesophytic forests of Japan, Korea, and Central China. One species is native to Pacific North America. The records from reliably dated Parathethys floras are scattered and all not older than the Sarmatian. The earliest well-documented record comes from the Ukraine (Amvrosievka). Other localities in Slovakia—Heiligenkreuz/Ziar (Stur 1867)—and Hungary (Rozsaszentmarton)

TABLE 30.9 *Stratigraphic and Geographic Distribution of* Acer vindobonense

Series	Stages Mediterr.	Stages C.Parateth.	Mamm. Z.	Germany	Austria	CSFR	Hungary	Poland	Ukraine/ Russia	Romania	Croatia Bosnia Serbia	Italy
PLIOCENE	Piacenzian	Romanian	MN 16									
PLIOCENE	Piacenzian	Romanian	MN 15									
PLIOCENE	Zanclean	Dacian	MN 14						Berezinka			
PLIOCENE	Messinian	Pontian	MN 13							Chiuzbaia		? Gabbro
MIOCENE	Tortonian	Pannonian	MN 12									
MIOCENE	Tortonian	Pannonian	MN 11									
MIOCENE	Tortonian	Pannonian (F)	MN 10		Schneegattern Lohnsburg Großenreith Laaerberg	Moravska N.V.						
MIOCENE	Tortonian	Pannonian (B)	MN 9									
MIOCENE	Serravallian	Sarmatian	MN 7/8				Saly[x]		Mukatchevo[x] Amvrosievka			
MIOCENE	Serravallian	Badenian	MN 6									
MIOCENE	Langhian											
MIOCENE	Burdigalian	Karpatian	MN 5									
MIOCENE	Burdigalian	Ottnangian	MN 4									
MIOCENE	Burdigalian	Eggenburgian	MN 3									
MIOCENE	Aquitanian	Egerian	MN 2									
OLIGOC.	Chattian	Egerian	MN 1									

x) unfigured record, specimen lost

are possibly of similar age. *A. vindobonense* occurs later in Pannonian age horizons of the Vienna Basin (Austria, southern Moravia) and the Molasse zone north of the Alps (Austria). This species survived until the early Pliocene in Central and Southern Europe, as documented by rare occurrences including Berezinka, the Ukraine, and Domanski Wierch, Poland (Zastawniak 1972).

Byttneriophyllum tiliifolium (Al. Braun) Knobloch and Kvacek (pl. 30.1:9; tab. 30.10)   The leaf-shape is characterized by its typical asymmetry, entire margin, and cordate leaf-base. The venation is prominent and reminiscent of a spider's web. The leaves may attain a large size: up to 20 cm in length.

Remains of this species have been described under various names as *Ficus tiliifolia, Dombeyopsis tiliifolia, Buettneria aequalifolia,* and *Alangium tiliifolium. Byttneriophyllum tiliifolium* belongs most probably to an extinct member of either the Tiliaceae (family of lime trees) or Sterculiaceae. The winged fruits of *Banisteriaecarpum giganteum* are supposed to be derived from the same plant.

This species is confined to Europe and adjacent regions of Asia. In the Pannonian, most records of this species are associated with lignitic facies (Givulescu 1992). There, the leaf remains occur in great quantities, sometimes as mono-dominants. Characteristically, *B. tiliifolium* is associated with some of the following few taxa: *Glyptostrobus europaeus, Alnus, Myrica, Salix,* and *Spirematospermum.* Therefore, those associations are most typical of a swampy environment. *B. tiliifolium* also occurs as an accessory element in alluvial environments associated with deciduous broad-leaved forest taxa.

There are no reliably dated records older than the Badenian occurrences. *B. tiliifolium* seems to disperse broady, beginning with the Sarmatian, and is well documented from reliably dated floras in the Ukraine (Zaporozhe, Amvrosievka). Rich records from Slovakia and Hungary are also believed to be of Sarmatian age (Handlova-Novaky Basin, Mociar in Slovakia; Felsötarkany and Füzerradvany, Hungary) as well as those from Germany (Öhningen; Steinheim, only a single leaf, Gregor 1983) and Austria (Kainberg, Andritz near Graz; Knobloch and Kvacek 1965). Most numerous are the lignitic facies that broadly occurred during the Pannonian F and younger age horizons in the Vienna Basin (Zillingsdorf), southern Moravia (Postorna, Dubnany), Hungary (Dozmat; Hably and Kovar-Eder 1996), and Romania (Lugoj and Baia Mare Basins; Givulescu 1992).

TABLE 30.10 *Stratigraphic and Geographic Distribution of* Byttneriophyllum tiliifolium

Series	Stages Mediterr.	Stages C.Parateth.	Mamm. Z.	Germany	Austria	CSFR	Hungary	Poland	Ukraine/ Russia	Romania	Croatia Bosnia Serbia	Italy
P L I O C E N E	Piacenzian	Romanian	MN 16									
	Piacenzian	Romanian	MN 15									
	Zanclean	Dacian	MN 14						Velikiy Rakovets Il'nica Berezinka[x]			
	Messinian	Pontian	MN 13				Iharosbereny			Chiuzbaia		
M I O C E N E	Tortonian	Pannonian	MN 12							Turt[y], Racsa		
	Tortonian	Pannonian	MN 11							Odesti Sinersig		
		Pannonian	MN 10			Postorna	Rudabanya					
		Pannonian	MN 9		Lohnsburg Vösendorf[xy]							
	Serravallian	Sarmatian	MN 7/8	Steinheim					Mukatchevo[x] Zaporozhe Amvrosievka	Comanesti[x]		
		Badenian	MN 6									
	Langhian											
		Karpatian	MN 5					Belchatow[x]				
		Ottnangian										
	Burdigalian	Eggenburgian	MN 4									
		Eggenburgian	MN 3									
	Aquitanian	Egerian	MN 2									
O L I G O C.	Chattian	Egerian	MN 1									

x) unfigured record
y) could not be verified

Northward (Poland, Germany) records of *B. tiliifolium* are almost lacking in lacustrine/fluviatile environment except for the site of Stroza (Striese, Poland; Miocene/Pliocene age, S. Dyjor, pers. comm.). However, lignitic facies were not developed at these places.

Pliocene occurrences are well documented from the Ukraine and Romania (Balteni). The youngest reported assemblages occur in the western Balkan Peninsula. These assemblages are yet to be verified: Glogovac (Czernejavski 1933) and Podvinje (Engelhardt 1895) in Croatia. There are no other reliable Pliocene records of this species.

Vegetational Development

The fossil plant assemblages discussed here are derived from a large portion of Central, Eastern, and Southeastern Europe. Because of the restricted number of sites that are dated by the criteria we adhere to here, actual site density is still unsatisfactory in some regions, such as Germany. At the moment, we are dealing with few selected taxa demonstrating different distribution patterns. So far, the frequency has only partially been considered. Therefore,

we are still unable to trace finer climatic and vegetational oscillations. The fossil plant assemblages discussed herein document a variety of forest community types.

The Paleogene vegetation is dominated by evergreen thermophilous elements indicative of a paleotropical origin (evergreen Fagaceae, Lauraceae, Juglandaceae, Hamamelidaceae, Mastixiaceae, Symplocaceae, ancient conifers, and others). The first invasion of deciduous, Arctotertiary elements are known from middle Oligocene age horizons.

Differences in paleovegetation distribution (north and south, east and west) are certainly traceable back into the Oligocene. However, transition from the Oligocene to early Miocene vegetation in Central Europe (Germany, Bohemia, Hungary) is characterized by the extinction of paleotropical taxa, including *Eotrigonobalanus furcinervis* (Fagacea), and new deciduous immigrants such as *Fagus*. While the early Miocene vegetation here remained rich in Paleogene taxa (e.g., *Tetraclinis, Platanus neptuni, Daphnogene, Engelhardia*), the forests show a progressive domination by Arctotertiary species. By the middle Miocene, the occurrence of Arctotertiary elements became even more pronounced. The Ukrainian and Russian plant assemblages (Pistynka, Zalestsy, Krepkaya River, Krynka River) are clearly deciduous, and include few evergreen components, while to the south evergreens dominated until the Pannonian. In Central Europe, transitions connect both extremes. The withdrawal of the evergreen elements from Central Europe toward the south and southeast is successively recognizable into the Pliocene. The deciduous roburoid oaks apparently invaded in an east to west direction and, subsequently, to the southeast. Therefore, the southern regions, as well as the western Balkan Peninsula, remained rich in evergreen species until the early Pannonian. The Pliocene record is far less complete. This period's paleovegetation development remains unclear.

Conclusions

The taxa discussed here reveal differing paleogeographic distributions through time. Paleotropic elements (*Tetraclinis, Platanus neptuni, Daphnogene, Engelhardia*) can be traced back to the European Eocene. The ancestors of *Platanus neptuni* and *Tetraclinis salicornioides* are native to Europe. The paleotropical elements indicate gradual changes from prevailing evergreen forest communities to prevailing deciduous communities during the Miocene.

We can draw a number of general conclusions. First, regional occurrences of any single taxon are not necessarily synchronous. Second, paleotropical elements tend to survive longer in Southeastern Europe (Romania, Balkan Peninsula) and the Caucasian region (*Platanus neptuni* in the Sarmatian of Croatia and Serbia, and the Pannonian of Romania; *Tetraclinis salicornioides* and *Daphnogene* in the latest Miocene and early Pliocene of the Caucasian region). Third, the retreat of different taxa from Central Europe toward the south and southeast is not entirely synchronous. *Platanus neptuni* does not occur during the Badenian Stage (early middle Miocene) and at younger sites in Central Europe, while *Daphnogene* and *Tetraclinis salicornioides* still persist there. Fourth, different deciduous taxa (e.g., *Quercus pseudocastanea* and roburoid oaks, *Alnus ducalis, Acer vindobonense*), appear in Central Europe as late as the middle Miocene. At least the deciduous oaks of the *Quercus pseudocastanea* and roburoid group can be traced back into the late Oligocene of Kazakhstan, and the origins of *Alnus ducalis* and *Acer vindobonense* are not yet clear. Fifth, no living counterpart has yet been discovered for *Byttneriophyllum tiliifolium*; although it is most characteristic for a swampy habitat (lignite facies), and very common in Sarmatian and late Miocene time, it is missing in comparable facies of older strata. In that the processes described are complex and did not occur synchronously across the area considered here, we cannot expect a simple replacement of one vegetation type (Paleotropical flora sensu Engler) by the other (Arctotertiary flora sensu Engler) (see also Mai 1981).

Further stratigraphic evaluation of the European fossil plant record should take into consideration the following: (1) fossil plant assemblage correlations are only reliable within paleogeographically restricted and reasonably homogenous areas; (2) first occurrences and extinctions can contribute significantly to the temporal resolution of correlations (i.e., within Central Europe the last occurrences of *Platanus neptuni* are known from the Ottnangian, and the first reliable record of *Alnus ducalis* is correlative with the Sarmatian Stage); (3) only after further critical revision of as many taxa as possible do we hope to be able to differentiate assemblage zones.

Acknowledgments

We would like to thank our colleagues G. Daxner-Höck, C. Buzek, S. Dyjor, O. Fejfar, E. Knobloch, N. Pantic, M. Piwocki, J. Rutkowski, V. Sitar, L. Stuchlik, and S. Syabryaj for important contributions and discussion. The investigations were supported by the IGCP-Project 326 (Oligocene–Miocene transition in the Northern Hemisphere).

LITERATURE CITED

Andreanszky, G. 1959. *Die Flora der sarmatischen Stufe in Ungarn.* Budapest: Akademie Kiado.

Avakov, G. S. 1979. *Miotsenovaya flora Medzhudy.* Tbilisi: Metsnereba.

Bachmayer, F., F. Rögl, and R. Seemann. 1991. Geologie und Sedimentologie der Fundstelle miozäner Insekten in Weingraben (Burgenland, Österreich). In *Jubiläumsschrift 20 Jahre Geologische Zusammenarbeit österreich-Ungarn,* pp. 53–70. Vienna.

Ballesio, R., H. Meon, and E. Samuel. 1979. Un gisement á plantes des formations Pliocènes des environs de Rasteau près Vaison-la-Romaine (Vaucluse). *Géobios* 12:235–65.

Berger, W. 1951. Pflanzenreste aus dem tortonischen Tegel von Theben-Neudorf bei Pressburg. *Sitzungsberichte der österreichische Akademie der Wissenschaften, math.-nat. Klasse* 160:273–78.

———. 1952. Pflanzenreste aus dem miozänen Ton von Weingraben bei Drassmarkt (Mittelburgenland). *Sitzungsberichte der österreichische Akademie der Wissenschaften, math.-nat. Klasse* 161:93–101.

———. 1953. Pflanzenreste aus den obermiozänen Ablagerungen von Hernals. *Annalen des Naturhistorischen Museums, Wien* 59:141–54.

———. 1955a. Nachtrag zur altpliozänen Flora der Congerienschichten von Brunn-Vösendorf bei Wien. *Palaeontographica, B* 97:74–80.

———. 1955b. Die altpliozäne Flora des Laaerberges in Wien. *Palaeontographica, B* 97:81–113.

———. 1957. Untersuchungen an der obermiozänen (sarmatischen) Flora von Gabbro (Monti Livornesi) in der Toskana. *Palaeontographica Italica* 51:1–96.

Berger, W. and F. Zabusch. 1953. Die obermiozäne (sarmatische) Flora der Türkenschanze in Wien. *Neues Jahrbuch für Geologie und Paläontologie,* Abhandlungen 98:226–76.

Buzek, C., Z. Dvorak, Z. Kvacek, and M. Proks. 1992. Tertiary vegetation and depositional environments of the "Bilina delta" in the North-Bohemian brown-coal basin. *Casopis pro mineralogii a geologii* 37:117–34.

Buzek, C., F. Holy, and Z. Kvacek. In press. Early Miocene flora of the "Cypris Shale" (western Bohemia). *Sbornik geologickych ved paleontologie Praha.*

Buzek, C. and Z. Kvacek. 1988. Nove nalezy tretihorni flory v hlavacovskych sterkopiscich u Nesuchyne na Rakovnicku. *Zpravy Geolog. vyzkumech roce* 1986:22–24.

Czernejavski, P. 1933. Beiträge zur Kenntnis der Pliozänflora in der Umgebung von Glogovac in Kroatien. *Bulletin du service géologique du royaume Yugoslavie* 2:1–12.

Depape, G. 1922. Récherches sur la flore pliocène de la Vallée du Rhône. *Annales des Sciences naturelles* 10:73–201.

Engelhardt, H. 1895. Flora aus den Paludinenschichten des Caplagrabens bei Podvin in der Nähe von Brood (Slavonien). *Abhandlungen der Senckenbergischen Naturforschenden Gesellschaft* 18:169–207.

Ercegovac, M. 1963. Beitrag zur Kenntnis der fossilen Flora Ostserbiens (zwischen Brza Palanka und Negotin). *Bulletin du Muséum d'histoire naturelle de Belgrade* A18:151–59.

Fischer, O. and L. Hably. 1991. Pliocene flora from the alginite at Gerce. *Annales Historico Naturales Musei Nationalis Hungarici* 83:25–47.

Gaudin, C.-T. and M. Strozzi. 1858. Mémoire sur quelques gisements de feuilles fossiles de la Toscane. *Neue Denkschriften der Schweizerischen Gesellschaft der gesamten Naturwissenschaften* 16:1–47.

Givulescu, R. 1957a. *Flora pliocena de la Cornitel.* Biblioteca de Geologie si Paleontologie III. Academei Republicii Populare Romaine.

———. 1957b. Note paleobotanice III. *Studii cercetari studii Academie, Cluj* 8/3–4: 381–86.

———. 1960. Neue Untersuchungen über die pflanzenführenden Mergel von Gheghie (Bez.Oradea-Grosswardein-Rumänien). *Acta Botanica Academiae Scientiarum Hungarica* 6:35–44.

———. 1962. Die fossile Flora von Valea Neagra, Bezirk Crisana, Rumänien. *Palaeontographica,* B 110:128–87.

———. 1968. Date noi privind flora fosila a Bazinului Comanesti. *Studii si Cercetarii Geologie, Geofizicia, Geografie Series,* Geologie, 13:285–88.

———. 1975a. Fossile Pflanzen aus dem Pannon von Delureni (Rumänien). *Palaeontographica,* B 153:150–82.

———. 1975b. Untersuchung einer Sammlung fossiler Pflanzen von Daia und Sacadate (Kreis Sibiu, Rumänien). *Studii si comunicari Muzeul Brukenthal, studii naturale* 19:69–79.

———. 1975c. Fossile Pflanzen aus dem unteren Pannon von Valea de Cris (Kreis Bihor-Rumänien). *Acta Palaeobotanica* 16(1): 71–82.

———. 1990. *Flora fosila a Miozenului superior de la Chiuzbaia.* Bucuresti: Academei Romane.

———. 1992. Les fôrets marécageuses du Miocène supérieur de Roumanie: un paléobiotope d'exception et sa vegétation dans le Miocène supérieur de l'ouest de la Roumanie. In *Palaeovegetational Development in Europe,* ed. J. Kovar-Eder, pp. 147–51. Proceedings of the Pan-European Palaeobotanical Conference 1991, Vienna, Naturhistorisches Museum, Wien.

Givulescu, R., and O. Edelstein. 1981. Plante fosile din Tara Oasului. *Studii si Cercetarii Geologie, Geofizica, Geografie ser. Geologie* 26301–8.

Givulescu, R., O. Edelstein, V. Dragu, and D. Stan. 1986. Plantes fossiles du Pontien d'Odesti (department de Maramures). *Dari de Seama Institutul Geologie si Geofizica* 70–71/3:195–205.

Givulescu, R., O. Edelstein, and E. Hady. 1975. Flora fosila a Maramuresului (IV). Plante fosile din forajul 69 de la Turt (Oas). *Contributions Botanica,* pp. 59–61.

Givulescu, R. and N. Florei. 1960. Die fossile Flora von Sinersig (Rumänien). *Geologie* 9:799–813.

Grangeon, P. 1958. Contribution à l'étude de la paléontologie végétale du Massif du Coiron (Ardèche) (Sud-Est du Massif Central francais). *Mémoires de la Société de l'Histoire naturelle Auvergne* 6:1–302.

Gregor, H.-J. 1983a. Die miozäne Blatt- und Fruchtflora von Steinheim am Albuch (Schwäbische Alb). *Documenta naturae* 10:1–45.

———. 1983b. Die Flora aus dem Mergel I der Kiesgrube Massendorf. *Documenta naturae* 11:30–47.

Hably, L. 1985. Early Miocene plant fossils from Ipolytarnoc, N. Hungary. *Geologica hungarica ser. palaeontologica* 45:77–255.

———. 1989. In P. Müller and M. Szokony: Faciostratotype the Tihany-Feherpart (Hungary). In *Pontien. Chronostratigraphie und Neostratotypen, Neogen der Westlichen und Zentralen Paratethys* 8, ed. P. Stevanovic, L. A. Nevesskaja, Fl. Marinescu, A. Sokac, and A. Jambor, pp. 427–35. Zagreb-Beograd: Jazu and Sanu.

———. 1990. Egerian plant fossils from Vertesszölös, NW Hungary. *Studia Botanica hungarica* 22:3–78.

———. 1992a. A palaeoflora of zonal vegetation from the Egerian (upper Oligocene) of Pomaz, Hungary. In *Palaeovegetational Development in Europe,* ed. J. Kovar-Eder, pp. 153–751. Proceedings of the Pan-European Palaeobotanical Conference 1991, Vienna, Naturhistorisches Museum, Wien.

———. 1992b. Early and late Miocene floras from the Iharosbereny-I and Tiszapalkonya-I Boreholes. *Fragmenta Mineralogica et Palaeontologica* 15:7–40.

Hably, L. and J. Kovar-Eder. 1996. A representative leaf assemblage of the Pannonian Lake from Dozmat near Szombathely (West Hungary), upper Pannonian, upper Miocene. *Festschrift 1100 Jahre Ungarne-1000 Jahre Österreich,* Budapest.

Il'inskaya, I. A. 1964. Tortonskaya flora Svoshovitse. *Trudy Botanicheskovo Instituta V.L. Komarova Akademii NAUK SSSR, ser. 8, Paleobotanika* 5: 115–44.

———. 1968. *Neogene floras of the Transcarpathian region of the USSR* [in Russian]. Leningrad: Nauka.

Jähnichen H., W. Friedrich, and M. Takac. 1984. Engelhardioid leaves and fruits from the European Tertiary. Part II. *Tertiary Research* 6:109–34.

Jähnichen, H., D. H. Mai, and H. Walther. 1977. Blätter und Früchte von *Engelhardia* LESCH. ex BL. (Juglandaceae) aus dem europäischen Tertiär. *Feddes Repertorium* 88(5–6): 323–63.

Jung, W. 1968. Pflanzenreste aus dem Jungtertiär Nieder- und Oberbayerns und deren lokalstratigraphische Bedeutung 25. *Bericht des Naturwissenschaftlichen Vereins Landshut:* 43–71.

Knobloch, E. 1969. *Tertiäre Floren von Mähren.* Brno.

———. 1975. Die Makroflora des Egerien von der Fundstelle Krumvir. In T. Baldi and J. Senes: OM, Egerien. Die Egerer, Pouzdraner und Puchkirchner Schichtengruppe und die Bretkaer Formation. In *Chronostratigraphie und Neostratotypen*, ed. E. Brestenska, pp. 547–50. Bratislava: V. Veda.

———. 1986. Die Flora aus der Oberen Süsswassermolasse von Achldorf bei Vilsbiburg (Niederbayern). *Documenta naturae* 30:14–48.

———. 1988. Neue Ergebnisse zur Flora aus der Oberen Süsswassermolasse von Aubenham bei Ampfing (Kreis Mühldorf a. Inn). *Documenta naturae* 42:2–27.

———. 1990. Willershausen, 3. Teil. Die Flora. *Fossilien* 5(1990): 216–22.

Knobloch, E. and Z. Kvacek. 1965. *Byttneriophyllum tiliaefolium* (Al. Braun) Knobloch et Kvacek in den tertiären Floren der Nordhalbkugel. *Sbornik Geologickych Ved, paleontologie* 5:123–66.

———. In press. Miozäne Floren der südböhmischen Becken. *Sbornik Geologickych Ved, paleontologie*.

Knobloch, E. and E. Velitzelos. 1986. Die obermiozäne Flora von Likudi bei Elassona (Thessalien, Griechenland). *Documenta naturae* 29:5–20.

Kolakovskiy, A. A. 1958. Pervoe dopolnenie k duabskoy pliotsenovoy flore. *Trudy Sukhumskovo Botanicheskovo Sada* 11:311–97.

———. 1962. Ponticheskaya flora Pitsundy. *Trudy Sukhumskovo Botanicheskovo Sada* 14:37–57.

———. 1964. *Pliotsenovaya flora Kodora.* Sukhumskiy Botanicheskiy Sad Monografii 1.

Kolakovskiy, A. A. and N. K. Ratiani. 1967. Pliotsenovaya flora Malykh Shirak. *Trudy Sukhumskovo Botanicheskovo Sada* 16:30–71.

Kolakovskiy, A. A., L. P. Rukhadze and A. K. Shakryl. 1970. Meoticheskaya flora Kodora. *Trudy Sukhumskovo Botanicheskovo Sada.* 17:89–110.

Kolakovskiy, A. A. and A. K. Shakryl. 1978. Kimmeriyskaya flora Gul'ripsha (Bagazhishta). *Trudy Sukhumskovo Botanicheskovo Sada* 24:134–56.

Kolasa, K. 1982. Miocene fossil flora in the Wieliczka salt deposit (in Polish). *Studia i Materialy do Dziejow Zup Solnych w Polsce* 11:45–57.

Kordos-Szakaly, M. 1984. New data to the Miocene flora of Nogradszakal (Hungary). *Annales Historico-naturales Musei Nationalis Hungarici* 76:43–63.

Kovar, J. 1982. Eine Blätter-Flora des Egerien (Ober-Oligozän) aus marinen Sedimenten der Zentralen Paratethys im Linzer Raum (österreich). *Beiträge zur Paläontologie von Österreich* 9:1–209.

———. 1988a. Obermiozäne (Pannone) Floren aus der Molassezone österreichs. *Beiträge zur Paläontologie von Österreich* 14:19–121.

———. 1988b. Three dimensional distribution maps for fossil plants. Examples from middle to upper Miocene leaf-floras of Central Europe. *Tertiary Research* 9:213–35.

Kovar, J. and J.-P. Berger. 1987. Die oberoligozäne Flora von Unter-Rudling bei Eferding in Oberösterreich. *Annalen des Naturhistorischen Museums, Wien* 89:57–93.

Kovar-Eder, J., L. Hably, and T. Derek. 1995. Neuhaus/Klausenbach-eine miozäne (pannone) Pflanzenfundstelle aus dem südlichen Burgenland. *Jahrbuch der Geologischen Bundesanstalt* 138(2): 321–47.

Kovats, J. 1856. Fossile Flora von Erdöbenye. *Arbeiten der geologischen Gesellschaft für Ungarn* 1:1–37.

Kräusel, R. 1921. Nachträge zur Tertiärflora Schlesiens. III. *Jahrbuch der Preussischen Geologischen Landesanstalt* 40 (1919), 3. Teil: 363–433.

Kretzoi M., E. Krolopp, H. Lörincz, and I. Palfalvy. 1975. Flora, Fauna und stratigraphische Lage der unterpannonischen Prähominiden-Fundstelle von Rudabánya (NE-Ungarn). *Annales Geologici Instituti Publici Hungarici* 1974:365–94.

Krishtofovich, A. N. and T. N. Baykovskaya. 1965. *Sarmatskaya flora Krynki.* Moskva-Leningrad: Nauka.

Kvacek, Z. 1989. Fosilni *Tetraclinis* Mast. (Cupressaceae). *Casopis Narodniho Muzea v Praze* 155(1–2): 45–52.

Kvacek, Z. and L. Hably. 1991. Notes on the Egerian stratotype flora at Eger (Wind brickyard), Hungary, upper Oligocene. *Annales Historico-naturales Musei Nationalis Hungarici* 83:49–82.

Mädler, K. 1939. Die pliozäne Flora von Frankfurt a. M. *Abhandlungen der Senckenbergischen naturforschenden Gesellschaft* 446:1–202.

Mai, D. H. 1981. Entwicklung und klimatische Differenzierung der Laubwaldflora Mitteleuropas im Tertiär. *Flora* 171:525–82.

Mai, D. H. and H. Walther. 1985. Die obereozänen Floren des Weisselster-Beckens und seiner Randgebiete. *Abhandlungen des Staatlichen Museums für Mineralogie und Geologie zu Dresden* 33:5–260.

———. 1988. Die pliozänen Floren von Thüringen, Deutsche Demokratische Republik. *Quartärpaläontologie* 7:55–297.

Meller, B. 1989. Eine fossile Blattflora aus Sprendlingen. *Documenta naturae* 54:1–109.

———. 1992. Samen und Früchte aus dem Köflach-Voitsberger Braunkohlenrevier-erste Ergebnisse. In *Palaeovegetational Development in Europe*, ed. J. Kovar-Eder, pp. 181–87. Proceedings of the Pan-European Palaeobotanical Conference Vienna, 1991, Naturhistorisches Museum, Wien.

Mihajlovic, D. 1977. Contribution to the study of Tertiary flora of Beograd surroundings. *Annales Géologiques de la Péninsule Balkanique* 41:269–79.

———. 1978. Fossil flora from Slanci near Belgrade. *Bulletin du Muséum d'Histoire Naturelle de Belgrade* A33:199–207.

———. 1980. Miocene flora from white marls of Jancici (Dobrinje-Jezevica Basin). *Comptess Rendus des Séances de la Société Serbe de Géologie* 1979:159–64.

———. 1985. Palaeogene fossil flora of Serbia. *Annales Géologiques de la Péninsule Balkanique* 49:299–434.

Milovanovic, Lj., and D. Mihajlovic. 1984. Miocene flora from Zagubica Basin, eastern Serbia. *Annales Géologiques de la Péninsule Balkanique* 48:201–13.

Nemejc, F. and E. Knobloch. 1973. Die Makroflora der Salgotarjaner Schichtengruppe. In *Ottnangien. Chronostratigraphie und Neostratotypen*, ed. A. Papp, F. Rögl, and J. Senes, 3:695–759. Bratislava: VEDA.

Palvalfy, I. 1981. A magyarorszagi *Engelhardia* fjok retegtani, ökologiai es cönologiai szerepe. *Magyar Földtani Intezet Evkoenive* 1979:491–95.

Pantic, N. 1956. Biostratigraphie des flores tertiaires de Serbie. *Annales Géologiques de la Péninsule Balkanique* 24:199–321.

Pantic, N. and D. Mihajlovic. 1977. Neogene floras of the Balkan land areas and their bearing on the study of the paleoclimatology, paleobiogeography, and biostratigraphy (part 2). *Annales Géologiques de la Péninsule Balkanique* 41:159–73.

Pantic, N., D. Mihajlovic, and S. Vrabac. 1988. Badenian fossil flora from Tuzla area. *Annales Géologiques de la Péninsule Balkanique* 51:321–27.

Pimenova, N. V. 1954. Sarmatskaya flora Amvrosievki. *Trudy Instituta Geologicheskikh Nauk serija stratigrafii i paleontologii* 8:3–96.

Purtseladze Kh. N. and E. A. Tsagareli. 1974. Meotitcheskaya flora Yugo-Zapadnoy Gruzii. *Akademiya Nauk Gruzinskoy SSR, Geologicheskiy Institut im A.I. Dzhanelidze, Trudy, Novaya Seria.* 45:1–227.

Raniecka-Bobrowska, J. 1957. Kilka szczatkow roslinnych z tortonu Gornego Slaska. *Kwartalnik Geologiczny* 1:275–97.

Ratiani, N. K. 1970. Novye dannye o pleystotsenovoy flore Gumisty. *Trudy Sukhumskovo Botanicheskovo Sada* 17:81–88.

Rayushkina, G. S. 1979. *Oligotsenovaya flora Mugodzhara i juzhnowo Altaya.* Alma Ata: Nauka.

Reichenbach, E. 1919. Coniferen und Fagaceen. In *Die Pflanzen des Schlesischen Tertiärs,* ed. R. Kräusel, pp. 97–144. Jahrbuch Preussische Geologische Landesanstalt 38 (1917), 2. Teil.

Reid, E. M., and M. E. J. Chandler. 1926. The Bembridge flora. *Catalogue of Cainozoic plants in the Department of Geology,* 1:1–206. London: British Museum.

Roiron, P. 1979. *Récherches sur les flores plio-quaternaires méditerranéennes: La macroflore pliocène de Pichegu près de Saint-Gilles (Gard).* Académie de Montpellier, Université des Sciences et Techniques du Languedoc, Thèse 10.9.1979.

Rögl, F. and G. Daxner-Höck. This volume. Late Miocene Paratethys Correlations.

Rögl, F., H. Zapfe, R. L. Bernor, R. Brzobohaty, G. Daxner-Höck, I. Draxler, O. Feyfar, J. Gaudant, P. Herrmann, G. Rabeder, O. Schultz, and R. Zetter. 1993. Die Primatenfundstelle Götzendorf bei Bruck an der Leitha, Niederösterreich (Obermiozän des Wiener Beckens). *Jahrbuch der Geologischen Bundesanstalt* 136(2): 503–26.

Sadowska, A. 1991. The stratigraphical table of the Neogene floras from Poland. In *Proceedings of the Symposium "Floristic and Climatic Changes in the Tertiary,"* pp. 133–39. Bratislava: Dionyz Stur Institute of Geology.

Schweigert, G. 1992. Die untermiozäne Flora (Karpatium, MN 5) des Süsswasserkalkes von Engelswies bei Meßkirch (Baden-Württemberg). *Stuttgarter Beiträge zur Naturkunde,* Serie B 188:1–55.

Shtefyrtsa, A. G. 1974. *Rannesarmatskaya flora Bursuka.* Kishinev: Shtintsa.

Shvareva, N. 1983. *Miotsenovaya flora Predkarpat'ya.* Kiev: Naukova Dumka.

———. 1989. *Verckne-badenskaya flora Zalestsev.* Akademiya Nauk Ukrainskoy SSR. Kiev: Naukova Dumka.

1969. Die Paläoflora des Turiec-Beckens und ihre Beziehung zu den mitteleuropäischen Floren. *Acta Geologica et Geographica Universitatis Comenianae,* Geologica 17:99–173.

———. 1973. Die fossile Flora sarmatischer Sedimente aus der Umgebung von Mociar in der mittleren Slowakei. *Acta Geologica et Geographica Universitatis Comenianae,* Geologica 26:5–85.

Sitar, V. and Z. Kvacek. 1991. A review of Tertiary floras in the Western Carpathians. In *Proceedings of the Symposium "Floristic and Climatic Changes in the Tertiary,"* pp. 77–80. Bratislava: Dionyz Stur Institute of Geology.

Sitar, V., Z. Kvacek, and C. Buzek. 1989. New late Neogene floras of southern Slovakia. *Zapadne Karpaty, ser. paleontologia* 13:43–59.

Staub, M. 1890. Beiträge zur fossilen Flora der Umgebung von Munkacs. *Földtanii Közlöny* 20:68–73.

Steininger, F. F., R. L. Bernor, and V. Fahlbusch. 1989. European Neogene marine/continental chronologic correlations. In *European Neogene Mammal Chronology,* ed. E. H. Lindsay, V. Fahlbusch, and P. Mein, pp. 15–46. New York: Plenum.

Stevanovic, P., L. A. Nevesskaja, Fl. Marinescu, A. Sokac, and A. Jambor, eds. 1989. Pontien. *Chronostratigraphie und Neostratotypen, Neogen der Westlichen und Zentralen Paratethys 8.* Zagreb: JAZU, SANU.

Stevanovic, P. and N. Pantic. 1954. O sarmatskoj flori i fauni iz zhekznitchkikh useka kod Bozhdarevatsa. *Annales Géologiques de la Péninsule Balkanique* 22:1–26.

Stojanoff, N. and B. Stefanoff. 1929. Beitrag zur Kenntnis der Pliozänflora der Ebene von Sofia. *Zeitschrift der Bulgarischen geologischen Gesellschaft* 1:1–120.

Stuchlik, L., A. Szynkiewicz, M. Lancucka-Srodoniowa, and E. Zastawniak. 1990. Results of the hitherto palaeobotanical investigations of the Tertiary brown coal bed Belchatow (Central Poland). *Acta Palaeobotanica* 30:259–305.

Stur, D. 1867. Beiträge zur Kenntnis der Flora der Süsswasserquarze, der Congerien- und Cerithien-Schichten im Wiener und Ungarischen Becken. *Jahrbuch der kaiserlich königlichen Geologischen Reichsanstalt* 17:77–188.

Suraru, M., N. Suraru, and R. Givulescu. 1978. Sarmatianul din Valea Baita (com.Borod) si paleoflora lui. *Nymphaea* 6:65–92.

Szafer, W. 1961. Miocene flora from Stare Gliwice in Silesia. *Instytut Geologiczny, Prace* 33:1–205.

Takac, M. 1974. Die Miozänflora des oberen Nitra-Gebietes. *Acta rerum naturalium Musei nationalis Slovenici* 19:25–101.

Teslenko, Yu. V. 1975. Novye rasteniya miotsenovoy flory Vostochnovo Donbassa. *Paleontologicheskiy Sbornik* 12:137–41.

———. 1977. Novi materiali do vivchennya miotsenovikh flor pivnichnovo Priazov'ya. *Geologichiy Zhurnal* 37:81–87.

Ticleanu, N. and L. Artin. 1982. Date noi prvind flora sarmatiana de la Deva-Timpa. *Dari de Seama sedintelor Institutul Geologie Geofizica, 3. Paleontologie* 67:173–86.

Ticleanu, N. and R. Givulescu. 1982. Plantes fossiles dans les dépots du Sarmatien de Racsa (District Satu Mare). *Dari de Seama sedintelor Institutul Geologie Geofizica, 3. Paleontologie* 66:115–25.

———. 1978. Contributions to the knowledge of the upper Egerian palaeoflora-fossil flora of Corus II, Cluj District. *Courier Forschungsinstitut Senckenberg* 30:133–50.

Unger, F. 1869. *Fossile Flora von Radoboj.* Denkschriften der kaiserlichen Akademie der Wissenschaften, mathematisch-naturwissenschaftliche Classe 29. Wien.

Velitzelos, E. and E. Knobloch. 1986. Die pliozäne Flora von Skoura bei Sparta auf dem Peloponnes (Griechenland). *Documenta naturae* 29:21–28.

Wilde, V. 1989. Untersuchungen zur Systematik der Blattreste aus dem Mitteleozän der Grube Messel bei Darmstadt (Hessen,

Bundesrepublik Deutschland). *Courier Forschungsinstitut, Senckenberg* 115:1–212.

Yakubovskaya, T. A. 1955. Sarmatskaya flora Moldavskoy SSR. *Trudy Botanicheskovo Instituta im. V.L. Komarova Akademii Nauk SSSR*, Series 1, 11:7–108.

Zastawniak, E. 1972. Pliocene leaf flora from Domanski Wierch near Czarny Dunajec (Western Carpathians, Poland). *Acta Palaeobotanica* 13:1–73.

———. 1980. Sarmatian leaf flora from the southern margin of the Holy Cross Mts. (South Poland). *Prace Muzeum Ziemi* 33:39–108.

Zastawniak, E., ed. 1992. The younger Tertiary deposits in the Gozdnica region (SW Poland) in the light of recent palaeobotanical research. *Polish Botanical Studies* 3:3–129.

Zhilin, S. G., and A. G. Andrejev. 1984. New data on the late Oligocene flora of Altyn-Shokysy tableland (North Aralian region). *Botaniceskij Zhurnal* 69(12): 1603–11.

31

Provinciality, Diversity, Turnover, and Paleoecology in Land Mammal Faunas of the Later Miocene of Western Eurasia

M. FORTELIUS, L. WERDELIN, P. ANDREWS, R. L. BERNOR, A. GENTRY,
L. HUMPHREY, H.-W. MITTMANN, AND S. VIRATANA

The relationship between paleoenvironmental change and biogeographic development of later Neogene Western Eurasian faunas is this volume's central theme. Elaboration and scrutiny of this problem rests on consideration of the systematic and geochronologic bases of the record, rooted in decades of active scientific investigation of the relationship between European Neogene chronology, continental vertebrate and marine invertebrate evolution, and paleoenvironmental and biogeographic reconstruction. The material brought together for this volume has allowed us, for the first time, to qualitatively and quantitatively approach the outstanding issues of provinciality, biodiversity, turnover, and paleoecology. This volume's approach is nascent, and coarse-grained in some areas, but the results go beyond the obvious and expected in several respects.

Discrimination of the Western Eurasian Neogene record has followed two parallel paths, which we have subsequently merged here for analysis and interpretation: (1) refinement of its chronologic basis, which seeks correlations based on the biochronologic ranking and chronometric tie points of vertebrate localities; (2) teasing apart the paleoenvironmental and biogeographic parameters that bias this zonation scheme so as to cause false, "biofacies-based" correlations—a problem resolved by exacting more precision in the chronology of critical localities and the mammalian systematic records we have reviewed across several studies in this volume. Obviously, an essential dichotomy exists in these two scientific pursuits: mammalian Neogene units seek isochronometric and homotaxial means for correlation; biogeographic and paleoenvironmental inquiries seek the variability between faunal provinces and the underlying mechanisms that might bias interprovincial correlations. The results of these somewhat contradictory activities must be weighed independently before an attempt is made to merge their records.

The recognition of Western Eurasian later Neogene faunal provinciality has emerged gradually over the last 130 years. Gaudry, through his monographic descriptions and comparisons of the Pikermi (1865) and Mont Lubéron (1873) faunas, is properly credited as being the first individual to recognize the striking differences between these Mediterranean "Pontian" age faunas and the classic Central European fauna from Eppelsheim. Increased research activity in the later part of the nineteenth and early twentieth centuries substantiated Gaudry's observations and led Koenigswald (1929), and later Thenius (1951), to demonstrate the existence of two ecological associations: one (Central European) "Eppelsheim" fauna with a prevalent woodland character; one (peri-Mediterranean) fauna with a prevalent "steppe and/or savanna character." Crusafont and Villalta's (1954) discoveries in northeast Spain suggested that the "Eppelsheim" woodland fauna stratigraphically preceded the "steppic or savanna" faunas of the peri-Mediterranean realm. In a short, and largely unrecognized address of this problem, Tobien (1967) recognized that two distinct, provincial ecosystems might have occurred synchronously in the late Miocene: "The theoretical possibility of a facies boundary between Vallesian and Pikermian in the Spanish Neogene basins of Valles-Penedes and Calatayud-Teruel involves the existence of an identical and unchanged steppe fauna in the East during the whole Pontian (i.e., Vallesian plus Pikermian) time span, and a similarly unchanged and evolutionarily stable woodland fauna in the central part of Europe" (1967:3).

The existence of distinct, geographically bound, and environmentally contrasting faunas in Western Eurasia was reexamined as a consequence of studies on the three classic "Pikermian" faunas of Maragheh, Samos, and Pikermi (Bernor 1978, 1985, 1986; Solounias 1981a, b). In the development of Maragheh's biostratigraphy and radioisotopic chronology (Bernor 1978, 1985, 1986; Bernor et al. 1979b; Campbell et al. 1980; Bernor et al this volume; Swisher this volume), and attempts to correlate this sequence based on homotaxial comparisons, especially using hipparionine horses (Bernor et al. 1980), it was found that there were woefully few species comparisons possible outside of a very narrowly circumscribed geographic area including the former Southern USSR, Greece, and Afghanistan.

In an attempt to understand the relationship between these apparent provinces and other potential Western Eurasian provinces, Bernor (1978) selected 38 well-known later Miocene localities for quantitative analysis and identified a minimum of four identifiable provinces: (1) an Eastern Mediterranean–Southwest Asian zoogeographic province; (2) an East African zoogeographic province; (3) an Indopakistan zoogeographic province; (4) a Central (A) and Western (B) European province. Bernor et al. (1979a) followed these observations with an early attempt to reconstruct the evolution of these provinces. Their survey of Miocene Eurasian localities suggested that those ungulate and carnivore groups that dominate late Miocene savanna-like open country faunas (hyaenids, felids, bovids, and giraffids) actually first evolved and radiated in the Central Asian and North African early Miocene.

Revision of the MN time scale by Mein (1979) led to a provincially based reassessment of the chronology of fauna change by Bernor (1983; 1984) and the recognition of at least six Eurasian zoogeographic provinces: (1) Western and Southern Europe; (2) Eastern and Central Europe; (3) Romania–Western USSR; (4) sub-Paratethyan (Greece, Turkey, Iran, and Afghanistan); (5) North Africa; (6) Siwaliks; with China being a seventh province with too poorly detailed a stratigraphic/chronologic record to be characterized at that time.

Beyond the early quantitative analyses by Bernor (1978), qualitative reconstructions (1983, 1984) of whole faunas verified that later Miocene "savannalike" faunas had a long evolutionary history, that they appeared diachronously across Western Eurasia, and that they were highly variable in their species composition, abundance, and diversity. Bernor (1984) offered an early characterization of these provinces' evolution, corroborating earlier documentation of their existence. Bernor and Pavlakis (1987) undertook a quantitative investigation of the biogeographic relationships of a late Miocene/early Pliocene (Howell 1987) North African locality, Sahabi, that showed a mixture of Eurasian and African faunal elements. This work confirmed the provincial nature of late Miocene and Pliocene Old World faunas, and revealed episodic shifts in African biotopes toward modern savannalike faunas.

An attempt to develop biogeographic reconstructions through systematic revisions of superspecific groups took place in parallel with the work reviewed above. Bernor (1983) initially synthesized the provincial diversification of

Miocene hominoid primates, an analysis that revealed very short intervals of intercontinental exchange followed by long intervals of provincial autochthonous evolution. Bernor et al. (1988) further demonstrated that early Miocene catarrhine deployment from Africa to Eurasia was provincial and developed across a separate northern, more temperate, latitudinal track and a southern, more tropical latitudinal track. However, it was the broadly distributed, well-calibrated, and evolutionarily resolved hipparionine record that best exemplified Eurasian and African late Miocene and Pliocene provinciality (Woodburne and Bernor 1980; Bernor and Lipscomb 1991; 1995; Bernor et al. 1980, 1989, 1990, this volume b). Similar studies in this volume of the Carnivora (Werdelin; Werdelin and Solounias), Rhinocerotidae (Heissig), Suoidea (Fortelius et al.), and Pecora (Gentry and Heizmann) indicate that this provinciality was a phenomenon affecting the mammalian fauna as a whole.

Quantitative analyses of diversity and turnover patterns have been virtually absent for the Old World Neogene until the production of this volume. The main points of reference are studies of the fossil land mammal faunas of North America (e.g., Simpson 1947; Webb 1969, 1984; Lillegraven 1972; Gingerich 1984; Krause and Maas 1990; Stucky 1990, 1992; Van Valkenburg and Janis 1993; Maas et al. 1995) and South Asia (e.g., Barry et al. 1985; 1990; 1995).

Paleoecological characterization of Neogene Eurasian land mammal faunas has been a less neglected area, as indicated above, but the literature is scattered and a recent synthesis is lacking (see the taxonomic and synthetic articles of this volume). Janis (1989, 1993) and Potts and Behrensmeyer (1992) have recently reviewed the subject in a global context. Although few ecomorphological studies have been undertaken until recently, they do extend back at least to Kurtén's (1952) classic treatment of the Chinese *Hipparion* fauna, which confirmed and refined Schlosser's (1903) identification of geographically distinct "forest" and "steppe" faunas. In this survey we have approached the question from the standpoints of analysing entire fossil faunas, following the procedures developed by Andrews et al. (1979), and by analysis of ecomorphological change in various subsets of the available data base, disregarding the completeness of individual faunas (Fortelius et al., in press).

The current review discusses the biogeographic, faunal, and paleoecological development of Western Eurasian later Neogene faunas. Our analysis follows the assumption that meaningful quantitative patterns can be derived from Western Eurasian later Neogene faunas. This assumption is found to be essentially valid, despite some results that have been identified as being biased by probable sampling errors. Most geographic areas and time intervals are represented by at least one and frequently several rich localities, providing "Rosetta Stone" faunal lists (Alroy 1992) that account for the great majority of taxa. In Europe such faunas are used for characterizing MN units and have been recommended by Fahlbusch (1991) to delimit the lower boundary of the referred MN unit. The small, poorly sampled localities (the number of which varies greatly between regions and intervals) contribute only a small fraction to the summed diversity. It follows that the present analysis is not much different in practice from studies using only selected ("representative") localities (e.g., Van Valkenburgh and Janis 1993). Indeed, our own paleoecological analysis of faunas is also based on selected faunas rather than on the entire material available.

The main quantitative analyses undertaken here are species richness and turnover, analysis of faunal similarity between regions, and breakdown of diversity changes by ecomorphological criteria. All of these are based on simple presence/absence analyses on taxa in variously defined regions and time intervals. For practical reasons we have relied on MN units for defining time intervals and on countries rather than geographic coordinates for defining areas. The former decision entails the logical problem that biotic change is being analyzed in a biochronologic framework, i.e., partly in terms of itself. Although Bernor et al. (this volume a) have noted a divergence in the MN time scales during the late Miocene, with the boundaries of MN 11 and 12 probably being diachronous, the bias is conservative in that regional diachroneity and shifting tempo of change will both be minimized rather than exaggerated. The latter decision means that provinciality will be analyzed, absurdly, according to modern political boundaries. For broad, regional comparisons this matters little, however, while for the most part the scatter of the data will not support the analysis of areas smaller than a group of several countries. It also provides an arbitrary standard for long-term comparisons involving shifting geographic conditions.

The sampling adequacy and data quality are obviously critical. Completeness indices and similar quantitative parameters are highly useful but crude indicators at best. Indeed, it may never be possible to assess the quality of the data and the completeness of the sampling in any strict sense, simply for lack of a commonly applicable standard. Paleontology is a historical subject dealing with a unique and, quite literally, unimaginably complex chain of events. An alternative to analysis of the raw data is offered by the study of associated information, such as ecomorphology (Damuth 1992), or paleoecological interpretations in general. If, for example, a pattern of diachronous change in diversity can be seen to correspond to a shift in the trophic or spatial composition of the faunas, then the ecological reasons for, and the reality of, the change can be evaluated.

We have undertaken simple ecomorphological analyses using relative molar crown height, estimated body mass, and interpreted diet at the basic level of animal-eater, plant-eater, or omnivore (see Damuth 1993). All these parameters reflect ecology only indirectly, but they have the advantages of being relatively easy to obtain and broadly applicable across taxonomic groupings. Additionally, we have provided simple bivariate comparisons between spatial and trophic distributions for a selection of recent and fossil faunas (Andrews, in press). Based on both sets of data, inferences will be made in terms of relative openness of the landscape, seasonality, "harshness," or some such unidimensionally conceived gradient.

Materials and Methods

Materials

In advance of the Schloss Reisensburg workshop, the participants were requested to supply updated locality faunal lists and age referrals (in terms of MN units) for the more complete and better-studied Western Eurasian localities referable to MN 6–13 (ca. 15–5 Ma). Later, the following definitions of areas were specified: "Western Europe (principally France [and if necessary] Spain), Central Europe (Germany, Switzerland, Austria, Poland, Czechoslovakia, Hungary), Southeastern Europe (Romania, Bulgaria, Yugoslavia, Greece, Georgia, Turkey-Thrace), Southwestern Asia (Anatolia and Iran)." These strict limitations and instructions made the entire undertaking possible, but have, of course, inevitably limited the potential for comparison, especially as far as the Messinian crisis at the end of the interval is concerned. Fortunately, not all authors followed the instructions to the letter, especially in their written contributions.

For the analysis we first reduced the groupings to the lowest level possible (that of individual countries), and then analyzed a number of smaller assemblages defined a priori (see below). The most detailed comparisons were confined to a comparison of two major blocks, "West" and "East" (fig. 31.1). In these analyses, West consisted of Portugal, Spain, France, Italy, Germany, Switzerland, Austria, Poland, Czechia, and Slovakia. East consisted of Hungary, Slovenia, Serbia, Bosnia, Macedonia, Albania, Greece, Bulgaria, Romania, Moldova, Ukraine, Georgia, Turkey, Iran, Afghanistan, and Kazakhstan (no Belarussian or Russian localities were included). As far as our analyses indicate, these two blocks were well defined during the study period, and apparently had quite different histories.

All chronologic referrals are in terms of MN units, except for when comparisons are made with North America (see below). We follow Steininger et al.'s (this volume) updated time scale, which includes revisions of the magnetic time scale (Kent) and marine time scale (Berggren) as well as the geochronologic results of several researchers contributing original information to this vol-

FIGURE 31.1 Map of the Western Old World, showing West and East blocks compared in this study.

ume (Bernor et al., this volume c; Kappelman et al., this volume; Sen, this volume; Swisher, this volume; Woodburne et al., this volume a and b). All volume contributors further follow de Bruijn et al. (1992) in merging MN 7 and 8 into a single zone: MN 7 + 8 (or, alternatively, 7/8). In the analyses we have included within each MN unit all localities assigned either to that particular unit only, or to it and/or the preceding or following units. For example, our data for MN 10 include localities assigned to MN 10, MN 9/10, MN 10/11, and MN 9–11, but not MN 7/8–10, MN 10–12, etc. It is our opinion that this procedure, which admittedly "smears" the data considerably, is in better agreement with the actual temporal resolution available than would be the restrictive use of only localities confidently assigned to a single MN unit (which would also reduce the sample size drastically). It is also a conservative choice, since it will tend to moderate the abruptness of change between intervals. The same conservatism also applies to the logical paradox that biotic change is investigated partly in terms of itself, as noted in the introductory paragraphs.

The data that were delivered before, during, and after the meeting were compiled into a data base by Hans-Walter Mittmann at Karlsruhe. Independently, a second data base was compiled by Mikael Fortelius at Helsinki, drawn from the manuscripts submitted to the volume (rather than from the lists supplied by the authors), and supplemented through direct communication with the authors and from the literature. This data base is the essence of the "NOW" (Neogene of the Old World) research project, financed by the Academy of Finland (Fortelius et al., in press). Structurally, it is a simplified version of the ETE data base described by Damuth (1993).

Thus far, the NOW data base is principally limited to large mammals. Of the perissodactyls, only hipparionine horses and rhinoceroses are covered by this volume. Data for *Anchitherium* were therefore added from Abusch-Siewert (1983) and for chalicotheres from Zapfe (1979; 1989). Since the suoid and rhinocerotid chapters largely ignored Western Europe, data for that region (mainly Spain) were added from Van der Made (1992) and Cerdeño (1989), respectively. Data for several Anatolian localities were also taken from Sickenberg (1975). The only additional published sources used are Andrews (1990) and Sen (1994). Unlike the chapter presented by Kappelman et al. (this volume), which is current up through 1992, faunal data for the Central Anatolian Sinap Formation includes all identifications made up through the 1994 field season. Proboscideans, tubulidentates, and hyracoids were excluded from all analyses, since they are not included in the book. All three are taxa with low diversity, and their exclusion is unlikely to affect the results significantly. The total number of localities in the NOW data base at the time of analysis was 511, the number of taxa (including incomplete and uncertain identifications) 1,073, and the number of taxon-locality occurrences 2,984.

The macromammal families included are the following: Primates—Hominidae and Pliopithecidae; Creodonta—Hyaenodontidae; Carnivora—Nimravidae, Canidae, Mustelidae, Ursidae, Amphicyonidae, Viverridae, Hyaenidae, Percrocutidae, and Felidae; Perissodactyla—Equidae, Chalicotheriidae, and Rhinocerotidae; Artiodactyla—Suidae, Tayassuidae, Tragulidae, Palaeomerycidae, Cervidae, Giraffidae, and Bovidae. The micromammal families included in the Karlsruhe data base were the following: Insectivora—Erinaceidae, Soricidae, Talpidae, and Dimylidae; Rodentia—Sciuridae, Castoridae, Cricetidae, Spalacidae, Gerbillidae, Arvicolidae, Muridae, Gliridae, Hystricidae, Ctenodactylidae, Anomalomyidae, and Eomyidae.

Both data bases have been utilized in the syntheses presented here: the Karlsruhe data base for the micromammals (Rodentia and Insectivora), the NOW data base for the remainder. For micromammals, about 200 taxa from 150 localities, with about 500 taxon-locality occurrences, were used. For macromammals, the corresponding numbers were about 420, 490, and 1,800 (more precise numbers would be meaningless because of the criteria employed for the use of uncertain identifications, as explained below).

Sampling and Completeness

Sampling has been investigated by a number of methods. At a very general level, sampling will be reflected in the correlation between the number of taxa and the number of available localities, since the latter will circumscribe the number of specimens available and this in turn is known to be related to the number of species found in a sample (Fisher et al. 1943; May 1975). However, a strongly positive correlation between these parameters cannot be taken at face value to imply poor sampling, since the specimen abundance patterns may well be the reflection of species present rather than the vagaries of the fossil record. Therefore, other means of estimating the quality of the data sampling must be found.

Krause and Maas (1990) and Maas et al. (1995) discuss two completeness indices that can potentially be used to evaluate sampling. The first (CI_1) is simply the percentage of all taxa inferred to be present in a stratigraphic interval (in this case, MN unit) that have been actually found in that interval. It is calculated as:

$$CI_1 = [Np / (Np + Nrt)] \times 10$$

where Np is the number of taxa actually found in the interval and N_{rt} the number of range-through taxa, i.e., the number of taxa found before and after the interval, but not during it. The latter are inferred to have been present in the interval but have not actually been recovered. The second index (CI_2) is more conservative in that it is based only on taxa whose ranges extend through the interval, excluding first and last occurrences in the interval. However, our data were in some cases too limited for this index to be calculated and we have omitted it from further consideration.

Sampling can also be studied by investigating the number of taxa per locality in different regions and times. These can be plotted as rank diagrams that indicate obvious absences (e.g., taxon-rich or taxon-poor localities) in the region and temporal record. Absence probably reflects incomplete sampling. Other means of investigating completeness and sampling will be referred to below as necessary.

Similarity

A great number of indices have been applied for investigating faunal similarity and/or difference. These have recently been evaluated from a number of perspectives (Flynn 1986; Archer and Maples 1987; Maples and Archer 1988). Archer and Maples have used Monte Carlo simulations to show that certain of these indices tend to give spurious results when there are few variables, or sparse data, or both. By convention, the table of presences and absences is separated into four quadrants: A = present in both faunas; B = present in fauna 1, absent in fauna 2; C = absent in fauna 1, present in fauna 2; and D = absent in both faunas. The quantity D is unknown in fossil faunas since the size of the population from which the fossil record is sampled is an unknown quantity and indices that require this value cannot be used. Of the remaining indices, the Dice index (Sokal and Sneath 1963) is the one most highly recommended by Archer and Maples (1987) and Maples and Archer (1988). It is calculated as **2A / (2A + B + C)**, and is our first choice of index. It is monotonic with the commonly used Jaccard index (Sepkoski 1988). However, for historical reasons most studies of similarity of fossil mammal faunas have used the Simpson index (Simpson 1943; Bernor 1978; Flynn 1986; Bernor and Pavlakis 1987) and we have incorporated this index as well. It is calculated as **A / (A + E)**, where E is the smaller of B or C. Other indices have been used in recent times in studies similar to ours. Van Valkenburgh and Janis (1993) use an index attributed to Pielou by Cody (1993), which is calculated as: **AB + AC / 2BC**. De Bonis et al. (1992a), on the other hand, use the Pickford index of faunal difference calculated as: **BC / ((A+B)*(A+C))**. Our data indicate that this index and the previous one behave as each others' inverses, and in our diagrams we have not presented the Pickford index.

Diversity and Turnover

Diversity has been based throughout on counts of the number of species present in more or less arbitrarily defined geographic regions. It is not possible to distinguish between alpha diversity (within a community) and beta diversity (between communities) in this material. What is discerned is a combination of the two, or "gamma diversity" in the original sense of Whittaker (1977; see Sepkoski 1988). The gamma diversity of current usage (between regions) has been studied here in terms of faunal similarity,

as described above. As far as possible, we have attempted to distinguish between imprecise, uncertain, and anonymous identifications, respectively. Identifications to the genus level or above have been excluded from all analyses, except when known to signify the presence of a specific but as yet formally unrecognized taxon (see Damuth 1993). Uncertain identifications (*Progonomys* cf. *P. cathalai*) were counted only if a certain identification to the conferred taxon (*P. cathalai*) was not available for the region and interval in question. Thus, uncertain identifications were not allowed to inflate the number of taxa included, but were accepted when rejection would have meant an artificial reduction of apparent diversity.

We also calculated standing richness according to the formula of Maas et al. (1995) and Barry et al. (1995):

$$N_{sr} = (N_{rt} + N_{bda}) + (N_f + N_l + N_o) / 2$$

where for each interval N_{rt} is the number of range-through taxa; N_{bda} the number of taxa known before, during, and after the interval; N_f the number of first occurrences; N_l the number of last occurrences; and N_o the number of taxa that are known only from that interval. The trend in standing richness matches that using simple counts very closely.

Turnover has been studied by a number of means. First appearances in an MN interval were taken to occur at the base of that interval, while last appearances were taken to occur at the end of the interval. In our analyses, the raw counts of entries and exits were converted to a percentage of the number of species present in each interval. Finally, relative turnover was calculated as the number of entries plus the number of exits (i.e., the total number of changes) in an interval, divided by the number of species present in that interval. This will produce a value between 0 and 2, where 0 represents a situation where no species had either a first appearance or last appearance in the interval, and 2 represents a situation where all species present in an interval had their first and last appearances in that interval.

Ecomorphology

For these analyses, we used three ecomorphological variables: molar crown height, body mass, and interpreted diet at the most basic level ("diet-1" of the ETE design; Damuth 1993). Molar crown heights were only applied to non-rodent herbivores, and were scored subjectively as either brachydont, mesodont, or hypsodont, according to the criteria described in Damuth (1993). Of the taxa included, only ruminant and perissodactyl ungulates had values other than brachydont. The crown height for the ruminants was scored by Alan Gentry, for the perissodactyls by Mikael Fortelius.

The body mass estimates are derived from several sources: Fleagle (1988) for primates, Fortelius et al. (this volume) for suoids, and unpublished analyses by Mikael Forelius, Suvi Viranta, and Lars Werdelin for horses, rhinoceroses, and carnivores. These analyses are mainly based on regression equations given in Legendre and Roth (1988), Damuth and MacFadden (1990), Anyonge (1993), and Fortelius and Kappelman (1993). Ruminant body masses are based on nine subjective size classes, as assigned by Alan Gentry. Each class has a single mass value, based on the mean for a representative, living ruminant. Body masses were analyzed only in terms of group means, with groups defined variously in terms of time, region, taxonomy, and/or ecomorphology.

There are three problems with the use of body mass in reconstructing paleoecology: taphonomic bias, lack of statistical significance, and errors in size determinations for fossil species. One of the most common taphonomic biases in the fossil record relates to body size distribution in fossil faunas. Depositional conditions associated with large mammal accumulations are frequently very different from those associated with small mammal accumulations, and the result is that in many fossil faunas there is a bias in favor of one or the other. If this size bias is not detected, the size distribution of the faunas can give a misleading representation of the paleoenvironments from which the faunas were derived (Andrews 1990).

The second problem concerns the lack of clear correlation between body size distributions and paleoenvironments. In an analysis of four ecological and taxonomic categories for a series of recent and fossil African faunas, it was shown that body size was the one criterion that lacked significant association with any particular habitat type (Andrews et al. 1979). Faunas from a range of tropical African habitats were analyzed for taxonomic order, body size, spatial (locomotor) distribution, and dietary guild. The latter two showed highly significant differences (chi-square test, P >0.001) in faunal distributions in habitats ranging from tropical forest to woodland savanna and grassland; even the distribution by taxa was significantly different at >0.05, but the size distribution showed no significant differences from one habitat to another, although there was a low level of consistency within habitats. If faunas from recent habitats as different as tropical forest and grassland show no significant difference in their distribution of body size, the application of these criteria to fossil faunas is necessarily limited.

Third, body mass estimates obtained from most regression equations are imprecise, fundamentally because in most cases the underlying relationship is not in itself directly mass dependent, but also for several other reasons. The fact that a single adult individual's mass may vary substantially during its life (e.g., by season or reproductive phase) should caution against any attempt to obtain or apply very precise estimates—body mass studies simply

have an intrinsically low resolution, reflected in the 20% prediction error seen even in the "best" prediction equations (Van Valkenburgh 1990). As long as this fact is accepted, it does not matter much that individual estimates based on different equations may differ by 100% or more (e.g., Roth 1990). Seen in this perspective, the fact that the ruminant masses used here represent subjective class means rather than regression equations is not particularly worrying. A difference between 20 and 40 kg may be spurious, but the difference between 20 and 100 kg is almost certainly real, and that between 20 and 200 kg definitely so. The 352 body mass estimates used in this study range from about 200 g to about 1500 kg, with a mean of 134 kg (standard deviation 223, coefficient of variation 63). It is highly unlikely that individual estimate errors in the 100% range would bias the results in any serious or systematic way. For obvious reasons, species known from very incomplete material lack mass estimates. The proportion of macromammal taxa with body mass estimates in the material used in these analyses was 68%, with roughly equal coverage of the orders.

Dietary interpretation was the most difficult, and therefore we simply recognized only three categories: animal-eater, plant-eater, and omnivore. These were only assigned for macromammals. The category "omnivore" in these analyses comprises most suoids (essentially all except *Listriodon* and *Schizochoerus*) and a number of generalist carnivores (some ursids, some amphicyonids, and some mustelids). For unambiguous comparison with body mass analyses, only taxa for which body mass estimates were available were included.

Paleoecology

To supplement the ecomorphological analyses, we have attempted to infer past environments by taxon-free analysis of the more complete faunas in the data base. They are compared with recent faunas from Europe, Asia, and Africa (Andrews, in press) in terms of body size distribution, spatial distribution, and dietary guild. Body mass estimates are the same as those described earlier, but the size distribution was subdivided into eight categories following the criteria of Andrews et al. (1979). Spatial zoning criteria followed the classification of Harrison (1952; refined by Andrews et al. 1979). Dietary guilds were based again on Andrews et al. 1979) and are equivalent to the "diet-2" design of the ETE data base (Damuth 1993). Although body size distribution has been found to associate nonsignificantly with ecology in recent mammal faunas, it has been found to contribute usefully in multivariate analyses in conjunction with the other ecological criteria (Andrews, in press), and we have retained it here.

More detailed analysis will be forthcoming in future

publications, but in order to investigate some of the specific issues raised by the diversity analyses presented here, we have opted to present the ecological data as bivariate plots comparing distributions through time of various trophic and spatial criteria.

Results and Discussion

Sampling

The correlation between number of species and number of localities is very high (fig. 31.2a and b). For West it is 0.968 and for East 0.917, both highly significant (fig. 31.2a; df = 6, p <0.01). In part, these high correlations are due to the large range in number of species per interval. This is particularly true of East, where high species counts for the locality-rich MN 11 and MN 12 raise the correlation coefficient considerably (fig. 31.2b). On the face of it, these diagrams and correlations indicate that the number of species is almost entirely dependent on the

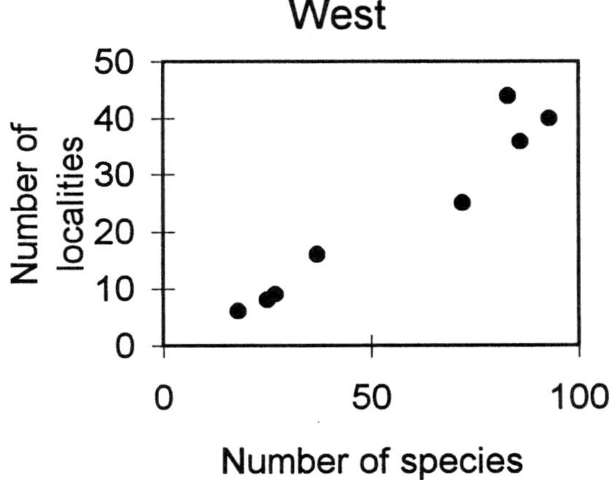

West

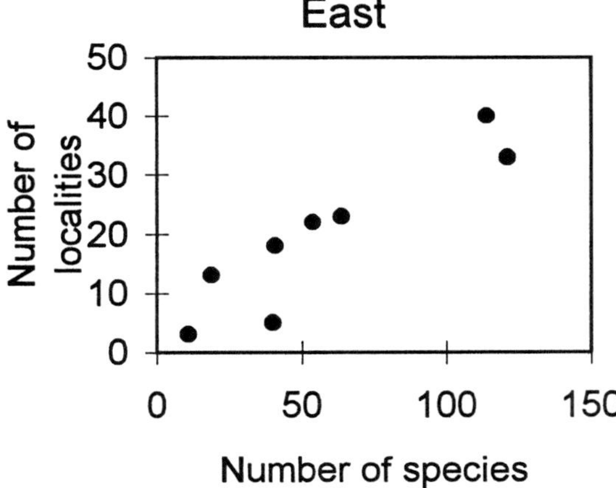

East

FIGURE 31.2 Correlation between number of species and number of localities. A: West. B: East.

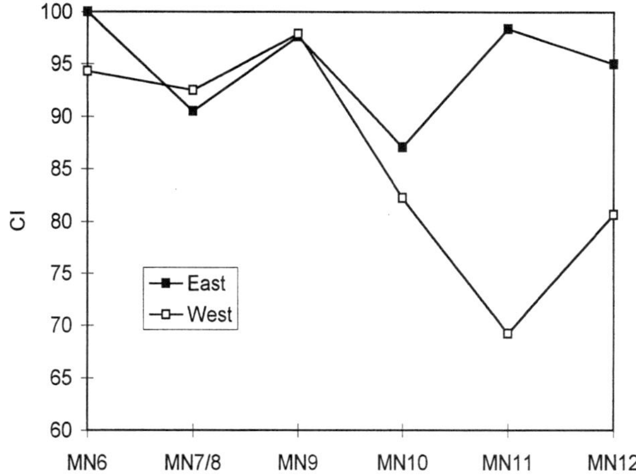

FIGURE 31.3 Plot of the Completeness Index (CI) for each block and MN unit.

number of localities sampled in an interval. At the same time, however, the number of species found is also dependent on the presence or absence of "Rosetta Stone" localities (Alroy 1992; cf. above). The presence and absence of such localities may in turn reflect a real presence or absence of taxon-rich environments. This means that the high correlations found between number of species and number of localities may equally well reflect a biological reality as a spurious effect of sampling and cannot be taken at face value.

In order to examine the question of sampling more closely we calculated the completeness indices of Krause and Maas (1990) and Maas et al. (1995). The results for CI_1 in West and East are shown in figure 31.3. It can be seen that the index is generally high (between 85 and 100) for East and somewhat more variable for West. If we use the (admittedly arbitrary) criterion employed by Maas et al. (1995) of considering intervals with a CI of less than 70 to be poorly sampled, only MN 11 in West stands out. Most significant, there is no consistent correlation between low diversity and poor sampling, which is clearly seen for East, where low pre-Vallesian diversity is nevertheless accompanied by high CI values.

A third way of viewing sampling is by looking at a rank diagram of the number of species per locality for each MN unit (figs. 31.4–31.5). This will give insights into whether a certain interval lacks localities with many taxa ("Rosetta Stone" localities) or lacks a long tail of localities with just a few taxa, or both. The diagram for West (fig. 31.4) shows that all intervals basically exhibit the same pattern. Intervals with low diversity have fewer localities, as noted above, and in particular they lack the one or two localities with many taxa that provide the greatest amount of the diversity in the taxon-rich intervals. In the diagram for East, on the other hand, there are some intervals that differ from the

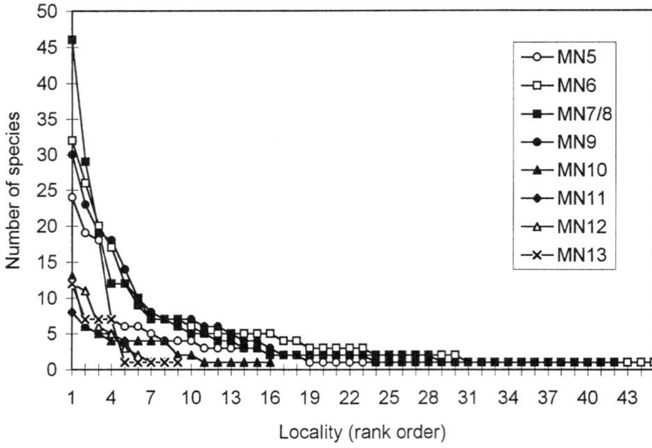

FIGURE 31.4 Rank diagram to illustrate the contribution of individual localities to the total species richness of the MN units in the West block.

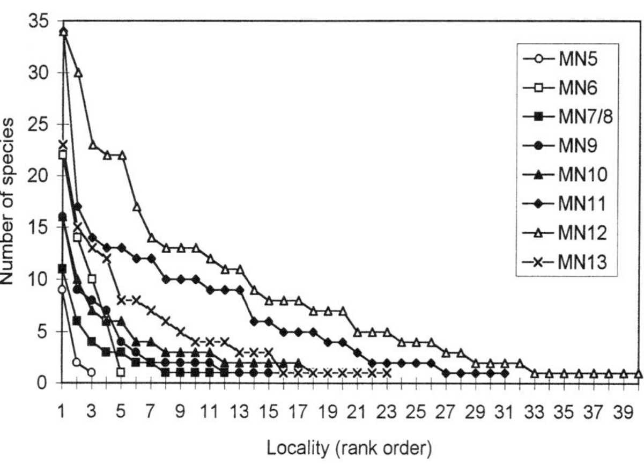

FIGURE 31.5 Rank diagram to illustrate the contribution of individual localities to the total species richness of the MN units in the East block.

others in their rank order pattern (fig. 31.5). These are especially MN 5 and MN 6. The first includes only three localities and must be considered inadequately sampled (MN 5 is in any case not a significant component of most analyses). There are only five localities known from MN 6, but they tend to be relatively diverse, especially the largest (Paşalar and Çandir). This contributes to making the rank order pattern different for MN 6 than for any other interval, as there is no indication of a tail to this distribution.

The fact that there are such species-rich localities in East in MN 6 might lead one to suspect a large increase in diversity in this interval compared to those surrounding it. There is such an increase (fig. 31.5), but it is not particularly great and does not obscure the overall pattern of post-Astaracian diversity increase in the eastern block. That the diversity signal obtained from the data is not obscured by the very different sampling regimes of MN 5, MN 6, and

MN 7/8 in East is an indication that sampling adequacy is not a problem with regard to the overall results obtained from the data base.

Provinciality

As a first step in the analysis, the original list of countries included in the data base was grouped into six regions. These were: Western Europe (WE) (Portugal, Spain, France, Italy); west Central Europe (WCE) (Germany and Switzerland); Austria (AUS); the Black Sea (BS) (Hungary, Romania, Moldova, Ukraine); the Balkans (BAL) (Slovenia, Croatia, Bosnia, Serbia, Macedonia, Albania, Greece except Samos, Turkish Thrace, and Bulgaria); and Anatolia (ANA) (Anatolian Turkey, Samos, and Georgia). The rest of Central Europe (Poland, Czechia, and Slovakia) and Western Asia (Iran, Afghanistan, Georgia, and Kazachstan) were too unevenly represented for meaningful comparisons with other regions.

Dice similarity indices were calculated between these regions. The comparisons that included adequate data are shown in figures 31.6 and 31.7. The first of these diagrams (fig. 31.6) shows some comparisons with western regions (WE and WCE) as baselines. The results indicate that WE, WCE, and AUS generally have higher similarities with each other than with other regions, which is an indication that they can be considered to belong to a single faunal block. The second diagram (fig. 31.7) shows comparisons using eastern regions (BAL and ANA) as the base-

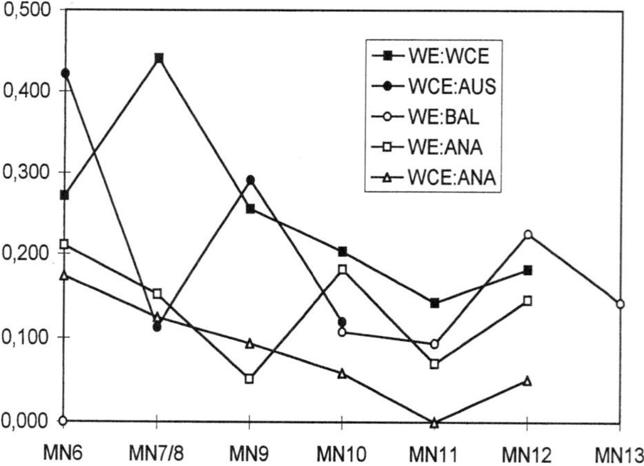

FIGURE 31.6 Plot of the Dice index of faunal similarity of various subregions compared with standards from West. The more distant the regions, the lower the index and the less noticeable the decrease in similarity during the early part of the interval. ANA: Anatolian Turkey, Samos, and Georgia; BAL: Slovenia, Croatia, Bosnia, Serbia, Macedonia, Albania, Greece except Samos, Turkish Thrace, and Bulgaria; BS: Hungary, Romania, Moldova, and Ukraine; AUS: Austria; WE: Portugal, Spain, France, and Italy; WCE: Germany and Switzerland.

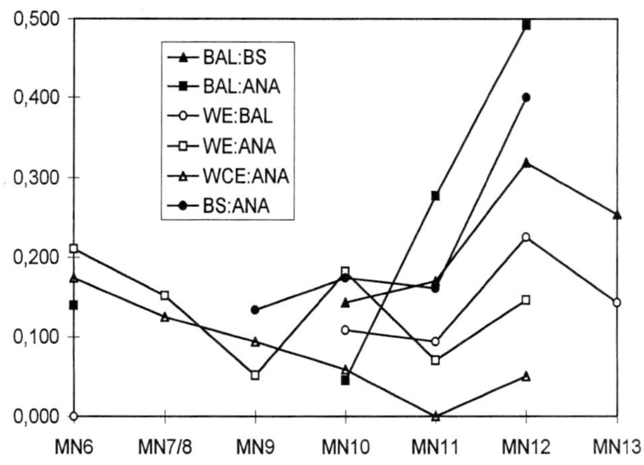

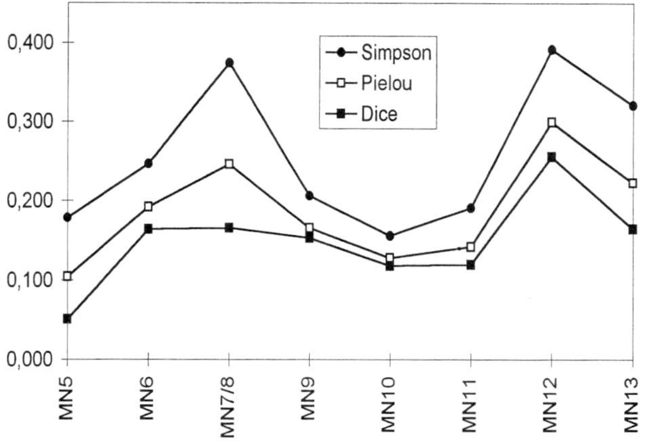

FIGURE 31.7 Plot of the Dice index of faunal similarity of various subregions compared with standards from East. The effect of increasing similarity seen in MN 11–12 is strongest between the geographically closest regions, reflecting a differential penetration of Turolian immigrants from East toward West. See figure 31.6 for abbreviations.

FIGURE 31.8 Mean similarity indices calculated from all the pairwise comparisons that were possible between subregions for each MN unit.

line. Few data are available before MN 10 and after MN 12, while during this interval there is a strong correlation between the similarity index and the geographic distance between the regions compared. The greatest difference was between WCE and ANA, and the smallest between BAL and ANA, with WE:ANA, WE:BAL, BAL:BS, and BS:ANA in that order between the extremes. Thus, the eastern regions showed greater similarity to each other than to more distant regions, just as did the western regions.

Figure 31.6 reveals a general decline in similarity of all faunas compared with WCE. Exceptions to this general decline include: WCE:AUS at MN 9, the base Vallesian "*Hippotherium* Datum" when a number of East Asian derived taxa extended their ranges into Central Europe; WE:ANA at MN 10, possibly due to the evolutionary radiation burst in Southeastern Europe–Southwestern Asia that extended some additional taxa into Central Europe; WE and WCE with all combinations during MN 12, when the Turolian open country faunas underwent their maximum geographic expansion, even in those areas that likely retained less seasonal habitats. Figure 31.7 reveals a general decline from MN 6 to MN 7/8, a general slow rise in MN 9 and MN 10, and a sharp rise in all regional relationships between MN 11 and MN 12. This again reflects the evolutionary and migratory pulse of East and Southwest Asian open country herbivore and carnivore faunas, which predominate in most of Eurasia in the later Miocene.

Overall, the mean inter-regional similarity index fluctuated without a clear trend over the period investigated, as shown in figure 31.8. While we have concentrated on the Dice index, it should be noted that the Simpson and Pielou indices do reveal the same trends, but more dramat-

ically. The lowest indices are seen at MN 5, an interval correlated with the late Burdigalian regression (Bernor in Steininger et al., this volume), when the first strong Neogene migratory pulse of African mammals into Western Eurasia faintly overprints a largely endemic European fauna. A distinct rise occurs in MN 6, accompanying a second strong interprovincial faunal extension of African and Asian elements into Western Eurasia. Similarities in faunas decrease through MN 9 and MN 10, with migrations of East Asian fauna into East, where immigrants and their descendant species replaced more archaic forms. Beginning in MN 10, and culminating in MN 11–12, these forms extended into West, so that faunal similarity reaches a maximum in MN 12. Whether the immigrants were replacing older taxa through competition or entering niches vacated by extinction is difficult to judge with the coarse time resolution available; in a few cases (e.g., replacement of *Listriodon* by *Schizochoerus*) replacement after extinction appears more likely (Fortelius et al., this volume). What occured in MN 13 is somewhat obscure (Bernor et al., this volume a), but the net result was decreased mean faunal similarity, probably related to biological fragmentation of the area by the extensive regressions of the Mediterranean and Paratethys systems (Rögl and Daxner-Höck, this volume; Bernor et al., this volume c). As a whole, these trends indicate that similarity decreases during intervals of high extinction (e.g., MN 7/8–MN 10, MN 13) and rises during intervals with high interregional immigration (e.g., MN 12)—another indication that replacement was not primarily through direct competition.

On the basis of inter-regional similarity comparisons and inspection of the geographic distribution of taxa, we separated the study area into the two main blocks (fig. 31.1) discussed earlier. The similarity indices between these blocks were calculated for each MN unit and are

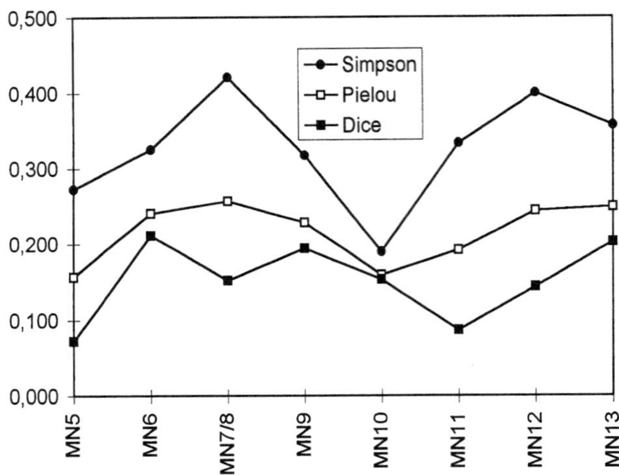

FIGURE 31.9 Similarity indices for West-East block comparisons for each MN unit.

shown in figure 31.9. These curves are very similar to those seen in figure 31.8, indicating that our grouping of the regions is correct and has not distorted the observed patterns in any significant way.

Diversity

Species richness of the two main blocks, East and West, exhibits two completely different trends (fig. 31.10). In West, species richness is high throughout the Astaracian and early Vallesian, slowly rising from MN 5 until MN 9. The mid-Vallesian crisis (Moyà-Solà and Agustí 1989) is seen as a sharp drop from MN 9 to MN 10, continuing into MN 11, after which species richness remains low, at about 25% of the precrisis level. The difference between MN 5–9 and MN 10–13 is significant (Mann-Whitney U-test, $P = 0.021$). East shows a pattern that is roughly opposite. Species richness is initially low, though slowly rising from MN 5 until MN 10, with a minor peak in the well-sampled MN 6. From MN 10 to MN 11 the curve for East rises abruptly to more than double richness, after which there is a decrease, especially from MN 12 to MN 13. This difference is also significant (Mann-Whitney U-test, $P = 0.021$). The differences between East and West are also significant at the same level for each of these intervals, but not for the entire period. The difference between MN 5–9 and MN 10–13 for the entire area is also not significant. Standing richness shows the same pattern (fig. 31.10).

For the individual regions defined above, the curves come in two types, corresponding to West and East (figs. 31.11 and 31.12). Western and west Central Europe both show versions of the West pattern (fig. 31.11), with a continuous and rapid decrease in richness from a peak at MN

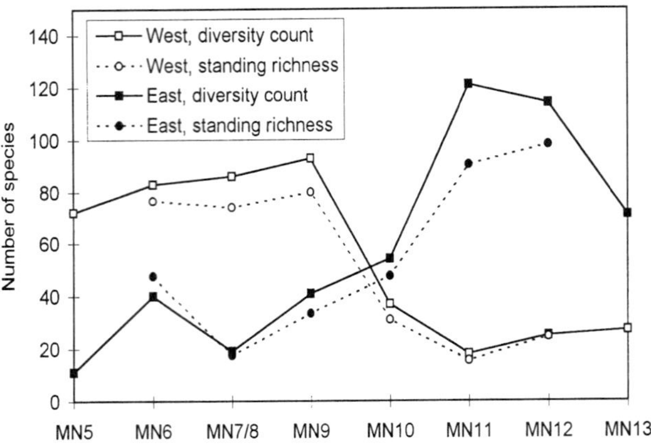

FIGURE 31.10 Diversity (species richness) and standing richness of the West and East blocks during the study interval.

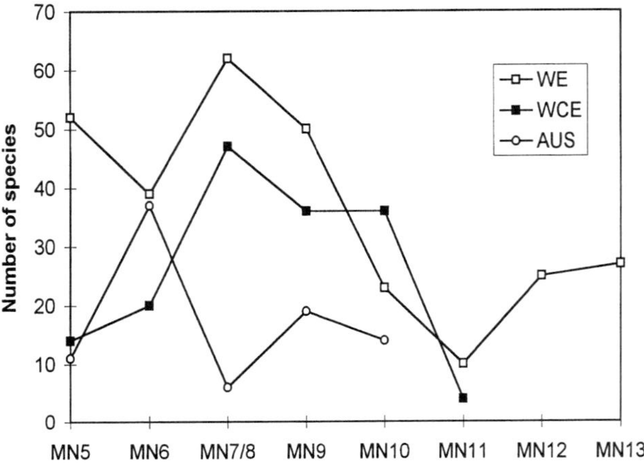

FIGURE 31.11 Diversity (species richness) for subregions of the West block. See figure 31.6 for abbreviations.

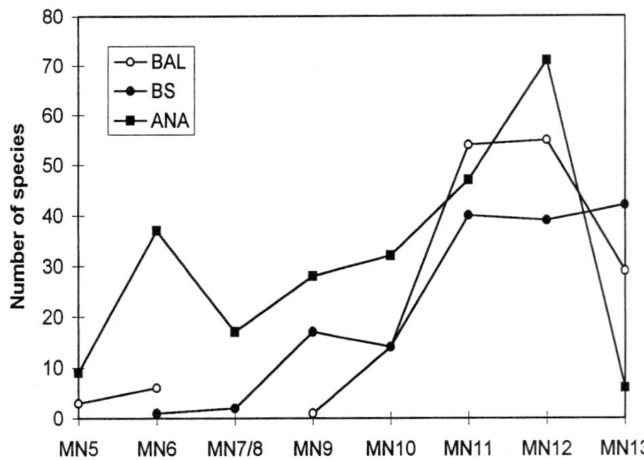

FIGURE 31.12 Diversity (species richness) for subregions of the East block. See figure 31.6 for abbreviations.

7/8 to a minimum at MN 11. The main drop occurs between MN 9 and MN 10 in Western Europe, while west Central Europe shows no change at this time. The meaning of the drop from MN 10 to the poorly sampled MN 11 is difficult to judge, as it is almost certainly at least partly an artifact. Austria shows a somewhat different pattern that is difficult to interpret owing to the absence of data after MN 10. The existing part actually resembles the Anatolian curve more than it does the European ones. The species richness curves for subregions of East are highly congruent (fig. 31.12), and the differences between them are due mainly to (or at least indistinguishable from) sampling error. The very low levels of the Black Sea and the Balkans before the Vallesian and the very low level of Anatolia in MN 13 both correspond to a few, poor fossiliferous localities. For the summed data matrix the completeness is quite high, however, as noted above. The artifacts thus clearly affect details but do not account for the pattern itself.

These results reflect the events already seen in faunal similarity: a general decline in species richness in West from MN 7/8 accompanying the contraction of subtropical habitats, with a sharp decline thereafter (MN 9–11), followed finally by an increase of open country large mammal diversity in Western Europe during MN 12–13. East, undergoing a steady immigration of East Asian elements, and the provincial evolutionary radiation of open country large herbivores and carnivores, exhibits a striking increase in species richness from MN 9 to MN 12, followed by a decline in MN 13, probably related to the regional regression of the Paratethys and the Mediterranean, and the consequently increased seasonality and decreased habitat (beta) diversity.

It would be illuminating to compare these patterns with what is known from elsewhere, but unfortunately very few studies using species-level analyses as justifications of whole faunas have been published. Van Valkenburgh and Janis (1993) do, however, provide roughly analogous diversity data for North America, covering most of the continent except for the east coast north of Florida. For purposes of comparison we assigned an absolute age to each MN unit, taken as the midpoint between the boundaries (based on the emerging consensus of this volume as interpreted by Bernor in October 1994).

Figure 31.13 shows that the North American curve is strikingly similar to that for West. In both areas, diversity is high and stable for an extended interval, until about 10–8 Ma, when it rapidly drops to a significantly lower level. Both events might well be variously delayed effects of the Serravallian global sea-lowering at 11 Ma, or rather exacerbated by the global sea-lowering events that occurred at 8 Ma (near base MN 12 correlative) and again later at 7.1 Ma (base MN 13 correlative). Regionally, these sea-lowering events are reflected in the progressive regression

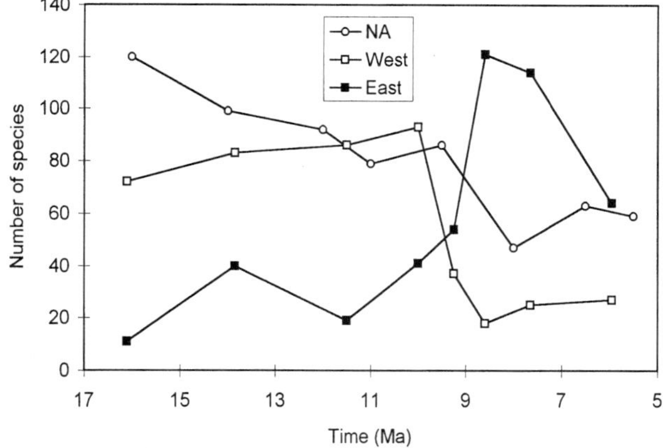

FIGURE 31.13 Diversity (species richness) of the West and East blocks compared with corresponding data for North America (from Van Valkenburgh and Janis 1993).

of the Paratethys and most probably coupled with orogenic pulses of the Alpine-Himalayan system that increased the relief and altered climates of the Southeast European and Southwest Asian areas. In East, late Miocene faunas were very evidently subjected to increased seasonality, which promoted the immigration of East Asian taxa as well as the provincial diversification of typical "Pikermian" faunas (Bernor et al., this volume a; Fortelius et al., in press). Whether this was the case cannot, however, be settled from diversity data alone. The (somewhat hypothetical) synchroneity of all or some of these changes could also be coincidental, and the causes in each case local, or the changes could even be partly artificial. For our immediate objective of evaluating and understanding the differences between East and West, the best (and perhaps only) option seems to be to investigate the details of turnover and ecology.

Turnover

Percent entries and exits are given in separate diagrams for West and East (figs. 31.14 and 31.15). Both blocks exhibit sharp entry peaks during MN 9, which, in association with the decreased faunal similarity, suggests that local speciation was the dominant source of new taxa at this time, at least in West. As noted, this period also corresponds to the terminal Serravallian sea-lowering event, which would have projected climatic effects over the greater portion of Eurasia. The second time of high entry in West is MN 12, and clearly corresponds to entry of taxa from East, as testified by elevated similarity. In West, percent exits shows a decrease from MN 6 to MN 7/8, an increase in MN 9 (at the *Hippotherium* Datum), a decrease in MN 10–11, and an increase in MN 12 (during explosive evolution of open country large herbivore and

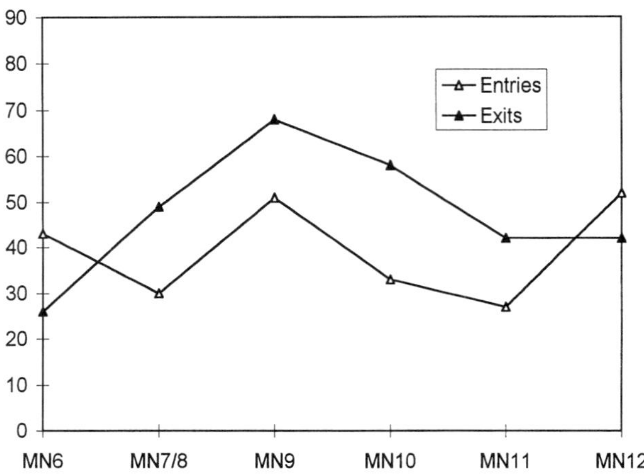

FIGURE 31.14 First and last appearances of taxa in the West block as a percentage of species richness.

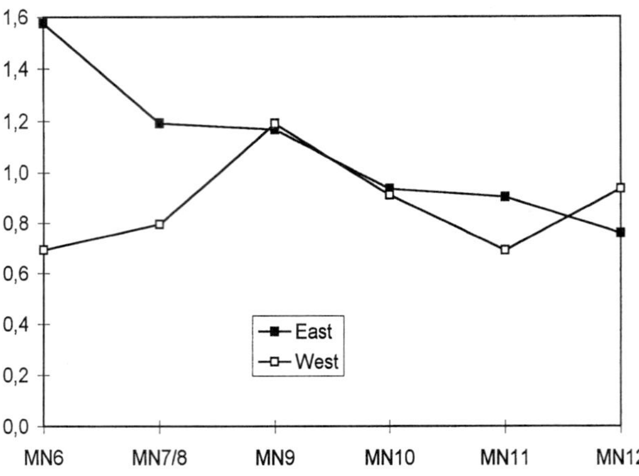

FIGURE 31.16 Relative turnover (summed first and last appearances divided by species richness) in West and East.

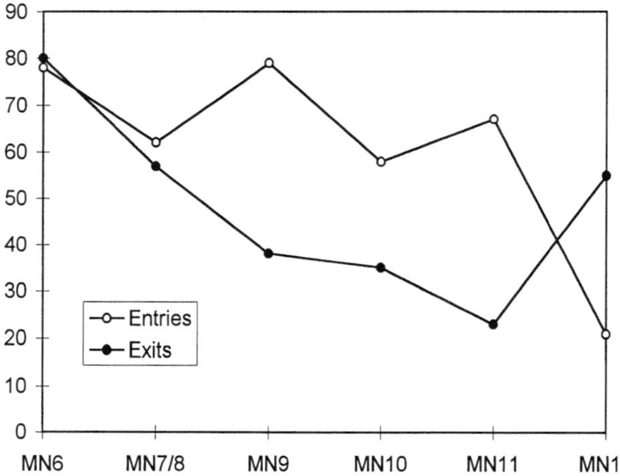

FIGURE 31.15 First and last appearances of taxa in the East block as a percentage of species richness.

carnivore faunas). Besides the MN 9 event, East shows a decline in entries during MN 10, and another entry pulse in MN 11, most probably related to further East Asian immigrations and the local evolutionary radiation of open country forms. East's general decline of exits from MN 6 to MN 11 is plausibly related to the progressive expansion of seasonal, open country habitats. The feature to note about East, however, is that there are more entries than exits throughout the interval when species richness increases in East (MN 7/8–MN 11). This reinforces the interpretation that increased species diversity was due to migration plus evolutionary radiation of open country taxa.

The patterns of entries and exits are summed in the diagram for relative turnover (fig. 31.16). Here again West and East are seen to exhibit very different patterns. Relative turnover in West shows an initial increase then declines, followed by an increase from MN 11 to MN 12. Seen over the entire period considered here, there is no significant change in relative turnover. In East, on the other hand, relative turnover decreases steadily from MN 6 to MN 13, and is on aggregate approximately halved during this time period, despite the diversity increase in this area during the same period. Thus, while one might expect that increased diversity should entail increased turnover, the opposite seems actually to be true. Furthermore, the higher turnover in East in MN 6 and MN 7/8 is brought about by high percentages of both entries and exits. The subsequent decrease in turnover follows the decrease in percent exits in East, suggesting that while diversity changes in East are entry-driven, turnover patterns are exit-driven. This might reflect, in an averaged and cumulative manner, the basic circumstance that the entry of new species into a local ecosystem, whether by immigration or speciation, is inherently a diffuse, successive phenomenon, while extinction due to an external event will take the form of a distinct pulse. Especially in an unconstrained, nonequilibrium situation, such as the expansive East, entries potentially provide volume, and exits, rhythm.

Entry-driven faunal aggregation with decreasing turnover is exemplified by the Paşalar fauna (sensu Bernor and Tobien 1990), which evolved in part by retaining a series of successful immigrants. Subsequent to their entry, these immigrants underwent evolutionary diversification and were further augmented by later immigrants. This process, continued into the late Miocene, resulted in an increased diversity, and decreased relative turnover in East: the open country "species-factory" proposed by Bernor et al. (this volume, b) .

Micromammals

The micromammal data are different in nature and coverage from the large mammal data, and are discussed separately here. There is a marked difference between the rich micromammal record of West and the much poorer one of East (fig. 31.17). There is no doubt that the difference is partly artificial, being due to the relatively short history and small volume of small mammal research in East. It should, however, not be dismissed out-of-hand as pure artifact. Indeed, the fact that the eastern species richness curve repeats the basic pattern of no diversity increase in the later Miocene seen also for the eastern brachydonts and the <30 kg class of macromammals suggests that it probably reflects the same differences.

Because of the uneven geographic coverage of the available material, detailed comparisons between micromammals and macromammals were only drawn from Central Europe (Germany, Switzerland, Austria, Czechia, Slovakia, and Poland). Adequate data were only available for MN 5–MN 11. The species richness pattern is quite similar for the macromammals of the same area, especially from MN 5 to MN 9; curiously, even the absolute numbers of species are closely matched (fig. 31.18). There is first a marked rise from MN 5 to MN 6, followed by a gradual decline into MN 9. After this the curves depart, macromammals showing the familiar mid-Vallesian drop into MN 10 while micromammal diversity increases. Both curves show a decrease from MN 10 to MN 11, but the micromammal curve now has higher absolute values than the macromammal curve. That the micromammals of Central Europe crashed in MN 10 rather than in MN 9, as the fauna of West generally, suggests this region responded differently. The macromammal fauna as a whole would not be expected to reflect the difference as clearly as the ecologically more sensitive small mammals.

The large number of MN 9 macromammal first and last appearances are reflected in a strong relative turnover

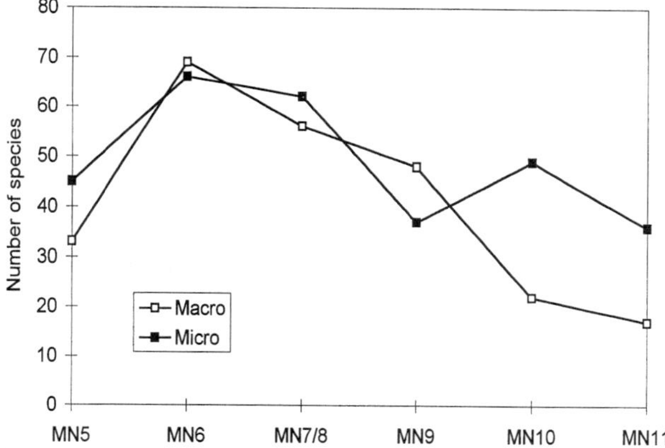

FIGURE 31.18 Micromammal and macromammal diversity in Central Europe (Germany, Switzerland, Austria, Czechia, Slovakia, and Poland). Note the increase in micromammal diversity after MN 9, in marked contrast to the universal diversity crash seen in West at this time.

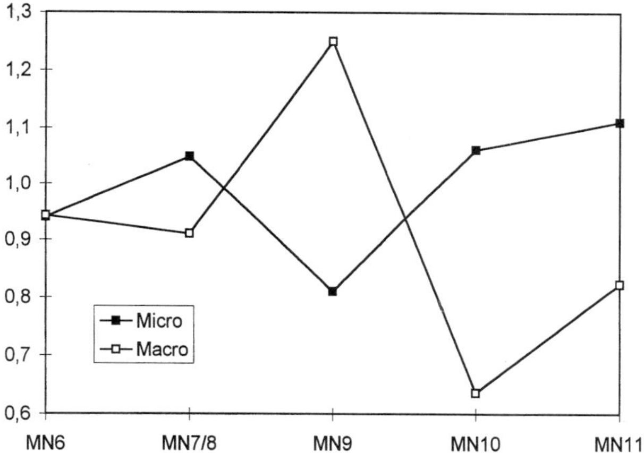

FIGURE 31.19 Relative turnover of micromammals and macromammals in Central Europe (see fig. 31.27). Note opposite overall trends and antiphasic disposition of the curves.

peak, followed by a greatly reduced relative turnover in MN 10 (fig. 31.19). This is essentially an exaggerated version of the turnover pattern for the entire West block (fig. 31.16). The pattern for micromammals is completely different, with lower relative turnover in MN 9 than in any other time period studied, followed by increase through MN 10 and MN 11. As a general statement, macromammal turnover decreases during the time interval (except for MN 9), while turnover increases for micromammals (again excepting MN 9).

Paleoecology

We have approached paleoecology mainly in terms of ecomorphology, i.e., of morphological properties that can

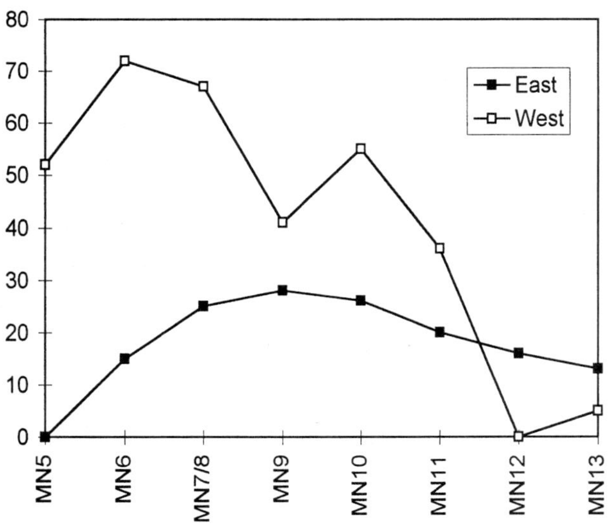

FIGURE 31.17 Micromammal diversity (species richness) in West and East.

be related to ecology directly, rather than via the uncertain route of taxonomy (Andrews et al. 1979; Damuth 1992). We present results of two somewhat different analyses, using slightly different data sets. The first analysis uses the same data base as the analyses of provinciality, diversity, and turnover presented above, and includes taxon/locality occurrences regardless of how complete or incomplete the faunal lists for individual localities are. The second set, which follows the basic methodology of Andrews et al. (1979), is restricted to selected faunas with reasonably complete faunal lists (for both small and large mammals). We refer to the former as "ecomorphology," the latter as "community paleoecology."

Ecomorphology

Ecomorphological study of the faunas has a dual purpose. First, it may provide ecological information about the faunas themselves. Second, it should function as a consistency check on the patterns seen in diversity, provinciality, and turnover. If the patterns make no ecological sense one may suspect that they are spurious. The opposite does not, of course, prove that the patterns are real, but it does mean they have at least passed one reasonable test. As the number of groups and independent (or, as is more often the case, semi-independent) parameters studied increases, the test becomes progressively more robust by congruence of pattern.

Body Mass

A first step in paleoecological analysis could be to explore changes within arbitrarily defined size classes. For this purpose, four size classes of macromammals were defined: <30 kg; 30–100 kg; 100–500 kg; and >500 kg. The boundaries were selected to provide classes with roughly equal numbers of species, except for the heaviest class, which was chosen specifically to include only very large mammals (rhinoceroses, giraffes, and a few giant suids). In West, the richest class by far until MN 10 is the lightest one, which also is the one most affected by the diversity crash from MN 9 to MN 10 (fig. 31.20). The 30–100 kg class is also strongly reduced at this time. The class least affected is the heaviest one, while the only western class to show some recovery in the Turolian is the 30–100 kg class. The eastern pattern differs in that all classes show a rather similar, slow increase from MN 5 to MN 10, after which all but the >500 kg class rise sharply, roughly double species richness in MN 11 (fig. 31.21). The greatest eastern increase is seen in the 100–500 kg class, the smallest in the 30–100 kg class. Both of these intermediate size classes show some decline in eastern species richness during the later Turolian, but the <30 kg class does not.

Pairwise comparison of these size classes by block re-

veals further patterning. Mean size in the <30 kg class is on average about 3.5 kg smaller in West (5.7 vs. 9.2 kg). The difference is consistent throughout the interval (Mann-Whitney U-test, P = 0.006; see fig. 31.26). The only exceptions to this rule are MN 5 and MN 7/8, which are poorly sampled for East (in MN 5 eastern taxa are much bigger, in MN 7/8 there is no difference between East and West). No significant difference between the blocks is seen in the 30–100 kg class, although the western taxa are larger in the middle of the study interval, eastern taxa at the beginning and the end. No directional change in size is seen in this class. For the two heaviest classes

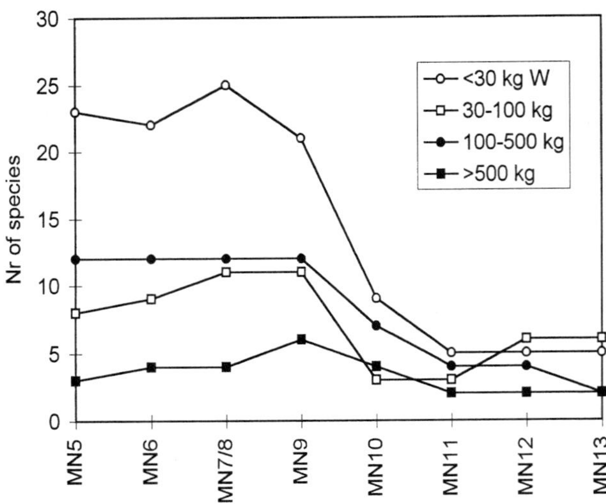

FIGURE 31.20 Diversity (species richness) in arbitrarily defined size classes in the West block. Note early predominance and strong reduction of the two smallest size classes included (macromammals <30 kg and 30–100 kg).

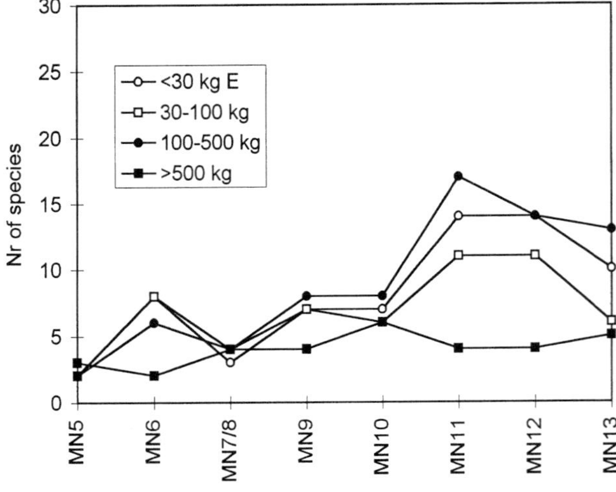

FIGURE 31.21 Diversity (species richness) in arbitrarily defined size classes in the East block. The main increase in MN 10–11 was in the two medium classes.

mean size is very similar in West and East, as is the common trend of minor increase in size (from about 160 to about 220 kg and from about 1,000 to about 1,250 kg, respectively).

Molar Crown Height

A breakdown of herbivores by molar crown height allows comparison of rough dietary categories, probably defined mainly by the relative fiber content of their food. The use of only the three states of brachydont, mesodont, and hypsodont provides a picture that is certainly very coarse, but relatively easy to interpret.

The proportions of the three crown height categories varies between the blocks. In West, the proportion of hypsodont herbivores is minuscule in MN 5 and zero in the later Astaracian (fig. 31.22). In the Vallesian of West, hypsodont forms constitute up to 10% of the large herbivore fauna (in MN 10), but no hypsodont forms are present in MN 11, and the proportion in the Turolian of West never exceeds the maximum of about 20% reached in MN 12. The proportion of mesodont forms in West is about 15% until MN 11 and increases sharply to about 40% in the Turolian, when the proportion of brachydont forms falls dramatically. In absolute terms, the number of mesodont and hypsodont forms decreases slightly, but there was a far greater reduction in numbers of brachydont species. This indicates that the diversity decrease from MN 9 to MN 11 (fig. 31.24) was mainly of animals dependent on low-fiber foods.

In East, the proportion of hypsodont herbivores is about 30% in MN 5 and MN 6, and then falls to less than 10% in MN 7/8 and MN 9, but regains the level of about 30% in the Turolian faunas (fig. 31.23). The Vallesian in East is characterized by a relatively high proportion of brachydont taxa, while the proportion of mesodont taxa generally increases throughout the interval, from about 10% in MN 5–6 to about 30% in the Turolian in East. In absolute terms, all dietary categories increased their species richness, roughly in proportion to the general diversity increase (fig. 31.25). In contrast to West, the diet-related composition of the East herbivore fauna shows only gradual changes during the interval studied.

Over the entire area, there is a trend of increasing molar crown height, indicating greater dependence on high-fiber foods (graze and/or fibrous browse). In these terms, the main differences between East and West is twofold: the western herbivores appear less adapted to high fiber diets throughout the interval, and the western low-fiber adapted herbivores suffered disproportionately in the general diversity crash in the mid-Vallesian. Similar results have been previously interpreted by Fortelius et al. (in press) to indicate that the eastern biota was to some extent preadapted to the environmental effects of the global climatic changes

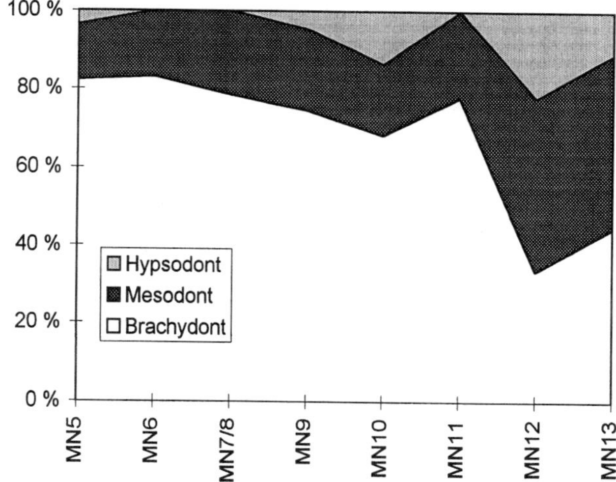

FIGURE 31.22 Composition of the ungulate fauna of West by molar crown height. Brachydont forms dominate until MN 12, when there is a sudden increase in the other groups, especially in the mesodonts. Compare with figure 31.21.

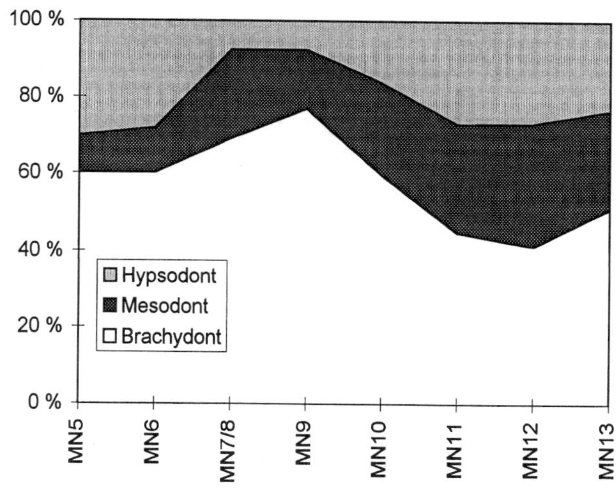

FIGURE 31.23 Composition of the ungulate fauna of East by molar crown height. Brachydonts are the largest group but relatively less common than in West except for MN 12, and hypsodonts are relatively more common. Compare with figure 31.22.

that occurred in the middle Vallesian and thrived on them, but that the western biota were not, and therefore experienced a crisis when the landscape became more open.

Crown Height and Body Mass Combined

Relating the crown height patterns to body mass serves to resolve the picture further (fig. 31.26). In West, the average body mass of brachydont herbivores was around 30 kg in the Astaracian, with an increasing trend from MN 7/

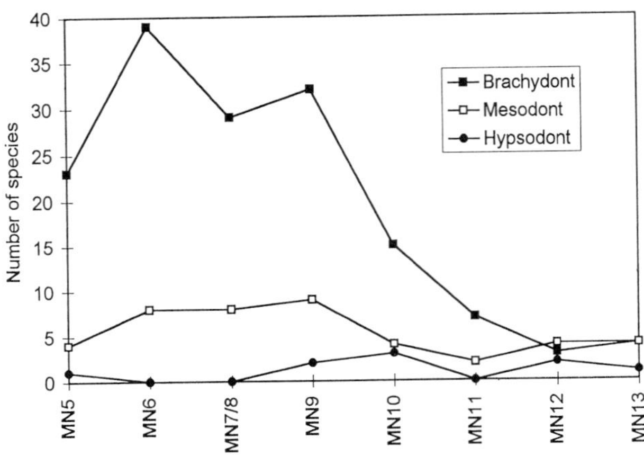

FIGURE 31.24 Diversity (species richness) of ungulates by molar crown height in West. Note absolutely large numbers of brachydonts until MN 9 and the subsequent crash of this group. Compare with figure 31.19.

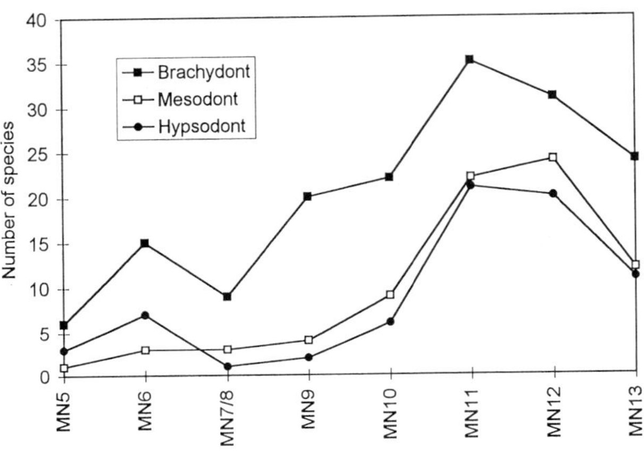

FIGURE 31.25 Diversity (species richness) of ungulates by molar crown height in East. All groups show broadly similar curves, and all increased in diversity from MN 7/8 to MN 11. Compare with figure 31.20.

8 to MN 11, concomitant with a decrease in the species richness of this group during the same period. This suggests disproportionate extinction of small brachydont forms in West from the Vallesian onward. The body mass curve rises gradually to a peak of about 63 kg in MN 11, and then falls quite steeply through MN 12 to about 30 kg in MN 13. In East, this figure was roughly constant, at approximately 100 kg, during the study interval, with some fluctuations. Note that eastern brachydonts remained larger than western ones throughout (mean size 108 kg vs. 55 kg; Mann-Whitney U-test, P = 0.006).

It follows from basic physiological considerations (especially the scaling of metabolism) that larger herbivores are better equipped, all else being equal, to obtain their energy from fibrous foods with long digestion times (e.g., Bell 1969, 1971; Janis 1976, 1982). It is also empirically well

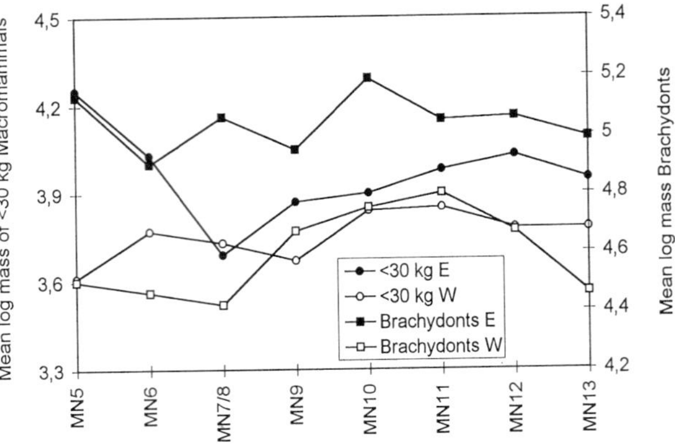

FIGURE 31.26 Mean body mass of small macromammals and brachydont non-carnivores in West and East. Throughout the interval, both groups maintained a significantly larger mean size in East.

established that forest communities are characterized by the presence of large numbers of small species, whereas this component is variable in open habitats (see Andrews, in press, for a review). The western increase in mean size of brachydont ungulates and the fact that eastern brachydonts were bigger than western ones throughout the interval (fig. 31.26), taken in conjunction with the drastic decrease in the diversity of brachydont herbivores (fig. 31.24) and of all mammals of less than 100 kg body mass (fig. 31.20), supports the interpretation that the environmental change was toward more open and probably more seasonal habitats.

Dietary Category and Body Mass

The rough dietary categories of animal-eater, plant-eater, and omnivore can be used in a similar manner to crown height. Indeed, the omnivore category overlaps to a major extent with "brachydont herbivore," but differs in excluding specialized folivores and including omnivorous carnivores. These categories are obviously further removed from the raw data than simple parameters like crown height, but are potentially more informative ecologically. Again, distinct differences are found between East and West, and between the categories.

The most obvious pattern is the familiar one of inverse diversity trends between East and West, seen for all groups except the omnivores of East (fig. 31.27). The western animal-eaters exhibit high species richness until MN 9, after which it drops precipitously, to less than half in MN 10. Western omnivores were highly diverse in the Astaracian, and did not recover from the reduction of 15 to 3 taxa between MN 7/8 to MN 10. In contrast, omnivore species richness in East never exceeded 5 taxa, and the increase in eastern plant-eaters and animal-eaters during the Vallesian and early Turolian is not reflected in this group. This pattern supports the idea, emerging from sev-

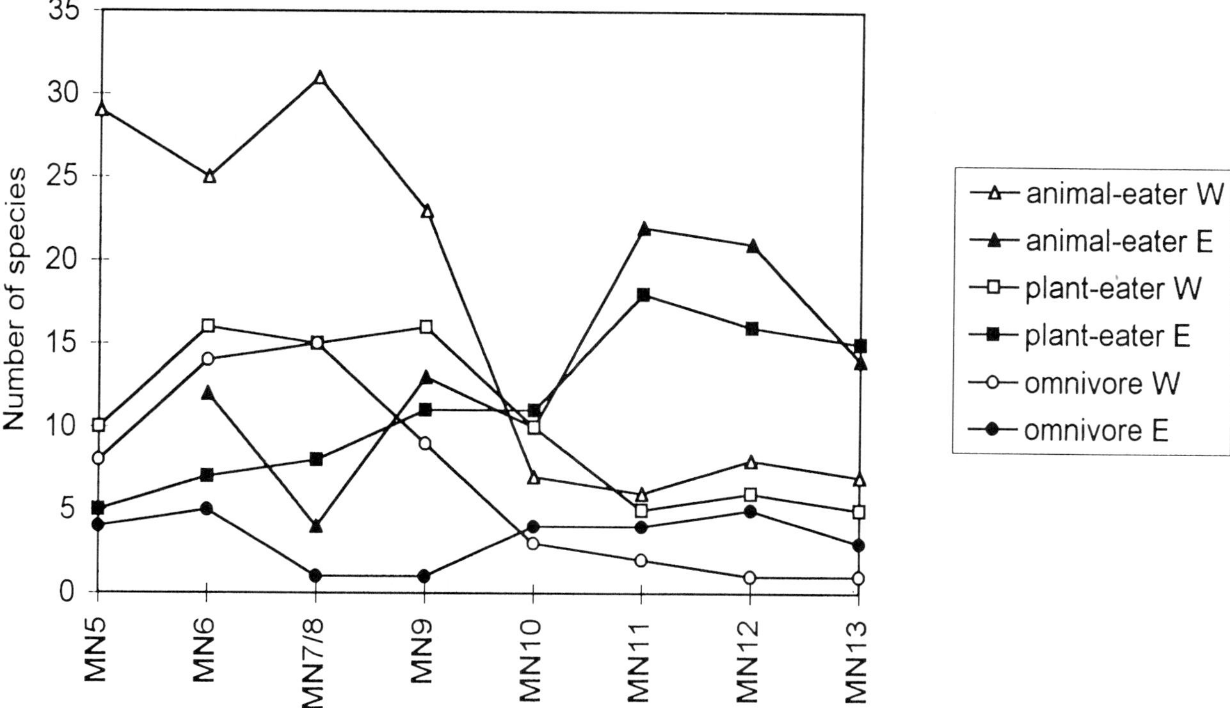

FIGURE 31.27 Diversity (species richness) of dietary categories in West and East. The groups that suffered worst in the crash of West were the animal-eaters, and this was also the group that responded most strongly to the diversity increase of East. The group that was least affected by the changes was the omnivores of East.

eral lines of evidence presented above, that the crash observed in West was linked to a disappearance of closed habitats providing a year-round supply of low-fiber (plant) foods, while the eastern diversity increase had a different, perhaps more complex, resource base.

Analysis of the body mass trends associated with these patterns reveals similar results to those obtained from previous analyses (fig. 31.28). In West, all groups show a size increase, preceding and overlapping with the diversity crash of the mid-Vallesian. In East the only group to show a sustained trend are the animal-eaters, which became gradually larger throughout the interval; no fluctuations are seen, as between MN 7/8 and MN 9 for the western animal-eaters. For all these categories, the eastern animals were larger than the western ones before the Vallesian. By MN 9 (animal-eaters and plant-eaters) or MN 10 (omnivores) there are no identifiable block differences, due to increasing size in West. During the Turolian no significant differences in mean size of these dietary groups are seen between the blocks. There is, here and elsewhere in the data, an impression of progressive east-west change: West, often abruptly, appears to "catch up" with East.

Community Paleoecology

As an addition to the ecomorphological study, we have attempted to investigate paleoenvironmental change by analyzing large and small mammals together for sites where both are known. The method we have used is the one developed by Andrews et al. (1979), and the comparative data base for this analysis has been extended to cover present-day faunas from habitats across Europe and Asia (Andrews, in press). Our intention is to compare community patterns of Miocene faunas in East and West through time, with the aim of investigating environmental patterns, and where change occurs, to see if the changes are synchronous. Changes within communities can be instructive in showing responses to ecological shifts, and beyond that it may be possible to draw inferences about more particular ecological variables. For example, herbivore feeding strategies may be inferred, carnivore guilds may be reconstructed, spatial distribution of the communities may indicate vegetation structure.

The same combination of the NOW and Karlsruhe data bases was used as for the other analyses (see above), but only a subset of the data was used (selected sites with micromammals as well as macromammals). The data include taxonomic revisions incorporated in various chapters of this book, so that the species lists differ in some respects from ones used previously for ecological reconstruction (Andrews, in press), and it also differs in the absence of certain mammalian groups, especially lagomorphs, hyraxes, and proboscideans. The absence of these groups leads to some loss of discrimination, but it has not

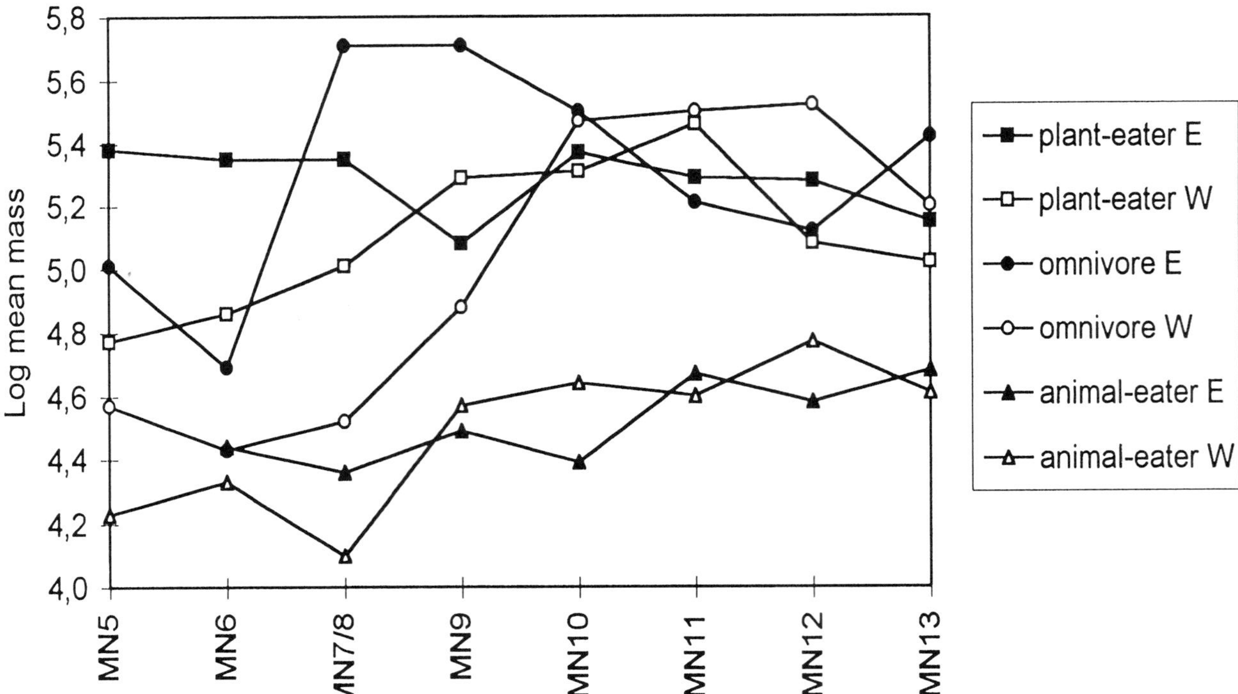

FIGURE 31.28 Mean body mass of dietary categories in West and East. Until MN 9–10 all categories were of smaller mean size in West, but later the difference disappeared. Note the very marked increase in the omnivore group, MN 6 to MN 7/8 in East and after MN 7/8 to MN 10 in West.

had any material effect on the results or on the environmental conclusions. The main biases are geographic and concern the nature of West. As a result of the scarcity of small mammal data for Western Europe in the Karlsruhe data base, West is mostly represented by Central European localities. Unlike in the other analyses, Hungary was here included in West (one of two faunas in MN 11 and the only fauna in MN 13). These biases serve to minimize the change seen in the Vallesian of West and pull the Turolian of West in an easterly direction.

Body size reconstructions are taken from the NOW data base and from the ones previously published (Andrews et al. 1979; Andrews, in press). The size categories range from small mammals (A = 0–100g; B = 101–1,000 g, etc.) and large mammals (H >360kg)—see the caption to figure 31.29. Spatial analysis is based on the concept of Harrison (1952), whereby the locomotor adaptations of the fauna are inferred from their postcranial adaptations, and these are interpreted in terms of the space occupied-for example, terrestriality or arboreality. The divisions are as follows: terrestrial, locomotion totally restricted to the ground and including both cursorial and springing species; semi-arboreal, originally referred to as SGM and includes mammals that can live both on the ground and in bushes and low trees; scansorial, large branch arboreality by clawed mammals; arboreal, mainly tree-living by mammals that can grip branches; aerial, mammals with adaptations

for flying; fossorial, mammals with adaptations for digging; and aquatic, mammals with adaptations for swimming.

Trophic guilds are very general in nature and are based on the major adaptations of the teeth: insectivores with high, pointed cusps (or lacking enamel in anteaters); frugivores with low, rounded cusps; brachydont browsing herbivores; hypsodont grazing herbivores; carnivores with cutting teeth; and omnivores. Species with mixed diets have been assigned to more than one guild except where so much variety is indicated that they should be assigned to the omnivore class.

Size Distribution

The size distributions for MN units 5 to 13 for West (fig. 31.29) and East (fig. 31.30) are mainly of use in detecting bias in the faunas. These figures show data from modern mammalian faunas on the left and the Miocene faunas averaged for MN unit on the right. There is little consistent difference between East and West. The latter is compared with seven modern faunas from Asian subtropical habitats from India through to China, and most of the MN units show a bias in favor of small mammals compared with the recent faunas, particularly MN 5. The opposite is seen in East, where several of the faunas have fewer small mammals, especially in the later time periods. These differences do not have any recognizable ecological

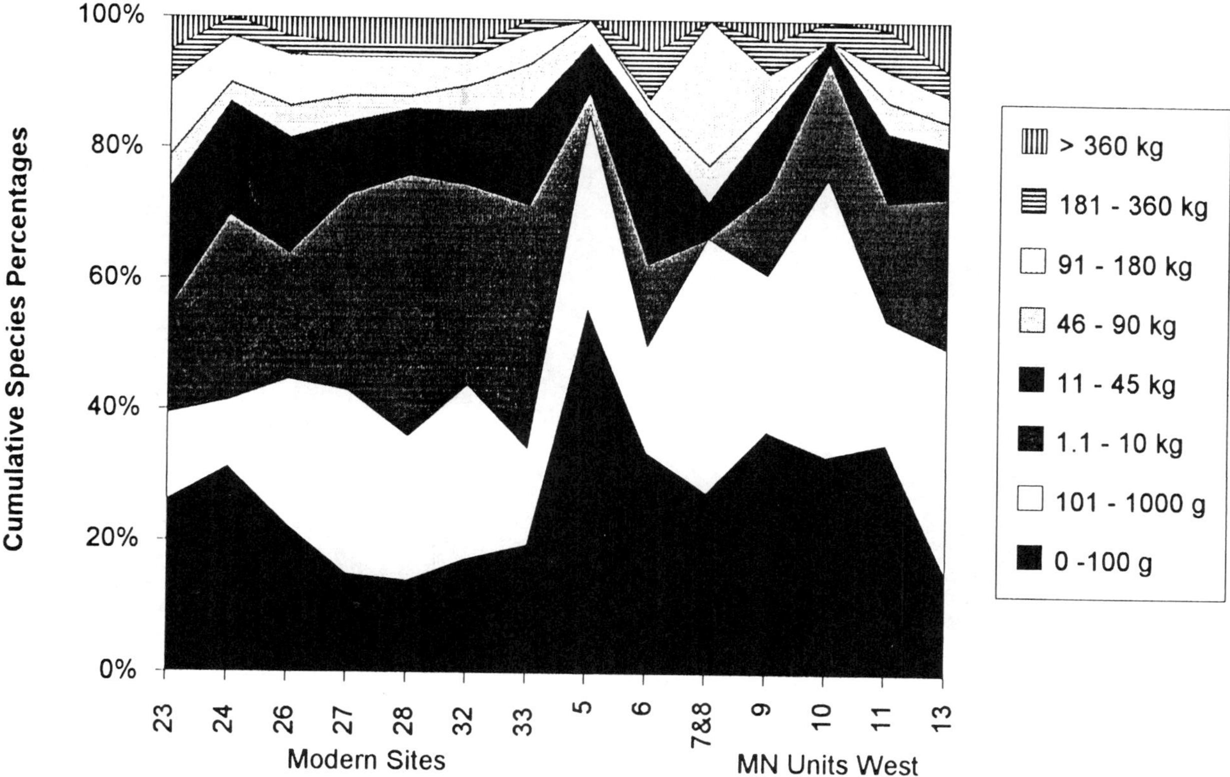

FIGURE 31.29 Cumulative distribution of body mass for West. The MN unit faunas are averaged for each unit. Modern sites are as follows: (23): Kanha, India, subtropical moist/dry deciduous forest, N = 39; (24): Lyallpur, Pakistan, subtropical xeric thorn scrub, N = 28; (26): Imphal, Assam, subtropical montane evergreen forest, N = 42; (27): Mokokchung, India, subtropical montane forest, N = 52; (28): Hkamti, Burma, subtropical montane evergreen forest, N = 56; (32): West Hsi-Shuan-Pan-Na, Yunnan, subtropical semi-deciduous forest, N = 47; (33): Kwangsi, China, subtropical semi-deciduous forest, N = 60.

meaning, for the variability in modern faunas, both within-habitat and between different habitats, is so great that size distribution has little significance. It should also be noted that in the faunal lists for the Miocene sites some are clearly small mammal sites, with few large mammals, and some are large mammal sites, with few small mammals, and in combining the faunas to produce average values for each unit there is an additional sampling bias introduced.

One result that may be drawn from these analyses is that there is no decrease in small mammal species diversity between MN 9 and MN 10 in (Central European) West. It was shown earlier (fig. 31.17) that the drop in species diversity in MN 9/10 affected mainly macromammal species less than 100 kg, but it is apparent from figure 31.29 that there is a slight *increase* in diversity of size classes A to C (up to 10 kg body mass). The consensus is that the greatest reduction in West diversity, as a whole, encompasses the medium-sized mammals from 10 to 100 kg.

Spatial Analysis

The spatial analyses have also been affected by taphonomic and sampling bias, as shown in the body size analy-

ses, but despite this some interesting patterns emerge. The modern subtropical forest faunas (figs. 31.31 and 31.32) exhibit a pattern of decreasing terrestriality and increasing arboreality from left to right, which corresponds to decreasing seasonality and change from deciduous to semi-deciduous or evergreen tree cover. In West (fig. 31.30), MN 5–9 correspond to the intermediate semi-deciduous pattern of the modern faunas, and in general terrestrial species are less well represented than in East (fig. 31.31). In East, MN 6–8 faunas correspond to the drier end of the modern range of subtropical forest, indicating deciduous forest at best for this period. In both West and East, proportions of arboreal, semi-arboreal, and scansorial decrease with time, starting from MN 10/11 in West and from MN 8/9 in East, and this indicates progressive loss of tree cover. At the same time the proportion of terrestrial species increases in East, further indicating opening of the habitat, while in West a more complex situation occurs.

There is a sharp decline in terrestrial diversity from MN 8 to MN 10 in West, with a later increase back to previous levels (fig. 31.31). At the same time there are moderate increases across the other spatial units. These changes can be correlated with the overall drop in species diversity in

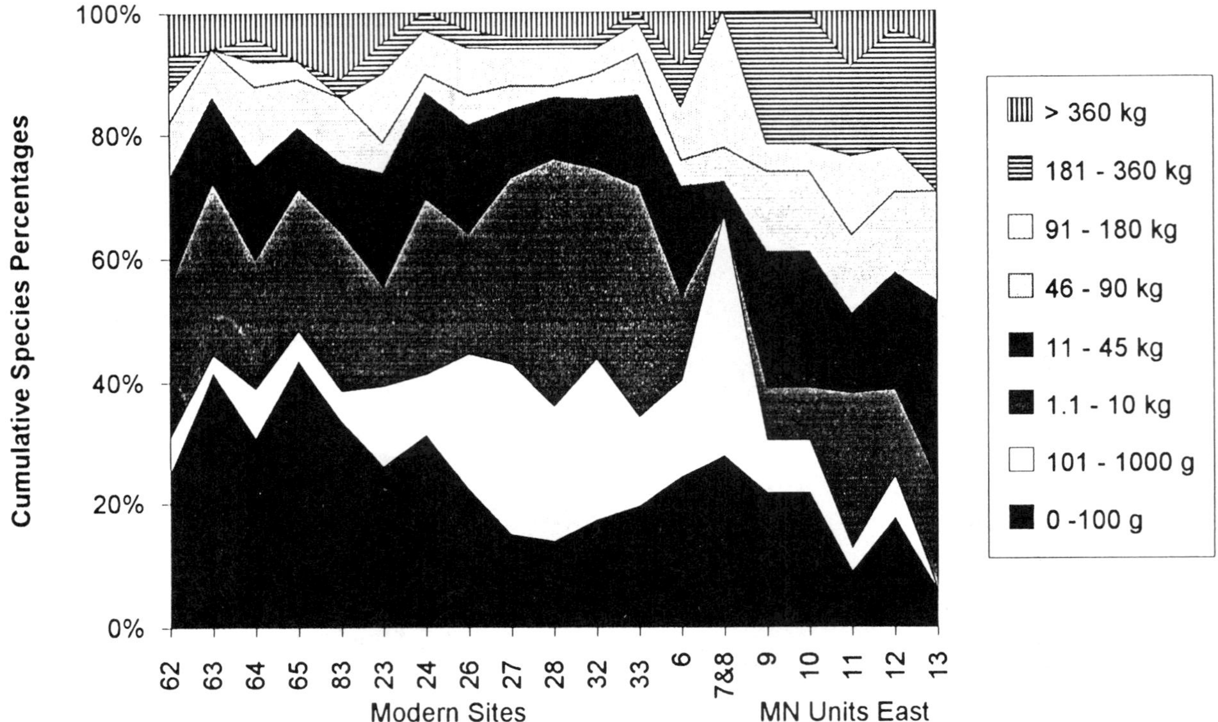

FIGURE 31.30 Cumulative distribution of body mass for East. Details as for figure 31.29 with the addition of the following modern sites: (62): Kruger, South Africa, subtropical deciduous bush-forest, N = 82; (63): Ingwaruma, South Africa, swamp forest, N = 64; (64): Ubombo, South Africa, moist evergreen forest, N = 48; (65): Manguzi, South Africa, swamp forest, N = 39; (83): Knysna forest, South Africa, temperate rain forest, N = 40.

MN 9/10 (see fig. 31.10) and the decrease in medium-sized species (10–90 kg) described above (fig. 31.29). It would appear, therefore, that in (Central European) West, the decline in medium-sized mammals during the transition from MN 9 to MN 10 is composed almost entirely of terrestrial mammals, and there is no corresponding decline in either nonterrestrial species or in small (<10kg) or large (>180kg) species. This may indicate a subtle shift in the nature of the forest environment at this time, but there is no evidence from these data for loss of tree cover and increasingly open habitats: quite the contrary. After MN 10, terrestriality rises in (Central European) West, as it does in East, and this compliments the decline in arboreal species, further supporting the decrease in tree cover, although the decrease is less than in East.

Dietary Guilds

The distribution of dietary guilds gives evidence of taphonomic bias against carnivorous species, which are underrepresented in many units (figs. 31.33–31.34). The West faunas are dominated by browsing herbivores, with grazing species at the lower limits of the modern subtropical faunas. There is little change in browsing/grazing proportions through time in the West sequence of faunas, but there

is a trend toward increasing insectivory and decreasing frugivory, particularly after MN 10 (fig. 31.33). This corresponds to the increasing terrestriality and decreasing arboreality after MN 10 described above. There is no marked decline in any dietary group during the MN 9/10 diversity change in West except for a drop in carnivores. Frugivores, insectivores, and browsing herbivores are as a whole more abundant than in East, indicating more closed habitats in West, with the trend toward opening of the tree cover never reaching the extent seen in East. These results are entirely consistent with the spatial analyses.

The pattern in East is more obscure. The analyses in figures 31.22–31.23 exhibited increases in grazers relative to brachydont species compared to the western patterns. Grazing proportions in East are higher throughout the middle and late Miocene than is seen in the subtropical Asian faunas except for those from the most open deciduous forests, and after MN 11 the proportion of grazers increases, so that the pattern of the fossil faunas is closest to those found in the South African woodland habitats (fig. 31.34). Browsing species reach their highest proportions during MN 7–10 and thereafter decline (with the possible exception of MN 13). Frugivorous species decrease at the same time that grazing species increase, indicating increased seasonality to at least the same extent as is found

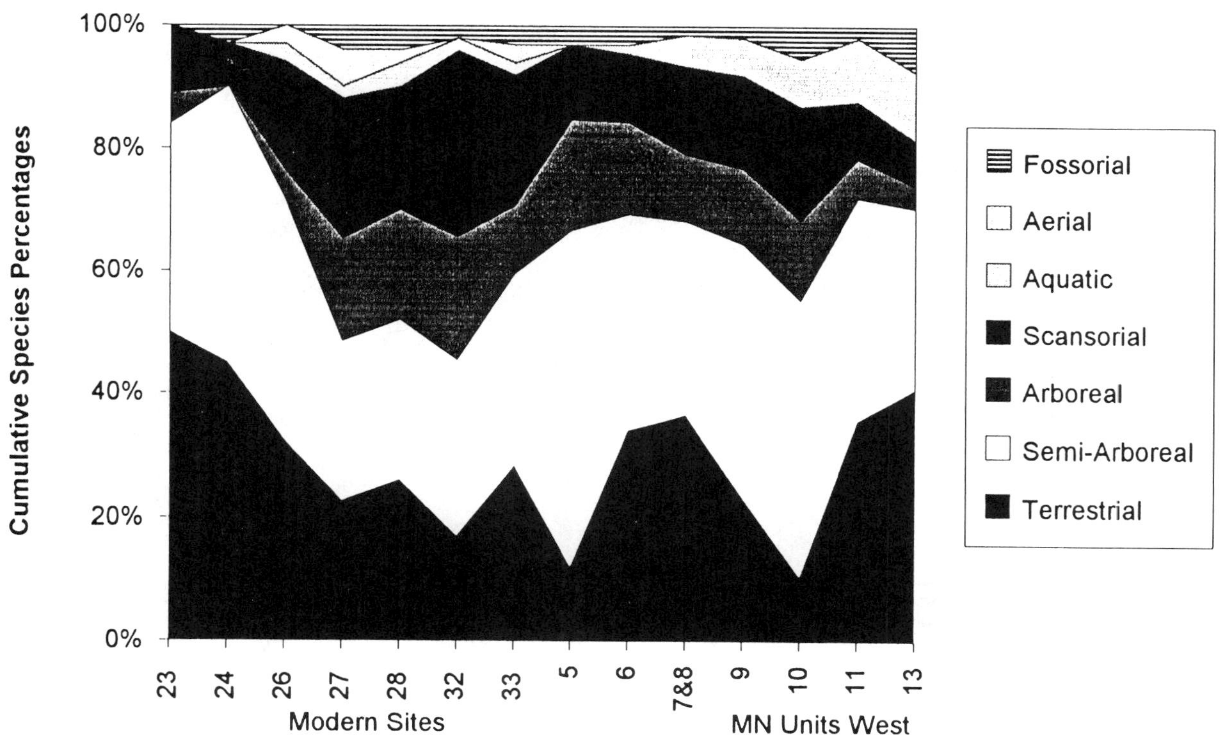

FIGURE 31.31 Cumulative distribution of locomotor spatial adaptations for West. Categories explained in the text and in figure 31.29 caption.

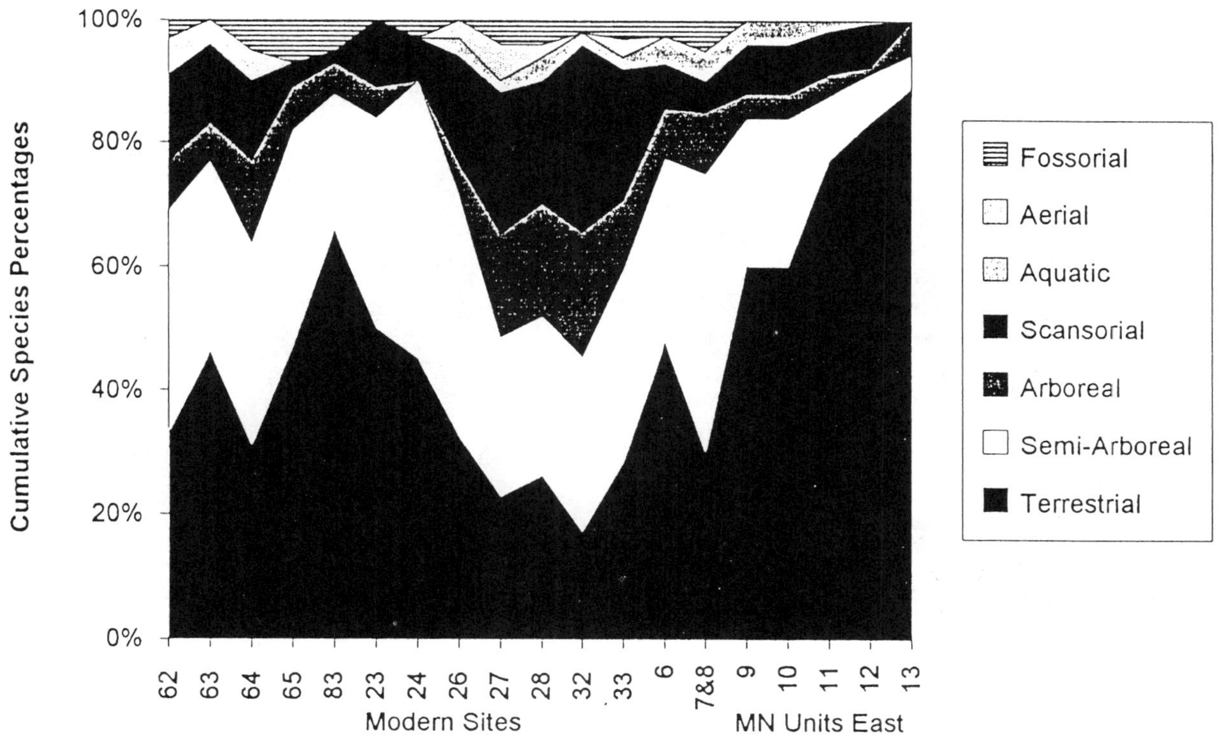

FIGURE 31.32 Cumulative distribution of locomotor spatial adaptations for East. Categories explained in the text and in figure 31.30 caption.

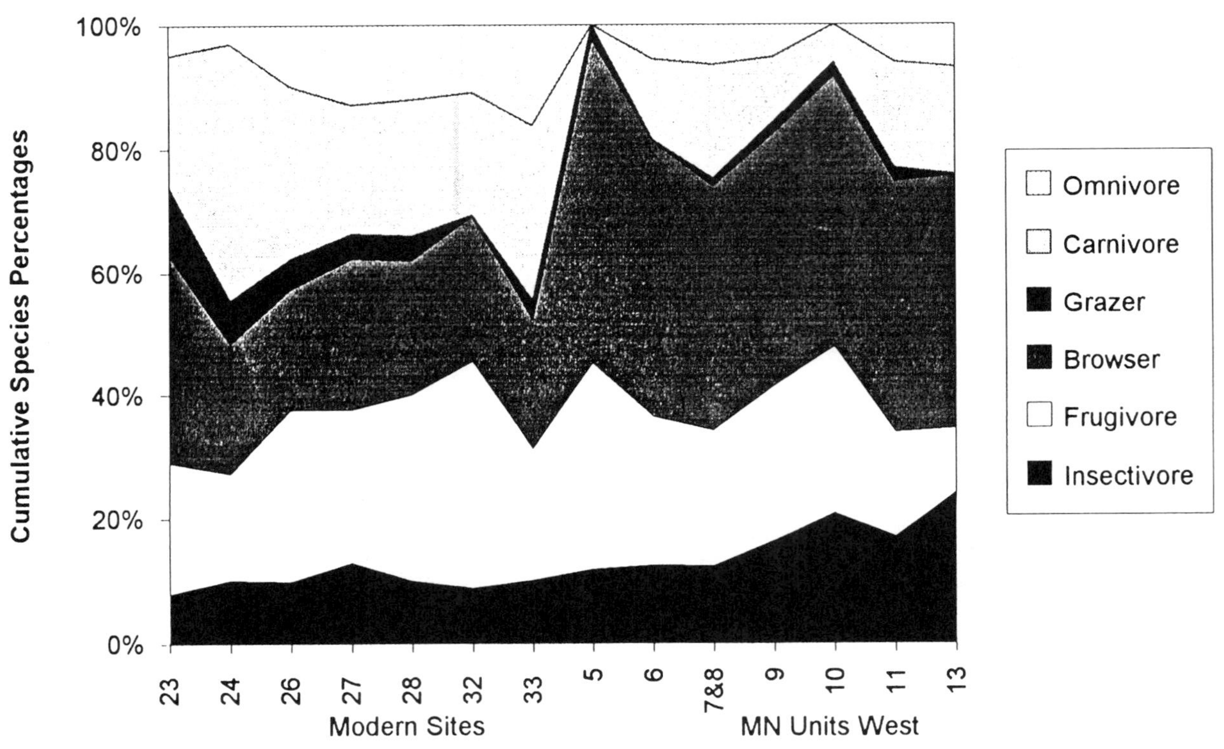

FIGURE 31.33 Cumulative distribution of dietary guilds for West. Categories explained in the text and in figure 31.29 caption.

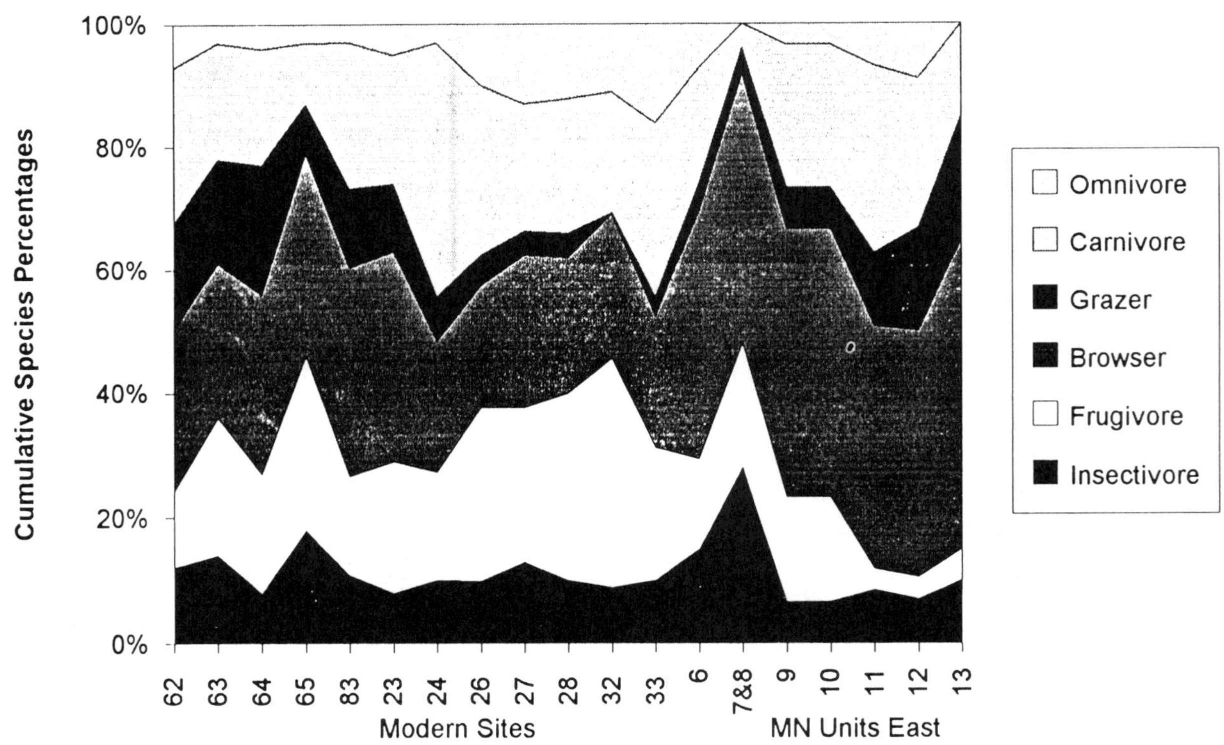

FIGURE 31.34 Cumulative distribution of dietary guilds for East. Categories explained in the text and in figure 31.30 caption.

today in the South African semi-arid woodlands, such as that in the Kruger National Park (site 62). Overall, the habitats indicated in East vary from deciduous seasonal forest in MN 6 to more seasonal conditions with decreased tree cover by MN 9.

Ecological Relationships

Comparisons between ecological variables have been made to gain further insights into the nature of the similarities and change in the Miocene faunal succession. Figure 31.35 shows percentages of species with frugivorous adaptations plotted against percentages with grazing adaptations for a number of modern faunas from tropical and subtropical habitats and for the means of the MN unit faunas. For West (fig. 31.35a), the faunas from MN 5–10 are shown grouped with modern tropical and subtropical forest faunas, while the MN 11 and MN 13 faunas (dominated by the "eastern" Hungary) are somewhat removed toward deciduous forest and savanna. There is no apparent differentiation between MN 9 and MN 10 in Central Europe, and in fact both faunas have higher proportions of frugivores and so are situated toward the more tropical forest faunas on figure 31.34a. For East (fig. 31.35b), only the MN 7/8 faunas are grouped with the subtropical forest faunas from Asia, at the bottom end of the tropical forest range, with MN 9/10 at the bottom limit, and the MN 11–13 faunas well removed and grouped with the tropical savanna modern faunas. These patterns reflect the changes in relative proportions of grazers and frugivores between East and West, and within these areas they show a change through time leading to more open environments. In the case of East, it appears likely that environments in MN 11–13 were highly seasonal, with open deciduous tree cover, probably lacking closed canopy.

Similar analyses of proportions of frugivores plotted against proportions of terrestrial species produces complimentary results (fig. 31.36a). Faunas from MN 5–10 in (Central European) West are within the tropical and subtropical forest faunal distribution, MN 11 in the tropical deciduous forest distribution, and MN 13 (Hungary) grouping at the upper limit of the savanna distribution (fig. 31.36a). The MN 9–10 faunas are again grouped more with modern tropical than with subtropical faunas. In East, the pattern is the same as in figure 31.35b, and this figure is not included here. When grazer proportions are compared with proportions of semi-arboreal species (fig. 31.36b), the distributions for East show MN 6–10 again grouping with modern tropical and subtropical forest faunas, with the MN 13 faunas grouped with the modern savanna faunas, and MN 11–12 offset but still closest to the savanna faunas.

These analyses strongly support the differences described earlier between East and West, with the former

having more open environments. Apparent also for both blocks during the interval under consideration is the marked opening of the environment: that is, a partial loss of tree cover. The dramatic loss of species diversity in West from MN 9 to MN 10 does not show up in these analyses, probably because the data are dominated by the environmentally stable Central Europe. Here, there was no opening up of the tree canopy, because the ecological trend is in the opposite direction (figs. 31.35a and 31.36a), and the closest modern analogies are with tropical forest faunas from both Africa and Asia, with no evidence of loss of diversity.

Ecomorphological evidence published by Köhler (1993) has shown similar trends. This study concentrated on two geographic areas, Spain and Turkey, which correspond (albeit to a more limited extent) to our West and East, respectively. Köhler based her work on a range of body parts of ruminants, establishing correlates between morphology and ecology for living ruminants before applying the same criteria to fossil faunas from both regions ranging between MN 6 and MN 12. During the Astaracian and the early Vallesian, MN 6–MN 9, the ruminant faunas from Spain (Manchones and Can Llobateres, supplemented by Steinheim because no MN 7 fauna was available from Spain) indicate humid woodland conditions, while for the same period the Turkish faunas from Çandir, Sofça, and Eşme Akçaköy appear to have been adapted to dryer and more open conditions (Köhler 1993). Similar conditions were found for the MN 10 fauna from Terrassa, in Spain, and no evidence of the massive diversity decline from MN 9 to MN 10 is evident, although the fauna is clearly impoverished with respect to the earlier sites, and this may indicate a difference. No MN 10 Turkish fauna is included, but the MN 11 faunas from both regions are dramatically different, indicating open woodland, although Köhler stresses that there is no evidence for any of the ruminants inhabiting open country. This drying trend is continued in MN 12, with both areas being dominated by open woodland species. It is evident from these ecomorphological reconstructions that there was a much greater change in the Spanish faunas (equivalent to our West) than was seen in Turkey (East), but the change was observed between MN 10 and MN 11, one MN unit later than our present results show. The change in East could only be placed between MN 9 and MN 11; our results indicate that it occurred mainly between MN 10 and MN 11.

Geographic Area

The Turolian localities of West are derived from a much smaller geographic area than the pre-Turolian ones. Since diversity is known to correlate strongly with area (McGuiness 1984; Rosenzweig and Abramsky 1993), this fact

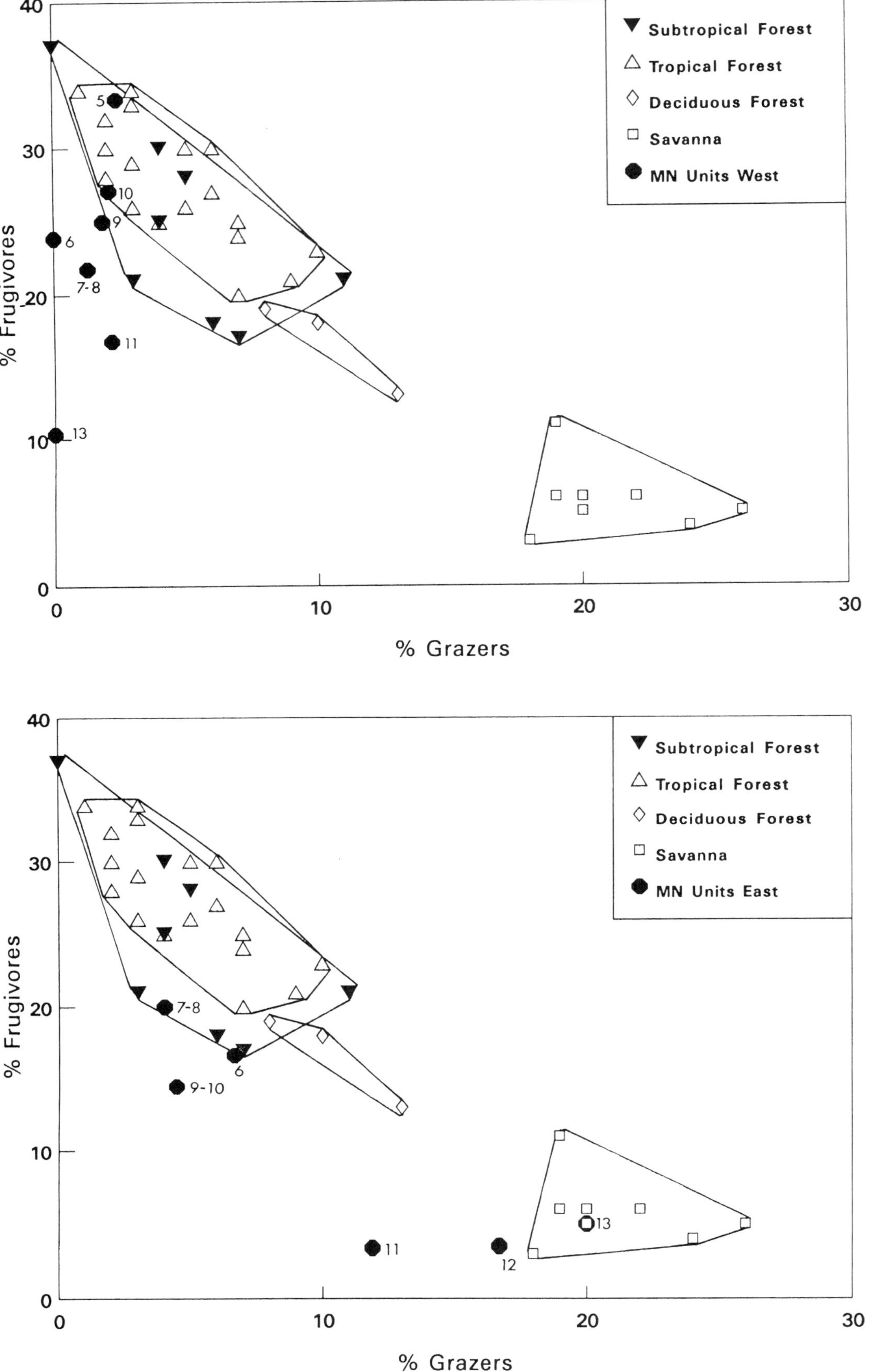

FIGURE 31.35 Bivariate plots of proportions of species with frugivorous adaptations against ones with grazing adaptations. Modern faunas are taken from subtropical forest habitats from Asia, tropical forest faunas from Africa and southeast Asia, tropical deciduous forest faunas from Africa, and tropical savanna faunas from East Africa. A: West MN units. B: East MN units.

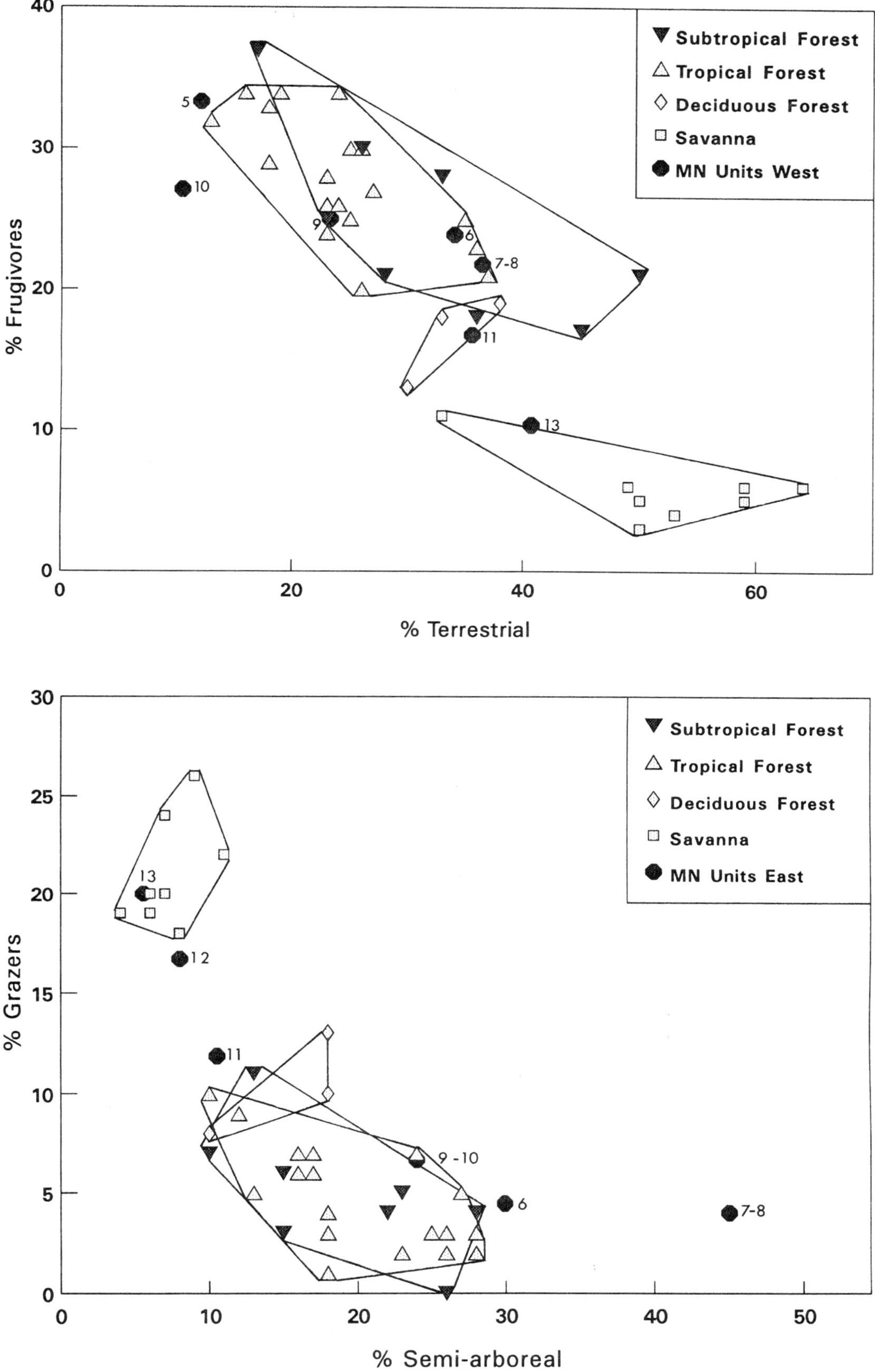

FIGURE 31.36 A: Bivariate plots of frugivore proportions against proportions of species with terrestrial adaptations. Miocene faunas from West. B: Plot of proportions of grazers against proportions of semi-arboreal species. Miocene faunas from East. The modern faunas are as listed in figure 31.35.

might explain the lower western diversity in the Turolian. It is observed, for instance, in South American habitats, that while mammal alpha diversity in tropical rain forest habitats is much higher than in dryland habitats, the total diversity is higher in the latter than in the former because it is more than four times larger in areal extent (Mares 1992).

For a rough evaluation of this hypothesis, we first divided all diversity values by total block area (fig. 31.37). The result is striking: West has a much higher diversity than East through MN 9, after which (base MN 10) the mid-Vallesian crisis depresses it to East's level. The Turolian diversity increase in East is reduced to a relatively minor feature, and by MN 13 West and East exhibit equivalent diversities because of concomitant West increase and East decrease. The principal feature of these diversity changes was the West's mid-Vallesian crash, which resulted in uniformly low diversity in the entire area. The difference between MN 5 and MN 9, and MN 10 and MN 13, is statistically significant for East as well as for West, however (Mann-Whitney U-test, P = 0.021 for both blocks). The difference between West and East is, of course, significant in MN 5 through MN 9 (P = 0.021), but not in MN 10 to MN 13.

This picture is probably grossly exaggerated by the enormous territories of East that are almost devoid of fossil localities of this age: Iran, Afghanistan, and Kazakhstan alone make up 72% of the block, and contribute only a few localities toward the end of the interval. Moreover, the localities are not evenly spaced, but clumped in small areas of those enormous countries. Nonetheless, it does provide a baseline against which to judge more realistic comparisons.

One such might be to sum the areas of all countries with fossil localities (one or more) in an MN unit, separately for each unit and block. It is not altogether obvious what this "political" area actually represents, but many factors (such as intensity and history of research or availability of exposed sediment) are obviously involved. Nevertheless, it may also reflect past reality in some meaningful sense, and is, at the very least, "closer to the data" than the use of total block area. The correlation between this summed area and species richness is significant for both blocks (West: r = 0.712, P <0.05; East: r = 0.757. P <0.05). For West, the diversity and area curves follow each other quite closely (fig. 31.38), except for MN 7/8 and MN 9, where diversity fails to fall with the drastically reduced area. What the delay of one MN unit means is difficult to assess, but at least on the face of it reduced area does not account for the mid-Vallesian crisis, since the area of West actually increases slightly from MN 9 to MN 10. There is little doubt, however, that the low diversity of the Turolian of West is related to the small area sampled. For East, the coupling between area and diversity appears very strong. The increase in diversity in the late Vallesian and Turolian corresponds to a marked increase in area, due to progressive inclusion of Iran, Afghanistan, and Kazakhstan (fig. 31.39). The diversity drop of MN 13 is an exception to this, since the area sampled for that interval is the largest of all.

When the species richness curves are plotted normalized for area, the basic pattern seen in the raw data remains but the changes in East are less marked, except for the peak in MN 6, which is more pronounced (fig. 31.40). The only difference that remains statistically significant is the dramatic change between pre-and postcrisis levels in West (Mann-Whitney U-test, P = 0.021).

The MN 6 peak for East could be either an artifact of anomalously good sampling (a "Paşalar-effect") or it could reflect genuinely high diversity, preceding an eastern crash similar to the mid-Vallesian crisis of West, especially since the low diversity of MN 5 is almost certainly an artifact of inadequate sampling. The paleoecological interpretation of the suoid material offered in this volume would favor the latter interpretation (Fortelius et al., this volume), and the timing of the increase in omnivore body mass also fits this pattern (fig. 31.27). An interesting detail of the normalized diversity curve for West is the very high value for MN 9, immediately preceding the crash of closed-habitat ecosystems. This, too, may be a real feature, as discussed below.

Regression analysis of species richness on area shows the relationship to be positively allometric: the larger the total area, the more species are found per unit area. For this reason, the residuals of a regression analysis of log-transformed data on area and diversity are shown in figure 31.41. The result once more repeats the familiar pattern,

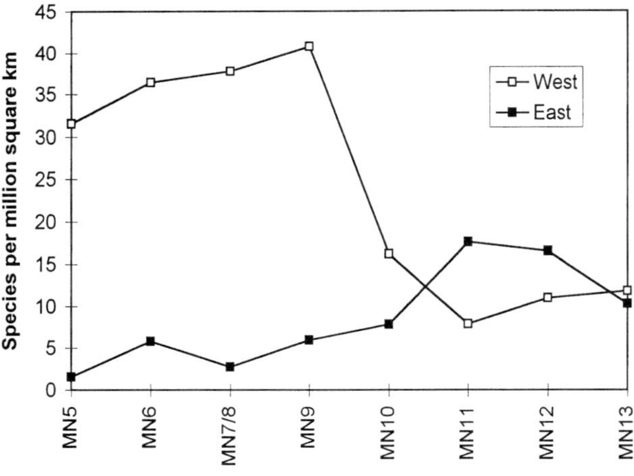

FIGURE 31.37 Diversity (species richness) of West and East normalized for total block area. The differences between pre- and post MN 9/10 boundary are significant for both blocks, but the difference between the blocks is not significant for the later part.

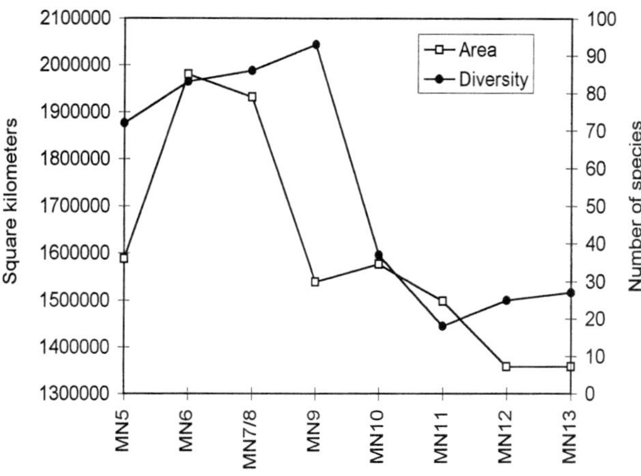

FIGURE 31.38 Plot of diversity (species richness) and sampled area for the West block. The correlation is significant (r = 0.712; P <0.05). Note great discrepancy in MN 9.

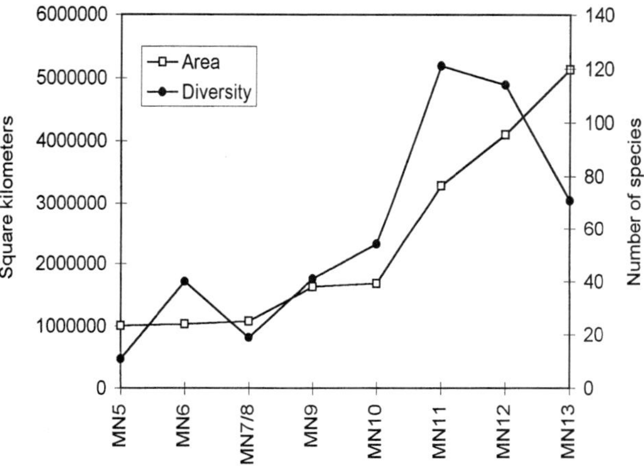

FIGURE 31.39 Plot of diversity (species richness) and sampled area for the East block. The correlation is significant (r = 0.757; P <0.05).

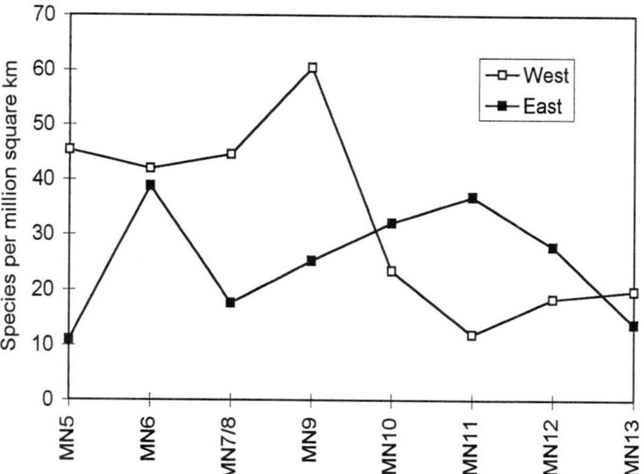

FIGURE 31.40 Diversity (species richness) of West and East normalized for sampled area. The only statistically significant difference is for West between pre– and post–MN 9/10 boundary (Mann-Whitney U-test, P = 0.021).

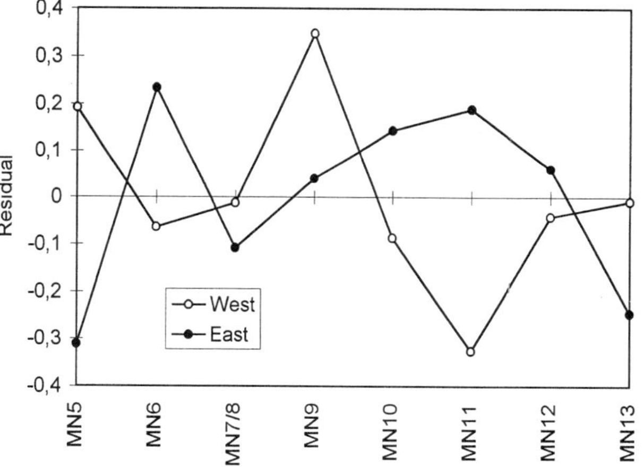

FIGURE 31.41 Plot of residuals of logarithmic regression of diversity (D) on area (A). Regression equations: West D = 3.318A-18.906, r^2 = 0.512; East D = 1.048A-4.936, r^2 = 0.686.

but with interesting differences of emphasis. Normalized diversity was higher than predicted by the regression equation for MN 6 and MN 10–12 in East and for MN 9 in West. The low points are MN 5 of East, MN 11 of West, and MN 13 of East, all relatively poorly sampled. The latter two are part of a general pattern that warrants serious consideration. Although the area sampled reflects past biogeography in indirect and somewhat obtuse ways, some indication of a real relationships may still remain in the data.

The fact that individual subregions of both blocks show roughly the same diversity changes as the block itself (figs. 31.11–31.12) reveals that it is not simply the area covered by the analysis that matters, but something less arbitrary, such as the degree of biogeographic continuity or the degree of contact with regions beyond the area studied. Such

factors might be the major determinants of diversity change, but this does not mean that the changes observed in individual regions are arbitrary. The apparent shrinkage of West probably reflects a real biogeographic process, a fragmentation of a previously uniform area. The fact that Central European large mammal sites are almost unknown after MN 11 (which is represented by the single controversial locality of Dorn Dürkheim) fits this interpretation, although the (presumably taphonomic) mechanisms involved remain obscure. The anomalously high diversity/area relationship of West might thus represent a phase of supersaturation, when diversity was briefly at a value above equilibrium. This would have increased the sensitivity of the system, and thereby contributed to the violence of the subsequent crash.

The progressive decline of East's diversity in the MN 11–13 interval is emphasized in the normalized diversity curve (fig. 31.40), and especially in the residual plot (fig. 31.41). This appears to presage the Messinian crisis. It is interesting that West exhibits no decline in diversity at this time, perhaps because it was already very low. It is necessary to recall that we have no data on abundance or biomass, only on species richness, and that low richness may reflect different ecological circumstances, depending on abundance.

Conclusions

The nature of the data places severe constraints on the scope of possible interpretation. In particular, the lack of species abundance estimates limits the ways in which the revealed patterns can be related to ecology, since the most informative parameters, such as equability and relative biomass, cannot be calculated (Andrews, in press). Species richness and mean body size estimates are better than nothing, but these do not provide particularly specific information about the environment or ecology of the faunas. Yet at the same time ecology emerges as the central theme, the one unifying aspect to which other data can be related and against which they can be evaluated. We have concentrated on some general questions that the data allow us to address: degree of similarity between regions, species richness, turnover, the timing and direction of changes, and beyond them whatever may be gleaned of the ecologic data.

The "faunas" analyzed here are all aggregates, derived from what is presumably a variety of environments, representing a mixture of community types and ecological settings, widely separated (in an ecological sense) in time and space. On the one hand, this means that the signal must inevitably be fuzzy, related only at a very general level to some average value of conditions in the selected area and interval. On the other hand, the likelihood that local environmental conditions would override the temporal or regional signals is minimized in this way, which, in the present case, is advantageous. In fact, it is a moot point to what extent fossil large mammals can, even potentially, provide high-resolution information about the environment. Many large mammal species move naturally between habitats, and a variety of taphonomic factors also operate to bring together the remains of individuals that lived in different local habitats. Thus even in an optimal case, the signal would not be expected to have particularly high resolution. Indeed, Andrews (in press) argues that the temporal and geographic signals in fossil mammal faunas generally do override the local ecological ones, and that the ecological signal that comes through is general rather than specific, distinguishing between such broad categories

as closed versus open country faunas, regardless of climatic or geographic setting.

The results we have presented above correspond in general outline to what has been suggested previously (e.g., Bernor 1983, 1984; de Bonis et al. 1992a, b; Köhler 1993). The mammal fauna of West showed only moderate change until the dramatic "crisis" of the middle Vallesian, when its western part underwent an abrupt change from closed to open environments and lost more than half of its species diversity, while its central part changed much less. In contrast, the open woodland mammal fauna of East developed incrementally, until it suddenly "exploded" and transgressed into the newly altered parts of West during the Turolian. We can now begin to elaborate on the ecological and evolutionary detail of the changes, to discuss their causes, and to relate them to a theoretical framework. We can also point to unexpected details and relationships that seem to merit attention beyond what has been possible in this coarse overview.

On a broad scale, Western Eurasian faunal provinciality shows a clear general relationship to the changes detailed above (fig. 31.8). Overall similarity increased from MN 5 through MN 7/8, reflecting progressive filtering of African and Central/East Asian taxa across the area. With rising extinction levels in West, similarity dropped in MN 9 and stayed low until MN 11, but rose sharply in MN 12 with the main wave of entry of eastern taxa into West. On a finer scale a similarity gradient is detected, with degree of similarity roughly inversely related to geographic distance between the regions compared (figs. 31.6 and 31.7). To what extent true faunal provinces can be defined is as yet unclear, and perhaps of secondary importance. The point to stress is that consistent and meaningful differences between arbitrarily assigned regions can be demonstrated.

Since the raw diversity curves of West and East are virtual mirror images of each other (fig. 31.10), it might be thought that the changes that took place in the two blocks between MN 9 and MN 11 were also, in some sense, opposite. This was not the case, however. Analysis of changes in community structure and ecomorphology clearly reveals that the direction of change was the same in both blocks, toward more open and more seasonal habitats. It is highly probable that the reason for the change was climatic forcing related to the terminal Serravallian global sea-lowering event, variously delayed and moderated by geographic and biotic filters. The markedly different initial conditions in West and East at the onset of the change seem the most likely explanation for the opposite diversity response: there was an abrupt change in West (except Central Europe), whereas in East a process that was already underway was further accelerated (Fortelius et al., in press; Bernor et al., this volume a).

Throughout the interval studied, turnover fluctuated at a high level in West, with peaks in MN 9 and MN 12, and a close coupling of fluctuations in extinction and origination. In East, turnover shows a clear decreasing trend, and the extinction rate remains low even when the origination rate rises, which produces diversity increase. The fact that the entry and exit curves follow each other in West suggests a closer relationship between origination and extinction there, perhaps reflecting more limited niche space or some such biotic constraint. The contrasting relationships between body mass and species richness that is seen between East and West may also reflect this pattern. In both blocks, the coming change is heralded by a marked increase in mean size of the macromammal species (plot not shown). Perhaps there really was less niche space available in West, where fewer large species could be accommodated than in East, where there seem to have been no ecological constraints on increase in either size or diversity? The West's diversity collapse could, for example, have been more a reduction of beta than alpha diversity, a loss primarily of habitat/community diversity. Increasing beta diversity is less likely to have been part of the diversity increase of East, which was primarily a matter of diversification of the open woodland fauna.

The larger ungulates exhibit a striking contrast in their diversity history. Hipparionine horses, for example, always remained relatively species-poor in West, as did the bovids, whereas both groups flourished in East (fig. 31.42). The relationship between hipparions and ruminants is not well understood. Hipparions were dietarily diverse (Hayek et al. 1991), and the lack of modern analogs for browsing horses complicates the issue further. Hipparions certainly must have had a dramatic impact on the Old World ungulate community, which they rapidly came to dominate, but the relationship was probably more complex than simple competition over food. It has even been suggested (Janis 1982) that the entry of hipparionine horses into the Old World would in itself have initiated the diversification of other ungulate groups, perhaps through direct impact on the vegetation and feeding successions (cf. Bell 1969; 1971). Although such a mechanism is not really needed to explain the pattern, it is congruent with the East's history, where the diversification of the bovid fauna increases with the entry of hipparions, with both bovid and hipparion curves peaking in MN 11–12, and where regional speciation is demonstrated by numerous cases of endemic sister taxa (Solounias 1981a, b; Bernor et al., this volume b; Gentry and Heizmann, this volume). The entry of hipparions never led to a comparable radiation in West, however, nor was there any diversification of the contemporaneous bovid fauna.

This points to an important and somewhat unexpected result of this survey: with the opening up of its habitats West (except Central Europe) became more like East in many respects, but failed to generate any equivalent of the eastern diversity increase, seen especially in the ungulate fauna. We suggest that this difference arises from basic biogeography. The Western Mediterranean is a relatively

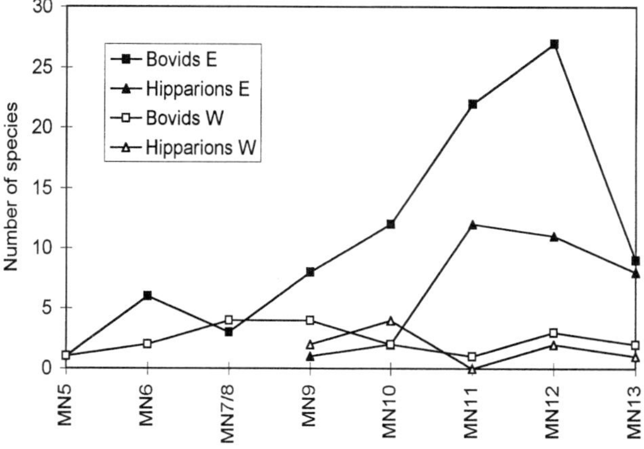

FIGURE 31.42 Diversity (species richness) of hipparions and bovids in the West and East blocks. No trace of the eastern radiation of these groups is seen in West.

isolated cul-de-sac compared to the Eastern Mediterranean, with its broad continental connections eastward to Central Asia and beyond. East, as used in this study, is an arbitrarily defined portion of the vast Eurasian landmass, while West is a more natural and much smaller area. The difference in turnover patterns fits this scenario, with the isolated West showing low origination and high extinction (fig. 31.14), while the opposite is true of East, the "species factory" of Bernor et. al. (this volume b; fig. 31.15).

Whether the lack of ungulate diversity in the Turolian of West corresponds to low abundance cannot be determined from the data presently available, but it is interesting to note that at least some of the carnivores that would have preyed on them did not experience the diversity crash seen in the rest of the fauna. Hyaenids maintained a high diversity (Werdelin and Solounias, this volume) and machairodontid sabertooths regained their diversity after a brief low in MN 10–MN 11. The felids in general showed a selective extinction of small forms during the process, so that by MN 12 only four large sabertooths remained. The possibility therefore exists that the ungulate fauna of West was species poor but still relatively abundant, providing sufficient food for large carnivores, rather than generally impoverished. Such low equability might indicate an environment that was in some sense "harsh" (Andrews, in press), or it might simply reflect low habitat (beta) diversity. The contrasting composition of entries in the Vallesian and Turolian of West (based on comparison with the level of similarity) may fit the latter alternative: in MN 9 speciation in the local communities was still the dominant source of new taxa, while in MN 12 most new taxa were immigrants from East, where the biota had a long history of adaptation to open habitats.

A major factor explaining the contrasting diversity histories of West and East is evidently geographic area, and its extension in the form of land connections to surrounding regions, probably augmented by the contrast between the uniform interior and a more fragmented periphery. The role of the strangely elusive Central Europe in this equation is unknown, because of an almost complete lack of data, but a reasonable guess would be that between MN 9 and MN 13 it was not, in ecological terms, part of the same "West" as the Western Mediterranean. Separation from the dynamic Mediterranean realm and proximity to the persisting Western Parathethys might be major factors responsible for an environmentally stable and biotically conservative interior. One of the few clues is the fact that micromammal diversity of Central Europe failed to crash in MN 9, and showed only a moderate drop in MN 10, so that micromammal diversity there was suddenly higher than macromammal diversity in MN 10 and MN 11 (fig. 31.18). The fact that the wide-ranging large mammals decreased in richness, while the more local small mammals did not, would support the interpretation that conditions were locally stable and different from those of West in general.

Paleoecological analysis allows us to fill in some more details of the emerging picture. The overall trend, in the entire area and during the entire interval, was from closed toward open habitats with increased seasonality. The faunas of West (represented mainly by Central Europe in this analysis) went from tropical to subtropical semi-deciduous forest in MN 5–10 to deciduous forest or savanna in MN 11–13 (mainly Hungary). East showed more open habitats throughout, with evidence of strong seasonality toward the end of the interval. The rise of the proportion of terrestrial species in the faunas of East from MN 7/8 to MN 13 was rapid and continuous, with the most rapid change seen at the beginning of the process (fig. 31.32). The proportion of frugivores to grazers (fig. 31.35) illustrates the relative degree of change within and between blocks clearly: the consistently more forested conditions of West, and the dramatic change to savannalike proportions in East between MN 10 and MN 11. It almost appears as though, in East, the locomotor response to changing conditions preceded the dietary response by one or two MN units.

The ecomorphological analysis (excluding small mammals) from the entire NOW data base (rather than selected faunas, as above) gives a similar result, with a few intriguing differences. The change in East appears, on the whole, to be more gradual than the change in West, and the principal ecomorphological changes in West are seen between MN 9 and MN 10, at the time of the main diversity crash, rather than between MN 10 and MN 11, which shows almost no change at all. Most notable of these changes is the combined drastic reduction in numbers and increase in mean size of western brachydont ungulates and omnivores and a reduction in the diversity of small- to medium-sized macromammals (mainly carnivores). The most plausible explanation for this difference is that the analysis of whole faunas reflects mainly the environment-

ally stable Central Europe, whereas the macromammal analysis includes the Western Mediterranean. This hypothesis is, of course, eminently testable.

Both data sets agree that the victims of the mid-Vallesian crisis of West were predominantly terrestrial forms in the 10–100 kg range, with a heavy emphasis on animal-eaters (of 110 large mammal species lost, 43 were carnivores, 20 of them mustelids). It should be noted that the large mammals lost between MN 7/8 and MN 9 were taxa traditionally associated with forests (ursids, cervids, primates, small suoids, tragulids, etc.), and the carnivores lost at that stage were invertebrate-eaters and omnivores. Taken together, these data suggest that the process, in West as well as in East, was prolonged and complex, not a simple response to an episodic change in external conditions. The nature of the crisis itself thus remains partly unresolved by our analysis.

It is interesting that brachydont non-carnivores and small (< 30 kg) macromammals remained significantly larger and the ungulates on average more hypsodont in East than in West throughout the study interval. This probably reflects an ever-present, general difference in seasonality or some such variable broadly related to continentality, and is not in retrospect surprising, especially as a similar size-difference is found today, for example in bovids and deer. That eastern brachydonts (and, with less confidence, micromammals) did not increase in diversity, unlike most other taxa in East, is probably part of the same pattern.

A somewhat unexpected discovery is that, especially for East, diversity appears to reflect primarily the level of entries. This plausibly reflects the inherently diffuse entry of taxa, by speciation or immigration, in contrast to the occasionally episodic extinction due to external events. It also supports the independently derived notion that diversity in West, where this does not apply, was constrained by availability of niche space or some such factor, probably related at least in part to habitat (beta) diversity, while diversity in East was well below the equilibrium (with the possible exception of MN 12). This has implications for the fragmentation of West, which appears to have followed the Vallesian crisis. Indeed, the crisis itself may have been aggravated by super-saturated diversity in MN 9, as indicated by analysis of diversity/area residuals (fig. 31.41).

The environmental changes in the later Miocene were many, their primary effects different, and their interactions consequently difficult to gauge. What really happened is mostly beyond analysis, but it appears that some of the tangle can be teased apart. The immediate effects of the Serravallian regression at about 11 Ma would, for example, have been different from the ones resulting from the entry of hipparions 0.5 Ma later (Kappelman et al., this volume; Swisher, this volume). The response would also be, and apparently was, different in regions with differing geographic relationships, ecological conditions, and history.

Despite the coarse chronology used in this study, and despite significant smearing of the data, there are several cases of what seems to be either delayed response or the result of some secondary process only indirectly related to the original, triggering event. The most obvious potential candidate for a delayed response is the mid-Vallesian crisis of West, if the biogeographic fragmentation of West really preceded the event, as suggested above. Indeed, on a broad scale the entire Vallesian emerges as a relatively brief phase of transition, a 1.5 Ma interlude between the middle Miocene and the full blown Turolian (fig. 31.13). The rapidity of the change is partly disguised by the use of formally equal MN units here.

Given the uncertainties involved in interpreting all but the most major features of the diversity data, and the complete lack of data on abundance, ecomorphological analysis emerges as the only key to the ecology behind the changes. Some of the ecomorphological plots presented above involve diversity and the same uncertainty applies to them, but as long as they are interpreted in relative terms this hardly matters. Whatever corrections are applied, it is still true that forms dependent on low-fiber foods suffered more than other groups in the mid-Vallesian crisis of West. Similarly, the contribution of bovids and hipparions, and their predators, to the increased Turolian diversity of East was disproportionately strong. For analyses using body mass or relative composition of the faunas, overall diversity is irrelevant (except in the trivial sense of affecting sample sizes). Ecomorphological analysis thus emerges as one of the more useful tools not only for describing ecological change but also for explaining patterns of diversity, turnover, and provinciality.

Acknowledgments

We wish to thank the organizers of the Schloss Reisensburg meeting and the VW Stiftung for inviting us to participate in this venture. Discussions with and comments from the following friends and colleagues helped us in the production of this chapter: J. Barry, M. Armour-Chelu, J. Damuth, G. Daxner-Höck, C. Janis, J. Jernvall, J. Kappelman, M. Maas, S. Sen, J. Sepkoski, and B. Weston. P. Haikonen drew the map of figure 31.1.

This is a contribution from the NOW project, supported by the Academy of Finland. Lars Werdelin is supported by the Swedish Natural Science Research Council. Formulation of this workshop occurred during the one-year tenure of a Humboldt Fellowship awarded to Bernor in 1988. Furthermore, Bernor's and Andrews's joint research in this effort has been largely supported by the National Geographic Society and L.S.B. Leakey Foundation.

LITERATURE CITED

Abusch-Siewert, S. 1983. Gebissmorphologische Untersuchungen an eurasiatischen Anchitherien (Equidae, Mammalia) unter besonderer Berücksichtigung der Fundstelle Sandelzhausen. *Courier Forschungsinstitut Senckenberg* 62:1–401.

Alroy, J. 1992. Conjunction among taxonomic distributions and the Miocene mammalian biochronology of the Great Plains. *Paleobiology* 18:326–43.

Andrews, P. (ed.). 1990. Special issue "The Miocene hominoid site at Paşalar, Turkey." *Journal of Human Evolution* 19:335–588.

——. In press. Palaeoecology and hominoid palaeoenvironments. *Biological Reviews (Cambridge)*.

Andrews, P., J. Lord, and E. M. N. Evans. 1979. Patterns of ecological diversity in fossil and recent mammal faunas. *Biological Journal of the Linnean Society* 11:177–205.

Anyonge, W. 1993. Body mass in large extant and extinct carnivores. *Journal of Zoology (London)* 231:339–50.

Archer, A. W. and C. G. Maples. 1987. Monte Carlo simulation of selected binomial similarity coefficients (I): Effect of number of variables. *Palaios* 2:609–17.

Barry, J. C., L. J. Flynn, and D. R. Pilbeam. 1990. Faunal diversity and turnover in a Miocene terrestrial sequence. In *Causes of Evolution: A Paleontological Perspective*, ed. R. Ross and W. Allman, pp. 381–421. Chicago: University of Chicago Press.

Barry, J. C., N. M. Johnson, S. M. Raza, and L. L. Jacobs. 1985. Neogene mammalian faunal change in Southern Asia: Correlations with climatic, tectonic, and eustatic events. *Geology* 13:637–40.

Barry, J. C., M. E. Morgan, L. J. Flynn, D. R. Pilbeam, L. L. Jacobs, E. H. Lindsay, S. M. Raza, and N. Solounias. 1995. Patterns of faunal turnover and diversity in the Neogene Siwaliks of Northern Pakistan. *Palaeogeography, Palaeoclimatology, Palaeoecology* 115:209–27.

Bell, R. H. V. 1969. The use of the herb layer by grazing ungulates in the Serengeti. In *Animal Populations in Relation to Their Food Resources*, ed. A. Watson, pp. 111–28. Symposium of the British Ecological Society. Oxford: Blackwell Scientific Publishers.

——. 1971. A grazing ecosystem in the Serengeti. *Scientific American* 225:86–93.

Bernor, R. L. 1978. *The Mammalian Systematics, Biostratigraphy, and Biochronology of Maragheh and Its Importance for Understanding Late Miocene Hominoid Zoogeography and Evolution.* Ph.D. diss., University of California, Los Angeles.

——. 1983. Geochronology and zoogeographic relationships of Miocene Hominoidea. In *New Interpretations of Ape and Human Ancestry*, ed. R. L. Ciochon and R. S. Corruccini, pp. 21–64. New York: Plenum.

——. 1984. A zoogeographic theater and biochronologic play: the time/biofacies phenomena of Eurasian and African Miocene mammal provinces. *Paléobiologie Continentale* 14:121–42.

——. 1985. Systematic and evolutionary relationships of the hipparionine horses from Maragheh, Iran (late Miocene, Turolian age). *Palaeovertébrata* 15:173–269.

——. 1986. The mammalian biostratigraphy, geochronology, and zoogeographic relationships of the late Miocene age Maragheh fauna. *Journal of Vertebrate Paleontology* 6:76–95.

Bernor, R. L., P. J. Andrews, N. Solounias, and J. A. Van Couvering. 1979a. The evolution of "Pontian" mammal faunas: Some zoogeographic, palaeoecologic, and chronostratigraphic considerations. *Annales Géologiques des Pays Helléniques*, hors série, 1:81–89.

Bernor, R. L., V. Fahlbusch, M. Fortelius, G. Daxner-Höck, F.

Rögl, F. F. Steininger, and L. Werdelin. This volume a. The evolution of Western Eurasian later Neogene mammal faunas: A chronologic, systematic, biogeographic, and paleoenvironmental synthesis.

Bernor, R. L., G. D. Koufos, M. O. Woodburne, and M. Fortelius. This volume b. The evolutionary history and biochronology of European and Southwest Asian late Miocene and Pliocene hipparionine horses.

Bernor, R. L., J. Kovar-Eder, D. Lipscomb, F. Rögl, S. Sen, and H. Tobien. 1988. Systematics, stratigraphic and paleoenvironmental contexts of first-appearing *Hipparion* in the Vienna Basin, Austria. *Journal Vertebrate Paleontology* 8:427–52.

Bernor, R. L., M. Kretzoi, H.-W. Mittmann, and H. Tobien. 1993a. Preliminary systematic assessment of the Rudabánya hipparions. *Mitteilungen der Bayerischen Staatssammlung für Paläontologie und historische Geologie* 33:195–207.

Bernor, R. L. and D. Lipscomb. 1991. The systematic position of "*Plesiohipparion*" aff. *huangheense* (Equidae, Hipparionini) from Gülyazi, Turkey. *Mitteilung der Bayerischen Staatsslammlung für Paläontologie und historische Geologie* 31:107–23.

——. 1995. A consideration of Old World hipparionine horse phylogeny and global abiotic processes. In *Paleoclimate and Evolution, With Emphasis on Human Origins*, ed. E. Vrba et al., pp. 00–00. New Haven: Yale University Press.

Bernor, R. L., D. Lipscomb, J.-P. Suc, and H. Tobien. 1990. A contribution to the evolutionary history of European late Miocene age hipparionines (Mammalia, Equidae). *Paléobiologie Continentale* 17:291–309.

Bernor, R. L., H.-W. Mittmann, and F. Rögl. 1993b. The Götzendorf hipparions. *Annalen des Naturhistorischen Museums, Wien* 95:101–20.

Bernor, R. L. and P. Pavlakis. 1987. Zoogeographic relationships of the Sahabi large mammal fauna (early Pliocene, Libya). In *Neogene Paleontology and Geology of Sahabi*, ed. N. T. Boaz, A. El-Arnauti, A. W. Gaziry, J. de Heinzelin, and D. D. Boaz, pp. 233–54. New York: Liss.

Bernor, R. L., N. Solounias, C. C. Swisher III, and J. A. Van Couvering. This volume c. The correlation of three classical "Pikermian" mammal faunas—Maragheh, Samos, and Pikermi—with the European MN unit system.

Bernor, R. L. and H. Tobien. 1990. The mammalian geochronology and biogeography of Paşalar (middle Miocene, Turkey). *Journal of Human Evolution* 19:551–68.

Bernor, R. L., H. Tobien, and J. A. Van Couvering. 1979b. The mammalian biostratigraphy of Maragheh. *Annals Geologica Pays Hellenica* 1979:91–99.

Bernor, R. L., H. Tobien, and M. O. Woodburne. 1989. Patterns of Old World hipparionine evolutionary diversification and biogeographic extension. In *European Neogene Mammalian Chronology*, ed. E. H. Lindsay, V. Fahlbusch, and P. Mein, pp. 263–319. New York: Plenum.

Bernor, R. L., M. O. Woodburne, and J. A. Van Couvering. 1980. A contribution to the chronology of some Old World Miocene faunas based on hipparionine horses. *Géobios* 13:25–59.

Bonis, L. de, G. Bouvrain, D. Geraads, and G. Koufos. 1992a. Diversity and paleoecology of Greek late Miocene mammalian faunas. *Palaeogeography, Palaeoclimatology, Palaeoecology* 91:99–121.

——. 1992b. Multivariate study of late Cenozoic mammalian faunal compositions and paleoecology. *Paleontologia i Evolucio* 24/25:93–101.

Bruijn, H. de, R. Daams, G. Daxner-Höck, V. Fahlbusch, L. Ginsburg, P. Mein, and J. Morales. 1992. Report of the RCMNS working group on fossil mammals, Reisensburg 1990. *Newsletters on Stratigraphy* 26:65–118.

Campbell, B. G., M. H. Amini, R. L. Bernor, W. Dickinson, R. Drake, R. Morris, J. A. Van Couvering, and J. A. H. Van Couvering. 1980. Maragheh: A classical late Miocene vertebrate locality in northwestern Iran. *Nature* 287:837–41.

Cerdeño, E. 1989. *Revision de la sistematica de los rinocerontes del Neogeno de España*. Ph.D. diss., University of Madrid.

Cody, M. L. 1993. Bird diversity components within and between habitats in Australia. In *Species Diversity in Ecological Communities: Historical and Geographical Perspectives*, ed. R. E. Ricklefs and D. Schluter, pp. 147–58. Chicago: University of Chicago.

Crusafont, M. and J. Villalta. 1954. Ensayo de sintesis sobre el Mioceno de la meseta castellana. *Boletín de la Real Sociedad Española de Historia Natural (Madrid)* 1954:215–27.

Damuth, J. 1990. Problems in estimating body masses of archaic ungulates using dental measurements. In *Body Size in Mammalian Paleobiology*, ed. J. Damuth and B. J. MacFadden, pp. 229–53. Cambridge: Cambridge University Press.

——. 1992. Taxon-free characterization of animal communities. In *Terrestrial Ecosystems Through Time*, ed. A. K. Behrensmeyer, J. D. Damuth, W. A. DiMichele, R. Potts, H. Sues, and S. L. Wing, pp. 183–203. Chicago: University of Chicago Press.

——. 1993. *ETE Database Manual*. Evolution of Terrestrial Ecosystems Consortium. Washington, D.C.: Smithsonian Institution.

Damuth, J. and B. J. MacFadden. 1990. *Body Size in Mammalian Paleobiology: Estimation and Biological Implications*. Cambridge: Cambridge University Press.

Fahlbusch, V. 1991. The meaning of MN zonation: Considerations for a subdivision of the European continental Tertiary using mammals. *Newsletters on Stratigraphy* 24:159–73.

Fisher, R. A., A. S. Corbet, and C. B. Williams. 1943. The relation between the number of species and the number of individuals in a random sample of an animal population. *Journal of Animal Ecology* 12:42–58.

Fleagle, J. G. 1988. *Primate Adaptation and Evolution*. San Diego: Academic Press.

Flynn, J. J. 1986. Faunal provinces and the Simpson coefficient. *Contributions to Geology, University of Wyoming Special Paper* 3:317–38.

Fortelius, M., P. Andrews, R. L. Bernor, S. Viranta, and L. Werdelin. In press. Preliminary analysis of taxonomic diversity, turnover, and provinciality in a subsample of large land mammals from the later Miocene of Western Eurasia. In *Neogene and Quaternary Mammals of the Palaearctic*, ed. A. Nadachowski and L. Werdelin. Acta Zoologica Cracoviensia.

Fortelius, M. and J. Kappelman. 1993. The largest land mammal ever imagined. *Zoological Journal of the Linnean Society* 108:85–101.

Fortelius, M., J. Van der Made, and R. L. Bernor. This volume. Middle and late Miocene Suoidea of Central Europe and the Eastern Mediterranean: Evolution, biogeography, and paleoecology.

Gaudry, A. 1865. *Animaux fossiles et Géologie de l' Attique.* Paris.
——. 1873. *Animaux fossiles du Mont Lebéron.* Paris.
Gentry, A. and E. Heizmann. This volume. Miocene ruminants of the Central and Eastern Paratethys.
Gingerich, P. D. 1984. Pleistocene extinctions in the context of origination-extinction equilibria in Cenozoic mammals. In *Quaternary Extinctions: A Prehistoric Revolution,* ed. P. S. Martin and R. G. Klein, pp. 211–22. Tuscon: University of Arizona Press.
Harrison, J. L. 1952. The distribution of feeding habits among animals in a tropical rain forest. *Journal of Animal Ecology* 31:53–64.
Hayek, L.-A. C., R. L. Bernor, N. Solounias, and P. Steigerwald. 1991. Preliminary studies of hipparionine horse diet as measured by tooth microwear. *Annales Zoologici Fennici* 3–4:187–200.
Heissig, K. This volume. The stratigraphical range of fossil rhinoceroses in the late Neogene of Europe and the Eastern Mediterranean.
Howell, F. C. 1987. Preliminary observations on Carnivora from the Sahabi Formation (Libya). In *Neogene Paleontology and Geology of Sahabi,* ed. N. T. Boaz, A. El-Arnauti, A. W. Gaziry, J. de Heinzelin, and D. D. Boaz, pp. 152–81. New York: Liss.
Janis, C. M. 1976. The evolutionary strategy of the Equidae and the origins of rumen and cecal digestion. *Evolution* 30:757–74.
——. 1982. Evolution of horns in ungulates: Ecology and palaeoecology. *Biological Reviews (Cambridge)* 57:261–318.
——. 1989. A climatic explanation for patterns of evolutionary diversity in ungulate mammals. *Palaeontology* 32:463–81.
——. 1993. Tertiary mammal evolution in the context of changing climates, vegetation, and tectonic events. *Annual Reviews in Ecology and Systematics* 24:467–500.
Kappelman, J., S. Sen, M. Fortelius, A. Duncan, B. Alpagut, J. Crabaugh, A. Gentry, J.-P. Lunkka, F. McDowell, N. Solounias, S. Viranta, and L. Werdelin. This volume. Chronology and biostratigraphy of the Miocene Sinap Formation of Central Turkey.
Koenigswald, G. H. R. von. 1929. Bemerkungen zur Säugetierfauna des rheinhessischen Dinotheriensandes. *Senckenbergiana* 11:267–79.
Köhler, M. 1993. Skeleton and habitat of recent and fossil ruminants. *Münchner Geowissenschaftliche Abhandlungen* 25:1–88.
Krause, D. W. and M. C. Maas 1990. The biogeographic origins of late Paleocene–early Eocene mammalian immigrants to the Western Interior of North America. In *Dawn of the Age of Mammals in the Northern Part of the Rocky Mountain Interior, North America,* ed. T. M. Bown and K. D. Rose, pp. 71–105. Geological Society of America Special Paper 243.
Kurtén, B. 1952. The Chinese Hipparion fauna. *Commentationes Biologicae Societatis Scientiarum Fennicae* 13:1–82.
Legendre, S. and C. Roth. 1988. Correlation of carnassial tooth size and body weight in recent carnivores (Mammalia). *Historical Biology* 1:85–98.
Lillegraven, J. A. 1972. Ordinal and familial diversity of Cenozoic mammals. *Taxon* 21:261–74.
Maas, M. C., M. R. L. Anthony, P. D. Gingerich, G. F. Gunnell, and D. W. Krause. 1995. Mammalian generic diversity and turnover in the late Paleocene and early Eocene of the Bighorn and Crazy Mountains Basins, Wyoming and Montana (USA). *Palaeogeography, Palaeoclimatology, Palaeoecology* 115:181–207

Maples, C. G. and A. W. Archer. 1988. Monte Carlo simulation of selected binomial similarity coefficients (II): Effect of sparse data. *Palaios* 3:95–103.
Mares, M. A. 1992. Neotropical mammals and the myth of Amazonian biodiversity. *Science* 255:976–79.
May, R. M. 1975. Patterns of species abundance and diversity. In *Ecology and Evolution of Communities,* ed. M. L. Cody and J. M. Diamond, pp. 81–120. Cambridge, Mass.: Belknap Press of Harvard University Press.
McGuiness, K. A. 1984. Equations and explanations in the study of species-area curves. *Biological Reviews* 59:423–40.
Mein, P. 1979. Rapport d'activité du groupe travail des vertébrés, mise à jour de la biostratigraphie du Néogène basée sur les mammifères. *Annales Géologiques des Pays Helléniques,* hors série, 3:1367–72.
——. 1989. Updating of MN zones. In *European Neogene Mammal Chronology,* ed. E. H. Lindsay, V. Fahlbusch, and P. Mein, pp. 73–90. New York: Plenum.
Moyà-Solà, S. and J. Agustí. 1990. Bioevents and mammal successions in the Spanish Miocene. In *European Neogene Mammal Chronology,* ed. E. H. Lindsay, V. Fahlbusch, and P. Mein, pp. 357–73. New York: Plenum.
Potts, R. and A. K. Behrensmeyer. 1992. Late Cenozoic terrestrial ecosystems. In *Terrestrial Ecosystems Through Time,* ed. A. K. Behrensmeyer, J. D. Damuth, W. A. DiMichele, R. Potts, H. Sues, and S. L. Wing, pp. 419–541. Chicago: University of Chicago Press.
Rögl, F. and G. Daxner-Höck. This volume. Later Miocene Paratethys correlations.
Rosenzweig, M. L. and Z. Abramsky. 1993. How are diversity and productivity related? In *Species Diversity in Ecological Communities,* ed. R. E. Ricklefs and D. Schluter, pp. 52–65. Chicago: University of Chicago Press.
Roth, V. L. 1990. Insular dwarf elephants: A case study in body mass estimation and ecological inference. In *Body Size in Mammalian Paleobiology,* ed. J. Damuth and B. J. MacFadden, pp. 151–79. Cambridge: Cambridge University Press.
Schlosser, M. 1903. Die fossilen Säugethiere Chinas nebst einer Odontographie der recenten Antilopen. *Abhandlungen der Bayerischen Akademie der Wissenschaften* 22:1–221.
Sen, S. (ed.). 1994. Les gisements de mammifères du Miocène supérieur de Kemiklitepe, Turquie. *Bulletin du Muséum Nationale d'Histoire Naturelle* C1:1–241.
Sepkoski, J. J., Jr. 1988. Alpha, beta, gamma: Where does all the diversity go? *Paleobiology* 14:221–34.
Sickenberg, O. 1975. Die Gliederung des höheren Jungtertiärs und Altquartärs in der Türkei nach Vertebraten und ihre Bedeutung für die internationale Neogen-Stratigraphie. *Geologisches Jahrbuch B* 15:1–167.
Simpson, G. G. 1943. Mammals and the nature of continents. *American Journal of Science* 241:1–31.
——. 1947. Evolution, interchange, and resemblance of the North American and Eurasian Cenozoic mammalian faunas. *Evolution* 1:218–20.
Sokal, R. R. and P. H. A. Sneath. 1963. *The Principles of Numerical Taxonomy.* San Francisco: Freeman & Co.
Solounias, N. 1981a. *The Turolian fauna from the island of Samos, Greece.* Contributions to Vertebrate Evolution 6. Basel: Karger.
——. 1981b. Mammalian fossils from Samos and Pikermi. Part 2.

Resurrection of a classic Turolian fauna. *Annals of the Carnegie Museum* 50:231–70.

Steininger, F. F., W. A. Berggren, D. V. Kent, R. L. Bernor, S. Sen, and J. Agusti. This volume. Circum-Mediterranean Neogene (Miocene and Pliocene) marine-continental chronologic correlations of European mammal units.

Steininger, F. F., R. L. Bernor, and V. Fahlbusch. 1989. European Neogene marine/continental chronologic correlations. In *European Neogene Mammal Chronology*, ed. E. H. Lindsay, V. Fahlbusch, and P. Mein, pp. 15–46. New York: Plenum.

Stucky, R. K. 1990. Evolution of land-mammal diversity in North America during the Cenozoic. In *Current Mammalogy*, ed. H. Genoways, pp. 2:375–432. New York: Plenum.

——. 1992. Mammalian faunas in North America of Bridgerian to Arikareean "Ages" (Eocene and Oligocene). In *Eocene-Oligocene Climatic and Biotic Evolution*, ed. D. R. Prothero and W. A. Berggren, pp. 464–93. Princeton: Princeton University Press.

Swisher, C. C., III. This volume. New ^{40}Ar/^{39}Ar dates and their contribution toward a revised chronology for the late Miocene of Europe and West Asia.

Thenius, E. 1951. Die jungtertiäre Säugetierfauna des Wiener Beckens in ihrer Beziehung zu Stratigraphie und ökologie. *Erdölztg* 5:52–54.

Tobien, H. 1967. Subdivision of Pontian mammal faunas. Committee Mediterranean Neogene Stratigraphy, Proceedings IV Session, Bologna 1967. *Giornale di Geologia* (2) 35:1–5.

Van der Made, J.. 1992. Iberian Suoidea. *Paleontologia i Evolucío* 23 (1989–1990): 83–97.

Van Valkenburgh, B. 1990. Skeletal and dental predictors of body mass in carnivores. In *Body Size in Mammalian Paleobiology*, ed. J. Damuth and B. J. MacFadden, pp. 181–205. Cambridge: Cambridge University Press.

Van Valkenburgh, B. and C. M. Janis. 1993. Historical diversity patterns in North American large herbivores and carnivores. In *Species Diversity in Ecological Communities*, ed. R. E. Ricklefs and D. Schluter, pp. 330–40. Chicago: University of Chicago Press.

Webb, S. D. 1969. Extinction-origination equilibria in late Cenozoic land-mammals of North America. *Evolution* 23:688–702.

——. 1984. Ten million years of mammal extinctions in North America. In *Quaternary Extinctions: A Prehistoric Revolution*, ed. P. S. Martin and R. G. Klein, pp. 189–210. Tucson: University of Arizona Press.

Werdelin, L. and N. Solounias. This volume. The evolutionary history of hyaenas in Europe and Western Asia during the Miocene.

Whittaker, R. H. 1977. Evolution of species diversity in land communities. *Evolutionary Biology* 10:1–67.

Woodburne, M. O. and R. L. Bernor. 1980. On superspecific groups of some Old World hipparionine horses. *Journal of Paleontology* 8(4):315–27.

Zapfe, H. 1979. *Chalicotherium grande (Blainv.) aus der miozänen Spaltenfüllung von Neudorf und der March (Devinska Nova Ves), Tscheckoslowakei*. Wien: Ferdinand Berger und Söhne.

——. 1989. *Chalicotherium goldfussi* Kaup aus dem Vallesien vom Höwenegg im Hegau (Südwestdeutschland). *Andrias* 6:117–26.

The Evolution of Western Eurasian Neogene Mammal Faunas: A Chronologic, Systematic, Biogeographic, and Paleoenvironmental Synthesis

R. L. BERNOR, V. FAHLBUSCH, P. ANDREWS, H. DE BRUIJN, M. FORTELIUS, F. ROGL, F. F. STEININGER, AND L. WERDELIN

This volume contains a revision of the Western Eurasian Neogene time scale, the systematics and evolution of most major small and large mammal groups, and interpretations of the relationship between environmental change and faunal turnover. This synthesis first discusses the principal changes in the Neogene time scale and where major discrepancies occur in their correlation. We follow with a summary of Western Eurasian later Neogene small and large mammal evolution and biogeography. Chronologic and mammalian evolutionary data are combined with the biogeographic and paleoecologic analysis given by Fortelius et al. (this volume b) to reveal the major faunal changes that occurred in the Western Eurasian later Neogene, and then to briefly examine the isochronous/diachronous nature of MN units, the extent to which faunal provinciality prevailed across Western Eurasia, and whether regional or global abiotic factors forced faunal change.

Chronologic Framework

Methods for correlating fossil mammal faunas vary greatly across the Old World. European, West Asian, and North African continental chronology is organized within the MN correlation system (Mein 1975, 1979, and 1989; Bernor 1983, 1984; Steininger et al. 1989; this volume). The North and East Asian regions have a less formalized mammalian biochronologic system, but recent studies have broadly correlated their sequences to the European MN unit system (Qiu 1989). South Asian Neogene mammal chronology has largely been based on a well-defined magnetostratigraphic framework (Barry et al. 1982; 1985). East African continental chronology differs from all other regions in predominately calibrating its biostratigraphic records using radioisotopic methods (see Harris 1983 and Brown and Feibel 1991 for a review).

Of the various correlation systems used, MN units (sensu Fahlbusch 1991) have the broadest regional application, but at the same time are the most complex in their development and usage. The considerable history of MN unit development is inextricably coupled with the development of contemporary Western Eurasian mammalian paleontology. We review this development as a background to our synthesis of the workshop's results.

MN Unit Development and Application

The first development of a European mammal chronology was presented by Thaler (1966). Thaler's initiative reflected vertebrate paleontologists' determination to depart from the existing makeshift correlation system, which inappropriately utilized the European marine stage system and estimated continental faunal correlation to it. Thaler's correlation scheme characterized well-known West European faunas using mammalian evolutionary grades. Stratigraphic superposition was not attempted, providing a truly biochronological system (Lindsay and Tedford 1989; Fahlbusch 1991). Thaler's system soon underwent a series of refinements by Sudre (1969; 1972), Hartenberger (1969), and Thaler (1972), and this set the stage for the next generation of revisions.

Mein (1975) revolutionized Thaler's nascent mammalian biochronologic framework by erecting sixteen Neogene "zones," which he characterized by three criteria: (1) characteristic taxa belonging to broadly distributed mammalian groups; (2) well-known generic first appearances; (3) characteristic associations of genera. Mein (1979; 1989) has since refined and revised this system so that seventeen Neogene zones, two subzones (2a, 2b), and Q (for Quaternary) are now recognized. These "zones" are applied by him to several biogeographic areas: Central Europe, Eastern Europe, Southeastern Europe, Western Asia, and North Africa. Mein (1989: tab. 2, pp. 81–89) further lists principal MN "zone" reference localities.

Daams and Freudenthal (1981) challenged the utility of European MN "zones," arguing that their lack of a biostratigraphic basis has often led to incorrect correlations. Bernor (1978; 1983; 1984) and Bernor and Pavlakis (1987) have demonstrated striking provincial diachroneity of MN unit faunas, and attribute these discrepancies to regional paleogeographic (re: Rögl and Steininger 1983) and paleoenvironmental differences. Steininger et al. (1989:20–21) have discussed the various approaches for developing a stratigraphic framework for MN "zones" and proposed the term "Neogene Mammal Faunal Zone" for those units that are truly biostratigraphically defined. So defined, these Neogene mammal faunal zones have been placed within larger biostratigraphic units (e.g., Turolic). However, as demonstrated by Fahlbusch (1991) for greater Europe, and in this volume for Western Eurasia, the application of MN "zones" is inappropriate since the concept is biochronologic rather than biostratigraphic; hence our usage of MN units (after Fahlbusch 1991).

Despite the various problems incurred by a transprovincial correlation system, the one based on European MN units has proven to be very useful. Research in the last ten years has included an increased application of radioisotopic and magnetostratigraphic technologies for MN unit boundary definition (re: Steininger et al. 1989). Furthermore, a new correlation tool, eustatic chronology, has allowed testable hypotheses to be posed on the relationship between short-term global eustatic events and unit-defining mammalian migration events (re: Rögl and Steininger 1983; Steininger et al. 1985; Barry et al. 1985; Bernor 1985; Bernor et al. 1988a, b; Bernor et al. 1989; Lindsay 1989; Opdyke et al. 1989; Bernor and Tobien 1990; Bernor and Lipscomb, 1995).

A Revised MN Chronology

A major impetus of this workshop was to refine MN unit chronology. Several geochronologists participated, forwarding new radioisotopic, magnetostratigraphic, marine-continental, and biochronologic correlations. Steininger et al. (this volume) integrated this work within the most recently revised Cenozoic time scale (Berggren et al. 1995), providing clear paleontologic, stratigraphic, and magnetostratigraphic definitions of marine stage boundaries. These correlations form the empirical bases for our MN correlations, and the obligatory starting point for the biogeographic and paleoenvironmental comparisons we summarize here.

Swisher's (this volume) single crystal $^{40}Ar/^{39}Ar$ dates on the early Vallesian locality of Höwenegg (Germany; also Woodburne et al., this volume a, b), and the Turolian

localities of Maragheh, Iran, and Samos, Greece, were critical for revising the later Miocene of Western Eurasia. These radioisotopic studies led to the reevaluation and correlation of several key magnetostratigraphic sequences in the circum-Mediterranean area. Höwenegg's 10.3 Ma date is critical from the following standpoints: (1) for providing a much needed Central European later Miocene radioisotopic tie point; (2) for providing a precise radioisotopic correlation for lower MN 9; (3) for providing a more accurate date for the Old World "Hipparion" Datum. While the Höwenegg hipparion exhibits a number of archaic morphologic features (Bernor et al., in press b), it is not the most primitive member of the Central European "Hippotherium" primigenium s.s. lineage. It is slightly derived compared to the Vienna Basin Pannonian C and D–E hipparions (Bernor et al. 1993a, 1993b; Bernor et al., this volume a; Woodburne et al., this volume a).

Swisher's (this volume) correlation of Höwenegg with other provincial hipparion first occurrences finds close agreement on the age of basal MN 9 with results derived from the Sinap section, Turkey (Kappelman et al., this volume): the "Hipparion" Datum is found to be isochronous and correlated with the middle portion of the Chron 5's long normal interval: 10.5 Ma. Pilbeam et al. (this volume) review the "Hipparion" Datum in the Potwar Plateau, Pakistan, and derive a slightly older age of 10.7 Ma. Rögl and Daxner-Höck (this volume) as well as Steininger (in Steininger et al., this volume) have forwarded an alternative correlation for the base of MN 9, ca. 11.2 Ma, based on Central–Eastern Paratethys correlations. Compared to former correlations of the peri-Mediterranean (ca. 12.5 Ma; Berggren and Van Couvering 1974) and the Siwaliks (ca. 9.5 Ma; Barry et al. 1982), the age discrepancy for Old World hipparion is now far less and most probably resolvable with further work.

Swisher's (this volume) reanalysis of volcanic tuffs in the Maragheh (Iran) and Samos (Greece) sections was further augmented by new systematic, stratigraphic, and biochronologic comparisons of these two localities and Pikermi (Greece; Bernor et al., this volume b). These three Turolian localities were formerly upheld as classic examples of "Pontian" faunas and were believed to represent the terminal Miocene extention of "savanna" biotopes across virtually all of Eurasia. Mein's (1975) earlier correlations originally placed Samos 1–4 and Pikermi in upper MN 12 and Samos 5 in MN 13. Radioisotopic dates from Maragheh (Campbell et al. 1980), coupled with biochronologic correlations between it, Samos, and Pikermi (Bernor et al. 1980; Bernor 1986; Solounias 1981), contradicted the notion that these open country faunas were constrained in time to the terminal Miocene. Bernor's (1978; 1983; 1984) and Bernor et al.'s (1979) view that these provincial faunas had a chronologically long evolutionary history that was tied to the paleogeographic and paleoclimatic development of Europe (Rögl and Steininger 1983) was later reflected in Mein's (1989) and Steininger et al.'s (1989) biochronologic and geochronologic revisions.

In the most recent correlation of these localities, de Bruijn et al. (1992) did not include the Maragheh correlations, placed Samos 1–4 in MN 11, Pikermi and the nearby locality of Chomateri in upper MN 12, and Samos 5 in MN 13. Bernor et al. (this volume b) now correlate these faunas as follows: Maragheh is basal MN 11 (9.0 Ma) to medial MN 12 (7.6 Ma) age; Pikermi is correlative with the MN 11/12 boundary (= Lower/Middle Maragheh; ca. 8.3 Ma); Samos' lowermost Old Mill Beds localities (such as Quarry X and locality 6) are correlative with basal Middle Maragheh (= lower MN 12, ca. 8.2 Ma), the Main Bone Beds (Samos 1–4, 5) are correlative with uppermost MN 12 (ca. 7.3–7.1 Ma), and only locality L is correlative with MN 13. These results are congruent with recent correlations of the Thessaloniki faunas (de Bonis et al. 1988; Koufos 1989), which record "Pikermian" open country faunas back into the late Vallesian (MN 10; re: de Bruijn et al. 1992) and perhaps MN 10 of Anatolia (Kappelmann et al., this volume; Fortelius, pers. comm.). Certainly, critical taxonomic elements of the "Pikermian" chronofauna occur by immigration, and others by evolutionary origin as early as MN 9. Indeed, the nascent community structure of this chronofauna is anticipated by the early diversification of the Hyaenidae, Bovidae, and Giraffidae in the Astaracian (MN 6–7/8).

Three papers attempted to draw together new lines of chronologic evidence and better resolve Western Eurasian Neogene correlations. Sen (this volume a) updated all available information on European and Anatolian magnetostratigraphic correlations, which were largely confined to sections measured in Spain, Greece, and Turkey. Rögl and Daxner-Höck (this volume), following a critical review by Rögl et al. (1993), significantly refined the Central and Eastern Paratethys time scale, affording a clear basis for correlations with the circum-Mediterranean and Western Eurasian realms. Kappelmann et al. (this volume) forwarded initial results on a key basal middle Miocene–Pliocene section at Sinap, Anatolia, which has proved critical for the east-west faunal comparisons that mammalian paleontologists pursue in this volume.

MN units have gained considerable precision as a result of these renewed efforts. While chronologic resolution is much improved, a general consensus of the age of all MN units is yet to be resolved. MN boundaries presently actively debated include: MN 5/6, variously late Burdigalian correlative (16.5 Ma; Steininger in Steininger et al., this volume) or late Langhian (15.2 Ma; Sen, this volume a; Bernor in Steininger et al., this volume); MN 8/9, variously 10.7/10.5 Ma, based on European and Asian correlations of the "*Hipparion*" Datum (Kappelman et al., this volume;

Pilbeam et al., this volume; Swisher, this volume; Woodburne et al., this volume a, b), or 11.2 (= late Bessarabian of the Eastern Paratethys; Rögl and Daxner-Höck, this volume) based on Paratethys correlations.

The MN Turolian boundaries are not in agreement with the West European localities and those of Central and Eastern Europe and Southwestern Asia. Sen (this volume a) has correlated these boundaries as: MN 10/11 = ?8.5 Ma; MN 11/12 = 7.2 Ma; MN 12/13 = 6 Ma. Bernor et al. (this volume b), Rögl and Daxner-Höck (this volume), and Steininger et al. (this volume) correlate these boundaries as: MN 10/11 = 9.0 Ma; MN 11/12 = 8.24 Ma; MN 12/13 = 7.1 Ma. Whereas Sen's correlation scheme relies on the Spanish sequence for this part of the record, the other authors rely on the Southeast European–Southwest Asian sequence. The differences in these various correlations may in part be due to the precision of various chronologic tools available, with Swisher's single crystal argon dates weighing heavily in favor of the "eastern chronology." However, studies of a number of the large mammal groups clearly show that the evolutionary radiation and succession of Turolian open country megafaunas was not isochronic, but distinctly provincial (Fortelius et al., this volume b), and most likely confounds correlations between Western Europe, Central Europe, and the Eastern Mediterranean–Southwest Asian provinces.

Mammalian Systematics, Biogeography, and Paleoecology

Lindsay (1989:3) has identified in the MN unit and other biochronologic systems a general failure to develop a well-defined and consistently applied methodology for stage-of-evolution correlations. This problem strikes at the heart of the MN correlation discrepancies. Old World MN correlations often employ genus-rank comparisons. Their use is inappropriate because these genus-groups are often paraphyletic, including multiple lineages united by symplesiomorphic characters (re: Bernor and Lipscomb, 1995). Definition of lineages by a character-based analysis can readily uncover these systematic problems and lead to: (1) greater precision in the lineage's characterization; (2) greater biochronologic precision by basing correlations on lineage-specific comparisons; (3) finer resolution in determining the chronology of biogeographic extension and vicariant divergence of provincially restricted lineages. In several instances, our sample of Neogene mammals exhibits very long, often intercontinental, geographic extensions. Therefore, a critical point for improving MN correlation precision is the employment of broad, interregional-, and even intercontinental-based reconstructions of mammalian evolutionary lineages.

The mammalian systematic chapters derived from the 1992 Schloss Reisensburg workshop were intended to present the participants' current understanding about the evolutionary history of the groups under consideration. In some cases these presentations followed years of detailed systematic and evolutionary research. In others, information was based on a very nascent systematic inquiry. In all cases the information is the most current for the various groups considered. The systematic chapters have laid the groundwork for the biochronologic, paleogeographic, and paleoecological interpretations set forth in the last chapters of this volume. Appendix 32.1 presents a summary of mammalian systematics, chronology, and biogeography of species distributions. This appendix is meant to serve as a quick reference to the interpretations that follow. For further details the reader is referred to the original articles on which this synthesis is based.

Biogeographic and Paleoecological Patterns

Fortelius et al. (this volume b) have taken the composite revised faunal data base developed from the systematic and geochronologic studies provided in this volume and exhaustively analyzed biogeographic and paleoecologic change from MN 5 to MN 13. We summarize the overall patterns below.

MN 5   MN 5 is the latest unit of the Orleanian faunal unit and correlative with the Mediterranean late Burdigalian and Central Paratethys late Karpathian Stage. Steininger prefers a 17.0 Ma correlation for the base of MN 5, while Bernor prefers a slightly later, terminal Burdigalian correlation of 16.5 Ma (re: Steininger et al., this volume). Either age is correlative with a major global sea-lowering event and is a time of extensive Eurasian-African biogeographic connection (Barry et al. 1985; Rögl and Steininger 1983; Steininger et al. 1985; Mein 1989; Bernor and Tobien 1990).

There were a number of small and large mammal lineages holding over from the earlier Miocene MN units. "Relictual" small mammals were particularly prevalent at this time in Central and Western Europe, including: a diverse suite of insectivores, chiropterans, and glirids, a relictual didelphid, eomyids, zapodids, and primitive cricetids (*Eumyarion*, *Megacricetodon*, and *Democricetodon*). Likewise, several carnivore families carried over, particularly in Central and Western Europe: viverrids, amphicyonids, mustelids, and ursids. Hyaenids first occurred in MN 4 of Western Europe, and by MN 5 there were two primitive insectivorous/carnivorous Western and Central European lineages: *Protictitherium* and *Plioviverrops*. The anomalomyid and spalacid rodents first occurred in the Eastern block as immigrants from the east. Eucatarrhine primates first emigrated from Africa during this interval with pliopithecids being represented by a diverse Western and Central European suite of species, and larger bodied

hominoids (*?Griphopithecus*) first occurred in Central Europe. The herbivore fauna carrying over from the early Miocene included the cosmopolitan rhinoceros *Brachypotherium* and broadly distributed hyotheriine *Hyotherium soemmeringi* (LAD MN 6 of Central Europe). The primitive bovid *Eotragus*, of probable Asian origin, carried over from MN 4, and the primitive giraffid *Giraffokeryx* immigrated into Eurasia from Africa/Arabia during this interval.

According to Fortelius et al. (this volume b), the Western block (hereafter West) was deeply forested, more than at any later time during our interval. Fortelius et al. (this volume b) did not look for regional differences within this block. The Eastern block (hereafter East) is poorly known, with only Inönü I (MN 5/6 boundary) and Mala Miliva to provide insight into regional environments. Ecologically, nothing indicates that MN 5 differed from MN 6 in East. The fact that faunal similarity was low despite a small sample effect pulling in the opposite direction is almost certainly significant. The ecologically distinct *"Listriodon"* fauna was yet to come.

MN 6   The base of MN 6 corresponds with the base of the Astaracian faunal unit and is alternatively correlative with the base of the Langhian Stage (Steininger, in Steininger et al., this volume) or terminal Langhian Stage (Bernor, in Steininger et al., this volume). MN 5–6 chronology pivots on whether one accepts de Bruijn et al.'s (1992) correlation of the Neudorf fissure fills as being an MN 6 fauna, or previous interpretations that it is correlative with MN 5 (Steininger et al. 1989). Steininger correctly indicates that the Devinska Nova Ves fissure fill is capped by a basal Langhian transgressive unit, requiring a pre-Langhian MN 6 correlation. Bernor, following Bernor et al. (1988a) and Steininger et al. (1989), continues to observe the MN 5 correlation of Devinska Nova Ves, further citing magnetostratigraphic correlations of the basal MN 6 localities of Sansan (Sen, this volume), Inönü (base of C5Bn, ca. 15.2 Ma; Kappelman et al., this volume), and the slightly later MN 6 locality of Paşalar (Bernor and Tobien 1990). Furthermore, the terminal Langhian-basal MN 6 correlation corresponds to a major global sea-level lowering event (Haq et al. 1987) and was a time of extensive Eurasian-African biogeographic connection (Barry et al. 1985; Mein 1989; Bernor and Tobien 1990). Moreover, with the MN 5 and MN 6 chronologies suggested by Bernor, there is a clear chronologic separation of these intervals, and the mammalian migratory events that define their bases would *both* be correlative with major short-interval global sea-lowering events.

During this interval didelphids became extinct. Insectivores, on the other hand, exhibited a tremendous diversity in Central Europe, and to a somewhat lesser extent in Western Europe from MN 6 through MN 7/8. Chiropterans exhibited a similar, albeit lower, diversity in these provinces. These two orders further showed a strong French–Central European biogeographic relationship, sharing many lineages in common. Sciurids were also diverse beginning in MN 6, with a number of taxa confined to the Central–West European provincial block (*Blackia, Forsythia,* and *Albanensia*). During MN 6 a second wave of anomalomyids invaded Southwestern Asia, Southeastern Europe, and Central Europe from the east. Also during this interval species of West European hyaenids, *Protictitherium* and *Plioviverrops,* extended their range eastward into Central Europe, Southeastern Europe, and Southwestern Asia. A characteristic Western Eurasian rhino fauna evolved that included species of *Plesiaceratherium, Hoploaceratherium,* and *Alicornops*. Elasmotherine rhinos extended their range in MN 6 from the east into Southwestern Asia only. The Central–West European lineage of tetraconodont pigs, *Parachleuastochoerus,* commenced its evolutionary record at this time. Two bovid species were common to Central European and the Southeast European–Southwest Asian provinces at this time: *Eotragus* and *Caprotragoides*. The Southeast European–Southwest Asian province records additional bovid origins during MN 6: *Hypsodontus serbicus* and *Hypsodontus pronaticornis, Miotragocerus* sp. and *Turcocerus gracilis.*

West would appear to have acquired slightly more open country environments. Andrews (in Fortelius et al., this volume b) has found that Western MN 5–9 faunas correspond most closely to the intermediate "semi-deciduous" forest pattern. MN 6–8 faunas in East similarly correspond to the drier end of the modern range of subtropical forest, having community structures most similar to deciduous forest at best, so that conditions were drier and probably more seasonal, and the environment was more open than in West. Faunal similarity rose during this time corresponding with the dispersion of the *Listriodon* fauna. Diversity remained static. Turnover was low in West and high in East.

MN 7/8   MN 7 and 8 have been combined into a single unit after the recommendations of de Bruijn et al. (1992). The correlation of the base of MN 7/8 has been changed here by Steininger et al. (this volume) from Steininger et al.'s (1989) estimation of 13.6 to 12.5 Ma based on a correlation with Central Paratethys Volhynian age sediments. However, Rögl still supports a 13.6 Ma age for the base Volhynian. This age estimation is correlative with the late Serravallian of the Mediterranean, middle Sarmatian s.s. of the Central Paratethys and late Volhynian of the Eastern Paratethys Stages.

MN 7/8 marks the initial crash in insectivore diversity. The sciurids *Heteroxerus, Atlantoxerus, Spermophilinus,* and *Tamias* established a broader East–West distribution at this time. Cricetids also showed some provinciality, with *Byzantinia* having evolved autochthonously in Southeast-

ern Europe–Southwestern Asia and *Hispanomys* in Southwestern Europe. Murines of the *Antemus* stage-of-evolution extended their provincial Indopakistan range westward into Southwestern Asia and North Africa during MN 7/8. The primitive early Miocene cricetid *Megacricetodon* became extinct during this interval in Western and Southeastern Europe, but extended its range well into MN 9 of Central Europe. The advanced hominoid primate *Dryopithecus* first occurred during this interval in both Western (*D. fontani*) and Central (*D. carinthiacus*) Europe. The first meat/bone eating hyaenids (*Thalassictis*) appeared in Western Europe. The rhino "*Dicerorhinus*" was prevalent in Central Europe and prevailed there well into MN 9. Tayassuids exhibited vicariance at this time with *Schizochoerus* (Southwestern Asia) and *Albanohyus* (Central and Western Europe) evolving provincially restricted ranges. The advanced suine *Korynochoerus palaeochoerus* first occurred and established a broad geographic range in Western and Central Europe and Southwestern Asia. The Bovidae continued to radiate in Southeastern Europe and Southwestern Asia with the addition of *Protoryx*.

Fortelius et al.'s (this volume b) analysis records very little change from the MN 6 interval in either West or East. Similarity continued to increase, and turnover values were converging: sinking in East and rising in West. Both blocks had fewer entries than in MN 6 or MN 9, but exits were rising quickly in West and sinking equally quickly in East. Faunal similarity peaks in MN 7/8, except for Dice, which shows a plateau (mean) or a minor second-order minimum (block). Of the carnivores in West, invertebrate-eaters (insectivores, chiropterans, but not primitive hyaenids) and omnivores decreased markedly in diversity, while other ecomorph groups continued to rise or level out. In the whole macromammal fauna of West, the lost taxa belonged to families usually associated with forest habitats (ursids, cervids, suoids, tragulids, etc.). Omnivores of East jumped to a much larger mean size while the small macromammals (<30 kg) became correspondingly smaller. East animal eaters also showed a diversity minimum here. In East the proportion of hypsodonts declined and that of brachydonts rose (beginning of a "forest episode" culminating in MN 9?).

MN 9   The base of MN 9 has been intensively studied by various authors in this volume, both from systematic (Bernor et al., this volume a; Woodburne et al., this volume a) and geochronologic stand points (Kappelman et al., this volume; Pilbeam et al., this volume; Sen, this volume a; Steininger et al. this volume; Swisher, this volume). From Paratethys correlations, Rögl and Daxner-Höck estimate an 11.2 Ma age of basal MN 9, whereas Pilbeam et al. argue for a 10.7 + Ma datum and estimates from other Western Eurasian sectors have suggested a 10.5 Ma correlation (Kappelman et al., this volume; Swisher, this volume) based on Central European, Anatolian, and Siwalik radio-isotopic or magnetochronologies. An 11.2 Ma age is correlative with the Serravallian–Tortonian boundary of the Mediterranean and early Pannonian Stage of the Central Paratethys and late Bessarabian of the Eastern Paratethys stage systems; the 10.5 Ma would be correlative with the early Tortonian, later early Pannonian and Khersonian, respectively. The 11.2 Ma correlation also corresponds to the terminal Serravallian sea-lowering event.

The lower boundary of MN 9 marked a second wave of Central and West European insectivore and chiropteran extinctions. Additionally, a number of Central European cricetid lineages, especially of the "*Neocricetodon*"–*Kowalskia*–group, radiated predominantly in Central Europe. Cricetids of the primitive *Megacricetodon*–group held on through MN 9 in Central Europe only. A number of Central European cricetid lineages were established in the Vallesian; some of these continued into the Turolian. Anomalomyid rodents underwent an extensive evolutionary radiation in Central and Western Europe. Murines of the *Progonomys* stage-of-evolution first occured in Southeastern and Southwestern Europe, but apparently not in Central Europe until MN 10. There were also a number of Western and Central European glirid origins at this time. Eucatarrhine primates underwent extensive extinctions by the end of this interval, with the notable rare exceptions of the derived Central European pliopithecid *Anapithecus* and the Greek hominid *Graecopithecus*.

The evolution of peri-Mediterranean "Pikermian"-like faunas is rooted in MN 9 (Bernor et al. 1988b). A new felid fauna evolved with the dispersion of *Machairodus* and later *Felis*. Of the hyaenids, *Thalassictis* continued and the whole group of doglike hyaenas began to radiate. MN 9 exhibits a major turnover in the large herbivore fauna with the first occurrence of the North American immigrant *Hippotherium* across all of Eurasia and Africa. The Southeast European–Southwest Asian realm underwent a complete turnover of the middle Miocene rhino fauna with *Chilotherium* having emigrated from Asia and *Ceratotherium* having emigrated from Africa. *Aceratherium* was yet another new rhino which occurred broadly across Western Eurasia. Quite to the contrary, the Central and West European rhino faunas underwent a gradual decline during the Vallesian. Bovids underwent an additional evolutionary explosion in Southeastern Europe and Southwestern Asia during MN 9.

Faunal similarity began to decrease during this interval, between blocks and within West (except that Austria is now more similar to Western Central Europe than it was in previous MN intervals). Diversity remained high, perhaps even rising in the West, where it became much higher than predicted by area, and definitely beginning to rise in East. There was a rise in the number of entries in West, but the number of exits also continued to rise, so that exits

outnumbered entries. East also exhibits an increased entry of taxa, but exits there continued to decrease. The turnover values for both blocks are identical in MN 9 (and, at lower levels, in MN 10). The animal eaters and <30 kg mammals of West decreased markedly in diversity. In West there was a noticeable carnivore species crash (forty-three species became extinct, almost half of them being mustelids). What happened? A matter of abundance rather than diversity of prey? Were carnivores somehow more sensitive to the fragmentation of West, the reduced area, or was there a critical change in forest environments? Other groups affected included the rhinos, bovids, and hominoids—not nearly as clear a forest guild as seen in the MN 7/8 exits. Andrews (in Fortelius et al., this volume b) has shown that the nature of the forest environment changed in an as yet unexplained manner, with loss of many medium-sized terrestrial species, but while equable conditions probably persisted in Central Europe, with closed semi-deciduous forests, there was probably some opening of the canopy in Western Europe. In East there was a great increase in terrestrial species with both browsing and grazing adaptations being represented, indicating the beginning of the break-up of the forest environments in the region.

MN 10   The base of MN 10 is variously estimated as being 9.5 or 9.3 Ma (Steininger et al., this volume); for the sake of continuity with most geochronology chapters we use the 9.5 Ma age here. A 9.5 Ma age would be correlative with the Mediterranean late early Tortonian, Central Paratethys medial Pannonian, and Eastern Paratethys early Maeotian Stages.

Murines of the *Progonomys* stage-of-evolution exhibit a late range extension into Central Europe, and underwent their initial radiation in Western Eurasia with the first occurrence of *Parapodemus*. Western Europe began to evolve an endemic murid fauna with the first occurrences of *Occitanomys* and *Huerzelerimys*. Glirids likewise exhibit a number of additional Central and West European origins. Other than the enigmatic form *Oreopithecus* (MN 12/13 of Italy), the last Miocene hominoid occured in Western Eurasia, as is recorded by the Macedonian form *Graecopithecus*. *Graecopithecus* is seen as being alternatively an African immigrant or a vicariant lineage of MN 8–9 *Dryopithecus*. The first bone-cracking hyaenid, *Adcrocuta eximia*, occurred in virtually all provinces. The equid lineages "*Hippotherium*" s.l. (Old World), *Cremohipparion* (Southeastern Europe/Southwestern Asia), and *Hipparion* s.s. (Western Europe–Southeastern Europe/Southwestern Asia) first radiated provincially during this interval. The large, open country suine *Microstonyx* dispersed across Western Eurasia corresponding with the LAD of *Korynochoerus*. The large giraffes *Decennatherium* and *Bohlinia*, as well as a diverse set of bovids, radiated in Southeastern Europe/Southwestern Asia.

Faunal similarity dropped to its lowest except for MN 5, with the exception of Dice East block, which is lowest in MN 11. Two areas contradicting the trend are Black Sea:Anatolia and, especially, Western Europe:Anatolia, which rises drastically from its all-time low in MN 9. We do not know which taxa are responsible, but for the future more detailed studies at the level of the primary data might be useful. Diversity in West fell drastically, while in East it continued to rise slowly. These diversity trends are similar in the subregions of East, but in West there is no reduction at all except for a very moderate one in Austria. Entries and exits both fall in West, while in East entries fell and exits remained at the level of MN 9. Turnover is still identical in both blocks, and lower than in MN 9. In West the small- to medium-sized macromammals decreased the most (Fortelius et al., this volume b), and Andrews has shown that this reduction mainly affected terrestrial species, continuing the trend seen in MN 9. In East the only increase was in the >500 kg class, so at some level the change affected size classes similarly in each block (but the initial conditions were very different!). The large mammal families that most declined in their species diversity in West include the equids, rhinos, hyaenids, cervids, and pigs. Moreover, brachydont equids and a suite of "forest" families (hominoids, peccaries, tragulids, and viverrids) became extinct. West brachydont diversity fell drastically, and western brachyodont taxa exhibit a trend toward the evolution of smaller-bodied forms. In East brachydont forms failed to diversify further but became larger. The proportion of brachydonts in East decreased during this interval. Otherwise, ecomorphological changes between MN 9 and MN 10 appear to be small in both blocks.

MN 11   The base of MN 11 is best calculated on the extrapolated age of basal Lower Maragheh, ca. 9.0 Ma. This age is correlative with the medial Tortonian, medial Pannonian, and lower Maeotian. A third-step extinction occurred, with several species of Soricidae and Heterosoricidae disappearing at the MN 10/11 boundary. Sciurids show some West–East connections as well as some Southeast European–Southwest Asian provinciality (*Miopetaurista, Hylopetes*). *Kowalskia* is first known to extend its range into the peri-Mediterranean region at this time. Cercopithecoid primates, also African immigrants, first occurred at the end of MN 11 in Southeastern Europe/Southwestern Asia. Typical "Pikermian" chronofaunal forms including hyaenids, hipparionine horses, the giraffes *Samotherium* and *Helladotherium*, and especially bovids underwent an extensive radiation in the Southeast European–Southwest Asian "superprovince."

Faunal similarity continued to be low, except Simpson block, which increased sharply, and Dice block, which decreased. By the Dice index, the only increased similarity is the sharp one seen between the Balkans and Anatolia,

while the rest exhibit a decrease or stasis, with the sharpest decrease being between Western Europe and Anatolia. Diversity fell in West to a level well below that predicted by area, and rose sharply in East, also more than predicted by area, with little difference between the subregions. West entries and exits both decreased, as did eastern exits. East entries rose, which is why diversity increased. Turnover decreased somewhat in West and remained almost unchanged in East (the smallest decrease of any interval during the unbroken descent from MN 6 to MN 13). The horses and bovids in East diversified: hipparions reached their peak diversity, while bovids continued their diversification into MN 12. In West, the fauna was more relictual, reflecting the retention of forested conditions. Andrews (in Fortelius et al., this volume b) has shown that, in terms of their community structure, these conditions were most similar to subtropical deciduous forests of today, while in East the climate was probably more seasonal, with open deciduous tree cover, lacking closed canopy but still far from the tropical savanna faunas seen today in tropical East Africa that are often used as a model for the "Pikermian" "savanna-like" faunas.

MN 12   Basal MN 12 is best correlated by the calibrated first occurrence of *Hipparion prostylum* at the base of Middle Maragheh: ca. 8.24 Ma. This age is in conflict with magnetostratigraphic correlations of basal MN 12 in Spain (7.2 Ma; Sen, this volume). An 8.24 Ma age is correlative with the upper half of the Tortonian, upper half of the Pannonian, and medial Maeotian Stages.

The West European endemic murine fauna continued to expand with the origin of the *Stephanomys* lineage. The enigmatic hominoid primate *Oreopithecus* is recorded in association with a strange endemic Italian fauna, apparently isolated from the rest of Western Eurasia for a long interval of time. Canids first immigrated from the east and the meat/bone-eating cursorial hyaenids *Lycyaena* and *Hyaenictis* first occurred in the Southeast European–Southwest Asian "superprovince." The first "bone-cracking" hyaenid, *Belbus*, occurred in Southwestern Asia. The more "grazing" form, *Hipparion* s.s., dispersed across a narrow latitudinal belt ranging from France, eastward as far as Indopakistan. The Bovidae continued their aggressive evolutionary radiation in Southeastern Europe and Southwestern Asia, bringing "Pikermian" faunas into full bloom.

The trends established in MN 11 were largely continued in MN 12, although data for West are poor. Faunal similarity increased sharply, while diversity was similar to MN 11. Turnover values are also similar, although somewhat elevated in West, where entries exceed exits. In East exits rose for the first time since MN 6, and quite sharply, while entries declined still further. Andrews (in Fortelius et al., this volume b) has shown the development for the first time in this region of faunas with savanna-like community structures, with further increases in terrestrial species with both browsing and especially grazing adaptations. In West brachydont ungulates continued to decrease in diversity and plant eaters became smaller: the general impression is one of unchanged ecology since MN 11.

MN 13 The majority of this volume's geochronology authors have adopted a base MN 13–base Messinian correlation, ca. 7.1 Ma (re: Steininger et al., this volume). At this time a long interval regional regression commenced that involved the Mediterranean, Central and Eastern Paratethys (= Pontian Stage of Rögl and Daxner-Höck, this volume) provinces. This broad regression is believed to have precipitated a strong shift to more seasonal, open country faunas in the circum-Mediterranean and West Asian regions. However, based on magnetostratigraphic calibration of the Spanish rodent succession, Sen (this volume) correlates the base of MN 13 as being 6.0 Ma.

Mean East–West block similarity declined during MN 13. Eastern block species diversity again declined. There was a provincial continuation of Southeast European/Southwest Asian "Pikermian" faunas, although there was a reduction in large mammal species diversity. One particularly striking species reduction was in bovid diversity, declining from 25+ species to 10 species, and reversing the trend of increasing diversity that occurred between MN 6 and MN 12. There is some evidence that many species typical of these "Pikermian" faunas extended their range into Western and Central Europe (Bernor 1983) and North Africa (Bernor and Pavlakis 1987).

Beyond the assimilation of some "Pikermian" elements, the Western block underwent its own evolutionary progression, documented in both the large- and small-mammal record. Western Europe would appear to have undergone a substantial ecological change that included the further adaptive radiation of terrestrial species and concomitant reduction in frugivorous species. As a result, the Western block's community structure evolved to resemble recent environments with warm, open deciduous woodlands.

The Mio–Pliocene boundary marks yet another major shift in environments and faunas from warm temperate seasonal climates to cooler temperate forest with a relatively impoverished megafauna.

Conclusions

Across the first thirty studies of this volume more than sixty authors have forwarded the results of our inquiry on the stratigraphic, chronologic, biogeographic, paleoenvironmental, and paleoecologic contexts of Western Eurasian later Neogene mammal evolution. The revised chronologic and systematic data bases have been combined in many of the chapters, and particularly the last two, to address the central themes of this workshop: (1) Are MN units isochronous? (2) Did Western Eurasian faunas turnover synchronously, or is there observable diachroneity? (3) Does Western Eurasia exhibit faunal, paleoenvironmental, and paleoecologic provinciality? (4) Was faunal change forced by regional or global abiotic agencies, or the coupling of the two? A careful assessment of the data presented here should convince the reader that these are complex issues with no clear-cut simple answers. However, as the result of this undertaking we do have a better understanding of these various issues.

First, MN unit boundaries have proven to be very difficult to correlate across Western Eurasia except in those limited circumstances when there are demonstrated interprovincial to intercontinental scale migration events such as the *Hipparion* Datum (basal MN 9), or an extensive immigration of extra-provincial forms such as is seen in MN 5 and MN 6 with the influx of African species. Still, even under the best of circumstances, the level of MN biochronologic precision falls far short of those regions where long magnetostratigraphic sequences (i.e., Siwaliks) or radioisotopic series (i.e., Höwenegg, Maragheh, Samos, and much of the East African record) have recently been applied.

Second, the issues of synchronous versus diachronous turnover and the existence of faunal and environmental provinciality are clearly related. Fortelius et al. (this volume b), analyzing results from the chronologically justified systematic studies, have demonstrated the existence of at least a Western "faunal block" (Central and Western Europe) and an Eastern (Southeastern Europe, Southwestern Asia) "faunal block" during the MN 5–13 interval. The Western block has been characterized as having retained a nonseasonal subtropical aspect with a distinct "forest" fauna later than the Eastern block. Within the Western block there is some evidence that Central Europe retained this more forested character throughout most of the late Miocene. The Eastern block was relatively precocious in its evolution of seasonal open country faunas, and further developed these faunas to a greater extent than the Western block did. Migration events of faunal elements clearly derived from outside of Western Eurasia had a substantial impact on the course of this region's faunal evolution. Whereas African and Asian emigrant species both influenced the development of Western Eurasian MN 5 and MN 6 ecosystems, later in the Miocene emigration from the east was a prevailing feature of provincial faunal change. As might be expected, later Miocene Central and East Asian immigrants, namely the large ungulates and carnivores, had their greatest impact on the relatively precocious late middle and late Miocene open country biotopes of Southwestern Asia and Southeastern Europe.

Western and Central Europe were apparently buffered from the effects of this change by provincial climates that retained a less seasonal character later in time (Bernor 1983; 1984).

Finally, in that the data presented here do not support synchronous faunal change across Western Eurasia during any interval of time, neither can they support a direct correlation between global abiotic events and global faunal change. However, the evidence advanced here does not dispel the plausibility of a global climatic influence on Western Eurasian faunal change. Our results did not include the identification of specific abiotic agencies that might have forced faunal change for any individual MN interval. Rather, the participants chose to focus on refining the stratigraphic, chronologic, and systematic data so as to document the diversity in provincial biogeography and paleoecology between MN 6 and MN 13. We believe that more basic scientific inquiry is needed before we can more pointedly address the relationships between abiotic processes and mammalian evolutionary response. However, it is not incidental that we have cited a number of correlations between global eustasis, regional paleogeographic change, environmental shifts, and faunal turnover: i.e., MN 6, MN 9, and MN 13.

This volume has assembled a current record of later Neogene mammal evolution in Western Eurasia. However, rather than a statement, we see this as a beginning for the study of this exciting multidisciplinary subject. As this final chapter was being written there was a flood of new chronologic and faunal data that may alter some of this volume's conclusions. These details are best left for future evaluation and will, with the accummulation of further data, lead to a constructive revision of this volume's contents. The most lasting contributions to this volume will likely be the new radioisotopic dates for key localities, the systematic revisions and interpretations derived from them, the methodologies applied to interpret faunal change, and of course the observation that the Western Eurasian later Neogene has an incredibly complex history that must be studied further.

Acknowledgments

We wish to thank the VW Stiftung for supporting the workshop, and the National Geographic Society and L. S. B. Leakey Foundation for funding the related Höwenegg and Rudabánya research projects. We further thank Professor Dr. Siegfried Rietschel for providing his unwavering support of this project. Bernor also thanks the Alexander Von Humboldt Stiftung for supporting his early formulation of this project. We wish to thank the township of Immendingen and the Schloss Reisensburg for extending every courtesy to us and the workshop participants. Finally, we thank all of those colleagues who agreed to participate in this project.

LITERATURE CITED

Andrews, P. 1992. Evolution and environment in the Hominoidea. *Nature* 360:641–47.

Andrews, P., T. Harrison, E. Delson, R. L. Bernor and L. Martin. This volume. Distribution and biochronology of European and Southwest Asian Miocene catarrhines.

Barry, J. C., E. H. Lindsay, and L. L. Jacobs. 1982. A biostratigraphic zonation of the middle and upper Siwaliks of the Potwar Plateau of Northern Pakistan. *Palaeogeography, Palaeoclimatology, Palaeoecology* 37:95–130.

Barry, J. C., N. M. Johnson, S. M. Raza, and L. L. Jacobs. 1985. Neogene mammalian faunal change in Southern Asia: Correlations with climatic, tectonic, and eustatic events. *Geology* 13:637–40.

Begun, D. 1992. Miocene fossil hominids and the chimp-human clade. *Science* 257:1929–33.

———. 1994. Relations among the great apes and humans: New interpretations based on the fossil great ape *Dryopithecus*. *Yearbook of Physical Anthropology* 37:11–63.

Berggren, W. A., J. Hilgen, D. V. Kent, C. G. Langereis, I. Raffi, J. D. Obradovich, M. E. Raymo, and N. Shakelton. In press. Late Neogene (Pliocene–Pleistocene) chronology: New perspectives in high resolution stratigraphy. In *Geochronology, Time Scales, and Stratigraphic Correlation: Framework for an Historical Geology*, ed. W. A. Berggren, D. V. Kent, and J. Hardenbol. Society of Economic Mineralogists and Paleontologists, Special Publication.

Berggren, W. A. and J. A. Van Couvering. 1974. The late Neogene: Biostratigraphy, geochronology, and paleoclimatology of the past 15 million years in marine and continental sequences. *Palaeogeography, Palaeoclimatology, and Palaeoecology* 16:1–216.

Bernor, R. L. 1978. *The Mammalian Systematics, Biostratigraphy, and Biochronology of Maragheh and its Importance for Understanding Late Miocene Hominoid Zoogeography and Evolution.* Ph.D. diss., University of California, Los Angeles.

———. 1983. Geochronology and zoogeographic relationships of Miocene Hominoidea. In *New Interpretations of Ape and Human Ancestry*, ed. R. L. Ciochon and R. S. Corruccini, pp. 21–64. New York: Plenum.

———. 1984. A zoogeographic theater and biochronologic play: The time/biofacies phenomena of Eurasian and African Miocene mammal provinces. *Paléobiologie Continentale* 14:121–42.

———. 1985. Neogene palaeoclimatic events and continental mammalian response: Is there global synchroneity? *South African Journal of Science* 81:261.

———. 1986. Mammalian biostratigraphy, geochronology, and zoogeographic relationships of the late Miocene Maragheh fauna, Iran. *Journal of Vertebrate Paleontology* 6:76–95.

Bernor, R. L., P. J. Andrews, N. Solounias, and J. A. Van Couvering. 1979. The evolution of "Pontian" mammal faunas: Some zoogeographic, palaeoecologic, and chronostratigraphic considerations. *Annales Géologiques des Pays Helléniques*, hors série. 1:81–89.

Bernor, R. L., L. J. Flynn, T. Harrison, S. T. Hussain, and J. Kelley. 1988a. *Dionysopithecus shuangouensis* (Catarrhini, Primates): A new element in the Kamlial fauna of Southern Sind, Pakistan. *Journal of Human Evolution* 17:339–58.

Bernor, R. L., G. D. Koufos, M. O. Woodburne. and M. Fortelius. This volume a. The evolutionary history and biochronology of European and Southwest Asian late Miocene and Pliocene hipparionine horses.

Bernor, R. L., J. Kovar-Eder, D. Lipscomb, F. Rögl, S. Sen, and H. Tobien. 1988. Systematics, stratigraphic, and paleoenvironmental contexts of first-appearing Hipparion in the Vienna Basin, Austria. *Journal of Vertebrate Paleontology* 8:427–52.

Bernor, R. L., M. Kretzoi, H.-W. Mittmann, and H. Tobien 1993a. A preliminary systematic assessment of the Rudabánya hipparions (Equidae, Mammalia). *Mitteilungen der Bayerischen Staatssammlung für Paläontologie und historische Geologie* 33:195–207.

Bernor, R. L. and D. Lipscomb. 1995. A consideration of Old World hipparionine horse phylogeny and global abiotic processes. In *Paleoclimate and Evolution, with Emphasis on Human Origins*, ed. E. Vrba et al. New Haven: Yale University Press.

Bernor, R. L., H.-W. Mittmann and F. Rögl. 1993b. Systematics and chronology of the Götzendorf "*Hipparion*" (late Miocene, Pannonian F, Vienna Basin). *Annalen des Naturhistorisches Museum, Wien* 95:101–20.

Bernor, R. L. and P. P. Pavlakis. 1987. Zoogeographic relationships of the Sahabi large mammal fauna (early Pliocene, Libya). In *Neogene Geology and Paleontology of Sahabi*, ed. N. T. Boaz, J. de Heinzelin, W. Gaziry, and A. El-Arnauti, pp. 349–84. New York: Liss.

Bernor, R. L., N. Solounias, C. C. Swisher III, and J. A. Van Couvering. This volume b. The correlation of three classical "Pikermian" mammal faunas—Maragheh, Samos, and Pikermi—with the European MN unit system.

Bernor, R. L. and H. Tobien. 1990. The mammalian geochronology and biogeography of Paşalar (middle Miocene, Turkey). *Journal of Human Evolution* 19:551–68.

Bernor, R. L., H. Tobien, and M. O. Woodburne. 1989. Patterns of Old World hipparionine evolutionary diversification and biogeographic extension. In *Topics on European Mammalian Chronology*, ed. E. H. Lindsay, V. Fahlbusch, and P. Mein, pp. 263–319. New York: Plenum.

Bernor, R. L., M. O. Woodburne, and J. A. Van Couvering. 1980. A contribution to the chronology of some Old World Miocene faunas based on hipparionine horses. *Géobios* 13:25–59.

Bolliger, T. This volume. A current understanding about the Anomalomyidae (Rodentia): Reflections on stratigraphy, paleobiogeography, and evolution.

Bonis, L. de, G. Bouvrain, D. Geraads, and K. D. Koufos. 1988. Late Miocene mammal localities of the lower Axios valley (Macedonia, Greece) and their stratigraphical significance. *Modern Geology* 13:141–47.

Brown, F. H. and C. S. Feibel 1991. Stratigraphy, depositional environments, and palaeogeography of the Koobi Fora Formation. In *Fossil Ungulates: Geology, Fossil Artiodactyls, and Palaeoenvironments*, pp. 1–30. Koobi Fora Research Project volume 3. Oxford: Clarendon.

Bruijn, H. de, R. Daams, G. Daxner-Höck, V. Fahlbusch, L. Ginsburg, P. Mein, and J. Morales. 1992. Report of the RCMNS working group on fossil mammals, Reisensburg 1990. *Newsletters on Stratigraphy* 26:65–118.

Bruijn, H. de and P. Mein. This volume. The middle and late Miocene record of the Sciuridae and Petauistidae in France, Central Europe, Southeastern Europe, and Anatolia.

Bruijn, H. de and E. Ünay. This volume. On the evolutionary history of the Cricetodontini from Europe and Asia Minor and its bearing on the reconstruction of migrations and the continental biotope during the Neogene.

Bruijn, H. de, J. A. Van Dam, G. Daxner-Höck, V. Fahlbusch, and G. Storch. This volume. The genera of the Murinae, endemic insular forms excepted, of Europe and Anatolia during the late Miocene and early Pliocene.

Campbell, B. G., M. H. Amini, R. L. Bernor, W. Dickenson, R. Drake, R. Morris, J. A. Van Couvering, and J. A. H Van Couvering. 1980. Maragheh: A classical late Miocene vertebrate locality in northwestern Iran. *Nature* 287:837–41.

Daams, R. and M. Freudenthal. 1981. Aragonian: The stage concept versus Neogene mammal zones. *Scripta Geologica* 62:152–82.

Daxner-Höck, G. This volume. Middle and late Miocene Gliridae of Western, Central, and Southeastern Europe.

Daxner-Höck, G., V. Fahlbusch, L. Kordos, and W. Wu. This volume. The late Neogene cricetid rodent genera *Neocricetodon* and *Kowalskia*.

Engesser, B. and R. Ziegler. This volume. Didelphids, insectivores, and chiropterans from the later Miocene of France, Central Europe, and Turkey.

Fahlbusch, V. 1991. The meaning of MN zonation: Considerations for a subdivision of the European continental Tertiary using mammals. *Newsletters on Stratigraphy* 24:159–73.

——. This volume. Middle and late Miocene common cricetids and cricetids with prismatic teeth.

Fahlbusch, V. and T. Bolliger. This volume. Eomyids and zapodids (Rodentia, Mammalia) in the middle and upper Miocene of Central and Southeastern Europe and the Eastern Mediterranean.

Fortelius, M., J. Van der Made and R. L. Bernor. This volume a. Middle and late Miocene Suoidea of Central Europe and the Eastern Mediterranean: Evolution, biogeography, and paleoecology.

Fortelius, M., L. Werdelin, P. Andrews, R. L. Bernor, A. Gentry, H.-W. Mittmann, and S. Viranta. This volume b. Provinciality, diversity, turnover, and paleoecology in land mammal faunas of the later Miocene of Western Eurasia.

Gentry, A. In press. Fossil pecorans from the Baynunah Formation, Abu Dhabi, United Arab Emirates. In *International Seminar on the Neogene Vertebrates of Arabia: The Rise of Modern Animal Communities and Climate Change in the Old World, with Special eference to the Miocene of the Emirate of Abu Dhabi*, ed. P. Whybrow and A. Hill.

Gentry, A. and E. Heizmann. This volume. Miocene ruminants of the Central and Eastern Paratethys.

Harris, J. M. 1983. Correlation of the Koobi Fora succession. In *Koobi Fora Research Project, vol. 2: The Fossil Ungulates: Proboscidea, Perissodactyla and Suidae*, ed. J. M. Harris, pp. 303–18. Oxford: Clarendon.

Hartenberger, J.-L. 1969. Les Pseudosciuridae (Mammalia, Rodentia) de l'Eocene moyen de Bouxwiller, Egerkinger et Lissieu. *Palaeovertébrata* 3:27–91.

Hayek, L., R. L. Bernor, N. Solounias, and P. Steigerwald. 1991. Methods in determining the dietary adaptations of extinct hipparionine equids. In *Bjorn Kurten: A Memorial Volume*, ed. A. Forstén, M. Fortelius, and L. Werdelin, pp. 187–200. Annales Zoologici Fennici 28.

Heissig, K. This volume. The stratigraphical range of fossil rhinoceroses in the late Neogene of Europe and the Eastern Mediterranean.

Kappelman, J., S. Sen, M. Fortelius, A. Duncan, B. Alpagut, J. Crabaugh, A. Gentry, J. P. Lunkka, F. McDowell, N. Solounias, S. Viranta, and L. Werdelin. This volume. Chronology and biostratigraphy of the Miocene Sinap Formation of Central Turkey.

Koufos, G. D. 1989. The hipparions of the lower Axios Valley (Macedonia, Greece): Implications for the Neogene stratigraphy and the evolution of hipparions. In *European Neogene Mammal Chronology*, ed. E. H. Lindsay, V. Fahlbusch, and P. Mein, pp. 321–38. New York: Plenum.

Lindsay, E. H. 1989. The setting. In *European Neogene Mammal Chronology*, ed. E. H. Lindsay, V. Fahlbusch, and P. Mein, pp. 1–14. New York: Plenum.

Lindsay, E. H. and R. Tedford. 1989. Development and application of land mammal ages in North America and Europe: A comparison. In *European Neogene Mammal Chronology*, ed. E.H. Lindsay, V. Fahlbusch and P. Mein, pp. 601–24. New York: Plenum.

Mein, P. 1975. Résultats du Groupe de Travail des Vertébrés. In *Report on Activity of the RCMNS Working Groups (1971–1975)*, ed. J. Senes, pp. 78–81. Bratislava: SAV.

———. 1979. Rapport d'activité du Groupe de Travail Vertébrés: Mise à jour de la biostratigraphie du Néogène basée sur les mammifères. *Annales Géologie Pays Hellénica* 1973 (3): 1367–72.

———. 1989. European mammal correlations. In *Topics on European Mammalian Geochronology*, ed. E. H. Lindsay, V. Fahlbusch, and P. Mein, pp. 73–90. Plenum: New York..

Moyà-Solà, S. and M. Köhler. 1993. Recent discoveries of *Dryopithecus* shed light on evolution of great apes. *Nature* 365:543–45.

Opdyke, N. D., P. Mein, E. Moissenet, A. Perez-Gonzales, E. H. Lindsay. and M. Petko. 1989. The magnetic stratigraphy of the late Miocene sediments of the Cabriel Basin, Spain. In *European Neogene Mammal Chronology*, ed. E. H. Lindsay, V. Fahlbusch, and P. Mein, pp. 507–14. New York: Plenum.

Pilbeam, D., M. Morgan, J. C. Barry, and L. J. Flynn. This volume. European MN units and the Siwalik faunal sequence of Pakistan.

Qiu, Z. 1989. The Chinese Neogene mammalian biochronology: Its correlation with the European Neogene mammalian zonation. In *European Neogene Mammal Chronology*, ed. E. H. Lindsay, V. Fahlbusch, and P. Mein, pp. 527–56. New York: Plenum.

Rögl, F. and G. Daxner-Höck. This volume. Late Miocene Paratethys correlations.

Rögl, F., H. Zapfe, R. L. Bernor, R. Brzobohaty, G. Daxner-Höck, I. Draxler, O. Fejfar, J. Gaudant, P. Herrmann, G. Rabeder, O. Schultz, and R. Zetter. 1993. Primatenfundstelle Götzendorf an der Leitha, Niederösterreich (Obermiozän, Pontien des Wiener Beckens). *Jahrbuch des geologischen Bundesanstalt* 136:503–26.

Rögl, F. and F. F. Steininger. 1983. Vom Zerfall der Tethys zur Paratethys. *Annalen des Naturhistorischen Museums, Wien* 85:135–63.

Sen, S. This volume a. Present state of magnetostratigraphic studies in the continental Neogene of Europe and Anatolia.

———. This volume b. Late Miocene Hystricidae in Europe and Anatolia.

Solounias, N. 1981. The Turolian fauna from the island of Samos, Greece. *Contributions to Vertebrate Evolution* 6:1–232.

Solounias, N., J. C. Barry, R. L. Bernor, E. H. Lindsay and M. Raza. In press. The earliest bovid from the Siwaliks. *Journal of Vertebrate Paleontology*.

Steininger, F. F., W. A. Berggren, D. V. Kent, R. L. Bernor, S. Sen, and J. Agusti. This volume. Circum-Mediterranean Neogene (Miocene and Pliocene) marine-continental chronologic correlations of European mammal units.

Steininger, F. F., R. L. Bernor, and V. Fahlbusch. 1989. European Neogene marine/continental chronologic correlations. In *European Neogene Mammal Chronology*, ed. E. H. Lindsay, V. Fahlbusch, and P. Mein, pp. 15–46. New York: Plenum.

Steininger, F. F., G. Rabeder, and F. Rögl. 1985. Land mammal distribution in the Mediterranean Neogene: A consequence of geokinematic and climatic events. In *Geological Evolution of the Mediterranean Basin*, ed. D. J. Stanley and F. C. Wezel, pp. 559–71. New York: Springer.

Sudre, J. 1969. Les gisements de Robiac (Eocene supérieur) et leurs faunes de mammifères. *Palaeovertébrata* 2:95–165.

———. 1972. Revision des Artiodactyles de l'Eocene moyen de Lissieu (Rhône). *Palaeovertébrata* 5:111–56.

Swisher, C. C., III. This volume. New ^{40}Ar/^{39}Ar dates and their contribution toward a revised chronology for the late Miocene nonmarine of Europe and West Asia.

Thaler, L. 1966. Les Rongeurs fossiles du Bas-Languedoc dans leurs rapports avec l'histoire des faunes et la stratigraphie du Tertiary d'Europe. *Mémoire du Muséum Nationale d'Histoire Naturelle* 17C:1–295.

———. 1972. Datation, zonation et mammifères. *Mémoire Bureau Recherche Geologie et Minérologie* 77:711–24.

Thenius, E. 1959. Tertiär. 2. Teil Wirbeltierfaunen. In *Handbuch der Stratigraphischen Geologie* 3(2): 1–328, ed. F. Lotze. Stuttgart: Enke.

Ünay, E. This volume. On fossil Spalacidae (Rodentia).

Werdelin, L. This volume. Carnivores, exclusive of Hyaenidae, from the later Miocene of Europe and Western Asia.

Werdelin, L. and N. Solounias. This volume. The evolutionary history of hyaenas in Europe and Western Asia during the Miocene.

Woodburne, M. O., R. L. Bernor, and C. C. Swisher III. This volume a. An appraisal of the stratigraphic and phylogenetic bases for the *"Hipparion"* Datum in the Old World.

Woodburne, M. O., G. Theobald, R. L. Bernor, H. König, C. C. Swisher III, and H. Tobien. This volume b. Advances in the geology and stratigraphy at Höwenegg, southwestern Germany.

Appendix 32.1

SUMMARY OF MAMMALIAN SYSTEMATICS,
BIOGEOGRAPHY, AND GEOCHRONOLOGY

Small Mammal Systematics and Biogeography

The small mammal groups evaluated in this volume included didelphids, insectivores, and chiropterans (Engesser and Ziegler, this volume); eomyids and zapodids (Fahlbusch and Bolliger, this volume); sciurids and petauristids (de Bruijn and Mein, this volume); "common" cricetids and cricetids with "prismatic teeth" (Fahlbusch, this volume); species belonging to *Neocricetodon* and *Kowalskia* (Daxner-Höck et al., this volume); "Cricetodontini" (de Bruijn and Ünay, this volume); anomalomyids (Bolliger, this volume); spalacids (Ünay, this volume); murines (de Bruijn et al., this volume); glirids (Daxner-Höck, this volume); and hystricids (Sen, this volume b).

Engesser and Ziegler's "non-rodent" small mammal chapter provides an account of didelphid, insectivore, and chiropteran biogeographic and chronologic distribution for most of the Western Eurasian middle and late Miocene. Didelphids are relictual during this period and are only known from Central Europe by a single species, *Amphiperatherium frequens*, which became extinct by the end of MN 6.

The insectivores were very diverse during this time, including no fewer than six families: Plesiosoricidae, Soricidae, Heterosoricidae, Talpidae, Dimylidae, and Erinaceidae. Most of the taxa included within these families are from Central Europe, followed in diversity by France, and with the lowest diversity being in Southwestern Asia and Southeastern Europe. The Plesiosoricidae included four species of *Plesiosorex*: *P. styriacus* (MN 6 of Central Europe), *P. germanicus* (MN 6 of Central Europe and France), *P. schaffneri* (MN 7/8–9 of Central Europe), and *P. sp.* (MN 9 of Central Europe and Southwestern Asia). The Soricidae were more diverse, including several genera: *Lartetium* (*L. dehmi*, MN 6 of Central Europe and MN 7/8 of France); *Miosorex* (*M. desnoyersianus*, MN 6 of France; *M. grivensis*, MN 7/8 of France and MN 9 of Central Europe); *Allosorex* (*A. gracilidens*, MN 6 of Central Europe; *A.* sp., MN 11 of France); *Paenelimnoecus* (*P. crouzeli*, MN 6–7/8 of France; *P. repenningi*, MN 9–10 of Central Europe; *P.* sp., MN 7/8 of Southwestern Asia); *Angustidens* (*A. exultus*, MN 9 of Central Europe); *Crusafontia* (*C. kormosi*, MN 9–11 of Central Europe, MN 10 of France); *Blarinella* (MN 6–7/8 of France); *Paracryptotis* (*P.* sp., MN 10 of Central Europe); *Amblycoptus* (*A.* sp., MN 10 of Central Europe; *A. oligodon*, MN 13 of Central Europe and Southwestern Asia).

The Heterosoricidae were not well known during this interval, but present two distinct lineages, *Heterosorex* and *Dinosorex*. *Heterosorex* includes a single species, *H. delphinensis*, known solely from MN 7/8 of France. *Dinosorex* was more diverse, including four species: *D. sansaniensis* (MN 6–7/8 of France, continuing up to MN 10 of Central Europe); *D. zapfei* (MN 6–9 of Central Europe); *D. pachygnathus* (MN 7/8 of Central Europe); and *D.* sp. (MN 7/8) of Southwestern Asia. The Talpidae exhibit great taxonomic diversity. The archaic talpid genus *Desmanella* was widely distributed throughout Europe during the Paleogene and appears to have diversified into a number of Western Eurasian vicariant lineages by MN 7/8: *D. stehlini* (MN 7/8 of Central Europe and MN 10 of France); *D. crusafonti* (MN 10 of Central

Europe); *D.* sp. (MN 7/8 of Southwestern Asia, MN 9–11 of Central Europe, and MN 11–13 of France); *D. sickenbergi* (MN 7/8 of Southwestern Asia); *D. cingulata* (MN 7/8 of Southwestern Asia); *D. dubia* (MN 12 of Southeastern Europe); *D. amasyae* (MN 13 of Southwestern Asia); "*D.*" *quinquecuspidata* (MN 9 of Central Europe).

Most talpid genera occurred entirely in Central Europe and France, including: *Talpa* (*T. minuta* MN 6–7/8 of France and Central Europe, continuing into MN 10 of Central Europe; *T. gilothi*, MN 10–11 of Central Europe; *T. vallesiensis*, MN 11 of Central Europe); *Proscapanus* (*P. sansaniensis*, MN 6–7/8 of France, continuing into MN 9 of Central Europe); "*Scaptonyx*" ("*S.*" *edwardsi*, MN 6–9, only MN 7/8 of France); *Urotrichus* (?*U. dolichochir*, MN 6–9 of Central Europe and MN 7/8 of France); *Asthenoscapter* (*A. meini*, MN 7/8 of France); *Mygalea* (*M. antiqua*, MN 6 of Central Europe and France; *M. jaegeri*, MN 6 of Central Europe); *Desmana* (*D.* sp., MN 10 of Central Europe); *Dibolia* (*D. pontica*, MN 10 of Central Europe; *D. vinea*, MN 11 of Central Europe; *D.* sp., MN 11 of France); *Mygalinia* (*M. hungarica*, MN 13 of Central Europe); *Desmanodon* (all taxa in Southwest Asian MN 7/8: *D. minor*, *D. major*, *D.* n. sp.).

The Dimylidae were represented by two genera: *Metacordylodon* (*M. schlosseri*, MN 7/8 of Central Europe and France; *M.* sp., MN 6 of Central Europe) and *Plesiodimylus* (*Pl.* sp., MN 6–7/8 of Central Europe and MN 7/8 of France; *Pl. chantrei*, MN 6–11 of Central Europe and France; *Pl. crassidens*, MN 7/8 of Southwestern Asia only). *Plesiodimylus* was a relatively conservative lineage from MN 4–11, except for the strongly divergent and highly specialized Turkish taxon *P. crassidens*.

The Erinaceidae were a diverse and successful group of insectivores. The genus *Lanthanotherium* is a rare and poorly understood taxon that ranges through the middle and most of the late Miocene: *L. sansaniensis* (MN 6–7/8 in France as well as Central Europe); *L.* n. sp. (MN 6 of Central Europe only); *L. sanmigueli* (MN 10–11 of Central Europe and MN 10 of France); *L.* sp. (MN 9–10 of Central Europe). *Galerix* is reported from Central Europe and France: *G. exilis* (MN 6 of both provinces); *G. stehlini* (MN 7/8 of France only); *G. socialis* (MN 7/8–10 of both provinces); *G.* cf. *symeonidisi* (MN 4 of Southeastern Europe); *G.* sp. (MN 6–7/8 of Central Europe). *Schizogalerix* first occurred in Anatolia during the early middle Miocene as two lineages quite distinct from European *Galerix* (*S. pasalarensis* [MN 6–7/8] and *S. anatolica* [MN 6–7/8]). Subsequent Southwest Asian lineages included *S. sinapensis* (MN 9) and *S.* n. sp. (MN 13). *Schizogalerix* first extended its range into Central Europe during MN 9 (*S. vosendorfensis*), continued into MN 10 (*S. zapfei*), and had a last occurrence in MN 11 (*S. moedlingensis* and *S.* sp.). Southeastern Europe records a single occurrence in MN 12 (*S. attica*).

Other erinaceids are poorly known: *Mioechinus* (*M. sansaniensis*, MN 6–7/8 of Central Europe and MN 7/8 of France; *M. oeningensis*, MN 7/8 of Central Europe; *M. tobieni* and *M.* sp. from Southwestrn Asia only); *Amphechinus* (*A. ginsburgi*, MN 6 of France only); *Postpalerinaceus* (*P. intermedius*, MN 7/8 of France only; *P. vireti*, MN 10 of France only).

The very high insectivore diversity recorded in the middle Miocene of Central Europe was unparalleled at other times and in other provinces of Western Eurasia. Although not as diverse as Central Europe, France maintained a diverse insectivore fauna during the middle Miocene. Insectivore diversity falls at the MN

8/9 boundary and again, precipitously, at the MN 9/10 boundary, and must reflect a sharp shift of environments from warm, moist conditions to more arid (= seasonal) ones. From the evidence provided here, Central Europe would appear to exhibit an extensive biogeographic connection with France, particularly in the middle Miocene. Southeastern Europe and Anatolia have autochthonous forms with limited biogeographic connection to the west. *Schizogalerix* is notable as a Southwest Asian lineage that appears to have arisen there during the Astaracian and undergone a westward extension into Central Europe at the base of the late Miocene.

Chiropterans are relatively poorly known. The Turkish record includes indeterminable material from five middle and late Miocene localities. While far better represented from Central and Western Europe, chiropterans are as of yet poorly understood evolutionarily. Western Eurasian middle and late Miocene chiropterans are represented by no fewer than five families with chronologic ranges extending back into the Oligocene: Hipposideridae, Rhinolophidae, Vespertilionidae, Molossidae, and Megadermidae.

The Hipposideridae had a limited diversity and distribution: *Asellia* (*A. mariatheresae*, MN 7/8 of France; *A.* sp., MN 6 of Central Europe); *Hipposideros* (*H. collongensis*, MN 7/8 of France).

The Rhinolophidae are another Central and West European family that was more diverse and had a longer chronologic range than the Hipposideridae: *Rhinolophus* (*Rh. delphinensis*, MN 6–10 of Central Europe, MN 7/8 of France; *Rh.* aff. *delphinus*, MN 6 of Central Europe; *Rh. grivensis*, MN 6 of Central Europe, MN 7/8 of France; *Rh. lissiensis*, MN 13 of France; *Rh.* sp., MN 7/8 of France, MN 11 of Central Europe; *Rh. similis*, MN 6 of Central Europe).

The Vespertilionidae included a number of taxa: *Vespertilio* (*V.* sp., MN 6 of Central Europe); *Plecotus* (*Pl. sanctialbani*, MN 7/8 of France, MN 10 of Central Europe; *Pl.* sp., MN 13 of Central Europe; *Pl. atavus*, ?MN); *Pareptesicus* (*P. priscus*, MN 6 of Central Europe and MN 7/8 of France); *Miniopterus* (*M. fossilis*, MN 6 of Central Europe, MN 7/8 of France; *M.* sp., MN 7/8 of France); *Eptesicus* (*E. campanensis* and *E. noctuloides*, both MN 6–7/8 of France); the solely French genus *Myotis* (*M. antiquus*, MN 7/8; *M. murinoides*, MN 6–7/8; *M. elegans*, MN 6; *M.* sp., MN 7/8; *M. boyeri*, MN 13); the Central European genus *Miostrellus* (*M. rigoviensis*, MN 6); the Greek genus *Samonycteris* (*S. majori*, MN 12).

The Molossidae included two genera: Central European *Mormopterus* (*M. helveticus*, MN 6–7/8; *M. kalorhinus*, MN 6) and *Tadarida* (*T. engesseri*, MN 6–7/8; *T. leptognatha*, MN 6; *T. monslapidis*, MN 6–7/8, all of Central Europe; *T.* sp., MN 7/8 of France).

The Megadermidae included two Central European/French genera: *Megaderma* (*M. lugdunensis*, MN 6–11 of Central Europe; *M. vireti*, MN 7/8–13 of France and MN 10 of Central Europe) and *Miomegaderma* (*M. gaillardi*, MN 6–7/8 of France).

The chiropterans showed an even more striking biogeographic and biochronologic pattern than the insectivores: there are strong biogeographic connections between Central Europe and France; there is a very sharp drop in diversity at the MN 8/9 boundary; of the thirty-five species reported by Engesser and Ziegler (this volume), only one appeared outside Central and Western Europe,

Samonycteris majori from MN 12 of Samos, Greece. Patterns of chiropteran diversity further support the interpretation of more forested environments in the middle Miocene of the west than the east and, furthermore, indicate a sharp environmental shift at the middle/late Miocene boundary.

Eomyid rodents are represented by limited fossil material. They are relatively well known from the Paleogene of the Western United States and Europe, and are more limited from China. They continued into the early Miocene of Europe. Fahlbusch and Bolliger (this volume) identify two principal eomyid evolutionary groups: the *Eomyops*–group and the *Keramidomys*–group.

Eomyops occured in Central Europe from MN 5 to MN 10, most frequently found in MN 9 localities. This genus also occurred in France (MN 7/8–MN 13) and had a limited occurrence in Southeastern Europe (MN 10 and MN 13) and Southwestern Asia (MN 6 and MN 12).

The *Keramidomys*–group had an extended evolutionary record in Central Europe (MN 5–MN 14) and France (MN 5–MN 13). In Turkey, other than a questionable occurrence from Çandir (MN 6), this group did not occur until the early Turolian (MN 11) and extended its range up into MN 14. In Southeastern Europe it is known from the late Turolian (MN 13) and early Ruscinian (MN 14). The Central European genus *Estramomys* first occurred in MN 14 and has an unknown origin.

Zapodids had a probable Holarctic distribution and a likely Asiatic occurrence since the Oligocene. The *Plesiosminthus*–group occurred in the earliest Miocene (Agenian) of Europe, but zapodids are unknown between MN 3 and MN 7/8. Thereafter, *Eozapus* is known to have occurred in Central Europe (MN 9–14), France (MN 10–13), and the Eastern Mediterranean in MN 12. A rarer zapodid, *Sminthozapus*, is known to have occurred in MN 14 of France and Central Europe.

De Bruijn and Mein (this volume) have summarized the chronologic and biogeographic distributions of the Sciuroidea, which included forms that exhibit flying (seven genera), ground (five genera), and arboreal (two genera) adaptations. Two families of sciuroid are included, the rarer Petauristidae and the more common Sciuridae.

Petauristid squirrels exhibit a flying adaptation and have extensive biogeographic distributions: *Aliveria* has a distribution restricted to MN 4 of Southeastern Europe. Central Europe and France appear to be uniquely related in their distributions of *Blackia* (MN 4–14) and *Forsythia* (MN 7/8); *Albanensia* is also found in Central Europe (MN 7/8–9) and France (MN 6–7/8); *Miopetaurista* occurs in France (MN 7/8–14), Central Europe (MN 4–11), Southeastern Europe (MN 13), and Anatolia (MN 7/8); *Pliopetaurista* occurs in France and Central Europe (MN 10–14), Southeastern Europe (MN 10–12), and Anatolia (MN 11); *Hylopetes* also occurs in all areas, including France (MN 10–14), Central Europe (MN 10–14), Southeastern Europe (MN 13/14), and Anatolia (MN 11).

Sciurid squirrels likewise exhibit a variety of biogeographic ranges. Those that are confined to Central Europe included: *Palaeosciurus* (Oligocene: MN 6), and only possibly in Anatolia (also MN 6); *Spermophilus* (MN 3–13/14); *Sciurus* (MN 14) during the period considered. *Heteroxerus* had a somewhat broader biogeographic distribution, including France (MN 6–10) and Central Europe (MN 6). The remaining taxa had broader distributions: *Atlantoxerus* (Central Europe, MN 13; Southeastern

Europe, MN 7/8; Anatolia, MN 7/8–14); *Spermophilinus* (in all four areas considered during MN 6–11, continuing into MN 12 of France, MN 13 of Southeastern Europe, and MN 14 of Anatolia); *Tamias* (known earliest in Anatolia [MN 6–14] with limited occurrences in France [MN 7/8], Central Europe [MN 14], and Southeastern Europe [MN 13]).

Cricetid rodents are a diverse and evolutionarily complex group. Fahlbusch (this volume) has reported on the chronologic and biogeographic distribution of nineteen genera, most of these being restricted to one or two geographic areas. He has recognized five broad evolutionary "groups" in his survey: (1) those common in the late Orleanean and Astaracian from Central and Western Europe (*Democrictodon, Megacricetodon, Kowalskia, Eumyarion,* and *Lartetomys*); (2) fossil relatives of extant dwarf hamsters (*Cricetus* s.l., *Cricetulus, Allocricetus, Mesocricetus,* and *Calomyscus*); (3) taxa with conservative tooth occlusal morphology but that increase crown height (*Collimys* and *Hypsocricetus*); (4) gerbellid-like cricetids (*Epimeriones* and *Pseudomeriones*); (5) microtoid cricetids and true arvicolids (*Microcricetus, Pannonicola, Baranomys/Microtodon, Bjornkurtenia,* and *Promimomys*).

Cricetid taxa with geographic distributions confined to Central Europe during the MN 5–14 interval are extensive and included: the primitive taxon *Eumyarion* (MN 4–10), *Lartetomys* (MN 6), *Cricetus* s.l. (MN 13–14), *Collimys* (MN 11), *Epimeriones* (MN 10–14), *Microtocricetus* (MN 9–10), *Pannonicola* (MN 13), *Baranomys/Microtodon* (MN 14), *Bjornkurtenia* (MN 14), and *Promimomys* (MN 14). Those taxa known from only one province other than Central Europe include: *Cricetulus* (MN 14 of the Eastern Mediterranean), *Allocricetus* (MN 13 of Southeastern Europe), *Calomyscus* (MN 14 of Southeastern Europe), and *Hypsocricetus* (MN 13 of Southeastern Europe). Two taxa are known only from Southeastern Europe and the Eastern Mediterranean: *Mesocricetus* (MN 14) and *Pseudomeriones* (MN 13–14 in Southeastern Europe and MN 12–14 in the Eastern Mediterranean). *Megacricetodon* occurred in all three areas considered between MN 4–7/8, continuing into MN 9. Likewise, *Democricetodon* occurred in all three areas, including MN 4–10 of Central Europe, MN 5 of Southeastern Europe, and MN 5–6 of the Eastern Mediterranean. *Kowalskia* is yet a third taxon distributed in all three areas, but later in time: Central Europe (MN 10–14); Southeastern Europe (MN 10–13); the Eastern Mediterranean (MN 12–14).

The cricetids "*Neocricetodon*" and *Kowalskia* have been considered separately by Daxner-Höck et al. (this volume). The "*Neocricetodon*"–*Kowalskia* complex clearly underwent a major evolutionary radiation in the Central and Eastern Paratethys, and exhibits a distinct biogeographic connection to the east with apparent limited connections to Western, Southern, and Southeastern Europe. A number of taxa are known only from the Central and Eastern Paratethys: *K. polonica* (MN 14); *K. magna* (MN 14); *K. schaubi* (MN 11); *K. fahlbuschi* (MN 10); *K. intermedia* (MN 15); *K. moldavica* (MN 10); *K. skofleki* (MN 11–12); *K. polgardiensis* (MN 13); *K. progressa* (MN 9). Additionally, one is known from Italy (*K. nestori*, MN 13/14), two are known from the latest Miocene of China (*K. neimengensis* and *K. similis*), two from Western Europe (*K. ?lavocati*, MN 13, and *K. occidentalis*, MN 11), and one, *K. browni*, from Southeastern Europe (MN 13/14).

De Bruijn and Ünay (this volume) have reviewed the chronologic and geographic distributions of cricetodontine rodents including: *Cricetodon* (also *Turkomys*), *Hispanomys, Ruscinomys* (including *Pseudoruscinomys*), and *Byzantinia* (including *Pararuscinomys*). The best records of this group are known from Southwestern Europe (MN 5–15) and Asia Minor (MN 1–12). Anatolian and Greek assemblages contain only species of *Cricetodon* from MN 1–6. *Cricetodon* is also reported from Central Europe in MN 6–7/8. These assemblages show some evolutionary progression (types 1–3 of de Bruijn and Ünay, this volume) but remain relatively conservative in their morphology. Between MN 7/8 and MN 12, Anatolian and Greek cricetodontines exhibit appreciable differences in their dental morphology that the authors refer to various species of *Byzantinia* (also, Cricetodontini "type 4"). A single West European species is referred to "group 4," "*Cricetodon*" *lavocati* (MN 7/8–9). Southwest European cricetodontines include the "group 5" forms that are time-successive: *Hispanomys* (MN 7/8–10) and *Ruscinomys* (MN 11–15).

Bolliger (this volume) has reviewed the evolutionary history of the Anomalomyidae. He presents evidence for two separate Western Eurasian anomalomyid radiations: an early Miocene radiation including a Central–West European clade of three taxa, and a more extensive middle Miocene–Pleistocene radiation of some seventeen East, Central, and West European taxa. Bolliger further suggests that the first radiation followed an Asian emigration, while the secondary radiation may have included from one to three separate Asian emigrations (MN 6, MN 12, and MN 14).

The initial evolutionary radiation of the Anomalomyidae (MN 4–5) included the oldest-known species, *Anomalomys aliveriensis* (MN 4 of Greece). This form was succeeded by the more widely distributed (in Europe) MN 4–5 species *A. minor* and the MN 5 species *A. minutus*. The next clade, the *A. gaudryi*–group, appears in MN 6 and includes two taxa, *A. gaudryi* and *A. kowalskii*. These taxa first occurred in MN 6 of Turkey and Central Europe, apparently as immigrants from the east, and continued in Central and Western Europe until MN 9. The *Anomalomys gaillardi*–group (*A. gaillardi* and *A. rudabanyensis*) would appear to have been derived from the *A. gaudryi*–group and occurred in Central and Western Europe between MN 9 and 11. *Prospalax* is yet another broadly distributed taxon which occurred over the eastern portion of Central Europe and is believed to have evolved from *A. gaillardi. Prospalax petteri* is another Central European lineage which occurred in MN 10–11 horizons and is apparently related to a series of local Ruscinian and Villanyan aged species: *P. kretzoii, P. rumanus,* and *P. priscus.* Alternatively, Bolliger argues that these later taxa may have been introduced by a separate Asian emigration.

The Spalacidae first occurred in Anatolia (MN 3) and are broadly related to muroid rodents, and apparently had a long independent evolutionary history (Ünay, this volume). As seen in species of Anomalomyidae, Rhizomyidae, and Tachyoryctoididae, spalacids were adapted for burrowing behavior. However, despite a number of recent finds, the spalacid fossil record is still limited.

The earliest spalacid, *Debruijnia*, exhibits more characters in common with cricetids than with its own family. From Anatolia it is believed that spalacids first extended their range into Southeastern Europe (Aliveri, MN 4), where they are represented by the derived genus *Heramys*.

The *Pliospalax*–group first occurred in Southwestern Asia (MN 6) and diversified into a number of local provincial taxa: *P. marmarensis* (the most primitive member, known from Paşalar); advanced *P. marmarensis* (Çandir); *P. primitivus* (Sariçay; MN 7/8);

P. canakkalensis (Bayraktepe 1; MN 7/8). *Pliospalax* is first documented in Southeastern Europe during MN 7/8 (Pismanköy) and is poorly known from MN 9–10 of Southwestern Asia. *Pliospalax composidontus* is present in two MN 10 southeastern localities, and two additional species (n. sp.) are documented from MN 11 of Anatolia. An advanced *Pliospalax, P. macovei*, is known from MN 13 of Southeastern Europe (Maramena) and MN 15 of Central Europe (Malusteni and Beresti) and Anatolia (Çalta). The last occurrence of *Pliospalax* is from the late Villanyian of Tourkobounia.

The oldest species of *Spalax, S. odessanus*, appeared in Southeastern Europe during MN 15 (Karaburun and Odessa). *Spalax minor* and *S. ehrenbergi* first appeared in the early Biharian of Southeastern Europe (Nogaisk, Ukraine) and Southwestern Asia (Ubeidiya). *Spalax* dispersed into Central Europe during the late Biharian (*Spalax advenus*; several localities) and Africa during the late Pleistocene.

With regard to Western Eurasian Neogene faunal history, spalacids clearly exhibit a Southwest Asian–Southeast European provincial origin and evolutionary radiation beginning by MN 3. Biogeographic extensions were evidently limited to Central Europe (MN 15; late Biharian) and Africa (late Pleistocene).

De Bruijn et al. (this volume) have given a brief yet thorough review of the major murine lineages. Murines evolved from an Asiatic myocricetodontine stock and underwent a number of east–west biogeographic extensions: first, in the late middle Miocene (MN 7/8; *Antemus* grade of evolution), they dispersed into Southwestern Asia and North Africa; second, in the early late Miocene (MN 9; *Progonomys* grade of evolution), they extended their range from Anatolia into Southeastern, Southwestern, and (in MN 10) Central Europe. Mein (pers. comm. to Rögl) has indicated that the first-occurring MN 10 Central European murine was of a different lineage than the first-occurring MN 9 circum-Mediterranean lineage.

Parapodeums (MN 10–13), *Apodemus* (MN 12–14+), *Rhagapodemus* (MN 13–14+), and *Micromys* (MN 13–14+) would appear to have first evolved in the Southwest Asian–Southeast European realm and variously extended their ranges principally into Southwestern Europe and, to a more limited extent, into Central Europe. *Pelomys* evolved endemically in Southeastern Europe (MN 14). "*Karnimata*" first evolved in the late Miocene of South Asia, extended its range initially into the Eastern Mediterranean in the late Miocene (MN 13), and later (MN 14) into Southwestern Europe. *Paraethomys* may be derived from *Karnimata* and established a circum-Mediterranean distribution in the late Miocene. *Castillomys* s.s. was an additional West European immigrant (MN 14), but has uncertain relationships to other late Miocene European members of *Castillomys* s.l.

Occitanomys (MN 10–13), *Huerzelerimys* (MN 10–13), and *Stephanomys* (MN 12–14+) apparently first evolved endemically in Southwestern Europe. Of these taxa, only *Occitanomys* is believed to have extended its range extraprovincially, and apparently it did so twice to Southeastern Europe–Southwestern Asia (MN 10 and MN 13 [based on occurrence at Chomateri, not referred here to MN 12 as presented by de Bruijn et al., this volume]). The possibility exists that the genus "*Occitanomys*" is paraphyletic, and included more than one evolutionary lineage.

Murine rodents underwent an initial South Asian radiation followed by a westward extension and evolutionary radiation in the Southeast European and Southwest Asian realm. There were subsequent evolutionary radiations in Southwestern Asia–Southeastern Europe, Western Europe, and Africa, which harbored a number of extant murine lineages. The Southwest Asian–Southeast European realm would appear to have been a "biogeographic hub" for a number of murine evolutionary radiations.

The Gliridae have a poorer fossil record in Southeastern Europe and Southwestern Asia than Central and Western Europe (Daxner-Höck, this volume). Glirid chronologic and geographic distribution suggest a greater degree of biogeographic unity between Central and Western Europe. From a high in the early Miocene, glirid diversity and abundance decreased sharply in the middle and late Miocene. In Western and Central Europe, a number of glirid genera continued from the early into the middle and late Miocene, and in some cases until the present: *Prodryomys* (LAD MN 6), *Bransatoglis* (LAD MN 7/8), *Glirudinus* (LAD MN 7/8), *Miodyromys* (LAD MN 10), *Myoglis* (MN 10), *Eomuscardinus* (MN 10), *Paraglirulus* (MN 10), *Microdyromys* (MN 11), *Vasseuromys* (MN 11), *Glis* (extant), and *Glirulus* (extant). Western and Central European glirid origins included: *Muscardinus* (MN 9–14+), *Eliomys* (MN 9–14+), *Myomimus* (MN 10–11), and *Graphiurops* (MN 10–11). Southeastern Europe had most genera in common: *Miodyromys* (LAD MN 9), *Eomuscardinus* (LAD MN 9), *Vasseuromys* (LAD MN 10 or 11), *Glis* (extant), *Muscardinus* (LAD MN 14+), *Eliomys* (LAD MN 10 or 11) and *Myomimus* (LAD MN 14+). *Ramys* (known only from MN 10) is the exception.

The Hystricidae are a poorly known group of rodents in the later Neogene of Western Eurasia (Sen, this volume b). The oldest Western Eurasian *Hystrix* comes from Yeni Eskihisar (MN 7/8), Turkey. *Hystrix* is known from the Vallesian by *H.* cf. *suevica* (MN 10 of Kohfidisch), and later *Hystrix primigenia* (and *Hystrix* sp.) became a relatively common, albeit unabundant, species of Turolian Central and Southeast European and Southwest Asian Turolian age faunas. Sen reports no *Hystrix* from the West European late Miocene. *Hystrix primigenia* continues its record in a few Central and West European Plio–Pleistocene localities.

Large Mammal Systematics and Biogeography

This volume has presented seven detailed evolutionary and biogeographic analyses of large mammals including: catarrhine primates (Andrews et al., this volume); carnivores exclusive of the Hyaenidae (Werdelin, this volume); Hyaenidae (Werdelin and Solounias, this volume); hipparionine horses (Bernor et al., this volume a); Rhinocerotidae (Heissig, this volume); Suoidea (Fortelius et al., this volume a); ruminants (Gentry and Heizmann, this volume).

Andrews et al. (this volume) have given a major revision of Western Eurasian catarrhine primate systematics, biogeography, and biochronology. The Western Eurasian catarrhine record began in MN 5 and is reconstructed through MN 14. Twenty-three taxa are recognized for this interval and mark several evolutionary radiations.

The initial catarrhine group to populate Western Eurasia is the Pliopithecidae, which first occurred in Western and Central Europe during MN 5. The group shows an early diversity in both these provinces, suggesting to Harrison (in Andrews et al., this volume) that more than one species immigrated from Africa into Southern Europe. Taxa unique to Central Europe included: *Pliopithecus platyodon* (MN 5–9), *Pliopithecus vindobonensis* (MN 6), *Plesiopliopithecus lockeri* (MN 6), and *Anapithecus hernyaki* (MN 9–?11). Taxa restricted to Western Europe included: *Pliopithecus priensis* (MN 9), *Plesiopliopithecus auscitanensis* (MN 6), and *Plesiopliopithecus rhodanica* (MN 7). Only *Pliopithecus antiquus* (MN 5–9) is known from both provinces, suggesting that it may be evolutionarily close to the founding pliopithecid stock. Clearly, there was a major adaptive radiation of the Pliopithecidae during the Astaracian, with several taxa evolving vicariantly in sectors of Western and Central Europe. The medial Vallesian incurred a major crisis for the Pliopithecidae, with only one very derived and relatively large taxon, *Anapithecus hernyaki*, continuing in Central Europe.

Until recently, hominoid primates were first reported from MN 6 of Europe. Heizmann (pers. comm.) has reported, and Andrews concurs, that the MN 5 locality of Engelswies (southern Germany) represents the earliest record of Western Eurasian hominoids. This may be a member of the same clade of hominoids found from early MN 6 of Paşalar (Turkey) and Neudorf Sandberg (Vienna Basin), *Griphopithecus*. *Griphopithecus* is a member of the East African middle Miocene Kenyapithecini. There is an apparent chronologic as well as morphologic hiatus between *Griphopithecus* and the succeeding MN 7/8 *Dryopithecus*, which leads Andrews et al. (this volume) to consider them as separate lineages. While it is currently unresolvable whether *Dryopithecus* evolved from a species of *Griphopithecus*, or as a separate immigration from Africa, it would appear that from an MN 8 Central and West European lineage, *Dryopithecus* diversified into provincially distinct Central (*D. carinthiacus*) and Western (*D. fontani, D. crusafonti*, and *D. laietanus*) European species in the MN 8–10 interval.

Sivapithecine primates, the sister taxon of modern Southeast Asian *Pongo*, predominately evolved in South Asia. However, Southwestern Asia (Anatolia) records a limited late MN 9 record of a member of this clade, *Sivapithecus meteai* (Kappelman et al., this volume). Sivapithecines are reported to have first occurred in the late middle Miocene of South Asia and it may be that the Turkish form evolved as a product of that initial radiation.

An enigmatic large hominoid from MN 10 of Greece, *Graecopithecus freybergi*, is recognized as a clade distinct from all other Western Eurasian hominoids and may indeed share relationships to the African human-ape lineage. It is recognized here as distinct from *Dryopithecus*, but Begun (1992; 1994) groups the two genera together in a single clade within the Hominidae. Moyà-Solà and Köhler (1993) propose the placement of *Dryopithecus* in the pongine clade, and they include *Graecopithecus* now in the same clade (Moyà-Solà, pers. comm. to Andrews). We (Andrews and Bernor) consider that *Dryopithecus* is a stem hominid predating the divergence of the great apes and humans and distinct from *Graecopithecus*, which we place in the hominine clade, but we recognize that it may prove to be convergent on that lineage. At issue here are both the relationships that can be recognized with the living great apes and the relationships of these fossil apes with each other, and whether they evolved independently from separate emigration events from Africa, or are linked in ancestral-descendant relationship.

Oreopithecus is yet another enigmatic hominoid primate restricted in its range to the late Miocene (MN 12–13) of Italy. Its relationship with other Western Eurasian hominoids (if any) remains unresolved.

It is currently not resolvable whether Western Eurasian hominoids are monophyletic, that is the evolutionary product of a single MN 5/6 migratory event (*Griphopithecus*), or polyphyletic and the product of multiple migrations of distinct East African lineages. Begun (1994) has forwarded a character state analysis of all Western Eurasian hominoids excluding *Oreopithecus*, and has defined the transformation series: *Hylobates–Proconsul–Kenyapithecus/Griphopithecus–Sivapithecus/Pongo–Dryopithecus/Ouranopithecus–Gorilla–Homo/Australopithecus–Pan*. Explicit in this series is the distinction of the three clades of Western Eurasian Miocene hominoids (and relevant sister taxa) *Kenyapithecus/Griphopithecus, Sivapithecus/Pongo*, and *Dryopithecus/Ouranopithecus* (= *Graecopithecus* in this volume). The distinction of these clades follows the scheme of Andrews (1992) except for the combination of *Dryopithecus* with *Graecopithecus* (*Ouranopithecus* in Begun's terminology). Ancestor-descendant relationship has not been established for any of these clades, and the possibility must be considered that there were three separate migratory events from Africa (MN 5/6, MN 7 + 8 [?separate *Sivapithecus* and *Dryopithecus* events]). This is a testable hypothesis, but we believe that there is presently too little fossil material, distributed through too much time and space, to choose between a single or multiple migratory events leading to the diversity of Western Eurasia Miocene hominoids.

Cercopithecoid primates were first abundant in MN 11/12 of Southeastern Europe, where they are represented by the colobine *Mesopithecus pentelicus*. *Mesopithecus* continues through the late Miocene and into the early Pliocene of all provinces under consideration. In late MN 13 and MN 14 a new colobine, *Dolichopithecus rusciniensis*, occurred in Western, Central, and Southeastern Europe, with the last recorded occurrence in MN 16/17. The cercopithecine *Macaca* is first recorded in latest MN 13 of Spain and is followed by a very broad Eurasian dispersion of the genus over the course of the Plio–Pleistocene interval.

The catarrhine primate record exhibits a repeated populating of greater Eurasia by African lineages. Pliopithecids and large-bodied hominoids underwent extensive middle Miocene provincial evolutionary radiations followed by precipitous extinction in the medial and late Vallesian; there are no known Turolian forms in Western Eurasia. Cercopithecines replaced these earlier Miocene pliopithecids and hominoids but never attained the level of diversity that the earlier lineages established in the Western Eurasian middle and late Miocene. The shift from a pliopithecid-hominoid fauna to a cercopithecoid fauna generally corresponded to the replacement of equable subtropical-warm temperate forest environments by more seasonal temperate woodlands in the early portion of the late Miocene.

Miocene to recent Carnivora have been a highly diverse mammalian order. Werdelin (this volume) has synthesized the carnivore evolutionary record exclusive of the Hyaenidae, while Werdelin and Solounias (this volume) have separately analyzed the better-understood hyaenid record for Western Eurasia. The carnivores reviewed in this volume include the following families: Hyaenidae, Percrocutidae, Canidae, Viverridae, Amphicyonidae,

Ursidae, Nimravidae, Felidae, Mustelidae, and Simicyonidae, as well as the enigmatic taxa *Schlossericyon viverroides, Parailurus anglicus,* and *Lophocyon carpathicus.*

Percrocutids appear to have originated either in the easternmost parts of the region investigated in this volume or still further east. They initially appeared in Southwestern Asia and Southeastern Europe in MN 6 and subsequently made only intermittent excursions further westward, first in MN 7/8 (France) and later in MN 12 (Spain). There are three genera currently placed in this family: *Percrocuta* (*P. abessalomi* [MN 6]; *P. miocenica* [MN 6]; and *P.* sp. [two separate forms, MN 6–8]); *Dinocrocuta* (*D. minor* [MN 8–9]; *D. senyureki* [MN 9–11]; *D.* sp. [MN 12]); *Allohyaena* (*A. kadici* [MN 11]).

Canids are a group of Nearctic origin that did not establish themselves in Eurasia until the end of the Miocene, when two genera appeared in Western Europe: "*Canis*" ("*C.*" *cipio* [MN 12] and "*C.*" sp. [MN 13]) and *Nyctereutes* (*N. donnezani* [MN 13]). These appearances represent a dilemma since there are no known canid taxa from the late Miocene in the vast area between Western Europe and the ancestral region of these forms, North America.

Western Eurasian Miocene viverrids are represented by two distinct groups: an early middle Miocene archaic group possibly stemming from Oligocene Eurasia Stenoplesictinae and a single later record of much more modern aspect. The first group reached its acme in MN 7/8 and rapidly became extinct at the end of that interval. It consists of a number of genera: *Semigenetta* (*S. elegans* [MN 4]; *S. sansaniensis* [MN 5–8]; *S. ripolli* [MN 9]); *Leptoplesictis* (*L. cadeoti* [MN 4]; *L. aurelianensis* [MN 4–8]); *Viverrictis* (*V. modica* [MN 5–8]); *Jourdanictis* (*J. grivensis* [MN 7/8]); "*Viverra*" *oenigensis* (MN 7); "*Herpestes*" *dissimilis* (MN 7). The latter record is an occurrence of *Viverra* sp. from Italy (MN 13).

Amphicyonids were remnants of a Paleogene radiation and, except for a minor peak in MN 7/8, underwent a steady decline in diversity and abundance during the MN 4–11 interval. The general pattern of amphicyonid evolution throughout the interval seems to be one of replacement of small forms by larger ones and of omnivorous/hypocarnivorous forms by hypercarnivorous ones. This may indicate a marginalization of the amphicyonid feeding niche correlated with the opening of the environment, a process eventually leading to their demise.

A number of genera are involved: *Amphicyon* (*A. major* [MN 4–9]; *A. giganteus* [MN 4–5]; *A. olisiponensis* [MN 4]; *A. steinheimensis* [MN 5–7/8]; *A. caucasicus* [MN 6]; *A. castellanus* [MN ?7/8]); *Pseudocyon* (*P. sansaniensis* [MN 4–6]); *Ysengrini* (*Y. vanentiana* [MN 4]); *Cynelos* (*C. helbingi* [MN 4]; *C. schlosseri* [MN 4]; *C. bohemicus* [MN 5]); *Agnotherium* (*A. grivense* [MN ?4–7/8]; *A. antiquus* [MN 9]); *Pseudarctos* (*P. bavaricus* [MN 5–7/8]); *Thaumastocyon* (*T. bourgeoisi* [MN 5]; *T. dirus* [MN 9]); *Amphicyonopsis* (*A. serus* [MN 7/8]); *Simamphicyon* (*S. batalleri* [MN 9]). In addition, there are a number of fragmentary and poorly dated amphicyonid records throughout the time interval.

Ursids are yet another diverse group that underwent a Western Eurasian adaptive radiation between MN 5 and MN 6, with a subsequent drop in diversity during MN 10, a level that is more or less maintained until the end of the Miocene. For convenience, two main groups within the Ursidae can be distinguished. The first includes genera such as *Plithocyon* (*P. stehlini* [MN 4–6]; *P. bruneti* [MN 5]; *P. armagnacensis* [MN 5–7/8]); *Ursavus*

(*U. brevirhinus* [MN 4–9]; *U. primaevus* [MN 6–9]; *U. intermedius* [MN 6–7/8]; *U. depereti* [MN 7/8–?12]); *Hemicyon* (*H. sansaniensis* [MN 5–6]; *H. goeriachensis* [MN 6–7/8]); and *Dinocyon* (*D. thenardi* [MN 7/8–9]). These are forms of archaic aspect and with the exception of *U. depereti,* all became extinct in MN 9 or earlier. The second group includes the agriotheriine genera *Indarctos* (*I. vireti* [MN 9]; *I. arctoides* [MN 10–11]; *I. atticus* [MN 11–13]) and *Agriotherium* (*A. roblesi* [MN 13]; *A.* sp. [probably two forms, MN 11 and MN 13]), which, as seen, all originated in MN 9 or later. Thus there was almost complete turnover of ursid taxa in the mid-Vallesian.

The Nimravidae exhibited a limited radiation in the middle Miocene, consisting of the genera *Prosansanosmilus* (*P. peregrinus* [MN 4]) and *Sansanosmilus* (*S. palmidens* [MN 5–6]; *S. jourdani* [MN 6–10]). The relationship between these forms and Paleogene nimravids is unclear at present. Nimravids appear to have been replaced ecologically by machairodont felids in the Vallesian.

The Felidae may have originated as early as the late Oligocene and underwent an extensive middle and late Miocene radiation. Their origin is likely to be in Western Eurasia, with subsequent frequent migrations out of this region. Felid diversity was approximately constant from MN 4 to MN 7/8, increasing with the beginning of the machairodont radiation in MN 8, and peaked with the broad dispersion of Southwest Asian–Southeast European "Pikermian" faunas beginning in MN 11. Werdelin (this volume) reports nine genera of felids from Western Eurasia: *Pseudaelurus* (*P. quadridentatus* [MN 4–9]; *P. turnauensis* [MN 4–?11]; *P. romieviensis* [MN 4–7/8]; *P. lorteti* [MN 4–?9]); *Miomachairodus* (*M. pseudaeluroides* [MN 8–9]; "*Felis*" ("*F».*" *antediluviana* [MN 9]; "*F.*" *attica* [MN 10–13]); *Machairodus* [*M. aphanistus* [MN 9–11]; *M. laskarevi* [MN 9]; *M. alberdiae* [MN 9]; *M. giganteus* [MN 11–13]; *M. copei* [MN 11]; *M. kurteni* [MN 13]; *M. irtyschensis* [MN 13]); *Paramachairodus* (*P. ogygia* [MN 9–11]; *P. orientalis* [MN 11–13]); *Metailurus* (*M. parvulus* [MN ?10–13]; *M. major* [MN 11–13]); *Epimachairodus* (*E. romeri* [MN 11]); *Stenailurus* (*S. teilhardi* [MN 11]); *Dinofelis* (*D.* sp. [MN 13]).

Mustelids were the most diverse yet the least understood carnivore group in the Miocene, in terms of both systematics and ecomorphology. Mustelid diversity underwent a marked rise from MN 4 to MN 5, followed by a plateau of high diversity until MN 10, when there was a sharp decline. Mustelid turnover was generally high throughout the middle and late Miocene of Western Eurasia.

The Hyaenidae are a phylogenetically and ecomorphologically diverse group of carnivores in the Miocene. Werdelin and Solounias (this volume) recognize three principal evolutionary phases in the Miocene hyaenid radiation: (1) a general increase in size and acquisition of a suite of fundamental hyaenidlike dental characters (genera *Ictitherium, Thalassictis, Hyaenotherium, Miohyaenotherium,* and *Hyaenictitherium*); (2) an evolutionary pathway toward greater adaptations to cursoriality and specialist meat-eating genera (*Lycyaena, Hyaenictis,* and *Chasmaporthetes*); (3) an evolutionary pathway toward scavenging, with generally less cursoriality (includes the modern genera *Hyaena, Parahyaena,* and *Crocuta*).

The most archaic hyaenid group, *Protictitherium,* first occurred in MN 4 of Western Europe. This group is dentally and postcranially only slightly modified from the generalized common ancestor of all feloid carnivores. *Proteletitherium* is surmised to have had an omnivorous/insectivorous diet (Werdelin and Solounias, this volume). From an origin in Western Europe, where several species have been recorded (*P. gaillardi* [MN 4–9]; *P. crassum* [MN 7/8–10]; *P. llopisi* [MN 10–11]), this genus extended its range eastward, with records from Central Europe (*P. sp.* [MN 6]; *P. crassum* [MN 9]), Western Asia (*P. gaillardi* [MN 6]; *P. cingulatum* [MN 8]; *P. crassum* [MN 8–11]), Hungary–Black Sea region (*P. crassum* [MN 10]; *P. sumegense* [MN 11]; *P. sp.* [MN 11]), and Southeastern Europe (*P. sp.* [two forms, MN 10 and MN 13]). This group clearly achieved its greatest diversity in the middle Miocene, retained a relatively high diversity during the Vallesian, and became more restricted in range and diversity in the early Turolian.

The closely related genus *Plioviverrops* is more mongooselike than *Proteletitherium*, and may have been more insectivorous. It too first occured in Western Europe at about the same time as *Proteletitherium* (*Plioviverrops gervaisi*) and underwent a limited middle and late Miocene evolutionary radiation in Western Europe (*P. gaudryi* [MN 7/8]; *P. sp.* [MN 9]; *P. guerini* [MN 11–12];, *P. faventinus* [MN 13–?14]), Southeastern Europe (*P. orbignyi* [MN 10–12]) and Western Asia (*P. orbignyi* [MN 12]). This suggests that *Plioviverrops* did not extend its range outside Western Europe until MN 10 in Southeastern Europe and somewhat later (MN 11/12) in Southwestern Asia. While *Proteletitherium* is well known from the Central and Eastern Parathethys, *Plioviverrops* is not known from these provinces. The general feature of *Plioviverrops*'s evolution was an increasingly hypocarnivorous dentition, which is in line with suggestions of increased insectivory in the group.

Hyaenid ecomorphological group 3, jackal- and wolflike meat and bone eaters, constitutes the backbone of the Miocene hyaenid radiation, with the genera *Ictitherium*, *Thalassictis*, *Hyaenotherium*, *Miohyaenotherium*, and *Hyaenictitherium*. *Thalassictis* is the first member of this group to appear in Western Eurasia with *T. certa* (MN 7/8) of Western Europe and current indications are that the group may have originated from *Proteletitherium*-like ancestors in this region. The subsequent late Astaracian–Vallesian radiation of *Thalassictis* is modest, including: *T. montadai* (MN 7/8 of Western Europe and MN 8 of Western Asia); *T. robusta* (MN 10 of Central Europe and MN 9 of the Hungary–Black Sea region); *T. spelaea* (MN 9 of the Hungary–Black Sea region); and *T. sarmatica* (MN 9 of the Hungary–Black Sea region).

Ictitherium (*I. intuberculatum*) first occurred in MN 9 of Western Asia. This genus subsequently occurred throughout Western Eurasia, with records from Western Europe (*I. viverrinum* [MN 10]; *I. pannonicum* [MN 11]; *I. sp.* [MN 12–13]), Central Europe (*I. viverrinum* [MN 10]; *I. sp.* [MN 10]), Southeastern Europe (*I. viverrinum* [MN 11–12]), Hungary–Black Sea (*I. tauricum* [MN 10]; *I. viverrinum* [MN 11–12]; *I. pannonicum* [MN 12–13]), and Western Asia (*I. sp.* [two forms, MN 12 and MN 13]; *I. viverrinum* [MN 12]; *I. ibericum* [MN 13]). This genus had a wide distribution in the Old World, and the provincial ages of the various taxa suggest a temporally staggered west to east dispersion.

The genus *Hyaenotherium* included a single common species, *H. wongii*. This species definitely occurred in Western Asia (MN 11–12), Southeastern Europe (MN 11–12), and Hungary–Black Sea (MN 11–12), but not further west. It is also very common in Turolian-equivalent aged localities from China, indicating an

eastern origin and dispersion into the eastern portions of Western Eurasia during MN 11.

Miohyaenotherium and *Hyaenictitherium* are morphologically constructed much like *H. wongii*, only larger. They are species poor, with one and four species, respectively. Both were limited in their distribution to the eastern regions, with only a single possible incursion into Western Europe of *Hyaenictitherium* sp. (MN 9). *Hyaenictitherium* is subsequently recorded from Western Asia (*H. hyaenoides* [MN 11–12]; *H. parvum* [MN 12]) and Hungary–Black Sea (*H. hyaenoides* [MN 11]; *H. sp.* [MN 11]; *H. parvum* [MN 11–12]), while *Miohyaenotherium* is known only from the latter region (*M. bessarabicum* [MN 12]). *Hyaenictitherium* is also known from China and Africa and may have originated in the latter region, with the MN 9 record from Western Europe representing an early attempt to migrate into Europe via a western route.

The cursorial meat and bone eaters include the genera *Lycyaena*, *Hyaenictis*, and *Chasmaporthetes*. Unlike the type 3 hyaenas, these forms are never abundant in the fossil material. *Lycyaena* is known from several regions: Western Europe (*L. sp.* [MN 11–13]), Southeastern Europe (*H. chaeretis* [MN 12]), Western Asia (*L. chaertis* [MN 13]), while *Hyaenictis* is known only from Western Europe (*H. almerai* [MN 12]) and Southeastern Europe (*H. graeca* [MN 12]). Both of these genera also occurred in Africa and may have had their origin there from *Hyaenictitherium*-like ancestors and subsequently dispersed into Eurasia. *Chasmaporthetes* had an apparent origin in Eurasia, but did not appear in the western parts of this region until the Ruscinian.

The transitional bone-crackers, the genera *Palinhyaena*, *Ikelohyaena*, *Belbus*, and *Leecyaena*, originated relatively late and for the most part had geographic distributions that lie outside of Western Eurasia. Only *Belbus* occurred in this area, with *B. beaumonti* from MN 12 of Western Asia.

The full scale bone-crackers include *Adcrocuta* and the modern genera *Hyaena*, *Parahyaena*, and *Crocuta*. Only a single species of this group occurred in our study area during the Miocene: *Adcrocuta eximia*. It first appeared in Western Eurasia in MN 10, presumably as an Asian emigrant and from the onset occurred in virtually all regions. Its provincial ranges were: Western Europe (MN 10–13), Western Asia (MN 10–11), Central Europe (MN 10–11), Southeastern Europe (MN 10–13), Hungary–Black Sea area (MN 11–13).

The hyaenid record exhibits an evolutionary radiation centered in Western Europe during the early and middle Miocene: evolutionary phases 1–4 of Werdelin and Solounias (this volume). The late Miocene radiation of larger taxa would seem to have shifted its locus eastward. Hyaenid diversity increased in Western Europe from MN 4 to MN 7/8 and was maintained until a moderate drop in MN 13. Central Europe never had a diverse hyaenid fauna, but reached a peak of four species in MN 10. Southeastern Europe and Western Asia exhibit a generally increasing diversity from the Astaracian into the Turolian, with a drop at the end of the Turolian (MN 12–13). The entire West Eurasian region exhibits an almost continuous rise, from one species in MN 4 to sixteen species in MN 11, followed by a drop to fourteen species in MN 12 and ten in MN 13. Of all the Miocene hyaenid species known, only one, *Plioviverrops faventinus*, survived into the early Pliocene. Werdelin (this volume) further notes that a number of the turnover peaks are broadly synchronous with abiotic events: MN 6 corresponding to the beginning of the Paratethys regression, MN 9 during the global-wide Serravallian regression and the MN 13 extinction crash with the global-wide late Miocene cooling event and Paratethys-Mediterranean regression.

As a whole, the Western Eurasian carnivore fauna exhibits a steady and high diversity from MN 5 to MN 9. From MN 9 to MN 10 there is a sharp drop in diversity. Carnivore diversity continues to drop steadily during the Turolian interval. At the end of the Turolian there is a mass extinction. Werdelin (this volume) perceives the total relative turnover as a two-phase pattern. The first is from MN 5 to MN 8, during which time little turnover occurred. The second is a high turnover pattern from MN 10 to MN 13, preceded in MN 9 by an initial elevation in turnover. The carnivore turnover pattern would generally appear to coincide with the progressive shift from equable forest environments to seasonal woodland-parkland environments.

Bernor et al. (this volume a) have given an evolutionary and biogeographic revision of Western Eurasian and North African hipparionine horses. This record supports the hypothesis that Old World hipparionines are monophyletic, likely derived from a member of the late middle/early late Miocene North American *Coromohipparion occidentale* s.l. evolutionary complex, and include at least four superspecific groups: the *Hippotherium* Complex, the *Hipparion* s.s.–group, the *Cremohipparion*–group, and the *Sivalhippus* Complex. Each one of these complexes/groups has its own evolutionary and biogeographic characteristics.

The *Hippotherium* Complex represents the initial radiation of Old World hipparionines (Bernor and Lipscomb, 1995), and included species of horses that appear to have been, for the most part, mixed browse-grazing species. Bernor et al. (this volume a) have identified an anagentic Central European evolutionary sequence ranging from MN 9–11, the *Hippotherium primigenium* stage I–V lineage. Additionally, Western Europe contained a number of vicariant derivatives of the *Hippotherium* Complex including: *H.* "koenigswaldi" (MN 9), "*H.*" *catalaunicum* (MN 9–10), "*H.*" aff. *catalaunicum* (MN 12), and "*H.*" *depereti* (MN 10). Koufos (in Bernor et al., this volume a) has reported *Hippotherium primigenium* III from Macedonia (MN 10). Two likely provincial derivatives, "*H.*" *brachypus* (MN 10–12) and "*H.*" *giganteum* (MN 11–13) evolved in the Southeast European–Southwest Asian "superprovince." North Africa had its own local derivation of the "*Hippotherium*" Complex, including "*H.*" *africanum* (MN 9/10) and "*H.*" *sitifense* (MN 13).

The *Hipparion* s.s. lineage is likely derived from the *Hippotherium primigenium* Complex during the late Vallesian. This lineage was apparently adapted to more open country habitats and incorporated more grass into its diet (Hayek et al., 1991; Bernor et al., this volume a). The earliest known occurrence of this group is from late MN 10 of Spain, where the autapomorphically derived form *H. melendezi* is known to first occur. Its sister taxon, *H. gettyi* (MN 11), occurred in Southeastern Europe and Southwestern Asia. The succeeding species, *H. prostylum*, established a West European–Southeast European/Southwest Asian–South Asian distribution at the base of MN 12, and within MN 12 apparently diversified into a number of provincial taxa: *H. concudense* (Spain), *H. dietrichi* (Southeastern Europe), *H. campbelli* (Southwestern Asia), and *H. antelopinum* (South Asia).

The *Cremohipparion*–group would appear to have originated in the Southeast European–Southwest Asian "superprovince" dur-

ing MN 10. Evolution proceeded along two evolutionary lines: the Southeast European–Southwest Asian *C. moldavicum–C. mediterraneum–C. proboscideum* lineage (MN 10–13) increased in size, developed hypertrophied facial fossae, and progressively deepened their nasal notch incision, indicating development of a pseudoprobosis for enhancing a diet with increased browse. The second is the *C. macedonicum–C. matthewi–C. nikosi*–"*C.*" *periafricanum* lineage that exhibits a progressive size decrease, reduction of facial fossae, and apparent grazing diet. This group arises in Southeast European–Southwest Asian MN 10 open country habitats, evolved in that province through MN 13, and apparently ranged into Western Europe during MN 13. The North African dwarf hipparion, "*H.*" *sitifense*, may eventually prove to be a member of this lineage.

The "*Plesiohipparion*"–group is a paraphyletic radicle of the predominantly Asian-African "*Sivalhippus*" Complex, which itself represents an extensive secondary evolutionary radiation of Old World hipparionines. This group of predominantly large, high-crowned cheek tooth horses evidently records two separate extensions from Asia into Western Eurasia: (1) an earlier Pliocene extension into Southwestern Asia ("*P.*" *longipes*), Southeastern Europe ("*P.*" *longipes* and "*P.*" *crassum*), Western Europe ("*P.*" *crassum* and "*P.*" *rocinantis*), Central Europe ("*P.*" *moritorum*); (2) a later Pliocene extension from East Asia into Southwestern Asia ("*P.*" aff. *huangheense*).

Hipparionine diversity was low at initial entry in MN 9. Diversity increased markedly in all provinces except Central Europe beginning in MN 10, when the "*Hippotherium*" *primigenium* s.l., *Hipparion* s.s., and *Cremohipparion* lineages all underwent their initial diversification. The maximum diversity in Western Europe was achieved in MN 10, with three taxa, and fluctuated thereafter between one and two taxa. The Southeast European–Southwest Asian "superprovince" reached its maximum diversity of eight taxa in MN 11 and thereafter underwent a slow decline until the terminal Turolian extinction of these three lineages. Turnover in hipparionine lineages is provincially antiphasic: in Southeastern Europe–Southwestern Asia diversity maxima are at MN 11 and MN 13, whereas in Western Europe diversity maxima are distinctly lower, and occur at MN 10 and MN 12. Other than the initial MN 9 extension of the *Hippotherium primigenium*, and later MN 14/15 "*Plesiohipparion*" extension into all provinces considered, as well as the more limited geographic extensions of the *Hipparion* s.s. (MN 12) and *Cremohipparion* (MN 11 and MN 13) groups, hipparionine species evolved mostly as provincial vicariant lineages.

The Rhinocerotidae exhibit a number of striking provincial changes during the middle Miocene–early Pliocene of Western Eurasia (Heissig, this volume). The Aceratherini and Teloceratini were the dominant middle Miocene rhinocerotid groups in Western Eurasia. Early Miocene holdover lineages included the small teloceratine *Prosantorhinus germanicus* and the aceratherine *Plesiaceratherium fahlbuschi*; Central Europe records the single occurrence of *P. mirallesi*.

The Aceratheriinae exhibit an extensive middle and late Miocene evolutionary record. *Brachypotherium brachypus* occurred throughout Western Eurasia during most of the middle Miocene, with its latest occurrence being MN 9 of Central Europe. The derived form *B. goldfussi* is yet another late-occurring (MN 9) Central European taxon. *Plesiaceratherium* is limited to an MN

5/6 Central European record. *Hoploaceratherium* first occurred in MN 6 and diversified into two taxa: *H. tetradactylum* (Western Europe [MN 6] and the Eastern Mediterranean [MN 6–7/8]) and *H. bavaricum* (Central Europe [MN 7/8–9]). *Alicornops* also diversified into two taxa: *A. simorrensis* (France [MN 6–7/8]; Central Europe [MN 6–7/8]; and the Eastern Mediterranean [MN 7/8]) and *A. alfambrensis* (Spain [MN 10]). *Aceratherium incisivum* occurred throughout Western Eurasia beginning at MN 9 but extended its range into MN 10 of Western Europe only. The *Chilotherium*-group originated in Asia and probably included multiple entries into Western Eurasia. The earliest-occurring Western Eurasian taxon is the MN 9 Eastern Mediterranean species *C. intermedium*. A related taxon, *C. zernowi*, is first recorded from MN 9 of Southeastern Europe and the Eastern Mediterranean. Another species diverse lineage of *Chilotherium* (*Chilotherium Chilotherium*) first occurred in MN 10 of the Eastern Mediterranean (*C. habereri*) and subsequently diversified into a number of Southeast European–Southwest Asian later Miocene lineages (*C. kiliasi*, *C. samium*, *C. habereri*, *C. kowalevski*, and *C. schlosseri*).

Rhinocerotine rhinos also exhibit an extensive middle and late Miocene evolutionary radiation. The influence of Asian rhino faunas is seen by the occurrence of two species of elasmotherini in Southwestern Asia during MN 6, *Begertherium tekkayai* and *B. grimmi* (extending into MN 7/8). *Diceros* has a limited occurrence in the Eastern Mediterranean during MN 6 (*D.* sp.) and MN 10–11 (*D. primaevus*). *Ceratotherium neumayri* was a successful immigrant from East Africa, first occurring in MN 9 of Southeastern Europe and Southwestern Asia, and subsequently occurring in Central Europe during MN 11. "*Dicerorhinus*" *steinheimensis* had a distribution limited to Western (MN 7/8) and Central Europe (MN 5–9). "*Dicerorhinus*" included four different species-lineages, each with a distinct biogeographic/chronologic distribution: "*D.*" *schleiermacheri* (in France [MN 10–11]; Central Europe [MN 9–11]; and Southeastern Europe [MN 11]); "*D.*" *megarhinus* "*D.*" *jeanvireti*, and "*D.*" *miguelcrusafonti* are Pliocene lineages of this group. *Stephanorhinus pikermiensis* first occurred at Pikermi (MN 11/12) and continued its provincial range up into MN 13. *S. etruscus* was a late Pliocene West European–Southeast European lineage of this genus.

Heissig (this volume) has identified a major rhinoceros MN 8/9 (middle/late Miocene boundary) faunal turnover. No rhino lineage is known to occur across the MN 7/8–9 boundary of the Eastern Mediterranean/Southwest Asian realm. Quite to the contrary, there is a more gradual decline of middle Miocene lineages during MN 9 and MN 10 of Central and Western Europe. Late Miocene Southeast European–Southwest Asian rhino faunas were dominated by diverse species of the Asian *Chilotherium* s.l. East Asian stock and the African derivative *Ceratotherium neumayri*. Western Europe was a refuge for a number of middle Miocene relictual species including *Alicornops*, *Aceratherium*, and "*Dicerorhinus*"; *Chilotherium* and *Ceratotherium* are not certainly known from Western Europe; and *Ceratotherium* has only a very limited occurrence documented in Central Europe. The differences in rhinoceros faunas are striking beginning in the late Miocene and certainly reflect the beginning of a paleoenvironmental and biogeographic divergence of Southeast European–Southwest Asian faunas from Central and West European faunas beginning in MN 9.

Fortelius et al. (this volume a) have provided an extensive revision of Miocene suoid systematics, chronology, and biogeography. The middle and late Miocene Western Eurasian suoid fauna includes two families, the Tayassuidae and Suidae. The Tayassuidae exhibit the more modest diversity and include the *Taucanamo–Schizochoerus*–group and *Albanohyus*. The Suidae are far more diverse and include four subfamilies: Hyotheriinae, Listriodontinae, Tetraconodontinae, and Suinae.

Tayassuids are first known from MN 4 of Western Europe by the occurrence of *Taucanamo sansaniensis* (MN 4–6). Closely related taxa occur broadly through Western Eurasia, including: *T. inonuensis* from Central Europe, Southeastern Europe, and Southwestern Asia (MN 5–6); cf. *Taucanamo* sp. from Southwestern Asia (MN 6); *Taucanamo grandaevum*, from Western and Central Europe (MN 6–9). *Schizochoerus* sp. is a Southwest Asian MN 7/8 derivative of this stock. The *Albanohyus* lineage is restricted to MN 7/8 of Central and Western Europe.

Hyotheriines are a predominately early Miocene paraphyletic group, broadly included within the Suinae, from which all suid lineages are believed to have been derived. *Hyotherium soemmeringi* is an MN 3–6 survivor known only from Central and Western Europe.

Listriodontine pigs included three major lineages in Western Eurasia: *Bunolistriodon*, *Kubanochoerus*, and *Listriodon*. *Bunolistriodon lockharti* has a likely African origin and first occurs in Western and Central Europe during MN 4, extending its range there through MN 5. *B. latidens* is an MN 5–6 Western Eurasian form that established a Central European, Southeast European and Southwest Asian distribution. *Kubanochoerus*, including the MN 6 taxa *K. robustus* and *K. khinzikebirus*, is known in our study area only from Southwestern Asia. These species are related to the East Asian radiation of "horned" non-lophodont listriodontines. *Listriodon splendens* is an immigrant from Asia or Africa that first occurs at the base of MN 6, and established a broad Western Eurasian distribution up into MN 9.

The Tetraconodontinae are a group that likely originated in Asia during the late early Miocene. Western Eurasia first records this group as the *Conohyus simorrensis* lineage in MN 5 and was maintained as a broadly distributed group up into MN 9. *Parachleuastochoerus* is a distinctly Western and Central European group of tetraconodonts that included two lineages: *P. steinheimensis* (MN 6–9) and *P. huenermanni* (MN 8–10).

The Suinae include a number of distinct lineages that likely originated in Asia: the *Korynochoerus* (= ?*Propotamochoerus*) *palaeaeochoerus* lineage, which occurred in Western–Central Europe and Southwestern Asia between MN 7/8 and 10; the *Microstonyx major* lineage, which occurred across all of Western Eurasia between MN 10 and 13; *Microstonyx erymanthius*, which occurred in Central Europe, Southeastern Europe, and Southwestern Asia between MN 11 and 12. The enigmatic form from Samos (late MN 12) *Potamochoerus hyotherioides* is the only occurrence of this taxon thus far reported from Western Eurasia.

Suoid provinciality was found to be marked, especially between Southwestern Asia/Southeastern Europe versus Central and Western Europe. Fortelius et al. (this volume a) characterize middle and late Miocene suoid faunal change as including a progressive decline in taxonomic and ecologic diversity accompanied by increasing regional uniformity. These phenomena apparently occur as a result of the successive replacement of autochthonous, ecologically diverse lineages by a diverse assemblage of eastern, ostensibly Asian, immigrations. In this regard, there are three critical intervals of suoid turnover: MN 6, a time of great Eurasian–African faunal exchange; MN 9, a time of major environmental change in circum-Mediterranean biotopes to more seasonal habitats; MN 12/13, marking a sharp increase in Western Eurasian seasonality resulting in the maximum extension of "Pikermian" open habitat faunas.

Gentry and Heizmann (this volume) have reviewed the evolutionary record of the dominant ruminant groups: the Bovidae and Giraffidae. Cervoids and giraffoids probably arose from vicariant origins in Eurasia and the African-Arabian-Indian faunal realm, respectively. By the late Miocene cervids were restricted largely to the forested areas of Central Europe. Giraffids first extended their range into Eurasia during the latest early Miocene, MN 5, but did not undergo their most extensive radiation in the Southeast European–Southwest Asian province during the late Miocene. Bovids may share a broad ancestry with the giraffoids but are first known from older, early Miocene South Asian horizons, ca. 18 Ma (Solounias et al., in press).

The first giraffe, *Giraffokeryx* cf. *punjabiense*, is known to have occurred in MN 5–6 of Southeastern Europe and MN 6 of Southwestern Asia. Giraffids underwent a major evolutionary burst in Southeastern Europe and Southwestern Asia beginning in MN 9, with a number of lineages diversifying: *Palaeotragus coelophrys* (MN 9–12), *P. roueni* (MN 9–12), *P. asiaticus* (MN 13), *Samotherium boissieri* (MN 11–12), *Decennatherium macedoniae* (MN 10), *Helladotherium duvernoyi* (MN 11–13), and *Bohlinia attica* (MN 10–13). Of these taxa, only *H. duvernoyi* is recorded outside this region, occurring in Central Europe during MN 13 when "Pikermian" faunas extended their range there during the Pontian s.s. (re: Rögl and Daxner-Höck, this volume).

After the bovids first arose in Asia during the early Miocene, they retained a low diversity including a few species of *Eotragus* in Eurasia and Africa until MN 5/6. The middle Miocene documented an initial diversification of Central Eurasian lineages including: *Eotragus clavata* (MN 5–7), *Protragocerus chanteri* (MN 7–9), *Austroportax latifrons* (MN 7), *Caprotragoides stehlini* (MN 6–10), and *Miotragocerus monacensis* (MN 7/8; Thenius 1959). The Southeast European–Southwest Asian middle Miocene faunas share *Eotragus clavata* (MN 6) and *Caprotragoides stehlini* (MN 6–10) in common with contemporaneous Central European faunas, but additionally exhibit a more extensive evolutionary radiation of bovid taxa, including: *Miotragocerus* sp. (MN 6–8; occurring later in Central Europe [MN 9–11]); *Hypsodontus serbicus* (MN 5–6); *H. pronaticornis* (MN 11); *Turcocerus gracilis* (MN 6); and *Protoryx enanus* (MN 7).

The late Miocene witnessed a virtual evolutionary explosion of bovid diversity, beginning in MN 9, increasing in MN 10, increasing even more in MN 11–12, and dropping in MN 13. Bovid genera are mostly species diverse. Bovid taxa which first occurred in the late Miocene (re: Gentry and Heizmann, this volume) include: *Tragoportax* (five species; MN 9–13); *Samokeras* (one species; MN 11–12); *Tyrrhenotragus* (one species; MN 12); *Gazella* (four species; MN 9–13); *Prostrepsiceros* (five or more species; MN 10–13 [of N. Africa and Arabia; Gentry, in press]); *Ouzoceros* (one species; MN 10); *Protragelaphus* (five species; MN 11–12); *Palaeoreas* (four species; MN 9–13); *Nisidorcas* (one species; MN 10–11); *Hispanodorcas* (one species; MN 11–13); *Oioceros* (five species; MN 10–13); *Samodorcas* (one species; MN 11–12); *Mesembriacerus* (one species; MN 10); *Plesiaddax* (one species; MN 11); *Urmiatherium* (one species; MN 12); *Parurmiatherium* (one species; MN 11–13); *Criotherium* (one species; MN 11–12); *Palaeoryx* (one species; MN 11–13); *Protoryx* (four species other than *enanus*; MN 9–13); *Pseudotragus* (two species; MN 11–12); *Procobus* (one species; MN 11—12); *Moldoredunca* (one species); *Maremmia* (two species MN 12–13); *?Redunca* (one species; MN 13). This bewildering radiation of late Miocene bovids was largely centered in the Southeast European–Southwest Asian realm. The notable exceptions to this were the radiation of the endemic late Miocene Italian "Baccinellö faunas (MN 12/13; *Tyrrhenotragus gracillimus, Oioceros occidentalis, Maremmia hauptii,* and *M. lorenzi*).

Locality Index

Illustrations are denoted by page numbers in *italics*.

Taxonomy Index

Illustrations are denoted by page numbers in *italics*.